PROTOZOA
IN BIOLOGICAL
RESEARCH

PROTOZOA
IN BIOLOGICAL
RESEARCH

Edited by
GARY N. CALKINS
and
FRANCIS M. SUMMERS

HAFNER PUBLISHING COMPANY, INC.

New York

1964

Originally published 1941
Reprint 1964

Printed and published by

HAFNER PUBLISHING COMPANY, INC.

31 East 10th Street

New York, N.Y. 10003

Library of Congress card catalog number 64-21056

PREFACE

NUMEROUS textbooks of varying degrees of excellence for the study of Protozoa are now on the market and should be consulted for a general treatment of these unicellular forms. This volume will not fill such a need, but has been prepared as a result of a discussion by a group of specialists assembled during the summer of 1937 at the Marine Biological Laboratory at Woods Hole, Massachusetts, for the purpose of ascertaining the best means to stimulate further research on these unicellular animals.

As a result of these discussions it was decided that one good way to attain our end would be to have a group of specialists in the field of protozoölogy prepare a work on research in this field, each specialist to provide a chapter on the subject in which he is best known, and about which he can speak with authority.

Our first real difficulty was to select a limited number of topics from a vast number of possibilities, and then to choose the biologists who, in our opinion, were the best men to write these chapters. As can be readily imagined, this opened up a long list of difficulties and led to many vexatious troubles, but the present work, finally, is the outcome.

To our very great regret, one of the men chosen—Professor Noland, of the University of Wisconsin—who was most enthusiastic over the project, has been forced by reason of continued illness, to drop out. The loss of others, on the plea of pressure of work, and so forth, has also depleted our ranks. But the remainder have completed their labors, have come through, and the results here speak for their continued interest and assiduity.

While the arrangement of our material does not make a great deal of difference, some order nevertheless is advantageous, and we have therefore arranged the chapters, according to the general character of their materials, into groups having a more or less common subject matter.

GARY N. CALKINS
FRANCIS M. SUMMERS

New York City
January 2, 1941

ACKNOWLEDGMENTS

We are indebted to the American publishers of the following books or periodicals for permission to use figures or other illustrations which are reproduced in this book. In a few cases these have been slightly modified by the contributors. Also we are under obligation to numerous foreign publishers who, for obvious reasons, we have been unable to ask for permission.

American Journal of Hygiene
  Figures 188, 191, 192

Biological Bulletin
  Figures 5, 10, 13, 74, 81, 82, 84, 96, 104, 181, 183, 185, 186, 187, 199A, 199B, 223

Biology of the Protozoa by G. N. Calkins, 1933. Lea and Febiger Co.
  Figures 1, 2, 3A, 3B, 9

Genetics
  Figures 166, 168

Journal of Animal Behavior, Henry Holt and Co.
  Figure 103

Journal of Cellular and Comparative Physiology, Wistar Institute of Anatomy and Biology
  Figures 119, 120, 121, 122, and Table 2

Journal of the Elisha Mitchell Scientific Society
  Figures 202A, 202B

Journal of Experimental Medicine, Rockefeller Institute for Medical Research
  Figures 189, 190

Journal of Experimental Zoology, Wistar Institute of Anatomy and Biology
  Figures 14, 94, 97, 98, 102, 105, 106, 149, 156, 162, 164, 179, 182

Journal of Morphology, Wistar Institute of Anatomy and Biology
  Figures 83, 91, 93, 154, 157, 160, 161, 163

Journal of Preventative Medicine, John McCormick Institute for Infectious Diseases
Figure 193

Light and Behavior of Organisms by S. O. Mast, 1911. John Wiley and Sons, Inc.
Figures 99, 107

Physiological Zoology, University of Chicago Press
Figures 95, 113, 126, 180

Publications of the Carnegie Institute of Washington
Figure 214

Science
Figure 123

Transactions of the American Microscopical Society
Figures 36, 37

University of California Publications in Zoology, University of California Press
Figures 15, 16, 39, 49, 50, 53, 54, 55, 70, 85, 86, 87, 88, 89, 146, 148, 150, 151, 196, 219, 225

United States Geological Survey of the Territories
Figure 221

Finally, it is a pleasure to recognize the care and patience on the part of the staff of the Columbia University Press in the preparation and distribution of this work. While all members have worked diligently and harmoniously with us, we are particularly indebted to Miss Georgia W. Read, Miss Ida M. Lynn, and Miss Eugenia Porter for the meticulous care with which they have sought to unify and, in some cases, to clarify the manuscripts on subjects unfamiliar to them.

THE EDITORS

New York City
January 2, 1941

CONTENTS

TABLES

ILLUSTRATIONS

FIGURES

PLATES

LIST OF ABBREVIATIONS

Abh. bayer. Akad. Wiss.: Abhandlungen der Kgl. Bayerischen Akademie der Wissenschaften. Math.-Phys. Kl. München.

Abh. naturw. Ver. Bremen.: Abhandlungen hrsg. vom Naturwissenschaftlichen Verein zu Bremen.

Abh. sencken b. natur f. Ges.: Abhandlungen hrsg. von der Senckenbergischen Naturforschenden Gesellschaft. Frankfurt a. M.

Acta Biol. exp.: Acta biologiae experimentalis. Warsaw.

Act. Zool. Stock: Acta Zoologica: Internationell Tidskrift för Zoologica. Stockholm.

Act. Sci. Indust.: Actualités scientifique et industriel. Paris.

Állatt. Közlem.: Állattani Közlemények. Budapest.

American Academy of Arts and Sciences, see Mem. Amer. Acad. Arts Sci.; Proc. Amer. Acad. Arts Sci.

Amer. J. Anat.: American Journal of Anatomy. Baltimore.

Amer. J. Bot.: American Journal of Botany. Lancaster.

Amer. J. Hyg.: American Journal of Hygiene. Baltimore.

Amer. J. Path.: American Journal of Pathology. Boston.

Amer. J. Physiol.: American Journal of Physiology. Boston.

Amer. J. Psychiat.: American Journal of Psychiatry. Baltimore.

Amer. J. publ. Hlth.: American Journal of Public Health. New York.

Amer. J. Syph.: American Journal of Syphilis. St. Louis, Mo.

Amer. J. trop. Med.: American Journal of Tropical Medicine. Baltimore.

Amer. Midl. Nat.: American Midland Naturalist. Notre Dame, Ind.

Amer. mon. micr. J.: American Monthly Microscopical Journal. Washington.

Amer. Nat.: American Naturalist. Boston.

Anat. Anz.: Anatomische Anzeiger. Jena.

Anat. Rec.: Anatomical Record. Philadelphia.

An. Fac. Med. Porto.: Anais da Faculdade de Medicina do Porto.

An. Inst. Biol. Univ. Méx.: Anales del Instituto de biologia, Universidade de México. México.

Ann. appl. Biol.: Annals of Applied Biology. Cambridge.

Ann. Bot., Lond.: Annals of Botany. London.

Ann. Fac. Med. S. Paulo.: Annaes da Faculdade de Medicina de São Paulo.

Ann. Igiene (sper.): Annali d'igiene (sperimentale). Torino.

Ann. Inst. océanogr. Monaco.: Annales de l'Institut océanographique de Monaco. Monaco.

Ann. Inst. Pasteur.: Annales de l'Institut Pasteur. Paris.

Ann. Mag. Nat. Hist.: Annals and Magazine of Natural History. London.

Ann. Microg.: Annales de Micrographie spécialement consacrée à la bactériologie au protophytes et aux protozoaires. Paris.

Ann. Mus. Stor. nat. Genova.: Annali del Museo cirico di storia naturale. Genova.

Ann. Mus. zool. polon.: Annales musei zoologici polonici. Warszawa.

Ann. Natal Mus.: Annals of the Natal Museum. Pietermaritzburg.

Ann. N. Y. Acad. Sci.: Annals of the New York Academy of Sciences. New York.

Annot. zool. jap.: Annotationes zoological jáponenses. Tokyo.

Ann. Parasit. hum. comp.: Annales de parasitologie humaine et comparée. Paris.

Ann. Physiol. Physicochim. biol.: Annales de physiologie et de physicochimie biologique.

Ann. Protist.: Annales de protistologie. Paris.

Ann. Sci. nat.: Annales des sciences naturelles. (*a*) Botanique. (*b*) Zoologie. Paris.

Ann. Soc. belge Micr.: Annales de la Société belge de microscopie. Bruxelles.

Ann. Soc. Sci. méd. nat. Brux.: Annales (et Bulletin). Société R. des sciences médicales et naturelles de Bruxelles.

Ann. trop. Med. Parasit.: Annals of Tropical Medicine and Parasitology. Liverpool.

Anz. Akad. Wiss., Wien.: Anzeiger der Kaiserlichen Akademie der Wissenschaften. Math. Kl. Wien.

Arb. Gesundh. Amt. Berl.: Arbeiten aus dem Kais. Gesundheitsamte. Berlin.

Arb. Staatsinst. exp. Ther. Frankfurt.: Arbeiten aus dem Staatsinstitut für experimentelle Therapie und dem Georg. Speyer-Hause zu Frankfurt a. M. Jena.

Arb. zool. Inst. Univ. Wien.: Arbeiten aus den Zoologischen Institut der Univ. Wien u. der Zoolog. Station in Triest. Wien.

Arch. Anat. micr.: Archives d'anatomie microscopique. Paris.

Arch. Anat. Physiol. Lpz.: Archiv für Anatomie und Physiologie. Leipzig.

Arch. Anat. Physiol. wiss. Med.: Archiv für Anatomie, Physiologie und wissenschaftliche Medicin. Leipzig.

Arch. argent. Enferm. Apar. dig.: Archivos argentinos de enfernedades del aparato digestivo de la nutrición. Buenos Aires.

Arch. Biol. Paris.: Archives de biologie. Paris.

Arch. exp. Zellforsch.: Archiv für experimentelle Zellforschung. Jena.

Arch. Hyg. Berl.: Archiv für Hygiene (und Bakteriologie). München u. Berlin.

Arch. Inst. Pasteur Afr. N.: Archives des Instituts Pasteur de l'Afrique du Nord. Tunis.

Arch. Inst. Pasteur Algér.: Archives de l'Institut Pasteur d'Algérie. Alger.

Arch. intern. Med.: Archives of Internal Medicine. Chicago.

Arch. int. Physiol.: Archives internationales de physiologie. Liége et Paris.

Arch. Méd. exp.: Archives de médecine expérimentale et d'anatomie patholo-gique. Paris.

Arch. mikr. Anat.: Archiv für Anatomie (und Entwicklungsmechanik). Bonn.

Arch. Mikrobiol.: Archiv für Mikrobiologie. Berlin.

Arch. Naturgesch.: Archiv für Naturgeschichte. Berlin.

Arch. parasit.: Archives de parasitologie. Paris.

Arch. Path. Lab. Med.: Archives of Pathology and Laboratory Medicine. Chicago.

Arch. physiol. norm. path.: Archives de physiologie normale et pathologique. Paris.

Arch. Protistenk.: Archiv für Protistenkunde. Jena.

Arch. russ. d'Anat.: Archives russes d'anatomie, d'histologie et d'embryologie.

Arch. russ. protist.: Archives russes de protistologie.

Arch. Schiffs- u. Tropenhyg.: Archiv für Schiffs- u. Tropenhygiene. . . . Leipzig.

Arch. Sci. biol., St. Pétersb.: Archives des sciences biologiques. St. Péters-bourg.

Arch. Sci. phys. nat.: Archives des sciences physiques et naturelles. Genève, Lausanne, Paris.

Arch. Tierernähr. Tierz.: Archiv für Tierernährung und Tierzucht. Berlin.

Arch. wiss. prakt. Tierheilk.: Archiv für wissenschaftliche u. praktische Tier-heilkunde. Berlin.

Arch. Zellforsch.: Archiv für Zellforschung. Leipzig.

Arch. zool. exp. gén.: Archives de zoologie expérimentale et générale. Paris.

Arch. zool. (ital.): Archivio zoologico italiano. Napoli. Torino.

Arhiva vet.: Arhiva Veterinara. Bucuresti.

Årsskr. Lunds Univ.: Årsskrift Lunds Universitet.

Atti Accad. gioenia.: Atti della R. Accademia gioenia di scienze naturali. Catania.

Atti della R. Accademia dei Lincei, see R. C. Acad. Lincei.

Atti Soc. Studi Malar.: Atti della Società per gli studi della malaria. Roma.

Aust. J. exp. Biol. med. Sci.: Australian Journal of Experimental Biology and Medical Science. Adelaide.

Bact. Rev.: Bacteriological Review. Baltimore, Md.

Beih. bot. Zbl.: Beihefte zum Botanischen Zentralblatt. Cassel.

Beitr. allg. Bot.: Beiträge zur allgemeinen Botanik. Berlin.

Beitr. Biol. Pfl.: Beiträge zur Biologie der Pflanzen. Breslau.

Ber. dtsch. bot. Ges.: Bericht der Deutschen Botanischen Gesellschaft. Berlin.

Ber. Ges. Wiss.: Berichte über die Verhandlungen der kgl. sächsischen Gesell-schaft der Wissenschaften.

Berl. klin. Wschr.: Berliner klinische Wochenschrift. Berlin.

Ber. naturf. Ges. Freiburg i. B.: Berichte der Naturforschenden Gesellschaft zu Freiburg i. Br.

Ber. senckenb. Ges.: Bericht der Senckenbergischen Naturforschenden Gesellschaft in Frankfurt a. M. Frankfort a. M.

Ber. wiss. Biol.: Bericht über die wissenschaftliche Biologie. Berlin.

Bibliogr. genet.: Bibliographia genetica. 'sGravenhage.

Bibliogr. zool.: Bibliographia zoologica. Lipsiae.

Biederm. Zbl.: Biedermanns Zentralblatt für Agrikultur-chemie und rationellen Landwirtschaftsbetrieb. Leipzig.

Bio-chem. J.: Bio-chemical Journal. Liverpool.

Biochem. Z.: Biochemische Zeitschrift. Berlin.

Biodyn.: Biodynamica. Normandy, Mo.

Biol. Bull. Wood's Hole: Biological Bulletin of the Marine Biological Laboratory, Wood's Hole, Mass.

Biol. Listy.: Biologické Listy. Prague.

Biol. Monogr.: Biological Monographs and Manuals. Edinburgh.

Biologe: Biologe; Monatsschrift zur Wahrung der Belange der deutschen Biologen. Munich.

Biol. Rev.: Biological Reviews and Biological Proceedings of the Cambridge Philosophical Society. Cambridge.

Biol. Zbl.: Biologisches Zentralblatt. Leipzig.

Biometrika.: Biometrika. Cambridge.

Bol. biol. Fac. Med. S. Paulo.: Boletim biologico, Laboratorio de parasitologia. Faculdade de medicina de São Paulo.

Boll. Lab. Zool. agr. Bachic. Milano.: Bollettino del Laboratorio di zoologia agraria e bachicoltura del R. Istituto superiore agrario di Milano.

Boll. Soc. eustach.: Bollettina della Società Eustachiana. Camerino.

Botaniste.: Le Botaniste. Caen, Poitiers, Paris.

Bot. Gaz.: Botanical Gazette. Chicago.

Bot. Rev.: Botanical Review. Lancaster, Pa.

Bot. Ztg.: Botanische Zeitung. Berlin and Leipzig.

Brazil-med.: Brazil-medico. Rio de Janeiro.

Brit. J. exp. Biol.: British Journal of Experimental Biology. Edinburgh.

Brit. J. exp. Path.: British Journal of Experimental Pathology. London.

Bronn's Klassen.: Bronn's Klassen und Ordnungen des Tierreichs. Leipzig.

Bull. Acad. Med. Paris.: Académie de Médecine, Bulletin. Paris.

Bull. Acad. Sci. St.-Petersbourg.: Bulletin de l'Académie Impériale des sciences de St.-Petérsbourg.

Bull. Acad. Sci. U.R.S.S.: Bulletin de l'Académie du sciences de l'U.R.S.S.

Bull. Amer. Mus. Nat. Hist.: Bulletin of the American Museum of Natural History. New York.

Bull. Bingham oceanogr. Coll.: Bulletin of the Bingham Oceanographic Collection, Yale University, New Haven.

Bull. biol. exp. med. U.R.S.S.: Bulletin de Biologique et de Medicine Experimentale de L'U.S.S.R. Moscow.

Bull. biol.: Bulletin biologique de la France et de la Belgique. Paris.

Bull. Inst. océanogr. Monaco: Bulletin de l'Institut océanographique de Monaco.

Bull. Inst. Pasteur.: Bulletin de l'Institut Pasteur. Paris.

Bull. int. Acad. Cracovie.: Bulletin international de l'Académie des sciences de Cracovie (de 'l'Académie polonaise des sciences).

Bull. int. Acad. Prag. Sci. Math. Nat.: Bulletin International. Česká akademie véd a umění v Praze. Sciences, mathématiques et naturelles. Prague. (Česká akademie césáre Františka Josefa pro védy, slovesnost a umění v Praze.)

Bull. Mus. Hist. nat. Belg.: Bulletin du Musée Royal d'histoire naturelle de Belgique. Bruxelles.

Bull. N. J. agric. Exp. Stas.: Bulletin of the New Jersey Agricultural Experimental Stations. New Brunswick.

Bull. sci. Fr. Belg.: Bulletin scientifique de la France et de la Belgique. Londres, Paris, Berlin.

Bull. Scripps Instn. Oceanogr. tech.: Bulletin Scripps Institution of Oceanography. Technical Series. La Jolla, Calif.

Bull. Sleep. Sickn. Bur.: Bulletin. Royal Society Sleeping Sickness Bureau. London.

Bull. Soc. bot. Fr.: Bulletin. Société botanique de France. Paris.

Bull. Soc. Hist. nat. Afr. N.: Bulletin de la Société d'histoire naturelle de l'Afrique du Nord. Alger.

Bull. Soc. imp. Natur., Moscow: Société impériale des naturalistes, Bulletin. Moscow.

Bull. Soc. méd.-chir. Indochine.: Bulletin de la Société medico-chiruricale de l'Indochine. Hanoï.

Bull. Soc. Path. exot.: Bulletin de la Société de pathologie exotique. Paris.

Bull. Soc. zool. Fr.: Bulletin de la Société zoologique de France. Paris.

Bull. U. S. Bur. Fish.: Bulletin of the Bureau of Fisheries. Washington.

Bull. U. S. nat. Mus.: Bulletin. United States National Museum. Smithsonian Institution. Washington.

Carnegie Institution, see Pap. Tortugas Lab.; Publ. Carneg. Instn.; Yearb. Carneg. Instn.

Camb. Phil. Soc. Proc.: Proceedings of the Cambridge Philosophical Society. Cambridge.

Cellule.: La Cellule. Recueil de cytologie et d'histologie générale. Louvain.

Cellulose-Chem.: Cellulose-Chemie. (Papierfabrikant. Suppl.) Berlin.

Chem. Abstr.: Chemical Abstracts. New York.

Clin. vet. Milano.: La clinica veterinaria. Milano.

Cold Spr. Harb. Monogr.: Cold Spring Harbor Monographs. Brooklyn, N.Y.

Cold Spring Harbor Symp. Quant. Biol.: Cold Spring Harbor Symposium on Quantitative Biology, 1933.

Coll. Net.: Collecting Net. Woods Hole, Mass.

Contr. zool. Lab. Univ. Pa.: Contributions from the Zoological Laboratory, Univ. of Pennsylvania. Philadelphia.

Copeia.: Copeia. Published to advance the science of cold-blooded Vertebrates. New York.

C. R. Acad. Sci. Paris.: Compte rendu hebdomadaire des séances de l'Académie des sciences. Paris.

C. R. Acad. Sci. U.R.S.S.: Compte rendu de l'Académie des sciences de l'U.R.S.S.

C. R. Ass. Anat.: Compte rendu de l'Association des anatomistes. Paris and Nancy.

C. R. Lab. Carlsberg.: Compte rendu des travaux du Laboratoire de Carlsberg. Copenhague.

C. R. XIIe Cong. Int. Zool.: Comptes rendues XIIe Congrès international de la Zoologie. Lisbon, 1935.

C. R. Soc. Biol. Paris.: Compte rendu hebdomadaire des séances et mémoires de la société de biologie. Paris.

Cytologia, Tokyo.: Cytologia. Tokyo.

Denkschr. Akad. Wiss. Wien.: Denkschriften der Kaiserlichen Akademie der Wissenschaften. Math.-nat. Kl. Wien.

Denkschr. med.-naturw. Ges. Jena.: Denkschriften der Medizinisch-naturwissenschaftlichen Gesellschaft zu Jena.

Dtsch. med. Wschr.: Deutsche medizinische Wochenschrift. Leipzig.

Dtsch. tierärzlt. Wschr.: Deutsche tierärztliche Wochenschrift. Hannover.

Dutch East Indies Volkksgesondheid.: Dutch East Indies. Dienst der Volksgesondheid in Nederlandsch-Indie. Mededeelingen.

Ecol. Monogr.: Ecological Monographs. Durham, N.C.

Ecology.: Ecology. Brooklyn.

Edinb. New phil. J.: Edinburgh New Philosophical Journal, 1826, 1854, 1855-'64.

Ergebn. Biol.: Ergebnisse der Biologie. Berlin.

Ergebn. Hyg. Bakt.: Ergebnisse der Hygiene, Bakteriologie, Immunitätsforschung u. experimentellen Therapie. Berlin.

Ergebn. inn. Med. Kinderheilk.: Ergebnisse der inneren Medizin u. Kinderheilkunde. Berlin.

Ergebn. Physiol.: Ergebnisse der Physiologie. Wiesbaden.

Ergeb. Zool.: Ergebnisse u. Fortschritte der Zoologie. Jena.

Fauna u. Flora Neapel.: Fauna u. Flora des Golfes von Neapel u. d. angrenz. Meeresabschnitte. Berlin.

Flora, Jena.: Flora, oder allgemeine botanische Zeitung. Jena, Regensburg.

Folia haemat.: Folia haematologica. Leipzig.

ForschBer. biol. Sta. Plön.: Forschungsberichte aus der Biologischen Station zu Plön. Stuttgart.

Gegenbaurs Jb.: Gegenbaurs morphologisches Jahrbuch. Leipzig.

Genetics.: Genetics: a Periodical Record of Investigations bearing on Heredity and Variation. Menasha, Wis.

Göttingen, mathem.-physical. Kl.: Nachrichten von der Königl. Gesellschaft der Wissenschaften zu Göttingen. Mathematisch-physikalische Klasse. Göttingen.

Grèce méd.: Grèce médicale. Syra.

Growth.: Growth. (A journal.) Series of numbered fasciculi at irregular intervals.

Handb. Vererbungsw.: Handbuch der Vererbungswissenschaft. Berlin.

Illinois biol. Monogr.: Illinois Biological Monographs. Univ. of Illinois. Urbana.

Indian J. med. Res.: Indian Journal of Medical Research. Calcutta.

Indian med. Gaz.: Indian Medical Gazette. Calcutta.

Indian med. Res. Mem.: Indian Medical Research Memoirs. Calcutta.

Indust. Engng. Chem.: Industrial and Engineering Chemistry. Easton, Pa.

Int. Rev. Hydrobiol.: Internationale Revue der gesamten Hydrobiologie u. Hydrographie. Leipzig.

Iowa St. Coll. J. Sci.: Iowa State College Journal of Science. Ames.

J. Acad. nat. Sci. Philad.: Journal of the Academy of Natural Sciences of Philadelphia.

J. agric. Res.: Journal of Agricultural Research. Washington.

J. Amer. med. Ass.: Journal of the American Medical Association. Chicago.

J. Amer. statist. Ass.: Journal of the American Statistical Association. Boston.

J. anat. Paris.: Journal de l'anatomie et de la physiologie normales et pathologique de l'homme et des animaux. Paris.

J. Anim. Behav.: Journal of Animal Behaviour. Boston.

J. appl. Physics: Journal of Applied Physics. Menasha, Wis.

J. Bact.: Journal of Bacteriology. Baltimore.

J. biol. Chem.: Journal of Biological Chemistry. Baltimore.

J. cell. comp. Physiol.: Journal of Cellular and Comparative Physiology. Philadelphia.

J. Coll. Sci. Tokyo.: Journal of the College of Science, Imp. University of Tokyo.

J. comp. Neurol.: Journal of Comparative Neurology (and Psychology). Philadelphia.

J. comp. Path.: Journal of Comparative Pathology and Therapeutics. Edinburgh, London.

J. Coun. sci. industr. Res. Aust.: Journal of the Council for Scientific and Industrial Research, Australia. Melbourne.

J. Dairy Sci.: Journal of Dairy Science. Baltimore.

J. Dep. Sci. Calcutta Univ.: Journal of the Department of Science of Calcutta University. Calcutta.

J. Elisha Mitchell sci. Soc.: Journal of the Elisha Mitchell Scientific Society, Chapel Hill, N.C.

J. exp. Biol.: Journal of Experimental Biology. Cambridge.

J. exp. Med.: Journal of Experimental Medicine. New York.

J. exp. Zool.: The Journal of Experimental Zoology. Philadelphia.

J. Fac. Sci. Tokyo Univ.: Journal of the Faculty of Science, Tokyo Imperial University. Tokyo. (4) Zool.

J. Genet.: Journal of Genetics. Cambridge.

J. gen. Physiol.: Journal of General Physiology. Baltimore.

J. Helminth.: Journal of Helminthology. London.

J. Immunol.: Journal of Immunology. Baltimore.

J. industr. Engng. Chem.: Journal of Industrial and Engineering Chemistry. Easton, Pa.

J. infect. Dis.: Journal of Infectious Diseases. Chicago.

J. Lab. clin. Med.: Journal of Laboratory and Clinical Medicine. St. Louis, Mo.

J. linn. Soc. (Zool.): Journal of the Linnean Society. (Zoology.) London.

J. Malar. Inst. India.: Malaria Institute of India, Journal.

J. Mar. biol. Ass. U. K.: Journal of the Marine Biological Association of the United Kingdom. Plymouth.

J. med. Res.: Journal of Medical Research. Boston.

J. Microbiol.: Journal de microbiologie. Petrograd. Moscou.

J. Morph.: Journal of Morphology (and Physiology). Philadelphia, Boston.

J. nerv. ment. Dis.: Journal of Nervous and Mental Diseases. New York.

J. Parasit.: Journal of Parasitology. Urbana, Ill.

J. Path. Bact.: Journal of Pathology and Bacteriology. London.

J. phys. Chem.: Journal of Physical Chemistry. Ithaca, N.Y.

J. Physiol.: Journal of Physiology. London and Cambridge.

J. Physiol. Path. gén.: Journal de physiologie et de pathologie générale. Paris.

J. prev. Med. Lond.: Journal of Preventive Medicine. London.

J. prev. Med., Oshkosh, Wis.: Journal of Preventive Medicine. Oshkosh, Wis.

J. R. Army med. Cps.: Journal of the Royal Army Medical Corps. London.

J. R. micr. Soc.: Journal of the R. Microscopical Society. London.

J. Sci. Hiroshima Univ. Journal of Science of the Hiroshima University. Hiroshima, Japan.

J. trop. Med. (Hyg.): Journal of Tropical Medicine (and Hygiene). London.

Jap. J. Zool.: Japanese Journal of Zoology. Tokyo.

Jber. Fortschr. Anat. EntwGesch.: Jahresberichte über die Fortschitte der Anatomie und Entwicklungsgeschichte. Jena.

Jb. wiss. Bot.: Jahrbuch für wissenschaftliche Botanik. Berlin.

Jena. Z. Naturw.: Jenaische Zeitschrift für Naturwissenschaft. Jena.

Jena. Z. Naturgesch.: Jenaische Zeitschrift für Naturgeschichte. Jena.

Kiev. obshch, estest. Zap.: Kievskve obshchestvo estestvoispytatelei Zapiski. Kief.

Kryptogamenfl. Mark Brandenb.: Kryptogamenflora der Mark Brandenburg und angrenzender Gebiete. (Botanische Verein der Provinz Brandenburg). Leipzig.

Matemat. termesz. estesitö.: Matematikai és természettudományi éstesito.

Math. naturw. Ber. Ung.: Mathematische und naturwissenschaftliche Berichte aus Ungarn. [Berlin: Budapest:] Leipzig.

Math.-nat. Kl., Heidelb. Akad. Wiss.: Mathematisch-naturwissenschaftliche Klasse, B. Biologische wissenschaften. Heidelberger Akademie der Wissenschaften.

Medicine.: Medicine. Baltimore, Md.

Med. Países calidos. Medicina de los países cálidos. Madrid.

Mém. Acad. roy. Belg.: Académie royale des sciences de Belgique, Mémoires. Brussels.

Mém. Acad. Sci., Paris.: Mémoires de l'Académie des sciences de l'Institut de France. Paris.

Mem. Amer. Acad. Arts Sci.: Memoirs of the American Academy of Arts and Sciences. Boston.

Mém. Cl. Sci. Acad. polon.: Mémoires de la classe des sciences mathématiques et naturelles. Académie Polonaise des Sciences et des Lettres. Cracovie.

Mem. Coll. Sci. Kyoto.: Memoirs of the College of Science, Kyoto Imp. University. Kyoto.

Mém. Inst. nat. genev.: Mémoires de l'Institut national genevois. Genève.

Mem. Inst. Osw. Cruz.: Memorias do Instituto Oswaldo Cruz. Rio de Janeiro.

Mém. Mus. Hist. nat. Belg.: Mémoires du Musée royal d'histoire naturelle de Belgique. Bruxelles.

Mém. Soc. Phys. Genève.: Mémoires de la Société de physique et d'histoire naturelle de Genève.

Mem. Soc. zool. tchec. Prague.: Československa společnost zoologická (Societas Zoologica čechoslovenica). Prague.

Microgr. prép.: Micrographe préparateur; journal de micrograprie générale, de technique micrographique et revue des journaux français et étrangers. Paris.

Microscope.: Microscope. Ann Arbor; Detroit; Washington.

Midl. Nat.: Midland Naturalist (Midland union of natural history societies). London; Birmingham.

Midl. Nat.: Midland Naturalist. Notre Dame, Ind.

Mikrokosmos.: Mikrokosmos (Deutsche mikrologische Gesellschaft, . . .) Stuttgart.

Mitt. zool. Inst. Univ. Münster.: Mitteilungen aus dem Zoologischen Institut der Westfälischen Wilhelms-Universität zu Münster i. W.

Mitt. Zool. Sta. Neapel.: Mitteilungen aus der Zoologische Station zu Neapel.

Monatsber. preuss. Akad. Wissensch.: Monatsberichte K. preussische Akademie der Wissenschaften zu Berlin.

Monit. zool. ital.: Monitore zoologico italiano. Sienna; Firenze.

Monogr. Inst. Pasteur.: Monographies de l'Institut Pasteur. Paris.

Morph. Jb.: Morphologisches jahrbuch. Leipzig.

Nature.: Nature. London.

Naturwissenschaften.: Naturwissenschaften. Berlin.

Nova Acta Leop. Carol.: Nova Acta Leopoldinisch-Carolinische deutsche Akademie der Naturforscher. Jena.

Occ. Publs. Amer. Ass. Adv. Sci.: Occasional Publications, American Association for the Advancement of Science.

Ohio J. Sci.: Ohio Journal of Science (Ohio State University . . . Ohio Academy of Science). Columbus.

Öst. bot. Z.: Oesterreichische botanische Zeitschrift. Wien.

Pap. Tortugas Lab.: Papers from the Tortugas Laboratory (Department of Marine Biology) of the Carnegie Institution of Washington. Washington.

Parasitology.: Parasitology, a supplement to the Journal of Hygiene. Cambridge.

Pasteur Institute, see Ann. Inst. Pasteur; Arch. Inst. Pasteur Afr. N.; Arch. Inst. Pasteur Algér.; Bull. Inst. Pasteur; Monogr. Inst. Pasteur.

Pflüg. Arch. ges. Physiol.: Pflügers Archiv für die gesamte Physiologie d. Menschen u. d. Tiere. Bonn.

Philipp. J. Sci.: Philippine Journal of Science. Manila.

Philos. Trans.: Philosophical Transactions of the Royal Society. London.

Physics.: Physics. Minneapolis.

Physiol. Rev.: Physiological Reviews. Baltimore.

Physiol. russe.: Physiologiste russe. Moscow.

Physiol. Zoöl.: Physiological Zoölogy. Chicago.

Planta: Planta. (Archiv für wissenschaftliche Botanik. Berlin. Abt. E.)

Poult. Sci.: Poultry Science. Ithaca, N.Y.

Proc. Acad. nat. Sci., Philad.: Proceedings of the Academy of Natural Sciences of Philadelphia.

Proc. Amer. Acad. Arts Sci.: Proceedings of the Academy of American Arts and Sciences. Boston.

Proc. Amer. phil. Soc. Proceedings of the American Philosophical Society. Philadelphia.

Proc. Boston Soc. nat. Hist.: Proceedings of the Boston Society of Natural History. Boston.

Proc. Calif. Acad. Sci.: Proceedings of the California Academy of Sciences. San Francisco.

Proc. Davenport Acad Sci.: Proceedings of the Davenport Academy of (Natural) Sciences. Davenport, Iowa.

Proc. Acad. Sci. Amst.: Proceedings of the Royal Academy of Sciences, Amsterdam.

Proc. Indian Acad. Sci.: Proceedings of the Indian Academy of Sciences. Bangalore.

Proc. K. Akad. Wetensch.: K. Akademie van Wetenschappen, Amsterdam. Proceedings of the Section of Sciences (translated).

Proc. Linn. Soc. N.S.W.: Proceedings of the Linnean Society of New South Wales. Sydney.

Proc. Minn. Acad. Sci.: Proceedings of the Minnesota Academy of Sciences. Minneapolis.

Proc. nat. Acad. Sci., Wash.: Proceedings of the National Academy of Sciences. Washington.

Proc. Pa. Acad. Sci.: Proceedings of the Pennsylvania Academy of Science. Harrisburg, Pa.

Proc. R. Irish Acad.: Proceedings of the Royal Irish Academy. Dublin. London.

Proc. roy. Soc.: Proceedings of the Royal Society. London.

Proc. roy. Soc. Edinb.: Proceedings of the Royal Society of Edinburgh.

Proc. R. Soc. Med.: Proceedings of the R. Society of Medicine. London.

Proc. Soc. exp. Biol. N.Y.: Proceedings of the Society for Experimental Biology and Medicine. New York.

Proc. zool. Soc. Lond.: Proceedings of the General Meetings for Scientific Business of the Zoological Society of London.

Protoplasma.: Protoplasma. Leipzig.

Protop.-Monog.: Protoplasma-Monographien. Berlin, 1928+

Protozoology.: Protozoology, a Supplement to the Journal of Helminthology. London.

Publ. Carneg. Instn.: Publications. Carnegie Institution of Washington. Washington.

Publ. Fac. Sci. Univ. Charles.: Spisy Vydávané Přírodovědeckou Fakultou Karlovy University. Praha.

Publ. Hlth. Rep. Wash.: Public Health Reports. Washington.

Quart. J. exp. Physiol.: Quarterly Journal of Experimental Physiology. London.

Quart. J. micr. Sci.: Quarterly Journal of microscopical Science. London.

Quart. Rev. Biol.: Quarterly Review of Biology. Baltimore.

Radiology.: Radiology. St. Paul.

Rasseg. faunist. Roma.: Ressegna faunistica. Roma.

R. C. Accad. Lincei.: Atti della R. Accademia dei Lincei. Rendiconti. Cl. di sci. fis. mat. e nat. Roma.

Rec. Malar. Surv. India.: Records of the Malaria Survey of India. Calcutta.

Rec. zool. suisse.: Recueil zool. suisse. Geneva.

Rep. Brit. Assoc. Adv. Sci.: Report of the British Association for the Advancement of Science. Bath.

Rep. N. Y. St. Conserv. Comm. Report of the New York State Conservation Commission. Albany.

Rep. Sleep. Sickn. Comm. roy. Soc.: Report of the Sleeping Sickness Commission. Royal Society. London.

Rep. vet. Res. S. Afr.: Report on (of Director of) Veterinary Research. Department of Agriculture, Union of South Africa.

Rev. biol. Nord Fr.: Revue biologique du Nord de la France. Lille.

Rev. Microbiol. Saratov.: Revue de microbiologie (et) d'épidémiologie, (et de parasitologie). Saratov.

Rev. sci. Instrum.: Review of Scientific Instruments. (Optical Society of America) Rochester, N.Y.

Rev. Soc. argent. Biol.: Revista de la Sociedad argentina de biología. Buenos Aires.

Rev. suisse Zool.: Revue suisse de zoologie et Annales de la Société zoologique suisse et du Muséum d'histoire naturelle de Genève. Geneva, 1893+

Rif. med.: Riforma medica. Napoli.

Riv. Fis. mat. Sci. nat.: Rivista di fisica, matematica e scienze naturali. Pavia.

Riv. Malariol.: Rivista dì malariologia. Roma.

Roux Arch. EntwMech. Organ.: Wilhelm Roux Archiv für Entwicklungsmechanìk der Organismen. Leipzig. (Abt. D, Zeitschrift für wissenschaftliche Biologie, Berlin.)

Russ. J. Zool.: Zoologicheskiĭ vîêstnik. Journal Russe de zoologie. Petrograd.

Russk. zool. Zh.: Russkii zoologicheskiĭ zhurnal. Revue zoologique russe. Moscow.

S. Afr. J. Sci.: South African Journal of Science. Capetown.

S. B. Akad. Wiss. Wien.: Sitzungsberichte der Kais. Akademie der Wissenschaften in Wien.

S. B. böhm. Ges. Wiss.: Sitzungsberichte der Kgl. Böhmischen Gesellschaft der Wissenschaften. Prag.-Math.-Nat. Klasse.

S. B. Ges. Morph. Physiol.: Sitzungsberichte der Gesellschaft für Morphologie und Physiologie in München. München.

S. B. Ges. naturf. Fr. Berl.: Sitzungeberichte der Gesellschaft Naturforschender Freunde zu Berlin.

S. B. preuss. Akad. Wiss.: Sitzungsberichte der Kgl. Preussischen Akademie der Wissenschaften zu Berlin.

Science. Science. New York.

Sci. Prog.: Science Progress; a quarterly review of current scientific investigation. London, 1894-98; Sci. Prog. Twent. Cent. Science Progress in the Twentieth Century, a quarterly journal of scientific thought. London, 1906.

Scientia, Bologna.: Scientia, rivista di scienza. Bologna.

Scientia, Sér. biol., Paris.: Scientia. Série biologique. Paris. 99-04

Sci. Mem. Offrs. med. san. Dept. Gov. India.: Scientific Memoirs by Officers of the Medical and Sanitary Department of the Government of India. Calcutta.

Sci. Mon., Lond.: Scientific Monthly. London.

Sci. Mon., N.Y.: Scientific Monthly. New York.

Sci. Proc. R. Dublin Soc.: Scientific Proceedings of the Royal Dublin Society. Dublin.

Sci. Rep. Tôhoku Univ.: Science Reports of the Tôhoku Imp. University. Sendai. S.4, Biology. Sendai.

Sci. Rep. Tokyo Bunrika Daig.: Science Reports of the Tokyo Bunrika Daigaku. Tokyo.

Scripta bot. Petropol.: Scripta botanica. Leningrad. (Glavnyi botanicheskii sad).

Smithsonian Institution, *see* Smithson. misc. Coll.; Bull. U. S. nat. Mus.

Smithson. misc. Coll.: Smithsonian Miscellaneous Collections. Washington.

Soc. geol. France.: Société géologique de France. Paris.

Special Rep. Ser. med. res. Counc.: Special report series, Medical research council. London.

Stain Tech.: Stain Technology. Geneva, N.Y.

Stat. Rep. N.J.: New Jersey. Statistical Report. Trenton, N.J.

Stazione Zoologica. Naples.: Fauna und Flora des Golfes von Neapel. 1880+

Sth. med. J., Nashville.: Southern Medical Journal. Nashville, Tenn.

Studies Inst. Divi Thomae.: Studies of the Institutum Divi Thomae, Cincinnati, Ohio.

Tabl. zool.: Tablettes zoologiques. Poïtiers.

Tijdschr. ned.: Nederlandsch Tijdschrift voor Geneeskunde. Amsterdam.

Trab. Lab. Invest. biol. Univ. Madr.: Trabajos del Laboratorio de investigaciones biológicas de la Universidad de Madrid.

Trans. Amer. Fish. Soc.: Transactions of the American Fisheries Society. New York.

Trans. Amer. micr. Soc.: Transactions of the American Microscopical Society. Menasha, Wis.

Trans. Amer. phil. Soc.: Transactions of the American Philosophical Society. Philadelphia.

Trans. Faraday Soc.: Transactions of the Faraday Society. London.

Trans. N. Y. Acad. Sci.: Transactions of the New York Academy of Science. New York.

Trans. roy. Soc. Edinb.: Transactions of the Royal Society of Edinburgh. Edinburgh.

Trans. R. Soc. trop. Med. Hyg.: Transactions of the Royal Society of Tropical Medicine and Hygiene. London.

Trav. Inst. biol. Peterhof.: Travaux de l'Institut des Sciences naturelles de Peterhof (Biologische Institut zu Peterhof). Peterhof [Petergof].

Trav. Lab. zool. Sebastopol.: Travaux lu Laboratoire zoologique et Station biologique de Sébastopol. St-Pétersbourg.

Trav. Soc. Nat. St.-Pétersb. (Leningr.): Travaux de la Société Imp. des naturalistes de St-Pétersbourg (Leningrad).

Trav. Sta. limnol. Lac Bajkal.: Travaux de la Station Limnologique du Lac Bajkal.

Trop. Dis. Bull.: Tropical Diseases Bulletin. London.

Univ. Cal. Publ. Bot.: University of California Publications in Botany. Berkeley.

Univ. Cal. Publ. Zool.: University of California Publications in Zoology. Berkeley.

Univ. Mo. Stud.: University of Missouri Studies. Columbia.

Unters. Bot. Inst. Tübingen.: Untersuchungen, Botanische Institut. Tübingen.

U. S. Geol. Surv. Terr.: U. S. Geological and Geographical Survey of the Territories. Washington.

U. S. Public Health Reports, *see* Publ. Hlth. Rep., Wash.

Verh. dtsch. zool. Ges.: Verhandlungen der Deutschen zoologischen Gesellschaft. Leipzig.

Verh. Ges. deutscher Naturf. Ärtzte: Verhandlung Gesellschaft deutscher Naturforscher und Ärtzte. Berlin. (Amtlicher Bericht über die zweiundzwanzigste Versammlung deutscher Naturforscher und Ärtzte in Bremen. Bremen.)

Verh. naturh.-med. Ver. Heidelberg.: Verhandlungen des Naturhistorisch-medizinischen Vereins zu Heidelberg.

Verh. naturh. Ver. preuss. Rheinl.: Verhandlungen des Naturhistorischen Vereins der Preussischen Rheinlande, Westfalens u. des Reg.-Bez. Osnabrück. Bonn.

Wiss. Arch. Landw.: Wissenschaftliches archiv für landwirtschaft. Berlin; Leipzig.

Yearb. Carneg. Instn.: Yearbook of the Carnegie Institution of Washington.

Z. allg. Physiol.: Zeitschrift für allgemeine Physiologie. Jena.

Z. Biol.: Zeitschrift für Biologie. München and Berlin.

Z. ges. Anat. 1. Z. Anat. EntwGesch. 2. Z. KonstLehre. 3. Ergebn. Anat. EntwGesch.; Zeitschrift für die gesamte Anatomie. Abt. 1. Zeitschrift für Anatomie u. Entwicklungsgeschichte. Abt. 2. Zeitschrift für Konstitutionslehre. Abt. 3. Ergebnisse der Anatomie u. Entwicklungsgeschichte. Berlin.

Z. ges. Neurol. Psychiat.: Zeitschrift für die gesamte Neurologie und Psychiatrie. Berlin.

Z. Hyg. Infektkr.: Zeitschrift für Hygiene. Leipzig, 1886-91; became (Z. Hyg. InfektKr.) Zeitschrift für Hygiene und Infektionskrankheiten. Leipzig.

Z. ImmunForsch.: Zeitschrift für Immunitätsforschung und experimentelle Therapie. Jena.

Z. indukt. Abstamm.- u. VererbLehre.: Zeitschrift für induktive Abstammungs- u. Vererbungslehre. Berlin.

Z. InfektKr. Haustiere.: Zeitschrift für Infektionskrankheiten, parasitäre Krankheiten u. Hygiene der Haustiere. Berlin.

Z. mikr.-anat. Forsch.: Zeitschrift für mikrospopische-anatomische Forschung. Leipzig.

Z. Morph. Ökol. Tiere.: Zeitschrift für Morphologie und Ökologie der Tiere. (Zeitschrift für wissenschaftliche Biologie. Berlin. Abt. A.)

Z. Naturw.: Zeitschrift für Naturwissenschaften. Leipzig.

Z. Parasitenk.: Zeitschrift für Parasitenkunde. (Zeitschrift für wissenschaftliche Biologie, Berlin. Abt. F.)

Z. phys. Chem.: Zeitschrift für physikalische Chemie, Stöchiometrie und Verwandtschaftslehre. Leipzig.

Z. Tierz. ZüchtBiol.: Zeitschrift für Tierzüchtung und Züchtungsbiologie. Berlin.

Z. vergl. Physiol.: Zeitschrift für vergleichende Physiologie. (Zeitschrift für wissenschaftliche Biologie. Berlin. Abt. C.) Berlin.

Z. wiss. Bot.: Zeitschrift für wissenschaftliche Botanik. Zurich.

Z. wiss. Mikr.: Zeitschrift für wissenschaftliche Mikroskopie und für mikroskopische Technik. Leipzig.

Z. wiss. Zool.: Zeitschrift für wissenschaftliche Zoologie. Leipzig.

Z. Zellforsch.: Zeitschrift für Zellforschung und Mikroskopische Anatomie. Berlin. (Zeitschrift für wissenschaftliche Biologie. Berlin. Abt. B [2].)

Z. Zell.- u. Gewebelehre.: Zeitschrift für Zellen- u. Gewebelehre. Berlin. (Abt. B [1] Zeitschrift für wissenschaftliche Biologie, Berlin.)

Zbl. Bakt.: Zentralblatt für Bakteriologie, Parasitenkunde u. Infectionskrankheiten. Jena.

Zool. Anz.: Zoologischer Anzeiger. Leipzig.

Zool. Anz. Suppl.: Zoologischer Anzeiger Supplement (Bibliographia zoologica of the Concilium Bibliographicum). Leipzig.

Zool. Jb.: Zoologische Jahrbücher. 1 Abt. für Anatomie. 2 Abt. für Systematik. 3 Abt. für allgemeine Zoologie und Physiologie der Tiere. Jena.

Zoologica.: Zoologica. Scientific Contributions of the New York Zoological Society. New York.

Zool. Zbl.: Zoologisches Zentralblatt. Leipzig.

PROTOZOA
IN BIOLOGICAL
RESEARCH

CHAPTER I

GENERAL CONSIDERATIONS

GARY N. CALKINS

ALTHOUGH Protozoa are known in all parts of the world, and evidence is at hand that they—in some cases even the same genera—have lived and thrived in all periods of the earth's history, little has been accomplished to show how this remarkable phenomenon of longevity has been brought about. Within the last century, however, the matter has been the subject of many studies, both theoretical and experimental, although the latter, it must be confessed, are nearly always combined with the former.

LIFE AND VITALITY

For many years it has seemed to the present writer that *life* and *vitality* are concepts which have so often been confused that at the present time they are held by many biologists, and by most philosophers, to be synonymous. There is reason, however, especially in connection with the protoplasm of Protozoa, for distinguishing between them, as I have maintained in my recent textbook, *The Biology of the Protozoa.*

It is generally recognized that life cannot be measured nor analyzed as such, except through its manifestations of *vitality.* It has long been known that each type of living thing has a specific organization which is carried on, subject to adaptations through reactions to the environment, or by inheritance, from generation to generation. At the present time we do not know what this finer organization is, but it is assumed that specific proteins form the basis of species differences, and that these, in combination with water, carbohydrates, fats, and salts of different kinds, provide the materials for metabolic activities. Thus we have the possibility of arriving at at least two concepts; first, the concept of the physical and chemical make-up, or, in general, the *organization;* and second, the concept of that same organization in action. I would apply this second concept to protoplasm during its activity and would limit the term vitality to this activity. It follows that life may be defined as specific organization having the possibility, or the potential of vitality. Such activity of protoplasm involves the interaction of its component parts with one another and with the environment. Life thus may be conceived as static, or analogous to an automobile in its garage, and dynamic, as an automobile in motion. A dried rotifer is almost, if not entirely static, but retains its organization; place it in water and it becomes dynamic within a few minutes. Similarly with an encysted protozoön, which, within its cyst walls is not freely exposed to water and oxygen but retains its specific organization and is apparently static and, like a dried rotifer, it may remain in this desiccated condition for years without losing its potential of vitality so that, when again placed in water, or culture medium, it soon emerges from its cyst, develops motile organs and other adult structures, and begins again its metabolic activities. The protoplasm within the cyst is undifferentiated, but soon after active metabolism begins, it becomes differentiated with structures peculiar to the species. After this the organization changes with each act of a metabolic nature.

FUNDAMENTAL AND DERIVED ORGANIZATION

We have reason, therefore, to speak of a *fundamental organization* of protoplasm, characterized by undifferentiated protoplasm of a cyst, or of an egg, and of a *derived organization* which comes from the fundamental through metabolic activity and is characteristic of the adult and all of its parts. It follows that *death* is not of necessity the absence of vitality but is due to the derangement or breakdown of the organization.

Thus all species of Protozoa have within themselves the potential of an endless existence, subject, of course, to the vicissitudes of the daily life of the adults, each of which is the custodian of a limited portion of the fundamental organization.

SOME ECOLOGICAL CONSIDERATIONS

As well known, Protozoa may be found wherever there is moisture without deleterious substances and the same species may be found in the littoral waters of the sea and in inland fresh-water lakes, ponds, and pools. Although many species are cosmopolitan, they tend to accumulate in certain places where the environments best suit their needs; hence it is possible to outline certain ecological limitations, although these must not be too closely limited. We speak, for example, of the water-dwelling forms, having reference to the great multitude of types which may be found in exposed waters of the earth's surface; or of an even greater number of species which live in the sea at all depths, from the surface down to 3,000 feet, while some Radiolaria are present in the most extreme depths of the ocean. Where salt and fresh waters mix we find special groups of brackish water fauna which are represented by thousands of species.

The fresh-water species are so numerous and so varied that some help is gained by grouping them in "habitat groups" which are adapted more or less to similar environmental conditions. An attempt to classify such fresh-water species on an ecological basis was made by Kolkwitz (1909), although this classification has never been widely used in protozoan taxonomy. It was based, in the main, on the amounts and conditions of organic matter and oxygen present in the water. The habitat groups were described by him under the terms katharobic, oligosaprobic, mesosaprobic, and polysaprobic types. The first are relatively rare types, being found in fresh-water springs, running rivers and streams, and wells which have little organic matter but are rich in oxygen.

Oligosaprobic types are characterized by a small amount of organic matter but are rich in minerals. Hence lakes and reservoirs become the haunts of chlorophyll-bearing forms in particular, which often accumulate to incredible numbers and frequently cause disagreeable odors and tastes in drinking waters and render them unpotable.

Mesosaprobic types are the most common of all fresh-water Protozoa, for here active oxidation is going on and organic matter, in the presence of sunlight, is in all stages of decomposition. Here the microscopist finds his richest collecting place.

Polysaprobic types, finally, live in waters with little or no free oxygen, but with an abundance of sulphureted hydrogen, carbon dioxide, and other products of putrefaction which are advertised by their foul odors, due to the gases which are formed. Here belong Lauterborn's (1901) "sapropelic fauna," which live, for the most part, as anaerobes, and which are often characterized by fantastic shapes and inability to live under aërobic conditions.

Another group have an almost terrestrial habitat and may be found in damp moss, sphagnum, or similar environments. A few types of ciliates and flagellates, and some testate rhizopods, may be found here, but one must be an optimistic collector, and an opportunist, to get good results.

It is somewhat the same with soil-dwelling Protozoa, among which it is to be expected that water-dwelling forms would occasionally be found and interpreted as casual soil-dwelling types. Sandon (1927), however, has shown that there is a characteristically well-defined group of forms living in this environment, although, as would be expected, there is a wide difference in soils, both as to depth of occurrence of Protozoa and chemical make-up. In arable soils, Sandon finds that not only are they most abundant, but that they live at a depth of four to five inches and, for the most part, are bacteria eaters. He, with other observers, has described some seventy-five species of flagellates as fairly common; and about eighty-five species of rhizopods and ciliates which are less common. There is not much evidence that life in the soil leads to any particular type of morphological adaptations, but there is a possibility that species adapted to life deep in the soil are already partly anaerobic, and such forms may more easily become parasitic.

Lackey (1925) enumerates no less than nineteen common forms of Protozoa which live in the sewage of Imhof tanks, while five common species of rhizopods, four of ciliates, and about twenty less common species of flagellates are occasionally found.

So-called coprozoic forms are rarely segregated, but may be found more or less sporadically almost anywhere on the earth's surface. These are Protozoa which are taken into the digestive tract by all animals, and, as encysted forms, become stored up in the intestine until they are passed out with the feces. In water, such cysts accompanied by nutrient material from the feces, develop into adult and active forms which may be mistaken by the unwary as entozoic parasites. So great and complete is the specificity of internal parasites, however, that such cysts, while not representing parasites of the host providing the feces, may nevertheless be cysts of active parasites of other hosts which now pass with disastrous results, into the digestive tract of a new definitive host by which they have been eaten with contaminated food and water.

In numbers of species there is little doubt that the free-living and water-dwelling Protozoa stand first, but the parasitic forms make a close second, for no type of animal is free from the possibility of infection by one or more species of parasitic Protozoa.

While parasitism will be dealt with by others in this volume (see Becker, *infra,* Chapter XVII; and Kirby, *infra,* Chapters XIX, XX), I will speak here merely of one or two types of adaptations which have arisen as a result of this mode of life. Ectoparasites and endoparasites have developed somewhat differently, as a result of their different modes of life and different needs. In common, they all possess the first great need of all parasites, viz. reproductive power, thus obeying a first law of nature to the effect that the number of offspring produced should vary according to the difficulties encountered in their youth and during development to maturity.

Adaptations of ectoparasitic types are mainly morphological, and here some striking structures may be developed, as a few illustrations will show. All may have a special feeding advantage by being transported from place to place; or when attached to gills or other structures of their hosts, they are continuously bathed by food-bearing currents of water. Attached forms live on algae, exoskeletons and appendages of arthropods, shells of molluscs, or gill filaments and gill bars of all kinds of water-dwelling animals. Thus we find species of *Zoothamnium* or of *Lagenophrys* on the legs of *Gammarus,* while species of *Spirochona* or *Dendrocometes* are on the gill filaments of the same hosts, or on *Asellus.* The suctorian *Trichophrya* adheres like a saddle to the gill bars of *Salpa,* while the vorticellid *Ellobiophrya* encircles a gill filament of the lamellibranch *Donax vittatus* by the union of two branch outgrowths of a more common adhesion disc (see Kirby, Fig. XXX). The latter is almost always provided with hooks or suckers, or both, to form the "scopula," as in species of *Trichodina* or *Cyclochaeta,* parasites on *Hydra.* A special thigmotactic reaction appears to keep the ciliate *Kerona pediculus* on the surface of *Hydra fusca.* Such forms seem to have no ill effects on their hosts and scarcely qualify as parasites. Schröder calls them "epibionts." Real ectoparasites are rare, as a matter of fact, and they are little different from free water-dwelling forms in structure. They do occur, however, especially on fish hosts. The flagellate *Costia necathrix* increases to such numbers that normal functions are impeded, and young fish are frequently killed. *Ichthyophthirius multifiliis* bores into fish skin and brings about distributed ulcerations which may be fatal.

Endoparasites may be more destructive, and, while they are relatively simple morphologically, they may be highly differentiated physiologically.

Knowledge of the life histories of endoparasites, particularly those of man, has grown amazingly, and prophylaxis has grown with it (see Becker, *infra,* Chapter XVII; and Kirby, *infra,* Chapter XIX-XX; and for immunity, see Taliaferro, *infra,* Chapter XVIII).

SOME HISTORICAL FACTS

Each of the myriads of species represented in this diversity of habitats has its own specific fundamental organization which carries the possibility of indefinite life in the future. We cannot visualize the conditions under which life came into being in times past, but we can observe, study, describe, and in part measure the manifestations—vitality—which have kept it going and enabled it to withstand all of the vicissitudes of nature, through upheavals, floods, droughts, and other symptoms of the might of nature which have played so prominent a part in the history of the earth's activity since the dawn of life.

We have as a basis for such study the vast mass of knowledge which has accumulated since Protozoa were introduced to science by the Dutch naturalist Leeuwenhoek in the latter part of the seventeenth century. Many different kinds were soon recognized, and this recognition led to classifications and logical groupings in general which facilitated the discovery and descriptions of new species, modes of life, adaptations to changing conditions, and the like.

Almost from the time of their discovery, the Protozoa have played an important part in problems of general biological interest. Spontaneous generation, for example, or origin of living things from nonliving matter, which was popularly and generally believed up to the sixteenth century, had received hard knocks from the experiments of Redi, Spallanzani, Harvey, and other scientific men. All life from life, all living things from eggs (*omne vivum ex ovo*) supplanting the common belief regarding generation.

The use of the microscope, revealing a novel and marvelous world of living things, gave a new lease of life to the theory of spontaneous generation. Appearing in containers of pure water, it was asked "How could such water become animated with living things if these had not arisen there by spontaneous generation?" The problem, thus reopened with the discovery of Leeuwenhoek's animalcula, was not solved until near the end of the nineteenth century by the careful work of Pasteur,

Lister, and a growing school of biologists who pushed back farther and farther the organisms supposed to arise *de novo*. Lower invertebrates, algae and Protozoa, and finally bacteria, one after the other forming the fighting lines of the army of ignorance, now fell back, little by little, before the slow but sure advance of science.

In connection with this theory of spontaneous generation, a novel conception sprang up with the observations of the French naturalist Buffon in 1749. This conception, expanded by Needham (1750), was a suggestive forerunner of the cell theory outlined by Schleiden and Schwann in 1839-40. The wealth of different forms of life and the numbers of each type in natural waters in which decomposition was under way, was interpreted by Buffon and Needham as the result of disintegration of animals and plants in water and the resulting liberation of myriads of "organic particles" (Buffon) of which such animals and plants were composed.

Protozoölogy, like cytology, owes its birth and development to the use of the microscope. There has been much discussion and much diversity of opinion over the discovery of the microscope, meaning the compound instrument, the significant principles of which have been adapted and improved until the beautiful microscopes of today are the outcome. Woodruff (1939) writes:

Galileo (1610) was the first to use the instrument,—but the first figures ever made with the aid of a compound microscope to appear in a printed book, were by Francesco Stelluti in 1625. But an Englishman, Robert Hooke, was the first to realize to the full the importance of using instruments which increase the powers of the senses in general and of vision in particular, and to express it convincingly in 1665, in a remarkable book: the "Micrographica." Here he described and emphasized for the first time, the "little boxes or cells" of organic structure, and indelibly inscribed the word "cell" in biological literature. (Woodruff, *loc. cit.*, p. 2.)

As stated above, the real discoverer of the Protozoa was a Dutch microscopist, Anton von Leeuwenhoek (1632-1723) who, using crude lenses of his own make, was one of the first to apply the microscope to scientific investigation. His contributions to microscopic anatomy and to physiology, inaugurating, as they did, the invaluable services of the microscope in biological research, marked an epoch in the history of science. In a letter to the Royal Society in 1675, Leeuwenhoek wrote that he had discovered living creatures in rain water which had stood but four days in a new earthen pot, glased blew within. This invited me [he continues] to view this water with great attention, especially those little animals appearing to me ten thousand times less than those represented by Mons. Schwammerdam and by him called *Water fleas* or *Water-lice,* which may be perceived in the water with the naked-eye. The first sort by me discover'd in the said water, I divers times observed to consist of 5, 6, 7, or 8 clear globuls, without being able to discern any film that held them together, or contained them. When these *animalcula* or living atoms did move, they put forth two little horns, continually moving themselves. The place between these horns was flat, though the rest of the body was roundish, sharp'ning a little towards the end, where they had a tayl, near four times the length of the whole body, of the thickness (by my microscope) of a spiders-web; at the end of which a globul, of the bigness of one of those which made up the body; which tayl I could not perceive, even in very clear water, to be mov'd by them. These little creatures, if they chanced to light upon the least filament or string, or other such particle, of which there are many in water, especially if it hath stood some days, they stook entangled therein, extending their body in a long round, and striving to dis-intangle their tayls whereby it came to pass, that their whole body lept back towards the globul of the tayl, which then rolled together Serpent-like, or after the manner of copper or iron-wire that having been wound about a stick, and unwound again, retains those windings or turnings. This motion of extension and contraction continued awhile; and I have seen hundreds of these poor little creatures, within the space of a grain of gross sand, lye cluster'd in a few filaments. (From Calkins, 1901, p. 5.)

This is the first description of a protozoön; and though the description is incomplete, it undoubtedly refers to a species of *Vorticella.* Leeuwenhoek observed several other forms, but their identity is uncertain.

Leeuwenhoek allowed his imagination to see what his eyes could not.

When we see [said he] the spermatic animalcula [spermatozoa] moving by vibrations of their tayls, we naturally conclude that these tayls are provided with tendons, muscles, and articulations, no less than the tayls of a dormouse or rat, and no one will doubt that these other animalcula which swim in stagnant water [Protozoa] and which are no longer than the tayls of the spermatic animalcula, are provided with organs similar to those of the highest animals. How marvellous must be the visceral apparatus shut up in such animalcula! (Quoted from Dujardin, 1841, pp. 21-22.)

In a letter a year later Leeuwenhoek further says: "The fourth sort of little animals . . . were incredibly small; nay, so small, in my sight that I judged that even if one hundred of these very wee animals lay stretched out one against another, they could not reach to the length of a grain of coarse sand." Later he discovered parasite Protozoa in man and beast, and bacteria in the human mouth. Of course Leeuwenhoek made many other discoveries during his long life—his studies were not confined to animalcules—but it is enough that he is justly regarded as the Father of Protozoölogy and Bacteriology (Woodruff, *loc. cit.*, p. 3).

Obviously an immense field awaited intensive study, and this was begun in a desultory way by many amateur and professional biologists during the closing decades of the eighteenth century, the outstanding contributions being made by O. F. Müller in 1773 and 1786. And then over half a century passed before Ehrenberg, in 1838, and Dujardin, in 1841, afforded a sufficiently broad view of the "simple" animals to justify the establishment of the phylum Protozoa by von Siebold in 1845 (Woodruff, *loc. cit.*, p. 4).

Müller, adopting the Linnæan nomenclature, described and named some 378 species, of which about 150 are retained today as Protozoa. His classification was the first successful attempt to bring order out of the heterogeneous collection of forms included under the name animalcula. He used Ledermüller's (1760-63) term Infusoria for the name of the entire group, which he placed as a class of the worms (see Bütschli, 1883, p. 1129). While he eliminated many inaccuracies, he confirmed the substantial observations of the earlier observers, extending many of them to all groups of the Protozoa. He ascertained the presence of an anus, showed that many Infusoria are carnivorous, and observed the process of conjugation, his description of the latter being more accurate than that of any of his followers until the time of Balbiani, in 1858-59.

Like his predecessors, Müller included among the Protozoa many other organisms, placing here diatoms, nematode worms, *Distomum* larvae, and larval forms of coelenterates and molluscs. The majority of these miscellaneous forms were, however, properly classified before 1840, while finally spermatozoa (discovered by Ludwig Hamm, who is said to have been a pupil of Leeuwenhoek) and which had been universally regarded as animalcula inhabiting the seminal fluid, were withdrawn during the last century.

Following John Hill (1752), Müller did not regard the Protozoa as complicated animals, but considered them as the simplest of all living beings, composed of a homogeneous gelatinous substance, a view in which he was followed by a majority of the "nature-philosophers," most of whom gave little or no attention to the Protozoa, but, accepting Müllers' work as final, based many of their speculations upon it.

It is rather remarkable that fifty years or so later, when the biological atmosphere was saturated with the idea of the cell theory, the justly famous microscopist of Europe, C. G. Ehrenberg (1795-1876), using much finer achromatic lenses, should have returned to the crude view of Leeuwenhoek, assigning to the Protozoa a system of minute but complete organs. His conclusions on Protozoa were brought together in one great work, the title of which alone shows his point of view (*Die Infusionsthierchen als vollkommene Organismen,* 1838). From the supposed possession of many stomachs he gave to one of his groups the name Polygastrica or Magenthiere, making it a sharply defined class of the animal kingdom. One of the sub-classes in which these supposed stomachs were apparent he called the Enterodela, while all other forms he included as Anentera. The red pigment spots of many forms were interpreted as true eyes, but as eyes could not be conceived without an accompanying nervous system, he sought for nerve ganglia in different organisms and found what he was looking for in a species of Astasia. He described the "eye" in this form as seated upon a "spherical granular mass" which he considered as equivalent to the suprapharyngeal ganglion of the rotifers. The myonemes in the stalks of *Vorticella,* in *Stentor,* and in many other ciliates he interpreted as muscles. Pigment spheres and protoplasmic granules were described as ovaries, the nucleus as a testis, while the contractile vacuole was at first regarded as a respiratory organ (see Weatherby, *infra,* Chapter VII).

A formidable opponent of Ehrenberg soon appeared in France—Felix Dujardin, who, influenced by long study of the Foraminifera, came to the conclusion in 1835 that these rhizopods, which up to that time had been classified with the cephalopod molluscs, were in reality the simplest of animal organisms, composed of a simple homogeneous substance to which he gave the name *Sarcode*.

Dujardin is best known by his systematic treatise on the Protozoa which he published in 1841, and in which he laid the basis for the modern classification of these unicellular forms. The first suggestion that Protozoa might be single cells was made by Meyen (1839), who compared the infusorian body with a single plant cell. But, according to Bütschli

(1883), the cell theory was first applied directly to the Protozoa by Barry (1843), who asserted that *Monas* and its allies among the flagellates were single cells, and that the nucleus found within them was the equivalent of the cell nucleus of higher forms. At the same time Barry expressed the view that cells increase only by division, and he compared the processes of multiplication in *Volvox* and *Chlamydomonas* with the cleavage of eggs (cf. Bütschli, 1887, p. 1152).

Barry's view was accepted, in part, by Owen, who thought, however, that the Infusoria could not be included with the flagellates as single cells because of their differentiation. It was von Siebold (1845) who finally asserted the unicellular nature of all Protozoa.

Balbiani's researches on the life history of Protozoa at first led him into a curious error, a reminiscence, apparently, of Ehrenberg's and the older point of view. O. F. Müller had observed and had correctly interpreted conjugation in different forms, but his successors down to Balbiani regarded this as incorrect, maintaining that Müller had seen only stages in simple division.

Balbiani (1861) returned to Müller's view, and clearly stated that, in addition to simple division, another and a sexual method of reproduction occurs. His interpretation of the sexual organs of the Protozoa was given in 1858, when he maintained that the larger of the two kinds of nuclei of Infusoria, the macronucleus, is the ovary, and that the smaller one, the micronucleus, is the testis. He saw and pictured the striped appearance of the micronucleus prior to its division and interpreted the stripes as spermatozoa. The eggs were said to be fertilized in the macronucleus, and then deposited on the outside where they developed into new ciliates. Stein at first (1859) opposed this assumption, but in the second volume of his work on the Infusoria (1867), misled by his own Acineta theory, in which certain Suctoria were thought to be stages in the life cycle of certain ciliates, he practically adopted it, maintaining however, that the young forms developed first in the nucleus and only later left the mother organism. Bütschli (1873) was apparently the first to point out Balbiani's error, and in his epoch-making work of 1876, after demonstrating the striped appearance of the nucleus in egg cells (mitotic figure) during division, he concluded that the "stripings" which Balbiani held to be spermatozoa were no other than this striated condition of the nucleus during division. Bütschli further showed that during conjugation the macronucleus disintegrates, and that the parts which Balbiani had considered eggs are resorbed in the protoplasm, the whole nucleus being replaced by one of the subdivisions of the *Nebenkern* (micronucleus). Bütschli's interpretation of the process of conjugation was equally happy. After observing that continued asexual division of certain forms resulted in decreased size and a general "lowering of the life energy," he concluded that the function of conjugation is to bring about a rejuvenescence (*Verjungung*) of the participants (see Turner, *infra*, Chapter XII).

In the same year Engelmann (1876) obtained very similar results. Quite independently of Bütschli he proved the error of Balbiani's view, and came to the conclusion not far different from Bütschli's: "The conjugation of the Infusoria," he said, "does not lead to reproduction through eggs, embryonic spheres, or any other kind of germ, but to a peculiar developmental process of the conjugating individuals which may be designated *Reorganization* (*loc. cit.*, p. 628).

THE USE OF CULTURES

Toward the end of the last century new methods of studying Protozoa gradually evolved; at first structures, or morphological considerations were predominant; then the uses of these structures led to a general treatment of physiology, and a vast literature on the functions of different parts of Protozoa grew up. All of this, in the main, was founded upon observations and little was done in the experimental field of protozoan research. Staining technique aided, and great strides were made in knowledge of the microchemistry of all classes of Protozoa. Variations in structures and in functional activities became apparent, and the conception of the life cycle, first definitely outlined by Schaudinn (1900) as a series of forms and activities consecutively produced and performed, was recognized as characteristic of every species (see Kofoid, *infra*, Chapter XI). The larger fields of study thus inaugurated have become the starting point for many lines of research, and the single individual has long since ceased to be the most important goal in study of Protozoa.

Any work on the life history, or portion of it, begins with observations on food and feeding habits of the protozoön in question. Experiments must first be undertaken to find a suitable culture medium upon which to grow the chosen form. The most universal of such media, in all probability, is the "hay infusion," in which hay in water, preferably after boiling, is allowed to stand for a certain time for bacteria to grow before the protozoön is placed in it. If this is a satisfactory medium, the organism will respond by dividing at a certain rate. After twenty-four hours the culture medium is replaced by fresh medium, made in the same way. In this manner the culture is inaugurated and carried on, and mass cultures are provided and sustained from the reserve individuals that are produced every day. In this manner material is produced for study of individual structures and for various activities characteristic of the different phases of the life cycle.

In many inaugural cultures made in this way, the outcome is different. No culture is started, or at best a very poor one. The reasons for this are manifold. Usually in such cases the medium turns out to be a suitable culture medium for some types of organisms, primarily bacteria, which are noxious to the protozoön under study. Hence a knowledge of bacteriological technique is valuable in determining the proper bacterial food to be used (see Kidder, *infra,* Chapter VIII).

A successful culture of a ciliated protozoön, for example, provides ample material for study of structures and functions; for encystment; or for the minutiæ of cell division, conjugation, sex phenomena, and the like. The appearance of derived structures of all kinds may be followed in sequence, from their origin in the fundamental organization as it appears in the recently encysted individual, to the active adult. Euplasmatic and alloplasmatic materials and their functional purposes in the cell may be determined, and the investigator proceeds with the confident expectation that plenty of material will be on hand for future study.

Very often the organism to be studied is carnivorous, and it is necessary to provide suitable food material, which must be cultivated for the purpose. Thus *Actinobolina radians* lives on *Halteria grandinella, Didinium nasutum* on *Paramecium,* and *Spathidium spathula* on *Colpidium,* and practically pure cultures of these food organisms must be kept on hand. For many purposes bacteria-free cultures must be prepared, and for this a knowledge of bacteriological technique is not only desirable but essential (see Hall, *infra,* Chapter IX; Kidder, *infra,* Chapter VIII).

FACTORS INFLUENCING LONGEVITY

The first factor in longevity is the set of functions which the isolated individual must perform in its daily life. It must react to stimuli, it must digest its food, it must excrete its waste matters (see Weatherby, *infra*, Chapter VII), and it must experience the processes involving katabolism. The euplasmatic structures having to do with these functions gradually lose their vitality, and, perhaps as a consequence of this, changes in activity are set up which lead to the second factor in lon-

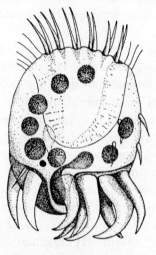

Figure 1. *Uronychia transfuga*, with giant cirri, membranelles for swimming, ten macronuclear segments, and single micronucleus. (After Calkins, 1933.)

gevity—reorganization through cell division. These changes have only within the last few decades been recognized and studied.

With every activity of the euplasmatic constituents of the protoplasmic make-up, the derived organization is changed. These changes may be studied day by day, and their significance in the life history of the organism under culture may be ascertained. Thus we learn that the macronucleus is not a portion of the fundamental organization, but is one of the derived organs of the ciliated Protozoa. These finer changes of the organization are difficult to note in the living animal, on any morphological basis, but physiologically it is possible to show that the organization is not equally responsive at all stages between divisions, and the implication is that protoplasmic changes must have taken place.

A merotomy experiment indicates this. A marine ciliate—*Uronychia transfuga* (Fig. 1)—is cut transversely through the center so that approximately half of the thirteen or fourteen beads which constitute the

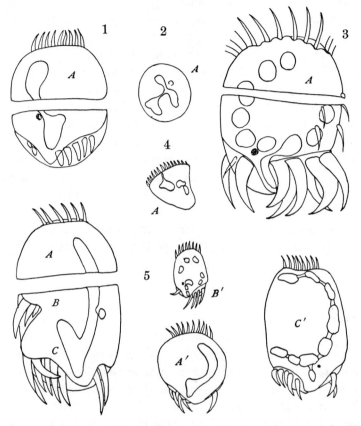

Figure 2. *Uronychia transfuga,* merotomy and regeneration. 1. Cell immediately after division, cut as indicated; 2, Fragment A of 1, three days after the operation, no regeneration; 3. cell cut five hours after division; 4, fragment A of 3, three days after operation, no regeneration; 5, cell cut at beginning of division as indicated into prospective fragment A, B, and C; A′, B′, C′, fragments A, B, and C twenty-four hours after the operation; fragment A regenerated into a normal but amicronucleate individual (A′); B, C, divided in the original division plane forming a normal individual C′ and a minute but normal individual B′. (After Calkins, 1933.)

macronucleus are left in each of the two halves, while the micronucleus is left undisturbed in one half. If the operation is made on an individual three to five hours, or less after the last previous division, the fragment

with the micronucleus regenerates perfectly, but the amicronucleate fragment, while it may live for a few days, never regenerates the missing
structures. The same result is obtained if the organism cut is from ten
to fifteen hours old. If, however, individuals older than this are cut in
the same manner, an increasing percentage of complete regenerations,
varying with increasing age, results (Fig. 2). At the time of the experiment the interdivisional period was twenty-six hours. If individuals
were cut after twenty-one hours of age, regeneration of the amicronucleate individuals was invariable. The experiment indicates a progressive differentiation in respect, at least to the power to regenerate and,
to that extent, a change in organization. Furthermore, if one individual
is similarly cut while the two daughter cells at division are still connected, or shortly afterwards, there is no regeneration of the amicronucleate fragment. This indicates that the condition which underlies the
power to regenerate is lost with processes of division and is not regained
until the young cell has undergone a considerable period of normal
metabolism (Calkins, 1911). This experiment was confirmed by Young
(1922) (see also Summers, *infra,* Chapter XVI).

CHANGES WITH METABOLISM

There is an accumulating amount of morphological evidence that a
derived structure, such as the macronucleus, is constantly changing with
continued metabolism. In *Uroleptus halseyi* reorganization after conjugation requires from four to five days for completion. At first there
is no chromatin in the young macronucleus, which stands out clearly
in the young organism, and attempts to stain it after fixation are futile.
By use of the Feulgen nucleal test, however, the chromatin reaction becomes increasingly intense and the chromatin granules more distinct
toward the end of the reorganization period; the nucleus disappears
from sight in the living organism, but now stains intensely with any
nuclear dye. The young macronucleus divides three times with the first
division of the ex-conjugant and each of the daughter cells receives
eight macronuclei, the chromatin staining deeply in all of them. This
chromatin is in the form of discrete granules of similar form and uniform size, but during the interdivisional period there appear a few
(three to five) larger granules, which are dissolved by the Feulgen

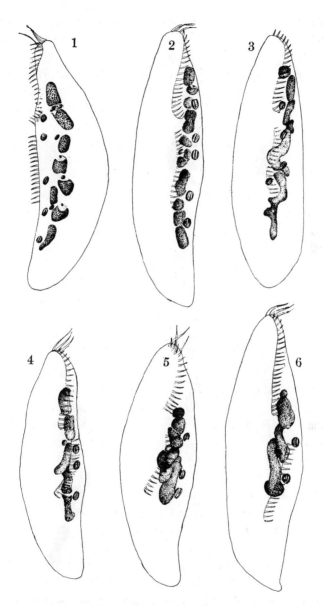

Figure 3A. *Uroleptus mobilis* Engelm. Stages in the fusion of the macronuclei prior to cell division. (After Calkins, 1933.)

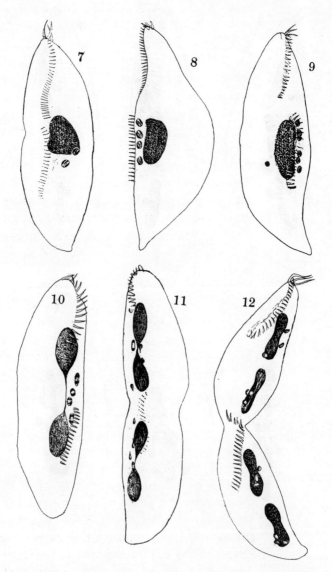

Figure 3B. *Uroleptus mobilis* Engelm. Further stages in preparation for division. (After Calkins 1933.)

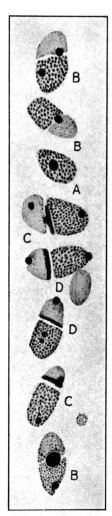

Figure 4. *Uroleptus halseyi* Calk. X-bodies, chromatin elimination, and nuclear cleft, in preparation for division of the macronucleus. (After Calkins, 1930.)

method, leaving vacuoles in the nuclear substance. Toward the beginning of the following period of division, these X-granules, as they were called, become larger and more numerous, and collect, forming a disc which extends across the nucleus nearer one end than the other. This disc stains with acid dyes and is entirely dissolved by the Feulgen hydrolysis. It forms the familiar *Kernspalt* of the ciliate macronucleus and its substance appears to *act as a catalyst,* for the smaller portion of the nucleus between the disc and the nearer end of the nucleus separates from the larger moiety (Figs. 3A, 3B; Fig. 4), and eight of these large fragments now unite in the cytoplasm to form a single division nucleus, while the eight smaller fragments, with their contained X-bodies and chromatin, are resorbed in the cell.

Here, then, is evidence of metabolic activity and change leading to the phenomena of cell division.

REORGANIZATION OF THE MACRONUCLEUS AND OTHER DERIVED STRUCTURES IN CILIATA

Analogous processes of chromatin elimination, which numerous observers have referred to as evidence of "nuclear purification," occur in different ways in other ciliates. Kidder (1933) described a core of modified chromatin accumulating in the center of the macronucleus of *Conchophthirus (Kidderia) mytili* (Fig. 5). This core condenses into a small deeply staining ball which, upon division of macronucleus, remains for a time in the connecting strand of the daughter nuclei, but ultimately disappears in the cytoplasm. A similar protrusion, referred to only incidentally by Rossolimo and

Jakimowitch (1929), occurs in *C. steenstrupii,* where it is in the form of a finely granular substance which comes from the macronucleus and remains for a time between the nuclear halves after division, but ultimately disappears in the cytoplasm.

There is reason to believe that this phenomenon has something

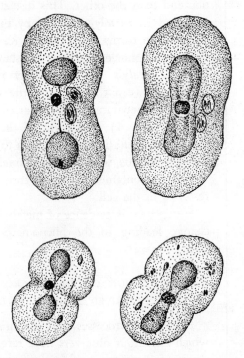

Figure 5. *Conchophthirus (Kidderia) mytili.* Extrusion of chromatin during division. (From a preparation made by Dr. G. W. Kidder.)

to do with restoring full metabolic powers of the macronuclear chromatin, possibly by the elimination of waste products of chromatin activity.

Another process of macronuclear reorganization during division which does not involve the visible elimination of nuclear substance, occurs in species of the families Aspidiscidae and Euplotidae of the Hypotrichida. At the approach of division in *Aspidisca* (Fig. 6), according to Summers (1935), the area of reorganization first appears as a wedge-shaped cleft at the approximate center of the convex side of the C-shaped macronucleus. A small granule in the center of the cleft represents the first of the reorganized nuclear material. As the cleft pushes across the diameter of the nucleus, the central reorganized chromatin increases in amount. Two separate "reorganization bands" result when the wedge-shaped cleft reaches the opposite side of the macronucleus. These bands then move in opposite directions, traversing the entire macronucleus and disappearing at the two ends. The increasing zone of staining chromatin granules, is quite different in appearance and in staining capacity from the chromatin in parts of the nucleus which have not been traversed

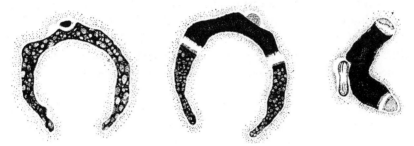

Figure 6. *Aspidisca lynceus.* Origin and further history of the reorganization bands. (After Summers, 1935).

by the bands. After disappearance of the bands at the ends, the nucleus condenses to form the typical division macronucleus of *Aspidisca.* Here again, therefore, there has been a physical change in the chromatin and a change that is brought about through activity of substances which form the nuclear cleft.

A similar process takes place in the genus *Euplotes,* the reorganization bands of which have been studied by numerous investigators. Griffin (1910) was the first to describe them fully as "reconstruction bands," but by later writers, beginning with Yocom (1918), they have been known as "reorganization bands." In *Euplotes* the bands begin one at each end of the macronucleus and proceed toward the middle, where they meet and disappear. There is no unanimity of opinion as to what takes place during this passage of the bands, but all agree that some re-organization of the chromatin occurs. Griffin (*loc. cit.*) expressed the opinion that all of the chromatin is disolved and later re-formed without the erstwhile impurities. Yocom (*loc. cit.*), on the other hand, holds that the chromatin does not go into solution in that part of the band known as the "solution plane," but appears there as closely packed granules and then passes to the part called the "reconstruction plane," not as precipitated granules, but as granules which have undergone some physical or possibly chemical change. Turner (1930), working on *Euplotes patella,* cites the absence of granules in the solution plane and regards the substance of this plane "as in the state of a colloidal solution." The chromatin reticulum in the center of the nucleus he interprets as in a continuous phase, while the karyolymph is dispersed. After action in the reorganization bands this condition is reversed, the chromatin granules being in the dispersed phase and the karyolymph continuous. Phenomena of an exactly similar nature have been observed in *Diophrys, Stylonychia,* and a host of other forms so that little doubt can be entertained as to the probability that this is a highly critical period in the daily life of a ciliated protozoan.

At the time of division, not only in the macronucleus but throughout the organism there is a wave of general house cleaning. Morphological parts which have been active in the metabolic reactions receive attention and all are built up afresh. This principle was recognized and applied to all parts of the cell by Wallengren in 1900 and again in 1901 when he wrote his illuminating paper in German.

In the Hypotrichida in particular, in which there is no covering of cilia but instead motile organs called cirri, of a more complex type, Wallengren was the first to show, in great detail, that the old cirri, just prior to cell division, are gradually resorbed, while new ones arising from the cortex at an adjacent spot, grow out slowly to take their places. In like manner cilia are also replaced, and thus it comes to pass that, inside and outside, the active organs of a ciliate are composed of new materials derived from the fundamental organization contained in every protozoön.

It is not only the ciliates that possess this apparent fountain of eternal youth; other groups of Protozoa manifest similar, if not identical phenomena.

With few exceptions, cell division in flagellates is longitudinal, beginning as a rule at the anterior or flagellar end, the cleavage plane passing down through the middle of the body. As the process continues, the two daughter cells separate and usually come to lie in one plane, so that final division appears to be transverse.

As there are few details in the structure of a simple flagellate on which to focus attention, descriptions of division processes are practically limited to the history of the nucleus, kinetic elements, and the more conspicuous plastids. Here, in the main, are fairly prominent granules of different kinds, which divide as granules, and, save for the chromatin elements of the nucleus, without obvious mechanisms (see MacLennan, *infra,* Chapter III).

In the simpler cases there is little evidence that can be interpreted to indicate reorganization at the time of division, and that little is confined to the motile organs. In the more complex forms, however, there is marked evidence of deep-seated changes going on within the cell.

The earlier accounts of cell division in the simpler flagellates described an equal division of all parts of the body, including longitudinal division of the flagellum, if there were but one, or equal distribution if there were more. One by one such accounts have been checked by use of modern methods, until today there remains very little substantial evidence of the division of the flagellum. The basal body and the blepharoplast usually divide, but the flagellum either passes unchanged to one of the daughter cells, as in *Crithidia, Trypanosoma,* and others, or is resorbed into the cell. In some doubtful cases it may be thrown off. If the old flagellum is retained in uniflagellate forms, the second flagellum develops by outgrowth of the basal body or the blepharoplast. If the old flagellum is resorbed, both halves of the divided kinetic body give rise to flagella by outgrowths. Similarly if there are two or more flagella, one or more may be retained by each daughter cell, while the others, making up the full number, are regenerated.

Reorganization is indicated, to some extent, by these cases in which the old flagellum is resorbed. It is still better indicated by a number of flagellates in which the cytoplasmic kinetic elements, as well as the flagella, are all resorbed and replaced by new combinations in each of the daughter cells. Thus in *Spongomonas splendida,* according to Hartmann and Chagas (1910), the old blepharoplasts and two flagella are resorbed and new ones are derived from centrioles of the nuclear division figure. The phenomenon cannot be regarded as typical of the simple flagellates, for in the great majority the kinetic elements are self-perpetuating, even the axostyles, according to Kofoid and Swezy (1915), dividing in *Trichomonas.* This, however, requires confirmation.

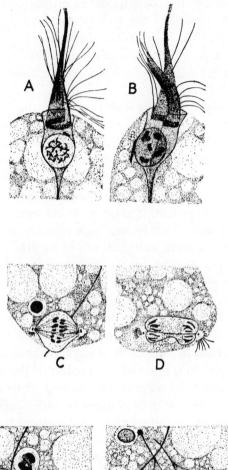

Figure 7. *Lophomonas blattarum* Janicki. Division of the nucleus and reorganization. (After Bělǎr, 1926.)

An extreme case of reorganization in flagellates is apparent in the two species of *Lophomonas* (*L. blattae* and *L. striata*) first described by Janicki (1915). Here the parental calyx, basal bodies, blepharoplasts, and rhizoplasts all degenerate and are resorbed during division (Fig. 7). At the beginning of division, a cytoplasmic centriole divides with a connecting fibril, which is retained throughout as a paradesmose. The nucleus emerges from the calyx in which it normally lies and moves with the spindle to the posterior end of the cell. The spindle takes a position at right angles to the long axis of the cell. Chromosomes,

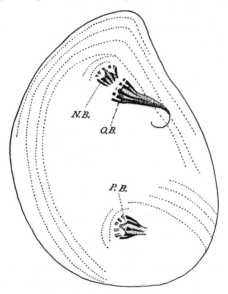

Figure 8. *Chilodonella uncinatus*. New pharyngeal basket and mouth, replacing old ones. (After MacDougall, 1925.)

probably eight in number, are formed and divided, and two daughter nuclei result, each of which is enclosed in a new calyx, while new basal bodies and new blepharoplasts apparently arise from the polar centrioles (Fig. 7).

This phenomenon in *Lophomonas* is strikingly similar to the reorganization processes occurring in one of the Chlamydodontidae—*Chilodonella*. Here, according to the observations of Enriques (1908), Nägler (1911), and MacDougall (1925), the old mouth of the cell and the oral basket of trichites are discarded and disappear in the cell, while two new oral aggregates are provided for the daughter individuals (Fig. 8).

In the majority of Sporozoa and Sarcodina simple division is not the usual mode of reproduction. Where it does occur, as in Amoebida, characteristic derived structures are absent and, if reorganization takes place, it is confined to the cytoplasm, where we have little evidence of change. A modified type of division, called "budding division," is widely spread in testacea. Here, water is absorbed by the old protoplasm, followed by lively cyclosis and the protrusion of a protoplasmic bud from the shell mouth. This bud grows, assumes the form of the parent organism, and secretes its own chitinous membrane on which foreign bodies (Arcellidae), or plates manufactured by the parent protoplasm (Euglyphidae), are cemented. Apart from withdrawal of old pseudopodia, there is little evidence of reorganization. In Heliozoa, however, pseudopodia, with their axial fiilaments, are drawn in and new ones are formed by the daughter cells.

In Radiolaria, Foraminifera, and Mycetozoa, indeed in the majority of the Sarcodina and in most Sporozoa binary fission is replaced by multiple division. The nuclei divide repeatedly, and a portion of the cytoplasm is finally parceled out to each of the nuclei. The minute cells thus formed leave the parent organism usually as swarm spores. Metabolic products, waste matters, and certain structures of the derived organization are left behind, and shells and skeletons alone mark the previous existence of living cells. There is often a small amount of protoplasm retained in these alloplasmatic products and it is not altogether fantastic to see in these remains what Weismann (1880-82) denied as occurring in Protozoa—viz., a corpse.

It might seem that these methods of restoring protoplasm to its full vitality by processes of division are adequate to account for indefinite longevity of Protozoa. This, however, is not the case, with the possible exception of the animal flagellates, in which division in some form is the only means of reproduction known. In ciliates the ability to divide, if other methods of reorganization are prevented, gradually weakens; the interdivisional periods are gradually lengthened, and the degeneration of the derived structures finally results. The micronucleus may hypertrophy (Calkins, 1904) or disappear (Maupas, 1888), and the motile organs may be lost (Maupas, *ibid.*). In *Uroleptus mobilis* the chromatin of the macronucleus ultimately disappears and only a few X-granules remain. The protoplasm apparently dies from "old age" (Fig. 9).

The effects of continued division in ciliates, if other means of reorganization are excluded, are clearly shown by the method of isolation

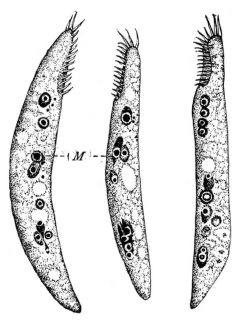

Figure 9. *Uroleptus mobilis* Engelm. Old-age specimens showing the degeneration of the macronucleus (M) and loss of micronuclei. (After Calkins, 1919.)

cultures. A single individual, preferably an ex-conjugant, is isolated in a suitable medium in a culture dish such as a Columbia isolation culture dish. The next day such an individual, for example a *U. mobilis,* is represented by four or eight daughter individuals, the number depending upon two or three divisions of the original one isolated. In our practice five of these are isolated in separate dishes, and five independent lines are started, all representing protoplasm which had been a part of the protoplasm of the original individual isolated. A single individual is isolated daily from each of these five lines and a daily record of generations is kept for all lines. The total number of divisions is recorded for all lines and these are averaged for ten-day periods. The division rate is taken as a measure of vitality and the history of the protoplasm is shown graphically by plotting the ten-day averages on a graph in which the ordinates represent the number of divisions and the abscissas the consecutive ten-day periods. Such a graph for *U. mobilis* is shown in Figure 10, which is a composite graph of 23 different series,

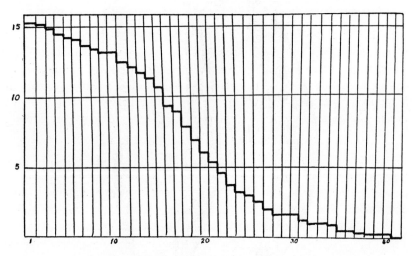

Figure 10. *Uroleptus mobilis* Engelm. Graph representing the life history by ten-day periods. (After Calkins, 1933.)

averaged for successive ten-day periods in isolation culture. It will be seen that there is a high initial vitality which gradually wanes through 360 divisions during 300 days. In these isolation cultures the individuals do not encyst nor conjugate; hence division is the only means of re-organization.

That division is not effective in checking waning vitality is shown by the fact that, in my experience, all such cultures of *Uroleptus* finally die, in some cases in from fourteen to sixteen months, in other cases in from three to ten months.

The macronucleus is probably the most important of all derived organs of *Uroleptus*. With each division of the cell, as shown above, it undergoes a process of reorganization whereby the chromatin is restored to a virile condition. This may be repeated upwards of 300 times, until reorganization, if it occurs, is ineffective and the protoplasm dies. Thus the macronucleus, like all other derived structures of the cell which come from the euplasmatic substances, apparently has a limited potential of activity. But the macronucleus may overcome this difficulty through the more deeply reaching phenomena in the life history, viz., endomixis and conjugation.

REORGANIZATION BY ENDOMIXIS AND BY CONJUGATION

In isolation cultures of any ciliate, if the extra individuals which remain over after the daily isolation is made on any day are put into a larger container with abundant food, a so-called "encystment test" or "conjugation test" is started. In my experiments these are begun regularly every ten days. Here the individuals multiply by division until there are thousands in the container. In the early stages of the life cycle of *U. mobilis* all such individuals die of starvation, but in a month or six weeks after the initial conjugation from which the series is started, such tests result in an increasing number of encystments. A type of breakdown which is not seen in the division phenomena is now manifested. The macronucleus is fragmented, and the fragments are distributed in the protoplasm where they ultimately disappear, while a new macronucleus is formed from a product of micronuclear division. Other structures of the derived organization are resorbed, waste matters and water are voided to the outside, and a cyst membrane is formed within which the organism may remain in a partially desiccated condition for months. When it is recovered from the cyst and reëstablished in isolation culture, it has an optimum vitality and passes through a complete life cycle exactly like that of an ex-conjugant. This phenomenon of endomixis, except for encystment, is the equivalent of endomixis in *Paramecium aurelia* as originally described by Woodruff and Erdmann (1914). Endomixis thus brings about a more far-reaching and more complete reorganization than does division; the new macronucleus arising from a micronucleus, is provided with a new potential of vitality. For an up-to-date account of endomixis, see Woodruff, *infra,* Chapter XIII.

An interesting phenomenon which I interpreted as analogous to endomixis occurs in the ciliate *Glaucoma (Dallasia) frontata* (Fig. 11). At an early stage in its life history, a stage corresponding to the period of encystment in *Uroleptus,* this organism loses its normal form which is assumed under certain feeding conditions (see Kidder, *infra,* Chapter VIII), becomes elongate and cylindrical, and darts about in the medium with unusual speed (Fig. 12). Soon its mouth and oral mem-

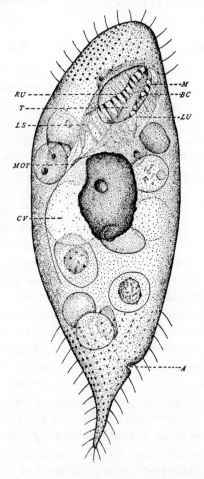

Figure 11. *Glaucoma (Dallasia) frontata* Stokes. Vegetative individual. *A,* anus; *BC* buccal system; *CV,* contractile vacuole; *LS* ladder system; *LU* left undulating membrane; *M* mouth of buccal cavity; *MOT,* region of motorium; *RU* right undulating membrane; *T,* "tongue" in buccal cavity. (After Calkins and Bowling, 1929.)

branes are resorbed and it divides into two cylindrical individuals. These divide into four, the four into eight, and the eight into sixteen, all without intervening growth. The last-formed individuals have no resemblance to the original parent form, but each has a macronucleus and a micronucleus which has undergone maturation divisions and is haploid as to chromosomes. The final sixteen are associated in pairs, which become surrounded by a cyst membrane within which they fuse (Calkins and Bowling, 1929). For other variations in the phenomena of endomixis, see Diller, 1936; and especially Woodruff, *infra*, Chapter XIII.

This unusual phenomenon in *Dallasia*, which was not encountered by Kidder in his studies on *Glaucoma*, finds its closest parallel in a group

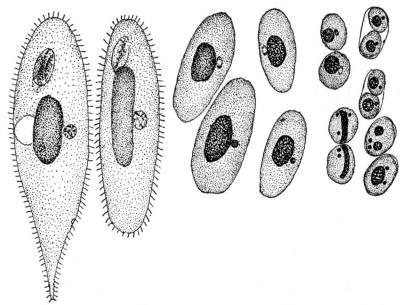

Figure 12. *Glaucoma (Dallasia) frontata* Stokes. History of normal vegetative individual with successive divisions resulting in copulating gametes and encysted zygotes. (After Calkins and Bowling, 1929.)

of ciliates in each of which the life history is unique. This group includes the ectoparasite *Ichthyophthirius multifiliis,* according to Buschkiel (1911) and Nerescheimer (1908), and *Opalina ranarum,* according to Metcalf (1909) and Nerescheimer (1907). With these forms we have good evidence, together with Dogiel's "transformation of a male pronucleus into a spermatozoön" in the Ophryoscolecidae, that the intracellular micronuclei forming pronuclei are a reminiscence of a brood of gametes (Dogiel, 1923; see also Turner, *infra*, Chapter XII).

The longevity of a ciliate's protoplasm in the past is shown by the fact that it is before us today and it has a possibility of indefinite longevity in the future. We know very little about the secrets of protoplasmic organization which underlie continued life, but we can analyse some, at least, of the conditions under which it maintains its animation. A proper environment and adequate food are essential factors. These lead to growth and to reproduction by division. We have seen that these latter bring about a reorganization and a return to full metabolic ac-

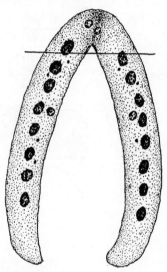

Figure 13. *Uroleptus mobilis* Engelm. Conjugation and merotomy. (After Calkins, 1921.)

tivity. The periodic restoration of these processes might well be enough to ensure protoplasmic longevity, as seems to be the case with the animal flagellates. At endomixis the ciliate macronucleus follows the fate of other parts of the derived organization; a new one is formed from the fundamental organization, and the result is, again, increased vitality of the protoplasm. This phenomenon, recurring every thirty days, ensured the longevity of *Paramecium aurelia* in Woodruff's hands for many years and through thousands of generations by division. As Woodruff states, it may be adequate, indeed, to ensure indefinite life in the future, as division alone is apparently adequate for animal flagellates.

In a conjugation test made with *Uroleptus,* the results showed that if the series is sixty or more days old, the individuals multiply by division until there may be thousands in the container. If conditions are appropriate, this massing may be followed by an epidemic of conjugations (see Sonneborn, *infra,* Chapter XIV). In this process two individuals fuse at the anterior ends and remain united for approximately twenty-four hours (Fig. 13). They then separate and each of the ex-conjugants begins a process of reorganization which lasts for four or five days. If such an ex-conjugant is isolated and fed, it will give rise to a new series and its progeny will pass through all the stages of vitality that the parent series passed through. At the outset its vitality will be approximately the same as the original vitality of the parent series, but it will be greater than that of the parent series at the time of conjugation. It passes through the same history of waning vitality, and its protoplasm finally dies, although this may be months after the death of the parent protoplasm.

These experiments with *Uroleptus,* continued for more than ten years and with 146 different series, always gave the same result. The length of life of each series varied from three months to over a year and all were descendants of a single bit of protoplasm making up the body of one *Uroleptus* cell.

There is little reason to doubt that some change in the protoplasmic make-up occurs with continued metabolic activity through many successive generations by division. In isolation cultures it is manifested by an increasing time interval and by a change in the physical properties of the protoplasm whereby fusion of cells, total or partial, is possible. A type of breakdown, not manifested early in the series, is set up. The cortex is liquified, in most cases in the region of the peristome, and two individuals fuse in conjugation. In some cases the entire cortex is thus modified, and individuals will fuse at any point on the periphery. I have often referred to one case where no less than nine *Paramecium caudatum* were thus fused into one amorphous mass.

The stimulus of normal fusion results in activities not manifested before. The micronuclei divide as they do in endomixis, and their division involves reduction in number of chromosomes and in the formation of pronuclei which meet and fuse. These copulating micronuclei, as pronuclei, are usually interpreted as a reminiscence of an ancestral gamete brood which is realized in *Dallasia* (*Glaucoma*) *frontata, Opalina,* and in gregarines. In the latter, as is well known, each individual of a syzygy forms a brood of gametes which copulate with similar gametes from the other individual of the pair (see Turner, *infra*, Chapter XII; and Kofoid, *infra*, Chapter XI).

Coming at the end of a vegetative period during which single or multiple divisions may have occurred, such gamete broods indicate a change in organization which, if copulation does not occur, results in death of the gametes. In other words gametes are so specialized that they require a complemental combination to ensure normal metabolism and life. In some gametes differentiation has been in the direction of greater constructive activity, resulting in food-stored macrogametes equivalent to egg cells; in others in the direction of greater katabolic activity, resulting in a brood of microgametes equivalent to spermatozoa, as in Coccidiomorpha.

The happenings at conjugation need only be recapitulated, for the phenomena are so well known that description here is unnecessary (see Turner, *infra*, Chapter XII, for detailed accounts). In *Uroleptus*, shortly after fusion of the anterior ends, the micronuclei of both individuals begin a series of maturation divisions, and the third division gives rise to the gametic nuclei. One of a pair of these remains *in situ*, while the other, or migrating pronucleus, passes through the protoplasmic bridge, unites and forms a fertilization nucleus by fusing with the quiescent pronucleus of the other conjugant. Thus a mutual fertilization occurs, the two migrating pronuclei in *Uroleptus* passing each other at the apex of the united pair of cells (Fig. 13). After this is accomplished, the two individuals separate, the amphinucleus of each conjugant divides twice, and one of the products forms a new macronucleus, which, after four or five days, is ready to divide. Another product forms a new micronucleus. In the meantime the old macronucleus begins to degenerate and is ultimately resorbed in the cytoplasm. The other structures of the derived organization of the old organism are resorbed and replaced by new ones. The young organism, with this new set up, starts a new life cycle with an optimum of vitality (Fig. 14).

Through these activities of conjugation, an old protoplasm with low vitality is made over out of its own contained substances, into a new protoplasm with high vitality. What explanation of this remarkable phenomenon can be given? The only apparent difference between the happenings at division and at conjugation are: (1) fertilization, and (2) loss and replacement of the old macronucleus, and so forth. Renewed vitality may be due to one or to both of these phenomena. Conjugation differs from endomixis mainly in gelatinization of the cortex and in

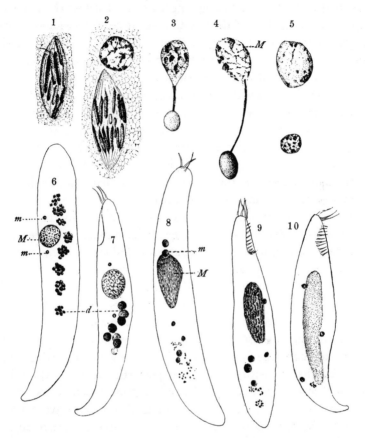

Figure 14. *Uroleptus mobilis* Engelm. Formation of new macronucleus after conjugation. 1. First metagamic or zygotic division of the amphinucleus; 2. One of the progeny of this division dividing again; 3, 4, 5, telophase stages of this divison resulting in a new macronucleus (above) and a degenerating nucleus (below); 6 to 10, stages in the differentiation of the new macronucleus and disintegration and resorption of the old macronucleus. In 10, two new micronuclei are in mitosis preparatory to the first division of the ex-conjugant. (After Calkins, 1919.)

amphimixis. Fertilization, with its sequelæ of hereditary possibilities, is a highly important result (see Jennings, *infra,* Chapter XV; and Sonneborn, *infra,* Chapter XIV). It is also assumed to be the *raison d'être* of all sorts of subsequent peculiarities, but for the phenomenon of increased vitality under consideration, a very simple experiment with *Uroleptus mobilis* shows that fertilization with amphimixis has little to do with the problem. A conjugating pair is cut across the apex with a scalpel. One individual of the two thus separated is fixed and stained, to show the stage of conjugation when cut, while the other one is cultivated in an isolation culture dish. The experiment is particularly striking when the two wandering pronuclei are cut off with the apical piece (Fig. 13). The cultivated individual goes through exactly the same processes as though conjugation had been completed in the usual manner, and a normal life cycle results. Amphimixis, however, had been prevented; conjugation apparently had been transformed into endomixis (Calkins, 1921). Here, however, the possibility remains that one of the supernumerary micronuclei of the amputated individual may divide and one part may unite with the pronucleus already present.

The loss and replacement of the old macronucleus, together with that of the old derived structures generally remains as a possible explanation. This phenomenon is common to endomixis, conjugation, and conjugation merotomy, and in all cases a renewed vitality results. This does not happen with division, although, as shown above, characteristic changes occur in the macronucleus. With its disintegration and resorption in the cytoplasm, a new supply of nucleoproteins and other chemical compounds useful in metabolic activities are distributed in the cell, and these may be potent factors in the new vitality. Added to these is the fact that an entirely new and powerful organ of the cell—the macronucleus—has been supplied from the fundamental organization.

The majority of the derived structures of the cell have a relatively short life, being resorbed and renewed at division (cirri and other motile organs; others, such as membranelles and oral structures, apparently have a longer life. Of all derived structures the macronucleus has the longest life; it divides at cell division and may remain functional through entire life cycles, including many hundreds of divisions. It is probably the chief metabolic agent of the cell, yet it apparently lacks the power of continued life which the micronucleus possesses. It is essentially somatic in character, and, like the soma in metazoa, it ultimately wears out; if not replaced, the rest of the cell, including the micronucleus depending on it, dies with it. It is quite possible, indeed

I regard it as probable, that the gradual deterioration of this important organ of the ciliate cell, is the underlying cause of waning vitality and ultimate death of protoplasm in isolation cultures (Fig. 10).

With parasitic forms there is probably the same underlying variation in vitality, but there is not the same possibility of measuring it, at least not in any direct way.

Remarkable as these phenomena are, they leave us cold so far as the matter of protoplasmic vitality is concerned. The interpretations presented here are, after all, essentially mechanistic, and even with the ciliates the phenomena described are by no means universal, while in some groups of Protozoa they are not shown at all, or else only in a vague and indefinite manner. In ciliates there is one organoid of the cell, the micronucleus, which transcends all other structures of the cell, and, although it is apparently not functional except in heredity and activities connected therewith, such as regeneration, and so forth, it does appear to be the most essential morphological element of the fundamental organization. Its agent in metabolism is the macronucleus, which is derived from it. For the secret of life and longevity in ciliates we must turn to this inconspicuous and often overlooked structure of the cell and of the cyst.

In other groups of the Protozoa, the homologue of this important element of the cell lies in the usually single nucleus. Furthermore, in the micronucleus it is probably the chromatin content that gives it its power; and, in other groups than the ciliates, it is the chromatin content that makes the nucleus what it is. The value and importance of chromatin is seen by the meticulous care with which it is distributed to daughter cells and to progeny generally, while the maturation divisions bespeak its significance in heredity. The secret of life and vitality must thus be sought not in the daily activities of living things, but in that enigmatical substance—chromatin—which is about us in all living things, including ourselves. That secret may never be disclosed.

Literature Cited

Balbiani, G. 1859. Du rôle des organes généateurs dans la division spont. des infusoires cilies. J. Physiol. Pathigén., 111: 71-87.

—— 1861. Recherches sur les phénomènes sexuelles des infusoires. J. Physiol. Pathigén.,   : 102-30, 194-220, 431-48, 465-520.

Barry, M. 1843. On fissiparous generation. Edinb. New phil. Jour., 35: 205-20.

Bělăr, K. 1926. Der Formwechsel der Protistenkerne. Ergebn. Zool., 6: 235-654.

Buffon, G. L. de. 1749. Histoire naturelle génèrale et particuliere.

Buschkiel, A. L. 1911. *Ichthyophthirius multifiliis*. Arch. Protistenk., 21: 61-102.

Bütschli, O. 1873. Einiges über Infusorien. Arch. mikr. Anat., 9: 657-78.

—— 1876. Studien über die ersten Entwicklungsvorgange der Eizelle, die Zelltheilung, und die Konjugation der Infusorien. Abh. senckenb. naturf. Ges., 10: 213-462.

—— 1887-89. Bronn's Klassen, und Ordnungen des Thier reichs. Erster Band. Protozoa. Leipzig. Pp. 1-2028.

Calkins, G. N. 1901. The Protozoa. New York, 1901.

—— 1904. Studies on the life history of Protozoa. IV. Death of the A-series. J. exp. Zool., 1: 423-59.

—— 1911. Regeneration and cell division in *Uronychia*. J. exp. Zool., 10: 95-116.

—— 1919. *Uroleptus mobilis* Engelm. I. History of the nuclei during division and conjugation. J. exp. Zool., 27: 293-357.

—— 1921. *Uroleptus mobilis* Engelm. Effect of cutting during conjugation. J. exp. Zool., 34: 449-70.

—— 1930. *Uroleptus halseyi*. II. The origin and fate of the macronuclear chromatin. Arch. Protistenk., 69: 151-74.

—— 1933. The Biology of the Protozoa. 2d ed., Philadelphia.

—— 1934. Factors controlling longevity in protozoan protoplasm. Biol. Bull. Woods Hole, 67: 410-41.

Calkins, G. N., and R. Bowling. 1929. Studies on *Dallasia frontata* Stokes. II. Cytology, gametogamy and conjugation. Arch. Protistenk., 66: 11-32.

Diller, Wm. F. 1936. Nuclear reorganization processes in *Paramecium aurelia*, with descriptions of autogamy and "hemixis." J. Morph., 59: 11-51.

Dogiel, V. 1923. The transformation of a male pronucleus into a spermatozoön. Zool. Lab. Univ. Petrograd, 1923.

Dujardin, F. 1835. Observations sur les rhizopodes et les infusoires. C. R. Acad. Sci. Paris, Nov. 1835, pp. 338-40.

—— 1841. Histoire naturelle des zoöphytes infusoires. Paris.

Ehrenberg, C. G. 1938-39. Die Infusionsthierchen als vollkommene Organismen. Leipzig.

Engelmann, Th. W. 1876. Über Entwicklung u. Fortpflanzung der Infusorien. Morph. Jb., 1: 573-635.

Enriques, P. 1908. Die Conjugation und sexuelle Differenzierung der Infusorien. Arch. Protistenk., 12: 213-74.

Griffin, L. E. 1910. *Euplotes worcesteri*, sp. nov. II. Division. Philipp. J. Sci., 5: 291-312.

Haas, G. 1933. Beiträge zur Kenntnis der Cytologie von *Ichthyophthirius multifiliis*. Arch. Protistenk., 81: 88-137.

Hartmann M., and C. Chagas. 1910. Flagellaten Studien. Mem. Inst. Osw. Cruz, 2, fasc. I.

Hill, John. 1752. History of Animals. 3d vol. of *a gen. Nat. Hist.*, 1748-52.

Hooke, Robt. 1665. The Micrographia. London.

Janicki, C. 1915. Parasiten Flagellaten, Teil 2. Z. wiss. Zool., 112: 573-691.

Kidder, G. W. 1933. Studies on *Concophthirius mytili* de Morgan. I. Morphology and division. Arch. Protistenk., 79: 1-24.

Kofoid, C. A., and O. Swezy. 1915. Mitosis in *Trichomonas*. Proc. nat. Acad. Sci., Wash., 1: 315-21.

Kolkwitz, R., and M. Marsson. 1909. Oekologie der tierischen Saprobien. Int. Rev. Hydrobiol., 2:

Lackey, J. B. 1925. Studies on the biology of sewage disposal. The fauna of Imhof tanks. Bull. N. J. agric., Exp. Stas., No. 417.

Lauterborn, R. 1901. Die "Sapropelische" Lebewelt. Zool. Anz., 24: 50-55.

Ledermüller, M. F. 1760-63. Mikroskopische Gemüths. u. Augenergörtzungen. Nurnberg, 1760: 88.

Leewenhoek, A. van. 1676. Observations concerning little animals by him observed in rain, well, sea, and snow water wherein pepper had lain infused. Philos. Trans., 12: 821-31.

MacDougall, M. S. 1925. Cytological observations on gymnostomatous ciliates, etc. Quart. J. micr. Sci., 69: 361-84.

Maupas, E. 1888-89. Le Rejeunissement karyogamique chez les ciliés. Arch. Zool. Exp. gén., (2) 7, 8.

Metcalf, M. M. 1909 *Opalina*. Its anatomy, etc. Arch. Protistenk., 13: 195-375.

Meyen, J. 1849. Einige Bemerk. über den Verdauungsapparat der Infusorien. Arch. Aanat. Physiol. Lp. z., 1839: 74-79.

Moody, J. 1912. Observations on the life history of two rare ciliates—*Spathidium spathula* and *Actinobolus radians*. J. Morph., 23: 349-99.

Müller, O. F. 1773. Vermium terrestr. et fluviatil. s. animal. infusor. etc. Hafniae u. Lipsiae, 1773.

——— 1786. Animalc. infusoria, fluviat. et marina, etc. (posth.) Hafniae et Lipsiae.

Nägler, K. 1911. Caryosom u. Centriol Beimteilungsvorgang von *Chilodon uncinatus*. Arch. Protistenk., 24: 142-48.

Needham, T. 1750. A summary of some late observations upon the generation composition and decomposition of animal and vegetable substances. Philos. Trans. n.s., 45: 615-66.

Nerescheimer, E. R. 1907. Der Zeugungkreis von *Opalina*. S.B. Ges. Morph. Physiol. München, 22:

——— 1907. Die Fortpflanzung der Opalinen. Arch. Protistenk., Festschr. Hertwig.

Rossolimo, L. I., and Frau K. Jakimowitsch. 1929. Die Kernteilung bei *Concophthirius steenstrupii* St. Zool. Anz., 84: 323-33.

Sandon, H. 1927. The composition and distribution of the protozoan fauna of the soil. Edinburgh.

Schaudinn, F. 1900. Untersuchungen über Generationswechsel bei Coccidien. Zool. Jb., Anat., *Abt.* 1, 13: 177-292.

Siebold, Th. von. 1845. Lehrb. der Verg. Anatomie d. Wirbellosen Thiere. Hf. 1, 1845.

Stein, F. von. 1859. Der Organismus der Infusionsthiere, Abth. I, Hypotrichida. Leipzig.

—— 1867. Der Organismus der Infusionsthiere, Abth. II, Heterotrichida. Leipzig.

Summers, F. M. 1935. The division and reorganization of the macronuclei of *Aspidisca lynceus* Müller, *Diophrys appendiculata* St. and *Stylonychia pustulata* Ehr. Arch. Protistenk., 85: 173-208.

Turner, J. P. 1930. Division and conjugation in *Euplotes patella* Ehr. Univ. Cal. Publ. Zool., 33: 193-258.

Wallengren, H. 1901. Zur Kenntnis der vergl. Morphol. der hypotrichen Infusorien. Zool. Jb., Anat. Abt. 1, 15: 1-58.

Weismann, August. 1880-83. Essays on life and death and heredity. London.

Woodruff, L. L. 1905. An experimental study on the life history of hypotrichus infusoria. J. exp. Zool., 2: 585-632.

—— 1939. Some pioneers in microscopy, with special reference to protozoölogy. J. Trans. N. Y. Acad. Sci., Ser. 2, 1: 1-4.

Woodruff, L. L., and R. Erdmann. 1914. A normal periodic reorganization process without cell fusion in *Paramecium*. J. exp. Zool., 17: 425.

Yocom, H. B. 1918. The neuromotor apparatus of *Euplotes patella*. Univ. Cal. Publ. Zool., 18:337-96.

Young, D. B. 1922. A contribution to the morphology and physiology of the genus *Uronychia*. J. exp. Zool., 36: 353-90.

Young, Dixie. 1939. Macronuclear reorganization in *Blepharisma undulans*. J. Morph., 64: 297-347.

CHAPTER II

SOME PHYSICAL PROPERTIES OF THE PROTOPLASM OF THE PROTOZOA

H. W. Beams and R. L. King

Introduction

Dujardin in 1835, a little over a century ago, was the first to carefully describe the physical properties of protozoan protoplasm which he termed "sarcode," although other earlier observers had seen and drawn living amoebae. For instance, Rösel von Rosenhof drew an amoeba in 1755, O. F. Müller described living amoeba in 1773 and Ehrenberg, a pioneer protozoölogist, undoubtedly observed living protozoan protoplasm. However, none of these observers emphasized the protoplasm as the living substance as did Dujardin. In addition, Dujardin's description of protoplasm was so accurate that his definition of it as a "living jelly, glutinous and transparent, insoluble in water, and capable of contracting into globular masses and of adhering to dissecting needles so that it can be drawn out like mucus" can be little improved upon today (see Faurè-Fremiet, 1935).

Perhaps no group of animals has served as the basis for so many and so extensive studies on the structure of protoplasm as the Protozoa. This is, no doubt, in part due to the fact that many of the earlier workers labored under the assumption that the "simplest type" of protoplasm should be looked for in the "lower forms" of animals. In addition many of the Protozoa are comparatively large and discrete cells, thus offering little mechanical difficulties to direct microscopic observations upon their living structure.

In a modern discussion of the physical properties of protoplasm, one must bear in mind the fact that the manifold chemical and physiological properties of living matter are intimately connected with the structure of protoplasm. That is, one must conceive of a system which is kept in existence, in spite of the fact that the katabolic phases of energy exchange tend to destroy its integrity; this involves a continuous automatic replacement of worn-out parts or the growth of new ones. Chemical processes must often be definitely localized in a small region within the framework of a single cell. However, visible differentiation usually does not take place within a single cell in the Metazoa to such an extent as in the Protozoa, where we have a whole series of special differentiated organelles for performing particular functions.

A protozoan is usually regarded as both an organism and a cell; in the Metazoa protoplasmic differentiation tends to be irreversible, and in Protozoa to be reversible, so that the most highly differentiated cells ever observed are found as individual Protozoa. This extreme differentiation is, of course, surprising to one accustomed to the usual structural simplicity of metazoan cells, and is probably the main reason for the position taken by Dobell (1911) and others who deny the cellular nature of the Protozoa. Several degrees of permanency of differentiation may be distinguished among the Protozoa, although of course the groups are not mutually exclusive. Thus we may distinguish: (1) temporary, completely reversible structures or differentiations of the protoplasm, such as pseudopodia and spindle elements; (2) differentiations usually irreversible and lasting throughout the life of the organism, which may be the seat of active chemical energy changes, such as cilia, flagella, and myonemes, or, with little chemical changes, such as morphonemes and pellicle; and (3) differentiations formed by protoplasm which are chemically distinct from it, such as shells and secretions.

Every time a protozoan divides, conjugates, or encysts there is a tendency toward dedifferentiation, followed by reorganization, so that the individual emerging from these processes often has developed a new set of structures derived from the fundamental structure. These changes must be accompanied or caused by physical realignments of protoplasmic elements, of which little is known. Perhaps in no other phylum of animals can such a wide range of type of mitosis and cytokinesis be found as exists in the Protozoa. The morphological details of the division process in these forms are fairly well known, and evidence is continually accumulating which shows that mitotic and cytokinetic mechanism does not differ fundamentally from that of higher forms.

In the Protozoa, as in other types of cells, little is actually known regarding the initial stimulus that starts the organism to divide, to undergo certain sexual phenomena, or to form cysts. However, environmental conditions, such as amount and kind of food present, temperature, and so forth, have been generally observed to affect the rate at which these phenomena take place. Since Protozoa under normal conditions seem to be limited to a more or less uniform maximum size for a given species, various suggestions have been made to the effect that the stimulus for division is in part controlled by definite ratio of volume to body surface, or to a possible ratio of nuclear volume to cytoplasmic volume. However, whatever may prove to be the explanation of this, it is an observed fact that in most cases conditions suitable for most rapid growth are the same as those best suited for most rapid division. On the other hand, relatively unfavorable conditions, such as scarcity of food or the concentration of metabolites, have been observed to induce sexual activity in the various Protozoa (see Giese, 1935, for literature). Practically nothing can be said of the physical changes undergone by the protoplasm of the organisms just before they start division, conjugation, or endomixis, except that before division they may shorten and thicken (*Uroleptus,* Calkins, 1919), or become spheroidal in form (*Amoeba,* Chalkley, 1935). Such observations possibly indicate a change in viscosity, the nature of which awaits investigation.

In certain of the hypermastigote flagellates Cleveland (1934, 1935) has found one of the best materials for the study of the cytology of cell division, not only in the Protozoa but of cells generally. Here the mitotic spindle, with its chromatic and achromatic parts, is clearly seen in the living condition; it may also be easily preserved and stained by ordinary techniques. In addition, experiments on pulling the centrosomes demonstrate the elasticity and contractility of the extranuclear chromosome fibers and those of the central spindle. Accordingly, in this material the various elements of the mitotic apparatus must be considered real and therefore demand careful consideration in any discussion of the mechanics of mitosis, not only in these forms but in astral types of mitotic divisions generally. Furthermore, a study of the physical problems involved in the origin and nature of the centrosomes, spindle fibers, asters, degeneration and reorganization of certain of the locomotor organs, as well as the correlated mechanics of cytokinesis in these forms would be most valuable. However, for the present we can only assume that the physical-chemical changes giving rise to these various structures are similar to those which occur in the mitotic divisions of cells of higher forms. For a discussion see Gray, 1931; Heilbrunn, 1928; and Chambers, 1938.

PROPERTIES OF PROTOPLASM AS EXHIBITED IN AMOEBA

To obtain a concept of protoplasmic structure no better course can be followed than to secure a microscope, some amoebae, and study them under relatively high power. Efforts to obtain this information from the literature alone often lead to confusion, because of the wide variety of terminology used by the various workers on the structure of amoebae. Careful microscopic examination of the protoplasm of *Amoeba proteus* in locomotion will reveal that it is composed of an outer, thin, colorless, hyaline layer which is for the most part optically structureless, and an inner granular region which makes up the greater portion of the animal. A detailed study of the granular portion will reveal that it is composed of a colorless continuous phase (hyaloplasm), in which are suspended crystals and granules of various sizes and shapes, nucleus, food vacuoles, contractile vacuole, refractive bodies, oil globules, and possibly other inclusions. Long intervals of observation on these structures will show that some of them seem to be permanent and self-perpetuating while others are only transitory. The question that naturally arises in this connection is, which are living and which are nonliving bodies? A further discussion of this point will be given elsewhere.

The question may now be asked, how are the various microscopically visible inclusions of *Amoeba* protoplasm kept in suspension? This may be because their density is only slightly different from that of the surrounding protoplasm and because of the viscosity of the protoplasm. Mast and Doyle (1935a) have presented evidence that the refractive bodies, which are the heaviest inclusions of the protoplasm, are found near the lower surface in living amoebae in which the protoplasmic viscosity is low; that they are distributed by protoplasmic flow has been demonstrated by Mast and Doyle (1935b). Upon centrifuging, the cytoplasmic constituents of amoeba are layered out, in order of their relative specific gravity, with the refractive granules at the centrifugal pole. Upon recovery pseudopods form first at or near the lighter end, and the heavy refractive bodies then flow forward followed by the other layers of constituents until all are thoroughly mixed.

The granular region of an amoeba is made up of an outer stationary layer, which forms a hollow cylinder through which an axial stream of protoplasm flows. The outer stationary layer has been called the ectoplasm and, more recently, because of its firm consistency, plasmagel; the axial flowing layer, endoplasm, and, more recently, because of its fluid consistency, plasmasol. More difficult to observe, because of its extreme thinness, is the surface layer, or plasmalemma, just external to the clear hyaline layer.

Ordinary methods of fixation often result in an apparent loss of the constituents which may be seen in the living cell; however, by the use of suitable fixing methods Mast and Doyle (1935a) were able to show that in *A. proteus* the cytoplasmic constituents may be preserved in a form scarcely distinguishable from the living condition, except for the nuclear granules and the water-soluble salts. Since protoplasm is an unstable intimate association of salts in solution, proteins, fats, and other materials which may give it a fibrillar or an alveolar appearance, we cannot hope to preserve it unchanged. Indeed, we may not even observe it in the living condition unchanged, as change is a universal characteristic of protoplasm. It must be constantly kept in mind that what is observed in a permanent fixed preparation is not living material, but likewise it must also be recognized that while the fixed material is not protoplasm, it is at least a significant artifact derived from protoplasm, since it has had its origin in the coagulation of proteins and other significant elements of living material. Most of the established facts of cytology have been first observed in fixed material, and later corroborated by a study of the living. However, it should be pointed out, from the works of Hardy and of Fisher, that probably many of the older theories of protoplasmic structure, such as the granular, fibrillar, alveolar, and reticular, have been based in part, at least, on artifacts induced by the methods employed.

Probably the most significant feature of the protoplasm of *Amoeba* is its ability to change from a relatively fluid state to a more solid jelly-like condition. Upon mechanical agitation, such as occurs in transferring an amoeba to a slide or by tapping on the slide, the organism may assume a globular form which may have short projecting pseudopodia. If the temperature be raised to 30° C., the projections may disappear (K. Gruber, 1912). This means that the protoplasm present has become relatively fluid, because only a liquid immiscible with the surrounding medium shows the globular form which is essential in the principle of minimal surfaces. This change in consistency may be brought about by mechanical agitation (Chambers, 1921) and has been experimentally studied by Angerer (1936), who presents data to show that mechanical agitation of amoebae causes a decrease in viscosity in the plasmagel (*A. proteus*) and an initial decrease followed by an increase in viscosity in the plasmasol (*A. dubia*). Extreme mechanical agitation according to Angerer, leads to a minimum viscosity of the plasmagel, which is eventually followed by the disintegration of the organism, the substance of which mixes with the surrounding medium.

If enough pressure is applied to the cover glass, the amoeba will burst, and it may be observed, providing the injury is not too extensive, that an effort is made on the part of the protoplasm at the region of rupture to form a water-insoluble membrane, thus inhibiting further mixing of the protoplasm with the water. This process has been recently termed "the surface precipitation reaction" by Heilbrunn (1928). If, however, the injury has been sufficient so that the interphase between the ruptured protoplasm and the surrounding medium is too great, the surface precipitation reaction is not sufficient to prevent a complete dissolution of the amoeba into the water.

Small hemispheres of clear protoplasm (the incipient pseudopodia) usually appear quickly upon the surface of a quiescent spheroidal amoeba in the form of liquid extrusions from the main mass of protoplasm. Into these extrusions, which gelate equatorially, a central flow of granular protoplasm may be seen streaming forward from the amoeba in the direction of the advancing pseudopodium, which constantly moves distally. As the pseudopodium advances, it may be seen to consist of an outer cylinder of motionless protoplasm and an inner streaming fluid protoplasm. That the cortical layer is gelated is indicated by the fact that its granules and other inclusions are stationary. As the central stream of flowing protoplasm (endoplasm or plasmasol) reaches the hyaline cap, it moves peripherally in all directions and gelates (i.e., becomes ectoplasm or plasmagel). At the other end of the amoeba the plasmagel becomes plasmasol and passes forward as a fluid core through the cylinder of plasmagel (Mast, 1926). The cause of the forward flow of protoplasm is obscure, but the suggestion has been made by

Hyman (1917) and many others that it is due to the tension exerted upon the fluid plasmasol by the elastic plasmagel.

There has been much discussion concerning the physical characteristics of the outermost layer, or plasmalemma, of different species of amoeba. In certain forms, such as *A. verrucosa,* the pellicle may be lifted with microdissection needles and stretched, apparently without injuring the organism in any way (Howland, 1924c). Seifriz (1936) also found a thin outer membrane on *A. proteus* which was resistant, elastic, and highly viscous except at the advancing tip of the pseudopodia. Likewise Chambers (1924) was able to lift the plasmalemma in *A. proteus* by injecting water beneath it, thus causing large blisters to form between the plasmalemma and the underlying surface which burst upon puncturing, leaving the pellicle collapsed. Mast (1926) caused blisters to appear by local pressure; further pressure caused disruption of the pellicle, the frayed ends of which were clearly observed.

Practically all investigators are agreed that the plasmalemma moves forward, at least on the upper surface of a moving amoeba. In addition, the method of formation of new food vacuoles requires a structure of the plasmalemma of such nature as to form new surfaces immediately by the replacement of large areas of the plasmalemma which have been used in forming the boundary of the food vacuoles (Schaeffer, 1920).

Immediately beneath the outer layer, or plasmalemma, is the clear hyaline layer. That this layer is fluid may be shown by the fact that it contains scattered granules which are in Brownian movement (Mast, 1926) and by the injection of a suspension of lamp black into it, the injected lamp-black particles spreading throughout this clear hyaline layer (Chambers, 1924). However, Schaeffer (1920) holds that sometimes it is more rigid than endoplasm, and at other times not.

COLLOIDAL NATURE OF PROTOPLASM

Any attempt to understand the processes occurring in protoplasm must depend for its success not only upon a knowledge of the chemical constitution of protoplasm but also of its physical structure. For this reason much ingenuity has been exercised in trying to discover the physical structure of the protoplasm in such relatively simple forms as *Amoeba,* in which permanently differentiated structures are at a minimum. For example, in *A. proteus* apparently only the nucleus is perman-

ently differentiated; the contractile vacuole and other structures are
evanescent, at least in part. We have already mentioned the more obvious physical characteristics of *Amoeba* protoplasm, which consists, as
does all protoplasm, of a complex heterogenous colloidal system, made
up of a suspension (granular, fibrillar, and alveolar) of many different
materials (fats, carbohydrates, and proteins) dispersed in a supporting
continuous liquid part. The greatest volume ingredient in protoplasm
is water; in solution in the water are various salts. Protoplasm may be
deprived of many of its visible inclusions without killing it, leaving
often a clear, colorless, optically homogenous hyaloplasm. Thus the
microscopic structure of protoplasm gives no direct evidence of the
finer submicroscopic structure. However, it is generally agreed that the
finer structure of protoplasm is dependent upon its colloidal nature,
which, because of the relatively large size of the particles present, results
in enormous intracellular surfaces.

Then too, the taking up of water by colloids is influenced both by the
solutes present and by the previous history of the colloid itself (hys-
teresis). Colloids often have the property of changing reversibly from
a relatively liquid (sol) to a relatively solid condition (gel). A gel
has many of the properties of a solid, among them elasticity, apparently
due to its structure; but differs from a solid in that diffusion in gels of
low concentration is often the same as in a simple solution, and in that
chemical reactions often can occur at velocities unaffected by the gel
condition. Every organism is dependent upon the temporal and spatial
coördination of its chemical reactions, and this depends largely on the
degree of dispersion and kinetic activity, because these regulate reaction velocities. Thus the organization of chemical events is due in
some way to the nature and architecture of the colloidal system in which
they occur. Therefore, biologists attempt to explain the physiological
action of various factors as influences on the colloids of protoplasm.

CONSISTENCY

It is a phenomenon of general observation that protoplasm flows,
but that it resists pressure. The former is a property commonly attributed
to liquids, the latter to solids. In liquids there is great internal mobility
of the molecules; this is essential to many physical activities, such as

movement, and to chemical reactions in protoplasm. On the other hand, solids have great internal cohesion of molecules, i.e., they retain their shape, show elasticity; this is essential to the maintenance of continuity and form of protoplasm. Colloids quite generally show changes from a relatively fluid to a firm jelly-like consistency, so that protoplasm as a colloidal system may partake of the nature of a solid (in the gel condition) or that of liquid (in the sol condition). Thus changes in viscosity are one of the essential factors in ameboid movement. In addition, the viscosity of the protoplasm as a whole, and particularly the changes in viscosity within a given portion of a protozoan, are very important factors to be considered in connection with a study of mitosis, cytokinesis, cyclosis, rate of diffusion of various substances, protoplasmic reorganization, and functioning of such organs as the contractile vacuole.

Besides the qualitative methods of estimating the viscosity of protoplasm by observing the presence or absence of movement (Mast, 1926b), presence or absence of Brownian movement (Bayliss, 1920; Pekarek, 1930; Brinley, 1928), microdissection studies (Kite, 1913; C. V. Taylor, 1920; Chambers, 1924; Howland, 1924c; and others), certain experimental methods, i.e., centrifugation (Heilbrunn, 1928; 1929b; Fetter, 1926), electromagnetic methods (Seifriz, 1936), and rate of diffusion of certain dyestuffs (Chambers, 1924; Needham and Needham, 1926) have been used to reveal data as to the relative and absolute viscosity of the protoplasm of various Protozoa. For a detailed discussion of these various methods the reader is referred to the works of Chambers (1924), Heilbrunn (1928), and Seifriz (1936).

The Protozoa show a wide range of viscosity values. In fact a variation in protoplasmic viscosity from 2 times that of water in *Amoeba* (Heilbrunn, 1929b) to over 8,000 times that of water in *Paramecium* (Fetter, 1926) has been reported. Undoubtedly a much higher viscosity exists in many protozoan cysts and other forms characteristic of resistant stages. Seifriz (1936) has reported that during the winter the plasmodium of a myxomycete becomes as hard and as brittle as a thin sheet of dry gelatin.

In addition to the information obtained about the consistency of various Protozoa from multilation experiments, for example those of Calkins (1911) and Peebles (1912) on *Paramecium caudatum*, the microdissection apparatus has proved to be a useful instrument in a study of the structure of Protozoa. By use of this method, Kite (1913) determined the ectoplasm of *A. proteus* to have a moderately high viscosity and cohesiveness. The substance forming the wall of the contractile vacuole seemed to possess a much higher viscosity than the surrounding endoplasm. The nucleus seemed to behave as a highly rigid granular gel. Kite further observed that the protoplasm of *Paramecium* was a soft, elastic, and somewhat glutinous gel; the surface seemed to be more viscous than the interior. More recent studies by Chambers (1924); Howland (1924a, 1924b); Howland and Pollack (1927); Seifriz (1936), and others on various species of *Amoeba* have confirmed in a general way the observations of Kite. In addition, C. V. Taylor (1920, 1923) has described the pellicle of *Euplotes patella* as a firm, fairly tough, rigid substance. The micro- and macronuclei were found to be highly gelatinous, rather rigid structures imbedded in a viscous hyaline matrix. Needham and Needham (1926) in attempting to inject *Opalina ranarum* made the observation that the outer membrane is composed of a thick, tough substance; the inner cytoplasm they found to be jelly-like, so that injected indicators would not spread within it. The suggestion was tentatively made that the failure of the injected indicators to spread within the *Opalina* might be due to its consistency or to membranes surrounding the numerous nuclei. Following injection, the organisms continued to swim about, and in a few moments the injected portions suddenly dropped out, often leaving the animals quite riddled with holes. See also Chambers and Reznikoff (1926); Reznikoff (1926), and Morita and Chambers (1929), on injection experiments in *Amoeba* with similar results.

Heilbrunn (1929b) has used the centrifuge method to study the absolute viscosity of *A. dubia,* which he found to be approximately 2 times that of water at 18° C. However, the viscosity was found to vary, with changes in temperature, from about 2 times water at 18° C. to 25 times water at $2\frac{1}{2}$° C. (1929a). Pekarek (1930), by studying the Brownian movement in *Amoeba,* has estimated its viscosity to be about 6 times that of water. More recently Seifriz (1936), by observing Brownian movement in *Amoeba,* estimated the viscosity in a quiet form to be 700 to 800 times that of water; in an active form it was much lower. Likewise Pantin (1924b) estimated the absolute viscosity of certain marine amoebae to be comparable to that of vaseline, i.e., over 1,000 times that of water.

Fetter (1926), using the same method as Heilbrunn, found the absolute viscosity of *Paramecium* to be 8,027 to 8,726 times that of water. Hyman (1917) has found that the heliozoön *Actinosphaerium eichhornii* can be cut into pieces as if it were solid. In addition to the viscosity and the changes in viscosity of the protoplasm which may take place within organisms under "normal" conditions, certain experimental conditions, such as abnormal salt concentration, acids, and alkalies, abnormally high or low temperature, mechanical agitation, changes in hydrogen-ion concentration, anesthetics and narcotics, radiation, sound waves, and so forth, may cause marked changes in the viscosity of the protoplasm.

The effect of various agents (chemical, mechanical, electrical, and so forth) upon the viscosity of protozoan protoplasm has been determined by studying their effect upon locomotion, Brownian movement, body form, pseudopod formation, rate of action of the contractile vacuole, as well as by centrifuging and by the microscopic appearance of the cytoplasm.

THE EFFECTS OF WATER

Since the principle solvent of protoplasm is water, any condition which tends to increase or decrease the water content of the organism also tends to change the viscosity or consistency of its protoplasm. Thus hypertonic solutions and desiccation usually cause the cells to shrink, this being accompanied by an increase in the viscosity of their protoplasm; a hypotonic solution or the injection of water directly into the organism tends to induce a swelling, accompanied by a decrease in the viscosity of the protoplasm (see Pantin, 1923).

THE EFFECTS OF SALTS

The effects of salts on the consistency of certain Protozoa, particularly *Amoeba,* have been rather extensively studied. Giersberg (1922), Edwards (1923), Chambers and Reznikoff (1926), Reznikoff and Chambers (1927), Pantin (1926a, 1926b), Brinley (1928), Heilbrunn and Daugherty (1931, 1932, 1933, 1934), Thornton (1932, 1935), Pitts and Mast (1934), Butts (1935), and others have studied the action of various salts and salt antagonisms upon the consistency of *Amoeba* protoplasm. In general it has been found that sodium and potassium ions tend to increase the viscosity of the internal protoplasm, while calcium and magnesium tend to lower it. However, Heilbrunn and Daugherty (1932, 1934), have found that this does not hold for the plasmagel (i.e., outer layer) of *A. proteus.* Here calcium produces a pronounced stiffening of the cortical gel, and this effect tends to be antagonized by Na, K, and Mg. Potassium has the strongest liquefying effect, Mg next, and Na the least action. The degree of reaction, particularly the antagonism of salts, seems to vary somewhat, depending upon the hydrogen-ion concentration (Pitts and Mast, 1934).

Greeley (1904) observed that KCl coagulates or increases the viscosity, whereas NaCl liquefies or decreases the viscosity of the protoplasm of *Paramecium.* Heilbrunn (1928) reports that unpublished work of Barth shows that lithium salts cause a coagulation or increase in viscosity of the protoplasm of both *Stentor* and *Paramecium.* Heilbrunn (1928) found that sodium, potassium, ammonium, and lithium chlorides all cause coagulation of *Stentor* protoplasm and that weak solutions of $HgCl_2$ produce a coagulation of the protoplasm of *Euglena.* The effect of salts upon reproduction in *Amoeba* has been studied by Voegtlin and Chalkley (1935) and by Butts (1935). Oliphant (1938) observed that potassium, lithium, sodium, and ammonium salts induce reversal in the direction of the effective beat of the cilia of *Paramecium,* whereas calcium and magnesium do not. The reversal in direction of the beat of the cilia is thought to be associated with an increase in viscosity of the cytoplasm. See also the work of Spek, 1921, 1923, and 1924, on the action of various salts on *Actinosphaerium, Opalina,* and other Protozoa.

Chambers and Howland (1930) have cut or torn *Spirostomum* in $CaCl_2$ solutions; the exposed protoplasm coagulates into a dense mass which the uninjured part of the organism pinches off. Injection of $CaCl_2$ produces localized coagulated regions which are pinched off. Potassium chloride and NaCl cause liquefaction. Ephrussi and Rapkin (1928), however, report that $CaCl_2$ facilitates "l'explosion" of this ciliate; KCl and NaCl render explosion more difficult.

Chambers and Howland (1930) have further performed injection and immersion experiments with *A. eichhornii,* a heliozoön with grossly vacuolated protoplasm. Immersion in NaCl or KCl dissolves the vacuolar membranes, with a dissolution of the intervacuolar protoplasm. Immersion in strong concentrations of $CaCl_2$ causes coagulation of the protoplasm; in weak solutions the coagulation may be local, and the living remnant rids itself of the coagulated regions. After immersion in strong concentrations of $MgCl_2$ the protoplasm coagulates into a flabby mass; in weaker solutions localized regions rupture. Injections result in similar but more localized effects, except in the case of $MgCl_2$.

THE EFFECTS OF ACIDS AND ALKALIES

There seem to be no general agreement on the effects of acids upon protozoan protoplasm. Some authors report that they cause an increase in viscosity, while others find that they produce a decrease in viscosity of the protoplasm.

Jacobs (1922), by bubbling CO_2 through the culture medium containing *Paramecium* and *Colpidium,* observed that short exposure of these organisms to the CO_2 caused a decrease in the viscosity of the protoplasm, while longer exposures increased it. Brinley (1928) found that CO_2 caused gelation of the ectoplasm and solation of the endoplasm of *A. proteus.* Reznikoff and Chambers (1927), after injecting bubbles of CO_2 into *A. dubia,* observed that it produced a decrease in the viscosity of the protoplasm and that the animal was not irreversibly injured unless the CO_2 destroyed the cell membrane. Hydrochloric acid has been observed to produce an increase in the viscosity of the protoplasm of *Amoeba* (Chambers, 1921; Edwards, 1923; Brinley, 1928). Heilbrunn (1937) finds that acids cause an increase in viscosity of both the plasmasol and plasmagel of *Amoeba.*

Chambers (1921) reports that basic dyes, which contain a relatively strong acid radical, jelly the protoplasm of *Amoeba,* whereas acid dyes, with a strong basic radical, liquefy it.

The action of alkalies has been reported to decrease the viscosity of *A. proteus* (Chambers, 1921; Edwards, 1923; Brinley, 1928). However, Heilbrunn (1937) reports that alkalies increase the viscosity of the plasmasol and decrease the viscosity of the plasmagel in *Amoeba.*

THE EFFECT OF TEMPERATURE

As pointed out by Brues (1927), the Protozoa are among the most resistant of all animals to high temperatures; they have been found living in hot springs at temperatures between 50° and 60° C. Motile forms seem to be much more susceptible to high temperature than do encysted forms (cysts of *Colpoda* are not killed at 100° C. dry heat for three days), a condition probably associated with the low water content of the organisms (Bodine, 1923). That Protozoa show ability gradually to acclimate themselves to increased temperatures has been shown by Jacobs (1919).

Since temperature changes are known to affect the rate of action of molecules in liquids, it is only reasonable to assume that their effect is somewhat similar upon protoplasm. In general it may be stated that a slight increase of temperature over that of the normal, causes a decrease in protoplasmic viscosity; a slightly higher temperature, a reversible coagulation, and a still higher temperature, an irreversible coagulation and death. Davenport (1897) has referred to these conditions as contraction, heat rigor, and death rigor respectively. See also the discussion of the action of temperature on protoplasm by Heilbrunn, 1928.

Pantin (1924a), working on two species of marine *Amoeba,* found that the viscosity was high near 0° C. and decreased with rise in temperature. The primary effect of temperature upon locomotion, he holds to be a direct effect upon the change of state of the protoplasmic sol-gel transformation.

Perhaps one of the most complete studies of the action of temperature upon protozoan protoplasm is that of Heilbrunn (1929a) on *A. dubia.* He found that at temperatures from about 3° to 10° C., the viscosity value was about 22 to 23 times that of water. From 10° to 18° C. there was a very rapid decrease in viscosity, approaching about 2 to 3 times that of water, while at 20° to 25° C. there was an increase in viscosity to about 8 to 9 times that of water, followed by a decrease to about 2 times that of water at 30° to 35° C. Although the range of Heilbrunn's experiments was not extended beyond 35° C., he states that higher temperatures cause a coagulation of the protoplasm. Thornton (1932) has observed that the maximum viscosity of the plasmagel of *A. proteus* occurs at 4.5° C. Between 4.5° and 30° C. the viscosity decreases progressively, with rise in temperature, until at 30° C. the decrease is more rapid. Thornton (1935) has further found that the action of certain salts does not alter the fundamental effect of temperature on the viscosity of the plasmagel of *A. proteus.* Daniel and Chalkley (1932) observed that the rate of mitosis, including nuclear and cytoplasmic division, of *A. proteus* varies inversely with temperature from about 4° to 30° C. From about 30° to 40° C. these processes vary directly with the temperature.

Greeley (1904) found that the consistency of the protoplasm of various Protozoa, like organic colloids, varies directly with the temperature, within certain limits. As the temperature is elevated above the normal, the protoplasm absorbs water, so that its fluidity and motility are greatly increased. These changes continue until a critical point is reached, at which coagulation occurs. The resistance of *P. caudatum* to a temperature of 40° C. has been shown to vary with the hydrogen-ion concentration of the medium and to exhibit two maxima of resistance: one on the alkaline and one on the acid side, with a region of minimal resistance at neutrality (Chalkley, 1930). In saline solutions the mechanism of death by heat seems to vary with different hydrogen-ion concentrations. At *p*H 6 or less the cell coagulates, at *p*H 8 or more the organism disintegrates; between these two extremes death occurs from rupture by swelling. Furthermore, Chalkley also found an increase in thermal resistance of *Paramecium* on the addition of Ca, and a decrease on addition of K. Oliphant (1938) found that the rate at which the cilia of *Paramecium* beat in reverse varies directly with temperature, a condition he implies is associated with changes in viscosity of the organisms. For a further discussion of the effects of high temperatures on organisms, including the Protozoa, see Bělehradek (1935).

It has been shown by Greeley (1901) that when *Stentor coeruleus* is suddenly lowered to the freezing point of water, it is usually killed. However, when the temperature is lowered slowly to 0° C., a remarkable dedifferentiation of the animal takes place. The resorption of the cilia and the gullet, and the throwing off of the ectosarc was observed, and there was finally formed a spherical cyst-like undifferentiated cell, which Greeley referred to as a "resting" cell. When returned to room temperature, a reverse process takes place, and the cyst-like organism becomes active. With the lowering of the temperature the organism was observed to lose water. Greeley (1903) further found that the method of reproduction in *Monas* could be controlled by temperature; at 20° C. the organisms reproduced sexually and by fission; at from 1° to 4° C. they reproduced by asexual spores.

According to Luyet and Gehenio (1938), Becquerel (1936) found that certain *Amoeba* in dry soil were not killed when subjected to —269° to —271° C. for 7.5 days, or when subjected to —190° C. for 480 hours. See the above-mentioned work of Heilbrunn (1929a) for absolute viscosity values of the protoplasm of *A. dubia* at low temperature.

Chambers and Hale (1932) observed that *Amoeba* exposed to —5° C. were not killed. However, by inserting an ice-tipped pipette into the interior of the organism, fine feathery crystals of ice were observed to appear immediately at —0.6° C.

In general it may be stated that low temperatures tend to increase the viscosity and thus to decrease the rate of locomotion, and to favor cyst formation in Protozoa. Motile forms are usually killed as the temperature of the water reaches 0° C. However, the Protozoa show some degree of acclimatization to low temperatures.

MECHANICAL AGITATION

Mechanical agitation may cause a marked change in the consistency of *Amoeba* protoplasm, apparently by causing a breakdown (thixotropic collapse) of its internal structure. Chambers (1921) has shown that churning *Amoeba* by microneedles caused a liquefaction of its protoplasm. Vigorous shaking (Angerer, 1936) caused at first a liquefaction of the plasmasol of *A. dubia*, followed by an increase in viscosity; continued agitation caused the complete dissolution of the organism. However, in *A. proteus* agitation caused a decrease in viscosity in the plasmagel to a minimum, without the subsequent increase observed in the plasmasol of *A. dubia*. High-frequency sound waves have been observed to produce whirling of the inclusions in the small vacuoles of *A. proteus* and *A. dubia;* higher intensities cause a mild whirling of the more liquid regions, followed by rupture of the organisms (E. N. Harvey, E. B. Harvey, and Loomis, 1928). A decrease in viscosity of the endoplasm was also observed.

HYDROGEN-ION CONCENTRATION

That Protozoa can live in wide ranges of hydrogen-ion concentrations is evident from the work of Alexander (1931) on *Euglena* which he found lived in ranges from pH 2.3 to pH 11. Studies on the effects of hydrogen-ion concentration on protoplasmic viscosity are complicated by the rate of entrance of the ions and by the fact that they may be neutralized by the buffers of the protoplasm (Pollack, 1928a).

Pantin (1926a) has found that the movement of certain marine amoebae takes place between pH 6 and 10. It is reversibly inhibited at the acid limit, but alkaline inhibition is reversible only after a brief immersion. The rate of movement he holds to depend upon the rate of change in sol⇌gel transformation. In addition, the water content of *A. proteus* has been shown to vary with the hydrogen-ion concentration of the medium (Chalkley, 1929). The effect of changes in hydrogen-ion concentration upon the action of cilia has been studied by Chase and Glaser (1930).

THE EFFECTS OF NARCOTICS

For a summary of the literature on the effects of narcotics in cells generally, the reader is referred to the work of Winterstein (1926) and of Henderson (1930). In general it may be stated that the action of narcotics upon cells is to change their permeability and viscosity, to inhibit enzyme action, and to affect the electric potential.

Alcohol has been reported to produce a lowering of the viscosity of the protoplasm of *A. proteus* (Edwards, 1923; Brinley, 1928). Ether and chloroform likewise have similar effects (Brinley, 1928). Certain paraffin oils have also been reported to have a narcotic effect on *A. dubia* (Marsland, 1933). More recently Daugherty (1937) found that the higher alcohols and ether in concentrations just below lethal cause liquefaction of the plasmagel of *Amoeba;* the same concentrations of the lower alcohols produce gelatin of the plasmasol; the higher alcohols and ether, liquefaction of the plasmasol. More dilute solutions of the higher alcohols and ether produce first liquefaction and then gelatin of the plasmasol (see also Frederikse, 1933a). Potassium salts and fat solvents liquefy the plasmagel in *A. proteus* (Heilbrunn, 1931).

Makarov (1935) has studied the effects of narcotics on various Infusoria, using vital stains in conjunction with the ultramicroscope. Narcotics cause a change in the dispersion of the colloids, which is reversible. Strong concentrations cause an irreversible coagulation.

THE EFFECTS OF RADIATION

The specific effects of radiation upon the cell are unknown. Furthermore, whether the nucleus or the cytoplasm is more susceptible to radiation is a debated question. According to Heilbrunn and Mazia (1936), Glocker and Reuss claim that isolated cells are less sensitive to Roentgen rays than are cells in mass.

The permeability of *Paramecium* and *Stylonychia* to NH_4OH has been shown to increase with increased exposure of the animals to radiation (Packard, 1923, 1924). After reviewing the literature on the biological effects of radiation, including numerous studies on Protozoa, Heilbrunn and Mazia (1936) reach the conclusion that ultra-violet rays, Roentgen rays, and radium all cause liquefaction of the protoplasm and, with an increase in exposure, coagulation. Coagulation of the protoplasm is frequently preceded by extensive vacuolization. Furthermore, Heilbrunn and Daugherty (1933), from their work on the effects of ultra-violet rays on *Amoeba,* offer the theory that the effect of radiation is to release the bound calcium from the cell cortex; it then enters the endoplasm, causing first liquefaction and then gelation. Likewise, ultra-violet radiation causes a release of fat in *Amoeba* (Heilbrunn and Daugherty, 1938). For the recent ingenius method of "microdissection" with ultra-violet rays, see Tchakhotine (1937).

THE EFFECT OF HEAVY WATER

E. N. Harvey (1934) has studied the effects of heavy water on *Paramecium, Amoeba, Euglena,* and *Epistylis* (see also Taylor, Swingle, Eyring, and Frost, 1933). Harvey found that paramecia were killed by 80 to 100-percent heavy water in from 6 to 10 hours. They first swim slowly, appearing bloated; the contractile vacuole stops functioning; blisters appear, followed by disintegration. However, they were not markedly affected by 0.2-percent heavy water. *Amoeba* rounds up and is killed in from 4 to 6 hours in 80 to 100-percent heavy water. *Euglena,* on the other hand, is not irreversibly injured by 90 to 97-percent heavy water. Gaw (1936b) found *Blepharisma* to become more spherical in 95-percent heavy water, which indicates a change (decrease in viscosity) in the physical nature of its cytoplasm.

HYDROSTATIC PRESSURE

High hydrostatic pressure (500 atmospheres) causes collapse of pseudopodia, and a rounding up of *A. proteus* (Brown and Marsland, 1936). This effect is apparently the result of liquefaction of the plasmagel, an effect which inhibits the normal sol-gel transformation. However, similar high hydrostatic pressures seemed to have little or no effect on the beating of flagella or cilia of other Protozoa (Marsland, 1939).

THE EFFECTS OF ELECTRIC CURRENT

The effects of electric current on many Protozoa have been studied (see Hahnert, 1932, for literature).

Bayliss (1920) has shown that on passage of an electric current of the proper intensity through *Amoeba,* immediate gelation of the protoplasm takes place and all Brownian movement stops. As the *Amoeba* recovers, the protoplasm solates and the particles again take up their active motion. Strong electric shocks caused irreversible coagulation. More recently Luce (1926), Mast (1931b), and Hahnert (1932) have studied the effects of electric current on *Amoeba.* In general, as stated by Hahnert, a constant electric current provokes responses in *Amoeba* by definite polar actions at the ends of the organism. Destruction or solation of the plasmagel occurs at the cathode end immediately, then contractive, and later, disintegrative processes occur at the anodal end. These responses are directly dependent upon the strength of the current.

IRREVERSIBLE COAGULATION

Acids and certain salts, such as mercuric chloride, in the proper concentrations, cause irreversible coagulation of protoplasm. Protozoa are therefore often "fixed" in such solutions for morphological studies. The causes of the irreversible coagulation which occurs at "normal" death are obscure.

SURFACE PROPERTIES

A discussion of the physical properties of the surface layer in Protozoa involves its structure, which has been analyzed by a study of its gross morphology, elasticity, contractility, extensibility, viscosity, and ultramicroscopic architecture. It is obvious that any conception of the physical structure of the surface membrane must be in harmony with its functions, such as preserving the integrity of the organism (by being immiscible with water), controlling the diffusion of materials in and out of the cell (permeability), acting as a seat of electromotive forces (membrane or diffusion potentials), maintaining form (possessing tension), forming secretions (for protection, adhesion, forming cyst walls, armor, and so forth) and for the reception of stimuli, and so on.

STRUCTURE AND ORIGIN

The problem of surface structure in the Protozoa is often complicated by a failure of many investigators to define clearly what they refer to as the "cell membrane." From a review of the literature it is quite obvious that the vital membrane essential to the cell may be reduced to only a very thin ultramicroscopic film, such as that which forms at the torn surface of an amoeba. In other cases the "cell membrane" is of microscopic dimensions and displays the physical properties of a thick, tough pellicle. In this connection the work of Nadler (1929) showed that the pellicle could be completely removed from *Blepharisma* without killing or even affecting the shape of the organism; after a few days a new pellicle may be formed in these animals. Could it not be, therefore, that in forms with thick surface layers, the physiologically active vital membrane is mainly limited to a thin film, and that the remaining part of the thick surface layer serves mainly for protection, support, giving anchorage to locomotor organs, secretion of slime for adhesion, forming cyst walls, or in the formation of other surface structures which possibly aid the animal in coping with its environment?

In addition to the vast literature describing definite morphological membranes at the surface of various Protozoa, both in the living and fixed condition, considerable experimental work has been done on the physical nature of the surface layers. Kite (1913), Chambers (1924), Howland (1924c), Howland and Pollack (1927), Taylor (1920), Needham and Needham (1925, 1926), and many others have actually established the presence of the surface membrane by puncturing it, tearing it, and in some cases actually removing it. In general it has been found to vary in thickness from that of a delicate film to that of a tough pellicle. Microdissection methods have also shown it to possess measurable elasticity and contractility, and to vary considerably in consistency from that of the underlying protoplasm.

The tension at the surface of *Amoeba* has been measured by E. N. Harvey and Marsland (1932). They injected drops of paraffin oil or olive oil into *A. proteus* and *A. dubia* and then subjected them to centrifugal force in the microscope-centrifuge. Because of the bouyancy of the oil, the organisms became stretched. The amount of distortion was photographed and, under certain assumptions, a value for the order of magnitude of the surface forces was calculated. In this way these authors found the tension at the surface of *A. dubia* to give values of one to three dynes per centimeter. In this form they concluded that "there can be no appreciable turgidity due to resisting surface layers." However, in *A. proteus* it was impossible to pull even large oil droplets out of the organism by the highest centrifugal force available. For this reason no tension at the surface of this organism could be calculated, but they estimated it to be about thirty times that for *A. dubia*. The surface of *A. proteus* was described as a firm, tough, external layer.

Since the surface membrane may be of ultramicroscopic dimensions, its physical properties are not easily determined, and for this reason much of our knowledge is of a theoretical nature. It has been held that the appearance of a new membrane at the surface of a torn bit of protoplasm is due to the accumulation at the surface of substances, chiefly lipoid, which tend to lower surface tension. However, Heilbrunn thinks of this process of new surface membrane formation as a "surface precipitation reaction," comparable in many ways to the clotting of blood. He has produced evidence to show that the presence of calcium is a prerequisite for the formation of new surface membranes. In addition, he holds that any factors which cause a release of calcium from its protein binding cause the "surface precipitation reaction" to take place within the cell interior, giving to it a froth-like appearance.

Whatever may be the exact mechanism of new membrane formation, the fact that a time factor is involved in its production from the cytoplasm, that it assumes increased tension over that of the underlying cytoplasm, that its consistency and durability depend both upon the environmental medium and the specific character of the protoplasm from which it is formed, as well as its semipermeable properties, all point to the surface membrane as being a definite, organized structure (Chambers, 1924).

Recent researches upon the optical properties of cell membranes other than those of Protozoa, by means of polarized light and X-ray diffraction methods, seem to indicate that they are constructed of lipoid and protein molecules, with their long axes arranged perpendicular and parallel, respectively, to the surface of the cell (Schmitt, 1938). E. N. Harvey and Danielli (1936) also hold the cell membrane to be composed of lipoid and protein substances. Furthermore, Langmuir has shown that the structure and surface properties of certain nonliving membranes are frequently determined by the orientation of the individual molecular layers; these may, under certain conditions, undergo almost instantaneous reversal or reorientation. This overturning may markedly alter the chemicophysical properties of the membrane, a fact which may prove to be of considerable interest in the study of the permeability of cells generally. As pointed out by Harvey (1936), one would seem justified in conceiving of the surface membrane in different cells as composed of a monomolecular layer and a polymolecular film, or as a polymolecular oil film with oriented adsorbed protein molecules varying from a rather liquid to solid consistency.

The remarkable film-forming properties of the ciliate *Spirostomum* have been studied by Fauré-Fremiet, Ephrussi, and Rapkine (1926). This organism explodes when it comes into contact with the air-water interface, and its solid constituents spread over the surface of the water. By first dusting the water with talc, it was determined that the surface film was 4.2 to 5.7 μ μ thick, and probably monomolecular.

PERMEABILITY

1. *Cell membrane.*—In Protozoa as in all other types of cells one of the most important properties of the protoplasm is its ability to form new surface membranes which have a selective permeability. In addition to preserving the integrity of the cell, the surface membrane regulates to a large extent the passage of dissolved foodstuffs and oxygen into the cell and the diffusion of waste materials from it. The rate of this exchange depends in part upon the degree of the permeability of the surface membrane and in part upon the osmotic concentration of the cell and of the surrounding medium.

It is not our purpose to discuss generally here the physiological problems of permeability and diffusion. However, many fresh-water Protozoa possess an interesting mechanism for controlling their osmotic state. The protoplasm of these organisms possesses a much higher osmotic concentration than that of the surrounding medium. For this reason water tends continually to enter them, and, were it not for the controlling mechanism of the surface membrane and the continuous bailing out process of the contractile vacuole, the animals would swell up and burst. Adolph (1931) has reviewed the literature dealing with the rate of water exchange in Protozoa and has found that the fastest turnover is in *Cryptochilum,* which excretes its body volume in 2 minutes; the slowest water exchange was in *Amoeba,* which requires 31.5 hours, while *P. caudatum* eliminates its own body volume of water in from 15 to 20 minutes.

Water, when injected in small amounts into an amoeba, readily diffuses throughout the cytoplasm, causing a temporary cessation of ameboid movement (Chambers, 1924). However, if the injection be great in amount i.e., equal to half the body volume, the water tends to collect on one side in the form of a large blister, which is eventually pinched off, and within a short time the amoeba resumes its normal activities (Howland and Pollack, 1927). In addition, a marked increase in contraction rate and water output of the contractile vacuole was noticed.

In contrast to the above experiment, Mast and Doyle (1934) caused *A. proteus, A. dubia, A. radiosa,* and *A. dofleini* to lose water by placing them in 3-percent egg albumen, hypertonic salt solutions, or solutions of calcium gluconate. The organisms decreased in size and their surfaces became wrinkled and covered with protuberances, folds, and crevices. The crests of some of the adjoining protuberances and folds fuse, giving rise to tubes open at one end. The region of the fused folds push out, forming a pseudopodium, thus extending the tubule, the mouth of which was observed to expand and contract, drawing the fluid in. Later the tubule disintegrates, releasing the fluid to the inside of the organism. This "drinking" of water by means of tubules seems to be an important mechanism, to compensate for the rapid loss of water in the organism. Further experiments by Mast and Fowler (1935) showed that when *A. proteus* was placed in Ringer solution containing various concentrations of lactose, it would decrease as much as 88 percent in total volume without injury. The rate at which the water leaves the cell was found to be approximately 0.026 cubic micra per minute through each square micron of surface. These authors concluded that the permeability to water was regulated by the plasmalemma.

It is well known that many marine Protozoa, especially Sarcodina, do not possess contractile vacuoles, but, when transferred to fresh water, contractile vacuoles may appear (Schaeffer, 1926). The contractile vacuoles of fresh-water forms either work very slowly or disappear entirely upon the organism being transferred to various concentrations of sea water. Furthermore, some Protozoa of the same species, for example *Actinophrys,* may be found in both fresh and salt water. The cytoplasm of the fresh-water form is greatly vacuolated and possesses a contractile vacuole, while the cytoplasm of the salt-water form is relatively free of vacuoles, including the contractile vacuole, and the general appearance of the cytoplasm has changed to a granular condition. Thus a gradual acclimatization of fresh-water Protozoa to salt water seems to reduce the difference between the external and internal osmotic pressures by a loss of water from the cell. This is probably accompanied by changes in the physical state of the protoplasm, particularly its consistency and specific gravity. It is of interest to recall here the extraordinary case of *Noctiluca,* in which the specific gravity of the organism is less than that of the surrounding sea water, owing to lower concentration of salts. Here the water, instead of entering the organism, tends to diffuse out of it, so that osmotic work must be done to retain its constant state.

Experiments designed to test the selective permeability of the surface membrane of the Protozoa have been carried out by utilizing various dyestuffs (Chambers, 1922; Ball, 1927). For example, according to Chambers (1922) an aqueous solution of eosin does not stain *Amoeba* from the exterior. However, if injected into the interior of the cell, it readily diffuses throughout the protoplasm.

Attempts to visualize the mechanism of permeability control in surface membranes have been made by assuming a solubility of the permeating substance in the membrane inself or by assuming the membrane to have a sieve-like structure similar to a filter, but much smaller. In any case, whatever may prove to be the final answer to this problem, it will undoubtedly involve a change in molecular aggregation, organization, and polarity of the elements of the physiological membrane.

2. *Nuclear membrane.*—The interphase nucleus, like the cytoplasm of

the Protozoa, is surrounded by a limiting surface membrane which may be demonstrated in many forms by the usual cytological techniques (see, for instance, Chalkley, 1936). In addition, experimental studies on the living nucleus by Kite (1913), Chambers (1924), C. V. Taylor (1920), and many others have substantiated the view of the presence of a definite nuclear membrane, often extremely thin, but usually composed of a moderately tough, solid substance, which, upon rupture of the surface of the organism, may preserve the integrity of the nucleus for a considerable time. Seifriz (1936) reports that the nuclear membrane of *Amoeba* may be readily removed by microneedles, following coagulation of the nucleus as a whole.

Morita and Chambers (1929) have shown that the nuclear membrane of *Amoeba* is permeable to acid, while the general body surface of the animal is not.

King and Beams (1937) report that the macronucleus of *P. caudatum* in the vegetative stage was greatly stretched by centrifugation. In some specimens it was separated into a relatively heavy chromatic portion and a relatively light achromatic portion. Animals with their macronucleus separated in this way were able to live and carry on apparently normal metabolic processes. Here it is evident that a mechanical disruption of the macronucleus did not cause a marked physical change of the cytoplasm, as often happens when the nucleus is punctured or cut by a needle (Kite, 1913; Peebles, 1912). King and Beams were unable to differentiate a limiting membrane surrounding the two separated portions of the macronucleus, and, if present, it must have been of ultramicroscopic dimensions.

Luyet and Gehenio (1935) were unable to demonstrate a definite membrane surrounding the macronucleus of *P. caudatum* by means of ultra-violet absorption methods.

Whatever, in the final analysis, the physical structure of such ultramicroscopic nuclear membranes may prove to be, it will probably involve a special molecular behavior, characteristic of surfaces much like that which is thought to occur between nonliving immiscible fluids.

3. *Contractile vacuole.*—For the early literature dealing with the long-disputed question of the presence or absence of a permanent membrane surrounding the contractile vacuole in Protozoa, the reader is referred to the works of C. V. Taylor (1923), Lloyd (1928), Howland (1924a),

King (1935), and Kitching (1938). It is well known that the contractile vacuole often arises from the coalescence of smaller vacuoles.

C. V. Taylor (1923) studied the structure of the contractile vacuole of *Euplotes* and found that it appeared to be composed of a definite "wall" of measurable thickness. However, he later concluded that the apparent "wall" was an optical illusion and that careful study revealed only the internal surface of the "wall" to be sharply delimited; externally it merges not abruptly, but gradually, into the surrounding medium. By moving the needle point against and about the contractile vacuole, he found it could be displaced and that its boundary was tolerably durable and its viscosity distinctly higher than the surrounding endoplasm. With completion of contraction, the vacuole wholly disappears.

In a similar study upon the contractile vacuole of *A. verrucosa* and *P. caudatum,* Howland (1924b) and Howland and Pollack (1927) have been able to dissect the contractile vacuole out of the organisms and observe it floating freely in the water. Here it may be stained by alizarin blue and manipulated by microneedles. Upon puncture, the surface of the membrane was observed to wrinkle. In other experiments, when the contractile vacuole was forced into contact with the plasmalemma, a fusion took place and, because of this fact, Howland and Pollack were led to suggest that the surface membranes of both the cell and the contractile vacuole must possess similar physical properties. These authors further found the contractile vacuole to lie in a region of gelated endoplasm, a condition they think necessary for its functioning.

Mast (1938) has described the membrane at the surface of the contractile vacuole in *A. proteus* as a well-differentiated structure about 0.5 micron in thickness.

King (1935) has described the permanent components of the contractile vacuole system of *P. multimicronucleata* as including the pore with its discharging tubule, and the feeding canals, each made up of a distal excretory portion, an ampulla, and an injection tubule. The membrane of the contractile vacuole itself is considered a temporary structure, disappearing at systole and closing the pore, which ruptures at the next systole. The new membrane of the contractile vacuole appears by the coalescence of the membranes of vesicles which lie just under the pore, and becomes continuous with that of the pore. In centrifuged *P. caudatum,* King and Beams (1937) observed that in some cases the vacuole seems to have been moved out of its position in contact with the membrane closing the pore. In spite of this, the feeding canals continue to form other main vacuoles, each of which expels its contents through the pore in normal fashion. The displaced vacuole is free in the endoplasm, and, if not too large, may move about in the protoplasmic stream. Such vacuoles have been observed to be present in the cytoplasm twenty-four hours after centrifuging. These observations were interpreted in further support of the view that the main contractile vacuole and its membranes are purely temporary, forming anew before each systole by the fusion of feeding vesicles formed at the vacuolar ends of the feeding canals. The feeding canals are markedly osmiophilic, and for this reason they have been described by Nassonov (1924) as homologous to the Golgi apparatus. However, this view is not supported by Beams and King (1932) and King (1935). Gelei (1928) thinks of the cytoplasm surrounding the feeding canals as "nephridial-plasma" and suggests a parallelism between the nephridial system in *Paramecium* and that of higher organisms. Metcalf (1910) observed, in the cytoplasm surrounding the contractile vacuole of *A. proteus,* small round granules which he termed "excretory granules." These bodies, which are permanent structures, are thought by Metcalf to be functionally connected with secretion. However, Mast (1938a) has been unable to substantiate this view.

It is interesting to note that the permeability of the cell membrane and that of the contractile vacuole system are in some ways similar. For instance, Morita and Chambers (1929) report that in *A. dubia* both the surface membrane and the contractile vacuole membrane are impermeable to HCl. Kitching (1936) reports that the vacuolar surface, like the cell surface, is relatively impermeable to salts in the peritrich ciliates. He further presents arguments to show that the vacuolar system in these forms actively secretes water.

4. *Food vacuoles.*—In many free-living Protozoa the ingested food particles are surrounded by a distinct vacuole. It is here that the digestive enzymes collect and act upon the solid food particles, converting them into a dissolved form suitable for use by the organism. The so-called food vacuolar membranes, in forms such as *Amoeba* and *Paramecium,* must perform much the same function as the cell membranes surrounding the cells in the intestine of higher organisms. In other words, the semiperme-

able membrane delimiting the food particles from the surrounding cytoplasm prevents the diffusion of undigested substances, such as proteins and starches, until they have been broken down into diffusable compounds of much smaller dimensions.

The formation of food vacuoles by means of food cups (Kepner and Taliaferro, 1913; Schaeffer, 1916) is of interest in connection with a consideration of their surface structure. In *Amoeba* the membrane surrounding the food particles has been observed to be derived directly from the surface of the organism i.e., the plasmalemma. Schaeffer (1916) reports that *A. proteus* may form several hundred such food vacuoles a day, a fact which would necessitate considerable replacement of the plasmalemma. If this observation be correct, one need only assume a structure for the food vacuole membrane like that of the plasmalemma. Furthermore, it has been shown that the membrane of a food vacuole which has completed its function in the cell may fuse upon egestion with the cell membrane from which it has been derived (Howland, 1924c). In addition, some vital stains, such as neutral red, readily penetrate the surface membranes of many Protozoa and stain the food vacuoles, which suggests similar permeable properties for both the membranes surrounding the surface and those surrounding the food vacuoles. In contrast to its penetration of the surface and food-vacuole membranes, neutral red does not readily diffuse through the membranes of the nucleus or that of the contractile vacuole.

Dissection and multilation studies upon a number of Protozoa have shown that the food vacuoles are capable of existing free of the cytoplasm for relatively long intervals of time. For instance, King and Beams (1937) have observed food vacuoles in water to retain their form for over one-half hour, after which time the vacuolar membrane was observed to wrinkle, followed by a breakdown of the vacuole.

Dogiel and Issakowa-Keo (1927) immersed *Paramecium* in various salt solutions and India ink. In solutions of $MgSO_4$, $MgCl_2$, and $FeSO_4$ the food vacuoles are much elongated. These sausage-like food vacuoles may swell up or may be extruded through the gullet. In $BaCl_2$ the food vacuoles are small and spindle-shaped. Whether this effect is produced on the membrane of the food vacuole or upon the cytoplasm is not clear.

Mast (1938b) has recorded for *A. proteus* that a food vacuole may divide several times, forming a number of vacuoles. In other cases food vacuoles have been observed to fuse (Mast and Hahnert, 1935).

From the comparatively few accounts of the physical structure of the food vacuole membrane, it is not possible to give an analysis of its structure. However, it seems reasonable to conclude that it is usually of molecular dimensions, capable of resisting deformation, and that it is permeable to enzymes, water, certain dyes, and digested food materials. Its architecture is probably much the same as that of the surface membrane, except perhaps for its thickness.

5. *Other types of vacuoles.*—In addition to the contractile vacuolar system and the food vacuoles, many other types of vacuoles may be found in the cytoplasm of the Protozoa. Such vacuoles may be characteristic of the cytoplasm, as the crystal vacuoles in *Amoeba* (Mast and Doyle, 1935a), transitory vacuoles (Hopkins, 1938), large acid-filled vacuoles associated with changes in specific gravity of *Noctiluca* (E. B. Harvey, 1917; Lund and Logan, 1925); or they may be induced by certain experimental methods, such as change in the salt content of the environment (Schaeffer, 1926), exposure to dyestuffs, X-rays, poisons, and so forth (Heilbrunn, 1928). These vacuoles, too, frequently have been reported to fuse; normally as in the coalescence of vacuoles in *Noctiluca* (Lund and Logan, 1925), and experimentally in *Paramecium* in which King and Beams (1937) observed the crystal vacuoles to fuse when centrifuged.

Little is known concerning the actual physical structure of the membranes of such vacuoles, but there is no reason for believing that their structure differs greatly from that of other surface membranes surrounding vacuoles, such, for instance, as the feeding vesicles of the contractile vacuole of a form like *Euplotes* or *Paramecium*.

ADHESIVENESS OR STICKINESS

Among the physicochemical properties of the protoplasm-liquid medium interface, adhesiveness is of importance in considering such subjects as ameboid movement, tissue culture, leucocyte activity, immunity, and cell movements in embryology. Pfeiffer (1935) has listed the literature on adhesiveness, which is scattered very widely.

It is well known that amoebae creep on vertical surfaces, even on the under side of the surface film of water, creeping on this as though it were a solid body, but Bles (1929) denies that *Arcella* can move on a clean surface film. Specimens of *Amoeba* on the under side of the cover glass swell up when disturbed, but may still stay attached to the "ceiling" (Dellinger, 1906). Adhesion to the substratum, according to Mast (1926b), is due either to the secretion of an adhesive substance or to a state of the plasmalemma. According to Rhumbler (1898) and Jennings (1904), an amoeba probably adheres to the substratum by a mucus-like secretion. Many observers have reported that if the surface of an amoeba is touched with a fine glass rod, it adheres to the glass rod, so that sometimes a bit of the organism may be pulled off (Mast, 1926b) or the whole organism may be dragged about by a thread of mucus from the amoeba which has become attached to the glass rod (Rhumbler, 1898; Jennings, 1904). According to Chambers (1924), a stationary amoeba accumulates a considerable amount of slime, by which it is attached to the substratum; an amoeba dragged out of position and then released will be pulled back toward its original position. However, Schaeffer (1920) has not been able to convince himself that amoebae secrete mucus; nevertheless, whatever the method, the tips of the pseudopods often adhere so firmly to the substratum that strong squirts from a pipette are necessary to dislodge them (Dellinger, 1906). Hyman (1917) applied a needle to the posterior end of an amoeba and pulled the animal in two, in spite of the tensile strength of the ectoplasm.

It is commonly held that attachment is one of the important factors in ameboid movement: when the organism is not attached there is no locomotion, although protoplasmic flow and the gel-sol and the sol-gel processes may be observed. Mast (1929) has studied the factors involved in attachment of *Amoeba* to the substratum and finds that simple agitation of the dish in which the amoeba are cultured may cause an increase in the firmness of adhesion. There was slight attachment in pure water (see also Parsons, 1926), but strong attachment over a great range of hydrogen-ion concentration, pH 4.6 to 7.8). All the salts tested by Mast caused a decrease in time for attachment, and an increase in firmness of attachment; since non-electrolytes have no effect, Mast believes that charges on the ions may change the surface charge of the amoeba or may change unknown internal forces. Bles (1929) has shown that under conditions of low oxygen tension, *Arcella* releases its attachment and floats to the surface.

According to Chalkley (1935), locomotor activity and attachment to the substratum are important factors in cytoplasmic fission in *Amoeba*. If one of the daughter cells is detached while the other remains attached, fission is not completed, but the unattached daughter flows back into the attached. However, if both amoebae are detached, as in distilled water, fission may be completed by the pushing of the pseudopodia of one animal against those of the other, thus breaking the narrow cytoplasmic bridge between the daughter cells.

In Testacea with lobose pseudopods, such as *Difflugia* and *Centropyxis,* the organism is pulled along by means of contraction of pseudopods, the tips of which have become attached to the substratum (Dellinger, 1906). That this attachment is of special nature and of considerable strength has been shown by Mast (1931a), who prevented the shell from being dragged along by the contraction of the pseudopods, under which conditions the pseudopods were torn loose and snapped back toward the shell. Testacea, such as *Difflugia, Lesquereusia* and *Pontigulasia,* attach to *Spirogyra* and devour the cell contents (Stump, 1935). According to Penard (1902), the filose pseudopods of such forms as *Pseudodifflugia* and *Cyphoderia* have as their principle function the fastening of the animal to the substratum, to which they adhere with extraordinary tenacity. *Acanthocystis ludibunda* (Helizoa) moves by adhesion of its axopods to the substratum, after which they contract, rolling the animal along over a distance twenty times its diameter in one minute (Penard, 1904). Schaeffer (1920) has estimated that the axopods of this form must adhere, contract so as to pull the animal along, and relax their hold, all in the short time of two seconds. The reticulose pseudopods of the Foraminiferan, *Astrorhiza limicola,* attach and contract much as do the lobopodia of the Testacea. Here, however, the organism leaves a trail of slime and bits of pseudopodia behind it, according to Schultz (1915). In *Gromia squamosa* the reticulose pseudopods play a very slight part in locomotion, but serve mainly as organs of attachment and food capture (Penard, 1902).

The whole outer layer of many rhizopods is sticky: it is well known that certain shelled rhizopods collect foreign particles to be included in their tests. According to Stump (1936), *Pontigulasia* will not reproduce unless shell materials are present in the cultures in which the organisms are grown; shell materials are collected just previous to division, but before nuclear changes have begun. Verworn (1888) has found that only after mechanical irritation of the pseudopodia of *Difflugia* do they become sticky enough for glass particles to adhere to them, to be later drawn into the shell. Foreign particles may adhere to the surface of a moving amoeba and be carried forward by the outer layer, sometimes making many complete revolutions. Parsons (1926) has observed that carmine granules adhere to and move over the surface of an amoeba floating in distilled water, although there has been a definite loss of capacity for adherence to substratum. Then, too, the outer layer of Foraminifera and Heliozoa is quite fluid, serving to capture food organisms which adhere to the surface. Certain of these organisms (i.e., *Actinophrys*) have been reported to form temporary colonies and to capture large objects of prey; such temporary colonies may be induced by mechanical means (Looper, 1928). However, it has been shown by Dawson and Belkin (1929) and Marsland (1933) that the adhesion between *A. dubia* and an oil surface is distinct from the process of ingestion, since a cap of oil may adhere to and spread over the tip of a pseudopod without ingestion taking place. The relation of adhesion to phagocytosis has been extensively studied in the amebocytes of the invertebrate Metazoa (see Fauré-Fremiet, 1930; Loeb, 1927, 1928).

A remarkable example of the stickiness of protoplasm is shown by the Choanoflagellata, in which the flagellum is surrounded by a protoplasmic collar; food particles, brought to the collar by the movements of the flagellum, adhere and are carried to the point of ingestion by the flowing protoplasm of the collar. Some flagellates (*Heteromita* and *Oikomonas*) have pseudopodia which are primarily adhesive in function. While flagella are primarily organs of locomotion, they frequently are used as organs of attachment. This is commonly observed in *Costia, Chilomonas, Heteromita, Pleuromonas, Anisonema,* and *Petalomonas.* The nature of this adhesion is not known.

Certain of the developmental forms of trypanosomes become attached by their flagella to the walls of the organ of the invertebrate host in which they are found. Lwoff (1934) has mentioned that *Strigomonas oncopelti* and *S. fasciculata* may swim free in cultures or may fasten to the glass side of the culture dish; they release upon a few seconds heating at 55° C. On the other hand, in *Leptomonas ctenocephali* the attached forms are held in position by the slimy enlarged tip of the flagellum, which is much shortened; they are killed at 55° C. before they can release their hold.

Adhesion phenomena have recently been introduced for diagnosing trypanosomiasis. In adhesion tests a drop of *Trypanosoma* suspension is added to one drop of equal parts of blood and 2-percent sodium citrate. If the blood comes from an infected animal, the red blood cells, and occasionally blood plates as well as bacteria, adhere to the trypanosomes. The immune serum contains an antibody that apparently acts upon the surface of trypanosomes to make it sticky. This test has been recently used by Taliaferro and Taliaferro (1934) in connection with other tests in equine trypanosomiasis; the adhesion phenomenon may persist for more than two years in animals that have been infected.

Fauré-Fremiet (1910) has emphasized the fact that while motility is the more general, it is not the only property of cilia. They may become immobile and serve as rigid stalks, or may be reduced to short rods and serve for fixation or for protection to the cell that bears them. It has frequently been observed that the movements of ciliates often cease temporarily when they come in contact with a firm surface. According to Jennings (1906) "the cilia that come in contact with the solid cease moving, and become stiff and set, seeming to hold the *Paramecium* against the object." Saunders (1925), however, holds that partially extruded trichocysts with sticky tips serve as temporary attachments in *Paramecium*. The hypotrichs generally can creep along vertical surfaces and on the under surfaces of the cover glass. The hypotrich *Ancystropodium maupasi* may attach itself by its posterior cirri (Fauré-Fremiet, 1908), and the holotrich *Hemispeira* by a bundle of fixative cilia (Fauré-Fremiet, 1905). Kahl (1935) thinks that such adhesion depends upon the ability of cilia to become sticky, at least at the tips. Cilia are associated with attaching organs in *Trichodina* and in *Ellobiophrya* (Chatton and Lwoff, 1923a). Chatton and Lwoff (1923b) have also described organs of attachment derived from the posterior ciliated region in the Thigmotricha. Fauré-Fremiet (1932) has studied the fixatory apparatus of *Strombidium calkinsi* which is constituted by two dorsal membranelles nearly as long as the body and made up of coalesced cilia which separate from each other at the adhesive distal extremity. At times the ciliate seems to be walking on the solid surface, the two membranelles moving one after the other, somewhat like the cirri of the hypotrichs. The membranelles are not contractile but very elastic; the *Strombidium* may unfasten itself all at once and swim hastily away. Similar fixing organs have been described for other oligotrichs (*S. urceolare* and *S. clavellinae*). *Metacystis lagenula* retracts by means of a large filament (a modified cilium glazed with sticky protoplasm), which adheres to the interior of the test, while *M. recurva* and *Vasicola gracilis* seem to be kept in their tests by a specialized cilium, a veritable bristle, which adheres to the inside of the test (Penard, 1922). *Mesodinium pulex* may adhere by means of tentacle-like structures, and various species of *Stentor* attach themselves to the substrate by means of cilia (Fauré-Fremiet, 1910) which form together with ectoplasmic extensions the so-called pseudopodia of the attaching disk of this form. The presence of slime-like or mucus secretions for attachment have been described for *Spirostomum* (Jennings, 1906), *Urocentrum turbo* and for *Strombilidium gyrans*. In the latter the slime, which is secreted by an attaching organ, the scopula, derived from modified cilia, congeals upon contact with the water into a very resistant thread which is attached to some structure in the medium (Fauré-Fremiet, 1910; Penard, 1922). The *Strombilidium* may then swing back and forth like a pendulum, held in position by the thread of slime which can be seen only because small particles of debris in the culture medium adhere to it. Certain observations of Chambers and Dawson (1925) on *Blepharisma* seem to show that the ability of cilia to combine into composite organelles, such as undulating membranes, membranelles, cirri, and so forth, may be dependent upon the presence of a slime-like secretion which spreads over the cilia and joins them into what looks like a homogeneous structure. At any rate these composite organelles are often seen to break down into their constituent cilia, after which they may be recombined. This separation of composite motor organelles into their components often takes place upon fixation.

According to Mast (1909), the prey of *Didinium* is held by means of the seizing organ, which in some way adheres to the surface when contact is made. Capture and ingestion of food depend upon the adherence of the seizing organ of *Didinium* and the strength of the ectoplasm of its prey, which is usually *Paramecium*. There are two different explanations of how the tentacles of the Suctoria adhere to their prey: by the presence of a viscid secretion, or by active suction of the hollow tentacles. However, the character of the outer surface of the prey is an important factor in its capture (Root, 1914).

SPECIFIC GRAVITY OR DENSITY

Whole organisms.—Since their specific gravity is slightly greater than that of water, most unicellular organisms can remain suspended in fresh or salt water only by the use of special locomotor organs such as cilia or flagella. Among the advantages of this slightly greater specific gravity is that the organisms are not caught in the surface film, nor in the congealing water on the surface when their locomotor activities are depressed because of low temperature; then, too, the simple methods of locomotion of Protozoa would hardly suffice to move bodies of very great density or to keep them suspended against the pull of gravity. In this connection the work of Jensen (1893) should be mentioned. In spite of the inaccuracy of his absolute measurements, it is clear that a *Paramecium* can lift 9 times its own weight in water. Probably a *Paramecium* with a specific gravity above 1.35 could not keep itself in suspension because a very insignificant amount of energy $\frac{1}{100} - \frac{1}{1000}$ of the total is available for locomotion in this form (Ludwig, 1928a, 1930).

Certain Protozoa of floating habit frequently have hydrostatic devices which aid in flotation, such as the gas bubbles secreted in the protoplasm of *Arcella* and *Difflugia;* special layers of vacuolated protoplasm, such as the calymma of Radiolaria; or very highly vacuolated protoplasm, as in *Noctiluca* and in Heliozoa.

According to Bles (1929) the gas bubbles of *Arcella* are formed in the marginal protoplasm and are filled with oxygen. They are secreted when the oxygen tension is reduced experimentally and are adaptive, in that they reduce the specific gravity of the organism so that when oxygen tension is low the organism may float to the surface, where the oxygen tension is always somewhat higher. When *Arcella* is turned upside down so that the external pseudopods cannot adhere to the substratum, gas bubbles appear in from three to six minutes, before the animal begins to right itself and aid in this process by lowering the specific gravity. The bubbles disappear rapidly after the righting process is completed. When there is more than one bubble present all grow at the same time and decrease at the same time. In the Radiolaria periodic migrations take place to and from the surface layers of the sea; these are brought about by changes in the vacuolar contents of the hydrostatic layer, which, according to Brandt (1885), is lighter than water. Schewiakoff (1927) has described the presence of a clearly defined gelatinous hydrostatic layer in the Acantharia (Radiolaria). Although the specific gravity of *Noctiluca* (1.014) is less than that of sea water (1.026) in which it floats, according to Massart (1893), E. B. Harvey (1917) has shown that this form can lessen and increase its specific gravity in a regulatory fashion. Lund and Logan (1925) have shown that the increase in specific gravity, following strong mechanical shock or electrical stimulus, is caused by the coalescence of large vacuoles and the liberation of their contents, which diffuse through the pellicle. The density of the solution in the vacuoles is less than that of sea water; according to Ludwig (1928b) this is because of its lower salt content, its osmotic pressure being about half that of sea water. Most marine animals possess body fluids almost isotonic with the external fluid, but *Noctiluca,* together with the marine teleosts, are hypotonic and, to a great extent, osmotically independent. Such organisms must be impermeable to water, absorb water in some way without salts, or take in sea water and excrete salts. Marine teleosts apparently take in sea water and excrete the excessive salts, and, since the membrane of *Noctiluca* is permeable to water, osmotic work (negative osmotic force) must be done by the membrane (E. B. Harvey, 1917) in maintaining this steady state instead of osmotic equilibrium.

The protoplasm of marine Protozoa frequently becomes much vacuolated upon transfer to fresh water. The marine variety of *Actinophrys sol,* according to Gruber (1889), has thick, granular protoplasm poor in vacuoles and entirely lacking a contractile vacuole; during gradual transfer to fresh water the protoplasm becomes foamy with bubbles and a contractile vacuole appears, so that the organism is indistinguishable from the fresh-water variety. The formation of vacuoles and the entrance of water into them undoubtedly lowers the specific gravity in this form, when transferred from salt to fresh water. The spine-like pseudopodia of Heliozoa, Radiolaria, and other floating forms also serve as a protection against sinking.

The first estimation of the specific gravity of a protozoan apparently was that of Jensen (1893), who attempted to determine the energy relations of the movement of *P. aurelia*. He obtained the value of 1.25 by suspending the organisms in solutions of potassium carbonate, a procedure which gave too high values because of excessive shrinkage, due to osmotic pressure. Later Platt (1899) suspended killed or anaesthetized *Paramecium* and *Spirostomum* in solutions of gum arabic and found their specific gravity to be 1.017. Lyon (1905) centrifuged living *Paramecium* in solutions of gum arabic and obtained 1.048 or 1.049. This was repeated by Kanda (1914, 1918), who finally arrived at a value of 1.0382 to 1.0393 for *Paramecium* and 1.028 for *Spirostomum*. Fetter (1926) utilized approximately the same value, 1.038, which she obtained by centrifuging *Paramecium* in sugar solutions, in calculating the protoplasmic viscosity of that form.

Leontjew (1927) has determined the density of various Protozoa (*Fuligo, Stemonitis,* slime molds; *Naegleria,* an *amoeba;* and *Dunaliella,* a flagellate) to be 1.020 to 1.065. Some of his readings on *Fuligo varians,* obtained with a micropyknometer, are interesting enough to be mentioned in detail: in moist weather the density was 1.016; in dry, 1.040, and 11 hours before spore formation, 1.065.

Heilbrunn (1929a, 1929b) used 1.03 as the specific gravity of the protoplasm in his studies on viscosity of *A. dubia*. Motile amoebae (*Naegleria*) have a density of 1.043, according to Leontjew (1926a), and cysts 1.060 to 1.070 (Joschida, 1920, cited by Leontjew, 1927); cysts of *Hartmanella hyalina,* a soil amoeba, have a specific gravity of 1.084 (Allison, 1924).

It is, of course, generally recognized that the protoplasm of encysted Protozoa contains less water than that of active forms. Allison (1924) determined the specific gravity of cysts of *Colpoda* by the time required to fall through water. He finds that four-day cysts averaging 40.1 microns in diameter, have a density of 1.042; while twenty-day cysts, averaging 25.1 microns, have a density of 1.061. Similar results were found for cysts of *Gonostomum*. The decrease in size and increase in specific gravity are apparently caused by water loss.

The specific gravity of protoplasm other than that of Protozoa has been found to vary from about 1.02 to 1.08, with average values about 1.045. The publication of Pfeiffer (1934) gives a résumé of the methods and results of such studies.

Relative specific gravity of cell inclusions and components.—It has long been known that the various inclusions of protozoan cells are of different specific gravities. In centrifuging cultures to obtain large numbers of organisms for fixation previous to morphological studies, it is often noticed that certain crystals have been displaced centrifugally. Mc-Clendon (1909) was one of the earliest workers to fix and stain *Paramecium* after long-continued centrifuging; he found that the crystals and nucleus were displaced centrifugally. Heilbrunn (1928) mentions that a centrifuged *Euglena* loses its spindle-shaped contour and becomes spherical, with the granular inclusions packed at the centrifugal end. The same author (Heilbrunn, 1929b) has used the speed of movements of crystals centrifugally through the cytoplasm of *A. dubia* to estimate the absolute viscosity of the protoplasm; the specific gravity of the crystals was estimated to be approximately 1.10. E. N. Harvey (1931) records that the crystals of *A. dubia* fall down so rapidly that their velocity can hardly be determined in the microscope-centrifuge; and that the crystals of *Paramecium* were rapidly thrown down, as was the nucleus. He was also able to cleave living *Stentor* into two parts in the microscope-centrifuge; the lighter, oral half contained none of the *Zoöchlorellae* which had been moved into the basal part. E. N. Harvey and Marsland (1932) observed the movement of cytoplasmic particles through the protoplasm of *A. dubia* and found them to be layered out in the following order: coarse granules and crystals, most centrifugal; nucleus, a visibly empty zone, a zone of fine granules, and, most centripetal, the contractile vacuole. Mast and Doyle (1935b) have recorded as follows the relative specific gravities of the various cytoplasmic components in *A. proteus,* from centrifugal to centripetal: refractive bodies, *beta* granules (mitochondria) and food vacuoles containing little or no fat, nucleus and food vacuoles containing much fat, hyaline protoplasm, contractile vacuole, crystal vacuoles without crystals, and fat globules. The position of the crystal vacuoles varies with the size of the included crystals: those with large crystals are heavy and move centrifugally in the centrifuge; those with small crystals are lighter. The small *alpha* granules, which are about 0.25 micron in diameter, are not layered out. All the refractive bodies, a large proportion of the crystals, and all the fat may be removed, with no injurious effects, from a centrifuged amoeba by cutting off the light and heavy ends. However, removal of the *beta* granules (mitochondria) resulted in the death of the amoeba. Singh (1939) has also centrifuged *A. proteus* 'Y,' and found the order of layering to be: nutritive spheres, nucleus, crystals, neutral red bodies, mitochondria, cytoplasm, contractile vacuole, and fat.

Patten and Beams (1936) centrifuged *Euglena* and found that the chloroplasts form a middle belt, having on the centrifugal side paramylum and neutral red bodies, while the clear cytoplasm containing the small spherical mitochondria is at the centripetal pole. In *Menoidium* the heaviest inclusions are the paramylum and neutral red bodies; in *Chilomonas* the starch and neutral red bodies are heaviest. Johnson (1939) has confirmed the results of Patten and Beams; in *Euglena rubra,* however, hematochrome is present and is displaced to the centripetal pole with mitochondria.

King and Beams (1937) ultracentrifuged *Paramecium;* in this form the various components and inclusions were layered in the following order from centrifugal to centripetal: crystals in vacuoles, compact chromatin of the macronucleus, food vacuoles and neutral red inclusions, achromatic matrix of the macronucleus, endoplasm, large clear vacuoles, and fat. Here the chromatin may be removed from the achromatic matrix of the macronucleus; the chromatin regenerates a macronucleus; the achromatic matrix persists for some time and apparently interferes with subsequent divisions. Browne (1938) has ultracentrifuged *Spirostomum* and has found the contents of the cell to be layered as follows: centrifugally located are the mitochondria, food vacuoles, and macronucleus; cytoplasm; Golgi bodies; and centripetally, vacuoles.

Daniels (1938) has used the ultracentrifuge in a study of gregarines; here the paraglycogen and chromidial granules are heaviest; next, the mitochondria and nucleus; cytoplasm; then the larger Golgi bodies; and lightest the smaller Golgi bodies and fat globules. In the gregarines studied the karyosome moved centrifugally in the nucleus, and the contents of the deutomerite layered independently of those in the primite because of the presence of the transverse septum.

It is obvious that the centrifuge may serve as an important research tool for the identification and study of the form, relative volume, and other characteristics of the components and inclusions found in protoplasm. For example, Holter and Kopac (1937), by cutting amoebae in half after centrifuging, were able to demonstrate that the enzyme dipeptidase is apparently associated with the cytoplasmic matrix, independent of all cytoplasmic constituents which could be stratified by centrifugal acceleration.

OPTICAL PROPERTIES

The ordinary optical characteristics of protoplasm, such as its transparency, color, and refractive index, do not seem to be of very great importance except that the observation of living protoplasm is conditioned by these properties. Too many have assumed that because a structure cannot be seen in living protoplasm, it is therefore nonexistent.

TRANSPARENCY

The protoplasm of the protozoan cell is generally transparent or translucent, but in the presence of granular or other inclusions it may appear to be opaque or nearly so. Many of the differentiations are so nearly of the same index of refraction that special fixing and staining methods are necessary in order to study them. The state of aggregation of the colloids of the general protoplasm seems to be dependent, at least to some extent, on the salt content of the surrounding medium. Thus *Actinophrys sol* in sea water is densely granular, while in fresh water it is alveolar and translucent (Gruber, 1889). Spek (1921) has shown that *Actinosphaerium* becomes relatively opaque in certain salt solutions, and that *P. bursaria,* which is glass-clear, becomes dark brown in artificial salt solutions, owing to the collection of albuminoid substances into large aggregates. Certain observations of Schaeffer (1926) on marine *Amoebae* are of interest here: *Flabellula pellucida,* a most transparent marine amoeba, becomes densely granular in 25-percent sea water, while *F. citata,* another marine amoeba, is unusually transparent in 364-percent sea water.

COLOR

Protoplasm is usually observed to be colorless or grayish; many of the shades of blue, green, or yellow described for Amoebae are merely diffraction phenomena or subjective in nature. The color of the endoplasm of various amoebae has been described by Schaeffer (1926) as pale bluish-green, yellowish-green, bluish-gray; *Hyalodiscus elegans* has endoplasm which is orange-yellow centrally and ashen-gray peripherally. The contents of the contractile vacuole are often slightly pink. This may be regarded as an optical illusion, since the color observed is complementary to the usual bluish-green of the endoplasm. Very frequently the color is obscured by the presence of colored inclusions of various kinds, or caused by colored inclusions of the same index of refraction as the protoplasm and therefore difficult to differentiate.

Ciliates may be colorless, gray, pink, blue, or violet. The blue color of *S. coeruleus* is caused by a coloring matter, called stentorin, diffused through the cytoplasm, but in *Blepharisma* the color, which may vary from none through pink and violet to purple, and varies with the cultural conditions and from individual to individual in the same culture, is apparently concentrated in the pellicle (Nadler, 1929). According to Jennings (1906), most colorless Infusoria do not react at all to a light of ordinary intensity; this has not been tested with forms such as *Blepharisma*, in which the color varies.

In the plant-like flagellates color is usually caused by chromatophores, which may be green, blue (Lackey, 1936), brown, or yellow. The most interesting colored inclusion is the hematochrome, found in such forms as *E. rubra* as red granules from 0.3 to 0.5 microns in diameter. These euglenae form a green scum in shaded places; the green chloroplasts mask the hematochrome, which is centrally located; in direct sunlight the scum is red, the hematochrome being peripherally located and masking the chloroplasts. Control of the distribution of the hematochrome is so delicately balanced that if the euglenae are shaded for fifteen minutes, they change from red to green (Johnson, 1939). The mechanics of this control needs to be investigated.

REFRACTIVE INDEX

Even the finest strands of protoplasm can be seen in water, in spite of the fact that they may be transparent and colorless. This is because of their relatively high index of refraction. It is surprising that so little is known about the optical characters of protoplasm which may be seen to change during cell division. Schaeffer (1926) has shown that the nuclei of certain marine amoebae become much more prominent by dilution of the sea water with fresh water, and Chalkley (1935) has shown that there is a change in the refractive index of the nuclei of *A. proteus* during division: the interkinetic nuclei can easily be seen at a magnification of 200 diameters; those in division cannot. Very few measurements of the refractive index of protoplasm have been made; Frederikse (1933b) has reported a value of 1.40 to 1.45 for *A. verrucosa;* Mackinnon and Vlès (1908) 1.51 for the cilia of *Stentor,* and 1.56 for the flagellum of *Trypanosoma (Spirochaeta) balbiani.* Mackinnon and Vlès made their determinations by immersing the organisms in media of different refractive indices; double refraction, due to depolarization, disappears in media of the same refractive index as the cilia and flagella. Fauré-Fremiet (1929) found the index of refraction for entire amoebocytes of *Lumbricus* to be 1.400, for the hyaloplasm 1.364; of *Asterias* to be 1.446 and 1.385 respectively (for methods see Pfeiffer, 1931).

STRUCTURAL PROPERTIES

It has long been known that the polarity of a cell may persist after the relative positions of its various visible constituents have been changed; this has led to the idea that polarity has its basis somehow or other in the structure of homogenous cytoplasm, which remains unchanged in spite of exposure to high centrifugal forces (Conklin, 1924). Polarity and symmetry are generally present in the Mastigophora and Infusoria, in which an anterior-posterior axis is usually persistent throughout active life, and has been described as present in cysts (Lund, 1917, in *Bursaria*). However, in Sarcodina such as *Amoeba,* polarity may be thought of as continually changing, being bound up with the gel-sol process at the temporary posterior end, the flow of protoplasm forward, and the sol-gel process at the temporary anterior end. Hyman (1917) has demonstrated that the temporarily differentiated anterior end of *Amoeba* is the region of highest susceptibility to cyanide. Mast (1931) and others have reported that electrical currents have a solating action on the plasmagel, on the side directed toward the cathode.

Recently Chalkley (1935) has studied the process of cytokinesis in *A. proteus.* He observed, with the onset of prophase, a loss in sensitivity, a swelling up of the organism, a decrease in activity of the contractile vacuole, and an increase in movement of the granules in the region of the nucleus. With the separation of the daughter chromosome plates, he observed a flow of the cytoplasm from the equator, in the same direction as the separating daughter plates. As the daughter plates approach the surface of the cell and the new nuclei begin to form, a solation of the plasmagel takes place, resulting in the formation of numerous pseudopodia which become attached and undergo active ameboid movement. At the same time, because of the flow of the cytoplasm from the equator toward the two poles, the region of the *Amoeba* at the equator has become narrowed to a thin neck. Presumably a solation of the plasmagel at the equator, together with a pull exerted by the two actively dividing ameboid daughter cells, produces the final separation of the organisms. The temporary polarity becomes immediately lost after the completion of division.

The physical-chemical factors involving the change in polarity of the protoplasm at the equator and the forces responsible for the flow of the protoplasm from this point are unknown. However, Chalkley thinks the fundamental principles involved in these processes are the same as those described by Mast for ameboid movement, namely sol-gel transformation. Further, it has been shown by Chambers (1938) that if the nucleus is moved toward the cell surface, pseudopodia are induced in that region. Becker (1928) has demonstrated that the factor which determines the direction of streaming and hence polarity is located in *Mastigina,* a flagellated amoeba, in the region of the nucleus. If the nucleus is moved posteriorly, streaming ceases and is then resumed toward the nucleus.

It should be pointed out that in the Protozoa nuclear division and cytoplasmic fission may be closely correlated or widely separated in tempo, and that they often exhibit a considerable degree of independence. However, the plane of separation of the nucleus usually determines the plane of cytokinesis, in that they usually take place at or nearly at right angles to each other. In organisms like the Mastigophora, the plane of nuclear division is parallel to the anterior-posterior axis and coincides with the plane of cytokinesis of the organism. In Ciliata the plane of nuclear division is perpendicular to the anterior-posterior axis and coincides with the plane of cytokinesis.

In some ciliates there appears to be a definite and permanent division zone laid down early in the life of the organism, which is not disturbed by diverse mutilations of the body (Calkins, 1926). Furthermore, in *Frontonia* this zone differs so markedly from the surrounding cytoplasm that it can be easily seen in the living condition (Popoff, 1908).

Multilation studies by Calkins (1911) and by Peebles (1912) on *Paramecium* have resulted in the production of numerous monsters. Peebles describes this condition as due to an upset in nuclear and cytoplasmic division tempo; thus when the nucleus is ready to divide, the cytoplasm is not, and vice versa. If this be true, the mitotic and cytokinetic phenomena in this form must be closely integrated, and the division mechanism of the organism as a whole be dependent upon the proper coördination of both the nuclear and the cytoplasmic division processes.

Child and Deviney (1926) and Child (1934), have shown that in ciliates generally there is an anteroposterior gradient, due to the existence of a physiological gradient in the longitudinal axis. The anterior end is more susceptible to many agents, and there is also an axial differential in the rate of reduction of methylene blue. Child is of the opinion that this metabolic gradient is the only basis of physiological polarity. Lund (1917, 1921) has found that reversal of polarity often occurs in cut halves of *Bursaria* undergoing regeneration; it may also occur in normal animals. An indication of this change in polarity was a reversed beat of the cilia. He further found that *Paramecium* showed a reversed beat of cilia, in direct electrical currents of proper strength. Verworn (1899) has shown that paramecia and other ciliates orient themselves with the anterior end of their bodies toward the cathode to which they swim. On the other hand, many flagellates show an opposite behavior.

Schaeffer (1931) has presented evidence that the protoplasm of the amoebae, and presumably of other organisms, consists primarily of specific molecules which are organized into definite patterns, and that most or all of the characteristics of the organisms are due to or correlated with positional relationships of the molecules.

It is generally thought at the present time that adjacent protein molecules, because of their multipolar character, have an orienting effect upon one another and that the resulting configuration may be equivalent to a net-like structure, extended in three dimensions. That this is true may be inferred from the anomalous viscosity of solutions of proteins and protoplasm: they show non-Newtonian flow, i.e., their viscosity varies with the stress applied, although they may outwardly conform to true fluids in being free from rigidity.

Bensley (1938) has recently isolated from the cytoplasm of liver cells a material called plasmosin, which he thinks is constituted of linear micelles.

The formation of a fiber results from the end-to-end orientation of these linear micelles. In protoplasm in the liquid state these micelles are probably independent and irregularly arranged but in flowing protoplasm they would be oriented parallel to the axis of flow. . . . From this state by simple end-to-end combination all changes in viscosity are possible up to the formation of a fibrous gel . . . or even discrete fibers.

Bresslau (1928) has also shown that "tektin," a material extruded by ciliates, has anisotropic properties somewhat similar to those of plasmosin.

The fact that the chromosomes probably represent gene-strings has been of enormous importance in determining our ideas of significant protoplasmic structure. The chromosomes apparently reproduce themselves at each cell division, so that their individuality is retained in all cell generations. The demonstration of the presence of these linear aggregates of visible size, which are self-perpetuating, cleared the way for the micellar theory of protoplasm structure.

Much of the evidence for the presence of linear aggregates in protoplasm has been obtained by microdissection, by the use of dark-field and polarization microscopes, by studies on cohesion and swelling, and by X-ray diffraction methods. Some of these have already been considered and others will be discussed below.

ELASTICITY

According to Seifriz (1936), "elasticity is the best indication we have of the structure of living matter" and is evidence for the presence of linear aggregates. A body is said to be elastic if after having been strained it tends to return to its original form when the stress is removed. Volume elasticity is characteristic of fluids and solids; shape elasticity (rigidity) of solids and colloids in the gel condition generally. The form assumed by the bodies of various flagellates and ciliates is characteristic and offers means of identification in many instances. Relative rigidity is of common occurrence among those Protozoa, such as *Euplotes,* which have a differentiated pellicle and a firm ectoplasm. C. V. Taylor (1920) demonstrated this elasticity by applying pressure with a microneedle; the body bent conspicuously over the needle but returned to normal shape upon release of the pressure. Animals cut two-thirds across may keep their shape; this argues for a stiff consistency of the ectoplasm, as well as a tough pellicle. Any apparent modification in the shape of E. *patella* occurs only from outside pressure, since the animal is unable to vary its shape. Other hypotrichs may be similarly armored, as is shown by the fact that they are broken in the ultracentrifuge (King and Beams, unpublished work) so that fragments are found swimming around as though the whole animal were brittle rather than plastic. In other forms, such as *Paramecium,* the body may be constricted when the animal forces its way through obstacles. Upon ultracentrifuging in gum solutions paramecia become much elongated and thin, because of the presence of materials of different specific gravities in the cell. Such elongated animals may survive and return to their normal shape unless the pellicle and ectoplasm have been strained beyond the limit of their elasticity, in which case the structure responsible for return to normal shape has been destroyed and they die, permanently deformed (King and Beams, 1937). That the form of *Paramecium* is determined by the relatively firm outer layers has been shown by some observations of Chambers (1924), who tore the ectoplasm with a microneedle. The fluid interior pours out into the surrounding water and the ectoplasm soon disintegrates; but occasionally the fluid endoplasm forms a delicate surface film which maintains the integrity of the extruded mass. Merton (1928) has studied these so-called autoplasmic paramecia which, deprived of pellicle, cilia, trichocysts, and ectoplasm, take on the form of a fan-shaped amoeba, which may live for some days, divide, and exhibit locomotor activities. Under unfavorable conditions a rayed stage, with long pseudopodia-like extensions reminiscent of a A. *radiosa,* may be assumed. That the pellicle is not the only element involved in rigidity has been shown for *Blepharisma* by Nadler (1929). The pellicle of this form may be shed after immersion in weak solutions of strychnine sulphate. The "naked" animal emerges from the old pellicle with the shape and elasticity characteristic of the species. Eventually the pellicle is reformed, and the process may then be repeated.

That amoebae have elasticity of form to a considerable degree has been shown by Jennings (1904), who bent a pseudopod with a glass rod; when released the pseudopod sprang back into its original position. Whole amoebae were also bent, with subsequent return to original form after release; Dellinger (1906) and Hyman (1917) have repeated these experiments, as indeed may be done by anyone. Howland (1924c) has stretched the outer layer of *A. verrucosa* with microneedles; upon release the animal recovers its normal shape, apparently unharmed. The protoplasm of plasmodia of slime molds (Seifriz, 1928) is at times poorly elastic, and at other times it may be stretched into very fine, long threads which snap back a goodly distance when released.

Seifriz has also determined elastic values by inserting minute nickel particles into the protoplasm of slime molds and attracting these particles electromagnetically. On release of the current the metal particles return to their original position; the distance traveled is measured and used as an indicator of elasticity. A maximum stretching value of 4.4 microns was obtained for liquid, previously streaming protoplasm of myxomycetes, a maximum value of 292 microns for quiescent, highly viscous exuded masses of protoplasm from plasmodia. This latter value is slightly greater than that for gelatin solutions and slightly greater than that of fresh egg albumen.

The long thread-like pseudopodia of Foraminifera, which usually pull the organism along by adhering to the surface terminally and then contracting, have been shown to be elastic by Schultz (1915), who cut these and observed them to snap back like a rubber band.

The reticulose pseudopodia (myxopodia) of the Foraminifera are very different from the lobose and filose forms in other Rhizopoda. The former have a soft miscible outer protoplasm which leads to fusion on contact with one another and a relatively rigid inner axial structure which shortens without wrinkling when the pseudopod is withdrawn. As this denser core is formed as an elongation "in the direction of growth, strains will be set up during the process which will give rise to ordered and preferential arrangement tending toward the crystalline state" (Ewles and Speakman, 1930). Thus it will be seen that the axial solid protoplasm of these myxopodia, although it is not in the form of a fiber, serves the same function as the axial filament of the Heliozoa and Radiolaria.

Up to this point elasticity has been considered principally in connection with the protoplasm itself, or in its temporary completely reversible structures. There remain for consideration those differentiations which last the whole life of the organism and are usually irreversible, such as flagella, axial filaments, cilia, myonemes, and supporting fibers (morphonemes) of various kinds. The fact that these are elastic is too well known to need more than mention here. Spindle fibers, on the other hand, are often thought to be artifacts, but Cleveland (1935), working on the hypermastigote flagellates, has pulled the centriole out of position; the chromosomes were also displaced, but both centriole and chromosomes immediately sprang back into position when released. This argues not only for the reality of the chromosomal and spindle fibers, but also that they are structures of considerable elasticity. If one considers the aphorism of Needham (1936) "that biology is largely the study of fibers," these fibrillar structures of the Protozoa are of great interest because they consist of parallel aggregations of the submicroscopic elongated particles (micelles) of protoplasm (see Taylor, *infra,* Chapter IV).

CONTRACTILITY

We may distinguish between active contraction, as in muscle fibers, and elastic shortening after having been stretched, as in elastic fibers in the higher animals. Although there are many examples of contractility and elasticity, in the Protozoa, associated with differentiated myonemes or morphonemes respectively, there are also many instances of active contraction in the absence of any optically differentiated structure. According to Lewis (1926), theories of contractility must be based on the presence of a contractile molecule, because the fibrillae seen in heart muscle in tissue culture are not "true" cytological structures but are due to reversible gelation. Fauré-Fremiet (1930) also holds that the gelified condition is often bound up with the existence of internal fibrillar structures, which disappear when solution occurs. That such fibrillae appear and disappear may, of course, be caused by aggregation and disaggregation of smaller invisible fibrillae, or may even be due to changes of refractive index. It is well known that objects that are of the same refractive index, transparency, and color cannot be seen, even in the dark field (Schmidt, 1929).

It is generally assumed that contraction of the gelled ectoplasmic cylinder in *Amoeba* forces the more fluid endoplasm forward (Schaeffer, 1920; Pantin, 1923; Mast, 1926b). The contraction is thought to be caused by the fact that the gel-sol process at the posterior end of the amoeba results in an increase of volume and so stretches the gelled cylinder of protoplasm; the resulting elasticity forces the sol forward where a decrease in volume has occurred upon gelation of the plasmasol. However, in shelled forms, such as *Difflugia* (Dellinger, 1906; Mast, 1931a) and *Centropyxis*, there is an active longitudinal contraction of the plasmagel cylinder which results in locomotion by pulling the shell along. That the circular elastic contraction in *Amoeba* and the longitudinal active contraction in *Difflugia* are different mechanisms may be doubted. However, Mast (1931a) has shown that *Difflugia* deprived of their shells move much as does *Amoeba*. It is to be noted that the source of energy is in the ectoplasm, so that the streaming of the endoplasm in these forms must be of a different nature from cyclosis in other forms such as *Paramecium* or *Frontonia*, where protoplasmic streaming is extremely difficult to explain in terms of contraction.

There are many examples of local contractility in the literature of ameboid movement. Swinging and revolving movements of lobose pseudopods when not in contact with the substrate have been described by Penard (1902), Jennings (1906), Hyman (1917), Kepner and Edwards (1917), and many others. These differential local contractions of the ectoplasm often approximate muscular activity, according to Kepner and Edwards. Schaeffer (1926) has described corkscrew-shaped pseudopodia in *Astramoeba flagellipoda*, with from two to eight spirals which wave about quite like flagella, often making a complete revolution in three seconds.

The most spectacular instances of local contraction are those in which an *Amoeba* pinches a large ciliate in half. Mast and Root (1916) describe this process as taking ten seconds for *Paramecium* and show that it cannot be explained in terms of the surface tension of the *Amoeba*. Beers (1924) describes the constriction of *Frontonia* by *Amoeba* until the former was dumbbell-shaped, and ascribes the pinching to centripetal pressure exercised by an extending collar of protoplasm which pinched the prey in half in eight minutes. Kepner and Whitlock (1921) saw a partly ingested *Paramecium* constricted much as described by Beers for *Frontonia*, except for the loss of the cilia from the ingested part. The figures and descriptions of the two latter instances are very similar to those of Grosse-Allermann (1909) for ingestion by invagination in

A. terricola, and of Mast and Doyle (1934) for the ingestion of water by various species of *Amoeba* in albumin solutions. Penard (1902) has described and figured the pinching off of an extensive injured portion by *A. terricola* as has Jennings (1906) for *A. limax.* The process of egestion, as figured by Howland (1924c) for *Amoeba verrucosa,* seems also to involve extensive local contraction.

The lobopodia and filopodia of the Amoebida and Testacea are solid peripherally with a central fluid region, while the pseudopods of the Radiolaria and Heliozoa (axopodia) and those of the Foraminifera (myxopodia) are more fluid peripherally and more solid axially. In axopods and myxopods there can be no flow of endoplasm caused by the elastic contraction of a gelled ectoplasmic cylinder. Roskin (1925) describes the origin of the axopods of *Actinosphaerium* by the flowing together and alignment of fibrillae, which eventually fuse into a hollow tube, filled with fluid which is associated with rigidity and contractility. That these axopods do contract rapidly has been shown for *Acanthocystis,* which, according to Penard (1904), may traverse twenty times its own diameter by rolling along the tips of the axopods, which must adhere, contract, and then release very rapidly. In the foraminiferan *Astrorhiza,* according to Schultz (1915), pseudopods stretch out five to six times the length of the body, make "feeling" movements and may finally either adhere to the substratum or contract and be withdrawn. The organism is usually fastened by three bundles of pseudopods; if one of these be torn loose, the others contract rapidly and the animal is pulled forward. The axial, more solid stereoplasm of these pseudopodia is distinctly fibrillar. Schmidt (1929) has also described the formation of contractile pseudopodia in the Foraminiferan *Rhumblerinella* by the alignment and coalescence of fibrillae.

The minute size of cilia and flagella makes it extremely difficult to determine whether their movements are due to active contraction or to changes occurring in the cell which bears them.

In addition to their usual method of locomotion by means of flagella, many Mastigophora such as the euglenoids show euglenoid movement, or metaboly. In these forms there are present in the pellicle more or less spiral striations, which apparently are elastic in nature and tend to preserve the form of the organism during and after the contraction of the superficial layers of the body. In some cryptomonads (e.g., *Chroömonas pulex*) springing movements are brought about by the strong contraction of the outer layer of the body, the resulting locomotion being independent of flagellar movement. Certain flagellates have an extraordinary superficial resemblance to medusae both in appearance and in method of locomotion: (Cystoflagellata: *Leptodiscus, Craspedotella;* Phytomonadida: *Medusochloris;* Dinoflagellata: *Clipeodinium*). In these types according to Pascher (1917), the movements represent a special form of metaboly and two mechanical systems should be present, one radial and dilating, the other peripheral and contracting. The former could of course be elastic only and bring about passive return after the contraction of the latter.

In the Infusoria there are many examples of specialized retractile organelles: the tentacles of Suctoria may be retracted and extended much as may the axopods of the Heliozoa. The structure of these tentacles is quite similar to that of the axopods(Roskin, 1925). The remarkable tentacles of the ciliate *Actinobolina* may be extended to a length twice the diameter of the body or may be completely retracted. They are associated internally with two groups of fibrils which seem to wind up to retract and unwind to extend them (Wenrich, 1929). In many ciliates, such as *Stentor* and *Spirostomum,* there are actively contractile myonemes in the ectoplasmic layer. In other related forms, such as *Climacostomum,* the corresponding structures are elastic only and have been referred to as morphonemes. The myonemes may even appear to be striated (Dierks, 1926, and others) but this has been denied (Roskin, 1923).

A most unusual case of extension and retraction occurs in the ciliate *Lacrymaria olor.* Here the "neck" may be extended to fifteen times the length of the body, the form of which remains unchanged. This "elasticity" is associated with the presence of what appears to be a series of spiral striations. However, upon complete extension of the neck, Penard (1922) has observed that there is only a single continuous spiral. No experimental work has been done upon this form, so that the nature of the extension and retraction is not understood. However, Verworn (1899) has cut the neck of *L. olor* free from the head and body. The neck retains its extensile and contractile properties exactly as when in connection with the body.

The finer structure of the contractile stalks of the Vorticellidae has been studied by Koltzoff (1912), Fortner (1926), and many others.

The stalk consists of an external wall, an inner liquid, and a spirally wound contractile cord. Within the contractile cord is an excentrically placed myoneme often called a spasmoneme, the contraction of which causes the stalk to become coiled like a spring. The distal end of the stalk is usually attached to the substratum, but in *Vorticella natans* and *V. mayeri* the organisms are never attached but swim, stalk first, through the water. In the former the stalk rolls up into narrow spirals, but in the latter the stalk swings in a wide loop on contraction somewhat like a flagellum. Bĕlehradek and Paspa (1928) have reconstructed a myogram from moving pictures of the stalk of a vorticellid and find that the spasmoneme does not function like a true muscle but like a modified flagellum. Various attempts have been made to explain the contraction and extension of the stalk of vorticellids as caused by the complex action of two opposing bundles of fibers, or in terms of internal pressure against coiling brought about by elastic fibers.

Myonemes are also well developed in the gregarines, where longitudinal and circular myonemes are apparently responsible for bending and peristalsis-like movements. Myonemes called myophrisks are also present in certain Radiolaria. Here they are associated with the spreading out of the gelatinous cortical layer, previous to their decrease in specific gravity and subsequent rise to the surface (Schewiakoff, 1927).

ROPINESS, OR THREAD FORMATION

Living material is said to be ropy if it can be drawn out into threads; ropiness is thought to be due to the micellar structure of the material. The ability of a drop of a pure liquid to resist distortion is due to its surface tension only, and in heterogenous mixtures the surface film may approximate a solid consistency. A column of fluid breaks up into a number of smaller spherical bodies. The formation of pseudopodia, especially of filopodia, axopodia, and myxopodia, demonstrates the presence of solid structures in protoplasm. Then, too, protoplasm may be drawn out into fine strands of considerable elasticity and tensile strength: the ectoplasm of *Amoeba* has been drawn out into fine strands by Hyman (1917) and many others; Schultz (1915) was able to draw out the protoplasm of the foraminiferan *Astrorhiza* into long threads; this was more marked in the outer layers than in the inner mass of protoplasm.

Many other observers, using microdissection methods, have confirmed this for a whole series of animal and plant cells. The spinning of threads from protoplasm is generally assumed to be dependent upon the submicroscopic fibrillar structure of the protoplasm, the latter being responsible for its elasticity and high degree of extensibility. The tensile strength of a strand of myxomycete protoplasm has been found by Pfeffer (quoted by Seifriz, 1936) to be 50 mgm. per square millimeter. The ropiness of protoplasm is of course conditioned by temperature, viscosity, hydrogen-ion concentration, and other factors which affect protoplasm (review by Jochims, 1930).

Further evidence of the micellar nature of protoplasm may be deduced from the experiments of Seifriz (1936), who has examined the protoplasm of slime molds with a Spierer lens. When the protoplasm is quiet it presents a mosaic appearance, but when in active flow or when formed into a thread is presents a striated appearance. Harvey and Marsland (1932) noted that the crystals of *Amoeba* fall in jerks when moving under centrifugal force; this may be due to the presence of structural elements in the protoplasm. Moore (1935) has found that slime molds in the plasmodial stage will flow through pores 1 μ in diameter. He also forced this living material through bolting cloth of various mesh sizes. The plasmodia survived after being forced through pores 0.20 mm. in diameter or larger, but died when forced through smaller pores. Moore thinks that there are fibrillar elements in slime mold protoplasm which are destroyed if forced unoriented through a pore through which they could flow if the micelles were properly oriented.

DOUBLE REFRACTION

The great advantage of the use of the polarizing microscope is that characteristic structure which is otherwise imperceptible may be revealed, even in the living organism, without any alteration of the specimen. Amorphous and pseudoamorphous materials (i.e., materials in which the orientation of the particles is a random one) are dark when viewed between crossed nicol prisms. True double refraction is always the result of an orientation of optically anisotropic elements; many substances, such as glass, become doubly refractive if deformed by external forces. Many sols when in movement or when placed in an electric or a magnetic field show double refraction, as do many gels when under pressure or when drawn out into threads. Mechanical stress may produce double refraction in gelatin even of 0.01 percent.

Phenomena of double refraction in protoplasm often indicate the presence of mechanically and optically anisotropic elements oriented in some definite way, and are among the best evidences for the presence of micelles in protoplasm. See Schmidt, 1937, for a complete discussion.

Valentin saw double refraction in the cilia of *Opalina ranarum* in 1861 and Rouget in the stalk muscle of *Carchesium* in 1862. Engelmann (1875) carried out very extensive pioneering work on double refraction on the Protozoa and other forms. He saw anisotropy in the pellicle of large ciliates, such as *Opalina,* but was unable to distinguish double refraction in the myonemes of *Stentor* because the entire outer layer is doubly refractive. Mackinnon (1909) mentioned that the protoplasm of *Actinosphaerium* is quite generally anisotropic, and Schmidt (1937) saw faint but unmistakable traces of double refraction at the edge of the surface layers of *Amoeba,* which became distinctly greater as the amoeba became rounded up before encysting, the cyst wall showing it very distinctly.

Engelmann (1875) also observed that the axopods of *Actinosphaerium* were doubly refractive in the living condition, the anisotropy being coextensive with the axial filaments which extend deep into the protoplasm. With the withdrawal of the pseudopods the condition vanishes. Mackinnon (1909) confirmed the findings of Engelmann. The structure of the axial filaments of these pseudopods is known to be fibrillar (Roskin, 1925; Rumjantzew and Suntzowa, 1925). Schultz (1915) found the fibrillae in the rhizopods of the foraminiferan, *Astrorhiza,* to be doubly refractive, and Schmidt (1929) found weak anisotropy in the axopods of the radiolarian, *Thalassicolla.* Schultz (1915) and Schmidt (1929) report that the stereoplasmic axis of the pseudopods of the Foraminifera are formed by the parallel alignment of fibers.

Cilia have been found to be doubly refractive by Valentin, Engelmann (1875), and Mackinnon and Vlès (1908). The latter authors consider this to be caused by the depolarization of the light by reflection from surfaces of the cilia, because of the difference of refractive index. If air penetrates into the axis of a cilium, upon drying the anisotropy is greatly increased (Schmidt, 1937). The so-called rootlets of the cilia are doubly refractive, but the basal granules are not (Engelmann, 1880.) Myonemes have been quite generally shown to be anisotropic, especially those in the stalk muscle of the vorticellids (Engelmann, 1875; Wrzesniowski, 1877; Mackinnon and Vlès, 1908). The myoneme of the stalk spreads out into fibrillae in the base of the animal. These, too, are doubly refractive. Engelmann (1875) found that the extensile neck of *Trachelocerca* (*Lacrymaria*) *olor,* when stretched out, was positively anisotropic in relation to its longitudinal axis. It is to be recalled that the neck may be extended as much as fifteen times the length of the body. Associated with this is a single spiral thread which becomes straight upon extension of the neck (Penard, 1922).

Brandt (1885) has shown that the isospore nuclei of the Radiolaria are anisotropic, but not the vegetative nor the anisospore nuclei. This has been confirmed by Schmidt (1932) on living and preserved material of the same form. Schmidt (1929) had also observed double refraction of the nuclear membrane in living nuclei of a foraminiferan and of *Amoeba;* in the latter he also observed weak anisotropy of small visible granules (chromatin?) in the living nucleus. Finally, Kalmus (1931) has recorded that certain elements of the division figure of the nuclei of *Paramecium* show slight traces of double refraction during fission and conjugation.

X-RAY DIFFRACTION AND ULTRACENTRIFUGATION

Early studies on cohesion and swelling relations of organic fibrillar structures indicated that the finer structures of which they are composed are micellar in nature. Recently X-ray diffraction methods have substantiated this view and have made it possible actually to measure the dimensions of these structural units. Much of this work has been done on keratin, elastin, chitin, myosin, cellulose, and other nonliving substances. Some observations have been made on living nerve fibers (Schmitt, Bear, and Clark, 1935).

The evidence from X-ray diffraction shows that animal fibers owe their anisotropic properties to the fact that they are composed of longitudinally oriented protein chains. However, in most protoplasm the configuration of such chains must be such that it may be altered rapidly and reversibly. For a review of X-ray diffraction, see Frey-Wyssling, 1938.

Beams and King (1937) and King and Beams (1938) have shown that the complicated process of karyokinesis may occur in eggs of *Ascaris*, although cytokinesis does not usually take place at very high centrifugal forces. They believe that if stratification occurred, it would involve a breakdown of the normal, submicroscopic spatial relations, which are of importance for the maintenance of life.

LITERATURE CITED

Adolph, E. F. 1931. The regulation of size as illustrated in unicellular organisms. Springfield.

Alexander, G. 1931. The significance of hydrogen ion concentration in the biology of *Euglena gracilis* Klebs. Biol. Bull., 61: 165-84.

Allison, R. V. 1924. The density of unicellular organisms. Ann. appl. Biol., 1: 153-64.

Angerer, C. A. 1936. The effects of mechanical agitation on the relative viscosity of *Amoeba* protoplasm. J. cell comp. Physiol., 8: 329-45.

Ball, G. H. 1927. Studies on *Paramecium*. III. The effects of vital dyes on *Paramecium caudatum*. Biol. Bull., 52: 68-78.

Barnes, T. C. 1937. Textbook of general physiology. Philadelphia.

Bayliss, W. M. 1920. The properties of colloidal systems. IV. Reversible gelation in living protoplasm. Proc. roy. Soc. B., 91: 196-201.

Beams, H. W., and R. L. King. 1932. The architecture of the parietal cells of the salivary glands of the grasshopper. J. Morph., 53: 223-41.

—— 1937. The suppression of cleavage in *Ascaris* eggs by ultracentrifuging. Biol. Bull., 73: 99-111.

Becker, E. R. 1928. Streaming and polarity in *Mastigina hylae* (Frenzel). Biol. Bull., 54: 109-16.

Beers, C. D. 1924. Observations on *Amoeba* feeding on the ciliate *Frontonia*. Brit. J. exp. Biol., 1: 335-41.

Bělehradek, J. 1935. Temperature and living matter. Protop. Monog., vol. 8. Berlin.

Bělehradek, J., and K. M. Paspa. 1928. Courbe myographique de la Vorticelle. Arch. Int. Physiol., 30: 70-72.

Bensley, R. R. 1938. Plasmosin. The gel- and fiber-forming constituent of the protoplasm of the hepatic cell. Anat. Rec., 72: 351-69.

Bles, E. J. 1929. *Arcella*. A study in cell physiology. Quart. J. micr. Sci., 72: 527-648.

Bodine, J. H. 1923. Excystation of *Colpoda cucullus*. Some factors affecting excystation of *Colpoda cucullus* from its resting cysts. J. exp. Zool., 37: 115-25.

Brandt, K. 1885. Die koloniebildenden Radiolarien (Spaerozoen) des Golfes van Neapel. Fauna u. Flora Neapel, Monogr. 13.

Bresslau, E. 1926. Die Stäbchenstruktur der Tektinhüllen. Arb. Staatsinst. exp. Ther. Frankfurt, H. 21 (Festschr. f. W. Kolle): 26-31.

Brinley, F. J. 1928. The effect of chemicals on viscosity of protoplasm as indicated by Brownian movement. Protoplasma, 4: 177-82.

Brown, D. E. S., and D. A. Marshland. 1936. The viscosity of *Amoeba* at high hydrostatic pressure. J. cell. comp. Physiol., 8: 159-65.

Browne, K. M. R. 1938. The Golgi apparatus and other cytoplasmic bodies in *Spirostomum ambiguum*. J. R. micr. Soc., 58: 188-99.

Brues, C. T. 1927. Animal life in hot springs. Quart. Rev. Biol., 2: 181-203.

Butts, H. E. 1935. The effect of certain salts of sea water upon reproduction in the marine *Amoeba, Flabellula mira* Schaeffer. Physiol. Zoöl., 8: 273-89.

Calkins, G. N. 1911. Effects produced by cutting *Paramecium* cells. Biol. Bull., 21: 36-72.

—— 1919 *Uroleptus mobilis*, Engelm. I. History of the nuclei during division and conjugation. J. Exp. Zool., 27: 293-355.

—— 1933. The biology of the Protozoa. 2d ed., Philadelphia.

Chalkley, H. W. 1929. Changes in water content in *Amoeba* in relation to changes in its protoplasmic structure. Physiol. Zoöl., 2: 535-74.

—— 1930. On the relation between the resistance to heat and the mechanism of death in *Paramecium*. Physiol. Zoöl., 3: 425-40.

—— 1935. The mechanism of cytoplasmic fission in *Amoeba proteus*. Protoplasma, 24: 607-21.

—— 1936. The behavior of the karosome and the "peripheral chromatin" during mitosis and interkinesis in *Amoeba proteus* with particular reference to the morphologic distribution of nucleic acid as indicated by the Feulgen reaction. J. Morph., 60: 13-29.

Chalkley, H. W., and G. E. Daniel. 1934. The effect of certain chemicals upon the division of the cytoplasm in *Amoeba proteus*. Protoplasma, 21: 258-67.

Chambers, R. 1921. The effect of experimentally induced changes in consistency on protoplasmic movement. Proc. Soc. exp. Biol. N. Y., 19: 87-88.

—— 1922. A micro injection study on the permeability of the starfish egg. J. gen. Physiol., 5: 189-93.

—— 1924. The physical structure of protoplasm as determined by microdissection and injection. Sec. V, General cytology, 234-309 (edited by E. V. Cowdry). Chicago.

—— 1938. Structural aspects of cell division. Arch. exp. Zellforsch. 22: 252-56.

Chambers, R., and J. A. Dawson. 1925. The structure of the undulating membrane in the ciliate *Blepharisma*. Biol. Bull., 48: 240-42.

Chambers, R., and H. P. Hale. 1932. The formation of ice in protoplasm. Proc. roy. Soc. B, 110: 336-52.

Chambers, R., and P. Reznikoff. 1926. Micrurgical studies in cell physiology. 1. The action of the chlorides of Na, K, Ca, and Mg on the protoplasm of *Amoeba proteus*. J. gen. Physiol., 8: 369-401.

Chase, A. M., and O. Glaser. 1930. Forward movement of *Paramecium* as a function of the hydrogen ion concentration. J. gen. Physiol., 13: 627-36.

Chatton, E., and A. Lwoff. 1923a. Sur l'évolution des Infusoires des Lamelle-branches, Les formes primitives du phylum des Thigmotriches: le genre Thigmo., Phrya C. R. Acad. Sci. Paris, 177: 81-83.

—— 1923b. Un Cas remarquable d'adaptation: *Ellobiophrya donacis*. n. g. n. sp., etc. C. R. Soc. Biol., Paris, 88: 749-52.

Child, C. M. 1934. The differential reduction of methylene blue by *Paramecium* and some other ciliates. Protoplasma, 22: 377-94.

Child, C. M., and E. Deviney. 1926. Contributions to the physiology of *Paramecium caudatum*. J. exp. Zool., 43: 257-312.

Cleveland, L. R. 1935. The centriole and its role in mitosis as seen in living cells. Science, 81: 598-600.

Cleveland, L. R., in collaboration with S. R. Hall, E. P. Sanders, and J. Collier. 1934. The wood-feeding roach *Cryptocercus*, its Protozoa ,and the symbiosis between Protozoa and Roach. Mem. Amer. Acad. Arts Sci., 17, No. 2; 187-342.

Conklin, E. G. 1924. Cellular differentiation. Sec. IX, General cytology, 539-607 (ed. by E. V. Cowdry). Chicago.

Daniel, G. E., and H. W. Chalkley. 1932. The influence of temperature upon the process of division in *Amoeba proteus* (Leidy). J. cell, comp. Physiol., 2: 311-27.

Daniels, M. L. 1938. A cytological study of the gregarine parasites of *Tenebrio molitor* using the ultra-centrifuge. Quart. J. micr. Sci., 80: 293-320.

Daugherty, K. 1937. The action of anesthetics on *Amoeba* protoplasm. Physiol. Zoöl., 10: 473-83.

Davenport, C. B. 1897. Experimental morphology. New York.

Dawson, J. A., and M. Belkin. 1929. The digestion of oils by *Amoeba proteus*. Biol. Bull., 56: 80-86.

Dellinger, A. P. 1906. Locomotion of amoebae and allied forms. J. exp. Zool., 3: 337-58.

—— 1909. The cilium as a key to the structure of contractile protoplasm. J. Morph., 20: 171-209.

Dierks, K. 1926. Untersuchungen über die Morphologie und Physiologie des *Stentor coeruleus*. Arch. Protistenk., 54: 1-91.

Dobell, C. C. 1911. The principles of protistology. Arch. Protistenk., 23: 269-310.

Dogiel, V., and M. Issakowa-Keo. 1927. Physiologische Studien an Infusorien. II. Der Einflusz der Salzlösungen auf die Ernahrung von *Paramaecium*. Biol. Zbl., 47: 577-86.

Edwards, J. G. 1923. The effect of chemicals on locomotion in *Amoeba*. I. J. exp. Zool., 38: 1-43.

Engelmann, T. W. 1875. Contractilität und Doppelbrechung. Pflüg. Arch. ges. Physiol., 11: 432-64.

—— 1880. Zur Anatomie und Physiologie der Flimmerzellen. Pflüg. Arch. ges. Physiol., 23:: 505-35.

Ephrussi, B., and R. Rapkin. 1928. Action des différents sels sur le *Spirostomum*. Protoplasma, 5: 35-40.

Ewles, N. A., and J. B. Speakman. 1930. Examination of the fine structure of wool by X-ray analysis. Proc. roy. Soc. B., 105: 600-7.

Fauré-Fremiet, E. 1905. L'appareil fixateur chez les Vorticellidae. Arch. Protistenk., 6: 207-26.

—— 1908. *L'Ancystropodium maupasi*. Arch Protistenk., 13: 121-37.

—— 1910. La fixation chez les Infusoires cilies. Bull. Sci. Fr. Belg., 44: 27-50.

—— 1929. Caracteres physico-chimiques des choanoleucocytes de quelques invertébrés. Protoplasm, 6: 521-609.

—— 1930. The kinetics of living matter. Trans. Faraday Soc., 26: 779-93.

—— 1932. *Strombidium calkinsi*, a new thigmotactic species. Biol. Bull. 62: 201-4.

—— 1935. L'Oeuvre de Félix Dupardin et la notion de protoplasma. Protoplasma, 23: 250-69.

Fauré-Fremiet, E., B. Ephrussi, and L. Rapkine. 1926. Lames minces formées par la diffluence du cytoplasma cellulaire. C. R. Soc. Biol. Paris, 94: 442-43.

Fetter, D. 1926. Determination of the protoplasmic viscosity of *Paramecium* by the centrifuge method. J. exp. Zool., 44: 279-83.

Fortner, H. 1926. Zur Morphologie und Physiologie des Vorticellenstieles. Z. wiss. Zool., 128: 114-32.

Frederikse, A. M. 1933a. Viskositätsänderungen des protoplasmas während der Narkose. Protoplasma. 18: 194-207.

—— 1933b. Der Brechungsindex des Protoplasmas. Protoplasma, 19: 473-48.

Frey-Wyssling, A. 1938. Submikroskopische Morprologie des Protoplasmas und seiner Derivate. Protop.-Monog. 15, Berlin.

Gaw, H. Z. 1936a. Physiology of the contractile vacuole in ciliates. 2. The effect of hydrogen ion concentration. Arch. Protistenk., 87: 194-200.

—— 1936b. Physiology of the contractile vacuole in ciliates 4. The effect of heavy water. Arch. Protistenk. 87: 213-24.

Gelei, J. V. 1928. Nochmals über den Nephridialapparat bei den Protozoen. Arch Protistenk., 64: 479-94.

Ginsberg, H. 1922. Untersuchungen zum Plasmabau der Amöben, im Hinblick auf die Wabentheorie. Roux Arch EntwMech. Organ., 51: 150-250.

Giese, A. C. 1935. The role of starvation in conjugation of *Paramecium*. Physiol. Zool., 8: 116-25.

Gray, J. 1931. A textbook of experimental cytology. New York.

Greeley, A. W. 1901. On the analogy between the effects of loss of water and lowering of temperature. Amer. J. Physiol., 6: 122-28.

—— 1902. The artificial production of spores in *Monas* by a reduction of the temperature. Biol. Bull., 3: 165-71.

—— 1903. Further studies on the effect of variations in temperature on animal tissues. Biol. Bull., 5: 42-54.

—— 1904. Experiments on the physical structure of the protoplasm of *Paramoecium*, etc. Biol. Bull., 7: 3-32.

Grosse-Allermann, W. 1909. Studien über *Amoeba terricola* Kreeff. Arch. Protistenk., 17: 203-57.

Gruber, A. 1889. Biologische Studien an Protozoen. Biol. Zbl., 9: 14-23.

Gruber, K. 1912. Biologische und experimentelle Untersuchungen an *Amoeba proteus*. Arch. Protestenk., 25: 316-76.

Hahnert, W. F. 1912. A quantitative study of reactions to electricity in *Amoeba proteus*. Physiol. Zoöl., 5: 491-526.

Harvey, E. B. 1917. A physiological study of specific gravity and of luminescence in *Noctiluca*, with special reference to anesthesia. Publ. Carneg. Institn. No. 251: 235-53.

Harvey, E. N. 1931. Observations on living cells, made with the microscope-centrifuge. J. exp. Biol., 8: 267-74.

—— 1934. Biological effects of heavy water. Biol. Bull., 66: 91-96.

—— 1936. The properties of elastic membranes with special reference to the cell surface. J. cell. comp. Physiol., 8: 251-60.

—— 1938. Some physical properties of protoplasm. J. appl. Physics, 9: 68-80.

Harvey, E. N., and J. F. Danielli. 1936. The elasticity of thin films in relation to the cell surface. J. cell. comp. Physiol., 8: 31-36.

Harvey, E. N., E. B. Harvey, and A. L. Loomis. 1928. Further observations on the effect of high frequency sound waves on living matter. Biol. Bull., 55: 459-69.

Harvey, E. N., and D. A. Marsland. 1932. The tension at the surface of *Amoeba dubia* with direct observations on the movement of cytoplasmic particles at high centrifugal speeds. J. cell. comp. Physiol., 2: 75-97.

Heilbrunn, L. V. 1928. The colloid chemistry of protoplasm. Berlin.

—— 1929a. Protoplasmic viscosity of *Amoeba* at different temperatures. Protoplasma, 8: 58-64.

—— 1929b. The absolute viscosity of *Amoeba* protoplasm. Protoplasma, 8: 65-69.

—— 1931. Anasthesia in *Amoeba*. Anat. Rec., 51: 27.

—— 1937. An outline of general physiology. Philadelphia.

Heilbrunn, L. V., and K. Daugherty. 1931. The action of the chlorides of sodium, potassium, calcium, and magnesium on the protoplasm of *Amoeba dubia*. Physiol. Zoöl., 4: 635-51.

—— 1932. The action of sodium, potassium, calcium, and magnesium ions on the plasmagel of *Amoeba proteus*. Physiol. Zoöl., 5: 254-74.

—— 1933. The action of ultraviolet rays on *Amoeba* protoplasm. Protoplasma, 18: 596-619.

—— 1934. A further study of the action of potassium on *Amoeba* protoplasm. J. cell. comp. Physiol., 5: 207-18.

—— 1938. Fat release in *Amoeba* after irradiation. Physiol. Zoöl., 11: 383-87.

—— 1939. The electric charge of protoplasmic colloids. Physiol. Zoöl., 12: 1-12.

Heilbrunn, L. V., and D. Mazia. 1936. The action of radiations on living protoplasm. Biological effects of radiation, (edited by B. M. Dugger) : 625-76. New York.

Henderson, V. E. 1930. The present status of the theories of narcosis. Physiol. Rev., 10: 171-220.

Holter, H., and M. J. Kopac. 1937. Studies on enzymatic histo-chemistry XXIV. Localization of peptidase in *Ameba*. J. cell. comp. Physiol., 10: 423-37.

Hopkins, D. L. 1938. The vacuoles and vacuolar activity in the Marine amoeba, *Flabellula mira* Schaeffer and the nature of the natural red system in Protozoa. Biodyn., No. 34: 1-22.

Howland, R. B. 1924a. On excretion of nitrogenous waste as a function of the contractile vacuole. J. exp. Zool., 40: 231-50.

—— 1924b. Experiments on the contractile vacuole of *Amoeba verrucosa* and *Paramecium caudatum*. J. exp. Zool., 40: 251-62.

—— 1924c. Dissection of the pellicle of *Amoeba verrucosa*. J. exp. Zool., 40: 263-70.

Howland, R. B., and H. Pollack. 1927. Micrurgical studies on the contractile vacuole. I. Relation of the physical state of the internal protoplasm to the behavior of the vacuole. II. Micro-injection of distilled water. J. exp. Zool., 48: 441-58.

Hyman, L. 1917. Metabolic gradients in *Amoeba* and their relation to the mechanism of amoeboid movement. J. exp. Zool., 24: 55-99.

Jacobs, M. H. 1919. Acclimatization as a factor affecting the upper thermal death points of organisms. J. exp. Zool., 27: 427-42.

—— 1922. The effects of carbon dioxide on the consistency of protoplasm. Biol. Bull., 42: 14-30.

Jennings, H. S. 1904. Movements and reactions of *Amoeba*. Publ. Carneg. Instn., No. 16: 129-234.

—— 1906. Behavior of the lower organisms. New York.

Jensen, P. 1893. Die absolute Kraft einer Flimmerzelle. Pflüg. Arch. ges. Physiol., 54: 537-51.

Jochims, J. 1930. Das Fadenziehen biologischer Substanzen. Protoplasma, 9: 298-317.

Johnson, L. P. 1939. A study of *Euglena rubra* Hardy 1911. Trans. Amer. micr. Soc., 58: 42-48.

Jordan-Lloyd, D. 1932. Colloidal structure and its biological significance. Biol. Rev., 7: 254-73.

Joyet-Lavergne, Ph. 1931. La Physico-Chimie de la sexualité. Protop.- Monog., Vol. V. Berlin.

Kahl, A. 1935. Wimpertiere order Ciliata (Infusoria), in Dahl: Die Tierwelt Deutschlands. 18 Teil, Jena.

Kalmus, . 1931. *Paramecium*. Jena.

Kanda, Sakyo. 1914. On the geotropism of *Paramecium* and *Spirostomum* Biol. Bull., 24: 1-24.

—— 1918. Further studies on the geotropism of *Paramecium caudatum*. Biol. Bull., 34: 108-19.

Kepner, W. A., and J. G. Edwards. 1917. Food reactions of *Pelomyxa carolinesis* Wilson. J. exp. Zool., 24: 381-407.

Kepner, W. A., and B. D. Reynolds. 1923. Reactions of cell-bodies and pseudopodia fragments of *Difflugia*. Biol. Bull., 22-47.

Kepner, W. A., and W. H. Taliaferro. 1913. Reactions of *Amoeba proteus* to food. Biol. Bull., 24: 411-22.

Kepner, W. A., and C. Whitlock. 1921. Food reactions of *Amoeba proteus*. J. exp. Zool., 32: 397-425.

King, R. L. 1928. The contractile vacuole in *Paramecium trichium*. Biol. Bull., 55: 59-64.

—— 1935. The contractile vacuole of *Paramecium multimicronucleata*. J. Morph., 58: 555-71.

King, R. L., and H. W. Beams, 1937. The effect of ultracentrifuging on *Paramecium*, with special reference to recovery and macronuclear reorganization. J. Morph., 61: 27-49.

—— 1938. An experimental study of chromatin diminution in *Ascaris*, J. exp. Zool., 77: 425-43.

Kitching, J. A. 1936. The physiology of contractile vacuoles. II. The control of body volume in marine Peritricha. J. exp. Biol., 13: 11-27.

—— 1938. Contractile vacuoles. Biol. Rev., 13: 403-44.

Kite, G. L. 1913. Studies on the physical properties of protoplasm. I. The physical properties of the protoplasm of certain animal and plant cells. Amer. J. Physiol., 32: 146-64.

Koltzoff, N. K. 1912. Untersuchungen über die Kontraktilität des Vorticellenstiels. Arch. Zellforsch, 7: 344-423.

Lackey, J. B. 1936. Some freshwater Protozoa with blue chromatophores Biol. Bull., 71: 492-97.

Leontjew, H. 1926a. Zur Biophysik der niederen Organismem. II. Mitteilung. Die Bestimmung des spezifischen Gewicht und der Masse von *Naegleria*. Pflug. Arch. ges. Physiol., 213: 1-4.

—— 1926b. Über das spezifische Gewicht des Protoplasmas Biochem. Z. 170: 326-29.

—— 1927. Über das spezifische Gewicht des Protoplasmas. III. Protoplasma, 2: 59-64.

Lewis, W. H. 1926. Cultivation of embryonic heart muscle. Publ. Carneg. Instn., 363: 1-21.

Lloyd, F. E. 1928. The contractile vacuole. Biol. Rev., 3: 329-58.

Loeb, L. 1927. Amoeboid movement and agglutination in amoebocytes of *Limulus* and the relation of these processes to tissue formation and thrombosis. Protoplasma, 2: 512-53.

—— 1928. Amoebocyte tissue and amoeboid movement. Protoplasma, 4: 596-625.

Looper, J. B. 1928. Cytoplasmic fusion in *Actinophrys sol,* with special reference to the karyoplasmic ratio. J. exp. Zool., 50: 31-49.

Luce, R. H. 1926. Orientation to the electric current and to light in *Amoeba*. (Abs.) Anat. Rec., 32: 55.

Ludwig, W. 1928a. Der Betriebsstoffwechsel von *Paramecium caudatum*. Arch. Protistenk., 62: 12-40.

—— 1928b. Permeabilität und Wasserwechsel bei *Noctiluca miliaris* Suriray. Zool. Anz., 76: 273-85.

—— 1930. Zur Theorie der Flimmerbewegung. Z. vergl. Physiol., 13: 397-504.

Lund, E. J. 1917. Reversibility of morphogenetic processes in *Bursaria*. J. exp. Zool., 24: 1-33.

—— 1921. Experimental control of organic polarity by the electric current. I. J. exp. Zool., 34: 471-93.

—— Lund, E. J., and G. A. Logan. 1925. The relation of the stability of protoplasmic films in *Noctiluca* to the duration and intensity of an applied electric potential. J. gen. Physiol., 7: 461-72.

Luyet, B. J., and P. M. Gehenio. 1935. Comparative ultra-violet absorption by the constituent parts of protozoan cells (*Paramaecium*). Biodyn., No. 7: 1-14.

—— 1938. The lower limit of vital temperatures, a review. Biodyn. No. 33: 1-92.

Lwoff, A. 1934. Die Bedeutung des Blutfarbstoffes für die parasitischen Flagellaten. Zbl. Bakt., 130: 498-518.

Lyon, E. P. 1905. On the theory of geotropism in *Paramaecium*. Amer. J. Physiol., 14: 421-32.

McClendon, J. F. 1909. Protozoan studies. J. exp. Zool., 6: 265-83.

Mackinnon, D. L. 1909. Optical properties of contractile organs in Heliozoa. J. Physiol., 38: 254-58.

Mackinnon, D. L., and F. Vlès. 1908. On the optical properties of contractile organs. J. R. micr. Soc., 553-58.

Makarov, P. 1935. Experimentelle Untersuchungen an Protozoen mit Bezug auf das Narkose-Problem. Protoplasma, 24: 593-606.

Marsland, D. 1933. The site of narcosis in a cell; the action of a series of paraffin oils on *Amoeba dubia*. J. cell. comp. Physiol., 4: 9-30.

Marsland, D. A. 1939. The mechanism of protoplasmic streaming, etc. J. cell, comp. Physiol., 13: 23-30.

Massart, J. 1893. Sur l'irritabilité des Noctiluques. Bull. sci. Fr. Belg., 25: 59-76.

Mast, S. O. 1909. The reactions of *Didinium nasutum*. Biol. Bull., 16: 91-118.

—— 1926a. The structure of protoplasm in *Amoeba*. Amer. Nat., 60: 133-42.

—— 1926b. Structure, movement, locomotion and stimulation in *Amoeba*. J. Morph., 41: 347-425.

—— 1929. Mechanics of locomotion in *Amoeba proteus* with special reference to the factors involved in attachment to the substratum. Protoplasma, 8: 344-77.

—— 1931a. Movement and response in *Difflugia* with special reference to the nature of cytoplasmic contraction. Biol. Bull., 61: 223-41.

—— 1931b. The nature of the action of electricity in producing response and injury in *Amoeba proteus* (Leidy) and effect of electricity on the viscosity of protoplasm. Z. vergl. Physiol., 15: 309-28.

—— 1938a. The contractile vacuole in *Amoeba proteus* (Leidy). Biol. Bull., 74: 306-13.

—— 1938b. Digestion of fat in *Amoeba proteus*. Biol. Bull., 75: 389-94.

Mast, S. O., and W. L. Doyle. 1934. Ingestion of fluid by *Amoeba*. Protoplasma, 20: 555-60.

—— 1935a. Structure, origin and function of cytoplasmic constituents in *Amoeba proteus*. I. Structure. Arch. Protistenk., 86: 155-80.

—— 1935b. II. Origin and function based on experimental evidence; effect of centrifuging on *Amoeba proteus*. Arch. Protistenk., 86: 278-306.

Mast, S. O., and C. Fowler. 1935. Permeability of *Amoeba proteus* to water. J. cell. comp. Physiol., 6: 151-67.

Mast, S. O., and W. F. Hahnert. 1935. Feeding, digestion, and starvation in *Amoeba proteus* (Leidy). Physiol. Zoöl., 8: 255-72.

Mast, S. O., and F. M. Root. 1916. Observations on *Ameba* feeding on rotifers, nematodes and ciliates, and their bearing on the surface-tension theory. J. exp. Zool., 21: 33-49.

Merton, H. 1928. Untersuchungen über die Enstehung amöbenahnlicher Zellen an absterbenden Infusorien. Math.-nat. Kl. Heidelb. Akad. Wiss., 5 B. 1928, Nr. 3: 51-29.

Metcalf, M. M. 1910. Studies upon *Amoeba*. J. exp. Zool., 9: 301-31.

Moore, A. R. 1935. On the significance of cytoplasmic structure in *Plasmodium*. J. cell. comp. Physiol., 7: 113-29.

Morita, Y., and R. Chambers. 1929. Permeability differences between nuclear and cytoplasmic surfaces in *Amoeba dubia*. Biol. Bull., 56: 64-67.

Nadler, E. J. 1929. Notes on the loss and regeneration of the pellicle in *Blepharisma undulans*. Biol. Bull., 56: 327-30.

Nassonov, D. 1924. Der Exkretionsapparat (kontraktile vakuole) der Protozoa als Homologon des Golgischen Apparats der Metazoazellen. Arch. mikr. Anat., 103: 437-82.

Needham, J. 1936. Order and life. New Haven.

Needham, J., and D. M. Needham. 1925. The hydrogen-ion concentration and oxidation-reduction potential of the cell interior: a micro-injection study. Proc. roy. Soc. B., 98: 259-86.

—— 1926. Further micro-injection studies on the oxidation-reduction potential of the cell-interior. Proc. roy. Soc. B., 99: 383-97.

Noland, L. E. 1931. Studies on the taxonomy of the genus *Vorticella*. Trans. Amer. micr. Soc., 50: 81-123.

Oliphant, J. F. 1938. The effect of chemicals and temperature on reversal in ciliary action in *Paramecium*. Physiol. Zoöl., 11: 19-30.

Packard, C. 1923. The susceptibility of cells to radium radiations. Proc. Soc. exp. Biol., N.Y., 20: 226-27.

—— 1924. The susceptibility of cells to radium radiations. Biol. Bull., 46: 165-77.

—— 1925. The effect of light on the permeability of *Paramecium*. J. gen. Physiol., 7: 363-72.

Pantin, C. F. A. 1923. On the physiology of amoeboid movement. I. J. Mar. Biol. Assoc., 13: 24-69.

—— 1924a. Temperature and the viscosity of protoplasm. Jour. Mar. biol. Ass. U.K., 13: 331-39.

—— 1924b. On the physiology of amoeboid movement. II. The effect of temperature. Brit. J. exp. Biol., 1: 519-38.

—— 1926a. On the physiology of amoeboid movement. III. The action of calcium. Brit. J. exp. Biol., 3: 275-95.

—— 1926b. On the physiology of amoeboid movement. IV. The action of magnesium. Brit. J. exp. Biol., 3: 297-312.

—— 1930. On the physiology of amoeboid movmeent: a.V. Anaerobic movement. Proc. roy. Soc. B. 105: 538-55. b.VI. The action of oxygen. Proc. roy. Soc. B., 105: 555-64. c.VII. The action of anaesthetics. Proc. roy. Soc. B., 105: 565-79.

Park, O. 1929. The differential reduction of osmic acid in the cortex of *Paramecium*, and its bearing upon the metabolic gradient conception. Physiol. Zoöl., 2: 449-58.

Parsons, C. W. 1926. Some observations on the behavior of *Amoeba proteus*. Quart. J. micr. Sci. 70: 629-48.

Pascher, A. 1917. Von der merkwurdigen Bewegungsweise einiger Flagellaten. Biol. Zbl., 37: 241-429.

108 PROTOPLASM OF PROTOZOA

Patten, R., and H. W. Beams. 1936. Observations on the effect of the ultracentrifuge on some free-living flagellates. Quart. J. micr. Sci., 78: 615-35.

Peebles, F. 1912. Regeneration and regulation in *Paramecium caudatum*. Biol. Bull., 23: 154-70.

Pekarek, J. 1930. Absolute Viskositätsmessung mit Hilfe der Brownschen Molekularbewegung. II. Viskositätsbestimmung des Zellsaftes der Epidermiszellen von *Allium cepa* und des Amöben-Protoplasmas. Protoplasma, 11: 19-48.

Penard, E. 1902. Faune Rhizopodique due Basin du Leman. Genève.

—— 1904. Les Héliozoaires d'eau douce. Genève.

—— 1922. Études sur les Infusoires d'leau douce. Genève.

Pfeiffer, H. 1931. Über mikro-refraktometrische methoden im Dienste der Protoplasma-Forschung. Z. wiss. Mikr., 48: 47-62.

—— 1933a. Beiträge zur quantitativen Bestimmung von molekularkräften des Protoplasmas. I. Eine Methode zur messung der Adhäsionsarbeit plasmatischer Oberflächen. Protoplasma, 19: 177-92.

—— 1933b. Beiträge zur quantitativen Bestimmung von molekularkräften des Protoplasmas. II. Eine modifikation der methode zur bestimmung des Reinbungswiederstandes nackter. Protoplasten. Protoplasma, 20: 73-78.

—— 1933c. III. Die Bestimmung des Randwinkels aus der Form nackter Protoplasten. Protoplasma, 20: 79-84.

—— 1934. Versuche zur Bestimmung des spezifischen Gewichts nackter Protoplasten. Protoplasma, 21: 427-32.

—— 1935. Literature on adhesiveness (stickiness) of protoplasm and related topics. Protoplasma, 23: 270-81.

—— 1936. Further tests of the elasticity of protoplasm. Physics, 7: 302-5.

—— 1937. Experimental researches on the non-Newtonian nature of Protoplasma. Cytologia, Tokyo. (Fujii Jubilaei vol.) 701-10.

Pitts, R. F., and S. O. Mast. 1934. The relation between inorganic salt concentration, hydrogen ion concentration and physiological processes in *Amoeba proteus*. III. J. cell. comp. Physiol., 4: 435-55.

Platt, J. B. 1899. On the specific gravity of *Spirostomum, Paramaecium*, and the tadpole in relation to the problem of geotaxis. Amer. Nat., 33: 31-38.

Pollack, H. 1928a. Intracellular hydrion concentration studies. III. etc. Biol. Bull., 55: 383-85.

—— 1928b. Micrurgical studies in cell physiology. VI. Calcium ions in living protoplasm. J. gen. Physiol., 11: 539-45.

Popoff, M. 1908. Experimentelle Zellstudien 1. Arch. Zellforsch., 1: 245-379.

Pütter, A. 1903. Die Flimmerbewegung. Ergebn. Physiol., 2: 1-102.

Reznikoff, P. 1926. Micrurgical studies in cell physiology. II. The action of the chlorides of lead, mercury, copper, iron, and aluminum on the protoplasm of *Amoeba proteus*. J. gen. Physiol., 10: 9-21.

Reznikoff, P., and R. Chambers. 1927. Micrurgical studies in cell physiology. III. The action of CO_2 and some salts of Na, Ca, and K on the protoplasm of *Amoeba dubia*. J. gen. Physiol., 10: 731-55.

Reznikoff, P., and H. Pollock. 1928. Intracellular hydrion concentration studies. II. The effect of injection of acids and salts on the cytoplasmic pH of *Amoeba dubia*. Biol. Bull., 55: 377-82.

Rhumbler, L. 1898. Physikalische analyse von Lebenserscheinungen der Zelle. Roux Arch. EntwMech. Organ., 7: 103-350.

Root, F. M. 1914. Reproduction and reactions to food in the suctorian, *Podophrya collini*, n. sp. Arch. Protistenk., 35: 164-96.

Roskin, G. 1923. La Structure des myonemes des Infusoires. Bull. biol., 57: 143-51.

—— 1925. Über die Axopodien der Heliozoa und die Greiftentakeln der Ephelotiden. Arch Protistenk., 52: 207-16.

Rumjantzew, A., and E. Suntzowa. 1925. Untersuchungen über den Protoplasmabau von *Actinosphaerium Eichhornii*. Arch. Protistenk., 52: 217-64.

Saunders, J. T. 1925. The trichocysts in *Paramecium*. Biol. Rev., 1: 249-69.

Schaeffer, A. A. 1916. On feeding habits of *Ameba*. J. exp. Zool., 20: 529-84.

—— 1920. Amoeboid movement. Princeton.

—— 1926. Taxonomy of the amebas. Publ. Carneg. Instn., No. 345.

—— 1931. On molecular organization in ameban protoplasm. Science, 74: 47-51.

Schewiakoff, W. 1927. Die Acantharia des Golfes von Neapel. Fauna u. Flora neapel. Monogr., 37.

Schmidt, W. J. 1929. Rheoplasma und Steroplasma. Protoplasma, 7: 353-94.

—— 1932. Der submikroskopische Bau des Chromatins. II. Über die Doppelbrechung der Isosporenkerne von *Sphaerozoen*. Arch. Protistenk., 77: 463-90.

—— 1937. Die Doppelbrechung von Karyoplasma, Zytoplasma und Metaplasm. Protop. Monog., 11. Berlin.

Schmitt, F. O. 1938. Optical studies of the molecular organization of living systems. J. Appl. Physics, 9: 109-17.

Schmitt, F. O., R. S. Bear, and G. L. Clark. 1935. X-ray diffraction studies on nerve. Radiology, 25: 131-51.

Schultz, E. 1915. Die Hyle des Lebens. I. Beobactungen und Experimente an *Astrorhiza limicola*. Roux Arch. EntwMech. Organ., 41: 215-36.

Seifriz, W. 1928. The physical properties of protoplasm. Colloid Chemistry (edited by J. Alexander), 11: 403-50.

—— 1936. Protoplasm. New York.

Singh, B. N. 1939. The cytology of *Amoeba proteus* "Y" and the effects of large and small centrifugal forces. Quart. J. micr. Sci., 80: 601-35.

Spek, J. 1921. Der Einfluss der Salze auf die Plasmakolloide von *Actinosphaerium eichhornii*. Acta Zool. Stock., 2: 153-200.

—— 1923. Über den physikalischen Zustand von Plasma und Zelle der *Opalina ranarum.* (Purk. et Val.) Arch. Protistenk., 46: 166-00.

—— 1924. Neue Beitrage zum Problem der Plasmastrukturen. Z. Zell. u. Gewebelehre, 1: 278-326.

Spek, J., and R. Chambers. 1934. Neue experimentelle Studien über das Problem der Reaktion des Protoplasmas. Protoplasma, 20: 376-406.

Stump, A. B. 1935. Observations on the feeding of *Difflugia, Pontigulasia* and *Lesquereusia.* Biol. Bull., 69: 136-42.

—— 1936. The influence of test materials on reproduction in *Pontigulasia vas* (Leidy) Schouteden. Biol. Bull., 70: 142-47.

Taliaferro, W. H., and L. G. Taliaferro. 1934. Complement fixation, precipitin, adhesion, mercuric chloride and Wassermann tests in equine trypanosomiasis of Panama (murrina). J. Immunol., 26: 193-213.

Taylor, C. V. 1920. Demonstration of the function of the neuromotor apparatus in *Euplotes* by the method of microdissection. Univ. Cal. Publ. Zool., 19: 403-70.

—— 1923. The contractile vacuole in *Euplotes:* an example of the sol-gel reversibility of cytoplasm. J. exp. Zool., 37: 259-89.

—— 1935. Protoplasmic reorganisation and animal life cycles. Biol. Rev., 10: 111-22.

Taylor, H. S., W. W. Swingle, H. Eyring, and A. A. Frost. 1933. The effect of water containing the isotope of hydrogen upon fresh water organisms. J. cell. comp. Physiol., 4: 1-8.

Tchakhotine, S. 1937. Radiations, cell permeability and colloidal changes. Trans. Faraday Soc., 33: 1068-72.

Thornton, F. E. 1932. The viscosity of the plasmagel of *Amoeba proteus* at different temperatures. Physiol. Zoöl., 5: 246-53.

—— 1935. The action of Sodium, Potassium, Calcium, and Magnesium ions on the plasmagel of *Amoeba proteus* at different temperatures. Physiol. Zoöl., 8: 246-54.

Tiegs, O. W. 1928. Surface tension and the theory of protoplasmic movement. Protoplasma, 4: 88-139.

Verworn, M. 1888. Biologische Protisten-Studien. Z. wiss. Zool., 46: 455-70.

—— 1899. General physiology. London.

Voegtlin, C., and H. W. Chalkley. 1935. The chemistry of cell division. IV. The influence of H_2S, HCN, CO_2 and some other chemicals on mitosis in *Amoeba proteus.* Protoplasma, 24: 365-83.

Wenrich, D. H. 1929. The structure and behavior of *Actinobolus vorax.* Biol. Bull., 56: 390-401.

Winterstein, H. 1926. Die Narkose. Berlin.

Wrzesniowski, A. 1877. Beiträge zur naturgeschichte der Infusorien. Z. wiss. Zool., 29: 267-323.

CHAPTER III

CYTOPLASMIC INCLUSIONS

RONALD F. MacLennan

ALL ACTIVE CELLS possess a large number of cytoplasmic granules which change in number, size, shape, and composition in accordance with the changes in the activities of the cell of which they are a part.[1]

The fact that small granules are so constantly present in the living substance is an indication that such a fine suspension of material represents a colloidal condition favorable for the life process. It seems certain that as the physiology of the cell becomes more clearly understood there will be shown to be a definite dependence of vital phenomena on the granular nature of protoplasm, on the properties which it possesses by virtue of the fact that it is a suspension (Heilbrunn, 1928, p. 20).

The cytoplasmic granules are a visible part of the fundamental organization of the cell, and the elucidation of their functions contributes not merely to a specialized branch of cytology but contributes directly to a solution of the fundamental problem of protoplasmic organization.

The richness of the granular complex early attracted the interest of cytologists, and many studies were made on their chemical composition. The report of Bütschli's discovery that the carbohydrate granules of gregarines differ from those in vertebrates is one of the classic papers in the group. During the first thirty years of this century the emphasis shifted from the earlier cytochemical methods to an interest in certain of the granules as permanent, self-perpetuating cytoplasmic organelles, which could be classified by certain empirical reactions such as osmic reduction or the segregation of janus green and neutral red. Dissatisfaction with the specificity of these methods has resulted recently in a renewed emphasis on methods which yield specific information on the chemical and physical nature of the cytoplasmic granules and their cyclic changes. Too often, however, there has been a tendency to carry

[1] This paper is a contribution from the departments of Zoölogy of the State College of Washington and Oberlin College.

on these two types of investigation separately, with the result that in many cases the morphological and functional studies of the cytoplasmic granules have become separated. The one group emphasizes the classification of granules into hard and fast categories of Golgi bodies, mitochondria, vacuome and so forth, with little specific consideration of function, while those engaged in functional studies tend to group all the granules together or to devise entirely new systems, which hinder the comparison of granules in several species. This review is an attempt to coördinate these two angles of approach, so that both may contribute to our understanding of the rôle of the cytoplasmic granules in the cell.

Since summaries are available of the characteristics of single groups of granules, this review is not intended to provide an exhaustive catalogue of the facts of any one group. Particular emphasis is placed on those granules which have been described with sufficient completeness to furnish evidence as to reactions, classification, and function, as well as their relationship to other granules in the same cell. Specific directions on standard techniques for demonstrating the various granules are available in the various books on microtechnique and histochemistry and so are not described in detail here. The more recent publications will be emphasized, since the specificity of methods has improved greatly and summaries of the earlier papers are available in the works of Calkins, Doflein-Reichenow, and others. The Protophyta have been omitted in most cases, since their inclusion would complicate the picture unnecessarily.

MITOCHONDRIA

Undoubtedly many granules described in early cytological studies of Protoza were actually mitochondria, but their status as a separate group of cytoplasmic constituents in the unicellular organisms was not recognized until the publication of the monograph of Fauré-Fremiet (1910), which emphasized the concept that mitochondria are universal, self-perpetuating cytoplasmic constituents.

The identification of mitochondria is not yet entirely satisfactory, since it depends upon stains and fixatives of the lipoid component, a material not restricted to mitochondria alone, or upon vital dyes which are not as effective in the Protozoa as in the Metazoa and which in certain cases stain other organelles as well. Typical mitochondria are refractile in life, become grey brown, or black in osmic techniques, but usually bleach faster than the Golgi bodies, reduce pyrogallol, take basic stains after fixation in lipoid preservatives, stain weakly or not at all after fixatives containing acetic acid, and are stained vitally with janus green B. This last method is often considered to be the final criterion, but unfortunately in many cases the mitochondria of Protozoa stain only a pale green (Subramaniam and Ganapati, 1938, and others), not the dark green described in metazoan cells. In addition, the stain is not always specific, since Lynch (1930) found that any concentration from 1:2000 to 1:500,000 tints the entire organism (*Lechriopyla*), although the mitochondria can be distinguished by their darker color. Hayes (1938) found that none of the granules in *Dileptus* stained electively with Janus green B. An additional source of difficulty is the fact that in flagellates, the parabasal bodies are often stained as darkly as the mitochondria. Although the parabasal bodies and mitochondria show additional similarities in staining reactions and composition, Volkonsky (1933) points out that the former are derivatives of the neuromotor system and cannot be considered as homologous with the mitochondria.

Mitochondria present a wide variety of shapes, but most commonly they are spherules (Fig. 17), chains of spherules (Fig. 26), short rods (Figs. 22, 23), or dumb-bells. The filamentous structures found so often in metazoan cells are found rarely in the Protozoa, but a good example has been described in the phytoflagellate *Polykrikos* by Chatton and Grassé (1929). Lynch (1930), in his studies on the ciliate *Lechriopyla,* found a compound structure (Figs. 15. 16), composed of several discs. Each disc is composed of chromophobic material, with a rim of chromophilic material which stains with Janus green and the other mitochondrial dyes. In cases of secretion (Fig. 17), the secretion granules often appear as a chromophobic center in the mitochondria (MacLennan, 1936), but the mitochondrial material itself usually appears to be homogeneous either in the living unstained ciliate or after any of the mitochondrial stains. It is probable that the chromophobic center of the discs of *Lechriopyla* represents material secreted by the rim, which is the mitochondrial part of the complex discs. In some cases, however, mitochondria may have a true duplex structure, since Mast and Doyle (1935b) showed that the outer surface of the mitochondria of *Amoeba* stain more deeply than the center. This differentiation may be explained either as a definite

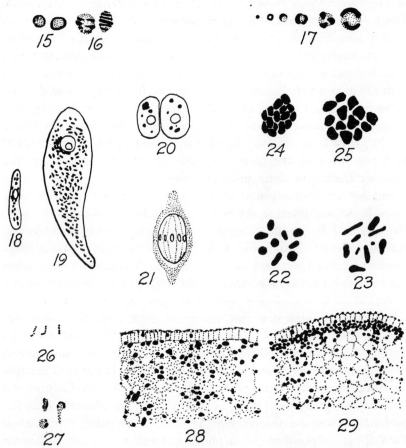

Figure 15-16. From *Lechriopyla mystax:* Figure 15, end view after Hirschler's mitochondrial technique; Figure 16, lateral view, after Champy-iron haematoxylin. (After Lynch, 1930.) Figure 17. From *Ichthyophthirius multifiliis,* series showing mitochondria and the secretion of paraglycogen, vital stain with janus green. (After MacLennan, 1936.) Figures 18-21. From *Monocystis* showing *de novo* origin of mitochondria: Figure 18, sporozoite; Figure 19, trophozoite; Figure 20, conjugating gametes from a cyst; Figure 21, spores, Champy or Flemmings—iron haematoxylin. (After Horning, 1929.) Figures 22-25. From *Amoeba proteus:* Figures 22-23, free in cytoplasm; Figure 22, normal: Figure 23, fixed in modified Regaud's fluid; Figures 24-25, on surface of contractile vacuole. (After Mast and Doyle, 1935.) Figures 26-27, From *Aggregata eberthi:* Figure 26, mitochondria proper; Figure 27, mitochondria associated with protein reserves. (After Joyet-Lavergne, 1926.) Figures 28-29. From *Bursaria truncatella:* Figure 28, section of early conjugant; Figure 29, section of later conjugant. (After Poljansky, 1934.)

In all cases, material which responds to mitochondrial stains is drawn in solid black; associated granules are stippled.

localization of stainable materials or as a dense surface with less dense centers, with the type of material the same in both places.

The mitochondria in most Protozoa are fairly evenly distributed through the cytoplasm, sometimes alone and sometimes associated with various types of storage granules (see the discussion of function). Fauré-Fremiet (1910) found that if any localization occurs, the mitochondria tend to concentrate beneath the pellicle and occasionally around the contractile vacuole. Horning (1927) extended these observations and contends that mitochondria tend to concentrate near all membranes—particularly around the food vacuoles during active digestion, beneath the pellicle, and around the nucleus. Hall and Nigrelli (1930) criticized Horning's identification of mitochondria, which was based largely on dark-field observations, and showed that in *Vorticella* sp. the mitochondria are not associated with the food vacuole. Volkonsky (1934) likewise rejected Horning's identification and showed that the granules associated with the digestive vacuoles, in a large number of species, are stainable only with neutral red. MacLennan (1936) found a similar situation in *Ichthyophthirius*. In other Protozoa, however, the accumulation of mitochondria near membranes has been confirmed. Mast and Doyle (1935b) find that the mitochondria in *Amoeba* are occasionally associated with the gastriole (food vacuole) and they were able to correlate the association with the type and stage of digestion. They also demonstrated an association with the contractile vacuole, confirming the earlier work of Metcalf (1910). Volkonsky (1934) found a small aggregation of mitochondria around the food vacuole of *Campanella* and *Paramecium* during the alkaline phase of digestion. Chatton and Grassé (1929) showed that the filamentous mitochondria of *Polykrikos* tend to accumulate near the pellicle, but instead of being parallel to the surface, as in the ciliates described by Horning, are perpendicular to the surface. Poljansky (1934) studied the changes occurring in the life cycle of *Bursaria* and found that the mitochondria of neutral individuals are uniformly dispersed throughout the cytoplasm, but during conjugation (Figs. 28, 29) the mitochondria migrate to the periphery and form a definite zone under the ectoplasm, and also around the micronuclear derivatives. The chondriosomes again scatter during the growth of the macronuclear primordium. There seems to be neither universal nor permanent localization of mitochondria near membranes, as is to be expected according to Horning's theory that mitochondria accumulate at the intracellular surfaces in accordance with the Gibbs-Thompson Law. Doyle (1935) suggests that mitochondria tend to collect at regions of active interchange, since he found that mitochondria are the only granules which flow out into the pseudopodia of the foraminiferan *Iridia*. While this theory is attractive in certain cases, it is difficult to see how this would apply to the concentrations which occur in conjugating *Bursaria*.

Mitochondria are more widely accepted as universal, permanent, and self-perpetuating granules than any other cytoplasmic component, and for this reason it is worth while to consider in detail the proof upon which such statements rest. Rigid proof of this theory requires a demonstration of mitochondria only, with, of course, a lack of evidence of any *de novo* origin. Furthermore, it is obvious that proof of the continuity of any cytoplasmic component cannot be based on a study of only one stage in the life cycle, but must rest upon adequate studies of the whole life cycle.

Mitochondria have been identified in a multitude of Protozoa of all groups by Fauré-Fremiet (1910) and later authors, and as a result these components are usually considered to be present in all Protozoa. However, recent evidence shows that this assumption is unjustified. An extreme example is *Trypanosoma diemyctyli*, in which Nigrelli (1929) was not able to demonstrate any pre-formed mitochondria, although granules which satisfy the general criteria of mitochondria are induced by exposure of the organisms to Janus green B. These induced granules are not permanent, but disappear after about two hours. The marine amoeba *Flabellula* is another species in which mitochondria are normally lacking. Hopkins (1938b) found that the normal amoeba possessed no granules which can be classified as mitochondria, but that when the amoeba is disturbed in a variety of ways, granules are precipitated in small pre-formed cytoplasmic vacuoles, the contents of which are normally a homogeneous fluid. After the recovery of the amoeba from the disturbing conditions, the granules are resorbed. These temporary granules possess the staining reactions of mitochondria, including the ability to segregate Janus green B, and are the only granules in this organism which can be classed in that group. As a contrast to these cases of induced mitochondria, Kirby (1936) reports the experimental destruction of the mitochondria which are a normal component in the cytoplasm of flagellates from termites. In normal *Pseudodevescovina,* large numbers of mitochondria can be demonstrated by the Flemming-Regaud method but, after feeding for three days on filter paper soaked in one-percent Janus green B, no mitochondria can be demonstrated. Additional examples of the lack of mitochondria are furnished by studies of the whole life cycle of certain Protozoa. Horning (1929) was able to demonstrate mitochondria in most stages of *Monocystis* (Figs. 19, 20), but found that these granules disappear in the sporozoite stage (Fig. 21). Beers (1935) and MacLennan (1936) also found that mitochondria disappear in the later encysted stages of ciliates, although the same methods give positive results in other stages of the cycle. These observations under both normal and experimental conditions demonstrate that mitochondria are neither universal nor permanent cytoplasmic constituents.

The crucial point in the classification of mitochondria as autonomous organelles is whether they always arise from preëxisting mitochondria. The occurrence of dumb-bell-shaped mitochondria and other possible division stages have been found so often in fixed and stained preparations that it is unnecessary to quote this evidence here. The observations on living material are few and perhaps the clearest is that of Horning (1926), who reported division stages of mitochondria in a living heterotrich and was able to confirm the descriptions based on fixed material. Thus division is a factor in the increase in numbers of mitochondria, but returning again to the studies of the whole life cycle, we find that division is not the only method, since mitochondria must be formed *de novo* (Figs. 18-21) in those species in which the mitochondria have disappeared during the quiescent phases of the life cycle. Mitochondria are not self-perpetuating organelles, but are differentiations which may endure for a longer or a shorter period during the cycle of the cell.

In the cases considered above, mitochondria are cytoplasmic in their origin, but this is by no means the only possibility. Joyet-Lavergne (1926) states that the group of mitochondria attached to protein granules (Fig. 27) are derived from the nucleus along with the protein reserves (for discussion of this point, see p. 163), but Daniels (1938) in the same or related species could find no mitochondria attached to the protein granules. Calkins (1930) found in *Uroleptus* one set of cytoplasmic granules, which he traced back to macronuclear fragments during conjugation, interpreting these as mitochondria. Since the staining methods were not very specific and these granules do not stain with Janus green, the statement that these are mitochondria must be accepted with caution. Miller (1937) has endeavored to prove that mitochondria, Golgi bodies, and other cytoplasmic inclusions in *A. proteus* are "bacteria spores, fungi, or yeasts, together with indigestible material of certain food organisms." This idea is, of course, similar to that of Wallin and has been so thoroughly criticized by Cowdry and others that it need not be considered here in detail. Miller was not able to culture these cytoplasmic "bacteria" and his main argument seems to be based on the observation that mitochondrial stains and Golgi type impregnations will demonstrate granules in the culture medium. This merely shows that the stains used are not always specific under all conditions, a fact which has been pointed out many times. Miller does not present any evidence which can stand up against the observations and experiments of Mast and Doyle (1935a, 1935b), Holter and Kopac (1937), and Holter and Doyle (1938) on the same species. The mitochondria are specializations of the cell itself, probably in all cases from the cytoplasm, and are neither artifacts nor invaders.

The composition of mitochondria is still incompletely known in any exact sense, in spite of the large amount of work done on these components. They have long been thought to be composed of both lipoids[2] and proteins, because of their staining reactions and solubilities (Fauré-Fremiet, 1910; Hirschler, 1924, 1927). Unfortunately no one has yet repeated in Protozoa the work of Bensley (1937), who isolated mitochondria from liver and was able to make both qualitative and quantitative analyses. These analyses confirmed the cytochemical analysis of lipoid and protein, but instead of the large amounts of phosphatids and so forth, which were predicted from the cytological reactions, he showed that the lipoid is largely neutral fat. Bensley also found that many reactions (for example the osmic-acid reaction) are much weaker

---

[2] Lipoid is used here, as in most cytological works, in a very general sense. It includes all materials which are soluble in ether, absolute alcohol, and so forth, and which stain with Nile blue sulfate, the sudans, and other fat soluble dyes. Neutral fat, fatty acid, phosphatids, and the like respond to these tests. Lison (1936) suggests the rather awkward term sudanophil material, in order to emphasize the cytological side and to avoid false implications as to chemical nature.

in mitochondria in the cytoplasm than in mitochondria isolated from the cytoplasm. These observations confirm the validity of the interpretation of mitochondrial reactions in the Protozoa as indicative of a lipoid-protein mixture, but emphasize the need of caution in more specific interpretations before methods as specific as Bensley's are applied to the Protozoa. It is clear that the evidence now available as to the nature of the lipoids and of the proteins in the mitochondria of Protozoa is significant largely as a lead for further work.

Horning (1927) adheres to the view that the lipoid component of mitochondria is a phosphatid, but presents no conclusive evidence for this statement. Wermel (1925) found that the mitochondria (or liposomes) of *Actinosphaerium* react with Ciaccio's method, but according to Lison the only valid interpretation is that unsaturated lipoids are present. MacLennan (1936) stated that the lipoid material in the mitochondria of *Ichthyophthirius* is a fatty acid, on the basis of a blue stain with Nile blue sulphate, used according to Lorrain Smith's method. However, since Lison (1936) presents evidence against the specificity of this method, the above interpretation is perhaps too strict, but it is interesting to note that the original interpretation is in accord with Bensley's analysis of the mitochondria of liver. There is likewise a lack of information on the nature of the proteins present. Hayes (1938) demonstrated a positive reaction to fuchsin-sulfurous acid reagent and claimed that nucleic acid is present. However, since lipoids may react in this manner in the "plasmal reaction," it is possible that this test was concerned with the lipoid component rather than the protein portion.

The metallic impregnation of mitochondria, or the depth of stain taken after the use of lipoid solvents, varies between the species of Protozoa and has usually been considered to be a rough indication of the proportion of lipoids present. Thus Scott and Horning (1932) find in *Opalina* a large amount of lipoid; Lynch (1930) in *Lechriopyla,* Patten (1932) in *Nyctotherus,* MacLennan and Murer (1934) in *Paramecium,* find some lipoid; while Beers (1935) in *Didinium* finds little if any evidence of lipoids in the mitochondria. Although this data is very crude, foundation is provided for the working hypothesis that there is a wide variation in the probable composition of mitochondria, ranging from practically pure lipoid to almost pure protein. Much of this difference represents constant differences between species and could be detected by Bensley's mass technique. However, Fauré-Fremiet (1910) found staining differences within the mitochondria of a single individual, and Peshkowskaya (1928) reports that the ectoplasmic chondriosomes of *Climacostomum* are resistent to fixatives which usually dissolve mitochondria, although the endoplasmic mitochondria are much more typical in their reactions. Pellissier (1936) found similar differences, not within the cell but between various individuals, and was able to show that all the mitochondria impregnate more deeply in the vegetative stages than in the stages just before reproduction. By the selection of species in which all granules are in the same stage at the same time, Bensley's mass technique could be used very profitably in exploring the changes in mitochondrial composition.

MacLennan and Murer (1934) found heavy deposits of ash in the typical mitochondrial rods, as well as in the other cytoplasmic granules of *Paramecium*.

The presence of enzymes in mitochondria have been indicated indirectly in many cases by the morphological association of these bodies with structures in which digestive or synthetic activity is going on. The only direct demonstration of the localization of cytoplasmic enzymes is due to Holter and Kopac (1937) and Holter and Doyle (1938), who showed that dipeptidase is not present in mitochondria, but that amylase is. The method used was a combination of centrifugal localization of granules and micro-methods for the measurement of enzymatic activity. The mitochondria were concentrated in one end of an *Amoeba*, which was then cut and the enzymatic activity of the mitochondria-rich and the mitochondria-poor portions of the cytoplasm compared. Since both the centripetal and the centrifugal portions had the same amount of dipeptidase (measured by the ability to split alanylglycine) per unit volume of cytoplasm, Holter and Kopac concluded that the enzyme is in the matrix. They point out that this proves nothing as to the origin of the enzyme, which might diffuse out from a granule as fast as it is formed. Holter and Doyle found that the middle region of the centrifuged *Amoeba* had the most amylase (measured by the digestion of starch). The nucleus, crystals, digestive vacuoles, mitochondria, and matrix are found in this zone of the centrifuged *Amoeba*. The enzyme could not be localized in the nucleus, since non-nucleated fragments show no significant diminution in amylase, nor in the crystals since most of these are in the centrifugal end which would thus have the highest enzymatic activity. They also ruled out a localization in the digestive vacuoles by a demonstration that the enzyme content of hungry and feeding *Amoeba,* with a resultant difference in the number of vacuoles, is the same. The only structures the distribution of which after centrifuging corresponds with the distribution of amylase are the mitochondria. The study of the centrifuged *Amoeba* presents many difficulties, since the stratification is never complete and there is always some mixing between the finishing of centrifuging and cutting the amoeba in parts, but the further use and development of these methods and their use in species in which stratification is complete will undoubtedly aid in the complete analysis of mitochondria and other cytoplasmic granules.

The theory that mitochondria are concerned with cellular respiration has led to attempts to identify in the mitochondria the materials known to be active in this respect. One of these is glutathione, in which the physiologically active group is sulfhydril, demonstrable cytochemically by the sodium nitroprusside reaction. Joyet-Lavergne (1927-29) found that the mitochondria of Sporozoa give a positive reaction with sodium nitroprusside, and this was confirmed by Cowdry and Scott (1928) in *Plasmodium.* Chalkley (1937), however, found that the strongest reaction in vegetative *Amoeba* is in the nucleus and that at the metaphase this material is poured into the cytoplasm. Some granules in the nucleus give a particularly strong reaction, but in the cytoplasm the coloration is diffuse. These results of Chalkley's extensive work on glutathione in *Amoeba* suggest the desirability of a reinvestigation of the Sporozoa, and certainly indicate that the materials containing the sulfhydryl group are not always localized in the mitochondria. Joyet-Lavergne (1934) has also shown that the mitochondria give a strong reaction with the antimony trichloride test for vitamin A, and concludes that this is a part of the respiratory mechanism along with glutathione. Although respiration cannot be discussed in detail here, it should be pointed out that the glutathione-vitamin A theory presents many difficulties and the system more usually accepted is glutathione-ascorbic acid (Holmes, 1937). Bourne and Allen (1935) and Bourne (1936) have demonstrated the concentration of ascorbic acid in cytoplasmic granules by the acetic-silver-nitrate method, but unfortunately have not correlated these with any particular type of granules. Daniels (1938) used this method in gregarines, but although she was able to demonstrate granular accumulations in the cells of the gut, the gregarines remained clear.

The functions ascribed to mitochondria in the Protozoa may be grouped under two main headings: synthesis (or segregation), and respiration. The first group includes a number of activities, all of which involve the accumulation of materials and in some cases the synthesis of new products from these raw materials. Examples of this type of function for which there is definite evidence are the secretion of reserve bodies, a digestive function in connection with the gastrioles, excretory function in connection with the contractile vacuoles, and the transport of materials from one organelle to another. Respiration also might be considered to fall in this category, since it would depend upon the accumulation in the mitochondria of the substances responsible for the oxidation-reduction processes of the cell.

The secretion of reserve bodies is cytologically the easiest phase of accumulation to demonstrate. Fauré-Fremiet described the formation of deutoplasmic granules by direct transformation of mitochondria, as well as by the more common method of segregation adjacent to or within the mitochondria which retain their identity. Joyet-Lavergne (1926a) occasionally found a relationship of this type between the mitochondria and paraglycogen and always found that protein granules possessed a mitochondrial cap (Fig. 27), but denied that this indicated anything more than a casual relationship. Horning (1925) described mitochondrial rims around the protein granules in *Opalina* (this identification of mitochondria is denied by Kedrowsky, 1931b) and accepts it as evidence of secretory activity. MacLennan (1936) described the origin of paraglycogen in the center of spherical mitochondria (Fig. 17). The identification of these granules was made not only with the usual permanent stains for mitochondria, but with specific microchemical stains (Sudan III, Nile blue sulfate, iodine, chlor-zinc-iodide), and their growth observed in live specimens stained with Janus green. The fact that the paraglycogen first appears as a center in a solid mitochondrial sphere refutes the usual suggestions that the secretion merely happens to be in contact with the mitochondria and that there is no real connection between the two. There is no evidence as yet to show whether these visible secretions are actually synthesized by the mitochondria or are the result of the segregation by the mitochondria of materials synthesized at other points. Wermel (1925) found that certain mitochondria of *Actinosphaerium* have a high lipoid content, as shown by Ciaccio's lipoid methods, i.e., a so-called liposome rather than an ordinary type of mitochondria, and concludes that they secrete the lipoid reserves. Except for the fact that these granules stain weakly

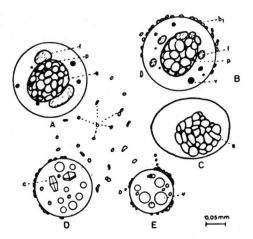

Figure 30. The association of mitochondria with the gastriole in *Amoeba proteus*. (From Mast and Doyle, 1935b.) A, 2-6 hours; B, 6-8 hours; C, 8-16 hours; D and E, 16-30 hours; b and b₁, mitochondria; f, fat; s, starch; v, vacuole refractive bodies; c, crystals; p, pellicle of *Chilomonas*.

with Janus green, they are like the intermediate lipoid bodies and could perhaps be classed more conveniently with them. Zinger (1928) counted the "spherical inclusions" and the mitochondria in *Ophryoglena* and found that they are roughly proportional in number. He concluded that the spherical inclusions are derived from mitochondria. Since many granules of entirely different origin will increase in number under favorable conditions, this conclusion cannot be considered as proved.

The digestive function of mitochondria rests upon two types of evidence: the actual demonstration of enzymes in these granules (see p. 125), and the correlation between the periodic aggregation of mitochondria around the gastriole and the type of digestion taking place within. Both types of evidence are available for *A. proteus*, so that the results of observations may be checked against a direct knowledge of the enzymes actually present. In *A. proteus,* Mast and Doyle (1935b) find that the mitochondria accumulate around the food vacuoles six to eight hours after its origin and then again at sixteen to thirty hours (Fig. 30). During the first contact, digestion begins, starch is changed to erythro-dextrine, and fat leaves the vacuole. In the second contact, vacuole re-fractive bodies and crystals disappear. These authors never found the mitochondria actually entering the vacuole. The relationship between mitochondria and starch digestion was directly demonstrated by Holter and Doyle, who showed that these granules contain amylase. Their function is restricted with respect to digestion, since they lack dipeptidase, according to similar studies by Holter and Kopac (1937). This situa-tion is not universal, since Hopkins (1938b) also working with Mast, showed that no formed granules are associated with digestion in the marine amoeba, *Flabellula,* although a material is dissolved in the vacuoles which when precipitated by disturbance, and so forth, forms granules which stain with Janus green B and other mitochondrial stains. Again, in many other Protozoa the mitochondria have no direct con-nection with digestion. Thus mitochondria may contain enzymes in some cases, but this is not a necessary association.

Excretory granules associated with contractile vacuoles have been described many times, but only recently have mitochondria been proved to be associated with the excretory process. Mast and Doyle (1935a) have shown that the excretory granules of *A. proteus* (Figs. 24, 25, 47) correspond in their staining reactions to mitochondria. By centrifuging the majority of these bodies into one end of the amoeba and removing this part, these authors showed that the formation of the vacuole is dependent upon these granules and that if most of the mitochondria are removed, death follows. In one experiment, most of the mito-chondria were left, with the result that the average interval between pulsations was 3.46 minutes. If few mitochondria were left, the time was correspondingly longer, and when very few granules were left, the average time between pulsations increased to twenty-five minutes. Mast and Doyle interpreted these experiments as indicating that some excretory material, toxic to *Amoeba,* is eliminated by the vacuole, and that the mitochondria function as the means of transport to the vacuole.

Doyle (1935) harmonized the apparently discrepant functions of digestion and excretion by pointing out that they may be united under the general heading of transport, and pointed out further that mitochondria tend to accumulate wherever exchanges are taking place, both within the individual and at the outer surface. Mast and Doyle (1935b) observed that the mitochondria accumulate around the crystal vacuoles while the crystals decrease in size, and with the surface of the refractive bodies while the latter are increasing in size. The mitochondria of *Amoeba* seem to be a mechanism for intracellular transport and for carrying amylase to the food vacuoles, digested material from the food vacuoles and crystal vacuoles to the refractive bodies, and metabolic wastes to the contractile vacuoles. However, since Holter and Kopac (1937) have shown that dipeptidase is not associated with mitochondria, and Volkonsky (1933) and MacLennan (1936) have shown cases in which vacuome alone touches the food vacuole, and Hopkins (1938b) cases in which no preformed granules are associated with the vacuole, it is clear that the mitochondria are not necessary in all cases either for transport function or a support for enzymes.

The supposed universality and permanence of mitochondria have led to many suggestions that they are concerned with some vital part of cellular activity, some function more universal than secretion and storage. Some evidence to this effect has been presented by Joyet-Lavergne (1927-35), both from the standpoint of the presence of materials active in respiration (see p. 121) and from the standpoint of a direct demonstration of respiratory activity. He finds that vital dyes are reduced most strongly near mitochondria and that individuals which have large amounts of mitochondria reduce the dyes faster than individuals with less mitochondria. He was able to demonstrate glutathione and vitamin A in the mitochondria and attempted to show that these two form an oxidation-reduction system. Rey (1931a, 1931b) repeated Joyet-Lavergne's staining experiments and obtained the same results, but criticized the latter's interpretation of his findings. Rey also repeated the experiments using an electrometric method for determining rH and found no significant differences. Wurmser (1932) likewise criticized Joyet-Lavergne's interpretations of the stain reactions as indications of oxidation-reduction differences. Bles (1929) found that the oxidation-reduction reactions which he studied in *Arcella* were associated with the hyaloplasm, rather than with any granules. Since Joyet-Lavergne found morphological continuity between the mitochondria and two types of reserve granules, the existence of a secretory function is possible, and it is not necessary to invoke respiration in order to find a function for these mitochondria. The respiratory function must be regarded as unproved, either from the standpoint of the proof of the presence of materials which could act as an oxidation-reduction system, or from the standpoint of the direct measurement of localized oxidation-reduction potentials. At best mitochondria as morphological entities cannot be necessary for respiration, since many species lack them at one time or another in the life cycle, and since in other cases they can be eliminated experimentally without fatal results (Kirby, 1936).

All the various types of evidence—staining reactions, composition, function, and tracing through the life history—show that mitochondria in the Protozoa do not form a homogeneous group, but are actually a heterogeneous assortment which are associated merely by their ability to segregate Janus green B or by even less specific staining reactions. No one type is found in all Protozoa, and in all cases which have been carefully studied mitochondria are not self-perpetuating but arise *de novo* at some time during the life cycle.

THE VACUOME HYPOTHESIS

According to the vacuome hypothesis as applied in the Protozoa, there are only two fundamental cytoplasmic components in the Protozoa—the chondriome and the vacuome, since the Golgi bodies and vacuome are merely different aspects of the same thing. The term vacuome was substituted for the earlier term segregation granule as an indication of the supposed homology between the neutral red bodies in animal cells and the vacuoles of plant cells. Volkonsky (1929 on), Kedrowsky (1931-33), Hall and his associates (1929 on), Lynch (1930), and others have upheld the general conclusion that the granules stainable with neutral red are identical with the Golgi bodies; but Kirby (1931), MacLennan (1933, 1940), Bush (1934), Kofoid and Bush (1936), Daniels (1938), and others have demonstrated many cases of neutral red granules which are not osmiophilic (for a more detailed discussion of this point, see p. 140). It should be pointed out that the acceptance of the vacuome hypothesis is by no means universal in the Metazoa or Metaphyta, according to Weier (1933) and Kirkman and Severinghaus (1938).

The ability to segregate neutral red has been ascribed to a specific reaction of the dye with a single material in the granule, the process being called the "neutral red reaction" by Koehring (1930) and included as a part of the "ferment theory of the vacuome" by Kedrowsky (1932b). The substance involved is supposed to be a proteolytic enzyme, a conclusion based on the work of Marston (1923), who showed that these enzymes are precipitated by combination with neutral red, Janus green, and other azine dyes. Le Breton (1931) has reviewed the rather voluminous literature resulting from this suggestion and concludes that the reaction is not specific, since ordinary proteins (an important constituent of most segregation bodies) also are precipitated. From the standpoint of cytology, the theory fails to explain how neutral red and Janus green have such different staining reactions in the granules of living cells. Hopkins's (1938a) experiments with *Flabellula* show that precipitates are formed in vacuoles after either Janus green or neutral red. If Janus green was first used, and then neutral red added, a red precipitate formed around the original green one. "The small neutral red vacuoles are, then, the same as the vacuoles in which the condensation granules are formed, but it appears that Janus green B stains a different component of these vacuoles than does the neutral red."

Not one factor, but many are responsible for the segregation of neutral red by cytoplasmic granules. The rôle of pH in neutral red staining has been demonstrated by Chambers and Pollack (1927) and Chambers and Kempton (1937), who showed that neutral red tends to go from an alkaline region to an acid region so that "segregation granules" would be those which are acid relative to the hyaloplasm. Kedrowsky (1931) demonstrated in *Opalina* that normally the segregation granules have this relationship with the hyaloplasm and that the staining reactions can be changed by altering the pH of the cytoplasm. He also showed that the granules will take up acid dyes in the presence of albumoses, which was confirmed by Volkonsky (1933) and included under the term "chromopexie." Since neutral red has long been known as a lipoid stain (Fauré-Fremiet, Mayer, and Schaeffer, 1910) it is possible that this may play a rôle in the staining of bodies which contain lipoids, such as the dictyosomes of gregarinida and the digestive granules of *Ichthyophthirius*. From these brief examples it is clear that the segregation of neutral red and other vital dyes is influenced by many internal and external factors and that it cannot be considered specific in the sense that it combines with a single definite substance.

Since many factors are involved in the segregation of neutral red, it is to be expected that more than one type of granule will be revealed by the use of this and similar dyes. Conclusive evidence of this lack of specificity is furnished by several Protozoa in which two or more types of granules are able to segregate neutral red at the same time, i.e., under identical conditions. Dangéard (1928) stained two types of granules in *Euglena* with neutral red—the vacuome and the mucous apparatus. The latter group may be extruded to form a mucous envelope or mucous hairs—an interesting example of true external secretion, and comparable to the staining of secretion granules in Metazoa by neutral red. However, Dangéard rejected the mucous apparatus as vacuome because it retains the neutral red after the death of the cell, while the other granules—the true vacuome—do not. Finley (1934) found four different groups of granules which segregate neutral red in *Vorticella*: pellicular secretions, pellicular tubercles, thecoplasmic granules, and refractile granules. He was careful to control the staining to avoid overstaining, so that his results cannot be questioned on that ground. Bush (1934) found in *Haptophyra* two sets of granules which are discontinuous in size and distribution. Mast and Doyle (1935b) complete the picture by showing that three groups of granules stain in *Amoeba*: vacuole refractive bodies, refractive bodies, and blebs on crystals, and they proved experimentally that these bodies are different in origin and in fate. The experiments of Kedrowsky (1931) on *Opalina* show the reverse picture—under certain feeding conditions, the growing segregation vacuoles lose their ability to take up neutral red, just as they do in vertebrate eggs. Hall and Loefer (1930) showed that the granules in *Euglypha* may vary in the same specimen from pink to bluish red or red violet.

Since neutral red stains several different groups of granules and also does not stain all stages of the same granule, it does not of itself reveal fundamental homologies, and it seems to me to be unjustifiable to group all of these bodies as vacuome or under any other catchall term. As a preliminary step toward an accurate classification of this group, the digestive granules—i.e., those neutral red bodies associated with the gastrioles—are separated from the segregation granules which correspond to the "vacuome de reserve" of Volkonsky and the segregation apparatus of Kedrowsky. The segregation granules of necessity still include a heterogeneous assortment and probably include granules other than those associated with synthesis and storage, but at present they cannot be classified properly because too many of them are known only by their ability to segregate neutral red. Kedrowsky and Volkonsky regard both types as carriers of enzymes, in one case acting to digest proteins and in the other case to synthesize them. However, each group appears *de novo* when the necessity arises, and the two groups show no direct continuity with each other. Further evidence of the independence of these two groups of granules is furnished by *Opalina,* in which only the segregation bodies are present, and by *Ichthyophthirius,* in which only the digestive granules are present, these latter having no connection with the storage of the numerous protein granules.

<h3 style="text-align:center">DIGESTIVE GRANULES</h3>

The digestive granules may be briefly defined as cytoplasmic granules stainable with neutral red, which become associated with the newly formed vacuoles containing food. This does not include all neutral red granules in the food vacuole, since such granules as the vacuole refractive bodies of *Amoeba* are derived from the food (Mast and Doyle, 1935b) and thus are not cytoplasmic components. Volkonsky (1934) points out that the term vacuole has been applied to so many structures that it is a source of confusion, and he has substituted the term gastriole. The fluid vacuole which contains ingested food is a progastriole, and with the addition of the digestive granules (vacuome in Volkonsky's terminology) becomes a gastriole. In addition to these terms it is convenient to use postgastriole for the structures containing undigested remnants.

There is no single pattern of the gastriole in the Protozoa. In *A. proteus,* mitochondria alone aggregate periodically around the gastriole (Mast and Doyle, 1935b); in *Paramecium* and *Campanella,* the digestive granules enter the gastriole, and mitochondria cluster around the membrane in the alkaline phase; in *Flabellula,* the materials stainable with Janus green or neutral red are normally dissolved in the fluid vacuole, which later forms the gastriole (Hopkins, 1938), and a somewhat similar situation is found in hypermastigote flagellates (Duboscq and Grassé, 1933); in *Ichthyophthirius* (MacLennan, 1936), the digestive granules (Fig. 31) enter the gastriole, but the mitochondria are never associated in any visible manner with the digestive mass. A more detailed description of digestion will not be given in this section, since attention is here centered on the granules.

All careful descriptions agree that the digestive granules vary in size and shape during the gastriolar cycle, ranging from minute spherules to relatively large rods. Both vital stains and metallic impregnation show a homogeneous structure, and the deformation of these granules either

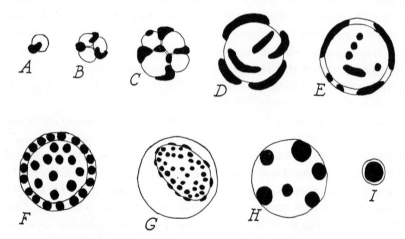

Figure 31. The association of the gastrioles and the digestive granules in *Ichthyophthirius multifiliis.* (From MacLennan, 1936.)

from other cytoplasmic granules or from outside pressure show that they have a soft, semifluid consistency. Volkonsky (1934) has found that the morphology of the gastriole varies in the same species with the food used. In *Acanthamoeba* he was able to induce the formation of large granules, small granules, or a homogeneous rim stainable with neutral red, by varying the food used. No digestive granules were formed around vacuoles which contained only starch.

The penetration of the digestive granules into the vacuole was not observed by Koehring (1930), Hall and Dunihue (1931), Dunihue (1931), or Hall and Nigrelli (1930), and the suggestion was made that the granules in the food vacuole are derived from food particles rather than from the cytoplasm, a suggestion which later was proved to be true, in the case of vacuole refractive bodies of *A. proteus,* by Mast and Doyle (1935b). Volkonsky, however, checked his observations on many ciliates and ruled out exogenous granules in the gastrioles of *Glaucoma* by the use of bacterial free medium; he still observed digestive granules in the vacuoles and was able to trace them from the cytoplasm. The migration of the digestive granules into the gastriole of *Ichthyophthirius* was found by MacLennan (1936), but it was shown that the granules do not penetrate any membrane. Instead, a new membrane is formed (Fig. 31) around the whole gastriole and the inner membrane, which is the original one, then disappears. The end result is exactly the same as in the cases described by Volkonsky, but the mechanism is somewhat different.

Volkonsky's interpretation of the digestive granules as vacuome is based upon their impregnation by the various Golgi-type methods. The question as to the identification of the granules as shown by entirely different methods is not present in this case, as it was in the case of the scattered cytoplasmic granules (segregation granules), since the digestive granules can be recognized independently of their staining reactions by their relationship to the gastriole. Hall and Nigrelli (1937) claim that the digestive granules are less consistent in impregnation than the scattered cytoplasmic granules and dispute Volkonsky's claim that they can be considered as vacuome. MacLennan (1940) showed that the various types of osmiophilic granules could not be distinguished on the basis of impregnation alone; in particular the digestive granules of *Ichthyophthirius* show 100-percent impregnation.

Few of the materials which occur in the digestive granules are known from direct evidence. The digestive granules of *Paramecium* are high in ash (MacLennan and Murer, 1934) and those in *Ichthyophthirius* contain lipoids (MacLennan, 1936). These latter bits of information are not as yet particularly useful and emphasize the need for more specific knowledge. The presence of enzymes is suggested by the morphological evidence and this would be a fertile field for the use of the microenzymatic methods.

The digestive granules are not permanent self-perpetuating structures, but appear to rise in the cytoplasm in response to the stimulus of feeding. Volkonsky (1934) found that when the preëxisting granules are utilized in the formation of gastrioles, new granules are formed in the cyto-plasm. The digestive granules in *Ichthyophthirius* were observed by MacLennan (1936) to arise *de novo* in direct contact with the ingested particles of food. Final evidence of their *de novo* origin was furnished by the fact that no digestive granules are present in the encysted stage, after the gastrioles which were formed during the feeding stage have disappeared. Volkonsky suggests that the materials of the digestive granule exist in the cytoplasm in a diffuse state. In the hypermastigote flagellates this material is concentrated in dissolved form in the gastrioles; in *Flabellula* the materials are first concentrated in vacuoles, which con-tribute the fluid part of the gastrioles; in the more common cases, the materials are condensed to form digestive granules. We must not for-get, however, that not all the digestive reactions result in granules, since dipeptidase in *Amoeba* is independent of any granules. Whatever the specific morphology of the reaction, whenever the cytoplasmic equilib-rium is disturbed by the ingestion of food there is an effective mobiliza-tion of this material to cope with the ingested food. Volkonsky calls these varied changes the "vacuolar reaction," an extremely useful term, but since we reject the term vacuome, "gastriolar reaction" would be more appropriate.

SEGREGATION GRANULES

The segregation granules are bodies which are able to concentrate, accumulate, and store within themselves vital dyes, proteins, and other materials. Unlike the definitions of most cytoplasmic granules, this definition is based upon function rather than upon morphology or staining reactions. The evidence for this definition is due largely to the work of Kedrowsky (1931-33) on *Opalina, Spirostomum,* and other ciliates. Since the functions of most of the granules stainable with neutral red have not been demonstrated, the work on *Opalina* will be discussed first and granules in other Protozoa considered in the light of this work. The accumulation of vital dyes in higher concentration than they occur in the medium is, of course, one piece of evidence of the segregating ability, even if this has not been demonstrated with the materials which enter into the metabolism of the cell.

The segregation bodies of *Opalina* (Figs. 32-35) are the external layer of granules or ectosomes and have been identified as Golgi bodies or as mitochondria by various authors, since they respond to some of the Golgi and mitochondrial techniques, although they are not stained specifically with Janus green B. Kedrowsky described four main morphological types—fine dispersed granules, large granules, alveoli, and heteromorphic granules (Figs. 32-35). In some cases the dispersion of granules is so accentuated that they lose their identity as granules. These changes in morphology are associated under natural conditions with the seasons of the year; for instance, the heteromorphic types are common in the spring and early summer, and the large granular type is found in the early spring. There is some variation between populations of different frogs in the same season, but all members of the same popula-

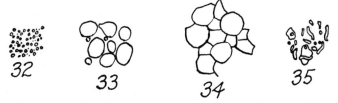

Figures 32-35. Basic morphological variations in the segregation granules of *Opalina ranarum*. Semischematic. (After Kedrowsky, 1931e): Figure 32. Dispersed type; Figure 33, homomorphic granular type; Figure 34, alveolar type; Figure 35, heteromorphic granular type.

tion have essentially the same type of granules. These different forms can be produced experimentally by changing the culture medium. The colloidally dispersed type is typical of amino-acid cultures. In distilled water, each ectosome swells and becomes a watery vacuole. The heteromorphic type is found in cultures which contain defibrinated and hemolysed blood. These changes, which occur both naturally and in artificial media, may well account for the disagreement among cytologists both as to the morphology and the identification of these granules.

The segregation bodies of *Opalina* are not permanent organelles. As indicated above, they may disperse homogeneously through the ectoplasm, which then takes a general pale stain with neutral red, and no method applied at this stage shows any indication of a remnant of the originally discrete granules. If the bodies are loaded with protein compounds of metals, such as silver, they may finally be extruded from the surface and be replaced by granules which arise *de novo*.

The composition of the segregation bodies is as varied as their morphology and results from the same causes—changes in the environment. Under natural conditions they may contain proteins (Millon's reaction), glycoproteins (Fischer's reaction), cholesterin (Schultz reaction, digitonin reaction), or be stained with bile pigments. The alveolar type of course contains mostly water. The segregation granules will store basic dyes in salt solution, acid dyes in the presence of proteins, silver in the form of kollargol or other similar compounds, iron albuminates, cholesterin, and so forth. According to the "ferment theory of the vacuome" of Kedrowsky (1932b), enzymes are present in the

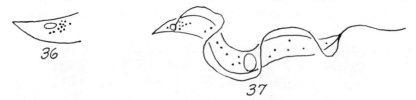

Figures 36-37. Segregation granules in *Trypanosoma diemyctyli,* neutral red stain. (After Nigrelli, 1929.) Figure 36. Preformed granules around the blepharoplast; Figure 37, preformed granules and granules induced by prolonged exposure to neutral red.

segregation bodies, but, as is the case with the digestive granules, no direct proof of this is available.

The segregation bodies obviously function in the concentration and storage of various materials, particularly proteins. This may be accounted for on a purely physical basis, as, for instance, the tendency of basic dyes to migrate to a more acid region, or the tendency of molecules to migrate toward bodies of opposite charge. The exact mechanism is, of course, a complex problem and cannot be taken up here, but the essential point at present is that the granules may play a purely passive rôle—the accumulation and storage of materials which originate elsewhere.

Accumulation is merely one part of the function, according to the enzyme hypothesis of the segregation granules, advanced at nearly the same time in slightly varying forms by Koehring (1930), Kedrowsky (1931 on), and Volkonsky (1929 on). Koehring based her conclusions on the supposed specificity of the neutral red reaction for proteolytic enzymes (see p. 178). Kedrowsky considers that both the synthesis of proteins and deaminization may occur in the segregation bodies. The evidence of synthesis may be summarized simply as the increase in size or number of segregation bodies in *Opalina* when immersed in various nutrient solutions, and the subsequent identification in the granules of materials from this culture medium. This certainly proves the segregating ability, but the actual synthesis might take place almost anywhere and the increase in size be due simply to the segregation of these pre-formed materials. The evidence of deaminization is based on the appearance of glycoprotein in the late stages of the segregation bodies and the fact that the segregation granules are able to oxidize Rongalit white vitally. This evidence could hardly be called more than suggestive, but the processes which are indicated by these tests seem to be localized in the segregation granules.

The neutral red granules of Protozoa other than *Opalina,* to be considered in the rest of this section, excludes only the group which were discussed above as digestive granules. It is thus essentially the group called segregation apparatus by Kedrowsky (1931), or the vacuome of Hall (1929 on). Since neither the history nor the function of most of these granules is known, this is doubtless a heterogeneous group, but I believe that further splitting at this time would merely add names without increasing understanding.

The segregation bodies are normal cytoplasmic constituents and are not induced by vital dyes, since they have been observed by many investigators in normal, unstained specimens (Hall, 1929 on; Finley, 1934; MacLennan, 1933, 1936; Volkonsky, 1929 on; Kedrowsky, 1931 on, and others). Prolonged staining (Fig. 23), it is true, may induce the formation of new granules (Kedrowsky, 1931; Cowdry and Scott, 1928; Nigrelli, 1929), but this is not a universal phenomenon, since many species, in my own observations, show the general diffuse staining of the cytoplasm and nucleus characteristic of severe overstaining, without the appearance of new granules. The normal origin of all types of segregation granules seems to be *de novo.* Mast and Doyle (1935b) removed most of the refractive bodies from *A. proteus* and observed the formation of new bodies several hours after feeding. The "blebs" on the crystals likewise clearly originate *de novo.* These cases are too few to justify any certain statements for all the many segregation granules, but it is indicative that in all the many descriptions of these granules, there has been no clear case of division described—all the evidence indicates a *de novo* origin.

Small homogeneous spherules are the commonest type of segregation body and they are found in all classes of Protozoa. There are often more than one group in a single individual, as shown by differences in size (Bush, 1934) or differences in localization (Finley, 1934). Complex bodies similar to those found by Kedrowsky in *Opalina* have been found in flagellates from termites (Kirby, 1932). In some cases such complex

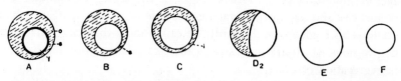

Figure 38. Segregation granule (refractive body) of *Amoeba proteus* in successive stages of resorption. (From Mast and Doyle, 1935.) A, normal body; B-F, stages in resorption, optical sections; D, surface view. Outer layer cross-hatched, shell black, vacuole in outline.

bodies may be the result of precipitation of dyes or other materials in an aqueous vacuole, but in the cases of the denser granules entirely different methods show the duplex structure. In the refractive bodies of *Amoeba* (Fig. 38), microchemical tests show that the duplex structure represents an actual difference in composition (Mast and Doyle, 1935a).

The segregation granules are most often scattered at random throughout the endoplasm (Hall and his associates, 1929 on), but occasionally are more definitely localized. In *Trypanosoma* (Figs. 36, 37) they are concentrated around the blepharoplast (Nigrelli, 1929), in the Ophryoscolecidae they are most common in the operculum and spines (MacLennan, 1933), and in *Lechriopyla* a marked concentration is found just under the pellicle (Lynch, 1930). In *Vorticella* a group of discrete globular inclusions is found scattered in the cytoplasm, and another group is found in the stalk (Finley, 1934). The tubercles of the pellicle also stain, as do the secretions of the pellicle, but the latter have been considered in the section on secretion granules and the former do not seem to me to fulfill the definition of cytoplasmic granules, or at least are clearly not in the same class as the ectosomes of *Opalina*. As a contrast to these Protozoa with several types of segregation granules, the ciliate *Ichthy-*

ophthirius has none at all at any time in its life cycle (MacLennan, 1936), and the function of synthesis and storage of the protein reserves is taken over by the macronucleus.

The composition of the segregation granules of Protozoa other than *Opalina* is known only in the case of the refractive bodies and blebs of *Amoeba.* The osmiophilic shell of the refractive bodies of *Amoeba* has a protein stroma impregnated with a lipid substance (Mast and Doyle, 1935a). This portion stains with Sudan III only after Ciaccio's method for "unmasking" the lipoids, and is intensely blue in Nile blue. These granules are not dissolved in alcohol in twelve hours, but they do lose their positive reaction to fat soluble dyes. They respond to the methylene-blue-sulphuric test for metachromatin and give a faint reaction with Millon's reagent. Within this layer a brittle carbohydrate shell and an unknown fluid are found. The stainable blebs on crystals in the same protozoan, when first formed, contain only lipoids, but as they grow larger, protein is added so that their final composition is similar to the shell of the refractive bodies of the cytoplasm. These granules are therefore quite different from the segregation granules of *Opalina,* since Kedrowsky showed that whatever else might be stored in these bodies, they do not segregate lipoids. However, since the refractive bodies do segregate proteins and stain with neutral red, they are included as segregation bodies. This is, of course, arbitrary, since they overlap on the Golgi granules and on the reserve granules. Sufficient mineral ash is present to mark the vacuome in incinerated specimens of *Opalina* (Horning and Scott, 1933) and *Paramecium* (MacLennan and Murer, 1934). The only striking fact about the segregation bodies in most Protozoa is that with the exception of those in *Opalina* and *Amoeba,* not even a sketchy outline of their composition is available.

The function, in so far as it is known, agrees closely with the storage function described for the granules in *Opalina.* Variations of the numbers of the segregation granules have been found in *Paramecium* (Dunihue, 1931) and in *Vorticella* (Finley, 1934). Dunihue was able to correlate the decrease in numbers with starvation, thus indicating a storage function. Mast and Doyle (1935a) showed that both protein and lipoid materials are found in the blebs on the crystals and on the refractive bodies. Since the blebs appear shortly after feeding and since they in turn disappear as the refractive bodies are increasing, the blebs seem to be temporary reserves, while the refractive bodies are the final reserves. In the blebs, the protein portion appears later than the lipoid portion. There is no evidence in these cases of anything more than the accumulation of materials.

GOLGI BODIES

The term Golgi body has been applied in Protozoa to organelles which differ fundamentally in both composition and function. These structures include contractile vacuoles (Ramon y Cajal, 1904-5), granules (Hirschler, 1914), specialized regions of cytoplasm around the contractile vacuoles (Nassonov, 1924), segregation granules (Cowdry and Scott, 1928), osmiophilic nets (Brown, 1930), and the parabasal apparatus (Duboscq and Grassé, 1925). The controversies which have arisen between advocates of one or another of these structures have been due not so much to disagreement upon the actual facts involved as to disputes concerning criteria for the identification of Golgi material. The selection of criteria is thus a crucial point in coördinating the investigations of Golgi bodies; yet even after years of work on representatives of all the major groups of animals and plants, and notwithstanding periodic reviews of the field, few criteria seem to have unanimous approval—few, indeed, have majority approval. The most recent review (Kirkman and Severinghaus, 1938) after failing to demonstrate any universal and objective basis of identification, quotes as follows from Gatenby (1930): "modern workers in general have experienced no difficulty in identifying Golgi bodies." This is very satisfactory as long as only one school of cytologists is considered, but such statements lose their attractive ring of authority when one tries to correlate the results presented by such experienced cytologists as Bowen, Parat, Gatenby, Canti, or Ludford, to mention only a few.

This is particularly true when seeking criteria on which to base a reasonable identification of Golgi bodies in the highly specialized cells of the Protozoa. In the following paragraphs the major objective criteria which have been used are discussed with particular reference to their applicability to the Protozoa. In general these criteria involve two points—consistent impregnation, and occurrence in all types of cells.

The first Golgi structures were discovered by the use of metallic impregnation methods, and ever since these methods have remained as the primary criteria (Gatenby, 1930), particularly since the discovery that the supposedly typical reticular network is a relatively rare type and in many cases is due to a temporary aggregation of granules (Hirschler, 1927). Both the osmic and the silver methods involve the same principle—the adsorption of the reduced metal on particular structures, although the actual reduction of the metallic compound may take place either where the deposits are located or in other parts of the cell (Owens and Bensley, 1929). Since the osmic acid methods have been most thoroughly investigated, this discussion will be restricted to these techniques, but the fact should be emphasized that the same general principles are involved in the silver techniques as well.

The preliminary treatment of the cell strongly affects the subsequent impregnation (Lison, 1936). Thus direct exposure of the protozoan to osmic fumes (Hall, 1929) is not equivalent to the full technique which involves preliminary fixation in such mixtures as Champy's fluid. Likewise, the blackening of granules by exposure to osmic fumes after they have been stained with neutral red is not the equivalent of the standard Golgi methods, since it has been shown in the case of *Flabellula* (Hopkins, 1938b) that the blackening is produced only in the presence of neutral red.

The Golgi methods are not specific for any single type of material, since a series of different lipoids extracted from echinoderm eggs give a typical Golgi reaction (Tennent, Gardiner, and Smith, 1931). Furthermore both lipoid and non-lipoid bodies of various types in the Protozoa react typically and identically to the Golgi techniques (MacLennan, 1940). Although the method is not specific in the sense that it reveals a single known substance, or some unknown Golgi material, the method is consistent in the sense that under proper conditions the results can be reproduced. In general, the reputation of osmic acid methods as carpricious and erratic is due to the slow penetration of tissues by osmium tetroxide, with the result that the cells in a block vary in exposure to the unreduced reagent because of their varying distances from the free surfaces. Another source of difficulty is the fact that the composition of cytoplasmic granules changes during the growth of that granule (Kedrowsky, 1931e; Mast and Doyle, 1935a, 1935b; MacLennan, 1936), and it is to be expected that the reducing power will also vary. The Protozoa are admirably adapted to solve both of these difficulties. The individual cells are either separate or in such small aggregates that all cells and all parts within the cell are exposed to essentially the same concentration of reagents. The variations in the stage of the Golgi granules can be observed directly in living Protozoa, so that the impregnation of the granules can be compared at equivalent stages, which can be identified with or without impregnation. In metazoan cells the Golgi nets are extremely difficult to demonstrate in the normal living cell, so that it is at present impossible in these cells to check the structures independently of impregnation. When standard conditions are achieved both around the cell and within the cell, absolutely consistent results are attained, even with osmic acid (MacLennan, 1940). Under these conditions, 100 percent of the digestive granules, fatty acid granules, excretory granules, and so forth, impregnate, with the result that these types cannot be separated on the basis of the Golgi-type reactions. The results of Hall and Nigrelli (1937), which seem to show that excretory granules and digestive granules are erratic in impregnation, is due to failure to allow for changes in composition and aggregation or for the occasional lack of these granules at certain times in the life cycle. All the stages have been lumped together, rather than equivalent stages compared.

In view of the fact that osmic acid is reduced by many different substances, it is highly important to prove that the impregnated bodies are not some other component (Hirschler, 1927; Bowen, 1928). Resistance to bleaching by turpentine or hydrogen peroxide is more pronounced in Golgi bodies than in most mitochondria, but unfortunately, even with this method, there are too many border-line cases in which the individual judgment of the observer is the determining factor. However, this individual judgment can be eliminated in comparing Golgi bodies and mitochondria by the use of such methods as the Altmann aniline acid fuchsin stain after osmic impregnation. Such a comparison of Golgi bodies and segregation granules is not possible, since the major criteria of the one is a fixation method and of the other is a vital staining method. If the two components have a different distribution, a comparison of different cells is sufficient to establish the difference (MacLennan, 1933), but too often there is in both cases merely a random distribution. The most famous case of this sort is that of the gregarines, in which Joyet-Lavergne (1926b) described the Golgi

bodies as being identical with the neutral red bodies because of similarity in form and distribution, and this has been widely accepted in
spite of objections presented by Tuzet (1931) and Subramaniam and
Ganapati (1938). The question was not settled conclusively until
Daniels (1938) applied the centrifuge to these species. She found
that Golgi bodies always moved to the centripetal pole of the cell, but

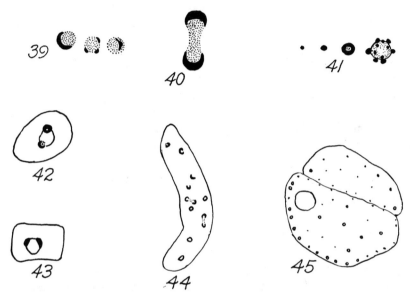

Figures 39-45. Dictyosomes: Figure 39, Dictyosomes from *Haptophrya michiganensis,*
Champy-osmic (after Bush, 1934); Figure 40, diagram of a dividing dictyosome of
Lecudina brasili (after Subramanian and Ganapati, 1938); Figure 41, stages in the
secretion of neutral fat in *Ichthyophthirius multifiliis,* Lorrain Smith Nile blue-sulphate
method, black represents blue stain, stippling represents pink (after MacLennan, 1934);
Figures 42-45, dictyosomes during the life cycle of *Lecudina brasili;* Figures 42 and 43,
intracellular stages; Figure 44, growing trophozoite; Figure 45, "association" stage,
either daFano or Nassonow impregnations (after Subramanian and Ganapati, 1938).

that the bodies stainable with neutral red were never displaced. Thus
in this case the Golgi bodies and the neutral red granules are not
identical. This does not mean that no osmiophilic granules segregate
neutral red, since, for example, the digestive granules react to both
impregnation and vital staining, but it does mean that the ability to
segregate neutral red is not a characteristic of all the Golgi bodies of
the Protozoa.

The theory that the Golgi apparatus is a universal organoid of the cell, as constant in its characteristics as is the nucleus, has given rise to a series of criteria requiring permanence during the whole cycle of the cell, as well as similarity in form and intracellular distribution. The presence of Golgi bodies in all stages of the life cycle has been demonstrated in Sporozoa (Joyet-Lavergne, 1926a) as well as their origin by the division of preëxisting Golgi bodies (Subramaniam and Ganapati, 1938, Figs. 42-45). However, this is not universal, since neither

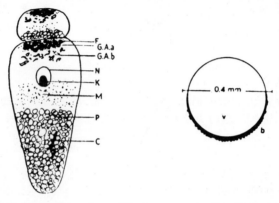

Figures 46-47. The effect of centrifuging upon the distribution of cytoplasmic granules. Figure 46, diagram of a centrifuged gregarine, F, fat, GAa granular Golgi material, GAb larger Goli elements, N nucleus, K karyosome, M mitochondria, P paraglycogen, C "chromidia," neutral red bodies not shown; the paraglycogen mass marks the centrifugal pole. (From Daniels, 1938.) Figure 47, contractile vacuole of a centrifuged amoeba, v vacuole, b mitochondria; the mitochondria are thrown to the centrifugal surface of the vacuole. (From Mast and Doyle, 1935b.)

of the types of Golgi bodies in *Ichthyophthirius* (MacLennan, 1936) nor *Amoeba* (Mast and Doyle, 1935a) are self-perpetuating or even present in all stages of the cycle, but arise *de novo*. Thus the Golgi bodies are not universally self-perpetuating and permanent.

The criterion of similarity in form has received considerable support, but the evidence as to what this form is has been discordant. Nassonov (1924) described a net-like structure around contractile vacuoles and homologized this with the Golgi net. Hirschler (1927) finds that the typical Golgi bodies have an osmiophil cortex and an osmiophobe center, this duplex structure being called a dictyosome. The complex nets around the contractile vacuoles are, according to Hirschler, aggregations of the dictyosomes. Subramaniam and Ganapati (1938) insist on this dictyosome structure, although they describe a homogeneous spherule in one stage of the Golgi cycle:

a Golgi granule when it enlarges becomes differentiated into a vesicle having chromophile and chromophobic regions. Rupture of the vesicle gives rise to batonettes in which the chromophobic part is in relation with the cytoplasm.

Since all of these structures—spherules, dictyosomes, aggregations—may be included in the cycle of a single granule (Kedrowsky, 1931; MacLennan, 1936; Mast and Doyle, 1935a) or as the result of the periodic aggregration of granules (Hirschler, 1927; MacLennan, 1933), it is impossible to insist on one type to the exclusion of the others, and we must conclude that morphology is not a criterion for the identification of various types of granules.

Similarity in distribution within the cell has also been urged as a criterion. Hall (1931) holds that a random distribution throughout the cytoplasm is the typical configuration, while Subramaniam and Gopala-Aiyar (1937) consider that an excentric juxtanuclear position, similar to that found in spermatids or gland cells, is typical. In metazoan cells both types have been found, and both may be typical of the same cell if all stages of the life cycle are considered. Similar periodic aggregations and dispersal of Golgi granules have been described in connection with the pulsatory cycle of the contractile vacuoles of ciliates (MacLennan, 1933, 1936). These movements are comparable to the migration of mitochondria and of digestive granules, which have likewise been associated with functional changes in the cell.

The chief characteristic either of form or of distribution of Golgi bodies is a variability which is associated with functional changes, and even this variability is not a criterion since it is characteristic of all granules which are actively concerned with the metabolism of the cell.

The criteria based upon impregnation and upon separation from other granules which can be identified by more specific methods are the only truly objective criteria, while the criteria of universality in form, distribution, and permanence are indefinite and are based upon various theories of the function, form, or derivation of Golgi bodies. The term Golgi body, being based upon nonspecific criteria (nor would retention of specific form or distribution make the definition more exact), includes a heterogeneous group of structures, including scattered endo-plasmic granules and granules associated with contractile vacuoles or fused to form heavy and permanent vacuolar membranes. A summary of the known functions of these granules includes the excretion of ma-terials through the contractile vacuoles, the formation of granules of neutral fat, the storage of lipoids other than neutral fat, a secretory cycle which does not involve lipoids, as well as many functions as yet unknown. These functions are different, but in a broad sense they are all varieties of secretion and in this respect conform to the Nassonov-Bowen theory of the relationship between Golgi bodies and secretion. The restriction of criteria for Golgi bodies to that of impregnation alone thus does not do violence to the concept of Golgi bodies as originally developed in verte-brate tissues.

The Golgi bodies are simply those secretory bodies (exclusive of mitochondria and segregation granules) which synthesize or store ma-terials which can be preserved by Golgi-type fixatives, and after this treatment are able to reduce (or adsorb the reduced metal) OsO_4 or silver nitrate. This is not a natural grouping, since on the one hand it includes several specific types of secretion, and on the other hand it does not include all types. In several cases the Golgi bodies can be classified according to function or composition, and in this discussion they are referred to as excretory granules, intermediate lipoid bodies (see p. 151), and, when the bodies are secretory in nature but the type of secretion body unknown, secretory Golgi bodies. This leaves a miscel-laneous group of Golgi bodies which are known only by their ability to reduce osmium or silver and for which there is no evidence as to either composition or function. Since the impregnation reactions them-selves do not reveal homologies which must be based on composition and function, the retention of the term Golgi body is merely a con-venience to bridge the change from reliance on the nonspecific osmic techniques alone to reliance upon specific cytochemical and physiological criteria.

EXCRETORY GRANULES

Ramon y Cajal (1903-4) was the first to suggest that the contractile vacuole is equivalent to the Golgi reticulum of the cells of the Metazoa. The first confirmation of this view was the demonstration in several ciliates by Nassonov (1924) that this vacuolar region is osmiophilic.

Further work has extended the number of such cases in ciliates and flagellates, but at the same time it has been definitely proved that the vacuoles in several species of both classes are never osmiophilic. An examination of these cases shows that they form a closely graded series ranging from no impregnation at any time to complete impregnation at all times. *Fabrea* has no ectoplasmic Golgi bodies and no osmiophilic contractile vacuoles (Ellis, 1937). *Lechriopyla* has many ectoplasmic Golgi bodies, but they never form an aggregation around the contractile vacuole (Lynch, 1930). *Epidinium, Eudiplodinium* (Fig. 54), and others show an accumulation of granules only during diastole (Krascheninnikow, 1929; MacLennan, 1933). This same type is found in *Ichthyophthirius* in the parasitic stages (Fig. 48), but neither free osmiophilic ectoplasmic granules nor accumulations around the contractile vacuoles during encystment (MacLennan, 1936) are present. *Metadinium* (Fig. 55) has a permanent granular nephridioplasm which waxes and wanes during the pulsatory cycle (MacLennan, 1933). *Paramecium caudatum* and *P. nephridiatum* have a permanent osmiophilic shell around the radiating canals, but not around the contractile vacuole itself (Nassonov, 1924; von Gelei, 1928). *Haptophrya possesses* a vacuolar apparatus which consists of a permanent, homogeneous osmiophilic tube (Bush, 1934). This nicely graded series shows that the impregnation of parts of the vacuolar apparatus is due to aggregations of osmiophilic granules around the fluid vacuoles and their membranes. The only cases in which the membranes themselves impregnate are those extremely specialized cases in which the osmiophilic material forms a permanent shell around the fluid vacuole. The highly complex osmiophilic apparatus in either *Paramecium* or *Haptophrya* is fundamentally no different from those vacuoles with a granular layer, and the homogeneous osmiophilic shells are merely the result of the aggregation and specialization of the ordinary undifferentiated ectoplasmic Golgi bodies. This view is similar to that of Nassonov (1924) and Hirschler (1927), except that these authors view the osmiophilic portion as the outer portion of the dictyosomes, and the fluid vacuole as the inner portion of the dictyosome structure. Both views are, of course, the same, if the osmiophobic portion of the dictyosome be accepted as a secretion droplet and not as an essential part of the dictyosome itself, as held by Gatenby and Subramaniam and Ganapati.

The demonstration that the osmiophilic granules are an addition to the simple contractile vacuoles which may exist independently of the granules, even in the same species, is a close parallel to the union of the progastriole and digestive granules. The only difference is that no gastriole yet discovered presents a permanent, highly developed mem-

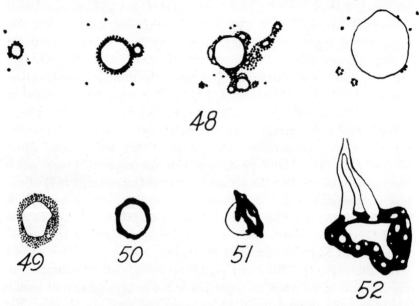

Figures 48-52. Excretory granules and contractile vacuoles. Figure 48, aggregation and disappearance of excretory granules during the pulsatory cycle, from *Ichthyophthirius multifiliis*, Champyosmic impregnation (after MacLennan, 1934) ; Figures 49-50, from *Polyplastron multivesiculatum;* Figure 49, cold impregnation; Figure 50, warm impregnation (after MacLennan, 1933) ; Figure 51, from *Dogielella sphaerii,* Champyosmic (after Nassonov, 1925) ; Figure 52, "nephridialplasm of *Campanella umbellaria,* Flemming-glychémalum (after Fauré-Fremièt, 1925).

brane or a permanent granular region, as in the contractile vacuoles of *Metadinium, Paramecium,* and *Haptophrya.*

The origin of a new vacuolar apparatus from the original structure of the parent has been described and compared with dictyokinesis in germ cells (Nassonov, 1924; von Gelei, 1928). The clearest case is that of *Haptophrya,* in which the vacuolar apparatus is a tube extending the full length of the ciliate. Studies of both live and fixed animals show that this tube is permanent and that the transverse fission of the cells divides the tube in two parts, each of which continues to function in the daughter cells (Bush, 1934). The only possible exception would be the stages in which the parasite is transferred from one host to another. The new vacuolar apparatus of *Paramecium* is said to arise by the multiplication of canals and by the division of the whole vacuolar apparatus just prior to fission (Nassonov, 1924). The formation of extra vacuoles has been noted many times in living ciliates, but there is no recorded observation of the actual division in a living *Paramecium,* and the interpretation of Nassanov's figures of fixed material is susceptible to the difficulties inherent in building any cycle from fixed material alone.

The majority of vacuolar systems, however, do not possess the thick, permanent wall similar to that in *Haptophrya,* but a temporary aqueous vacuole which certainly arises *de novo* (Taylor, 1923; Day, 1927; Mac-Lennan, 1933, 1936). These vacuoles and their membranes are not osmiophilic, the impregnation of the vacuolar system being due to the aggregation of the osmiophilic excretory granules. The fundamental question with respect to the origin of Golgi bodies is in these cases not the origin of the vacuoles, but the origin of the individual excretory granules. These granular aggregations in the Ophryoscolescidae have been observed in living specimens (Figs. 53-55) to be ectoplasmic Golgi bodies which migrate into the region of the vacuole during systole and the earliest stages of diastole (MacLennan, 1933). In specimens fixed during division of the ciliate, the newly arising vacuolar regions sometimes overlap the old ones and give the appearance of a division of the old one as described in *Didinium* (von Gelei, 1938), but a study of similar stages in living ciliates shows that they originate independently. These granules are continually migrating toward the vacuoles and dissolving there, and no granules migrate outward, so the question of origin is shifted to the granules at the time they are scattered in the ectoplasm. No cases of division were observed, either in fixed or in living material, at any place nor at any stage of the life cycle. This negative proof is not entirely satisfactory, since these granules are small (0.25—0.50 μ) and a very rapid division might escape notice. This problem does not occur in *Ichthyophthirius,* since all ectoplasmic granules (whether scattered or around the vacuoles) are absent in the encysted stage (MacLennan, 1936), so that in this ciliate they must originate *de novo* in the young parasites, whether or not they continue from them by division as the ciliate grows during the feeding stages. Persistence and genetic continuity is restricted to a few very highly specialized types of vacuoles, and in most cases there is no continuity either in whole or in part.

The osmiophilic reaction of the differentiated cytoplasm around the contractile vacuoles has given rise to statements that this is a lipoid structure (Nassonov, 1924; Volkonsky, 1933; Haye, 1930), and to the interpretation of the vacuolar system as due to the accumulation of lipoids at the vacuole-cytoplasm interface, which is similar to Parat's theory of the Golgi region around the vacuome. It is true that these structures are partially destroyed by lipoid solvents, but such evidence is not to be completely trusted and before acceptance must be corroborated by more specific methods. In *Ichthyophthirius* these granules are negative to Sudan III and Nile blue sulphate. In *Paramecium* (unpublished work) I have used Ciaccio's long unmasking process and find only a very faint reaction—only slightly more than in the hyaloplasm—and negative results with Nile blue sulphate. These experiments show that there is no concentration of lipoids, either in the excretory granules or in the cytoplasm around the vacuoles.

The variations in impregnation during the pulsatory cycle and during the life cycle suggest that the osmiophilic reaction is due to some reducing agent (not a lipoid) which is poured into the vacuole during diastole. The excretory theory of the contractile vacuole is indicated by the name nephridialplasm (Fauré-Fremiet, 1925). The similarity between the cytological changes during the pulsatory cycle and in the glandular epithelium or renal tubules has been used in support of this theory (Nassonov, 1924, 1925). Further support was given by the demonstration that the osmiophilic granules dissolve in the vacuolar fluid which is then discharged (MacLennan, 1933). In *Amoeba* the activity of the contractile vacuole is roughly proportional to the number of *beta* granules around it (Mast and Doyle, 1935b). It is certain in these cases that water plus some other material is being excreted.

The question, then, is not whether excretion in a broad sense takes place, since water plus dissolved materials is certainly being excreted, but what are the dissolved substances which are carried to the vacuole in granular form? Nitrogenous excretion has often been assumed, but Weatherby (1927, 1929) found that the vacuolar fluid extracted by an application of the microdissection technique showed upon analysis too

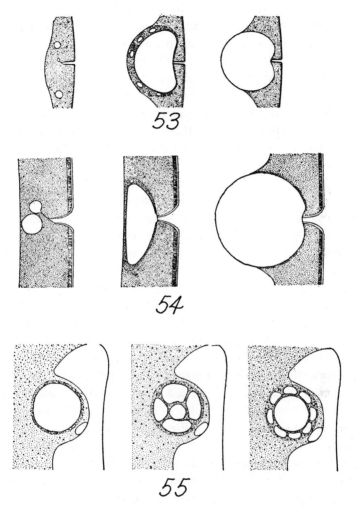

Figures 53-55. Excretory granules in living ciliates. Figure 53, from *Polyplastron multivesiculatum;* Figure 54, from *Eudiplodinium maggii;* Figure 55, from *Metadinium medium* (from MacLennan, 1933).

little nitrogenous materials—probably not more than one percent—to account for any significant part of the total excretion of *Paramecium*. The application of these results to all Protozoa is by no means certain, since I have shown previously (1933) that the pellicle of the Ophryoscolecidae is highly impermeable and the only pathway for the excretion of katabolic wastes is through the contractile vacuoles. Frisch (1938) has demonstrated a similar impermeability of the pellicle in *Paramecium* itself, and suggests that Weatherby's experiments be repeated, with the use of more delicate tests which have been devised recently.

The materials concentrated in the osmiophilic structures of the contractile vacuole are not salts, since no significant accumulation is demonstrated in the nephridialplasm by microincineration (MacLennan and Murer, 1934). These materials may possibly be incidental in some species, but *Amoeba* dies if this elimination is prevented (Mast and Doyle, 1935b), demonstrating that these wastes are toxic and that in this respect they have the known properties of the nitrogenous wastes of metabolism. Evidence that these materials are the result of metabolism is found in *Ichthyophthirius* (MacLennan, 1936). The vacuoles are osmiophilic only during the active feeding phase, when large amounts of food are being used and converted into storage bodies; but during encystment, when these activities cease, the vacuoles are not osmiophilic. The food of this protozoan consists entirely of epithelial cells, which are largely proteins, but not more than a third of the reserves of the ciliate are proteins, the rest being carbohydrate and fat. During the feeding stage a large amount of the ingested protein would be deaminized to form other reserves, with the result that much larger amounts of nitrogenous wastes would be formed during the feeding stage than in the encysted stage.

Frisch (1938) suggests that the contractile vacuoles also function in respiration. This function cannot be correlated with the variations in the nephridialplasm in various Protozoa and thus, if this is a function of the contractile vacuoles, it is probably independent of the granules which are being considered here.

LIPOID RESERVES

The lipoid materials considered in this section include all lipoids which are visible as granules and which are laid down during active feeding stages and are used during hunger or encystment. Zinger (1933) included all sudanophil particles as lipoid reserves, but digestive granules, mitochondria, as well as other bodies respond to Sudan III because of their lipoid content. Zinger pointed this out, for in his conclusion he states that the sudanophil bodies are more than reserve materials. However, until more is known of the functions of the intracellular lipoids, it is impossible to indicate accurately the boundary between reserve lipoids and those active directly in the metabolism of the cell.

Lipoid reserves have been found in a large number of Protozoa (for a detailed list see von Brand, 1935). Usually, if not always, these granules are in the endoplasm, either distributed at random, as in *Ichthyophthirius* (MacLennan, 1936), or concentrated at one end, as in *Anoplophrya* (Eksemplarskaja, 1931). Although these visible lipoid granules occur in many Protozoa, they are not universal. *Trypanosoma evansi* lacks all lipoid reserves, a fact which is correlated with a lack of lipase (Krijgsman, 1936). The Ophryoscolecidae and Cycloposthiidae, noted for their tremendous glycogen reserves, have no important lipoid reserves. *Mesnilella multispiculata* has no lipoid reserves, although five other species of the same genus have many fat globules (Cheissin, 1930).

The formation of droplets of neutral fat inside a granule of fatty acid has been demonstrated in *Opalina* (Kedrowsky, 1931) and *Ichthyophthirius* (MacLennan, 1936) by the Nile blue sulphate method (Fig. 41). Since, after staining with Nile blue sulphate, very small quantities of fatty acid dissolved in neutral fat result in an intense blue color rather than the pink which is characteristic of pure neutral fats, the pink stain observed in the cases above indicates that there are no free fatty acids in the neutral fat granules, as would be expected if the fats were synthesized on the surface of these granules. The fatty acids and glycerine dissolved in the endoplasm are first segregated into granules, and in these granules the neutral fat is synthesized. Then this fat is segregated into the visible droplets of pure neutral fat inside the active granules. These latter granules are typical Golgi bodies (MacLennan, 1936, 1940), as indicated by the name endoplasmic Golgi bodies. However, since these are functionally an intermediate stage in the development of the fat reserves, the descriptive term "intermediate lipoid body" is more appropriate in a functional classification.

In *A. proteus* the fat droplets grow in the cytoplasm without any intermediate granules being visible. These cytoplasmic fat droplets are not derived directly from fat in the food vacuoles, but the ingested fat is absorbed as free fatty acid and glycerine, synthesized into neutral fat in the cytoplasm and then stored as granules (Mast, 1938). Free fatty acids were demonstrated in the food vacuoles when fat was being digested, but none were demonstrated in the cytoplasmic fat droplets or on their surface when they were being formed. In the cases in which this process is not visible, either the synthesis is carried on elsewhere and the fat transported to the granules as such, or the fatty acids are never allowed to accumulate sufficiently to show under the microscope.

The visible lipoids, i.e., those which are found in definite globules and demonstrable by the ordinary fat-staining technique, include only a part of the total lipoids of the cell. "The pathologists have known for many years that the fats and fat-like substances of protoplasm are so bound or united to proteins as to be for the most part non-recognizable in the living or stained cell" (Heilbrunn, 1936). Besides the factors of food and the formation of fat from other substances such as carbohydrate or protein, changes from bound lipoids to free globules must be considered in any estimation of the reserve lipoids. Heilbrunn demonstrated an increase in lipoid globules in specimens of *A. proteus* kept in a dilute solution of ammonium salts. Three types of amoebae were found —those which show lipoids in culture, those in which lipoid globules appear after treatment with NH_4Cl, and those in which no free lipoid appears even after treatment. In similar experiments with *Arbacia* eggs, Heilbrunn showed that the total lipoids of the protoplasm remained constant; therefore the newly visible bodies are derived from bound lipoids, not from new fat formation. Since ammonium salts in the culture medium raise the pH of the immersed cells, the results were attributed to alkalinization of the protoplasm. The fact that CO_2 bubbled in the medium (which would tend to lower the protoplasmic pH), inhibits the formation of visible lipoids, confirms this hypothesis. Old cultures of *Paramecium* show larger amounts of fat than new cultures, although the paramecia divide and show no ill effects (Zinger, 1933); and since such cultures contain ammonia (Weatherby, 1927), the presence of abnormal amounts of visible fats may be due to the resulting alkalinization of the protoplasm.

Ultra-violet radiation causes a release of lipoid in *Amoeba* (Heilbrunn and Daugherty, 1938). The release of lipoids was greatly increased by a preliminary immersion in ammonium chloride solutions. "Further study is necessary in order to determine whether this fat release is due to a direct action of the radiation on the protein-lipoid binding or whether it may not be due indirectly to an alkalinization of the protoplasm" (Heilbrunn and Daugherty, 1938). In the same publication it is stated that any stimulus in which localized increases in temperature occur is efficient in the release of fat. This is also shown by the experiments of Sassuchin (1924), who compared the protoplasm of *Opalina* kept at room temperature with the protoplasm of those kept at 35-38° C. In the first group he found elongate mitochondria in the endoplasm (Kedrowsky's endosomes), but in the heated group only fat spherules and protein spherules, and these results were interpreted as due to the separation of mitochondria into their two components. These latter experiments should be repeated in individuals with little or no fat, and in species as to which there is more agreement on the identification of mitochondria.

In *Paramecium* (Zweibaum, 1921) and *Stentor* (Zhinkin, 1930) fat is stored under conditions of low oxygen tension and lost when the oxygen tension is restored. The rate of loss in this case is dependent upon the temperature.

Pathological conditions are often marked by fatty degeneration in the Protozoa. Degenerating coccidial oocysts show an increase in fat globules (Thélohan, 1894), and in *Aulacantha* fatty vesicles are formed and the nucleus is finally replaced by fatty bodies (Borgert, 1909). Individuals of *Actinophrys* which show depression by a lowered division rate and otherwise, have an abnormal number of lipoid bodies, and in extreme cases show typical fatty degeneration. In the macronucleus of *Paramecium* parasitism by bacteria also results in tremendous quantities of visible lipoids in the cytoplasm and also of crystals (Fiveiskaja, 1929).

CARBOHYDRATE RESERVES

Granules containing carbohydrates are found in most Protozoa, although in a few species this reserve is in a diffuse form which is precipitated as granules or irregular masses by fixation. The lack of any carbohydrate reserve at all has been proved in only a few species, such as *Trypanosoma evansi*.

The carbohydrate reserves in *Paramecium* (Rammelmeyer, 1925) and in the cysts of *Bursaria* (Poljansky, 1934) are probably dissolved in the protoplasm, since they are visible in fixed specimens only as cloudy masses, not in regular granules. Homogeneous vacuoles, granules, or platelets visible in the living normal Protozoa are very common. They are well known in *Iodamoeba* and other intestinal amoebae. Large numbers of these granules are found in the flagellates from termites and wood-eating roaches (Cutler, 1921; Kirby, 1932; Cleveland, 1934; Yamasaki, 1937a). The carbohydrate granules of *Stentor* tend to be localized in a peripheral sheath of the endoplasm (Zhinkin, 1930) and just beneath the pellicle. In *Arcella* these granules are embedded in the chromidial net. In *Ichthyophthirius* these smaller granules are always associated with mitochondria (MacLennan, 1936). Glycogen granules are often associated with the parabasal bodies in flagellates (Duboscq and Grassé, 1933).

Carbohydrate granules with definite internal structure are by no means uncommon. The granules of Sporozoa (Fig. 60) have a cross or star-shaped center (Joyet-Lavergne, 1926a; Daniels, 1938), the general appearance of which and ability to accumulate iodine suggest vacuoles. Identification of a lipoid center (Erdmann, 1917) is based on insufficient evidence and, in view of the later work quoted above, seems unlikely. Vacuolated bodies are also found in *Balantidium* (Fig. 57) with the added feature of crystals floating in some of the vacuoles (Jirovec, 1926). Two types of granules are found in *Difflugia,* small homogeneous spherules and larger elliptical bodies with a center granule which stains a pale blue after hematoxylin and a rim which is rose-colored after Best's stain (Rumjantzew, 1922).

In the Ophryoscolecidae, the granules possess a spherical center (Fig. 56) denser than the rest of the granule (MacLennan, 1934). The most spectacular of the carbohydrate reserves are the skeletal plates of the Cycloposthiidae, Ophryoscolecidae, and related families. The plates themselves are probably supporting structures, but in their meshes are platelets of the same type as the scattered cytoplasmic granules. The platelets in the Cycloposthiidae (Fig. 61) are roughly spool-shaped with slender strands connecting the flanges of adjacent granules (Strelkow, 1931), but in the Ophryoscolecidae (Fig. 62) the polygonal plates are unconnected (MacLennan, 1934).

The formation of paraglycogen bodies has been followed in only a few cases. The bodies of *Pelomyxa* behave like permanent bodies with a protein stroma and with the paraglycogen being built up or released as the case demands (Leiner, 1924). The paraglycogen granules of *Polyplastron* are likewise independent of other formed components and are apparently self-perpetuating (Fig. 56), since they show regular division (MacLennan, 1934). The dense centers may be naked or, more often, surrounded by an envelope of varying thickness. In the largest of these compound granules, the centers are dumb-bell-shaped or double, and in the latter the envelope also is constricted. These stages probably represent growth or utilization stages and division stages, although this was not confirmed by following a single granule in live ciliates. The complex granules of *Amoeba hydroxena* (Fig. 58, 59), in which a varying number of glycogen granules are imbedded in glycoproteid (Wermel, 1925), suggests a conversion of glycogen into glycoprotein for storage and the reversal of this process in utilization.

Some paraglycogen granules are formed in association with mitochondria instead of being independent bodies. The paraglycogen in *Ichthyophthirius* first appears as a minute vacuole (Fig. 17) in the center of a sphere of mitochondrial material (MacLennan, 1936). As the granules grow, this mitochondrial shell breaks into short rods fused to the surface of the paraglycogen. The mitochondria disappear after the granule has attained full size. Joyet-Lavergne (1926b) also noticed a morphological relationship between mitochondria and paraglycogen of gregarines, but says "il y a là un simple rapport de contact et nous n'avons aucune raison de supposer une intervention dans da génèse du paraglycogène." However, in the case of a granule in the center of an unbroken sphere, as in *Ichthyophthirius,* it is difficult to list the relationship as merely an incidental contact.

Duboscq and Grassé (1933) show that the glycogen granules of *Cryptobia helicis* are not found scattered in the cytoplasm, but are formed in close contact with or in the strands of the parabasal bodies (Fig. 65). The glycogen is not laid down in the summer, but only in the winter, a fact which, they point out, would explain the negative results of other authors. This formation of glycogen by the parabasal body parallels the secretion of protein granules by the macronucleus of ciliates—a part of the segregation function which, in most species, is performed by isolated

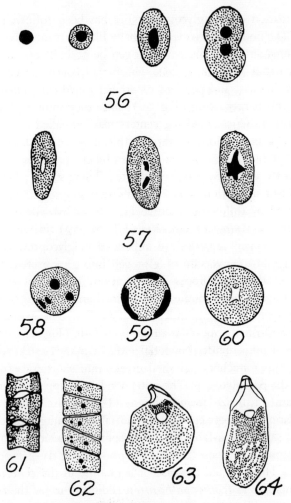

Figures 56-64. Carbohydrate reserves. Figure 56, stages in the paraglycogen granules of *Polyplastron multivesiculatum*, Champy-osmic impregnation followed by Sudan III in hot paraffin (after MacLennan, 1934) ; Figure 57, growth of crystals in paraglycogen granules of *Balantidium elongatum*, Zenkers-dahlia (after Jirovec, 1926) ; Figures 58-59, glycogen droplets in a glycoproteid granule, from *Amoeba hydroxena*, Carnoy-Best (after Wermel, 1925) ; Figure 60, vacuolated paraglycogen body from Sporozoa (after Joyet-Lavergne, 1926) ; Figure 61, skeletal platelets of *Cycloposthium edentatum*, Lugol (after Strelkow, 1929) ; Figure 62, skeletal platelets of *Polyplastron multivesiculatum*, Champy-osmic-Sudan III in hot paraffin (after MacLennan, 1934) ; Figures 63-64, glycogen reserves in *Trichonympha agilis*, Best's stain, 63 normal, 64 showing loss of glycogen just before death under conditions of lowered temperature and raised oxygen pressure (after Yamasaki, 1937).

cytoplasmic granules, is performed in the one case by a nuclear structure and in the other case by a neuromotor structure.

The differentiation between the various carbohydrates found in the Protozoa is based on their staining reactions and solubility, since the exact nature of the sugars involved in the formation of protozoan poly-saccharides is unknown. Zhinkin (1930) and von Brand (1935) pointed out that this is unsatisfactory and contend that no separation should be made from glycogen until this is known. However, the differences are so pronounced that it is convenient to retain the name paraglycogen.

Soluble glycogen as found in vertebrate liver cells is relatively rare. The diffuse materials found in *Paramecium* and *Bursaria* are probably of this type. The commonest carbohydrate is paraglycogen, distinguished by Bütschli (1885) from glycogen on the basis of its relative insolubility in water as compared with true glycogen. It is digested by ptyalin and diastase and the sugar produced reduces Fehling's solution. It stains a light brown in iodine and brown or brown purple in iodine-sulphuric acid or chlor-zinc-iodide. Probably all of the granular reserves of carbo-hydrate in Protozoa are paraglycogen or some similar relatively insoluble compound. The reserve granules of the flagellates of termites have been identified as glycogen (Yamasaki, 1937a; Kirby, 1932); but in the re-lated flagellates of the wood roach, since the Protozoa contain no enzyme capable of breaking down glycogen, it has been suggested that the gran-ules which stain with iodine consist of some other product which results from the breakdown of cellulose (Cleveland, 1934). The material in the platelets of the Ophryoscolecidae has been named *ophryoscolecin* on the ground that it is unique in this family and is more like cellulose than paraglycogen (Dogiel and Fedorowa, 1925). It was later identified as a hemicellulose (Strelkow, 1929). This interpretation is based on slight variations in solubility and color reactions, but other authors, using some of the same methods and some different methods, were not able to find any difference between the reactions of paraglycogen and the platelets (Schulze, 1922, 1924, 1927; Weineck, 1931, 1934; MacLennan, 1934). However, such arguments cannot be settled, as von Brand suggests, until the exact structure of these polysaccharides is known, and the term para-glycogen in this discussion is used in a rather general sense for carbo-hydrates more insoluble in water than glycogen and differing in color reactions from starch and cellulose.

The presence of more than one type of material in the same granule has been demonstrated in several cases, in spite of the relative crudity of the cytochemical methods for the demonstration of carbohydrates. The oval carbohydrate bodies of *Difflugia* are not completely dissolved in ptyalin, and their staining reactions suggest the presence of a glyco-proteid (Rumjantzew, 1922). Two types of carbohydrate reserves have been reported from Sporozoa by Dobell (1925). Chakravarty (1936)

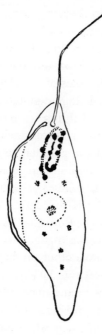

Figure 65. The association between glycogen and the parabasal body in *Cryptobia helicis,* winter forms stained with iodine. (After Duboscq and Grassé, 1933.)

also differentiated two sets of granules by differences in the speed of destaining after treatment in iodine. These authors refer to one set as glycogen, the other as paraglycogen. Two types of carbohydrate have been found in *Actinosphaerium* (Rumjantzew and Wermel, 1925), and were identified as glycogen and a glycoprotein on the basis of their re-action to Best's and Fischer's stains. *Pelomyxa* loses its paraglycogen dur-ing prolonged starvation, but since the remnants of these granules may be stained with haematoxylin (Leiner, 1924), it is probable that the carbohydrate is here associated with a protein. The carbohydrate bodies of *Amoeba hydroxena* contain two different materials (Figs. 58, 59) and on the basis of Best's and Fischer's methods have been interpreted as granules of glycogen embedded in glycoprotein.

The decrease in glycogen or paraglycogen during hunger or encystment, and its storage during the feeding stages, has been noted so often that detailed descriptions of observations under controlled conditions are rare. The glycogen in *Stentor* is deposited during low temperatures and utilized at higher temperatures, and this process is accelerated by starvation (Zhinkin, 1930). Fat, rather than glycogen, is deposited, if the oxygen tension is lowered. Under such conditions some of the carbohydrate is probably converted into fat. In *Trichonympha* also, available food, oxygen tension, and temperature affects the amount of glycogen present (Yamasaki, 1937b). The cytoplasm of this species is divided into two parts by a fibrillar basket, which suspends the nucleus from the anterior cone of the body and separates this portion from the rounded posterior part in which the food vacuoles are formed. Both regions normally contain glycogen (Fig. 63), but during starvation the glycogen in the posterior part disappears first, the glycogen in the anterior part then diminishes and disappears, and the death of the organism follows shortly. Similar results are observed at high temperatures, or with oxygenation at room temperature. However, when the termites are oxygenated at low temperatures (Fig. 64), the glycogen in the posterior portion often shows little change, but the portion anterior to the nucleus disappears rapidly. As soon as the glycogen in the corbula disappears the protozoan dies, even though glycogen remains in the body region. Yamasaki states that the posterior region is simply one of synthesis and storage, while the anterior region is the region of consumption. He concludes that defaunation by oxygen is due not only to toxicity but also to a depletion of the glycogen available for the nucleus and motor organelles.

Trypanosoma evansi possesses no glycogen and, since it possesses no amylase, is not able to synthesize it (Krijgsman, 1936). Other trypanosomes do deposit glycogen, but at best it forms an insignificant reserve, since trypanosomes may use three times their body weight in sugar in twenty-four hours (von Brand, 1938). In this case the glycogen reserves of the trypanosomes are the liver glycogen of the host.

PROTEIN RESERVES

This term is one of convenience and, as in the case of the term lipoid, cannot be taken in a strict sense, but is used here to include, besides true proteins, bodies which contain lipoids or carbohydrates, as well as proteins, amino acids, nucleic acid, and so forth. Since most of the fixing agents precipitate at least the protein portions of such granules, many have been described, although relatively few have been identified by acceptable microchemical methods. For this reason they have been described under a variety of names, many of which mention incidental staining properties. Some of the names which are most securely embedded in the literature are chromidia, volutin, metachromatic granules, basophilic granules, chromatoidal bodies, and albuminoid reserves. The confusion in these terms is best illustrated by chromidia. This was originally used to designate chromatin bodies which are extruded into the cytoplasm from the nucleus (Hertwig, 1902) and which have the ability to reaggregate to form new nuclei. Although this interpretation has been disproved, the name may be retained to designate these granules (Meyers, 1935). In other cases it is used to designate nonchromatin material which is supposed to be extruded from the nucleus (Daniels, 1938). Other authors use it even more loosely to designate basophilic and metachromatic cytoplasmic bodies which are secretory in nature (Campbell, 1926). The elimination of the original meaning was due to the improvement of both cytoplasmic and nuclear methods, accompanied by detailed studies of life cycles. The last stronghold of this theory—the Foraminifera—was eliminated by the tracing of the nuclear history in live *Patellina* throughout the vegetative and sexual stages, with a complete demonstration of the cycle with moving pictures (Meyers, 1935). The exclusion of chromidia in the original sense, with respect to the cells of the Metazoa, has already been accepted (Wilson, 1928).

Many chromidia are actually mitochondria (Fauré-Fremiet, 1910) which contain a high percentage of protein and are therefore resistant to routine fixatives. This probably led to one revival of the chromidial theory, according to which all cytoplasmic structures are formed from mitochondria, which in turn originate from the nucleus as chromidia.

Alexeieff in a series of works on the Flagellata strives to prove that all cell structures are formed at the expense of mitochondria. The latter, according to Alexeieff, in their turn are not autonomic, as the majority of investigators suppose, but originate from the nucleus as chromidia. . . . In cases where the autonomy of the mitochondria and of the blepharoplast is indisputable, this author always attributes to them a nuclear origin though phylogenetic.

After this summary of Alexeieff's theory, Milovidov (1932) rejects it. Certain cytoplasmic granules of *Uroleptus* are derived from the nuclei during reorganization (Calkins, 1930), and these granules were described as mitochondria, but since they do not stain with janus green they do not seem to be typical mitochondria.

The term chromidia, as now accepted, includes cytoplasmic granules supposed to be derived from the nucleus (but not necessarily chromatin), particularly in the Sporozoa. It also includes granules in the rhizopods, at one time supposed to be examples of the chromidial theory, but now retained without any such implication.

The chromidial net, characteristic of many of the rhizopods with shells, is a definite morphological entity which may be recognized independently of particular staining methods. The net itself is negative to Feulgen's stain in *Arcella* and *Chlamydophrys,* either with or without hydrolysis, and is digested more rapidly than the nucleus by pepsin or trypsin (Reichenow, 1928). Since the net in *Difflugia* gives a positive reaction with Ciaccio's lipoid method (Rumjantzew, 1922), it probably has a lipoid component in addition to the protein component in this species. Although basophilic, it is not directly related to the nuclear material. On the other hand, the net of *Patellina* is positive to Feulgen's method, but complete studies show that it is independent of the nuclei (Meyers, 1935). In both cases the net is a specialized mass of reserve protein, and within it may be found two other types of reserve, volutin and glycogen granules. This is not true in all species, since no glycogen is found in the net of *Difflugia* (Rumjantzew, 1922). The chromidia of gregarines are similar in ordinary staining reactions to the karyosome and to the protein reserves (Daniels, 1938).

The chromidia of several Sporozoa (Joyet-Lavergne, 1926a) are positive to Millon's reagent and are therefore certainly protein and they appear to be associated with mitochondria. In gregarines from mealworms, on the other hand, these granules are negative to both Millon's reagent and Feulgen's reagent, and show no morphological relationship with mitochondria (Daniels, 1938). Daniels found chromidia and volutin similar in shape, distribution, and so forth, but found fewer black granules in haematoxylin preparations than blue granules after the methylene blue method. She concludes that they are separate types of granules, although there is a close relationship. Joyet-Lavergne was not able to decide whether chromidia and volutin are really separate.

Daniels observed buds on the surface of the karyosome, then bodies in the nuclear sap, and finally in the cytoplasm near the nucleus, but she found no direct evidence as to how they penetrate the nuclear membrane. On the basis of these suggestive observations, she concludes that these bodies are derived from the karyosome. With regard to the validity of this conclusion, the comment of Wilson (1928, p. 96) with regard to a similar case in oogenesis is highly pertinent: "To the writer none of these cases yet seems to be satisfactorily demonstrated, and the question is a most difficult one to be settled by studies on fixed material alone." Joyet-Lavergne (1926a) calls these protein granules albuminoid reserves, a name far more appropriate than chromidia, which at least implies a nuclear origin.

Volutin granules are basophilic granules which are also metachromatic. Because of their pronounced basophilia, volutin granules have often been linked with chromatin. However, they are negative to Feulgen's stain after hydrolysis, but give a positive reaction when the preliminary hydrolysis is omitted (Reichenow, 1928), a characteristic of free nucleic acid. The full Feulgen reaction apparently dissolves this type of volutin granule, so that in *Arcella* there results a diffuse Feulgen reaction in the chromidial net. This is a possible explanation of the positive Feulgen test by the chromidial net of *Patellina*. The volutin granules of *Trypanosoma melophagium* contain no nucleic acid (van Thiel, 1925), while those of *T. equinum* do (Reichenow, 1928). The volutin bodies of *T. evansi* were not tested in this respect (Krijgsman, 1936), although they are listed as containing nucleic acid. Since reserve bodies are not the same in all species of the genus (some trypanosomes are able to store glycogen while others are not, according to von Brand, 1938), both analyses of the basophilic granules may be correct.

Volutin granules increase and nuclear granules decrease in trypanosomes which have been treated with atoxyl (Swellengrebel, 1908). This fact in conjunction with the staining reactions of volutin, were interpreted as indicating a direct nuclear origin—in other words, a type of chromidia. In these experiments the results are probably a degeneration phenomenon, since in *Pelomyxa* the expulsion of chromatin into the cytoplasm is found just prior to death (Schirch, 1914). Hindle (1910) thought of this as a degeneration phenomenon in *T. gambiense*. In various phytoflagellates division stops when volutin is lost, and the volutin was interpreted as a nuclear reserve (Reichenow, 1928), although no direct connection between the two was demonstrated. A dehydrase has been demonstrated in *Trypanosoma* by the leucomethylene blue method and localized in the volutin granules (Krijgsman, 1936). Krijgsman, however, holds to Reichenow's views of volutin as a nuclear reserve.

Protein bodies in *Oxymonas dimorpha,* which are negative to Feulgen's stain and stain with either basic or acid dyes (i.e., not metachromatic), have been called volutin granules (Connell, 1930), although they are not volutin in the sense used by Reichenow. However, in *Oxymonas,* as in the phytoflagellates of Reichenow's experiments, division ceases when these granules are exhausted. Since the division stages of *Oxymonas* are also the flagellated stage, the protein granules could be explained as reserve bodies for the expenditure of energy by these organelles. Neither explanation has adequate proof, since each merely correlates obvious phenomena.

Volutin is thus a term which has no standard usage, but wherever microchemical tests have been made volutin has been found to contain proteins, nucleic acid, or other similar materials. Since the available evidence shows that it behaves as a reserve material, it seems to me to be convenient to include it as one of the various types of protein reserves and to eliminate the terms volutin and metachromatin, neither of which seems to have been used consistently by protozoölogists.

The macronuclei of many ciliates contain one or more large, intensely basophilic bodies lodged in vacuoles among the closely packed granules of chromatin (Chakravarty, 1936; MacLennan, 1936). Since their number and size vary, it has been suggested that these are reserve materials (Kazancev, 1928). In *Ichthyophthirius* these granules have been traced in living ciliates from the macronucleus through temporary breaks in the macronuclear membrane into the cytoplasm (Fig. 66) where they are stored until resorption and utilization occurs in the encysted stages (MacLennan, 1936). These bodies take both acid and basic dyes even more strongly than chromatin and, unlike the chromatin, are negative to Feulgen's reaction and Macallum's tests for iron. These granules first appear as minute bodies at the lower limits of visibility, embedded in the chromatin net during the feeding stages of the ciliate but not during encystment. The macronucleus is positive to Feulgen's reagent without preliminary hydrolysis during the formation of these granules, but at no other time. It seems probable that food materials are built up in the macronucleus into chromatin, which is then split into a group containing iron and nucleic acid and another protein group which lacks these substances. The first group is used to rebuild more chromatin and the latter group is segregated into the granules which are ejected into the

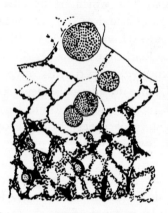

Figure 66. The formation and release of protein granules from the macronucleus of *Ichthyophthirius multifiliis,* Feulgen-light green. (After Mac-Lennan. 1936.)

cytoplasm. Since granules of this type are found in both the macronucleus and the cytoplasm of so many ciliates, this is probably quite a general phenomenon. The Protociliata lack macronuclei, but perform this same function by the segregation apparatus.

The balls of chromatin and other macronuclear fragments which are extruded during the various types of macronuclear reorganization are only incidentally reserve material, if at all, and will be considered in detail in the chapters on nuclear phenomena.

The crystals which are common in various Protozoa are often considered to be excretory products, and in some cases have been identified as uric acid (for a discussion of this work, see Reichenow, 1929). Recent work (Mast and Doyle, 1935b) shows, however, that some of the crystals must be regarded as reserve material. *A. proteus* contains two types of crystals, a bipyramidal type and a plate-like type, which are suspended in vacuoles containing an alkaline fluid. A careful study of spectroscopic analysis, solubility, and form shows that the bipyramidal type probably consists of a magnesium salt of a substituted glycine. The plate-like crystals are insoluble in a saturated solution of leucine, and in structure resemble leucine crystals. If the crystals are removed by centrifuging and the *Amoeba* is then put in a solution which contains amino acids and egg albumin, the platelets are formed in the vacuoles which contain leucine, while the bipyramidal crystals are formed in all solutions. "Crystals are normally formed from amino acids derived from food during digestion" (Mast and Doyle, 1935b). These crystals decrease in number just before the refractive bodies increase in number, indicating that the crystals are an intermediate stage in the transfer of food from the food vacuoles to the lipoprotein refractive bodies.

The chromatoidal bodies of various parasitic amoebae, are intensely basophilic structures, the fixing and staining reactions of which suggest a protein composition. They possess neither chromatin nor free nucleic acid (Reichenow, 1928), so are not volutin; but they are similar in their reactions to the protein bodies of ciliates. Since they disappear during encystment they are reserve bodies.

The reserve proteins are often found in combination with other materials. In *Amoeba,* lipoids and proteins are bound together in the refractive bodies (Mast and Doyle, 1935a). In *Actinosphaerium,* granules of glycoproteid are present (Rumjantzew and Wermel, 1925). Similar inclusions are found in *Ophryoglena* (Zinger, 1928). In the Foettingeriidae the protein reserves have the characteristics of the vitellin of the hen's egg and in *Polyspira* there is a single central mass consisting of protein associated with a carotenoid (Chatton, Parat, and Lwoff, 1927). The protein portion alone is used, the carotenoid remaining in the old cyst, and finally disappearing during encystment.

The protein bodies which are found throughout the Protozoa vary greatly in their specific structure and composition. This variation, with the resulting variation in staining reactions, has resulted in a complicated nomenclature with the usage of terms proposed by each author. The term chromidia is so definitely bound up with disproved theories that it should be dropped. Volutin should either be dropped or definitely restricted to metachromatic granules which respond to Feulgen's stain when used without hydrolysis. Whenever the function is that of a reserve, as in the majority of known cases, I believe these granules should be called simply protein reserve bodies, a usage found convenient by Joyet-Lavergne (1926a). At the same time, it should be recognized that a reserve function is the one most easily identified by morphological methods, and that other functions must be investigated. The presence of dehydrogenase (Krijgsman, 1936) in protein granules of *Trypanosoma evansi* is one definite lead.

The protein reserve bodies are as catholic in origin as in structure and may result from the activities of the segregation granules, macronucleus, mitochondria, or food vacuoles, or may be independent of other formed bodies. The crudity of our knowledge of cytoplasmic granules is illustrated by the fact that no suggestion of the significance of these differences can be made.

EXTERNAL SECRETION

This important cytological subject has been greatly neglected in the Protozoa and is generally ignored in a discussion of the protozoan cytoplasm, except as the vacuolar apparatus is considered to be a secretory organelle. It is an important subject in itself and is most nearly comparable to secretion studied in the Metazoa.

The attaching organs, or at least the cementing portion, of sessile Protozoa are secreted structures. Just preceding the formation of the peduncle of *Campanella,* granules are found in the basal region (Fauré-Fremiet, 1905). *Tintinnopsis nucula* is cemented to the lorica by a mucus secretion which is derived from basophilic granules in the stalk (Campbell, 1926).

The lorica of *Favella* is likewise derived from cytoplasmic granules (Campbell, 1927). Granules which are to form the new lorica accumulate near the mouth, in dividing animals. After division these are forced out through the cytostome, expand, fuse, and harden. At the same time fecal pellets are molded into this secreted material and the whole lorica is shaped by the activities of the motor organelles. There is in this form a local zone of secretion, as in gland cells, not a general secretion over the whole surface.

The shell of *Euglypha* is formed from separate shell plates, which are secreted within the cytoplasm (Hall and Loefer, 1930). They appear first as small refractive spheres in vacuoles, then enlarge and elongate to become typical shell plates. (Fig. 67). The finished plates lie free in the cytoplasm. It was demonstrated that the reserve plates have no connection with either the mitochondria or the neutral red globules.

The cyst of *Ichthyophthirius* is secreted in two parts: first a homogeneous clear membrane is formed (Fig. 68) and then individual fibrils are extruded, apparently between the bases of the cilia (MacLennan, 1937). These sticky fibrils are stroked into rope-like fibers, which adhere to the under side of the outer membrane (Fig. 69) by the activities of the cilia. Although seven types of granules were demon-

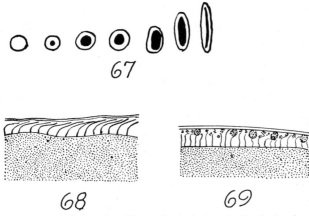

Figures 67-69. External secretion. Figure 67, inclusions in *Euglypha alveolata*, probably representing the formation of reserve shell plates (after Hall and Loefer, 1930); Figures 68-69, secretion of the cyst wall in *Ichthyophthirius multifiliis;* Figure 68, section of early stage showing only the homogeneous layer; Figure 69, section of later stage showing the addition of the fibrillar layer (from MacLennan, 1936).

strated, no granules could be associated with the secretion of this cyst. In *Nyctotherus,* variations in the thickness of the secreted cyst may be correlated with the distribution of ectoplasmic structures (Rosenberg, 1937), but no granules responsible for the secretion were noted in this form either. The lorica of *Folliculina ampulla* is likewise secreted in the form of a clear fluid, which hardens to form a membrane just beyond the tips of the somatic cilia (Fauré-Fremiet, 1932).

The secretion of vacuoles of oxygen in *Arcella* is not associated with granules, but with nongranular regions of hyaloplasm, and is probably a result of oxidative and reductive processes of the cell (Bles, 1929).

The most striking aspect of these examples of protozoan secretion is that none of them has been traced to Golgi bodies, mitochondria, or segregation bodies—i.e., they do not react with osmic acid, Janus green B, nor neutral red. It is evident that not all important segregation nor synthesis is revealed by these stains.

The mucus granules of *Euglena* (Dangéard, 1928) and the pellicular secretions of *Vorticella* (Finley, 1934) stain with neutral red and thus might be classed the segregation granules of *Opalina* and others, the only difference being that in the former is segregated mucus, which is not used within the cell but is extruded in the normal functioning of the protozoön, while in the latter are segregated proteins, which are normally used within the cell. It is interesting, however, that Kedrowsky found that when the segregation granules were filled with foreign materials, such as the organic silver compounds, the granules are extruded. The expulsion of droplets containing neutral red may be induced in *Paramecium* (Frisch, 1938) and other ciliates. These examples indicate that the formation of the segregation granules and the secretion granules is comparable, the only difference being that in one the material is used internally and in the other externally.

THE GRANULAR COMPLEX

The detailed consideration of each of the types of cytoplasmic granules has resulted in the conclusion that there are no universal cytoplasmic components and that each of the terms mitochondria, Golgi bodies, neutral red granules, and so forth has been applied to a heterogeneous assortment of granules of widely different functions. This conclusion, derived from a consideration of the types of granules separately, becomes inescapable if we consider the whole granular complex. The problem is on the surface one of classification, but fundamentally it is one of function—what functions are performed by cytoplasmic granules, and is the same function always performed by the same type of granule in different Protozoa? Since these granules are not independent units but are part of a granular complex which in turn is a part of the whole cell, this whole complex must be considered in seeking an answer to these problems of function. The investigations which seem to be suitable for this comparison are those of Mast and Doyle (1935a, 1935b), Holter and Kopac (1937), Holter and Doyle (1938), all on *Amoeba proteus;* Hopkins (1938a, 1938b) on *Flabellula mira;* Kedrowsky (1931-33) on *Opalina ranarum;* MacLennan (1936, 1937) on *Ichthyophthirius multifiliis;* and Joyet-Lavergne (1926 on) on several Sporozoa, supplemented by the observations and experiments of Daniels (1938) on similar species. Since a summary of the individual granules has been given in the previous sections, only a general account of each granule will be given in this comparison.

The number of types of granules ranges from two in the marine amoeba *Flabellula* to at least six in *Ichthyophthirius* and some of the Sporozoa. In *Flabellula* there are only digestive granules and small granules of unknown composition and function. This small number contrasts sharply with *A. proteus,* which has four types of granules of cytoplasmic origin: refractive bodies (dictyosomes), alpha granules unknown in composition and function, mitochondria (beta granules), and neutral fat granules. In addition to these granules there are two types of crystals, blebs on these crystals, and vacuole refractive bodies, all of which arise in connection with the food vacuole. The two ciliates *Opalina* and *Ichthyophthirius* also show marked differences in number of granules—the former with only four types and the latter with seven. *Opalina* has segregation bodies, endosomes (mitochondria?), intermediate lipoid bodies (endoplasmic Golgi), and neutral fat, while *Ichthyophthirius* has intermediate lipoid bodies (endoplasmic Golgi), neutral fat, excretory granules (ectoplasmic Golgi), mitochondria, paraglycogen, and protein bodies. Gregarines and Coccidia have at least one type of Golgi body, neutral fat, one or two types of mitochondria, paraglycogen, one or two types of protein reserves, and neutral red bodies, a total of six to eight types of granules, allowing for differences in the accounts of Daniels and Joyet-Lavergne. Not all the Sporozoa present such a complicated picture, since there are probably not more than three types of granules in *Plasmodium:* mitochondria, segregation granules, and pigment. These marked differences, which appear even with a crude comparison based only on number of types of granules, show clearly that at best only very few granules could be universal. Furthermore, the number of granules varies independently of the relationships of the Protozoa involved, since both large and small numbers of granules are found in Protozoa of the same class.

The immediate facts which stand out with respect to the staining reactions of these five species of Protozoa are that in each species are granules or vacuoles which are stained specifically by Janus green B and other mitochondrial stains, in each are granules which segregate neutral red, and in each, with the exception of *Flabellula,* are granules which may be specifically impregnated by the Golgi methods. If we do not press the comparison any further, we can say that chondriome and vacuome are universal cell constituents but, even with the very crude definition based on impregnation alone, Golgi bodies are lacking in one of the five Protozoa discussed. However, this apparent uniformity is reduced if these granules are compared with respect to other characteristics than the so-called specific staining methods, which are in reality quite crude in spite of brilliant contrasts. With respect to mitochondria, the bodies range from one extreme of temporary induced granules in *Flabellula* to two separate types in some of the Sporozoa. The neutral red bodies are even less comparable: in *Ichthyophthirius* they are lipoid-containing bodies, which are found only in association with gastrioles. In *Opalina* they range from watery vacuoles to dense bodies of proteins and are obviously not connected with any gastrioles. In *Flabellula* they are vacuoles in which is also dissolved the material stainable with Janus green. In *Amoeba* there are two types of neutral red bodies which are of cytoplasmic origin, and both contain large amounts of lipoids and at some stages proteins as well. One of these types, the refractive bodies, are apparently built up from material derived from the blebs and the crystals. In a structural sense also the refractive bodies are unique and are much more complex than any of the other neutral red bodies. In the Sporozoa, lipoid dictyosomes are weakly stainable with neutral red, but in addition there are non-lipoid bodies which stain much more specifically with neutral red. With respect to Golgi bodies, even if we ignore *Flabellula,* harmony is not achieved. In *Amoeba* the two types of neutral red bodies, as well as the mitochondria, respond to impregnation and bleach with difficulty. A comparison between these lipoid bodies and the endoplasmic Golgi bodies (intermediate lipoid bodies) of *Ichthyophthirius, Opalina,* and the dictyosomes of Sporozoa seems quite logical, until we consider that in these latter three species, these fatty acid bodies are an intermediate step in the formation of neutral fat, while the refractive bodies of *Amoeba* are finished bodies, the granules of neutral fat being morphologically independent of them. None of these lipoid bodies is comparable with the non-lipoidal ectoplasmic Golgi bodies (excretory Golgi) around the contractile vacuoles of *Ichthyophthirius*.

These comparisons demonstrate that with respect to actual composition not even one single type of granule is found throughout these five species of Protozoa, and the apparent universal presence of certain types of granules is due to the lack of specificity in Janus green, neutral red, osmic acid, and other stains. This is demonstrated in spite of the fact that the known composition of these bodies can be stated only in qualitative terms which are actually very broad—i.e., lipoid, non-lipoid, protein, and so forth.

A comparison on the basis of composition alone is open to criticism if it is not checked from other angles. The segregation apparatus of *Opalina* may be aqueous vacuoles, dense protein bodies, or it may contain bile pigments, depending on the medium and the temperature. Using composition alone as a criterion (or the neutral red reaction, for that matter), these granules would be separately classified, actually Kedrowsky showed they are the same granules with the same function— segregation and synthesis. In this case the stain reactions and the composition are incidental, and they are important in the classification of the segregation apparatus only if they can be used to reveal the function. Can groups of granules be demonstrated if the granules are compared, not on the basis of structure nor of composition, but on function? If so, are any of these groups represented in all of these five species which we are considering? This comparison cannot be in any way as complete as the comparison based on staining reactions and composition, since for the most part this knowledge is restricted to those functions which have a definite morphological expression—digestion, storage, external secretion, and so forth. The apparent emphasis on these functions should not be considered as an implication that these are the only functions in which the cytoplasmic granules may play a rôle, but as an inverse expression of the difficulties of localizing functions which do not produce visible structures.

The segregation and storage of protein reserves is obvious morphologically and the materials which are stored can be identified by adequate cytochemical methods. All the five Protozoa, again with the exception of *Flabellula*, have visible stores of proteins or derivative substances.

A. proteus forms leucine and glycine crystals in the gastrioles (they are therefore not strictly speaking cytoplasmic bodies), and these are then separated from the gastriole and the materials are transported from the crystal vacuoles to the growing refractive bodies and are there stored in the form of the protein stroma of these lipoid-protein bodies. In *Opalina* proteins are stored in the ectoplasmic segregation bodies. In both these cases, the final structures are of cytoplasmic origin and although they are so different they could perhaps be harmonized on a functional basis. In *Ichthyophthirius,* on the other hand, the protein spherules are stored and utilized in the cytoplasm, but originate in the macronucleus by splitting from the chromatin a portion which contains nucleic acid and iron, leaving a reserve protein in the form of large granules which are then discharged as completed bodies into the cytoplasm. There is some evidence in the Sporozoa also of a nuclear origin of some of the protein reserves, although it is entirely possible that they are connected with mitochondria, since Joyet-Lavergne noted a morphological relationship between the two in the Sporozoa he studied. However, even if we disregard the somewhat questionable case of the Sporozoa, we find that an identical function—the storage in the cytoplasm of proteins—is accomplished in two cases by cytoplasmic structures, but in a third case by the macronucleus.

Digestion, except in the astomatous species, is accomplished by the gastriole, a structure formed by the union of a vacuole which contains the food particles with granules or vacuoles of cytoplasmic origin. In *Ichthyophthirius* the granules involved are cytoplasmic in origin, but become enclosed within the vacuole; the cytoplasmic vacuoles of *Flabellula* apparently furnish the fluid in the gastriole; while in *A. proteus* the granules merely aggregate around the gastriole. From morphological evidence, the granules in *Ichthyophthirius* and *Flabellula* are concerned with all types of digestion, but in *A. proteus* both morphological and microenzymatic studies show that the mitochondria are concerned with the digestion of carbohydrates and with the transport of digested materials from the gastriole to such bodies as the refractive granules. There is thus some variation in the digestive function, but clearer evidence of differences is the fact that in both *Ichthyophthirius* and *Flabellula* the diffusion of materials outward from the vacuole is accomplished without the intervention of any visible granules. The granules of *Amoeba,* therefore, have a transport function, but none of the granules of either *Ichthyophthirius* or *Flabellula* give evidence of a relationship with the gastriole which would permit such a function.

The storage of lipoids is accomplished in almost identical fashion in *Opalina, Ichthyophthirius,* and *Gregarina* by the formation of intermediate lipoid bodies which are converted into neutral fats. *A. proteus* also stores neutral fats, but in this case no visible intermediate bodies are formed. This is not necessarily in conflict with the facts observed in the other species, since it is very possible that the intermediate bodies of fatty acid might be present but never get as large as the lower limits of microscopic visibility. However, not all the visible lipoid reserves of *Amoeba* are in the form of neutral fat; some are in the form of masked lipoids in the refractive bodies, and none of the other Protozoa in this group have granules which are strictly comparable with these complex structures.

Carbohydrate reserves are found in the form of paraglycogen granules in *Ichthyophthirius* and the Sporozoa, and in both cases are secreted by mitochondria. In *Opalina,* glycoprotein is found in certain cases in the segregation apparatus, but according to Kedrowsky, there are no important stores of carbohydrates in this species. For *Amoeba* likewise this statement holds, the only carbohydrate being the shell between the fluid and the lipoid-protein rim of the refractive bodies. No carbohydrate reserves are found in *Flabellula.* From these cases it would appear that the function of carbohydrate storage is largely accomplished by mitochondria, but it must be remembered that this is not a general rule, since these reserves may be formed by independent bodies, as in the Ophryoscolecidae, or by the parabasal body, as in certain flagellates.

The vacuolar apparatus performs at least two functions—the excretion of water to maintain the proper water balance, and the excretion of other materials which are probably metabolic wastes. The first function may be performed without the intervention of granules, as in the encysted stages of *Ichthyophthirius,* but the excretion of other materials is accompanied by the periodic aggregation of granules around the vacuole in *Amoeba* and *Ichthyophthirius.* In the former function the granules conform to the definitions of mitochondria, and in the latter to the Golgi bodies. These bodies are certainly a group having to do, in *Ichthyophthirius,* only with the contractile vacuoles; but in *Amoeba* they are apparently also associated with the food vacuoles, the refractive bodies, and so forth, in a general transport function. The two groups are comparable, but the granules of *Ichthyophthirius* perform only a part of the functions assigned to them in *Amoeba*. It may be possible that the mitochondria or beta granules of *Amoeba* include more than one type, but in view of the detailed experiments which have been made on this form, this explanation is hardly more than a possibility.

This survey of functions which have a granular basis fails to reveal any general uniformity, even in this restricted group of five Protozoa. The most clear-cut case of two different mechanisms having the same function is that of the segregation of reserve protein in *Ichthyophthirius* and *Opalina*. No matter how broad the definitions are made, the fact remains that the identical result is attained through the mediation of two different cellular mechanisms—in the one case the mechanism is the macronucleus, in the other it is strictly cytoplasmic, the segregation bodies. The general concept of transport introduced by Doyle helps in several cases to group apparently diverse functions within a single functional concept; but here again none of Kedrowsky's published observations in the case of *Opalina,* nor my own observations in *Ichthyophthirius,* would support this. The stored fat, paraglycogen, and proteins merely decreased in size during encystment, and no intermediate bodies aid in the redistribution of this material.

The granules which are produced in a particular species are typical of that species, but in other species the same function may be accomplished by granules different in composition and relationships from those of the first species, or the same function may be accomplished without the formation of visible granules. The cell is not restricted in the accomplishment of its functions by any system of universal and invariable cytoplasmic components.

THE CONTINUITY OF CYTOPLASMIC GRANULES

The failure to find evidence of universal cytoplasmic components by the use of either composition or function as criteria, brings the discussion to a much more general concept—the distinction between granules as permanent organelles and as temporary reserve granules. This distinction can be traced back to Altmann's bioblast theory, but it has been applied more recently in a modified and refined form to the mitochondria,

Golgi bodies, and the vacuome. The rejection of the universality of any one of these components still leaves the possibility that in each cell there are two sets of granules—the group which is permanent in organization, and a group of temporary granules, usually "passive" reserve bodies, which may be derived from the activities of the first. This is not a restatement of Altmann's theory, in the sense that it implies that the first group are living units as such, nor even that these granules are regarded in any strict sense as independent, since their maintenance obviously depends upon their interaction with the other parts of the cell. The question is whether any set of granules are present during the whole life cycle, and further whether new granules of the same group arise directly from the old granules and never arise *de novo*.

The morphological studies of Fauré-Fremiet (1910), Joyet-Lavergne (1926a), MacLennan (1934), Subramaniam and Ganapati (1938), and others have shown that mitochondria, Golgi bodies, glycogen granules, and so forth, in various Protozoa, undergo division in such a manner as to retain the original organization of the granules, and that these bodies are found in all stages of the life cycle. On the other hand, Horning (1929), Volkonsky (1929 on), MacLennan (1936), and Kedrowsky (1931 on) find that one or more of the supposedly fundamental components arise *de novo* either continuously or at some stage of the life cycle. A *de novo* origin has been proved by Mast and Doyle (1935b) not only for granules which are simple morphologically, but even for the complex tripartite refractive granules of *A. proteus*. In *Ichthyophthirius* apparently none of the granules are retained through the life cycle, thus clearly eliminating in this protozoan any distinction based on continuity. These morphological studies show that there may be a genetic continuity with respect to some granules in some of the Protozoa, but that it is not a general thing.

The observation that in some Protozoa all of the granules arise *de novo* at some time or other, raises the question whether the observed divisions are significant or are merely incidental. Kedrowsky was able to induce typical division figures in the endosomes of *Opalina* by altering the culture medium. Horning (1929) showed that dividing mitochondria are found in the trophozoite of *Monocystis* but that these granules disappear completely during the spore stages and form *de novo* in the newly liberated sporozoite. The digestive granules of *Ichthyophthirius* divide inside the gastriole to form a larger number of small granules, and both the mitochondria and the intermediate lipoid bodies (endoplasmic Golgi) fracture and split into rods as the granule within the sphere grows; yet in none of these cases is the division more than an incident in the cycle of the original granule, which arises in all cases *de novo*. The division of cytoplasmic granules is merely an indication that the granule is unstable under the particular conditions of size, surface tension, and so forth.

The term "vacuolar reaction" was introduced by Volkonsky to describe the relationship between the formation of new digestive granules (his vacuome) and the presence of food. The pattern of the reaction depended both upon the species of cell and upon the type of food present. This formation of granules as a response to a specific stimulus is by no means restricted to the single case of the digestive granules. The formation of secretion granules in the Tintinnidae is a specific response to the factors which require a new lorica, and these granules are present at no other time. The formation of the complex refractive bodies in *Amoeba* is a specific response of that particular protozoön to excess food; when this condition no longer holds, the granules are resorbed. The excretory granules of *Ichthyophthirius* are the response of this protozoön to the presence of metabolic wastes, which result from active feeding and growth and which disappear in encystment, when the original condition no longer holds good. The segregation bodies of *Opalina* likewise exhibit changes which are specific responses to the particular food which is available. Horning (1929) points out that the disappearance and reappearance of mitochondria in *Monocystis* is correlated with the decrease and increase of metabolism resulting from encystment and excystment. Volkonsky's vacuolar reaction is one case of general response, or "granular reaction," of the cell to a host of stimuli. If the stimulus is always present, the particular granules which are characteristic for the stimulus and for the particular cell under consideration are always present, but if the stimulus is intermittent, the particular granules involved are present only for the corresponding period.

Continuity is of no significance in the evaluation of the granules, but is rather a criterion of the continuity of the stimulus which induces the formation of the granule. This, together with the demonstration that the division of the granules is purely incidental, shows that it is not possible to distinguish between permanent organelles and temporary components, nor between active and passive granules.

THE CLASSIFICATION OF CYTOPLASMIC GRANULES

The cytoplasmic granules are a visible reaction of the cell to various stimuli, with the result that they show as great a variety as do the functions of which they are the visible expression and as the cells which form these granules. Any final classification must be based, then, on function, composition, and origin, rather than on a few nonspecific stains which give the impression of universal components, or on a distinction between permanent organelles and temporary reserves. Since function, composition, and relationship vary widely from one cell to another, the cytoplasmic granules, even of the Protozoa alone, cannot be divided into three or four sharply defined types, but must be separated into more types, with a classification sufficiently flexible to allow for the combination of several functions in the same granule. Such an ideal classification may be defined briefly as functional.

A functional classification is impossible at the present time, since the usual cytological or cytochemical methods reveal only those functions which result in the accumulation of visible masses of material—the segregating functions. The general type of material which is segregated has been identified in many cases, but usually there is insufficient evidence to determine whether this is a simple segregation process or whether there is actual synthesis involved. A functional classification on such a narrow base would lack permanent value, but it is necessary to readjust the present classification, in order to separate granules which are obviously unlike, even on the relatively scanty evidence now available. This separation has been outlined in the previous sections with the detailed evidence, but it is worthwhile to assemble these suggested changes here in one place.

The *mitochondria* are those granules which respond to mitochondrial methods, such as those employed by Regaud, Benda, and so forth, and which usually segregate Janus green. This is admittedly a heterogeneous group, but there is insufficient information at present to separate any groups on a logical basis.

The term *Golgi body* is used to designate granules or structures which impregnate specifically with the classic reduction methods, but excepts those bodies which are mitochondria, or which segregate neutral red vitally. The term as used here is thus merely a convenient way to indicate briefly certain techniques. Included in this group of granules are the fatty acid bodies, which are simply a stage in the formation of neutral fat granules and which have been called *intermediate lipoid bodies*. The granules or membranes which are associated with the contractile vacuoles are a separate group in composition and function, and are called *excretory granules*. There is a third group which display a characteristic secretory cycle but which are neither lipoidal in composition nor excretory in nature (Ellis, 1937). For these bodies and other unknown granules, the term *Golgi body* is appropriate, since it merely designates them according to the techniques used and implies nothing as to their composition or functions.

The term *neutral red granule* refers to any body which segregates neutral red or similar basic dyes in the living normal protozoön. Again, it is a term which indicates only the technique used and is a convenience when there is no evidence as to function. In this group are the *segregation granules,* which accumulate and perhaps synthesize proteins and similar materials, as in *Opalina*. The refractive bodies of *Amoeba* may be included here because of the neutral red reaction and because of the protein stroma, or they could be listed in the lipoid reserves because of their high lipoid content. The *digestive granules* are a separate group and are associated with the gastriole. One of the problems in this connection is how these digestive granules differ in function and composition from the mitochondria, which may also be associated with digestion.

Many different types of granules, some of them with the power to segregate neutral red, are expelled from the cell in the formation of shells, cysts, cement, and so forth, and are named *secretion granules.* These granules should be given more attention, since they indicate a situation similar to the secretion granules of gland cells.

The *reserve bodies* have been separated on the basis of the material stored—protein, lipoid, and carbohydrate—which also allows for the various combinations which do occur, and which will permit further subdivision when justified by an increase in the precision of cytochemical methods. This is convenient in summarizing the reserves, but for final classification it is unsatisfactory, since it ignores the differences in origin— whether they are independent bodies, as the segregation granules of *Opalina* and the paraglycogen granules of *Polyplastron,* or whether they are products of the macronucleus, mitochondria, and so forth. This is a difficulty which cannot be overcome until the fundamental processes which are involved have been worked out.

In addition to these granules are a heterogeneous group of *unknown granules* such as the alpha granules of *Amoeba* and the accessory bodies formed by the neuromotor system of *Haptophrya* (Fig. 70); various *pigment granules,* which may in some cases be part of the lipoid reserves, or in some cases residues of food, as in *Plasmodium* and *Ichthyophthirius,* this latter type of course not being true cytoplasmic granules. *Crystals* also are often present, the ones in *Amoeba* being classed as a part of the protein reserves (although here there is a question, since they orig-

Figure 70. Accessory bodies being formed from the neuromotor ring in *Haptophrya michiganensis,* Zenker's haematoxylin. (After Bush, 1934.)

inate in the food vacuoles) on the evidence of Mast and Doyle. Other crystals, according to Reichenow, may be excretory granules.

COMPARISON WITH CELLS OF THE METAZOA

Cytological investigations in the Protozoa have always been influenced by the transfer of concepts originally developed from a study of the cells of the Metazoa, particularly the vertebrates, with the result that the division of granules into mitochondria, Golgi bodies, vacuome, and passive reserve bodies are as common in the literature of the Protozoa as in that of the Metazoa. In spite of Dobell's denial of the cellular nature of the Protozoa, any consideration of the granules of the Protozoa necessitates a comparison with the cytoplasmic granules in other animal cells.

The lack of a "typical" reticular net of Golgi in the Protozoa, the infrequency of filamentous mitochondria, and other striking morphological differences between the protozoan and metazoan cells have been stressed so often that it is well to present several cases of equally striking similarities in both structure and function. Volkonsky (1934) found that the digestive granules and gastrioles of Protozoa, choanocytes, and leucocytes are entirely comparable, and included all of these cells in his vacuolar reaction. Fauré-Fremiet (1909) and later Kedrowsky (1932, 1933) compared the segregation granules of Protozoa and vertebrate tissues and found they were similar in appearance, staining reactions, and function. Chatton, Parat, and Lwow (1927), on the basis of specific microchemical reactions, compared the protein reserves in certain of the Foettingeriidae with the vitellin of hen's eggs. Kedrowsky (1931) and MacLennan (1936) have given figures of the formation of neutral fat bodies in the Protozoa which are almost identical with the description and figures of Bowen (1929) in the relationship between Golgi bodies and lipoid secretion in cells of the mouse. Examples could be multiplied, but these are sufficient to emphasize the fact that there are similarities as well as differences between protozoan and metazoan cells, with respect to their cytoplasmic granules.

The same difficulties with respect to the so-called specific staining reactions arise in both Protozoa and Metazoa. In a study of echinoderm eggs, Tennent, Gardiner, and Smith (1931) showed that not one material, but many reduce osmium in the Golgi techniques. The presence of more than one type of osmiophilic granules has been proved by Mast and Doyle (1935) and MacLennan (1936) in Protozoa. Although Kirkman and Severinghaus (1938) hold to the idea of a particular Golgi substance, they bear witness to the occurrence of additional osmiophilic materials: "One often finds small osmiophilic granules of uncertain significance in Kolatchev sections, but they are present in addition to the Golgi apparatus and appear to bear no relation to the latter structure." In neither group is there any evidence that there is a particular Golgi substance, any more than a particular Golgi structure.

There is also at times an embarrassing overlapping in the results obtained from the use of Janus green and neutral red. These dyes were found by Hopkins (1938) to stain the same vacuoles in the marine amoeba, *Flabellula,* and Uhlenhuth (1938) reports similar results in the thyroid cells of amphibia. The mitochondria turn out to be not a simple group but a complex group, as indicated by the distinction between mitochondria and active mitochondria, or mitochondria proper, by Parat (1927) and by Joyet-Lavergne (1926).

Evidence is accumulating in both Protozoa and Metazoa that no type of cytoplasmic granule (with the the exception of the centriole) can be considered as permanent, self-perpetuating structures. Wilson and Pollister (1937), in connection with an investigation on sperm formation in scorpions, review the division and distribution of mitochondra, Golgi bodies, and vacuome, and show that the supposed accurate division is actually an incidental fragmentation of large masses and that the distribution is random during the division of the cell. They state: "There is, however, little ground for the contention that either Golgi bodies or chondriosomes can be regarded as permanent individuals having the power of self perpetuation by growth and regular division." Exactly the same situation has been shown in this discussion with respect to the cytoplasmic granules of the Protozoa, from studies of both fixed material and of living normal cells. The permanence of mitochondria in metazoan cells has been summed up recently by Bensley (1937): "The disappearance and reappearance of mitochondria in living cells under observation, as described by Chambers, however repugnant the idea may be to those who would elevate mitochondria to the dignity of living, self-reproducing units, must be definitely entertained as probable." With respect to the *de novo* origin of Golgi apparatus, Kirkman and Severinghaus (1938) assert that "there is little to favor such a view," although they quote at least a dozen authors who have advanced evidence of a *de novo* origin in one form or another, admitting in several cases that the evidence presented has not been refuted. For a detailed discussion, the reader is referred to the review of the subject by these authors and to the original publications, but it is clear that not all cytologists agree with these authors on the permanence of the Golgi bodies.

The concept that the cytoplasmic granules arise or are resorbed as a result of specific conditions of metabolism in the cell—in other words that there is a granular reaction—is a logical result of the evidence that the granules are neither permanent nor self-perpetuating, and it is therefore no surprise to find that this interpretation has been made with respect to the cytoplasmic constituents of the Metazoa as well as of the Protozoa. One of the clearest statements of this concept has been made by Tennent, Gardiner, and Smith (1931): "The results of this research have been the conviction that neither Golgi bodies nor Chondriosomes are structural elements in the cellular architecture, but that both are the chemical products of physiological processes." Nahm (1933) likewise states that "they are the visible products of chemical reactions that occur in the cell."

182   CYTOPLASMIC INCLUSIONS

LITERATURE CITED

Beers, C. D. 1935. Structural changes during encystment and excystment in the ciliate *Didinium nasutum*. Arch. Protistenk., 84: 133-55.

Bensley, R. R. 1937. On fat distribution in mitochondria of the guinea pig liver. Anat. Rec., 69: 341-54.

Bles, E. J. 1929. *Arcella*, a study in cell physiology. Quart. J. micr. Sci. n. s., 72: 527-648.

Borgert, A. 1909. Ueber erscheinungen fettiger Degeneration bei tripyleen Radiolarien. Arch. Protistenk., 16: 1-25.

Bourne, G. 1936. The role of vitamin C in the organism as suggested by its cytology. Physiol. Rev., 16: 442-49.

Bourne, G., and R. Allen. 1935. The distribution of vitamin C in lower organisms. Aust. J. exp. Biol. med. Sci., 13: 165-74.

Bowen, R. H. 1928. The methods for the demonstration of the Golgi apparatus. VI. Protozoa. The vacuome. Plant tissues. Aust. J. exp. Biol. med. Sci., 40: 225-76.

—— 1929. The cytology of glandular secretion. Quart. Rev. Biol., 4: 484-519.

Brand, T. von. 1935. Der Stoffwechsel der Protozoen. Ergebn. Biol., 12: 161-220.

—— 1938. The metabolism of pathogenic trypanosomes and the carbohydrate metabolism of their hosts. Quart. Rev. Biol., 13: 41-50.

Brown, V. E. 1930. The Golgi apparatus of *Pyrsonympha* and *Dinenympha*. Arch. Protistenk., 71: 453-62.

Bush, M. 1934. The morphology of *Haptophrya michiganensis* Woodhead. Univ. Cal. Publ. Zool., 39: 251-76.

Bütschli, O. 1885. Bemerkungen ueber einen dem glikogen verwandten Koerper in Gregarinen. Z. Biol., 21: 603-12.

Cajal, Ramon y, Santiago. 1904. Variaciones morfologicas del reticulo nervioso de invertebrados y vertebrados sometidos á la acción de condiciones naturales. Trab. Lab. Invest. biol. Univ. Madr., 1: 3.

Calkins, G. N. 1930. *Uroleptus halseyi* n. sp. II. The origin and fate of the macronuclear chromatin. Arch. Protistenk., 69: 151-74.

Campbell, A. S. 1926. The cytology of *Tintinnopsis nucula* (Fol) Laakman. Univ. Cal. Publ. Zool., 29: 179-236.

—— 1927. Studies on the marine ciliate *Favella*, with special regard to the neuromotor apparatus and its rôle in the formation of the lorica. Univ. Cal. Publ. Zool., 29: 429-52.

Chakravarty, M. 1936. On the morphology of *Balantidium depressum* Ghosh from a mollusc *Pila globosa*, with a note on its nucleal reaction and cytoplasmic inclusions. Arch. Protistenk., 87: 1-9.

Chalkley, H. W. 1937. The chemistry of cell division VII. Protoplasma, 28: 489-97.

Chambers, R., and R. T. Kempton. 1937. The elimination of neutral red by the frog's kidney. J. cell. comp. Physiol., 10: 199-221.

Chambers, R., and H. Pollack. 1927. Micrurgical studies in cell physiology. IV. Colorimetric determination of the nuclear and cytoplasmic pH in the starfish egg. J. gen. Physiol., 10: 739-55.

Chatton, E., M. Parat, and A. Lwow. 1927. La formation, la nature et l'evolution des réserves chez les *Spirophrya*, les *Polyspira*, et les Gymnodinioides (Infusoires Foettingeriidae). C. R. Soc. Biol. Paris, 96: 6-8.

Chatton, E., and P. P. Grassé. 1929. Le chondriome, le vacuome, les vésicules osmiophiles, le parabasal, les trichocystes, et les chidocystes du Dinoflagellé *Polykrikos Schwartzi* Bütschli. C. R. Soc. Biol. Paris, 100: 281-85.

Cheissin, E. 1930. Morphologische und systematische Studien über Astomata aus dem Baikalsee. Arch. Protistenk., 70: 531-618.

Cleveland, L. R. 1934. The wood-feeding roach *Cryptocercus*, its Protozoa, and the symbiosis between Protozoa and roach. Mem. Amer. Acad. Arts Sci., 17: 185-331.

Connell, F. H. 1930. The morphology and life cycle of *Oxymonas dimorpha* sp. nov. from *Neotermes simplicicornis* (Banks). Univ. Cal. Publ. Zool., 36: 51-66.

Cowdry, E. V. 1924. General cytology. Chicago.

Cowdry, E. V., and G. H. Scott. 1928. Études cytologiques sur le paludisme. III. Mitochondries, granules colorable au rouge neutre et appareil de Golgi. Arch. Inst. Pasteur Afr. N., 17: 233-52.

Cutler, M. 1921. Observation on the Protozoa parasitic in *Archotermopsis wroughtoni* Desn. Part III. *Pseudotrichonympha pristina*. Quart. J. micr. Sci., 65: 248-64.

Dangéard, P. 1928. L'Appareil mucifère et le vacuome chez les Eugléniens. Ann. Protist., 1: 69-74.

Daniels, M. L. 1938. A cytological study of the gregarine parasites of *Tenebrio molitor*, using the ultra-centrifuge. Quart. J. micr. Sci., 80: 293-320.

Day, H. C. 1927. The formation of contractile vacuoles in *Amoeba proteus*. J. Morph. 44: 363-72, 6 figs.

Dobell, C. C. 1925. The life history and chromosome cycle of *Aggregata eberthi*. Parasitology, 17: 1-136.

Doflein, F., and E. Reichenow. 1929. Doflein, Lehrbuch der Protozoenkunde. 5th ed., Jena.

Dogiel, V., and T. Fedorowa. 1925. Über den Bau und die Funktion des inneren Skeletts der Ophryoscoleciden. Zool. Anz., 62: 97-107.

Doyle, W. L. 1935. Distribution of mitochondria in the foraminiferan, *Iridia diaphana*. Science, n.s., 81: 387.

Duboscq, O., and P. Grassé. 1925. Notes sur les protistes parasites des Termites de France. Appareil de Golgi, mitochondries, et vésicules sous-flagellaires de *Pyrsonympha vertens* Leidy. C. R. Soc. Biol. Paris, 93: 345-48.

—— 1933. L'Appareil parabasal des flagellés, avec remarques sur le trophosponge, l'appareil Golgi, les mitochondries, et le vacuome. Arch. zool. exp. gén., 73: 381-621.

Dunihue, F. W. 1931. The vacuome and the neutral red reaction in *Paramoecium caudatum*. Arch. Protistenk., 75: 476-97.

Eksemplarskaja, E. V. 1931. Morphologie und Cytologie von *Anoplophrya sp.* aus dem Regenwurmdarm. Arch. Protistenk., 73: 147-63.

Ellis, J. M. 1937. The morphology, division and conjugation of the salt marsh ciliate *Fabrea salina* Henneguy. Univ. Cal. Publ. Zool., 41: 343-88.

Erdmann, R. 1917. *Chloromyxum leydigi* und seine Bezeichnung zu anderen myxosporidien. Arch. Protistenk., 37: 276-326.

Fauré-Fremiet, E. 1905. Sur l'organisation de la *Campanella umbellaria*. C. R. Soc. Biol. Paris, 58: 215-17.

—— 1909. Vacuoles colorables par le rouge neutre chez un infusoire. C. R. Ass. Anat., 11: 286-88.

—— 1910. Étude sur les mitochondries des protozoaires et des cellules sexuelles. Arch. Anat. micr., 11: 457-648.

—— 1925. La Structure permanente de l'appareil excréteur chez quelques vorticellides. C. R. Soc. Biol. Paris, 93: 500-3.

—— 1932. Division et morphogenèse chez *Folliculina ampulla* OFM Bull. biol., 66: 79-109.

Fauré-Fremiet, E., A. Mayer, and G. Schaeffer. 1910. Sur la microchimie des corps gras. Arch. Anat. micr., 12: 19-102.

Finley, H. E. 1934. On the vacuome in three species of *Vorticella*. Trans. Amer. micr. Soc., 53: 57-64.

Fiveiskaja, A. 1929. Einfluss der Kernparasiten der Infusorien auf den Stoffwechsel. Arch. Protistenk., 65: 275-98.

Frisch, A. 1938. The rate of pulsation of the contractile vacuole in *Paramecium multimicronucleatum*. Arch. Protistenk., 90: 123-61.

Gatenby, J. B. 1930. Cell nomenclature. J. R. micr. Soc., 50: 20-29.

Gelei, J. von. 1928. Nochmals über den Nephridialapparat bei den Protozoen. Arch. Protistenk., 64: 479-94.

—— 1938. Das Exkretionsplasma von *Didinium nasutum* in Ruhe und Teilung. Arch. Protistenk., 90: 368-82.

Hall, R. P. 1929. Modifications of technique for demonstration of Golgi apparatus in free-living Protozoa. Trans. Amer. micr. Soc., 48: 443-44.

—— 1931. Vacuome and Golgi apparatus in the ciliate, *Stylonychia*. Z. Zellforsch., 13: 770-82.

Hall, R. P., and F. W. Dunihue. 1931. On the vacuome and food vacuoles in *Vorticella*. Trans. Amer. micr. Soc., 50: 196-205.

Hall, R. P., and J. B. Loefer. 1930. Studies on *Euglypha*. I. Cytoplasmic inclusions of *Euglypha alveolata*. Arch. Protistenk., 72: 365-76.

Hall, R. P., and R. F. Nigrelli. 1930. Relation between mitochondria and food vacuoles in the ciliate, *Vorticella*. Trans. Amer. micr. Soc., 49: 54-57.

—— 1937. A note on the vacuome of *Paramecium bursaria* and the contractile vacuole of certain ciliates. Trans. Amer. micr. Soc., 56: 185-90.

Haye, Ans. 1930. Über den Exkretionsapparat bei den Protozoen, nebst Bemerkungen über einige andere feinere Strukturverhältnisse der untersuchten Arten. Arch. Protistenk., 70: 1-87.

Hayes, M. L. 1938. Cytological studies on *Dileptus anser*. Trans. Amer. micr. Soc., 57: 11-25.

Heilbrunn, L. V. 1928. The colloid chemistry of protoplasm. Berlin.

—— 1936. Protein lipid binding in protoplasm. Biol. Bull., 71: 299-305.

Heilbrunn, L. V., and K. Daugherty. 1938. Fat release in *Amoeba* after irradiation. Physiol. Zoöl., 11: 383-87.

Hertwig, R. 1902. Die Protozoen und die Zelltheorie. Arch. Protistenk., 1: 1-40.

Hindle, E. 1910. Degeneration phenomena of *Trypanosoma gambiense*. Parasitology, 3: 423.

Hirschler, J. 1914. Über Plasmastrukturen (Golgi'scher Apparat, Mitochondrien, u.a.) in den Tunicaten-, Spongien, und Protozoen-zellen. Anat. Anz., 47: 289-311.

—— 1924. Sur les composants lipoidifères du plasma des Protozoaires. C. R. Soc. Biol. Paris, 10: 891-93.

—— 1927. Studien über die sich mit Osmium schwärzenden Plasmakomponenten (Golgi-apparat, Mitochondrien) einiger protozoenarten, nebst Bemerkungen über die Morphologie der ersten von ihnen im Tierreiche. Z. Zellforsch., 5: 704-86.

Holmes, E. 1937. The metabolism of living tissues. Cambridge.

Holter, H., and W. L. Doyle. 1938. Über die lokalisation der Amylase in der Amöben. C. R. Lab. Carlsberg (Sér. Chim.), 22: 219-25.

Holter, H., and M. J. Kopac. 1937. Studies in enzymatic histochemistry. XXIV. Localization of peptidase in the *Amoeba*. J. cell. comp. Physiol., 10: 423-37.

Hopkins, D. L. 1938a. The mechanism for the control of the intake and output of water by the vacuoles in the marine amoeba, *Flabellula mira* Schaeffer. Biol. Bull., 75: 353.

—— 1938b. The vacuoles and vacuolar activity in the marine amoeba *Flabellula mira* Schaeffer and the nature of the neutral red system in Protozoa. Biodyn., No. 34, 22 pp.

Horning, E. S. 1925. The mitochondria of a protozoön (*Opalina*) and their behavior during the life cycle. Aust. J. exp. Biol. med. Sci., 2:

—— 1926. Studies on the mitochondria of *Paramoecium*. Aust. J. exp. Biol. med. Sci., 3: 91-94.

—— 1927. On the orientation of mitochondria in the surface cytoplasm of infusorians. Aust. J. exp. Biol. med. Sci., 4: 187-90.

—— 1929. Mitochondrial behavior during the life cycle of a sporozoön (*Monocystis*) Quart. J. micr. Sci., 73: 135-43.

Horning, E. S., and G. H. Scott. 1933. Comparative cytochemical studies by micro-incineration of a saprozoic and an holozoic infusorian. J. Morph., 54: 389-94.

Jirovec, O. 1926. Protozoenstudien I. Arch. Protistenk., 56: 280-90.

Joyet-Lavergne, P. 1926a. Recherches sur les cytoplasmes des Sporozoaires. Arch. Anat. micr., 22: 1-128.

—— 1926b. Sur la coloration vitale des elements de Golgi des gregarines. C. R. Soc. Biol. Paris, 94: 830-32.

—— 1927. Sur les rapports entre le glutathion et le chondriome. C. R. Acad. Sci. Paris, 184: 1587.

—— 1928. Le Pouvoir oxydo-reducteur du chondriome des Gregarines et les procédés de recherches du chondriome. C. R. Soc. Biol. Paris, 98: 501.

—— 1929. Glutathion et chondriome. Protoplasma, 6: 84-112.

—— 1931. Le Potentiel d'oxydo-reduction et la sexualisation cytoplasmique des Gregarines. C. R. Soc. Biol. Paris, 107: 951-52.

—— 1932a. Sur le pouvoir oxydant du chondriome dans la cellule vivante. C. R. Soc. Biol. Paris, 110: 552-53.

—— 1932b. Sur la mise en evidence des zones d'oxydation dans la cellule animale. C. R. Soc. Biol. Paris, 110: 663-64.

—— 1934. Une Théorie nouvelle sur le mécanisme des oxydo-réductions intracellulaires. C. R. Acad. Sci. Paris, 199: 1159.

—— 1935. Réchèrches sur la catalyse des oxydo-reductions dans la cellule vivante. Protoplasma, 23: 50-69.

Kazancev, V. 1928. Beitrag zur Kenntnis der Grosskerne der Ciliaten. Trav. Lab. zool. Sebastopol, Ser. II, 13: 1-30.

Kedrowsky, B. 1931a. Die Stoffaufnahme bei *Opalina ranarum*. I. Methodik der Kultur in künstlichen medien, pH regulieren und Ionen Gleichgewichte im Kulturmedien. Protoplasma, 12: 356-79.

—— 1931b. II. Struktur. Wasseraufnahme und Wasserzustand in Protoplasma von *Opalina*. Protoplasma, 14: 192-255.

—— 1931c. III. Aufnahme und Speicherung von Farbstoffen. Z. Zellforsch., 12: 600-65.

—— 1931d. IV. Die synthetische Fettspeicherung. Z. Zellforsch., 12: 666-714.

—— 1931e. V. Der Segregationsapparat. Z. Zellforsch., 13: 1-81.

—— 1932a. Vitalfärbungsstudien an Infusorien. Z. Zellforsch., 15: 93-113.

—— 1932b. Über die Natur des Vakuoms. Z. Zellforsch., 15: 731-60.

—— 1933. Neue Probleme im Studium des Eiweissstoffwechsels der Zelle. Arch. exp. Zellforsch., 14: 533-54.

Kirby, H. 1931. Trichomonad flagellates from termites. II. *Eutrichomastix* and the subfamily Trichomonadinae. Univ. Cal. Publ. Zool., 36: 171-262.

—— 1932. Flagellates of the genus *Trichonympha* in termites. Univ. Cal. Publ. Zool., 37: 349-76.

—— 1936. Two polymastigote flagellates of the genera *Pseudodevescovina* and *Caduceia*. Quart. J. micr. Sci., n.s., 79: 309-35.

Kirkman, H., and A. E. Severinghaus. 1938. A review of the Golgi apparatus. Anat. Rec., 70: 413-32, 557-74; 71: 79-104.

Koehring, V. 1930. The neutral red reaction. J. Morph., 49: 45-130.

Kofoid, C. A., and M. Bush. 1936. The life cycle of *Parachaenia myae*, gen. nov., sp. nov., a ciliate parasitic in *Mya arenaria* Linn. from San Francisco Bay, Calif. Bull. Mus. Hist. nat. Belg., 12: 1-15.

Kraschenninikow, S. 1929. Über den Exkretionsapparat einiger Infusorienarten der Familie Ophryoscolecidae. Z. Zellforsch., 8: 470-83.

Krijgsman, B. J. 1936. Verleichend physiologische Untersuchungen über den stoffwechsel von *Trypanosoma evansi* in Zusammenhang mit der Anpassung an das Wirtstier. Z. vergl. Physiol., 23: 663-711.

LeBreton, E. 1931. Mitochondries et ferments protéolytiques. Examen de l'hypothèse de Robertson-Marston. Arch. Biol. Paris, 42: 349-63.

Leiner, M. 1924. Das Glycogen in *Pelomyxa palustris* Greef mit Beiträgen zur kenntnis des Tieres. Arch. Protistenk., 47: 253-307.

Lison, L. 1936. Histochimie Animale. Paris.

Lynch, J. E. 1930. Studies on the ciliates from the intestine of *Strongylocentrotus*. II. *Lechriopyla mystax*, gen. nov., sp. nov. Univ. Cal. Publ. Zool., 33: 307-50.

MacLennan, R. F. 1933. The pulsatory cycle of the contractile vacuoles in the Ophryoscolecidae, ciliates from the stomach of cattle. Univ. Cal. Publ. Zool., 39: 205-50.

—— 1934. The morphology of the glycogen reserves in *Polyplastron*. Arch. Protistenk., 81: 412-19.

—— 1936. Dedifferentiation and redifferentiation in *Ichthyophthirius*. II. The origin and function of cytoplasmic granules. Arch. Protistenk., 86: 404-26.

—— 1937. Growth in the ciliate *Ichthyophthirius*. I. Maturity and encystment. J. exp. Zool., 76: 423-40.

—— 1940. A quantitative study of the osmic acid reaction in Protozoa. Trans. Amer. micr. Soc., 59: 149-59.

MacLennan, R. F., and H. K. Murer. 1934. Localization of mineral ash in the organelles and cytoplasmic components of *Paramecium*. J. Morph., 55: 421-33.

Marston, H. R. 1923. The azine and azonium compounds of the proteolytic enzymes. Bio-chem. J., 17: 850-59.

Mast, S. O. 1938. Digestion of fat in *Amoeba proteus*. Biol. Bull., 75: 389-94.

Mast, S. O., and W. L. Doyle. 1935a. Structure, origin, and function of cytoplasmic constituents in *Amoeba proteus*. I. Structure. Arch. Protistenk, 86: 155-80.

—— 1935b. Structure, origin, and function of cytoplasmic constituents in

Amoeba proteus with special reference to mitochondria and Golgi substance. II. Origin and function based on experimental evidence; effect of centrifuging on *Amoeba proteus*. Arch. Protistenk., 86: 278-306.

Metcalf, M. M. 1910. Studies upon *Amoeba*. J. Exp. Zool., 9: 301-32.

Meyers, E. H. 1935. The life history of *Patellina corrugata* Williamson, a foraminifer. Bull. Scripps Instn. Oceanogr. tech., Ser. 3: 355-92.

Miller, E. D. 1937. A comparative study of the contents of the gelatinous accumulations of the culture media and the contents of the cytoplasm of *Amoeba proteus* and *Arcella vulgaris*. J. Morph., 60: 325-54.

Milovidov, P. F. 1932. Independence of chondriosomes from nuclear matter. Cytologia, 4: 158-73.

Nahm, L. J. 1933. A study of the Golgi elements. J. Morph., 54: 259-301.

Nassonov, D. 1924. Der Exkretionsapparat (kontraktile Vakuole) der Protozoa als Homologen des Golgischen Apparats der Metazonzellen. Arch. Mikr. Anat., 103: 437-82.

—— 1925. Zur Frage über den Bau und die Bedeutung des lipoiden Excretionsapparates bei Protozoa. (*Chilodon, Dogielella*). Z. Zellforsch., 2: 87-97.

Nigrelli, R. F. 1929. On the morphology and life history of *Trypanosoma diemyctyli* and the relation of trypanosomiasis to the polynuclear count. Trans. Amer. micr. Soc., 48: 366-87.

Owens, H. B., and R. R. Bensley. 1929. On osmic acid as a microchemical reagent, with special reference to the reticular apparatus of Golgi. Amer. J. Anat., 44: 79-110.

Parat, M. 1927. A review of recent developments in histochemistry. Biol. Rev., 2: 285-97.

Patten, R. 1932. Observations on the cytology of *Opalina ranarum* and *Nyctotherus cordiformis*. Proc. R. Irish Acad., 41: 73-94.

Pellissier, M. 1936. Sur certains constituants cytoplasmiques de l'infusoire cilié, *Trichodinopsis paradoxa* Clap. et Lach. Arch. zool. exp. gén, 78: 32-36.

Peshkowskaya, L. 1928. On the biology and the morphology of *Climacostomum virens*. Arch. russ. protist., 7: 205-35.

Poljansky, G. 1934. Geschlechtsprozesse bei *Bursaria truncatella* OFM. Arch. Protistenk., 81: 420-546.

Rammelmeyer, H. 1925. Zur Frage ueber die Glykogendifferenzierung bei *Paramecium caudatum*. Arch. Protistenk., 51: 184-88.

Reichenow, E. 1928. Ergebnisse mit der nuclealfaerbung bei Protozoen. Arch. Protistenk., 61: 144-66.

—— 1929. Doflein, Lehrbuch der Protozoenkunde. 5th ed., Jena.

Rey, P. 1931a. Potentiel d'oxydo-reduction et sexualité chez les Gregarines. C. R. Soc. Biol. Paris, 107: 611-14.

—— 1931b. Coloration vitale et potentiel d'oxydo-reduction chez les gregarines. C. R. Soc. Biol. Paris, 107: 1508-11.

Rosenberg, L. E. 1937. The neuromotor system of *Nyctotherus hylae*. Univ. Cal. Publ. Zool., 41: 249-76.

Rumjantzew, A. 1922. Über den Bau der Chromidial substanz bei *Difflugia pyriformis*. Arch. russ. protist., 1: 87-105.

Rumjantzew, A., and E. Wermel. 1925. Untersuchungen ueber den Protoplasmabau von *Actinosphaerium Eichhornii*. Arch. Protistenk., 52: 217-64.

Sassuchin, D. 1924. Zur Kenntnis der Plasmaeinschlüsse bei den Opalinen. Arch. russ. protist. 3: 147-54.

Schrich, P. 1914. Beiträge zur Kenntnis des Lebenscyclus von *Arcella vulgaris* und *Pelomyxa palustris*. Arch. Protistenk., 33: 247-71.

Schulze, P. 1922. Über Beziehungen zwischen tierischen und pflanzlichen Skelettsubstanzen und über Chitinreaktionen. Biol. Zbl., 42: 389-94.

—— 1924. Der Nachweis und die Verbreitung des Chitins mit einem Anhang über das komplizierte Verdauungssystem der Ophryoscoleciden. Z. Morph. Ökol. Tiere, 2: 643-66.

—— 1927. Noch einmal die "Skelettplatten" der Ophryoscoleciden. Z. Morph. Ökol. Tiere, 7: 670-89.

Scott, G. H., and E. S. Horning. 1932. The structure of Opalinids as revealed by the technique of microincineration. J. Morph., 53: 381-88.

Strelkow, A. 1929. Morphologische Studien über Oligotriche Infusorien aus dem Darme des Pferdes. Arch. Protistenk., 68: 503-54.

—— 1931. II. Cytologische Untersuchungen der Gattung *Cycloposthium* Bundle. Arch. Protistenk., 75: 191-220.

Subramaniam, M. K., and P. N. Ganapati. 1938. Studies on the structure of the Golgi apparatus. I. Cytoplasmic inclusions in the gregarine *Lecudina brasili* n.sp. parasitic in the gut of *Lumbriconereis*. Cytologia, 9: 1-16.

Subramaniam, M. K., and R. Gopala Aiyar. 1937. The Golgi apparatus and the vacuome in protozoa—some misconceptions and the question of terminology. Proc. Indian Acad. Sci., Sec. B., 6: 1-18.

Swellengrebel, N. H. 1908. La Volutine chez les trypanosomes. C. R. Soc. Biol. Paris, 64: 38-40.

Taylor, C. V. 1923. The contractile vacuole in *Euplotes* an example of the sol-gel reversibility of cytoplasm. J. exp. Zool., 37: 259-82.

Tennent, D. H., M. S. Gardiner, and D. E. Smith. 1931. A cytological and biochemical study of the ovaries of the sea-urchin *Echinometra lucunter*. Pap. Tortugas Lab., 27: 1-46.

Thélohan, P. 1894. Nouvelles réchérches sur les coccidies. Arch. zool. exp. gén., 3: 541-73.

Tuzet, O. 1931. Une Grégarine parasite de *Bythinia tentaculata*. *Gonospora duboscqui* nov. sp. Arch. zool. exp. gén., 71: 16-21.

Uhlenhuth, E. 1938. A quantitative approach to the secretion process of the thyroid. Coll. Net, 13: 76-87.

Van Thiel, P. H. 1925. Was ist *Rickettsia melophagi?* Arch. Protistenk., 52: 394-403.

Volkonsky, M. 1929. Les phénoménes cytologiques au cours de la digestion intracellulaire de quelques ciliés. C. R. Soc. Biol. Paris, 101: 133-35.

—— 1933. Digestion intracelluaire et accumulation des colorants acides. Étude cytologique des cellules sanguines des Sipunculides. Bull. Biol., 67: 135-286.

—— 1934. L'Aspect cytologique de la digestion intracellulaire. Arch. exp. Zellforsch., 15: 355-72.

Weatherby, J. H. 1927. The function of the contractile vacuole in *Paramecium caudatum*; with special reference to the excretion of nitrogenous compounds. Biol. Bull., 52: 208-22.

—— 1929. Excretion of nitrogenous substances in the Protozoa. Physiol. Zoöl., 2: 375-94.

Weier, T. E. 1933. A critique of the vacuome hypothesis. Protoplasma, 19: 589-601.

Weineck, I. 1931. Die chemische Natur der Skelettsubstanzen bei den Ophryoscolec iden. Jena. Z. Natur., 65: 739-51.

—— 1934. Die Celluloseverdauung bei den Ciliaten des Wiederkäuermagens. Arch. Protistenk., 82: 169-202.

Wermel, E. 1925. Beiträge zur Cytologie der *Amoeba hydroxena* Entz. Arch. russ. protist., 4: 95-120.

Wilson, E. B. 1928. The Cell. 3d ed., New York.

Wilson, E. B., and A. W. Pollister. 1937. Observations on sperm formation in the centrurid scorpions with especial reference to the Golgi material. J. Morph., 60: 407-44.

Wurmser, M. 1932. Sur l'emploi de certain colorants pour l'évaluation des proprietés oxydantes du cytoplasme. C. R. Soc. Biol. Paris, 111: 690.

Yamasaki, M. 1937a. Studies on the intestinal Protozoa of termites. III. The distribution of glycogen in the bodies of intestinal flagellates of Termites *Leucotermes* and *Coptotermes*. Mem. Coll. Sci. Kyoto, B 12: 212-24.

—— 1937b. IV. Glycogen in the body of *Trichonympha agilis* v. *japonica* under experimental conditions. Mem. Coll. Sci. Kyoto, 12: 225-35.

Zhinkin, L. 1930. Zur frage der Reservestoffe bei Infusorien (Fett und Glykogen bei *Stentor polymorphus*). Z. Morph. Ökol. Tiere, 18: 217-48.

Zinger, J. A. 1928. Morphologische Beobachtungen an *Ophryoglena flava* Ehr. Arch. russ. protist., 7: 179-204.

—— 1929. Beiträge zur Morphologie und Cytologie der Süsswasserinfusorien. Arch. russ. protist., 8: 51-90.

—— 1933. Beobachtungen an Fetteinschlüssen bei einigen Protozoen. Arch. Protistenk., 81: 57-87.

Zweibaum, I. 1921. Ricerche sperimentali sulla conjugazione degli Infusori. II. Influenza della coningazione sulla produzione dei materiali di riserva nel *Paramecium caudatum*. Arch. Protistenk., 44: 99-114.

CHAPTER IV

FIBRILLAR SYSTEMS IN CILIATES

C. V. Taylor

Introduction

The essential nature of Leeuwenhoek's "little animals" remained obscure for more than 150 years, evidently because the methods of observation which characterized that ingenious microscopist of Delft were replaced largely by fruitless speculation. Otherwise, man's epochal discovery of the cellular nature of living things might have been realized sooner.

Meantime, it is true, a prodigious diversity of macroscopic forms had been examined and classified. But the disclosure of cellularity, which eventually unified all of this diversity in organic form, had to depend upon the detailed analysis of organic structure.

During the hundred years that have now intervened, that common denominator of organic form and function has come to be regarded, for multicellular plants and animals, as a sort of master key to the solution of their fundamental problems. And for the major advances in biology during that memorable century, we are surely indebted primarily to this cellular concept of the organization of living things.

For the microörganisms, however, the concept of cellularity, although generally conceded, has encountered not infrequently some confusing difficulties. With von Siebold's pronouncement in 1845 of the unicellularity of the Protozoa, the way at first seemed clear toward simplifying and unifying all forms of life, in terms of the cell as the universal unit. Eventually, however, it was evident that, for the Protozoa, this concept did not simplify matters so satisfactorily. The chief difficulty here arose in trying to equate the protozoon cell with a tissue cell of the Metazoa. And even in recent times this comparison has again been challenged by Dobell (1911) and others, who would maintain that Protozoa are not cells at all and so should be regarded as non-celluar organisms.

But similar difficulties in comparing microscopic with macroscopic forms of life had confronted investigators several years before von Siebold's pronouncement and, in fact, before the concept of cellularity had been definitely formulated. As is well known, this all culminated in the Ehrenberg-Dujardin controversy, beginning in 1835. Obviously for these investigations, the issue was not one of cellularity, but it had to do with complexity versus simplicity in the organization of the Infusoria. It seems probable that Ehrenberg defended his thesis of "complete organisms" partly in refutation of the theory of spontaneous generation, then vigorously championed for microörganisms. At any rate, he sought to identify in the Infusoria all the organs common to other animals. Much of his adduced evidence, it will be recalled, was successfully refuted by Dujardin, who described among other things, his newly discovered "sarcode" in support of his contentions for uniqueness and simplicity in the organization of the Infusoria.

The essentials of these contrasting views of Ehrenberg and Dujardin on the nature of infusorian organization have recurred, in varied guise, many times in the literature since their day. These opposing viewpoints have, of course, become translated into terms of the concept of cellularity, so that now the nature of unicellular organization, or "protoplasmic differentiation," is commonly contrasted with "cellular differentiation" of multicellular organisms.

Accordingly, in the following review of literature on fibrillar systems in ciliates, it will become evident that some discrepancies in both the analysis of structure and the interpretation of functions may owe their origin largely to contrasting points of view on the essential nature of "protoplasmic differentiation" in the Protozoa and "cellular differentiation" in the Metazoa.

Before beginning that review, however, the fact should be emphasized that, as Maupas (1883) has pointed out, the Ehrenberg-Dujardin controversy marks a turning point in protistological investigations. Not only did it enlist a wider interest in these microörganisms, but it made clear the necessity of a critical structural analysis of their greatly diversified types of organization and of a comparative study of such types before any satisfactory interpretations were possible.

The literature resulting from those analyses is so voluminous that when one undertakes to review the accounts of a given system of organelles, such as the fibrillar system, and to condense that review within reasonable bounds, the difficulties soon become evident. For this reason it has seemed advisable in the review that follows, in the interests of students and laity as well as of specialists, to present a fairly detailed account of the structural analysis, together with interpretations of the fibrillar differentiations of a well-known representative of each of four major groups of ciliates. This is followed by a brief review of other published work, mostly since 1920, on fibrillar systems in other ciliates, with some suitable illustrations; and finally, a few paragraphs of general discussion are added under the caption "Conclusions."

The discussion of the structural analysis of the fibrillar systems of the four representative ciliates, *Paramecium, Stentor, Euplotes* and *Vorticella,* is offered first and separate from the interpretations for these four ciliates, whose order is then, for convenience, reversed. This separate treatment was decided upon primarily for the sake of accuracy and clarity. Often in the literature the author's interpretations are so intermixed with his factual descriptions that it is sometimes very difficult to make certain just what he observed and undertook to describe.

EXAMPLES OF FIBRILLAR SYSTEMS

A. STRUCTURAL ANALYSIS

1. *Paramecium.*—This familiar representative of the holotrichs has doubtless been more generally used in both teaching and research laboratories than has any other of the numerous kinds of ciliates. Probably its apparent simplicity, more than its smaller size, tended to discourage a search for a fibrillar system, such as had been found in *Stentor* and other forms.

In 1905, however, Schuberg described for both *Paramecium* and *Frontonia* fibrillar differentiations which, running close under the pellicle, united the basal granules in the longitudinal rows of cilia. By means of a bichromate-osmic fixative and Loeffler's stain, not only was this relationship of fibril and basal granules clearly defined, but also, because of their staining properties, they could be well differentiated from the hexagonal, or rhomboidal, pattern of the pellicle, as was well illustrated in Schuberg's several figures.

In 1925, J. von Gelei described in *Paramecium nephridiatum* a peripheral network of fibrils which was not connected with the familar polygonal pattern observable in the living organism. For fixation, he used Apathy's sublimate-osmium and stained with toluidin blue.

In the following year Klein (1926a) reported the results of his studies on a peripheral fibrillar complex in certain ciliates by means of a new silver-nitrate technique which involved no previous fixation. The method, thus employed, is now well known as the "dry method," in contrast to von Gelei's (1932a) "wet method," and the resulting silver-impregnated fibrillar complex is quite commonly referred to as the "silverline system."

The several subsequent publications of these two authors on the fibrillar system of *Paramecium*, using especially the silver-nitrate technique but also other methods, admit of useful comparisons for this brief review, so that their results will now be considered together.

In most of these various articles, the author's account of the structures that were clearly observed is at times so involved with his avowed interpretations of their functions that it has been found difficult to sift out the essential data for which this review is intended.

Their structural analyses of the fibrillar system of *Paramecium* have, nevertheless, several important points in common which may now be fairly, and as simply as possible, presented. On the basis of these common points, certain discrepancies will then be indicated.

To this end, it will be convenient to recall the findings of Schuberg (1905). He observed (1) a differentially stainable *pellicular pattern,* which was hexagonal over the body and rhomboidal on each side of the mouth, and *below* this (2) *a longitudinal fibril* connecting (3) the *basal granules* in each row of cilia—each such granule appearing below the center of each pellicular polygon (Fig. 71).

In outline, the descriptions of both von Gelei and Klein present this same general picture, which may now serve to simplify a brief comparison of their essential findings. In their later papers, both of these authors agree that the fibrillar system of *Paramecium* is entirely *subpellicular.* Bearing this in mind, we may note that:

1. Schuberg's pellicular pattern corresponds *in general outline* to von Gelei's "Stützgitter System" and to Klein's "Indirekt verbindung System" ("Meridiaan II. Ordnung").

2. Both the Stützgitter System and the Indirekt verbindung System, lying *under* the pellicle, comprise each: (1) a longitudinal fibril, between

the rows of cilia; (2) cross   fibrils, connecting adjacent longitudinal fibrils; (3) opening for a trichocyst midway on the cross fibril; and (4) a suture line at the anterior and the posterior poles. The resulting lattice, therefore, has its counterpart in Schuberg's pattern.

3. Beneath von Gelei's Stützgitter System lies his "Neuronem System."

4. Beneath Klein's Indirekt verbindung System lies his "Direkt verbindung System."

5. Both Neuronem System and Direkt verbindung System comprise

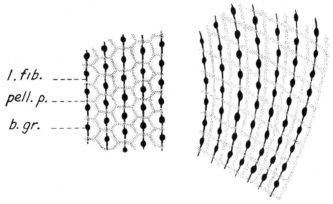

Figure 71. Pellicular Pattern and Longitudinal Fibrils Connecting Basal Granules in *Paramecium*. (Modified from Schuberg, 1905.)
b. gr.—basal granule   l. fib.—longitudinal fibril   pell. p.—pellicular pattern

each: (1) an *interciliary fibril,* connecting the "Basalapparaten" in each longitudinal row of cilia; (2) cross fibrils, connecting the interciliary fibrils; and (3) *Relationskörner,* which include the *Basalapparat* and the "Trichocystenkorn."

The *interciliary fibril* and the *Basalapparaten,* therefore, find their counterpart essentially in Schuberg's *longitudinal fibril* and its connected *basal granules.*

For both von Gelei and Klein, accordingly, it is evident that the fibrillar system of *Paramecium* includes two subsystems, or fibrillar complexes. For present convenience and especially for later discussion, I shall refer to these as von Gelei's outer fibrillar complex and inner fibrillar complex and Klein's outer fibrillar complex and inner fibrillar complex respectively.

TABLE 1: PARAMECIUM

Von Gelei

I. Stützgitter System

A "statisch-mechanisch" network of fibrils directly under the pellicle; ectoplasmic differentiation
1. Langsfasern—between rows of cilia—1 strand
2. Querbalken—between longit. fibrils, separating 1 ciliary insertion—1 strand
   a. With opening for Trichocyst
   b. With Gitterkorn at each end
3. Nahtlinie
   a. Anterior
   b. Posterior

II. Neuroneme System

Network of fibrils not connected with I
1. Interciliarfaser—with trichocyst **gran.**
   a. Praciliares } but continuous
   b. Postciliares } throughout
2. Querverbindungen
3. Vorderpol Naht—commissural system of **cross**-fibrils between Langsneuronemen
4. Hinterpol Naht Geflecht—of 2 or 3 fibr.
5. Relationskörner—divide during fission or arise anew on neuroneme
   a. Basalapparat
      (1) Basalkorn
      (2) Basal ring um Basalkorn Anlage
      (3) Nebenkorn
   b. Trichocystenkorn on interciliarneuroneme
6. Neuroplasma surrounding Basalapparat and Interciliarfaser

III. Intraciliare Gittersystem (of G. von Gelei)

An ectoplasmic complex of fibrils located beneath and at the same level with basal granule; fibres of different grades of thickness; in continuity at crossings; contractile in funtion (?)
1. Langsfibrillen
2. Querfibrillen
3. Diagonale Fibrillen

IV. Intraplasmatic fibrillar system (Rees, 1922)

Klein

I. Meridian II. Ordnung (indirekt verbind. System)

A nonstatic network under the pellicle; derived from Meridian I.1Ordnung
1. Langfaser—between rows of cilia; more than 1 strand
2. Querbalken—between longit. fibrils; more than 1 strand
   a. With opening for trichocyst
   b. Rudimentare protrichocystenkorn
3. Nahtlinie
   a. Vordere Pol—Circumpolar Linie
   b. Hintere (direkt u. indirekt fibrils merge)

II. Meridian I. Ordnung (direkt verbind. System)

Fibrils in continuity with Mer. II. Ord.
1. Interciliarfaser, with
2. Zircularfibrille (derived from 1. by splitting)
3. Querverbindungen
4. Elementarefibrille
5. Relationskörner—arise from silverline at junction of fibrils (Stosspunkten)
   a. Basalapparat (Dreiergruppen)
      (1) Basalkorn } Within
      (2) Two Nebenkörner} Zircularf.
   b. Trichocystenkorn—a granule at outer end of Trichocyst—in I, 2 (above) (Zwischen Stift u. Korp.)

III. Unkown

The several components of these fibrillar complexes, which were listed above, constitute the main structural features shared by Klein and von Gelei in their various accounts of the fibrillar system of *Paramecium*. But certain discrepancies appear in their descriptions of these and especially of some other components.

Space does not permit a discussion of all of these discrepancies, but most of them will be found listed in Table 1, which comparatively summarizes all the structural components ascribed to *Paramecium's* fibrillar system by von Gelei and by Klein in their various publications. Some of

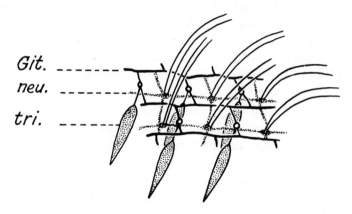

Git.

neu.

tri.

Figure 72. Diagram of *Gitter* (lattice) with attached trichocysts and of the neuronemes connecting bases of the cilia. (Von Gelei, 1925.)
Git.—Gitter   neu.—neuroneme   tri.—trichocyst

the discrepancies may very possibly be attributable to species differences, since von Gelei studied *P. nephridiatum* and *P. caudatum,* while Klein's descriptions are of *P. aurelia.*

A few discrepancies should, however, here be pointed out. The first and most important of these has to do with any structural integration between the two fibrillar complexes. Klein especially emphasizes the fact that his outer fibrillar complex and inner fibrillar complex are a *continuum* through interfibrillar connectives. Von Gelei, on the contrary, denies that any connection exists and so states that his outer and inner fibrillar complexes are only *contiguous* (Fig. 72).

A further discrepancy concerns the basal apparatus of the cilium. According to von Gelei (1932a), this apparatus consists of (1) a basal ring surrounding (2) the basal granule of the cilium, and (3) the "Nebenkorn" occurring at the right of the junction of the basal ring and the interciliary fibril (p. 158). According to Klein (1931), the elements of the basal apparatus include (1) a ring (Zirkularfibrille) within which are usually three granules (Drierkörner). The central granule is the basal granule, and the other two are *Nebenkörner*. The discrepancy here is not only in the relative number of granules in the basal apparatus but in their relationship, since von Gelei regards his "Nebenkorn" as identical with Klein's basal granule, and accounts for Klein's third granule as being only a thickening of the basal ring.

Finally, mention may be made in this connection of an additional system of fibrils described by Gabor von Gelei (1937) in three species of *Paramecium,—P. caudatum, P. multimicronucleata,* and *P. trichinum.* This third fibrillar complex, in addition to the two noted above, was found at the level of and below the basal granules. Its fibrils spread throughout the entire body surface, including the vestibule, where it sends a thickened fibril into the cytopharynx between the membranelles.

The general pattern of this complex, made up of longitudinal and cross fibrils, resembled that of the outer fibrillar complex, the meshes of the former being smaller and more numerous, however, than those of the latter. Also, the course of the fibrils of this third complex were more irregular. Its longitudinal fibrils become fewer in the middle region of the body and at times are directed diagonally, even spirally. A splitting of fibrils was occasionally observed, as well as variations in their thickness. They may also anastomose and form a "Schaumgiter."

The author could discover no connection between this fibrillar complex and either of the other two.

Apparently the most recent detailed account of the fibrillar system of *P. caudatum,* including that of its cytostome, was made by Lund (1933), working in Kofoid's laboratory. After comparing the descriptions of earlier workers (Engelmann, 1880; Maupas, 1883; Schuberg, 1905; Rees, 1922; von Gelei, 1925-32; Klein, 1926-31; Jacobsen, 1931) with his own findings, Lund concluded that previous investigators had confused "parts of at least two and possibly three quite different aggregations of structures, namely, the pellicle, the trichocysts and the peripheral portion of the neuromotor system. In addition" they had failed "completely to demonstrate the great pharyngeal complex," which is an integral part of this system.

Lund was able to differentiate between these "different aggregations of structures" by means of the silverline technique (Klein's and von Gelei and Horváth's) on the one hand and, on the other, by the use of iron-haematoxylin and Mallory's stain.

The former method demonstrated von Gelei's "Stützgitter System" and Klein's "Indirekt verbindung System," i.e., their "outer fibrillar complex" noted above. It also revealed essentially their "inner fibrillar complex," also as noted above.

But, according to Lund, these are separate and distinct "aggregates." The "outer fibrillar complex" is not subpellicular, as both von Gelei and Klein maintain, but represents rather the sculptural polygonal pattern of the pellicle itself. A similar interpretation was made by Brown (1930).

The "inner fibrillar complex" of von Gelei and of Klein comprises the basal granules, their connecting longitudinal body fibrils, transverse fibrils connecting the longitudinal fibrils, and others which include the "radial fibrils." These last mentioned "originate as longitudinal fibrils in the cytopharynx and oesophagus, spread radially out from the oral opening over the body surface and terminate a short way from the cytostome."

This inner fibrillar complex may be clearly demonstrated by the silverline techniques, especially by the wet method. There is, however, a portion of the fibrillar complex within the cytopharynx and cytoesophagus which is not wholly demonstrable by these techniques. This was well differentiated by the iron-haematoxylin and Mallory's methods, and described as "seven major parts, namely: (1) the pharyngo-esophageal network, (2) the neuromotorium, (3) the penniculus, (4) the oesophageal process, (5) the paraesophageal fibrils, (6) the posterior neuromotor chain, and (7) postesophageal fibrils."

For the descriptive details of this very elaborate complex of fibrils and associated parts, obviously the original account must be read. It is evident, however, from this brief review of the results of these several workers on the fibrillar system of *Paramecium* that a number of discrepancies need to be cleared up and perhaps further structural analysis of this system made before we can hope to have a complete understanding of the parts that are, or are not, structurally integrated.

In recent publications Chatton and Lwoff (1935, 1936) have described a fibrillar complex in several ciliates, which has long been known (Chatton and Lwoff, 1936) but has not been clearly distinguished from Klein's silverline system. The fibrils are visible *in vivo* and may be clearly differentiated in preparations fixed in Bouin's or Champy's solution and stained in iron-haematoxylin.

Each fibril (*cinétodesme*) has connected, *always along its left side,* the basal granules (*cinétosomes*) of a longitudinal row of cilia. The fibrils, together with their adjoined basal granules (the so-called *infraciliature*), are each essentially an independent entity. They are never united by anastomosis or otherwise at either body pole, and so include no transverse or other fibrillar connectives throughout their course.

The fibrils of this *infraciliature* are entirely superficial and adhere to the pellicle as rectilinear (never sinuous) threads. Other granules, as well as the ciliary basal granules, appear likewise attached, and these represent successive stages of the multiplication of the basal granules. Fibrils and granules stain alike, but in some species the fibrils cannot be impregnated with silver by the usual techniques. After fixation (osmic acid, Da Fano, Champy) and covering with gelatin or gelose, the fibrils may show, upon silver-nitrate treatment, the basal granules connected to a sinuous thread which, with its various connectives, represents Klein's silverline "plexus." This plexus, according to Chatton and Lwoff (1935), is acid labile and cannot be stained.

The selective staining properties and relations of the infraciliature show that it is quite distinct from the silverline fibrils, and is comparable with the flagellar ridges of the Hypermastigidae, marking the place of formation and of the insertion of the cilia.

It would appear that Chatton and Lwoff's *infraciliature* may be identified with the longitudinal fibrils and basal granules of the inner fibrillar complex reviewed above. The left lateral attachments of the *cinétosomes* to the *cinétodesmes* is evidently a new finding.

2. *Stentor.*—In his search for organs that would account for the well-known contractile behavior of *Stentor,* Ehrenberg (1838) saw in its conspicuous longitudinal bands the seat of that contractility. This interpretation of its contractile mechanism was accepted by several later

investigators including Kölliker (1864) and, according to Neresheimer (1903), Haeckel's (1873) "Myophanen" should be so construed.

It was between these longitudinal stripes, within the clearer non-pigmented meridians (Bütschli's "Zwischenstreifen"), that Lieberkühn, in 1857, found a distinct contractile fibril coursing from the basal disc forward to the adoral zone. Greeff (1870) confirmed these findings and Engelmann (1875) made detailed studies on the refractive and contractile properties of the fibers which have come to be commonly referred to as myonemes.

Four authors may be cited, among many others, for the descriptive details of the fibrillar system of *Stentor:* Schuberg (1890), Johnson (1893), Neresheimer (1903), and Dierks (1926). These have been the main sources for the following brief review of this system.

Schuberg (1890) made several important observations on the arrangement of the myonemes of *Stentor coeruleus* and an analysis of the basal apparatus of the membranelles. He found that the course of the body myonemes, from the basal disc to the peristome border, was not constant. Instead, some showed bifurcations, with occasional re-branching. This branching of myonemes followed consistently a corresponding branching of the longitudinal rows of cilia and their adjacent, non-pigmented bands. Similar relations of bands, ciliary rows, and myonemes obtained also for the peristome field.

Schuberg further observed that the double row of cilia, comprising a membranelle, was seated in an ectoplasmic *basal platelet* ("Basalsaum"), itself bipartite, below which appeared a triangular *lamella* ("Basallamelle"). The inwardly directed apex of this triangle was continued as a fibril ("Endfädchen") which was, in turn, united to all other such *end fibrils* by a *basal fibril*. The latter then ran rather deep below and parallel to the entire series of membranelles.

Schuberg's account of Stentor's myonemes and his analysis of its membranelles have been generally confirmed, with the exception of the *basal fibril*. The latter was identified by Johnson (1893), Maier (1903), and Schröder (1906). But Neresheimer (1903) and Dierks (1926) are certain that, as such, it does not exist. It is worth noting that Schuberg's "Basalfibrille" has been widely cited in the literature.

Johnson's (1893) work is not concerned primarily with a structural analysis of *Stentor's* fibrillar system, but his observations were thorough and critical and, for the most part, they have remained valid. Reference now may be made to his search for the so-called myoneme canal, described by Bütschli and Schewiakoff (1889, p. 1297).

Beneath the "Zwischenstreifen" they found a fairly spacious fluid-filled canal which surrounded the myoneme throughout its course. Johnson looked in vain for this canal, finally deciding that he was "unable to find the least evidence of such a structure, either in optical or actual sections." The majority of authors—including Delage and Herouard (1896), Maier (1903), Neresheimer (1903), and especially Dierks (1926)—agree with Johnson that the canal does not exist except as an artefact. Schröder (1907), on the contrary, affirms its form to be oval or circular in cross-section, its shape and position varying with the degree of body constriction. Roskin (1918) and von Gelei (1929) claim also to have definitely identified it. The latter regards it as an "organic part" of the myoneme, "solid and elastic."

Neresheimer's chief contribution to the microanatomy of Stentor was his discovery of another complex of fibrils to which he gave the name "Neurophanen." These were associated contiguously with the myonemes, but coursed usually peripheral to them. In suitable preparations which had been differentiated with Mallory's triple stain, the myonemes were distinctly red, whereas the neurophanes were colored a dark violet. The Zwischenstreifen remained unstained. Schröder (1906) maintained that these fibrils were rather only a structural feature of the "Zwischenstreifen," which, according to his results, also with Mallory's stain, did show an intense purple color. More recently, however, von Gelei (1925) and Dierks (1926) have identified similar fibrils, as will be noted further on.

Neresheimer traced these neurophanes as coursing, each fibril directly over a myoneme, from the aboral plate to about halfway up the body. Here some ended in a knob and all others disappeared before reaching the peristome border. While the myonemes became shorter and thicker in fully contracted Stentors, the neurophanes appeared sinuous but otherwise remained unchanged. It is not clear, however, how Neresheimer could make sure of the changed or unchanged appearance of these fibrils, since he stated that he was not able to fix Stentor in an uncontracted state. Evidently the myonemes may be visible in the living organism (Bütschli, 1889; Johnson, 1893), which may have been

Neresheimer's means of observing a "three-fold increase" in the thickness of the contracted myonemes.

Perhaps the most complete structural analysis of the fibrillar differentiations of *S. coeruleus* is that by Dierks (1926). His work, which was carried out in Korschelt's laboratory, considerably revised and extended earlier accounts of the fibrillar system of this heterotrich.

He noted a gradual thickening of the myonemes from the peristome border down to the aboral pole, where the fibers do not end abruptly,

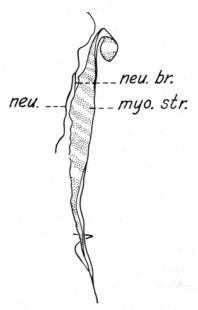

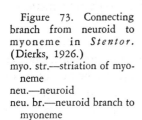

Figure 73. Connecting branch from neuroid to myoneme in *Stentor*. (Dierks, 1926.)
myo. str.—striation of myoneme
neu.—neuroid
neu. br.—neuroid branch to myoneme

as Johnson (1893) thought, but bend sharply inward and revert anteriorly to form a pencil-like bundle (see also Schröder, 1906a). This bundle soon becomes fimbriated, its component fibrils branch, and their tapering ends disappear in the cytoplasm "near the center of the contracted animal."

Dierks confirmed Neresheimer's (1903) findings of a second fibril coursing parallel and usually peripheral to the myoneme, both of which also stained differentially by Mallory's method. But the relationship of these two sorts of fibrils was found by Dierks to be evidently more intimate than Neresheimer had observed. In various sectioned and stained preparations, the smaller fibril, or "neuroid," gave off one or more branches (Fig. 73) to its adjacent myoneme, with which it apparently united (cf. von Gelei, 1929b).

He observed also the knob-like endings of these neuroids, as described by Neresheimer for his neurophanes, but Dierks apparently could account for such knobs as being merely the cross sections of fibrils that happened to be bent near the plane of section.

The cross striations of the myonemes, described by Bütschli and Schewiakoff (1889), were observed by Dierks in the living organism as well as in his preparations. Johnson (1893) had regarded these as artifacts due possibly to wrinkling of the myonemes, but the regularity of their recurrence and spatial relations seemed to preclude this. The myonemes were usually elliptical in cross section, with the longer axis of the ellipse directed toward the center of the body. This cross section revealed definitely an outer cortex (Plasmahülle) and a medulla (Plasmamark) (cf. Roskin, 1918).

Dierks' analysis of *Stentor's* membranelle apparatus differs in several points from most earlier descriptions. The membranelle platelet, supporting each membranelle, represented essentially the aggregate of basal granules of the component cilia. Continuing from these granules into the cytoplasm was a basal lamella, the outline of which was clearly rectangular and not triangular as Schuberg (1890) had claimed for his "Basallamelle." Dierks' rectangular lamella could appear as a triangle, whose apex might be directed either toward or away from the basal granules, depending upon their position when viewed. For these lamellae were as ribbons, each about three times as long as broad, and each alike was slightly twisted on its long axis. This, according to Dierks, accounted for the erroneous interpretation of Schuberg (1890), Schröder (1906), and others. Not only might the lamellae appear as triangles, but also their inwardly directed "apexes" might then seem to be continued as a fibril ("Endfädchen"). To account for the basal fibril, which Schuberg thought united all of the end fibrils ("Endfädchen"), Dierks observed that his basal lamellae overlapped in such a way that their ends could give the impression of a continuous fiber, comparable in appearance, direction, and extent to Schuberg's described "Basalfibrille."

3. *Euplotes.*—As a major group of ciliates, the hypotrichs probably mark the acme of highly differentiated motor organelles (undulating membranes, membranelles, and cirri) the related fibrillar system of

which may appear correspondingly specialized. Another structural differentiation, the pellicle, assumes in this connection a significant importance in maintaining the flattened bodily form so characteristic of the hypotrichs. The often remarkable rigidity of this pellicle has long been recognized.

Euplotes, the representative whose fibrillar system will now be reviewed, suitably illustrates these well-known characteristics of the hypotrichous ciliates.

Maupas (1883) was apparently the first to identify fibrillar differentiations in this genus. He described in *Euplotes patella* var. a fibril extending anteriorly from the basal plate of each of the five anal cirri. These five fibrils united into a single fiber, which continued anteriorly and disappeared near the bases of the adoral membranelles.

Maupas's findings were essentially confirmed by Prowazek (1903) in his brief account of protoplasmic reorganization in *E. harpa.* He further observed the "solide und fest" nature of fixed anal cirri fibrils, as indicated upon sectioning, when they might be pulled and bent thread-like by the microtome knife. Similar fibrils were seen to extend radially from the bases of the other cirri. Prowazek also described and figured still finer fibrillar lines ("Fibrillenzüge") going in parallel to the adoral membranelles. These finer fibrils have apparently not been identified as such by later workers.

Some years later Griffin (1910) gave a fairly detailed description of the fibrillar system which he discovered in *E. worcesteri.* From the base of each of its five anal cirri, he observed a fiber extending anteriorly. All five fibers converged toward the adoral membranelles, near which they disappeared "close to each other." Unlike similar fibers described by Maupas (1883) and by Prowazek (1903) for other species of *Euplotes,* these of *E. worcesteri* apparently did not unite to form a common strand and were not traceable to the membranelles. Several finer fibers were found associated with the bases of some of the other cirri. Their number and direction varied, however, and they had no connection with those of the anal cirri. Griffin suggested that all of these fibers might be comparable to myonemes, the number of which had become reduced with a reduction in rows of cilia, as postulated for the hypotrichs generally; but he noted also that some of the fibers may be directed even transverse to the hypothetical original ciliary rows.

Yocom (1918), working in Kofoid's laboratory, found and described in *E. patella* a fibrillar system much more extensive than that delineated in other *Euplotes* by Maupas, Prowazek, and Griffin, as noted above. In addition to the anal cirri fibers, such as they had found, Yocom discovered in *E. patella* "a *fiber* connecting the inner ends of the cytostomal membranelles" ("anterior cytostomal fiber"), and a *"motorium"* (after Sharp, 1914) which united the membranelle fiber with those from the anal cirri. A structural integration was traced, therefore, between the cytostomal membranelles and the anal cirri. *Similar fibers,* radiating from the bases of the other cirri, were also described, but no connection was found between them and the others above mentioned. In the "anterior lip" of this species, Yocom depicted a *fibrillar lattice-work* which was united by "short rodlike projections" to the membranelle fiber.

The intimate contiguity between this fibrillar system of *E. patella* and its motor organelles was clearly detailed by Yocom. Certain minor modifications and additions to his account were made by Taylor (1920), from studies especially of dissected and slowly disintegrating organisms. Following Sharp's (1914) terminology for a comparable fibrillar system which he had found and elaborately described in *Diplodinium ecaudatum,* Yocom designated this system in *E. patella* a "neuromotor apparatus."

The neuromotor apparatus discovered by Yocom is to be distinguished from an additional fibrillar system in this same species, which was carefully worked out by Turner (1933) by means of his modification of Klein's (1926) and von Gelei and Horváth's (1931) methods. His technical procedure is here worthy of note. After fixing the organisms in osmic acid vapor for about three seconds, and before the material was quite dry, Turner added two or three drops of 2-percent silver nitrate. Within four to eight minutes the nitrate was poured off and the slide placed in distilled water, barely covering the preparation. Over a white background, the slide was then exposed to the sun until the reduction of the nitrate had progressed as desired, according to occasional microscopic examinations. The preparation was then thoroughly washed in distilled water, dehydrated, and mounted. "The method gives strikingly clear-cut results." For this study, various other techniques were also employed, on both whole mounts and sections.

By these several methods Turner was able to disclose an "external fibrillar network," which included:

1. The "dorsal network," comprising (1) seven to nine longitudinal rows of granular rosettes, from each of which protruded a central bristle; (2) seven to nine longitudinal fibrils, uniting the bristles of the rosettes and designated *primary fibrils;* (3) longitudinal *secondary fibrils,* running midway between the primary fibrils; (4) commissural fibrils connecting, midway between the rosettes, the *primary fibril* and on either side its *secondary fibril,* "pulling the latter slightly out of line." The square meshes of this dorsal network, which appeared remarkably constant, averaged about four microns across.

2. The "collar," anterior to this dorsal network, comprising, (1) the inclined row of parallel *basal plates* of the adoral membranelles, the lower ends of which "rest" on the oral lip; (2) a *posterior membranelle fibril,* connecting the "upper" ends of the membranelle *basal plates;* (3) an anterior

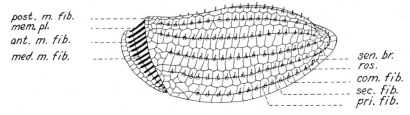

Figure 74. *Euplotes patella:* dorso-lateral view of external fibrillar system. (Turner, 1933.)

ant. m. fib.—anterior membranelle fibril	com. fib.—commissural fibril
med. m. f.—median membranelle fibril	mem. pl.—membranelle plate
post. m. fib.—posterior membranelle fibril	pri. fib.—primary fib.
ros.—rosette sec. fib.—secondary fibril	sen. br.—sensory bristle

membranelle fibril, connecting short commissures at the "lower" ends of these basal plates; (4) a *median membranelle fibril,* attached on the *basal plates* between the anterior and posterior membranelle fibrils.

These *anterior, median,* and *posterior membranelle fibrils* continue this same relationship with the *membranelle basal plates* throughout the course of the membranelles along the ventral "lapel" down to their ending in the cytopharynx (Fig. 74).

Of primary importance is Turner's observation that his *anterior membranelle fibril* is identical with Yocom's *anterior cytostomal fiber,* because this obviously integrates structurally the neuromotor apparatus described by Yocom and the external fibrillar system described by Turner.

3. The "ventral network," which is entirely comparable, in its general features, with the dorsal network, and includes also *primary* and *secondary longitudinal fibrils* and rosettes, with their central bristles.

Its pattern, although "constant and characteristic," is less regular and "reminds one of badly treated chicken wire."

Sectioned material showed the fibrils of both *dorsal* and *ventral networks* "to be immediately *under* the pellicle *and in contact with it.*"

The fibrils stained *intra vitam* were distinctly more delicate than those impregnated with the silver.

Turner confirmed Yocom's observations on the neuromotor apparatus, excepting the motorium. In the *E. patella* which he studied, he was unable to detect this cited organelle. Instead, the single fiber, formed by fusion of the five anal cirri fibrils, was traceable to the "collar," without a break, where it continued as the *anterior membranelle fibril* (Turner's designation) noted above.

4. *Vorticella.*—The vorticellids, by their size and quick reactions, caught the eye of the earliest microscopists, including, of course, Leeuwenhoek. The sudden contraction of the spiraling stalk along with the inversion and closure of the adoral membranelles naturally invited speculations on the kinds of mechanisms that might account for such reactions. Geza Entz (1893) cites Wrisberg (1765) as among the first to describe "mit recht treffenden Worden" this surprising behavior and to point out its elastic nature.

To Ehrenberg (1838), however, apparently should go the credit for the earliest detailed studies of the fibrous nature of the contractile stalk and the detection of longitudinal and circular fibers in the body of several vorticellids. He attributed to all of these fibers a contractile function and described in the stalk "muscle" cross striations comparable to those of the striated muscles of other animals.

This fibrillar complex in these peritrichs came to be a favorite object of investigation by many able workers, especially during the latter half of the past century: Dujardin (1841), Czermak (1853), Lachmann (1856), Lieberkühn (1857), Kühne (1859), Rouget (1861), Cohn (1862), Haeckel (1863), Metschnikoff (1863), Köllicker (1864), Greeff (1871), Everts (1873), Engelmann (1875), Wrzesniowski (1877), Forrest (1879), Maupas (1883), Brauer (1885), Bütschli (1889), Schewiakoff (1889), and Entz (1893). The literature for this period has been reviewed by Greeff (1871), Wrzesniowski (1877), and Bütschli (1889). Similar investigations on the vorticellids have been relatively meager during the present century, and the most detailed and careful analysis of the finer structure of their fibrillar system, so far as I have found, is that of G. Entz (1893) on "Die elastischen und contractilen Elemente der Vorticellen." In the brief review here presented of this system, I have followed chiefly this excellent account.

In this review of the fibrillar system of *Vorticella* and its relatives,

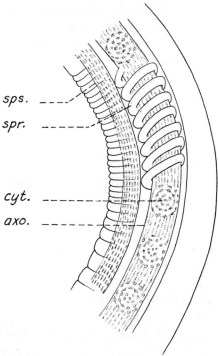

Figure 75. Stalk of *Zoothamnium arbuscula*. (Entz, 1893.)
axo.—Axonem   cyt.—Cytophane   spr.—Spironem   sps.—Spasmonem

it will be more convenient to consider first the fibrils of its contractile stalk, then those of its body and its peristome.

Previous to the critical investigations of Entz (1893), the *"Stielstrang"* in the contractile stalk of the vorticellids, which had been identified by Ehrenberg (1838) as the "Stielmuskel," was found by later analysts to be composed of two parts: (1) the "Stielmuskel," a cylindrical, or band-like, strongly refractive fiber, and (2) an adjacent, granular *"Protoplasmastrang,"* which accompanied the former throughout its course. Both were surrounded by a delicate membrane, the "Strangscheide," in the same manner that the whole *Stielstrang* is enclosed within the outside membrane, "Stielscheide," of the stalk.

The *Stielmuskel,* so most authors (e.g., Engelmann, 1875; Wrzesniowski, 1877) agreed, was made up of distinct fibrils running variously transverse or parallel to its long axis. For Bütschli (1889), however, this composition of the Stielmuskel represented rather (and probably more in line with his alveolar hypothesis) an attenuated meshwork. And Ehrenberg (1838), Leydig (1883), *et al.* could apparently see cross striations in the *Stielmuskel* of some vorticellids, comparable with those of metazoan muscle.

Entz (1893) described for the giant stalk of *Zoothamnium arbuscula,*

Figure 76. Spasmonem (cross section) in *Zoothamnium.* (Entz, 1893.)

and as typical for all the Contractilia, a *Stielstrang* comprising *three* well-defined fibers: a *"Spasmonem,"* a *"Spironem,"* and an *"Axonem"* (Fig. 75), all enclosed within the *Strangscheide.* He identified the *Spasmonem* with the *Stielmuskel* (noted above), and the *Spironem* and *Axonem* with the *Protoplasmastrang.*

In the smaller branches of *Z. arbuscula,* the *Spasmonem (Stielmuskel)* is a round fiber which, in the main stalk, becomes compressed by the adjacent *Spironem* so that a cross section of the former appears crescent-shaped or, since one edge of the crescent is swollen, rather like the form of a comma (Fig. 76). As shown by Engelmann (1875), it is birefringent and may frequently be tinted a steel-gray or appear faintly greenish or yellowish. It is stainable in carmine and disintegrates in alkali and in mineral salts—reactions which are more rapid in the younger stems, and which, according to Entz, would indicate that the *Spasmonem* is not cellulose; neither is it chitin nor keratin, although least unlike chitin.

This age difference shows itself also in the finer structure of the

Spasmonem. In the younger branches it is fairly homogeneous, but in the older stems and in the main trunk one may distinctly observe, in its hyaline ground substance, parallel longitudinal fibrils that gradually disappear toward the distal end. Nodal interruptions in these parallel fibrils apparently account for the cross striations cited by earlier workers. Outside these longitudinal fibrils is a single, or composite, spiraling fibril, and centrally placed along the *Spasmonem* are ovoid discs, each containing a central granule or "nucleus."

Entz's *Spironem* and *Axonem* constitute, as noted above, the *Proto-*

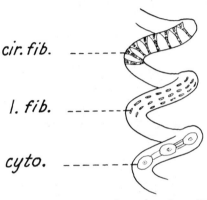

cir. fib.

l. fib.

cyto.

Figure 77. Three components of Spironem of *Zoothamnium.* (Entz, 1893.)
cir. fib.—circular fibril   cyto.—cytophanes   l. fib.—longitudinal fibrils

plasmastrang. This structural duality had been previously overlooked, perhaps because the *Spironem* is wound closely around the *Axonem.* Also, the spirals of the former, which are contiguous when the stalk is fully contracted, separate increasingly as the stalk is extended, so that the two fibers may easily appear as one in the completely extended stalk.

In its finer details, the *Spironem* (Fig. 77) shows beneath its investing membrane a spirally wound fibril, under which are several longitudinal fibrils. All these fibrils are birefringent and, as noted for those in the *Spasmonem,* nodes of less refringence give here also the effect of cross striations. Along the axis of the *Spironem* are oval bodies (*Cytophanes*) containing each a central granule (*Caryophane*). These *Cytophanes* are connected by a longitudinal fibril (Fig. 75) similar to a string of pearls.

In the *Axonem* (Fig. 75) longitudinal fibrils comparable to those in the *Spironem* also occur, but their course is apparently completely interrupted at regular intervals by *Cytophanes* which are relatively much larger than those found in the *Spironem*.

It seems to be generally agreed that the stalk of all the vorticellids, both the Contractilia and the Acontractilia, is a direct continuation of the body. Bearing this in mind, we may now briefly review the fibrillar

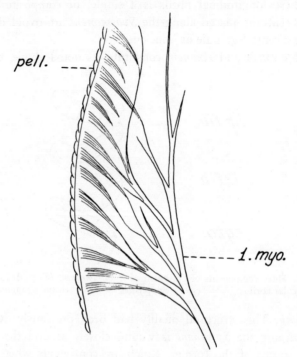

Figure 78. Pellicular structure, and branching of longitudinal myonemes in *Zootham-nium*. (Entz, 1893.)
lon. myo.—longitudinal myoneme   pell.—pellicle

system of the body proper and later note how the parts of this system are related to those in the stalk.

For convenience in description and with special reference to the genus *Vorticella*, we may regard the body as divisible into three fairly well-defined regions: (1) the *funnel*, lying between the stalk and the ciliary ring; (2) the *bell*, that part of the body above the ciliary ring; and (3) the *disk*, which includes the peristomal border, adoral zone, and cytostome.

The entire body, like the stalk, is covered by a pellicle which, according to Entz (1893), is not homogeneous but distinctly sculptured (Fig. 78) as if composed of "Stäbchen" that overlap, somewhat as tile on a roof.

The fibrillar system of the body lies immediately beneath the pellicle and comprises an outer and an inner complex of fibrils, or myonemes. Each such complex is in turn composed of (1) an outer circular layer, and (2) an inner longitudinal layer, making in all, then, four fibrillar layers.

Lachmann (1856) was first to describe the outermost circular layer.

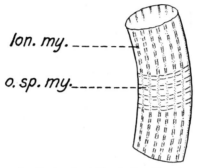

Figure 79. Myonemes of stalk sheath of *Vorticella*. (Entz, 1893.)
lon. my.—longitudinal myoneme   o. sp. my.—outer spiral myoneme

It was later recognized also by Stein (1867), but apparently overlooked by other investigators previous to Entz (1893). As a single fibril, it spirals directly beneath the pellicle and may be followed from the attachment of the stalk uninterruptedly to the center of the disc. Its spiral course accounts for the annular appearance of the pellicle. Entz (1893) thinks also that the birefringence of the pellicle may be due to this underlying fibril, since, as shown by Engelmann (1875), all the fibrils are refractive in polarized light.

The fibrils of the next layer, the longitudinal fibrils, lie immediately below layer (1), noted above, and likewise pass from the style's attachment to the center of the peristomal disc. On this disc they of course are radially arranged. This layer was found by Greeff (1871), but Bütschli (1889) questioned its existence.

It should further be noted also that this outer fibrillar complex is continued uninterruptedly into the protoplasmic lining of the style sheath (Fig. 79). This will be referred to again in later discussion.

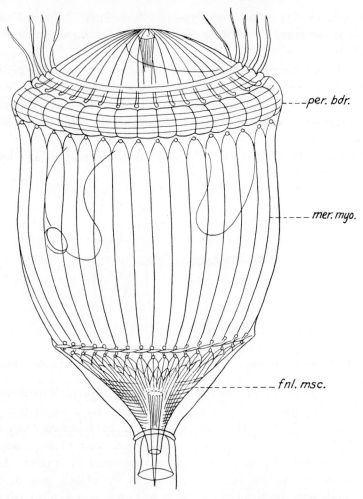

Figure 80. Arrangement of second complex of body fibrils in *Epistylis*. (Entz, 1893.)
fnl. msc.—funnel muscle   mer. myo.—meridional myoneme
per. bdr.—peristome border

Of the second complex of fibrils, its outer component, unlike the outer
single fibril of the first complex, behaves rather as a large fiber that may
become split into several fibrils. This set of fibrils courses spirally around
only the lower half of the *funnel* portion of the body (Fig. 80). Toward
the ciliary ring it disappears and, according to Entz (1893), reappears
actually to form this ciliary ring "out of a number of fine fibrils (Myone-

men)." This would account for the interlaced appearance of the ciliary ring. Above the ring, in the *bell* part of the body, this set of fibrils could not be detected except in the peristome border, where it forms a fairly strong spiral, narrowly wound. Its occurrence there was described by earlier authors as a "sphincter ring" of the peristome border (Fig. 80).

On the *disk,* this fiber appears as a fairly thick strand, one end of which originates in the center of the disk and, after following at their base the several turns of the adoral membranelles along the outer margin of the disk and through the mouth into the gullet, it continues down the inner surface of the gullet the whole length. Along the gullet wall Entz thought that, in some preparations, he could observe this fiber branching into many increasingly fine fibrils.

The inner, longitudinal layer of the second fibrillar complex is a direct continuation of the *Spasmonem* (Stielmuskel) of the contractile stalk. Upon entering the body this *Spasmonem* breaks up into its component fibrils, which diverge and so form the *"funnel muscle"* (Fig. 80) of earlier authors. Higher in the funnel, these fibrils may rebranch, then anastomose into a network. At the ciliary ring, the fibrils curve round it as crescent-like spans, which are somewhat larger and apparently more dense, and continue directly up the wall of the *bell* to the peristome border. These meridional myonemes (Fig. 80) may end singly on this border or, as Engelmann (1875) noted, they may bifurcate and unite, each of a pair with its adjacent neighbor, to form "arcades" just below the peristome border. The fibrils then span this projected border, much as they curved over the ciliary ring, and thereupon proceed as radii to the center of the *disc.* Toward this center, these fibrils may sometimes branch and anastomose to form varied and striking patterns.

Finally, it may be mentioned that, in addition to the four layers of fibrils noted above, Entz (1893) describes a fairly thick strand of other fibrils which originate from the middle of the *disc* and project downward through the endoplasm and finally disappear in the region of the cytostome.

B. INTERPRETATION

The essential structural components of the fibrillar systems of several representative ciliates have been briefly reviewed in the foregoing paragraphs. It remains now to consider the several functions that have been ascribed in the literature to each of these fibrillar mechanisms, as illustrative for similar systems that have been studied thus far in a large number and variety of other ciliates.

It will be recalled that Ehrenberg (1838) saw in the "stalk muscle" of *Vorticella* cross striations which he regarded as comparable to those of the striated muscles of other animals. Also, he detected both longitudinal and circular fibers in the body of several other vorticellids. Quite in line with his search for such comparisons, he attributed to all of these fibrils a contractile function. For him they were literally miniature muscles which had their structural and functional counterparts in the muscles of macroscopic organisms (Maupas, 1883). Evidently, Ehrenberg's comparison between microörganisms and macroörganisms led him to look for a one-to-one correspondence between microörgans and macroörgans.

It will be recalled that what Ehrenberg regarded as the stalk muscle (Stielmuskel) was analyzed into several components by later investigators. These, according to Entz (1893), included: a cross-striated, rod or band-like *Spasmonem,* and another rod-like strand, the *Axonem,* around which was coiled the *Spironem.* Each of these was, in turn, analyzable into longitudinal fibrils, whose properties apparently accounted for their birefringence and cross striations. The *Spasmonem* and *Spironem* showed also, spiraling within their own spiral strand just outside its longitudinal fibrils, a fine fibrillar coil coursing throughout their length. Also, both *Spironem* and *Axonem* had, centrally located, the "cytophanes" which, with their longitudinal interconnections, appeared like a string of pearls. Thus Ehrenberg's "Stielmuskel" turns out to be a highly complex mechanism which is structurally quite different from the "muscles of other animals" to which he had compared it.

Finally, mention was made also of the circular and longitudinal myonemes of the protoplasmic lining of the style sheath, which are continuations of the *outer* fibrillar complex of the body myonemes.

Taking into account this assemblage of highly differentiated components within the stalk of the Contractilia, together with a few rather inconclusive experimental results, various interpretations were advanced to account for both the contraction and the extension of this stalk. These may be summarized as follows:

1. The majority of investigators agree that the quick spiraling con-

traction of the stalk is referable to contractile-properties of some one or more of its components. Dujardin (1841) was apparently the only one to regard the stalk sheath as the seat of that contractility. Ehrenberg's *Stielmuskel* was for him the contractile component, but also the only component that he had recognized within the stalk sheath. Of the three fibers into which this *Stielmuskel* was later analyzed, viz., the *Spasmonem* and *Spironem* and *Axonem,* Kühne (1859) considered the latter two, or *Protoplasmastrang,* as the contractile organelle. This he likened to the sarcoglia of metazoan muscle. Other proponents of this contractile theory (Czermak, 1853; Engelmann, 1875; Wrzesniowski, 1877; and others) regarded the *Spasmonem* as the contractile fiber which, for later authors, represented a bundle of fibrils or myonemes that continued without interruption into the body of the vorticellid.

2. Apparently there is general agreement among all the investigators that the mechanism of *stalk extension* inheres in the elastic properties of the pellicle (Bütschli, 1889; Entz, 1893). Only Kühne (1859), who alone regarded the *Spironem* and *Axonem* as the contractile organelles, attributed elasticity to both the pellicle and the *Spasmonem,* thus to account for the stalk's extension.

3. While the majority, as noted, support the contractile theory of stalk contraction, there is a minority group who maintained that the vorticellid stalk, in its quick spiraling retraction, actually does not contract as such but instead recoils somewhat as a coiled spring.

A chief proponent of this elastic theory is Entz (1893), who sees the *Spasmonem* not as a contractile fiber but, like Kühne, as an elastic fiber. This organelle is, according to Entz, primarily responsible for the sudden recoil of an extended stalk. He likens this elastic fiber to a curly hair which when stretched and then released will resume its spiral form.

The opposing force, tending to "stretch" this normally spiral *Spasmonem,* is inherent in the elastic pellicle. Thus pellicle and *Spasmonem* constitute a pair of "antagonistic elements."

Associated with this pair is another pair of antagonistic elements, viz., (1) the longitudinal myonemes in (a) the stalk sheath and (b) the *Spironem;* and (2) the circular myoneme in (a) the stalk sheath and (b) the *Spironem.* These opposing pairs of fibrils are relatively weak, but of a strength sufficient to determine by their antagonistic contractions whether the stalk "contracts" with the recoil of the elastic

Spasmonem or extends by means of the elastic pellicle. Accordingly the longitudinal fibrils of the stalk sheath and the *Spironem* tend to reinforce the elasticity of the *Spasmonem,* and the spiral myoneme of both sheath and *Spironem,* upon relaxation of the longitudinal pair of myonemes, tend to reinforce the elasticity of the pellicle.

Entz's rather elaborate thesis at least makes evident the elaborate differentiation of the vorticellid's stalk, and so provides one working hypothesis which might be experimentally tested. With modern techniques, this should not prove very difficult. And, in the writer's opinion, the vorticellids offer extraordinary possibilities for some clear-cut and fruitful experiments which should help toward a better understanding of their fibrillar mechanisms.

It seems to be more generally agreed that the fibrillar complexes of the *funnel, bell,* and *disc* are primarily contractile. If so, and if the *Spasmonem* of the stalk is, as Entz and others claim, continued without interruption into the funnel and bell as the longitudinal fibrils of the "inner complex," then it should follow that if these fibrils of the inner complex are contractile, one should expect the *Spasmonem* also to be contractile. Entz (1893) apparently does not discuss this point.

Finally, it may be noted that Entz's thesis attributes no specific function to the *Axonem.* Both he (1893) and Stein (1854) suggest that it might function as a neuroneme.

Observations on the fibrillar system of *Stentor* have provided a basis for some interpretations which appear more plausible than those cited above for the vorticellids. These interpretations have to do exclusively with the myonemes. The other fibrils are poorly understood.

Before discussing these, however, the essential components of this system should be recalled to mind. As was pointed out, and as will doubtless be familiar to the reader, the myonemes of *Stentor* are band-like fibers which course beneath and slightly lateral to the longitudinal rows of body cilia and the curved ciliary rows on the oral disc.

Associated with these band-like myonemes, but lying peripheral to and mostly contiguous to them, are other fibrils which Neresheimer (1903) described as "neurophanes" and which Dierks (1926) recently further described and named "neuroids," since he could not fully identify these with Neresheimer's neuronemes.

Dierks added one very significant observation, which, if verifiable, would definitely suggest a conductive function for his neuroids. He saw, in several preparations, that the smaller fibril gave off one or more branches to its adjacent myoneme. As will be noted further on, this claim has recently been challenged by von Gelei (1929b).

Another fibril, found and described by Schuberg (1890), connected the basal lamellae of the entire series of membranelles. This basal fibril was noted also by Johnson (1893), Maier (1903), and Schröder (1906), but Neresheimer and Dierks maintain that such does not exist. Dierks (1926) does describe a fibril coursing along the platelets of the membranelles, which he thinks both Schuberg (1890) and Meyer (1920) may have seen but misinterpreted.

The evidence supporting the interpretation that the myonemes are contractile organelles has been supplied from many sources, beginning with their discoverer, Lieberkühn (1857). Ehrenberg (1838) claimed to have seen in a living *Stentor* that in the extended state the myonemes were serpentine, while with the contraction of the body they became shortened and straight. But since he misidentified the pigmented meridians as myonemes, the significance of his observation is uncertain.

Probably the most significant evidence for the contractile nature of *Stentor's* myonemes was provided by Johnson (1893) and verified by Dierks (1926). The former investigator compressed the living organism beneath the cover slip and could then observe the myonemes "alternately to extend and contract," concluding that "no one who has once observed them under these conditions can doubt that they are responsible for the contractions of the animals." Bütschli (1889) apparently made similar observations on living *Stentor* upon applying an electric stimulus. Merton (1932) has confirmed these former evidences of the contractile properties of *Stentor's* myonemes.

Dierks (1926) states further that "these fibrils are definitely shorter and thicker in the contracted animal and longer and shorter" when the organism is extended. He also noted, however, that this behavior was lacking in the myonemes toward the posterior end of the body, and supposed that here the body protoplasm contracted more or less independently, suggesting a progressive differentiation in the myonemes anteriorly along their course.

The assigned contractile function of the myonemes of *Stentor* appears to be agreed upon without exception. Schröder (1906) added to the knowledge of their contractile function probably also that of conducting. But for the other fibrils of this ciliate, evidence regarding their actual function or functions is mostly wanting. Interpretations based on morphological evidence concern chiefly the relations of these fibrils (neurophanes or neuroids) to other organelles.

Neresheimer's observation on the parallel course of the neurophanes with the myonemes and the differential stainability of these two kinds of fibrils, suggested for the neurophanes a conductive function. By means of a drop-weight apparatus (Fallmachine) he tested the responses of *Stentor* to a variety of narcotics: morphine, strychnine, atropine, caffein, and so forth, and to other chemicals which are known to affect the nerves of higher animals. On the basis of an assumed selective action of these reagents, he derived evidence which seemed to favor assigning a nervous function to the neurophanes. The technique for these tests was ingeniously devised and might prove useful also for others. But owing to lack of adequate controls and because of other possible interpretations, his results do not seem very convincing.

Schröder (1906) considered Neresheimer's interpretation of the neurophanes invalid. He believed that Neresheimer had mistaken a structural feature of the "mid-stripes" for his neurophanes, as indicated upon comparing the latter's Figures 7 and 8 with Schröder's Figures 1-5.

From Dierks' more extensive and detailed studies, however, there can be no doubt about the identity of another fibril coursing with each of the myonemes. The occurrence of these was consistent and their relationship with the myonemes apparently significant. As formerly noted, Dierks found that in good preparations branches were given off from the neuroids to the myonemes which they evidently joined. He considered several possible functions which this connection between neuroids and myonemes might indicate, and finally, as a plausible hypothesis, suggested for them a conductive function.

Von Gelei (1926c) rejects this interpretation. From his own previous studies (1925) on *Stentor,* he concludes that Neresheimer's neurophanes and Dierks's neuroids are real and are identical structures. But, according to von Gelei, these neuroids are not fibers but bands composed of fibrils, which are quite wide, especially in the aboral region. They are not a structural feature of the pellicle (Schröder, 1906), but are *subpellicular* and *fused* to the pellicle. This was indicated in contracted *Stentors* by the body groove overlying the neuroid, as produced by the contraction of the myoneme under the neuroid. The attachment of the pellicle to the neuroid and the latter, in turn, to the myoneme would, upon contraction of the myoneme, cause an ingrooving of the pellicle.

Studies on the hypotrichous ciliates yielded apparently the first recorded example of fibrils directly associated with ciliary locomotor organelles. Engelmann (1880) traced such fibrils from the bases of the marginal cirri of *Stylonychia* "nach der Mittellinie des Leibes." He postulated for these fibrils a conductive function in transmitting impulses, as in nerves of higher animals, from the ventral region of this hypotrich to the cirri, the movements of which might thereby be regulated.

Engelmann's interpretation was formulated entirely by analogy and he offered no substantial evidence to support it. It seems to have been generally rejected by his contemporaries. Maupas, who three years later (1883) described similar fibrils in *E. patella* var., said, in a brief footnote (p. 622) concerning his own findings: "Quant a la signification physiologique de ces racines ciliaires, j'avoue ne pas la connaitre." And with reference to Engelmann's discovery of "fibrilles nerveuses" in *Stylonychia mytilus,* Maupas adds: "Je ne sais comment concilier des interpretations aussi divergents." Bütschli regarded Engelmann's interpretation as untenable and proposed, on equally meager evidence, a contractile function for these fibrils. Maier (1903), who attributed a contractile function to the basal fibril associated with the basal lamellae of membranelles in *Stentor,* concluded that the lateral cirri fibers in *Stylonychia* were "required" for the support of these cirri.

Prowazek (1903), on the other hand, suggested that the anal cirri fibers which he found in *E. harpa* might perform the dual functions of contractility and conductivity. He noted, however, that *E. harpa* could move each anal cirrus independently and could flex the tip of these cirri quite at will. He observed, moreover, that a detached anal cirrus might continue its contractions for a time. These important observations have apparently been overlooked by later investigators.

It will be recalled that Griffin (1910) compared the anal cirri fibrils of *E. worcesteri,* which he found and described, with myonemes whose number might have become reduced phylogenetically during reduction in the hypotrichs's rows of cilia. He pointed out a difficulty in this interpretation, viz., that some of the few fibrils of other cirri in this species were not aligned longitudinally, but might be even transverse to the longitudinal axis.

Griffin based his concept of contractility for these fibrils in E. *worcesteri* only partly, however, on this comparison with myonemes in other ciliates. He accepted Bütschli's interpretation of a contractile function for the marginal and anal cirri fibrils in *Stylonychia* as a much more reasonable view, and also noted that "Every detail of arrangement and structure indicates that the fibrils are, principally at least, contractile in function." He observed, also, that the fibrils were developed around the bases of the cirri in such a way as to assist in producing the ordinary motions. "As the anal cirri have only a single strong motion, a vigorous kick directed backward, each needs but a single strong fibril."

Evidently Griffin had not seen any reversal in the effective stroke of these anal cirri. In several other species of *Euplotes* this reversal is not uncommon in both swimming and creeping movements. Should that be the case in E. *worcesteri*, then a contractile function for the anal cirri fibers is scarcely conceivable. The effective function of a contractile fibril is obviously a pulling but not a pushing function. Moreover, Prowazek's account of the contractile behavior of the anal cirri in E. *harpa*, cited above, largely vitiates any claims for contractility in the cirri fibrils of this species, and similar behavior has been observed by the writer in the anal cirri of E. *patella*. It was also noted (Taylor, 1920) that cutting these fibrils apparently did not impair the effective stroke of the anal cirri, whether that stroke was directed backward or forward.

The latter evidences were thus cited against the assumption that these fibrils in E. *patella* were contractile. It was further shown by those experiments that cutting the anal cirri fibrils interrupted the coördinated movements between anal cirri and adoral membranelles. Also, severing the membranelle fiber likewise interrupted the coördinated movements of the membranelles on opposite sides of the incision. Incisions in other parts of the body did not impair the coördination of these organelles. It should be pointed out, however, that those incisions which did interrupt coördinated movement of organelles cut not only the anal cirri fibrils or the membranelle fiber, but also the peripheral fibrils which have since been described especially by Turner (1933). Whatever rôle, if any, these peripheral fibrils may have in E. *patella's* coördinated behavior was not demonstrated by the writer's (1920) experiments. Reinvestigation of this problem, especially on a more favorable form such as *Lichnophora* (Stevens, 1891), ought, therefore, to be undertaken in order to determine what relative rôles the so-called introplasmic fibrils and the peripheral fibrils each perform in the coördinated movements of the organelles with which such fibrils are demonstrably associated.

Several investigators (Bělař, 1921; Jacobson, 1931; Peschowsky, 1927) have maintained that all such cirri fibrils in the hypotrichous ciliates and the fibrillar systems in various other ciliates are primarily or exclusively supporting in function. Jacobson (1931), for example, studied by means of various techniques, including the silver-nitrate methods, the fibrillar systems of some twenty-seven ciliates. These comprised representatives of all the major groups and included among the hypotrichs *E. patella* and *E. charon*. She concluded from the results that no evidence was found in support of a conductive function for any of the fibrillar systems studied. It was pointed out that in the hypotrichs whose motor organelles are localized, fibrils are nevertheless present, as on the dorsal side where cilia are wanting. Reference on this point should be made to Turner's (1933) studies on *E. patella,* which showed, as previously mentioned, that longitudinal fibers connect the bases of the dorsal and ventral rows of bristles, whose function is not known. But they may have a function and if that function is, as has been suggested, sensory in nature, it is surely not inconceivable that the associated fibrils may facilitate its performance.

Another and more significant observation was made by Jacobson, viz., that in *Sciadostoma difficile,* where three ciliary rows surround the anterior body pole, no silverline connection exists between the basal granules. An impulse, therefore, originating at the anterior ciliary ring would need to pass nearly to the posterior body pole and back again in order to affect adjacent cilia. Since our assumptions should be, first of all, plausible, one would justifiably regard this morphological evidence contradictory to the thesis that these silverlines in *S. difficile* are conductive in function. It may be pointed out, however, that Chatton and Lwoff (1935) apparently demonstrated that the fibrils (cinétodesme) described for several holotrichous ciliates, which alone were connected with the basal granules, could not be stained by silver impregnation, whereas other adjacent fibrils (interpreted by them as silverline fibrils) could be silver impregnated but apparently were not in contact with the basal granules.

The thesis that all fibrils of all ciliates are only supportive in function is, of course, not tenable. One would at once except myonemes. But why limit the exceptions to myonemes? If, in the eons of time, protoplasmic fibrils have become differentiated so as to facilitate contractility in protistan organisms, who can deny them the capacity to have become differentiated also to facilitate conductivity or some other function in these unicellular forms of life? All our assumptions should be both plausible and reasonable assumptions, the validity of which may, in the last analysis, be demonstrated only by experiment.

In conformity with the less specialized differentiation of its motor organelles, the fibrillar system of *Paramecium* is also relatively less specialized, as compared with those of the other three representative ciliates reviewed in foregoing paragraphs.

From the accounts especially of Schuberg (1905), von Gelei (1925-31), and Lund (1933), it appears that two separate and distinct complexes have been described in the literature, which may be represented by the outer fibrillar complex of von Gelei and of Klein (1926-32), and by their inner fibrillar complex. For these latter authors, both complexes are *subpellicular.*

According to Schuberg (1905) and Lund (1933), however, the above-mentioned outer fibrillar complex is not subpellicular, but actually represents the polygonal pattern of the pellicle itself. Lund emphasizes the fact, therefore, that the essential fibrillar system of *Paramecium* is exclusively the complex which is associated with the basal granules of the entire motor mechanism, including the cilia of the mouth, cytopharynx, and cytoesophagus, as well as the body cilia. Lund's fibrillar system would essentially include, therefore, the inner fibrillar complex of von Gelei and of Klein.

The discrepancies just noted in the structural interpretations of *Paramecium's* fibrillar system are obviously crucial, since they go hand in hand with discrepancies in the functional interpretations of that system. This holds, of course, not only for the investigators cited above, but for various others also.

Referring now to these diverse functional interpretations, Schuberg (1905) suggested that the fibrils connecting the basal granules, such as he had described for *Paramecium* and *Frontonia,* might function in the well known metachronism of ciliary movement. He was not inclined, however, to compare these fibrils with Neresheimer's neurophanes nor with the neurofibrils of the Metazoa. Obviously, since it had not yet been proved that the neurofibrils were the conductive elements of nerves, then an analogy between these neurofibrils and the fibrils found in *Paramecium* would not add to our understanding of either the one or the other.

Mention may here be made of the system of fibrils described for *Paramecium* by Rees (1922). These have an arrangement and relationship quite different from any of the fibrils referred to above. According to Rees, all of the fibrils he observed connected the basal granules of the cilia of the body and cytopharynx with the motorium, located just anterior to the cytostome, by coursing through the cytoplasm in several graceful whorls.

From the results of his few experiments, Rees concluded that these fibrils were conductive in function. By severing with a microneedle the fibrils connecting the cytopharyngeal membranelles with the motorium, the coördinated movements of these membranelles was interrupted. Likewise, the coördinated movement of body cilia was interrupted when the neuromotor center was destroyed.

In view of the later descriptions of the more peripheral system of fibrils in *Paramecium,* Rees's experiments should be repeated. It may be noted also that Jacobson (1931) reinvestigated this fibrillar complex described by Rees, and concluded that he had observed not fibrils but internally discharged trichocysts, as effected by the killing agents used. Her figures of fixed *Paramecium* illustrating such trichocysts are comparable to some of Rees's figures, except for certain regularities in the "fibrillar whorls" depicted by Rees. One could suppose that the internally discharging trichocysts might tend to follow the course of these whorls of fibrils and so reproduce that course in fixed material, which, through differential staining, revealed the trichocysts but not the fibrils. Only further careful investigation can clarify this discrepancy.

The interpretations of von Gelei (1925-31) and of Klein (1926-32) agree in ascribing a conductive function to the inner fibrillar complex of *Paramecium,* but their views on the relations and functions of its outer fibrillar complex are not in accord. The basis for their interpretations is confined primarily to morphological evidence. This evidence is derived, however, not only from their studies on the fibrillar system of *Paramecium,* but also from comparative studies of a considerable number and variety of other ciliates.

According to von Gelei's interpretation, Klein's strongly birefringent "indirekt System" (Meridian II. Ordnung) is nothing more than the well known "pelliculäre Gittersystem" of the ciliates, which functions as a supporting network to maintain body form. As such, it has no differentiated structural connections with the less refractive *inner fibrillar complex* (von Gelei's neuroneme system), which alone, therefore, is the peripheral conductive mechanism. Von Gelei sees in the longitudinal and transverse fibrils of his neuroneme network a mechanism which structurally integrates the ciliary basal granules, the trichocysts, and even the contractile vacuole pore into a coördinated whole ("einer koordinierten Einheit").

This integrated mechanism coördinates automatically the metachronous effective strokes of successive cilia. The direction of the stroke of one cilium activates conditionally that of the next by way of the basal apparatus (basal granule, basal ring, and "Nebenkorn"), which represents a sensory organelle. For the coördinated activity of the organism as a whole, however, von Gelei recognizes in his peripheral neuroneme system the absence of a neuromotor center. Accordingly, he regards the "intraplasmatic" fibrillar complex described by Rees as a centralized mechanism which may complement the peripheral neuroneme complex, thus to provide reasonably a unified neuromotor system. Von Gelei (1929b) observed by means of his silver-osmium-formol method a "platte" beneath the basal apparatus, which, he thought, might serve as a sort of end plate connecting Rees's intraplasmic fibrils with the peripheral neuroneme complex.

As has been previously indicated, Klein's (1926-32) interpretation of his inner fibrillar complex (Meridian I. Ordnung) agrees essentially with von Gelei's concept of a conductive function for his peripheral neuroneme network. Among their minor differences, Klein would account for the commonly observed reversal in the effective stroke of cilia, not by means of Rees's intraplasmic complex as von Gelei supposed, but by way of a "primitive reflex arc." This reflex arc includes the axial filament of the cilium as the receptor, the basal apparatus as the "relator," and the protoplasmic sheath as the effector. The basal apparatus comprises the basal granule and two "Nebenkörner." The latter serve to spread the proximal part of the reflex arc and, with the basal granule, may function as a kind of commutator that regulates the direction and change in direction of the effective stroke of the cilium.

We may note that while such a mechanism might conceivably account for a reversal in the effective stroke of a given cilium, obviously it does not, as such, provide for the synchronous reversal of the many other cilia, which is an essential part of the problem of coördinated movement.

In this connection, reference may be made to a fairly recent paper by J. C. Hammond (1935), who would refer the phenomenon of synchronous and metachronous ciliary behavior to an anterior-posterior physiological dominance in the organization of the cell, as opposed to the concept of a neuromotor mechanism. This thesis might help to account for this more general coördinated behavior (Rees, 1921), but it would presuppose a reversal in anterior-posterior dominance of the physiological axis in order to explain the reversal of ciliary stroke so common in the swimming or creeping behavior of ciliates. Moreover, the well known localized reversal of a few or many of the cytostomal membranelles in the intake or ejection of solids would obviously require a similar presupposition.

The chief discrepancy between Klein's and von Gelei's interpretations of the fibrillar system of *Paramecium* concerns the structural and functional relations between their outer fibrillar complex and their inner fibrillar complex. Klein (1928, p. 203) regards these two fibrillar complexes as a continuum. His "Meridian II. Ordnung" is a derivative of his "Meridian I. Ordnung," as are also all cilia, basal granules, "Nebenkörner," trichocyst granules, and protrichocyst granules (secretory, or "Tektin" granules). In its re-genesis, observed during reorganization, many more fibrils (Profibrille) are formed than are retained, varying with the species and genus, and of those that persist some may unite to form composite fibrils (Bündelfibrillen), as occurs in *Paramecium* (Klein, 1932).

Klein's silverline system incorporates as a unit, therefore, both structurally and functionally, the outer and inner fibrillar complexes. In its re-genesis, it functions as a "form-building system," and some of its fibrils become more resistant and rigid by the addition of a secondary substance, the "fibrilläre Komponente" (1928, p. 255). By virtue of its integral relation with the basal apparatus of the cilia and with the trichocysts, he attributed to his silverline system also a specifically conductive, coördinative function.

Chatton and Lwoff's (1935-36) criticisms of Klein's interpretations bear mainly upon the structural relations of the fibril to which are adjoined, *always* on its *left side,* the basal granules. The former authors think Klein's silverline system is quite separate and distinct from their *infraciliature.* It appears to the writer, however, that this discrepancy might be completely resolved by identifying Chatton and Lwoff's *infraciliature* with the longitudinal fibrils and basal granules of Klein and von Gelei's inner fibrillar complex. Further critical investigation would be needed, of course, to establish such identity.

FIBRILLAR SYSTEMS OF OTHER CILIATES

A. HOLOTRICHA

Ancistruma (Kidder, 1933).—The fibrils in *Ancistruma mytili* and *A. isseli* are of three types—the *longitudinal fibrils* of the ciliary rows, the irregularly distributed *transverse fibrils,* and the *net complex* of fine fibrils in the peristomal region. The latter fibrils seem to connect directly with the basal bodies of the peristomal cilia, and they are coarser and less numerous in *A. isseli* than in *A. mytili.* The longitudinal fibrils in *A. mytili* are continuous around the posterior end, but in *A. isseli* they center about a posterior suture.

A number of fine fibrils connect the inner dorsal row of peristomal cilia to the outer dorsal row. These seem to be distinct from the *net complex* and probably are in the nature of concentrated transverse fibrils. In *A. isseli* such fibrils are absent, and instead the fibril of the outer ciliary row joins that of the inner dorsal row.

Fibrils resembling the interstrial fibrils of *Boveria* (Pickard, 1927) are sometimes seen in *A. mytili,* but these may represent a deep-lying network of the same type regularly seen in the peristomal region.

METHODS

Fixatives: Schaudinn, sublimate-acetic, Bouin's, Zenker's, Champy's.
Stains: Heidenhain's and Delafield's haematoxylin, crystal violet-sulphalizarinate (Benda's).

Klein's silver method, modified to include fixation with osmic acid fumes and impregnation period of one to three hours in 2-percent solution of silver nitrate.

Boveria teredinidi Nelson (Pickard, 1927).—Surrounding the cytostome is an *oral ring* which begins and ends in the *motorium.* From the *motorium* arise the anterior and the posterior *adoral fibrils,* which bound the adoral zone. The *posterior fibril* joins the *anterior fibril* distally, and the latter continues as the pharyngeal fibril. This enters the endoplasm in the region of the pharynx and spirals around the potential gullet. A fibril from a point on the *ring* opposite the *motorium* enters the *pharyngeal fibril* near the margin of the peristome. Within the peristomal field delicate fibrils connect the *anterior adoral fibril* with the *ring.*

The longitudinal lines of the body surface consist of contractile *myonemes* and basal granules of the ciliary rows. *Myonemes* arise at the anterior end directly from the posterior *adoral fiber,* or indirectly from fibrils of the posterior granular line. The myonemes pass posteriorly and in somewhat oblique parallel lines. They gradually converge in the posterior field. Basal granules of cilia rest on *myonemes, oral ring,* and the *anterior* and *posterior adoral fibrils* (except the free end of *anterior adoral fibril*). A *deep nerve net,* consisting of longitudinal interstrial and transverse fibrils, interconnect the area "between the myonemes and their basal granules."

METHODS

Fixatives: Schaudinn's (60° C.), Bouin, Zenker, formalin, osmic acid, Da Fano.

Stains: Delafield's iron haematoxylin, Mallory's triple, alum carmine, and Yabroff's silver-gold method.

Clearing: Xylol, oil of cedar; equal parts of bergamot, oil cedar, and phenol, and sometimes before imbedding, in synthetic oil of wintergreen.

Chlamydodon sp. nov. (MacDougall, 1928).—The structural analysis of this new species of *Chlamydodon* disclosed "a complex neuromotor apparatus, including a coördinating center, and systems of fibers connected with cilia, the mouth opening, the pharyngeal basket, and the 'railroad track.' "

The *motorium* was identified as a bilobed mass located just below the anterior end of the large pharyngeal basket. This mass and all the fibrils of the neuromotor system stained bright red by Mallory's method. When dislodged along with the basket by means of microdissection, the *motorium* was "a refringent body." In stained sections, however, it appeared granular.

"Fans of fibrils" join the *motorium* at both its anterior and posterior ends. Those at the anterior end mark the convergence of *longitudinal fibrils* (Fig. 81A) of the body cilia, of *ventral fibrils* from the posterior region of the body and of fibrils from the mouth. The *circular myonemes* of the mouth are traversed by many fine fibrils that continue on toward the "railroad track." A fibrillar fan from the posterior end of the *motorium* continues as the *dorsal fibers* to the "railroad track," which, in turn, show a very complex system of fibrils associated with its trichites.

The basal bodies of the cilia are connected not only by the *longitudinal fibrils* but also by *cross fibrils* (Fig. 81B).

By methods of microdissection it was shown that (1) after destruction of the *motorium* "there is a marked disturbance in the action of the cilia, in no way comparable to the disturbance of the cilia if other parts of the body are injured"; (2) the cilia may still exhibit a wave-like motion, but they do not reverse after the *motorium* is destroyed. This seemed to suggest a coördinating function for the fibrillar system of this *Chlamydodon,* whose inconspicuous motor organelles are not favorable for an experimental study of modifications of their coördinated activity.

METHODS

Fixatives: Schaudinn's, Bouin's, and strong Flemming's.
Stains: Iron-haematoxylin, Mallory's (after Zenker's or picromercuric fixation).
Microdissection.

Conchophthirus mytili De Morgan (Kidder, 1933).—The peripheral longitudinal fibrils linking the ciliary basal bodies originate from a *transverse fibril* in the anterior ventral region. These almost parallel rows of fibrils pass around the posterior end of the organism uninterrupted, continue over the anterior end, and again return to the *transverse fibril* of the ventral surface. Each basal body is furnished with a *ciliary rootlet* extending toward the endoplasm.

An elongate mass of homogeneous material, the *motorium,* follows the posterior line of the cytopharynx and continues into the endoplasm as a

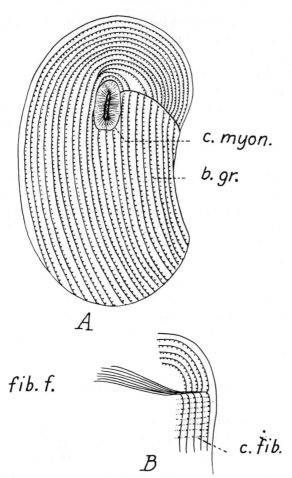

c. myon.

b. gr.

A

fib. f.

c. fib.

B

Figure 81. *Chlamydodon* sp. (MacDougall, 1928.)
A. c. myon.—circular myoneme and traversing fibrils   b. gr.—basal granule
B. fib. f.—fibrillar fan   c. fib.—cross fibril

strand which appears to be made up of fibrils. This *strand* follows the posterior margin of the gullet and frays out in the endoplasm near the left side of the organism. From the outer end of the *motorium* originate two fibrils: one joins the basal elements of the posterior brush, while the other is the fiber of the dorsal peristomal row of cilia. The latter proceeds anteriorly to join the *anterior fibril* and thus to the *transverse fibril* mentioned above.

The fused basal elements of the large *oral brush* of cilia form a deeply staining band or plate at right angles to the long axis of the *motorium*. From its anterior end two fibrils are given off which join the anterior and middle *oral brush plates*. *Fine fibrils* arise from the basal elements of the large oral brush line of the cytopharynx and connect with the *motorium* along its anterior side. Finally, a *ventral peristomal fiber* runs under the *basal plate* of the long ventral peristomal cilia, curves slightly to the outside, and ends just anterior to the oral brush.

The brushes of cilia about the mouth seem to connect directly with the *motorium* but not with the *peristomal fibers*. This lack of connection is noted in the movements of the cilia, the peripheral and peristomal cilia beating regularly, continuously and metachronously while the beating of the oral brushes is non-continuous and synchronous.

METHODS

Fixatives: Flemming's, Zenker's (for whole mounts), Bouin's, Zenker's, strong Flemming's (for sections).
Stains: Heidenhain's haematoxylin, Mallory's triple.

Conchophthirus (Kidder, 1934).—The "well integrated and closely interconnected neuromotor systems" of three species of *Conchophthirus* —*C. anodontae* Stein, *C. curtus* Engl., and *C. magna* sp. nov.—are quite comparable. The description is given of the external, internal, and peristomal fibrillar complexes, with reference especially to *C. magna*.

Most of the numerous, closely set ciliary rows originate in an *antero-ventral suture* and terminate in a *dorsal suture* near the posterior end. The *ventral suture* comprises two fibrils which are united at their ends and are connected irregularly by cross fibrils. This ventral suture is continued posteriorly as the *pre-oral connecting fiber,* from which arise two fibrils: (1) connecting the rows of basal granules of the dorsal lip, and (2) connecting the basal granules of the pharyngeal ciliary row.

The *pre-oral connecting fiber* itself becomes the *peristomal net fiber*. The latter gives off secondary fibrils to the peristomal field. On the left, these secondary fibrils are bounded by a longitudinal *inner net fiber* (Fig. 82) from which arise numerous fine fibrils that line the ventral side of the *peristomal basket*. On the floor of this *basket,* these fine fibrils join the *inner basket fiber,* which, in turn, gives off many branches that line the dorsal surface of the *basket*. These branches then unite with the fiber (noted above) that connects the basal granules of the dorsal lip.

At the anterior, inner end of the pharynx is a large *pharyngeal ring fiber* that fuses with the *inner basket fiber,* previously noted, to form the *fibrillar bundle.* This point of fusion is regarded as comparable to the motorium of *C. mytili.* From the *fibrillar bundle* a *gullet fiber* extends inward and courses throughout the floor of the gullet, finally fraying out at the posterior end.

The *inner basket fiber* is united posteriorly with the *posterior basket connecting fiber.* The latter, bending dorsally and to the right, comes to join the *post-oral connecting fiber.* Thus a direct connection is made between the peristomal region and the *dorsal suture.* From this *post-oral fiber,* numerous cross fibrils connect with adjacent ciliary rows.

"This neuromotor system is thought to be mainly conductive but some parts of it may possibly be contractile or even supportive."

METHODS

Fixatives: Klein's (1926), von Gelei-Horváth's (1931), strong Flemming's for four hours.

Stains: Heidenhain's haematoxylin, destained with hydrogen peroxide.

Dallasia frontata Stokes (Calkins and Bowling, 1929).—The most conspicuous part of the neuromotor system of *Dallasia frontata* was found to be the complicated apparatus of the mouth. This is composed of a tongue running through the buccal cavity, supported by bars which are anchored in long strands of dense material lying on the floor of the buccal cavity. There are two of these *longitudinal strands* and two series of bars from the tongue, one on each side. On the right and left sides are undulating membranes. A *ladder-like organ* originates anteriorly just below the membrane of the buccal cavity and at the right side of the mouth, and runs into the gullet.

A discoidal mass on the left side of the gullet is interpreted as the *motorium.* It is connected by fibers directly to the proximal end of the tongue and strands. Similar fibrils connect the outer and the inner margins of the *ladder-like structure* with the *motorium,* and these fibers appear to form the outer and inner margins of this organ. Minute granules are present at the ends of each bar, at the points where the bars join with the longitudinal fibers. Posteriorly *two fine fibers* run from the *motorium*

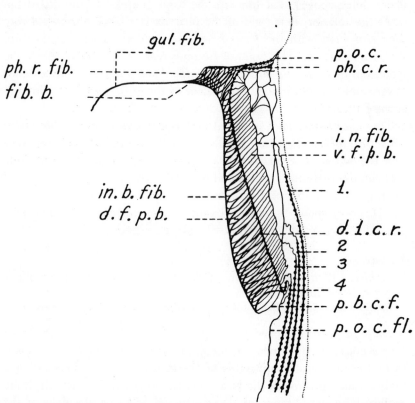

Figure 82. *Conchophthirus magna.* (Kidder, 1934.) Diagram of the fibrillar system of peristomal region.

d. f. p. b.—dorsal fibers of peristomal basket   d. l. c. r.—dorsal lip ciliary row
fib. b.—fibrillar bundle   gul. fib.—gullet fibril   i. n. fib.—inner net fibril
in. b. fib.—inner basket fibril   1, 2, 3, 4—rows of body cilia
p. b. c. f.—posterior basket connecting fibril   p. o. c. f.—pre-oral connecting fibril
p. o. c. f¹.—post-oral connecting fibril   ph. c. r.—pharyngeal ciliary row
ph. r. fib.—pharyngeal ring fibril   v. f. p. b.—ventral fibril of peristomal basket

deep into the endoplasm, where they are lost, one in the vicinity of the macronucleus, the other near the contractile vacuole.

Basal granules are described, but no connecting fibrils mentioned.

The function of this enigmatical organ is purely conjectural; possibly it has something to do with the opening and closing of the mouth. . . . Whatever the specific function may be there is little reason to doubt that it is intimately connected with the irritability of the mouth region.

METHOD (whole mounts and sections)
   Fixatives: Schaudinn's, made up in 95-percent ethyl alcohol, followed by
   prolonged treatment with turpentine.
   Stains: Iron haematoxylin; solution of acid fuchsin and methyl green.

Dileptus gigas (Visscher, 1927).—From a rod-like *motorium* found
near the base of the gullet, several sets of fibrils extend to different
parts of the body, as follows: (1) one set supplies the wall of the fun-
nel-shaped gullet, (2) two more distinct fibrils course anteriorly along
the proboscis, one on either side of a row of trichocysts; and (3) a set
of branching finer fibrils spreads posteriorly over the body, where they

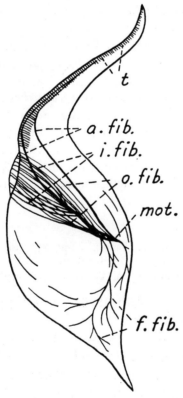

Figure 83. *Dileptus gigas.* Neuromotor apparatus (camera drawing). (Visscher,
1927.)

a. fib.—anterior fibril	f. fib.—fine fibrils	i. fib.—inner fibrils
mot.—motorium	o. fib.—outer fibrils	t.—trichocysts

are probably associated with the longitudinal rows of cilia running parallel to the contractile fibrils (Fig. 83).

METHODS

    Fixative: Schaudinn's.
    Stains: Iron-haematoxylin, acid borax carmine.

Entodiscus borealis (Powers, 1933).—The following distinct groups of fibrils and associated structures characterize the fibrillar system of *Entodiscus borealis*: (1) the *stomatostyle,* with its dorsal and ventral anterior horns and their *adoral fibrils;* the *labial fibrils;* the *pharyngeal fibrils;* and the *circumpharyngeal rods;* (2) the anterior fibrillar center, or *motorium,* with its anterior, posterior, and marginal strands; the posterior auxiliary fibrillar center with its associated strands; and (3) the peripheral *transverse commissural fibrils* interconnecting the basal granules. Closely associated with the *commissural fibrils* are the distal branching ends of most of the internal fibrils. The ventral peripheral layer is further complicated by the long (6-12 μ) *ciliary rootlets,* which extend into the endoplasm from the basal granules of the cilia.

    The pellicle of the ventral surface is thicker than elsewhere and is highly differentiated owing to *pellicular fibrils* accompanying each longitudinal ciliary row. These fibrils are conspicuous only at the anterior end. They are interpreted as either supporting or contractile in nature (Fig. 84).

METHODS

    Fixatives: Schaudinn's, Da Fano's, 25-percent osmic acid, Flemming's
      without acetic acid (the two latter fixatives for sections).
    Stains: Iron alum haematoxylin, Yabroff's silver method.

Entorhipidium echini (Lynch, 1929).—One fibrillar system of *Entorhipidium echini* consists of a *motorium* which is connected to a network of *peripheral fibrils* linking all the basal granules. The other set of fibrils is developed in the pellicle, chiefly on the ventral surface.

    The *motorium* is located to the left of the buccal cavity. It is composed of five heavy strands united into a rod-like body by *longitudinal fibrils* which appear to be continuous with the *neurofibrils* of the basal granules. These delicate *longitudinal fibrils* unite the basal granules of each peripheral row of cilia, and other fibrils encircle the body, uniting

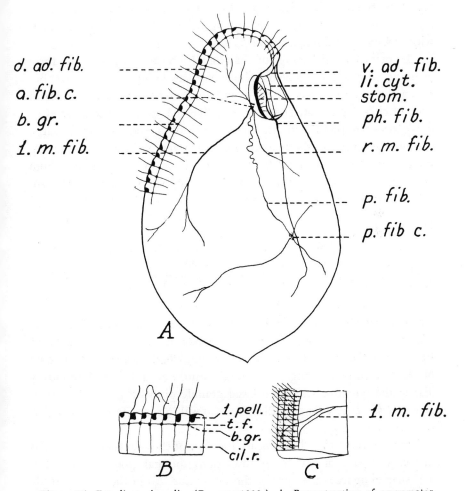

d. ad. fib.
a. fib. c.
b. gr.
l. m. fib.

v. ad. fib.
li. cyt.
stom.
ph. fib.
r. m. fib.

p. fib.
p. fib c.

A

l. pell.
t. f.
b. gr.
cil. r.

B

l. m. fib.

C

Figure 84. *Entodiscus borealis.* (Powers, 1933.) A. Reconstruction of neuromotor system, dorsal surface (modified from author). B and C. Relations of body cilia and their basal granules with fibrils of neuromotor system.

a. fib. c.—anterior fibrillar center
cil. r.—ciliary rootlet
l. m. fib.—left marginal fibril
li. cyt.—lips of cytostome
p. fib. c.—posterior fibrillar center
r. m. fib.—right marginal fibril
t. f.—transverse fibril

b. gr.—basal granule
d. ad. fib.—dorsal adoral fibril
l. pell.—longitudinal pellicular thickening
p. fib.—posterior fibril
ph. fib.—pharyngeal fibril
stom.—stomatostyle
v. ad. fib.—ventral adoral fibril

the basal granules with perfectly regular *transverse commissures.* In the region posterior to the frontal lobe, the *commissural fibrils* branch, forming numerous collaterals extending in various directions. These unite with *transverse* and *longitudinal fibrils.*

The pellicular thickenings of the frontal lobe are present in the form of heavy, deeply staining fibers which alternate with the ciliary rows. Between the anterior ends of these and the anterior ends of the dorsal rows of cilia is a delicate fretwork, or polygonal area. The boundaries of polygons correspond with the boundaries of a double row of large vacuoles. Posterior to the frontal lobe, the *heavy fibrils* become fine and lie so close to the *neurofibrils* that they are almost indistinguishable from them. Similar fibrils are evident on the dorsal surface of the organism only at the anterior end (Fig. 85).

METHODS

> *Fixatives:* Schaudinn's, Da Fano's, 2-percent osmic acid, Flemming's without acetic acid (the two latter for sections).
> *Stains:* Iron-haematoxylin, Yabroff's silver method.

Eupoterion pernix (MacLennan and Connell, 1931).—Most of the *longitudinal ciliary fibrils* of the body take their origin from a heavy bar, the *anterior connective fibril,* that lies dorsoventrally across the anterior tip, whence they extend to the posterior end, where they fuse. Additional fibrils arise in pairs along the pre-oral suture line and these fuse mostly also at the posterior end. Each basal granule gives off one or two *commissural fibrils* to adjacent ciliary rows, resulting in a fairly regular latticework (Fig. 86).

The *neuromotorium* lies beneath the wall of the cytostome. The fibrils of the four pairs of oral ciliary rows along the oral groove are fused in a V-shaped figure at the apex of the suture line. All of these end directly in the *motorium* or are closely connected to it by the *transverse fibril* or by the *longitudinal ciliary fibrils.* The *transverse fibral* lies across the end of the *neuromotorium;* its right end joins the *pharyngeal fibrils,* thus forming a *pharyngeal strand;* and its left end fuses with the two outermost peristomal ciliary fibrils and ends farther left in a connective fibril of adjacent outer rows of cilia. The two rows of cilia arising from the anterior ends of the outer peristomial rows (of the ordinary peripheral

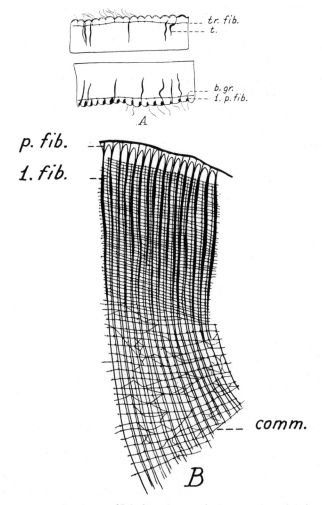

Figure 85. A. *Entorhipidium echini*. (Lynch, 1929.) Cross section, showing periplast of the anterior dorsal fibril. B. *Entorphidium echini*, tangential section of anterior ventral surface.

A. b. gr.—basal granule
   l. p. fib.—longitudinal pellicular fibril
   t.—trichocyst
   tr. fib.—transverse fibril

B. comm.—commissural neuro-fibril
   l. fib.—longitudinal fibril
   p. fib.—pellicular fibril

type) are connected to the rest of the body rows by *commissural fibrils*, thus uniting the peristomial cilia with the rest of the body cilia.

METHODS

Fixative: Schaudinn's (60° C.).
Stain: Heidenhain's iron haematoxylin.

Haptophrya michiganensis Woodhead (Bush 1934).—*Haptophrya michiganensis* has an integrated system of fibrils that center in a moto-

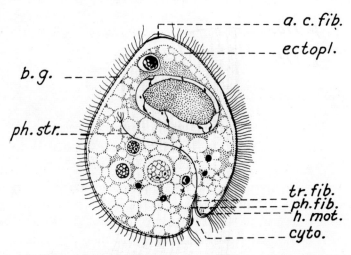

Figure 86. *Eupoterion pernix*, optical section. (MacLennan and Connell, 1931.)

a. c. fib.—anterior connective fibril	n. mot.—neuromotorium
b. gr.—basal granule	ph. fib.—pharyngeal fibril
cyto.—cytostome	ph. str.—pharyngeal strand
ectopl.—ectoplasm	tr. fib.—transverse fibril

rium. Within the sucker at the anterior end is a *fibrillar ring*, homologous with the esophageal ring of stomatous ciliates. The *motorium* is located in the center of this ring and is connected with it by *radial connectives*. Accessory bodies of the *motorium* are suspended from various points on the inner edge of the ring. (These are not evident during fission.) Numerous myonemes radiate from the inner edge of the *fibrillar ring* and extend to the opposite walls, posteriorly and laterally, dividing into fine fibrils at their outer ends. Equally spaced *peripheral myonemes* arise from the external edge of the *fibrillar ring*, adhere to the inner layer of the ectoplasm, and extend to the posterior end of the animal. These myonemes are "closely associated with the basal granules." Commissures connecting the basal granules form a close network over the entire body. Supporting fibrils from the nuclear membrane, the endoplasmic cone, and the contractile canal extend to the peripheral ectoplasm (Fig. 87).

The deeply staining mass was interpreted as a *motorium* because "(1) it is connected, directly or indirectly, with all parts of the fibrillar system; (2) it is near the anterior end of the ciliate; (3) if, with this mass, the sucker is removed, the animal loses its power of worm-like forward movement even though the cilia continue to beat; and (4) a toxic substance acts first upon the anterior part, particularly the sucker, whereupon the animal ceases its forward movement."

METHODS

Fixatives: Schaudinn's and Zenker's.
Stains: Delafield's and Heidenhain's haematoxylin; also Kolatschev's osmic impregnation, as outlined by Bowen; Yabroff method.

Ichthyophthirius (MacLennan, 1935).—The *longitudinal fibrils* connecting the basal granules beneath the ciliary rows of the body surface are linked together anteriorly and to some extent posteriorly by small centers, the anterior and the posterior fields. The centers are connected by a "suture fibril" which marks the ventral side. The *suture fibril* is interrupted by the oral region, thus dividing it into pre- and post-oral segments. Concentric *accessory suture fibrils* lie on the sides of the oral region and terminate anteriorly and posteriorly in the *suture fibrils*. The lip of the oral opening is bounded by (1) the outer peristomal fibrils, which are also linked to the *suture fibril*. (2) *Circular fibrils* line the walls of the oral cavity and *radial fibrils* intersect the two sets of fibrils at right angles. These transverse connections between the ciliary fibrils are present only in the oral region.

Two heavy *basophilic rods*, each attached to a heavy *esophageal fibril* are located near the esophageal plug. An *inner peristomal fibril* runs from this bilobed *neuromotorium* to the basal granules.

Ciliary rootlets are developed in the region of the *inner peristomal fibrils;* they are less well developed in the region of the *outer peristomal fibrils,* and not found in the region of ordinary body cilia. About 50-100 individual ciliary rootlets combine to form numerous esophageal strands,

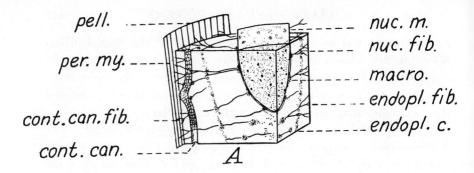

pell. nuc. m.
nuc. fib.
per. my. macro.
endopl. fib.
cont. can. fib. endopl. c.
cont. can.

A

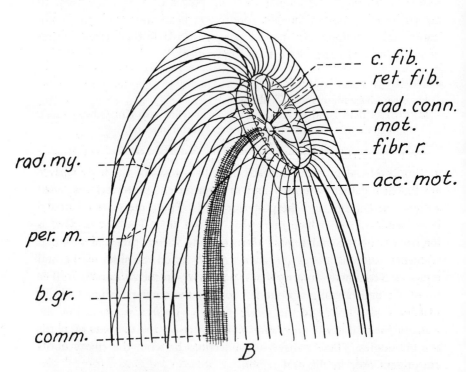

c. fib.
ret. fib.
rad. conn.
mot.
fibr. r.
rad. my. acc. mot.

per. m.

b. gr.

comm.

B

Figure 87. *Haptophrya michiganensis.* (Bush, 1934.) A. Diagrammatic section showing one-fourth of the animal body with part of the macronucleus. B. Diagram of anterior part of neuromotor system.

acc. mot.—accessory motorium
b. gr.—basal granule
comm.—commissures
c. fib.—coarse fibrils
cont. can.—contractile canal
cont. can. fib.—contractile canal fibrils
endopl. c.—endoplasmic cone
endopl. fib.—endoplasmic fibrils
fibr. r.—fibrillar ring

mot.—motorium
macro.—macronucleus
nuc. fib.—nuclear fibrils
nuc. m.—nuclear membrane
pell.—pellicle
per. my.—peripheral myoneme
rad. my.—radial myonemes
rad. conn.—radial connectives
ret. fib.—reticulate fibrils

each of which turns sharply and continues to the endoplasm, parallel to the main axis of the oral pit.

METHOD

Fixatives: Schaudinn's, Zenker's.
Stains: Heidenhain's iron-haematoxylin, Delafield's (for fibrils), Klein's (1926) (for ciliospores), Lund's wet silver method (for adults).

Lechriopyla mastax (Lynch, 1930).—The cilia of *Lechriopyla mastax* can be divided into three areas: (1) the cilia of the general surface of the body, (2) the cilia of the peristome, and (3) a transverse band of cilia known as the supraoral band. The basal granules of all these cilia are connected by delicate longitudinal fibrils without transverse connectives. Below the ciliary lines of the peristome and vestibule are vertical pellicular lamellae, which are fused to the furcula described below.

A crescent-shaped neuromotorium lies beneath the pellicle at the left end of the peristome. Its ends are continued as long fibers, the anterior and *posterior adoral fibers,* which form a complete (or nearly complete) ring about the peristome. From the ring arise the ciliary lines which extend over the surface of the body, or pass into the peristome and the pharyngeal involution. A variable number of fibrils arising from the outer border of the *motorium* extend through the cytoplasm for varying distances, to fuse with the pellicle just beneath a ciliary line. They may branch or anastomose, and they become more delicate distally.

The *furcula,* shaped much like a heavy tuning fork, partly surrounds the vestibule. The ends fuse with the walls of the pharynx, and delicate fibrils from the pharyngeal wall extend to the *furcula.* This organelle may be an additional element of the neuromotor system.

A long *pellicular fiber* extends from the left end of the internal opening of the cytopharynx to the middle of the posterior end of the organism and occasionally curves along the right side for varying distances. This fibril is not included in the author's description of the neuromotor system.

METHODS

Fixatives: Bouin's, Schaudinn's, Da Fano's cobalt nitrate-formalin.
Stains: Iron alum haematoxylin, carmines, cochineals, Yabroff's silver-nitrate method (no success), Klein's silver method.
Sections prepared in a variety of ways.

Ptychostomum chattoni Rossolimo (Studitsky, 1932).—The mouth of this parasitic ciliate is at its posterior end. At the anterior end is a horse-shoe-shaped sucker with a projecting rim for attachment.

The sucker (*Fixationsapparat*) is provided with a system of fibrils. The largest fibril, the *peripheral cord,* borders the sucker and gives off at its ends fine fibrils that extend into the cytoplasm. The sucker's disc has four sets of fibrils: (1) the deeper set comprises two groups of parallel fibrils that cross each other at a sharp angle as they traverse the disc, both groups coming to adhere to the *peripheral cord;* (2) the uppermost set, visible in the living organism, courses from right to left and from anterior to posterior; (3) a third set of fine fibrils are attached to the *peripheral cord* by their anterior ends; and (4) the fourth set, composed of sixteen or seventeen strands that run from right to left and from anterior to posterior, frays into four or five fine fibrils at the anterior end of each strand, to become attached to the *peripheral cord* and to other adjacent fibrils; the posterior ends of these strands become fimbriated also into fine fibrils. All apparently serve for support.

METHODS

> *Fixatives:* Schaudinn's, Carnoy's, Bouin's, Champy's and Benda's (for whole mounts), Altmann-Kull's (for sections).
> *Stains:* Heidenhain's haematoxylin, safranin, von Gelei's toluidin blue, Mallory's triple.

B. HETEROTRICHA

Balantidium coli Malmsten and *B. suis* sp. nov. (McDonald, 1922).— The *motorium* of *Balantidium suis* lies within the ectoplasm of the apical cone, close to the right ventral wall of the esophagus. A fibril encircling the esophagus arises and ends at the anterior end of the *motorium,* where the *adoral ciliary fiber* also arises. The *circumesophageal fibril* has irregular enlargements, from which fibers pass both posteriorly and anteriorly into the ectoplasmic mass of the anterior end. These fibrils appear to fade out in the ectoplasm. The *adoral fiber* connects the basal granules of the adoral cilia. The remainder of the fibrils are not directly connected with the *motorium.*

The basal granules of the peripheral longitudinal spiral rows of cilia are so closely set that it has been impossible to see a fibrillar connection. No transverse fibrils connecting the rows were observed. A *ciliary rootlet* extends from each *basal granule,* and a *second small granule* is found at the junction of the ectoplasmic and endoplasmic layers. The ectoplasmic layer is quite thin, except at the anterior end of the organism, where it is deep and the distance between the granules of each cilium correspondingly long. The rootlets of the row of adoral cilia around the margin of the peristome, the "radial fibrils," are exceedingly long, ending in about the posterior third of the body without connection or attachment. The cilia immediately posterior also have long rootlets, but they become shorter as they approach the base of the apical cone (Fig. 88).

METHODS

Vital stain: Neutral red (differentiates the neuromotor apparatus).
Fixatives: Schaudinn's, Zenker's, Formalin, osmic acid, picrocuric, 60-80° C.
Whole mounts and sections
Stains: Iron haematoxylin, Mallory's triple stain (particularly for sections).

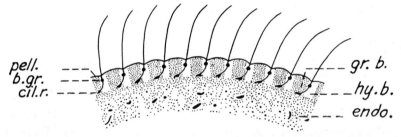

Figure 88. *Balantidium coli,* cross section of peripheral region. (McDonald, 1922.)
b. gr.—basal granule   gr. b.—granular band
cil. r.—ciliary rootlet   hy. b.—hyaline band
endo.—endoplasm   pell.—pellicle

Balantidium sushilii (Ray, 1932).—Fibrils associated with the contractile vacuoles have been described in *Balantidium sushilii.* Each of the two lateral vacuoles has a fibril running from the wall of its outer half to the neighboring pellicle. The neck of the terminal vacuole is surrounded by a diaphragm of fibrils running from the wall of the neck to the surrounding pellicle. These unusual fibrils are described and figured in the extended as well as the contracted condition.

An axial and a peripheral system of fibrils can be seen also in the living organism. Stained preparations show that the former consists of three or four fibrils, parallel or twisted after the manner of a rope, which originate below the pellicle at the anterior end and extend posteriorly a short distance beyond the mouth. The individual fibrils gradually become thinner toward the posterior end, where they come in contact with the limiting membrane formed by some of the fibrils of the peripheral system. At the anterior termination of these fibrils is found a knob-like structure which may be imbedded in the epithelium of the host.

The peripheral system of fibrils is arranged in two conspicuous arches along the left anterior border of the organism. Anteriorly, fibrils extend both to the right and to the left peristomal lips. Posteriorly, the two arches converge. Some of the fibers continue mesially and come to form a kind of limiting membrane beyond which no fibrils are traceable.

The group of fibers forming the axial system, together with the borer attached at its anterior end, is called the boring apparatus, as the first of its kind to be noted. It may be compared with the axostyle of some flagellates.

The peripheral system is believed to represent morphonemes.

METHODS

> *Fixatives:* Brasil's modification of Bouin-Duboscq's (for whole mounts), Bouin's alcoholic, twenty-four hours (for sections 5 μ).
> *Stain:* Heidenhain's haematoxylin (whole mounts and sections).

Fabrea salina Henneguy (Ellis, 1937).—In *Fabrea* there are numerous longitudinal rows of closely set body cilia (in pairs). These are interrupted on the ventral side by a coiling adoral zone. The basal granules are connected by fine longitudinal fibrils. No transverse connections or ciliary rootlets were observed. Each membranelle consists of two rows of basal granules whose ciliary rootlets fuse into a single plate, the *basal lamella*. Each longitudinal fibril of the dorsal and ventral surfaces, with the exception of those that merge with one another, is connected at the adoral zone with the basal lamella of a membranelle. The basal lamella is connected with the *adoral fibril* by fibril running across the peristomal groove. The *adoral fibril* starts at the anterior tip and follows the course of the inner border of the adoral zone. The *peristomal fibril* is continued beyond the end of the adoral zone on the wall of the funnel and ends in a ganglion-like body on the left wall of the ventral lobe. From this *motorium* arise several fibrils—the *adoral fibril*, which follows the course of the adoral zone, and other fibrils which appear to end blindly in the endoplasm of the ventral lobe. Anteriorly the fibrils of the frontal field tend to converge and end very obliquely on the *adoral fibril*.

No pellicular pattern was demonstrated by any of the silver methods. but they did show the longitudinal fibrils connecting the cilia.

METHODS

Fixatives: Schaudinn's, Bouin's and Flemming's.
Stains: Iron-haematoxylin, Mallory's triple.
Silver method: Yabroff's modification of Da Fano's.

Metopus circumlabens (Lucas, 1934).—The *motorium* of *Metopus circumlabens* lies posterior to the cytostome. From its left side it gives rise to a pair of *ventral adoral fibrils* which follow the peristomal curvature outward to the oral margin and there end in a sort of arborization. Each row of peristomal membranelles arises immediately in contact with a connective between these two fibrils. A *dorsal adoral fibril* extends from the right side of the *motorium*. At slightly irregular intervals is gives rise to from ten to twenty heavy connectives, which may partially fuse in pairs as they curve beneath the dorsal wall of the peristome to its left side. There they turn ventrad and unite with the *ventral adoral fibers*. A fibrillar pharyngeal strand arises from the posterior end of the *motorium*. Its course varies in different organisms, but it usually lies along the right lateral wall of the organism to the right of the cytopharynx. Near the posterior end of the latter structure it forms a large spiral coil.

The entire body surface, except the right lateral margin, is covered with rows of cilia. Longitudinal ciliary fibrils are present, but no commissural fibrils were observed. Each basal body in the most dorsal of the five rows of the crest cilia gives rise to *rootlets* which end freely in the cytoplasm. A relation between these *peripheral fibrils* and the fibrils of the *motorium* seems to be suggested by the numerous fine branches which arise from the *ventral adoral* fibrils. These extend indefinitely into the cytoplasm toward the longitudinal ciliary rows of the ventral surface.

In view of the contrastingly striking and obvious specialization in the fibrillar structure of the neuromotor system about the peristomal, pharyngeal, and central endoplasmic regions of the cell, one is inclined to believe that the neuromotor system of this ciliate is vitally, though not exclusively, concerned in the conductile functions related to the metabolic activities of the organism. It is possible that, because they are located within the mobile cytoplasm of the protozoan, the stouter of these various fibres may serve in addition some function in the nature of support.

METHODS

Fixatives: Schaudinn's, Bouin's, Jorgensen's, Van Rath's.

Stains: Heidenhain's iron haematoxylin, Regaud's haematoxylin, Mallory's triple (whole mounts and sections).

Nyctotherus hylae (Rosenberg, 1937).—A detailed description of the neuromotor system of Nyctotherus hylae is given, including two centers and a group of special fibrils believed to control the reversal of ciliary action. An incomplete account of some of these structures was given by Kirby (1932) for N. sylvestrianus. The movements of N. hylae were studied by cinematographic methods.

From the main motorium located at the distal end of the cytopharynx arise two sets of unbranching fibrils. One set extends along the anterior border of the pharynx, eventually terminating at the end of the peristome. The other set follows the posterior border and arm of the ectoplasmic thickening. Some of the latter fibrils unite at the cytostomal border with the peripheral peristomal fibril. The so-called "reversal fibrils" originate from the posterior part of the motorium, radiate through the endoplasm, and at their distal ends unite with the ciliary lines at several points, not including the presutural ciliary lines.

The membranelles have each a basal plate and two rows of basal granules, from which fine fibrils connect with the circumpharyngeal fibrils. The latter become the transverse peristomal fibrils, of which there are at least two for each membranelle of the series.

A number of fibrils from the motorium directly connect with the anterior neuromotor center. From this structure arise many ciliary lines that connect rows of basal granules. Commissural fibrils between these lines are present only in the apical post-sutural region. The lateral and sagittal sutures which divide the ciliation into definite regions were interpreted as probable conductors between ciliary lines.

The pharyngeal terminus, a deeply staining structure, gives rise to a post-pharyngeal bundle of fibrils that have no apparent distal attachment. Kirby (1932) described a similar "band formed structure" in N. silvestrianus which may extend beyond the cytopharynx, its course in the endoplasm varying in different individuals. He considers it "homologous with the 'continuation tube' of the 'subpharyngeal canal' described by Higgins (1929) in *N. cordiformis"* (Kirby, 1932, p. 298). Whether or not this strand is a part of the neuromotor system is an open question.

Fibrils interpreted as *morphonemes* are as follows: (1) those extend-

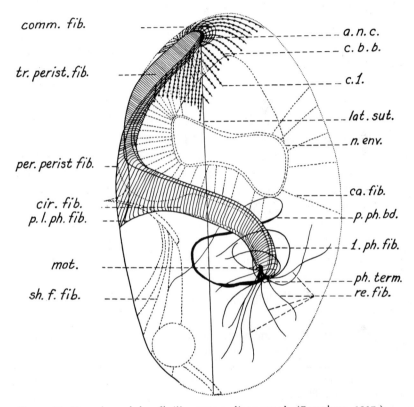

Figure 89. *Nyctotherus hylae,* fibrillar system, diagrammed. (Rosenberg, 1937.)

ant. neur. cen.—anterior neuromotor center
c. b. b.—ciliary basal body
c. l.—ciliary line
ca. fib.—caryophore fibril
cir. fib.—circumpharyngeal fibril
comm. fib.—commissural fibril
l. ph. fib.—longitudinal pharyngeal fibril
lat. sut.—lateral suture
mot.—motorium

n. env.—nuclear envelope
p. l. ph. fib.—posterior longitudinal pharyngeal fibril
p. ph. b.—post-pharyngeal bundle
ph. term.—pharyngeal terminal
per. perist. fib.—peripheral peristomal fibril
re. fib.—reversal fibrils
sh. s. fib.—shelf supporting fibrils
tr. perist. fib.—transverse peristomal fibril

ing from the right to the left of the body, (2) the caryophore fibrils, and (3) the shelf-supporting fibrils (Fig. 89).

METHODS

Fixatives: Schaudinn's (5-percent acetic), Flemming's.
Stain: Heidenhain's iron haematoxylin, aqueous and alcoholic.
Silver techniques: Klein, Gelei-Horváth, Yabroff (negative).

Spirostomum ambiguum Ehrbg. (Bishop, 1927).—The ridges and furrows in the ectoplasm of *Spirostomum ambiguum* follow a sinistral spiral course from the anterior to the posterior end of the body. Beneath the furrows lie thread-like *myonemes,* somewhat beaded in appearance, but without light and dark alternating bands. The *myonemes* taper gradually as they approach either end of the body and finally disappear from view. They are not attached to any structure. *Longitudinal myonemes* were found on either side of and running parallel to the band of peristomal membranelles, except along those membranelles nearest the cytostome. No evidence was obtained confirming the presence of other fibrils such as neurophanes.

On the anterior side of each *myoneme* lie the basal granules of the body cilia. The rows of granules are parallel to and slightly above the level of the *myonemes.* No ciliary rootlets nor connections between basal granules or myonemes were discovered.

The system of fibrils underlying the membranelles (Fig. 90) of *S. ambiguum* includes:

an anterior basal fibril extending from the anterior end of the body to the beginning of the peristomial depression; a middle fibrillar system which varies in its course in different individuals, but which collects the end-threads of the membranelles lying on the left side of the peristomial depression; and a posterior basal fibril into which the end-threads of the membranelles at the posterior end of the peristomial depression and in the cytopharynx join. A connection between the posterior basal fibril and the middle fibrillar system is seldom found, and there is always a break between the middle fibrillar system and the anterior basal fibril.

A central body to which the fibrils join was found in no case.

METHODS

Fixatives: Schaudinn's, picro-mercuric (hot).
Stains: Iron-haematoxylin (alcoholic and aqueous), Mallory's triple (Sharp's modification), Fuchsin S.

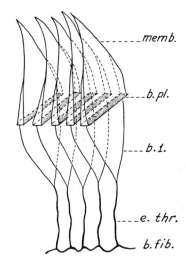

Figure 90. *Spirostomum ambiguum.* Diagram of membranelles and their intracytoplasmic structures. (Bishop, 1927.)
b. fib.—basal fibril
b. l.—basal lamella
b. pl.—basal plate
e. thr.—end thread
memb.—membranelle

C. OLIGOTRICHA

Diplodinium ecaudatum (Sharp, 1914).—The motorium is located in the ectoplasm above the base of the left skeletal area and between the left extremities of the dorsal and the adoral membranelle zones. Dorsally it is connected with the bases of the dorsal membranelles by a *dorsal motor strand,* which also sends a branch along the base of the inner dorsal lip. A *ventral motor strand* connects the *motorium* with the bases of the adoral membranelles and its branch and passes along the base of the inner adoral lip. Numerous fibers from the *motorium* follow the contour of the operculum and disappear near the base of the right skeletal structure. A *circumesophageal ring* surrounds the esophagus at the level of the outer adoral furrow, from which a fibril connects with the *motorium.* Certain fibrils in the wall of the esophagus appear to unite with the *circumesophageal ring;* others are attached to skeletal structures and are considered contractile fibrils. Rootlets from the oral cilia end in, or close to, the ring. A coördinating (conductive) function was ascribed to this fibrillar system of D. *ecaudatum,* because of its inititate relationships with the motor organelles and its complete structural integration through the *motorium.*

METHODS

Fixatives: Schaudinn's (alcoholic, hot), Zenker's, Flemming's, Worcester's, Bouin's, formalin 4 percent, osmic acid one percent (formalin 36° C.)
Stains: Heidenhain's haematoxylin and Mallory's triple.

Diplodinium medium (Rees, 1931).—The structure of *D. medium* is compared with *D. ecaudatum* as described by Sharp. The disagreement between Rees's interpretation and the interpretation of Sharp has to do primarily with the ectoplasmic layer directly under the pellicle, except in the region of the adoral lip and the inner boundary layer of the ectoplasm. In *D. medium* this layer, according to Rees, is more prominent than in *D. ecaudatum* and consists, instead of fine alveoli, of an *interwoven network,* or complex system of fibrils. Serial cross and longitudinal sections, 3 μ in thickness, made it possible to trace the ectoplasmic layers.

Cross sections of *D. medium* show a fold of this middle layer of the ectoplasm which corresponds in its position to Sharp's *motorium.* Furthermore, the *esophageal ring* described by Sharp is interpreted as section of the inner boundary layer of ectoplasm. The fibers connecting the *mortorium* with the membranelles and esophagus could not be differentiated.

The ciliary rootlets are attached to membranes composed of sheets of the fused middle and inner layers of ectoplasm, which in turn are attached to the fibrillar system of the ectoplasmic layers The structures are membranes, according to Rees, because "one or the other of them occurs in all longitudinal and oblique sections, whether cut with reference to the parasagittal plane or to a plane at right angle or at any other angle to it." The *esophageal tractor strands* are considered to be a part of the ectoplasmic network of fibrils. The membranes, instead of strands, are believed by Rees to function in the retraction of the adoral and the dorsal cilia.

The occurrence of fine fibrillae in the non-ciliated ectoplasm of *Diplodinium* is of interest in connection with other papers on the neuromotor system. It is obvious that in the latter ciliate the fibrillae of the ectoplasmic layers have no relationship to a neuromotor system.

METHODS

Fixatives: Not listed.

Stains: Iron-haematoxylin, Zirkle's N-butyl alcohol method of dehydration.

Diplodinium Schbg. (Kofoid and MacLennan, 1932).—In addition to the neuromotor apparatus as described by Sharp (1914) for *D. ecaudatum* and incidentally confirmed in this systematic investigation, a fibrillar complex also was found to occur in the caudal spines of *D. dentatum,* in contrast to the apparently structureless spines in *Entodinium.* Along the bases of the caudal spines appeared a heavy *marginal fibril* from which finer anchoring fibrils extended into each spine, terminating under the cuticle of its outer margin. A very heavy *main anchoring fibril* bordered the inner edge of each spine. From the *anchoring fibril* of the ventral spine smaller fibrils branched toward the anus, where they ended, one on each side. Small branches from these coursed in the wall of the rectum, parallel to its main axis.

No connection was evident between this fibrillar complex of the caudal spines of *D. dentatum* and its neuromotor system. The location and relationships of the former suggested a supporting function similar to that of the longitudinal surface fibrils. Also, since these spines undergo a change in their curvature such as might obviously be facilitated especially by the *main anchoring fibril* together with the *marginal fibrils,* these caudal fibrils were considered to be myonemes.

A similar fibrillar system had been described by Bělař (1925) in *Epidinium caudatum,* on the basis of which Reichenow (1929) denied a neuromotor function for all fibrils of the Ophryoscolecidae. The clear difference in the morphological relationships of the two fibrillar systems in *D. dentatum,* however, indicated that these systems have quite different functions:

The caudal fibrils are admirably situated to serve as supporting and contractile structures. The motor fibrils are so situated as to be of little or no use either as supporting or as contractile fibrils. The caudal fibrils show no connection to the motor organelles. The motor fibrils link together (through the neuromotorium) all the motor organelles of the individual.

METHODS

Fixative: Schaudinn's.
Stain: Iron-haematoxylin.

Favella jorgensen (Campbell, 1927).—The *neuromotorium* is a spindle-shaped body in the ventral ectoplasmic wall in the mid-region of the gullet. This organelle gives rise to five intracytoplasmic fibrils as follows: (1) the *adoral fibril,* extending to and interconnecting the membranelles; (2) the *circumesophageal fibril,* with branches surrounding the gullet; (3) a *dorsal fibril* which appears to connect with the striations of the oral plug; and (4, 5) two *ventral fibrils* extending downward and ending freely in the endoplasm. It was observed that the membranelles not only serve the organism in feeding and in locomotion, but, during periods of binary fission, function in the building of the lorica. In addition to their fibrillar connection with the *motorium,* each membranelle is supplied with three large basal bodies.

METHOD

    Fixative: Schaudinn's (aqueous and alcoholic), 90° C.
    Stain: Iron-haematoxylin (whole mounts and sections).

Tintinnopsis nucula (Campbell, 1926).—The somatic ciliation is confined to the column and forms in longitudinal rows along the *myonemes.* These showed basal granules, but without fibril connection. The *myonemes* are ectoplasmic structures, arranged longitudinally and unbranched. Anteriorly, they extend to the reflexed margin of the collar and possibly connect with basal granules of the adoral membranelles; posteriorly, they fade out.

    At the base of each adoral membranelle are three basal granules connected by fibrils. Through this triangular base passes the *adoral motor fibril.* Three oral membranelles (flat plates of fine cilia) follow the spiral of the gullet. Each oral membranelle ends in a distinct basal body. The ciliary membrane (undulating membrane of unusual construction), which functions in house-building and repair, is connected through its basal granules to the adoral fiber. A retractile tentaculoid is found between each adoral membranelle. Tentaculoids, accessory combs, and trichocysts have no known fibrillar connection.

    The *motorium* is located in the ectoplasm of the ventral wall of the column. From it arise directly (1) the *adoral fiber* (granular), which connects with the adoral membranelles, the oral membranelles, and the ciliary membrane; (2) two *dorsal fibers,* extending into the ectoplasm adorally, where they end freely; and (3) the *ventral fiber,* which extends downward and also ends freely in ectoplasm. The circumesophageal ring is connected to the *motorium* indirectly by a single fiber. Short fibers from the *ring* surround the gullet.

METHODS

    Fixatives: Schaudinn's, 90° C.
    Stain: Heidenhain's iron-alum haematoxylin, aqueous and alcoholic.

D. HYPOTRICHA

Oxytricha (Lund, 1935).—The parts of the neuromotor system of *Oxytricha* apparently are confined to the more specialized organelles. No fibrils were found in connection with most of the ventral cirri.

A long *membranelle* fibril connects the inner ends of the membranelle plate. The frontal membranelles, in addition, have *ciliary rootlets* which arise only from the most proximal basal granules of each of the three rows. These combine into a stouter fibril for each membranelle, which ends free in the endoplasm.

Along the ventral margin of the peristome are numerous fibrils (Fig. 91). One of these connects the basal granules of the undulating mem-

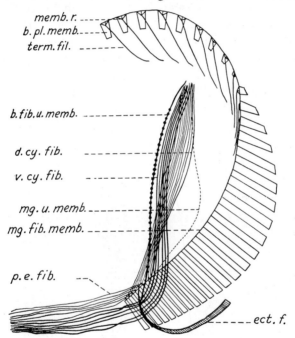

Figure 91. *Oxytricha.* Diagram of fibrillar complex of cytostome, ventral view. (Lund, 1935.)
b. fib. u. memb.—basal fibril of undulating membrane
b. pl. memb.—basal plate of membranelle   d. cy. fib.—dorsal cytostomal fibril
ect. f.—ectoplasmic fold   memb. r.—membranelle rootlet
mg. fib. memb.—marginal fibril of membranelles
mg. u. memb.—position of marginal undulating membrane
p. e. fib.—post-esophageal fibril   term. fil.—terminal filament
v. cy. fib.—ventral cytostomal fibril

brane. Twenty-two originate near the anterior end of this fibril. Their posterior destinations are as follows: ten terminate in ten small granules attached to the marginal fibril in the region of the posterior seven or eight membranelles; six pass along the dorsal wall of the gullet and extend into the endoplasm to a point near the right side of the body; and the other six continue into the gullet to form the ventral *post-esophageal fibrils*. These fibrils along the right side of the peristome were observed to be lax, apparently nonelastic, and capable of individual movement.

A delicate fibril extends anteriorly from each of the five anal cirri. In the region at the left of the posterior macronucleus they disappear from view. Their position suggests, however, that they may join the other parts of the neuromotor system, in the region of the posterior *undulating membrane* fibril.

METHODS

Fixatives: Schaudinn's (with 5-percent acetic), Zenker's.

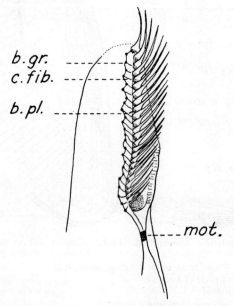

Figure 92. *Uroleptus halseyi,* section through anterior end. (Calkins, 1930.)
b. gr.—basal granule   c. fib.—coördinating fibrils
b. pl.—basal plate   mot.—motorium

Stains: Iron-haematoxylin, Mallory's triple (Sharp's modification).
Microdissection

Uroleptus halseyi Calkins (Calkins, 1930).—The conspicuous parts
of this kinetic system are the *motorium* near the right side of the gullet
and, leading from it, the longitudinal *anterior fibril* which links a row of
endoplasmic granules. Each basal plate of the membranelle series is connected by a short fibril to one of these basal granules, in regular order,
and additional connectives unite the basal granules of each frontal cirrus
with the chain. One short *anterior fibril* from the *motorium* extends to
the margin of the peristome; the other leads to the undulating membrane,
and two *posterior fibrils* are soon lost in the endoplasm. (Fig. 92).

METHODS

> *Fixatives:* HgCl$_2$ saturated in 95-percent alcohol.
> *Stain:* Iron-haematoxylin.

CONCLUSIONS

It is evident from the review presented in foregoing paragraphs that
the differentiation of fibrils in ciliates has been established for various
representatives of their major groups beyond any doubt. Such differentiations, as revealed in fixed and stained material, are not artifacts,
for many may be seen in living or in slowly disintegrating organisms
(Worley, 1933).

It is not certain, however, that all of the structures thus identified are
actually fibrillar. Some may represent rather a sculptural, fibrillar-like
pattern in the pellicle.

It is also clear that the fibrils are not all alike, either structurally or as
related to other protoplasmic differentiations of the cell. This was well
illustrated in the several complexes of fibrils in the contractile stalk
of the vorticellids. Here the structural elements composing the *Spironem,*
for example, differed partly in kind, but especially in arrangement,
from those of the *Axonem.* Also, their relations to the fibrillar complexes of the *bell* were found to be different.

Again, it is known that many fibrillar complexes of various ciliates
are intimately associated with the basal apparatus of the motor organelles. This was shown for the inner fibrillar complex, or the *infracilia-*

ture, of holotrichs such as *Paramecium;* for the membranelle fibril of *Stentor* and of *Euplotes;* and for most of the fibrillar systems of the other ciliates, the accounts of which were more briefly reviewed.

Also, the literature contains numerous records of fibrils, in a variety of ciliates, which are structurally integrated into a so-called fibrillar system. It is with such fibrillar systems that this review has been chiefly concerned.

Having established the identity of these fibrils and fibrillar systems and, for many, their structural continuity or contiguity with other organelles of the cells, especially the motor organelles, the investigators' further interest has, of course, been concerned with the function or functions which may be performed by such definitely related and integrated fibrillar systems.

It was previously pointed out that the interpretation of the function or functions of these fibrillar systems has been based largely on the evidences of their structural integration and their relationship to other organelles. Relatively few of the interpretations have been made from experimental evidence. From both kinds of evidence, it was noted that at last four elementary functions have been ascribed by the many investigators to these various fibrils or fibrillar systems: (1) elasticity, (2) mechanical support, (3) contractility, and (4) conductivity.

Some examples of these included (1) *Elasticity,* the *Spasmonem* and pellicle in the contractile stalk of the vorticellids (Entz, 1893), the axial filament of cilia (Koltzoff, 1912); (2) *Mechanical support, Stützgitter system of Paramecium* (von Gelei, 1929); fibrils generally (Jacobsen, 1931); (3) *Contractility,* myonemes of *Stentor* (Johnson, 1893; and other authors), and of *Boveria* (Pickard, 1927); (4) *Conductivity,* neuromotor system of many ciliates (Sharp, 1914; Yocom, 1918; and other authors).

A fifth function, "metabolic influence," not previously noted, has recently been proposed by Parker (1929) for the fibrillar complex in *Paramecium* (Rees, 1922) and in other ciliates, comparable to the function of fibrils in nerve cells. The neurofibrillar hypothesis for conductivity in nerves was regarded as untenable by Parker (1929). After an extensive review of the evidences for and against this thesis, including Bethe's (1897) experiment on the brain neurones in the crab *Carcinus,* which showed that the nerve impulses did not have to traverse the fibrils of the cell body, Parker suggested that neurofibrils generally, and possibly the fibrils described for certain ciliates, may serve to transmit, from the metabolic center or nucleus, metabolic influences "essential for the continued life of the whole neurone." How these transmissions might be made was not clear. He thought they might involve "chains of ionic readjustment such as have been proposed as an explanation of the nerve impulse." Aside from whether or not such might apply to the function of the fibrils in ciliates, however, he rightly observed that these fibrils may not be intimately associated with the nucleus, as seems to be the case in neurones.

Entz's (1893) interpretation of an elastic function for the *Spasmonem* in the recoiling stalk of the vorticellids was discussed under the caption "Interpretations." In addition to this, reference may be made briefly to Koltzoff's (1903, 1906, 1912) similar interpretations for elastic fibrils in cilia and in cells generally. He would ascribe elasticity to all fibrils in maintaining all organic form other than spherical. Since protoplasm is liquid, as shown by the sphericity of its enclosed vacuoles, then elastic elements must be postulated to counteract the physical forces of inner and outer osmotic pressure and surface tension, which tend always to effect spherical form. Such elements are fibrillar, as observed in the many kinds of cells investigated. The amount of evidence adduced by Koltzoff **is** impressive, but his interpretation obviously cannot apply exclusively **to** all fibrils.

Similar claims for a supporting function for fibrils are rather widespread in the literature. Thus Jacobson (1931), as already stated, is disposed to attribute a supporting function to all noncontractile fibrils.

These few citations, together with many others previously noted, may serve to indicate the diversity of functions that have been variously attributed to fibrillar differentiations in ciliates. In so far as they suggest that these fibrils and fibrillar systems may differ in their structure, functions, and relationships among the manifold kinds of ciliated Protista, certainly no one could present conclusive evidence to the contrary. But when, in the absence of proof, an investigator seriously contends that in these unicellular organisms any and all fibrillar differentiations perform only one elementary function, whether it be that of elasticity, mechanical support, contractility, conductivity, or "metabolic transmission," or when he assigns to these fibrils or fibrillar systems one or two functions to the exclusion of another possible function or functions, then surely that investigator thereby adopts a point of view which is inconsistent and indefensible as well.

In such instances we begin to sense a recrudescence of the opposing claims advanced in the Ehrenberg-Dujardin controversy and of the non-cellular theory of protistan organization proposed by Dobell and others. Once having denied the validity of Ehrenberg's extreme contention that the organs of the Infusoria are essentially miniature counterparts of those of macroscopic organisms, a comparably extreme viewpoint is substituted, which would maintain that the Protista represent a complete departure in the organization of living things and so belong in the wholly exclusive category of non-cellular organisms. Thus the claims of these counter extremists would have us search for identities in organization, on the one hand, or only for differences in protistan and multicellular organization on the other hand.

In Ehrenberg's day similar extreme points of view were quite irreconcilable, but in our day they can scarcely represent anything less than rash inconsistencies. Obviously the thesis of non-cellular organization tends to place exclusive emphasis on *differences* between protozoan and metazoan organization and, if one is still inclined to accept that thesis, one might well refer to Bělař's (1926) excellent monograph on the protistan nucleus. Variable as are the nuclei, in form and behavior, of the many kinds there described and illustrated—where they appear to differ from one another more than some differ from metazoan nuclei—surely one cannot fail to recognize that their numerous modifications do not represent discrete differences, but clearly betray the indelible marks of a common origin. They are like the musical variations of some great motif. They demonstrate irrefutably that living nature has been both labile and stable in its evolutionary history, so that we are amply justified in searching out and emphasizing not merely differences but also, and more fundamentally, similarities, in both the structural and functional processes of protoplasmic differentiation.

And since all cellular differentiation is referable in its last analysis to protoplasmic differentiation, then certainly the fibrils and fibrillar systems of multicellular tissues, such as those described by Grave and Schmitt (1925), may belong in the same category, both structurally and functionally, as *some* fibrillar differentiations that have been described and some that we may afford further to search for, in unicellular organisms. Knowing today the general properties and behavior of the long-chain protein molecules, if such fibrils are proteinaceous, as evidently they are, then fibrillar differentiation is one of the most likely kinds of protoplasmic differentiation that might be expected.

But by the same token, we would not expect all such proteinaceous fibrils to be alike, either structurally or functionally. Both by virtue of their intrinsic properties and their relations to other organelles, some fibrils of protistan cells, or of metazoan cells, may serve for support, others for contraction, and still others for conduction of impulses to and from motor or other organelles. Or any one fibrillar complex may perform more than one of these, or of other yet unkown, functions. And this duality or plurality of fibrillar functions may obtain for protistan cells and for tissue cells of multicellular organisms. Certainly we know of no evidences contradicting this *possibility*.

The actual function or functions of most fibrils or fibrillar systems are not as yet finally known. There can be no doubt about the contractile properties of the myonemes of *Stentor,* and some experimental evidences indicate a coördinating (conductive) function for some fibrils in several ciliates and in epithelial tissue. The outer fibrillar systems of *Paramecium* and other ciliates may be fibrillar, or only apparently fibrillar, as an integral part and pattern of the pellicle. If of the pellicle, then at least one of its functions would evidently be that of support.

Much more study and more critical analysis of these fibrillar systems are greatly needed by both improved old and newly devised observational methods, perhaps such as that of the recently developed electron microscope. Then complementing these observational and comparative studies will be required indispensably, as the crucial test of all of our hypotheses, exceedingly refined and precise tools and methods of experimentation. Even today there are many devices suitable for this purpose, if properly adapted and fully utilized by the ingenious well-trained hands and eyes of thoroughly informed, exceptionally endowed minds. It is a mistake to suppose that such microtechniques are peculiarly difficult and that such problems are really unapproachable. It is rather that these techniques are different and that their use requires special training. With such training, it may be easier to transect a ciliate or a marine ovum, with much more accuracy, than to perform "free hand" some of the disections on macroscopic organisms.

By micromanipulative methods and with the aid of other modern devices, such as the ultracentrifuge, and by micromethods of irradiation, and the like, we may expect for the future, once the world has recovered its sanity, notable advances in protistological investigations such as may not have been dreamed of, even by the most sanguine of our predecessors. No other group of organisms may offer more than the unicellular forms toward the solution of some of our most fundamental problems. The results, therefore, will provide a better understanding not only of these unicellular organisms, but of all forms of life. In their last analysis, however, all such problems must surely depend for any final solution upon unique exacting methods of biological experimentation. "Belief unconfirmed by experiment is vain" (Francesco Redi, of Florence, 1668).

Literature Cited

Alverdes, F. 1922. Untersuchungen über Flimmerbewegung. Pflüg. Arch. ges. Physiol., 195: 243-49.

Bělař, K. 1921. Protozoenstudien III. Arch. Protistenk., 43: 431-62.

—— 1925. Hartmann, Allgemeine Biologie. Jena.

—— 1926. Die Formwechsel der Protistenkerne. Ergbn. Zool., 4: 235-664.

Belehradek, Jan., 1921. Rozbor pohybu Vorticell. Biol. Listy, 8: 49-53.

Beltran, Enrique. 1933. *Gruberia calkinsi* sp. nov., a brackish-water ciliate (Protozoa, Heterotrichida) from Woods Hole, Mass. Biol. Bull., 64: 21-27.

Bethe, A. 1897. Das Nervensystem von *Carcinus maenas*. I. Arch. mikr. Anat., 50: 589-639.

Bishop, Ann. 1927. The cytoplasmic structures of *Spirostomum ambiguum* (Ehrbg.). Quart. J. micr. Sci., 71: 147-72.

Brauer, A. 1885. *Bursaria truncatella* unter Berücksichtigung anderer Heterotrichen und der Vorticellinen. Jena. Z. Naturw., 19: 489-519.

Bresslau, E. 1921. Die Gelatinierbarkeit des Protoplasmas als Grundlage eines Verfahrens zur Schellanfertigung gefärbter Dauerpräparate von Infusorien. Arch. Protistenk., 43: 469-80.

Bretschneider, L. H. 1931. Beiträge zur Strukturlehre der Ophryoscoleciden. I. Ekto- und Entoplasma. Fibrillen. Zool. Ans. Suppl., 5: 324-30.

—— 1934. Beiträge zur Strukturlehre der Ophryoscoleciden. II. Arch. Protistenk., 82: 298-330.

Brown, V. E. 1930. The neuromotor apparatus of *Paramecium*. Arch. zool. exp. gén., 70: 469-81.

Bush, Mildred. 1934. The morphology of *Haptophrya michiganensis* Woodhead, an astomatous ciliate from the intestinal tract of *Hemidactylium scutatum* (Schlegel). Univ. Cal. Publ. Zool., 39: 251-76.

Bütschli, O. 1887-89. *Protozoa*. Bronn's Klassen, Bd. 1-3.

Calkins, Gary N. 1930. *Uroleptus halseyi* Calkins, III. The kinetic elements and the micronucleus. Arch. Protistenk., 72: 49-70.

Calkins, Gary N., and R. Bowling. 1929. Studies on *Dallasia frontata*. II. Cytology, gametogamy, and conjugation. Arch. Protistenk., 66: 11-32.

Campbell, A. S. 1926. The cytology of *Tintinnopsis nucula* (Fol) Laackmann, with an account of its neuromotor apparatus, division, and a new intranuclear parasite. Univ. Cal. Publ. Zool., 29: 179-236.

—— 1927. Studies on the marine ciliate *Favella* (Jorgensen), with special regard to the neuromotor apparatus and its rôle in the formation of the lorica. Univ. Cal. Publ. Zool., 29: 429-52.

Chakravarty, M. 1936. On the morphology of *Balantidium depressum* (Ghosh) from a mollusc, *Pila globosa,* with a note on its nuclear reactions and cytoplasmic inclusions. Arch. Protistenk., 87: 1-9.

Chatton, E., and S. Brachon. 1935. Discrimination, chez deux Infusoires du genre *Glaucoma,* entre systeme argentophile et infraciliature. C. R., Soc. Biol. Paris, 118: 399-402.

Chatton, E., and A. Lwoff. 1930. Impregnation, par diffusion argentique, de l'infraciliature des ciliés marins et d'eau douce, apres fixation cytologique et sans desiccation. C. R., Soc. Biol. Paris, 104: 834-36.

—— 1935. I La Constitution primitive de la strie ciliare des Infusoires. La desmodexie. C. R. Soc. Biol. Paris, 118: 1068-71.

—— 1936. Les Remaniements et la continuité du cinétome au cours de la scission chez les Thigmotriches Ancistrumidés. Arch. zool. exp. gén., 78: 84-91.

Cohn, F., and C. V. Siebold. 1862. Ueber die contractilen Staubfäden der Disteln. Z. wiss. Zool., 12: 366-71.

Czermak, J. 1853. Über den Stiel der Vorticellen. Z. wiss. Zool., 4: 438-50.

Delage, Y., and Ed. Herouard. 1896. Traité de zoologie concrète. 1. La Cellule et les protozoaires. Paris.

Dierks, K. 1926. Untersuchungen über die Morphologie und Physiologie des *Stentor coereuleus* mit besonderer Berücksichtigung seiner kontraktilen und konduktilen Elemente. Arch. Protistenk., 54: 1-91.

D'Udekem, M. J. 1864. Description des Infusoires de la Belgique. I. Ser. Les Vorticelliens. Mém. Acad. roy. Belg., 35: 1-52.

Dujardin, F. 1835. Sur les prétendus estomacs des animalcules infusoires et sur une substance appelée sarcode. Ann. Sci. nat., 2: 4.

—— 1841. Histoire naturelle des (Zoophytes) Infusoires. Suites à Buffon. Paris.

Ehrenberg, C. G. 1838. Die Infusionsthierchen als Vollkommene Organismen. Leipzig. Atlas, 64 pls.

Ellis, John. 1937. The morphology, division, and conjugation of the salt-marsh ciliate *Fabrea salina* Henneguy. Univ. Cal. Publ. Zool., 41: 343-88.

Dobell, C. C. 1911. Principles of protistology. Arch. Protistenk., 23: 269-310.

Doflein, F., and E. Reichenow, 1929. Doflein, Lehrbuch der Protozoenkunde. 5th ed., Jena.

Engelmann, T. W. 1875. Contractilität und Doppeltbrechung. Pflüg. Arch. ges. Physiol., 11: 432-64.

—— 1879. Physiologie der Protoplasma und Flimmerbewegung. Hermann's Handbuch d. Physiol., 1: 343-408.

—— 1880. Zur Anatomie und Physiologie der Flimmerzellen. Pflüg. Arch. ges. Physiol., 23: 505-35.

Entz, G. 1893. Die elastischen und contractilen Elemente der Vorticellen. Math. naturw. Ber. Ung., 10: 1-48.

Everts, E. 1873. Untersuchungen über Vorticella nebulifera. Z. wiss. Zool., 23: 592-622.

Fabre-Doumergue, P. 1888. Recherches anatomiques et physiologiques sur les infusoires ciliés. Ann. Sci. nat. (b) Zool., 1: 1-140.

Forest, H. E. 1879. The natural history and development of the Vorticellidae. Midl. Nat., 2:

Gelei, G. von. 1937. Ein neues Fibrillensystem im Ectoplasma von Paramecium. Arch. Protistenk., 89: 133-62.

Gelei, J. von. 1925. Ein neues Paramaecium aus der Umgebung von Szeged, Paramaecium nephridiatum n. sp. Allatt. Közlem., 22: 121-59 (résumé in German, 245-48).

—— 1926a. Zur Kenntnis des Wimperapparates. Z. Anat. Entw. Gesch., 81: 530-53.

—— 1926b. Cilienstruktur und Cilienbewegung. Zool. Anz. Suppl., 2: 202-13.

—— 1926c. Sind die Neurophane von Neresheimer neuroide Elemente? Arch. Protistenk., 56: 232-42.

—— 1927. Eine neue Osmium-Toluidinsmethode für Protistenforschung. Mikrokosmos, 20: 97-103.

—— 1929a. Sensorischer Basalapparat der Tastborsten und der Syncilien bei Hypotrichen. Zool. Anz., 83: 275-80.

—— 1929b. Über das Nervensystem der Protozoen. Allatt. Közlem. (Zool. Mitt.), 26: 186-90.

—— 1932a. Die reizleitenden Elemente der Ciliaten in nass hergestellten Silber-bwz. Goldpräparaten. Arch. Protistenk., 77: 152-74.

—— 1932b. Ein Neue Goldmethode für Ciliatenforschung und eine neue Ciliate: Colpidium pannonicum. Arch. Protistenk., 77: 219-30

—— 1933. Über den Bau die Abstammung und die Bedeutung der sog. Tastborsten bei den ciliaten. Arch. Protistenk., 80: 116-27.

—— 1934. Der Cytopharynx der Paramecien. Matemat. és termész. értesitö, 51: 736-50.

—— 1934. Die Differenzierung der Cilienmeridiane der Ciliaten und der Begriff des Richtungsmeridians. Matemat. és termész. értesitö, 51: 632-44.

—— 1934. Das Verhalten der ectoplasmatischen Elemente des Parameciums während der Teilung. Zool. Anz., 107: 161-77.

—— 1934. Die Vermehrung der Sinneshaare von *Euplotes* während des Teilungsprozesses. Zool. Anz., 105: 258-66.

—— 1934. Eine mikrotechnische Studie über Färbung der subpellicularen Elemente der Ciliaten. Z. wiss. Mikr., 51: 103-78.

—— 1934. Der feinere Bau des Cytopharynx von *Paramecium* und seine systematische Bedeutung. Arch. Protistenk., 82: 331-62.

—— 1935. Eine neue Abänderung der Klein'schen trockenen Silbermethode und das Silberliniensystem von *Glaucoma scintillans*. Arch. Protistenk., 84: 446-55.

—— 1935. Der Richtungsmeridian und die Neubildung des Mundes während und ausserhalb der Teilung bei den ciliaten. Biol. Zbl., 55: 436-45.

—— 1935. Historisches und Neues über die interciliaren Fasern und ihr morphologische Bedeutung. Z. Zellforsch., 22: 244-54.

—— 1935. *Colpidium glaucomaeformae* n. sp. (Hymenostomata) und sein Neuronensystem. Arch. Protistenk., 85: 289-302.

—— 1936-37. Der schraubige Körperbau in der Ciliatenwelt im Vergleich zu Symmetrieverhältnissen der vielzelligen Tiere. Arch. Protistenk., 88: 314-38.

—— 1937. Pori secretorii am Ciliatenkörper. Biol. Zbl., 57: 175-87.

—— 1938. Schraubenbewegung und Körperbau pei Paramecium. Arch. Protistenk., 90: 165-77.

—— and P. Horváth. 1931. Eine nasse Silber-bzw. Goldmethode für die Herstellung der reizleitender Elemente bei den Ziliaten. Z. wiss. Mikr., 48: 9-29.

Grave, C., and C. O. Schmitt. 1925. A mechanism for the coördination and regulation of ciliary movement as revealed by micro-dissection and cytological studies of ciliated cells of molluscs. J. Morph., 40: 479-515.

Greeff, R. 1870-71. Untersuchungen über den Bau und die Naturgeschichte der Vorticellen. Arch. Naturgesch., I-II Abt., 36: 37.

Griffin, L. E. 1910. *Euplotes worcesteri* sp. nov. I. Structure. Philipp. J. Sci., 5: 291-312.

Haeckel, E. 1873. Zur Morphologie der Infusorien. Jena. Z. Naturgesch., 7: 516-60.

Hall, R. F. 1923. Morphology and binary fission of *Menoidium incurvum*. Univ. Cal. Publ. Zool., 20: 447-76.

Hammond, D. M., and C. A. Kofoid. 1936. The continuity of structure and function in th neuromotor system of *Euplotes patella* during its life cycle. Proc. Amer. phil. Soc., 77: 207-18.

Hammond, J. C. 1935. Physiological dominance as a factor in ciliary coördination in the Protozoa. Ohio J. Sci., 35: 304-6.

Hartmann, M. 1925. Allgemeine Biologie. Eine Einleitung in die Lehre vom Leben. Jena.

Heidenhain, M. V. 1899. Beiträge zur Aufklärung des wahren Wesens der faserformigen Differenzierungen. Anat. Anz., 16: 97-131.

—— 1911. Plasma und Zelle. Jena.

Heidenreich, E. 1935. *Ptychostomum lumbriculi* n. sp. Arch. Protistenk., 85: 303-5.

Higgins, H. T. 1929. Variations in *Nyctotherus* found in frog and toad tadpoles and adults. Trans. Amer. Micr. Soc., 48: 141-57.

Hofker, J. 1928. Das neuromotorische Apparat der Protozoen. Tijdschr. Nederlandsch. 1:34-38.

Horváth, J. 1938. Eine neue Silbermethode für die Darstellung des Stützgitters und der erregungsleitenden Elemente der Ciliaten. Z. wiss. Mikr., 55: 8-122.

Jacobson, Irene. 1931. Fibrilläre Differenzierungen bei Ciliaten. Arch. Protistenk., 75: 31-100.

Johnson, H. P. 1893. A contribution to the morphology and biology of the stentors. J. Morph., 8: 467-562.

Kate, C. G. B. ten. 1926. Über das Fibrillensystem der Ciliaten. Dissertation Univ. Utrecht. Zutphen.

—— 1927. Über das Fibrillensystem der Ciliaten. Arch. Protistenk., 57: 362-426.

—— 1928. Über das Fibrillensystem der Ciliaten. 2. Das Fibrillensystem der Isotrichen. Arch. Protistenk., 62: 328-54.

Kidder, G. W. 1933a. On the Genus *Ancistruma* Strand (*Ancistruma* Maupas). I. The structure and division of *A. mytili* Quenn. and *A. isseli* Kahl. Biol. Bull. 64: 1-20.

—— 1933. Studies on *Conchophthirus mytili* De Morgan. I. Morphology and division. Arch. Protistenk., 79: 1-24.

—— 1934. Studies on the ciliates from fresh water mussels. I. The structure and neuromotor system of *Conchophthirus anodontae* Stein, *C. curtus* Engl., and *C. magna* sp. nov. Biol. Bull. 66: 69-90.

Kirby, Harold. 1932. Protozoa in termites of the Genus *Amitermes*. Parasitology, 24: 289-304.

Klein, B. M. 1926a. Über eine neue Eigentümlichkeit der Pellicula von *Chilodon uncinatus* Ehrbg. Zool. Anz., 67: 160-62.

—— 1926b. Ergebnisse mit einer Silbermethode bei Ciliaten. Arch. Protistenk., 56: 243-79.

—— 1927. Die Silberliniensysteme der Ciliaten. Ihr Verhalten während Teilung und Conjugation, neue Silberbilder, Nachträge. Arch. Protistenk., 58: 55-142.

—— 1928. Die Silberliniensysteme der Ciliaten. Weitere Resultate. Arch. Protistenk., 62: 177-260.

—— 1929. Weitere Beiträge zur Kenntnis des Silberliniensystems der Ciliaten. Arch. Protistenk., 65: 183-257.

—— 1930. Das Silberliniensystem der Ciliaten. Weitere Ergebnisse. IV. Arch. Protistenk., 69: 235-326.

—— 1931. Über die Zugehörigkeit gewisser Fibrillen bzw. Fibrillenkomplex zum Silberliniensystem. Arch. Protistenk., 74: 401-16.

—— 1932. Das Ciliensystem in seiner Bedeutung für Locomotion Coordination und Formbildung usw. Ergebn. Biol., 8: 75-179.

—— 1933. Silberliniensystem und Infraciliatur. Eine kritische Gegenüberstellung. Arch. Protistenk., 79: 146-69.

—— 1935. Die Darstellung des Silberlinien oder neuroformativen Systems nebst Gründsätzlichem zur Silbermethodik. Z. wiss. Mikr., 52: 120-57.

—— 1936. Wirkung von Schlangengiften auf Leben und Silberliniensystem von Infusorien. Arch. Protistenk., 87: 299-313.

—— 1936. Beziehungen zwischen Maschenweite und Bildungsvorgängen im Silberliniensystem der Ciliaten. Arch. Protistenk., 88: 1-22.

—— 1937. Über die Eigenkörplichkeit des Silberliniensystems. Arch. Protistenk., 88: 188-91.

—— 1937. Regionäre Reaktionen im Silberlinien- oder neuroformativen System der Ciliaten. Arch. Protistenk., 88: 192-210.

—— 1938. Miss-bzw. Doppelbildung am Silberliniensystem von Ciliaten. Arch. Protistenk., 90: 292-98.

Kofoid, C. A., and R. F. MacLennan 1932. Ciliates from *Bos indicus* Linn. II. A revision of *Diplodinium* Schuberg. Univ. Cal. Publ. Zool., 37: 53-152.

Kölliker, A. 1845. Die Lehre von der tierischen Zelle. Z. wiss. Bot., 1: 2.

—— 1864. Icones histiologicae 1. Leipzig.

Koltzoff, N. K. 1903. Über formbestimmende elastische Gebilde in Zellen. Biol. Zbl., 23: 680-96.

—— 1906. Die Spermien der Decopoda, als Einleitung in das Problem der Zellengestalt. Arch. mikr. Anat., 67: 364-571.

—— 1909. Studien über die Gestalt der Zelle. II. Untersuchungen über das Kopfskelett des tierschen Spermiums. Arch. Zellforsch., 2: 1-65.

—— 1912. Studien über die Gestalt der Zelle. III. Untersuchungen über die Kontraktilität des Vorticellenstiels. Arch. Zellforsch. 7: 344-423.

Kühne, W. 1859. Untersuchungen über Bewegung und Veränderungen der contractilen Substanz. Arch. Anat. Physiol., 1859: 748-835.

—— 1859. Über sogenannte idiomuskuläre Kontraktion. Arch. Anat. Physiol., 1859: 418-20.

Lachmann, J. 1856. Über die Organization der Infusorien, besonders der Vorticellen. Arch. Anat. Physiol., 1856: 340-98.

Lang, A. 1901. Lehrbuch der vergleichenden Anatomie der wirbellosen Tiere. (Protozoa). 2. Aufl., Abt. 1. Berlin.

Lebedew, W. 1909. Über *Trachelocerca phoenicopterus* (ein marines Infusor). Arch. Protistenk., 13: 70-114.

Leiberman, P. R. 1929. Ciliary arrangement in different species of *Paramecium*. Trans. Amer. Micr. Soc., 48: 1-11.

Leydig, F. 1883. Untersuchungen über Anatomie und Histologie der Tiere. Bonn.

Lieberkühn, N. 1857. Beiträge zur Anatomie der Infusorien. Arch. Anat. Physiol., 1857: 20-36.

Lucas, M. S. 1934. Ciliates from Bermuda sea urchins. I. *Metopus.* J. R. micr. Soc., 54: 79-93.

Lühe, M. 1913. A. Lang, Handbuch der Morphologie der wirbellosen Tiere. 1. *Protozoa.* Berlin.

Lund, E. E. 1933. A correlation of the silverline and neuromotor systems of *Paramecium.* Univ. Cal. Publ. Zool., 39: 35-76.

—— 1935. The neuromotor system of *Oxytricha.* J. Morph., 58: 257-77.

Lynch, J. E. 1929. Studies on the ciliates from the intestine of *strongylocentrotus.* I. *Entorhipidium* Gen. nov. Univ. Cal. Publ. Zool., 33: 27-56.

—— 1930. Studies on the ciliates from the intestine of *Strongylocentrotus.* II. *Lechriopyla mystax,* Gen. nov., sp. nov. Univ. Cal. Publ. Zool., 33: 307-50.

McDonald, J. Daley. 1922. On *Balantidium coli* (Malmsten) and *Balantidium suis* (sp. nov.), with an account of their neuromotor apparatus. Univ. Cal. Publ. Zool., 20: 243-300.

MacDougall, Mary S. 1928. The neuromotor system of *Chlamydodon* sp. Biol. Bull., 54: 471-84.

MacLennan, R. F. 1935. Dedifferentiation and redifferentiation in *Ichthyophthirius.* I. Neuromotor system. Arch. Protistenk., 87: 191-210.

MacLennan, R. F., and F. H. Connell. 1931. The morphology of *Eupoterion pernix* Gen. nov., sp. nov. Univ. Cal. Publ. Zool., 36: 141-56.

Maier, H. N. 1903. Über den feineren Bau der Wimperapparate der Infusorien. Arch. Protistenk., 2:73-179.

Maupas, E. 1883. Contribution a l'etude morphologique et anatomique des infusoires ciliés. Arch. zool. exp. gén., (2) I: 427-644.

Merton, H. 1932. Gestalterhaltende fixierungsversuche an besonders kontraktilen Infusorien nebst Beobachtungen über das verhalten der lebenden Myoneme und Wimpern bei *Stentor.* Arch. Protistenk., 77: 491-521.

Metschnikoff, E. 1863. Untersuchungen über den Stiel der Vorticellen. Arch. Anat. Physiol., 1863: 180-86.

—— 1864. Nachträgliche Bemerkungen über den Stiel der Vorticellen. Arch. Anat. Physiol., 1864: 291-302.

Meyer, A. 1920. Analyse der Zelle. Jena.

Neresheimer, E. R. 1903. Die Höhe histologischer Differenzierung bei heterotrichen Ciliaten. Arch. Protistenk., 2: 305-24.

—— 1907. Nochmals über *Stentor coeruleus.* Arch. Protistenk., 9: 137-38.

Parker, G. H. 1929. What are neurofibrils? Amer. Nat., 63: 97-117.

Pensa, A. 1926. Particolarita strutturale di alcuni protozoi cigliati in rapporto con la contratilia. Monit. zool. ital., 37: 165-73.

Peschkowsky, L. 1927. Skelettgebilde bei Infusorien. Arch. Protistenk., 57: 31-57.

Pickard, E. A. 1927. The neuromotor apparatus of *Boveria ternedinidi* Nelson, a ciliate from the gills of *Teredo navalis*. Univ. Cal. Publ. Zool., 29: 405-28.

Powers, P. B. A. 1933. Studies on the ciliates from sea urchins. II. *Entodiscus borealis* (Hentschel). Behavior and morphology. Biol. Bull., 65: 122-36.

Prowazek, S. 1903. Protozoenstudien. III. *Euplotes harpa*. Arb. zool. Inst. Univ. Wien., 14: 81-88.

Pütter, A. 1904. Die Reizbeantwortung der ciliaten Infusorien. Z. allg. Physiol., 3: 406-54.

Raabe, Zdz. 1932. Untersuchungen an einigen Arten des Genus *Conchophthirus* Stein. Bull. Int. Acad. Cracovie, Cl. Sci. Math et Nat. Ser. B. Sci. Nat. (II) (Zool.) 1932 (8/10) : 295-310.

Ray, Harendranath. 1932. On the morphology of *Balantidium sushilii* n. sp., from *Rana tigrina* Daud. J. R. micr. Soc., 52: 374-82.

Rees, C. W. 1921. The neuromotor apparatus of *Paramecium*. Amer. Nat., 55: 464-68.

—— 1922. The neuromotor apparatus of *Paramecium*. Univ. Cal. Publ. Zool., 20: 333-64.

—— 1930. Is there a neuromotor apparatus in *Diplodinium ecaudatum?* Science, 71: 369-70.

—— 1931. The anatomy of *Diplodinium medium*. J. Morph., 52: 195-215.

Reichenow, E. 1929. Doflein, Lehrbuch der Protozoenkunde. 5th ed., Jena.

Rosenberg, L. E. 1937. The neuromotor system of *Nyctotherus hylae*. Univ. Cal. Publ. Zool., 41: 249-76.

Roskin, G. 1915. La Structure des myonèmes contractiles de *Stentor coeruleus*. Tirage à part des memoires scientifique des Chaniavsky Université de Moscou. 1:

—— 1918. Sur la structure de certains elements contractiles de la cellule. Arch. russ. d'Anat.,

—— 1923. Die Cytologie der Kontraktion der glatten Muskelgellen. Arch. Zellsforsch., 17: 368-81.

Rouget, C. 1861. Sur les phenomenes de polarisation qui s'observent dans quelques tissus vegetative et des animaux. J. Physiol. Path. gén., 5:

Schewiakoff, W. 1889. Beiträge zur Kenntnis der holotrichen Ciliaten. Bibliogr. zool., 1: H5, 1-77.

Schmidt, W. J. 1924. Die Bausteine des Tierkörpers im polarisierten Lichte. Bonn.

Schröder, O. 1906a. Beiträge zur Kenntnis von *Campanella umbellaria*. Arch. Protistenk., 7: 75-105.

—— 1906b. Beiträge zur Kenntnis von *Epistylis plicatilis*. Arch. Protistenk., 7: 173-85.

—— 1906c. Beiträge zur Kenntnis von *Vorticella monilata*. Arch. Protistenk.. 7: 395-410.

—— 1907. Beiträge zur Kenntnis von *Stentor coeruleus* und *Stentor roeselii* Ehrbg. Arch. Protistenk., 8: 1-16.

Schuberg, A. 1890. Zur Kenntnis des *Stentor coeruleus.* Zool. Jahrb., Abt. Anat. Ontog., 4: 197-238.

—— 1905. Über Cilien und Trichocysten einiger Infusorien. Arch. Protistenk., 6: 61-110.

Seifriz, W. 1929. The contractility of protoplasm. Amer. Nat., 63: 410-34.

Sharp, R. G. 1914. *Diplodinium ecaudatum* with an account of its neuro-motor apparatus. Univ. Cal. Publ. Zool., 13: 42-122.

Siebold, C. Th. von. 1849. Über einzellege Pflanzen und Tiere. Z. wiss. Zool., 1: 270.

Siebold, C. Th. von, and H. Stannius. 1845. Lehrbuch der vergleichenden Anatomie, H. 1. Berlin.

Stein, F. 1854. Die infusionsthiere auf ihre Entwicklung untersucht. Leipzig.

—— 1859. Der Organismus der Infusionsthiere nach einigen Forschungen in systematischer Reihenfolge bearbeitet. I. Abtheilung Allgemeiner Theil und Naturgeschichte der hypotrichen Infusionsthiere. Leipzig.

—— 1867. *Ibid.,* II. Abtheilung Naturgeschichte der heterotrichen Infusorien, pp. 140-352. Leipzig.

Stevens, N. M. 1891. Studies on ciliate Infusoria. Proc. Calif. Acad. Sci., 3: 1-42.

Studitsky, A. N. 1932. Über die Morphologie, Cytologie u. Systematik von *Ptychostomum chattoni* Rossolimo. Arch. Protistenk., 76: 188-216.

Taylor, C. V. 1920. Demonstration of the function of the neuromotor apparatus in *Euplotes* by the method of microdissection. Univ. Cal. Publ. Zool., 19: 403-70.

—— 1929. Experimental evidence of the function of the fibrillar system in certain Protozoa. Amer. Nat., 63: 328-45.

Turner, J. P. 1933. The external fibrillar system of *Euplotes,* with notes on the neuromotor apparatus. Biol. Bull., 64: 53-66.

Verworn, Max. 1889. Protisten-Studien. Jena.

Visscher, J. P. 1927. A neuromotor apparatus in the ciliate *Dileptus gigas.* J. Morph., 44: 373-81.

Worley, L. G. 1933. The intracellular fibre systems of Paramecium. Proc. nat. Acad. Sci. Wash., 19: 323-26.

Wrisberg, H. A. 1765. Observationum de animalculis infusoriis satura. Göttingae.

Wrzesniowski, A. 1877. Beträge zur Naturgeschichte der Infusorien. Z. wiss. Zool., 29: 267-323.

Yabroff, S. W. 1928. A modification of the Da Fano technique. Trans. Amer. micr. Soc., 47: 94-95.

Yocom, H. B. 1918. The neuromotor apparatus of *Euplotes patella.* Univ. Cal. Publ. Zool., 18: 337-96.

Zon, Leo. 1936. The physical chemistry of silver staining. Stain Tech., 11: 53-65.

CHAPTER V

MOTOR RESPONSE IN UNICELLULAR ANIMALS

S. O. Mast

Responses consist of changes in structure, composition, form, or movement in organisms, which, in turn, are correlated with changes in the constituents of the environment or the organisms. Responses are found in all living systems and are among the most fundamental distinguishing characteristics of life. Motor responses consist of changes in rate or direction of movement of organisms or their constituents. They facilitate the control of the environment by the organisms involved and are consequently of great importance to them. Knowledge concerning these responses and their relation to the factors correlated with them makes it possible to control the activities of organisms, it throws light on the distribution of pain and pleasure (consciousness) which profoundly affects the attitude of man toward his fellow creatures, and it illuminates the processes involved in instincts and in learning. Such knowledge is therefore very valuable.

The unicellular organisms are in many respects extraordinarily favorable for the study of the more fundamental characteristics of these responses. They are relatively simple in structure. Many of them can be readily procured and maintained in great numbers, and the factors in their environment can be accurately controlled or changed as desired; and, in addition, details concerning the responses can be readily seen. Moreover, the motor responses in these organisms are very favorable for the study of adaptation, as pointed out long ago by Jennings (1906).

In the following pages are presented the more important facts in hand concerning the motor responses of the rhizopods, flagellates, ciliates, and colonial forms to light, electricity, and chemicals. These are presented with the view of encouraging the use of these organisms in further work on various biological problems. Important results have also been obtained on responses to contact, temperature, and gravity, but limitation in space prevents the consideration of these.

RESPONSES TO LIGHT

A. RHIZOPODS

Many of the rhizopods respond to light and some orient if they are exposed in a beam of light. Some of the responses are correlated with the rate of change in intensity, others are not. They are fundamentally the same in all the species which have been investigated, but they have been more thoroughly studied in *Amoeba proteus* than in any of the other species. The following considerations therefore refer largely to this species.

A. proteus consists of a thin elastic outer membrane, the plasmalemma, a central relatively fluid granular mass, the plasmasol, surrounded by a

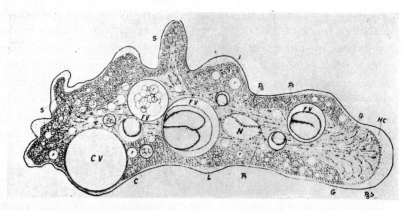

Figure 93. Camera sketch of horizontal optical section of *Amoeba proteus*. Ps, plasmasol; Pg, plasmagel; Pl, plasmalemma; HC, hyaline cap; Pgs, plasmagel sheet; L, Liquid layer; S, region of solation; G, region of gelation. (After Mast, 1926.)

relatively solid granular layer, the plasmagel, and a thin fluid hyaline layer between the plasmagel and the plasmalemma (Fig. 93). During locomotion, the plasmalemma is attached to the substratum and to the adjoining plasmagel; the plasmagel at the posterior end is transformed into plasmasol, which flows forward to the anterior end and is there transformed into plasmagel. The forward flow of the plasmasol is due to contraction of the plasmagel at the posterior end and expansion at the anterior end, owing to difference in its elastic strength in these two regions. In some species surface tension is probably also involved in locomotion. In *A. mira,* for example, the anterior portion consists of a thin sheet of hyaline cytoplasm in close contact with the substratum (Hopkins, 1938). The surface tension at the water-substrate interface is probably greater than the combined surface tensions at the cytoplasm-substrate and cytoplasm-water interfaces. This would result in a spreading of the cytoplasm over the substrate (like oil over water) wherever the two are in contact, i.e., at the anterior end of the organism.

With reference to a marine amoeba, Pantin (1923-31) contends that the cytoplasm at the anterior end swells and extracts water *"from the posterior protoplasm of the amoeba itself,* and that this will cause a stream from the posterior to the anterior end." Presumably he holds that the cytoplasm shrinks there and gives it off. However, that would necessitate absorption of water during gelation at the anterior end, and elimination during solation at the posterior end, which is contrary to what is ordinarily observed in the process of gelation and solation of gels.

Marsland and Brown (1936) suggest that the forward flow is due to increase in volume during solation at the posterior end, and decrease in volume during gelation at the anterior end. They give no direct evidence in support of this suggestion, but simply say that "The magnitude of these volume changes in relation to the observed rate of flow is problematical."

Responses of *Amoeba* to light are therefore probably due to localized changes in (1) the elastic strength of the plasmagel, (2) the rate of transformation of plasmasol to plasmagel and vice versa, or (3) the firmness, extent, or region of attachment to the substratum (Mast, 1923, 1926a, 1931a).

Shock-reactions.—Engelmann (1879) long ago observed that if strong light is flashed on an amoeba, movement stops suddenly, but that if the intensity is gradually increased, movement continues. This response therefore depends upon the rate of change of light intensity. Such responses are usually designated "Schreckbewegungen," or shock-reactions. They are closely correlated with adaptation. The shock-reaction in *Amoeba* produced by light varies greatly. It may consist merely of momentary retardation in streaming in a localized region in a pseudopod, of total cessation throughout the entire animal with reversal in direction of streaming after recovery, or of any one of an endless number of modifica-

tions between these extremes. The character of the response is correlated
with the amount of light received, as well as with the rate of reception.
There is no fixed threshold and the "all-or-none" law does not apply
(Mast, 1931a).

If an amoeba is intensely illuminated for a very short time only, movement does not cease until some time after the light has been cut off.
The period between the beginning of illumination and the response is

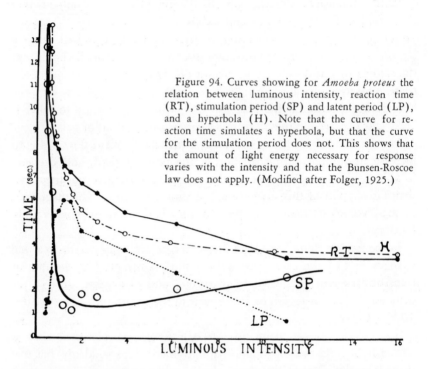

Figure 94. Curves showing for *Amoeba proteus* the
relation between luminous intensity, reaction time
(RT), stimulation period (SP) and latent period (LP),
and a hyperbola (H). Note that the curve for reaction time simulates a hyperbola, but that the curve
for the stimulation period does not. This shows that
the amount of light energy necessary for response
varies with the intensity and that the Bunsen-Roscoe
law does not apply. (Modified after Folger, 1925.)

known as the "reaction time"; the time illumination must continue, the
"stimulation period"; and the time it need not continue, the "latent
period" (Folger, 1925). There are therefore two processes involved in
producing this response. The first occurs only in light, the other in light
or in darkness. The action of light probably results in the formation of
a substance which acts to produce, independent of light, another substance which induces the response.

After an amoeba has responded to rapid increase in illumination, some time must elapse before it will respond again to the same increase in illumination. There is therefore a refractory period, a period during which the amoeba recovers from the effect of the stimulation. During a part of this period the amoeba may remain either in light of the same intensity, such as that which induced the response, in light of lower intensity, or in darkness; but during the remainder of the period it must be in light of lower intensity or in darkness. There are therefore two processes which occur during the refractory period, one (1 to 2 minutes) which proceeds with or without any change in luminous intensity, and another (10 to 20 seconds) which proceeds only if the intensity is decreased. These processes result in the production of the physiological state which existed before the exposure; that is, in recovery (Folger, 1925).

The latent period and the amount of light energy required to induce cessation of movement vary with the intensity of the light used (Fig. 94). Figure 94 shows that as the intensity increases, the latent period increases rapidly from about one second at 500 $\pm$ meter-candles to a maximum of about 6 seconds at 1,000 $\pm$ meter-candles, and then decreases gradually to about 0.75 seconds at 11,000 $\pm$ meter-candles; and that the light energy required to induce cessation of movement decreases from about 7,000 $\pm$ meter-candle seconds at 500 $\pm$ meter-candles to a maximum of about 24,000 $\pm$ meter-candle seconds at 1,500 $\pm$ meter-candles, and then increases to about 30,000 $\pm$ meter-candle seconds at 11,000 $\pm$ meter-candles. These results are, however, only rather crude approximations. They were obtained by a method of calculation which yields results with a large probable error and they have not been confirmed. The data are, however, sufficiently accurate to substantiate Folger's conclusion that the Bunsen-Roscoe law does not hold.

This work should be repeated, and the latent period established by direct observation in all luminous intensities, instead of by calculation. This is especially desirable since recent experience makes it possible to select specimens of *A. proteus* in which the responses are much more consistent than they were in those used by Folger.

No explanation has been offered for the mode of variation in the latent period, with variations in luminous intensity during the period of stimulation. However, it has been suggested that the variation in the amount of light energy required to induce cessation of movement is due, at least in part, to adaptation (Mast, 1931a). For if the light is rapidly increased and then held, streaming soon begins again, i.e., the organism recovers from the effect of the increase in light. In other words, it becomes adapted (Mast, 1939; Folger, 1925). This shows that the effect of rapid increase in light is eliminated while the organism is continuously

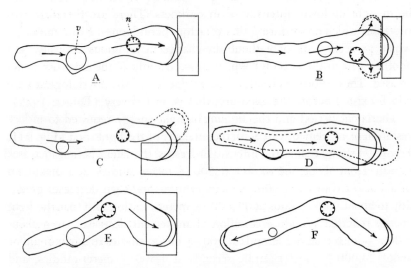

Figure 95. Camera drawings of *Amoeba* sp. illustrating the response to localized illumination. Rectangular areas, regions of high illumination; arrows, direction of protoplasmic streaming; dotted lines in *B*, *C* and *D*, positions and forms shortly after the illumination of the parts indicated; *n*, nucleus; *v*, contractile vacuole. *E* and *F*, same specimen; *F*, form and direction of streaming assumed by *E* after the anterior end had been illuminated for a few minutes. (After Mast, 1932.)

exposed to the light. It also indicates that there are two opposing processes involved, i.e., that increase in light induces certain changes in the organism and that internal factors tend continuously to oppose and to eliminate these changes. If this is true, the more rapidly a given amount of light is received, the less time there is for recovery, and consequently the greater will be the effect of a given quantity of light. This probably accounts for the increase in the amount of light energy required (with decrease in intensity) when observations are made in weak light; but it does not account for the increase in the amount required (with increase in intensity) if the observations are made in strong light.

The quantity of light energy required to induce cessation of movement depends upon the chemical composition of the surrounding medium. Increase in HCL, for example, causes an increase in the quantity of light required. On the other hand, an increase in CO_2 causes decrease in the quantity required. In solutions of KCl, $CaCl_2$, and $MgCl_2$, respectively, the quantity of energy required increases as the salt concentration decreases, but in solutions of NaCl there is no consistent correlation between the quantity of energy and the concentration of the salt. In general, the quantity of energy required appears to vary directly with the viscosity of the cytoplasm (Mast and Hulpieu, 1930). The observations made by these authors extended, however, over only a very limited range of environmental variation. The conclusions reached are therefore not applicable to wide ranges of variations in the environment (Mast and Prosser, 1932).

Increase in the illumination of any localized region of an amoeba results in an increase of the thickness of the plasmagel in this region (Fig. 95). Increase in the illumination of the entire amoeba results in an increase in the thickness of the plasmagel at the tip of the advancing pseudopods. In turn, this causes a cessation of movement (shock-reaction).

The shorter waves of light are more efficient in inducing this response than the longer waves (Harrington and Leaming, 1900; Mast, 1910). According to Inman, Bovie, and Barr (1926), ultra-violet light is probably more efficient than visible light. Although the distribution in the spectrum of stimulating efficiency has not been precisely ascertained, Folger (1925) maintains that it is not closely correlated with temperature. He did not thoroughly investigate the problem, however.

Kinetic responses.—If an amoeba is kept for some time in very weak light it becomes inactive; if the light is then increased, the organism gradually becomes active again. This response is similar to the response to change in temperature. It is primarily correlated with the magnitude of the change, not with the rate of change in intensity. It is probably due to the effect of light on the rate of transformation of gel to sol and vice versa. This type of response occurs also in *Difflugia* (Mast, 1931c), but

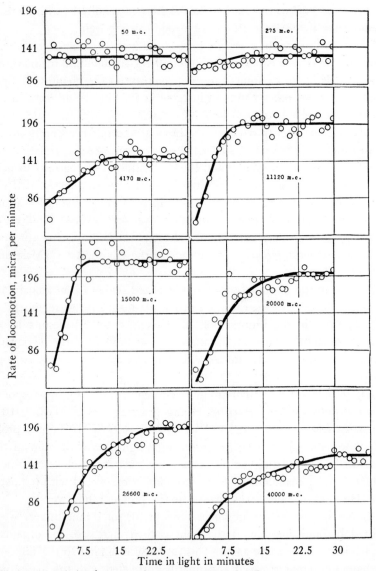

Figure 96. Relation between adaptation to light of different intensities and rate of locomotion in *Amoeba proteus*. Each point in the figure represents the average for one measurement on each of from fourteen to twenty-three specimens. The time in light is the time from the beginning of movement after exposure until the measurements were made. (After Mast and Stahler, 1937.)

the observations on it should be repeated and extended under carefully controlled conditions.

Mast and Stahler (1937) made a thorough study of the relation between luminous intensity and rate of locomotion in *A. proteus*. They found that if dark-adapted amoebae are exposed to light, the rate of locomotion gradually increases to a maximum and then remains constant; that the time required to reach the maximum decreases from 15 minutes at 225 meter-candles to a minimum of 7 minutes at 15,000 meter-candles, and then increases to 30 minutes at 40,000 meter-candles; and that the rate of locomotion at the maximum increases from 128.8 $\pm$ 10.8 micra per minute at 50 meter-candles to 219.3 $\pm$ 11.4 micra per minute at 15,000 meter-candles, and then decreases to 150.2 $\pm$ 8.5 micra per minute at 40,000 meter-candles (Fig. 96). They present evidence which indicates that the increase in rate of locomotion with increase in light intensity is due to the action of the longer waves, and that the decrease in rate in intensities beyond the optimum is due to the action of the shorter waves. This action of light on rate of locomotion is similar to the action of temperature. It is probably due to changes in the rate of sol-gel and gel-sol transformations. If this is true, both of these transformations must be augmented by the longer waves and retarded by the shorter.

Orientation.—Davenport (1897) found that *A. proteus* orients fairly precisely in a beam of direct sunlight and that it is photonegative, but he did not ascertain the processes involved in orientation.

Mast (1910) demonstrated that if an amoeba is unilaterally illuminated, pseudopods develop more freely on the shaded side than on the illuminated side, and that this results in gradual turning from the light (Fig. 97). He concludes that orientation is due to retardation in the formation of pseudopods on the more highly illuminated side, owing to increase in the thickness of the plasmagel on this side caused by the gelating effect of light.

There is some evidence which indicates that *A. proteus* is photopositive in very weak light (Schaeffer, 1917; Mast, 1931a). More carefully controlled observations concerning this are highly desirable.

It is possible that the kinetic responses in *Amoeba* are due to changes in the rate of sol-gel transformations at the anterior end, and gel-sol transformations at the posterior end. Moreover, the shock reactions appear to be associated with rapid local increases in the sol-gel transformation. If the views concerning the process of orientation as presented

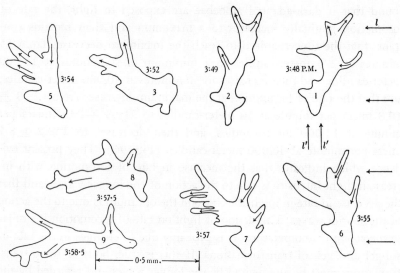

Figure 97. Camera outlines representing different stages in the process of orientation in *Amoeba proteus*. 1, *Amoeba* oriented in light *l'*; 2-9, successive positions after exposure to light *l*, time indicated in each. Arrows represent the direction of streaming of protoplasm in the pseudopods. In those which do not contain arrows there was no perceptible streaming at the time the sketch was made. *l* and *l'* direction of light; *mm*, projected scale. (After Mast, 1910.)

above are correct, it is obvious that orientation is the result of shock reactions rather than kinetic reactions.

B. FLAGELLATES

Response to rapid changes in the intensity of light is very widespread among the flagellates, and many of them orient fairly precisely. The processes involved are essentially the same in all. *Euglena* is representative of those which orient, and *Peranema trichophorum* is representative of those which do not.

Euglena rotates continuously on its longitudinal axis as it swims. The flagellum extends backward along the ventral or abeyespot surface. This causes continuous deflection of the anterior end toward the opposite

surface, resulting in a spiral course. Its direction of movement is changed
by the shifting of the distal end of the flagellum from the surface of the
body so as to increase the angle between it and the surface. This increases the deflection of the anterior end (Fig. 98).

Shock reaction and aggregation.—Engelmann (1882) observed that if
the intensity of the light in a field in which euglenae are swimming about
at random is rapidly decreased, they stop suddenly, then turn and pro-

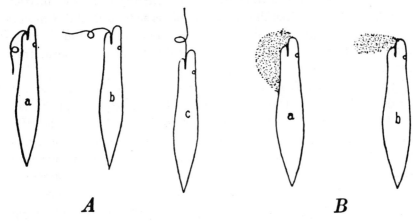

Figure 98. A. Diagrams showing the position of the flagellum as seen in a viscid
medium; a, when *Euglena* is swimming forward in a narrow spiral; b, when swerving
sharply towards the dorsal side; c, when moving backwards. B. Dotted area, shows the
position of the moving India-ink particles. a, when *Euglena* is swimming forward in a
narrow spiral; b, when swerving toward the dorsal side during a shock-movement.
(After Bancroft, 1913.)

ceed in various directions. He designated the response as a "Schreck-
bewegung" (fright movement, or shock reaction), because the re-ori-
ented organisms gave the impression of having been frightened. It was
found that if the intensity is slowly changed this response does not
occur. It is therefore dependent upon the rate of change in intensity.

He says that if there is a spot of relatively strong light in the field,
it acts just like a trap; owing to random movements, the euglenae get
into this spot, but as they reach the boundary on the way out, the rapid
reduction in intensity induces the shock reaction and consequently prevents their exit.

Under some conditions the euglenae respond to rapid increase in intensity and aggregate in a spot of relatively weak light in the field (Mast, 1911).

Orientation.—If euglenae are exposed in a beam of light, they usually swim toward or away from the source of light, i.e., they may be either photopositive or photonegative.

Verworn (1895) postulated that if the euglenae are not directed toward or away from the light, so that one side is more strongly illuminated than the other, the flagellum beats more effectively in one direction than in the opposite, that this causes the euglenae to turn until both sides are equally illuminated, and that the flagellum then beats equally in opposite directions and the organism moves directly toward or away from the source of light.

The above hypothesis is, in principle, essentially the same as that formulated by Ray (1693), in reference to orientation in plants, and later accepted by de Candolle (1832). According to this idea, the effect of light on the activity of the motor mechanism, or upon the photoreceptors connected with it, is dependent upon the intensity (not upon change of intensity) of the illumination. The light acts continuously after orientation has been attained, as well as during the process of orientation. During the process of orientation, the illumination on opposite sides is unequal, which results in quantitatively unequal action in the motor mechanism; but after orientation, it is equal on opposite sides, and consequently the action of the mechanism on opposite sides is equalized. Verworn applied this theory to ciliates as well as flagellates. In his earlier work, Loeb (1890) strenuously opposed the theory outlined above, accepting Sachs's "ray-direction theory" as the alternative. He adopted it later (1906), however, and applied it to higher animals, introducing the idea that the action of the locomotor appendages is quantitatively proportional to the intensity of the light on the photoreceptors connected with them. He maintained that this is due to the effect of light on muscle tonus. This theory has been designated the "difference of intensity theory," the "continuous-action theory," the "tropism theory," and "Loeb's muscle-tonus theory" (Mast, 1923).

Engelmann (1882) demonstrated that only the anterior end of *Euglena* is sensitive to changes in luminous intensity. Jennings (1904) contends that because of this, all turning from the light results in a reduction of illumination, whereas all turning toward the light results in an increase in illumination of photosensitive substance. The photopositive specimens consequently turn until they face the light, whereas photonegative specimens turn until they face in the opposite direction. When the stimulus which induces turning ceases, the organisms continue either directly toward or from the light.

Mast (1911) made a very intensive study of the process of orientation in a species of *Euglena* which crawls on the substratum but continuously rotates on the longitudinal axis as it proceeds. This *Euglena* orients very precisely in light, it has a well-developed eyespot, and it moves so slowly that the different phases of its responses can readily be followed in detail. It is therefore very favorable for the study of the process of orientation.

If the intensity of the light is rapidly decreased in a beam in which specimens are proceeding toward the source of light, they stop suddenly and bend in the middle toward the abeyespot surface until the two halves form nearly a right angle; then they begin again to rotate on the longitudinal axis; and, while rotating, they gradually straighten and proceed once more toward the light source. If the intensity of the light is increased, or if it is slowly decreased, there is no perceptible response. The cessation of movement and the bending are therefore dependent upon the rate of decrease in the intensity of the light in the field, i.e., it is a shock reaction. The decrease in the intensity of light in the field necessarily results in decrease in intensity of light on all the substance in the field; it therefore must cause decrease in the illumination of the photosensitive substance. The response, then, is dependent upon the rate of decrease in the light on the photosensitive substance.

If the direction of the beam of light is changed through 90° without alteration in intensity, the specimens oriented in it are illuminated laterally. Those in which the eyespot surface faces the light after the direction of the rays has been changed, stop at once. They bend in the middle toward the abeyespot surface, then rotate, and gradually straighten to resume their crawling movements. Those in which the eyespot surface does not face the light after the direction of the rays has been changed, do not respond to the changed direction of the rays, until, in the process of rotation on the longitudinal axis, this surface faces the light; then they also stop, bend, rotate, straighten, and proceed. Thus they continue until, in the process of rotation, the eyespot surface again faces the light, when they again respond. The gradual straightening during rotation results in greater deflection of the anterior end toward,

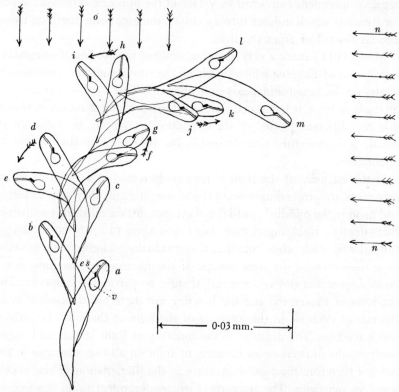

Figure 99. *Euglena* sp. in a crawling state, showing details in the process of orientation; *v*, contractile vacuole; *es*, eyespot; *n*, *o*. direction of light; *a-c*, positions of *Euglena* with light from *n* is intercepted; *c-m*, positions after light from *n* is turned on and that from *o* cut off, so as to change the direction of the rays. If the ray direction is changed when the *Euglena* is in position *c*, there is no reaction until it reaches *d*. Then it suddenly reacts by bending away from the source of light to *e*, after which it continues to rotate and reaches position *f*, where it gradually straightens to *g*, and rotates to *h*, when the eyespot again faces the light and the organism is again stimulated and bends to *i*, from which it proceeds to *j*, and so forth. If the ray direction is changed when the *Euglena* is at *d*, it responds at once and orients as described above. If the intensity from *n* is lower than that from *o* the organism may respond at once when the ray direction is changed, no matter in which position it is. (After Mast, 1911.)

rather than away from the light. Thus the anterior end becomes directed more and more nearly toward the light source, until an axial position is reached in which changes in illumination of the eyespot surface, owing to rotation, disappear. The organism is then oriented (Fig. 99). The response induced by changing the direction of the rays, or by rotation in lateral illumination of uniform intensity, is precisely the same as the response induced by a decrease in the intensity of the light in the field

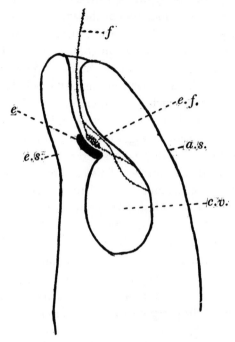

Figure 100. Side view of anterior end of *Euglena viridis*. *e*, pigmented portion of eyespot; *f*, flagellum; *e.f*, enlargement in flagellum; *c.v*, contractile vacuole; *e.s.* eyespot surface of the organism; *a.s.* abeyespot surface of the organism. (After Wager, 1900.)

without a change in the surface illuminated. The change from illumination of the anterior end or the abeyespot surface to illumination of the eyespot surface therefore must, in some way, result in a rapid decrease in the illumination of the photosensitive substance. How is this brought about?

Wager (1900) demonstrated that the eyespot in *Euglena* consists of a spoon-shaped portion containing red pigment and a small globular enlargement of one of the roots of the flagellum in the concavity of the pigmented portion (Fig. 100). The eyespot is situated near the eyespot surface, a short distance from the anterior end, with the convex surface directed outward and backward. When the anterior end, or the abeyespot surface, is directed toward the light, the enlargement in the eyespot is fully exposed; but when the eyespot surface faces the light, the enlargement is in the shadow cast by the pigmented portion. It is evident, then, that rapid change from illumination of the anterior end, or the abeyespot surface, to illumination of the eyespot surface causes rapid decrease in illumination of the enlargement in the eyespot, and that if the enlargement is photosensitive, change in the direction of the rays or rotation on the longitudinal axis has the same effect as decrease in the intensity of the light in the field. It is therefore highly probable that the enlargement in the eyespot is photosensitive and that the pigmented portion functions in producing changes in intensity of light on it, when the axial position of the organism changes and when it rotates on the longitudinal axis in lateral illumination. This contention is supported by the facts that the region of maximum stimulating efficiency in the spectrum is in the blue for *Euglena* (Mast, 1917) and that blue is absorbed by the yellowish-red pigmented portion of the eyespot (Fig. 102).

In photonegative specimens the responses to changes in light intensity in the field and to changes in the surfaces illuminated are precisely like those of photopositive specimens, except that the responses are induced by (1) increase rather than decrease in light intensity and (2) by change from illumination of the eyespot surface to illumination of the abeyespot surface.

The process of orientation in free-swimming specimens is, in principle, precisely the same as it is in crawling specimens.

Orientation in *Euglena* is, then, clearly due to a series of responses dependent upon the rate of change in the intensity of the light on the photosensitive substance, which is probably situated in the concave surface of the pigmented portion of the eyespot. The light does not act continuously, and there is no evidence whatever indicating anything in the nature of balanced or antagonistic action of locomotor appendages on opposite sides, in accord with the Ray-Verworn theory.

The evidence in hand indicates, in short, that the photosensitive substance is confined to the concavity in the pigmented portion of the eyespot; that rotation on the longitudinal axis results in alternate shading and exposing of this substance, if the organisms are not directed toward or from the light; that this induces shock reactions which result in orientation; and that the organisms remain oriented and proceed directly toward or away from the light, because, after they have attained either of these two axial positions, rotation no longer produces changes in the illumination of the photosensitive substance in the eyespot, and they therefore continue in the direction assumed. In other words, the orienting stimulus ceases after the organism has become oriented. The organism then continues directly toward or away from the light because (1) owing to internal factors, it tends to take a straight course, and because (2) if for any reason it is turned from this course, the orienting stimulus immediately acts, and induces shock reactions which bring it back on its course.

Bancroft (1913) presented evidence against the contention that photic orientation in *Euglena* is due to shock reactions and concluded that it is due to tonus effects brought about by "the continuous action of the light," in accord with his conception of Loeb's tropism theory. Mast (1914) demonstrated, however, that if the evidence presented by Bancroft is valid, it proves that his explanation of orientation in *Euglena* is not correct. Moreover, the fact that after *Euglena* is oriented, the rate of locomotion is practically independent of the luminous intensity (Mast and Gover, 1922) also militates against his explanation.

Orientation in light from two sources.—In a field of light consisting of two horizontal beams crossing at right angles, *Euglena* orients and goes toward or away from a point between the two beams. The location of this point is related to the relative intensity of the two beams in such a way that the tangent of the angle between the direction of locomotion and the rays in the stronger beam is approximately equal to the intensity of the weaker divided by that of the stronger (Fig. 101) (Buder, 1917; Mast and Johnson, 1932). Buder maintains that this demonstrates that there is a quantitative proportionality between the stimulus and the response. Mast and Johnson conclude that "it has no bearing on the problem concerning the quantitative relation between stimulus and response," but that it can be explained on the assumptions that the eyespot is a photoreceptor and that the stimulating efficiency of light varies with the angle of incidence.

Wave length and stimulating efficiency.—The shorter waves in the visible spectrum are more efficient than the longer in stimulating *Euglena*

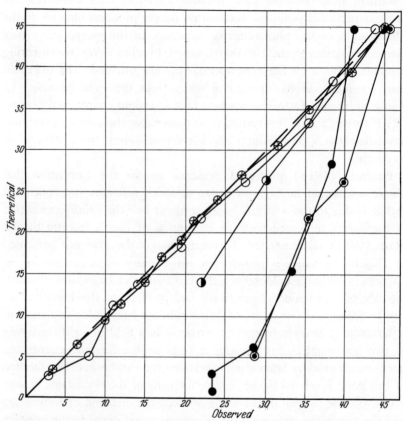

Figure 101. Graphs showing the relation between the direction of locomotion observed in a field of light produced by two horizontal beams crossing at right angles and that demanded by the "Resultantengesetz." Abscissae, angles between the direction of locomotion and the direction of the rays in the stronger beam observed with different ratios of intensities in the two beams, ranging from 0 at 0° to 1 at 45°; ordinates, angles between the direction of locomotion and the direction of the rays in the stronger beam demanded by the "Resultantengesetz." ◑ *Euglena rubra;* ○ *Gonium pectorale;* ● *Volvox minor;* ⊙ *Volvox globator.* (After Mast, 1907.) Note that if the observed direction of locomotion were the same as the theoretical all the points would fall on the broken line, and that this practically obtains for *Euglena* but not for *Volvox* and *Gonium.* Note also that for the latter, as the ratio between intensity in the two beams decreases from 1, the difference between the theoretical and the observed results increases to a maximum, then decreases to zero, after which it increases in the opposite direction. (After Mast and Johnson, 1932.)

and other flagellates. Strasburger (1878) concluded that stimulation is confined to violet, indigo, and blue in the solar spectrum, with the maximum in the indigo. Engelmann (1882) maintains that for *Euglena* the maximum is in the blue between 470 *m*μ and 490 *m*μ, and Loeb and

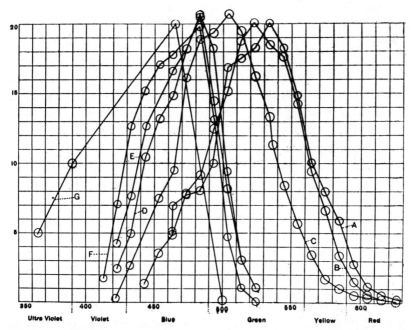

Figure 102. Curves representing the distribution in the spectrum of stimulating efficiency, constructed from data given in Table 15 (See Mast 1917). A, *Pandorina* (negative); B, *Pandorina* (positive); C, blowfly larvae; D, *Euglena viridis* (negative); E, *Euglena viridis* (positive) F, *Euglena tripteris* (negative); G, *Avena sativa* (oat seedlings). (Constructed from data obtained by Blaauw.) The circles represent points experimentally established abscissae, wave lengths; ordinates, relative stimulating efficiency on the basis of equal energy. The curves for *Eudorina* and *Spondylomorum*, not represented in the figure, are in position and form essentially like those for *Pandorina*; the curve for *Chlamydomonas* is much like that for blowfly larvae; those for *Euglena gracilis, E. minima, E. granulata, Phacus, Trachelomonas, Gonium, Arenicola,* and *Lumbricus* are nearly like those for *E. viridis* and *E. tripteris*. (After Mast, 1917.)

Maxwell (1910) assert that in the carbon-arc spectrum it is between 460 and 510 *m*μ. The unequal distribution of energy in the spectrum was not considered in these conclusions. Mast (1917) made corrections for unequal distribution of energy and ascertained the relative stimulating efficiency of negative and positive orientation at intervals of 10 *m*μ throughout the visible spectrum. He found that as the wave length increases, the stimulating efficiency also increases very rapidly from zero at about 410 $m\mu$ to a maximum of 21 arbitrary units at 485 $m\mu$, and then decreases equally rapidly to zero at about 540 $m\mu$ (Fig. 102). He holds, however, that the limits of the stimulating region depend upon the luminous intensity.

Kinetic responses.—If *Euglena* is subjected for long periods to low illumination or to darkness, it gradually becomes less active; and if the illumination is then increased, it gradually becomes more active. The rate of change in activity varies with the magnitude of the change in intensity. But this response is never so sudden and abrupt as the shock reaction. There are therefore two types of responses to light in *Euglena,* one depending primarily upon the rate of change in luminous intensity, the other primarily upon change in the amount of light received. The one results in orientation and aggregation, the other in change in activity.

Mast and Gover (1922) measured the rate of locomotion in several different flagellates in different intensities of light and found very little correlation between rate and intensity. The environmental factors were, however, not accurately controlled, and adaptation was not considered. The measurements should therefore be repeated, with the methods used by Mast and Stahler (1937) in their observations on *Amoeba.*

Reversal in response.—*Euglena* is ordinarily photopositive in weak light and photonegative in strong light. The orienting response therefore tends to keep it in light of moderate intensity, indicating that these responses are fundamentally adaptive. This has not been demonstrated, however, because the direction of orientation is not specifically correlated with luminous intensity. For example, euglenae which are strongly photopositive in a given intensity of light at room temperature may become equally strongly photonegative if the temperature is rapidly decreased 10 to 15 degrees, the extent of the requisite decrease depending upon the state of adaptation (Mast, 1911).

This problem is much in need of thorough investigation. It is a very important problem because it concerns the biological significance of response to light in these organisms.

II. *Peranema tricophorum.*—*Peranema* is a colorless flagellate without an eyespot. It is usually in contact with the substratum and moves slowly with the flagellum extending forward (Fig. 103). If the luminous intensity is rapidly increased, it stops suddenly and then deflects the anterior end sharply to one side. If the intensity is slowly increased, or if it is decreased, there is no response. The entire organism is sensitive to

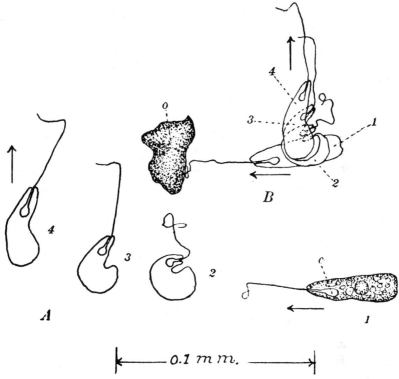

Figure 103. Camera drawings illustrating the response of *Peranema* to contact or to rapid increase in luminous intensity. A1, normal locomotion; 2, immediately after response; 3, 4, recovery from response; B, response to contact with grain of sand, o, and recovery. Note that response results in a change in the direction of motion of approximately 90 degrees. (Mast, 1912.)

light, but the flagellum is most sensitive and the posterior end least sensitive (Shettles, 1937).

Dark adaptation.—Mast and Hawk (1936) demonstrated that if light-adapted peranemae are subjected to darkness, the time required in light of 2,000 meter-candles to induce the response decreases from 30.95 seconds after 15 minutes in darkness to a minimum of 4.54 seconds after one hour in darkness, then increases to a maximum of 63.46 seconds after 6 hours in darkness, and then remains nearly constant. As the time in darkness increases, the sensitivity to light rapidly increases to a maximum, then decreases to a minimum, before it becomes constant (Fig. 104).

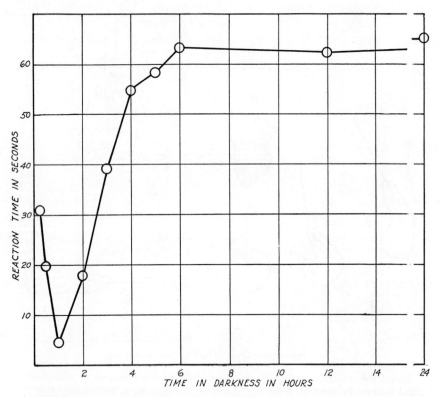

Figure 104. Graph showing the effect of dark-adaptation on sensitivity to light in *Peranema trichophorum*. Each point on the curve, except the last two, represents the average reaction time for from fifteen to seventeen tests. (After Mast and Hawk, 1936.)

Light adaptation.—Shettles (1937, 1938) made a much more extensive and thorough investigation of this response. He confirmed the conclusions reached by Mast and Hawk (1936). A brief summary of other results obtained, and the conclusions reached, follows:

If dark-adapted peranemae are subjected to light, their sensitivity to light increases rapidly to a maximum, then decreases considerably, after which it remains nearly constant (Fig 105). The reaction time consists of an exposure period and a latent period. With increase in intensity of illumination, the exposure period decreases, at first rapidly, then more slowly, until it becomes nearly constant; the latent period increases to a maximum and then decreases, and the amount of light energy required during the exposure period increases from 22,970 meter-candle seconds at 538 meter-candles to 54,315 meter candle seconds at 2,152 meter-candles, and then decreases to 13,498 meter-candle seconds at

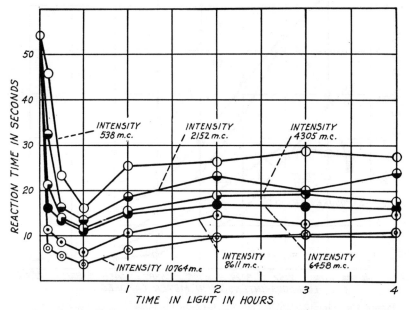

Figure 105. Graphs showing rate of light-adaptation. Dark-adapted peranemas were exposed to light (intensity given in the graph) for the time indicated, then subjected to darkness one half hour, then exposed to 2,152 m.c., and the reaction time measured. Each point on the curves represents the average reaction time for ten tests, different individuals being used in each test. (After Shettles, 1937.)

6,458 meter-candles (Fig. 106). He concludes that "the amount of light energy required to induce a shock-reaction in *Peranema* varies greatly with the intensity of the light and that the Bunsen-Roscoe law consequently does not hold."

The latent period decreases from 39.68 seconds at 10° C. to 24.3 seconds at 30° (15.38 seconds), but the exposure period decreases from 27.87 to 22.97 seconds (only 5.8 seconds). This indicates that there

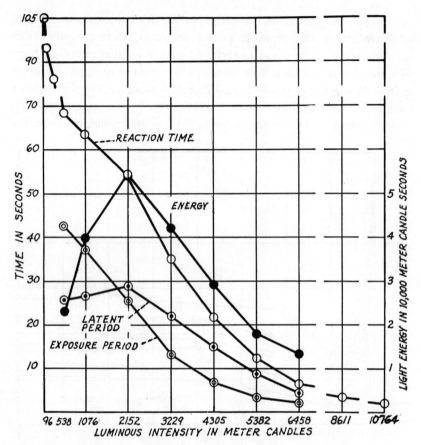

Figure 106. Graphs showing relation between luminous intensity and reaction time, exposure period, latent period, and energy. Each point on the curves of the reaction time and the exposure period represents the average for ten tests, different individuals being used in each test. Each point on the curve of the latent period represents the difference between the reaction time and the exposure period. Each point on the energy curve represents exposure period × luminous intensity. (After Shettles, 1937.)

are at least two processes involved in the response of *Peranema* to light. One of these is nearly independent of temperature and is therefore probably photochemical. The other is closely correlated with temperature and is therefore not photochemical.

After *Peranema* has responded in light of a given intensity, it must be subjected to light of a lower intensity or to darkness before it will again respond. The time required for recovery varies directly with the luminous intensity in which the response occurred, and inversely with the temperature.

Wave length and stimulation.—Stimulating efficiency of light is closely correlated with wave length. There are two maxima in the spectrum, one in the ultra-violet at 302 $m\mu$ and one in the visible at 505 $m\mu$. The latter is nearly twice as great as the former.

The absorption of light by *Peranema* in the violet remains nearly constant as the wave length decreases from 450 $m\mu$ to 325 $m\mu$, then increases rapidly and extensively to a maximum at 253 $m\mu$. The maximum injuring efficiency is also at 253 $m\mu$. Injury is therefore closely correlated with the amount of light absorbed, but stimulating efficiency is not, for the maximum is at 302 $m\mu$ in place of 250 $m\mu$. The processes involved in stimulation therefore differ from those involved in injury. Injury, in *Peranema,* is due to coagulation of the protoplasm, whereas stimulation is probably not due either to coagulation or to increased viscosity.

Peranema responds very precisely and very consistently. Its movements are very slow and its reaction time long. Since it can be grown under fairly accurately known environmental conditions in total darkness, it is well suited for quantitative work of a high order.

C. CILIATES

Very few of the ciliates respond to light. Only one of these, *Stentor coeruleus,* has been investigated extensively.

If the luminous intensity is rapidly increased, this organism stops, turns toward the aboral surface, and then proceeds. This is a shock reaction, because if the intensity is slowly increased there is no response. If the intensity is decreased, no matter whether rapidly or slowly, there is no response. If *Stentor* is exposed in a beam of light, it orients fairly precisely and swims away from the light, i.e., it is photonegative. It rotates on the longitudinal axis as it swims, so that if it is not oriented, the oral and the aboral surfaces are alternately shaded and illuminated. The oral surface is much more sensitive than the aboral; therefore every time that this surface is carried from the shaded to the illuminated side, the result is the same as an increase in the illumination of the entire organism, and it consequently responds, i.e., it turns toward the aboral surface. This continues until it is directed away from the light, and

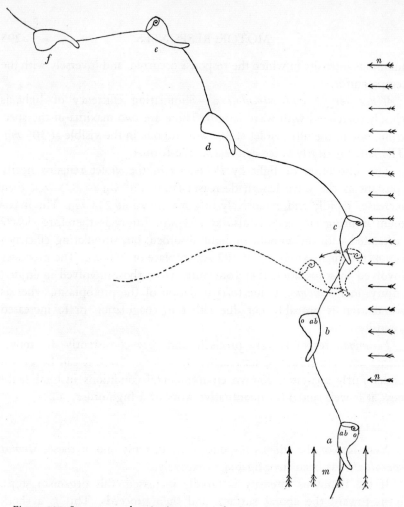

Figure 107. *Stentor coeruleus* in the process of orientation. Curved line, spiral course; arrows *m* and *n*, direction of light from two sources; *a-f*, different positions of *Stentor* on its course; *o*, oral surface; *ab*, aboral surface. At *a* the *Stentor* is oriented in light from *m*, *n* being shaded. If *n* is exposed and *m* shaded simultaneously when the *Stentor* is in position *b*, there is usually no reaction until it reaches *c* and the oral side faces the light; then the organism may respond by suddenly stopping, backing, and turning sharply toward the aboral side (dotted outline), and become oriented at once; or it may merely swerve toward the aboral side without stopping. At *e* the oral side is again exposed, and the organism is again stimulated and it again swerves from the source of light. This process continues until the oral side is approximately equally exposed to the light in all positions on the spiral course. If the Stentor is at *c* when *n* is exposed, it responds at once and orients as described above. If the light from *n* is more intense than that from *m*, or if the organism is very sensitive when *n* is exposed and *m* shaded, it responds at once, no matter in which position it is. If it is at *b*, it turns toward the source of light, but now repeats the reaction, successively turning in various directions until it becomes oriented. (After Mast, 1911.)

rotation no longer produces changes of intensity on the opposite surfaces (Fig. 107). Photic orientation in *Stentor* is therefore the result of a series of shock reactions, as is the case in *Euglena*. There is no evidence in support of the view that it is the result of a continuous quantitative difference in the activity of the cilia on opposite sides, in proportion to the difference in the illumination of these sides. The process of orientation in *Stentor* is therefore not in accord with the Ray-Verworn theory.

If stentors are exposed in a field of diffuse, non-directive light which contains a dark spot, they aggregate in this spot. The process of aggregation is, in principle, precisely the same as the process of aggregation of photonegative euglenae in a dark spot. They reach the dark spot by random movements. No reaction occurs when the organisms enter the unilluminated area. However, at the periphery on the way out, as the light intensity rapidly increases, they stop suddenly, turn sharply toward the aboral surface, and then proceed in a different direction. The dark spot therefore acts like a trap (Jennings, 1904; Mast, 1906, 1911).

The relative stimulating efficiency of different regions in the spectrum has not been investigated; no observations have been made on the quantitative relation between the different phases of the shock reaction, the state of adaptation, and the extent of change in luminous intensity. Indeed, very little is known about the body processes involved in stimulation and response.

D. COLONIAL ORGANISMS

Response to light is essentially the same in all of the colonial forms in which it has been studied, but it has been more intensively investigated in *Volvox globator* than in any of the other species.

Volvox is a slightly elongated, globular colonial organism somewhat less than one mm. in diameter. It consists of numerous cells (zoöids), each of which contains two flagella and an eyespot. The zoöids are arranged in a single layer at the surface of the colonies. The eyespot in each zoöid is directed toward the posterior end of the colony, but those at the anterior end are much larger than the rest (Fig. 108).

Mast (1927a) presented evidence which demonstrates that the eyespots consist of a cup-shaped pigmented portion, a lens-like structure near the opening of the cup, and photosensitive substance between this and the inner surface of the cup. The evidence also indicates that the lens-like structure brings the longer incident waves of light to focus in the wall of the cup; and that the shorter wave lengths, after being reflected from the inner surface of the cup, are focused in the photosensitive substance (Fig. 109).

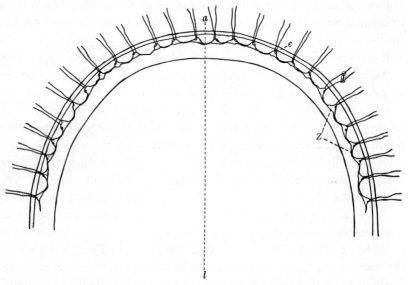

Figure 108. Camera drawing showing the zoöids in about one half of an optical section through the longitudinal axis of a colony of *Volvox*. *l-a*, longitudinal axis of colony; *a*, anterior end; *z*, zoöids, *f*, flagella; *e*, eyes. Note that the eyes are located at the outer posterior border of the zoöids and that they become larger as the anterior end of the colony is approached. (After Mast, 1927.)

Movement, response, and orientation in *V. globator* have been thoroughly studied by Mast (1907, 1926b, 1927b, 1932b). The more important of the results obtained in this study lead to the following conclusions.

Shock reaction.—*Volvox* colonies rotate on the longitudinal axis as they swim. This is due to the diagonal stroke of the flagella. In a beam of light they usually orient and go almost directly either toward or away from the light, i.e., they may be photopositive photonegative, or neutral.

If, while the colonies are swimming toward the light, the intensity is rapidly decreased without any change in the direction of the rays, rotation on the longitudinal axis stops and forward movement increases greatly. On the other hand, if the intensity is rapidly increased, the forward movement stops and the rate of rotation increases. If the colonies are swimming away from the light, the reverse occurs, i.e., forward movement decreases if the intensity is increased, and increases if it is decreased. If the colonies are neutral, there are no such responses to changes of intensity. These responses consist chiefly, if not entirely, of rapid changes in the direction of the stroke of the flagella. In other words, a rapid decrease in the illumination of photopositive colonies changes the

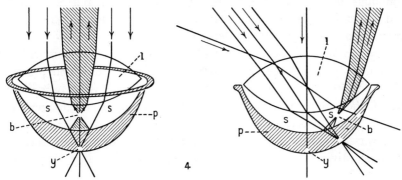

Figure 109. Sketches showing the structure of the eyespot in *Volvox* and its action on light entering the pigment-cup at different angles. *p*, pigment-cup; *l*, lens; *y*, yellow focal spot; *b*, bluish green focal spot; *ss*, photosensitive substance; large arrows, incident rays of light. Note that the longer waves of the incident light are brought to focus in the wall of the pigment-cup and that the shorter waves are brought to focus in the cup, after being reflected from the inner surface, and then continue in the form of a concentrated beam of bluish-green light. Note also that the more obliquely the incident light enters the pigment-cup, the nearer the edge of the cup the yellow focal spot is located. (After Mast, 1927.)

stroke of the flagella from diagonally backward to straight backward. An increase in the illumination causes it to change from diagonally backward to sidewise. In photonegative colonies precisely the reverse obtains. These responses continue for only a few seconds, although if the luminous intensity is slowly changed they do not occur at all. They are therefore dependent upon the rate of change in intensity, i.e., they are shock-reactions which are somewhat similar to those observed in *Euglena*.

Kinetic responses.—If *Volvox* is kept in weak illumination or in darkness for several hours, it becomes inactive; but if the illumination is afterwards increased, it gradually becomes active again. These responses consist chiefly, if not entirely, in changes in the rate or the efficiency of the stroke. Changes in the direction of the stroke of the flagella are not involved. They are relatively slow responses which occur, even if the luminous intensity is gradually changed. The responses are primarily dependent upon change in luminous intensity, not upon the rate of change. Consequently, there are, in *Volvox,* two different types of response: (1) typical shock reactions, and (2) responses which consist merely in changes in activity.

Holmes (1903) maintains there is no consistent correlation between luminous intensity and rate of locomotion in *Volvox.* But his methods did not exclude the effect of adaptation. Further work concerning this correlation is therefore highly desirable.

Orientation.—If a colony of *Volvox* in a beam of light is laterally illuminated, it turns gradually until it is oriented, and then proceeds either toward or away from the light source. When it is laterally illuminated, the zoöids, owing to rotation of the colony on the longitudinal axis, are continuously transferred from the light side to the dark side, and vice versa. As the zoöids pass from the light side to the dark side, the photosensitive substance in the eyespots becomes shaded by the pigment cup. As they pass from the dark side to the light side, this substance becomes fully exposed. A rapid decrease in the illumination of the sensitive substance on the dark side of photopositive colonies induces shock reactions on this side, and the flagellar stroke increases in its backward phase. A rapid increase in illumination of the sensitive substance on the light side of a colony induces shock reactions. The latter consist of increase in the lateral phase of the stroke of the flagella (Fig. 110). This difference in the direction of the stroke of the flagella causes the colonies to orient gradually, until they are directed toward the light, after which all sides are equally illuminated. Rotation on the longitudinal axis then no longer produces changes in the illumination of the photosensitive substance, and the shock reactions cease. The *Volvox* colonies continue directly toward the light because, in the absence of external stimulation, they tend to take a straight course. Furthermore, if they are forced out of their course, opposite sides immediately become unequally illuminated, the intensity of the illumination of the photosensitive substance in the eyespots changes, and consequently reorientation occurs.

If photopositive colonies are exposed in a field of light consisting of two horizontal beams which cross at right angles, they orient and swim toward a point between the two beams. The location of this point de-

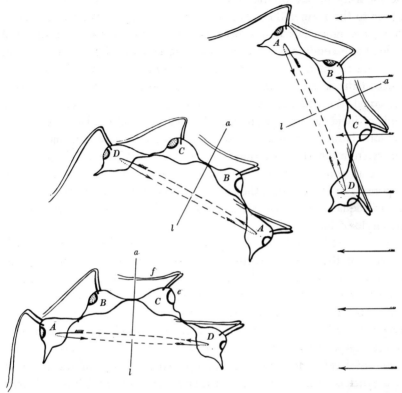

Figure 110. Diagrammatic representation of the process of orientation in *Volvox*.
A, B, C, D, four zoöids at the anterior end of the colony; *l-a*, longitudinal axis; large
arrows, direction of illumination; small arrows, direction of locomotion; curved arrows,
direction of rotation; *f*, flagella; *e*, eyes, containing a pigment-cup represented by a
heavy black line and photosensitive tissue in the concavity of the cup. Note that when
the colony is laterally illuminated, the photosensitive tissue in the eyes on the side
facing the light is fully exposed and the flagella on this side beat laterally. Those on the
opposite side, shaded by the pigment-cup and the flagella on this side, beat directly
backward. The difference in the direction of the beat of the flagella on these two sides is
due to alternate decrease and increase in the luminous intensity to which the photosensitive tissue in the eyes is exposed, owing to the rotation of the colony on its longitudinal axis—an increase causing, in photopositive colonies, a change in the direction
of the stroke of the flagella from backward or diagonal to lateral; and a decrease, a change
from lateral or diagonal to backward. In photonegative colonies, precisely the opposite
obtains. In photopositive colonies, this results in turning toward and in photonegative
colonies turning from the source of light. In both, the turning continues until opposite
sides are equally illuminated, when changes of intensity on the photosensitive tissue are
no longer produced by rotation and the orienting stimulus ceases. (After Mast, 1926a.)

pends upon the relative intensity of the light in the beams. The higher the intensity in one of the beams in relation to that in the other, the nearer to the former the point is. If the intensity in the two beams is equal, the point is halfway between them. The colonies are oriented under these conditions, opposite sides are equally illuminated, both in reference to intensity and direction of the rays, i.e., the angle of incidence at the surface of the colony. If the intensity in the two beams is not equal, the illumination of the oriented colonies is higher and the angle of incidence greater on one side than on the other. However, when a colony is oriented in a field of light, no matter how unequal the intensity from different directions may be, transfer of the zoöids from side to side in consequence of rotation on the longitudinal axis causes no responses. In other words, the effect of unequal illumination on opposite sides is equal. This obviously must be correlated with the difference in the angle of incidence.

Mast (1927a) and Mast and Johnson (1932) demonstrated that the location of the point of focus in the eyespot varies with the angle of incidence. By ascertaining the location of these points in the eyespots on opposite sides of the colonies, in relation to the relative intensity of the two beams, they calculated the distribution of sensitivity and found that the photosensitive substance is much more sensitive in the central regions of the eyespot than at the periphery (Fig. 109). The stimulating efficiency of light, therefore, depends upon the location of the point of focus; this, in turn, depends upon the angle of incidence. The equal effect of light on the sides of colonies which are unequally illuminated on opposite sides when they are oriented, is therefore due to the fact that the point of focus in the eyespots is more nearly centrally located on the side which receives the least light than on that which receives most.

In photonegative colonies the process of orientation is precisely the same as it is in photopositive colonies, except that decrease in intensity causes increase in the lateral phase, and increase in the light intensity increases the backward phase of the stroke of the flagella. In consequence, the illuminated side moves more rapidly than the shaded side. The colonies therefore turn away from the light source.

Orientation of *Volvox* in light is the result of qualitative differences in the action of the locomotor appendages on opposite sides. These differences are due to shock reactions induced by rapid change in the intensity or the location of the light in the photosensitive substance in the eyespot, by virtue of colony rotation on the longitudinal axis. It should be noted that the responses observed are not the result of quantitative differences due to continuous action of the light. The explanation offered is therefore not in accord with the Ray-Verworn theory.

Wave length and response.—The distribution of stimulating efficiency in the spectrum for *Volvox* (Laurens and Hooker, 1920) and *Gonium* (Mast, 1917) is essentially the same as it is for *Euglena;* but for the closely related forms *Pandorina* and *Spondylororum* (Mast, 1917) the maximum is at 535 $m\mu$ in place of 485 $m\mu$, and the effective region extends from this wave length much farther in either direction than it does for *Euglena, Gonium,* and *Volvox* (Fig. 102). The orange, pigmented portion of the eyespot in these forms is opaque in reference to the light of all those regions of the spectrum which have the highest stimulating efficiency. The distribution of stimulating efficiency for these forms consequently supports the conclusions reached concerning the structure of the eyespots, the distribution of photosensitive substance in them, and their function in the process of orientation.

Threshold.—Mast (1907), on the basis of quantitative results, concludes that the minimum difference in light intensity on opposite sides of a colony ·which is necessary to induce a response varies greatly with the physiological state of the colony; but that with colonies in a given physiological state, the response varies directly with the intensity, and the ratio is nearly constant, *i.e.,* nearly in accord with the Weber-Fechner law. His observations, however, covered such a small range (2-27 meter-candles) and the probable error in the results is so large that further observations concerning this relation are highly desirable.

Reversal in response.—*Volvox* is usually positive in weak, and negative in strong light. However, the reverse obtains under some conditions. It may be positive, negative, or neutral in every condition of illumination in which orientations occurs. If it is positive, a shadow on the photosensitive substance in the eyespots causes a change in the direction of the stroke of the flagella of the zoöids from diagonal to backward. A flash of light on this substance causes a change of stroke from diagonal to sidewise. If the colony is photonegative, the reverse obtains; and if it is neutral, there is no response unless the changes in luminous intensity are great.

Reversal in the direction of orientation from positive to negative is therefore due to internal changes of such a nature that shock reactions which were produced by decrease are produced by increase in the illumination of the photosensitive substance. The shift from negative to positive is due to the reverse. The nature of the response to light in *Volvox* depends upon the state of adaptation and upon the intensity of the illumination. If *Volvox* is fully adapted in a given intensity, it becomes positive if the intensity is increased or negative if it is decreased. If the colony is not fully adapted, it becomes negative if the intensity is increased or positive if it is decreased.

The time required for colonies of *Volvox* to become negative or positive after the luminous intensity has been changed (the reaction time) depends upon the degree of adaptation and the extent of the change. If colonies which have been subjected first to strong light (1-2 hours) and then to a variable period in darkness are exposed to strong light, the time required to become positive (reaction time) increases with increase in the length of the period in darkness (dark adaptation) from 0.04 minutes (with 2 minutes in darkness) to a maximum of 0.52 minutes (with 16 minutes in darkness) and then decreases to 0.18 minutes (with 25 minutes in darkness). If the colonies are kept longer in strong light and are then subjected to darkness, the reaction time decreases to a minimum and then increases as the time in darkness increases. If they are left in darkness until they are fully dark-adapted, and are then exposed to light of different intensities, the reaction time (as the intensity increases) decreases from 29 minutes in 5.24 meter-candles to a minimum of 0.098 minutes in 7.5 meter-candles, and then increases to 0.358 minutes in 62,222 meter-candles. The energy required to make the colonies positive varies directly with the light intensity, over the whole range tested. Over most of the range this variation is nearly proportional to the variation in intensity. No satisfactory explanation of this relation is available.

If colonies are kept in a given intensity or in darkness, they become adapted, *i.e.,* they lose the ability to respond to light. Their responsiveness is regained if, after dark adaptation, the intensity is changed. The processes associated with adaptation and those induced by change in illumination are therefore antagonistic. The rate of these antagonistic processes varies greatly, depending upon the magnitude of the change in intensity. For example, if dark-adapted colonies are exposed to light of 22,400 meter-candles for 0.05 minutes then returned to darkness, it takes 20 minutes or more in darkness to eliminate the effect of the light. This indicates that, under these conditions, the processes which occur in light proceed at least 400 times as fast as the reverse processes which occur in darkness.

To account for the phenomena described, it is necessary to postulate at least three interrelated processes, some of which must be directly correlated with light in such a way that change in illumination of very short duration can cause complete reversal in the nature of the response. It is altogether probable (1) that some of these processes are photochemical reactions; (2) that others are dependent upon the results of these; and (3) that all are closely correlated with the physiological state of the organism as a whole (Mast, 1932b). The evidence now available clearly indicates that such simple processes as those postulated by Mast (1907) in his first publication dealing with this problem, and those postulated by Luntz (1932) are very inadequate (Mast, 1932b).

A considerable number of other facts have been established concerning reversal in *Volvox* and related forms. For example, increase in temperature of hydrogen-ion concentration, and some anesthetics (especially chloroform) cause photonegative colonies to become strongly photopositive. However, they usually remain positive only a few moments, then become negative again (Mast, 1918, 1919). There is also a very interesting correlation between reversal in light and response to electricity, in that photopositive colonies always swim toward the cathode and photonegative colonies toward the anode (Mast, 1927c). These facts show that reversal in light is not due to direct action of environmental factors. They also indicate that it is correlated with the rate of metabolism; but there is no clue to the nature of the processes involved.

RESPONSES TO ELECTRICITY

A. RHIZOPODS

All the rhizopods which have been investigated (*Amoeba, Pelomyxa, Difflugia, Arcella, Actinosphaerium,* and others) respond to electricity. Kühne (1864) and Engelmann (1869) observed that if they are subjected to a series of induction shocks (alternating current), streaming in them stops and they then round up. Verworn (1895), from observations on rhizopods in a direct current, maintains that immediately after the circuit is closed, there is marked contraction at the anodal side and then movement toward the cathode, and that if the current is strong enough, disintegration begins on the anodal surface of the organism.

Greeley (1904), in referring to *Amoeba,* says that "on the anodal side of the cells the protoplasm is coagulated . . . and on the cathode side it is liquefied." Bayliss (1920) maintains, however, that the current causes only gelation. According to the careful observations of Luce (1926), hyaline blisters appear on pseudopods oriented with their longitudinal axis perpendicular to the direction of the current. With the aid of superior optical apparatus, he observed the transformation of these blisters into pseudopodia. Since there was no indication of gelation, the phenomenon must have been due to a liquefaction of the plasmagel at the cathodal surface.

More details concerning response of rhizopods to electricity were obtained by Mast (1931b) in observations on *A. proteus* in direct and alternating currents of various intensities. The results obtained are as follows.

Direct current.—In direct current of low density, movement continues no matter how the amoebae are oriented in the field, but the formation of pseudopods is inhibited on the anodal side, resulting in gradual turning toward the cathode. In stronger currents movement ceases immediately after the circuit is closed, then in a few moments one or more pseudopods appear on the cathodal side and movement continues directly toward the cathode. In still stronger currents there is marked contraction on the anodal side immediately after the circuit is closed. This is soon followed by disintegration which begins at this side.

If the anterior end of the amoeba faces the cathode when the current is made, there is, in the lowest density that produces an observable effect, merely a slight momentary increase in the rate of flow in the plasmasol immediately back of the hyaline cap. No change in the rate of flow is seen elsewhere. If the current is stronger, this increase extends back farther, the hyaline cap disappears, the plasmasol extends to the tip, the anterior end becomes distinctly broader, and the plasmagel becomes very thin (Fig. 111). If the current is strong enough, this is followed by violent contraction at the posterior end, slight contraction at the an-

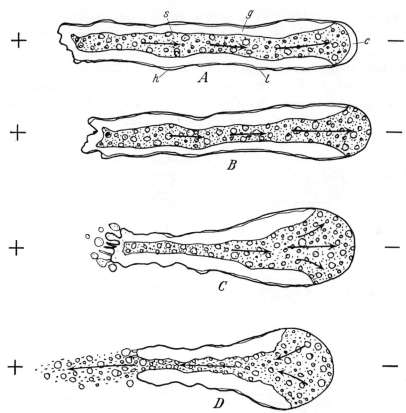

Figure 111. Sketches illustrating the effect of a galvanic current on a monopodal *Amoeba* moving toward the cathode. *g*, plasmasol; *s*, plasmasol; *l*, plasmalemma; *c*, hyaline cap; *h*, hyaline layer; —, cathode; +, anode; arrows, direction of streaming. A, very weak current, B, C, D, progressively stronger current. Note that in a current of moderate density the hyaline cap disappears and the plasmasol extends to the plasmalemma and that in stronger current the cathodal end expands and the anodal end contracts and finally breaks, after which the granules in the plasmasol flow toward the anode, indicating that they are negatively charged. If the surrounding medium is acid, the amoebae do not break. (After Mast, 1931a.)

terior end, and, finally, by disintegration beginning at the posterior end, i.e., that which is directed toward the anode.

If the anterior end of the amoeba is directed toward the anode and the current is weak, there is merely a momentary retardation of streaming at the posterior (cathodal) end. With successively stronger currents, the streaming at this end (1) stops a few moments, then begins again, and continues in the original direction; (2) it stops a bit longer, then begins, and continues in the reverse direction, i.e., the plasmasol then streams toward the cathode at one end and toward the anode at the other (Fig. 112); (3) the reversal extends to the anodal end, and a new hyaline cap forms at the original posterior end, which now becomes the anterior end; (4) the reversal is followed by marked contraction at the

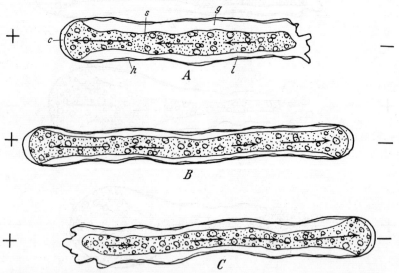

Figure 112. Sketches illustrating the effect of a galvanic current on monopodal amoebae moving toward the anode in a weak current. Labels are the same as in Figure 111. Note that the direction of streaming reverses, that it begins at the cathodal end and proceeds toward the anodal end, and that this results in movement in opposite directions at the two ends during one phase in the process of reversal. (After Mast, 1931a.)

anodal end, which is followed by partial or complete disintegration, always beginning at the anodal end. The fact should be emphasized that no matter how extensive a reversal in the direction of streaming may be, it always begins at the cathodal end of the amoeba.

When the contraction at the anodal end begins, the hyaline layer in this region becomes thicker here and there, resulting in the formation of several small blisters and in numerous minute papilla-like foldings in the plasmalemma (Fig. 111). As contraction continues, some of the blisters, containing fluid and a few granules in violent Brownian movement, round up and are pinched off; and others together with the plasmagel break, after which granules in the plasmasol stream out and proceed rapidly toward the anode. This continues until frequently there is nothing left intact except a crumpled membranous sac, the plasmalemma. The plasmagel changes entirely into plasmasol and is carried away. The fact that the granules are carried toward the anode shows that they are negatively charged.

McClendon (1910) came to the same conclusion concerning the granules in the eggs of frogs and the cells in root tips of onions. But Heilbrunn (1923) says: "Particles in the interior of living cells bear a positive, whereas the particles in the surface layer have a negative charge."

If a small amount of HCl is added to the culture solution, the amoeba does not disintegrate, regardless of the current strength. Its plasmagel turns distinctly yellow at the anodal end immediately after contraction begins, after which it increases in thickness until the entire amoeba has solidified and is dead. If the current is broken before more than about one-fourth of the amoeba has gelated, the gelated portion is usually pinched off. The rest of the amoeba then proceeds normally.

If the amoeba is moving toward the anode when the current is made, streaming of the plasmasol reverses before it stops at the anodal end. This behavior demonstrates conclusively that the effect of the current begins at the surface directed toward the cathode. The fact that before the reversal occurs, the thick plasmagel at the cathodal end is replaced by a very thin plasmagel sheet and a hyaline cap strongly indicates that the first effect of the current is solation of the plasmagel at the cathodal surface. This conclusion is supported by the facts that if the anterior end is directed toward the cathode when the current is made, the plasmagel sheet disappears entirely, the anterior end enlarges, and the plasmasol extends to the plasmalemma. It is also true that if the current passes through the amoeba in a direction perpendicular to the longitudinal axis, the forward streaming stops, and pseudopods are formed on the cathodal side.

The contraction of amoeba at the anodal end, and the increase in the thickness of the plasmagel—especially in specimens directed toward the anode—seem to show that the current causes gelation at the anodal surface.

The facts that the end directed toward the cathode enlarges, that the opposite end decreases in size, breaks up, and becomes yellow just as do amoebae killed in an acid solution, and the granules in the amoeba then move toward the anode, all indicate (1) that the granules bear a negative charge, (2) that water in the amoeba is carried toward the cathode, and (3) that the plasmagel at the anodal surface becomes acid.

In very accurately controlled observations on *A. proteus,* in direct current, Hahnert (1932) found that there is at first a momentary increase in rate of locomotion and then a gradual decrease. He also noted

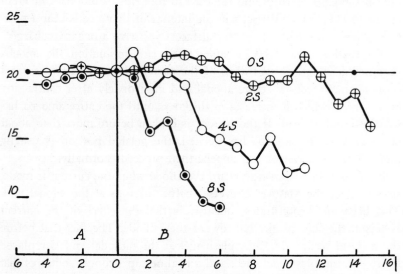

Figure 113. Graphs showing comparative effect of different densities of current on the rate of locomotion. Abscissae, time in minutes, (A) before and (B) during the passage of current; ordinates, apparent rate of locomotion. To obtain actual rate in millimeters per minute, divide the apparent rate by 85. (After Hahnert, 1932.)

that the rate of decrease varies directly with the current density (Fig. 113), and that the time required to induce cessation of movement in specimens directed toward the anode varies nearly inversely with the square of the current density, "the time-intensity relation being nearly in accord with the equation $i\sqrt{t} - (xi) = K$ in which i is the intensity of the current and t its duration."

Alternating current.—If an active *Amoeba* mounted in culture fluid is subjected to a weak alternating current, locomotion ceases at once. The pseudopods are then retracted partially or entirely, and the animal becomes somewhat rounded. It remains in this condition a few moments, then a pseudopod appears on one of the sides between the two surfaces facing the poles and projects at right angles to a line connecting the two poles. Almost immediately after the first pseudopod begins to form, another usually appears on the opposite surface and extends in the opposite direction. Thus the amoeba becomes oriented perpendicularly to the direction of the current, i.e., at right angles to the direction in which orientation occurs in a direct current. These two pseudopods usually continue to stretch out in opposite directions until the amoeba becomes greatly elongated. Then one is withdrawn, and the amoeba continues in the opposite direction, soon moving out of the field (Fig. 114).

These pseudopods usually contain no hyaline cap and no plasmagel at the tip, and the plasmagel elsewhere is very thin. Sometimes such a large portion of the distal end is without plasmagel that the plasmasol very definitely streams back at the surface. If the circuit is opened shortly after these pseudopods have begun to form, no change in movement is seen; if it is closed again, there is still no response. This indicates that pseudopods in which the plasmagel is very thin, or absent, do not respond to electricity; and that the response to electricity consequently is due to its action on the plasmagel.

If the current is stronger, movement ceases and the pseudopods retract just as they do in weak currents. The contraction which follows the retraction of the pseudopods is much more marked, especially on the surfaces directed toward the poles. Here the plasmagel fairly shrinks up and becomes yellowish in color. This is apparently precisely what occurs at the anodal end of an amoeba subjected to the action of a direct current. If the circuit is opened immediately after the violent contraction has occurred, the amoeba soon recovers. But the two masses of plasmagel that have become yellowish are usually pinched off. An irreversible transformation takes place in them, which results in the death of a portion of the cytoplasm. Occasionally, however, these masses, especially if they are relatively small, are taken into the plasmasol and are there digested.

If observations are made under an oil immersion objective on a specimen in an alternating current, the plasmagel can be seen to contract and become yellowish. Fluid is squeezed out on either side of the organism and the adjoining plasmalemma is thrown into folds and papillae of

various sizes. The latter are filled with hyaline substance containing a
few scattered granules which exhibit violent Brownian movement, showing that the substance in which they are suspended is a fluid with low
viscosity. Some of the folds and papillae round up and pinch off, to

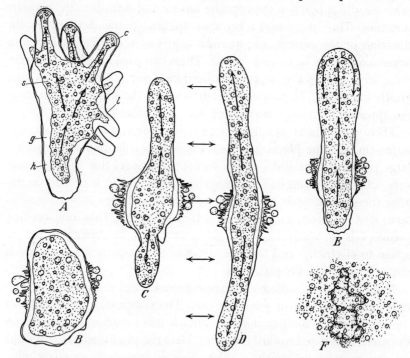

Figure 114. A series of camera sketches of an *Amoeba*, showing the effect of an
alternating current. *A*, before current was made; *B-F*, successive stages after it was
made; *g*, plasmagel; *s*, plasmasol; *l*, plasmalemma; *h*, hyaline layer; *c*, hyaline cap;
arrows, direction of streaming; double headed arrows, direction of the current. Note
that the *Amoeba* orients perpendicularly to the direction of the current, that the plasmagel
in the pseudopods is at this time very thin or absent, that the plasmagel contracts violently
at the surface directed toward the poles, that blisters are formed on these surfaces, and
that the *Amoeba* eventually breaks here and then disintegrates. If the surrounding medium
is acid, the *Amoeba* does not break and disintegrate. (After Mast, 1931a.)

form spherical bodies filled with granule-containing fluid observed in
the folds (Fig. 114).

If the current is strong enough, the plasmagel usually breaks after
it has thus contracted. The plasmasol flows out and the entire amoeba
soon disintegrates and dissolves. Sometimes breaks occur in the pseudo-

pods before there is much contraction, and then the plasmasol flows out through these. As the plasmasol flows out, it collects about the amoeba; and the granules and fluid in it do not stream toward the poles as they do in direct current. Cataphoresis and electroendosmosis are thus neutralized, owing to the reversal in the direction of flow of the current.

There is no change in the responses if acid is added to the culture fluid. However, instead of disintegrating as usual, the plasmasol coagulates after the amoeba breaks up.

The essential phenomena observed in the effect of the alternating current on *Amoeba* appear to be: (1) mild contraction in extended pseudopods, beginning at the tip; (2) violent contraction on the two surfaces facing the poles, with the formation of blisters in these regions; (3) formation of highly fluid pseudopods between these two surfaces; and, finally, (4) rupture at the surfaces directed toward the poles, followed by disintegration of the organism.

Contraction in the pseudopods, as was repeatedly observed, occurs first in those directed toward the poles and last in those in which the longitudinal axis is perpendicular to the direction of the current. Indeed, there is some indication that there is no contraction at all if the axis is actually perpendicular to the direction of the current. Contraction in the pseudopods under these conditions results in retraction which, in all respects, appears to be the same as retraction of pseudopods in normal locomotion. The retraction appears to be due to increase in the elastic strength of the plasmagel, owing to reversible gelation of adjoining plasmasol. It may be concluded, then, that the first effect of the alternating current is reversible gelation of the plasmasol adjoining the plasmagel at the tip of the pseudopods directed toward the poles. This results in an increase in elastic strength of the plasmagel in this region, and in a retraction of the pseudopods. The contraction at the surfaces facing the poles is, in the beginning, doubtless due to the same phenomena; but the facts (1) that the plasmagel in this region later changes in color, (2) that it does not become thicker, (3) that fluid is squeezed out of it, and (4) that it is killed and then breaks, show that contraction here is associated with profound changes in the plasmagel itself. These changes result in such marked decrease in the strength and elasticity of this structure that it breaks readily. These changes are also associated with simultaneous increase in fluidity of the plasmasol, as indicated by the structure of the pseudopods which form at this time. The evidence presented above shows that the described changes are brought about by the action of the current.

Alsup (1939) measured the time required in alternating current and in light respectively to cause cessation in streaming (reaction time) and that required for recovery after this response (recovery period). He found in both that the reaction increased as the recovery period decreased. This indicates that after the plasmasol has gelated, owing to the action of electricity or light, and then solated, owing to the recovery processes, it no longer gelates so readily as it did.

Alsup also found that subminimum exposure to an alternating current followed by a subminimum exposure to light, or vice versa, may induce a response, indicating that the effects of these two agents are additive.

In order to account for the essential phenomena observed in amoebae when subjected to the action of an alternating current, it is then necessary to explain reversible gelation of the plasmasol adjoining the plasmagel on the sides of the organisms facing the poles. It should be noted that the gelation of the plasmasol is followed by changes in the plasmagel in the adjoining regions—changes which result in violent contraction, loss of fluid, decrease in elasticity, and rupture, and by increase in the fluidity of the plasmasol.

Mechanics of response.—In rhizopods all of the responses to direct current appear to be due primarily to solation at the cathodal surface, followed by gelation at the anodal surface. The question then arises as to what causes this.

If an electric current is passed through a culture solution containing amoebae, the negative ions in the solution and in the amoebae move toward the anode, whereas the positive ions migrate toward the cathode. If the surface layers of the amoebae are semipermeable, as they undoubtedly are, there will be an accumulation of positive ions (e.g., Na) on the inside, and negative ions (e.g., Cl) on the outside of the surface of the amoebae directed toward the cathode. Positive ions will accumulate on the outside and negative ions on the inside of the surface directed toward the anode (Ostwald, 1890). The positive ions will, however, unite with the hydroxyl ions of the water, forming bases (e.g., NaOH); and the negative ions will unite with the hydrogen ions of the water, forming acids (e.g., HCl). The cathodal surface layer of the amoebae should therefore become alkaline on the inside and acid on the outside. The anodal surface layer should react in the opposite manner.

Numerous observations were made on specimens stained with neutral red and subjected to direct current under various conditions. No difference whatever was observed in the color of different regions in any of these specimens. The neutral red staining in *A. proteus* is, however, confined to granules and vacuoles. It was observed that if the specimens are crushed, the color of these granules and vacuoles does not immediately change, in accord with the hydrogen-ion concentration of the solution in which they are immersed. It is obvious, then, that the fact that no difference in color was observed in the vacuoles and granules does not prove that the hydrogen-ion concentration of the cytoplasm was the same. Moreover, there is indirect evidence which indicates that it was not the same.

Kühne (1864) long ago observed in certain epidermal cells of the leaves of *Tradescantia* subjected to a galvanic current that the ends of the cells directed toward the cathode become alkaline, and that those directed toward the anode become acid. These cells contain a natural indicator which is bluish in neutral solutions, red in acid solutions, and green in alkaline solutions.

Habenicht (1935) came to the same conclusion in experiments on the effect of the galvanic current on cylinders of egg white.

Mast (1931b) repeated and extended Kühne's experiments and obtained results which confirm his contentions. It may then be assumed with considerable confidence that when an amoeba is subjected to a direct current, the hydrogen-ion concentration in the cytoplasm decreases at the cathodal end and increases at the anodal end.

Edwards (1923) demonstrated that if an alkaline solution is locally applied to the surface of an amoeba, the plasmagel in this region disintegrates; and that if acid is applied, it becomes thicker, owing to gelation of the adjoining plasmasol. This has been confirmed indirectly by Pantin (1923), Chambers and Reznikoff (1926), and others. If, then, the direct current produces a decrease in hydrogen-ion concentration at the cathodal end and an increase at the anodal end, one would expect the plasmagel to become thinner at the cathodal end and thicker at the anodal end. This is precisely what was observed. And if the elastic strength of the plasmagel varies directly with its thickness, as is doubt-less true, this would result in formation of pseudopods at the cathodal surface. This has been confirmed by observation. Since streaming toward the cathode begins at the cathodal surface before it does at the anodal surface, movement toward the cathode must be due primarily to the solation of the plasmagel at the cathodal surface. The accumulation of positive ions at this surface therefore must produce the solation. But this obviously does not account for the disintegration of the entire organism. Neither does it account for the violent contraction preceding disintegra-tion at the anodal side. It will be remembered that violent contraction and disintegration beginning at the anodal surface were observed only in relatively strong currents. Furthermore, the reactions were observed to occur only after large pseudopods develop and begin to advance to-ward the cathode.

An amoeba disintegrates only if the direct current applied is suffi-ciently strong. After the circuit is closed, there is, on the inner surface of the plasmagel or in the plasmagel, an accumulation of positive ions at the cathodal side, and of negative ions at the anodal side. The former produces a decrease in the elastic strength of this layer, which results in the formation of a pseudopod directed toward the cathode—a pseudopod in which the plasmagel extends to the plasmalemma. Local disintegra-tion occurs at first; but, as the current continues, more and more of the plasmagel in this pseudopod disintegrates. The accumulation of nega-tive ions and consequently of hydrogen ions at the anodal end causes a thickening of the plasmagel, as well as gelation of the adjoining plas-masol. This results in violent contraction and finally in the rupture ob-served at the anodal end.

Cataphoresis and electroendosmosis are probably also involved. The granules in *Amoeba* are negatively charged in relation to the fluid. The fluid consequently tends to flow from the anodal toward the cathodal end. This would facilitate contraction at the former and expansion at the latter end, which is precisely what was observed. There is, however, some evidence to indicate that transfer of water is of little importance in the rupture and disintegration at the anodal end. This will be presented later.

It is therefore fairly clear how, in a direct current, substances accumu-late locally; and how this can produce most of the processes associated with the responses of amoebae in it. But in an alternating current the situation is quite different. There is an equal movement of all substances in opposite directions. Consequently there can be no accumulation of different substances in different parts of the organism, unless there is some process which makes the movement in one direction greater than that in the other.

The essential phenomena observed in *Amoeba* as a consequence of exposure to alternating current may be summarized in the order of their appearance. A mild contraction begins first at the tips of extended pseudopodia. This is followed by violent contraction on the two surfaces facing the poles (blisters appear on these surfaces). Then highly fluid pseudopods form between these two surfaces. Finally, the surfaces directed toward the poles rupture and the organism disintegrates.

Numerous observations with the best lens system obtainable were made on the movement of microscopic particles, both in the field of the alternating current and in the amoebae in this field. There was no indication of a drift of these particles nor of their accumulation in any part of the organisms. It therefore is evident that cataphoresis and electroendosmosis cannot be involved in the observed contraction. It seems necessary, then, to conclude that the phenomenon of contraction is associated with the movement of the ions produced by the electric current; and, further, that ion movements are accompanied by processes which result in the accumulation of ions in certain regions of the organism.

Dixon and Bennet-Clark (1927) and others maintain that alternating current causes increase in the permeability of the plasmamembrane in cells. If so, then may not the contraction observed in the plasmagel be due to the action of substances which enter from the surrounding medium, since localized accumulation of ions at the surface of the amoeba increases its permeability owing to the action of the current?

Two facts suggest that the contraction of *Amoeba* during exposure to alternating current cannot be due to the entrance of substances from the outside. In the first place, the contractions are known to occur in both alkaline and acid solutions. Secondly, alkaline solutions tend to produce solation in the plasmagel, thus decreasing its elastic strength. Possibly the accumulation of ions in or near the plasmagel causes the contraction.

It is well known that the positive ions, Na, K, Ca, and others ordinarily pass through membranes more readily and more rapidly than the negative ions, SO_4, PO_4, NO_3, and others. In a structure like the plasmagel, the movement of ions is undoubtedly hindered. Nevertheless, the positive ions may move farther from their initial positions toward the pole than do the negative ions. It may be assumed that the negative ions tend to remain in the plasmagel, whereas the positive ions tend to leave it and return again as the direction of the current reverses. If the return movement of the positive ions is inhibited, there may be a momentary preponderance of negative ions within the plasmagel, and of positive ions in the adjoining substances, i.e., in the plasmasol, the hyaline layer, and the plasmalemma. If the movement of the positive ions away from the negative ions is extensive enough, the negative ions remaining in the plasmagel will unite with the hydrogen ions of the water surrounding them, to form acids. The hydroxyl ions thus liberated will, owing to the fact that they pass rapidly and freely through tissues, move out and unite with the positive ions, to form bases in the substance adjoining the plasmagel. It is possible that this union retards the return movement of the positive ions during the next reversal in the direction of the current. Consequently the effect would be cumulative, gradually increasing the acidity within the plasmagel and the alkalinity of the substance on either side.

The increase in acidity in the plasmagel would produce gelation in this layer and probably also in closely applied plasmasol. This would increase the thickness and elastic strength of the plasmagel. Contraction would be the result. If the accumulation of negative ions were great enough, the increase in acidity in the plasmagel would cause irreversible coagulation (death), accompanied by violent contraction and dehydration, thus making the coagulated plasmagel so brittle that it would break readily. The increased alkalinity in the plasmasol and plasmalemma would tend to make the former more fluid, and it would tend to break up the latter. The postulated action of the current is precisely in accord with the observations. It also accounts for the fact that no response was observed in pseudopods which contained no plasmagel. According to the explanation offered, an accumulation of ions occurs in or near the plasmagel; there could be no action in structures which have none.

Moreover, on the basis of this hypothesis, it is possible to account for the well-known fact that in many organisms the effect of a current varies inversely with the frequency of reversal. It is necessary only to assume that the higher the frequency of reversal, the more restricted the movements of the ions toward the poles, and the shorter the period of separation of positive from negative ions; and that the shorter this period, the more restricted the union of the negative ions with hydrogen ions, and the positive ions with hydroxyl ions. The more restricted these unions, the less the increase in acidity in the plasmagel, and the less the increase in alkalinity in the plasmasol, and the less the stimulating and the injurious effect. The hypothesis, then, that the action of the electric current on organisms is due to localized increase in acidity and alkalinity in different regions of the cell is in full accord with the fact that the effect in alternating currents varies inversely with the frequency.

What bearing has all this on the problem concerning the observed contraction at the anodal side of amoebae subjected to direct current?

Carlgren (1899), as previously stated, holds that this is due to electroendosmotic extraction of water by the current, owing to negative charge of the solid substance. However, the fact that the same phenomenon occurs in alternating current, in which electroendosmosis is neutralized, strongly indicates that Carlgren's conclusion is not valid. It also seems to show that the anodal contraction in direct current must be due, as appears to be the case in alternating current, to the action of the current on the movement of ions.

The evidence presented in reference to the effect of both direct and alternating current on *A. proteus* indicates that the assertion of Bayliss (1920), Weber and Weber (1922), Taylor (1925), and others that electricity gelates cytoplasm, is misleading, for it shows that if an electric current causes gelation in a cell, it probably always causes simultaneous solation, each being confined to a portion of the cell.

It was demonstrated above that this does not obtain for light. The implication frequently found in the literature that the action of electricity on protoplasm is the same as the action of light, appears therefore to be erroneous.

Heilbrunn and Daugherty (1931) found that if ammonium hydroxide or chloride is added to the culture fluid, *A. proteus* becomes anopositive. They maintain that the "protoplasmic granules" are ordinarily positively charged and are consequently carried (cataphoretically) toward the cathode, and that ammonium hydrate or chloride causes a change in the charge to negative and a consequent reversal in the direction in which they are carried. They contend that the contact of these granules with the inner surface of the plasmagel causes it to liquefy, and that this results in the formation of a pseudopod which is directed toward the cathode if the granules are positively charged and toward the anode if they are negatively charged. These authors say:

As the granules move either toward cathode or anode, they must tend to break down the thixotropic gel on the side toward which they move. . . . If this gel is liquefied in any local region, such a region becomes pushed out to form an advancing pseudopod.

These are interesting views, but they obviously do not account for the direction of movement of pseudopods in alternating current (Fig. 114), nor for the direction of movement of pseudopods under some conditions in direct current (Fig. 112). For this would require cataphoretic movement of the granules perpendicularly to the direction of the current in the former, and in opposite directions at opposite ends of the amoeba in the latter. Moreover, the fact that the granules stream toward the anode after the amoeba disintegrates, indicates that they are normally negatively (not positively) charged, and that this charge is consequently not involved in the cathopositive response. How, then, can the observed reversal in the direction of orientation be explained?

The cathopositive response is probably brought about as described above. The ammonium hydrate or chloride added to the culture fluid probably results in the liquefaction of all the plasmagel, and consequently in free movement toward the anode of the negatively charged granules suspended in it. According to this view, the reversal in the direction of galvanic orientation is due to liquefaction of the plasmagel caused by the ammonium compounds used, not to change in the electric charge on the "protoplasmic granules."

B. FLAGELLATES

Some of the flagellates orient very precisely in a direct current. Some are cathopositive, others anopositive, and still others both or neutral, depending upon the environmental conditions (Verworn, 1889; Pearl, 1900; Bancroft, 1913). Verworn maintains that orientation is brought about by differences in the effective stroke of the flagellum in opposite directions. Pearl believes that it is the result of typical avoiding reactions, whereas Bancroft thinks that it is identical with the process of orientation in light. Further details concerning the processes involved are much desired. Moreover, it is noteworthy that some of the flagellates have proved to be excellent material for quantitative study of the relation between stimulus and response.

C. CILIATES

The responses to electricity have been more intensively and extensively studied in the ciliates than in any of the other groups of Protozoa.

Jennings (1906) presents an excellent review of all the earlier investigations concerning these responses. He maintains that the results obtained show the following:

The principal feature in the response of all of the different species studied consists of reversal in the direction of the effective stroke of the cilia on the cathodal surface. In those species in which other ciliary actions are only slightly or not at all involved (e.g., *Paramecium*), this results in direct turning until the anterior end is directed toward the cathode. The organism then moves toward this pole. The extent of the cathodal surface affected varies directly with the strength of the current. If the current is strong enough to produce reversal over more than half of the surface of the *Paramecium*, it swims backward toward the anode. A still stronger current causes marked swelling of the anterior end and contraction of the posterior end, changes which are followed by disintegration beginning posteriorly (Fig. 115). The effect of the reversal on the cathodal surface is variously modified by the normal action of the cilia in other regions, in such a way that "with different strengths of current, and with infusoria of different action systems, this results sometimes in movement forward to the cathode; sometimes in movement forward to the anode; sometimes in cessation of movement, the anterior end continuing to point to the cathode; sometimes in a backward movement to the anode; sometimes in a position transverse to the current, the animal either remaining at rest or moving across the current."

Jennings holds that all the responses of the ciliates to electricity are due to stimulation at the cathodal surface, resulting in local reversal in the direction of the effective stroke of the cilia on this surface.

Bancroft (1906), in his experiments with *Paramecium*, observed that if certain salts (especially potassium, sodium, or barium salts) are added to the solution, the organisms will swim forward toward the anode. He maintains that this is due to stimulation at the anodal surface, and that consequently it is an exception to Pflüger's law.

It is well known that some salts cause paramecia to swim backward,

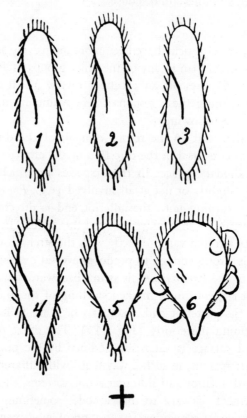

Figure 115. Progressive cathodic reversal of the cilia and change of form in *Paramecium* as the constant electric current is made stronger. The cathode is supposed to lie at the upper end. The current is weakest at *1*, where only a few cilia are reversed; 2-6, successive changes as the current is gradually increased. (After Statkewitsch, 903.)

owing to the forward stroke of all the cilia (Jennings, 1899; Mast and Nadler, 1926; Oliphant, 1938). Obviously, if under such conditions there is reversal in the direction of the stroke at the cathodal surface, the paramecia will turn and swim forward toward the anode (Mast,

1927c). If this reaction obtains, the forward swimming toward the anode is in full accord with Pflüger's law.

Kamada (1929) made a study of the correlation between the effect of many different salts on reversal from forward to backward swimming, and the direction of orientation in a direct current. He maintains that some salts which induce forward swimming toward the anode do not induce backward swimming. He consequently supports Bancroft's views.

Kamada (1931) also maintains that in paramecia which are ano-positive there is ciliary reversal at the anodal surface, in place of the cathodal, and that with increase in current density this is modified in various ways. The evidence he presents is, however, by no means conclusive. Further observations are therefore needed.

Paramecia are most sensitive if the anterior end is directed toward the cathode, less sensitive if it is directed toward the anode, and least sensitive if the longitudinal axis is perpendicular to a line connecting the two poles (Statkewitsch, 1903; Kinosita, 1936). The same holds for *Spirostomum* (Kinosita, 1938a).

Statkewitsch (1907) subjected to direct and to alternating currents paramecia which had been stained with neutral red. He maintains that the stained structures in them became violet (acid) in weak currents and distinctly yellowish (alkaline) in strong currents. He apparently did not observe any difference in the color at the two ends. In Kinosita's (1936) experiments with paramecia stained with either neutral red or Nile blue sulphate, the color of opposite ends differed. He says that the changes in color observed show that the paramecia become acid at the cathodal and alkaline at the anodal end, but that the alkaline portion rapidly extends forward and soon includes the entire body.

There is consequently a diversity of opinion concerning the effect of the electric current on the hydrogen-ion concentration of the cytoplasm in *Paramecium*. This also obtains for other cells, since, as previously stated, Kühne maintains that cells of *Tradescantia* become acid at the anodal end and alkaline at the cathodal end. Mast confirmed this, but could observe no difference in this respect between the two ends in *Amoeba*. It is therefore obviously desirable to have further observations concerning this problem, for it is theoretically very important.

Kinosita measured the time required to make the anodal end alkaline in different current densities, extending over a wide range. He maintains that the results indicate that $i\sqrt{t-a} = K$. This equation is similar to the one obtained by Kamada in observations on the destruction of the surface membrane in paramecia by a direct current. It is also related to the one obtained by Hahnert (see above) in his observations on the cessation of streaming in *Amoeba* in a constant current.

Internal processes involved in response.—Several theories have been formulated to account for the responses of ciliates to an electric current. Loeb and Budgett (1897) contend that these are in reality the result of responses to changes in the chemical constitution of the environment produced by the electric current. Pearl (1901) asserts that the direction of the stroke of the cilia is specifically correlated with the direction of protoplasmic streaming directly below the surface. Coehn and Barratt (1905) hold that the movements are purely cataphoretic; Bancroft (1906) maintains that the galvanic responses are due to local changes in the calcium content of the tissue in relation to that of other ions, especially monovalent cations. Although Carlgren (1899) lays especial stress on localized changes in water content within the organisms, due to endosmotic streaming. Some assert that the movements are due to direct action of the electric current on the cilia, others that they must be due to action on a coördinating center. Ludloff (1895), Verworn (1895), and Koehler (1925) postulate functional division of the organisms into anterior and posterior halves, such that one responds in one way that the other responds in another way. The views of Nernst (1899), Lucas (1910), Lillie (1923), and others regarding stimulation in higher organisms would lead to the idea that local changes in permeability of the surface membrane is the all-essential in controlling the movements of the lower organisms in an electric field.

The relation between these hypotheses and the facts established has been very illuminatingly discussed by Jennings (1906) and Koehler (1925). Both conclude that the facts in hand are not adequately accounted for by any of the hypotheses presented. Jennings contends, as previously stated, that the most important perceptible characteristic of the response is reversal in the direction of the stroke of the cilia on the cathodal side. Koehler holds that the processes involved cannot be solely dependent upon surface phenomena, that somehow the current results in a division of the organism into cathodal and anodal portions which function differently, owing to different internal factors. But neither Jennings nor Koehler offers any explanation of how the responses are regulated.

Among the most important of the known facts concerning the responses in Protozoa to an electric current are those discovered by Ludloff in observations on *Paramecium*. Ludloff (1895) found, as has been abundantly confirmed, that when the circuit is closed, the direction of the stroke of the cilia on the surface of the paramecia directed toward the cathode reverses; but that if the longitudinal axis of the organisms is directed obliquely to the direction of the current, reversal occurs on

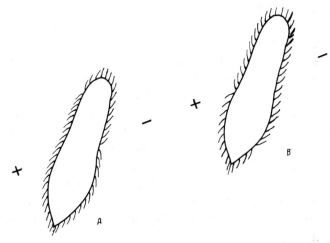

Figure 116. *Paramecium* showing reversal in the direction of the stroke of the cilia in a galvanic current. A, weak current; B, strong current; +, anode, —, cathode. (After Ludloff, 1895.)

all sides of the end of the body nearest the cathode, extending to a line around the body produced by passing a plane through it at right angles to the direction of the current. The extent of the portion of the body on which such reversal occurs depends upon the strength of the current: the stronger the current, the larger the portion affected (Fig. 116).

The fact that the cilia in different regions on the same side of the paramecia are not always affected equally by the current seems to show that the responses observed cannot, as Koehler points out, be due to direct action on the cilia or to surface phenomena alone; for if they were, all the cilia on either side should act alike, with the possible exception of those in the oral groove. If this is true, it is evident that the responses in this form must be associated with internal changes.

The fact that in specimens with the longitudinal axis directed obliquely to the direction of the current, the entire end nearest the cathode is affected, and the fact that the size of the portion affected varies directly with the strength of the current, indicates not only that the current results in a functional division of the organism (as maintained by Ludloff, Verworn, and Koehler), but also that whatever the factors involved may be, they act on a structure which is well distributed through the entire body and which is located some distance below the surface—probably the neuromotor apparatus. For only in a structure which is some distance from the surface, could a current produce the same changes in the distribution of substances on the anodal and the cathodal sides of the portion affected, resulting in reversal in the direction of the stroke in all the cilia on this portion.

It is well known that momentary reversal in the stroke of the cilia can be induced in *Paramecium* by almost any sudden environmental change (Jennings, 1906), and that more prolonged reversal can be induced by transfer from culture fluid to distilled water or from distilled water to solutions of monovalent cation salts, but not usually by transfer to solutions of bivalent cation salts (Mast and Nadler, 1926; Oliphant, 1938). These changes, therefore, produce the same result as is produced by an electric current on the cathodal surface, indicating similarity in action.

Greeley (1904) maintains that paramecia drift toward the anode, i.e., that they are negatively charged, indicating that there is a negative layer at the surface. Statkewitsch (1903) observed that if one end is directed toward the anode and the other toward the cathode, the former shrinks and the latter swells, indicating that the more solid substance is negative in relation to the more fluid substance. If all this obtains, a direct current will result in a decrease in the concentration of the positive, or an increase in the concentration of the negative ions on the cathodal side of the surface of the organism and on each semipermeable structure within. There will also be a decrease in the concentration of the negative, or an increase in the concentration of the positive ions on the opposite side. As a result, water will drift toward the cathodal side, and the solid particles will drift in the opposite direction.

Since reversal in the stroke of the cilia begins on the cathodal side, it would seem that it must be associated either with an increase in the concentration of the negative ions or with a decrease in the concentration of the positive ions, on the cathodal side of the semipermeable structures on this side of the organisms. Either a decrease of polarization or an increase in the water content of the cytoplasm may be involved. If this is true, then transfer from culture fluid to distilled water and from distilled water to solutions of monovalent cation salts should, since this results in reversal of the stroke of all the cilia, produce similar changes in the concentration of ions or water. In other words, the transfer from culture fluid to distilled water should produce a decrease in the concentration of positive ions on the outside of the semipermeable structures, or an increase in the water content at the surface. A transfer from distilled water to solutions of monovalent cation salts should produce like changes. Whether or not this obtains is at present unknown, but one would expect it to obtain, if, as is frequently asserted, permeability is increased by monovalent cation salts.

D. COLONIAL ORGANISMS

No detailed observations have been made on the response to electricity of any of the colonial organisms except *Volvox*. *Volvox* orients very precisely in direct current. It swims toward the cathode under some conditions and toward the anode under others. Carlgren (1899) maintains that reversal in the direction of orientation is correlated with the duration of exposure to the current. Terry (1906) and Bancroft (1907) contend that it is correlated with the intensity of the light received and the duration of exposure to it. Mast (1927c) found that *Volvox* swims toward the cathode when it is photopositive, and toward the anode when it is photonegative, i.e., that the response to electricity is specifically correlated with the response to light.

In photopositive colonies in which rotation on the longitudinal axis is inhibited by means of pressure, the flagella on the cathodal side stop beating immediately after the circuit is closed. They remain inactive 4.5 to 6 seconds and then begin to beat again (Fig. 117). If the circuit is now opened, those on the anodal side stop for a few moments and then beat again. If the colonies are swimming and rotating on the longitudinal axis in the normal way, but are not proceeding directly toward either pole when the current is made, flagellar inactivity on the cathodal side is continuous, owing to the continuous transfer of the zoöids from the anodal to the cathodal side. The colonies therefore turn toward the cathode until they face it directly, and the transfer of zoöids from side to side ceases.

Orientation.—In photonegative colonies precisely the opposite occurs. The flagella on the anodal side stop beating after the circuit is closed,

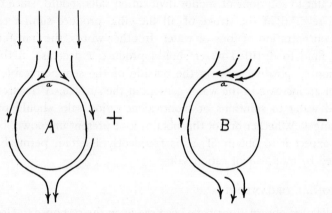

Figure 117. Sketch showing in a stationary photopositive colony of *Volvox* the effect of a galvanic current on the currents of water produced by the flagella. A, outline of colony oriented in light; B, same colony immediately after the circuit was closed; *a,* anterior end; straight arrows, direction of illumination; curved arrows, currents produced by the flagella; +, positive pole; —, negative pole. (After Mast, 1927.)

and the colonies turn toward the anode until they face it and then swim toward it (Mast, 1927c).

Galvanic orientation is consequently correlated with photic orientation, but the processes involved differ, for, as previously stated, photic orientation is due to a change in the *direction* of the stroke of the flagella on opposite sides, while galvanic orientation is due to *decrease* or *cessation* in the activity of the flagella on one side.

Electric charge on the colonies.—Galvanic orientation in *Volvox* also differs from that in *Paramecium,* for while the one is due to decrease or cessation in activity on one side, the other, as previously stated, is due to reversal in the direction of the effective stroke of the cilia on one side.

Since a given colony of *Volvox* may be either photopositive or photonegative in the same environment, the difference in response to the light in this environment must be due to changes in the colony itself. Since photopositive response is specifically correlated with cathopositive response, and photonegative response with anopositive response, the difference in the response to electricity must be due to like changes in the colony. The only difference observed in the colonies in connection with the response to light concerns the electric charge. Referring to this, Mast (1927c) says:

> Most of the photopositive colonies observed drifted toward the anode and most of the photonegative ones drifted toward the cathode, indicating that the former were negatively and the latter positively charged. However, owing to the negatively charged glass bottom of the aquarium in which the observations were made, there was produced in the solution near the bottom an endosmotic current of water toward the cathode and this current produced at the upper surface a current in the opposite direction, i.e., toward the anode. In making the observations it was impossible to ascertain precisely the location of the colonies in relation to these currents, resulting frequently in uncertainty as to whether the drift was due to cataphoresis or to endosmosis. The results obtained are consequently somewhat equivocal.

These observation should therefore be repeated under more favorable conditions, for the results are of fundamental importance in the analysis of the mechanics of the response to electricity, as will be shown presently.

Mechanics of response.—The outstanding characteristics of the responses to the electric current of the colonial forms, exemplified in *Volvox,* consist in momentary decrease in the action of the flagella, correlated with the direction and the density of the current and the nature of the response to light. In photopositive colonies this occurs in such a way that the flagellar activity decreases on the cathodal side after the current is made. In photonegative colonies the activity decreases on the anodal side, continues a few seconds, and then increases again. But if the current is broken, it decreases on the opposite side, continues a few seconds, then begins again. The extent of the region affected under all conditions varies directly with the density of the current.

The action of the current must be due to movement of ions, particles, or fluid in the colonies or the surrounding solution, and to differences in the responses in photonegative and photopositive colonies to differences in the effects produced by the movements of these substances.

There is considerable evidence which indicates that when colonies are photopositive and cathopositive, they are negatively charged; and when they are photonegative and anopositive they are positively charged (Mast, 1927c). If this obtains, the movement of ions in photopositive colonies in a direct current would result in decrease in the negative ions

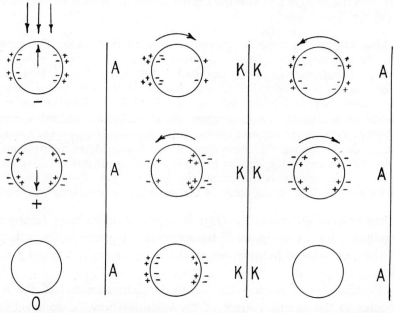

Figure 118. Diagrams illustrating the effect of direct current on the distribution of ions in colonies of *Volvox* and their response. —, negative charge on colony; +, positive; *o,* no charge; *A,* anode; *K,* cathode; large arrows, beam of light; small arrows, direction of movement of colonies in the beam of light; curved arrows, direction the colonies turn after the current is made and broken respectively. Note that there is in negatively charged colonies, after the circuit is closed, a decrease in potential at the cathode, and a decrease at the anode after it is opened, that the opposite obtains for positively charged colonies, and that there is no change in potential in neutral colonies. (Modified after Mast, 1927b.)

on the cathodal side. In turn, this would cause decrease in potential and increase in permeability here. It would also result in increase in the negative ions at the anodal side and increase in potential and a decrease in permeability there (Fig. 118). The movement of ions in photonegative colonies would result in increase in positive ions, in the potential, and increase in permeability on the cathodal side. The reverse would occur on the anodal side. The ion movement in photoneutral colonies would result in increase in positive ions, and increase in potential and decrease in permeability on the cathodal side. On the anodal side there would be equivalent increase in negative ions and increase in potential and decrease in permeability. After the current is broken, the change in distribution of ions, and its effect, would be precisely opposite in all respects (Fig. 99).

If the decrease in flagellar activity is due to local decrease in polarization and increase in permeability, it accounts for the observed direction of movement of photopositive and photonegative colonies in a direct current. And if the decrease in flagellar activity is correlated with rate of change in these characteristics, it also accounts for the fact that the decreased activity induced by making or breaking the current continues only a few seconds, for the change in polarization and permeability undoubtedly lasts but a few seconds.

How does the fact that strong currents cause decrease in the activity of the flagella simultaneously on all sides of the colonies, harmonize with this view?

It is well known that a galvanic current will produce cytolysis if it is strong enough, and that cytolysis is associated with increase in permeability and decrease in polarization. It is therefore not difficult to see that such a current could cause decrease in polarization simultaneously on all sides.

It is probable, however, that the processes involved in galvanic stimulation, resulting in orientation, are not the same in all organisms. For example, Ludloff (1895) and Statkewitsch (1903) found that in a galvanic current the fluid in the body of *Paramecium* is carried endosmotically toward the surface, on which the stroke of the cilia reverses, i.e., in the specimens which swim forward toward the cathode, there is, if the current is strong enough, contraction at the anode and expansion at the cathode end. In *Volvox* precisely the opposite obtains. In the one, stimulation appears to be associated with increase, in the other, with decrease in water content. In both, however, as pointed out above, it appears to be associated with decrease in polarization.

In general, it may be said that a galvanic current usually induces in the lower organisms chemical and physical changes which differ at opposite sides, the one facing the anode, the other the cathode; and that either or both of these two sets of changes may result in what is usually called a response, although not necessarily an orienting response. Thus while the motor responses in these organisms usually occur at the cathodal side when the circuit is closed, Kühne (1864), Verworn (1895), and McClendon (1911) observed in *Amoeba* and other rhizopods contraction at the anodal side. In some of these organisms, the anodal contraction appears to be involved in streaming toward the cathode; but there are other anodal responses which obviously have nothing to do with locomotion. For example, Loeb and Budgett (1897) assert that there is, in *Amblystoma,* copious secretion of mucus on the anodal side. Moore (1926) obtained bioluminescence and contraction on the anodal side of the Ctenophores, *Mnemiopsis,* and *Beröe.* Lyon (1923) and Lund and Logan (1925) observed, in *Noctiluca,* a sort of contraction first at the anodal side and later also at the cathodal side, and sometimes the reverse.

In all of the organisms referred to above, except *Noctiluca,* the anodal responses differ radically from the cathodal responses. This is very evident from the results of observations on *Amoeba,* in which it can be clearly seen that, after the circuit is closed, there is first local liquefaction of the plasmagel on the cathodal side, then contraction, and finally cytolysis on the anodal side. In *Volvox,* however, the response on the cathodal side in colonies which are positive to the cathode, is, in all perceptible characteristics, precisely the same as the response on the anodal side in those which are positive to the anode. Here, then, is an actual reversal in the action of the current, i.e., an actual reversal of Pflüger's law. Closing the circuit apparently produces the same effect on the anodal side of colonies in certain physiological states as it does on the cathodal side of colonies in other physiological states.

These physiological states are, as set forth above, specifically associated with those involved in reversal in the direction of photic orientation. As demonstrated above, these are dependent upon illumination, temperature, and chemicals in the environment, which apparently control the electric charge carried by the colonies and the bodies within them. What is much needed now is a more comprehensive study of these charges, in relation to the chemical and the physical content of the environment as well as the character of the responses of the colonies.

RESPONSES TO CHEMICALS

A. RHIZOPODS

None of the rhizopods except *Amoeba* has been studied with reference to motor responses to chemicals. Many of the observations made on *Amoeba* are so indefinite that further work under more carefully controlled conditions is highly desirable. In this work the large form, *Pelomyxa carolinensis,* which can now be readily procured, would doubtless be very favorable.

Using a capillary pipette, Edwards (1923) applied various chemicals locally to the surface of active specimens of *A. proteus.* With a few of these chemicals he obtained fairly definite results, which lead to the following conclusions:

If an alkali comes in contact with the side of an active amoeba, streaming stops and a local protuberance is formed at the point of contact. If the solution is weak, the protuberance develops into a normal pseudopod, which continues indefinitely toward the source of the solution. If it is strong, the protuberance breaks at the tip and the central portion of the amoeba flows out, leaving nothing but a crumpled membrane.

If an acid is applied, streaming stops and a similar protuberance is formed but it does not become large and does not develop into a normal pseudopod. If the acid is weak, the streaming soon begins again and the protuberance gradually disappears. If it is strong, the protuberance is very small and pseudopods form on the region opposite the point of application.

If acid is applied after a rupture in the surface of an amoeba has been produced by local application of an alkali and after the central portion begins to flow out, the flow immediately stops and the amoeba soon proceeds normally. These and other facts show that alkalies cause the cytoplasm of *Amoeba* to solate and that strong acids cause it to gelate; but what is involved in the formation of a protuberance by weak acid at the region of application is not clear.

A strong solution of sodium chloride results in formation of pseudopods opposite the region of application; a weak solution results only in the formation of a protuberance in the region to which it is applied. Thus *Amoeba* is negative to a solution of this salt if it is strong, and positive if it is weak. This indicates that the former induces gelation of the cytoplasm, and the latter, solation. Whether or not this obtains for other salts has not been ascertained.

Strong alcohol produces a blister at the point of application, followed by formation of pseudopods on the opposite surface.

The effect of all these substances is correlated with the kind of chemi-

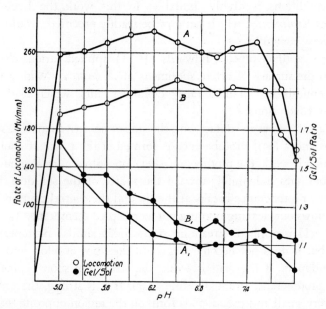

Figure 119. The relation between rate of locomotion, gel/sol ratio, and hydrogen-ion concentration in a balanced salt solution. A, A₁, rate of locomotion and gel/sol ratio in solutions containing salts in concentrations given in Table 2; B, B₁, rate of locomotion and gel/sol ratio in salts in concentration five times as great as those given in Table 2. Curve A, mean rates of movement of images of amoebae given in Table 2; B, means of the rates of movement of seventeen to twenty-four individuals for from five to seven minutes each in each hydrogen-ion concentration; A₁ and B₁, mean gel/sol ratios calculated from results obtained in measurements made on seventeen to twenty-four individuals in each hydrogen-ion concentration as described above. (After Pitts and Mast, 1933.)

cals, with their concentration in the medium surrounding the amoebae, and with the length of exposure to them.

If these conclusions are valid, one would expect amoebae in a culture medium to aggregate in regions which are alkaline or which contain relatively little salt. Hopkins (1928) made some observations concerning the former. He put specimens into a drop of solution which was pH 7.1 and then joined this drop with other drops which were respectively pH 6 and 8. He found that the amoebae at the border between the drops formed pseudopods which protruded toward the drop added, no matter whether it was alkaline or acid. This indicates that amoebae in

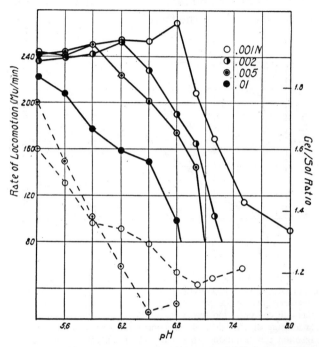

Figure 120. The relation between rate of locomotion, gel/sol ratio, hydrogen-ion concentration, and sodium-ion concentration. Each of the curves represents a series of experiments conducted at constant sodium-ion concentration. The four solid curves are based on the measurement of the rate of locomotion of an average of 14.8 different specimens for an average of 88.1 minutes in each hydrogen-ion concentration tested. The two broken curves are based on the measurement of the gel/sol ratio in an average of 19 different individuals in each hydrogen-ion concentration tested. (After Pitts and Mast, 1934b.)

a neutral solution tend to aggregate, either in acid or alkaline regions. The observations made were, however, not extensive enough to warrant any definite conclusions. There is, then, nothing definite known concerning the relation between differences in the chemical composition of the medium and aggregation in any of the rhizopods, and but little concerning the action of chemicals in relation to changes in direction of movement. On the other hand, there are some very definite results concerning the relation between the rate of locomotion and the chemicals in the surrounding medium, especially hydrogen ions.

Rate of locomotion and H-ion concentration.—Hopkins (1928) observed that *A. proteus,* in an ordinary hay culture fluid, is inactive if the fluid is neutral, but active if the fluid is either acid or alkaline, and that the rate of locomotion is maximum and nearly equal at about pH 6.5

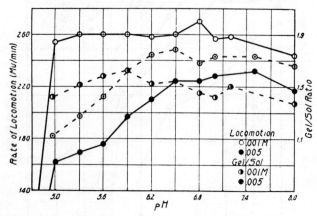

Figure 121. The relation between rate of locomotion, gel/sol ratio, hydrogen-ion concentration, and calcium-ion concentration. Curve ○, mean rate of locomotion in 0.001 M, based on measurements on an average of 22.1 different specimens for 11.6 minutes in each hydrogen-ion concentration tested; curve ●, mean rate of locomotion in 0.005 M, based on an average of 18.7 different specimens for a total average of 96.6 minutes in each hydrogen-ion concentration tested; curve ◐, mean gel/sol ratio in 0.001 M, based on measurements on an average of 21.3 different specimens in each hydrogen-ion concentration tested; curve ⊙, mean gel/sol ratio in 0.005 M, based on measurements on a total average of 14.6 different specimens in each hydrogen-ion concentration tested. (After Pitts and Mast, 1934b.)

and pH 8 respectively. Mast and Prosser (1932) confirmed these results. Pitts and Mast (1933), in a much more thorough study on *A. proteus,* demonstrated that the inactivity at neutrality is correlated with the relation between the amount of sodium or potassium and calcium present; and that the relation between activity and the concentration of hydrogen ions varies greatly with the kind, the concentration, and the proportion of salts in the surrounding medium. Some of the results obtained in this study are presented in figures 119, 120, 121, and 122. These figures show the following:

In a balanced salt solution the activity is minimum at neutrality and maximum on either side; but the activity at any given hydrogen-ion concentration varies with the salt concentration (Fig. 119). In sodium or calcium salt solutions (Table 2) the maximum rate of locomotion is nearly as high as in a balanced salt solution. Furthermore, the rate at

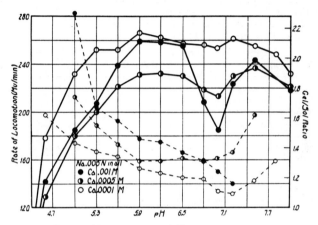

Figure 122. The effect of adding calcium in different concentrations to 0.005 N sodium solutions, on the relation between hydrogen-ion concentration, rate of locomotion, and gel/sol ratio. The solid curve for the solutions containing calcium 0.001 M is based on measurements of the rate of locomotion, on an average of 14.75 different specimens for a total average of 89.9 minutes; that for calcium 0.0005 M, on an average of 18.6 different specimens for a total average of 95 minutes; and that for calcium 0.001 M, on an average of 20.5 different specimens for a total average of 104.6 minutes in each hydrogen-ion concentration tested. The broken curve for the solution containing calcium 0.001 M is based on measurement of the gel/sol ratio, on an average of 19.8 different specimens; that for calcium 0.0005 M, on an average of 17.9 different specimens; and that for calcium 0.0001 M, on an average of twenty different specimens in each hydrogen-ion concentration tested. (After Pitts and Mast, 1934c.)

any given hydrogen-ion concentration varies with the salt concentration, although the relation between the rate and the hydrogen-ion concentration in the one differs greatly from that in the other, and there is no indication of inactivity at neutrality in either (Figs. 120, 121). If calcium salt is added to a solution of sodium salt, the activity decreases greatly at neutrality (Fig. 122). The ratio between the amount of plasmagel and the amount of plasmasol in *Amoeba* varies with the hydrogen-ion concentration; but the rate of locomotion is not specifically correlated with this ratio.

TABLE 2: RATE OF LOCOMOTION OF A. PROTEUS IN SODIUM AND
CALCIUM SALT SOLUTION*

Composition of the solution used to ascertain the effect of hydrogen-ion concentration
on rate of locomotion and gel/sol ratio in a balanced salt solution. To obtain different
hydrogen-ion concentrations, the acid and the alkaline components were mixed in different proportions.

Note that the concentration of Na, K, Ca, and Mg remains constant, no matter what
the proportion of the two components is.

In the second series of experiments the concentration of all the salts was increased
five times (after Pitts and Mast, 1933).

Acid Component		Alkaline Component		Molar Ratio	
NaH₂PO₄	0.00150N	NaOH	0.00150N	Na	60
KH₂PO₄	0.00010N	KOH	0.00010N	K	4
CaH₄(PO₄)₂	0.00010N	Ca(OH)₂	0.00010N	Ca	2
MgCl₂	0.00005N	MgCl₂	0.00005N	Mg	1

* After Pitts and Mast, 1933, by permission of the *Journal of Cellular and Comparative Physiology*,
Wistar Institute of Anatomy and Biology.

Mechanics of response.—Concerning the action of chemicals producing these responses on *Amoeba*, Pitts and Mast make the following statements:

It is obvious that a substance in the environment may influence processes
which occur in a cell either by entering the cell and acting directly on
substances in the cell, or by acting on the surface of the cell in such a way
as to retard or facilitate the passage into or out of the cell of other substances which, owing to their presence or absence, induce alterations in
internal processes. . . .

It may be assumed, then, either (1) that the gel/sol ratio depends upon
the entrance of salts into the cell and reaction between these and internal
substances, and that the rate of entrance of salts varies with the hydrogen
ion concentration; or (2) that the gel/sol ratio varies with the rate of entrance of hydrogen and hydroxyl ions into the cells and reaction between
these and internal substances, and that the rate of entrance of these ions
varies with the concentration of the salts and the hydrogen ions; or (3) that
the gel/sol ratio depends upon the exit of substances from the cell, e.g.,
water, and that this depends upon the hydrogen ion concentration, and the
salt concentration and the kinds of salts present. Let us now attempt to
ascertain if the processes in question are in accord with any of these groups
of assumptions. . . .

If they are in accord with the first of these assumptions, the gel/sol
ratio must vary directly or indirectly with the amount of salt that enters the
cell and this must vary directly or indirectly with the hydrogen ion con-

centration. If it varies indirectly with the amount of salt that enters and this varies either directly or indirectly with the hydrogen ion concentration practically none of the results obtained are in accord with the assumptions. If it varies directly with the amount of salt that enters and this varies directly with the hydrogen ion concentration the results obtained with balanced salt solutions and some others are in full accord with the assumptions; but the assumption that the entrance of salts varies directly with hydrogen ion concentration is not in harmony with the results obtained by practically all who have investigated this problem. Moreover, the assumptions do not account for the independence or inverse variation between hydrogen ion concentration and gel/sol ratio in calcium solutions. . . .

If they are in accord with the second of the three assumptions made above, the gel/sol ratio must depend upon the hydrogen ion concentration within the cell and this must vary with the concentration of the salts and the hydrogen ions in the surrounding medium. If it varies indirectly with the hydrogen ion concentration of the surrounding medium and directly or indirectly with the salt concentration, few if any of the results obtained are in accord with the assumptions. If it varies directly with the hydrogen ion concentration and the salt concentration of the surrounding medium, the results obtained with balanced salt solutions and some others are in accord with the assumptions, but those obtained with calcium solutions are not. There is, moreover, no evidence which indicates that the hydrogen ion concentration within *Amoeba* varies appreciably with variation in the hydrogen ion and the salt concentrations of the surrounding medium (Chambers, 1928). . . .

In accord with the third group of assumptions, the gel/sol ratio must vary with the rate of exit of substances from the cell, and this must vary directly or indirectly with the salt and the hydrogen ion concentration, and it must also vary with the kind of salts present in the environment. Without entering upon a detailed analysis of the correlation between these assumptions and the results under consideration, it is evident that, no matter what combination is selected, there are between them and the results inconsistencies of the same nature as those presented above. . . .

It is consequently obvious that the results in hand cannot be consistently explained by any one of the three groups of assumptions made, and that there must be a fairly complicated interaction between the various factors involved. If this is true, the statement, without qualification, that any given factor facilitates gelation or solation is obviously so incomplete that it is without value. This conclusion is in full harmony with that reached by Mast and Prosser (1932). . . .

In reference to the relations between rate of locomotion and kind of salts, salt concentration and hydrogen ion concentration, the results obtained are, on the basis of any one of the groups of assumptions considered above, even less explicable than are those in reference to the gel/sol ratio. The outstanding difficulty here concerns the remarkable decrease in rate of locomotion as neutrality is approached either from the acid or from the alkaline side. The results presented show that this decrease occurs in balanced solutions; that is that it is specifically correlated with the Na/Ca ratio; that the higher this ratio within the limits of the concentrations tested, the greater the decrease; and that it does not occur in solutions containing only one salt. . . .

The results show that if calcium is added to a solution containing only sodium salts, the rate of locomotion in the alkaline range increases greatly, with but little change in the acid range and in the region of neutrality; and that if sodium is added to solutions containing calcium salts, the rate decreases greatly in the region of neutrality, with but little change elsewhere. . . .

The questions now arise as to why addition of calcium to solutions containing only sodium salts causes great increase in rate of locomotion in the alkaline range, and why addition of sodium to solutions containing only calcium salts causes great decrease in the rate in the region of neutrality.

Similar questions have arisen in reference to the bimodal curves obtained by a number of other investigators in plotting the rate of various physiological processes against hydrogen ion concentration, e.g., by Robbins (1926) and Farr (1928) in various processes in plants; by Ephrussi and Neukomm (1927) in the resistance to heat in the eggs of a sea urchin; by Hopkins (1928) in the rate of locomotion in *Amoeba;* by Mast (1928) in the rate of assumption of stellate forms in *Amoeba;* by Eisenberg-Hamburg (1929) in the rate of increase in water content in infusoria; by Chalkley (1929) in water content and gel/sol ratio in *Amoeba;* and (1930a, 1930b) thermal death rate in *Paramecium;* by Chase and Glaser (1930) in rate of locomotion in *Paramecium;* and by Mast and Prosser (1932) in rate of locomotion in *Amoeba.* . . .

Only a few of these investigators attempted to elucidate the phenomenon. Mast and Prosser (1932), as previously stated, concluded that it is correlated with salt concentration. We have already considered this view. Robbins (1926) contends that the hydrogen ion concentration, at which the median minimum in the plant processes studied occurs, coincides with the isoelectric point of the principal proteins in the plant. Farr (1928) maintains, however, that this view is not tenable. He found in observations on the relation between hydrogen ion concentration and rate of growth in root hairs of collards that the hydrogen ion concentration at which the median minimum occurs varies greatly with salt concentration and he concludes that it therefore cannot be specifically correlated with the isoelectric point of any given protein in the organism. . . .

In reference to *Amoeba,* the constancy of a median minimum rate of locomotion at pH 7.0 in various solutions indicates some fixity of mechanism determining this median minimum. This mechanism might be correlated with the behavior of the membrane in the neighborhood of an ampholyte isoelectric point near neutrality in accord with the view of Robbins. (1926). But arguing against such an isoelectric point are the pertinent facts that, (1) the median minimum is lacking in solutions of single salts, though locomotion in the dilute solutions occurs at hydrogen ion concentrations in which it is usually found; (2) the difference between the maximum and the median minimum rate of locomotion depends on the sodium/calcium ratio and only slightly if at all on the total salt concentration; (3) the cations have marked effect on the rate of locomotion on the acid side of the neutral point as well as on the alkaline side; (4) the anions (so far as chloride and phosphate are concerned) have little effect down as far in the acid range as the observations were made. . . .

We are at present unable to suggest a satisfactory explanation for this median minimum. Whatever the cause of it may be, the relation between the rate of locomotion in *Amoeba* and the factors in its environment is doubtless fairly complex, for it is probable that these factors influence locomotion in it by their action on the surface, affecting adhesiveness and other properties of the surface, as well as by their action on permeability of the surface membrane (Mast, 1926a).

B. MARINE AMOEBAE

Pantin (1923-31) made observations on the relation between the rate of locomotion in a marine amoeba and various chemicals. He found that as the hydrogen-ion concentration increases, the rate increases rapidly from zero at pH 10 to a maximum at about pH 8, and then decreases rapidly to zero at pH 5.5. He observed no indication of decrease in activity at neutrality. He maintains that the rate is closely correlated with the relative concentration of sodium, potassium, magnesium, and calcium salts, that more than one of these salts is necessary for locomotion, and that calcium is required in all combinations. He maintains that calcium functions primarily in the contractile mechanism, and the others in the regulation of permeability. The evidence presented in support of these conclusions is, however, not unequivocal.

Oxygen is necessary for locomotion in *Amoeba* (Hulpieu, 1930; Pantin, 1930) but only in very low concentrations. Pantin maintains that none is immediately necessary, and that it functions in recovery somewhat as it does in the contraction of muscles. This, however, has not been demonstrated.

C. CILIATES

The most prominent response of the ciliates to chemicals consists in reversal in the direction of the effective stroke of the locomotor cilia, and consequent backward swimming. This reversal may be so brief that it results in scarcely perceptible backward movement, or it may continue for several minutes. It has been investigated in some detail in *Paramecium*, but not in any of the other ciliates. The rate of locomotion in all the ciliates is doubtless correlated with the chemical composition of the surrounding medium, but no measurements concerning this correlation have been made in any of them.

The reversal in the direction of the stroke of the cilia of paramecia in response to chemicals extends to the entire surface of the body, except the oral groove. It consequently results in backward swimming. It is the same as the reversal induced by contact and by rapid changes in temperature or osmotic concentration, and it is usually followed by turning toward the aboral surface and forward movement in a new direction, i.e., it usually constitutes the first stage in the avoiding reaction.

Jennings (1906) describes in detail how this response results in aggregation of paramecia in regions which contain ineffective chemicals in relation to those in adjoining regions, e.g., regions which are slightly acid, surrounded by regions which are slightly alkaline. If this obtains, he says, paramecia do not respond as they enter the acid region, but do when they reach the edge of this region and are about to enter the alkaline region. They consequently remain in the acid region. As more paramecia enter, owing to random movements, an aggregation is formed.

He maintains that these responses usually result in aggregations in regions which are favorable for the organisms, but that there are exceptions. That is, he holds that these responses are, in general, adaptive. However, he has made no suggestions concerning the processes in the organism involved in producing these responses.

Merton (1923) found that sodium and potassium salts induce reversal in *Stentor* and that calcium and magnesium do not, but he offers no suggestions concerning the nature of the action of the former nor the cause of inaction of the latter.

Mast and Nadler (1926) ascertained the effect of fifty-six different chemicals on the direction of the effective stroke of the cilia in *Paramecium*. They maintain that all of the monovalent cation salts and hydrates tested (thirty-one), except $(NH_4)_2SO_4$ and $NH_4C_2H_3O_2$, induce reversal; but that none of the bivalent and trivalent cation salts tested (nineteen), except $CaHPO_4$ and $MgHPO_4$, induce it. Also that $Ba(OH)_2$, H_3PO_4 and $H_2C_2O_4$ induce reversal, while HCL and lactose do not.

These authors contend that the duration of reversed action is closely correlated with the concentration of the salts; that it varies with the kind of salt at any given concentration; and that bivalent and trivalent cation salts neutralize the effect of monovalent salts.

Concerning the physiological processes involved in reversal in ciliary action, they make the following statement:

Copeland (1919, 1922) and Grave and Schmitt (1925) demonstrated that the cilia in higher forms are closely connected with nerve fibers and that their action is in all probability controlled by nerve impulses. The results obtained in the investigations of Yocum (1918), Taylor (1920), Rees (1922), Visscher (1927) and others on the neuromotor apparatus in the protozoa, indicate that the action of the cilia in these forms is similarly controlled. If this is true, the question arises as to how environmental changes which have been observed to induce reversal in ciliary action influence the neuromotor apparatus. This may be conceived to be either through chemical changes produced in the receptors or elsewhere in the organism or through changes produced in the electric potential at the surface, or in the permeability and the consistency of the surface layer. . . .

The results obtained by Mast and Nadler indicate that reversal in ciliary action is largely dependent upon the cations, that it is induced by monovalent but usually not by bi- and tri-valent cation salts, and that it depends upon the concentration of the salts. . . .

It is well known that adsorption of the cations is usually relatively greater in bi- and tri-valent than in monovalent cation salt solutions and that the adsorption varies with the concentration of the salts in the solution. This seems to indicate that the difference in the action of the monovalent and the bi- and tri-valent cation salts and the difference in the action of different concentrations of the monovalent cation salts is at least in part associated with differential adsorption of the ions in the various solutions, resulting in changes in the electrical potential at the surface of the paramecia, which directly or indirectly produce impulses in the neuromotor apparatus which pass to the cilia and influence their action. The facts, however, that $Ba(OH)_2$ produces reversal while $BaCl_2$, $CaCl_2$, $MgCl_2$ do not, that $H_2C_2O_4$ and H_3PO_4 produce reversal while HCl does not, that $CaHPO_4$ and $MgHPO_4$ produce reversal while $Ca_3(PO_4)_2$ and $Mg(PO_4)_2$ do not, all indicate that there must be other factors involved in reversal aside from differential adsorption of the cations.

Oliphant (1938) made a much more extensive and thoroughly controlled study of the effect of monovalent and bivalent cation salts on ciliary reversal in *Paramecium*.

The results obtained support the contentions of Mast and Nadler (1926) that the monovalent salts induce reversal, whereas bivalent cation salts do not, and that the duration of the reversed action varies with the kind and the concentration of the salts; but they do not support their contentions as to the nature of this variation. The results show that the effect of the salts is due primarily to the action of the cations, and that the anions have little, if any, effect. They show that the order of effectiveness of the cations is $K > Li > Na > NH_4$, and that the duration of their effect varies inversely with the temperature.

Oliphant (1938) cites work which indicates that in *Paramecium, Amoeba, Actinosphaerium, Spirogyra,* root hair of *Trianea,* and eggs of *Arbacia,* monovalent cations induce increase, and that bivalent cations decrease the viscosity of the cytoplasm (Spek, 1921; Heilbrunn, 1923, 1931, Cholodnyj, 1923; Weber, 1924). He concludes that this indicates that reversal in ciliary action is correlated with increase in viscosity of the cytoplasm, and contends that this conclusion is supported by the fact that "reversal in response to temperature occurs only at temperatures almost immediately lethal," i.e., at temperatures which cause marked increase in viscosity. He holds that the action of the cilia is controlled by the neuromotor apparatus and that increase in viscosity produces impulses in this structure which cause reversal in ciliary action, but he thinks that changes in electric potential, in permeability, in the consistency of the surface layer, or in the chemical composition of the receptors or other structures in the organism may be involved.

It is obvious from the above discussion that there is still much to be learned concerning the processes involved in the responses of the ciliates to chemicals.

LITERATURE CITED

Alsup, Fred W. 1939. Relation between the responses of *Amoeba proteus* to alternating electric current and sudden illumination. Physiol. Zool., 12: 85-95.

Alverdes, F. 1923. Der Sondercharakter der von den Ciliaten gezeigten Galvanotaxis. Pflüger's Archiv ges. Physiol., 198: 513-42.

Bancroft, F. W. 1906. The control of galvanotropism in *Paramecium* by chemical substances. Univ. Cal. Publ. Physiol., 3: 21-31.

——— 1907. The mechanism of the galvanotropic orientation in *Volvox*. J. exp. Zool., 4: 157-63.

——— 1913. Heliotropism, differential sensibility, and galvanotropism in *Euglena*. J. exp. Zool., 15: 383-428.

Bayliss, W. M. 1920. The properties of colloidal systems. IV. Reversible gelation in living protoplasm. Proc. roy. Soc., B, 91: 196-201.

Buder, J. 1917. Zur Kenntniss der phototaktischen Richtungsbewegungen. Jb. wiss. Bot., 58: 105-220.

Carlgren, O. 1899. Über die Einwirkung des konstanten galvanischen Stromes auf niedere Organismen. Arch. Anat. Physiol. Lpz., Physiol. Abth., 49-76.

Chalkley, H. W. 1929. Changes in water content in Amoeba in relation to changes in its protoplasmic structure. Physiol. Zoöl., 2: 535-74.

———1930a. Resistance of *Paramecium* to heat as affected by changes in hydrogen ion concentration and in inorganic salt balance in the surrounding medium. Pub. Hlth. Rep., Wash., 45: 481-89.

——— 1930b. On the relation between resistance to heat and the mechanism of death in *Paramecium*. Physiol. Zoöl., 3: 425-40.

Chambers, R. 1928. Intracellular hydrion concentration studies. I. The relation of the environment to the pH of the cytoplasm and of its inclusion bodies. Biol. Bull., 55: 369-76.

——— Chambers, R., and P. Reznikoff. 1926. Micrurgical studies in cell physiology. I. The action of the chlorides of Na, K, Ca, and Mg on the protoplasm of *Amoeba proteus*. J. gen. Physiol., 8: 369-401.

Chase, A., and O. Glazer. 1930. Forward movement of *Paramecium* as a function of the hydrogen-ion concentration. J. gen. Physiol., 13: 627-36.

Cholodnyj, N. 1923. Zur Frage über die Beeinflussung des Protoplasmus durch mono und bivalente Metallionen. Beih. Bot. Zbl., 39: 231.

Coehn, A., and W. Barratt. 1905. Über Galvanotaxis vom Standpunkte des physikalischen Chemie. Z. allg. Physiol., 5: 1-9.

Copeland, M. 1919. Locomotion in two species of the gastropod genus *Alectrion* with observations on the behavior of pedal cilia. Biol. Bull., 37: 126-38.

——— 1922. Ciliary and muscular locomotion in the gastropod genus *Polinices*. Biol. Bull., 42: 132-42.

Davenport, C. B. 1897. Experimental Morphology. Vol. 1, New York, 280 pp.

De Candolle, A. P. 1832. Physiologie vegetale. Paris.

Dixon, H. H., and T. A. Bennet-Clark. 1927. Responses of plant tissues to electric currents. Sci. Proc. R. Dublin Soc., 18: 351-72.

Edwards, J. G. 1923. The effect of chemicals on locomotion in *Amoeba*. J. exp. Zool., 38: 1-43.

Eisenberg-Hamburg, E. 1929. Recherches comparatives sur le fonctionnement de la vacuole pulsatile chez les infusoires parasites de la grenouille et chez les infusions d'eau douce. Influence de la pression osmotique, des electrolytes et du pH. Arch. Protistenk., 78: 251-70.

Engelmann, T. W. 1869. Beiträge zur Physiologie des Protoplasm. Pflüg. Arch. ges. Physiol., 2: 307-22.

—— 1879. Über Reizung contraktilen Protoplasmus durch plötzliche Beleuchtung. Pflüg. Arch. ges. Physiol., 19: 1-7.

—— 1882. Über Licht- und Farbenperception niederster Organismen. Pflüg. Arch. ges. Physiol., 29: 387-400.

Ephrussi, Boris, and Alex Neukomm. 1927. Resistance a la chaleur des oeufs d'oursin (Paracentrotus lividus, Lk). Protoplasma, 2: 34-44.

Farr, C. H. 1928. Studies on the growth of root hairs in solutions. IV. The pH molar rate relation for collards in calcium chloride. Amer. J. Bot., 15: 6-31.

Folger, H. T. 1925. A quantitative study of reactions to light in Amoeba. J. exp. Zool., 41: 261-91.

Grave, C., and F. O. Schmitt. 1925. A mechanism for the coördination and regulation of ciliary movement as revealed by microdissection and cytological studies of ciliated cells in mollusks. J. Morph., 40: 479-513.

Greeley, A. W. 1904. Experiments on the physical structure of the protoplasm of Paramecium and its relation to the reactions of the organism to thermal, chemical and electrical stimuli. Biol. Bull., 7: 3-32.

Habenicht, Wilhelm. 1935. Die Wirkung des galvanischen Stromes auf den Eiweisszylinder nach Du Bois-Reymond. Protoplasma, 22: 321-36.

Hahnert, W. F. 1932. A quantitative study of reactions to electricity in Amoeba proteus. Physiol. Zoöl., 5: 491-526.

Harrington, N. R., and E. Leaming. 1900. The reactions of Amoeba to light of different colors. Amer. J. Physiol., 3: 9-16.

Heilbrunn, L. V. 1923. The colloid chemistry of protoplasm. I. General considerations: II. The electrical charges of protoplasm. Amer. J. Physiol., 64: 481.

Heilbrunn, L. V., and K. Daugherty. 1931. The action of sodium, potassium, calcium, and magnesium on the protoplasm of Amoeba dubia. Physiol. Zoöl., 4: 635.

Holmes, S. J. 1903. Phototaxis in Volvox. Biol. Bull., 4: 319-26.

Hopkins, D. L. 1928. The effects of certain physical and chemical factors on locomotion and other life processes in Amoeba proteus. J. Morph., 45: 97-119.

—— 1929. The effects of the substratum, divalent and monovalent cations on locomotion in Amoeba proteus. J. Morph., 48: 371-83.

—— 1938. The vacuoles and vacuolar activity in the marine Amoeba, Flabellula mira Schaeffer and the nature of the neutral red system in Protozoa. Biodyn., No. 34: 1-22.

Hulpieu, H. R. 1930. The effect of oxygen on *Amoeba proteus*. J. exp. Zool., 56: 321-61.

Inman, O. L., W. T. Bovie, and C. E. Barr. 1926. The reversal of physiological dominance in *Amoeba* by ultraviolet light. J. exp. Zool., 43: 475-84.

Jennings, H. S. 1899. Studies on reactions to stimuli in unicellular organisms. II. The mechanism of the motor reactions of *Paramecium*. Amer. J. Physiol., 2: 311-41.

—— 1904. Contributions to the study of the behavior of lower organisms. Publ. Carneg. Instn., No. 16: 256 pp.

—— 1906. Behavior of the Lower Organisms. New York, 366 pp.

Kamada, T. 1928-31. Control of galvanotropism in *Paramecium*. J. Fac. Sci. Tokyo Univ., 2: 29, 123-39, 285-307.

—— 1934. Some observations on potential differences across the ectoplasm membrane of *Paramecium*. J. exp. Biol., 11: 94-102.

Kinosita, H. 1936. Electric stimulation of *Paramecium*. Jour. Fac. Sci. Tokyo Univ., 4: 137-94.

—— 1938a. Electric stimulation of *Spirostomum*. Jour. Fac. Sci. Tokyo Univ., 5: 71-105.

—— 1938b. Electric stimulation of *Paramecium* with two successive subliminal current pulses. J. cell. comp. Physiol., 12: 103-17.

—— 1939. Electrical stimulation of *Paramecium* with linearly increasing current. J. cell. comp. Physiol., 13: 253-61.

Koehler, O. 1925. Galvanotaxis. Handb. d. Norm. u. Pathol. Physiol., 11: 1027-49.

Kühne, W. 1864. Untersuchungen über das Protoplasma und die Contractilität. Leipzig, 158 S.

Laurens, H., and H. D. Hooker, 1920. Studies on the relative physiological value of spectral lights. II. The sensibility of *Volvox* to wave-lengths of equal energy content. J. exp. Zool., 30: 345-68.

Lillie, R. S. 1923. Protoplasmic Action and Nervous Action. Chicago, 417 pp.

Loeb, J. 1890. Der Heliotropismus der Tiere und seine Uebereinstimmung mit dem Heliotropismus der Pflanzen. Würzburg, 118 pp.

—— 1906. The Dynamics of Living Matter. New York, 233 pp.

Loeb, J., and S. P. Budgett. 1897. Zur Theorie des Galvanotropismus. IV. Mitt. Über die Ausscheidung elektropositiver Ionen an der äusseren Anodenfläche protoplasmitischer Gebilde als Ursache der Abweichungen vom Pflügerschen Erregungsgesetz. Pflüg. Arch. ges. Physiol., 66: 518-34.

Loeb, J., and S. S. Maxwell. 1910. Further proof of the identity of heliotropism in animals and plants. Univ. Cal. Publ. Physiol., 3: 195-97.

Lucas, K. 1910. An analysis of changes and differences in the excitatory process of muscles and nerves based on the physical theory of excitation. J. Physiol., 40: 225-49.

Luce, R. H. 1926. Orientation to the electric current and to light in *Amoeba*. Anat. Rec., 32: 55.

Ludloff, K. 1895. Untersuchungen über den Galvanotropismus. Pflüg. Arch. ges. Physiol., 59: 525-54.

Lund, E. J., and G. A. Logan. 1925. The relation of the stability of protoplasmic films in *Noctiluca* to the duration and intensity of an applied electric potential. J. gen. Physiol., 7: 461-71.

Luntz, A. 1931-32. Untersuchungen uber die Phototaxis. Z. vergl. Physiol. I, 14: 68-92; II, 15: 652-78; III, 16: 204-17.

Lyon, E. P. 1923. Effects of Electricity on Noctiluca. Proc. Soc. exp. Biol. N.Y., 20: 284-85.

McClendon, J. F. 1910. On the dynamics of cell division. I. The electric charge on colloids in living cells in the root tips of plants. Archiv. f. Entw. mech. d. Organismen. 31: 80-90.

McClendon, J. J. 1911. Ein Versuch amöboide Bewegung als Folgeerscheinung wechselnden elektrischen Polarisationszustandes der Plasmahaut zu erklären. Pflüg. Arch. ges. Physiol., 140: 271-80.

Marsland, D. A., and D. E. S. Brown. 1936. Amoeboid movement at high hydrostatic pressure. J. cell. comp. Physiol., 8: 167-78.

Mast, S. O. 1906. Light reactions in lower organisms. I. *Stentor coeruleus.* J. exp. Zool., 3: 359-99.

—— 1907. Light reactions in lower organisms. II. *Volvox.* J. comp. Neurol., 17: 99-180.

—— 1910. Reactions in *Amoeba* to light. J. exp. Zool., 9: 265-77.

—— 1911. Light and the Behavior of Organisms. New York, 410 pp.

—— 1912. The reactions of the flagellate *Peranema.* J. Anim. Behav., 2: 91-97.

—— 1914. Orientation in *Euglena* with some remarks on tropisms. Biol. Zbl., 34: 641-74.

—— 1916. The process of orientation in the colonial organism, *Gonium pectorale,* and a study of the structure and function of the eyespot. J. exp. Zool., 20: 1-17.

—— 1917. The relation between spectral color and stimulation in the lower organisms. J. exp. Zool., 22: 471-528.

—— 1918. Effects of chemicals on reversion in orientation to light in the colonial form, *Spondylomorum quaternarium.* J. exp. Zool., 26: 503-20.

—— 1919. Reversion in the sense of orientation to light in the colonial form, *Volvox globator* and *Pandorina morum.* J. exp. Zool., 27: 367-90.

—— 1923. Photic orientation in insects with special reference to the dronefly, *Eristalis tenax* and the robber fly, *Erax rufibarbis.* J. exp. Zool., 38: 109-205.

—— 1926a. Structure, movement, locomotion and stimulation in *Amoeba.* J. Morph., 41: 347-425.

—— 1926b. Reactions to light in *Volvox,* with special reference to the process of orientation. Z. vergl. Physiol., 4: 637-58.

—— 1927a. Structure and function of the eyespot in unicellular and colonial organisms. Arch. Protistenk., 60: 197-220.

—— 1927b. Reversal in photic orientation in *Volvox* and the nature of photic stimulation. Z. vergl. Physiol., 5: 730-38.

—— 1927c. Response to electricity in *Volvox* and the nature of galvanic stimulation, Z. vergl. Physiol., 5: 740-61.

—— 1928. Factors involved in changes in form in *Amoeba*. J. exp. Zool., 51: 97-120.

—— 1931a. The nature of response to light in *Amoeba proteus* (Leidy). Z. vergl. Physiol., 15: 139-47.

—— 1931b. The nature of the action of electricity in producing response and injury in *Amoeba proteus* (Leidy) and the effect of electricity on the viscosity of protoplasm. Z. vergl. Physiol., 15: 309-28.

—— 1931c. Movement and response in *Difflugia* with special reference to the nature of cytoplasmic contraction. Biol. Bull., 61: 223-41.

—— 1932a. Localized stimulation, transmission of impulses, and the nature of response in *Amoeba*. Physiol. Zoöl., 5: 1-15.

—— 1932b. The rate of adaptation to light and to darkness in *Volvox globator*. Z. vergl. Physiol., 17: 644-58.

—— 1936. Motor responses to light in the invertebrate animals. Biological Effects of Radiation (ed. by B. M. Dugger). New York. Pp. 573-623.

—— 1939. The relation between kind of food, growth, and structure in *Amoeba*. Biol. Bull. 77: 391-98.

Mast, S. O., and Mary Gover. 1922. Relation between intensity of light and rate of locomotion in *Phacus pleuronectes* and *Euglena gracilis* and its bearing on orientation. Biol. Bull. 43: 203-9.

Mast, S. O., and Brainard Hawk. 1936. Response to light in *Peranema trichophorum*. I. Relation between dark-adaptation and sensitivity to light. Biol. Bull. 70: 408-12.

Mast, S. O., and H. R. Hulpieu. 1930. Variation in the response to light in *Amoeba proteus* with special reference to the effects of salts and hydrogen ion concentration. Protoplasma, 11: 412-31.

Mast, S. O., and P. L. Johnson. 1932. Orientation in light from two sources and its bearing on the function of the eyespot. Z. vergl. Physiol., 16: 252-74.

Mast, S. O., and J. E. Nadler. 1926. Reversal of ciliary action in *Paramecium caudatum*. J. Morph., 43: 105-17.

Mast, S. O., and C. L. Prosser. 1932. Effect of temperature, salts, and hydrogen ion concentration on rupture of the plasmagel sheet, rate of locomotion, and gel/sol ratio in *Amoeba proteus*. J. cell. comp. Physiol., 1: 333-54.

Mast, S. O., and N. Stahler. 1937. The relation between luminous intensity, adaptation to light, and rate of locomotion in *Amoeba proteus* (Leidy). Biol. Bull., 73: 126-33.

Merton, H. 1923. Studien über Flimmerbewegung. Pflüg. Arch. ges. Physiol., 198: 1-28.

Moore, A. R. 1926. Galvanic stimulation of luminescence in *Pelagia noctiluca*. J. gen. Physiol., 9: 375-79.

Nernst, W. 1899. Zur Theorie der electrischen Reizung. Nachr. v. d. Kgl. Gesell. Wiss. Göttingen, Math. physik. Kl., 104-8.

Oliphant, J. F. 1938. The effect of chemicals and temperature on reversal in ciliary action in *Paramecium*. Physiol. Zoöl., 11: 19-30.

Ostwald, W. 1890. Elektrische Eigenschaften halbdurchlässiger Scheidewände. Z. phys. Chem., 6: 71-82.

Pantin, C. F. A. 1923-31. On the physiology of amoeboid movement. J. Mar. biol. Ass. U. K., 13: 24-69; Brit. J. exp. Biol., 1: 519; 3: 275, 297; 8: 365-78; Proc. roy. Soc., 105: 538-79.

Pearl, R. 1900. Studies on electrotaxis. I. On the reactions of certain infusoria to the electric current. Amer. J. Physiol., 4: 96-123.

—— 1901. Studies on the effects of electricity on organisms. II. The reactions of Hydra to the constant current. A. Jour. Physiol., 5: 301-20.

Pitts, R. F., and S. O. Mast. 1933. The relation between inorganic salt concentration, hydrogen ion concentration and physiological processes in *Amoeba proteus*. I. Rate of locomotion, gel/sol ratio and hydrogen-ion concentration in balanced salt solutions. J. cell. comp. Physiol., 3: 449-62.

—— 1934a. II. Rate of locomotion, gel/sol ratio and hydrogen ion concentration in solutions of single salts. J. cell. comp. Physiol., 4: 237-56.

—— 1934b. III. The interaction between salts (antagonism) in relation to hydrogen-ion concentration and salt concentration. J. cell. comp. Physiol., 4: 435-55.

Ray, J. 1693. Historia Plantarum, 2: 985-1944. London.

Rees, C. W. 1922. The neuromotor apparatus of *Paramecium*. Univ. Cal. Publ. Zool., 20: 333-64.

Robbins, W. J. 1926. The isoelectric point for plant tissues and its importance in absorption and toxicity. Univ. Mo. Stud., 1: 3-60.

Schaeffer, A. A. 1917. Reactions of *Amoeba* to light and the effect of light on feeding. Biol. Bull. 32: 45-74.

Shettles, L. B. 1937. Response to light in *Peranema trichophorum* with special reference to dark-adaptation and light-adaptation. J. exp. Zool., 77: 215-49.

—— 1938. Effect of ultraviolet light and x-rays on *Peranema trichophorum*. J. cell. comp. Physiol., 12: 263-72.

Spek, J. 1921. Der Einfluss der Salze auf die Plasmokolloide von *Actinosphaerium Eichhornii*. Acta Zool. Stock., 2: 153.

Statkewitsch, P. 1903. Über die Wirkung der Induktionsschläge auf einige Ciliata. Physiol. russe, 3: 1-55.

—— 1907. Galvanotropismus and Galvanotaxis der Ciliata. Z. allg. Physiol., 6: 13-43.

Strasburger, E. 1878. Wirkung des Lichtes und der Wärme auf Schwärm-sporen. Jena. Z. Naturw., n. f., 12: 551-625.

Taylor, C. V. 1920. Demonstration of the function of the neuromotor apparatus in *Euplotes*, by the method of microdissection. Univ. Cal. Publ. Zool., 19: 403-470.

—— 1925. Cataphoresis of ultramicropic particles in protoplasm. Proc. Soc. exp. Biol. N.Y., 22: 533-36.

Terry, O. P. 1906. Galvanotropism of *Volvox*. Amer. J. Physiol., 15: 235-43.

Verworn, M. 1889. Psycho-physiologische Protistenstudien. Jena, 218 pp.

—— 1895. Allgemeine Physiologie. Jena, 615 pp.

Visscher, J. P. 1927. A neuromotor apparatus in the ciliate *Dileptus gigas*. J. Morph. and Physiol., 44: 373-81.

Wager, H. 1900. On the eyespot and flagellum of *Euglena viridis*. J. linn. Soc. (Zool.), 27: 463-81.

Weber, E. B., and I. Weber. 1922. Reversible Viskositätserhöhung des Cytoplasmas unter der Einwirkung des elektrischen Stromes. Ber. dtsch. bot. Ges., 40: 254-58.

Weber, F. 1924. Plasmolyseform und Protoplasmaviskosität. Öst. Bot. Z., 73: 261.

Yocum, H. B. 1918. The neuromotor apparatus of *Euplotes patella*. Univ. Cal. Publ. Zool., 18: 337-96.

CHAPTER VI

RESPIRATORY METABOLISM

THEODORE LOUIS JAHN

STUDIES OF THE RESPIRATION of the Protozoa have, for the most part, been fragmentary, and our information on the subject resembles an accumulation of a number of more or less isolated data, rather than a unified body of knowledge. This situation is due to a variety of causes, chief among which is probably the fact that most studies of respiration have been made by physiologists who chose, among the members of the animal kingdom, the organisms which seemed to be the most suitable for a particular type of experiment. From this viewpoint the study of protozoan respiration has suffered a severe handicap, in that a considerable mass of protozoan protoplasm, free from bacteria and other organisms, has not always been easy to obtain, and in that our methods for measuring very small rates of respiratory exchange have not been nearly as accurate or as convenient as we might desire. However, with gradual technical advances, it seems probable that in the near future we shall see the development of an organized account of protozoan respiration, and it also appears probable that this development will take place among investigators who are primarily interested in the Protozoa. Therefore, it seems advisable to combine a review of data on protozoan respiration with a discussion of the general problems of respiratory metabolism in other biological materials, and to outline for the student not thoroughly trained in the lore of respirometry some of the purposes, methods, and possible interpretations of such a study.

Among the Protozoa, the intake of oxygen does not require complicated respiratory mechanisms. Apparently diffusion, high rate of water exchange, and protoplasmic movements (cyclosis, amoeboid streaming, and "metabolic" movements) are sufficient to maintain a suitable level of O_2 tension in the protoplasm and to prevent the accumulation of toxic amounts of CO_2. The mechanisms which are responsible for protoplasmic movements and the high rate of water exchange are more properly treated under the subjects of movement, permeability, and excretion. Therefore the problem of respiratory metabolism of a protozoan organism, at least for the present discussion, is easily resolved into a problem comparable to that of cellular respiration in the Metazoa.

PURPOSES OF STUDYING RESPIRATION

One of the first questions to be considered is: "What can be learned by studying respiration?" The measurement of gaseous exchange is not an end in itself, but it is a tool which, when used singly or in combination with other tools, may help us to obtain answers to the following types of questions.

1. What is the rate of energy expenditure of the organism? How does the metabolic rate (basal and otherwise) of one species compare with that of another? How does it change during starvation? Or conjugation? Is this rate dependent upon the O_2 or CO_2 tension of the environment? How is it affected by narcotics? How does it vary with temperature? Or with other chemical and physical factors of the normal environment?

2. What is the source of this energy? Is it obtained by oxidation of fats, carbohydrates, or proteins? Or by anaërobic oxido-reductions? Can this source be shifted by changing the chemical or physical environment of the organism? What are the intermediate products formed during oxidation of the substrate? Is oxidation of the substrate complete (i.e., to CO_2 and H_2O), or may these intermediate products be excreted by the cell?

3. What is the mechanism by which energy is obtained from the substates? Is this brought about through the intermediary action of dehydrogenase, cytochrome, and Warburg's respiratory enzyme? Or is it brought about through dehydrogenase and reversible oxidation-reduction systems, such as yellow pigment or pyocyanine? Or by glutathione? Can these substances be replaced by artificial oxidation-reduction systems? Can a shift in this mechanism of respiration be induced by changing the available substrate? How is this mechanism related to the degree of anaërobiosis which the organisms can endure? Is the organism capable of synthesizing the respiratory enzymes from simple substances, or must they be obtained from complex outside sources, i.e., from vitamins?

Measurements of the respiratory metabolism of an organism may be used as indices of the rate at which it uses energy, the substrate from which this energy is derived, and the mechanism by which it is obtained from the substrate. In addition to these factors, there is a possibility of a direct relationship between certain metabolic processes and the pathogenicity of parasitic forms.

METHODS OF MEASURING AËROBIC RESPIRATION

Methods that have been or might be used for measurement of aërobic protozoan respiration fall naturally into two groups—those applicable

TABLE 3: SENSITIVITY OF RESPIROMETERS

Type of Respirometer	Nearest Unit to Which Meniscus Can Be Read*	Approximate Sensitivity in Terms of Scale Divisions
Standard Warburg (Warburg, 1926)	0.2 mm.	1 mm.=1.0–2.0 mm.³ O_2
Microsemidifferential (Duryée, 1936)	0.2 mm.	1 mm.=0.5 mm.³ O_2
Microdifferential (Fenn, 1928)	0.1 mm.	1 mm.=0.3 mm.³ O_2
Microdifferential (Described in text)	0.2 mm.	1 mm.=0.2 mm.³ O_2
Straight capillary tubes (Howland and Bernstein, 1931)	0.01 mm.	0.01 mm.=0.001 mm.³ O_2
Straight capillary tubes in closed air chamber (Gerard and Hartline, 1934)	0.006 mm.	0.006 mm.=0.0013 mm.³ O_2
Microdifferential (Schmitt, 1933)	1 micron	1 micron=0.0005 mm.³ O_2
Cartesian diver (Needham and Boell, 1938)	0.2 mm.	1 mm.=0.001 mm.³ O_2

* It is assumed that the meniscus is read to the nearest 0.2 mm., unless otherwise stated by the authors. In those cases where the author claims greater reading accuracy, the smallest unit of change which can be detected is given in this column. If stability is adequate, the sensitivity of several models may be considerably increased by the use of special reading devices.

to concentrated suspensions of organisms, and those applicable to a few or to single cells. For concentrated suspensions, titration, gas analysis, and standard manometer methods have been used; and for studies of single protozoan cells, micromanometric methods have been devised (Kalmus, 1927; Howland and Bernstein, 1931). More recently, still better micromanometric methods have come into existence, but these have not yet been applied to the respiration of protozoa. Table 3 gives the sensitivity of various types of manometers, some of which have not yet been used to measure the respiration of Protozoa. It should be remembered that whenever a respirometer is made more sensitive to changes in gas volume produced by organisms, it simultaneously becomes more sensitive to slight changes produced by the environment (thermal and barometric effects) and to inaccuracies arising from imperfect design and construction (ground-glass connections and stopcocks, surface phenomena at the meniscus of manometer fluid, inaccuracies of capillary bore and so forth). Therefore, stability of the apparatus, on which final accuracy must depend, becomes more and more difficult to obtain. For that reason the most sensitive types should be reserved solely for those problems in which concentrated suspensions are undesirable or unobtainable. By comparison of Tables 3 and 4, it should be possible to determine approximately the type of respirometer necessary for any one of a variety of problems.

1.TITRATION METHODS

a. Dissolved O_2 determinations. For any aquatic animal it is possible to measure O_2 consumption by placing the organisms in a closed chamber filled with water of known O_2 content and by measuring the amount of O_2 left after a definite period of time. For this purpose a modified Winkler titration method is usually used (*Standard Methods of Water Analysis*, 1936). Special precautions are necessary whenever the animals cannot be removed from the solution, or if iron is present. This method has been used by Lund (1918a, 1918b, 1918c) and Leichsenring (1925) on *Paramecium* and *Colpoda*.

b. Measurement of CO_2 production. Production of CO_2 may be measured by placing the organisms in a small amount of solution in a small open container. This is placed inside of a larger closed container in which an alkali, preferably $Ba(OH)_2$, is present. The CO_2 given off by the organisms is absorbed by the alkali, which can then be titrated with acid in the presence of an indicator. This method was recommended by Lund (1918d) for use with *Paramecium*.

The above titration methods are applicable only to quite large numbers of organisms. It seems possible that these methods could be improved by the use of accurately controllable microburettes, smaller volumes of liquid, and so forth, but it is doubtful whether they could be made as accurate and reliable as some of the manometric procedures discussed below. Also, respiratory quotients are not easily obtained by titration methods.

2. GAS ANALYSIS

Soule (1925), Amberson (1928), and Root (1930) have applied the standard Haldane-Henderson methods of gas analysis to respiration of the Protozoa. These methods are adequate for use with rather concentrated suspensions and pressures, and possess certain definite advantages when O_2 or especially CO_2 tension is being varied experimentally and would have to be determined separately if manometric methods were used. Whenever gases other than CO_2 are evolved by an organism, gas analysis seems to be the only satisfactory method of measurement. The details of gas analysis methods are discussed by Peters and van Slyke (1932). Soule (1925), for studies of the metabolism of *Leishmania tropica* and *Trypanosoma lewisi,* used gas analysis, supplemented by readings of an insensitive manometer, the purpose of which was principally to indicate when gas exchange was taking place (method described in detail by Novy, Roehm, and Soule, 1925).

3. STANDARD MANOMETRIC METHODS

The principle of the manometric method is somewhat as follows: the organisms, in a suitable immersion medium, are placed in a closed flask large enough so that a considerable air space is present. The flask is connected to a capillary manometer tube partially filled with a liquid. Alkali may be present in a separate small container inside of the flask. If so, then CO_2 is absorbed, and the amount of O_2 consumed may be measured by means of the movement of fluid in the manometer tube as changes in volume (Haldane, Thunberg, Winterstein, Duryée, and Dixon types), or in pressure at a given volume (Warburg), or as the resultant of simultaneous changes in both (Barcroft differential type). Manometric methods, although very simple in outline, are filled with pitfalls for the inexperienced investigator, and a careful reading of the excellent treatise of Dixon (1934) is recommended. In this publication the theory and the more common forms of the apparatus are described in detail.

The Barcroft differential type can be made sufficiently small and sensitive for the study of respiration of Protozoa, when moderately concentrated mass cultures are available. A manometer designed by Dr. T. C. Evans (modified from that of Bodine and Orr, 1925), which has been in use in this laboratory for some time, seems to be quite suitable. It resembles the standard Barcroft (Dixon, 1934, p. 37) except that the two stopcocks are replaced by one double stopcock, which insures the simultaneous opening and closing of both flasks. The cups may be relatively small (about 5 cc.), and the U-shaped portion of the capillary (0.3 mm. bore) is placed at an angle of about twenty degrees from the horizontal. This design combines a high degree of sensitivity with ruggedness and dependability. The Duryée (1936) modification of the Thunberg-Winterstein principle and the Fenn (1928) form of microdifferential respirometer also seem to have a sensitivity adequate for moderately concentrated suspensions. Schmitt (1933) has devised an extremely sensitive form of microdifferential manometer, in which the gain in sensitivity and accuracy is due chiefly to an elaborate reading device and a system of temperature control which is stated to make meniscus movements of as little as one micron both detectable and significant.

4. CAPILLARY MANOMETER

The use of capillary tubes for measuring respiration of single protozoan cells was introduced by Kalmus (1927, 1928a). This method was improved by Howland and Bernstein (1931), who by means of a microinjection device drew small amounts of oil, air, KOH solution, and water containing an animal into small capillary tubes, so that they were finally arranged in the following order: oil, KOH, air, animal in water, oil. As the animal consumed O_2, the distance between the oil-water and the oil-KOH interfaces decreased. This change was measured microscopically by means of an ocular micrometer and a calibrated mechanical stage. By using control tubes made in a similar manner but without an animal, it was possible to correct for slight movements due to thermobarometric changes and to osmotic differences between the water and

KOH. Gerard and Hartline (1934) have improved the method by enclosing the tubes in an air tight chamber to eliminate barometric disturbances, and by using a screw micrometer to increase accuracy of reading.

5. CARTESIAN DIVER ULTRAMICROMANOMETER (NEEDHAM AND BOELL, 1938)

This is an application of the principle of the Cartesian diver for use as a constant-volume manometer. The "diver" chambers are constructed from capillary tubing and consist of a bulb partially filled with gas, an open capillary neck, and a solid glass tail to ensure that the diver floats upright. The diver is placed in a closed chamber partially filled with a strong salt solution, the specific gravity of which is such that the diver maintains a position below the surface of the salt solution, and that a small amount of this solution enters the neck. If the amount of gas in the diver is changed by a reaction, more salt solution will be drawn in or forced out of the neck, the specific gravity of the diver will change, and the level of flotation will also change. By changing the pressure on the salt solution, the diver may be brought back to any given level. Therefore, the diver may serve as a constant-volume manometer, and changes in the amount of gas in the chamber may be calculated from the changes in the external pressure which are necessary to maintain the diver at a definite level. This application of the Cartesian diver was suggested by Linderstrøm-Lang (1937), and has been used for parts of amphibian embryos by Needham and Boell (1938), who describe the use of this instrument for measurement of O_2 consumption, anaërobic glycosis, and respiratory quotient. From Table 3 it may be noted that this instrument when read only to 0.2 mm. has a sensitivity as great as those which employ special reading devices, and that it therefore has the possibility of being made more sensitive.

AËROBIC RESPIRATION

1. THE NORMAL RATE OF RESPIRATION

One fundamental essential in measuring the respiration of any biological material is that other material, also capable of respiratory activity, be absent or very well controlled. This means that bacteria must be absent, or at least must contribute only a negligible amount to the measured respiration. Various workers have tried removing bacteria by washing and filtering, or have corrected the figures for O_2 consumption by running controls of bacteria without Protozoa. Data obtained by these methods are extremely difficult for a reviewer to evaluate critically, and usually one may either accept them at face value until they can be checked with bacteria-free cultures or ignore them entirely. In the present discussion the tendency has been to accept all data in which the magnitude of the error is not obviously large, and to point out possible difficulties involved. In view of the fact that some bacteria have a respiratory rate per gram many times that of other types of cells (the rate for *Azotobacter* is stated by Burk [1937] to be equivalent to that of a 200-pound man consuming one ton of glucose per hour), the present viewpoint may be considered far from conservative. In some cases (e.g., the papers on cyanide insensitivity of *Paramecium*) the data on Protozoa seem quite adequate to prove the principal conclusions of the author, but are not accurate enough to afford detailed comparisons of respiratory rate. In such cases only the main points (e.g., insensitivity to cyanide) are given serious consideration. In only a few cases have investigators used bacteria-free cultures for measurement of respiration (cf. Table 4).

For comparative purposes in work with metazoan tissues, it is customary to express oxygen consumption in cubic millimeters (at normal temperature and pressure denoted as N.T.P.) per hour per milligram of dry weight of the tissue (symbolized by Q_{O_2}). In the protozoan literature, where the rate of O_2 consumption is expressed in absolute units, this unit is sometimes the Q_{O_2} but is more usually mm^3 per hour per organism, principally because the counting of organisms is simpler than measuring dry weight. Some authors use the symbol Q_{O_2} for O_2 consumption per 1,000,000 or per 100,000,000 organisms (e.g., von Fenyvessy and Reiner, 1928; Hall, 1938), but it seems preferable to avoid confusion by retaining this symbol for its original meaning and using a new symbol for consumption per 1,000,000 organisms, perhaps Q_{O_2} as used in Table 4. However, since not dry weight, nor wet weight, nor number of organisms affords the possibility of comparing the oxygen consumption per unit of respiring protoplasm, these discrepancies are not as important as one might at first suppose. In the case of flagellates such as *Astasia* and *Chilomonas,* in which a high percentage of the weight may be in the form of carbohydrate reserves, either dry or wet weight would be a poor index of the amount of respiring protoplasm. If one wishes to compare the rate per unit of protoplasm, the best index is probably the nitrogen content, because protein and other nitrogen-containing compounds are not ordinarily stored as reserve food. However, in the case of Protozoa which secrete nitrogen-containing tests, this criterion might also be very poor. Therefore any comparison of the absolute rate of different species, even after all differences in technique, immersion fluid, and physiological condition of the animal have been overcome, must usually be made with reservations, or at least with an adequate understanding of the limitations involved. Some authors have chosen to calculate O_2 consumption in terms of cubic millimeters of organisms, but the errors inherent in the methods of packing the animals for measurement (usually centrifuging) or in calculating volume from linear dimensions are too great to allow close comparison of data for different types of organisms. However, for an extended series of experiments on the same or very similar organisms, the use of dry weight (A. Lwoff, 1933) or of volume measurements (Elliott, 1939) seems to be entirely satisfactory.

If concentrated suspensions of organisms are used, the rate of shaking should be carefully controlled. The importance of this factor is demonstrated by the data of Hall (1938). If ammonia is produced by an organism, it is necessary to maintain acid within the respiring chamber in order to obtain true values of O_2 consumption. This very important procedure is discussed by Specht (1935).

Another question which arises in expressing results in absolute form is that of measuring basal metabolism, i.e., the metabolism of rest. In a mammal, for instance, there are certain well-defined limitations of conditions under which O_2 consumption may be termed a measurement of basal metabolism. In a protozoan it is more difficult, if at all possible, to apply these criteria, and in all known measurements we have a sum of the total metabolic processes, i.e., of those to which we refer as "basal," those due to movement of the organism, and, if the medium is nutrient, those due to the manufacture of reserve food material and to growth. The energy expended for each of these purposes will probably vary with the species, the physiological state, and the environmental conditions.

If the metabolic substrates of an organism undergo complete oxidation, we are able, by means of measurements of O_2 consumption, to determine directly the amount of energy available to the organism. If oxidation is incomplete, a further knowledge of the oxidation products is necessary. It is usually assumed, unless we have knowledge to the contrary, that oxidation is complete in the aërobic Metazoa (cf. intestinal nematodes, von Brand and Jahn, 1940). Among the bacteria and also among the Protozoa this is not always true, even in the presence of normal O_2 tension. However, since carbohydrate cleavage and intramolecular oxidation, even in the presence of O_2, may be considered an anaërobic process, that question will be discussed under anaërobiosis.

From studies on the heat of combustion, we know that complete oxidation of glucose yields 677,000 calories per gram molecule, or about 3,700 calories per gram. Complete combustion of protein yields 5,700 calories per gram, and fats yield 8,000-9,000 calories per gram. Therefore, if we know the substrate being oxidized and the rate at which O_2 is consumed, we can calculate the energy made available by oxidation. According to the equation

$$C_6H_{12}O_6 + 6 O_2 \rightarrow 6 CO_2 + 6 H_2O + 677,000 \text{ cal.}$$

one gram molecule of glucose requires six gram molecules of O_2. The volume of O_2 consumed (at N.T.P.) is 6×22.4 liters, or 134.4 liters. The ratio of O_2 consumed to calories released is 134 liters/677,000 calories, or about one calorie for each 200 mm^3 of O_2 consumed. Similar calculations may be made for fats and proteins.

Table 4 contains most of the known data for respiratory rates for the Protozoa which can be expressed in absolute terms—either as mm^3 O_2 per organism per hour, or mm^3 O_2 per gram dry or wet weight per hour. Similar tables are given by von Brand (1935) and Hall (1938).

2. THE EFFECT OF O_2 TENSION ON O_2 CONSUMPTION

For many types of biological material it has been quite well established that, under usual experimental conditions, O_2 consumption is independent of O_2 tension, within very wide limits (exceptions cited by Tang, 1933, McCoy, 1935). Recently, however, Kempner (1936, 1937) demonstrated that this is not true for several species of bacteria, for human leucemic leucocytes, for red blood cells of man, fowl, and alligator, and for pine needles if CO_2 is present or if the temperature is above 25° C. With these same materials in CO_2-free alkaline media below 25° C., O_2 consumption was independent of O_2 tension. The effect of O_2 tension apparently varied with pH, CO_2 tension, salt con-

TABLE 4: MEASUREMENTS OF PROTOZOAN RESPIRATION

Species	$Mm^3 O_2$ per Hour per Million Q_{O_2}	$Mm^3 O_2$ per Hour per mg. Dry Weight Q_{O_2}	Temperature, C.	Bacteria-free	Investigator
Paramecium caudatum	120 (CO_2)		21°	No	Barratt (1905)
	140		21°	No	Lund (1918c)
	2,250		23°	No	Zweibaum (1921)
	3,900		19°	No	Necheles (1924)
	5,600		22°	No	Kalmus (1928b)
	500		21.2°	No	Howland and Bernstein (1931)
Paramecium multimicronucleatum	1,021		25°	No	Mast, Pace, and Mast (1936)
Colpidium campylum	200		24.0°	No	Pitts (1932)
	112.5		19.8°	Yes	Hall (1938)
Colpidium colpoda	200		17.0°	No	Wachendorff (1912)
	200			No	Peters (1929)
Colpoda sp.	600–1,200		19.7°	No	Adolph (1929)
Glaucoma piriformis		35	22.0°	Yes	M. Lwoff (1934)
Blepharisma undulans		(0.5)*	20.8°	No	Emerson (1929)
Spirostomum ambiguum	2,590		25.0°	No	Specht (1935)
Strigomonas oncopelti	0.4	62	28.0°	Yes	A. Lwoff (1933)
Strigomonas fasciculata	0.4	55	28.0°	Yes	A. Lwoff (1933)
Leptomonas ctenocephalus	0.3	40	28.0°	Yes	A. Lwoff (1933)
Trypanosoma equiperdum	0.05		37.0°	Yes	Von Fenyvessy and Reiner (1928)
Chilomonas paramecium	17–26		25.0°	Yes	Mast, Pace, and Mast (1936, 1937)
Astasia sp.	2,400			Yes	Jay (1938)
Khawkinea halli	2,050			Yes	Jay (1938)
Actinosphaerium eichhornii	1,100		20.0°	No	Howland and Bernstein (1931)
Amoeba proteus		(0.2)*	20.0°	No	Emerson (1929)

* Not Q_{O_2} but $mm^3 O_2$ per hour per mm^3.

tent, and temperature. Since O_2 consumption of the yellow pigment of respiration (see below) varies with O_2 tension, it would not be surprising to find a similar relationship in organisms with this mechanism. However, this is not always the case (Schlayer, 1936). Clarification of the relationship between O_2 tension and O_2 consumption in Protozoa will apparently require much more data than is now available. A discussion of the theoretical relationship between oxygen tension and oxygen consumption is given by Marsh (1935).

For Protozoa, the available evidence indicates that within wide limits O_2 tension has little or no effect on the rate of O_2 consumption for *Paramecium* and *Colpoda*, and that it does have an effect on *Spirostomum*. Lund (1918a) found that the rate of O_2 consumption for *Paramecium* was independent of O_2 tension between 0.04 cc. and 2.2 cc. O_2 per 137 cc.—a 55-fold range. This was determined by placing thick suspensions of *Paramecium* in stoppered bottles and measuring the dissolved O_2 content of the water by the Winkler method, until the animals died. Lund's conclusion was confirmed by Amberson (1928), who placed the organisms in a closed vessel, in contact with an atmosphere of known O_2 content. By gas analyses he demonstrated a uniform rate of O_2 consumption, with O_2 partial pressures which varied from 50 to 220 mm. Hg, and only a slight decrease (about 20 percent) at pressures as low as 11 mm. Hg. Adolph (1929) found that the O_2 consumption of *Colpoda* did not vary significantly with O_2 tension between 155 and 750 mm. Hg. In a single experiment at 4-8 mm. Hg, O_2 consumption decreased to 31 percent of its previous value. However, Adolph did find that low O_2 tension (40 mm.) was correlated with smaller size of the progeny of cultures. Specht (1935) measured the respiration of *Spirostomum* in pure oxygen, in air, and in 0.5 percent O_2 in N_2. He found that O_2 consumption in these gases was in the ratio of 151 to 100 to 71, and that CO_2 production was in the ratio of 175 to 100 to 70.

When considering the effect of low O_2 tensions on O_2 consumption for any of the larger Protozoa, one should consider the O_2 tension at various points within the organism as well as at the surface. This can be calculated by the diffusion equations of Harvey (1928) and others, on the assumption that the rates of cyclosis and water exchange are low. The O_2 tension at the center of an ellipsoid which is consuming O_2 uniformly throughout its substance, will be zero when the shortest radius

$$a = \frac{5Dc}{A}$$

where D is the diffusion coefficient of O_2, c is the O_2 tension at the surface, and A is its rate of O_2 consumption. In the case of a cylinder (e.g., *Spirostomum*) the factor 5 should be 4. For *Colpoda*, Adolph (1929) calculated the value of *"a"* to be 148 μ at atmospheric O_2 tension, and 72 μ at 40 mm. Hg partial pressure of O_2. In large ciliates this factor might be important at low O_2 tensions, even with a rather low rate of O_2 consumption.

3. EFFECT OF CO_2 TENSION ON O_2 CONSUMPTION

Root (1930) showed that when CO_2 tension was raised from 1 mm. Hg to 15-20 mm. Hg, the respiration of *Paramecium* increased slightly (less than 15 percent), and it was believed that this increase might be caused by increased activity of the organisms. As the CO_2 tension was increased above 60 mm. Hg, O_2 consumption decreased continuously to about 40-60 percent of the control when CO_2 tension reached 220-360 mm. Hg. Similar experiments on fertilized *Arbacia* eggs did not show increase at low concentrations, and all CO_2 tensions greater than 30 mm. Hg produced a decrease to less than 40 percent of normal. *Paramecium* apparently was much more resistant to increase of CO_2 tension than *Arbacia*. For both *Paramecium* and *Arbacia* only slight effects were obtained with HCl, at pH values comparable to those present during the CO_2 experiment (4.5 to 7.5 for *Paramecium*).

From the information available, it is not possible to determine the mechanism of action of CO_2 on protozoan respiration, and results with other organisms are few and variable. Apparently CO_2 is not involved as an inhibitor or accelerator of any of the known mechanisms of respiration (to be discussed below), and at present we can only say that the effect on respiration seems to be indirect, and that the results are not due to pH changes in the external fluid. However, internal pH changes, as suggested by Root (1930), might account for the effect. This is an explanation comparable to that given by Jahn (1936) for the effect of the lack of CO_2 on growth of *Chilomonas* and bacteria. It is well established that certain bacteria will not grow in the absence of CO_2. Jahn (1936) studied the effect of CO_2-free media on growth of *Chilomonas* and *Colpidium* and found a distinct inhibition with *Chilomonas* and none with *Colpidium*. The explanation was offered that the inhibition of growth might be caused by inadequate intracellular buffering in those species which were affected by lack of CO_2, and that organisms whose normal environment is high in CO_2 might depend more on CO_2 buffering than those the normal environment of which is low in CO_2. (For possible application of this idea to culture of intestinal forms, see Jahn, 1934).

4. THE EFFECT OF THE PHYSIOLOGICAL STATE ON O_2 CONSUMPTION

It seems as if the effect of various factors which influence the physiological state of an organism may be reflected in measurements of O_2 consumption. Factors which have been investigated for the Protozoa are starvation, age of the culture, and conjugation.

Lund (1918c) starved *Paramecium* in tap water and noted an appreciable decrease in respiration during the first twenty hours. This was simultaneous with the disappearance of deutoplasmic food reserves from the protoplasm. Upon feeding starved animals with boiled yeast suspensions, the rate of oxygen consumption could be increased two to three times. This increase was independent of cell division. Leichsenring (1925) demonstrated a decrease of 23 percent after twenty-four hours of starvation and 29 percent after seventy-two hours.

The effect of the age of the culture on O_2 consumption was first studied by Wachendorff (1912), who found that for *Colpidium colpoda* the O_2 consumed per organism diminished from 191 mm³ per hour the first day to 151 mm³ the tenth day, and to 59 mm³ on the thirtieth day. Reidmuller (1936) reported a higher rate of O_2 consumption for young cultures of *Trichomonas foetus* than for older cultures. Twenty-four to forty-eight-hour-old cultures consumed 4.3 mm³ O_2 per mm³ of organisms, sixty-hour cultures, 2.8 mm,³ and three-day-old cultures only 0.81 mm³ O_2. Andrews and von Brand (1938) reported a decrease in sugar consumption per organism for this species, with increasing age of the culture, and their data indicate that the differences observed by Reidmuller were real, in spite of objections to the method used because of the possibility of hydrogen or methane evolution.

Zweibaum (1921) studied the rate of O_2 consumption of *Paramecium caudatum* in relation to conjugation. He found that the rate just before conjugation was about 0.73 mm³ O_2 per thousand organisms per hour. During conjugation this rate rose to 3.4 mm³/1,000/hour, and immediately after conjugation decreased to about 0.73. During the first eight or nine days following conjugation, the rate rose slowly to 2.0 mm³/1,000/hour, and remained at this value from four to five months.

5. THE EFFECT OF TEMPERATURE ON O_2 CONSUMPTION

Data concerning the effect of temperature on O_2 consumption are not numerous, and in most cases are incomplete. Barratt (1905) determined the CO_2 production of *Paramecium* at various temperatures and found that the rate at 27°-30° C. was more than twice that at 15° C. Wachendorff (1912) found that *C. colpoda* respired about four times as fast at 17° as at 7° C. Leichsenring (1925) demonstrated that *Paramecium*, when transferred from a temperature of 20° C. to one of 35° C., showed a respiratory increase of 35 percent, and that when transferred to a temperature of 15°, 10°, 5°, 0° C. respiration was decreased 30 percent, 34 percent, 50 percent, and 58 percent respectively. These effects were not completely reversible. The data of Kalmus (1928b) showed a Q_{10} value (temperature coefficient) of 1.5 for *Paramecium* respiration between 23° and 32° C. A. Lwoff (1933) found a Q_{10} value of 2.1 between 13° and 23° C., and about 1.5 between 23° and 32° C. The temperature characteristic (μ value) was 9,830 calories for the range 13° to 34° C.   Lwoff also calculated a μ value of 21,350 for the synthesis of respiratory enzyme (oxidase, see below) by the organism between 18° and 31.5° C., and a value of —52,000 between 31.5° and 34.0° C.

6. THE EFFECT OF ANESTHETICS AND POISONS ON O_2 CONSUMPTION

The effect of various toxic agents (e.g., KCN, CO, N_3H, arsenite, urethanes, and so forth) which are supposed to exert a specific effect on the normal functioning of certain respiratory enzymes, will be discussed in connection with the mechanism of respiration. The effect of other anesthetics on respiration has not been extensively studied. Leichsenring (1925) found that ethylene and nitrous oxide had no effect on the respiration of *Paramecium*, and that ether and chloroform produced decreases of as much as 25 percent (after two and one hours respectively). The effect was reversible with ether, but not with chloroform if the exposure was more than one-half hour. *Colpoda* was more sensitive to these substances than *Paramecium*. Von Fenyvessy and Reiner (1928) found no decrease in respiration of *Trypanosoma equiperdum* when one-percent Germanin was added. This was surprising, in view of the fact that Germanin is very toxic for trypanosomes *in vivo*.

7. THE EFFECT OF NUTRITIVE SUBSTANCES AND OTHER MATERIALS

The effect of various nutritive substances on Protozoa has been demonstrated by a number of investigators, but the criteria used are usually growth or the accumulation of food reserves, and not respiration. In some cases it is possible to determine that the substance is oxidized directly (e.g., glucose), or that it contributes toward the synthesis of the respiratory enzyme (A. Lwoff, 1933). These data will be discussed below. The effect of various substances on the respiration of *Paramecium* was studied by Leichsenring (1925), who found that caprine, glutamic acid, peptone, and aminoids increased respiration 12-18 percent; that glycocoll and succinic acid increased respiration 8-9 percent; and that tyrosine and cystine produced little effect. Lactose gave an increase of 16 percent, and other sugars and polysaccharides gave increases of 3-10 percent. Thyroxin gave an increase of 13 percent. No explanation was made of the mechanism of these effects.

Mast, Pace, and Mast (1936) found that *Chilomonas* grew well, formed considerable starch but little fat, and consumed 0.17 mm³ of O_2 per 10,000 organisms per hour, in a solution of $MgSO_4$, NH_4Cl, K_2HPO_4, Na-acetate, and silicon. When sulphur was omitted, the starch remained constant, fat accumulated, O_2 consumption decreased, and the animals finally died. When acetate was omitted, the organisms decreased in size, starch and fat decreased, and O_2 consumption decreased to 0.07 mm³/10,000/hour. When both sulphur and acetate were omitted, starch decreased to zero, fat accumulated, O_2 consumption decreased, and the organisms died. These authors conclude that starch is normally changed to fat, that sulphur induces oxidation of fat, thereby increasing respiration and preventing accumulation of fat, and that fat oxidation is probably associated with a cystine-cysteine mechanism (see glutathione, below). Mast and Pace (1937) reported a 25 percent increase in respiration of *Chilomonas* when Na_2SiO_3 was added to inorganic media. This was supposedly caused by the catalytic action of Si on organic syntheses.

8. EVOLUTION OF GASES OTHER THAN CO_2

The possibility that Protozoa evolve gases other than CO_2 was first shown by Cook (1932) for the flagellates of termites (*Termopsis nevadensis*). A gas which was not absorbable by hydroxide was evolved by normal termites, under anaërobic conditions. However, this gas was not evolved if the intestinal flagellates had been removed by oxygenation, and it was suggested that it might have been evolved by the flagellates. The evidence is incomplete, however, in that oxygenation probably also changed the bacterial flora. Witte (1933) observed the production of gas bubbles by *Trichomonas foetus*. This was confirmed by Andrews and von Brand (1938), who found that this gas was not absorbable by alkali and that when mixed with O_2 it burned with indications of explosiveness. Final identification was not made. The formation of gas vacuoles has been reported for several organisms, but in most cases there is little evidence regarding the identity of the gas. Bles (1929) believed that the gas vacuoles of *Arcella* contained O_2.

The chlorophyll-bearing flagellates, of course, might give off O_2 in the presence of strong light, because of photosynthesis, and this might also be true of the ciliates which harbor zoöchlorellae. There is considerable indirect evidence that this is true, but no direct measurements are available.

INVESTIGATIONS WHICH CONCERN THE SOURCE OF ENERGY

Whenever an oxidizable material is subjected to complete combustion, the ratio of CO_2 given off to O_2 consumed will vary with the type of material. This ratio (CO_2/O_2) is called the respiratory quotient (R.Q.). Carbohydrates (relatively rich in oxygen) have an R.Q. of 1.0; fats (relatively poor in oxygen) have an R.Q. of about 0.71; and proteins have an R.Q. of 0.83 if the nitrogen is eliminated as urea, and 0.93 if it is eliminated as ammonia. By measurements of the respiratory quotient we may obtain an index of the type of material which is being oxidized by the organism, at least as to whether it is predominantly carbohydrate or predominantly fat. If the excreted nitrogenous material can be identified and measured, the amount of protein and consequently the amounts of carbohydrates and fats consumed can be calculated. Under these conditions the measurement of R.Q. becomes more significant.

These interpretations are based on the assumptions that complete oxidation of metabolic substrates is the sole cause of gaseous exchange, and that gases other than O_2 and CO_2 are not involved. If these assumptions are unsound, then any interpretation of the R.Q. is necessarily more difficult. If carbohydrate is being converted into fat within the organism for purposes of storage, the R.Q. may reach a value of 1.4. If fat or protein is being converted to carbohydrate, there will be a corresponding tendency for the R.Q. to be lowered. A high value for the R.Q. can also arise whenever CO_2 is removed from a compound without the consumption of O_2, or whenever an oxygen debt accumulates (during heavy exercise). Whenever an oxygen debt is being removed, the R.Q. may fall to extremely low values (during rest after exercise). Unusual values of R.Q. may be obtained if substrates other than carbohydrates, fats, or proteins are being consumed. In the case of very rapid protozoan growth on substrates of organic acids, the R.Q. should vary with the oxygen content of the molecule being oxidized (acetic acid, 1.0; proprionic acid, 0.85; butyric acid, 0.80).

The respiratory quotient of an organism may be calculated with almost any of the manometric methods described above, if separate measurements are made with and without a CO_2-absorbing alkali in the respiratory chamber. In this case, one set of readings (with KOH) will give a measure of the O_2 consumed, and the other set (without KOH) will be an index of the difference between CO_2 given off and O_2 consumed. From this the R.Q. may be calculated, provided no NH_3 is evolved and no CO_2 is retained in the immersion fluid. If these complications arise, suitable modifications may be introduced. The use of somewhat more complicated manometer chambers allows measurements of simultaneous O_2 consumption and CO_2 production to be made on the same material (Dixon, 1934). For the Protozoa this method seems preferable because, in addition to its usual advantages, it prevents the results from being affected by the possible secretion of ammonia (Specht, 1935) and other bases (e.g., sodium carbonate from oxidation of Na-acetate, Jahn, 1935a). However, even this method does not correct for the possible evolution of hydrogen or methane (cf. *Trichomonas foetus*, Andrews and von Brand, 1938).

For the ciliates, several measurements of R.Q. have been made. Wachendorff (1912) reported R.Q. values of about 0.3 for *Colpidium*, but in view of later developments it seems as if this material should be reexamined with more modern methods. Emerson (1929) studied the respiration of *Blepharisma undulans* and found an R.Q. slightly less than 1.0. Daniel (1931) obtained an R.Q. of 0.84 for *Balantidium coli*, but the possible effects of bacteria were not well controlled. Amberson

(1928) reported an R.Q. of 0.69 for *Paramecium,* and Root (1930) in a number of experiments obtained an R.Q. value of 0.62. Root also found that the R.Q. varied somewhat irregularly with changes in CO_2 tension. However, there was a definite trend toward high R.Q. values in media of high CO_2 tension, and the average R.Q. at 238-423 mm. Hg. was 1.43. This apparently was caused by a decrease in O_2 consumption (see above) without a corresponding decrease in CO_2 production, thereby giving a high R.Q. According to Root, "It is possible that the suppression of oxidations under these conditions results in the production of acid metabolites which drive out carbon dioxide from the bicarbonate contained in the cells and in the surrounding medium." Similar experiments on *Arbacia* eggs did not show an increase in R.Q., and it was assumed either that acid substances were not produced or that they were rapidly converted into a non-acid form and did not accumulate in appreciable amounts. Apparently CO_2 tension is a factor which should be considered when making measurements of R.Q. However, if experiments are conducted with standard manometric techniques, this factor is probably not important.

Specht (1935) measured the R.Q. of *Spirostomum* in manometers, both with and without the presence of acid in a side arm of the manometer flask. He found an R.Q. of 0.24 without the acid and 0.84 when acid was present. This discrepancy was explained as being caused by the elimination of NH_3 by the organisms, and the value of 0.84 is therefore accepted as more nearly correct. However, it was also demonstrated that the R.Q. was 0.98 in an atmosphere of O_2, and that this value was not affected by the presence or absence of acid. Apparently NH_3 is not produced at high O_2 tension. These experiments indicate very clearly that ammonia secretion is a possible source of error in measurements of protozoan respiratory quotients, and therefore one should be suspicious of the validity of low R.Q. values unless adequate precautions have been taken against the ammonia error.

For the free-living flagellates, very low values of R.Q. have been reported. Jay (1938) found R.Q. values of 0.34 and 0.56 for *Astasia* and *Khawkinea,* respectively. Mast, Pace, and Mast (1936) reported R.Q. values of 0.28 to 0.37 for *Chilomonas,* and 0.72 for *Paramecium multimicronucleatum* under similar conditions. The possible explanations mentioned by Jay for the low R.Q. value are conversion of protein to carbohydrate, or incomplete oxidation of carbohydrate. The explanation of Mast et al. is that carbohydrate was being synthesized from carbon dioxide. This explanation is based on previous nutritional studies, but these are open to question (review, Hall, 1939). One obvious possibility is that CO_2 may be retained in the immersion fluid, but Mast and his coworkers obtained only a slightly higher R.Q. value when the bound CO_2 was liberated by acid (single experiment only). In these cases the explanations offered must be considered as only tentative, until the possibilities of NH_3 production and CO_2 retention are positively eliminated. Mast and his coworkers also reported for *Chilomonas* that under certain conditions starch was converted to fat, and that fat oxidation could be decreased by depriving the organism of sulphur. However, values obtained for the R.Q. were variable and showed no definite correlation with these conditions.

Values of the R.Q. reported for members of the family Trypanosomidae are within the normal range. Soule (1925) obtained an R.Q. of 0.84-0.91 for *Leishmania tropica* and 0.74-0.89 for *Trypanosama lewisi* in blood agar medium. When glucose was present, the R.Q. rose to 0.95 for *L. tropica* and 0.94 for *T. lewisi*. Novy (1932) reported respiratory quotients of 0.93 to 1.0 for *T. lewisi, L. tropica, L. donovani, L. infantum, Strigomonas oncopelti, S. culicidarum, S. culicidarum var. anophelis, S. lygaeorium, S. media, S. muscidarum,* and *S. parva,* when grown on glucose-blood agar. When grown on glycerol-blood agar or plain blood agar, the R.Q. was about 0.8 to 0.87 for the four species of *Leishmania*. Von Fenyvessy and Reiner (1924) found an R.Q. of 0.60 for *Trypanosoma equiperdum* in diluted blood. A. Lwoff (1933) obtained R.Q. values of 1.0 for *Strigomonas oncopelti* and *S. fasciculata,* and a value of 0.88 for *Leptomonas ctenocephali*.

Apparently the only R.Q. measurement on a rhizopod is that of Emerson (1929) on *Amoeba proteus,* which gave a value slightly less than 1.0.

Another method, in addition to that of the respiratory quotient, which might be used as an index of the source of energy in an organism is the calorific quotient. Since the ratio of heat produced to oxygen consumed differs with carbohydrate, fat, and protein (3.5, 3.3, and 3.2, respectively), it is possible to measure heat production and O_2 consumption and to use this ratio as an index of the substrate being utilized.

However, because of the small differences in the ratios and because of the complications of the technique of heat measurement, the calorific quotient has not been found to be very useful in determining the energy source for Metazoa (Needham, 1931), and apparently has not been tried for Protozoa.

INVESTIGATIONS WHICH CONCERN THE MECHANISM OF RESPIRATION

1. GENERAL THEORY

For a general consideration of the mechanism of respiration, the reader is referred to the monographs of Meldrum (1934) and Holmes (1937), to standard textbooks of general physiology, to several excellent discussions in recent volumes of the *Annual Review of Biochemistry,* and to the forthcoming volume of the Cold Spring Harbor *Symposia in Quantitative Biology* (Vol. VII). The present discussion of the mechanism of respiration will include only those portions of a bare outline which are necessary for an understanding of the data and interpretations which are to follow.

The first step in oxidation of a substrate is the removal of hydrogen from the substrate molecule, and the addition of this hydrogen to any other molecule which will serve as a hydrogen acceptor. After dehydrogenation is accomplished, the resulting molecule is supposed to be very unstable and easily undergoes oxidation by molecular oxygen, to form CO_2 and water. The enzymes necessary for these final stages in respiration are not well known, but the enzymes and respiratory pigments responsible for dehydrogenation and the subsequent transfer of the hydrogen to O_2 with formation of water are listed below. (Any distinction between respiratory enzymes and pigments is purely arbitrary.)

(1) *Dehydrogenases* are enzymes which bring about activation of the substrate, so that it may be oxidized by oxygen or intermediate hydrogen acceptors such as cytochrome. These enzymes are highly specific, in that they react with only one or a few substrates. Dehydrogenases are divided into two groups: anaërobic dehydrogenases, which cannot reduce molecular O_2 in the presence of their substrates, and aërobic dehydrogenases which can do so. Cytochrome and cytochrome oxidase are important factors in the completion of oxidation by anaërobic dehydrogenases.

(2) *Cytochrome* is a group of pigments or enzymes, which in the living cell are oxidized under aërobic, and reduced under anaërobic conditions, but which cannot be oxidized directly by molecular O_2. These serve as hydrogen acceptors for anaërobic dehydrogenase systems.

(3) *Oxidase.*—The term oxidase includes all enzymes which are capable of performing oxidations in the presence of molecular oxygen. To this group belongs the respiratory enzyme of Warburg, which is perhaps identical with the oxidase of cytochrome, which in turn is also referred to as indophenol oxidase because of one method of detecting its activity. This enzyme brings about the oxidation of reduced cytochrome by molecular oxygen, and water is supposed to be oxidized to hydrogen peroxide during the process. Aërobic dehydrogenases are sometimes classified as oxidases.

(4) *Catalase* is an enzyme present in aërobic organisms and usually absent in anaërobes. This enzyme converts hydrogen peroxide to water and molecular oxygen, and its place in the respiratory chain is given below.

(5) *Peroxidases* are enzymes which in the presence of an oxidizable substrate convert hydrogen peroxide to water and activated oxygen, thereby causing oxidation of the substrate. The exact rôle of peroxidases in cellular respiration is not understood. The peroxidases are iron compounds, and other iron compounds, such as cytochrome and methemoglobin, exhibit some peroxidase-like activity.

(6) *Yellow respiratory pigment, or enzyme,* is a flavo-protein capable of reversible oxidation and reduction which may be reduced in a reaction involving oxidation of substrate (through the intermediary action of a co-enzyme) and which can then be reoxidized in the presence of molecular O_2 (with formation of H_2O_2) or other hydrogen acceptors.

(7) *Glutathione* is an amino-acid complex capable of reversible oxidation and reduction, and which may act as a hydrogen acceptor through the reduction of an -S-S- group to two -SH groups (cysteine to 2 cysteine) which are auto-oxidizable in the presence of molecular O_2.

The details of how these substances function in the living cell are subject to considerable controversy, but for our present purpose we may regard the general outline for the first four items listed above as follows:

(1) substrate + 2 oxidized cytochrome $\underset{\text{dehydrogenase}}{\rightleftharpoons}$
oxidized substrate + 2 reduced cytochrome

(2) 2 reduced cytochrome $+ O_2 \underset{\text{oxidase}}{\rightleftharpoons} 2$ oxidized cytochrome $+ H_2O_2$

(3) $H_2O_2 \underset{\text{catalase}}{\rightleftharpoons} H_2O + \frac{1}{2} O_2$

The substrate may be activated by "anaërobic" dehydrogenase, and it is then oxidized by cytochrome, the cytochrome itself being reduced in the process (equation 1). Cytochrome is, in turn, oxidized by an oxidase system which may be identical with Warburg's respiratory enzyme (equation 2). During this process H_2O_2 is formed and is then broken down to water and molecular O_2 by catalase (equation 3).

The oxidase and catalase systems are inhibited by the presence of HCN and H_2S, and the oxidase system is also inhibited by CO. In the presence of any of these reagents, reaction (1) can proceed but not reactions (2) or (3). Therefore all of the cytochrome becomes reduced, and respiration by means of this mechanism is stopped. The dehydrogenase systems are inhibited by narcotics (e.g., the urethanes), by warming and cooling, and these agents leave all of the cytochrome in the oxidized state. These two general methods of treatment, therefore, may be used as tools in studying the above respiratory mechanisms. There are also aërobic dehydrogenases which, in addition to activating the substrate, can react directly with molecular oxygen without the mediation of cytochrome and oxidase. Respiration which is brought about by this type of system is not supposed to be affected by HCN.

It is possible to demonstrate that in some systems certain reversible oxidation-reduction indicators (e.g., methylene blue) can replace the cytochrome-cytochrome oxidase system, and that in this capacity the action of these indicators may or may not be affected by HCN and CO (e.g., grasshopper embryos, Bodine and Boell, 1937; *Escherichia coli*, Broh-Kahn and Mirsky, 1938). If HCN and methylene blue are added to such a respiratory system and inhibition does not occur, the ensuing reactions might be visualized as follows:

(4) substrate $+$ methylene blue $\underset{\text{dehydrogenase}}{\rightleftharpoons}$ oxidized substrate
$+$ leuco-methylene blue

(5) leuco-methylene blue $+$ oxygen $\rightleftharpoons$ methylene blue $+ H_2O_2$

Since catalase is inactivated by HCN, the hydrogen peroxide presumably accumulates and in *E. coli* cultures can be measured experimentally. In anaërobes which normally do not possess catalase, this mechanism might explain the bacteriostatic effect of oxidation-reduction indicators.

It is also known that there are certain pigments of bacteria and yeasts (e.g., yellow enzyme, or pyocyanine) which are capable of bringing about a similar result, and other naturally occurring oxidation-reduction indicators have been described (echinochrome, hermidin, and pigments from *B. violaceus* and *Chromodoris zebra*) which apparently might function in a similar fashion. The reactions involving yellow pigment and its coenzyme may be indicated as follows:

(6) substrate + coenzyme $\xrightleftharpoons[\text{dehydrogenase}]{}$ oxidized substrate + reduced coenzyme

(7) reduced coenzyme + yellow pigment $\rightleftharpoons$ coenzyme + leuco-yellow pigment

(8) leuco-yellow pigment + oxygen $\rightleftharpoons$ yellow pigment + H_2O_2

(9) $H_2O_2 \xrightleftharpoons[\text{catalase}]{} H_2O + \frac{1}{2} O_2$

In this case only the action of catalase is prevented by HCN, and therefore H_2O_2 accumulates. In the absence of O_2 the leuco-yellow enzyme may be oxidized by other substances (e.g., by methylene blue). The yellow enzyme has been found to be a combination of protein and vitamin G, and it is believed that while this sort of system is present in aërobic organisms, it assumes its greatest importance in anaërobic species. In anaërobic organisms (yellow enzyme can be prepared from bottom beer yeast or lactic acid bacilli) we have, then, a respiratory system which is quite independent of cytochrome and Warburg's oxidase, and which therefore is insensitive to HCN and CO. Perhaps when we say that the respiration of a given species is cyanide insensitive, we may be inferring that that species has a respiratory system more suited to anaërobic conditions (temporary or otherwise). Under anaërobic conditions the leuco-yellow pigment is probably oxidized by substances other than molecular oxygen, and H_2O_2 is not formed. The known respiratory enzymes of bacteria are summarized by Frei (1935) and Stephenson (1939).

The relationships between the various respiratory enzyme systems are not so well known nor so clear-cut as, for the sake of clarity and brevity, they have been made to appear in the above outline. Whenever we conclude, on the basis of the action of certain reagents on respiration, that one respiratory mechanism is very important and that another is not, we should do so only with certain mental reservations, and the conclusions should not be considered final, but merely indicative.

2. EXPERIMENTS WHICH CONCERN THE CYTOCHROME-CYTOCHROME OXIDASE SYSTEM OF HYDROGEN ACCEPTORS

We have, through the action of HCN and CO on respiration, a tool for determining how much of the respiratory activity of a given organism is carried on by means of the cytochrome-respiratory enzyme system and how much is not. Respiration which is not cyanide and CO sensitive may be due to aërobic dehydrogenases, or to anaërobic dehydrogenases plus an enzyme of the yellow-pigment type or perhaps to the action of peroxidases. Such analyses have been made for several types of biological material. It has been determined that respiration of some cells is extremely sensitive to HCN (e.g., yeast, *B. coli,* most bacteria, and mammalian tissues), while that of others is quite resistant (*Chlorella, Paramecium, Sarcina, Pneumococcus, B. acidophilus, Streptococcus, Staphylococcus*); also, the same organism may differ in sensitivity at different periods during its life history (grasshopper eggs, Robbie, Boell, and Bodine, 1938). One technical precaution which should be observed in cyanide experiments is the use of a KOH-KCN absorbing fluid for CO_2 (van Heyningen, 1935). By the selection of the proper KOH-KCN mixture, the osmotic transfer of KCN through the air from the experimental material to the KOH solution can be prevented. This is apparently one possible source of error in all work on the effect of cyanide on protozoan respiration—that none of the authors has used balanced KOH-KCN solutions.

Among the Protozoa the effect of cyanide has been studied on several ciliates and flagellates. It is quite well established that the respiratory mechanism of *Paramecium* is insensitive to cyanide (Lund, 1918b; Shoup and Boykin, 1931; Gerard and Hyman, 1931), and the work of Shoup and Boykin (1931) shows that the addition of iron salts does not increase respiration and that very little or no iron is present in *Paramecium*. These results may be interpreted to mean that the cytochrome-respiratory enzyme system plays no part in the respiration of *Paramecium*. Peters (1929) obtained no inhibition with M/500 KCN on *Colpidium colpoda*. The data of Pitts (1932) for *C. campylum* shows that less than 20 percent of the respiration is cyanide sensitive and that this depression is only temporary, and that while still in cyanide the respiration may rise to a rate which is as much as 25 percent above normal and drop again to 80 percent of the normal rate. M. Lwoff (1934) found that the respiration of *Glaucoma piriformis* in peptone solution was depressed as much as 80 percent during the first half hour, but that by the third hour the rate had returned to normal (M/1,000 KCN), or almost normal (15 percent below in M/450 KCN), and then decreased. In a weaker KCN solution (M/4,500) this latter decrease did not occur. The organisms were able to live twenty-four hours in M/450 KCN and eight days in M/1,000, but they did not divide. In M/2,000 and M/5,000 KCN multiplication of the organisms occurred slowly. In glucose-Ringer solution, M/450 KCN did not inhibit, but produced an acceleration of as much as 36 percent during the first half hour. The conclusion to be drawn from these data is that the ciliates, as far as we know, are relatively insensitive to the action of cyanide, and we might consider the temporary inhibitions produced in some cases as secondary effects rather than direct effects upon the respiratory mechanism (cf. another explanation mentioned below). It would be interesting to reinvestigate the effect of KCN on some of these ciliates by the use of balanced KCN-KOH solutions. It seems possible that the data, especially in the case of the temporary effects, may be complicated by the loss of KCN from the experimental solution to the KOH solution.

Among some of the flagellates, however, there seems to be quite a different respiratory mechanism. A. Lwoff (1933) found that M/3,000 KCN inhibited respiration of *Strigomonas oncopelti* 90 percent, of *S. fasciculata* 83 percent, and of *Leptomonas ctenocephali* 95 percent. With M/1,000 KCN, both species of *Strigomonas* were inhibited 90 percent, and *L. ctenocephali* 95 percent. The latter species was extremely sensitive and was inhibited 92 percent by M/20,000. Growth was also decreased in those concentrations which inhibited respiration, and the organisms were killed only by much greater concentrations. M. Lwoff (1934) reported an inhibition of 90 percent for *Polytoma uvella,* and Jay (1938) an inhibition of 60-65 percent for *Khawkinea* and *Astasia* at a concentration of M/100 KCN. Von Fenyvessy and Reiner (1928), however, reported no effect with 0.1 percent KCN (M/65) on either oxygen consumption or acid production of *Trypanosoma equiperdum* in glucose-bicarbonate-Ringer solution.

These results demonstrate that the respiratory mechanisms of various Protozoa are probably not the same. The respiratory mechanisms of some Protozoa seem to resemble those of *Chlorella* and *Sarcina,* while those of other species resemble the mechanisms of yeast and mammalian and other tissues. This question is one which should be studied carefully in a wide variety of organisms, and with a wide concentration range of cyanide solutions. The taxonomic position of the Protozoa should make such an investigation doubly interesting. It would also be of interest to know if the cyanide insensitivity of *Paramecium* is still maintained in the presence of glucose and other substances, or if an apparent change in the respiratory mechanism is brought about by the presence of glucose. Emerson (1929) found that the respiration of *Chlorella* was cyanide sensitive only in the presence of glucose; Gerard (1931) found that glucose had no effect on the cyanide sensitivity of *Sarcina,* but M. Lwoff (1934) found that the respiration of *Glaucoma* was accelerated by HCN when glucose was absent. These divergent results should have a final explanation in terms of the respiratory or other metabolic mechanisms involved.

An alternative theory to the supposed coexistence of cyanide sensitive and insensitive fractions in the normal cell is that all normal respiration is CN sensitive, and that in the presence of CN an entirely new respiratory mechanism is called into existence. This interpretation would indicate that among many bacteria, algae, and ciliates (but not among certain flagellates) there is a greater adaptability of the respiratory mechanism than among the more specialized cells of the Metazoa. Such generalizations are probably premature, but it does seem possible that aërobic protozoa which can live anaërobically for considerable periods of time might have a dual respiratory mechanism.

It has been demonstrated for several biological materials that respiration sensitive to cyanide is also sensitive to CO (because of CO-inhibition of cytochrome oxidase), and that respiration insensitive to cyanide is not depressed and may even be stimulated by CO (literature cited by Bodine and Boell, 1934). A. Lwoff (1933) showed that the KCN sensitive fraction (90 percent) of *S. fasciculata* respiration was also sensitive to CO. In an 80/20 mixture of CO/O_2 inhibition was 61 percent, in 95/5 mixtures 85 percent, and in 98/2 mixtures 90 percent. Values of K for the Warburg-Negelein equation

$$K = \frac{A}{A_0 - A} \times \frac{CO}{O_2}$$

where A is O_2 consumption in the CO/O_2 mixture and A_0 is O_2 consumption in the control, varied from 2.58 in 80 percent CO to 5.3 in 98 percent CO. M. Lwoff found that carbon monoxide (2-5 percent O_2 in CO) produced the same effect on *Glaucoma* as KCN in both peptone and in glucose-Ringer solutions. In peptone there was a marked inhibition for the first half hour and then a return to normal or almost normal, and in glucose-Ringer there was an increase of 20 percent. Reidmuller (1936) reported no appreciable effect of CO on O_2 consumption for *Trichomonas foetus* in 95/5 mixtures of CO and O_2. The effect of KCN was not investigated. The effect of CO on other Protozoa should be investigated together with the effect of cyanide. Recently azide (HN_3) has been found to have an effect on respiration which is similar to but not identical with that of HCN and CO (Keilin and Hartree, 1936; and others), and it would be interesting to make comparisons of these reagents on protozoan material.

For the purpose of inhibiting the cytochrome-cytochrome oxidase system, CO is apparently much more specific then HCN or azide. This is especially true if inhibition accurs in the dark but not in the presence of bright light, because the inactive compound formed by CO and cytochrome oxidase is dissociated upon illumination into CO and active oxidase. The reversibility upon illumination of CO inhibition has not been investigated for Protozoa.

The distribution of cytochrome among the Protozoa is a relatively untouched subject. A. Lwoff (1933) found two absorption bands in *Strigomonas fasciculata,* one at 530, and another rather broad band at 555 mμ. These bands disappeared upon passage of O_2 through the solution. Upon addition of KCN no other bands became visible, and the question arises as to whether the 555 band was the b and c bands of cytochrome or the b band of the hemochromogen, as has been found for various bacteria. Lwoff found the 555 band also in *S. oncopelti, Glau-*

coma piriformis, and *Euglena gracilis.* By treating *G. piriformis* and *Polytoma uvella* with sodium hydrosulphite and pyridine, he obtained the bands of pyridine-hemochromogen. The observations on *Glaucoma* are especially interesting because respiration is KCN and CO insensitive, and the explanation of the KCN and CO experiments therefore needs further clarification. Perhaps these organisms contain both cytochrome and a KCN insensitive system (e.g., yellow pigment) which may function interchangeably. This would explain their adaptability to both aërobic and anaërobic conditions, the presence of cytochrome and KCN and CO insensitivity, and perhaps also the somewhat oscillatory character of *Colpidium* and *Glaucoma* respiration in the presence of KCN (Pitts, 1932; M. Lwoff, 1934). This, of course, is pure speculation. However, the possibility of any discrepancies in the supposed parallelism between CN, CO, and HN_3 insensitivity and the absence of cytochrome should warrant an intensive investigation. Reidmuller (1936) was unable by spectroscopic methods to find either cytochrome or hemochromogen in *Trichomonas foetus,* and this result should be expected because of the CO-insensitivity of *Trichomonas* respiration.

3. Experiments Which Concern Other Systems of Hydrogen Acceptors

If we assume that the Warburg-Keilin system is not present in the ciliates, then we must seek another respiratory mechanism. Is this to be found in the action of glutathione? According to M. Lwoff (1934), the effect of arsenious acid on respiration offers a tool for detecting the action of glutathione because it is not supposed to affect the Warburg-Keilin respiratory system and because it does combine with -SH groups, thereby inhibiting the normal functioning of glutathione. In *Glaucoma piriformis* M. Lwoff found that M/1,900 arsenious acid (neutralized sodium arsenite) inhibited 75-80 percent of the respiration and that M/1,150 inhibited 90 percent. The organisms moved slowly in M/400 to M/2,000 and remained alive more than thirteen days in M/6,000, but did not multiply. The inhibition of respiration was entirely reversible (recovery in 1 1/2-2 hours from M/2,000). Monoiodoacetic acid, at least in some cases, is supposed to be similar in its action to arsenious acid, that is, it combines with -SH. (In other cases its action may be different, e.g., in the prevention of lactic-acid production from glycogen in muscle extracts in which glutathione is not present.) Therefore we may use it as an additional indicator in detecting the action of glutathione in respiration. M. Lwoff found that monoiodoacetic acid produced 61-82 percent inhibition of respiration of *Glaucoma* in concentrations of one part in 121,000 to 77,000, while the ciliates appeared normal and moved slowly. These results are interpreted to indicate that glutathione seems to be quite important in the respiration of *Glaucoma*. Another interpretation which has been used for work on other material (Korr, 1935; Cohen and Gerard, 1937) is that arsenites inhibit dehydrogenases (Szent-Gyorgyi and Banga, 1933). It is interesting that the accepted dehydrogenase inhibitors (urethanes, see below) do not result in as great an inhibition with *Glaucoma* as arsenites and monoiodoacetic acid.

Another respiratory mechanism which might exist among the cyanide-insensitive Protozoa is the yellow pigment found in yeast and other anaërobic organisms. It seems as if an investigation of the distribution of enzymes of this type should be made among the Protozoa, especially with those species in which respiration proves to be cyanide insensitive. Since a large number of Protozoa are presumably facultative anaërobes, it might be possible to poison the normal aërobic mechanism and study the anaërobic mechanisms under various conditions, as has been done for *Escherichia coli* by Broh-Kahn and Mirsky (1938).

4. INHIBITION OF THE DEHYDROGENASE SYSTEM

The dehydrogenases are apparently a part of the respiratory chain involved in several of the enzyme mechanisms. Therefore one would expect any substance which inhibits the dehydrogenases to inhibit respiration. Such is the case with the urethanes. M. Lwoff (1934) found that in the respiration of *Glaucoma* one percent methyl urethane produced 9 percent inhibition of respiration, 2 percent inhibited 38 percent, 2.5 percent inhibited 52 percent, and 3.5 percent inhibited 55-63 percent. Ethylurethane in concentration of 1.66 percent inhibited 44 percent, and 2 percent inhibited 57-61 percent. Propylurethane in 0.5 percent solution produced an inhibition of 47 percent. The ciliates appeared normal, movement was slow, but the effect on respiration was reversible. Therefore we may conclude that dehydrogenase systems are probably involved in the respiration of *Glaucoma*. It might be interesting to try the combined effects of urethane and arsenious acid, in order to obtain evidence for or against the idea that both dehydrogenase and glutathione are part of the same respiratory chain.

5. SYNTHESIS OF RESPIRATORY ENZYMES—VITAMINS

After we have determined that a certain organism has a certain type of respiratory system, the question arises of how the enzymes are formed. Is the organism capable of synthesizing them from relatively simple compounds, or must certain prosthetic groups be present in the nutritive substrate? For certain flagellates this question has been answered very definitely by A. Lwoff (1933). *Strigomonas oncopelti, S. fasciculata,* and *Leptomonas ctenocephali* have respiratory systems which are 90 percent dependent upon cytochrome (as demonstrated above with KCN and CO). It was found that *S. oncopelti* could live indefinitely in peptone solutions without the addition of hematin compounds. For the other two flagellates, hematin compounds were found to be necessary. *L. ctenocephali* would not grow unless rabbit blood (or an equivalent amount of hematin) were present in concentrations of one part to 1,200. *S. fasciculata* showed growth in blood dilutions as great as 1/1,000,000, and within limits the amount of growth was directly proportional to the amount of blood. Hemoglobin disappeared rapidly from the culture medium, and it apparently was being used to form more respiratory enzyme. When small amounts of blood were added, the Q_{O_2} increased linearly for several hours, until apparently all of the hematin was converted into respiratory enzyme; then the Q_{O_2} remained constant. This constant level varied with the amount added. It was found that for each gamma of blood (between one and 5 gamma) added to 1 mgm. of flagellate (dry weight) the Q_{O_2} was increased about 3 units. After 8.5 hours 5 gamma of blood raised the Q_{O_2} from 19.5 to 37.0. Blood could be replaced with hematin, prohemin, and protoporphyrin, but not by cytochrome C nor by a wide variety of synthetic hematin and porphyrin compounds, chlorophyll, peroxidase, or active iron (Lwoff, 1938). Apparently only the porphyrin compound which contained the vinyl (-CH $=$ CH$_2$) radical (protoporphyrin) was effective. Deuteroporphyrin, which differs from protoporphyrin in having hydrogen in place of the vinyl groups, was not effective. More recent investigations of the chemical structure of cytochrome indicate that the vinyl groups may be necessary for linking the prosthetic group (iron porphyrin) to the protein portion of the cytochrome molecule through a pair of sulphur atoma. Lwoff calculated that each flagellate required 520,000 molecules of protoporhyrin in order to bring the Q_{O_2} to 55, and that each organism must contain about 700,000 molecules of protoporphyrin before division would take place.

Although cytochrome C was ineffective alone, the action of protoporphyrin was increased when cytochrome was present (A. Lwoff, 1936). A. Lwoff (1933) found that the organisms had absorption bands at 555 and 530 mμ, and this indicated the presence of cytochrome. Therefore we may conclude that protoporphyrin is necessary for the building of (1) cytochrome or (2) cytochrome oxidase. Since cytochrome C cannot be substituted for protoporphyrin, it seems as if the reaction protoporphyrin → porphyrin C is irreversible and that protoporphyrin is necessary for the synthesis of something other than cytochrome—probably cytochrome oxidase. On the assumption that all of the iron is used to build the respiratory enzyme, we may calculate the rate of catalysis: one gramatom of iron at 28° carries 4.83 grammolecules of O_2 per second. For yeast, Warburg obtained a value of 100. Therefore, on the basis of these assumptions, it seems as if the respiratory enzyme of yeast is 20.8 times as active as that of *Strigomonas*. It has been demonstrated by M. Lwoff (review, A. Lwoff, 1938) that hematin is necessary for the growth of *S. muscidarum, S. culicidarum* var. *anophelis, L. tropica, L. donovani, L. agamae, L. ceramodactyli,* and *Schizotrypanum cruzi,* as well as for the organisms discussed above. The mechanism of its action has not been intensively studied, but presumably it may serve a purpose similar to that it serves in *S. fasciculata.*

We may conclude from the above experiments that protoporphyrin is necessary for the normal metabolism and growth of *Strigomonas fasciculata* and that the organism is not capable of its synthesis. Therefore protoporphyrin may be considered a vitamin. Comparable examples are known for other organisms: e.g., cholesterol for *Trichomonas columbae,* T. *foetus,* and *Eutrichomastix coluborum; aneurine* (vitamin B_1) for *Glaucoma piriformis, S. oncopelti, S. fasciculata, S. culicidarum,* certain bacteria, and fungi; pyrimidine and thiazol (parts of the aneurine molecule) for *Polytomella caeca* and *Chilomonas paramecium;* ascorbic acid for *Schizotrypanum cruzi;* and lactoflavine, nicotinic acid, and phospho-pyridine-nucleotides for various bacteria. In all cases where the function of these essential compounds is known or indicated, they seem to be necessary for the formation or at least for the normal functioning of the respiratory enzymes (review, A. Lwoff, 1938). Recent work on the supposed function of most of these substances has been reviewed by Burk (1937) and Stern (1938).

6. THE DETECTION OF OXIDASE, PEROXIDASE, AND CATALASE

No extensive investigation of isolated enzyme systems of the Protozoa has been made. Certain enzymes, especially oxidase and peroxidase, are supposed to react with certain stains so that the position of the enzyme may be located in the cell. The methods used involve the Nadi, Dopa, benzidine-H_2O_2, and pyronine-αnapthol-H_2O_2 reactions (methods given by Roskin and Levinsohn, 1926; Guyer, 1936; McClung, 1937). These reactions are important in studying vertebrate blood and nerve cells, but apparently no correlation has been made of the presence of "oxidase" or "peroxidase" granules, detectable by these methods, and the respiratory mechanisms discussed above, and it has not been demonstrated that these reactions are specific for oxidase or peroxidase. Several observations have been made on the Protozoa (Roskin and Levinsohn, 1926; Bles, 1929), and when more is known about the respiratory mechanisms of the Protozoa it might be possible to correlate the results of staining and of manometric methods. However, such attempts are omitted in the present discussion. Perhaps a certain degree of localization of the enzymes could be obtained by centrifuging the organisms and making activity tests on cell fragments, as has been done for peptidase in marine ova (Holter, 1936).

The presence of catalase can be detected by adding hydrogen peroxide to a cell suspension and measuring (chemically or manometrically) the oxygen evolved. Burge (1924) studied the catalase action of *Paramecium* and *Colpoda* and found that it was decreased by ether and chloroform, but not by ethylene or nitrous oxide. A much more accurate method was used by Holter and Doyle (1938), who found that the average catalase activity per individual of *Frontonia, Paramecium,* and *Amoeba* was in the ratio 190:30:5. Considerable variation was found between different cultures and even between different individuals from the same culture. Reidmuller (1936) reported only a trace of catalase, no peroxidase, and no indophenol oxidase for *Trichomonas foetus.*

The Measurement of Anaërobic Metabolism and Glycolysis

The measurement of anaërobic metabolism is somewhat more complex than the measurement of aërobic. The standard criterion of O_2 consumption does not exist, and the auxiliary criterion of CO_2 production indicates only the carbon which is completely oxidized. Sometimes this may comprise only a small percentage of the total metabolic changes; and in some cases measurements may be complicated by the presence of hydrogen, methane, and other gases. Therefore we must usually attempt to trace anaërobic metabolism by measuring changes in the concentration of several substances in the liquid phase, instead of one or two substances in the gaseous phase; and this is more difficult. In some cases it is possible to give an organism a known substrate, for example, carbohydrate, and to measure the decrease in the quantity of the substrate and the increase in the amount of decomposition products at various intervals. From carbohydrate decomposition these may be alcohols, aldehydes, and organic acids. From protein decomposition we might expect a wide variety of nitrogen-containing amino-acid fragments, and by deaminization of amino acids a wide variety of organic acids may be produced. From decomposition of lipoids we may expect products somewhat similar to those from carbohydrates. Methods for the final identification of these compounds usually take one into the field of microanalytical biochemistry (see Peters and van Slyke, 1932; Friedemann, 1938; and publications by von Brand, Reiner, and others, cited below). The identification of the acid formed is important in any study of the energetics of anaërobic carbohydrate metabolism, because the processes which yield the various acids release quite different amounts of energy. For many purposes, however, it is customary, if not adequate, to measure acid production by manometric measurement of the amount of CO_2 which is released from a bicarbonate buffering system, as CO_2 or stronger acids are produced by the organism. This gives an index of acid production, but leaves us ignorant of the nature of the acid. Changes in total titratable acidity or alkalinity are also used, and it seems as if, under certain conditions, accurate titration curves might be obtained which would give a fair index of the kind and amount of acids present.

In metazoan metabolism it is usually assumed that oxidation of glucose is preceded by a molecular rearrangement which results in the formation of lactic acid.

$$\text{Glucose} \rightleftarrows 2 \text{ lactic acid} + 43{,}000 \text{ cal. } (\Delta H)$$
$$2 \text{ lactic acid} + 6O_2 \rightarrow 6CO_2 + 6H_2O + 634{,}000 \text{ cal. } (\Delta H)$$

The first reaction is referred to as glycolysis, or cleavage. Glycolysis is reversible, and it occurs under both aërobic and anaërobic conditions, but the rate of the reverse reaction (lactic acid → glucose) is very much less under anaërobic than under aërobic conditions. Consequently, in some tissues (or in the tissue medium) lactic acid may accumulate aërobically, but usually it accumulates only during anaërobiosis. If lactic acid does tend to accumulate, it can be measured by allowing it to displace CO_2 from a bicarbonate immersion medium (usually glucose-bicarbonate-Ringer). If it is assumed that the CO_2 given off by oxidation is equal to the O_2 consumed, then the amount of lactic acid can be calculated as the "excess CO_2," i.e., the CO_2 evolved in addition to that released by oxidation. This is expressed as $Q_G^{O_2}$ for the unit one cmm. of CO_2 per mg. dry weight of tissue per hour. (Older authors use $Q_{CO_2}^{O_2}$ which is easily confused with Q_{CO_2}, the respiratory CO_2, and recent German authors use $Q_M^{O_2}$ for the same quantity.) If O_2 is replaced by N_2, all of the CO_2 evolved must come from glycolysis, and the unit is expressed as $Q_G^{N_2}$ (or $Q_{CO_2}^{N_2}$ or $Q_M^{N_2}$). For comparative studies on various organisms, it has been found to be useful to calculate the Meyerhof quotient (M.Q.), which is defined as

$$\frac{Q_G^{N_2} - Q_G^{O_2}}{Q_{O_2}}$$

This is an index of the amount of lactic acid reconverted to glucose per unit of oxygen consumption, i.e., a measure of the resynthesis of glucose. Recent work on the interpretation of Meyerhof quotients is reviewed by Burk (1937). Recent investigations of glycolysis in vertebrate tissue indicate that glucose is converted into pyruvic acid, part of which is oxidized and part of which is resynthesized into glucose, and that lactic acid is a step in the resynthesis, rather than the end product of glycolysis. It seems as if this may also be true for Protozoa.

Occurrence of Anaërobiosis and Glycolysis

Many Protozoa live in media which are almost if not entirely devoid of oxygen. Examples are those which inhabit the bottom of stagnant ponds (especially if a considerable amount of decaying organic matter, and consequently hydrogen sulphide, is present), those which are found in sewage-disposal plants, those which appear near the bottom of putrid laboratory cultures, and those which inhabit the lumen of the lower intestine of Metazoa. These organisms, because of the characteristics of their environment, are deprived of one of the chief sources of energy available to other animals—the reduction of molecular oxygen, and they must be able to obtain energy by other methods, such as molecular rearrangements (e.g., glucose to lactic acid) or oxido-reductions (e.g., glucose to CO_2 and alcohol). A summary of the early theories of anaërobic fermentations is given by Slater (1928), and a review of the data pertaining to anaërobic life of Protozoa and other invertebrates is given by von Brand (1934). Some of the anaërobic Protozoa seem to be obligatory anaërobes and are quickly killed by aëration (e.g., *Trepomonas agilis*, Lackey, 1932). Therefore one might expect them to have a type of metabolism comparable to those of the anaërobic bacteria.

Other organisms, such as certain intestinal forms, are certainly not strict anaërobes, but are facultative, or amphibiotic. Measurements of the intestinal gases (reviewed by von Brand and Jahn, 1940) and of the oxidation-reduction potential of the digestive tract (Jahn, 1933a) indicate that the lumen of the intestinal tract is largely devoid of oxygen. However, organisms which live at the surface of the epithelium (e.g., *Giardia*) and within the villi, and especially those such as *Endamoeba histolytica* and *Balantidium coli* which invade the tissue, do have access to molecular oxygen. The O_2 tension of the environment of the rumen infusoria of ruminants must be extremely variable, but, for considerable periods of time, almost devoid of oxygen. The question then arises as to what kind or kinds of respiratory mechanisms are present in these facultative organisms. The same question arises with such organisms as *Paramecium*, which are normally aërobic but can withstand lack of oxygen for a relatively long period of time. Do the facultative anaërobes of the phylum Protozoa have respiratory mechanisms comparable to those of bacterial facultative anaërobes? This question, although interesting and suggestive, is unanswerable at present because we know nothing about the respiratory mechanisms of anaërobic Protozoa, and not very much about those of bacteria. However, recent investigations indicate that among bacteria the respiratory mechanism of the strict anaërobes is probably different from the anaërobic mechanism of the facultative anaërobes (Broh-Kahn and Mirsky, 1938).

There is considerable evidence that carbohydrate decomposition takes place in Protozoa under anaërobic conditions. It was found by Pütter (1905) that the glycogen content of *Paramecium* decreased under anaërobic conditions. He also found that *Paramecium* poor in glycogen could live anaërobically for a considerable length of time, probably at the expense of albumen. A. Lwoff (1932) found that *Glaucoma piriformis* could live three days without oxygen only if sugar were present. M. Lwoff (1934) obtained a value of 10 for the $Q_G^{N_2}$ of *G. piriformis* in peptone broth ($Q_{O_2} = 35$). Emerson (1929) found that under anaërobic conditions 80 mm^3 of *Blepharisma* released 12.5 mm.[3] CO_2 per hour from a bicarbonate buffer mixture; negative results were obtained with *Amoeba proteus*. Zhinkin (1930) demonstrated that the glycogen content of *Stentor* decreased under anaërobic conditions and that visible fat increased. Upon exposure to O_2 that fat disappeared. This apparent conversion of glycogen to fat was observed in experimental cultures and under natural conditions in winter when the O_2 content of ponds was negligible, but it did not occur in experimental cultures in the presence of light because of the photosynthetic action of zoöchlorellae. Some data of this type was also obtained for *Prorodon teres* and *Loxodes* (Zhinkin, 1930) and for *Paramecium* (Pacinotti, 1914). The possible changes which take place in the glycogen content of intestinal amoebae and ciliates should also be investigated in this connection (see discussion by von Brand, 1934).

The trypanosomes are a group of organisms which live under conditions of high O_2 tension, but they apparently have a high degree of anaërobic metabolism (glycolysis). At least, they use much more sugar than they could possibly oxidize with the O_2 which they consume, and apparently the amount of acid produced by glucose destruction does not differ much under aërobic or anaërobic conditions. According to the data of von Fenyvessy and Reiner (1924), the O_2 consumption for a billion trypanosomes (*T. equiperdum*) suspended in diluted blood was about 0.07 mg. per hour. The sugar consumed under similar conditions (Yorke, Adams, and Murgatroyd, 1929) was about 5 mg. Since complete oxidation of 5 mg. of sugar requires about 5 mg. of O_2, it appears as if the major portion of the sugar destruction was anaërobic. This is discussed by von Brand (1934, 1935). On the basis of an assumed R.Q. of one, von Fenyvessy and Reiner (1928) subtracted the Q_{O_2} from the rate of CO_2 evolution from a bicarbonate solution and found that the resulting value was equal to the $Q_G^{N_2}$. Therefore the rate of glycolysis (acid production) is the same in O_2 or N_2 and is independent of O_2 consumption. The data of von Fenyvessy and Reiner showed that the amount of CO_2 evolved when the organisms were in bicarbonate-glucose-Ringer was so high under aërobic conditions that apparent R.Q. values of 1.7 to 3.6 were obtained. Therefore we must conclude that glucose → acid conversion is very high in *T. equiperdum*. According to the experiments of von Brand, Regendanz, and Weise (1932), this acid production is apparently not a true glycolysis, because lactic acid could not be detected in the medium by chemical methods. However, if we consider lactic acid to be a step in the resynthesis of pyruvic acid to glucose (see above), the absence of lactic acid may merely mean that resynthesis does not occur. In this case the Meyerhof quotient should be zero. It was shown by Reiner and Smythe (1934) and Reiner, Smythe, and Pedlow (1936) that aërobic sugar destruction by *T. equiperdum* was as follows:

1 glucose → 1 glycerol + 1 pyruvic acid
1 glycerol + O_2 → 1 pyruvic acid + $2H_2O$

Apparently lactic acid and CO_2 were not produced. For *T. lewisi,* under aërobic conditions, the end products were identified as succinic, acetic, and formic acids, ethyl alcohol, and carbon dioxide.

It has been suggested that the large amount of acid produced by trypanosomes might be the mechanism by which toxic effects are produced. This possibility was investigated for *T. evansi* by Kligler, Geiger, and Comaroff (1929), who analyzed the blood of infected rats and concluded that death was caused by lactic acid acidosis. In subsequent publications (Kligler, Geiger, and Comaroff, 1930; Geiger, Kligler, and Comaroff, 1930) they reported the measurement of glycolysis of *T. evansi,* having obtained even higher values (9.2 mgm./billion/hour) than Yorke, Adams, and Murgatroyd had obtained for *T. equiperdum.* Von Brand, Regendanz, and Weise (1932) measured the glucose, lactic acid, and alkali reserve of animals infected with *T. gambiense, T. brucei,* and *T. equiperdum,* and found no evidence of low glucose, high lactic acid, or low alkali reserves, and therefore no support for the acidosis theory of death. Von Brand (1933) measured the rate of sugar destruction for various trypanosomes, and obtained high values (8.0 mgm./billion/hour) for the pathogenic trypanosomes *T. brucei, T. gambiense, T. rhodesiense,* and *T. congolense,* and very low values for the nonpathogenic *T. lewisi* (about 1.4), and still lower values for the pathogenic *Schizotrypanum cruzi.* The results of these investigators indicate that although sugar destruction and formation of acid may be a contributing factor, it will not explain all of the pathological effects of the trypanosomes. This is reviewed by von Brand (1938).

It is interesting to compare the high aërobic glycolysis rate of trypanosomes with that of malignant tumors. The Warburg quotient (aërobic glycolysis/Q_{O_2}) for normal tissues is usually less than 0.3 (except retina and placenta), while that for benign tumors is about one, and that for malignant tumors is 3.1-3.9 (review, Needham, 1931). The Warburg quotients calculated from the data of von Fenyvessy and Reiner (1928) for *T. equiperdum* are 0.78 to 2.67. It has been found that KCN may change the Warburg quotient of chick embryos from 0.1 to 3.4, but the quotient for *T. equiperdum,* in the presence of KCN, showed no significant change (0.78 and 0.80 in two experiments). These comparisons may be taken to indicate that the relative glycolytic rate of the trypanosomes is different from that of normal tissues and resembles in certain respects that of benign or malignant tumors, but the fact that pyruvic and other acids are formed by trypanosomes instead of lactic acid invalidates this comparison.

Meyerhof quotients were calculated by A. Lwoff (1933) for *Strigomonas fasciculata, S. oncopelti,* and *Leptomonas ctenocephali* and were found to be 1.20, 1.38, and 0.125 respectively. The first two are within the normal range of metazoan tissues (Needham, 1931), but that of *Leptomonas* is very low. Values of the M.Q. calculated from the data of von Fenyvessy and Reiner (1928) on trypanosomes are approximately zero, indicating no reversal of glycolysis, and this conclusion agrees with the chemical equations given above.

WHY ARE ANAËROBES ANAËROBES, AND AËROBES AËROBES?

One question which arises in any treatment of anaërobiosis is, "Why does oxygen prevent growth of obligatory anaërobes?" There are several explanatory theories:

1. Oxygen is directly lethal to the cell.

2. Anaërobes do not contain catalase and therefore are incapable of destroying the toxic H_2O_2 which is formed by reduction of oxygen (see equations given above).

3. Growth of anaërobes is dependent upon the presence of a low oxidation-reduction potential in the medium, the attainment of which is prevented by oxygen.

4. O_2 forms a loose chemical complex with the respiratory system of obligatory anaërobes, and thereby inhibits its activity.

The relative merits of these theories, as applied to bacteria, are discussed by Hewitt (1936) and Broh-Kahn and Mirsky (1938). The first theory is certainly not true for those anaërobic organisms which will grow under anaërobic conditions after exposure to oxygen. The second theory is supported by considerable evidence, in that most anaërobes do not contain catalase, and in that some bacteria (e.g., pneumococci) will grow aërobically until they are killed by the accumulation of H_2O_2 resulting from their metabolism (so-called "suicide" of cultures). However, some anaërobes do contain catalase, and apparently it has not been definitely demonstrated that strict anaërobes consume O_2 in order to produce H_2O_2, or even that obligatory anaërobes do produce H_2O_2 (Broh-Kahn and Mirsky). The theory, however, might still be applicable to organisms such as penumococci and hemolytic streptococci, and to *Escherichia coli* in the presence of HCN and methylene blue. In these cases appreciable amounts of H_2O_2 can be detected.

Among the Protozoa we have very little evidence of the relative merits of the first two of these theories. It is shown by the work of Cleveland on termites and on xylophagous cockroaches (Cleveland, Hall, Sanders, and Collier, 1934, include citations of earlier papers) that the symbiotic Protozoa which inhabit the digestive tracts of these organisms are probably strict anaërobes. At least the O_2 tension of their normal environment is extremely low, and they are rapidly killed by appreciable quantities of molecular O_2. Cleveland found that at 23° C. the time necessary for death of the symbionts of termites was an inverse function of oxygen tension (e.g., in *Termopsis*, all Protozoa were dead in 72 hours at one atmosphere, in 30 minutes at 3.5 atmospheres). This is apparently due to an increase in O_2 concentration in the digestive tract, with increased O_2 pressure in the atmosphere, and this could easily be explained on the basis of either of the above two theories. Additional evidence, however, might be gained for the second theory from the fact that one atmosphere of O_2 is more toxic at low temperatures (4-5° C.) than at high (23-25° C.), and that four atmospheres of O_2 are more toxic at high temperatures than at low. The greater solubility of O_2 at 4-5° C. can account for the greater toxic effect with one atmosphere pressure. However, the reverse effect at four atmospheres O_2 pressure must, as suggested by Cleveland, be connected in some manner with metabolic processes. Superficially, at least, these results seem to be explicable on the basis of the second of the above theories, i.e., the organisms grew more rapidly and produced more H_2O_2 at the higher temperatures. An examination of the protozoa for catalase, or of the digestive contents of oxygenated insects for H_2O_2, might yield pertinent information.

The data of Cleveland (1925) on the toxicity of oxygen for the intestinal Protozoa of earthworms, salamanders, frogs, and goldfish, are possibly open to this explanation, but here also we lack experimental evidence. Where such defaunation procedures failed, as in the rat, we might assume that the O_2 tension of the digestive tract was not raised in a manner comparable with that which occurred in the smaller organisms, or that the Protozoa present were more resistant. The former theory seems much more probable. (For review of the chemistry of the intestinal contents, see von Brand and Jahn, 1940.)

The theory that the growth of anaërobes is dependent upon a low oxidation-reduction potential in the medium was proposed by Quastel and Stephenson (1926), and has gained considerable support among bacteriologists (review, Hewitt, 1936) and some dissent (literature cited by Broh-Kahn and Mirsky, 1938). Positive evidence consists mainly of the facts that (1) during the growth phases of anaërobic cultures, especially the sporulating anaërobes, much lower oxidation-reduction potentials are produced than during the growth phases of cultures of aërobes; and (2) anaërobic forms do not start growing until the potential is quite low ($E_h < + 100$ mv.). The first type of evidence does not help to distinguish between cause and effect, and there are some exceptions to the general trend. Most of these, however, are due to the fact that the organism is only one factor which tends to determine the E_h of the medium; the chemical composition of the medium certainly determines, to a great extent, what potentials may be attained. The second type of evidence is well founded in fact—anaërobes do not grow in media of high E_h value. However, if the E_h value of a suitable medium is lowered through displacement of air with H_2 or N_2, or by various chemical reagents, or by the growth of an aërobic organism, then the anaërobic forms are capable of growth. According to this theory, anaërobes and aërobes differ in their ability to grow at various points along the E_h scale, in a manner comparable to that which is exhibited by various acidophilic and basiphilic forms in growing at various points along the pH scale. Of course, there are intermediate-range and wide-range forms in respect to both pH and E_h. The fact that the toxic effects of lack of O_2, or of supernormal O_2 tensions, are not equal in all species supports this idea, but these data, of course, are subject to other interpretations.

Investigations of the rôle of oxidation-reduction potentials among the Protozoa have never passed the preliminary stages. The possible importance of such a study was pointed out by Jahn (1933b, 1934), in connection with experiments on -SH compounds, on the toxic action of methylene blue and on possible relationships with auto- and allelocatalysis. Measurements of the E_h of *Chilomonas* cultures (Jahn, 1935b), of hay infusions (Efimoff, Nekrassow, and Efimoff, 1928), and of digestive contents (Jahn, 1933a) have been made. However, until more data become available, most of these results are difficult to interpret. It was determined that *Chilomonas* would grow in mixtures of NaSH and H_2O_2 only if the concentrations of these were balanced so that the medium just failed to reduce methylene blue. This might indicate a microaërophilic tendency for *Chilomonas*. Other experiments with *Chilomonas* indicated that it could live, but could grow only slowly, however, in media in which methylene blue was reduced. Neither of these ideas is contradictory to its known habits in laboratory cultures. It was also demonstrated (Jahn, 1935b) that casein-acetate broth cultures of *Chilomonas*, when exposed to the air, developed potentials of -20 mv. at pH 7.55, a point at which methylene blue is about half reduced. The chief difficulty in interpreting experiments pertaining to the effect of E_h on growth is that it is necessary to change O_2 tension in order to change E_h. This makes an experiment containing only one variable seemingly impossible to execute, and the theory, therefore, has not been amenable to experimental approach.

The fourth theory of the effect of O_2 on anaërobes—that of an inactivation of the respiratory mechanism of obligatory anaërobes by O_2—is mentioned by Broh-Kahn and Mirsky (1938), but at present is unsupported by experimental evidence. The bacteriostatic effect of dyes might also be interpreted to mean that these have an inactivating effect on the respiratory mechanism.

In connection with the effect of oxygen on anaërobes, it should be mentioned that the converse problem also exists. Why do aërobes die in the absence of oxygen? On this question there has been considerably less discussion in the literature than on the former. It can be seen from the respiratory equations given above that activation of the substrate and partial oxidation can proceed without O_2. Even in the presence of O_2 a considerable destruction of substrate, in some forms (e.g., trypanosomes, cited above), seems to be incomplete. Poisoning of the aërobic cell under anaërobic conditions is supposedly caused by the accumulation of toxic products of carbohydrate cleavage or of incomplete oxidation, or by the conversion of all of the respiratory pigment to the reduced state. In facultative anaërobes the former, and in obligatory aërobes the latter theory seems more probable. The observation of Fauré-Fremiet, Léon, Mayer, and Plantefol (1929) that *Paramecium* withstands lack of O_2 longer at $4°$ C. than at higher temperatures, is open to either interpretation. The data of Pütter (1905), which indicate that *Paramecium* can live longer under anaërobic conditions when the ratio of volume of medium to cells is higher, can be explained on the basis of accumulation of toxic products. What these toxic products are probably depends to a large extent upon the organism, the substrate, and the conditions of the experiment, and the most likely possibilities include lactic and lower fatty acids.

OXIDATION-REDUCTION POTENTIAL VERSUS RESPIRATION AND GROWTH

Another question which arises is why we might or might not expect the oxidation-reduction potential of the medium to affect the respiration and growth of the organism. It is obvious from the outline of the respiratory processes given above, and from many other types of data, that oxidation-reduction phenomena are involved in respiration. The respiratory pigments and perhaps also the respiratory enzymes are reversible oxidation-reduction systems. Therefore their action should be affected by the potential of the medium in which they are found, that is, by the oxidation-reduction potential of protoplasm, and they, in turn, must to a large extent determine this potential. If we were to mix numerous half-reduced, completely reversible oxidation-reduction indicators in a homogeneous solution, an equilibrium would be reached, and the potential attained would depend upon the E_o values of the substances and upon their relative amounts. Substances with E_o values far from the resulting E_h value of the solution would be either completely reduced or completely oxidized, and would be unable to contribute much toward oxidizing or reducing small amounts of added materials unless the E_h of the mixture were appreciably changed.

Obviously such a simple system is not present in protoplasm. Because of the presence of irreversible systems and of the continual introduction of new substrate and the removal of certain end products, a true equilibrium is never attained. Also, the colloidal nature of protoplasm makes possible the existence of different E_h values in different phases of the substance, and the differential adsorption of oxidized and reduced material at interfaces may produce a potential different from that in any of the phases. Therefore the term "oxidation-reduction potential of protoplasm" may be without any interpretable significance (for discussion, see Jahn, 1934; Korr, 1938). But there must certainly be a significance to the E_o values of the respiratory pigments, and the possibility of an individual expression of these values may be maintained by the polyphasic nature of protoplasm. The oxidation-reduction potential, which can be measured with indicators, is probably an index of the potential developed by one or more of these pigments (for summary of such measurements, see Chambers, 1933; Cohen, 1933). It is known that the apparent E_h value of protoplasm, as measured by indicators, varies with the E_h of the external medium when the external O_2 tension is changed. Therefore, why cannot the E_h of the external medium determine the degree of reduction of the respiratory pigments and therefore the rate of respiration? This mechanism might be used to explain the inhibition which is produced by oxidation-reduction indicators in cultures of bacteria (Dubos, 1929) and in cultures of *Chilomonas* (Jahn, 1933b). One difficulty in predicting what reactions would occur in protoplasm, even if we had a thorough knowledge of the oxidation-reduction systems involved, is the fact that such knowledge can tell us only what reactions might or might not occur if all of the reactions were reversible. Since many of the reacting substances are changed irreversibly and since the rates of reactions are dependent not only upon E_o values but upon enzymes, knowledge of E_o and E_h values cannot indicate what reactions will occur.

Much of this discussion of oxidation-reduction potentials is pure speculation, but it is the type of speculation (often unexpressed) which has spurred investigators to a study of the naturally occurring oxidation-reduction systems, of the apparent oxidation-reduction potential of protoplasm, and of the E_h values developed in bacterial and protozoan cultures (reviews, Needham and Needham, 1927; Wurmser, 1932; Chambers, 1933; Clark, 1934; Hewitt, 1936). Interpretations of the data have not always been as fruitful as one might expect, and one is led at times to suspect that the modes, if not the points of attack on the problem, are in need of revision. However, since the necessity of some such relationship as outlined above seems sound, it is more probable that merely the time for the harvest has not yet arrived. Clark (1934) estimated that another half century will be necessary for the solution of these problems.

Another means by which the oxidation-reduction potential of the medium is supposed to affect metabolism is described by the surface catalysis theory suggested by Quastel (1930), Kluyver (1931), and others (discussed by Hewitt, 1936). It is suggested that many oxidative processes of bacteria take place at the surface of the cell (Quastel, 1930), and it seems as if for these reactions the E_h of the medium would be more important than that of the protoplasm. It is also very probable that oxidation-reduction enzymes are merely surface catalysts, which produce their effect by nature of intense interfacial electrical fields (Kluyver, 1931). These fields might be affected by the potential of the medium, whether they occur at the cell surface or within the protoplasm. An intriguing speculation would be that the respiratory pigments are distributed among the various phases of protoplasm, and that the enzymes are actually the interfaces of the emulsion. Unfortunately, such ideas are difficult to check experimentally. However, since oxidation-reduction enzymes are proteins which in all probability exert their catalytic properties through surface action, it is possible that the catalytic interfaces of the cells, discussed by Kluyver (1931), are merely the surfaces of the protein molecules.

LITERATURE CITED

Adolph, E. F. 1929. The regulation of adult body size in the protozoan *Colpoda.* J. exp. Zool., 53: 269-312.

Amberson, W. F. 1928. The influence of oxygen tension upon the respiration of unicellular organisms. Biol. Bull., 55: 79-91.

Andrews, Justin, and Th. von Brand. 1938. Quantitative studies on glucose consumption by *Trichomonas foetus.* Amer. J. Hyg., 28: 138-47.

Barratt, J. O. W. 1905. Die Kohlensäureproduktion von *Paramecium aurelia.* Z. allg. Physiol., 5: 66-72.

Bles, E. 1929. *Arcella.* A study in cell physiology. Quart. J. micr. Sci., 72: 527-648.

Bodine, J. H., and E. J. Boell. 1934. Respiratory mechanisms of normally developing and blocked embryonic cells (Orthoptera). J. cell. comp. Physiol., 5: 97-113.

—— 1937. The action of certain stimulating and inhibiting substances on the respiration of active and blocked eggs and isolated embryos. Physiol. Zoöl., 10: 245-57.

Bodine, J. H., and P. R. Orr. 1925. Respiratory metabolism. Biol. Bull., 48: 1-14.

Brand, Th. von. 1933. Studien über den Kohlehydratstoffwechsel parasitischer Protozoen. II. Der. zuckerstoffwechsel der Trypanosomen. Z. vergl. Physiol., 19: 587-614.

—— 1934. Das Leben ohne Saurstoff bei wirbellosen Tieren. Ergebn. Biol., 10:37-100.

—— 1935. Der stoffwechsel der protozoen. Ergebn. Biol., 12:161-220.

—— 1938. The metabolism of pathogenic trypanosomes and the carbohydrate metabolism of their hosts. Quart. Rev. Biol., 13: 41-50.

Brand, Th. von, and T. L. Jahn. 1940. Chemical composition and metabolism of Nematode parasites of vertebrates, and the chemistry of their environment. *In* An introduction to Nematology (ed. by J. R. Christie). (In press.) Baltimore.

Brand, Th. von, P. Regendanz, and W. Weise. 1932. Der Milchsäuregehalt und die Alkalireserve des Blutes bei experimentellen Trypanosomeninfektionen. Zbl. Bakt., I. (Orig.) 125: 461-68.

Broh-Kahn, R. H., and I. A. Mirsky. 1938. Studies on anaerobiosis. I. The nature of the inhibition of growth of cyanide-treated E. coli by reversible oxidation-reduction systems. J. Bact., 35: 455-75.

Burge, W. E. 1924. The effect of different anaesthetics on the catalase content and oxygen consumption of unicellular organisms. Amer. J. Physiol., 69: 304-6.

Burk, Dean. 1937. On the biochemical significance of the Pasteur reaction and Meyerhof cycle in intermediate carbohydrate metabolism. Some Fundamental Aspects of the Cancer Problem. Occ. Publ. Amer. Ass. Adv. Sci.

Chambers, R. 1933. An analysis of determinations of intracellular reduction potentials. Cold Spring Harbor Symp. Quant. Biol., 1: 205-13.

Clark, W. M. 1934. The potential energies of oxidation-reduction systems and their biochemical significance. Medicine, 13: 207-50.

Cleveland, L. R. 1925. Toxicity of oxygen for Protozoa in vivo and in vitro. Biol. Bull. 48: 455-68.

Cleveland, L. R., S. R. Hall, E. P. Sanders, and Jane Collier. 1934. The wood-feeding roach *Cryptocercus*, its Protozoa, and the symbiosis between Protozoa and roach. Mem. Amer. Acad. Arts Sci., 17: 185-342.

Cohen, Barnett. 1933. Reactions of oxidation-reduction indicators in biological material, and their interpretation. Cold Spring Harbor Symp. Quant. Biol., 1: 214-23.

Cohen, R. A., and R. W. Gerard. 1937. Hyperthyroidism and brain oxidations. J. cell. comp. Physiol., 10: 223-40.

Cook, S. F. 1932. The respiratory gas exchange in *Termopsis nevadensis*. Biol. Bull. 63: 246-57.

Daniel, G. E. 1931. The respiratory quotient of *Balantidium coli*. Amer. J. Hyg., 14: 411-20.

Dixon, M. 1934. Manometric methods as applied to the measurement of cell respiration and other processes. London.

Dubos, R. 1929. The relation of the bacteriostatic action of certain dyes to oxidation-reduction processes. J. exp. Med., 49: 575-92.

Duryée, W. R. 1936. A modified microrespirometer. Z. vergl. Physiol., 23: 208-13.

Efimoff, W. W., N. J. Nekrassaw, and Alexandra W. Efimoff. 1928. Die Einwirkung des Oxydationspotentials und der H-Ionenkonzentration auf die Vermehrung der Protozoen und Abwechselung ihrer Arten. Biochem. Z., 197: 105-18.

Elliott, A. M. 1939. A volumetric method for estimating population densities of Protozoa. Trans. Amer. micr. Soc., 58: 97-99.

Emerson, R. 1929. Measurements of the metabolism of two protozoans. J. gen. Physiol., 13: 153-58.

Fauré-Fremiet, E., C. Léon, A. Mayer, and L. Plantefol. 1929. Recherches sur le besoin d'oxygène libre. L'oxygène et les mouvements des paramécies. Ann. Physiol. Physicochim. biol., 5: 633-41.

Fenn, W. O. 1928. A new method for the simultaneous determination of minute amounts of carbon dioxide and oxygen. Amer. J. Physiol., 84: 110-18.

Fenyvessy, B. von, and L. Reiner. 1924. Untersuchungen über den respiratorischen Stoffwechsel der Trypanosomen. Z. Hyg. Infektkr., 102: 109-19.

—— 1928. Atmung und Glykolyse der Trypanosomen. II. Biochem. Z., 202: 75-80.

Frei, W. 1935. Atmungssysteme der Bakterien. Zbl. f. Bakt., 134: 26-35.

Friedemann, T. E. 1938. Metabolism of pathogenic bacteria. J. Bact., 35: 527-46.

Geiger, A., I. J. Kligler, and R. Comaroff. 1930. The glycolytic power of trypanosomes (*Trypanosoma evansi*) in vitro. Ann. Trop. Med. Parasit., 24: 319-27.

Gerard, R. W. 1931. Observations on the metabolism of *Sarcina lutea*. II. Biol. Bull., 60: 227-41.

Gerard, R. W., and H. K. Hartline. 1934. Respiration due to natural nerve impulses. A method for measuring respiration. J. cell. comp. Physiol., 4: 141-60.

Gerard, R. W., and L. H. Hyman. 1931. The cyanide sensitivity of *Paramecium*. Amer. J. Physiol., 97: 524-25.

Guyer, M. F. 1936. Animal micrology. Chicago.

Hall, Robert H. 1938. The oxygen-consumption of *Colpidium campylum*. Biol. Bull., 75: 395-408.

Hall, R. P. 1939. The trophic nature of the plant-like flagellates. Quart. Rev. Biol., 14: 1-12.

Harvey, E. N. 1928. The oxygen consumption of luminous bacteria. J. gen. Physiol., 11: 469-75.

Hewitt, L. F. 1936. Oxidation-reduction potentials in bacteriology and biochemistry. 2d ed. (London County Council) London.

Heyningen, W. E. van. 1935. The inhibition of respiration by cyanide. Biochem. J., 29: 2036-39.

Holmes, Eric. 1937. The metabolism of living tissues. London.

Holter, H. 1936. Studies on enzymatic histochemistry. XVIII. Localization of peptidase in marine ova. J. cell. comp. Physiol., 8: 179-200.

Holter, H., and W. L. Doyle. 1938. Studies on enzymatic histochemistry. XXVIII. Enzymatic studies on protozoa. J. cell. comp. Physiol., 12: 295-308.

Howland, R. B., and A. Bernstein. 1931. A method for determining the oxygen consumption of a single cell. J. gen. Physiol., 14: 339-48.

Jahn, Theo. L. 1933a. Oxidation-reduction potential as a possible factor in the growth of intestinal parasites in vitro. J. Parasit., 20: 129.

—— 1933b. Studies on the oxidation-reduction potential of protozoan cultures. I. The effect of —SH on *Chilomonas paramecium*. Protoplasma, 20: 90-104.

—— 1934. Problems of population growth in the Protozoa. Cold Spring Harbor Symp. Quant. Biol., 2: 167-80.

—— 1935a. Studies on the physiology of the euglenoid flagellates. VI. The effects of temperature and of acetate on *Euglena gracilis* in the dark. Arch. Protistenk., 86: 251-57.

—— 1935b. Studies on the oxidation-reduction potential of protozoan cultures. II. The reduction potential of cultures of *Chilomonas paramecium.* Arch. Protistenk., 86: 225-37.

—— 1936. Effect of aeration and lack of CO_2 on growth of bacteria-free cultures of protozoa. Proc. Soc. exp. Biol. N.Y., 33: 494-98.

Jay, George, Jr. 1938. Respiration of *Astasia* sp. and *Khawkinea halli.* Anat. Rec., 72 (Suppl.) : 104.

Kalmus, H. 1927. Das Kapillar-Respirometer: Eine neue Versuchsanordnung zur Messung des Gaswechsels von Mikroörganismen. Vorläufige Mitteilung demonstriert an einem Beispiel: Die Atmung von *Paramecium caudatum.* Biol. Zbl., 47: 595-600.

—— 1928a. Die Messung der Atmung, Gärung und CO_2-Assimilation kleiner Organismen in der Kapilläre. Z. vergl. Physiol., 7: 304-13.

—— 1928b. Untersuchungen über die Atmung von *Paramecium caudatum.* Z. vergl. Physiol., 7: 314-22.

Keilin, D., and E. F. Hartree. 1936. On some properties of catalase hematin. Proc. roy. Soc. B., 121: 173-91.

Kempner, W. 1936. Effect of low oxygen tension upon respiration and fermentation of isolated cells. Proc. Soc. exp. Biol. N.Y., 35: 148-51.

—— 1937. Effect of oxygen tension on cellular metabolism. J. cell. comp. Physiol., 10: 339-64.

Kligler, I. J., A. Geiger, and R. Comaroff. 1929. Susceptibility and resistance to trypanosome infections. VII. Cause of injury and death in trypanosome infected rats. Ann. Trop. Med. Parasit., 23: 325-35.

— 1930. Effect of the nature and composition of the substrate on the development and viability of trypanosomes. Ann. Trop. Med. Parasit., 24: 329-45.

Kluyver, A. J. 1931. Chemical activities of microorganisms. London.

Korr, I. M. 1935. An electrometric study of the reducing intensity of luminous bacteria in the presence of agents affecting oxidations. J. cell. comp. Physiol., 6: 181-216.

— 1938. Oxidation-reduction potentials in heterogeneous systems. J. cell. comp. Physiol., 11: 233-45.

Lackey, J. B. 1932. Oxygen deficiency and sewage Protozoa: with descriptions of some new species. Biol. Bull., 63: 287-95.

Leichsenring, J. M. 1925. Factors influencing the rate of oxygen consumption in unicellular organisms. Amer. J. Physiol., 75: 84-92.

Linderstrøm-Lang, K. 1937. Principle of the Cartesian diver applied to gasometric technique. Nature, 140: 108.

Lund, E. J. 1918a. Relation of oxygen concentration and the rate of intracellular oxidation in *P. caudatum.* Amer. J. Physiol., 45: 351-64.

—— 1918b. Rate of oxidation in *P. caudatum* and its independence of the toxic action of KCN. Amer. J. Physiol., 45: 365-73.

—— 1918c. III. Intracellular respiration, relation of the state of nutrition of

Paramecium to the rate of intracellular oxidation. Amer. J. Physiol., 47: 167-77.

—— 1918d. A simple method for measuring carbon dioxide produced by small organisms. Biol. Bull., 36: 105-14.

Lwoff, A. 1932. Recherches biochimique sur la nutrition des Protozoaires. Monogr. Inst. Pasteur, 1932.

—— 1933. Die Bedeutung des Blutfarbstoffes für die parasitischen Flagellaten. Zbl. Bakt. Abt. 1 (Orig.), 130: 498-518.

—— 1938. Les Facteurs de croissance pour les microörganismes. Premier Congrès des Microbiologistes de Langue française, à Paris.

Lwoff, M. 1934. Sur la respiration du Cilié *Glaucoma piriformis.* C. R. Soc. Biol. Paris, 115: 237-41.

Marsh, G. 1935. Kinetics of an intracellular system for respiration and bioelectric potential at flux equilibrium. Plant Physiol., 10: 681-97.

Mast, S. O., and D. M. Pace. 1937. The effect of silicon on growth and respiration in *Chilomonas paramecium.* J. cell. comp. Physiol., 10: 1-14.

Mast, S. O., D. M. Pace, and L. R. Mast. 1936. The effect of sulfur on the rate of respiration and on the respiratory quotient of *Chilomonas paramecium.* J. cell. comp. Physiol., 8: 125-39.

McClung, C. E., Ed. 1937. Handbook of microscopical technique. New York.

Meldrum, N. U. 1934. Cellular Respiration. London.

Necheles, H. 1924. Unpublished experiments. Cited by Kestner and Plant, *in* Winterstein's Handbuch der vergleichenden Physiologie. Bd. 2/2.

Needham, Joseph. 1931. Chemical embryology, Vol. II. Cambridge Univ. Press, London.

Needham, Joseph, and E. J. Boell. 1938. Metabolic properties of the regions of the amphibian gastrula. Proc. Soc. Exp. Biol. N.Y., 39: 287-90.

Needham, J., and D. M. Needham. 1927. The oxidation-reduction potential of protoplasm: a review. Protoplasma, 1: 255-94.

Novy, F. G. 1932. Respiration of microorganisms. J. Lab. clin. Med., 17: 731-47.

Novy, F. G., H. R. Roehm, and M. H. Soule. 1925. Microbic respiration. I. The compensation manometer and other means for study of microbic respiration. J. Infect. Dis., 36: 109-67.

Pacinotti, G. 1914. Infusorien, welche Glykogen in Fett umwandeln. Boll. Soc. eustach. No. 3.

Peters, J. P., and D. D. Van Slyke. 1932. Quantitative clinical chemistry Vol. II. Methods. Baltimore.

Peters, R. A. 1929. Observations on the oxygen consumption of *Colpidium colpoda.* J. Physiol., 68: ii-iii.

Pitts, R. F. 1932. Effect of cyanide on respiration of the protozoan, *Colpidium campylum.* Proc. Soc. exp. Biol. N.Y., 29: 542.

Pütter, A. 1905. Die Atmung der Protozoen. Z. allg. Physiol., 5: 566-612.

Quastel, J. H. 1930. The mechanism of bacterial action. Trans. Faraday Soc., 26: 853-64.

Quastel, J. H., and M. Stephenson. 1926. Experiments on "strict" anaerobes. I. The relation of B. sporogenes to oxygen. Bio-chem. J., 20: 1125-37.

Reiner, L., and C. V. Smythe. 1934. Glucose metabolism of the Trypanosoma equiperdum in vitro. Proc. Soc. exp. Biol. N.Y., 31: 1086-88.

Reiner, L., C. V. Smythe, and J. T. Pedlow. 1936. On the glucose metabolism of trypanosomes. J. biol. Chem., 113: 75-88.

Riedmuller, L. 1936. Beitrag zum kulturellen Verhalten von Trichomonas foetus. Zbl. Bakt. (Orig.), 137: 428-33.

Robbie, W. A., E. J. Boell, and J. H. Bodine. 1938. A study of the mechanism of cyanide inhibition. I. Effect of concentration on the egg of Melanoplus differentialis. Physiol. Zoöl., 11: 54-62.

Root, W. S. 1930. The influence of carbon dioxide upon the oxygen consumption of Paramecium and the egg of Arbacia. Biol. Bull., 59: 48-62.

Roskin, Gr., and L. Levinsohn. 1926. Die Oxydasen und Peroxydasen bei Protozoa. Arch. Protistenk., 56: 145-66.

Schlayer, C. 1936. The influence of oxygen tension on the respiration of pneumococci (Type I). J. Bact., 31: 181-90.

Schmitt, F. O. 1933. The oxygen consumption of stimulated nerve. Amer. J. Physiol., 104: 303-19.

Shoup, C. S., and J. T. Boykin. 1931. The insensitivity of Paramecium to cyanide and effects of iron on respiration. J. gen. Physiol., 15: 107-18.

Slater, W. K. 1928. Anaerobic life in animals. Biol. Rev., 3: 303-28.

Soule, M. H. 1925. Microbic Respiration. III. Respiration of Trypanosoma lewisi and Leishmania tropica. J. Infect. Dis., 36: 245-308.

Specht, H. 1935. Aerobic respiration in Spirostomum ambiguum and the production of ammonia. J. cell. comp. Physiol., 5: 319-33.

Standard Methods of Water Analysis. 1936. New York.

Stephenson, M. 1939. Bacterial Metabolism. London.

Stern, Kurt G. 1938. The relationship between prosthetic group and protein carrier in certain enzymes and biological pigments. Cold Spring Harbor Symp. Quant. Biol. 6: 286-300.

Szent-Gyorgyi, A., and I. Banga. 1933. Über das Co-ferment der Milchsauer-oxydation. Z. phys. Chem., 217: 39-49.

Tang, Pei-Sung. 1933. On the rate of oxygen consumption by tissues and lower organisms as a function of oxygen tension. Quart. Rev. Biol., 8: 260-74.

Wachendorff, T. 1912. Der Gaswechsel von Colpidium colpoda. Z. allg. Physiol., 13: 105-10.

Warburg, O. 1926. Über den Stoffwechsel der Tumoren. Berlin.

Witte, J. 1933. Bakterienfreie Züchtung von Trichomonaden aus dem Uterus des Rindes in einfachen Nährböden. Zbl. Bakt. Abbt. I. Orig., 128: 188-95.

Wurmser, R. 1932. La Signification biologique des potentiels d'oxydoréduction. Biol. Rev., 7: 350-81.

Yorke, W., A. R. D. Adams, and F. Murgatroyd. 1929. Studies in Chemotherapy I. A method for maintaining pathogenic trypanosomes alive in vitro at 37°C. for 24 hours. Ann. trop. Med. Parasit., 23: 501-18.

Zhinkin, L. M. 1930. Zur Frage der Reservestoffe bei Infusorien. Z. Morph. Ökol. Tiere, 18: 217-48.

Zweibaum, J. 1921. Richerche sperimentali sulla coniugazione degli Infusori. I. Influenze della coniugazione sull' assorbimento dell' O_2 nel *Paramecium caudatum*. Arch. Protistenk., 44: 99-114.

CHAPTER VII

THE CONTRACTILE VACUOLE

J. H. WEATHERBY

INTRODUCTION

SINCE THE FIRST description of the protozoan contractile vacuole, probably made by Spallanzani in 1776, few structures in these organisms have received such intensive investigation. Unfortunately, solutions of many of the perplexing questions which have arisen as a result of these studies are not yet at hand. Indeed, much of the more recent work has given rise to entirely new questions which are no less insistent in their demands for answers than were the earlier ones. In the literature claims and counterclaims are abundant; important discoveries have been made only to be discarded because of lack of confirmation, or, in some instances, because of direct contradiction. In view of the somewhat confused state of the evidence concerning contractile vacuoles a re-survey of some of the more important questions seems to be in order.

Probably the first question asked by the first investigator to see a contractile vacuole was "What is its function?" Needless to say, this first investigator did not learn the answer, and, in the opinions of many, the most recent investigation probably does not supply the complete answer. Following this question there have been others hardly less interesting. Is it essential to life? Is it a permanent structure, or does it arise anew at the beginning of each new cycle? Does it always occupy the same position in the organism with respect to other structures? Is the vacuole surrounded by a permanent membrane? Is its discharge to the exterior through a permanent excretory pore? If there is no pre-formed excretory pore, how may one explain the formation even of a temporary pore, and once formed how is it closed again? What natural forces operate to expel the contents of the vacuole? There are many other equally interesting questions, but only a few can be considered at this time.

The most promising order for discussing these problems seems to be to deal first with those pertaining to origin of the vacuole—to see, if possible, just where this organelle comes from, if it does not exist in the cell as a permanent structure. Then, having traced its origin, questions dealing with structure and function will follow in more logical order. An attempt will be made to follow this general plan, but the very nature of the subject will necessitate digressions from time to time.

While a conscientious review of the literature has been attempted, it is quite possible that important publications have been overlooked. It is hoped that this will prove not to be true, not only for the sake of completeness, but also for the sake of giving credit where credit is due. The author herewith offers his apologies to anyone whose labors have not been acknowledged.

THE ORIGIN OF CONTRACTILE VACUOLES

Metcalf (1910) noticed in amoeba of the proteus type that the vacuole is surrounded by a layer of granules of the same approximate size and appearance as the "microsomes" of the general cytoplasm. When the vacuole is of moderate size, these granules form a layer on its surface one granule thick; when the vacuole is fully distended, as just before systole, there are spaces between the granules; but when the vacuole is small the layer may be several granules thick. At systole the vacuole usually collapses completely, and the granules may be seen clumped together in the region of the cytoplasm previously occupied by the vacuole. The new vacuole arises in the midst of these granules, and is formed by the fusion of several small vacuoles. According to Metcalf, who reported observations which sometimes lasted for as long as several hours on a single organism, the vacuole never arises in any other part of the body under normal conditions, except among the granules which surrounded it before its last contraction. From these observations he concludes that the granules are associated in some way with the origin and the function of the vacuole, and for this reason calls them "excretory granules." However, he decides that the granules are not essential for life, since most of them, together with the vacuole, may be removed from an *Amoeba* by operation without a fatal result. Under these conditions a new vacuole develops, although there are few if any granules to be seen surrounding it when it first appears. Metcalf reaffirmed his statement concerning these observations in 1926.

Mast (1926) agrees with Metcalf concerning the frequent presence of granules around the vacuole, but does not interpret this as indicating a physiological association between them. This opinion is based on his having observed vacuoles functioning perfectly normally without the presence of a single granule in the immediate vicinity of the vacuole. To these granules Mast applies the name "beta granules," to distinguish them from others of a different nature which are also present in the cytoplasm. Mast and Doyle (1935) reinvestigated the relationship between granules and vacuole. By centrifuging amoebae it is possible to cause stratification of various cytoplasmic constituents. Organisms treated in such a manner can be operated on so as to remove all or any desired portion of almost any one of the constituents, including these granules. It was found by Mast and Doyle that removal of all or most of the granules resulted in the death of the organism. Removal of fewer granules caused a decrease in pulsation frequency of the vacuole, which was directly proportional to the relative number of granules removed; that is, pulsation frequency was found to be directly proportional to the number of granules remaining. Removal of the contractile vacuole alone resulted in the prompt formation of another. Concerning this same question, Mast (1938, p. 312) more recently states:

The beta granules around the contractile vacuole vary greatly in number and the layer of substance in which they are embedded varies greatly in thickness, without any apparent variation in the function of the vacuole. These facts indicate that neither the granules nor the layer of substance is involved in the function of the contractile vacuole, at least not directly.

Howland (1924a) found that there is no concentration of granules on the surface of the vacuole in *Amoeba verrucosa,* but she considers it likely that the vacuole arises from the coalescing of small hyaline globules, which in turn are derived from the dissolving of granules. In any case, Howland traces the ultimate origin of the vacuole back to granules in somewhat the same manner that Metcalf does, although in *A. verrucosa* these granules are probably dispersed throughout the cytoplasm. On the other hand, Haye (1930) found in fixed and stained preparations of *A. verspertilio* essentially the same relationship between granules and vacuole as described by Metcalf and later by Mast; that is, the filled vacuole is more or less covered by granules, and after systole the new vacuole arises in the midst of these granules.

Hall (1930a) studied the cytoplasmic inclusions in *Trichamoeba* after osmic and silver impregnation. In a few instances he observed the adherence of blackened globules to the outer surfaces of vacuoles. At first glance these appeared to be vacuoles with heavily impregnated walls, but close observation revealed the granular or globular nature of the blackened material.

It must be remembered that these granules are not confined to the immediate vicinity of the contractile vacuole, but usually are scattered throughout the entire cytoplasm as well. If the origin of the vacuole is associated with and dependent on the presence of these granules, then one would expect other parts of the organism to be at least potentially capable of giving rise to vacuoles, since some granules are present in other parts. That such a phenomenon actually occurs in *Amoeba* has been observed by various authors, among whom are Day (1927), as well as Howland and Mast and Doyle. In this connection it is interesting to note that Dimitrowa (1928) was able to induce formation of extra vacuoles in *Paramecium caudatum* by interfering mechanically with the normal function of those already present. These extra vacuoles usually appeared to be entirely normal, although in a few instances there were no radial canals. The customary number of vacuoles was restored at fission by failure of the organism to form new ones if two extra ones had been induced, or by the formation of one new vacuole if only one had been induced artificially. In the event that there were three extra vacuoles, one daughter cell received three and formed a single new one when it in turn divided.

Haye (1930) investigated eight species from two orders of flagellates. In *Phacus pleuronectes* he found that the walls of contractile vacuoles contain lipoid granules which are arranged in a net-like fashion. In *Euglena pisciformis* and *Trachelomonas hispida* the surfaces of both the reservoir and the vacuoles show a granular structure. In *Peridinium steinii* no granules were observed, nor were accessory vacuoles seen, except in organisms obtained from a laboratory aquarium in which conditions were thought to have been abnormal. A differentiated plasma zone, reminding one of the "excretory plasma" of fresh-water Protozoa, was noted around the pusule. In the wall of the pusule of *P. divergens* were observed lipoid granules similar to those in the wall of the contractile vacuole of fresh-water Protozoa. Besides the two sac pusules, a collecting pusule with daughter vacuoles and numerous accessory vacuoles were observed in *Phalacroma* sp. In *Goniodoma* sp. there were, besides a sac pusule, a collecting pusule with daughter pusules and an accessory vacuole. Only one large pusule was noted in *Ceratium hirundinella*. In both orders of flagellates Haye believes that emptying of accessory vacuoles is accomplished by diffusion through the walls into the contractile vacuole, rather than by coalescence with it.

Hall (1930b) found that in *Menoidium,* stained according to the Da Fano silver method, the contractile vacuole is formed by the fusion of several smaller vacuoles arising near the gullet.

The mode of origin of contractile vacuoles has been studied in a greater variety of ciliates than in either rhizopods or flagellates, and information on this subject is proportionally more abundant. Taylor (1923) observed in *Euplotes* that the vacuole (V_1), in its final form immediately before contraction, is the result of the fusion of several smaller vacuoles, and that these smaller vacuoles (designated as group V_2) in turn are formed by the fusion of still smaller vacuoles (group V_3). The smallest vacuoles in the series are thought to arise as the result of the dissolving of granules, or to arise *de novo.* Thus Taylor suggests granules as a possible source of vacuolar fluid, and he observed formation of the vacuole by the fusion of several small accessory vacuoles. King (1933), who studied *Euplotes* after impregnation with osmic acid, found that the smallest visible accessory vacuoles (V_3) have their origin at the distal ends of a very large number of collecting canals, located just under the ectoplasm on the dorsal surface of the ciliate. These canals radiate like a sun-burst from the vicinity of the vacuoles, and seem to end blindly in the protoplasm of the organism. These canals have a diameter of approximately 0.5 micron at their distal ends, and become relatively much narrower as they pass away from the region of the vacuoles. The canals are not visible in living organisms, but may be clearly demonstrated by proper impregnation with osmic acid. On the basis of information now available, it is difficult to tell whether the canals described by King and the granules mentioned by Taylor represent different interpretations of the same structures, observed under different conditions, or whether the canals merely provide a means for the transport of fluids which have originated in more distant parts of the body as a result of the activity of granules.

Of particular interest are the observations of MacLennan (1933) on the Ophryoscolecidae, ciliates from the stomachs of cattle. The cycle of the contractile vacuole was studied in both living and fixed material, including the following genera: *Ophryoscolex, Epidinium, Ostracodinium, Polyplastron, Eudiplodinium,* and *Metadinium.* In all these genera the contractile vacuole is formed by the coalescence of small accessory vacuoles, just as in *Euplotes,* as described by Taylor and also by King, and in *Amoeba,* as described by Metcalf. These accessory vacuoles arise from the dissolving of granules which are found in sharply defined regions around the contractile vacuole in *Eudiplodinium* and *Metadinium,* in a narrow dorsal strip of the ectoplasm in *Ostracodinium,* and in the whole ectoplasm in *Ophryoscolex* and *Epidinium.* If one may be permitted to assume that the canals and granules in *Euplotes* are identical, then the mode of origin of the contractile vacuole in this form is quite similar to that described by MacLennan for the Ophryoscolecidae; and in its fundamental features it also resembles that reported for amoebae as well as for some of the flagellates.

In addition to the flagellates previously mentioned, Haye (1930) also studied representatives of thirteen genera from four orders of ciliates. In *Opalina dimidiata, Isotricha prostoma, Spirostomum ambiguum,* and *Nyctotherus cordiformis* little was observed which suggests the mode of origin of contractile vacuoles. Except for the fact that the walls of the canals were found to contain lighter and darker zones—probably because of the presence of lipoid granules—little was observed in *Paramecium caudatum* which may be associated with origin of vacuoles or their contents. Rod-shaped entosomes were found closely packed about the wall of the vacuole in *Lionotus fasciola.* In *Stentor polymorpha* the wall of the vacuole is very delicate and shows only here and there a granular structure; several secondary vacuoles are usually present. In *Blepharisma undulans, Balantidium entozoön, Polyplastron multivesiculatum, Ostracodinium gracile,* and *Ophrydium versatile* are to be found granules (entosomes) within the wall, or closely associated with the wall, of the contractile vacuole. The vacuole in *Epistylis plicatilis* is formed from numerous secondary vacuoles; no granules or entosomes are to be observed.

Von Gelei (1933) states that the vacuole system in *Spathidium* consists of a primary vacuole, located usually in the posterior end of the organism. Around this are one or two rows of smaller secondary vacuoles, which fuse and give rise to a new primary vacuole following systole. Whether or not these secondary vacuoles originate from granules was not ascertained by von Gelei. Essentially the same relationship between primary and secondary vacuoles in *Blepharisma* was described by Moore (1934), who made the further statement that excretory granules could not be observed. Both Wenrich (1926) and King (1928) found the vacuoles of *P. trichium* to be vesicle-fed, although neither author mentioned the origin of these vesicles. Day (1930) found that the vacuolar fluid reaches the elongated canal of *Spirostomum* by the fusion of small vacuoles with the canal throughout its entire length. A similar source of fluid in canals of *P. caudatum* was also reported, but in neither instance was the origin of the accessory vacuoles mentioned.

Fauré-Fremiet (1925) observed the filling of contractile vacuoles in several species of *Vorticella* by the discharge of small vesicles into the vacuole. These vesicles originate in the wall of the vacuole, and correspond to the "mural vacuoles" described by Haye (1930) for *Campanella, Chilodon, Dogielella,* and some of the Ophryoscolecidae. Nassonov (1925) also investigated *Chilodon* and *Dogielella,* and found structure and mechanism of filling to be somewhat different from that described by Haye. According to Nassonov, vacuoles in these forms do not appear to have the thick walls described by Haye, nor even to have any sort of membrane, but lie directly in the cytoplasm. However, there is a strongly osmiophilic structure closely associated with them, which, for *Chilodon* at least, and possibly also for *Dogielella,* may be mistaken for a thick wall or membrane under certain conditions. In both forms the osmiophilic structure remains essentially unaltered in appearance after collapse of the vacuole. Nassonov observed the origin of accessory vacuoles (the mural vacuoles of Haye) in these osmiophilic structures, and believes them to contribute to the filling of the contractile vacuole.

Many authors hold that in certain Protozoa, typified by *Paramecium caudatum,* the question of origin of the contractile vacuole does not arise, since in these forms the vacuole system is a permanent structure. This view is not universally accepted, as will be pointed out later. But, whether permanent or temporary, there still exists a no less fundamental question as to the origin of the fluid which finds its way into these organelles. If the origin of this fluid is associated with granules in many diverse organisms, as much of the evidence implies, then it would be somewhat un-expected if such granules are not to be found distributed generally throughout the Protozoa. MacLennan called attention to the fact that many investigators have demonstrated a more or less solid membrane surrounding vacuoles in a variety of Protozoa, these demonstrations having been made by osmic-acid impregnation. Most of these workers, according to MacLennan, used the warm method of impregnation advo-cated by Nassonov. Hirschler showed that this method tends to produce overimpregnation, resulting in the production of a heavy black band or membrane in what is actually a granular zone. MacLennan found this to be true in the Ophryoscolecidae, while impregnation by the cold method shows this same region to be granular, a condition which can be seen in living material. He further calls attention to the fact that figures showing solid impregnation of the vacuolar walls of *Chilodon* and *Dogielella,* published by Nassonov, of *Balantidium* by Bojewa-Petrus-chewskaja, and of the Cyclopostheiidae by Strelkow, indicate a marked granular roughening of the outer margin of the osmiophilic layer. He takes this to indicate that what has been interpreted by these authors as a solid membrane may, in fact, be only the result of overimpregnation of a granular zone. Among the Ophryoscolecidae alone MacLennan found various degrees of aggregation of these granules—from virtually none to a very pronounced aggregation—around the contractile vacuole. He fur-ther showed that localization of the origin of contractile vacuoles in these forms is correlated with the degree of aggregation of the granules.

Of interest in this connection are the observations of Lloyd and Scarth (1926) on the origin of vacuoles in *Spirogyra.* These authors found that in sufficiently high concentrations even the most innocuous plasmolytes may by themselves cause subsidiary vacuoles to arise in the cytoplasm. It is not only by plasmolytes that this effect is produced however, but also by other more readily penetrating substances such as the narcotics, chloro-form, and ether, and by very low concentrations of salts. But without any artificial influence, similar vacuoles may form in normal cells. Their constant occurrence was demonstrated in the gametes during conjugation in *Spirogyra,* and their excretory function in the taking up of water from the central vacuole and its discharge to the exterior in typical "contrac-tile" fashion was proved. The authors state that these vacuoles originate from peculiar "lecithin-like" bodies already present in the cytoplasm.

Scarth and Lloyd (1927) claim that the vacuolar wall arises from the "kinoplasm" of Strasburger. They observed a reciprocal quantitative relation between kinoplasm and mitochondria. The activity of kinoplasm resembles that of lecithin, which is abundant in mitochondria. On the basis of this resemblance, they conclude that water at least may accumulate in the vacuoles without the visible interaction of any other structure.

The observations and opinions reported in the foregoing pages, while somewhat contradictory at times, point to two general conclusions concerning the origin of contractile vacuoles. First, in the great majority of forms, perhaps in all forms, fluid reaches the contractile vacuole through the fusion of small vesicles, or accessory vacuoles, with the contractile vacuole or its filling canals. The vesicles arise within what often appears to be the wall of the vacuole; the accessory vacuoles usually originate at a greater or less distance from the contractile vacuole, and coalesce to form the latter. Second, vesicles originate within walls of vacuoles which have been shown in many instances to be granular in nature or to be intimately associated with granules; accessory vacuoles have been reported by various authors as originating among granules which may be closely associated with the vacuole, or occasionally removed some distance from it. Thus it appears that in spite of their great variety of shapes and general appearances under the microscope, contractile vacuoles originate in a remarkably similar manner in all forms so far investigated with this problem in mind.

In certain instances authors have reported the absence of granules in the vicinity of the contractile vacuole, and from this have concluded that granules are not concerned in the origin of vacuoles. In this connection it must be remembered that in certain forms, e.g., *Euplotes,* some of the Ophryoscolecidae, and apparently in others as well, the accessory vacuoles which ultimately give rise to the contractile vacuole originate at some distance from the ultimate site of the final vacuole. Also, granules frequently are visible only after osmium or silver impregnation. Keeping in mind the greater vulnerability of negative evidence, one is justified in the thought that perhaps a reëxamination of organisms for which the absence of granules in the vicinity of the vacuole has been reported, may reveal the presence of granules in other parts of the body, either scattered or in aggregates. Such scattered granules, which are known to exist in some amoebae, may be the site of origin of new vacuoles when the function of the original vacuole is disturbed by artificial means or removed by operation. While such granules have not been demonstrated to be scattered about in the cytoplasm of *P. caudatum,* their presence would explain the origin of extra vacuoles in this form, when function of the original vacuoles is interfered with mechanically, as reported by Dimitrowa. The origin of new vacuoles at fission would have a similar explanation, since, as proposed by Dimitrowa, during fission the greater abundance of metabolites would impose a necessity on the organism essentially similar to interference with normal function. After fission, when the daughter cells are smaller than the parent cell was immediately prior to fission, and the metabolic rate is lowered, there no longer exists a stimulus for the formation of extra vacuoles, and the daughter cells appear quite normal, with the usual number.

THE STRUCTURE OF CONTRACTILE VACUOLES

The question of the structure of the contractile vacuole and its associated parts has occupied the attention of protozoölogists for many years. As a result the main question has been broken up into several parts, each concerned with a limited phase of this main question. Is the vacuole surrounded by a permanent membrane? Is its discharge to the exterior through a permanent excretory pore? If there is no permanent pore, how may one explain the formation even of a temporary pore, and once formed how is it closed again? Is the vacuole a permanent structure, or does it arise anew at the beginning of each new cycle?

The dispute as to the presence or absence of a permanent membrane surrounding the vacuole began over a hundred years ago, and continues, with little to indicate that is will end within the near future. According to Taylor (1923), to whom we may refer for a more detailed account of the history of this question, the following investigators have written in support of the idea of a permanent membrane: Ehrenberg, Siebold, Claparède, Lachmann, Degen, and Stempell. Those who believe that the vacuole possesses no permanent wall are: Dujardin, Meyen, Stein, Wrzesniowski, Perty, Schmidt, Zenker, Maupas, Rhumbler, Bütschli, Lankester, and Khainsky. Taylor himself holds this view, at least for *Euplotes.* Without reflecting unfavorably in any way on the researches of those who worked on this subject prior to 1900, or possibly as late as

1920, one must admit that only limited importance can be attached to their opinions. Unquestionably most of these investigators were careful observers, and expressed opinions only after due consideration of all the factors which they were able to recognize. But the microscope of today is a far different instrument from that of a hundred years ago, or even fifty years ago; and chemical procedures, particularly those dealing with colloids, have undergone extensive development. However highly one may regard this or that early investigator, the fact remains that none could have been better than the tools with which he worked, and admittedly the tools were poor. Consequently, the author maintains that it is neither unkind nor unappreciative to propose that these various early opinions be considered mainly as of historical interest, and of little worth in settling the question as to the presence or absence of permanent membranes, or of any kind of membrane for that matter, around contractile vacuoles. The employment of the best of modern instruments and techniques leaves the question in an unsatisfactory state.

Before presenting the more recent evidence concerning this question of membranes, perhaps it will not be unwise to present briefly the more fundamental question of what constitutes a membrane. Most textbooks either avoid the issue more or less completely or describe the structure and properties of the artificial membranes so often used in the laboratory for experimental purposes. Although reliable information concerning living membranes is scant, there is sufficient evidence to justify the division of membranes into two types: morphological membranes and physiological membranes. Morphological membranes are permanent structures which are frequently visible in living material viewed through the microscope, and usually may be demonstrated more or less clearly by suitable staining techniques. Apparently they consist mainly of a reticulum, or framework, which is described by some authors as being composed largely of protein. Such membranes are usually thought to possess an appreciable amount of rigidity, and to serve primarily as supporting structures. Free permeability in both directions is usually assigned to them.

Physiological membranes are entirely different from morphological membranes in many important respects. Usually they are considered to be so thin as to be invisible even with the highest magnification. They may possess a certain degree of rigidity, but probably much less than morphological membranes with which they are often associated in living material. Semipermeability, or more properly selective permeability, is a property of all living physiological membranes. Colloid-chemists, as well as many physiologists, are agreed that a physiological membrane is simply a phase boundary, an interface between two different fluids. In order that such phase boundary may be more or less permanent, it is necessary that the two phases be only slightly miscible at most—the more complete the immiscibility the more nearly perfect and permanent the membrane. A very wide variety of molecules show polar phenomena; that is, the two ends of the molecules are electrically and chemically different. This results in orientation of molecules with respect to one another and to various other molecules, in much the same manner that a compass needle becomes oriented with respect to the magnetic poles of the earth. This phenomenon of orientation is associated with organic acids, alcohols, aldehydes, lipoids, fats, proteins, and many other so-called "physiological" compounds. Thus at the interface of the two-phase system, oil-water, the oil molecules (glycerol-esters of fatty acids) become oriented in such a manner that the hydrocarbon ends of the molecules project into the oil phase, whereas the glycerol ends project into the water. Fat molecules undergo much more nearly perfect orientation than water molecules, although with the latter there appears to be a certain degree of orientation. Such an aggregation and packing together of oriented polar molecules at an interface represents a physiological membrane. Since protoplasm contains a variety of polar molecules, the membrane formed between protoplasm and water is composed of various types of molecules apparently arranged in the form of a mosaic. The thickness of such a membrane has not been definitely established. Some authors maintain that it is only a single molecule thick, or at most only one or two milli-micra thick, but at least one author (Peters, see Clark, 1933, p. 40) has advanced a theory according to which the cell is composed of a three-dimensional protein mosaic, with the molecules in the interior of the cell oriented on the surface film. Since the interior of many cells is known to be fluid, the structure must be regarded as an orientation rather than as an anatomical skeleton. This theory of Peters's agrees fairly well with certain evidence concerning the action of drugs on cells. Without entering into the question as to how far orientation extends beneath the surface layer, suffice it to say that it is well established that orientation of surface molecules occurs at the interface between two different phases (provided, of course, that at least one phase contains polar molecules), whether the system is composed of oil and water or protoplasm and water. The converse of this is equally true; since the very existence of this oriented layer depends on the presence of *two* different phases, the removal of one phase necessarily results in disintegration of the membrane. Whether or not this surface layer of oriented molecules actually comprises the true physiological membrane may be subject to debate, but the importance of such a membrane is obvious, since it not only separates the organism from its surroundings but at the same time provides the only means of communication between the interior and the exterior of the cell.

If one accepts the idea of a physiological membrane as a layer of oriented molecules at a phase boundary, as outlined briefly above, then it necessarily follows that any cell vacuole which contains a fluid different from cytoplasm must be surrounded by such a membrane. In the light of the information available at the present time, membranes around protozoan contractile vacuoles probably should be considered as temporary; although it is not inconceivable that in some forms the new vacuole may be formed with such rapidity, in the midst of oriented molecules remaining after systole, that dispersion of the membrane does not have time to occur before the second phase is present again. If such condition obtains, then the membrane may be considered as having a greater or less degree of permanence. There are numerous references to such physiological membranes in the literature on contractile vacuoles, so it appears that the idea has gained rather wide acceptance. The controversy is not so much concerned with such membranes as with the presence or absence of morphological, and hence permanent, membranes.

Before leaving the subject of physiological membranes, there are several phenomena which may be discussed profitably with this concept in mind. Repeatedly authors speak of the coalescence of accessory vacuoles to form contractile vacuoles. Taylor (1923) refers to this, and considers coalescence to be due to a reversion of the gel state of the surrounding film to the sol state. He further states that vacuoles are surrounded by highly viscous boundaries of endoplasm, and that the consistency of the papilla pulsatoria strikingly resembles that of the endoplasmic boundaries. Without raising the question as to whether or not coalescence of accessory vacuoles and the simultaneous rupture of vacuole wall and papilla pulsatoria represent sol-gel reversibility, it is obvious that exactly these phenomena must be anticipated, on the basis of the concept of physiological membranes such as described above. When two accessory vacuoles lie touching each other, the cytoplasmic phase of the two-phase system is pushed aside, at the same time removing the basic forces on which the presence of these membranes depends. In the absence of these forces, the membranes disintegrate at the site of contact, and the two vacuoles fuse into one. Likewise, when the filled vacuole comes in contact with the papilla pulsatoria, one phase (the cytoplasmic phase again) is pushed aside, the membranes disintegrate at the site of contact, and the contents of the vacuole are discharged to the exterior. After discharge the two-phase system is again established, since there is cytoplasm on one side of the pore (within the organism) and water on the other (outside the organism), whereas prior to discharge there was water on the outside as well as a fluid composed chiefly of water within the vacuole. The papilla is formed in this manner from the vacuole wall, which readily accounts for the similarity noted by Taylor.

Other phenomena which can be explained in like manner by the presence of physiological membranes are easy to find in the Protozoa. The ingestion of food by *Amoeba* is essentially the result of fusion of the walls of the organism after they have been extruded around the food particle in such a manner as completely to enclose it. The two-phase system exists as long as there is water on one side of the cell membrane and cytoplasm on the other; but when the engulfing process is complete and cell membrane is in contact with cell membrane with no water separating the two portions, the membrane disintegrates at the site of contact, and continuity of cell structures as well as of the vacuole membrane is established. The food-vacuole membrane persists as long as there is water within to maintain the two-phase system. In a similar manner one may explain the readiness with which an amputated fragment of an amoeba unites with the parent body when the two portions come together, although it has no bearing on the fact that a fragment from a diverse strain is refused. It has also been observed that occasionally an amoeba attempts to engulf a relatively large organism, such as *P. caudatum*, but is unable to accomplish this completely. The *Paramecium* is squeezed in two, apparently, with half inside the amoeba and half outside. Calculations have been made of the physical force necessarily exerted by the amoeba to accomplish this, but they have not taken into account some of the properties of physiological membranes, such as spontaneous disintegration when membrane comes into direct contact with membrane. The adherence of conjugating organisms may be dependent, likewise, on these same properties of membranes. While it is interesting to speculate on such matters, it must be admitted that these remarks on feeding amoebae are purely speculative, with little other than superficial observation to support them.

Concerning the presence of a permanent (morphological) membrane surrounding the contractile vacuole, there are diverse opinions. These diverse opinions apply not only to different species but even to the same species. Howland (1924a) found that the contributory globules as well as the vacuole may be removed from the organism to the surrounding water, where they retain their identity for an indefinite period of time. This may be taken to indicate a considerable degree of permanence of the vacuole wall, such as would be possessed by a morphological membrane, although Howland is of the opinion that these vacuole membranes are temporary. A temporary physiological membrane, formed of oriented molecules in a compact layer, may be expected to retain its identity for an appreciable period of time before dispersion of the molecules occurs, but it hardly seems probable that this "appreciable period of time" can be more than a few minutes. Day (1927) expresses the opinion that the vacuole wall in *A. proteus* is a "condensation membrane," or gel, disappearing with each contraction. By "condensation membrane" Day very likely implies such a structure as has been described above as a physiological membrane, so that the two terms may be taken as synonymous. Concerning *A. proteus,* Mast (1938, p. 307) states: "At the surface of the contractile vacuole under the layer of substance containing the beta granules, there is a layer or membrane about 0.5 micron thick which is optically well differentiated from the adjoining substance on either surface, for under favorable conditions a line indicating an interface can be clearly seen at both these surfaces." Whether this membrane is a permanent structure or is formed anew with each successive vacuole was not suggested by Mast. After an examination of fixed, stained, and sectioned material, Haye (1930) concluded that vacuole membranes are lacking in *Amoeba,* although it is probable that this author referred to morphological and not to physiological membranes, since these latter cannot be demonstrated in this manner.

Among the flagellates, Nassonov (1924) found an osmiophilic membrane surrounding the vacuole in *Chilomonas paramecium*. Haye (1930) could distinguish no vacuole wall in the Euglenoidina, but in many other flagellates examined by him distinct walls were visible in stained material.

Among the ciliates, morphological membranes are reported in a great variety of organisms by many investigators. Nassonov (1924) reports osmiophilic walls for the vacuoles in *Paramecium caudatum, Lionotus folium, Nassula laterita, Campanella umbellaria, Epistylis gallea, Zoothamnium arbuscula,* and *Vorticella* sp. Fauré-Fremiet (1925) confirmed the findings of Nassonov, using several species of *Vorticella,* in which osmiophilic walls were observed, even after collapse of the vacuole. Young (1924) concludes from studies on *P. caudatum* stained with iron hematoxylin that the vacuole system is a permanent and continuous structure. King (1928) arrived at essentially the same conclusion concerning the vesicle-fed system of *P. trichium.* Wenrich (1926) observed definite vacuole walls in *P. trichium* stained with Mayer's hemalum or Heidenhain's iron-alum-hematoxylin. Concerning this Wenrich states (p. 89):

It was somewhat surprising to find how distinctly the vacuolar walls showed in fixed and stained specimens. The relative thickness of the wall is noteworthy and it usually appears to be laminated. In sectioned material the walls contained strands of more or less intensely staining material, suggesting the presence of contractile fibers.

Von Gelei observed osmiophilic walls in *P. caudatum* (1925, 1928) and in *Spathidium giganteum* (1935). Haye (1930) found thin vacuole walls in the following forms: *Blepharisma undulans, Lionotus fasciola, Ophrydium versatile, Stentor polymorphus, Spirostomum ambiguum, Balantidium entozoön,* and *Isotricha prostoma.* Thick walls were observed in *Campanella, Chilodon, Dogielella, Paramecium,* and the Ophryoscolecidae. As previously mentioned, Nassonov (1925) examined *Chilodon* and *Dogielella* after osmium impregnation and concluded that the vacuoles possess no membranes, but lie directly in the cytoplasm. In these latter organisms, structures considered by Nassonov to be the Golgi apparatus surround the vacuole in such a manner that they may be mistaken for vacuole walls in certain preparations.

The presence of definite vacuole walls or morphological membranes around the vacuoles of ciliates is not accepted by all authors. Thus, Taylor (1923) believes the vacuole in *Euplotes* to disappear completely at systole, and to be replaced by an entirely new structure. If a morphological membrane were present, this could hardly obtain, although Taylor was able to distinguish a "highly viscous boundary" of endoplasm surrounding the vacuole. Moore (1931) was unable to cause osmication of the vacuole wall in *Blepharisma undulans*, using the technique of Nassonov, and from this she concludes that the vacuole lacks a permanent wall. This opinion was again expressed (1934) after further observation. The findings of Moore are in opposition to those of Haye, who reported thin vacuole walls in the same species. Day (1930) concludes from his observations on *Paramecium caudatum, Spirostomum ambiguum,* and *S. teres* that vacuoles in these forms are temporary structures which disappear at systole. King (1935, p. 564) found that:

The permanent components of the contractile vacuole system in *Paramecium multimicronucleata* include the pore with its discharging tubule, and the feeding canals, each made up of a distal excretory portion, an ampulla and an injection tubule. . . . The membrane of the contracting vacuole is a temporary structure, disappearing at systole. The pore is closed by the remnant of the old vacuole which ruptures at the next systole.

Essentially the same was found in *P. aurelia*. These observations were made on material osmicated at 38° C. It is somewhat surprising that the vacuole proper of the contractile vacuole systems in *P. multimicronucleata* and *P. aurelia* is a temporary structure, replaced anew after each contraction, whereas that of the closely related species *P. caudatum* is commonly believed to be a permanent structure. Perhaps the stainable vacuole wall described by Young for *P. caudatum* represents the same kind of material which King believes closes the excretory pore following systole in *P. multimicronucleata* and *P. aurelia*.

After examining both living and stained material, Fortner (1926) concludes that the membrane of the vacuole in Protista is a temporary structure, which, after fulfilling its purpose, closes the excretory pore during the period of diastole. He further believes that all surface layers sharing in the excretion process have the property of fusing together again merely on contact.

The present state of the knowledge concerning the presence or absence of membranes around contractile vacuoles is exceedingly unsatisfactory. Aside from the fact that many investigators admit the probability of, or in some instances the necessity for, a physiological membrane of the type described above as composed of molecules oriented and more closely packed together at the phase boundary, very little of a positive nature is known. It is difficult to understand how a vacuole without any sort of membrane can retain its identity in cytoplasm with which its contents appear to be freely miscible. That most of the cellular contents are freely miscible with water is indicated by the fact that discharge of the cell contents into the surrounding water, following rupture of the cell wall, is soon followed by dispersion of most of the cytoplasm into the surrounding medium; usually only granules, of one sort or another, remain to indicate the original position of the extruded material. If the vacuole content is mostly water, a belief quite generally if not universally held, how can a vacuole ever be formed in the absence of any kind of membrane to prevent this water from flowing back into the cytoplasm as rapidly as it is mobilized?

As indicated above, a great many investigators have demonstrated structures which were interpreted as vacuole "walls." In some material these walls were visible in living organisms as layers of substance, optically different from substances on either side of it. In other material the walls were visible only after fixation and staining. One is more or less obliged to accept an author's description of structures in material examined by him; but an observation sometimes is subject to two entirely different interpretations. This is clearly shown by MacLennan (1933) in his work on the Ophryoscolecidae, as pointed out earlier. In several instances the presence of morphological membranes is claimed by various authors, on the basis of observations on what may have been overimpregnated material. Because of the extreme thinness of physiological membranes, it is doubtful if they ever can be demonstrated visually, but evidence obtained from the study of other colloidal systems indicates that they almost certainly exist.

The Function of Contractile Vacuoles

During the years that have intervened since the discovery of the contractile vacuole, an extensive literature concerning its function (or functions) has accumulated. Excellent reviews of this literature have been published from time to time (Howland, 1924b; Day, 1927; Lloyd, 1928). Therefore no attempt will be made to present another review at this time, except in so far as the works to be mentioned have a direct bearing on one or the other of the two functions generally conceded to be most probable.

Of the various functions assigned to the contractile vacuole those of excretion of metabolic waste products and regulation of hydrostatic pressure within the cell have received most frequent support. Some authors prefer to limit "metabolic waste products" to nitrogenous substances, although others include carbon dioxide as well. In view of the scarcity of evidence bearing directly on the subject, it hardly seems advisable at this time to distinguish between different kinds of metabolic wastes. On the other hand, if one is to understand excretion to mean the expulsion of any sort of waste material from the organism, then the function was definitely established as excretory when Stokes (1893), and later Jennings (1904) proved the discharge of the vacuole to the exterior. But such a generalization offers little satisfaction.

Probably the earliest suggestion that the vacuole is an excretory organelle was made by Stein and Schmidt (see Kent, 1880, p. 69), who stated that "the functions discharged by the contractile vacuole are excretory and correspond most nearly with that of the renal organs of the higher animals." Griffiths (1888) made the statement, based on his own experiments, that the vacuole performs the function of a kidney, and that its secretions are "capable of yielding microscopic crystals of uric acid." As material for these experiments he used *Amoeba, Paramecium,* and *Vorticella.* In describing these experiments, Griffiths says (p. 132):

After the addition of alcohol minute flakes could be distinctly seen floating in the fluid of certain vacuoles. Bearing in mind the murexide reaction, there is every reason to believe that these flakes are nothing more or less than minute crystals of uric acid.

These experiments were repeated many times, generally with positive results, indicating the presence of uric acid. At times, however, the vacuole was found not to contain the slightest trace of uric acid. Howland (1924b) repeated these experiments using *Paramecium, Centropyxis,* and *Amoeba,* but always with negative results. However, uric acid was found in cultures of *Paramecium* and *Amoeba* by Howland, and the concentration was observed to be roughly proportional to the age of the culture. From this she concludes that uric acid is excreted by these forms, though probably not by the vacuole.

Experiments of Nowikoff (1908), Shumway (1917), and Riddle and Torrey (1923), in which the effects of thyroid feeding and the response of *Paramecium* to thyroxin were observed, offer further though indirect evidence in favor of the excretory function. Flather (1919) found that epinephrine, posterior pituitary extract, and pineal gland extract produce similar results—an acceleration in pulsation frequency, and a dilatation of the vacuole. Since these drugs cause diuresis in vertebrates, the action on vacuoles may be interpreted as resembling stimulation of excretion.

Weatherby (1927) found that urea is excreted by *Paramecium caudatum*, but was unable to detect urea in the fluid of the contractile vacuole by means of the micro-injection of his own modification of the xanthydrol reagent of Fosse (1913). This reagent yields positive results with dilutions of urea as great as one part in 12,000. Calculations based on the volume of fluid eliminated by vacuoles and the quantity of urea excreted by known numbers of organisms in mass cultures indicate that the concentration in fluid of the vacuole would be of the order of one part in 2,000 or 3,000, if all the urea were excreted via this route. It therefore appears that at most only a small part of the total urea is excreted in this manner. After removal of the fluid from the contractile vacuole of *Spirostomum* by means of micro-manipulation apparatus, and subsequent hydrolysis with urease, Weatherby (1929) found urea to be present in the vacuolar fluid in a concentration of about one part in 100,000. Calculations of the rate of excretion of urea by known numbers of *Spirostomum* in mass cultures indicate that this amount of urea accounts for only about one percent of the total urea excreted.

Parnas (1926) concludes from observed differences in pulsation frequency that the vacuole is mainly excretory in marine Protozoa, and both excretory and osmotic-pressure-regulatory in fresh-water forms. The excretory function is accepted apparently without reservation by von Gelei (1925, 1928), who homologizes the various parts of the vacuole system in *Paramecium* with the vertebrate kidney, ureter, bladder, and urethra, although he admits the possibility that this system may aid in removing excess water from within the organism. In *Paramecium,* von Gelei states that the vacuole removes approximately ten times as much water as is taken in with food, a fact which he fails to correlate with his claim of a predominantly excretory function. Day (1927) suggests that vacuoles in *Amoeba* originate in "the fusion and coalescence of ultramiscroscopic droplets of soluble katabolic waste which may include water of osmosis." He observed that conductivity water increases size, number, and pulsation frequency of vacuoles. Essentially the same observations and conclusions were extended by him to *Paramecium* and *Spirostomum* (1930). MacLennan (1933) observed in the Ophryoscolecidae that granules accumulate around the vacuole during the early part of diastole and then are gradually reduced in number. The formation of accessory vacuoles in these granular regions involves a solution of granules in the vacuolar fluid. He suggests this as a possible method for the elimination of katabolic wastes. Since he found the pellicle of these organisms to be relatively impermeable, MacLennan believes an excretory function to be all the more probable in these forms, since the vacuole is the only visible means for the removal of wastes. Adolph (1926) found that no change of external conditions alters significantly the rate of elimination of fluid by vacuoles of *Amoeba,* and from this concludes that water is not eliminated merely because it has unavoidably diffused into the body.

Dimitrowa (1928) observed (as mentioned earlier) that mechanical interference, as by pressure on the cover glass, induces the development of extra vacuole systems in *Paramecium.* In most instances these vacuoles assumed normal structure, size, and pulsation frequency, although in some cases there were actively pulsating vacuoles with no radial canals. Dimitrowa explained the formation of extra vacuoles, as well as of those normally formed at fission, by the assumption that if for one reason or another the excretory organs become inadequate to remove wastes, extra organs are formed. If one vacuole is rendered ineffective by mechanical interference, then another is formed to take over its function. Likewise, since metabolism is thought to be increased during fission, new vacuoles are formed to care for the increased production of wastes. Extra vacuoles, induced artificially, obviate the necessity for the formation of a like number at fission, since an ample excretory function is already present.

Somewhat contradictory evidence has been presented by various authors concerning the nature of nitrogenous end products of metabolism in the Protozoa. As previously mentioned, Griffiths (1888) reported uric acid in the vacuolar fluid of several forms. Howland (1924b) was unable to confirm this, but found uric acid in mass cultures of *Amoeba* and *Paramecium*. Weatherby (1929) found urea to be excreted by *Paramecium* and *Spirostomum,* but detected no ammonia nor uric acid; ammonia, as well as a questionable trace of uric acid, were found to be excreted by *Didinium*. Specht (1934) found that *Spirostomum* excretes ammonia, the amount being augmented by lack of oxygen and minimized by abundance of it. Weatherby noticed that cold aqueous extracts of many substances commonly used in culture media (hay, wheat, barley, rye, oats, malted milk, beef extract, blood fibrin, and blood albumen) yield positive tests for uric acid, and suggested this as a possible source of the uric acid found by Howland in cultures of *Paramecium* and *Amoeba.* Lwoff and Roukhelman (1926) found amino-nitrogen as well as additional nitrogen, which they report as ammonia plus amide-nitrogen, in pure cultures of *Glaucoma*. No urea nor uric acid was present. Doyle and Harding (1937) analyzed the food (in the form of *Pseudomonas*) supplied *Glaucoma,* and found that most of the nitrogen present was excreted as ammonia approximately six hours after ingestion of food. No urea was detected.

If the contractile vacuole is active in excretion of nitrogenous wastes, as is frequently maintained, then one would expect it to be able to excrete certain dyes which had been injected into the cytoplasm. Many attempts doubtless have been made to demonstrate such a phenomenon, but few accounts of such experiments are to be found in the literature. Apparently negative results have discouraged publication. A personal communication from one investigator reports complete failure to demonstrate elimination of dyes by way of the contractile vacuole, although the dyes used in these experiments are known to be excreted readily by the kidney of higher forms. Howland and Pollack (1927) found that picric acid, injected into the cytoplasm of *Amoeba dubia,* is picked up and excreted by the contractile vacuole.

Ludwig (1928) studied gaseous metabolism in *Paramecium,* and found that the amount of oxygen dissolved in water taken with food is insignificant, compared with the respiratory requirement of the organism. For the satisfaction of the oxygen requirement, there must be a quantity of water, saturated with oxygen, equivalent to 260 to 30,000 times the amount taken in through the gullet. Oxygen intake must also occur through the cell surface. The amount of water expelled by the vacuole corresponds within reasonable limits to that necessary for the excretion of carbon dioxide, if it is assumed that this water is saturated with the gas. From this Ludwig concludes that the vacuole is of special significance not only in the regulation of osmotic pressure within the cell, but also in excretion of carbon dioxide.

Evidence bearing directly on the excretory nature of the vacuole function is exceedingly scant, and for the most part negative. The reason for the relatively few observations is immediately apparent to all who have attempted experiments of this nature. Perhaps a more thorough investigation of the nature of nitrogenous waste products in other Protozoa will suggest more effective methods for answering the question, by revealing other chemicals which may be detected more readily.

Hartog (1888), Degen (1905), Zuelzer (1910), Doflein (1911), and others maintain that the contractile vacuole is concerned primarily in the regulation of hydrostatic pressure within the cell, or the prevention of overdilution of the cell contents by water taken into the cell in feeding as well as through the cell membrane by osmosis. Harvey (1917) found that *Noctiluca*, which normally lives near the surface of sea water, sinks when transferred to diluted sea water, but ultimately rises to the surface again. Meanwhile, expansion takes place, owing to the taking up of water by the organism. This passage of water from exterior to interior is from a region of higher concentration to one of lower concentration, and therefore contrary to the laws of osmosis in simple systems. When organisms are transferred from diluted sea water to pure sea water, they shrink, vacuoles are formed, and these appear to discharge to the exterior in somewhat the same manner as contractile vacuoles. This appears to aid in reëstablishing the normal salt concentration within the organism. Hance (1917) made extensive observations on a race of *Paramecium* possessing extra contractile vacuoles. He found that these animals cannot withstand immediate immersion in water containing 0.5 percent sea salt, but can be acclimated gradually to this concentration. The number of vacuoles is not reduced by this treatment, but the pulsation frequency is reduced. This response may be taken to indicate a decreased rate of entry of water into the cell, presumably because of the higher external osmotic pressure; but Hance observed also an increased viscosity and toughness of the pellicle, which may indicate that a decreased permeability of the cell wall is partly responsible.

Herfs (1922) investigated the effects of changes in tonicity of the external medium on several kinds of organisms, both free-living fresh-water forms and parasitic forms. He found the pulsation frequency in *Paramecium* to be decreased to about one-fourth the normal when the organism is transferred from fresh water to 0.75 percent NaCl solution. Lower concentrations of salt produce less marked changes. *Gastrostyla steinii* showed essentially the same reaction, except that organisms kept for about fourteen days in one-percent NaCl solution were found to contain no contracting vacuoles. With *Gastrostyla* the vacuole seems to disappear at a NaCl concentration of 1.1 percent to 1.3 percent, and to reappear at a concentration of about 0.5 percent. The pulsation frequency of *Nyctotherus cordiformis,* an intestinal parasite of the frog, was found to vary between wide limits, presumably because of corresponding variations in the water content of the medium. Graded pulsation frequencies were observed *in vitro* when the exterior medium varied from tap water to one-percent NaCl solution. *Opalina ranarum,* which possesses no contractile vacuole, can adapt itself to relatively wide variations in tonicity of the exterior medium without developing a vacuole, if the changes are made gradually. From this Herfs was led to doubt whether or not a vacuole is necessary for the prevention of overdilution of the cytoplasm. He noted further that the lack of a vacuole in *Opalina* goes hand-in-hand with the lack of a cell mouth; whereas *Nyctotherus* possesses both a cell mouth and a vacuole. From this he assumes that it is the water taken in through the mouth that is pumped out by the vacuole. Herfs does not seem to be altogether consistent in this idea, since he further states, as his opinion, that in ordinary cases at least the water taken in through the entire cell surface is of decisive significance for the appearance of the vacuole. With respect to the adaptation of organisms possessing no vacuoles to variations in tonicity of the exterior medium, Herfs seems to have overlooked as an explanation, the possibility of an interchange of salts between cell and medium, a possibility which will be mentioned again later.

Eisenberg (1926), assuming the volume of *Paramecium* to be approximately equal to that of an ellipsoid of rotation having the same dimensions, found that the two vacuoles discharge a volume of liquid equal to that of the organism in 20 minutes, 51 seconds. The average amount of fluid expelled in three experiments, each of 10 minutes' duration, was 45,000 cubic micra; and the average amount of water taken into the organism with food was 11,700 cubic micra, or about one-fourth the total amount expelled. In animals not feeding, the entire amount of fluid expelled entered the body otherwise than with food. Basing his theory on the work of Nirenstein as well as on his own observations, Eisenberg concludes that water penetrates the body by way of the peristome, even when a food vacuole is not in process of formation. It was further observed that an increase in the osmotic pressure of the exterior medium results in a decrease in pulsation frequency, and that equi-osmotic solutions of different chemicals may cause different degrees of slowing.

Fortner (1926) concludes, largely on the basis of theoretical considerations, that the vacuole operates for the preservation of vital cell turgescence, since there must be an accumulation of water in the protoplasm because it is surrounded by a membrane impermeable to water and aqueous solutions.

Eisenberg (1929) investigated the relationship between the osmotic pressure and the pulsation frequency of the vacuole in *Balantidium entozoön*. He found that the frequency of the formation of vacuoles depends on the osmotic pressure, and is all the greater the more the pressure is reduced below that of the usual environment. A pulsation frequency accelerated by the removal of the organism to a medium of lower osmotic pressure does not remain accelerated, but returns to normal after a certain period of time. The rapidity and extent of this return to normal are proportional to the osmotic pressure of the medium.

Frisch (1935) was unable to adapt *Paramecium caudatum* and *P. multimicronucleata* to sea water, the organisms dying when the concentration reached 40 percent. However, among other marked changes in the organisms was a pronounced decrease in pulsation frequency of the vacuoles.

Day (1930) concludes from his observations on *Spirostomum* and *Paramecium* that the vacuole is a hydrostatic organelle, which functions also in elimination of metabolic wastes. He found conductivity water to increase the size, number, and rate of pulsation of vacuoles. The lowering of the temperature of the culture medium slows the organisms and retards the contraction rate of vacuoles, while the raising of the temperature increases movement and pulsation frequency.

Kitching (1934) found that the rate of output of fluid from the contractile vacuole of a fresh-water peritrich ciliate is decreased to a new steady value immediately, when the organism is placed in a mixture of tap water and sea water. The rate of output returns to its original value immediately, when the organism is replaced in tap water. Pulsation is stopped when the medium contains more than 12 percent of sea water. Transference of marine peritrich ciliates from sea water to mixtures of sea water and tap water leads to an immediate increase in body volume, to a new and generally steady value. Return of the organisms to pure sea water results in an immediate return of body .volume to normal, or less. When the concentration of sea water is less than 75 percent, the pulsation rate increases, and then generally falls off slightly to a new steady value which is still considerably above the normal in sea water. The maximum sustained increase in rate observed by Kitching was 80-fold. From these observations it is concluded that the vacuole is probably a regulator of hydrostatic pressure in the fresh-water Protozoa, but in those marine Protozoa which possess vacuoles the functions remain obscure.

Hyman (1936) believes that the vacuole in *Amoeba verspertilio* serves to discharge water which has necessarily entered the cell from a hypotonic medium.

One of the most remarkable instances of adjustment of a protozoan to abnormal media is shown in the experiments of Hopkins (1938) on the marine amoeba, *Flabellula mira*. He found that this amoeba can be cultured in any concentration, from sea water diluted twenty times with fresh water to sea water concentrated ten times by evaporation. It never forms contractile vacuoles such as are typical for fresh-water Protozoa. The food vacuoles, when extruded from the cell, contain large quantities of water as well as fecal material. The rate of elimination of fluid by means of these vacuoles is inversely proportional to the concentration of the medium, and directly proportional to the volume of the amoeba. When the concentration of the medium is decreased, the organism swells at first, and then shrinks to its original volume. During shrinkage, elimination of fluid by food vacuoles does not nearly account for the volume loss. If the concentration of the medium is increased, the amoeba shrinks at first, and then swells to its original volume. Only a small increase in concentration is necessary to cause shrinkage, indicating an osmotic value for the cytoplasm, after adjustment, only slightly above that of the medium. Hopkins concludes from these observations that when the medium is either diluted or concentrated, the organism automatically loses or gains osmotically active substances to or from the medium respectively, in such proportion that when adjustment is completed the osmotic value of the cytoplasm is but slightly higher than that of the medium, and that this is accomplished independently of the action of vacuoles. Herfs, whose observations on *Opalina* have been described previously, may find such an explanation applicable to the unexpected behavior of this organism. It is interesting to speculate as to whether or not such adjustment to external osmotic-pressure differences as postulated for *Flabellula,* and possibly *Opalina,* represents the most primitive type of mechanism for this type of adjustment with *Noctiluca* (see Harvey 1917), which develops contractile vacuoles when the tonicity of the external medium is greatly reduced, occupying a position intermediate between *Flabellula* and those forms which possess vacuole systems.

In spite of the quite extensive literature dealing with the question, one is obliged to admit that virtually nothing has been proved beyond question concerning the function or functions of contractile vacuoles. Carbon dioxide and nitrogenous wastes of one sort or another are undoubtedly excreted by Protozoa. It is reasonable to suppose that at least a part of these highly soluble wastes finds its way into the fluid of the vacuole and is excreted in this manner. Many authors hold that it is not only reasonable to suppose this, but that it is unreasonable to suppose that it does not occur. But, be that as it may, the contractile vacuole certainly has not been proved an organelle whose main function is excretion of metabolic wastes. Likewise, there is indisputable evidence that many fresh-water Protozoa show a decreased pulsation frequency when the tonicity of the exterior medium is increased; and there is equally valid evidence indicating that the reverse occurs when many marine and parasitic Protozoa are transferred to a medium having a decreased tonicity. One may regret the fact, but it is none the less true, that these observations prove nothing more than the bare statement which describes the observations. They strongly suggest that the vacuole operates to prevent excessive dilution of the cytoplasm, or to regulate osmotic pressure within the cell, but beyond this the interpretation is subject to criticism. As stated by Calkins (1926), these supposed functions are not necessarily exclusive, and the possibility still exists that other functions, as well as these, are performed by the contractile vacuoles.

CONTRACTILE VACUOLES AND THE GOLGI APPARATUS

Few publications within recent years on the general subject of contractile vacuoles have aroused as much interest or stimulated as much constructive research as that of Nassonov (1924), in which he suggests that the vacuole in Protozoa is homologous with the Golgi apparatus in metazoan cells. Neglecting for the moment the ultimate status of this proposed homology, one must admit that this article is responsible, either directly or indirectly, for valuable work which otherwise might have been delayed indefinitely.

Before attempting a discussion of the literature bearing on this proposed homology, a few words concerning the general nature of the Golgi apparatus may be of benefit. Its discovery in 1899 is attributed to the man whose name it bears. For approximately twenty-five years after the first description of such a structure its actual existence was doubted by many competent cytologists. Demonstration of the Golgi apparatus in most cells requires a somewhat rigorous treatment of the tissue with various chemical agents, some of which may reasonably be suspected of leaving in the cytoplasm chemical or physical changes of such a nature as to be visible after the Golgi technique, when in reality no such structures exist, pre-formed, in the cell. The problem is probably complicated even further by the multiplicity of forms and shapes which the Golgi apparatus is observed to assume in different cells. At present there seems to be little doubt but that such structures exist, pre-formed, in most cells. Many investigators go so far as to state that the·Golgi apparatus is one of two or three cytoplasmic constituents which are invariably present in all cells, both plant and animal. If this is true, then it is probable that the rôle of the Golgi apparatus in the life history of the cell is of very great importance.

Demonstration of the Golgi apparatus, or Golgi bodies as the structures are frequently called, depends on the reduction of certain metallic compounds to the free metals, the compounds most frequently used containing either osmium or silver. The reduced metal results in blackening of the structure. It has been observed that treatment of stained material with turpentine or hydrogen peroxide results in the bleaching of most structures other than the Golgi apparatus, which may have been blackened by the procedure; Golgi bodies resist even prolonged bleaching effects of these agents. Structures which normally are blackened by the Golgi technique, or any of its modifications, are not blackened if the cell or tissue is first subjected to alcohol or dilute acetic acid. For these reasons, and others which need not be mentioned here, the Golgi apparatus is thought to be composed largely of lipoid substances.

Together with the Golgi apparatus, mitochondria are generally conceded to be invariably present in all cells. These structures, variously called chondriome, chondriosomes, cytomicrosomes, and so forth, are frequently present in the form of short rods or ovoid granules, although the shape is not constant for different types of cells. Some authors maintain that besides the Golgi apparatus and mitochondria a third invariable cytoplasmic constituent, the vacuome, is also present. Whether invariably present or not, the vacuome is at least frequently found in cells. The literature dealing with these structures is exceedingly confusing, owing largely to the lack of a uniform nomenclature. Repeatedly several authors have written of the same structure under different names, or different structures under the same name. The lack of standard techniques also contributes to the confusion. One of the most commonly used techniques for differentiating between the Golgi apparatus and mitochondria is staining of the tissue with a mixture of neutral red and Janus green; mitochondria readily stain with the latter, and some authors maintain that the Golgi apparatus is stainable with neutral red. The vacuome also stains readily with neutral red, and on this basis it has been proposed that the two structures, Golgi apparatus and vacuome, are identical. Others have found within the same cell neutral-red stainable inclusions which are not osmiophilic, and osmiophilic inclusions which are not neutral-red stainable; so it appears that the two structures are not identical in all cells, but in some exist as separate entities. Furthermore, it is sometimes claimed that osmiophilic bodies (the Golgi apparatus) are derived from mitochondria (Janus-green stainable, but not osmiophilic). If this latter is true, then one might expect to find in occasional cells structures which are both osmiophilic and Janus-green stainable, although such a situation has not come to the attention of the author. Unfortunately, no method or group of methods has been devised for the identification of these structures, which is acceptable to all concerned. However, for the protozoan Golgi apparatus it is generally conceded that in most instances it exists as granules, globules, spherules, short rods, or ovoid structures; but there appear to be many exceptions. These bodies reduce certain osmium and silver compounds to the free metals, thereby causing a blackening of the structures which resists bleaching with turpentine and hydrogen peroxide. Most stains commonly used in cytological studies are ineffective, although in some instances neutral red is found to stain some structures which answer other requirements for the true Golgi apparatus. Except for the occasional positive reaction to neutral red, the protozoan Golgi apparatus reacts in a manner practically identical with that of the metazoan Golgi apparatus.

Nassonov (1924) demonstrated the presence of osmiophilic membranes around the vacuoles in *Paramecium caudatum, Lionotus folium, Nassula laterita, Campanella umbellaria, Epistylis gallea, Zoothamnium arbuscula, Vorticella* sp., and *Chilomonas paramecium.* These membranes he found to be permanent structures, merely collapsing at systole of the vacuole—not disappearing, to be reformed anew during the next period of diastole. In *Paramecium* the vacuole system was found to consist of a thin-walled reservoir and filling canals, the latter composed of the short injection canal, the ampulla, and the distal section. The distal section Nassonov found to be surrounded by a specially differentiated plasma, from which hypertonic fluid is secreted into the lumen of the canal. This hypertonicity results in the passage of water into the canal, and ultimately into the vacuole. The vacuole wall in *Paramecium* is considered not to take part in secretion, but to serve only as a temporary reservoir or bladder. In other forms which possess no filling canals, the osmiophilic vacuole wall is considered capable of performing the secretory function as well. The formation of small droplets of fluid within this wall was sometimes seen to occur, following partial systole of the vacuole.

Subsequent observations by Nassonov (1925) on *Chilodon* and *Dogielella* necessitated a modification of the original view so as to include conditions which were not observed in the organisms mentioned in the earlier article. In *Chilodon* the osmiophilic material appears as a heavy black ring, although this ring is not always complete. If fixation occurs at diastole, the vacuole appears to lie within the ring. The ring does not collapse at systole, but remains more or less unchanged. The vacuole is believed to be formed by the flowing together, or coalescing, of small droplets (Sammelvacuolen) which form within the substance of this ring. In *Dogielella* the osmiophilic material is in the form of a ring around the vacuole, resembling, as Nassonov describes it, the rings around the planet Saturn. On contraction of the vacuole, the ring remains essentially unaltered, showing a certain amount of elasticity. The vacuole seems to arise as a result of the coalescing of numerous droplets of fluid, just as in *Chilodon*. These two forms, as well as many others described by other authors, represent a separation of the Golgi apparatus from the vacuole, although the close functional association remains. Nassonov's conception of this close functional association is expressed in a third publication (1926), in which he states that the Golgi apparatus serves as a mechanism for collecting certain materials from the cell substance and preparing them in such a way that they can be discharged from the cell by the vacuole. To do this the Golgi apparatus need not be a part of the vacuole system, nor even in direct contact with it. This conception represents an important departure from the first, in so far as morphology is concerned, but does not alter the essential physiological relationship. Further evidence that such is the function of the metazoan Golgi apparatus was obtained from experiments in which the dye, Trypan blue, was injected into mice. On examination of sections taken from the livers and kidneys of these mice, it was found that the dye was concentrated in that region of the cells of the liver and of the convoluted tubules of the kidney in which the Golgi apparatus is situated. Distribution of mitochondria in these cells was found to be quite different, indicating that these structures are not intimately associated with the collection of the dye.

Some authors summarily reject the idea of a relationship between vacuole and Golgi apparatus, solely on the ground that the wall of the vacuole proves not to be osmiophilic. However, some of these same authors present evidence which supports the idea of a physiological relationship, even though the actual identity of the two structures is disproved. Nassonov himself was among the first to demonstrate that by no means all contractile vacuoles have osmiophilic walls, but this does not alter the possibility of such a functional relationship as he suggested.

Brown (1930) found that the Golgi apparatus of *Amoeba proteus* is the characteristic protozoan type of globules and spherules, with clear centers and dark rims. From a central focus these spherules appear under the microscope to be crescent-shaped structures. He suggests that the minute vacuoles which occur in the endoplasm of *Amoeba* are associated in some way with these crescent-shaped structures, and that they unite to form the contractile vacuole. Brown further suggests this as the reason that the vacuole in this form is not blackened by osmic acid, as it is in *Paramecium*.

Hall (1930a) found small globular inclusions in *Trichamoeba* which are osmiophilic, and which resist bleaching by either hydrogen peroxide or turpentine. These inclusions are similar in size and distribution to those which are stained vitally by neutral red. In material impregnated by the Kolatchev method, the contractile vacuoles are not blackened. In material prepared according to the Mann-Kopsch method, small globules, similar to those seen in the Kolatchev material, are blackened. These globules likewise resist bleaching by turpentine and hydrogen peroxide. In the Mann-Kopsch material, small vacuoles—two, three, or more in number—are blackened in many amoebae. In a few instances a number of blackened globules were seen adherent to the wall of the contractile vacuole, which, on casual examination, gave the appearance of a vacuole with blackened walls. Hall suggests that in material less effectively bleached, such a condition might easily be mistaken for heavily impregnated vacuoles. Nigrelli and Hall (1930) report the presence of small osmiophilic and neutral-red stainable granules in *Arcella vulgaris*.

Mast and Doyle (1935) apply the name "beta granules" to small structures, usually spherical but sometimes ellipsoidal or rod-like in shape, which have a diameter of about one micron. These granules are distributed more or less uniformly throughout the cytoplasm, except at the surface of the contractile vacuole, where they tend to become concentrated in a layer. Aggregation of granules on the surface was described by Metcalf (1910), as previously mentioned. These granules, according to Mast and Doyle, are stained vitally by Janus green, but only on the surface, indicating that they have a differentiated surface layer similar to that in mitochondria. In addition to beta granules, these authors investigated other more or less spherical cytoplasmic inclusions, which they call "cytoplasmic refractive bodies." The outer layers of these bodies are readily stained by neutral red and osmium, whereas the central portions react negatively to osmium and stain but faintly with neutral red. Apparently these are the same structures studied by Brown, who believes them to give rise to minute vacuoles which are precursors to contractile vacuoles. The beta granules are not blackened by osmium. Many of these granules are usually situated close to the contractile vacuole, while others are scattered throughout the entire cytoplasm. The pulsation frequency of the vacuole is proportional to the number of beta granules remaining after some have been removed by operation, indicating a close relationship between granules and vacuole function. Removal of most of these granules results in the death of the organism.

The relationship between the contractile vacuole and cytoplasmic inclusions in *Amoeba* is puzzling. One would be inclined to accept, at least tentatively, the idea of the origin of vacuoles in the beta granules which surround it, were it not for the fact that in a variety of other Protozoa the vacuole has been seen to originate as minute droplets in the region of the cell occupied by osmiophilic granules. Yet in *A. proteus* the granules among which the vacuole apparently arises are not osmiophilic nor stainable by neutral red, but are stainable by Janus green. It might be suggested that the situation in *Amoeba* is the reverse to what it appears to be in other Protozoa, but such a suggestion offers no satisfaction. A more likely explanation lies in the uncertainty of identification of these cytoplasmic inclusions. Some authors (e.g., Hall, 1930a) consider the vacuome, which is neutral-red stainable, identical with the Golgi apparatus, which is osmiophilic; this Hall observed to be true in *Trichamoeba*. Others (e.g., MacLennan, 1933) have identified both neutral-red stainable and osmiophilic granules as separate structures within the same organism. Apparently Dunihue (1931) finds the same in *Paramecium*. Further, MacLennan observed that the only granules in *Eudiplodinium* which can be impregnated with osmium are those found in the vacuolar region; yet in a study of living material, it was shown that this region is composed of granules which originate in the surrounding ectoplasm. Therefore, these granules, as they assemble in the vacuolar region, undergo some change, either chemical or physical or both, which makes them osmiophilic. It has been suggested at one time or another that the Golgi apparatus is derived from mitochondria. Until this puzzling situation is clarified, it seems necessary to assume that in some organisms neutral-red stainable granules (vacuome), osmiophilic granules (the Golgi apparatus), and Janus-green stainable granules (mitochondria) exist as separate and distinct entities, whereas in others the Golgi apparatus may be combined with one or the other of the two remaining types of granules. Hirschler (1924) found only one kind of lipoid body in *Gregarina* and *Spirostomum,* and suggested that these represent a primitive type of organism in which Golgi apparatus and mitochondria are combined in a single type of granule. Until the identity and function of the various types of granules in *A. proteus* have been investigated further, it is difficult to arrive at any reasonable conclusion concerning the relationship of the contractile vacuole to them.

Hirschler (1927) examined a variety of organisms after fixation and staining with several dyes, as well as impregnation with osmium and silver. From these studies he concludes that both Golgi apparatus and mitochondria are present in *Bodo lacertae, Lophomonas blattarum, L. striata, Trypanoplasma dendrocoeli, Entamoeba blattae, Monocystis agilis, Trypanoplasma helicis, Diplocystis phryganeae, Gregarina polymorpha,* and *Clepsidrina blattarum.* In these organisms the Golgi apparatus and the mitochondria were shown to have the same staining reactions as corresponding structures in metazoan cells.

Hall (1929) found osmiophilic granules which resist bleaching with hydrogen peroxide in *Peranema trichophorum;* in *Menoidium* and *Euglena* (1930b); in *Chromulina* sp., *Astasia* sp., and *Chilomonas paramecium* (1930c); and in *Stylonychia* (1931). Hall and Dunihue (1931) found similar granules, or globules, in *Vorticella.* In many of these experiments two or more methods of osmium impregnation, as well as silver impregnation, were used. In some of them the osmiophilic bodies were found to be stainable with neutral red also. In some species the wall of the contractile vacuole was found to be osmiophilic after prolonged osmication, but generally this was readily bleached by hydrogen peroxide or turpentine. Janus green and neutral red were used as vital stains for several organisms; in these the osmiophilic granules were identified as the neutral-red stainable material, whereas smaller granules were stained with Janus green.

Fauré-Fremiet (1925) observed in several species of *Vorticella* essentially the same type of structure as that described by Nassonov. The vacuole wall was found to be in the form of a ring deeply blackened by osmium. Following systole, the vacuole collapses, but the wall remains quite evident. Small vesicles or droplets appear within the thickness of the wall, fuse together, and thus give rise to the new vacuole. On the other hand, Finley (1934) demonstrated, by means of recognized osmium and silver-impregnation techniques, discrete globular inclusions in the cytoplasm of *Vorticella convallaria, V. microstoma,* and *V. campanula.* These globules were readily distinguishable from the rod-shaped mitochondria by staining with a mixture of Janus green and neutral red, the globules reacting positively to neutral red and negatively to Janus green, whereas with mitochondria the reverse was true.

Moore (1931) found distributed through the entire endoplasm of *Blepharisma* globules with osmiophilic cortices and osmiophobic centers. These structures resist bleaching with turpentine. Only in instances of overimpregnation is the contractile vacuole blackened in this form, although paramecia, mixed with the *Blepharisma* uniformly show blackened vacuole systems. Where impregnation of the vacuole is produced in *Blepharisma,* it is readily bleached with turpentine. No evidence was noted by Moore that in these osmiophilic globules lay the origin of the contractile vacuole. In a later investigation, Moore (1934) found that the secondary vacuoles do not empty their contents into the primary vacuole, and thus contribute to its filling; but as contraction of the primary vacuole occurs, the secondary vacuoles move into the place it had occupied, where they coalesce to form a new primary vacuole. No "excretory granules" were observed, but in the earlier work Moore described osmiophilic globules scattered throughout the cytoplasm. On the basis of these observations, Moore rejects the Nassonov homology for *Blepharisma.*

King (1933) found in *Euplotes* that the vacuoles termed group V_3 by Taylor (1923) have their origin at the distal ends of a very large number of collecting tubules, located just under the ectoplasm on the dorsal surface of the organism. The presence of these tubules was demonstrated by impregnation with osmium. King believes that these tubules, or canals, like those in *Paramecium,* are responsible for collection of fluid which ultimately reaches the contractile vacuole.

In a comprehensive series of observations on the Ophryoscolecidae,

MacLennan (1933) found that the osmiophilic granules contribute directly to the formation of accessory vacuoles, which in turn form the contractile vacuole. With respect to the possible function of the vacuole, he states (p. 236):

The vacuolar region found in these ciliates shows definite evidence of the elimination of materials by means of the vacuolar fluid and corresponds to the secretary region or "region of Golgi" in gland cells. The nature of the materials eliminated by the vacuolar region was not determined in this investigation. Since, however, the pellicle in the Ophryoscolecidae has been shown to be relatively impermeable and since the vacuolar region is the only demonstrable path by which materials are constantly being passed to the exterior, it is likely that the katabolic wastes of these ciliates are eliminated by this organelle rather than by direct diffusion through the pellicle.

Dunihue (1931) found that the vacuole system in *Paramecium caudatum* is osmicated only after the neutral-red stainable globules. These globules and Janus-green stainable elements, he believes, represent the vacuome and chondriome (mitochondria) respectively. King (1935) noted a "specialized excretory protoplasm" surrounding the feeding canals in *P. multimicronucleata,* but denied that this material is homologous with the Golgi apparatus of metazoan cells.

The opinions of von Gelei (1925, 1928) concerning the structure and function of the contractile vacuole in *Paramecium* are of special interest. He described essentially the same structures in stained *Paramecium* as those mentioned by Nassonov. The zone of specialized plasma around the distal portion of the canals, particularly, was described in detail, and an excretory function assigned to it. This specialized plasma von Gelei calls "nephridial plasma," its excretory function being implied by its name. This "excretory" function von Gelei believes is entirely different from the "secretory" function assigned by Nassonov, when the latter considers the specialized plasma to be the Golgi apparatus. This disagreement appears to be imaginary rather than real, since an analysis of their respective views indicates that the two authors observed structures which are identical in practically every respect; but to describe the function, they selected different words. Moreover, these different words, when translated into terms of physiological processes, are practically identical. Von Gelei pictures two different arrangements of the deeply staining material in this zone of specialized plasma; in one the stained elements are in the form of short rods, which lie with their long axes at right-angles to the long axis of the filling canal, in much the same manner as the bristles of a test-tube brush are arranged with respect to the wire handle to which they are attached; and in the other, these elements are in the form of a net surrounding the filling canal. It is interesting to note that this net-like arrangement is commonly seen in the Golgi apparatus of many metazoan cells, as well as in *Dogielella*. Von Gelei (1933) observed in *Spathidium giganteum* that not only the contractile vacuole and the smaller vacuoles in its immediate vicinity possess osmiophilic walls, but also others further removed. One can but wonder if this represents the origin of contractile vacuoles by the coalescence of secondary vacuoles, which have arisen in more or less remote parts of the organism.

In *Monocystis agilis* and *M. ascidiae*, Hirschler (1924) identified two kinds of lipoid bodies. The smaller of these he considers mitochondria, the larger the Golgi apparatus. In *Gregarina polymorpha*, *G. blattarum*, and *Spirostomum ambiguum* only one kind of lipoid body was observed. From this Hirschler concludes that the latter are representatives of a more primitive state, in which lipoid bodies are not yet differentiated into mitochondria and Golgi apparatus.

In most instances in which description of structures are given in sufficient detail, and in which organisms have been subjected to a variety of stains as well as osmium and silver impregnation, very strong evidence has been presented to indicate that contractile vacuoles derive the fluid which they expel to the exterior from granules which are osmiophilic, argentophilic, and sometimes neutral-red stainable. In some cells the osmiophilic granules are aggregated around the vacuole or in that part of the cell in which the vacuole ordinarily arises; this is usually associated with the origin of the vacuole in a restricted portion of the cell. In other cells the osmiophilic granules are dispersed to a greater or less extent, sometimes apparently uniformly throughout the cytoplasm; this is usually associated with at least the potential origin of vacuoles in almost any part of the cytoplasm. Evidence bearing on the subject indicates that these osmiophilic granules may represent at least one type of the "excretory granule" so frequently mentioned in the literature.

Several authors have reported osmiophilic substances in the form of relatively broad bands, or rings, which may or may not be in direct contact with the vacuole wall. Most authors seem to agree that the usual form for the protozoan Golgi apparatus is that of granules, globules, or short rods. Overimpregnation of a granular region, which occurs when the process is carried out at too high temperatures, has been shown to produce heavy black bands, or rings, in certain organisms. This fact suggests the possibility that the Golgi apparatus may be of the usual form, even in those organisms in which the band, or ring, type has been observed. Identification of osmiophilic substances answering the known criteria for the Golgi apparatus has been extended to include representatives of the four classes of Protozoa: Mastigophora, Sarcodina, Sporozoa, and Ciliata. This substantiates the idea that the Golgi apparatus is a cytoplasmic inclusion of all living cells.

CONCLUSION

In spite of the multiplicity of claims, counter claims, theories, and suggestions, a few generalizations seem to be established well enough to indicate at least some of the fundamental processes associated with activity of the contractile vacuole. It is not intended that these shall be accepted as proved beyond question, but rather that the evidence points in their direction more consistently than in any other. Further investigation may necessitate a complete revision of opinion concerning these processes, but in the light of the information available at the present time the following conclusions seem to be justified.

1. Contractile vacuoles originate as a result of the activity of certain cytoplasmic inclusions, which may be aggregated in the immediate vicinity of the vacuole in some species, or distributed more or less generally throughout the cytoplasm in others. Temporary contractile vacuoles are formed by the fusion or coalescence of small accessory vacuoles, which in turn originate by the fusion of still smaller accessory vacuoles, the last and smallest vacuoles being formed in or associated with the cytoplasmic inclusions mentioned above. More or less permanent contractile vacuoles (e.g., those of *Paramecium*) receive fluid as small droplets, or accessory vacuoles which fuse with some portion of the filling canals; these droplets originate in or on cytoplasmic inclusions in the same manner as those mentioned above.

2. On the basis of known physicochemical laws and processes, it is necessary to postulate the existence of a physiological membrane surrounding the contractile vacuole. In some organisms, particularly those

possessing more or less permanent vacuole systems, these organelles appear to be surrounded by morphological membranes.

3. Direct evidence concerning the function of contractile vacuoles is almost entirely lacking. Indirect evidence indicates that in fresh-water forms the vacuole protects the organism against excessive dilution of its cytoplasm. In marine and parasitic forms such a function would seem to be largely superfluous, although even in these the elimination of at least a small quantity of water by some mechanism appears to be necessary. Direct evidence indicating the presence of waste products of metabolism in the vacuolar fluid is very scant, although, in those forms possessing relatively impermeable surface structures, the vacuole is the only visible means by which such wastes may be passed to the exterior.

4. In some Protozoa three types of cytoplasmic inclusions have been identified, in others only two types. In all Protozoa so far examined with this in view, at least some of the inclusions are osmiophilic. In some others these osmiophilic inclusions are also stainable by neutral red, but not by Janus green. Osmication of certain Protozoa by one technique or another frequently shows more than one type of inclusion to be osmiophilic, but generally one of these resists bleaching by hydrogen peroxide or turpentine more completely than the others. Such inclusions are generally recognized as the Golgi apparatus. By comparing living organisms with those stained vitally with various dyes, as well as with others impregnated with osmium or silver, identity of the Golgi apparatus and the cytoplasmic inclusions concerned with the origin of the contractile vacuole has been established for many forms. The usual form of protozoan Golgi apparatus is granular, globular, or rod-like. In a few species (*Paramecium, Dogielella, Chilodon,* and others), it frequently appears as a network, while in others (*Lionotus, Nassula, Campanella,* and others) it is in the form of a thick ring, or membrane, surrounding part or all of the vacuole. Evidence has been presented which indicates that in some of these, if not all, a granular structure has been overimpregnated, this causing it to assume the appearances mentioned. It therefore appears that fluid which is expelled from the organism by the contractile vacuole originates as droplets in association with the Golgi apparatus, although the Golgi apparatus is not necessarily in intimate contact with the vacuole. Concerning the origin of secretions in metazoan gland cells, Bowen (1929, p. 511) states:

Secretion is in essence a phenomenon of "granule" or droplet formation. Starting with a single such secretory droplet about to be expelled from the cell, we find it possible to trace its origin step by step to a minute vacuole, which has thus from the beginning served as a segregation center for a specific secretion-material. The primordial vacuole is found to arise in that zone of the cell characterized by the presence of the Golgi apparatus, and the evidence indicates, if it does not demonstrate, that the primordial vacuole arises through the activity of the Golgi substance and undergoes a part at least of its development in contact with, or imbedded in, the Golgi apparatus.

The idea of Nassonov, as developed in 1925 and 1926, as well as that of MacLennan (1933), concerning the origin of protozoan vacuoles could hardly be expressed more exactly.

The outstanding features of contractile vacuoles, taken collectively, then, do not lie in differences among them, but rather in similarities, both morphological and physiological. Another fundamental link in the kinship between all cells seems to be established by the apparent homology, both structural and functional, between the protozoan and the metazoan Golgi apparatus.

LITERATURE CITED

Adolph, E. F. 1926. The metabolism of water in *Amoeba* as measured in the contractile vacuole. J. exp. Zool., 44: 355-81.

Bowen, R. H. 1929. The cytology of glandular secretion. Quart. Rev. Biol., 4: 299-324 and 484-519.

Brown, V. E. 1930. The Golgi apparatus of *Amoeba proteus* Pallas. Biol. Bull. 59: 240-46.

Bütschli, O. 1887. Die Contraktilen Vacuolen. Bronn's Thierreich, Protozoa, 1: 1411-59.

Calkins, G. N. 1926. The biology of the protozoa. Philadelphia.

Clark, A. J. 1933. Mode of action of drugs on cells. London.

Day, H. C. 1927. The formation of contractile vacuoles in *Amoeba proteus*. J. Morph., 44: 363-72.

—— 1930. Studies on the contractile vacuole of *Spirostomum* and *Paramecium*. Physiol. Zoöl., 3: 56-71.

Degen, A. 1905. Untersuchungen über die Contractile Vacuole und die Wabenstruktur des Protoplasmas. Botan. Ztg., 63: 160-202.

Dimitrowa, Ariadne. 1928. Untersuchungen über die überzähligen pulsierenden Vakuolen bei *Paramecium caudatum*. Arch. Protistenk., 64: 462-78.

Doflein, F. 1911. Lehrbuch der Protozoenkunde.

Doyle, W. L., and J. P. Harding. 1937. Quantitative studies on the ciliate *Glaucoma*. Excretion of ammonia. J. exp. Biol., 14: 462-69.

Dunihue, F. W. 1931. The vacuome and the neutral red reaction in *Paramecium caudatum*. Arch. Protistenk., 75: 476-97.

Eisenberg, E. 1926. Recherches sur le fonctionnement de la vesicule pulsatile des Infusoires dans les condition normales et sous l'action de certains agents experimentaux. Arch. Biol. Paris, 35: 441.

—— 1929. Recherches comparatives sur le fonctionnement de la vacuole pulsatile chez les Infusoires parasites de la grenouille et chez les Infusoires d'eau douce. Influence de la pression osmotique, des electrolytes et du pH. Arch. Protistenk., 68: 451-70.

Fauré-Fremiet, E. 1925. La Structure permanent de l'appareil excréteur chez quelques vorticellides. C. R. Soc. Biol. Paris, 93: 500-3.

Finley, H. E. 1934. On the vacuome in three species of *Vorticella*. Trans. Amer. micr. Soc., 53: 57-65.

Flather, M. D. 1919. The influence of glandular extracts upon the contractile vacuoles of *Paramecium caudatum*. Biol. Bull., 37: 22-39.

Fortner, H. 1926. Zur Frage der diskontinuierlichen Excretion bei Protisten. Arch. Protistenk., 56: 295-320.

Fosse, R. 1913. Sur l'Identification de l'urèe et sa précipitation de solution extrêment diluèes. C. R. Acad. Sci. Paris, 137: 948-57.

Frisch, J. A. 1935. Experimental adaptation of fresh water ciliates to sea water. Science, 81: 537.

Gelei, J. von. 1925. Nephridialapparat bei Protozoen. Biol. Zbl., 45: 676-83.

—— 1928. Nochmals über den Nephridialapparat bei den Protozoen. Arch. Protistenk., 64: 479-94.

—— 1933. Wandernde Exkretionsvakuolen bei den Protozoa. Arch. Protistenk., 81: 231-42.

Griffiths, A. B. 1888. A method of demonstrating the presence of uric acid in the contractile vacuoles of some of the lower organisms. Proc. roy. Soc. Edinb., 16: 131-35.

Hall, R. P. 1929. Reaction of certain cytoplasmic inclusions to vital dyes and their relation to mitochondria and Golgi apparatus in the flagellate, *Peranema trichophorum*. J. Morph., 48: 105-18.

—— 1930a. Cytoplasmic inclusions of *Trichamoeba* and their reaction to vital dyes and to osmic and silver impregnation. J. Morph., 49: 139-51.

—— 1930b. Cytoplasmic inclusions of *Menoidium* and *Euglena*, with special reference to the vacuome and "Golgi apparatus" of Euglenoid flagellates. Ann. Protist., 3: 57-68.

—— 1930c. Osmiophilic inclusions similar to Golgi apparatus in the flagellates, *Chromulina, Chilomonas,* and *Astasia.* Arch. Protistenk., 69: 7-22.

—— 1931. Vacuome and Golgi apparatus in the ciliate, *Stylonychia*. Z. Zellforsch., 13: 770-82.

Hall, R. P., and F. W. Dunihue. 1931. On the vacuome and food vacuoles in *Vorticella*. Trans. Amer. micr. Soc., 50: 196-205.

Hance, R. T. 1917. Studies on a race of *Paramecium* possessing extra contractile vacuoles. J. exp. Zool., 23: 287-333.

Hartog, M. M. 1888. Preliminary notes on the functions and homologies of the contractile vacuole in plants and animals. Rep. Brit. Assoc. Adv. Sci., Bath, pp. 714-16.

Harvey, E. B. 1917. A physiological study of specific gravity and of luminescence in *Noctiluca*, with special reference to anesthesia. Pap. Tortugas Lab., 11: 237-53.

Haye, Ans. 1930. Über den Exkretionsapparat bei den Protisten, nebst Bemerkungen über einige andere feinere Strukturverhältnisse untersuchten Arten. Arch. Protistenk., 70: 1-86.

Herfs, A. 1922. Die pulsierende Vakuole der Protozoen, ein Schutzorgan gegen Aussüssung. Arch. Protistenk., 44: 227-60.

Hirschler, J. 1924. Sur une Méthode de Noircissment de l'Appareil de Golgi. C. R. Soc. Biol. Paris, 90: 893.

—— 1927. Studien über die sich mit Osmium schwärzenden Plasmakomponenten einiger Protozoenarten. Z. Zellforsch., 5: 704-86.

Hopkins, D. L. 1938. Adjustment of the marine amoeba, *Flabellula mira* Schaeffer, to changes in the total salt concentration of the outside medium. Biol. Bull., 75: 337 (Abstr.).

Howland, R. B. 1924a. Experiments on the contractile vacuole of *Amoeba verrucosa* and *Paramecium caudatum*. J. exp. Zool., 40: 251-62.

—— 1924b. On excretion of nitrogenous waste as a function of the contractile vacuole. J. exp. Zool., 40: 231-50.

Howland, R. B., and H. Pollack. 1927. Expulsion of injected solute by the contractile vacuole of *Amoeba*. Proc. Soc. exp. Biol. N.Y., 25: 221-22.

Hyman, Libbie H. 1936. Observations on Protozoa: The impermanence of the contractile vacuole in *Amoeba verspertilio*. Quart. J. micr. Sci., 79: 43-56.

Jennings, H. S. 1904. A method of demonstrating the external discharge of the contractile vacuole. Zool. Anz., 27: 656-58.

Kent, W. S. 1880. A Manual of the Infusoria. London.

Khainsky, A. 1910. Zur Morphologie und Physiologie einiger Infusorien (*Paramecium caudatum*) auf Grund einer neuen histologischen Methode. Arch. Protistenk., 21: 1-60.

King, R. L. 1928. The contractile vacuole in *Paramecium trichium*. Biol. Bull., 55: 59-69.

—— 1933. Contractile vacuole of *Euplotes*. Trans. Amer. micr. Soc., 52: 103-6.

—— 1935. The contractile vacuole of *Paramecium multimicronucleata*. J. Morph., 58: 555-71.

Kitching, J. A. 1934. The physiology of contractile vacuoles: osmotic relations. J. exp. Biol., 11: 364-81.

Lachmann, K. F. 1891. Über einiger neu entdekte Infusorien und über Contractilen Blasen bei den Infusorien. Verh. Naturh. Ver. preuss. Rheinl., 16: 66-68, 91-93.

Lloyd, F. E. 1928. The contractile vacuole. Biol. Rev., 3: 329-58.

Lloyd, F. E., and G. W. Scarth. 1926. Origin of vacuoles. Science, 63: 459-60.

Ludwig, W. 1928. Der Betriebsstoffwechsel von Paramecium caudatum. Arch. Protistenk., 62: 12-40.

Lwoff, André, and Nadia Roukhelman. 1926. Variations de quelques formes d'azote dans une culture pure d'infusoires. C. R. Acad. Sci. Paris, 183: 156-58.

MacLennan, R. F. 1933. The pulsatory cycle of the contractile vacuoles in the Ophryoscolecidae, ciliates from the stomach of cattle. Univ. Cal. Publ. Zool., 39: 205-50.

Mast, S. O. 1926. Structure, movement, locomotion, and stimulation in Amoeba. J. Morph., 41: 347-425.

—— 1938. The contractile vacuole in Amoeba proteus (Leidy). Biol. Bull., 74: 306-13.

Mast, S. O., and W. L. Doyle. 1935. Structure, origin, and function of cytoplasmic constituents in Amoeba proteus with special reference to mitochondria and Golgi substance. Origin and function based on experimental evidence; effect of centrifuging on Amoeba proteus. Arch. Protistenk., 86: 278-306.

Maupas, E. 1883. Contribution a l'Étude Morphologique et Anatomique des Infusoires Ciliés. Arch. zool. exp. gén., 2 serie, 1: 427-664.

Metcalf, M. M. 1910. Studies upon Amoeba. J. exp. Zool., 9: 301-31.

—— 1926. The contractile vacuole granules in Amoeba proteus. Science, 63: 523-24.

Moore, Imogene. 1931. Reaction of Blepharisma to Golgi impregnation methods. Proc. Soc. exp. Biol. N.Y., 28: 805-6.

—— 1934. Morphology of the contractile vacuole and cloacal region of Blepharisma undulans. J. exp. Zool., 69: 59-104.

Nassonov, D. 1924. Der Exkretionsapparat der Protozoen als Homologen des Golgischen Apparats der Metazoazellen. Arch. mikr. Anat. 103: 437-82.

—— 1925. Zur Frage über den Bau und die Bedeutung des lipoiden Exkretionsapparates bei Protozoa. Z. Zellforsch., 2: 87-97.

—— 1926. Die physiologische Bedeutung des Golgi Apparats im Lichte der Vitalfärbungsmethode. Z. Zellforsch., 3: 472-502.

Nigrelli, R. F., and R. P. Hall. 1930. Osmiophilic and neutral-red-stainable inclusions of Arcella. Trans. Amer. micr. Soc., 49: 18-25.

Nowikoff, M. 1908. Über die Wirkung des Schildrusen Extrakts und einiger anderer Organstoffe auf Ciliaten. Arch. Protistenk., 11: 309-26.

Parnas, J. K. 1926. Allgameines und Vergleichendes des Wasserhaushalts. Handbuch. d. norm. u. pathol. Physiol., 17, correlationem 3: 137-60.

Riddle, M. C., and H. B. Torrey. 1923. The physiological response of *Paramecium* to thyroxin. Anat. Rec., 24: 396.

Scarth, G. W., and F. E. Lloyd. 1927. The rôle of kinoplasm in the genesis of vacuoles. Science, 65: 599-600.

Shumway, W. 1917. Effects of thyroid diet upon paramecia. J. exp. Zool., 22: 529-62.

Specht, Heinz. 1934. Aerobic respiration in *Spirostomum ambiguum* and the production of ammonia. J. cell. comp. Physiol., 5: 319-33.

Stempell, W. 1914. Über die Funktion der Pulsierende Vacuole und einen Apparat zur Demonstrationen derselben. Zool. Jahrb., Abt. Allg. Zool. Physiol. Tiere, 34: 437-78.

Stokes, A. C. 1893. The contractile vacuole. Amer. mon. micr. J., 14: 182-88.

Taylor, C. V. 1923. The contractile vacuole in *Euplotes:* an example of the sol-gel reversibility of cytoplasm. J. exp. Zool., 37: 259-90.

Weatherby, J. H. 1927. The function of the contractile vacuole in *Paramecium caudatum;* with special reference to the excretion of nitrogenous compounds. Biol. Bull., 52: 208-18.

—— 1929. Excretion of nitrogenous substances in Protozoa. Physiol. Zoöl., 2: 375-94.

Wenrich, D. H. 1926. The structure and division of *Paramecium trichium* Stokes. J. Morph., 43: 81-103.

Young, R. A. 1924. On the excretory apparatus in *Paramecium*. Science, 60: 244.

Zülzer, Margaret. 1910. Über den Einfluss des Meerwassers auf die pulsierende Vacuole. Roux Arch. Entw. Mech. Organ., 29: 632-40.

CHAPTER VIII

THE TECHNIQUE AND SIGNIFICANCE OF CONTROL IN PROTOZOAN CULTURE

GEORGE W. KIDDER

INTRODUCTION

DURING THE LAST FEW YEARS there has come to be an appreciation of methods of culturing Protozoa which will permit the investigator to determine the conditions under which his study is being made. Studies of populations and the various interesting and important factors involved, mass physiology, nutrition, and numerous other phases of cellular activity may be profitably dealt with by the student of the Protozoa only when he can be sure that the effects noted are due to the conditions under investigation. The science of protozoölogy has passed through the phase of "pure-mixed" methods of culture. This term simply means that a single strain of Protozoa is grown in association with a chance combination of other microörganisms, usually bacteria. Many valuable and thought-provoking contributions, based upon this method, have been made, and these contributions have paved the way to the more precise evaluations of the present.

In the culture of practically any species of Protozoa, it may be safely said that the bacteria as a group offer the most serious obstacle to controlled conditions. Experimental modifications of factors such as nutritive materials, temperature, oxygen or carbon dioxide tensions, oxydation-reduction potentials, and so forth, may produce effects, but whether these effects are the result of changes in protozoan activity *per se,* or are secondary through the change of activity of the bacteria, is usually nearly or totally obscure. These facts are recognized, and there has been built up a body of literature reporting progress in methods which will allow for the control or, better still, the elimination of bacteria. Numerous investigators have succeeded in sterilizing various species of Protozoa and have made great strides in advancing our knowledge of cellular activities through the use of pure cultures.

It is the purpose of this section to devote some time to methods or techniques of protozoan sterilization, in order to bring before the reader some of the many problems which must be dealt with in work of this kind. As may be supposed, the nature of such a discussion makes it necessary to assume at least a rudimentary knowledge of bacteriological technique. And above all there must be a thorough appreciation of the potentialities of many different types of bacteria to resist even the most careful methods of irradication, potentialities which express themselves in some cases only after prolonged periods of apparent sterility.

In addition to an outline of techniques for sterilization, the question of acceptable tests for the sterility of cultures will be considered, and finally some of the problems and conditions arising from the establishment of sterile Protozoa in culture.

THE PROBLEM OF PROTOZOAN STERILIZATION

1. GENERAL MATERIAL

Protozoa from natural waters, soil, and so forth, are, and have been throughout their existence, in association with bacteria. This does not mean that the bacterial flora of their surroundings has remained constant either as to numbers or types. The flora is probably continually changing. This very change is one of the most important factors in the succession of microscopic animals in ponds and streams. The variety of bacterial types one would expect to encounter in any extended survey of natural ponds is practically limitless. Therefore it is impossible to do more than discuss the general factors to be taken into account in dealing with bacteria.

To attain successful sterile cultures of Protozoa, it is, of course, desirable to have rather large numbers of healthy organisms with which to work. It is usually possible to isolate single organisms into fresh infusion and obtain from fair to good growth. If they are bacteria-feeders (the great majority of free-living ciliates are), enough food organisms are brought over in the isolation to insure, at least in a high percentage of cases, against starvation. As the bacteria multiply they, in turn, are utilized by the Protozoa.

In dealing with ciliates from the wild, a partial substitution method may be attempted. For a number of species, this method facilitates later sterilization. If the ciliate to be used will feed on living yeast (and this can be determined only by experimentation), then cultures may be established by suspending yeast cells in spring water, distilled water, or balanced salt solutions, and introducing the desired ciliate. Often abundant growth will result and after a few subcultures have been made, the ratio of bacteria to Protozoa will be reduced. When serious attempts at sterilization are then carried out, the yeast cells will be found relatively easy to eliminate. It should not be supposed that it will be possible to eliminate the contaminating bacteria in this way, as they will be multiplying slowly all the while. In fact, long-continued cultures of this type are apt to show a decided increase in non-nutritive bacteria over those which were present at the start (see Kidder and Stuart, 1939). Therefore it is advisable to make from one to three subcultures only, and then to start the sterilization procedure.

In general it can be said that the larger the protozoan the more difficult it will be to sterilize. This fact becomes apparent from an examination of the literature, and was noted by Hetherington (1934). Physical properties likewise play a rôle in ease of sterilization. Holotrichous and heterotrichous ciliates, possessing large numbers of closely set cilia, are apt to retain a few of their associated bacteria even after repeated washing, while hypotrichous ciliates may be washed free of bacteria more readily. Flagellates, being for the most part smooth in surface, are relatively easy to sterilize. Activity is also important, both as to movement and metabolism. Highly motile forms may usually be freed of bacteria more readily than sluggish types. Those possessing a high rate of metabolism tend to utilize or defecate the contents of their food vacuoles more rapidly than the slow-growing types, and do not tend to carry over viable spores to contaminate later cultures. Many other characters which will influence the facility with which sterilization may be accomplished might be mentioned, but these will become apparent when we examine some of the procedures.

2. GENERAL METHODS OF STERILIZATION

In order to rid the Protozoa of their associated bacteria, workers have made use of three principles. The first and most generally useful is simple washing in sterile fluid. This is the dilution method whereby the bacteria are diluted out of the solution. The Protozoa must be retained of course, and a number of different manipulations have been devised to insure this.

The principle of dilution takes for granted that the bacteria either are suspended in the fluid or that they may be caused to become suspended.

The second principle is one of migration. The Protozoa to be sterilized are allowed or caused to swim through sterile fluid or semisolid medium or over the surface of solid medium, leaving the bacteria behind. This method has been used with success on a number of different types of ciliates and flagellates and a few amoebae. The Protozoa may be induced to migrate laterally by introducing them into one side of a flat dish of sterile fluid. Or, if they happen to be negatively geotropic, they may be introduced into the bottom of a vessel of sterile fluid and taken off at the top. Those that are positively geotropic may be introduced at the top and taken off at the bottom. Extremely active types may be able to migrate through a semisolid medium and literally scrape off their adhering bacteria.

Combinations of the above two principles have been used with marked success, and a number of ingenious pieces of apparatus have been designed to facilitate the manipulations and to reduce the chance of extraneous contamination. These will be described in detail later.

The third principle that has been applied to this problem is that of bactericidal agents. This method has met with questionable success and then usually only in cases where resistant phases (cysts) could be obtained. As might be expected, any agent which would kill the bacteria in a culture would most surely kill trophic Protozoa. It has been shown many times that the various species of Protozoa are much more susceptible to the usual toxic agents than many of the common bacteria. (An exception to the foregoing statement is indicated in the work of Brown, et al., 1933, using X-rays as a sterilizing agent.)

3. SPECIAL METHODS AND MANIPULATIONS

This section will be devoted to a description of the procedures which have been used by various investigators to rid the different types of Protozoa of their associated bacteria. Considerable pains will be taken to describe the apparatus used, the manipulations performed, and the results obtained. It is hoped that by so doing the reader will be able to gain constructive ideas which will allow him either to utilize one of the described methods or to formulate a modification which will meet his needs.

A. *Dilution*.—One of the first reports of the sterilization of Protozoa by this method is that of Hargitt and Fray (1917), using *Paramecium aurelia* and *P. caudatum*. They experimented with a number of modifications of the washing technique and followed their results by plating on nutrient agar. Their first procedure was dilution by centrifugation, wherein they centrifuged down the paramecia and then quickly withdrew the supernatant fluid with a sterile pipette. The paramecia were then covered again with sterile fluid, and the process repeated five times. At the end of the fifth wash they found that the number of bacteria "per drop" had decreased from 500 colonies (per plate) in the first wash to 3 colonies in the fifth. These results were not satisfactory, however, and the method was abandoned. The authors offer the following objections to the method:

a great deal of time was consumed, the wash waters had to be drawn off immediately after the centrifuge stopped or the paramecia rose in a body and prevented the removal of the wash water. . . . Another serious drawback to the centrifuge method is the difficulty of keeping the wash waters free from contamination by bacteria from the air. The air may contain such enormous numbers of bacteria that instruments and media which are sterile to start with will be contaminated unless precautions are taken to prevent the contact of air bacteria. A sterile pipette laid down on the table is no longer sterile, a wash water left unprotected is soon contaminated by air bacteria [p. 435].

It will become obvious from later discussions of these points that success of the centrifuge method in the hands of Hargitt and Fray was prevented by two principal faults in their manipulations—too few washes, and failure to keep their centrifuge and wash tubes plugged at all times.

These authors next attempted to reduce the chances of outside contamination by the transfer method, using watch crystals enclosed within Petri dishes, and transferring single ciliates through five separate dishes. They again failed to effect sterility, but succeeded in reducing the number of colonies "per drop" from 2,500 in the first wash to one colony in the fifth wash. They blame their lack of success with this procedure upon the fact that the amount of wash fluid was so large that considerable time was required to locate the ciliate between transfers.

Successful sterilization was accomplished by transferring individual paramecia through five successive washes of sterile tap water in sterile depression slides. The transfer pipettes were sterilized by dry heat before using, as were the Petri-dish-contained depression slides. The tap water was autoclaved. They claim to have effected sterilization in a high percentage of their trials, with a total time consumption per ciliate of not more than five minutes. If their bacteriological tests are accepted (development of bacterial colonies on agar plates, time of incubation not given) then we can only conclude that they were extremely fortunate in avoiding ciliates with ingested viable spores.

A chief criticism of the work of Hargitt and Fray was given by Parpart (1928). He pointed out that most of their sterility tests were confined to the washing fluids and not to the supposedly sterilized paramecia. In reinvestigating their results, he found that five washes gave sterile fluid at the end, but that even after ten washes in six out of eight trials the animals themselves were contaminated. He ascribed this fact to the probability of the carrying over of viable spores within the vacuoles of the ciliates. He suggested a simple modification of the method of Hargitt and Fray, which yielded fifty sterile paramecia out of fifty trials. Instead of five washes he employed ten, thereby increasing the dilution factor. Time was allowed (five hours) for the ciliates to void their vacuoles of possible spores in the fifth wash.

His method is essentially as follows: A single *Paramecium* was transferred with a sterile pipette from a wild culture to a sterile Petri-dish-enclosed depression slide containing about six drops of washing fluid. After about one minute, the animal was transferred to the next similar bath. At the fifth bath the *Paramecium* was allowed to swim about for five hours and was then carried through five further washes. All of the manipulations were carried out under a rather elaborate hood to minimize the possibility of contamination from the air.

This modification of the simple washing method has probably been more generally used than any other, owing principally to its simplicity of manipulation. It is admirably adapted to large ciliates which can be followed with ease under the low powers of the dissecting microscope. The smaller the organism the more difficult this method becomes. Of course failure will most surely follow any deviation from absolutely aseptic technique.

In an attempt further to simplify the technique as outlined by Parpart, especially regarding the hood under which the transfers were carried on, the following procedure has yielded extremely satisfactory results in our laboratory (Kidder, Lilly, and Claff, 1940). Syracuse watch glasses are enclosed in cellophane bags, the ends of which are folded over, and the whole sterilized in the autoclave. After cooling, the bags are carefully opened and 5 ml. of sterile wash fluid is placed in each watch glass by means of sterile serological pipettes. The protozoan to be sterilized is placed in the first bath by means of a micro-pipette inserted through the open end of the bag. There are three obvious advantages in this modification, aside from simplicity of apparatus. The opening of the bag is at the side of the dish and at some distance from it. The top of the dish and therefore the fluid is never exposed to the air from above. The same situation is here repeated as obtains when making tube inoculations in ordinary bacteriological technique, where the tube is always held at a slant. This system is less dangerous than one in which the top of a Petri dish must be removed, and obviates the necessity for a hood or drape. The second advantage is that the observer may follow the movements of the protozoan at all times and may then readily draw it up in the transfer pipette. This is usually impossible or difficult when using a Petri dish with the cover in place, as the water of condensation reduces the visibility markedly. Water does not condense on the cellophane, and observations are therefore not hampered. The third advantage is simply one of choice of containers. The Syracuse watch glasses holding 5 ml. of fluid raise enormously the all-important dilution factor.

When it is possible to obtain large numbers of Protozoa in heavy concentrations, sterilization may be accomplished by centrifugation. This method, although unsuccessful in the hands of Hargitt and Fray (1917), has been used to advantage recently (Kidder and Stuart, 1939). It is recommended for use with those species of Protozoa which are so small as to make them difficult to follow under the powers of a dissecting microscope. By choosing a washing fluid favorable for the species to be used and carrying the number of washes far enough, it is usually possible to recover large numbers of sterile Protozoa after the final wash. The method which was finally adopted for the sterilization of *Colpoda steinii* (see Burt, 1940, for species designation) is quoted below:

After excystment had occurred the ciliates were concentrated by slow centrifugation and the concentrate removed to a single, sterile, cotton-plugged centrifuge tube. This ciliate concentrate was diluted with 10 ml. of sterile distilled water and recentrifuged at a speed which would just throw down the majority of ciliates in 3 minutes. It was found that the most satisfactory speed for this purpose was 2000 revolutions per minute. As soon as the centrifuge stopped the supernatant fluid (9 ml.) was immediately withdrawn with a sterile 10 ml. pipette and the tube was allowed to stand for about two minutes in order that the ciliates might swim to the top of the remaining milliliter of water. With a sterile 1 ml. pipette, 0.5 ml. of ciliate suspension was withdrawn and placed in an empty sterile centrifuge tube. This suspension was again diluted, and the process repeated until the ciliates had gone through an average of fifteen such transfers with the accompanying dilutions (a dilution factor of approximately 10^{14}). This method entails a great loss of ciliates but was found necessary inasmuch as, without removal to fresh tubes with the consequent discarding of the residue, contaminations were invariable. It was demonstrated that the contaminations resulted from the fact that ciliates died or became immobilized during centrifugation and were carried passively to the bottom of the tube with their adhering bacteria. However, by discarding the dead forms we were able to completely sterilize several hundred ciliates at each attempt and these gave us the necessary organisms with which to work [Kidder and Stuart, 1939, p. 332].

The washing fluid which was used in this case was sterile Pyrex-distilled water. All pipettes were paper-wrapped and autoclaved. The centrifuge tubes were closed with large cotton plugs and autoclaved. During centrifugation the cotton was folded over and fastened with a rubber band, to prevent the plug from being drawn into the tube.

B. *Migration.*—Probably the first report of a technique for obtaining Protozoa free from bacteria by the utilization of migration was that of Ogata (1893). He reports that he was able to recover as many as fifty-two sterile phytomonads (*Polytoma uvella*) within five to thirty minutes after the start of the migration. The apparatus he employed was a capillary tube 10 to 20 cm. in length, with a 0.3 to 0.5 mm. bore. This tube was filled to within one to 2 cm. from the end with a sterile fluid, and then inserted into a culture of the flagellates and allowed to fill completely. Care was taken to avoid aid bubbles between the layers. Both ends of the capillary were sealed by heat and the whole allowed to stand for from five to thirty minutes. Eventually some of the flagellates were found to have migrated away from their associated bacteria, and when a number had collected in the upper end of the tube, this end was broken off and the flagellates inoculated into nutrient media.

A refinement of the same technique is reported by Stone and Reynolds (1939) for the sterilization of the parasitic flagellate *Trichomonas hominis.* Their capillary tube was made from a piece of 6 mm. Pyrex tubing about eight inches in length, which, before sterilization, had been plugged at both ends with cotton. The capillary was then drawn from one end of the tube, its tip broken off with sterile forceps, and a series of

Figure 123. Capillary tube used for the sterilization of *Trichomonas hominis.* The whole tube is filled with sterile fluid, the lower end sealed in a flame, and the Protozoa to be sterilized are layered on to the fluid at the large end. The Protozoa eventually migrate through the capillary portion, but the associated bacteria are trapped at the first or second bend. (Redrawn from Stone and Reynolds, 1939.)

loops constructed (Fig. 123). All of these manipulations were carried on with care not to contaminate the outside of the tube, for after the loops were made the whole tube was filled to within one inch of the top with sterile fluid (in this case, one part Ringer's, eight parts horse serum) by suction, applied to the large end. The capillary end was then sealed off and the tube, in a vertical position, was incubated forty-eight hours as a check on sterility. If no turbidity developed, contaminated *Trichomonas* were layered onto the fluid in the large end of the tube. Within forty-eight hours many flagellates had migrated down the tube and could be seen in the last inch or two of the capillary portion. The authors state that the bacteria failed, for the most part, to migrate past the first loop, and never passed the second. The last portion of the capillary was cut off and sealed by means of a flame and the cut-off portion was submerged in tincture of iodine (7 percent) for one hour. Then one end was grasped in the fingers and the tube held upright to drain. When dry, pieces of the tube were broken off with sterile forceps and dropped into selected culture media. The authors state that they have repeatedly isolated *T. hominis* bacteria-free by this method, but have not tested it with other Protozoa.

The above method appears to be applicable to many types of Protozoa, and should receive serious consideration. The manipulations offer some difficulty, however, and extreme care will have to be exercised to insure against outside contamination, especially during the filling of the tube with the sterile fluid and again during the breaking of the sections of capillary into nutrient culture media.

Probably the simplest method which takes advantage of the migration of Protozoa in fluid media is the Petri-dish method. A sterile Petri dish is partially filled with sterile fluid and placed on the stand of the dissecting microscope so that one edge is under the objective. After all motion of the fluid has ceased, the Petri dish cover is raised and a drop of concentrated protozoan culture is placed very near the edge opposite the one under the lens. This manipulation must be done with great care, so that the fluid is disturbed as little as possible. The cover is then gently lowered and sufficient time (five to ten minutes) is allowed for the Protozoa to swim to the opposite edge of the dish. The cover is again raised, and single organisms are picked out with sterile pipettes and transferred to selected media. Minimum time for the migration is important, so that none of the highly motile bacteria will reach the area from which the Protozoa are being taken. Enough Protozoa should be separated singly in this way to allow for the law of averages. The greater the motility of the Protozoa, the smaller their size, and the smoother their bodies, the greater the chance for successful sterilization by this method.

We have used the above method to sterilize a number of flagellates (*Euglena, Astasia, Chilomonas*). With a single migration across a Petri dish and the selection of twenty-five organisms at each trial, the percentage of sterile to contaminated cultures was very high (80 to 90 percent). We used a Plastocoel (transparent) shield over the microscope and always worked in a draft-free room. *Tetrahymena geleii* (Furgason, 1940) was also sterilized by a single migration, but with about only 10-percent efficiency. Although the *Tetrahymena* were more motile than the flagellates, they proved harder to rid of their bacteria, probably because of the tendency of the bacteria to become lodged among the cilia.

Oehler (1919) states that he was able to free various Protozoa of bacteria by allowing them to migrate over the surfaces of agar in Petri dishes. This technique may be applicable to a few types which are able to swim in a very thin film of moisture.

The utilization of large tubes of sterile fluids for migration purposes was first mentioned by Purdy and Butterfield (1918). In their studies on the growth of bacteria in sewage, they state that they obtained on one occasion sterile *Paramecium* after allowing the ciliate to swim through thirty feet of sterile water. They call this the "marathon bath," but give no details of its construction, use, or (and this appears to be important from the practical standpoint) means of sterilization of so long a tube.

Glaser and Coria (1930) carried out a rather exhaustive study on methods of sterilization. One of the methods they used with success was the large-tube migration, used with negatively geotropic Protozoa. Their apparatus consisted of a tube fourteen or more inches long, with one-fourth inch bore and a fine tapering point. The large end was plugged with cotton and the whole sterilized in a container. Sterile fluid was drawn up to within two inches of the top by applying suction to the large end through a rubber tube. About 2 ml. of contaminated protozoan culture was then drawn up, and this formed a layer beneath the sterile fluid. The fine end was sealed in the flame and the tube mounted upright in a rack (Fig. 124). After periods of time varying from five minutes to eighteen hours, depending on the species of Protozoa, samples from the top of the tube contained many organisms which had "washed themselves free of most other microorganisms" (p. 790). It was usually necessary to repeat the migration, and this was accomplished by filling a second tube as before, and then drawing up two inches of fluid from the top of the first tube. Occasionally a third wash was necessary to render the Protozoa bacteriologically sterile. Glaser and Coria were successful in sterilizing three species of ciliates and three species of flagellates by this method, and later (1935) three other species of ciliates were added to this list. The identifications of the organisms sterilized are uncertain, except for the well-known types, *Paramecium* and *Chilomonas*.

Another method of migration described by Glaser and Coria in their

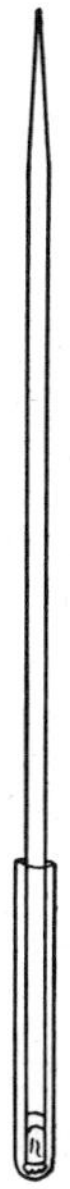

Fig. 124. Migration tube. The tube is filled with a sterile fluid to within about two inches from the top. Protozoa to be sterilized are drawn up under the sterile fluid and the small tip sealed in a flame. The Protozoa migrate upward and are taken off at the top. (Redrawn from Glaser and Coria, 1930.)

1930 paper was one employing a V shaped tube (Fig. 125), filled with Noguchi's semisolid medium. The larger arm of the tube measured 12 cm. in length and had an inner diameter of 28 mm. The smaller arm was 9 cm. in length with an inside diameter of 8 mm. After sterilization, the tube was filled with 15 ml. of sterile melted medium and this was allowed partially to solidify. Then the contaminated culture was placed at the bottom of the tube by injection through the small arm with a long, fine pipette. Air bubbles were excluded. The tube was then allowed to stand at room temperature for a sufficient time for some of the Protozoa to reach the top of the large arm, from the surface of which they were recovered.

This method, employing the use of semisolid media, appears to be applicable for many types of Protozoa. The consistency of the medium through which the Protozoa migrate seems to favor the removal of bacteria, in that vigorous motion is necessary. The bacteria, on the other hand, would be largely prevented from dispersing far from the point of inoculation, at least for some time. This method, or some modification of it, should receive serious consideration from future investigators, in-

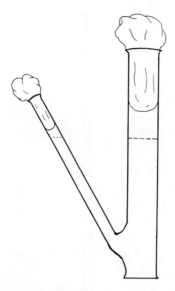

Figure 125. V migration tube for semisolid media. The Protozoa to be sterilized are injected through the small arm and deposited at the bottom of the V. They migrate up through the semisolid medium and are removed at the top of the large arm. (Redrawn from Glaser and Coria, 1930.)

terested in the problems arising from the use of bacteriologically sterile protozoa.

C. *Combinations of dilution and migration.*—Cleveland (1928) described in detail the various manipulations which he employed to effect the sterilization of *Tritrichomonas fecalis,* a flagellate parasitic in the human intestine. The method which yielded consistently satisfactory results was a combination of washing and migration. The flagellates to be sterilized were concentrated by centrifugation, and the supernatant fluid drawn off with sterile pipettes. The packed flagellates were than layered onto the surface of sterile fluid (serum-saline) in centrifuge tubes, and the centrifuging process repeated. This procedure was continued through twenty sets of tubes, at which time Cleveland states that the ratio of *Tritrichomonas* to bacteria was about fifty to one. This constituted the dilution part of the technique, and by this part of the procedure alone he was able to obtain many sterile flagellates. Higher percentages of sterile flagellates resulted, however, when he added a final migration to the above washing. After washing, a drop of the packed flagellates was placed in the center of a large Petri dish filled with sterile fluid. The Protozoa migrated in all directions, and after various intervals of time loops of medium, taken two to three inches from the center, were found to contain ten to fifteen trichomonads. These loops were inoculated directly into tubes of media and the majority proved to be sterile.

Hetherington (1934) described a much simpler technique, wherein he alternated the dilution method with migration. He employed Columbia culture dishes in Petri dishes, micropipettes for the transfer of the organisms, and 10 ml. serological pipettes for filling the dishes. The procedure was as follows: A drop of concentrated protozoan suspension was placed in the left margin of the first dish, care being taken not to disturb the one ml. of sterile fluid. By observing the activity of the Protozoa, it was seen that numbers of them migrated to the right edge of the fluid. Fifteen to twenty-five Protozoa were picked up in sterile micropipettes and transferred to the left side of a second dish. These were allowed to migrate. Those Protozoa which migrated were then placed in a third dish and allowed to remain there for three hours. They were transferred to the left side of the fourth dish and allowed to migrate, then transferred to the fifth dish for a second three-hour period. The sixth and seventh dishes were again used for migration, the Protozoa from the seventh being placed in culture medium.

By exercising care in the handling of the fluids, the pipettes, and the covers of the Petri dishes, this method gave excellent results. Spores were defecated during the two three-hour periods. Especially adherent bacteria were lost during the migrations and washing, according to Hetherington, owing to the fact that the medium was nutritive (Bacto yeast extract in Peter's medium), resulting in the heightened activity of the bacteria.

Recently Claff (1940) has described an apparatus designed to sterilize negatively geotropic Protozoa, which employs both the principles of dilution and migration and at the same time reduces the chances of air contamination. This apparatus consists of six flasks in series (Fig. 126). The Protozoa are injected into the bottom of flask 1 through a rubber vaccine cap with a hypodermic needle. From this point on, until they are finally recovered from the sixth flask, they are in an entirely closed system. After injection, the Protozoa are allowed to migrate to the narrow top of the flask and, when large numbers have collected, they are forced over into the bottom of flask 2 by a volume (1-2 ml.) of sterile medium from the liter reservoir. After each migration the process is repeated, and the fluid drained from the system at the top of flask 6 is kept in

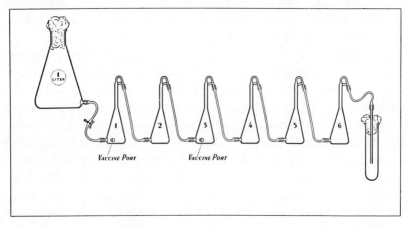

Figure 126. Migration-dilution apparatus drawn to show construction. All flasks are filled with fluid and sterilized. After cooling, the Protozoa to be sterilized are injected through the vaccine port of flask 1. Successive migrations result, and the Protozoa are finally collected in the test tube from flask 6. (From Claff, 1940.)

sterile test tubes and used as bacteriological controls on the fluid going before the Protozoa. The first Protozoa collected in the sixth tube are then separated into the various culture media.

The chief advantages of this apparatus are the simplicity of operation and the reduction of chance contamination. The whole apparatus is assembled in a compact unit and can be sterilized partially full of medium. Completion of the filling of the flasks is carried out after cooling, extreme care being taken to expel all air bubbles. No air bubbles may be allowed to enter during the injection of the contaminated Protozoa, as these will practically always rise ahead of the migration and contaminate every flask in order.

Claff gives experimental evidence for the sterilization of *Paramecium, Tillina, Tetrahymena* and *Glaucoma* in this apparatus. It has been used on numerous occasions in this laboratory and found to be very satisfactory. Perhaps its chief drawbacks are that it is limited to negatively geo-tropic organisms and to those types which are relatively powerful swimmers, and that the flasks must be specially built, as well as the carriage, and that the whole is rather expensive. However, in those laboratories in which an extensive program of investigation requiring sterile Protozoa is being carried on, the purchase or the building of this apparatus will be found advantageous.

D. *Bactericidal agents.*—Many methods have been tried in which agents were employed to kill bacteria without killing the Protozoa under investigation. Inhibition of bacterial overgrowth has been reported on numerous occasions by the use of various chemicals (Zumstein, 1900; Kofoid and Johnstone, 1929; and others) but sterility has rarely been obtained. Cleveland (1928) states that he used numerous chemicals in the hope that some would prove less injurious to *Tritrichomonas* than to the associated bacteria. His results were entirely negative, and he states that this type of investigation "appears to be almost a hopeless undertaking" (p. 256).

We do have, however, a few reports which indicate the possibility of obtaining sterile Protozoa after treatments with various chemicals which are more toxic to the bacteria than to the Protozoa. In all cases the bactericidal agents were used on protozoan cysts, not on the trophic forms. Frosch (1897) was the first to report the sterilization of cysts by chemical means. By the immersion of old cysts of *Amoeba nitrophila* in saturated sodium carbonate for a period of three days, Frosch claims to have killed all the associated non-spore-forming bacteria and to have recovered some of the amoebae. This method was repeated by Walker (1908) and the results confirmed. Oehler (1924) examined the possibility of treating ciliate and amoebae cysts with a variety of disinfectants, acids, alkalies, and salts but these results were far from encouraging. Severtzoff (1924) investigated the action of toluene, chlorine, and calcium sulphide on the cysts of "soil amoebae" and found that the cysts were able to withstand the deleterious effects of these chemicals better than the associated bacteria (non-spore-forming types). He claims to have effected complete sterility by the use of calcium sulphide, but was unable to establish pure cultures from the resulting cysts. Glaser and Coria (1930) were unable to obtain sterile *Euglena proxima* by their washing methods, but succeeded by treating "round or encysted stages" (p. 803) with a solution of potassium dichromate (1.25 to 2.5 percent) for from fifteen to thirty minutes. Perhaps the most exhaustive study on the relative effects of chemicals on bacteria and protozoan cysts was carried out by Luck and Sheets (1931). They investigated the lethal concentrations of some eighteen substances on two-day-old cysts of *Euplotes taylori* and their associated bacteria. They appear to have obtained sterile ciliates in some cases, when silver nitrate was used in high concentrations of glucose and sucrose. All other substances either were more toxic to the protozoan cysts than to the bacteria or were uniformly lethal to both.

This line of approach to protozoan sterility does not seem to be too encouraging. As might well be supposed, the investigator is limited in any case to cyst-forming types. Even these are not uniformly resistant to chemical action, so that it appears that little can be learned at this time from the experiences of others. It seems that the methods referred to above, while they may have produced the desired results in some cases, are not to be recommended for routine work. There is always the probability that any type of protozoan collected from the wild will have associated with it spore-forming bacteria, which are always highly resistant to disinfectants. If a cyst-forming type of protozoan is first sterilized by washing or some other method and established in culture with a single known bacterium, then some of these disinfectants might be used to advantage later for obtaining large numbers of sterile organisms for experimentation.

Of the physical bactericidal agents which have been employed, heat, used in different ways but always upon cysts of various species, has been most generally used. Walker (1908) reports that he was able to obtain sterile *"Amoeba intestinalis"* by inoculating an agar plate, first with concentric rings of *B. coli* and then, in the center, with amoebae cysts. The plate was heated to 70°-75° C. for one hour, which was sufficient to kill the *Bacillus coli* but did not kill the cysts. He then added fresh bacteria to the center of the plate and the amoebae excysted, fed, and migrated through the rings of dead *B. coli*. Walker states that in their migration through the dead bacteria they freed themselves of all living bacteria (by scraping them off?). He was able to recover sterile amoebae at the periphery of the plate. This report is surprising, in that moist heat was used, and cysts of Protozoa in general, under these conditions, are usually killed.

Oehler (1924) reports that he was able to obtain sterile *Colpoda* from cysts which had been heated either at 37° C. for six weeks or at 60°-64° C. for several hours. While it is true that certain types of bacteria are killed by desiccation for considerable lengths of time, it would be nothing short of a miracle if only these delicate forms happened to make up the associated flora from the wild. In this laboratory we have employed a modification of Oehler's method for obtaining large numbers of *C. steinii,* but a short description of the results will indicate that the method is not to be relied upon for initial sterilization.

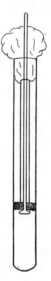

Figure 127. Details of construction of the sponge-and-glass plunger for the collection of cysts. The center rod is solid, and a piece of tubing holds the sponge down.

We maintain cultures of *C. steinii,* sterilized by centrifugation, as described in a preceding section, in suspensions of the common nonspore-forming coliform bacterium, *Aërobacter cloacae.* From time to time we have needed large numbers of sterile ciliates for experimental work, and these were obtained from cysts collected as follows. A ring of sponge is placed at the bottom of a glass plunger which will reach well into a culture tube (Fig. 127). These sponge and glass plungers, with the top of the rods wrapped in cotton, are placed in test tubes of water and autoclaved. When cool, the plunger is transferred aseptically to a culture of *Colpoda* in *Aërobacter.* After the food organism is largely depleted, the ciliates encyst on and within the sponge. The plunger is then withdrawn and placed in a dry sterile test tube and set aside to dry. At first it was hoped that prolonged desiccation, which we knew would not harm the cysts, would cause the death of the bacteria. Tests were made after one, two, three, and six months (duplicate preparations being placed by desiccators) by placing the plungers into tubes of sterile excysting fluid (yeast extract). While the length of time required to show turbidity increased with the time of drying, viable *Aërobacter* were present in every case. Heating the dry sponge to 80° C. for three hours did not kill the cysts, nor did it kill the *Aërobacter*. The heat resistance is very different in the dry and the wet states, as evidenced by the fact that one hour of heating at 50° C. is enough to kill a suspension of this bacterium.

It was found, however, that a simple manipulation could be employed with eight-months'-old sponge-glass plunger preparation to obtain large numbers of sterile ciliates. U-shaped tubes were partly filled with yeast extract and sterilized, and the sponge-glass plunger was inserted aseptically into one arm. The tube was placed in an upright position and the minimum time was allowed for the ciliates to excyst. Then, without disturbing the fluid, the plug from the other arm was lifted and some of the fluid containing the freshly excysted ciliates was withdrawn. These ciliates were sterile, provided not too long a time had elapsed since the sponge was introduced and provided the fluid was not unduly disturbed. What appears to happen is that the dry bacteria take a considerable time to become active and to disperse through the medium, while a few seconds after the ciliates leave their cyst walls they migrate to the top of the fluid. It should be remembered that this method has been used only when the associated bacteria have been reduced to a single species. By careful manipulation and by having the conditions just right, it might be possible to use this method on wild, cyst-forming material, but this is only a conjecture as it has not been done to date.

The possibility of using radiation for sterilization was investigated by Brown, *et al.* (1933). They found that *E. taylori,* trophic or cysts, withstood a much longer exposure to X-rays (2,110 Roentgen units per second) than did the associated bacteria (in this case *Pseudomonas fluorescens* and *B. coli*). They were able to obtain many sterile ciliates by this method. Here again, the success of the method appears to be due to the specific type of bacteria present, and it is extremely doubtful if, under the conditions of wild cultures, sterile Protozoa could be obtained with any regularity by this method.

THE IMPORTANCE OF ADEQUATE STERILITY TESTS

It is, of course, obvious that any method for ridding Protozoa of bacteria must be carried out under the rigid rules of bacteriological technique. Bacteria are so varied in form and activity that special pains must be taken to check the results of any method before the treated Protozoa may be pronounced sterile. Microscopic examination is of little or no use, at least until time has been given for any accompanying bacteria to multiply. We must therefore give any possible contaminant every conceivable chance to multiply, thereby revealing its presence.

Sterility tests are usually made in two ways, by inoculation into fluid media and by spreading on nutrient solid media. In the fluid media (broth) contamination shows itself when the broth becomes turbid. The turbidity test is sufficient, when the contaminant is such a one as will distribute itself through the media. In other words, turbidity denotes contamination. But lack of turbidity does not always denote sterility. Some bacteria grow very slowly in broth and form small clumps which sink to the bottom of the test tube, leaving the broth clear. This experience was reported by Hetherington (1933, 1934) and by Stuart, Kidder, and Griffin (1939). A macroscopic examination of a tube was not enough to reveal the presence of these organisms.

Plating, either by pipetting fluid from the culture to be tested or by streaking the surface of solid media (nutrient agar) with a needle dipped into the culture, is usually more satisfactory than the turbidity test. The plates are allowed to incubate, and the surface is examined for bacterial colonies. The usual contaminants from wild infusions will appear in from twenty-four to forty-eight hours. But this method has its limitations. Some bacteria grow very slowly at room temperatures, but well at higher temperatures. Others are the reverse. Duplicate sets of plates should always be made, one set to be incubated at room temperature and the other at temperatures from 30° to 37° C. The time factor should be carefully considered. The slow-growing types (such as the *Mycobacterium* reported by Stuart, Kidder, and Griffin, 1939) may not appear until many days after the inoculation. It is necessary in all tests with agar plates to keep the plates for at least ten days and it is safer to keep them two weeks. The plating method is usually a better criterion of conditions within a culture than the turbidity test, for another reason. When dealing with ciliates, normally bacteria-feeders, it is often the case that the bacteria are eaten out of the media almost as fast as they multiply. This is more likely to happen when the ciliate is a voracious feeder and multiplies rapidly and when the contaminant is one of the slow-growing variety. It should not be supposed, however, that all of the bacteria will be eaten, although two such cases are on record (Elliott, 1933; Johnson, 1935). Some of the bacteria will almost invariably escape and be carried along from transplant to transplant. On the solid media, however, the ciliates do not move about, and colonies of bacteria develop unhampered.

Although they are not prevalent in wild infusions, tests should always be conducted for anaërobic bacteria. The simplest test and one which will usually determine their presence or absence is the following: Tubes containing not over 3 ml. of nutrient broth plus a two to two-and-a-half-inch layer of paraffin oil, are plugged with cotton and autoclaved for twenty minutes at fifteen pounds' pressure. Rubber stoppers are sterilized at the same time. Immediately after sterilization the rubber stoppers are fitted into the tubes, which are then allowed to cool. When the broth is cool, inoculations are accomplished by injecting the material to be tested through the paraffin oil into the broth, the rubber stopper being immediately replaced. This type of culture will allow even obligatory anaërobes to multiply, although not necessarily to the height of their capacity. Enough growth is obtained, however, to determine the presence of anaërobic contaminants. The obligatory anaërobic bacteria, it must be admitted, do not form an important group for our consideration, as they occur so infrequently. The facultative anaërobes may be detected by more common procedures.

ESTABLISHMENT OF STERILIZED PROTOZOA IN CULTURE

The sterilization of Protozoa is, after all, only a means to an end. It is of very little value to the investigator if, after going to the trouble to rid a species of Protozoa of their associated bacteria, the Protozoa fail to live. For the most perfect control of a protozoan culture for experimental work, pure cultures are necessary. This means that the protozoan under investigation must be established in a medium containing no other living organism. This is easily accomplished in a number of cases, even to the establishment in media containing only dissolved proteins. A large number of flagellates probably exist in nature by the utilization of dissolved substances and, when sterilized, continue to employ this type of nutrition. Among the free-living ciliates, all species are known to possess oral openings into which solid foods are drawn. Some of these, however, are able to live on dissolved proteins in pure culture, e.g., *Tetrahymena*. Other types may be able to obtain only a small amount of nutriment from the dissolved proteins, but are able to feed on nonliving particulate matter. This was found to be the case in this laboratory with *Glaucoma scintillans* (unpublished work). When sterilized ciliates were placed in a wide variety of media containing dissolved proteins, very little multiplication took place. However, good growth resulted in unfiltered Yeast Harris, containing quantities of broken-down yeast cells. Still other types of free-living ciliates appear to be unable to exist without living organisms as food. This may be the case with the true carnivores, and here there is an excellent opportunity to make some interesting studies on the "Zweigliedrige Kulture" without employing bacteria. It is only necessary to be able to grow the food Protozoa in pure culture and to supply it to the sterile carnivores. Some work along these lines has already been started in our laboratory, using pure cultures of *Tetrahymena* as the food organism and studying the effects of such a diet on *G. vorax* (Kidder, Lilly, and Claff, 1940). Similar cultures of *Stylonychia pustulata* are being studied by D. M. Lilly (Lilly, 1940), and the nutritional requirements of a *Euglena*-feeding *Perispira* is being investigated by the author and V. C. Dewey in this laboratory. Oehler (1919) stated that *Colpoda steinii* was unable to live on dissolved nutrients but would live on particulate matter, including dead bacteria. Kidder and Stuart (1939) were unable to confirm Oehler's results, but found that *C. steinii* was dependent upon living organisms. They remark, however, that the possibility does exist that some combination of food substances and conditions, as yet not known, may possibly allow this important ciliate to reproduce in the absence of living organisms.

It may be inferred from what has already been said that the establishment of a sterile protozoan in culture is not a routine matter. Of course the goal is a pure culture, for more precise control is then possible. With some types it has been found that growth follows when they are placed immediately in a medium containing dissolved proteins (tryptone, proteose-peptone, yeast extract, and so forth). These types are the true saprozoic forms. Others must have particulate matter, so it is best, especially when dealing with a new type, to inoculate into a wide variety of media. In this laboratory it is the practice to start our newly sterilized Protozoa in five different types of media, usually ten isolations into each. Our standard five types for first tests are 0.1-percent proteose-peptone; 5-percent yeast autolysate; 0.5-percent yeast extract; 0.5-percent malted milk; and 0.5-percent unfiltered Yeast Harris. It is sometimes necessary to have quite a range of pH values within the different media, in order to obtain growth in even one or two of the tubes.

A number of species of Protozoa appear to be dependent upon living organisms as a source of food. This is true not only of the carnivores, but seems to hold for a number of bacteria-feeders as well. With the carnivores it is usually sufficient to observe their diet in nature to decide upon a suitable food animal. If the food animal can be grown in pure culture, then the chances are good that it will be possible to establish the carnivore in "Zwiegliedrige Kultur." Some carnivores have been found to be very selective, while others are able to feed on any one of a number of organisms. Occasionally a natural bacteria-feeder will turn carnivorous and then can be established without bacteria.

With the obligatory bacteria-feeders the best that can be done, as far as we now know, is to establish them on a single species of favorable food bacteria. Here again it is absolutely necessary to start with sterile Protozoa, as even in the so-called non-nutritive fluids (salt solutions, distilled water, and so forth) many extraneous bacteria, which are not favorable as food, will multiply and be continually present from transplant to transplant. Results may be entirely misleading under these conditions, as a number of common bacteria prove to be deleterious to many Protozoa (see Kidder and Stuart, 1939). One method of setting up cultures containing a single protozoan species in a suspension of a single species of bacteria is simply to try out a number of known bacteria until one is found which will support growth. However, some ciliates prove to be extremely selective, even as to specific bacteria. If poor growth or no growth results after all the known bacteria have been used, then the investigator must try to isolate from the wild culture the type of bacteria upon which the ciliate was originally feeding. This procedure is tedious, but sometimes necessary. A wild culture is selected in which the Protozoa under investigation are multiplying rapidly. From the fluid of this culture, surface-streak plates are prepared and pure cultures of all the different types of bacteria are obtained. From these pure cultures, suspensions are systematically prepared and inoculated with the sterile Protozoa. This was the method used by Johnson (1933) in his work on *Oxytricha.* Johnson states that his selection was made on the basis of prevalence, in a thriving wild culture. In other words, the type of bacteria found in the greatest abundance he supposed to be the type upon which the Protozoa were most likely to be feeding. We have found that this is not always the case. In our work on the ciliate *Tillina* (*T. canalifera,* obtained from Dr. J. P. Turner and described by him in 1937) we were able to obtain growth on one species only, out of twenty-six types isolated from a thriving culture. This one species (a *Zopfius*) was the least prevalent of all on our plates. The reason for this appears to be that *Tillina* being so very selective, the *Zopfius* were eaten out of the culture by the time we took our samples, while the other twenty-five species were left to multiply. We have found this situation to hold in a number of cases, so that we are of the opinion that the results obtained by Johnson were due to the fact that he was dealing with a ciliate which was not rigidly selective.

The work tending to show that supplementary factors (viz., thiamin and the like) are necessary for the growth of several Protozoa in pure culture has been reviewed in a subsequent chapter, but several observations regarding the same theme may be given here, as they apply to "Zwiegliedrige Kultur." These supplements may make the difference between success and failure to establish a protozoan in bacteria-free culture. Investigations are now going on in this laboratory on the supplement question, but they are as yet far from complete. Therefore little can be said as to the exact nature of the substances or factors to be described. In order to present this problem clearly to the reader, a description of a typical example will be given.

D. M. Lilly, working in this laboratory, has studied the nutritional requirements of two hypotrichous ciliates, *Stylonychia pustulata* and *Pleurotricha lanceolata.* Both of these forms are bacteria-feeders in nature, but will also become carnivorous in the presence of other small ciliates. Sterilization was carried out, with the use of the dilution method, in Syracuse watch glasses enclosed in cellophane bags, as previously described. In the case of *Stylonychia,* it was found possible to establish them on living yeast cells, suspended in distilled water, in the absence of any other food material. Sterile ciliates would not live on autoclaved yeast, however. Sterile ciliates would eat quantities of living *Tetrahymena* (taken from agar slants and suspended in distilled water), but would not divide. Sterile ciliates, placed in suspensions of autoclaved yeast, and living *Tetrahymena* grew well and established flourishing cultures. Sterile ciliates in dead yeast and dead *Tetrahymena,* failed to multiply. Additions of none of the known water-soluble vitamins changed the situation. The inference is, as Lilly points out (1940), that *Stylonychia* requires, among other things, two unknown factors—one found in yeast (even after autoclaving), but not present, at least in sufficient quantities, in *Tetrahymena;* the other what might be called a living factor, present in living yeast and *Tetrahymena.* Both of these factors are present in certain favorable species of bacteria when the bacteria are alive, but the "living factor" is destroyed with the death of the bacteria. The so-called living factor is not a surprising requirement among Protozoa, as experience has shown that many different types will not live without being supplied with some type of living organism. The yeast factor seems to belong to the water-soluble, heat-stabile group, but is not identifiable with any one of the known B complex. While this factor is present in dried and pasteurized yeast (Brewer's Yeast Harris), it is not present in sufficient quantities in Difco dehydrated yeast extract. Concentration and partial purification of the yeast factor have been carried out, but until this work is further along we must content ourselves with these few facts.

This example is one of many similar cases and serves to point out that several conditions must be recognized and fulfilled, if the investigator is to be successful in establishing sterile Protozoa in culture. The possibility of supplementary factors must be considered before it can be said of any type that it cannot be grown bacteria-free. Somewhat the same situation was encountered by Glaser and Coria (1933) in their work on *Paramecium.* They finally announced a complicated medium which proved to be successful, and this medium contained pieces of fresh rabbit kidney (possibly supplying the living factor during the early growth phases of the ciliate).

It is not the purpose of this chapter to consider in detail all of the interesting work which has been done regarding accessory growth factors and nutritional supplements. A large number of these are considered in the chapter on pure cultures (Chapter IX). It might be suggested, however, that one of the most fertile fields of protozoan investigation has been opened up with the development of bacteria-free techniques, and our knowledge of unsuspected requirements in the nutrition of carnivores should be extended greatly in the near future. The possibilities are many along these lines, and therefore considerable time has been devoted to the methods which will have to be employed in the beginning of any such studies.

LITERATURE CITED

Brown, M. G., J. M. Luck, G. Sheets, and C. V. Taylor. 1933. The action of X-rays on *Euplotes taylori* and associated bacteria. J. gen. Physiol., 16: 397-406.

Burt, R. L. 1940. Specific analysis of the genus *Colpoda* with special reference to the standardization of experimental material. Trans. Amer. Micros. Soc. (in press.)

Claff, C. L. 1940. A migration-dilution apparatus for the sterilization of Protozoa. Physiol. Zoöl. (in press.)

Cleveland, L. R. 1928. The suitability of various bacteria, molds, yeasts and spirochaetes as food for the flagellate *Tritrichomonas fecalis* of man as brought out by the measurement of its fission rate, population density, and longevity in pure cultures of these microorganisms. Amer. J. Hyg., 8: 990-1013.

Elliott, A. M. 1933. Isolation of *Colpidium striatum* Stokes in bacteria-free cultures and the relation of growth to pH of the medium. Biol. Bull., 65: 45-56.

Frosch, P. 1897. Zur Frage der Reinzüchtung der Amöben. Zbl. Bakt., Orig., 21: 926-32.

Furgason, W. H. 1940. The significant cytostomal pattern of the "*Glaucoma-Colpidium* group," and a proposed new genus and species, *Tetrahymena geleii*. Arch. Protistenk. (In press.)

Glaser, R. W., and N. A. Coria. 1930. Methods for the pure culture of certain Protozoa. J. exper. Med., 51: 787-806.

—— 1933. The culture of *Paramecium caudatum* free from living microorganisms. Jour. Parasit., 20: 33-37.

—— 1935. The culture and reactions of purified Protozoa. Amer. J. Hyg., 21: 111-20.

Hargitt, G. T., and W. W. Fray. 1917. The growth of *Paramecium* in pure cultures of bacteria. J. exper. Zool., 22: 421-54.

Hetherington, A. 1933. The culture of some holotrichous ciliates. Arch. Protistenk., 80: 255-80.

—— 1934. The rôle of bacteria in the growth of *Colpidium colpoda*. Physiol. Zoöl. 7: 618-41.

Johnson, D. F. 1935. Isolation of *Glaucoma ficaria* Kahl in bacteria-free cultures, and growth in relation to pH of the medium. Arch. Protistenk., 86: 263-77.

Johnson, W. H. 1933. Effects of population density on the rate of reproduction in *Oxytricha*. Physiol. Zoöl., 6: 22-54.

Kidder, G. W., D. M. Lilly, and C. L. Claff. 1940. Growth studies on ciliates. IV. The influence of food on the structure and growth of *Glaucoma vorax*, sp. nov. Biol. Bull., 78: 9-23.

Kidder, G. W., and C. A. Stuart. 1939. Growth studies on ciliates. I. The role of bacteria in the growth and reproduction of *Colpoda*. Physiol. Zoöl., 12: 329-40.

Kofoid, C. A., and H. G. Johnstone. 1929. The cultivation of *Endameba gingivalis* (Gros) from the human mouth. Amer. J. publ. Hlth., 19: 549-52.

Lilly, D. M. 1940. Nutritional and supplementary factors in the growth of carnivorous ciliates. (MS.)

Luck, J. M., and Grace Sheets. 1931. The sterilization of Protozoa. Arch. Protistenk., 75: 255-69.

Oehler, R. 1919. Flagellaten- und Ciliatenzucht auf reinem Boden. Arch. Protistenk., 40: 16-26.

—— 1924. Weitere Mitteilungen über gereinigte Amöben- und Ciliatenzucht. Arch. Protistenk., 49: 112-34.

Ogata, M. 1893. Über die Reinkultur gewisser Protozoen (Infusorien). Zbl. Bakt., Orig., 14: 165-69.

Parpart, A. K. 1928. The bacteriological sterilization of Protozoa. Biol. Bull., 55: 113-20.

Purdy, W. C., and C. T. Butterfield. 1918. Effect of plankton animals upon bacterial death rates. Amer. J. publ. Hlth., 8: 499-505.

Severtzoff, L. B. 1924. Method of counting, culture medium and pure cultures of soil amoebae. Zbl. Bakt., Orig., 92: 151-58.

Stone, W. S., and F. H. K. Reynolds. 1939. A practical method of obtaining bacteria-free cultures of *Trichomonas hominis*. Science, n.s., 90: 91-92.

Stuart, C. A., G. W. Kidder, and A. M. Griffin. 1939. Growth studies on ciliates. III. Experimental alteration of the method of reproduction in *Colpoda*. Physiol. Zoöl., 12: 348-62.

Turner, J. P. 1937. Studies on the ciliate *Tillina canalifera*, n. sp. Trans. Amer. micr. Soc., 56: 447-56.

Walker, E. L. 1908. The parasitic amebae in the intestinal tract of man and other animals. J. med. Res., 17: 379-459.

Zumstein, H. 1900. Zur Morphologie und Physiologie der *Euglena gracilis* Klebs. Jb. wiss. Bot., 34: 149-98.

CHAPTER IX

FOOD REQUIREMENTS AND OTHER FACTORS INFLUENCING GROWTH OF PROTOZOA IN PURE CULTURES

R. P. HALL

IT IS OBVIOUS that the growth of Protozoa is influenced by many different factors. The importance of some of these is well recognized and the relationships to growth are partially understood in a few instances, but there is little or no detailed information bearing on other factors. Here and there, investigations have suggested possible solutions to certain problems, but just as frequently have uncovered new problems which in turn must be solved in the approach to an understanding of protozoan growth. The present lack of information extends to such questions as the list of essential elements, the nature of the simplest organic foods adequate for various species, "growth factor" or vitamin requirements, and the combined effects of various environmental factors on growth. Furthermore, Protozoa in cultures constitute populations and presumably are subject to general laws of population growth. Hence the final interpretation of many experimental results demands further knowledge of the behavior of populations.

From the experimental standpoint, several types of protozoan populations may be distinguished. (1) The pure culture contains a single protozoan species with no other microörganisms. In most cases such cultures have been started from pure lines and are thus genetically homogeneous. The number of bacteria-free strains now in existence is uncertain, although an estimate of 100 may be fairly accurate. Many strains of Phytomastigophora are maintained by Pringsheim (1930), while additional species belonging to various groups of Protozoa are to be found in several other laboratories. (2) The species-pure culture contains a single protozoan species, usually in pure line, with bacteria, algae, or other microörganisms as sources of food. Populations of this type have been maintained on known species of microörganisms (e.g., Oehler,

1916, 1919; Philpott, 1928; Geise and Taylor, 1935; D. F. Johnson, 1936; W. H. Johnson, 1933, 1936; Loefer, 1936d) or on mixtures of bacteria. (3) Mixed populations, as described in the work of Gause (1935), contain two species of Protozoa feeding on other microörganisms, or perhaps one upon the other. This technique presents interesting possibilities. (4) Wild populations are mixtures of species as obtained from natural sources. Such populations have been studied particularly in relation to succession of species in cultures (e.g., Woodruff, 1912).

The present discussion deals primarily with investigations on pure cultures, which, with their obvious advantages, afford favorable material for the study of many problems. With the exclusion of other microörganisms, it is possible to control the food supply and to determine, more accurately than by other methods, the relation of environmental factors to growth. Detailed investigation of metabolic activities is possible with pure cultures, whereas allowance must be made for other microörganisms when bacteria-free material is not used. The pure-culture technique and scrupulous cleanliness of glassware are essential in studies on food requirements. This is true particularly of investigations on autotrophic nutrition, since protein contamination, to the extent of one part in millions, may influence growth. Likewise, pure cultures are a prerequisite to investigations on specific growth factors, or vitamins. Some of the methods used by various investigators have been described elsewhere (Pringsheim, 1926; Hall, 1937a). The technique is not particularly difficult and, while the preparation of glassware is somewhat laborious and constant precaution against contamination must be exercised, the results more than justify the additional time and effort.

Food Requirements of Protozoa

Food requirements of the various groups of Protozoa differ in certain general respects. The chlorophyll-bearing species may utilize carbon dioxide, while other types require a more complex carbon source. Nitrogen requirements also vary. Some forms thrive on ammonium salts or on nitrates; growth of other species is supported by nothing simpler than an amino acid; while that of a third group is dependent upon peptones or comparable protein-cleavage products. On the basis of such criteria, a number of different methods of protozoan nutrition have been recognized (Lwoff, 1938a; Pringsheim, 1937d). A somewhat simplified classification (Hall, 1939b) is presented below:

I. *Phototrophic nutrition* is characteristic of chlorophyll-bearing species, which utilize the energy of light in photosynthesis. Some appear to be obligate phototrophs, while others may be grown in darkness under suitable conditions. On the basis of nitrogen requirements, several varieties of phototrophic nutrition may be recognized:

(1) *Photoautotrophic nutrition* is characteristic of species which can grow in inorganic media; *Chlorogonium euchlorum* (Loefer, 1934; Hall and Schoenborn, 1938a) is typical. No obligate photoautotroph is known.

(2) *Photomesotrophic nutrition* is that in which one or more amino acids serve as nitrogen sources. In *Euglena deses* (Dusi, 1933b) this seems to be the simplest possible method of nutrition. Photomesotrophic nutrition may also be carried on by facultative photoautotrophs.

(3) *Photometatrophic nutrition* is characteristic of species which grow in peptone solutions or comparable protein media. *Euglena pisciformis* (Dusi, 1933b) has been described as an obligate photometatroph. This type of nutrition may also be carried on by facultative photoautotrophs and photomesotrophs.

II. *Heterotrophic nutrition* is characteristic of species which have no chlorophyll and hence require an organic carbon source. Some chlorophyll-bearing species have been grown in darkness and may, in this sense, be considered facultative heterotrophs. On the basis of nitrogen requirements, three varieties of heterotrophic nutrition may be distinguished.

(1) *Heteroautotrophic nutrition* involves utilization of inorganic nitrogen compounds in the presence of an organic carbon source. *Polytoma uvella* (Pringsheim, 1921; Lwoff and Dusi, 1938a) and *Astasia* sp. (Schoenborn, 1938) are examples.

(2) *Heteromesotrophic nutrition:* growth requirements may be satisfied by one or more amino acids as sources of nitrogen and carbon. Growth is usually much more vigorous with an additional carbon source, such as acetate. *Polytomella caeca* (Pringsheim, 1937a, 1937c) is representative.

(3) *Heterometatrophic nutrition* is characteristic of organisms which grow in peptone solutions or similar media. Obligate heterometatrophs, such as *Hyalogonium klebsii* (Pringsheim, 1937b) and *Glaucoma piriformis* (A. Lwoff, 1932), cannot be grown in amino-acid solutions or simpler media. This type of nutrition is exhibited by various holozoic Protozoa (ciliates, amoebae) which have been grown in pure culture. Among the parasitic flagellates, certain Trypanosomidae (M. Lwoff,

1930, 1933a, 1936) have been grown under comparable conditions; other parasitic flagellates (M. Lwoff, 1929a, 1929b, 1929c, 1929d, 1933a, 1933b, 1937, 1938a; Glaser and Coria, 1935b; Cailleau, 1936a, 1936b, 1937a, 1937b, 1938a, 1938b) apparently require, in addition, blood, serum, tissue extracts, or special growth factors.

PHOTOAUTOTROPHIC NUTRITION

Photoautotrophic nutrition is generally attributed to the chlorophyll-bearing plant-like flagellates and is, by definition, limited to this group of Protozoa. On the other hand, there is no evidence to support the assumption that all chlorophyll-bearing species are photoautotrophic, since several green flagellates have been grown only in amino acid or peptone media. Furthermore, in the absence of pure cultures, there is no conclusive evidence that any member of the Chrysomonadida, Heterochlorida, Cryptomonadida, Dinoflagellida, or Chloromonadida is capable of carrying on photoautotrophic nutrition. While it may be expected that such flagellates will be found in each of these orders, speculation must remain subject to experimental verification.

The known facultative photoautotrophs are: *Chlamydomonas agloëformis* (M. Lwoff and A. Lwoff, 1929), *Chlorogonium elongatum* (Loefer, 1934), *C. euchlorum* (Loefer, 1934; Hall and Schoenborn, 1938a), *Haematcoccus pluvialis* (M. Lwoff and A. Lwoff, 1929), and *Lobomonas piriformis* (Osterud, 1938, 1939), representing the Phytomonadida; and *Euglena anabaena* (Dusi, 1933b; Hall, 1938b), *E. gracilis* (Pringsheim, 1912; Dusi, 1933a; Hall and Schoenborn, 1939a), *E. klebsii* and *E. stellata* (Dusi, 1933b) and *E. viridis* (Hall, 1939a), representing the Euglenidae.

The establishment of autotrophic strains has often encountered difficulties, and conflicting results have sometimes been reported for the same species. Some of the apparent contradictions may be the result of differences in culture media and in technique. In addition, the technical difficulties may sometimes be augmented by a selective action of inorganic media, as observed in *Euglena* (Hall and Schoenborn, 1938b).

The present knowledge of food requirements in photoautotrophic nutrition is far from complete. In fact, it is not yet possible to list all the elements which are essential to growth, and little or nothing is known about quantitative food requirements. However, the following elements, which are found as general constituents of protoplasm, may be listed as probably essential to growth: C, H, O, N, P, S, Ca, Fe, K, Na, Mg, Cl. Additional possibilities include Cu, Sr, Al, Mn, Zn, Ni, B, Rb, Ba, Si, Ti, V, As, Co, and Cr, since these have all been demonstrated in plant or animal tissues and some appear to be essential to the growth of higher organisms.

By a process of successive eliminations, it should be possible to determine which elements are and which are not essential to growth. Such investigations, however, are entirely dependent upon adequately purified chemicals. In certain investigations (Hall, 1938b, 1939a; Hall and Schoenborn, 1939a; Osterud, 1938) analyzed reagents have been used in the preparation of culture media and, within such limits, the composition of each medium is known. One of these media (EF) contains the following elements: C, H, O, N, P, K, Mg, S, Ca, and Cl in appreciable amounts, and traces (1×10^{-6} to 1×10^{-11} gm. per cc.) of Cu, Ba, Fe, As, Mn, Na, Zn, and Pb. This medium has supported growth of *Euglena gracilis, E. viridis, E. anabaena,* and *Lobomonas piriformis.* Another medium (EC) has supported growth of *E. gracilis, E. viridis,* and *Chlorogonium euchlorum.* So far as the component elements are concerned, this medium differs from EF in the absence of Ba, in lower concentrations of Ca, Cl, Mg, and Mn, and in higher concentrations of P and K. Media EA and EAB, which have supported growth of *E. gracilis, E. viridis,* and *C. euchlorum,* contain a trace of Al, but no Ba; except for concentrations, the list of elements is otherwise the same as in EC and EF. Just how many of the "trace" elements are actually essential to growth has not been determined. The omission of Ba from three media and of Al from two media seems to be of little significance, and the status of these two elements as essential substances is questionable. By comparable methods of elimination, it may be possible to determine whether various other elements are actually essential in photoautotrophic nutrition.

In a few cases there is evidence that particular elements exert significant effects on growth. Calcium requirements of *Euglena stellata* (Dusi, 1933b) are much greater than those of other Euglenidae investigated, and manganese (Hall, 1937c) has been found to accelerate growth of *E. anabaena.* In addition, a few similar observations on heteroautotrophic flagellates have been reported. For instance, A. Lwoff (1930) has reported that Fe is essential to growth of *Polytoma uvella.* Similarly, Mast and Pace (1935) found that *Chilomonas paramecium* survived for only a few transfers in media without S. For example, one S-free line died on the seventh day and others on the third, while several lines in media containing S were maintained for from twenty to twenty-four days. Another example is that of *Hyalogonium klebsii,* which requires relatively large amounts of calcium (Pringsheim, 1937b).

It is possible, of course, that the action of certain elements may not be specific; in other words, comparable effects on metabolism may be exerted by several different elements, one of which may be substituted for another. This possibility should be considered in investigations on food requirements of photoautotrophs and heteroautotrophs.

Photomesotrophic Nutrition

Euglena deses (Dusi, 1933b) may be considered an obligate photomesotroph, a flagellate which has lost the primitive photoautotrophic ability characteristic of various other green flagellates. In addition to this species, several facultative photoautotrophs among the Euglenidae are known to carry on photomesotrophic nutrition: *E. anabaena* (Dusi, 1933b; Hall, 1938b), *E. gracilis* (Dusi, 1933a), *E. klebsii,* and *E. stellata* (Dusi, 1933b). An interesting feature of these Euglenidae is that a particular amino acid may support growth of one species but not another (Dusi, 1931). For example, phenylalanine was satisfactory for *E. anabaena, E. gracilis,* and *E. stellata,* but not for *E. deses and E. klebsii,* while serine was adequate for growth of all except *E. anabaena.* Comparable differences were noted for several other amino acids.

Among the Phytomonadia, photomesotrophic nutrition has been demonstrated in *Chlamydomonas agloëformis* and *Haematococcus pluvialis* (A. Lwoff, 1932), and also in *Lobomonas piriformis* (Osterud, 1939). In addition, Loefer (1935b) observed, in *Chlorogonium elongatum* and *E. euchlorum,* acceleration of growth by glycocoll and several other amino acids, added separately and in mixtures, to an inorganic medium and to a salt solution containing sodium acetate.

The growth of photomesotrophic species may be accelerated by the addition of various carbon sources (e.g., sodium acetate) to an amino-acid medium. Concerning mineral requirements in photomestrophic nutrition, nothing is known beyond the fact that amino acids have often been added to salt solutions comparable, except for the omission of inorganic nitrogen, to the media used for photoautotrophic nutrition.

PHOTOMETATROPHIC NUTRITION

Photometatrophic nutrition can be carried on by all of the chlorophyll-bearing flagellates which have been established in pure culture. Certain species, such as *E. pisciformis* (Dusi, 1933b), may prove to be obligate photometatrophs, although recent observations (Dusi, 1939) indicate that *E. pisciformis* should not be so classified. Peptones of one type or another have usually furnished the food supply, and in at least a few cases a solution of peptone in distilled water has supported growth. In addition to peptones, gelatin (Hall, 1938b) may also support growth, and certain species are known to produce proteolytic enzymes (Mainx, 1928; Jahn, 1931; Hall, 1937b). Many of the flagellates grow well on agar slants, provided the agar is enriched with a suitable peptone and sometimes with an additional source of carbon; such cultures are convenient for the maintenance of laboratory stocks.

Although various peptone media are satisfactory for all the species which have been studied, growth may be accelerated by the addition of salts of certain fatty acids, various carbohydrates, and several alcohols. Acceleration of growth by carbohydrates has been noted in *E. gracilis* (Jahn, 1935b) and in two species of *Chlorogonium* (Loefer, 1935a). Fermentation of dextrose by *E. proxima* was reported by Glaser and Coria (1930, 1935a), but other workers have failed to note such changes in cultures of Euglenidae. Furthermore, Loefer (1938b), using Benedict's colorimetric method, failed to detect utilization of dextrose by *C. elongatum* and *C. euchlorum,* in spite of the accelerating effect on growth. Acceleration of growth by ethyl alcohol has been reported by Loefer and Hall (1936) in *E. deses* and *E. gracilis,* and similar effects of several alcohols on the latter species have been described by Provasoli (1938c). Acceleration of growth by fatty acids, in cultures exposed to light, has been reported for *E. gracilis* (Jahn, 1935d) and *E. stellata* (Hall, 1937d). Furthermore, macroscopic observations on cultures in various stock-culture media have indicated such effects in approximately thirty species maintained in our laboratory. However, most of the quantitative studies on carbon sources have been based upon cultures maintained in darkness (heteromesotrophic and heterometatrophic nutrition), as described below.

Hutner (1936) failed to note acceleration of growth by fatty acids or carbohydrates in *E. anabaena* or by carbohydrates in *E. gracilis*. Since Hutner's conclusions apparently were based upon the macroscopic appearance of his cultures, he may have overlooked effects comparable to those reported by other workers.

Heteroautotrophic Nutrition

The utilization of inorganic nitrogen compounds in the presence of acetate or another organic carbon source, has been attributed to several colorless Phytomastigophora: *Chilomonas paramecium* (Mast and Pace, 1933), *Polytoma uvella* (Pringsheim, 1921; Lwoff and Dusi, 1938a), *P. obtusum* (Lwoff, 1929b, 1932), and *Astasia* sp. (Schoenborn, 1938, 1940). The results of Mast and Pace have not been duplicated by Loefer (1934) nor by Hall and Loefer (1936). Pringsheim (1935a) reported growth of *C. paramecium* in an ammonium-salt and acetate medium, but Lwoff and Lederer (1935) and Pringsheim (1935b) have pointed out that Pringsheim's medium contained "extract of soil," without which the flagellates failed to grow. Hence, Pringsheim did not confirm the observations of Mast and Pace. More recently, Lwoff and Dusi (1937a, 1938a, 1939b) have grown a strain of this species in an ammonium acetate medium, but only in the presence of either thiamine or thiazole and pyrimidine. Since Lwoff and Dusi added organic nitrogen compounds to their medium, application of the term *heteroautotrophic* to *C. paramecium* may be inappropriate. The contradictory results obtained by various workers with this species have not yet been explained. It is possible that different strains may vary in their nutritional requirements. Or it is conceivable that the strain of Mast and Pace was established through a selective process, similar to that reported in several species of *Euglena* (Hall and Schoenborn, 1938b).

The first known instance of heteroautotrophic nutrition in Euglenida is that described by Schoenborn (1938, 1940) in *Astasia* sp. This strain has now passed the nineteenth transfer, so that the peptone carried over from the original stock culture has been reduced, through serial dilution alone, to a calculated concentration of less than 1.8×10^{-12} gm. per cc.

According to Pringsheim (1937b), *Chlorogonium euchlorum* may be grown in darkness as a facultative heteroautotroph, provided glucose caramel is added to the medium. Osterud (1939) has reported growth of *Lobomonas piriformis* in an ammonium-nitrate and acetate medium for three transfers (twelve weeks) in darkness. Likewise, in a medium similar to that of Osterud, growth of *Euglena gracilis* has been observed (Schoenborn, 1939) through four successive transfers, covering a period of eighteen weeks. These suggestive observations indicate that certain chlorophyll-bearing flagellates may retain the ability to grow in inorganic nitrogen media, even after suppression of photosynthesis.

Heteromesotrophic Nutrition

Heteromesotrophic nutrition has been demonstrated in several color-less Phytomastigophora. The Cryptomonadida are represented by *Chilomonas paramecium,* which has been grown in an amino-acid and acetate medium by Mast and Pace (1933) and by Hall and Loefer (1936). This type of nutrition has not yet been demonstrated in colorless Euglenida, and *E. gracilis* has been grown in darkness for only a few transfers in a medium containing asparagin and acetate (A. Lwoff and Dusi, 1929, 1931). Such results are modified by the addition of thiamine, as described below. Several heteromesotrophs have been identified among colorless Phytomonadida. Pringsheim (1921) found glycocoll an adequate nitrogen source for *Polytoma uvella,* and comparable results were later obtained for *P. obtusum* (A. Lwoff, 1929b, 1932; A. Lwoff and Dusi, 1934). In addition, *P. caudatum* var. *astigmata* (A. Lwoff and Provasoli, 1935), *Polytomella agilis* (A. Lwoff, 1935b), and *P. caeca* (Pringsheim, 1935, 1937c; A. Lwoff and Dusi, 1937a) seem to be capable of heteromesotrophic nutrition. On the other hand, the related species, *Hyalogonium klebsii* (Pringsheim, 1937a), appears to be an obligate heterometatroph. The chlorophyll-bearing phytomonad, *Chlorogonium euchlorum,* has been grown in darkness in an asparagin medium (A. Lwoff and Dusi, 1935b); likewise, *Lobomonas piriformis* is capable of growth under similar conditions in a glycocoll and acetate medium (Osterud, 1939).

Heterometatrophic Nutrition

All of the colorless Phytomastigophora which have been investigated appear to thrive in simple peptone media, although growth is always accelerated by the addition of a suitable organic carbon source. In addition, a few of the chlorophyll-bearing species have been maintained in darkness in such media (Jahn, 1935c, 1935d; A. Lwoff and Dusi, 1929, 1931, 1935a, 1935b; M. Lwoff and A. Lwoff, 1929; Loefer, 1934; Provasoli, 1938b), especially with added acetate. Certain Trypanosomidae, such as *Strigomonas oncopelti,* have been grown in peptone media (M. Lwoff, 1930, 1933a, 1936), but growth of other Trypanosomidae (M. Lwoff, 1929a, 1929b, 1929c, 1929d, 1933a, 1933b) and of Polymastigida (Glaser and Coria, 1935b; Cailleau, 1935, 1936a, 1936b, 1937a, 1937b, 1938a, 1938b, 1939) seems to be supported only by more complex organic media containing blood, serum, tissue extracts, or particular growth factors. Several of the ciliates—*Colpidium campylum* and *C. striatum* (Elliott, 1933, 1935b), *Glaucoma ficaria* (D. F. Johnson, 1935a), *G. piriformis* (Lwoff, 1924, 1925, 1929a), *Paramecium bursaria* (Loefer, 1934b, 1936b, 1936c), and certain Sarcodina—*Acanthamoeba castellanii* (Cailleau, 1933, 1934), *Mayorella palestinensis* (Reich, 1935, 1936), have been grown in peptone media comparable to those which support growth of the heterometatrophic Phytomastigophora. On the other hand, Glaser and Coria (1930, 1933, 1935a) have used somewhat more complex media for several free-living ciliates.

At present little is known of the nitrogen requirements in heterometatrophic nutrition, and definite conclusions regarding the saprophilic or saprogenic nature of particular species are not always possible. Enzymes which hydrolyze gelatin and casein are produced by *C. striatum* (Elliott, 1933), by *G. piriformis* (A. Lwoff and Roukhelman, 1929; Lawrie, 1937), and by *Saprophilus oviformis, Trichoda pura,* and *Chilodon cucullus* (Glaser and Coria, 1935a); hence these ciliates cannot be considered saprophilic organisms. A. Lwoff (1924, 1925) found that complete peptones were satisfactory for growth of *G. piriformis,* whereas silk peptone, gelatin, and fibrin, each supposedly lacking certain amino acids, were inadequate. Recently, however, several strains of *C. campylum* have been grown in the writer's laboratory for twenty-four transfers in silk peptone media and for eighteen transfers in gelatin media. In a comparison of various peptones, Elliott (1935b) noted that *C. striatum* and *C. campylum* grew most rapidly in the peptones containing high percentages of free amino N. and Van Slyke amino N. Growth of the same ciliates (Hall and Elliott, 1935) was also accelerated by certain amino acids, added singly to a medium which supports slow multiplication. As in

Colpidium, Loefer (1936c) found for *P. bursaria* that the least satisfactory of several peptones were those containing the smallest amounts of amino N. Hence preliminary partial hydrolysis of peptones appears to be advantageous, especially in the early growth of ciliate populations.

Nitrogen metabolism of *G. piriformis* has been investigated by A. Lwoff and Roukhelman (1929), who have traced the quantitative changes in total N, peptone N, amino N, ammonia N, and amide N. In Witte peptone medium, peptone N decreased steadily. Amino N increased for the first two weeks, and then gradually decreased for two or three weeks; later, a secondary increase sometimes followed the death of many ciliates. Somewhat comparable results have been reported for *Acanthamoeba castellanii* (Cailleau, 1934), although hydrolysis was always less extensive than in cultures of *G. piriformis* and much less ammonia N was produced.

Fermentation of carbohydrates and the acceleration of growth by carbohydrates and other carbon sources have been reported in many species. Colas-Belcour and A. Lwoff (1925) observed fermentation of dextrose and levulose by *Leptomonas ctenocephali, Leishmania tropica,* and *L. donovani* (var. *infantum*). More recently, M. Lwoff (1936) has reported fermentation of fourteen carbohydrates by *Strigomonas muscidarum* and of a smaller number by *S. media* and *S. parva.* These three flagellates showed specific differences in their fermentation reactions. Likewise, Cailleau (1937b) has described fermentation of several monosaccharides and disaccharides by *Eutrichomastix colubrorum,* and the fermentation of dextrin, starch, and inulin, as well as some of the simpler carbohydrates, by *Trichomonas foetus* and *T. columbae.* Utilization of dextrose by trypanosomes (for review, see von Brand, 1938) has been known for some years. Recently, utilization of dextrose by *T. foetus* — the strain of Glaser and Coria (1935b)—has been measured by Andrews and von Brand (1938), who found that rate of utilization was correlated with growth rate. Although growth of *C. paramecium* is accelerated by dextrose (Loefer, 1935a), utilization of the sugar could not be detected (Loefer, 1938b) by means of Benedict's colorimetric method. Acceleration of growth by starch has been reported for *Polytoma caudatum* (A. Lwoff and Provasoli, 1935), *P. obtusum* (A. Lwoff and Provasoli, 1937), and *Polytomella agilis* (A. Lwoff, 1935b).

Few observations have been reported for the Sarcodina. Fermentation of carbohydrates by *Acanthamoeba castellanii* was not observed by Cailleau (1933), who also obtained no evidence that sugars are actually consumed by this species. On the other hand, dextrose accelerates growth of *Mayorella palestinensis* (Reich, 1936), maximal effects being produced by concentrations of 0.5 to 1.0 percent.

In *Colpidium campylum* and *Glaucoma piriformis,* Loefer (1938b) has measured dextrose consumption over a short pH range. In general, the rate of utilization followed the growth-pH relationship. Fermentation of carbohydrates has been demonstrated previously in several species. *G. piriformis* produces acid from dextrose, levulose, galactose, and maltose (Colas-Belcour and A. Lwoff, 1925), and also from dextrin and soluble starch (D. F. Johnson, 1935b); *G. ficaria* (D. F. Johnson, 1935b) ferments the same carbohydrates, with the apparent exception of levulose. The reactions of *C. campylum* and *C. striatum* (Elliott, 1935a) are similar to those of *G. piriformis,* except for fermentation of mannose and failure to ferment galactose. Growth of *C. campylum* is accelerated by several carbohydrates, in addition to those which are fermented. According to Glaser and Coria (1935a), dextrose and maltose are fermented, and starch and cellulose are hydrolyzed, by *Saprophilus oviformis* and *Trichoda pura;* starch and cellulose are attacked also by *Chilodon cucullus, Paramecium caudatum,* and *P. multimicronucleatum.* Growth of *P. bursaria* (Loefer, 1936c) is increased by dextrose, mannose, maltose, dextrin, and melizitose, while little or no effect is produced by other carbohydrates. No marked change in pH occurred in any case.

Other carbon compounds known to accelerate growth of heterometatrophs include various alcohols and salts of certain organic acids. The effects of several fatty acids on the growth of *C. paramecium* have been compared quantitatively by Loefer (1935a): the greatest acceleration was produced by acetate, butyrate, and valerate. Recently, Provasoli (1937a, 1937b, 1938a, 1938b, 1938c) has completed more extensive investigations on nine colorless flagellates (*C. paramecium, Hyalogonium klebsii, Polytoma obtusum, P. caudatum, P. uvella, P. ocellatum, Polytomella caeca, Astasia quartana, A. chattoni*). Acetate and butyrate accelerated growth of all, while the effects of other fatty acids varied with the species. Provasoli has pointed out that the negative results previously obtained with certain salts probably resulted from their use in toxic concentrations. The effects of sodium acetate on the growth of several colorless species had been described previously by A. Lwoff (1929b,

1931, 1932, 1935a, 1935b, 1938a), who proposed a class of "Oxy-trophes" to include organisms showing marked acceleration with acetate. Acceleration of growth by several alcohols has been reported by Provasoli (1938c) in *Astasia chattoni*, *A. quartana*, *Polytomella caeca*, *Polytoma ocellatum*, and *Chilomonas paramecium*.

For chlorophyll-bearing flagellates maintained in darkness, the effects of various carbon sources on growth seem to be much the same as for the colorless species. The growth of several species of *Euglena* (A. Lwoff and Dusi, 1929, 1931; Dusi, 1933a; 1933b; Jahn, 1935c, 1935d; Hall, 1937d) is accelerated by salts of certain fatty acids, particularly acetic, just as in the case of *Astasia* (A. Lwoff and Dusi, 1936). Jahn (1935d), in comparing the effects of certain salts on growth of *E. gracilis* in darkness and in light, found that butyrate and acetate were most effective in either case, while lactate produced a much greater acceleration in darkness than in light. Succinate was toxic in light, but produced a slight acceleration in darkness. Comparable effects of fatty acids have been reported in several Phytomonadida—*Chlamydomonas agloëformis* and *Haematococcus pluviatis* (M. Lwoff and A. Lwoff, 1929), and *Chlorogonium elongatum* and *C. euchlorum* (Loefer, 1935a). Accelerating effects of carbohydrates on the growth of *Euglena anabaena* in darkness have also been noted (Hall, 1934).

Elliott (1935b) has described acceleration of growth by acetate and butyrate in the ciliates *Colpidium campylum* and *C. striatum*, the effects being limited to the pH range 6.5 to 7.5, approximately. The increases ranged from about 15 percent to 300 percent at different pH values. Accelerating effects of pimelic acid on *C. campylum* also have been reported (Hall, 1939c), but the substance was used in low concentrations and may have been important as a catalyst, rather than as a carbon source.

TROPHIC SPECIALIZATION

In this brief survey of the food requirements of Protozoa, it has been pointed out that different methods of nutrition may be exhibited by different members of the same family or even of the same genus. Such varying degrees of specialization are particularly interesting, in that they afford a basis for speculation concerning the evolution of the more animal-like Protozoa from the plant-like flagellates.

Theoretically, the evolution of animal-like flagellates from chlorophyll-bearing facultative photoautotrophs may have proceeded as follows: (1) Certain flagellates lost the ability to use inorganic compounds as the sole source of nitrogen, except in the presence of a suitable organic carbon compound. This type of specialization may or may not have involved the loss of chlorophyll in the beginning. (2) The ability to grow in inorganic-nitrogen media was lost completely, so that a single amino acid represented the simplest adequate nitrogen source. (3) The ability to grow in an amino-acid medium was lost, as the degree of specialization approached that of the animal-like flagellates.

On the other hand, the existence of heteroautotrophic flagellates, which can utilize inorganic nitrogen sources without carrying on photosynthesis, suggests the possibility that primitive colorless flagellates may have appeared before the origin of chlorophyll. Chlorophyll, with the attendant power of photosynthesis, would thus have been acquired during the evolution of plant-like flagellates. This hypothesis would gain additional support from the demonstration of chemoautotrophic nutrition in flagellates, and the suggestive report of such a phenomenon in *Chilomonas paramecium* (Mast and Pace, 1933) is particularly interesting. Further evidence may eventually necessitate revision of the current view that the chlorophyll-bearing flagellates are the most primitive of all the Protozoa.

Even in adhering to the concept of a primitive chlorophyll-bearing stock, it must be admitted that in the known cases of heteroautotrophic nutrition, the presumed "loss" of chlorophyll has introduced only one new food requirement, a simple organic carbon source (e.g., acetate). A second stage of specialization is represented by such types as the chlorophyll-bearing *Euglena deses* and the colorless *Polytomella caeca*, each of which requires a simple organic nitrogen source. The third step in specialization also appears in the Phytomastigophora, and is illustrated by the chlorophyll-bearing *Euglena pisciformis*, described by Dusi (1933a) as an obligate photometatroph, and the colorless *Hyalogonium klebsii*, reported to be an obligate heterometatroph (Pringsheim, 1937a). If the types of nutrition described for these various species are taken for granted, it must be admitted that the presence of chlorophyll is no handicap to progressive physiological specialization. Furthermore, the mere absence of chlorophyll has not necessitated specialization beyond the first degree, although it may be accompanied by the assumption of heteromesotrophic or heterometatrophic nutrition. Hence the case for primitive chlorophyll-bearing forms as the ancestors of all the Protozoa may not be so strong as is generally assumed.

The plant-like flagellates as a group, however, furnish a logical starting point for the evolution of other groups of Protozoa. Primitive methods of nutrition are not the only methods to be observed in Phytomastigophora, and it is obvious that some of these flagellates approach in their growth requirements the Zoömastigophora, Sarcodina, and Ciliata, representatives of which have been grown in peptone media comparable to those required by *E. pisciformis* and *H. klebsii*. Accordingly, it seems that, so far as physiological modifications are concerned, the evolution of animal-like flagellates and other groups of Protozoa from an ancestral stock of plant-like flagellates could have presented few problems.

Specific Growth Factors, or Vitamins

A concise definition of the term, *growth factor,* is not yet available. The term is now usually restricted to an essential substance which the organism in question cannot synthesize, or perhaps cannot synthesize rapidly enough to meet the normal requirements for growth. Such a growth factor may exert its characteristic effects, even when present in low concentration. By general agreement, the concept excludes the essential food substances and elements necessary for synthesis of protoplasm. While a growth factor may in itself accelerate growth, it is to be distinguished from nonessential *growth stimulants,* which also produce noticeable effects when present in low concentrations.

A survey of the rapidly growing literature reveals that growth-factor requirements may differ among the species of a single protozoan genus, and that some species can be grown in media apparently containing no growth factors, while related types are much more exacting. At present, there is no sound basis for generalization. Lack of information concerning food requirements makes it impossible in some cases to decide whether or not a specific growth factor is necessary, and occasionally a postulated need for growth factors has disappeared after further investigation. Thus Dusi (1936) suggested that growth of *E. viridis* in inorganic media might be impossible without a growth factor, but the species has since been grown as a photoautotroph (Hall, 1939a). Similarly, Hutner (1936) concluded that a vitamin-like substance is a necessary constituent of media for the growth of *E. anabaena* and *E. gracilis* in light. The former has been grown in inorganic media by Dusi and by Hall; the latter, by Dusi and by Hall and Schoenborn. It is obvious, therefore, that the inorganic food requirements of a given species should be satisfied, as a prerequisite to the evaluation of specific growth factors.

Aneurin, or thiamine (vitamin B₁).—It must be admitted that the facultative photoautotrophs and heteroautotrophs are capable of synthesizing aneurin, if this substance is actually essential to the growth of such organisms. Various other Protozoa, however, apparently show a definite need for aneurin, or for one or both of its constituents.

Among the Cryptomonadida, the thiamine requirements of *C. paramecium* have been investigated by A. Lwoff and Dusi (1937b, 1938a). In their first publication these workers stated that growth of the flagellate in asparagin medium is supported by thiamine, or by thiazole alone. In their later article, they have concluded that for growth in an ammonium acetate medium, thiamine can be replaced by thiazole and pyrimidine, but not by either one separately. Without the growth factor, growth in the control medium was always negative. Recently, A. Lwoff and Dusi (1938b, 1938c) have shown that these substances are not specific; so far as *C. paramecium* is concerned, several thiazoles and pyrimidines are satisfactory for growth.

In the Euglenida, the existence of photoautotrophic species (*E. gracilis,* and others) and the occurrence of heteroautotrophic nutrition (*Astasia* sp., Schoenborn, 1938, 1940) seem to belie a need for thiamine in certain species. Furthermore, Elliott (1937a) observed no accelerating effect of this substance on the growth of *E. gracilis* in light. However, it has been assumed that such flagellates are capable of synthesizing thiamine in light, and this hypothesis receives indirect support from reports that the growth of *E. gracilis* in darkness is possible in an asparagin and acetate medium only when thiamine (Lwoff and Dusi, 1937c) or pyrimidine (Lwoff and Dusi, 1938a) is present. On the other hand, Dusi (1939) has concluded that *E. pisciformis* requires such a growth factor even in light, since an asparagin medium containing thiamine (or both thiazole and pyrimidine) supported growth, while the same medium without a growth factor was unsatisfactory.

In this connection, it has been noted (Hall, 1938b) that growth of *E. anabaena* in light is little, if any, better in an asparagin medium than in an ammonium-nitrate medium, the two media differing only with respect to the nitrogen source. Hence asparagin may actually be a poor nitrogen source for Euglenidae, even in light, and any growth stimulant might produce an effect comparable to that noted by Dusi in *E. pisciformis*. It should be noted, also, that the serial-transfer technique, in which Dusi apparently used two-drop inocula, might require an increase of as much as a hundred times in each transfer, if the original density of population is to be maintained. In the writer's experience with E. *anabaena,* the increase in asparagin medium was never greater than twenty-five times in any tranfer, and was often less. Hence the possibility exists that Dusi's rate of dilution in serial transfers was much more rapid than the growth of his flagellates in media without growth factors, and that the use of larger inocula might reveal *E. pisciformis* to be capable of slow growth in asparagin media.

If growth factors are actually essential, all the chlorophyll-bearing Phytomonadida which have been investigated appear to synthesize such substances from the constituents of suitable inorganic media. The colorless species, *P. uvella* and *P. obtusum* (Lwoff and Dusi, 1938a), show the same synthetic ability in salt solutions to which acetate has been added. *Polytoma ocellatum* and *P. caudatum* (Lwoff and Dusi, 1937b, 1937c) apparently require thiazole for growth in such media, while *Polytomella caeca* (A. Lwoff and Dusi, 1937a, 1938a, 1938b, 1938c) requires both thiazole and pyrimidine. A. Lwoff and Dusi (1937a) have shown that *P. caeca* grows fairly well in an asparagin medium, and much more rapidly after the addition of thiamine or of thiazole and pyrimidine. They have assumed, accordingly, that the growth in asparagin alone was dependent upon a trace of thiamine in the asparagin itself. The same interpretation is also applied to several other flagellates, on the basis of similar evidence. Just as in the case of *C. paramecium,* several different pyrimidines and thiazoles accelerate the growth of *P. caeca* (Lwoff and Dusi, 1938b, 1938c), and several thiazoles are also effective with *P. ocellatum.*

Among the Zoömastigophora, *Strigomonas oncopelti* (M. Lwoff, 1937), *S. culicidarum,* and *S. fasciculata* (M. Lwoff, 1938b) appear to require thiamine, which cannot be replaced by thiazole and pyrimidine. The last two species require hematin in addition to thiamine.

Of the Sarcodina, *Acanthamoeba castellanii* (A. Lwoff, 1938b) apparently requires thiamine, or both pyrimidine and thiazole. Accordingly, Lwoff has concluded that this species is capable of synthesizing this growth factor from the two components, although such a synthesis has not been demonstrated. With respect to growth-factor requirements, *A. castellanii* thus seems to resemble *Polytomella caeca* and *Chilomonas paramecium*.

So far, only a few investigations have been completed on the ciliates. Hall and Elliott (1935) noted that the addition of yeast extract in low concentration to a gelatin medium would support growth of *Colpidium campylum* and *C. striatum,* whereas gelatin medium alone was unsatisfactory. A. Lwoff and M. Lwoff (1937) have since found that *Glaucoma piriformis* will grow in a silk-peptone-dextrose medium containing thiamine, while the control cultures failed in the second or third transfer. Likewise, Elliott (1937a, 1939) noted a marked acceleration of growth in *C. striatum* when thiamine was added to a standard peptone solution and to a peptone medium autoclaved at pH 9.6. In the latter case, the controls showed very little growth in the first transfer. Observations of the Lwoffs (1937, 1938) indicate that *G. piriformis* requires the entire thiamine molecule, and is presumably unable to synthesize the substance from the thiazole and pyrimidine constituents. Various other related compounds cannot be substituted for thiamine.

Other Growth Factors.—Vitamin B_2 (riboflavin) apparently will not replace thiamine in meeting the growth requirements of *Colpidium striatum,* although a moderate acceleration of growth by this factor has been noted (Elliott, 1939). A vitamin B_6 concentrate has produced even less noticeable effects on the growth of the same species (Elliott, 1939).

Vitamin C (ascorbic acid) requirements have been investigated in several species. M. Lwoff (1938a, 1939) has reported that ascorbic acid is one of the factors essential to growth of *Trypanosoma cruzi, Leishmania tropica,* and *L. donovani* in cultures; and Cailleau (1938a, 1938b, 1939) has reached the same conclusion for *Trichomonas foetus, Eutrichomastix colubrorum,* and *T. columbae.*

Nicotinic acid and nicotinamide both seem to serve as growth factors for certain bacteria (for review, see Koser and Saunders, 1938). None of the investigations on Protozoa has yet been completed.

Hematin has been found essential for the growth of certain Trypanosomidae—for example, *Trypanosoma cruzi, Leishmania donovani,*

L. tropica (M. Lwoff, 1938a, 1939), *Leptomonas ctenocephali,* and *Strigomonas fasciculata* (M. Lwoff, 1933a). Protohemin and protoporphyrin have been substituted for hematin in the case of *T. cruzi.* The significance of such growth factors has been discussed by A. Lwoff (1934, 1936), who suggested that these substances may enter into the composition of respiratory catalysts (cytochrome).

Cholesterol.—The investigations of Cailleau (1936a, 1936b, 1937a, 1937b, 1938a, 1938b) indicate that cholesterol and certain other sterols serve as grown factors for the parasitic Polymastigida, *Trichomonas columbae, T. foetus,* and *Eutrichomastix colubrorum.* The physiological significance of these substances has not yet been determined for Protozoa.

Extract of soil.—An aqueous extract of soil has been used extensively by Pringsheim, who found it to accelerate the growth of a number of the plant-like flagellates and also to facilitate the growth of certain species in simple media. Accordingly, Pringsheim has considered this extract a source of unknown growth factors. The accelerative action has been verified by A. Lwoff and Lederer (1935), whose results suggest that soil extracts contain organic nitrogen in concentrations sufficient for growth of *Polytomella agilis.* Hence the status of soil extracts as a source of growth factors is yet to be evaluated.

GROWTH STIMULANTS

Growth stimulants differ from growth factors in that they are not essential to life. Like growth factors, however, they may be effective in low concentrations. So far as their relation to Protozoa is concerned, pantothenic acid and the plant "hormones" (auxins) may be placed in this category. Elliott (1935c) has shown that pantothenic acid accelerates the growth of *Colpidium campylum,* the maximal effect being noted at pH 6.0. Above 7.0 there was either no acceleration, or else a slight decrease in growth rate. Similar experiments with *Haematococcus pluvialis,* within the pH range 4.5 to 8.5, showed no acceleration. Addition of pantothenic acid to gelatin, gliadin, and zein media, which in themselves did not support the growth of *Colpidium* (Hall and Elliott, 1935), was without effect. These results indicate that pantothenic acid is not a substitute for thiamine.

The effects of several plant hormones, or auxins, on the growth of *Euglena gracilis, Khawkinea halli,* and *C. striatum* have also been investigated by Elliott (1938). The growth of *E. gracilis* was markedly accelerated at pH 5.6, while effects at lower and higher pH values were much less significant. In the colorless euglenoid, *K. halli,* and in *C. striatum* no acceleration of growth was observed at any pH. Thus the effects of the auxins may be correlated with the presence of chlorophyll, as well as the pH of the medium. Elliott (1937b) has shown further

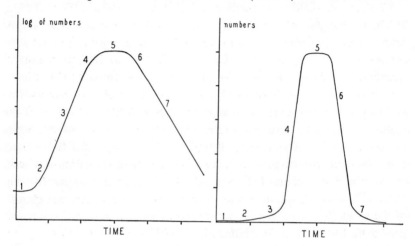

Figure 128. Growth phases in a hypothetical population. In the curve at the left, logarithms of numbers are plotted against time; on the right, numbers are plotted against time for a comparable population. Successive growth phases are numbered from 1 to 7.

that such acceleration may also be dependent upon light, since there was no effect on growth of *E. gracilis* in darkness.

An accelerating effect of pimelic acid upon the growth of *Colpidium campylum* has been noted by Hall (1939c); concentrations ranging from 10^{-10} to 10^{-4} gm. per cc. were effective in gelatin and in peptone media. These results are comparable to the findings of Mueller (see Koser and Saunders, 1938) with the diphtheria bacillus. Certain preliminary observations (Hall, 1938a) may indicate a possible growth-factor status for pimelic acid, but a definite conclusion is not yet warranted and pimelic acid may be considered, at least for the present, a growth stimulant for *C. campylum.*

Glucose caramel, as used by Pringsheim (1937b, 1937c), may also be classified as a growth stimulant. Pringsheim insists that this substance does not serve as a carbon source in his cultures of plant-like flagellates, and that it should be considered a "growth factor." On the other hand, it has not yet been demonstrated that glucose-caramel is essential to the life of Protozoa, and until such evidence is available the substance should not be classified as a growth factor.

GROWTH IN CULTURES AS A POPULATION PROBLEM

The growth of microörganisms in cultures has been described by Buchanan (Buchanan and Fulmer, 1928) in terms of seven phases: (Fig. 128): (1) Initial stationary phase, during which there is no increase in population; (2) lag phase (phase of positive growth acceleration), in which the division rate increases to a maximum; (3) logarithmic growth phase, during which the maximal rate is maintained; (4) phase of negative growth acceleration, in which the division rate decreases steadily; (5) maximum stationary phase, in which the population remains practically constant; (6) phase of accelerated death, in which the total population begins to decrease; and (7) a so-called logarithmic death phase, during which the population decreases at a more or less constant rate.

Little is known about the history of protozoan populations, and complete growth curves seem to have been traced for only two species in pure cultures—*Paramecium bursaria* (Loefer, 1936b) and *Polytoma* (Provasoli, 1938c). Loefer's growth curves, comparable to the numbers curve in Figure 128, show in general the phases recognized by Buchanan. Since counts were made at intervals of twenty-four hours or more, an initial stationary phase was not detected in several of the cultures. Total population histories covered from twenty to forty days in different media. Provasoli's curve for *Polytoma* also shows the general growth phases. Phelps (1935, 1936) traced *Glaucoma piriformis* well into the maximal stationary phase and observed in most cases the first five of the conventional growth phases. On the basis of such evidence, it may be assumed that the growth of Protozoa in pure cultures follows the general trends observed in populations of bacteria and yeasts.

More information concerning growth of protozoan populations is needed, since interpretation of experimental results may depend upon such knowledge. For example, the addition of a given substance to a

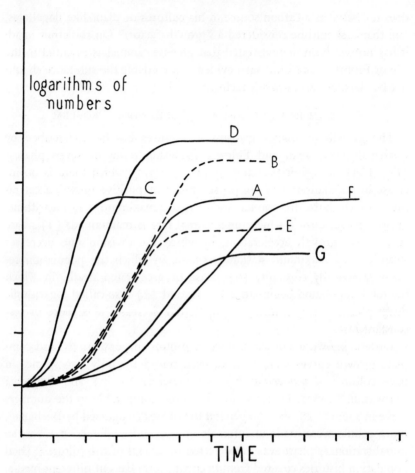

Figure 129. Hypothetical modifications (curves B-G) of the normal growth of a population (curve A), from the initial stationary to the maximal stationary phase.

certain medium might produce any one of several effects: (1) The maximal population might be increased without any appreciable effect on the growth rate; that is, the length of the logarithmic phase might be increased without any change in the division rate (curve B, Fig. 129), as compared with that in the control medium (curve A). (2) The growth rate might be increased without a change in density in the maximal stationary phase (curve C). (3) Both the growth rate and the maximal density of population might be increased (curve D). (4) The maximal density of population might be decreased, with no appreciable effect on the early growth rate (curve E). (5) The growth rate might be decreased, without any effect on the maximal density of population (curve F). (6) Both the growth rate and the maximal density of population might be decreased (curve G). These possibilities will serve as illustrations. Curves A, B, and E would show no significant differences in the early histories of the cultures; yet each reaches a different maximal stationary phase, one higher and one lower than that of the control (curve A). On the other hand, early examination of cultures A, C, and F would show significant differences in population density and in growth rate, although each culture eventually reaches the same maximal density. Hence without detailed information concerning the behavior of populations, there is obvious need for caution in interpreting experimental results.

In an analysis of the conditions which might influence the population-growth curve, the concentration of available food, the general condition and density of the initial population, the pH of the medium, the temperature of incubation, the accumulation of waste products, the oxygen tension, and the redox potential of the medium—to mention some of the more apparent factors—all seem to be significant. Although the importance of such factors may seem obvious, their detailed relationships to growth are mostly unknown.

The Initial Population

Both the physiological condition and the size of the initial population may influence the rate of growth. Phelps (1935) has reported that the length of the initial stationary phase and the lag period bears a relation to the age of the inoculum. With inocula taken from the logarithmic-growth phase of a stock culture, the length of the combined lag and initial stationary phases was usually reduced to zero. With inocula from older cultures, these two phases were usually detectable and were often quite pronounced. Obviously, therefore, age and other qualities of the inoculum must be considered in comparative studies on population growth and on the effects of environmental factors.

The relation between the initial density of population and the growth rate in bacterized cultures has been disputed for many years. Robertson and others have described an allelocatalytic effect, in which the rate of population growth is correlated more or less directly with the size of the initial population. A number of investigators, however, either have noted no significant correlation, or else have found that the growth rate varies inversely with the initial density of population (e.g., Woodruff, 1911). W. H. Johnson (1936, 1937) has pointed out that these results must be interpreted in relation to the concentration of bacteria in the cultures. Jahn (1929), who partially eliminated bacteria by growing *Euglena* sp. in inorganic media and washing the flagellates before inoculation, found that the growth rate varied inversely with the initial density of the population. Most of the literature on bacterized cultures has been reviewed by Jahn (1929), W. H. Johnson (1933, 1936, 1937), and Petersen (1929).

In contrast to the studies on bacterized cultures, very little work has yet been done with bacteria-free strains. Evidence bearing on the growth of such populations has been presented by Phelps (1935), Mast and Pace (1938), Reich (1938), and Hall and Schoenborn (1939b). Phelps concluded, for *Glaucoma piriformis,* that the density of population at the end of the logarithmic phase is, within wide limits, independent of the initial density of population. This may mean that the logarithmic phase is prolonged in the cultures with small inocula; or that the growth rate is higher in the cultures with low initial density; or perhaps that both the growth rate and the length of the logarithmic phase are increased. Phelps did not consider this question in detail, but some of his data (ser. II, Figs. 4, 5) indicate, at the end of the logarithmic phase, an average generation time of about four hours in the low-initial-density cultures and approximately five hours in the cultures started with larger inocula. Such data suggest an inverse relationship between the initial density of population and the rate of population growth.

In *Chilomonas paramecium* (Mast and Pace, 1938) the rate of reproduction varies directly with the initial density of population under some conditions, but inversely under others. Some sort of optimal relationship is indicated by the findings of Mast and Pace, since the growth rate increased to a maximum with decreasing volumes of medium per flagellate and then decreased to zero.

Reich (1938) observed that in *Mayorella palestinensis* the division rate varies directly with the initial density of population in cultures started with less than 3,000 amoebae per cc., although the population "eventually attained is largely independent of the quantity of inoculum."

These results are interpreted as supporting Robertson's concept of allelo-catalysis, although Reich does not subscribe to the theory of an auto-catalyst of growth.

In observations on *Euglena,* Hall and Schoenborn (1939b) have noted that the population tends to reach a concentration which is more or less independent of the initial density of population under the conditions described, and thus that the increase in the population varies inversely with the initial density of population. These conclusions were based upon counts made after specified periods of incubation, and population curves were not traced.

At present it is impossible to correlate the results which have been obtained with bacteria-free cultures, although the cited observations all indicate that the initial density of population influences the rate of population growth. Since the experiments on various species have been carried out under different conditions and in different media, it is pos-sible that some of the puzzling contradictions may eventually be traced to differences in technique, rather than differences in the nature of popu-lation growth. In fact, different relationships between the density of population and the rate of growth might reasonably be expected under different experimental conditions.

GROWTH IN RELATION TO WASTE PRODUCTS

Investigations on so-called waste products of Protozoa have led to con-flicting opinions. Woodruff (1911, 1913) concluded that waste prod-ucts inhibit growth of the homologous species, although growth of a different species may be relatively unaffected by the same substances. The other extreme is represented by Dimitrowa's (1932) observation that the growth of *Paramecium caudatum* is accelerated by the addition of small amounts of old medium to the experimental cultures. More recently, W. H. Johnson and Hardin (1938) have observed no significant effects of old culture medium on the growth of *P. multimicronucleata.*

Very little work on this problem has been carried out with pure cultures. The observations of A. Lwoff and Roukhelman (1929), that growth of *G. piriformis* ceases long before the food supply is exhausted, has lent some support to the view that growth may be inhibited by accumulated waste products. The results of later investigations are not so readily interpreted. Mast and Pace (1938) have noted that old culture medium, in high concentrations, inhibits the growth of *Chilomonas paramecium,* while small amounts accelerate growth. Reich (1938) found that the addition of culture filtrate produced no effect on the growth of *M. palestinensis.* However, Reich's filtrate was obtained from young cultures (twenty-four-hour cultures in one experiment, for example), and his technique is not entirely comparable to that of Mast and Pace. Hall and Loefer (1940), working with *C. campylum* in peptone medium, found that the addition of old culture filtrates (one part in ten, to five parts in ten) markedly increased the population yield, as compared with that in control cultures. Furthermore, the growth rate, after the first or second day of incubation, was noticeably higher in the cultures containing old-culture filtrate. Acceleration of comparable magnitude was also produced by the addition of aged sterile medium to fresh peptone medium. In view of the latter observation, it now seems impossible to attribute the effects of old-culture filtrates solely to a "biological conditioning" of the medium, or entirely to a product or products elaborated by the organisms growing in the medium. A basic explanation for these various phenomena is not yet available, and it is possible that any single explanation may be inadequate. Thus the "factor" of Mast and Pace is said to be heat-labile, whereas the effects noted by Hall and Loefer were produced by culture filtrates and aged sterile medium which had been sterilized in the autoclave. At any rate, these results are not only interesting in themselves, but they may also furnish important clues in untangling the conflicting opinions concerning allelocatalytic and autocatalytic phenomena. For instance, some preliminary observations of the writer have already shown that the "accelerating factor" in old cultures may have a definite bearing on the growth of *C. campylum,* in relation to initial density of population.

GROWTH IN RELATION TO FOOD CONCENTRATION

It seems obvious that, within reasonable limits, the density of a protozoan population should vary more or less directly with the concentration of available food until an optimal concentration is reached, although the relationship might not be evident in the early history of the culture. Such a generalization is supported by studies on pure cultures.

Cailleau (1933) noted that peptone concentrations of 3.0 percent supported abundant growth of *Acanthamoeba castellanii,* while lower concentrations were much less favorable. In *Colpidium campylum,* Bond (1933) observed comparable relationships between growth rate and concentration of yeast autolysate. Optimal concentrations of peptone for *C. striatum* (Elliott, 1935b) lie between 1.0 and 3.0 percent, while for *Glaucoma ficaria* (D. F. Johnson, 1935a) the optimum is about 1.5 percent. In both species of *Colpidium* the effects were apparent after twenty-four hours of incubation and became more marked in older cultures. Loefer's (1936b) observations on *Paramecium bursaria* indicate similar relationships, although the limits are somewhat narrower than for *Colpidium* and *Glaucoma.* For example, one of the peptones tested was optimal in 0.5-percent solution, whereas no growth of *P. bursaria* occurred in a concentration of 1.4 percent. Phelps (1936) observed that in the logarithmic phase the growth rate of *G. piriformis* was, within wide limits, practically independent of the food concentration. In later history of the cultures, however, the relationships were comparable to those observed by Elliott and D. F. Johnson. Rottier (1936b) described, in *Polytoma uvella,* a direct relation between the growth rate and the concentration of peptone (0.2 to 1.0 percent), and of asparagin (0.2 to 2.0 percent), significant differences being noted after about five days of incubation.

The effect of a substance added to an adequate medium also varies with concentration, as would be expected. Johnson (1935a) reported for *Glaucoma ficaria* in peptone medium, maximal acceleration by 0.5-percent yeast extract and inhibitory effects of concentrations above 2.0 percent. In *Paramecium bursaria* (Loefer, 1936b), the optimal concentration of the same yeast extract was 0.03 percent, and growth was definitely inhibited in a 0.5-percent solution. Dextrose was most effective for *P. bursaria* in a concentration of 0.5 percent. Reich (1936) has obtained similar results with *Mayorella palestinensis.* The effects of added sodium acetate on *P. uvella* (Rottier, 1936b) vary in the same fashion, the maximal effect being produced by 0.8-percent acetate. The optimum is much lower in *Euglena stellata* (Hall, 1937d), in which 0.05-percent sodium acetate was most effective in both light and darkness.

Growth in Relation to pH of the Medium

It has been known for many years that the acidity or the alkalinity of the medium bears some relation to the growth of Protozoa, and investigations on bacterized cultures have determined the optimal pH and pH range for a number of species (for reviews, see Loefer, 1935c; D. F. Johnson, 1935a). More recently, similar investigations have been carried out with bacteria-free material.

Dusi (1930) has shown that each of six species of *Euglena* has a characteristic pH range in certain media, and that the optimum varies somewhat for the different species. Jahn (1931), using quantitative methods, has studied that pH relationships of *E. gracilis* in detail, and similar relationships have been determined for *E. anabaena* and *E. deses* (Hall, 1933a) and for two species of *Astasia* (Schoenborn, 1936). Among the Cryptomonadida and Phytomonadida, the pH-growth relationships of *Chilomonas paramecium*, *Chlorogonium elongatum*, and *C. euchlorum* have been investigated by Loefer (1935c). Growth of the two phytomonads, with an optimum slightly above pH 7.0, was more or less comparable to that of several Euglenidae; *C. paramecium*, on the other hand, showed a bimaximal pH-growth curve with peaks at pH 4.9 and 7.0 and an intermediate low point at pH 6.0.

Relationships between growth and pH have also been determined for several ciliates. Elliott (1933, 1935b) has described the *p*H ranges and optima for *Colpidium campylum* and *C. striatum*, and has pointed out that the pH relationships vary with the type of medium. In one peptone medium (Difco tryptone) a bimodal curve, with peaks at pH 5.5 and 7.5, was noted; in certain other peptone media, a unimaximal pH-growth curve was observed. The addition of sodium acetate or a carbohydrate (e.g., maltose) to tryptone medium changed the shape of the curve from bimaximal to unimaximal. The extent of the pH range also varied with the type of medium. D. F. Johnson (1935a), in similar fashion, has compared the pH-growth curves of *Glaucoma ficaria* and *G. piriformis* in different types of media. Appreciable differences between the two species were noted, and the pH range and general form of the growth curves were found to vary with the type of medium, much as in *Colpidium*. More recently, Loefer (1938a) has studied the growth rate and general morphology of *Paramecium bursaria* in relation to the pH of the medium. Conditions known to be optimal for the symbiotic *Chorella* (Loefer, 1936a) did not coincide with those most favorable to the growth of *P. bursaria* containing the algae. The pH optimum for the ciliate was approximately 6.8, and growth occurred within the range 4.9 to about 7.8. The size of the ciliates varied with the pH, but independently of the growth rate.

In addition to the presumably direct influence upon growth rate, the pH of the medium has been found to modify the effects of other factors. For example, Elliott (1935a) has shown that the maximal accelerating effect of certain carbohydrates on the growth of *C. striatum* is exerted below pH 7.0, with little or no acceleration above that point. Some of Jahn's (1935b) results with *E. gracilis* in inorganic medium also seem to show a correlation between pH and the effect of several carbohydrates. Elliott noted also that the effects of sodium acetate and butyrate varied with the pH. The former inhibited growth of *C. striatum* more or less completely at pH 6.0 and lower, but produced moderate acceleration near the neutral point. Butyrate was toxic at pH 6.5 or below, but showed an accelerating effect at pH 7.0-7.5. Jahn (1934) has suggested that such effects of acetate and butyrate may be explained on the basis that only the undissociated organic-acid molecule is toxic. A. Lwoff (1935a), in reviewing Elliott's work, stated that acetate and butyrate inhibit the growth of *Colpidium;* this is true for only a certain pH range. Another indirect effect of the pH is the influence on temperature relationships, as indicated in Jahn's (1933a) observation that the susceptibility of *E. gracilis* to relatively high temperature is lowest at pH 5.0 and greatest above pH 7.0. The growth-accelerating effects of pantothenic acid on *C. striatum* (Elliott, 1935c) and of plant auxins on *E. gracilis* (Elliott, 1938) are also dependent upon the pH of the medium.

The evidence already accumulated shows that the pH relationships of Protozoa are exceedingly complex, and that they vary not only with the individual species but also with the composition of the medium and with other environmental conditions. Furthermore, such relationships may vary with time, since Jahn (1931) has observed that the optimal pH differs in young and in old cultures of *E. gracilis*. To some extent, the pH-growth relationships may be correlated with the activities of enzymes, which may show characteristic pH optima; for example, the protease of *G. piriformis* (Lawrie, 1937) shows maximal activity at pH 6.0. But this may represent only one of many ways in which growth is related to the pH of the medium.

OXYGEN RELATIONSHIPS

That oxygen tension of the medium influences growth of Protozoa is obvious, but relatively little detailed information has been accumulated in experimental studies. Observations on the natural occurrence of Protozoa indicate definite differences in oxygen requirements. Some species appear to be strict aërobes, others are perhaps comparable to the microaërophiles among the bacteria, and many intestinal parasites are probably to be regarded as facultative anaërobes. Investigations on proto-zoan respiration are discussed in another chapter of this volume (Chapter VI).

Growth in relation to oxygen requirements has been investigated for only a few bacteria-free strains, and no attempt has been made to correlate definite oxygen tensions with growth rate. *G. piriformis,* according to Lwoff (1932), is incapable of growth under anaërobic conditions. Likewise, Hall (1933b) found that under reduced oxygen tension (Buchner pyrogallol method), growth of *C. campylum* was approximately 50 percent less than in aërobic controls in peptone medium. With added dextrose, however, growth was greater than in the aërobic controls in peptone medium. The results suggest a certain degree of similarity between *C. campylum* and the facultative anaërobes among the bacteria. Phelps (1936) demonstrated that aëration of flask cultures in yeast autolysate produces a much heavier population of *G. piriformis* than in unaërated flasks. These results are somewhat comparable to the findings of Jahn (1936), who compared *G. piriformis* and *Chilomonas paramecium* with respect to effects of aëration. Growth of the ciliates was most rapid at first in unaërated flasks, but after three days of incubation the aërated flasks showed heavier populations. In *C. paramecium,* however, growth was consistently more rapid in the unaërated flasks. Rottier (1936a) has reported that the growth of *Polytoma uvella* in flasks is more rapid than in tubes, after approximately forty hours of incubation. Likewise, aërated tube cultures showed heavier growth than unaërated ones.

THE REDOX POTENTIAL

As applied to culture media, the redox potential may be considered an indication of the oxidizing or reducing power of such an oxidation-reduction system. In other words, the more positive the redox potential, the more highly oxidized is the medium; the more negative the potential, the more highly reduced will be the medium. In effect, the potential is a measure of intensity rather than of oxidizing or reducing capacity, and hence is somewhat analogous to the pH, which gives no indication of the amount of acid or alkali necessary to change the reaction by a given amount. In addition, the redox potential varies with the pH of the medium. A number of investigators have correlated the redox potential of culture media with the growth and metabolism of bacteria, but very little work along this line has yet been done in protozoölogy.

So far, the only detailed investigations are those of Jahn (1933b, 1935a), who has studied growth of *Chilomonas paramecium* in relation to the redox potential of the medium. In his first publication, Jahn found that growth is accelerated by NaSH, while the addition of H_2O_2 to a peptone and acetate medium inhibited growth. On the other hand, relatively rapid growth occurred when both peroxide and a high concentration of -SH were added to the medium. These results were explained on the basis of the redox potential. In his second article, Jahn traced the continuous changes in the pH and Eh in cultures of *C. paramecium*. The Eh of different media was found to drop as much as 300-460 mv. during the first few days of incubation, and Jahn suggested that such changes may involve not only a lowering of the oxygen tension but also the accumulation of reducing substances in the medium. After three to five days, depending upon the type of medium, the Eh began to rise; this change was attributed to a sharp decrease in the growth rate of the flagellates, with a corresponding decrease in oxygen consumption.

So far as the Protozoa are concerned, Jahn's results indicate that there is much to be learned concerning detailed relationships between growth and the redox potential of culture media. The exact effects of changes in the redox potential are still unknown, and the relative importance of the redox potential and the oxygen tension in different cases is yet to be determined. Possible relationships to growth have been discussed by Jahn (1934).

GROWTH IN RELATION TO TEMPERATURE

The importance of temperature relationships is obvious, and rigid control of temperature is essential in many types of experimental investigations. The actual relationships between growth and temperature are undoubtedly complex, since changes in temperature may not only affect metabolic activities of the organism directly, but may also modify the action of other environmental factors. Conversely, changes in various environmental conditions may modify the temperature relationships of a given species.

In one of the few investigations carried out on pure cultures, Jahn (1935c) has demonstrated an interesting temperature relationship in *Euglena gracilis*. In darkness, the optimal temperature for this species in a peptone medium was about 10° C. When sodium acetate was added to the medium, not only was growth accelerated, but the optimal temperature was shifted to about 23°, a point approaching the optimum for growth in light. Another instance in which the temperature relationships vary with other environmental conditions is represented by the thermal death time of *E. gracilis* (Jahn, 1933a), which appears to be a function of the pH, the greatest resistance to a temperature of 40° C. being noted at pH 5.0, and a greater susceptibility above pH 7.0 than below.

GROWTH IN RELATION TO LIGHT AND DARKNESS

Little or nothing is known concerning the relation between light and the growth of colorless Protozoa. On the other hand, the importance of light is obvious in the case of the chlorophyll-bearing species, and the relation to photosynthesis probably accounts for a number of the known effects of light. Dusi (1937) has noted certain interesting peculiarities of several Euglenidae. In constant light, *E. gracilis* grows well in peptone medium, but poorly in inorganic media. On the other hand, *E. klebsii* grows perfectly in inorganic medium under constant illumination, while *E. viridis* is incapable of growing under such conditions, even in peptone medium. No explanation for such specific differences in light relationships is yet available. An apparent relationship between light and the optimal temperature for growth has been noted by Jahn (1935c), who reported that the optimal temperature for growth of *E. gracilis* in darkness lies near 10° C., whereas the optimum in light for the same species is approximately 25°. The presence or absence of light is also a factor which must be considered in interpreting the effects of carbon compounds on growth. Thus Jahn (1935d) has observed that the accelerating effects of several organic acids on the growth of *E. gracilis* are relatively much greater in darkness than in light. Succinate, on the other hand, produced a slight acceleration in darkness, but was mildly toxic in light, while oxalate exerted no effect in darkness and a slight acceleration in light. Hall (1937d) observed also that in *E. stellata* tolerance to concentrations of acetate above 0.2 percent was much less in light than in darkness. Furthermore, Jahn (1936b) has obtained some evidence that intensity of light may influence the effects of carbohydrates on the growth of *E. gracilis* in an inorganic medium. Another instance involving light relationships is the effect of plant "hormones" on *E. gracilis,* in which Elliott (1937b) has shown that growth is accelerated in light, but not in darkness.

ACCLIMATIZATION

Acclimatization of Protozoa to various experimental conditions has been reported in many instances, ranging from acclimatization to toxic chemicals to the development of resistance to antibodies. A few cases have been described in bacteria-free cultures of free-living species. Such a process may occasionally be involved in the establishment of pure cultures, as reported by Elliott (1933) for *Colpidium striatum* and by Johnson (1935a) for *Glaucoma ficaria.* More recently, Loefer (1938c) has studied the acclimatization of several species (*C. campylum, G. piriformis, Chlorogonium euchlorum, Euglena gracilis,* and *Astasia* sp.) to progressively increased salt concentrations. *E. gracilis* developed no appreciable tolerance, but, after a series of transfers, the other species all showed the ability to grow in salt concentrations which were lethal in the initial exposures. The salinity finally tolerated by *C. campylum* was higher than that of ordinary sea water. Further investigations on acclimatization should prove interesting, and may throw some light on various experimental results which at present seem very puzzling.

LITERATURE CITED

No attempt has been made to include all the literature on bacteria-free cultures of Protozoa. Among the papers cited, those which contain good bibliographies are indicated by an asterisk.

Andrews, J., and T. von Brand. 1938. Quantitative studies on glucose consumption by *Trichomonas foetus.* Amer. J. Hyg., 28: 138-47.

Bond, R. M. 1933. A contribution to the study of the natural food cycle in aquatic environments. Bull. Bingham oceanogr. Coll. 4 (Art. 4), 89 pp.

Brand, T. von. 1938.* The metabolism of pathogenic trypanosomes and the carbohydrate metabolism of their hosts. Quart. Rev. Biol., 13: 41-50.

Buchanan, R. E., and E. I. Fulmer. 1928. Physiology and biochemistry of Bacteria. Vol. I. Baltimore.

Cailleau, R. 1933. Culture d'*Acanthamoeba castellanii* sur milieu peptone. Action sur les glucides. C. R. Soc. Biol. Paris, 114: 474-76.

—— 1934. Utilization des milieux liquides par *Acanthamoeba castellanii*. C. R. Soc. Biol. Paris, 116: 721-23.

—— 1935. La Nutrition de *Trichomonas columbae* en culture. C. R. Soc. Biol. Paris, 119: 853-56.

—— 1936a. Le Cholestérol, facteur de croissance pour le flagellé *Trichomonas columbae*. C. R. Soc. Biol. Paris, 121: 424-25.

—— 1936b. L'Activité de quelques stérols envisagés comme facteurs de croissance pour le flagellé *Trichomonas columbae*. C. R. Soc. Biol. Paris, 122: 1027-28.

—— 1937a. Nouvelles Recherches sur l'activité de quelques stérols considerées comme facteurs de croissance pour le flagellé *Trichomonas columbae*. C. R. Soc. Biol. Paris, 124: 1042-44.

—— 1937b.* La Nutrition des flagellés Tetramitides. Les stérols, facteurs de croissance pour les Trichomonades. Ann. Inst. Pasteur, 59: 137-293.

—— 1938a. Le Cholestérol et l'acide ascorbique, facteurs de croissance pour le flagellé tetramitide *Trichomonas foetus* Riedmüller. C. R. Soc. Biol. Paris, 127: 861-63.

—— 1938b. L'Acide ascorbique et le cholestérol, facteurs de croissance pour le flagellé *Eutrichomastix colubrorum*. C. R. Soc. Biol. Paris, 127: 1421-23.

—— 1939. L'Acide ascorbique, facteur de croissance pour le flagellé *Trichomonas columbae*. C. R. Soc. Biol. Paris, 130: 319-21.

Colas-Belcour, J., and A. Lwoff. 1925. L'Utilisation des glucides par quelques Protozoaires. C. R. Soc. Biol. Paris, 93: 1421-22.

Dimitrowa, A. 1932. Die Fördernde Wirkung der Exkrete von *Paramecium caudatum* Ehrbg. auf dessen Teilungsgeschwindigkeit. Zool. Anz., 100: 127-32.

Dusi, H. 1930. Limites de la concentration en ions H pour la culture de quelques euglènes. C. R. Soc. Biol. Paris, 104: 734-36.

—— 1931. L'Assimilation des acides aminés par quelques eugléniens. C. R. Soc. Biol. Paris, 107: 1232-34.

—— 1933a. Recherches sur la nutrition de quelques euglénes. I. *Euglena gracilis*. Ann. Inst. Pasteur, 50: 550-97.

—— 1933b. Recherches sur la nutrition de quelques euglènes. II. *Euglena stellata, klebsii, anabaena, deses* et *pisciformis*. Ann. Inst. Pasteur, 50: 840-90.

—— 1936. Recherches sur la culture et la nutrition d'*Euglena viridis*. Arch. zool. expr. gén., 78 (M. et R.): 133-36.

—— 1937. Le Besoin de substances organiques de quelques eugléniens à chlorophylle. Arch. Protistenk., 89: 94-99.

—— 1939. La Pyrimidine et le thiazol, facteurs de croissance pour le flagellé à chlorophylle, *Euglena pisciformis*. C. R. Soc. Biol. Paris, 130: 419-22.

Elliott, A. M. 1933.* Isolation of *Colpidium striatum* Stokes in bacteria-free culture and the relation of growth to pH of the medium. Biol. Bull., 65: 45-56.

—— 1935a. Effects of carbohydrates on growth of *Colpidium*. Arch. Protistenk., 84: 156-74.

—— 1935b.* Effects of certain organic acids and protein derivatives on the growth of *Colpidium*. Arch. Protistenk., 84: 225-31.

—— 1935c. The influence of pantothenic acid on growth of Protozoa. Biol. Bull., 68: 82-92.

—— 1937a. Vitamin B_1 and growth of Protozoa. Anat. Rec., 70 (Suppl.): 127.

—— 1937b. Plant hormones and growth of *Euglena* in relation to light. Anat. Rec., 70 (Suppl.): 128.

—— 1938.* The influence of certain plant hormones on growth of Protozoa. Physiol. Zoöl., 11: 31-39.

—— 1939. The vitamin B complex and the growth of *Colpidium striatum*. Physiol. Zoöl., 12: 363-373.

Gause, G. F. 1935. Experimentelle Untersuchungen über die Konkurrenz zwischen *Paramecium caudatum* und *Paramecium aurelia*. Arch. Protistenk., 84: 207-24.

Geise, A. C., and C. V. Taylor. 1935. Paramecia for experimental purposes in controlled mass cultures on a single strain of bacteria. Arch. Protistenk., 84: 225-31.

Glaser, R. W., and N. A. Coria. 1930. Methods for the pure culture of certain Protozoa. J. exp. Med., 51: 787-806.

—— 1933. The culture of *Paramecium caudatum* free from living micro-organisms. J. Parasite., 20: 33-37.

—— 1935a. The culture and reactions of purified Protozoa. Amer. J. Hyg., 21: 111-20.

—— 1935b. Purification and culture of *Tritrichomonas foetus* (Riedmüller) from cows. Amer. J. Hyg., 22: 221-26.

Hall, R. P. 1933a. On the relation of hydrogen-ion concentration to the growth of *Euglena anabaena* var. *minor* and *E. deses*. Arch. Protistenk., 79: 239-48.

—— 1933b. Growth of *Colpidium campylum* with reference to oxygen relationships. Anat. Rec., 57 (Suppl.): 95.

—— 1934. Effects of carbohydrates on growth of *Euglena anabaena* var. *minor* in darkness. Arch. Protistenk., 82: 45-50.

—— 1937a.* "Growth of free-living Protozoa in pure cultures." *In* Culture Methods for Invertebrate Animals. Ithaca. Pp. 51-59.

—— 1937b. Certain culture reactions of several species of Euglenidae. Trans. Amer. micr. Soc., 56: 285-87.

—— 1937c. Effects of manganese on the growth of *Euglena anabaena, Astasia* sp. and *Colpidium campylum*. Arch. Protistenk., 90: 178-84.

—— 1937d. Effects of different concentrations of sodium acetate on growth of *Euglena stellata*. Anat. Rec., 70 (Suppl.) : 127.

—— 1938a. Pimelic acid as a growth factor for the ciliate, *Colpidium campylum*. Anat. Rec., 72 (Suppl.) : 110.

—— 1938b. Nitrogen requirements of *Euglena anabaena* var. *minor*. Arch. Protistenk., 91: 465-73.

—— 1939a. The trophic nature of *Euglena viridis*. Arch. zool. expr. gén., 80 (N. et R.) : 61-67.

—— 1939b. The trophic nature of the plant-like flagellates. Quart. Rec. Biol., 14: 1-12.

—— 1939c. Pimelic acid as a growth stimulant for *Colpidium campylum*. Arch. Protistenk., 92: 315-19.

Hall, R. P., and A. M. Elliott. 1935. Growth of *Colpidium* in relation to certain incomplete proteins and amino acids. Arch. Protistenk., 85: 443-50.

Hall, R. P., and J. B. Loefer. 1936. On the supposed utilization of inorganic nitrogen by the colorless cryptomonad flagellate, *Chilomonas paramecium*. Protoplasma, 26: 321-30.

—— 1940. Effects of culture filtrates and old medium on growth of the ciliate, *Colpidium campylum*. Proc. Soc. exp. Biol. N. Y., 43: 128-33.

Hall, R. P., and H. W. Schoenborn. 1938a. Studies on the question of autotrophic nutrition in *Chlorogonium euchlorum, Euglena anabaena* and *E. deses*. Arch. Protistenk., 90: 259-71.

—— 1938b. The selective action of inorganic media in bacteria-free cultures of *Euglena*. Anat. Rec., 72 (Suppl.) : 129-30.

—— 1939a. The question of autotrophic nutrition in *Euglena gracilis*. Physiol. Zoöl., 12: 76-84.

—— 1939b. Fluctuations in growth rate of *Euglena anabaena, E. gracilis* and *E. viridis* and their apparent relation to initial density of population. Physiol. Zoöl., 12: 201-08.

Hutner, S. H. 1936. The nutritional requirements of two species of *Euglena*. Arch. Protistenk., 88: 93-106.

Jahn, T. L. 1929.* Studies on the physiology of the euglenoid flagellates. I. The relation of the density of population to the growth rate of *Euglena*. Biol. Bull., 57: 81-106.

—— 1931. Studies on the physiology of the euglenoid flagellates. III. The effect of hydrogen-ion concentration on the growth of *Euglena gracilis*. Biol. Bull., 61: 387-99.

—— 1933a.* Studies on the physiology of the euglenoid flagellates. IV. The thermal death time of *Euglena gracilis* Klebs. Arch. Protistenk., 79: 249-62.

—— 1933b.* Studies on the oxidation-reduction potential of protozoan cultures. I. The effect of -SH on *Chilomonas paramecium*. Protoplasma, 20: 90-104.

—— 1934. Problems of population growth in the Protozoa. Cold Spring Harbor Symp. Quant. Biol., 2: 167-80.

—— 1935a. Studies on the oxidation-reduction potential of protozoan cultures. II. The reduction potential of cultures of *Chilomonas paramecium*. Arch. Protistenk., 86: 225-37.

—— 1935b. Studies on the physiology of the euglenoid flagellates. V. The effect of certain carbohydrates on the growth of *Euglena gracilis* Klebs. Arch. Protistenk., 86: 238-50.

—— 1935c. Studies on the physiology of the euglenoid flagellates. VI. The effects of temperature and of acetate on *Euglena gracilis* cultures in the dark. Arch. Protistenk., 86: 251-57.

—— 1935d. Studies on the physiology of the euglenoid flagellates. VII. The effect of salts of certain organic acids on growth of *Euglena gracilis* Klebs. Arch. Protistenk., 86: 258-62.

—— 1936. Effect of aeration and lack of CO_2 on growth of bacteria-free cultures of Protozoa. Proc. Soc. Exper. Biol. N. Y., 33: 494-98.

Johnson, D. F. 1935a.* The isolation of *Glaucoma ficaria* in bacteria-free cultures, and growth in relation to pH of the medium. Arch. Protistenk., 86: 263-77.

—— 1935b. Fermentation of carbohydrates by *Glaucoma* and effects of carbohydrates on growth of two species. Anat. Rec., 64 (Suppl.) : 106-07.

—— 1936.* Growth of *Glaucoma ficaria* Kahl in cultures with single species of other microorganisms. Arch. Protistenk., 86: 359-78.

Johnson, W. H. 1933.* Effects of population density on the rate of reproduction in *Oxytricha*. Physiol. Zoöl., 6: 22-54.

—— 1936.* Studies on the nutrition and reproduction of *Paramecium*. Physiol. Zoöl., 9: 1-14.

—— 1937.* Experimental populations of microscopic organisms. Amer. Nat., 71: 5-20.

Johnson, W. H., and G. Mardin. 1938. Reproduction of *Paramecium* in old culture medium. Physiol. Zoöl., 11: 333-46.

Koser, S. A., and F. Saunders. 1938. Accessory growth factors for bacteria and related microorganisms. Bact. Rev., 2: 99-160.

Lawrie, N. R. 1937. Studies in the metabolism of Protozoa. III. Some properties of a proteolytic extract obtained from *Glaucoma piriformis*. Biochem. J., 31: 789-98.

Loefer, J. B. 1934. The trophic nature of *Chlorogonium* and *Chilomonas*. Biol. Bull., 66: 1-6.

—— 1935a.* Effect of certain carbohydrates and organic acids on growth of *Chlorogonium* and *Chilomonas*. Arch. Protistenk., 84: 456-71.

—— 1935b. Effects of certain nitrogen compounds on growth of *Chlorogonium* and *Chilomonas*. Arch. Protistenk., 85: 74-86.

—— 1935c.* Relation of hydrogen-ion concentration to growth of *Chilomonas* and *Chlorogonium*. Arch. Protistenk., 85: 209-23.

—— 1936a. Isolation and growth characteristics of the "zoochlorella" of *Paramecium bursaria*. Amer. Nat., 70: 184-88.

—— 1936b. Bacteria-free cultures of *Paramecium bursaria* and concentration of the medium as a factor in growth. J. exp. Zool., 72: 387-407.

—— 1936c. Effect of certain "peptone" media and carbohydrates on the growth of *Paramecium bursaria*. Arch. Protistenk., 87: 142-50.

—— 1936d. A simple method for maintaining pure-line mass cultures of *Paramecium caudatum* on a single species of yeast. Trans. Amer. micr. Soc., 55: 254-56.

—— 1938a. Effect of hydrogen-ion concentration on the growth and morphology of *Paramecium bursaria*. Arch. Protistenk., 87: 142-50.

—— 1938b.* Utilization of dextrose by *Colpidium, Glaucoma, Chilomonas* and *Chlorogonium* in bacteria-free cultures. J. exp. Zool., 79: 167-83.

—— 1938c. Effect of osmotic pressure on the motility and viability of fresh-water Protozoa. Anat. Rec., 72 (Suppl.) : 50.

Loefer, J. B., and R. P. Hall. 1936. Effect of ethyl alcohol on the growth of eight protozoan species in bacteria-free cultures. Arch. Protistenk., 87: 123-30.

Lwoff, A. 1924. Le Pouvoir de synthèse d'un protist hétérotrophe: *Glaucoma piriformis*. C. R. Soc. Biol. Paris, 91: 344-45.

—— 1925. La Nutrition des infusoires au dépens des substances dissoutes. C. R. Soc. Biol. Paris, 93: 1272-73.

—— 1929a. Milieux de culture et d'entretien pour *Glaucoma piriformis* (cilié). C. R. Soc. Biol. Paris, 100: 635-36.

—— 1929b. La Nutrition de *Polytoma uvella* Ehrenberg (flagellé Chlamydomonadinae) et le Pouvoir de synthèse des protistes hétérotrophes. Les protistes mésotrophes. C. R. Acad. Sci. Paris, 188: 114-16.

—— 1930. Le Fer, élément indispensable au flagellé *Polytoma uvella* Ehr. C. R. Soc. Biol. Paris, 104: 664-66.

—— 1931. La Nutrition carbonée de *Polytoma uvella*. C. R. Soc. Biol. Paris, 107: 1070-72.

—— 1932.* Recherches biochimiques sur la nutrition des protozoaires. Le pouvoir synthèse. Monogr. Inst. Pasteur.

—— 1934. Die Bedeutung des Blutfarbstoffes für die parasitischen flagellaten. Zbl. Bakt., Orig. 130: 498-518.

—— 1935a. L'Oxytrophie et les organisms oxytrophes. C. R. Soc. Biol. Paris, 119: 87-90.

—— 1935b. La Nutrition azotée et carbonée de *Polytomella agilis* (Polyblépharidée incolore). C. R. Soc. Biol. Paris, 119: 974-76.

—— 1936. La Fonction de la protohémin pour les protozoaires et les bactéries parahémotrophes. C. R. Soc. Biol. Paris, 122: 1041-42.

—— 1938a. Remarques sur la physiologie comparée des protistes eucaryotes. Les Leucophytes et l'oxytrophie. Arch. Protistenk., 90: 194-209.

—— 1938b. La Synthèse de l'aneurine par le protozoaire, *Acanthamoeba castellanii.* C. R. Soc. Biol. Paris, 128: 455-58.

Lwoff, A., and H. Dusi. 1929. Le Pouvoir de synthèse d'*Euglena gracilis* cultivée à l'obscurité. C. R. Soc. Biol. Paris, 102: 567-69.

—— 1931. La Nutrition azotée et carbonée d'*Euglena gracilis* en culture pure à l'obscurité. C. R. Soc. Biol. Paris, 107: 1068-69.

—— 1934. L'Oxytrophie et la nutrition des flagellés leucocophytes. Ann. Inst. Pasteur, 53: 641-53.

—— 1935a. La Supression expérimentale des chloroplastes chez *Euglena mesnili.* Ann. Inst. Pasteur, 119: 1092-95.

—— 1935b. La Nutrition azotée et carbonée de *Chlorogonium euchlorum* à l'obscurité; l'acide acétique envisagé comme produit de l'assimilation chlorophylliene. C. R. Soc. Biol. Paris, 119: 1260-63.

—— 1936. La Nutrition de l'euglénien *Astasia chattoni.* C. R. Acad. Sci. Paris, 202: 248-50.

—— 1937a. La Pyrimidine et le thiazol, facteurs de croissance pour le flagellé *Polytomella caeca.* C. R. Acad. Sci. Paris, 630-32.

—— 1937b. Le Thiazol, facteur de croissance pour les flagellés *Polytoma caudatum* et *Chilomonas paramecium.* C. R. Acad. Sci. Paris, 205: 756-58.

—— 1937c. Le Thiazol, facteur de croissance pour *Polytoma ocellatum* (Chlamydomonadine). Importance des constituants de l'aneurine pour les flagellés leucophytes. C. R. Acad. Sci. Paris, 205: 882-84.

—— 1938a. Culture de divers flagellés leucophytes en milieu synthétique. C. R. Soc. Biol. Paris, 127: 53-56.

—— 1938b. L'Activité de diverses pyrimidines, considérées comme facteurs de croissance pour les flagellés *Polytomella caeca* et *Chilomonas paramecium.* C. R. Soc. Biol. Paris, 127: 1408-11.

—— 1938c. Influence de diverses substitutions sur l'activité de thiazol considéré comme facteur de croissance pour quelques flagellés leucophytes. C. R. Soc. Biol. Paris, 128: 238-41.

Lwoff, A., and E. Lederer. 1935. Remarques sur l'"extrait de terre" envisagé comme facteur de croissance pour les flagellés. C. R. Soc. Biol. Paris, 119: 971-73.

Lwoff, A., and M. Lwoff. 1937. L'Aneurine, facteur de croissance pour le cilié *Glaucoma piriformis.* C. R. Soc. Biol. Paris, 126: 644-46.

—— 1938. La Specificité de l'aneurine, facteur de croissance pour le cilié *Glaucoma piriformis.* C. R. Soc. Biol Paris, 127: 1170-72.

Lwoff, A., and L. Provasoli. 1935. La Nutrition de *Polytoma caudatum* var. *astigmata* (Chlamydomonadine incolore), et la synthèse de l'amidon par les leucophytes. C. R. Soc. Biol. Paris, 119: 90-93.

—— 1937. Caractères physiologiques du flagellé *Polytoma obtusum.* C. R. Soc. Biol. Paris, 126: 279-80.

Lwoff, A., and N. Roukhelman. 1929. Variations de quelques formes d'azote dans une culture pure d'infusoires. C. R. Acad. Sci. Paris, 183: 156-58.

Lwoff, M. 1929a. Culture de *Leptomonas ctenocephali* var. *chattoni* Laveran et Franchini, en milieu privés de sang frais: milieux liquides au sange chauffé. Bull. Soc. Path. exot., 22: 247-51.

—— 1929b. Milieu d'isolement et d'entretien pour *Schizotrypanum cruzi* Chagas. Bull. Soc. Path. exot., 22: 909-12.

—— 1929c. Action favorisante du sang sur la culture due *Leptomonas ctenocephali* (flagellé trypanosomide). C. R. Soc. Biol. Paris, 99: 472-74.

—— 1929d. Culture de *Leptomonas ctenocephali* Fanth. (flagellé trypanosomide) en milieu privé de sang frais: les organes stérilisés. C. R. Soc. Biol. Paris, 99: 1133-35.

—— 1929e. Influence du degré d'hydrolyse des matières protéiques sur la nutrition de *Leptomonas ctenocephali* (Fantham) *in vitro*. C. R. Soc. Biol. Paris, 100: 240-43.

—— 1930. Une Flagellé parasite hétérotrophe: *Leptomonas oncopelti* Noguchi et Tilden (Trypanosomidae). C. R. Soc. Biol. Paris, 105: 835-37.

—— 1933a.* Recherches sur la nutrition des trypanosomides. Ann. Inst. Pasteur, 51: 55-115.

—— 1933b. Remarques sur la nutrition des trypanosomides et des bactéries parahémotrophes. Le "fer actif" de Baudisch. Ann. Inst. Pasteur, 51: 707-13.

—— 1936. Le Pouvoir de synthèse des trypanosomides des muscides. C. R. Soc. Biol. Paris, 121: 419-21.

—— 1937. L'Aneurine, facteur de croissance pour le flagellé trypanosomide *Strigomonas oncopelti* (Noguchi et Tilden). C. R. Soc. Biol. Paris, 126: 771-73.

—— 1938a. L'Hématine et l'acide ascorbique, facteurs de croissance pour le flagellé *Schizotrypanum cruzi*. C. R. Acad. Sci. Paris, 206: 540-42.

—— 1938b. L'Aneurine, facteur de croissance pour le *Strigomonas* (flagellés Trypanosomides). C. R. Soc. Biol. Paris, 128: 241-43.

—— 1939. Le Pouvoir de synthèse des leishmanies. C. R. Soc. Biol. Paris, 130: 406-8.

Lwoff, M., and A. Lwoff. 1929. Le Pouvoir de synthèse de *Chlamydomonas agloëformis* et d'*Haematococcus pluvialis* en culture pure, à l'obscurité. C. R. Soc. Biol. Paris, 102: 569-71.

Mainx, F. 1928.* Beiträge zur Morphologie und Physiologie der Eugleninen. II. Teil. Untersuchungen über die Ernährungs-und Reizphysiologie. Arch. Protistenk., 60: 355-414.

Mast, S. O., and D. M. Pace. 1933. Synthesis from inorganic compounds of starch, fats proteins and protoplasm in the colorless animal, *Chilomonas paramecium*. Protoplasma, 20: 326-58.

—— 1935. Relation between sulphur in various chemical forms and the rate of growth in the colorless flagellate, *Chilomonas paramecium*. Protoplasma, 23: 297-325.

—— 1938. The effect of substances produced by *Chilomonas paramecium* on rate of reproduction. Physiol. Zoöl., 11: 359-82.

Oehler, R. 1916. Amöbenzucht auf reinem Boden. Arch. Protistenk., 37: 175-90.

—— 1919. Flagellaten- und Ciliatenzucht auf reinem Boden. Arch. Protistenk., 40: 16-26.

Osterud, K. L. 1938. The nitrogen requirements of *Lobomonas piriformis*. Anat. Rec., 72 (Suppl.) : 128-29.

—— 1939. The nitrogen and carbon requirements of *Lobomonas piriformis*. Anat. Rec., 75 (Suppl.) : 150-51.

Petersen, W. A. 1929. The relation of density of population to rate of reproduction. Physiol. Zoöl., 2: 221-54.

Phelps, A. 1935. Growth of Protozoa in pure culture. I. Effect upon the growth curve of the age of the inoculum and of the amount of the inoculum. J. exp. Zool., 70: 109-30.

—— 1936. Growth of Protozoa in pure culture. II. Effect upon the growth curve of different concentrations of nutrient materials. J. exp. Zool., 72: 479-96.

Philpott, C. H. 1928. Growth of *Paramecium* in pure cultures of pathogenic bacteria and in the presence of soluble products of such bacteria. J. Morph., 46: 85-129.

Pringsheim, E. G. 1912. Kulturversuche mit chlorophyllführenden Mikroorganismen. II. Zur Physiologie der *Euglena gracilis*. Beitr. Biol. Pfl., 12: 1-48.

—— 1921. Zur Physiologie saprophytischer Flagellaten (*Polytoma, Astasia* und *Chilomonas*). Beitr. allg. Bot., 2: 88-137.

—— 1926. Kulturversuche mit chlorophyllführenden Mikroorganismen. V. Mitt. Methoden und Erfahrungen. Beitr. Biol. Pfl., 14: 283-312.

—— 1930. Algenreinkulturen. Eine Liste der Stämme welche auf Wunsch abgegeben werden. Arch. Protistenk., 69: 659-65.

—— 1935a. Über Azetatflagellaten. Naturwissenschaften, 23: 110-14.

—— 1935b. Wuchstoffe im Erdboden? Naturwissenschaften, 23: 197.

—— 1937a. Assimilation of different organic substances by saprophytic flagellates. Nature, 139: 196.

—— 1937b.* Beiträge zur Physiologie saprophytischer Algen und Flagellaten. 1 Mitt.: *Chlorogonium* und *Hyalogonium*. Planta, 26: 631-64.

—— 1937c.* Beiträge zur Physiologie saprophytischer Algen und Flagellaten. 2 Mitt.: *Polytoma* und *Polytomella*. Planta, 26: 665-91.

—— 1937d.* Beiträge zur Physiologie saprotropher Algen und Flagellaten. 3 Mitt.: Die Stellung der Azetatflagellaten in einem physiologischen Ernährungssystem. Planta, 27: 61-92.

Provasoli, L. 1937a. La Nutrition carbonée du flagellé *Polytoma uvella*. C. R. Soc. Biol. Paris, 126: 280-82.

—— 1937b. La Nutrition carbonée du flagellé *Polytoma ocellatum*. C. R. Soc. Biol. Paris, 126: 847-49.

—— 1938a. La Nutrition carbonée de l'euglénien *Astasia quartana* (Moroff). C. R. Soc. Biol. Paris, 127: 51-53.

—— 1938b. Remarques sur la nutrition carbonée des eugléniens. C. R. Soc. Biol. Paris, 127: 190-92.

—— 1938c. Studi sulla nutrizione dei Protozoi. Boll. Lab. Zool. agr. Bachic. Milano, 9 (rpr.), 124 pp.

Reich, K. 1935. The cultivation of a sterile amoeba on media without solid food. J. exp. Zool., 69: 497-500.

—— 1936. Studies on the physiology of *Amoeba*. I. The relation between nutrient solution, zone of growth and density of population. Physiol. Zoöl., 9: 254-63.

—— 1938. Studies on the physiology of *Amoeba*. II. The allelocatalytic effect in *Amoeba* cultures free of bacteria. Physiol. Zoöl., 11: 347-58.

Rottier, P. B. 1936a. Recherches sur les courbes de croissance de *Polytoma uvella*. L'influence de l'oxygénation. C. R. Soc. Biol. Paris, 122: 65-68.

—— 1936b. Recherches sur la croissance de *Polytoma uvella*. L'influence de la concentration des substances nutritives. C. R. Soc. Biol. Paris, 122: 776-80.

Schoenborn, H. W. 1936. Growth of two species of *Astasia* in relation to pH of the medium. Anat. Rec., 67 (Suppl.): 121.

—— 1938. Growth of *Astasia* sp. and *Euglena gracilis* in media containing inorganic nitrogen. Anat. Rec., 72 (Suppl.): 51.

—— 1939. Growth of *Euglena gracilis* on inorganic nitrogen sources in the absence of light. Anat. Rec., 75 (Suppl.): 151.

—— 1940. Studies on the nutrition of colorless euglenoid flagellates. I. Utilization of inorganic nitrogen by *Astasia* in pure cultures. Ann. N. Y. Acad. Sci., 40: 1-36.

Woodruff, L. L. 1911. The effect of excretion products of *Paramecium* on its rate of reproduction. J. exp. Zool., 10: 557-81.

—— 1912. Observations on the origin and sequence of the protozoan fauna of hay infusions. J. exp. Zool., 12: 205-64.

—— 1913. The effect of excretion products of Infusoria on the same and on different species, with special reference to the protozoan sequence in infusions. J. exp. Zool., 14: 575-82.

CHAPTER X

THE GROWTH OF THE PROTOZOA

Oscar W. Richards

GROWTH is a fundamental attribute of living organisms, manifested by a change of size of the individual, or in the number of organisms in a unit of environment. Negative growth may occur during adverse conditions or in certain dimensions when growth involves change of form. The analysis of population growth requires knowledge of the environment, the individuals, and the interactions of each on the other.

Growth is determined by measurement, and the information gained from any single measure is delimited by the nature of the measuring unit chosen. Rarely is a single measure adequate for the study of growth, even though it may be useful for practical application. Analytical studies require the simultaneous use of as many different measures as are necessary to give a picture sufficiently complete for the analysis. When there is no change in form, certain dimensions may be related directly, as length with volume or weight, but in allometric growth the conversion constants may change during the course of the growing period. These problems will be illustrated and discussed in this chapter, in so far as numerical data are available.

METHODS FOR THE MEASUREMENT OF GROWTH

Individual Protozoa have been measured to show growth changes in length and breadth, but these two dimensions may not permit very exact calculation of volume if the shape of the animal departs much from that of a sphere, cube, ellipsoid or other simple geometrical form. The area of the animal may be calculated by the use of a planimeter, from an enlarged photomicrograph or a tracing of the outline of the animal. The softer animals may be gently compressed between a slide and a cover glass, and the area measured. Multiplication of this figure by that of the thickness of the preparation gives the volume. The three-halves power of the area obtained from planimetric measurement may give the volume of some species fairly accurately. This is true for only a few solids, such as the cube. The method has an error of about 33 percent when used with spherical organisms. Chalkley (1929) measured the volume of *Amoeba* by gently drawing it into a capillary tube of known diameter and calculating the volume from the length of tube filled plus the two hemispherical ends.

Populations of Protozoa are usually measured by counting a sample of the population in a Sedgewick-Rafter cell with a Whipple disc in the eyepiece of the microscope, or with a hemocytometer (cf. Woodruff, 1912; Hall *et al.*, 1935). The chief source of error of this method depends on how closely the sample represents the population. Care must be used that none of the animals are lost by sticking to the transfer pipette and to make sure that all are counted once only. Berkson *et al.* (1935) has given a quantitative treatment of the errors of counting red blood cells with a hemocytometer, and their evaluation might be applied to estimates of protozoan populations.

Tippett (1932) has suggested that the mean number may be estimated by counting the squares containing 0, 1, 2, and so forth animals and using the tables prepared for the Poisson distribution. With Protozoa, greater precision may be obtained by killing the animals before making the count. Many killing fluids are hypertonic, and animals may be lost from the osmotic effects of the killing fluid. Jennings (1908) and others have found that Worcester's fluid causes little change with paramecia when a sufficient amount is used to overwhelm the animals. Hardy's (1938) method of estimating numbers by comparing with standards containing a known number of dots might be used when high precision is not required.

Protozoa may be centrifuged into a tube with a calibrated capillary bottom, similar to that used by Carlson (1913) for yeast. Elliott (1939) obtains greater precision and convenience by fusing a hematocrit tube to a 10 ml. centrifuge tube. When the animals are killed before centrifuging, it is necessary that the volume of the animals not be changed by an anisotonic killing fluid. Commercially made tubes should be carefully calibrated, as errors as great as 12 percent have been reported for some makes of Hopkins vaccine tubes. Solid packing may not be possible with the usual laboratory centrifuges, but for given conditions constant packing may be obtained in equal time intervals. If the values are to be used for other than intercomparison, the centrifugal force used should be stated. The nomogram of Shapiro (1935a) simplifies this computation. If the distribution of animals of different sizes changes, e.g., just after a large proportion of them have divided or during endomictic reorganization, the total volume may not indicate the number present. Size changes of yeast cells and failure to obtain constant packing have been reported by Richards (1934). Shapiro (1935b) has discussed the validity of the centrifuge method with respect to marine ova. Simultaneous counts and volume determinations of *Colpidium campylum* have been made by Bond (1933).

The population density of pigmented forms may be estimated from the optical density of the suspension, by means of a nephelometer (Richards and Jahn, 1933). A beam of light is passed through the suspension and the amount of light absorbed by the organisms is measured by a photoelectric cell and a microammeter. When I is the microammeter reading with a given tube and medium and I_t is the reading of the suspension at time t, then the optical density, $D = log\ I_t - log\ I$. In this way the small variations in transmission of different test tubes may be canceled out, and it is not necessary to open the tube, a factor which may be important if the organisms are reared in a bacteria-free culture.

The optical density depends on the number of organisms present, the distribution of organisms of various sizes, and their metabolic condition. It is sometimes difficult to relate measurements with this criterion to the number of organisms present, because changes in internal cell structure (e.g., storage products) may alter their transparency. With proper care and control, the nephelometer may give a useful measure of the amount of protoplasm present in the population. For technical information the following may be consulted: Kober and Graves (1915), Mestre (1935), Russell (1937), and Müller (1939). Difficulties in the use of the method have been summarized by Loofbourow and Dyer (1938) and by Stier, Newton, and Sprince (1939). Miss Wright (1937) has measured the turbidity of bacterial suspensions by passing the light beam through the suspension at right angles to the axis of the photoelectric cell.

The dry weight of Protozoa may be obtained by filtering them from the culture medium with filter paper of fine porosity, a sintered glass filter, an alundum crucible, or an asbestos mat; washing them rapidly to remove the culture fluid, but not to burst the cells, and drying them to a constant weight. A vacuum desiccator containing sulphuric acid or phosphorous pentoxide at room temperature may give better results than a drying oven. This is a difficult method to control, so as to get consistent results.

Bacteriological methods of diluting and plating are not often applicable to Protozoa, but may be useful for testing the culture medium to make certain that it is bacteria-free. Standard texts should be consulted for methods. The ordinary nutrient agar is not a certain medium for estimating the bacteria found in water, and special media must be used. The number of colonies on an incubated plate may be less than the number of bacteria unless care has been used to prevent clumping of the bacteria. The errors of plate counts have been evaluated by Mattick *et al.* (1935) and by Ziegler and Halvorson (1935). Gordon (1938) has questioned these probability tables.

Other methods which might be useful to protozoölogists are the measurement of the suspension in terms of viscosity (Shapiro, 1937) and the determination of the velocity of sedimentation (Nielsen, 1933). These would be used with killed or nonmotile animals.

THE GROWTH OF INDIVIDUAL PROTOZOA

Simpson (1902) measured the length and breadth of *Paramecium caudatum* with an ocular micrometer at a few and at many hours after fission. Jennings (1908) supplied the first detailed measurements, and the data from his summary table are plotted in Figures 130 and 131. At division the animal decreases in breadth and increases in length. After the separation the increase in both dimensions is increasingly rapid, then proceeds at a nearly constant relative rate, and finally slows until the cycle is repeated.

The graphs of the growth are made on arithlog, or semilogarithmic paper, to facilitate analysis. This equivalent to plotting the logarithm of the size against a linear time axis. The slope of the growth curve at any point is the relative rate of growth (dy/ydt). Two growth curves, parallel to each other, are changing at the same relative rates. When no change in form occurs, the curves for area and volume will be correspondingly above and have slopes two and three times as great as a linear dimension.

A year later Popoff (1908) measured the growth of *P. caudatum* by killing a sister cell immediately and the other member of the pair at a given time after fission. His results are expressed in micrometer units which have been converted into microns. The average of the cells killed

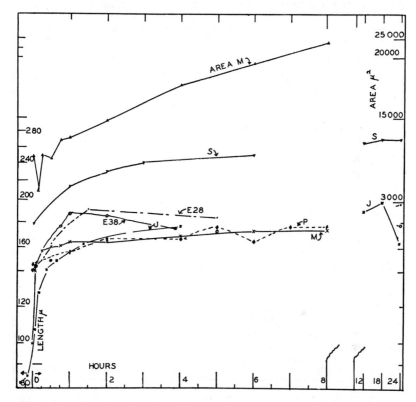

Figure 130. Growth in length and in area of *Paramecium caudatum*. Data: M from Mizuno (1927), S from Schmalhausen and Synagajewska (1925), E28 and E38 from Estabrook (1910), J from Jennings 1908), and P from Popoff (1909).

at fission was taken as the standard size at zero time, and the average differences for the intervals were added successively to obtain the data plotted in Figures 130, 131, and 132. He measured length (L), breadth (B), and thickness (T) and computed the volume as $= 4\pi LBT/24$, assuming the form to be ellipsoid, and he also gives similar measurements of the growth of the nucleus and of the nucleocytoplasmic ratio.

Estabrook (1910) investigated the effect of various chemicals on the

growth of *P. caudatum* and some of his control series, with more frequent measurements, are plotted. Measurements of growth in length by
Schmalhausen and Syngajewskaja (1925) and by Mizuno (1927) are
also available for *P. caudatum*. Mizuno also determined the area of the
animal with a planimeter, from a camera-lucida tracing.

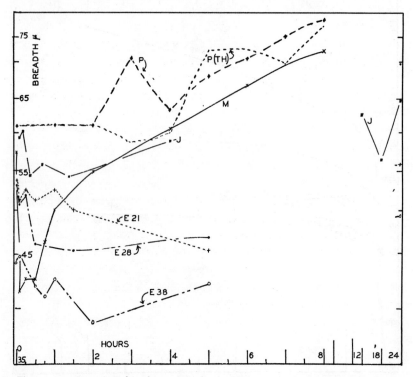

Figure 131. Growth in breadth and thickness (TH) of *Paramecium caudatum*. Data:
P from Popoff (1909), M from Mizuno (1927), J from Jennings (1908), E21, E28,
and E38 from Estabrook (1910).

All of the investigators took precautions to prevent any change in the
dimensions of the organism from killing the organism. The curves for
growth in length are quite similar. All were grown in hay infusion
medium and the temperature, when stated, was 24-26° C. Mizuno's animals grew at first at a more rapid relative rate than the others. Estabrook's and Schmalhausen's animals were appreciably larger than
Jennings's, Mizuno's, and Popoff's *Paramecia*. The decrease in the early

logarithmic growth is more rapid in the data from Jennings's measurements; however, his animals continued to grow for a longer period and reached a slightly greater length than did those measured by Mizuno and Popoff. The variations may be due to differences in nutrition or to race. The results are all in terms of averages from different animals, and do not show the continuous change of size of a given animal. This type of averaging of cross-sectional data is known to give variation.

The measurements of breadth (Fig. 131) are not as consistent as those of length. With the exception of one of Estabrook's series, the breadth decreases following fission, and growth in this dimension commences later, in the measurements given. Thickness and breadth measurements, plotted from Popoff's observations, show no change in size for the first two hours. The lack of agreement of the different series of measurements suggests differences in the pattern of growth for the different races. Growth in thickness occurred later than growth in breadth with Popoff's *P. caudatum*.

Growth in area (Fig.´ 130) is negative during the time that the breadth is decreasing, after which the increase continues for most of the cycle at about the same relative rate. Growth in volume, from Popoff's calculations (Fig. 132), does not show the early negative phase, because his animals apparently did not change form during division as the others did. With minor fluctuations the increase in volume continues in a suitable environment until the time of the next fission.

The growth of *Paramecium aurelia* (Fig. 132), plotted from the measurements of Erdmann (1920), is quite similar to that of *P. caudatum*. Erdmann's three races differed from one another in size. No decrease in breadth was reported for this species.

The growth in volume of the first individual and its progeny is given in Figure 132 for the soil amoeba, *Hartmanella hyalina* (Cutler and Crump, 1927). The growth curve is very similar to that of *P. caudatum*. Ten drawings were made rapidly at each time, and the volume was obtained from the average areas, on the assumption that the animals were 1 µ thick. The temperature was 21° C. The growth in volume of *Amoeba proteus* was measured by Chalkley (1929). The animals were pipetted until they assumed a spherical form, and the diameters of the cell and nucleus were then measured. The growth of the *Amoeba* is slow, and Chalkley's observations were not continued until the equilibrium size

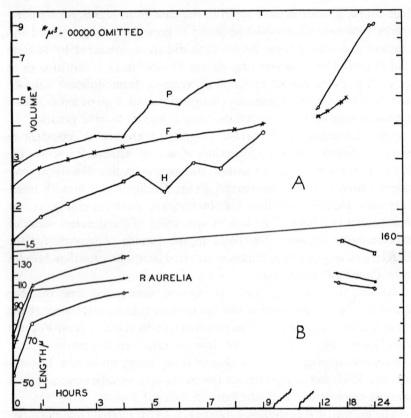

Figure 132. A. Growth in volume of *Paramecium caudatum* (P), *Frontonia leucas* (F), from Popoff (1907, 1908), and of *Hartmanella hyalina* (H) from Cutler and Crump (1927). B. Growth in length of *P. aurelia*. Data from Erdmann (1920).

was reached. The rate of growth is influenced by the number of nuclei present in the animal.

Popoff (1908) measured the growth of *Frontonia leucas* in length, breadth, and thickness, and estimated the volume as the product of these three factors. The growth of the nucleus was measured similarly. His data are given in arbitrary size units, which were converted into micra. The growth of *Frontonia* was much more variable than Popoff's growth data indicated to be the case for *Paramecium*, except for the volume changes. The average volume of the cytoplasm and the nucleus of the control animals killed at fission, was used as the value for zero time. To these values were added successively the average increments, in order to obtain the values at each time. The sum of the nuclear and the cytoplasmic volumes is plotted in Figure 132. The increments were read from Popoff's summary graph.

The growth in volume of *F. leucas* is more rapid during the first hour; the rate then decreases, until the relative rate is nearly linear, that is, until about one hour before the next division. The rate increases before division, mainly owing to the increase in nuclear volume. The volume of the cytoplasm increases continually. The volume of the nucleus decreases for the first two hours after division to about 86 percent of its size at fission, then increases slowly, until it shows the customary rapid increase in the three hours preceding a new fission.

Entz (1931) has followed the growth of populations and of individual dinoflagellates, *Ceratium hirudinella,* in their natural habitats. The largest animals were found in April. The highest rate of division occurred in late June, July, and August, which coincides with the period of maximum temperature. The growth was measured in three dimensions, and the time, in hours, for the growth stages is: nucleal and cytoplasmic division, one; horn regeneration, 2; slim stage, 27; indifferent stage, 72; compacted stage, 18; total, 120 hours.

Extensive studies have been made of the variation in the sizes of Protozoa, and this information has been collated by Adolph (1931). A change in form during growth is reported by MacLennan (1935) for *Ichthyophthirius.* This and other Protozoa would furnish excellent material for the study of allometric growth. Cf. Huxley (1932), Needham (1934), Teissier (1934), Richards (1935), and Huxley and Teissier (1936). The size of *Paramecium bursaria* was modified by changing the pH of the culture medium (Loefer, 1938a). Cell size and nuclear size in *Oxytricha fallax* was found by Woodruff (1913a) to be least during the periods of rapid reproduction, and to become larger as the division rate decreases. The nucleocytoplasmic ratio was highest during the period of greatest reproductive activity, and this was interpreted as an incidental result rather than as a cause of the rate of division.

The first evidence of division is a slight groove encircling the animal, and separation occurs in *Paramecium* about one-half hour later (Jennings, 1908). The nucleocytoplasmic ratio increases for the first three hours, and then decreases for the next three hours (Popoff, 1909).

During the first hour, Paramecia are little affected by chemicals, and food plays no rôle (Estabrook, 1910). Increase in length is predominant at first, and then the animals fill out in breadth. Negative correlations of length and breadth were found by Mizuno at one, four, and six hours after fission. Erdmann (1920) found changes in the sizes of Paramecia which were related to the endomictic rhythm.

Individual growth has been measured in relatively few species, mostly Infusioria. Data on other forms would be useful for comparison. Growth studies should be made successively on the same living animals, to avoid the difficulties of averaging and of using information from different individuals. A promising method for obtaining the measurements would be to take pictures of the animal in isolation culture, at frequent and regular intervals, through the microscope, with a motion-picture camera. The images could be measured from the film, which would provide a permanent record for the analysis of the growth.

The growth of individual Protozoa is quite similar to that observed with other plants and animals. If we disregard the first decrease in breadth, the growth curves are sigmoid. Sufficient information is not available to indicate to what extent there is a general change of form during growth. Many of these studies were not made with a constant culture fluid. In order to compare the growth patterns of different species, the culture conditions should be known to be optimal and reproducible. Such studies would establish standards for the further study of environment on the individual growth pattern. The application of geometrical and metabolic considerations, similar to those used by Schmalhausen and Syngajewskaja (1925) with bacteria, will contribute to the theory of growth.

The Growth of Colonial Protozoa

The colonial Protozoa are believed by Fauré-Fremiet (1930) to constitute an intermediary step between the unicellular and the multicellular organisms. The colony grows by regular division from a free-swimming cell, until a size characteristic for the species is reached. When the environment is unfavorable, the growth is restricted. In favorable cultures two types of growth are found. In certain species, e.g., *Epistylis arenicolae* and *E. Perrieri,* the first divisions are dichotomous and equal, and the mass growth of the colony follows in geometrical progression. Later, the sister cells divide unequally, and the growth becomes arithmetical, until the final size of the colony is reached. The time interval between cell divisions also gradually increases.

The growth of *Zoöthamnium alternans* depends on the common stalk, according to Fauré-Fremiet (1925). The cell initiating the colony has a given component of granular material which is distributed unequally during division and, as it is used up, the growth of the colony ceases through transformation of material into the stalk. This may be analogous to the accumulation of nonliving material in multicellular organisms, a phenomenon which some biologists believe results in the limited growth of such organisms. The growth curves are sigmoid (Fauré-Fremiet, 1925, 1930), and the growth is believed to be auto-catalyzed by the granular material.

Pedigree Isolation Culture and Life Cycles

The culture of a single cell in a drop of a suitable food and the isolation of one of the divided cells, shortly after division, into a new drop of medium, effectively maintains the environment constant, and the growth is then potentially unlimited. With a constant medium and suitable bacterial food, the rate of growth is very nearly constant, as has been demonstrated by Woodruff and Baitsell (1911), Darby (1930, 1930a, 1930b), Beers (1929), and others. If the growth curve for the sum of the individuals be plotted on arithlog paper, a straight line will result, because the growth (y) is exponential, $y = y_0 e^{kt}$, when y_0 is the amount of the seeding, or the growth at time, $t = O$ and e, is the Naperian base. The proportionality constant $k = (ln\ 2)/G.T.$, or $0.639/G.T.$ The generation time ($G.T.$) is the time between divisions. Tabel 5 summarizes the growth rates under fairly constant conditions for certain Protozoa.

The study of the variation in cell division has been obscured by the unfortunate practice of plotting the division rates in the form of a histogram. The bars of the plot commonly give the average division rate for ten-day periods. The histogram is used properly to show unit events which have no intermediate values, e.g., the results of tossing dice which cannot assume a position intermediate between two of the numbered sides. The average division rate is no such discrete, mutually exclusive attribute, but may take any fractional value within the limits of the experiment. An example showing how information on the division

TABLE 5. DIVISION RATES OF PROTOZOA WITH CONSTANT CONDITIONS

Organism	Divisions	° C	Remarks	Reference
Chilomonas paramecium	3.5/day 0.14/hour	24 26–30.5	NaAc-mineral salts	Mast and Pace (1934) Smith (1938)
Didinium nasutum	3.6/day	21	Fed on *Paramecia*	Beers (1929)
Euglena gracilis	.031* 0.47* 3.5 day	10 23 25	In dark no NaAc In dark with NaAc In wheat infusion	Jahn (1935) Sweet (1939)
Glaucoma pyriformis	6.86/day 7.65 to 8.02/day 5.62/day	24.2– 25.2	Yeast extract Whole yeast+yeast extract or peptone	Hetherington (1936) Phelps (1934)
Paramecium aurelia	0.72/day 1.2±/day 1.4/day 2.02/day	20± 26.8 28	Lettuce and bacteria	Phelps (1934) Woodruff and Baitsell (1911a) Phelps (1934) Phelps (1934)
Paramecium caudatum	2.1/day 1.8/day 2.3/day	25? 25–28 26	Over 200 days 51-day av.Min. salt +B. subtilis Oaten medium+bac- teria	Darby (1930b) W. H. Johnson (1936) Gause (1934)
Polytoma uvella	4.4/day 1.85/day	22 22	Aërated peptone me- dium Unaërated peptone me- dium	Rottier (1936) Rottier (1936)
Stentor coeruleus	2.1–0.7/hour 0.65/hour	18–20 22	Modified Peters' me- dium+ciliates Hetherington medium +Blepharisma	 Hetherington (1932) Gerstein (1937)
Stylonychia pustulata	4.5–5/day 3.2/day 3.7/day	25? 24 25.2		Darby (1930) Baitsell (1912)† Maupas†

* Divisions per day, per organism.   † From Darby (1930a).

rate may be obscured by incorrect histogram plots was given by Richards and Dawson (1927). The changes in division rate may be plotted to advantage as a running average. The three-day running average is readily calculated. If $x_1, x_2, x_3, \ldots x_n$ are the daily rates for n days, the value for the first is $\bar{x}_1 = (2x_1 + x_2)/3$; the second, $\bar{x}_2 = (x_1 + x_2 + x_3)/3$;

and the last is $(x_{n-1} + 2x_n)/3$. If further smoothing is desired, a five-day instead of a three-day running average may be used, and it is computed in a corresponding manner. Phelps (1934) gives an example of the running average plot.

In many of the earlier studies, the culture medium was inadequate, and after a time the division rate approached zero. Unless the animals were transferred to a favorable medium, the strain then died out. Such a growth period has been termed a "cycle" by Calkins. During a cycle, or, with some Protozoa during periods of nearly constant growth, small fluctuations in the growth rate occur. These minor variations are termed "rhythms" (Woodruff, 1905; Woodruff and Baitsell, 1911). Rhythms are associated with cellular reorganization (endomixis). The constant culture of *Didinium nasutum* without rhythms led Beers (1928) to believe that rhythms were due to food, temperature, and the condition of the culture medium. Rhythms may have a function in some species and be merely effects of the environment in others.

That the Calkins *cycle* depends on the adequacy of the culture medium has been demonstrated by a number of experimenters, e.g., Woodruff and Baitsell (1911a), Mast (1917), Beers (1928b), Darby (1930a), and Gerstein (1937). A medium that may be adequate for a few weeks may not be suitable for long periods. Dawson's *Paramecium* and *Blepharisma* showed gradual negative trends during the three years of the culture. At this rate the cycle would not end for several years (Richards and Dawson, 1927) and, in the meantime, some slight change in the medium might reverse the trend by supplying the cultural inadequacy, thus prolonging the cycle. Competition may bring out more rapidly the effects of the environment with populations than with individuals. The study of Protozoa, maintained for some time in an effectively constant culture medium, should add materially to our knowledge of growth. Peters (1901) gives useful methods; yeast techniques are summarized by Richards (1934).

Dawson kept pedigreed isolation cultures of *Histrio complanatus, Blepharisma undulans,* and a mutant *P. aurelia* for three years. A statistical analysis of the division rates removed the long-time trends, and established a seasonal cycle, with a maximum division rate in the summer and a minimum rate in winter (Richards and Dawson, 1927). The statistical methods used were those used in economics in the study of cyclic phenomena. Further study suggested that the seasonal cycle was associated with sunlight (Richards, 1929). The pigmented *Blepharisma* followed, more closely than the others, the seasonal variation of radiant energy. A recent graphic method of Spurr (1937) could be used to advantage in the analysis of seasonal cycles in the division rate of Protozoa. Properly controlled studies should be made to determine just how much effect light has, over a considerable period of time, on the growth of Protozoa. Such seasonal effect appears reasonable, as it is known that the reproductive cycles of some birds and other animals are initiated by the increased amount of light during the early part of the year.

Conjugation restricts variation, which aids in survival during adverse conditions, according to Pearl (1907). Endomixis occasioned large variations in size, which Erdmann (1920) believed aided in survival. She advised that attempts at selection be made during or immediately after endomixis. Changes that aid in survival of a species through an unfavorable period are important in population studies; they might even effect the growth of the individual, and they deserve further investigation by protozoölogists. Selection of rapidly dividing *Amoeba proteus* by Halsey (1936) did not produce a permanent race of rapidly dividing individuals. Burnside (1929) failed to change the size of *Stentor coeruleus* by fragmenting animals with large and small amounts of nuclear material. When the animals regenerated, the regulatory processes produced the same sized biotype.

Variations in the cell, at the time of division or during periods of intercellular reorganization, may aid in the adaptation of the cell to a new environment and may account for the success or failure of investigators in acclimating an organism to life in a synthetic liquid culture medium. Hegerty (1939) has shown that young *Streptococcus lactis*, at the end of the lag period and just before the period of logarithmic growth, can produce new enzymes which permit the use of a new substrate, to which the bacteria could not adapt themselves at any other period of their life cycle. Do comparable changes occur in Protozoa? If so, pure culture methods would be facilitated.

Thus the nature of the life cycles, as demonstrated by the earlier investigators (M. Robertson, 1929), may now be studied effectively by physiological methods, as well as by post-mortem cytology. The detailed discussion of reproduction must be left for consideration in the other chapters. The critical periods of binary fission, conjugation, and intercellular reorganization are important to the study of growth, and further information on these phenomena will facilitate our understanding of growth.

The variety of Protozoa and the numbers of each vary in time, and the abundance of individuals is usually inversely correlated with the diversity of kinds. At Geneva, Roux (1901) found the largest variety of species in January and in October and found that in the same locations there was considerable variation at the corresponding time in two successive years.

The sequence of Protozoa on sewage filtration beds (New Jersey) was followed by Crozier (1923) and Crozier and Harris (1923). A maximum number of rhizopods was found in August and of ciliates in May-June and November-December. *Paramecium* had a sharp maximum in December-January, *Vorticella* in late December and in May, and *Colpoda* in the first third of the year. The sequence was attributed to the amount of anaërobiosis and to the formation and sloughing of the film. In this environment the abundance was directly correlated with the diversity of types.

Noland (1925) found the sequence of Protozoa related to the temperature, oxygen, and carbon-dioxide concentrations in natural ponds. The hydrogen-ion concentration was not believed to be a controlling factor. Most of the Protozoa found were not those usually studied in the laboratory, but when samples were transferred to the laboratory, *Colpoda cucullus, Glaucoma pyriformis,* and *Paramecium caudatum* appeared, showing that these animals may thrive better in the laboratory than in natural habitats.

Changes in the concentration of Protozoa in a Philadelphia pond were followed for a year by Wang (1928), who measured also the temperature, oxygen concentration, pH, and relative amount of dissolved gases. The surface forms showed the greatest variation, which was believed due to the dissolved oxygen, depending on temperature and on the activity of the plants. A marked increase of acidity could be a limiting condition. The maximum number of forms was found in September-October. Since the amount of sunlight was greatest at this time and the temperature was not at a maximum, Richards (1929) has suggested that sunlight may have had more effect than temperature on numbers. The kinds of Mastigophora and Infusoria were inversely correlated with the abundance of individuals during the seasonal variations.

Coe (1932) found Protozoa attached to cement blocks suspended in the Pacific Ocean at La Jolla, from June to October. Protozoan sequences and numbers have been used by Lackey (1938a) for the study of sewage pollution of streams.

Sufficient information is not yet available to explain the sequence of protozoan population growths, or the declines and succession by other species in nature. Many of the factors are interrelated, since the solubility of dissolved gases is a function of temperature, and the oxygen production of aquatic plants depends on the amount of light. The solution of these ecological problems promises to be of considerable practical value to man, as well as an aid in the elucidation of the growth processes.

PROTOZOAN SUCCESSIONS: LABORATORY

Cultures maintained in the laboratory are more readily followed than those in natural habitats, and there are many records of the growth of populations of Protozoa, their decline and succession by a comparable growth of another species. Woodruff (1912) reported that near the surface of mass cultures the sequence was monads, *Colpoda,* Hypotrichida, *Paramecium, Vorticella,* and then *Amoeba.* The sequence of increase and disappearance was identical with appearance, except that the *Amoeba* advanced from the sixth to the fifth, and then to the fourth place. A definite succession was not apparent at the center or the bottom of the cultures, and a second cycle was rarely observed. The maximum rise and fall was about equal, but the final disappearance might be long delayed. The differences in the relative potential of division were believed to establish the sequence, which was determined by the food and waste products secreted by the animals. The waste products were shown to be toxic, and the toxicity was species-specific and did not effect other species (Woodruff, 1913b). No relation was found between the titrable acidity and the sequence of the Protozoa (Fine, 1912). The acidity was related, rather, to the activities of the bacteria present.

Fifteen series of two-liter cultures, made to imitate natural conditions, were followed by Eddy (1928). Counts were made with a Sedgewick-Rafter cell, but the results were not published, beyond general statements of sequence and dominance. Light had no effect on the sequence. Temperature exerts its influence by way of the bacteria serving as food for the Protozoa. Oxygenation of the culture increased the growth, especially at the bottom of the culture. Too great concentrations of carbon dioxide were deleterious and could be buffered by including soil in the culture. The sequence was effected by the quantity and type of the infusion material. Dominance of a species was believed to depend on favorable growing conditions for that species, rather than on the rate of reproduction (cf. Woodruff, above).

Unger (1931) has listed the sequence of Protozoa for two years in five cultures started from five different plants.

Laboratory cultures, not restricted to a single species, show a regular series of population growths and declines for different species. The nature of the culture, its bacterial flora, and the reproductive potential of each species regulate the period of intensive growth, and the accumulating excretion products of the animals bring about the decline of the population. The growth cycle of a species may modify the medium so that it becomes favorable for the growth of the next following species. Limiting conditions are oxygen and carbon-dioxide concentrations, pH, and temperature, and these will be discussed later. Remarkable flowerings of Algae and Protozoa in the ocean and in lakes have been reported and are apparently on a more intense scale than occurs in laboratory cultures. Some bacteria are inadequate as food sources; others are poisonous for some Protozoa; and the rise of a population of these bacteria would eliminate the susceptible Protozoa in the culture. Poisonous bacteria have been reported by Hargitt and Fray (1917) and by Kidder and Stuart (1938). The Protozoa commonly studied in the laboratory are apparently less frequently found in natural habitats. The growth of protozoan populations in mixed mass cultures is different from that of most other organisms, as no equilibrium is reached and maintained; instead, extinction seems to be the rule.

Autocatalysis and Allelocatalysis

The theories of T. B. Robertson have greatly influenced the study of growth, and the first of these has been concerned primarily with the growth of Protozoa. Robertson (1923) believed that the growth of an organism, or of a population of organisms, was *autocatalytic,* because the growth curves were sigmoid and could be fitted by the equation for a monomolecular, autocatalytic, chemical reaction. The slowest chemical reaction in the growth process was believed by Robertson to be the controlling *master* reaction for the process which established the form of the growth curve, and this could be discovered from the shape of the growth curve. His particular choice of chemical reaction was not satisfactory, and later he and other investigators have found difficulties which have, for the most part, led to the abandonment of the autocatalytic theory. Cf. Robertson (1923), Snell (1929), Jahn (1930), Kavanagh and Richards (1934).

The sigmoid nature of the growth curve is the inevitable result of the regular geometrical increase during the time that the environment is favorable, and the slowing of this increase when the environment becomes unfavorable as a result of the growth in it (decrease of foodstuffs and accumulation of excretion products). As long as the environment is maintained effectively constant, the rate of growth is constant and the growth curve is exponential. However, it eventually becomes impossible to maintain this constancy, and the growth is thus ultimately arrested. In this sense Bernard's "milieu interieur" is part of the environment. The granular material which Fauré-Fremiet believes to limit the growth of some colonial Protozoa is one of the few reported examples of limitation in growth which apparently follows the appearance of a single substance. Such a substance might be considered a catalyst in the Robertson sense. Teissier (1937) has questioned this conclusion, and Snell's (1929) objections are also applicable. Such substances, however, are rare.

Allelocatalysis, according to Robertson (1924a), is "the acceleration of multiplication by the contiguity of a second organism in a restricted volume of medium." Robertson reported (1921b) that two *Enchelys farcimen,* or two *Colpidium colpoda* in a drop of culture medium divided more rapidly than twice the division rate of one individual in an environment of equal volume. It was shown later that his *Colpidium* was *Colpoda cucullus.* Other publications followed, reporting that some unknown substance, the allelocatalyst, stimulated cell division, and Robertson believed this was formed during nuclear division and effected the permeability of the cells.

Cutler and Crump (1923), using *Colpidium,* were unable to confirm Robertson; and Greenleaf (1924) failed to demonstrate allelocatalysis with *Paramecium aurelia* and *P. caudatum,* and with *Pleurotricha lanceolata.* Peskett (1924) could not demonstrate allelocatalysis with yeast. Robertson (1924) attributed their failures to the fact that they had not washed their cultures free from the catalyst present in the medium from which the cells were removed for inoculation. Cutler and Crump (1925) and Peskett (1925) repeated their work, but were unable to demonstrate allelocatalysis either with washed or unwashed cultures.

Yocom (1928) found the division rate of *Oxytricha* higher in cultures of four drops of medium than in ten-drop cultures, and attributed the difference to an allelocatalyst. Petersen (1929) found that division of *P. caudatum* was accelerated in volumes of culture of 0.83 ml., but not in volumes of less than 0.21 ml. Dimitrowa (1932) obtained better growth in "conditioned" medium which had previously supported the growth of *Paramecium* than in medium which had not been "conditioned." *Colpidium campylum* grew better when some sterile filtrate from an old culture was added to a synthetic medium, according to Hall and Loefer (1938). Garrod (1936) reported that small inocula of *Staphylococcus aureus* did not grow in broth, but that large inoculations would grow. Mast and Pace (1937, 1938b) give evidence in support of an unknown substance produced in cultures, which, in low concentrations, stimulates the growth of *Chilomonas paramecium,* but which in high concentrations retards the growth of the animals. A soil amoeba, *Mayorella,* grown bacteria-free in mass cultures by Reich (1938), divided less when the initial populations were small. His data, replotted in the form of Figure 134, shows that the populations were proportionate to the seeding in rate of growth, within the large errors of observation, and do not support the allelocatalytic theory.

Yeast populations grew at the same rate when the inoculation was varied from 5 to 8×10^6 cells per ml. (Clark, 1922); and from 12 to 1,200 cells per cu. mm. (Richards, 1932). Peskett (1927) found no difference when one yeast cell was introduced into volumes from 0.008 to 40 cu. mm. Meyers (1927) failed to demonstrate allelocatalysis with *P. caudatum* and found that conditioning the medium lessened the growth of the animals. Increasing inoculations of *Glaucoma* up to

70,000 times gave no allelocatalysis (Phelps, 1935). Darby (1930) maintained that allelocatalytic effects were due to the pH of the medium, and Jahn (1933) believed them due to the oxidation-reduction poising of the medium. When the medium was optimum, there would be no increased rate of reproduction; but if the medium was suboptimum, two or more organisms might modify it enough to permit growth whereas one organism could not do so and would grow slowly or fail to survive.

Johnson and Hardin (1938) reported that medium conditioned by the growth of *Pseudomonas fluorescens* inhibits the reproduction of *Paramecium micronucleatum*. With the saline medium, used old-culture medium was as efficient as fresh medium. The difference between these and Woodruff's conclusions may be due to the effects of mixed bacteria in the natural medium used by Woodruff. Kidder (1939) studied the effect of conditioning with a bacteria-free *Colpidium campylum* culture in proteose-peptone, dextrose broth. He believes that there is an accelerator and an inhibitor in the conditioned medium for growth. These were separated by absorption and filtration. Caution should be exercised in the use of filtered media, as some kinds of filters make the filtrate toxic (Richards, 1933).

Sweet (1939) reinvestigated the volume seeding relation, using *Euglena gracilis,* and found that seedings of one and two individuals grew better in four drops of about 0.05 ml. each and inoculations of four and eight individuals in slightly larger, five-drop environments. This author's methods and technique illustrate survival, rather than growth, and while a volume effect of the environment was found, her results did not support the Robertson theories.

The observations of these investigators and others focused attention on the suitability of the culture medium and suggested that the allelocatalytic effects found by some biologists and discredited by others might be explained on this basis. Woodruff's (1911) demonstration that the waste products limited growth was recalled and clarified some of the volume effects on growth, wherein the yield of cells depended on the volume of the culture medium rather than on the size of the inoculation. Johnson (1933) explained allelocatalytic effects on the relation of the bacterial food concentration to the number of Protozoa in the culture. An allelocatalytic effect on *P. caudatum* and on *Moina macrocopa* was found with a high nutrient concentration, and the reverse of this with media of low nutrient concentrations (McPherson, Smith, and Banta, 1932).

Another possible interpretation depends on the presence or absence of essential elements, both organic and inorganic, or on vitamin or hormone-like effects. This field has hardly been touched, and investigations here may clear up many problems concerning the nutritional requirements and the responses of the organisms to various culture fluids. The present tendency is to look in this direction for an understanding of variations in the reproductive rate, rather than to attribute them to special allelocatalysts. Cf. Elliott, (1936), Hammond (1938), Koser and Saunders (1938), Hall (1939), and other chapters of this book.

Another explanation of the effect of the volume of the culture on the reproduction rate of the organisms might be that in larger volumes the organisms use more energy swimming about, which would leave less energy for reproduction. This view could be tested by the use of cinephotomicrographic films in measuring the amount of activity of animals in large and small isolation cultures, and correlating this figure with the rate of multiplication. The relation might be different in rich and in poor nutrient media and, if so, this would elucidate some of the contradictory observations in the literature.

NUTRITION AND GROWTH

Protozoa (Ciliophora) feed naturally on bacteria, and with mixed population of both it is difficult to analyze the growth. Maupas recognized this difficulty in the nineteenth century and recommended that Pasteur's methods be applied to the pure culturing of Protozoa. However, for some time little was done, other than to insure a uniform and adequate supply of bacteria in the medium by cross culturing.

Hargitt and Fray (1917) isolated and identified a number of bacteria from protozoan cultures and endeavored to grow *Paramecium* on pure cultures of bacteria, but found that no single species of bacteria was as satisfactory food as mixed cultures. *Bacillus subtilis* was the nearest satisfactory single species. Some species of bacteria were found to be toxic to the paramecia, and other poisonous bacteria have been reported by Kidder and Stuart (1938). Phillips (1922) extended the work of Hargitt and Fray and was unable to find a single species of bacteria suitable for the maintenance of *P. aurelia*. She concluded also that the paramecia could not live on dissolved substances, but were dependent on particulate food. *Glaucoma ficaria* was grown on a number of single species of bacteria, yeast, and flagellates by D. E. Johnson (1936). *B. prodigiosus* was the most satisfactory food organism. The results depended largely on the food being small enough for ingestion.

Recent studies have been directed toward determining the food elements required by Protozoa and toward devising synthetic media in which the Protozoa could be grown in bacteria-free, pure cultures. While it it not possible to separate studies on growth and nutrition except for convenience, this chapter will be limited to studies occupied primarily with the analysis of growth. The broad problem of nutrition will be covered elsewhere. Different species have different nutritional requirements, and the failure of some protozoölogists to realize this fully has led to confusion in the literature on growth. Very few data are available which give the growth of the bacteria, as well as that of the Protozoa, present in mixed cultures. Considerably more labor would be involved in securing this information, but the methods have been worked out and the information gained would justify the work. It is now possible to grow pure cultures of a variety of Protozoa in bacteria-free synthetic media. Some of the nutrient conditions limiting growth will be examined briefly.

Tolerance to changes of osmotic pressure was found by Loefer (1938), in attempts to adapt fresh-water Protozoa to artificial sea water, to be limited. Yocom (1934) was more successful. Loefer (1939) found that tolerance to diluted Van Hoff solution developed over several generations. Changes in the oxidation-reduction potential have been measured in *Chilomonas paramecium* cultures by Jahn (1933), and his results suggest that when the medium is poised at the optimum rH, growth will be most rapid.

The increased growth of Protozoa at the surface of mass cultures shows their sensitivity to oxygen. Aëration will often extend the growth to deeper levels. Inadequate amounts of oxygen limit the growth of *Polytoma uvella,* and sufficient oxygen must be provided before the effects of other nutrients may be evaluated (Rottier, 1936; Mond, 1937). Reich (1936) believes oxygen concentration more important in *Amoeba* cultures than acidity. Jahn (1936) aërated bacteria-free cultures of *Glaucoma pyriformis* and *Chilomonas paramecium* in a hydrolyzed casein medium with air, and air freed of carbon dioxide.

The *Chilomonas* grew best in unaërated cultures and not so well in the cultures aërated with CO_2 free air. The *Glaucoma* grew equally well with and without CO_2, but better than in unaërated cultures. Jahn believes that CO_2 is necessary to some organisms to avert the weakening of the buffer systems within the cell. The anaërobes are believed less sensitive to CO_2 removal because the amino acids and other weak acids may replace the carbonic acid. The lag period in bacterial growth varies with the CO_2 concentration (Walker, 1932), and increased production is associated with physiological changes in the bacterial cells (Huntington and Winslow, 1937; Gladstone et al., 1935). Similar effects should be watched for in protozoan populations.

Temperature has long been known to affect growth. Woodruff and Baitsell (1911b) found that the Q_{10} for the cell division of *P. aurelia* was 2.7, over a range of 21.5° to 31.5° C., and that the optimum range for them was 24° to 28.5° C. Individual pedigree cultures and mass cultures were measured by Mitchell (1929) over a range of 12° to 27° C., and the thermal increment (μ) for cell division was found to be 23,000 calories. A lag was found in the isolation cultures, and a method is given for calculating the division rates from data covering several days. Possibly with a different culture medium the lag might have been avoided or changed. Daniel and Chalkley (1933) found μ to equal 16,500 for the whole division process of *Amoeba proteus* (4° to 30° C). For nuclear division μ equals 16,600 (4° to 35° C.); for cytoplasmic division, 20,500 (11° to 21° C.), 7,300 (21° to 26° C.); prophase 11,700, and anaphase 20,200 (13° to 26° C.). The increments suggest that oxidative processes control cell division.

Jahn (1935) found a maximum growth rate for *Euglena gracilis*, grown in a hydrolyzed casein medium at 10° C., but the addition of sodium acetate changed the temperature of maximum growth to 23°. Motility and the occurrence of encystment and palmella stages were related to the temperature and food. Smith (1938) reported that *Chilomonas paramecium* grew in a sodium acetate-mineral salts medium from 9.5° to 35° C., with an optimum range of 26° to 30.5° C. Prolonged exposure to the lower temperatures decreased the resistance of the animals to the cold. Adaptation to changed temperature required at least forty-eight hours. The synthesis of fat and starch is a result of temperature and in turn may control the division rate.

The chlorophyll-containing Protozoa vary in their light requirements.

TABLE 6. THE EFFECT OF HYDROGEN-ION CONCENTRATION ON THE GROWTH OF PROTOZOA

Organism	Range	Optimum	Medium*	T°C	Remarks	Reference
Amphileptus sp.	6.8-7.5	7.1-7.3				Pruthi (1926)
Chilomonas paramecium	4.8-8.0	6.8	Inorganic salts acetate	24.4		Mast and Pace (1938)
Chlorogonium elongatum	4.4-8.6	7.5	Tryptone salts	28	70 hours in light	Loefer (1938c)
Chlorogonium euchlorum	4.4-8.6	7.5	Tryptone salts	28	70 hours in light	Loefer
Chlorogonium tetragonium	4.4-8.6	8.6	Tryptone salts	28	70 hours in light	Loefer
Colpidium striatum	4.0-8.6	5.8 and 7.4	Tryptone	25		Elliott (1936)
Colpidium sp.	6.0-8.5	—				Pruthi (1926)
Colpidium sp.	4.5-8.0	6.0				Mills (1931)†
Colpoda cucullus	5.5-9.5	6.5 and 7.5				Morea (1927)†
Didinium sp.	5.0-9.6					Beers (1927)
Euglena anabaena	5.5-8.0	6.8	MS+YE or P	29.5		Hall (1933)
Euglena deses	5.3-7.8	7.0	YE or P	29.5		Hall (1933)
Euglena gracilis					Enzyme opt. 6.6	Jahn (1931)
Gastrostyla sp.	6.0-8.5		YE+P			Pruthi (1926)
Glaucoma pyriformis		7.0	YE+P	24-25		Hetherington (1936)
Glaucoma ficaria	4.0-8.9	5.1 and 6.7	MS+YE+Tryptone	28		Johnson (1935)
Glaucoma pyriformis	4.0-8.9	5.4	MS+YE+Tryptone	28		Johnson (1935)

Species	pH range	pH optimum	°C	Medium	Remarks	Reference
Holophyra sp.	6.5-7.4					Pruthi (1926)†
Paramecium sp.		7.8-8.0				Pruthi (1926) Saunders (1924)
Paramecium aurelia	5.7-7.8 5.9-8.2	7.0 5.9-7.7			Range favorable no optimum	Darby (1929, 1930)† Phelps (1931, 1934)
Paramecium caudatum		7.0				Darby (1930)
Paramecium multimicronucleata	4.8-8.3	7.0				Jones (1930)†
Plagiopyla sp.	6.9-7.5					Pruthi (1926)†
Spirostomum sp.	6.5-8.0	7.5				Morea (1927)†
Spirostomum ambiguum	6.8-7.5	7.4				Saunders (1924)†
Stentor coeruleus	7.8-8.0		18-20	Modified Peters+c. campyl.	Calk ratio important	Hetherington (1932)
Stylonychia pustulata	6.0-8.0 6.1-	6.7 and 8.0 7.6				Darby (1929)† Darby (1930)
Uroleptus mobilis						Gregory (1926)

* MS=mineral salts, YE=yeast extract, P=peptone.   † From Johnson (1935).

Euglena gracilis can be grown in the dark for extended periods. Without light, some of the organisms require more complicated food substrates, and the experiments demonstrating this have been summarized by Hall (1939). The amount of sunlight may exert a seasonal effect on the division rate of Protozoa (Richards, 1929). Reflected light stimulated multiplication of *Paramecium caudatum* in the red, but had a depressing action in the violet, according to Zhalkovskii (1938). Filtered, transmitted light had a greater depressing effect than reflected light. The difference was believed to be due to the polarizing effect of the reflected light. Heritable changes in the size and form of *Chilodon uncinatus* have been produced by McDougall (1929) with ultra-violet radiation. Giese (1939) found that ultra-violet of 2,654Å injured the nuclear material of *P. caudatum* and that the damage was less readily repaired than was the damage to the cytoplasm caused by 2,804Å.

The ease with which the acidity of the culture medium may be measured is responsible for a considerable volume of information (Table 6). The early measurements of Peters (1904, 1907) and Fine (1912) were made by titration. The advent of simple methods for the measurement of the hydrogen-ion concentration was welcomed by the protozoölogists, and Bodine (1917) related the old and the new methods. Pruthi (1926) found a sequence in hay infusion of *Holphyra, Plagiopyla, Colpidium, Amphileptus,* and *Paramecium.* The first two do not persist beyond pH 7.5 and the paramecia did not appear before the pH reached 7.0. Mass cultures and some synthetic media change during the growth of the organisms, and the change in pH of mixed cultures is probably more the result of bacterial action than that of the Protozoa. Eddy (1928) believed that the changes in pH were not of importance in themselves, but rather the result of other effects. Phelps (1931) attributed the changes to the food supply, and Johnson (1935, 1936) stressed the effect of bacterial action. The accelerative effects of stimulants depend on the acidity of the medium, and there now seems no question but that there is an optimum pH range for different media and Protozoa, and that beyond the optimum range growth is less and may be entirely inhibited. Elliott (1935b) found that sodium acetate stimulated *Colpidium* at pH 6.8-7.5, and butyrate at a pH less than 7.0. The size of *P. bursaria* has been shown by Loefer (1938a) to depend on the acidity of the tryptone and proteosepeptone culture media.

A usable source of C, H, O, N, Ca, K, P, and Na is probably required by all Protozoa. Inorganic media have been used by Hall and Schoenborn (1938) to separate strains of flagellates, by choosing media in which one strain will survive and others perish. Some Protozoa can obtain nitrogen from nitrates or ammonium salts, while others require amino acids, proteoses, or peptones. Sodium acetate, glycerate, or glycerophosphate are among the simplest carbon sources required by the nonphotosynthetic organisms. Loefer (1935) has summarized the carbohydrate requirements. The growth of *Chilomonas paramecium* requires sodium acetate, magnesium, sulphur, and silicon (Mast and Pace, 1933), and vanadium and copper increase the rate of growth (Bowen, 1938). *Colpidium* needs phosphate and a minimum three-carbon source, according to Peters (1920). Potassium and magnesium may be omitted from glass cultures, but are required when quartz vessels are used. Uranium salts cannot be substituted for potassium (Peters, 1921). The addition of pimelic acid to a glycerine-dextrose medium permitted growth of *Colpidium* (Hall, 1938b). Bacteria-free *P. bursaria* grew in proportion to the concentration of the culture medium (Loefer, 1938d).

Polytoma grows better when aneurine (synthetic B_1) and thiazol compounds are present (Lwoff and Dusi, 1937, 1938); and trypanosomes need hematin and cholesterol. *Ameoba* and *Paramecium* grow better in the presence of sulfhydryl, and this may be a general requirement of Protozoa. Hammett (1929) obtained an increased growth of *Paramecium*, although it was not proportional to the SH content. Hall (1938) found that manganese stimulated the growth of *E. anabaena*, but failed to stimulate *Astasia* sp. and *Colpidium campylum*.

Culture media have been improved by the addition of yeast extract for *Uroleptus, Dallasia, P. bursaria, Pleurotricha,* and *Stylonychia,* and Gregory (1925-28) found that the stimulation or depression of the division rate of *Uroleptus mobilis* depended on the age of the culture. Beef extract has proved a suitable food for mixed cultures. Plant hormones, indoleacetic, indolebutyric, and indoleproprionic acids increase the growth of chlorophyll-containing Protozoa, while pantothenic acid stimulated those tested by Elliott (1935a, 1938) which did not have chlorophyll. Mottram (1939) reported that 3 : 4 benzpyrene is a growth stimulant for *Paramecium*.

Beers (1928a, 1928c) grew two parallel lines of *Didinium nasutum* on well fed and on starved *Paramecium,* and showed that the inadequate growth of the *Didinium* restricted to a diet of starved *Paramecium* was due to a qualitative deficiency, rather than to a shortage of food. Mond (1937) reported that Infusoria grown in known concentrations of *Bacillus coli* and *B. subtilis* grow in a linear relation to the available food. The same amounts of bacteria were used for each division of the Infusoria. Such studies will permit the determination of the amounts of energy used for the growth process and for maintenance of life. When enough data become available, Wetzel's (1937) methods may be used and the resulting data would aid in evaluating his theory of growth.

The knowledge of the nutritional requirements of the Protozoa is increasing rapidly and suitable methods for growing bacteria-free pure culture of a number of species are now available. It will be difficult to decide what is the optimum culture medium for a given species. The lack of trace elements may appear only after a period of years. Superoptimal media will give an increased rate of growth which may not be best for the species (McCay, 1933). Pearl's (1928) generalization that the length of life is inversely correlated with the rate of living must be remembered when experimental conditions are devised either to yield a maximum amount of Protozoa in a given time, or to provide an opportunity for the study and perpetuation of the species under the most favorable conditions.

The Growth of Population

Adequate measurement of population growth should include the following information, as well as the number of organisms present at a given time, per unit of environment: food concentration; the concentration of excretion products, pH, rH, oxygen, and carbon-dioxide concentrations; temperature; the amount of light, when light-sensitive organisms are used; and the effects of other species, when mixed populations are used, on the species measured and on its environment. Few studies approach this degree of completeness. The earlier studies made no attempt to measure the food concentration, when bacteria were the main source of organic food.

A few Protozoa inoculated into a limited amount of an adequate culture medium, soon begin to increase in numbers and continue to do so until a maximum yield is produced. The course of the population growth may be divided into the following phases: (1) a stationary period, (2) a lag period of increasing rate of growth, (3) a logarithmic period of constant relative rate of growth, (4) a period of declining rate of growth, (5) a period of equilibrium of numbers, and finally (6) a period of declining numbers. The duration of these phases, and even the presence or absence of some of them, depends on the age of the inoculation and the nature of the environment. The stationary and the lag phases may be eliminated when the inoculation has been taken from a culture during the logarithmic period. Very large inoculations may exhaust the food supply or make the environment toxic from the excretion products, before any appreciable growth can take place.

The understanding of population-growth studies may be clarified by the aid of a hypothetical example, Figure 133. If ten organisms were seeded into a limited amount of a suitable culture medium from a population in the logarithmic phase of population growth, they would grow at a constant rate, doubling their number at the end of each generation time (curve A). If the conditions of growth were identical, the rate of growth of the inoculum would be the same as it was in the parent population. After a time the environment will no longer be effectively constant, and the rate of growth will decrease. This may come about by the lapse of more time between generations (curve B), or by only a part of the animals being capable of reproducing (curve C). It is apparent that if there had been different periods of increasing generation time, or if different numbers had been permitted to reproduce, it would have been possible to make curves B and C coincide at all points. Therefore, it is not possible to decide from the shape of the growth curve alone, the cause of the slowing of the growth rate. A third possibility, which would give a curve of the same general shape, would be a selective encystment or the death of some of the animals, which would reduce the number of individuals capable of reproduction. Such considerations emphasize again the necessity of information from the use of more than one criterion for the analysis of growth.

The stationary or equilibrium period, when the population does not change in numbers, is indicated by the curve D and the period of declining population by curve E. The stationary period is usually a dynamic equilibrium wherein the birth and death rates balance each other, but it could be static if all of the cells encysted or became otherwise inactive.

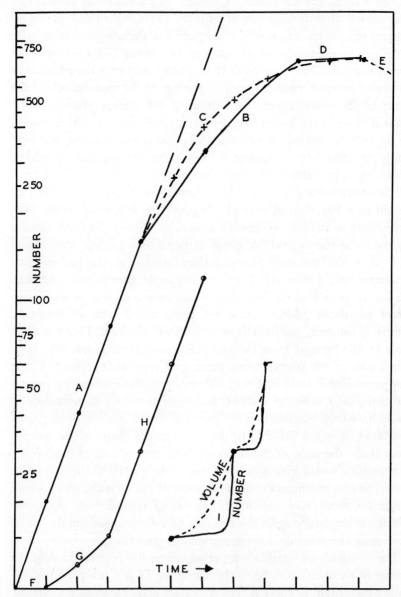

Figure 133. Hypothetical curves to illustrate phases of population growth. *Cf. text.*

The period of decline will depend on the nature of the environment and the rate of growth. It may show a phase of increasing death rate, a logarithmic death rate, or a decreasing death rate, or these phases may follow one another.

Had the inoculum been taken from an old culture which had reached the equilibrium or the period of decline, then there might have been a stationary period (F), followed by a period of increasing rate of growth (G), which would be followed by a constant relative rate (H), shown by the curve becoming parallel with the A curve. It is advantageous to know and to take into account these phases, in experiments with populations of unicellular organisms. The duration of the stationary and the lag phases will vary with the age of the inoculum and the effect of the previous unfavorable environment of them. Populations from old cells often provide considerable variation. Whenever possible, experiments should be made during the logarithmic period, to insure uniformity of material.

The detailed shape of the growth curve is often not known, because of infrequent measurements. If the Protozoa divided synchronously at the end of the generation time the curve would be like curve I.

The difference between the number of organisms in a population, shown by curves B or C, and the number theoretically possible, shown by the extension of curve A, is a measure of the inadequacy of the culture medium. The difference between the expected maximal number, curve D, and the number at a given time measures the potential growth yet to be achieved. The environmental resistance may be expressed as one minus (the potential growth divided by the expected number). This type of analysis, in terms of the logistic equation, has been made by Gause (1934) for the population growth of *P. caudatum,* and his instructive graph should be examined by all students of population growth. For information on the mathematics of growth, Pearl (1925), Jahn (1930), Richards and Kavanagh (1937) may be consulted. Protozoölogists have not used mathematical methods to any extent. Park (1939) also reviews Gause's analysis. Similar growth studies of other protozoan populations, besides presenting local data, should contribute to the general understanding of growth.

Buchanan and Fulmer (1928) have reviewed the literature of the growth of bacterial populations; Richards (1934) yeast populations;

Jahn (1934) protozoan populations; and for other animals, Pearl (1925), Gray (1929), Gause (1932), Johnson (1937), Hammond (1938), and Park (1939) may be consulted for reviews and bibliography.

Jones (1928) followed the population growth of an inoculation of 200 *P. multimicronucleatum* in 70 cultures at 80° F., with counts of 0.5 ml. samples made periodically. The pH of the medium was also measured. A maximum crop of 10^6 paramecia were obtained in 700 ml. cultures. Growth stopped when the pH decreased to 5. With hay-flour infusions, two cycles of growth were found (1930). The first was terminated by the high acidity; when the acidity returned to about pH 7, the second growth cycle commenced. During a three-day period Jones (1937a) found that the number having died at the close of the first growth cycle exceeded the maximum number present during the second period. The death of the animals was believed to be due to toxic excretion products, which were neutralized by the materials liberated from the cytolysis of the dead animals. Death was apparently disruptive, as no intact dead animals were observed. With large one-gallon cultures, the decline of the populations was related to the decline of food; and, by periodically renewing the food, the cultures could be maintained for four years.

The growth of *Euglena gracilis* in mass cultures was used by Jahn (1929) to test the allelocatalytic theory of Robertson. The organisms were derived from a single cell isolation and grown in an autotrophic mineral-salts medium, with temperature and light controlled. Jahn's larger inoculations gave a population growth with two cycles (Fig. 134). No evidence for allelocatalysis was obtained. The relative rates of growth were computed (Jahn, 1930) and found to give a decreasing sigmoid curve. Jahn emphasized the difference between the absolute rate of growth (dy/dt) of the total number and the relative rate of growth (dy/ydt), or division rate, of the organisms, without entering into the discussion of the relative growth rate as such.

Phelps (1935) measured the population growth of bacteria-free *G. pyriformis* in 700 ml. cultures of a mineral salts-yeast extract medium in one-liter flasks. The length of the stationary and the lag phases were proportional to the age of the seeding, and seeding from populations in the logarithmic phase gave no stationary or lag phases. Increasing

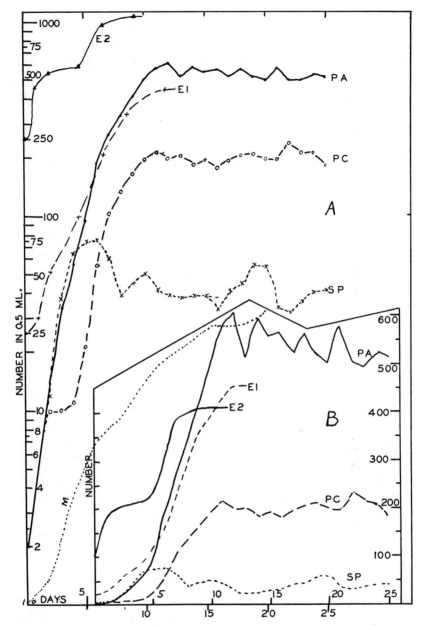

Figure 134. A. Population growth curves plotted on arithlog coordinates of *Euglena* (E₁ and E₂) from Jahn (1929); *Paramecium aurelia* (PA), *P. caudatum* (PC), *Stylonychia pustulata* (SP) from Gause (1934); and M. *Mayorella palestinensis* from Reich (1938). B. Population growth curves plotted on Cartesian coördinates (same data).

initial densities up to 70,000 times failed to show any allelocatalytic effect. The number of animals present at the end of the logarithmic phase was independent of the number in the seeding. In comparing the phases of *Glaucoma* population growth with those of bacterial and yeast populations, he found the following differences: the initial stationary and lag phases in *Glaucoma* populations are much shorter, in proportion to the optimum generation time; the stationary phase is independent of the size of the seeding; and the change from the logarithmic to the equilibrium phase of growth is more abrupt. No period of decreasing population size appeared within 120 hours.

Changing from yeast extract to yeast autolysate increased the yield (Phelps, 1936). The rate of growth was found to be independent of the food concentration within wide limits, but the total number of animals was proportional to the amount of food. The concentration of excretion products did not inhibit the growth until very great population densities were reached. This again is quite different from yeast cells, which are adversely effected by low amounts of excretion and fermentation products. A more favorable food medium and the use of aëration flasks, as well as differences in the species of animals used, may account for the lesser effect of waste products observed by Phelps than by Woodruff (1911). The *G. pyriformis* used by Phelps is identified now as *Tetrahymena glaucomiforma*.

The growth of populations of *Colpidium campylum* was measured by Bond (1933). With small amounts of yeast autolysate, the growth was slight and the lag period was greatly prolonged. With greater amounts of food, the equilibrium population was greater, the logarithmic phase was longer, the rate of growth greater, and the transition from the logarithmic phase to the equilibrium phase of the growth curve less abrupt. Bond's evidence suggests that the yield of animals depends more on the amount of food available than on an inhibitory effect of excretion products.

Gause (1934) presented the growth of a population of *P. caudatum* on an oatmeal infusion, with bacteria. He fitted the S-shaped growth curve with the logistic equation, and his analysis of the curve has been mentioned before. The growth curve of *Stylonychia pustulata,* Fig. 134, illustrates rapid growth, with a short equilibrium phase, followed by a period of negative growth leading to a lower equilibrium level. The second equilibrium level decreased slightly from the eighth day to the sixteenth day, when a second and shorter growth cycle commenced. The second cycle passed through a brief equilibrium period and then declined to about the same level as that which followed the first growth cycle. Population growth curves are given for *S. mytillus, P. aurelia* and, in a later monograph (1935), for *Glaucoma scintilans, Didinium nasutum, Bursaria truncatella,* and *P. bursaria.* Some of these will be discussed in the next section. One set of data is interesting from the viewpoint of population growth, that for *P. aurelia* and *P. caudatum,* grown separately in a standardized medium which was changed every twenty-four hours (Fig. 134). The equilibrium phases showed that there were over twice as many *P. aurelia* produced as *P. caudatum.* Gause then measured the sizes of the animals and computed the mean volume of each and the total volume of population. The volume curves showed that very nearly the same volume of protoplasm was produced by each species, with the same medium and conditions of culture.

P. caudatum was grown in a balanced salt medium, with one unit of concentration, and with five units' concentration of bacteria, by Johnson (1935). The growth curves are sigmoid and show no stationary phase and only a short lag phase. The equilibrium number was maintained with no decline for seven days. The number of animals produced in the greater concentration was more than five times the number in the lesser concentration. In the lower concentration a single animal divided more times than did a group of animals, while in the greater concentration a group divided more rapidly, for about three days, when the population figure from the single animal seeding passed the group curve to reach a higher equilibrium level.

Mond's (1937) estimates of *both* the bacteria and the infusorian populations point the way to more adequate studies of protozoan growth. Populations of *Colpoda duodenaria* were maintained in aëration flasks for four months by Taylor and Strickland (1938). By continuous feeding, densities of 6×10^6 per milliliter were produced. The size of the population fluctuated with the amount of food available and could be modified as the experimenters wished. Over the whole period the number of Protozoa produced from a given amount of food was constant. Excretion products did not limit the growth, but the continuous aëration may have ameliorated the effects of the waste products, so that the conditions are not comparable with unaërated cultures.

The growth curves of protozoan populations are sigmoid and resemble closely in form those of other populations. The growth curves of some Protozoa show all phases. The growth of different Protozoa depends on environmental conditions, and for details the reader should consult the original publications. The size of a protozoan population depends primarily on the amount of available food. Waste products do not limit the growth, as they do with yeast populations, and are inhibitory only in very dense populations. However, yeast populations contain more organisms than the protozoan populations—*Paramecium* (Jones, 1928) 10^5; *Glaucoma* (Phelps, 1936) 7.25×10^5; *Colpoda* (Taylor and Strickland, 1938) 60×10^5; yeast (Richards, 1932) 335×10^5 per ml.; bacteria (Steinhaus and Birkeland, 1939) to 2.5×10^9 —and laboratory populations of yeast are far less dense than those produced in aërated and cooled commercial fermenters. The total volume of protoplasm (number of individuals, $\times$ mean size) should be considered, and metabolic rates known, when comparing populations of different organisms. Under identical conditions *P. aurelia* and *P. caudatum* produced nearly the same total volume, although there were over twice as many of the smaller *P. aurelia*.

So far no selective mortality has been reported for protozoan population growth, although this is well known in yeast populations. The decline may occur because fewer of the Protozoa reproduce or it may be due to a slowing of the rate of cell division. Jahn (personal communication) believed the latter true for his *Euglena* populations. Jones reported a disruptive mortality in his *P. multimicronucleata* populations. No evidence of differences in the sizes and their distribution among Protozoa —which would reveal how homogeneous the populations are from time to time during the population growth—has been given in recent studies, with carefully controlled conditions (e.g., bacteria-free cultures, on synthetic media). Can Protozoa become resistant to an unfavorable medium and remain able to reproduce? Is encystment always governed by food concentration (Taylor and Strickland, 1938), or do other factors have a rôle? To what extent can an equilibrium population be maintained by en- and excystment? The lack of information on these and many other problems should attract more students of physiology and of growth to protozoölogy.

The Struggle for Existence

The mathematical analysis of the question of survival by Volterra, Lotka, Haldane, and others has established certain principles. Gause (1934, 1935) has contributed to both the experimental and the theoretical advancement of the subject. The mathematical analyses are complicated, even though in the state of first approximations, and the interested reader should consult the original articles. Cf. Lotka (1925, 1934), Kostitzin (1934), Gause (1934, 1935). Chapman (1931) gives a translation of part of Volterra's work. Protozoan populations have been used to test the hypothesis, and some of the experiments of Gause are here summarized to illustrate the beginning of a quantitative attack on the problems of struggle for existence and survival of the fittest.

Separate and mixed populations of *Paramecium caudatum* and *Stylonychia mytilus* were grown on an oatmeal infusion inoculated with *B. subtilis*. Neither species grew as well in mixed populations, but the influence of *Stylonychia* on *Paramecium* is about forty times as great as the effect of the latter on the former. With more food, provided by mixed, wild bacteria, *Paramecium* grew to about the same level in mixed populations as it did in pure population. *Stylonychia* grew only to about half the number when competing in the same environment with *Paramecium* as it would have alone, and its population soon declined, while that of the *Paramecium* maintained itself despite the competition.

Paramecium caudatum and *P. aurelia* may be grown together, and will compete for the same food. It is necessary to make comparisons in terms of volume of protoplasm, as discussed in the previous section. In mixed populations the growth curves for the two populations are quite similar for the first eight days, after which the *P. aurelia* population continues to grow, while that of the *P. caudatum* declines, reaching the point of extinction in about sixteen days. *P. caudatum* has an advantage in a greater coefficient of geometrical increase, but requires 1.64 times as much food as *P. aurelia*. Consequently, the greater rate of growth is a liability in competition. *P. aurelia* is less affected by excretion products, as it can live twice as long in the presence of a strong concentration of waste excretion products as *P. caudatum*. With the amount of food available and the medium used, only the *P. aurelia* could survive the competition of the mixed population. *Glaucoma scintillans,* growing in competition with *P. aurelia,* will survive when the latter perishes.

A more complicated series of experiments was made on *P. aurelia* or *P. caudatum* and *P. bursaria* with food supplied by bacteria and yeast. The *P. busaria* could eat the yeast, but the two other species could not. Varying equilibria of populations could be established, depending on the initial concentrations of the four organisms. In this case the competition is in different niches.

Populations of predators and prey are interesting and have been studied in epidemiology, notably by Ross and Lotka working with the malarial parasite. A simpler case, of less personal interest to man, is the competition of mixed populations of bacteria, *Paramecium,* and *Didinium nasutum.* The latter consumes a *Paramecium* every three hours. In such a mixed population, Gause found that at first both the *Paramecium* and the *Didinium* populations grew, but later the didinia ate all of the paramecia and then promptly starved. With medium with sediment in which some of the paramecia moved about and thus were not available as food for the didinia, the didinia ate the available paramecia and then starved while the remaining paramecia grew. Another experiment utilized *Bursaria truncatella,* which preyed on *P. bursaria.*

The experiments may be grouped in three classes: (1) two species in the same ecological niche, competing for the same food; (2) two species in different niches, competing for the same food; or (3) two species, one eating the other. Gause (1935) has given mathematical analyses of the equilibria, depending on the variables involved. Much progress has been made in this phase of biological science, even though it is less than a quarter of a century old, and well-planned experiments or heuristic theoretical analysis may be expected to contribute to an understanding of the growth of the Protozoa, to ecology, and to historical (evolutionary) biological science.

LITERATURE CITED

Adolph, E. F. 1931. The regulation of size. Springfield, Ill. 235 pp.
Allee, W. C. 1934. Recent studies in mass physiology. Biol. Rev., 9: 1-48.
Beers, C. D. 1928a. The regulation of dietary insufficiency to vitality in the ciliate *Didinium nasutum.* J. exp. Zool., 51: 121-33.
—— 1928b. Rhythms in infusoria with special reference to *Didinium nasutum.* J. exp. Zool., 51: 485-93.
—— 1928c. Some effects of dietary insufficiency in the ciliate *Didinium nasutum.* Proc. Nat. Acad. Sci., Wash., 14: 132-37.

—— 1929. On the possibility of indefinite reproduction in the ciliate *Didinium nasutum* without conjugation or endomixis., Amer. Nat., 63: 125-29.

Berkson, J., T. B. Magath, and M. Hurn. 1935. Laboratory standards in relation to chance fluctuations of the erythrocyte count as estimated with the haemocytometer. J. Amer. statist. Ass., 30: 414-26.

Bodine, J. H. 1917. Hydrogen ion concentration of protozoan cultures. Biol. Bull., 41: 73-77.

Bond, R. M. 1933. A contribution to the study of the natural food-cycle in aquatic environments. Bull. Bingham oceanogr. Coll., 4: 1-89.

Bowen, W. J. 1938. The effects of copper and vanadium on the frequency of division. Biol. Bull., 75: 361. (*Cf.* also Amer. J. Physiol., 1939, 126: 439.

Buchanan, R. E., and E. I. Fulmer. 1928. Physiology and biochemistry of bacteria. Vol. I. Baltimore.

Burnside, L. H. 1929. Relation of nuclear size to body size in *Stentor coeruleus.* J. exp. Zool., 54: 473-83.

Carlson, Tor. 1913. Über Geschwindigkeit und Grösse der Hefevermehrung in Würze. Biochem. Z., 57: 313-35.

Chalkley, H. W. 1929. Changes in water content of *Amoeba* in relation to change in its protoplasmic structure. Physiol. Zoöl., 4: 535-74.

—— 1931. The chemistry of cell division. II. The relation between cell growth and division in *Amoeba proteus.* Publ. Hlth. Rep. Wash., 46: 1736-54.

Chapman, R. N. 1931. Animal ecology. New York.

Clark, N. A. 1922. Rate of formation and yield of yeast in wort. J. phys. Chem., 26: 42-60.

Coe, W. R. 1932. Season of attachment and rate of growth of sedentary marine organisms at the pier of the Scripps Institution of Oceanography, La Jolla, Calif. Bull. Scripps Instn. Oceanogr. tech. 3: 37-86.

Crozier, W. J. 1923. On abundance and diversity in the protozoan fauna of a sewage "Filter." *Science,* 58: 424-25.

Crozier, W. J., and E. S. Harris. 1923. Animal population of a sewage sprinkling filter. Stat. Rep. N. J., 503-16.

Cutler, D. W., and L. M. Crump. 1923. The rate of reproduction in artificial culture of *Colpidium colpoda.* Bio-chem. J., 17: 878-86.

—— 1924. The rate of reproduction in artificial culture of *Colpidium colpoda.* Bio-chem. J., 18: 905-12.

—— 1925. The influence of washing upon the reproduction rate of *Colpidium colpoda.* Bio-chem. J., 19: 450-53.

—— 1927. The qualitative and quantitative effects of food on the growth of soil *Amoeba.* Brit. J. exp. Biol., 5: 155-65.

Daniel, G. E., and H. W. Chalkley. 1933. The influence of temperature on the process of division in *Amoeba proteus* (Leidy). J. cell. comp. Physiol., 2: 311-27.

Darby, H. H. 1930a. The experimental production of life cycles in ciliates. J. exp. Biol., 7: 132-42.

―― 1930b. Studies on growth acceleration in Protozoa and yeast. J. exp. Biol., 7: 308-16.

Dimitrowa, A. 1932. Die fördernde Wirkung der *Paramaecium caudatum*. Ehrbg. auf dessen Teilungsgeschwindigkeit. Zool. Anz., 100: 127-32.

Eddy, S. 1928. Succession of Protozoa in cultures under controlled conditions. Trans. Amer. micr. Soc., 47: 283-339.

Elliott, A. M. 1935a. The influence of pantothenic acid on growth of Protozoa. Biol. Bull., 68: 82-92.

―― 1935b. Effects of certain organic acids and protein derivatives on the growth of *Colpidium*. Arch. Protistenk., 84: 472-94.

―― 1936. Nutritional studies on Protozoa. Proc. Minn. Acad. Sci., Rept. 6 pp.

―― 1938. The influence of certain plant hormones on growth of Protozoa. Physiol. Zoöl., 11: 31-39.

―― 1939. A volumetric method for estimating population densities of Protozoa. Trans. Amer. micr. Soc., 58: 97-99.

Entz, G., Jr. 1931. Analyse des Wachstums und der Teilung einer Population sowie eines Individuums des Protisten *Ceratium hirudinella* unter den natürlichen Verhältnissen. Arch. Protistenk., 74: 311-61.

Erdmann, R. 1920. Endomixis and size variations in pure bred lines of *Paramecium aurelia*. Roux Arch. EntwMech. Organ., 46: 85-148.

Estabrook, A. H. 1910. Effect of chemicals on growth in *Paramecium*. J. exp. Zool., 8: 489-543.

Fauré-Fremiet, E. 1925. La Cinétique du développement. Paris.

―― 1930. Growth and differentiation of the colonies of *Zoothamnium alternans*. Biol. Bull., 58: 28-51.

Fine, M. S. 1912. Chemical properties of hay infusions with special reference to the titratable acidity and its relation to the protozoan sequence. J. exp. Zool., 12: 265-81.

Garrod, L. P. 1936. Allelocatalysis. J. Path. Bact., 42: 535-36.

Gause, G. F. 1932. Ecology of populations. Quart. Rev. Biol., 7: 27-46.

―― 1934. The struggle for existence. Baltimore.

―― 1935. Verifications expérimentales de la théorie mathématique de la lutte pour la vie. Act. Sci. Indust., No. 277. Paris.

Gerstein, J. 1937. The culture and division rate of *Stentor coeruleus*. Proc. Soc. exp. Biol. N. Y., 37: 210-11.

Giese, A. C. 1939. Ultraviolet radiation and cell division. I. Effects of λ 2654 and 2804 Å upon *Paramecium caudatum*. J. cell. comp. Physiol., 13: 139-50.

Gladstone, G. P., P. Fildes, and G. M. Richardson. 1935. Carbon dioxide as an essential factor in the growth of bacteria. Brit. J. exp. Path., 16: 335-48.

Gordon, R. D. 1938. Note on estimating bacterial populations by the dilution method. Proc. nat. Acad. Sci. Wash., 24: 212-15.

Gray, J. 1929. The kinetics of growth. Brit. J. exp. Biol., 6: 248-74.

Greenleaf, W. E. 1924. Influence of volume of culture medium and cell proximity on the rate of reproduction of Protozoa. Proc. Soc. exp. Biol. N. Y., 21: 405-6.

Gregory, L. H. 1925. Direct and after effects of changes in medium during different periods in the life history of *Uroleptus mobilis*. Biol. Bull., 48: 200-8.

—— 1926. Effects of changes in medium during different periods in the life history of *Uroleptus mobilis*. Biol. Bull., 51: 179-88.

—— 1928. The effects of changes in medium during different periods in the life history of *Uroleptis mobilis* and other Protozoa. Biol. Bull., 55: 386-94.

Hall, R. P. 1933. On the relation of hydrogen-ion concentration to the growth of *Euglena anabaena* var. *minor* and *E. deses*. Arch. Protistenk., 79: 239-48.

—— 1938a. Effects of manganese on the growth of *Euglena anabaena, Astasia* sp. and *Colpidium campylum*. Arch. Protistenk., 90: 178-84.

—— 1938b. Pimelic acid as a growth factor for the ciliate, *Colpidium campylum*. Anat. Rec., 72 (Suppl.) : 110.

—— 1939. The trophic nature of the plant-like flagellates. Quart. Rev. Biol., 14: 1-12.

Hall, R. P., D. F. Johnson, and J. B. Loefer. 1935. A method for counting Protozoa in the measurement of growth under experimental conditions. Trans. Amer. micr. Soc., 54: 298.

Hall, R. P., and J. B. Loefer. 1938. Effect of the addition of old culture medium on the growth of *Colpidium campylum*. Anat. Rec., 72 (Suppl.) : 50.

Hall, R. P., and H. W. Schoenborn. 1938. The selective action of inorganic media in bacteria-free cultures of *Euglena*. Anat. Rec., 72 (Suppl.) : 129.

Halsey, H. R. 1936. The life cycle of *Amoeba proteus* Pallas, Leidy and of *Amoeba dubia* Schaeffer. J. exp. Zool., 74: 167-203.

Hammett, F. S. 1929. Chemical stimulants for growth by increase in cell number. *Protoplasma*, 7: 297-322.

Hammond, E. C. 1938. Biological effects of population density in lower organisms. Quart. Rev. Biol., 13: 421-38; 14: 35-59.

Hardy, A. C. 1938. Estimating numbers without counting. Nature, 142: 255-56.

Hargitt, G. T., and W. W. Fray. 1917. The growth of *Paramecium* in pure cultures of bacteria. J. exp. Zool., 22: 421-55.

Hegerty, C. P. 1939. Physiological youth as an important factor in adaptive enzyme formation. J. Bact., 37: 145-52.

Hetherington, A. 1932. The constant culture of *Stentor coeruleus*. Arch. Protistenk., 76: 118-29.

—— 1936. The precise control of growth in a pure culture of a ciliate *Glaucoma pyriformis*. Biol. Bull., 70: 426-40.

Huntington, E., and C.-E. A. Winslow. 1937. Cell size and metabolic activity at various phases of the bacterial culture cycle. J. Bact., 33: 123-44.

Huxley, J. S. 1932. Problems in relative growth. London.

Huxley, J. S., and G. Teissier. 1936. Terminology of relative growth. Nature, 137: 780-81.

Jahn, T. L. 1929. Studies on the physiology of the Euglenoid flagellates. I. The relation of the density of population to the growth rate of *Euglena*. Biol. Bull., 57: 81-106.

—— 1930. Studies etc. II. The autocatalytic equation and the question of an autocatalyst in the growth of *Euglena*. Biol. Bull., 58: 281.

—— 1933. Studies on the oxidation-reduction potential of protozoan cultures. I. The effect of -SH on *Chilomonas paramecium*. Protoplasma, 20: 90-104.

—— 1934. Problems of population growth in the Protozoa. Cold Spring Harbor Symp. Quant. Biol., 2: 167-80.

—— 1935. Studies etc. VI. The effects of temperature and of acetate on *Euglena gracilis* cultures in the dark. Arch. Protistenk., 86: 251-57.

—— 1936. Effect of aeration and lack of CO_2 on growth of bacteria-free cultures of Protozoa. Proc. Soc. exp. Biol. N. Y., 33: 494-98.

Jennings, H. S. 1908. Heredity, variation and evolution in Protozoa II. Proc. Amer. phil. Soc., 47: 393-546.

Johnson, D. E. 1936. Growth of *Glaucoma ficaria* Kahl in cultures with single species of other organisms. Arch. Protistenk., 86: 359-78.

Johnson, W. H. 1933. Effects of population density on the rate of reproduction in *Oxytricha*. Physiol. Zoöl., 6: 22-54.

—— 1935. Isolation of *Glaucoma ficaria* Kahl in bacteria-free cultures and growth in relation to the pH of the media. Arch. Protistenk., 86: 263-77.

—— 1936. Studies in the nutrition and reproduction of *Paramecium*. Physiol. Zoöl., 9: 1-14.

—— 1937. Experimental populations of microscopic organisms. Amer. Nat., 71: 5-20.

Johnson, W. H., and G. Hardin. 1938. Reproduction of *paramecium* in old culture medium. Physiol. Zoöl., 11: 333-46.

Jones, E. P. 1928. Population curves of *Paramecium*. Proc. Pa. Acad. Sci., 2: 1927-28.

—— 1930. *Paramecium* infusion histories. I. Hydrogen ion change in hay and hay-flour infusions. Biol. Bull., 59: 275-84.

Kavanagh, A. J., and O. W. Richards. 1939. The autocatalytic growth curve. Amer. Nat., 68: 54-59.

—— 1937a. The unusual mortality which characterizes a *Paramecium* culture. Anat. Rec., 70 (Suppl.) : 39.

—— 1937b. The potential longevity of a *Paramecium* culture. Anat. Rec., 70 (Suppl.) : 39.

Kidder, G. W. 1939. The effects of biologically conditioned medium on the growth of *Colpidium campylum*. Biol. Bull., 77: 297-98.

Kidder, G. W., and C. A. Stuart. 1938. The rôle of chromogenic bacteria in ciliate growth. Biol. Bull., 75: 336.

Kober, P. A., and S. S. Graves. 1915. Nephelometry (Photometric Analysis). I. History of method and development of instruments. J. industr. Engng. Chem., 7: 843-47.

Koser, S. A., and F. Saunders. 1938. Accessory growth factors for bacteria and related microorganisms. Bact. Rev., 2: 99-190.

Kostitzin, V. A. 1934. Symbiose, parasitisme et évolution. Act. Sci. Indust., No. 96. Paris. 47 pp.

Lackey, J. B. 1938a. Protozoan plankton as indicators of pollution in a flowing stream. Publ. Hlth. Rep. Wash., 53: 2037-58.

—— 1938b. A study of some ecologic factors affecting the distribution of Protozoa. Ecol. Monogr., 8: 501-27.

Loefer, J. B. 1935. Effects of certain carbohydrates and organic acids on growth of *Chlorogonium* and *Chilomonas*. Arch. Protistenk., 84: 456-71.

—— 1938a. Effects of hydrogen ion concentration on the growth and morphology of *Paramecium bursaria*. Arch. Protistenk., 90: 185-93.

—— 1938b. Effect of osmotic pressure on the motility and viability of fresh-water Protozoa. Anat. Rec., 72 (Suppl.) : 50.

—— 1938c. Growth of *Chlorogonium tetragonium* at different hydrogen-ion concentrations. Anat. Rec., 72 (Suppl.) : 129.

—— 1938d. Bacteria-free culture of *Paramecium bursaria* and concentration of the medium as a factor in growth. J. exp. Zool., 72: 387-407.

—— 1939. Acclimatization of fresh-water ciliates and flagellates to media of higher osmotic pressure. Physiol. Zoöl., 12: 161-72.

Loofbourow, J. R., and C. M. Dyer. 1938. Relative consistency of weights and counts in determining microorganisms by photoelectric nephelometers. Studies Inst. Divi Thomae, 1: 129-35.

Lotka, A. J. 1925. Elements of physical biology. Baltimore.

—— 1934. Théorie analitique des associations biologiques. Act. Sci. Indust., No. 187. Paris.

Lwoff, A., and H. Dusi. 1937. La Pyrimidine et le thiazol, facteurs de croissance pour le Flagellé *Polytoma caeca*. C. R. Soc. Biol. Paris, 126: 630-32.

—— 1938. Culture de divers flagelles leucophytes en milieu synthétique. C. R. Soc. Biol. Paris, 127: 53-56.

McCay, C. M. 1933. Is longevity compatible with optimum life? Science, 77: 410-11.

MacDougall, M. S. 1929. Modifications in *Chilodon uncinatus* produced by ultraviolet radiation. J. exp. Zool., 54: 95-109.

MacLennan, R. F. 1935. Dedifferentiation and redifferentiation in *Ichthyophthirius*. I. Neuromuscular system. Arch. Protistenk., 86: 191-210.

McPherson, M., G. A. Smith, and A. M. Banta. 1932. New data with possible bearing on Robertson's theory of autocatalysis. Anat. Rec., 54 (Suppl.): 23.

Mast, S. O., and D. M. Pace. 1933. Synthesis from inorganic compounds of starch, fat, proteins and protoplasm in the colorless animal *Chilomonas paramecium*. Protoplasma, 20: 326-58.

—— 1935. Relation between sulphur in various chemical forms and the rate of growth in the colorless flagellate *Chilomonas paramecium*. Protoplasma, 23: 297-325.

—— 1937. The relation between the number of individuals per volume of culture solution and rate of growth in *Chilomonas paramecium*. Anat. Rec., 70(Suppl.): 40.

—— 1938a. The relation between the hydrogen ion concentration of the culture medium and the rate of reproduction in *Chilomonas paramecium*. J. exp. Zool., 79: 429-31.

—— 1938b. The effects of substances produced by *Chilomonas paramecium* on the rate of reproduction. Physiol. Zoöl., 11: 360-82.

—— 1938c. The relation between the age of the cultures from which *Chilomonas* is taken and the rate of reproduction in fresh culture fluid. Anat. Rec., 72(Suppl.): 62.

Mattick, A. T. R., J. McClemont, and J. O. Irwin. 1935. The plate count of milk. J. Dairy Sci., 6: 130-47.

Mestre, H. 1935. A precision photometer for the study of suspensions of bacteria and other microorganisms. J. Bact., 30: 335-58.

Meyers, E. C. 1927. Relation of density of population and certain other factors to survival and reproduction in different biotypes of *P. caudatum*. J. exp. Zool., 49: 1-43.

Mitchell, W. H., Jr. 1929. The division rate of *Paramecium* in relation to temperature. J. exp. Zool., 54: 383-410.

Mizuno, F. 1927. Sur la croissance du *Paramecium caudatum*. Sci. Rep. Tôhuku Univ., 4th Ser., 2: 367-81.

Mond, J. 1937. Réaction d'entretien et ration de croissance dans les populations bactériennes. C. R. Acad. Sci. Paris, 205: 1456-57.

Mottram, J. C., 1939. An increase in the rate of growth of *paramecium* subjected to the blastogenic hydrocarbon 3:4-benzpyrene. Nature, 144: 154.

Müller, R. H. 1939. Photoelectric methods in analytical colorimetry. Industr. Engng. Chem., 11: 1-17.

Needham, J. 1934. Chemical heterogony and the ground-plan of animal growth. Biol. Rev., 9: 79-109.

Nielson, N. 1933. A method for determining the velocity of sedimentation of yeast. C. R. Lab. Carlsberg, Nr. 19.

Noland, L. E. 1925. Factors influencing the distribution of fresh water ciliates. Ecology, 4: 437-52.

Park, T. 1939. Analytical population studies in relation to general ecology. Amer. Midl. Nat., 21: 235-55.

Pearl, R. 1907. A biometrical study of conjugation in *Paramecium*. Biometrika, 5: 213-97.

—— 1925. Biology of population growth. New York.

—— 1928. The rate of living. New York.

Peskett, G. L. 1924. Allelocatalysis and the growth of yeast. J. Physiol., 59: xxxiii.

—— 1925a. Studies on the growth of yeast. I. Influence of the volume of culture medium employed. Bio-chem. J., 19: 464-73.

—— 1925b. Studies etc. II. A further note on allelocatalysis. Bio-chem. J., 19: 474-76.

—— 1927. Studies etc. III. a further study on the influence of volume of media employed. Bio-chem. J., 21: 104-10.

Peters, A. W. 1901. Some methods for use in the study of Infusoria. Amer. Nat., 35: 553-59.

—— 1904. Metabolism and division in Protozoa. Proc. Amer. Acad. Arts. Sci., 39: 441-516.

—— 1907. Chemical studies on the cell and its medium. Amer. J. Physiol., 17: 443-77; 18: 321-46.

Peters, R. A. 1920. Nutrition of the Protozoa. J. Physiol., 54: L.

—— 1921. The substances needed for the growth of a pure culture of *Colpidium colpoda*. J. Physiol., 55: 1-32.

Petersen, W. A. 1929. Relation of density of population to rate of reproduction of *Paramecium caudatum*. Physiol. Zoöl., 2: 221-54.

Phelps, A. 1931. Effects of H-ion concentration on the division rate of *Paramecium aurelia*. Science, 74: 395-96.

—— 1934. Studies on the nutrition of *Paramecium*. Arch. Protistenk., 82: 134-63.

—— 1935. Growth of Protozoa in pure cultures. I. Effect upon the growth curve of the age of the inoculum and of the amount of the inoculum. J. exp. Zool., 70: 109-30.

—— 1936. Growth etc. II. Effect upon the growth curve of different concentrations of nutrient materials. J. exp. Zool., 72: 479-96.

Phillips, R. L. 1922. The growth of *Paramecium* in infusions of known bacterial content. J. exp. Zool., 36: 135-83.

Popoff, M. 1908. Experimentelle Zellstudien. I. Arch. Zellforsch., 1: 246-379.

—— 1909. Experimentelle Zellstudien. II. Arch. Zellforsch., 3: 124-80.

Pruthi, H. S. 1926. On the hydrogen ion concentration of hay infusions with special reference to its influence on the protozoan sequence. Brit. J. exp. Biol., 4: 292-300.

Reich, K. 1936. Studies on the physiology of *Amoeba*. I. The relation between nutrient solution zone of growth and density of population. Physiol. Zoöl., 9: 254-63.

—— 1938. Studies etc. II. The allelocatalytic effect in *Amoeba* culture free of bacteria. Physiol. Zoöl., 11: 347-58.

Richards, O. W. 1929. The correlation of the amount of sunlight with the division rate of ciliates. Biol. Bull. 56: 298-305.

—— 1932. The second cycle and subsequent growth of a population of yeast. Arch. Protistenk., 78: 263-301.

—— 1933. The toxicity of some metals and Berkefeld filtered sea water to *Mytilus edulis*. Biol. Bull., 65: 371-72.

—— 1934. The analysis of growth as illustrated by yeast. Cold Spring Harbor Symp. Quant. Biol., 2: 157-66.

—— 1935. Analysis of the constant differential growth ratio. Pap. Tortugas Lab., 29: 173-83.

Richards, O. W., and J. A. Dawson. 1927. The analysis of the division rate of ciliates. J. gen. Physiol., 10: 853-58.

Richards, O. W., and T. L. Jahn. 1933. A photoelectric nephelometer for estimating population density of microörganisms. J. Bact., 26: 385-91.

Richards, O. W., and A. K. Kavanagh. 1937. The course of population growth and the size of seeding. Growth, 1: 217-27.

Robertson, M. 1929. Life cycles in the Protozoa. Biol. Rev., 4: 152-79.

Robertson, T. B. 1921a. Experimental studies on cellular multiplication. II. The influence of mutual contiguity upon reproductive rate and the part played therein by the "x-substance" in bacterial infusions which stimulates the multiplication of Infusoria. Bio-chem. J., 15: 612-19.

—— 1921b. The multiplication of isolated Infusoria. Bio-chem. J., 15: 595-611.

—— 1922. Reproduction in cell communities. J. Physiol., 56: 404-12.

—— 1923. The chemical basis of growth and senescence. Philadelphia.

—— 1924a. The influence of washing on the multiplication of isolated Infusoria and upon the allelocatalytic effect in cultures initially containing two Infusoria. Aust. J. exp. Biol. med. Sci., 1: 151-73.

—— 1924b. Allelocatalytic effect in cultures of *Colpidium* in hay infusion and in synthetic media. Bio-chem. J., 18: 1240-47.

—— 1927. On some conditions affecting the viability of cultures of Infusoria and the occurrence of allelocatalysis therein. Aust. J. exp. Biol. med. Sci., 4: 1-23.

Rottier, P.-B. 1936a. Recherches sur les courbes de croissance de *Polytoma uvella*. L'influence de l'oxygénation. C. R. Soc. Biol. Paris, 122: 65-67.

—— 1936b. Recherches sur la croissance de *Polytomella uvella*. L'influence de la concentration des substances nutritives. C. R. Soc. Biol. Paris, 122: 776-79.

Roux, J. 1901. Faune infusorienne des eaux stagnantes des environs de Genève. Mém. Inst. nat. genèv., Vol. 19. 142 pp.

Russell, J. 1937. A photoelectric cell circuit with a logarithmic response. Rev. sci. Instrum., 8: 495-96.

Schmalhausen, I., and E. Syngajewskaja. 1925. Studien über Wachstum und Differenzierung. I. Die Individuelle Wachstumskurve von *Paramecium caudatum*. Roux Arch. Entwkmech. Organ., 105: 711-17.

Shapiro, H. 1935a. Nomogram for centrifugal speed. Industr. Engng. Chem., 7: 25.

—— 1935b. The validity of the centrifuge method for estimating aggregate cell volume in suspensions of the eggs of the sea-urchin *Arbacia punctulata*. Biol. Bull., 68: 363-77.

—— 1937. The viscosimeter method for determination of cell concentration in suspensions of living cells. Anat. Rec., 70 (Suppl.) : 110.

Simpson, J. Y. 1902. The relation of binary fission to variation. Biometrika, 1: 400-7.

Smith, J. A. 1938. Some effects of temperature on the reproduction of *Chilomonas paramecium*. Biol. Bull. 75: 336-37.

—— 1939. Effects of temperature on starch & fat in *Chilomonas*. Coll. Net, 14: 35-36.

Snell, S. D. 1929. An inherent defect in the theory that growth rate is controlled by an allelocatalytic process. Proc. nat. Acad. Sci. Wash., 15: 274-81.

Spurr, W. A. 1937. A graphic method for measuring seasonal variation. J. Amer. statist. Ass., 32: 281-89.

Steinhaus, E. A., and J. M. Birkeland. 1939. The senescent phase in ageing cultures and the probable mechanisms involved. J. Bact., 38: 249-61.

Stier, T. J. B., M. I. Newton, and H. Sprince. 1939. Relation between the increase in opacity of yeast suspensions during glucose metabolism and assimilation. Science, 89: 85-86.

Sweet, H. E. 1939. A micro-population study of *Euglena gracilis* Klebs in sterile, autotrophic media and in bacterial suspensions. Physiol. Zoöl., 12: 173-200.

Taylor, C. V., and A. G. R. Strickland. 1938. Reactions of *Colpoda duodenaria* to environmental factors. I. Some factors influencing growth and encystment. Arch. Protistenk., 90: 396-409.

Teissier, G. 1928. Croissance des populations et croissance des organismes. Ann. Physiol. Physicochim. biol., 4: 342-86.

—— 1934. Dysharmonies et discontinuités dans la croissance. Paris.

—— 1937. Les Lois quantitatives de la croissance. Paris.

Tippett, L. H. C. 1932. A modified method of counting particles. Proc. roy. Soc., A137: 434-46.

Unger, W. B. 1931. The protozoan sequence in five plant infusions. Trans. Amer. micro. Soc., 50: 144-53.

Walker, H. H. 1932. Carbon dioxide as a factor affecting lag in bacterial cultures. Science, 76: 602-4.

Wang, C. C. 1928. Ecological studies of the seasonal distribution of Protozoa in a fresh water pond. J. Morph., 46: 431-78.

Wetzel, N. C. 1937. On the motion of growth. XVII. Theoretical foundations. Growth, 1: 6-59.

Woodruff, L. L. 1905. An experimental study on the life history of hypotrichous Infusoria. J. Exp. Zool., 2:585-632.

—— 1911. The effect of excretion products of *Paramecium* on its rate of reproduction. J. exp. Zool., 10: 557-81.

—— 1912. Observations on the origin and sequence of the protozoan fauna of hay infusions. J. exp. Zool., 12: 205-64.

—— 1913a. Cell size, nuclear size and the nuclear cytoplasmic relation during the life of a pedigreed race of *Oxytricha fallax*. J. exp. Zool., 15: 1-22.

—— 1913b. The effect of excretion products on the same and on different species with special reference to the protozoan sequence in infusions. J. exp. Zool., 14: 575-82.

Woodruff L. L., and G. A. Baitsell. 1911a. The reproduction of *Paramecium aurelia* in a "constant" culture medium of beef extract. J. exp. Zool., 11: 135-42.

—— 1911b. The temperature coefficient of the rate of reproduction of *Paramecium aurelia*. Amer. J. Physiol., 29: 147-55.

—— 1911c. Rhythms in the reproductive activity of the Infusoria. J. exp. Zool., 11: 339-59.

Wright, E. V., and H. Kersten. 1937. An apparatus for measuring turbidity of bacterial suspensions. J. Bact., 34:581-83.

Yocom, H. B. 1928. The effect of the quantity of culture medium on the division rate of *Oxytricha*. Biol. Bull., 54: 410-16.

—— 1934. Observations on the experimental adaptation of certain fresh water ciliates to sea water. Biol. Bull., 67: 273-76.

Zhalkovskii, B. G. 1938. The differences in biological action of transmitted and reflected light. I. Experiments with *Paramecium caudatum*. Bull. biol. exp. med. U. R. S. S. 493-95. From Chem. Abstr., 1939, 33: 2544.

Ziegler, N. R., and H. O. Halvorson. 1935. Application of statistics to problems in bacteriology. IV. Experimental comparison of the dilution method, the plate count, and the direct count for the determination of bacterial populations. J. Bact., 29: 609-34.

CHAPTER XI

THE LIFE CYCLE OF THE PROTOZOA

CHARLES ATWOOD KOFOID

INTRODUCTION

THE ORGANISM has the fourth dimension of time. In the course of its life cycle, its three spatial dimensions change. The fourth changes also, interacting with the three. It may be measured by metabolic rate, by structural results of growth, or by organismal cyclic changes which follow one after the other in sequences. These may be regular, interrupted, repeated, or in some other way responsive to or dependent upon internal environmental conditions, or to external conditions, such as changing quantity or quality of food supply; rise or fall in temperature, of seasonal origin, or due to migration; inciting or deterrent chemical or physical factors, such as pH, intensity and duration of light, and changes of host.

The Protozoa* differ from the Metazoa because of their smaller size and the resulting more highly significant and potent surface-volume relations, as these affect the rate and intensity of the impact of environmental factors upon the organism and the changes they initiate and induce. It is therefore to be expected that the Protozoa will be relatively more susceptible to the modification of the individual and to the distortion and interruption of its normal life cycle than are the Metazoa, thus obscuring and complicating the evidence of the existence of life cycles among them. The factors of time, volume, and season enter more or less definitely into the life cycles of Metazoa such as hydroids, flukes, tapeworms, crustaceans, insects, and tunicates. Among the Protozoa, on the other hand, the time units required for the various cyclic changes may be very brief, and these changes very often have little or no dependence upon cosmic cycles, with the result that the evidence of their

* Assistance in preparing this chapter, rendered by the personnel of Work Projects Administration Official Project No. 65-1-08-113, Unit C1, is acknowledged.

occurrence and continuity is more difficult to organize than is that for cycles in the larger, longer-lived Metazoa.

Furthermore, a certain hesitancy about life cycles in the Protozoa has arisen historically because of the fact that skilled workers in this field have been caught in error by reason of the difficulties above noted. Following upon Bütschli's (1876) and Hertwig's (1889) fundamental analyses of conjugation in the Ciliata and its resemblance to maturation and fertilization in other organisms, there arose a Munich school of protozoölogists whose labors brought forth an array of protozoan life cycles fitted to the metazoan pattern. Under the brilliant leadership of Fritz Schaudinn, most of the major groups of Protozoa were subjected to this pattern of analysis, with resulting marvelous conformity to type. Some of these, notably those of *Trypanosoma, Endamoeba,* and *Mastigella,* have not stood the test of subsequent critical reëxamination. Others, such as those of *Plasmodium, Coccidium,* and *Paramecium,* have, on the other hand, survived and have proved the validity of the basic assumption that there are life cycles in the Protozoa, though not necessarily all of the same type.

The life cycle in the Metazoa starts with the diploid or polyploid zygote, a unicellular stage whose genes, derived from the haploid gametes, determine the characters of all of the varied subsequent stages unfolded in the ensuing life cycle. This cycle in many instances is marked by indirect development with one or more larval stages, followed by metamorphosis into the adult, sexual maturity, gametogenesis, senescence, and death. In other instances the development is direct, with adolescence replacing metamorphosis. In both types asexual reproduction may intervene at different periods in embryonic, larval, and even adult life, giving rise by budding, binary and multiple fission, and sporulation to two or many different functional individuals, all with the original genetic constitution. Parthenogenesis may also intervene and alternate with normal sexual reproduction. There is often considerable change in the external appearance of the successive stages, as in larva, pupa, and imago of the Lepidoptera, though a striking similarity, even continuity, may occur in various organ systems from stage to stage.

The stages occurring in the metazoan life cycle are brought about by the processes of cleavage, gastrulation, organogenesis and histogenesis, growth, adolescence or metamorphosis, gametogenesis, senescence, and death. Asexual reproduction may be interjected into the midst of any one of these processes, resulting in from one to many repeated generations of functional individuals. Not infrequently these individuals are heterogonous, with marked differences in structure from the parent, as for example in the larval stages of the Trematoda.

This alternation of sexually and asexually produced generations is widely distributed in the living world, ranging from some of the lower algae to the Quints. The ease with which regeneration occurs after mutilation and with which experimental asexual multiplication of functional individuals may be imposed upon the genetic individual is indicative of the fundamental organic basis of asexual reproduction, perhaps as a corollary of the still more fundamental capacity of growth on the part of the organism.

The Protozoa, from the evolutionary point of view, are of exceptional interest among the phyla, since it is among these primitive organisms that most of the basic biological properties, structures, and functions of the organism have had their evolution. Within these microcosms all of the basic functions of living must be performed. As one surveys their diversities and complexities of pattern, one is impressed with the evidence that among these minute organisms, adapted to so many ecological niches and exhibiting so many types of behavior, a vast deal of evolutionary experimentation has been enacted. It is among the Protozoa and Protophyta that the following have been evolved: nuclear structure, sex, sexual dimorphism, sexual reproduction, mitosis, chromosomes, gametogenesis, histogenesis, multicellularity, sex and somatic cells, asexual reproduction by the various methods of binary and multiple fission, budding and sporulation, and the beginnings of the organization of organ systems. Varying combinations and sequences of these evolutionary accomplishments are exhibited in the diverse patterns of life cycles to be detected among the Protozoa. Cycles of comparable type, in some instances apparently independently of one another, have emerged to a varying degree in the different classes and orders of Protozoa.

These cycles fall into two major groups. The first is the simpler and the more primitive. It consists merely of recurrent rhythms of homogeneous asexual reproduction, in which mitosis produces a multicellular ($=$ multinuclear) body of from two to many cells, forming a plasmodium, coenobium, sporocyst, or cyst. Fission of binary, multiple, or budding type breaks up this body into functional individuals of the ancestral type. Although nuclear division is essential in the accomplishment of this cycle, it does not initiate it. This is shown strikingly in the Polymastigophora, in which the entire neuromotor complex of centrosome, blepharoplast, flagella, undulating membrane, and axostyle of the individual is duplicated by new growth, accompanied by extensive dedifferentiation of the parental equipment before the nuclear phenomena of mitosis ensue. Asexual reproduction is thus profoundly an organismal phenomenon involving a rejuvenation of the organelle systems of the body of the individual.

This type of life cycle seemingly exists without any evidence of sex, sexual reproduction, or sexual dimorphism. Efforts to establish sexuality on the basis of the relative size of supposedly male and female individuals and upon interpretations of behavior are will-o'-the-wisps of wishful thinking. The only basis is gametogenesis, verified by reduction of the diploid to the haploid number of chromosomes, and fertilization, with the resulting return to the diploid state.

The juxtaposition or even fusion of motile individuals among flagellates and rhizopods may occur when adverse conditions or internal states induce an adhesive periphery; sometimes cannibalistic feeding of rhizopods resembles fusion; and changes in position from divergence to lateral contact in sister schizonts among Mastigophora resemble conjugation, all of which evidence is never to be accepted as sexual behavior unless confirmed by critical cytological evidence.

The not uncommon opinion that sex is an inherent characteristic of organisms and that sexual reproduction is to be expected in all animals and plants and even in bacteria, is as yet without convincing cytological evidence among the more primitive forms. It has, however, been clearly demonstrated in the Sporozoa, Ciliata, Foraminifera, and the Volvocidae, all representatives of the more highly evolved Protozoa. The present evidence, negative though it be, lends support to the view that sex was evolved in the Protozoa, perhaps independently in the different classes. It may well be that its origin rests ultimately on differential metabolism within the species, leading in time to more favorable conditions for permanent fusion of gametes, though this alone makes no provision for gametogenesis. The fact that some flagellates and rhizopods have an odd number of chromosomes suggests that they are not zygotes nor derived from zygotes, but primitive haploids. While it is to be expected that the cases of critically proved instances of sexual reproduction will increase in both number and systematic range with further investigation, even this will be far from establishing the universality of sexual differentiation among the Protista. Haploid (odd) numbers of chromosomes in primitive species will still require a solution. There are three chromosomes in *Trichomonas buccalis* (Hinshaw, 1926) and five in *Iodamoeba bütschlii,* according to unpublished observations made in my laboratory by Dr. Dora P. Henry.

In the absence of sex and sexual reproduction among primitive Protozoa, this first type of a merely asexual life cycle is the only one feasible. It is, however, incorporated into the second type of cycle, in which it alternates in varying irregularity with sexual reproduction and may even exhibit several forms with structurally different functional individuals within the same cycle, as in *Plasmodium.*

Asexual Reproduction in Alternating Binary and Multiple Fission (Type I)

An example of the first type of the protozoan life cycle among the Mastigophora is seen in *Trichomonas augusta,* in which asexual reproduction by binary fission prevails, but is interrupted at unknown intervals by the formation of an eight or sixteen-celled somatella with a common cytoplasm, each cell of which has its own neuromotor apparatus. Within this plasmodium paired schizonts, temporarily joined to each other by the paradesmose, ceaselessly tug at this tether until they are disunited except by the common cytoplasm. Serial plasmotomy releases each schizont, to start again the cycle with binary fission. There is in this type of cycle no clue to sexual reproduction.

Another example from the Rhizopoda is found in *Councilmania lafleuri,* usually called *Endamoeba coli,* in which there is an alternation between a unicellular free motile phase and a multicellular encysted one. During the motile phase binary fission prevails, and reversion to the unicellular condition follows each mitosis. This is interrupted from time to time by the encysted phase, in which, following reduction in volume, the body rounds out and secretes about itself an impervious membrane or cyst wall of elastin, with a differentiated exit pore closed by a plug. Encystment follows feeding and the accumulation of food reserves, which

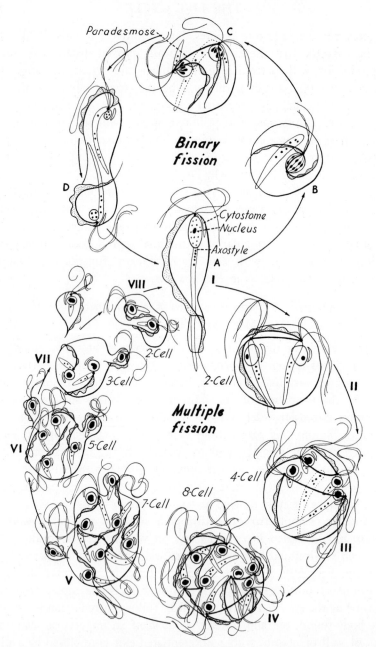

Figure 135. Diagram of the life cycle of *Trichomonas augusta* from the intestine of the frog, *Rana boylei,* including an alternation of binary fission of the two-cell somatella and of multiple fission of the eight-cell plasmodium.

in the cyst are stored in a large central glycogen vacuole. The glycogen is transformed into chromatoidal structures of unknown composition, staining deeply and formed on the surface of the glycogen vacuole. These progressively disappear as mitotic divisions ensue.

Soon after encystment is completed, a series of mitotic nuclear divisions occur, resulting in two, four, eight, and sixteen-cell stages, rarely thirty-two-cell, and in one observed instance approximately a sixty-four-cell stage, thus running the rhythm of normal cell division in a metazoan egg. Plasmotomy, however, does not attend the nuclear divisions. Cyst formation, in this instance, serves the function of assimilation and growth. Measurements of cysts in the one, two, four, eight, and sixteen-cell stages show a slight progressive increase in diameter.

Excystment occurs normally in the bowel, as shown by the occurrence of cysts free from glycogen or chromatoidals, with reduced numbers of nuclei from fifteen down. It can also be followed in fresh stools, as the small mononucleate amoebulae escape singly out of the exit pore. Excystment is a form of asexual reproduction, of budding, or progressive multiple or serial fission. In this type of life cycle we find an alternation of a unicellular free phase with reproduction by binary fission, with the formation of a multicellular encysted somatella, with reproduction by multiple fission and a return to the unicellular motile phase.

ALTERNATION OF ASEXUAL AND SEXUAL REPRODUCTION (TYPE II)

The second major type of the protozoan life cycle is that in which asexual and sexual reproduction alternate. It may or may not be accompanied by sexual dimorphism, as exhibited by differential reaction to aniline stains in *Nina,* by structural differentiation of gametes in *Eimeria,* of gametocytes in *Plasmodium,* or of conjugants in *Vorticella.* It seems probable that sex has become a genetic characteristic of the individual throughout the whole cycle, in all life cycles having sexual reproduction, even though structural features indicative of sexual dimorphism cannot be detected.

From the biological point of view, it is unfortunate that the life cycles of parasitic Protozoa have been arranged, in illustrations, in sequences as parasitic cycles, rather than biological life cycles. They are usually designated as beginning with the infection of the host, or in the case of a parasite with two hosts with that of the primary host, or

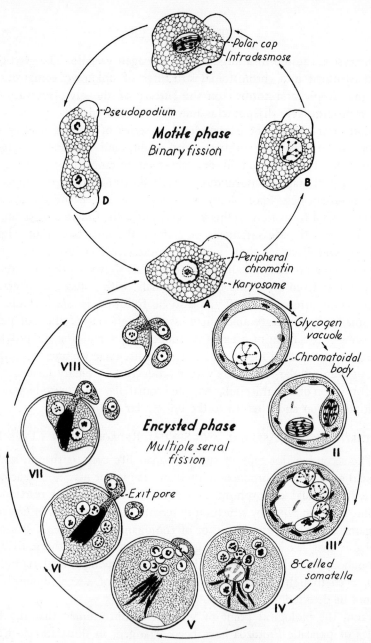

Figure 136. Diagram of the life cycle of *Endamoeba coli* (= *Councilmania lafleuri*, Kofoid and Swezy, 1921) from the intestine of man, including an alternation of binary fission in the motile phase and of budding, or serial multiple fission, of the eight-cell encysted somatella.

with that of the more significant host, for example, with the infection of man in the case of *Plasmodium*. The psychological effect of this is to deflect interest from the significant biological aspects of such cycles.

In order to follow these life cycles in their true biological sequences, we have rearranged them and will now proceed to discuss three of the most widely known ones, viz., those of *Eimeria, Plasmodium,* and *Paramecium.*

THE LIFE CYCLE OF *Eimeria schubergi*

The life cycle of *Eimeria schubergi*, a parasite in the intestinal epithelium of *Lithobius forficatus,* is a typical one with an alternation of asexual and sexual reproduction, and of a sexual phase with asexual ones. In this cycle no less than five different structural types of functional individuals appear, each with a distinctive pattern of shape, size, structure, and activity. Four of the five appear but once, but one is subject to numerous repetitions under favorable conditions.

As rearranged, the biological cycle begins with the zygote formed in the lumen of the intestine of the host by the fusion of a flagellated spermatozoan with a yolk-laden egg, recently emerging from an intestinal epithelial cell of its host. Even before the pronuclei fuse, the fertilized egg forms a fertilization membrane and secretes a cyst wall around its spherical body (Fig. 137, I). Two nuclear divisions bring the organism to the cleavage stage of a four-celled somatella, the sporoblast. Thereupon there ensues the first asexual reproduction, when this somatella divides into four unicellular spores. Unlike their spherical parent, these functional individuals are ellipsoidal, and they, too, secrete about their respective bodies a resistant ellipsoidal spore wall.

There then ensues the second asexual reproduction, when each spore cell divides into two spindle-shaped, naked unicellular sporozoites, retained within the spore case and the enveloping cyst wall of the sporoblast. At about this stage of the cycle, the sporocyst with its eight sporozoites in four spores, is discharged from the intestine of its host, and further development ceases until this infective stage is eaten by a *Lithobius*. Here the digestive fluids unstopper the cyst, the sporozoites are released from the spores, each escapes singly through the pore and enters an epithelial cell of the intestine of the host where it develops as a trophozoite, changing in pattern from a spindle shape to a spherical one. The organism at this stage is devoid of any special protecting cover and

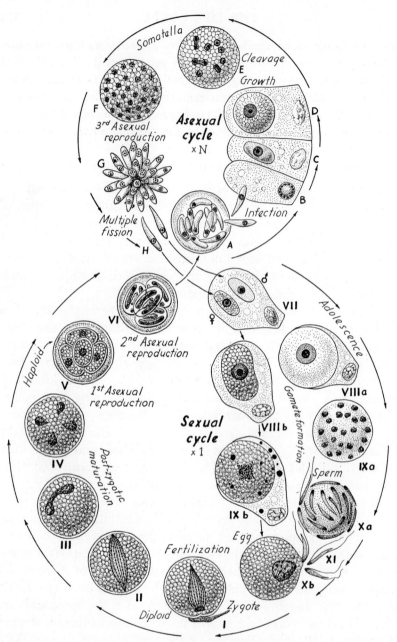

Figure 137. Diagram of the alternating sexual and asexual reproduction in the life cycle of *Eimeria schubergi*, from the intestine of *Lithobius forficatus*. For convenience, the first two asexual divisions within the sporoblast cyst wall are figured in the sexual cycle. They are transitional to the repeated multiplicative multiple fissions attending the infection of the intestinal cells of the host. Owing to post-zygotic reduction, both phases are haploid except for the diploid zygote. (Modified from Schaudinn, 1900.)

grows into a somatella of about sixty-four cells, in a rhythm of repeated mitoses.

Then follows the third asexual reproduction, in which the organism divides by plasmotomy into motile unicellular spindle-shaped merozoites, similar in size and pattern to the sporozoites. These in turn infect other intestinal epithelial cells, and this phase of the cycle is repeated an unknown number of times so long as susceptible host cells are available for infection. This phase ends the asexual part of the life cycle, as gametogenesis approaches, except in the male.

The sexual phase is marked by developing sexual dimorphism among the merozoites. Presumably sex is determined at fertilization, and all functional individuals derived from one zygote will accordingly be of one sex only, and the myriapod host must have acquired an infection by spores of each sex, in order that both male and female gametes of *Eimeria,* fertilization, and spore formation may ensue in the intestine. The sexual dimorphism of the gametocytes is determined by two factors, the metabolic rate and probably also the chemical nature of the food reserves on the one hand, and the structure and number of gametes produced on the other. Both male and female gamete mother cells grow to the size and spherical form of the trophozoite, but do not run its type of rhythm of cleavage, mitoses, and plasmotomy. The female is early differentiated from the male by the internal elaboration of spherical granules of food reserves or yolk, whereas in the male none appear. This functional and structural dimorphism is accompanied by a difference in nuclear behavior. In the male there appear to be as many as six successive mitoses, as in the trophozoite, producing up to sixty-four gametes. These are elongated slender, deeply staining bodies, largely of nuclear substance, with one trailing flagellum and a second one laterally attached to the anterior half of the body and free posteriorly.

The female gamete mother cell, on the other hand, undergoes no divisions, and transforms directly into the egg, though indications of metabolic activity appear in deeply staining spherules adjacent to the parasite in the host's cytoplasm. This absence of divisions in the female gametocyte, and their superabundance in the male, not only emphasizes a metabolic contrast, but also on cytological grounds offers cytological difficulties to the existence of maturation in these phases. These obstacles, which Schaudinn (1900) left unresolved, were removed by the discovery by Jameson (1920)—later extended by Dobell (1925), Naville (1931), Yarwood (1937), and Noble (1938)—that the maturation division takes place in the first division of the zygote and that, aside from the diploid zygote itself, the rest of the cycle is a haploid one.

It is obvious that in the case of the male there is a fourth asexual reproduction by multiple fission of a sixty-four-celled somatella, and that this does not occur in the female.

This life cycle is typical in having an alternation of sexual and asexual reproduction upon which are superposed certain features, in part adaptive and in part more fundamentally a part of the cycle. The first of these features is the building up of multicellular somatellas numbering respectively four, two, $\pm$ sixty-four (x n), and $\pm$ sixty-four in male cells only, prior to multiple fission. The body thus formed is temporary, lacking both nervous and hormonal mechanisms of integration to insure the maintenance of interacting relations. The adaptive aspect lies in the fact that these multiplicative reproductions make possible, with the least expenditure of individuals, the quick utilization of the food supply in the host's intestinal cells.

This cycle from the cytological point of view, as well as the general biological one, is atypical in the animal kingdom, though less so in the plant kingdom, in that only the zygote is diploid and all of the rest of the cycle is haploid. The fact that other Coccidiomorpha are known to have the same limitation and that this subclass has affinities with the flagellates, in some of which an odd number of chromosomes are known, suggests that the primitive Protozoa are haploid and that the diploid phase, like the polyploid, is a secondary evolutionary acquisition, dependent, in part at least, on the union of individuals or gametes in sexual reproduction. Thus both sex and sexual reproduction have had their origin in the Protozoa. The limitation of the diploid phase to the zygote only in *Eimeria* thus has a basic evolutionary significance.

Another feature of this life cycle which also has a basic significance is the fact that every one of the four phases of asexual reproduction results in the formation of a somatella of from two to sixty-four cells, and that the sexual phase also leads to a four-celled somatella. This evidence clearly indicates that these Protozoa are as truly multicellular, as are the early stages of the Metazoa. They undergo asexual reproduction as do Metazoa, from Porifera to Quints, but with this difference: that the

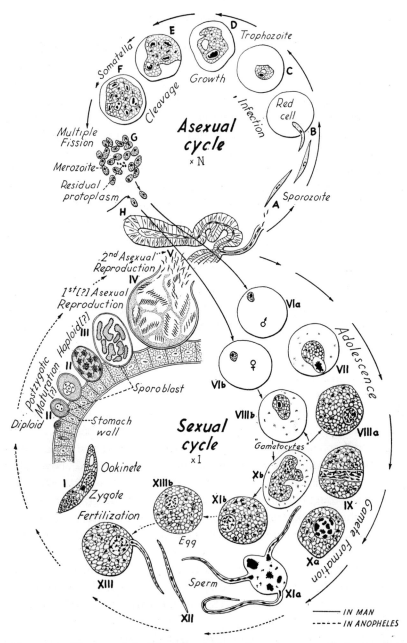

Figure 138. Diagram of the life cycle of *Plasmodium vivax*, parasitic in *Anopheles* and in the red cells of man. For convenience, the asexual reproductions of the sporoblast are figured in the sexual cycle, since they are preparatory to the repeated multiple fissions of the asexual cycle in man. By analogy with *Eimeria*, maturation is post-zygotic and both sexual and asexual phases are haploid except for the zygote. There is possibly an asexual reproduction. (Modified from Schaudinn, 1902.)

units into which they split are single cells, instead of flagellated chambers or axial organizers.

THE LIFE CYCLE OF *Plasmodium vivax*

A second example of the same general pattern, with added specializations due to parasitism in two hosts, is found in the malarial parasite, *Plasmodium vivax,* with the sexual and one (or two?) asexual phases in the mosquito, *Anopheles,* and oft-repeated merogony in the red cells of the blood of man, and a second asexual phase in the gamete mother cells of the male only on transfer of these cells to a lower temperature than that of the blood of man, as on a microscope slide or in the stomach of the mosquito.

THE LIFE CYCLE OF *Paramecium caudatum*

The life cycle of *Paramecium caudatum* makes a definite evolutionary advance in the Ciliata in two mutually interdependent features. The first is the differentiation of sex and somatic cells in the same individual, and the second is a permanent multicellular condition of two cells, derived from an undifferentiated eight-celled cleavage stage.

The original description of the cell, the selection of its name, the focusing of attention on total cleavage, with plasmotomy in embryology, rather than upon mitosis, all have combined to emphasize the separation of one cell from another by a wall or dividing structural boundary. These are all minor considerations. On the other hand, the significance of derivation, continuity in time, physiological functions, and above all of genetics, focus attention on the nucleus *and the cytoplasm* associated with it or brought in in the normal sequence of growth, fertilization, appropriation, or experiment under its control. These are all major considerations. In this modern sense it is biologically medieval to refer, as do many textbooks and other works, to *Paramecium* as a unicellular organism. It is biologically quite as logical to call a whale unicellular. Both start their cycles as one cell and both achieve multicellularity. No great biological significance attaches to the particular number of cells in the multicellular body, except during maturation. The significant achievement is the differentiation of sex and somatic cells. One of the primary distinctions in function, as well as in embryological origin, in the multicellular metazoan is this differentiation.

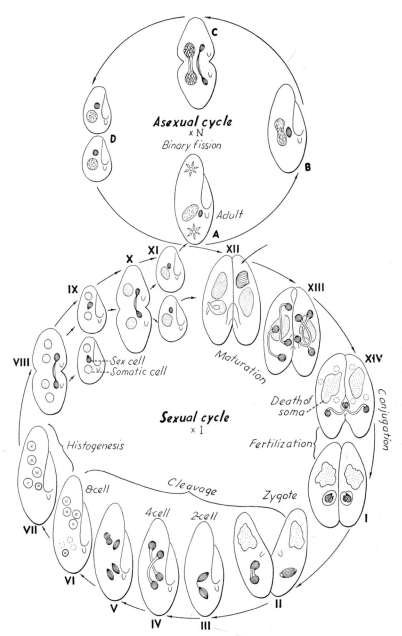

Figure 139. Diagram of the life cycle of *Paramecium caudatum* exhibiting an alternation of asexual reproduction, or binary fission of the two-cell somatella, and sexual reproduction with mutual fertilization of conjugants. The first and second divisions of the eight-cell somatella, or exconjugant, are for convenience included in the diagram of the sexual cycle, since they are preparatory to the asexual phase.

The life cycle of *Paramecium* is further complicated by the fact of conjugation and mutual fertilization of the conjugants. Sexual dimorphism is not evident between the conjugants as a whole, but appears in the behavior of the gamete nuclei. The migrant one is assumed to be male because of its motility, and the resident one female because of the lack of this quality. Dimorphism of the conjugants is structurally evident in *Vorticella,* in which the males are small and the females large.

The biological life cycle in *Paramecium* starts with the zygote, formed from the body of a conjugant by the fusion of the two haploid nuclei, one from the immigrant male gamete and the other the resident nucleus of the egg. The cleavage nucleus thus formed utilizes the cytoplasm of the egg, as in the Metazoa, with only a small amount from the male gamete. The old macronucleus in each continues (Fig. 139, I-IV) to disintegrate and is soon entirely metabolized into cytoplasm as food. This is the death of the soma of the conjugant, the future of which is henceforth under a new genetic control. There then ensue three successive mitotic divisions (Fig. 139, II-VI), representing the cleavage of the egg to an eight-celled somatella, when cleavage abruptly stops (Fig. 139, VI) and differentiation into four somatic and four sex cells occurs by the enlargement of the nuclei of the former and an increase in their chromatin. The four sex cells do not all survive. Three of them disintegrate at once, leaving a somatella of five cells, four somatic and one sex cell. Then begins asexual reproduction which in two peculiar binary fissions distributes the four macronuclei among the four daughter schizonts, with an accompanying division of the sex cell or micronucleus at each of the two fissions. In the diagram these two asexual fissions have been included in the sexual cycle, since they are necessary to restore the organism to the pattern in which regular asexual reproduction prevails. They otherwise belong in the asexual period.

The precise period in which maturation occurs in the sexual cycle is perhaps undetermined. It has been generally assumed that it occurs in the first two divisions of the micronucleus in the conjugant, in which case its third division would be an asexual reproduction of the gamete. This view does not rest upon exact chromosome count. The occurrence of post-zygotic maturation in the Sporozoa suggests the possibility of its occurrence in the Ciliata also. This view is further supported by the death of three of the post-zygotic sex nuclei and in *Paramecium* by the unnecessary third mitosis of the pronuclei in the conjugants. The evidence for chromosome reduction in the conjugants (Calkins and Cull, 1907) is inconclusive, because of the small size and the large number of the chromosomes in the three divisions prior to the formation of the zygote. A cytological examination of chromosome number during conjugation, in some ciliate with a small number of large chromosomes, may throw light on this problem. Both pre and post-zygotic alternatives should be explored.

The asexual cycle proper of *Paramecium* is one of oft-repeated simple binary fission, prior to which the two-celled somatella may grow, and by nuclear division become a four-celled one for a brief period. This cycle is one of indefinite duration.

This survey and interpretation of life cycles among the Protozoa exhibit the basic similarity of this fundamental characteristic of organisms among them to those emergent among the Metazoa. Peculiar to the Protozoa is the absence of sexual reproduction in the life cycles of the more primitive forms, among which it appears that they live a haploid life and that sexual reproduction has not as yet been evolved. Rare cases among Metazoa and the Metaphyta of the seeming absence of sexual reproduction are obviously secondary phenomena, but this interpretation is less defensible for the primitive Protozoa.

In the higher Protozoa, as in the Metazoa, the life cycle includes maturation, fertilization, cleavage to a multicellular stage, histogenesis of organelles, asexual reproduction with resulting functional individuals of differing structure in the different asexual phases, sexual dimorphism, adolescence, gametogenesis, senescence, and death.

The emphasis so generally placed upon the unicellular phase of the Protozoa, as against all Metazoa, tends to obscure their basic similarity in life cycle to that of the Metazoa, and thus to minimize the biological significances of the varied evolutionary accomplishments which have occurred in this primitive phylum. Similarities in biological phenomena are the bases on which an integrated concept of the evolution of life can be erected.

LITERATURE CITED

Bütschli, Otto. 1876. Studien über die ersten Entwicklungsvorgänge der Eizelle, die Zelltheilung und die Conjugation. Abhl. senckenb. naturf. Ges., 10: 213-452, 15 pls.

Calkins, G. N., and Sarah W. Cull. 1907. The conjugation of *Paramoecium aurelia (caudatum)*. Arch. Protistenk., 10: 375-415, pls. 12-18.

Dobell, Clifford. 1925. The life-history and chromosome cycle of *Aggregata eberthi*. Parasitology, 17: 1-136, 6 pls., 3 figs. in text.

Hertwig, Richard. 1889. Über die Conjugation der Infusorien. Abh. bayer. Akad. Wiss., Math.-Natur. Kl., 17: 151-234, 4 pls.

Hinshaw, H. C. 1926. On the morphology and mitosis of *Trichomonas buccalis* (Goodey) Kofoid. Univ. Cal. Publ. Zool., 29: 159-74, 1 pl., 2 figs. in text.

Jameson, A. P. 1920. The chromosome cycle of gregarines, with special reference to *Diplocystis schneideri* Kunstler. Quart. J. micr. Soc., London, 64: 207-66, pls. 12-15.

Kofoid, C. A., and Olive Swezy. 1915. Mitosis and multiple fission in trichomonad flagellates. Proc. Am. Acad. Arts and Sci. Wash., 51: 290-371, 8 pls., 7 figs. in text.

—— 1921. On the free, encysted, and budding stages of *Councilmania lafleuri*, a parasitic amoeba of the human intestine. Univ. Cal. Publ. Zool., 20: 169-98, pls. 18-22, 3 figs. in text.

Naville, André. 1931. Les Sporozoaires (cycles chromosomiques et sexualité). Mem. Soc. Phys. Genève, 41: 1-223, 3 tables, 150 figs. in text.

Noble, E. R. 1938. The life cycle of *Zygosoma globosum*, sp. nov., a gregarine parasite of *Urechis caupo*. Univ. Cal. Publ. Zool., 43: 41-66, pls. 7-10, 3 figs. in text.

Schaudinn, Fritz. 1900. Untersuchungen über Generationswechsal bei Coccidien. Zool. Jb. Abt. Anat., 13: 199-292, pls. 13-16.

—— 1902. Studien über krankheitserregende Protozoen. II. *Plasmodium vivax* (Grassi and Feletti) der Erreger des Tertianfiebers beim Menschen. Arb. GesundAmt. Berl., 19: 169-250, pls. 4-6.

Yarwood, Evangeline A. 1937. The life cycle of *Adelina cryptocerci* sp. nov. a coccidian parasite of *Cryptocercus punctulatus*. Parasitology, 29: 370-90, pls. 15-19, 1 fig. in text.

CHAPTER XII

FERTILIZATION IN PROTOZOA

John P. Turner

Much has been written on the phenomena which accompany fertilization in the Protozoa. For detailed analyses of this literature the reader is referred to texts by Minchin (1912), Doflein-Reichenow (5th Ed. 1929), and Calkins (1933). The aim of this chapter is to give a bird's-eye view of the subject, to present some of the more significant facts already discovered, and last and most important to point out the need for investigation to determine those facts and principles still awaiting discovery.

If we consider sex to be essentially the formation of gametes and the fusion of those gametes in the fertilization process, we are using the term sex in a somewhat broader sense than if we limit it to the difference or distinction between the two sexes. In the Metazoa these phenomena seem to be fairly uniform for all groups; consequently, when the basic principles of the process are understood for one animal, those same principles may be applied to all the higher animals. Until very recently, however, the Protozoa were thought to belong in a different category, and one did not apply to them the general laws which were considered applicable to all other animals.

With recent discoveries, more and more of these preconceived differences have disappeared and we are now faced with the question of how close we can draw the analogies in sex phenomena between the Protozoa and the Metazoa. In other words, are the fundamentals of sex, i.e., maturation of gametes and fertilization, common to all animals, both metazoan and protozoan? If so, how similar or how dissimilar are the processes, and if not, just how do they differ? No final answer can be given to these questions in our present state of knowledge, but considerable evidence may be pointed out that is extremely significant.

In the Metazoa fertilization is accomplished by a small, active microgamete (spermatozoön) penetrating and fusing with a large, nonmotile macrogamete (ovum). The difference in appearance between the male and the female gametes, as well as between the two kinds of animals which produce them, is so clearly recognizable in most cases that we have come to think of sex in terms of the differences between maleness and femaleness. Among the Protozoa we find some species which also show a clear differentiation between male and female gametes, even though the sex differences between the organisms producing them are not so apparent. The protozoan organism is unicellular, and in many cases this single cell produces both male and female gamete nuclei in a kind of hermaphroditism (e.g., wandering and stationary pronuclei in ciliates). This complication makes the homologies between Metazoa and Protozoa less easily understandable.

In a great many Protozoa there is no apparent differentiation between gametes, yet their formation and fusion are accompanied by the same fundamental processes as is the case with differentiated gametes. Isogamous reproduction, therefore, is considered a sexual process.

The difference between the individual and the gamete is not always clear in Protozoa. Perhaps the most primitive kind of sexual phenomena is exemplified by two Protozoa, apparently identical to the vegetative forms, coming together and fusing in a fertilization process. According to Dobell (1908), this occurs in the flagellate *Copromonas subtilis*. Nuclear "reduction" occurs after partial fusion of the cell bodies and before nuclear fusion (Fig. 140).

COPULATION

GAMETIC MEIOSIS AND FERTILIZATION

Copulation, the complete and permanent fusion of gamete cells, is the type of sexual activity found generally in the Plasmodroma. In cases in which the parent organism gives rise to specialized cells which perform in fertilization, the process is known as fertilization by union of gametes, or simply gamogamy. Both isogamy, the union of similar gametes, and anisogamy or heterogamy, the union of dissimilar gametes, are found in this group. In cases in which gametes are as extremely dissimilar as spermatozoa and ova, the union is sometimes referred to as oögamy. In cases in which the organism itself fuses with another organism in permanent union, the whole organism functions as a gamete and the process is called hologamy.

In all of these cases, presumably, maturation of nucleus occurs to prepare it for union with its mate. In most cases maturation takes place in the last two divisions of the nucleus, before formation of the fusion nuclei or pronuclei. These two meiotic divisions are similar to those characteristic of spermatogenesis and oögenesis in Metazoa (see Sharp, 1934). The result of meiosis is the halving of the chromosome number to the haploid condition, so that fusion of the two haploid

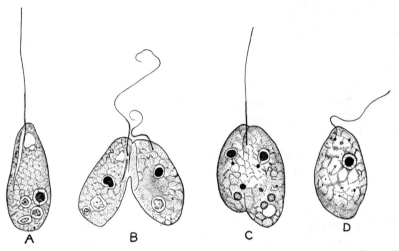

Figure 140. *Copromonas subtilis* in hologamous copulation. A, vegetative form; B, two individuals beginning to fuse anteriorly; C, cytoplasmic fusion well under way, nuclei in heteropolar, second "reduction" division; D, zygote with synkaryon and single flagellum. (After Dobell, 1908.)

gametes will reëstablish the diploid number which is characteristic of the species.

In the Mastigophora, syngamy has been described in very few forms except among the Phytomastigina, in which it seems to be the general rule. A typical case of hologamous copulation was described by Dobell (1908) in the colorless phytomonad *Copromonas subtilis* (Fig. 140). Two organisms which appear identical to each other and to ordinary vegetative forms come together and partially fuse. This partial union evidently acts as a stimulus to the nucleus of each gamont, for it proceeds to undergo two "maturation divisions" before fusing with its mate. Of the two products of the first progamic division, one degenerates and the other divides again into two very unequal parts. The smaller degenerates and the larger is the functional pronucleus, which fuses with the pronucleus of the other member of the pair. Superficially, this type of reduction is strikingly similar to polar body formation in metazoan oögenesis, although the actual reduction in chromosome number was not established by Dobell. The assumption that the two nuclear divisions preceding syngamy are reduction divisions seems reasonable, considering the almost universal occurrence of two divisions in the maturation of gamete nuclei. However, we are not justified in concluding that reduction in chromosome number occurs, unless positive determinations can be made of the chromosome number before and after reduction. In *Copromonas* there is obvious reduction in the amount of chromatin when all but one product of the two divisions disintegrate; but reduction in chromosome number has not been demonstrated, although it must occur somewhere in the life cycle, if chromosomes exist in this species. From the genetic angle, this is a critical point and should be determined if possible. After syngamy the zygote may encyst or it may develop directly into a vegetative form.

Hologamous fertilization has been reported in a few other members of the Plasmodroma, but is not confined to this group if Brumpt's (1909) description of fertilization in the parasitic ciliate *Balantidium coli* is correct. In this case two individuals come together and are invested by a common membrane, as in pseudoconjugation of gregarines. But the two balantidia then fuse completely and permanently. Other workers have not supported Brumpt's description and, although the details differ with species and author, Jameson (1927), Scott (1927), and Nelson (1934) agree that conjugation and not copulation is the form of sexual union found in this ciliate.

The occurrence of fertilization has been reported for a number of flagellates, but in very few cases has the evidence been convincing except for the Phytomonadida. Goldschmidt (1907) gave a detailed description of a sexual cycle in the animal flagellate *Mastigella,* similar in type to those described for *Arcella* and other Rhizopoda. If substantiated, another close link between the Pantostomatida and the Rhizopoda will be established. According to Goldschmidt, vegetative forms develop into macrogametocytes and microgametocytes, the nuclei of which give off chromatin into the cytoplasm. These chromidia in turn produce secondary nuclei, each of which appropriates some cytoplasm, forms a gamete, and undergoes reduction. Active flagellated macrogametes seek out and fuse with nonmotile microgametes, in contrast to the usual method. The zygote retains the flagellum and reproduces its monad-like self for several generations by fission. Then the offspring develop into the adult vegetative *Mastigella.*

Heterogamy seems to be clearly established in this amoeboid flagellate by the differences in the activity of the gametocytes as well as in the differences in size and motility of the gametes. Another point of interest is the pedogenic reproduction of the zygote. The interpolation of this asexual cycle into the life cycle bears a strong similarity to sporogony in *Plasmodium* and other Sporozoa. In both cases it is the asexual multiplication of the zygote before the adult stage, or trophozoite, is formed.

A point of particular interest to cytologists and geneticists alike is the origin of the gamete nuclei from chromidia. Not only here but also in a number of the Sarcodina, the origin of gamete and vegetative nuclei from chromidia has been reported. If the chromidial origin of nuclei is a fact, what of the genetic and structural continuity of the chromosomes and of the genes? Must we accept Hartmann's (1911) "polyenergid" interpretations that the nucleus is really an aggregate of many small nuclei, each with its sphere of influence, and that chromidia represent the scattered little nuclei or energids? At present we can only speculate. The problem is one of fundamental significance and is in great need of further investigation.

Chatton (1927) described a case of gametic meiosis in the flagellate *Paradinium poucheti* which is "exactly comparable in its progress and complexity" to spermatogenesis in certain insects. Included in his outline of this process are leptotine and pachytine stages, and diakinesis with tetrad rings and crosses. It is surprising that such highly developed processes should be found in such a primitive flagellate, although similar stages are not uncommon in ciliates and also occur in some Sporozoa and Sarcodina. The wide distribution of typical meiotic phenomena indicates that they are fundamental in nature, and it is probable that processes of comparable nature also occur in all forms in which fertilization takes place. Meticulous examination with improved techniques will throw much light on this question.

In the Dinoflagellata a few cases of syngamy have been reported, but most of them are too fragmentary or are supported by too little evidence to be discussed here.

An extremely interesting case has recently been reported by Diwald (1938) in *Glenodinium lubiniensiforme*. In this form, four flagellated isogametes are produced by the subdivision of each parent protoplast, and these gametes will copulate only with gametes of certain other clones. The obvious question is, of course, what is the nature of the difference between these "+" and "−" strains, a difference which inhibits their copulating among themselves or stimulates them to copulate with gametes of the other strain? This problem is yet to be solved, but it is similar to that found in *Paramecium aurelia* by Sonneborn (1937) and in *P. bursaria* by Jennings (1938). Diwald states that after fertilization the zygotes rest, then germinate and undergo two reduction divisions to form a "tetrad" of four potential individuals, only one of which, however, usually persists. This is the only described case of zygotic reduction outside the Telosporidia, with the possible exception of the amoeba *Sappinia diploidea* (see p. 595 below). A reasonable doubt remains, however, as to Diwald's interpretations. He gives no chromosome counts that would support his contention, and any assumption of chromosome reduction not based on determinations of chromosome numbers before and after reduction, especially in such an unusual case, is open to serious question. Dinoflagellates are not popular subjects for cytological investigation at the present time, but perhaps the work of Diwald will stimulate further research in this group.

In *Ceratium hirundinella* fertilization is accomplished in a way similar to that of the filamentous algae, according to the description of Zederbauer (1904). Two flagellates come together, the protoplasm of each extrudes from the lorica and makes contact with that of the other. The two masses now copulate, forming a zygote outside the loricas. Zederbauer observed these protoplasmic fusions only in the living state, so his account leaves much to be desired in the way of cytological details on which to base sound conclusions.

Chatton and Biecheler (1936) have more recently reported fertilization by slightly anisogamous gametes in the parsitic form *Coccidinium mesnili*.

In *Noctiluca scintillans* (*miliaris*) gamete formation has been repeatedly reported. In recent accounts Pratje (1921) could find no conclusive evidence for copulation, while Gross (1934) described copulation of isogametes. It seems odd that the life history of this abundant and spectacular species should still be a matter of such uncertainty.

Sexual reproduction is widespread among the Phytomonadida. These plant-like flagellates illustrate so nicely the gradations in sexual development and differentiation that they have been favorite material for classroom instruction.

Chlamydomonas is a non-colonial genus among the species of which both isogamy and heterogamy are found. In *C. steinii,* according to Goroschankin (1891), the flagellate divides into many isogametes within a cyst. The gametes fuse, beginning at their flagellated ends, and zygotes are formed which develop into resistant cysts. In *C. braunii,* the same author (1890) describes gametes of different sizes fusing in anisogamous copulation. Besides being definitely though not pronouncedly smaller, the microgamete is slenderer and more pyriform than the macrogamete. In still another species the differentiation is still more striking. Some individuals of *C. coccifera* (Goroschankin, 1905) are transformed directly into large nonmotile, egg-like macrogametes, while others divide into relatively very small, flagellated, sperm-like microgametes.

Chlamydomonas, therefore, illustrates possible stages in the evolution of sex from isogamy, in which the gametes are smaller than vegetative individuals but otherwise similar; through early differentiation of gametes, wherein the gametes are only slightly though clearly differentiated in size and therefore exhibit the very beginnings of anisogamy, to extremely well-differentiated heterogamy or oögamy, in which the microgametes and macrogametes are almost typical sperms and eggs. Only one more fundamental advance has been made in the evolution of sex in the Metazoa, and that is the differentiation of the adult forms into male and female individuals. Structural developments for the production and care of offspring belong to a different category.

It is unfortunate that so little is known of the maturation processes in *Chlamydomonas.* It is not known where reduction occurs, much less what the nature of the chromosomes and their behavior in reduction are. Some investigators assume that because of the way in which the zygotes of *Chlamydomonas, Gonium, Pandorina,* and so forth, behave at division, reduction is zygotic. Dangeard (1898) found no nuclear reduction taking place before fertilization in *Chlamydomonas* and suggested that it occurs during the division of the zygote. Pascher (1916), in his work on Mendelian inheritance in *Chlamydomonas,* presented genetic evidence that reduction is zygotic—that is, reduction in chromosome number occurs in the first two divisions of the zygote, which produce four swarmers. This would mean that only the zygote is diploid and that all other stages in the life cycle are haploid (see p. 611 below).

Here, indeed, is a peculiarly promising opportunity for the correlation of cytological and genetic evidence of chromosome behavior, if only the cytological data were available. Meiosis, not complicated by subsequent fusion of gametes, and the attendant bringing together of homologous chromosomes would offer some interesting possibilities.

Another excellent example of a series of organisms exhibiting progressively advancing stages in the evolution of sex is found in the colonial phytomonad flagellates. This series is so well known that it is usually discussed even in textbooks on general biology.

At one end of the series is *Gonium pectorale.* At certain times the sixteen cells making up this flat colony function as gametocytes by producing isogametes, which copulate in pairs to form zygotes. The gametes may vary somewhat in size, but the manner in which they copulate— apparently at random with any of the others—indicates that the slight variation in size of the gametes is without significance. In *Stephano-sphaera pluvialis,* a colony of eight cells, the gametes are all identical. The chief advance which these isogamous Volvocidae exhibit over the *Copromonas* type of fertilization is that the gametes are different from the vegetative form. In other words, vegetative or asexual forms may be distinguished from sex cells.

In *Pandorina morum* (Pringsheim, 1869), a subspherical colony of sixteen cells, two distinct sizes of gametes are produced, and two combinations are possible. Small gametes may fuse with other small gametes, and small gametes may fuse with large ones. Large gametes, however, never fuse with other large ones. Here, then, in a single species is exhibited both isogamy and heterogamy, for the failure of the large gametes to fuse with each other indicates that the size difference is significant. The critical factor may be the size itself, or it may be some less obvious factor associated with size.

It might be argued from this that primitive heterogamy is associated with hermaphroditism. The same colony produces both large and small

gametes; so, if the size difference means a step in the direction of maleness and femaleness, then the colony is monoecious or hermaphroditic.
This obviously leads to the conclusion that, in the evolution of sex in
these forms, gametes became differentiated into male and female types
before the parent organisms did. The fact that a larger percentage of
primitive Metazoa exhibit hermaphroditism than do the higher forms
lends weight to the assumption that this is a general truth.

In *Eudorina elegans* differentiation of gametes has become very
marked, and in *Volvox* the series is climaxed by such extreme differentiation between the microgametes and the macrogametes as is seen in metazoan sperms and eggs. Furthermore, the vegetative cells of the *Volvox*
colony are comparable to the somatic cells of the Metazoa, while relatively few cells of the colony carry on the germ line. Another advance
seen in *Volvox* is that some species have developed the dioecious
condition, wherein some colonies produce only microgametes and others
produce only macrogametes.

Among the Infusoria are found a few cases in which gametes are
formed that unite in complete and permanent fusion. This process is
therefore copulation, rather than the usual ciliate conjugation (see p.
617). In most cases of copulation in ciliates, the gamonts undergo repeated divisions, which result in the production of numerous small
"microgametes" which copulate with each other. Thus copulation of
gametes has been described for *Trachelocerca phoenicopterus* (Lebedew, 1909), in the Opalinidae (Neresheimer, 1907; Metcalf, 1923),
and in *Glaucoma* (*Dallasia*) *frontata* (Calkins and Bowling, 1929).
While these gametes may differ a little in size and be called "microgametes" and "macrogametes" by some, the differences do not appear
to be very significant. In general, they bear considerable resemblance
to the trophic individuals except in size. In the Vorticellidae, however,
gametes are formed which are truly anisogamous and which fuse permanently, although cytologically they more nearly resemble anisogamous
conjugants (see p. 621 below).

In the ciliate *Metopus sigmoides,* Noland (1927) described a sexual
process which is somewhat intermediate between copulation and conjugation. Conjugants come together and join anteriorly, but instead of
exchanging pronuclei as conjugants usually do, most of the cytoplasm
and both pronuclei of one member of the pair pass over into the body

of the other member, leaving behind only a shrunken remnant of the donor, which then detaches itself from the recipient and dies. This process is functionally very similar to that which occurs in the Vorticellidae, while the differences in structural details serve as a connecting link to typical ciliate conjugation.

Sexual phenomena seem to be fairly common in the Sarcodina, but they are not so characteristic of the group as was thought by many of the earlier workers.

In spite of the many reports of sexual stages in *Amoeba proteus,* several recent investigators have failed to observe any type of reproduction other than fission. Johnson (1930), whose article includes a review of the literature, believes that parasites and aquatic fungi have led to many misinterpretations of the life cycle of *Amoeba proteus.* Liesche (1938) carried *A. proteus* through 800 generations and observed no sexual stages and no cysts. It is quite possible that there is a sexual stage that occurs only at long intervals, or only under conditions which ordinarily do not obtain in the laboratory. It is obvious that if sexual stages occurred very often in this form, it would be reported more frequently and more convincingly than it has been, in view of the fact that this species is cultivated and studied so constantly in scores of biological laboratories. For instance, at the University of Minnesota *A. proteus* has been cultivated continuously for nine years, during which period thousands of observations have been made and several hundred permanent slides have been prepared from time to time. In spite of this prolonged search, no sexual stages have ever been found. Binucleate forms, presumably early dividing stages, have been observed frequently, but nothing suggesting gamete formation has ever been noted. It is true that sexual stages could have occurred and escaped observation, for examinations have not been made daily, but it seems reasonable to suppose that they would have been discovered at some time if they occurred at any but the rarest intervals.

However, Jones (1928) confirms the earlier work of Calkins (1907) and others, with descriptions and photomicrographs of gamete formation by fragmentation of the primary nucleus. He further claims that fertilization is accomplished by means of flagellated gametes.

A skeptic might point out that his photomicrographs of gametes and zygotes (his Figs. 13, 15, Plate 11) are strikingly similar to the figures of the parasite *Sphaerita* in *Amoeba limax,* as pictured by Chatton and Brodsky (1909, Figs. 2, 3). The possibility of confusing sporulation of *Sphaerita* with gamete formation in *Amoeba* is not too remote to be considered; although, as Calkins has pointed out, a parasitologist is inclined to see parasites in everything, and the parasite explanation has probably been over emphasized. The contradictory reports leave us in the peculiar position of not being very sure of the life cycle of our best known and most widely used protozoön, "the common laboratory *Amoeba.*"

Fertilization processes have been described for a number of other amoebae, including *Pelomyxa palustris* (Bott, 1907) and *Sappinia* (*Amoeba*) *diploidea* (Hartmann and Nägler, 1908). Bott's account of fertilization in *Pelomyxa* is unusual indeed. The nuclei of this multinucleated plasmodium extrude vegetative and generative chromidia into the cytoplasm. The chromidia form secondary nuclei, which in turn cast out the vegetative chromatin. The secondary nuclei, which now contain only generative chromatin, undergo the first maturation division, in which the chromosome number is reduced from eight to four. In the second maturation division four chromosomes appear and split, so that four go to each pole. Now each granddaughter nucleus divides into two compact masses of chromatin and a vacuole is formed near-by. The chromatin of the two masses then migrates into the adjacent vacuole, in the form of minute granules. After receiving the chromatin, the vacuole forms a membrane and becomes the definitive pronucleus of the gamete. The pronuclei which have arisen in this unique manner appropriate some cytoplasm and wander out as heliozoön-like gametes, which copulate in pairs to form zygotes. Each zygote grows into a new multinucleate *Pelomyxa.*

Aside from the peculiar rôle played by the vacuole in this maturation process, which introduces a sort of modified autogamy into the cycle just before the regular fertilization process, the cycle is worthy of further examination. Formation of secondary nuclei from chromidia, which in turn have resulted from the extrusion of chromatin from primary nuclei, has been described in many Sarcodina, several Sporozoa, at least one flagellate (*Mastigella,* see above) and one ciliate (*Trachelocerca phoenicopterus,* Lebedew, 1909).

Many protozoölogists remain skeptical of the entire proposition of the chromidial origin of nuclei (Doflein, 1916). Kofoid (1921) says "The evidence thus far presented of the *de novo* chromidial origin of protozoan nuclei is wholly inadequate to establish this hypothesis." Intracellular parasites are held responsible for some of the misinterpretations. More recent investigations have clearly refuted at least some of the earlier reports of the chromidial origin of nuclei. The reports of Myers (1935, 1938) and of Le Calvez (1938) on Foraminifera are good examples of this. However, many other reports must be reinvestigated before we can establish any very firm basis for our views.

Calkins (1933, p. 70) points out that the chromidial net of *Arcella* stains green with the Borrel mixture and usually gives a negative reaction to the Feulgen treatment. This supports Hartmann's experiments, in which the chromidia were dissolved out by pepsin, while the chromatin of the secondary nuclei remained conspicuous. Bělař (1926) believes this is conclusive evidence that chromidia are not composed of chromatin. However, Calkins shows that by omitting the strong hydrolysis of the Feulgen reaction, the chromidia are positively stained and therefore are composed of chromatin, or at least that nucleic acid is present in them. Nucleic acid becomes more concentrated in the nuclei, and this may explain why the nuclei resist pepsin digestion while the residue is dissolved. The author can confirm Calkins's positive results in staining *Arcella* chromidia with Feulgen. This organism has been stained with Feulgen at many stages in its life history by omitting strong hydrolysis, and intense staining of both chromidia and nuclei has resulted. Chromidia are colored an intense purple in forms containing nuclei, as well as in forms in which no detectable nuclei are present. It is not impossible that in the latter forms, some of the larger chromidia are actually minute nuclei which are lineal descendants by mitosis of the original nuclei.

According to Elpatiewsky (1907) and Swarczewsky (1908), the life cycle of *Arcella vulgaris* is extremely complicated. In addition to several methods of asexual reproduction, both chromidiogamy and anisogamous syngamy occur. In chromidiogamy two *Arcella,* the nuclei of which are degenerating into chromidia, come together. The protoplasm of one passes over into the shell of the other and, after the intermingling of the chromidia, half of the protoplasm passes back into the first shell, and the two organisms pull apart. After separation, the chromidia of each individual give rise to the nuclei of amoebulae, which bud off and grow into new adults. Zuelzer (1904) described chromidiogamy in *Difflugia urceolata,* but in this case all the chromidia are said to fuse into a single mass, and the united protoplasmic bodies condense and form a cyst. New nuclei form from the chromidia.

The significance of chromidiogamy has never been satisfactorily explained; in fact, the existence of the process itself remains in considerable doubt. While it is true that specimens of *Arcella* may frequently be found in which no typical nuclei are visible and the cytoplasm of which may contain numerous chromidia, these may be degenerating forms, and only a thoroughgoing reinvestigation of the life history of this interesting organism will convince the skeptics or disillusion the credulous.

In Elpatiewsky's account of anisogamy, some individuals form macrogametes by repeated nuclear division, while others form microgametes. The gametes are amoebulae, and the difference between male and female is one of size. After copulation between large and small gametes, the zygotes grow up into adult arcellae.

A remarkable type of sexual process was described by Hartmann and Nägler (1908) in *Sappinia (Amoeba) diploidea,* a binucleate form (Fig. 141). The active organism contains two nuclei, derived originally from two parents. It is therefore a kind of adult prezygote. Two such binucleate amoebae come together and develop a common cyst, but their bodies do not fuse. In each amoeba the two nuclei now fuse in a long-delayed fertilization, or karyogamy, after first giving off "vegetative chromidia." The cytoplasms of the two amoebae now fuse completely. Each synkaryon undergoes two "reduction divisions," after which three products of each degenerate, leaving one reduced nucleus from each synkaryon. These two are the nuclei of the vegetative form. If these two divisions are in reality meiotic divisions, the organism lives a haploid existence, and constitutes the only known case of zygotic reduction in the Sarcodina. If not, some other interpretation must be found for the two divisions which follow syngamy. Since chromosome number and behavior are not known in this form, no satisfactory conclusions may be drawn. It may be argued, of course, that the two haploid nuclei, lying close together in the cytoplasm, are the equivalent of one diploid nucleus, but such speculation must await the positive determination of the chromosome behavior.

Another noteworthy point here is that if the two amoebae which encyst together are derived from the same parent, the process is a case of
autogamy or pedogamy; if not, it is delayed hologamy.

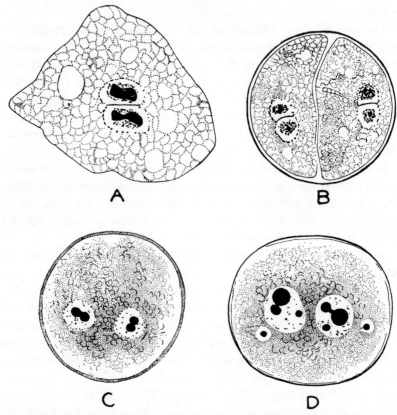

Figure 141. *Sappinia (Amoeba) diploidea.* A, the binucleate vegetative form; B, two
such individuals (sister cells ?) encyst within a common capsule and in each amoeba
the two nuclei fuse together; C, the bodies of the two amoebae now unite, and the two
fusion nuclei undergo two "reduction divisions," C, D, after which the nuclei lie side
by side, as in A, throughout the vegetative period. (After Hartmann and Nägler, 1908.)

The life history of the Foraminifera has been a subject of controversy
for many years. Since the pioneer researches of Lister (1895), which
were confirmed by Schaudinn (1903) and others, it has been generally
believed that the life history of *Polystomellina crispa* is fairly representative of the group. According to these investigators, alternation of

sexual (macrospheric generation $=$ gamont) and asexual (microspheric generation $=$ agamont) generations occurs, and the two generations may be distinguished morphologically, chiefly on the basis of the relative size of the original chamber of the shell. The protoplasm of the two adult generations was said to fragment, to produce flagellated isogametes from the gamont and agamete amoebulae from the agamont. Fertilization occurs free in the water, and the zygotes develop into agamonts, while the amoebulae develop directly into new gamonts. The nuclei of both the agametes and the gametes were said to arise from chromidia which are derived from the fragmentation of the primary nuclei.

In recent studies on the Foraminifera, Myers (1935, 1936, 1938) has confirmed the earlier work of Lister and Schaudinn, except for the origin of the gamete and agamete nuclei. In *Patellina corrugata, Polystomellina crispa, Spirillina vivipara* and *Discorbis patelliformis,* the nuclei of all stages, according to Myers, are derived by an orderly process of mitotic divisions from preëxisting nuclei. He believes that the chromidia are "concerned with feeding and metabolic activities" and in no case give rise to nuclei. This is another blow to those who hold to the chromidial origin of nuclei in Protozoa.

Myers (1935) further states that gametic reduction occurs in *Patellina corrugata* and that the haploid number of chromosomes is twelve. These observations differ from those of Schaudinn on the same species. In *P. corrugata* and in *S. vivipara,* the isogametes are amoeboid, but in *D. patelliformis* and *Polystomellina crispa* they are biflagellated, as indicated by the earlier workers.

In some forms, two or more gamonts become more or less closely associated in a kind of pseudoconjugation known as a syzygy, wherein the pseudopodia may temporarily fuse with those of close neighbors, while in other species they may encyst in a common capsule. This intimate association possibly has a synchronizing effect on gamete formation.

Le Calvez (1938) supports Myers's contention that gamete nuclei are not derived from chromidia, but arise by mitotic divisions from preexisting nuclei. In *Iridia lucida,* he states, the secondary nuclei "disintegrate" by rapid divisions which at first are typically mitotic. Later, because they are so small and the character so obscure, the mitoses are recognizable more by the centrosome than by the clarity of the chromosomes. Concerning the origin of the secondary nuclei, Le Calvez states that he "has not been able to discover the chain of processes which, from the disintegration of the vegetative nuclei, lead to the formation of a well defined micronucleus" (secondary nucleus). He believes that the hypothesis of generative chromatin ought to be completely abandoned.

In the Actinopoda, sexual reproduction has been reported for both Radiolaria and Heliozoa, but in only two forms has the process been reliably described. The classical case is that of *Actinosphaerium eichhornii* (Hertwig, 1898). The multinucleated vegetative individual forms a "mother cyst" and absorbs all but a few (up to 20) of its nuclei. The cytoplasm divides into as many primary cysts (cytospore number one) as there are nuclei. Each primary cyst divides into two distinct secondary cysts (cytospore number two), the nuclei of which undergo two successive "reduction" divisions, resulting in one pronucleus and two "polar bodies" each. The matured secondary cysts reunite with their sisters as gametes, and the nuclei fuse to complete fertilization. This is obviously a type of autogamy. Hertwig's claim that in both reduction divisions the chromosomes (numbering between 120 and 150) are divided in the metaphase seems open to question. If this were true, the divisions would not be reductional in character, so that the chromosome number would have to be reduced in some other manner than the usual gametic meioses.

According to Schaudinn (1896), *Actinophrys sol* undergoes isogamous macrogamy, or hologamy. He stated that two full-grown similar individuals come together and form a common cyst. The nucleus of each divides twice, and at both divisions one nuclear product degenerates and is expelled. The two cells, with their matured pronuclei, then fuse. The resulting zygote soon divides into two individuals, which later escape from the common cyst as vegetative animals.

The more recent and detailed investigations of Bělař (1923) have demonstrated in this species a type of sexual activity similar in many respects to that described by Schaudinn, except for the significant difference that the two original gametocytes within the cyst are sister cells, since they are derived by a progamous division of the original gamont (Fig. 142). The process, therefore, is a type of autogamy (pedogamy) similar to that occurring in *Actinosphaerium*, except that in the latter case the palmella produces several pairs of sister gametocytes. Incipient

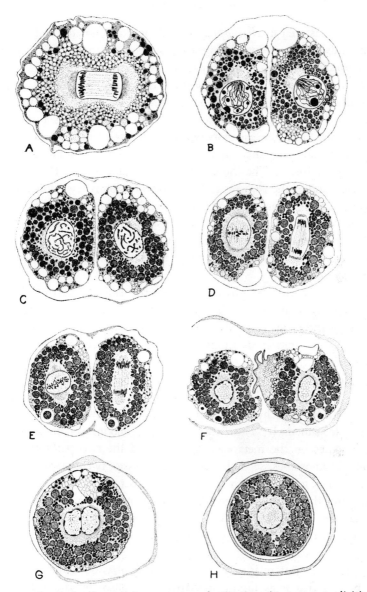

Figure 142. *Actinophrys sol* in autogamous fertilization. A, progamous division of
original gamont; B, the two daughters of this division within a common envelope, their
nuclei showing looping chromatin threads; C, pairing and twisting of thread-like chromosomes (left), and shortening and thickening of chromosomes (right); D, first maturation (reduction) division, with bivalent chromosomes on the equatorial plate (left),
and disjunction and separation of homologous chromosomes (right); E, second maturation (equational) division, with first "polar bodies" below; F, pseudopodium of ♂
gamete making contact with ♀ gamete; G, fusion of cell bodies; H, fusion of nuclei to
form zygote. (After Bělǎr, 1923. D is a composite.)

heterogamy is seen in *Actinophrys sol*. When the gametes unite, one of them sends out a pseudopodial process to the other, to initiate the fusion. This pseudopodium is formed by only one member of the pair, and the maturation processes in this one seem to occur a little ahead of those in the other. These slight differences between the two gametes are interpreted as the beginnings of differentiation toward maleness and femaleness. In rare cases the pseudopodium of the male fails to make contact with the female, and then the female sends out a pseudopodium which brings about fusion. The indication here is that whatever the degree of differentiation of the gametes is, this differentiation is reversible. Perhaps the potentiality for pseudopodial formation is retained in all gametes, but only the one completing maturation first ordinarily exhibits it. When neither gamete succeeds in connecting with its pseudopodium, no sexual differentiation is demonstrable. In such cases both gametes form parthenogenetic cysts.

The most noteworthy phase of gamete formation in *Actinophrys sol* is the striking similarity of the meiotic stages to those of the Metazoa. Following the progamous division of the gamont into the two gametocytes, two maturation divisions occur which reduce the chromosome number from the diploid forty-four to the haploid twenty-two. In the prophase of the first maturation division, the chromatin forms into slender looping threads (leptonema) which pair off (parasynapsis), become thicker (pachynema), and are obviously twisted around each other (strepsinema). Then they shorten (diakinesis) into compact chromosomes on the metaphase spindle, and the two parts of the bivalent chromosomes separate in the anaphase, twenty-two univalent chromosomes going to each pole. One product of this division degenerates, and the other undergoes the second maturation division, which is equational. The twenty-two chromosomes split longitudinally, so that the pronucleus and the two polar bodies of each gamete have twenty-two chromosomes.

It seems that this relatively simple heliozoön has developed a maturation process that is as highly specialized and clear-cut as any found in the Metazoa. It is probably safe to say that further diligent search will undoubtedly reveal other species of Protozoa with equally well developed meiotic phenomena.

SPOROZOA

Among the Sporozoa fertilization is almost universally present. As would be expected in such a heterogeneous group, all kinds of fertilization processes are known. Isogamy, heterogamy of all degrees of differentiation, pseudoconjugation, gametic meiosis, zygotic meiosis, and many other variations of the fertilization process have been described. Naturally only a few typical examples, illustrating the chief types of these phenomena, can be mentioned here.

Monocystis, the gregarine parasite in the seminal vesicles of the earth-

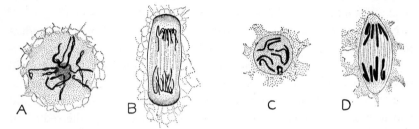

Figure 143. *Monocystis rostrata.* A and B, metaphase and anaphase of early progamous divisions of pseudoconjugant, eight chromosomes splitting, eight going to each pole; C and D metaphase and anaphase of last progamous (reduction) division, paired chromosomes disjoining, four going to each haploid pole. (After Mulsow, 1911.)

worm, illustrates typical pseudoconjugation, gametic meiosis, and isogamous fertilization. Two adult gregarines come together and are enclosed in a common cyst, but do not fuse. This intimate association without protoplasmic union is pseudoconjugation, and the members of this chaste betrothal are now gametocytes. The nucleus of each gametocyte divides again and again to form a large number of small nuclei, which migrate to the periphery and eventually become the gamete nuclei. According to Mulsow (1911), reduction occurs in the last of these divisions before formation of the pronucleus, in *Monocystis rostrata.* The earlier mitoses (Fig. 143) show eight thread-like chromosomes which split longitudinally, eight halves going to each daughter nucleus. In the last division the eight chromosomes associate in four pairs. In the anaphase that follows, members of the pairs separate and pass to different poles, thus reducing the number of chromosomes from eight to four. The surface of the gametocyte produces many small buds, each containing a pronucleus. These pinch off as gametes, and the walls between the associated gametocytes break down, allowing the gametes of one to fuse with those of the other. Thus cross-fertilization occurs and the process is isogamous, as there is no differentiation between gametes in this species.

Calkins and Bowling (1926), working on a species of *Monocystis,* have confirmed Mulsow's interpretations and have furnished additional critical evidence in support of Mulsow's belief. They found the early progamous divisions with the diploid number of chromosomes (ten) in each daughter plate and also the final progamous divisions with the haploid number (five) in each daughter plate.

Naville (1927a) has shown that in three types of *Monocystis* reduction is gametic. In types "A" and "B" early divisions of the pseudoconjugants show eight chromosomes and type "C" shows four as the diploid number. Anaphases of the last two divisions preceding gamete formation show four chromosomes going to each pole in types A and B, and two in type C. The next to the last division, therefore, is the reduction division. The first amphinuclear division is not reductional, and the sporoblast is diploid during its subsequent development.

Naville (1927b) also showed that in *Urospora lagidis* reduction of chromosome number from a diploid four to a haploid two occurs in the formation of its anisogamous gametes. A noteworthy occurrence here is that synaptic conjugation of chromosomes takes place in the synkaryon. The subsequent division of the synkaryon is equational, but the phenomenon serves to illustrate the possibility, in other forms, of an extremely precocious synapsis being prolonged throughout the life cycle until the next sexual stage appears, when the two members of the synaptic pairs would separate in a progamic division. Such a condition, if it exists, would explain in terms of gametic reduction the few known examples of zygotic reduction. This hypothesis seems worth investigating. Valkanov (1935) found pairing of chromosomes (synapsis) and condensation of chromosomes (diakinesis) into rings and crosses, in the zygotes of *Monocystella arndti;* and he believes that reduction occurs in the first zygotic division (see p. 613 below), although he was not able to follow the subsequent behavior of the chromosomes. His figures show eleven pairs in the zygote and eleven single chromosomes in the early divisions of the pseudoconjugants. These numbers indicate

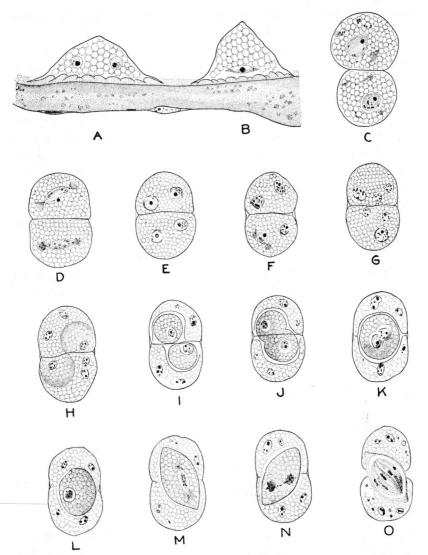

Figure 144. *Ophryocystis mesnili.* Isogamous gamete formation and fertilization. A and B, two trophic forms attached to ciliated cells of host; C, gamonts pairing in pseudoconjugation; D, E, F, two nuclear divisions, probably meiotic; G, mature gamont; H and I, formation of gametes by internal budding; J, K, L, fusion of gametes in fertilization; M, N, O, divisions of the zygote to form eight sporozoites in the single spore. (After Léger, 1907.)

zygotic reduction, but actual separation of chromosomes in the reduction division must be observed before the case is considered to be proved. *M. arndti* may have zygotic reduction, as the evidence indicates, but Naville's interpretation may apply to this case, so that gametic reduction remains a possibility.

Gamete formation by endogenous budding was found by Léger (1907) to occur in *Ophryocystis mesnili* (Fig. 144). In this form two gamonts adhere in pseudoconjugation, and the nucleus to each divides twice, one product of each division being destined to degenerate. These are presumably reduction divisions, although cytological evidence for this is lacking. One product of the two divisions becomes the pronucleus of the single large gamete which is formed inside the gamont as a loose internal bud. The walls between the gamonts break down and the two isogametes fuse. The zygote thus formed develops a spore wall, and eight sporozoites are produced by metagamic divisions.

While isogamy is most frequently observed in the gregarines, as for instance in the species already named and in *Diplocystis schneideri* (Jameson, 1920), *Gregarina cuneata* (Milojevic, 1925), *Actinocephalus parvus* (Weschenfelder, 1938), and others, several species show various degrees of anisogamy. Species other than *Urospora lagidis,* already mentioned, are *Echinomera hispida* (Schellack, 1907), *Stylocephalus longicollis* (Léger, 1903), and *Nini gracilis* (Léger and Duboscq, 1909), the last of which shows a marked degree of differentiation between the microgametes and macrogametes. This differentiation approaches that usually seen in the Coccidiomorpha.

An extreme differentiation of gametes (oögamy) is seen in *Eimeria* as well as in other Coccidia and in many Haemosporidia. The type of syngamy observed by Schaudinn (1900) in *Eimeria schubergi* will serve to illustrate this group. Here, as in other cases in which accurate chromosome determinations have not been made, it is assumed that reduction is gametic. The sexual phase starts with some of the merozoites developing into gametocytes, instead of repeating their asexual cycle. It is difficult to explain on a purely environmental basis why some merozoites repeat the asexual cycle while others, obviously in the same environment (intestinal epithelium) develop into gametocytes. If external factors play the chief rôle in determining whether a protozoan will continue asexual multiplication or enter a sexual phase, then it will be necessary to look further for the effecting stimulus in the Coccidiomorpha.

The phenomenon is more easily explained in these forms by the interpretation of Maupas, which has since been developed especially by Calkins, that internal factors play the determining rôle. This would mean that when the protoplasm had reached a certain degree of maturity in its cycle of development, the sexual phase would be initiated, even though the external conditions remained unchanged.

Whatever the cause, gametocytes appear and are differentiated into male and female gametocytes. The macrogametocytes are said to eliminate their karyosomes to accomplish reduction. By this process they are transformed into large, yolk-filled, egg-like macrogametes. The nuclei of the microgametocytes are said to give off chromidia and then degenerate. The chromidia condense into a number of clusters to form the nuclei of small, sperm-like, flagellated microgametes. A macrogamete is found and fertilized by a microgamete, and the resulting zygote forms an oöcyst. The synkaryon divides twice to produce four sporoblasts, each of which now develops two sporozoites.

A more thorough cytological study of the cycle may eventually reveal chromosome reduction taking place in the two divisions of the synkaryon, in which case meiosis would be zygotic; or in nuclear divisions prior to gamete formation, in which case it would be gametic. Karyosome extrusion and the formation of gamete nuclei from chromidia cannot be accepted today as conclusive evidence of meiotic reduction. If, indeed, no chromosomes are formed in *Eimeria schubergi,* then we shall be forced to modify our concept of meiosis. Here again we find urgent need for the application of improved techniques in cytological studies of a fundamental nature.

The sexual processes in the Adeleidea differ from those of the other Coccidia in several interesting respects. In *Adelina dimidiata,* according to Schellack (1913), two gametocytes of different sizes unite in a pseudo-conjugation process similar to that of the gregarines. The nucleus of the microgametocyte divides twice, and one of the nuclei enters and fertilizes the macrogamete. In this species only one macrogamete is formed and it is fertilized by one pronucleus of the microgamete in a way similar to anisogamous conjugation in the Vorticellidae. The peculiar behavior of the ciliate *Metopus sigmoides* (see p. 622 below) in conjugation also resembles the sexual union of *A. dimidiata.*

Fertilization in the Cnidosporidia is typically autogamous, and will be dealt with under the subject of autogamy. Little is known of the fertilization phenomena in the Acnidosporidia, but Crawley (1916) describes gamete formation in *Sarcocystis muris*, similar in general to oögamy in *Eimeria* except that the microgametocyte gives rise to the microgametes by a peculiar kind of nuclear fragmentation.

Autogamy

Autogamy, or self-fertilization, is accomplished in several ways in the Protozoa, but the result in all cases is the fusion of two gamete nuclei, both of which have been derived from the same parent cell. In some cases the two pronuclei have been separated by cytoplasmic divisions into separate cells which later fuse. In other cases the two pronuclei remain in the undivided cytoplasm and fuse after casting out part of the chromatin, with or without visible meiotic reduction.

Whatever benefit there may be to the individual or the race in exogamy, or cross-fertilization, in the way of renewing the vigor of the protoplasm and in propagating the race, this benefit is also a property of autogamy. There is no apparent reason why autogamy in these respects should not be as efficacious as exogamy. In two respects, however, the processes differ greatly. In the uniparental inheritance of autogamous individuals, meiosis will shuffle and sort out whatever genes are present; and if the allelomorphs are different, the resulting gametes and offspring will vary in their characteristics. However, the tendency in this kind of inbreeding would be overwhelmingly toward the production of homozygous races. Therefore the pronuclei and the resulting offspring would be less variable than in races in which exogamy brings together two sets of genes from two different parents, in the production of a heterozygous individual (see Jennings, 1920). From the genetic and the evolutionary standpoint, it is evident that exogamy would tend to produce a more heterozygous race; and, if natural selection is the critical factor in evolving organisms by the selection of favorable variants, then the advantage of exogamy over autogamy is apparent.

Another difference is the advantage the autogamous species has over the exogamous species in accomplishing fertilization. It is obvious that in exogamous species either the sexual organism or the gamete must seek out and find a mate before syngamy may occur; and if the organ-

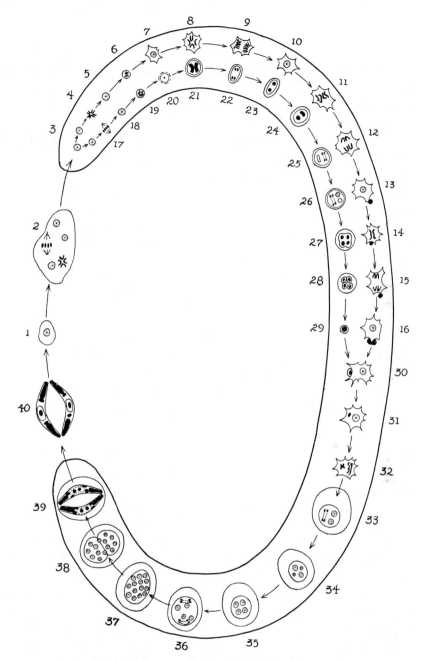

Fig. 145. Diagrammatic life cycle of *Sphaeromyxa sabrazesi*, 1, mononuclear zygote (sporozoite); 2, multinuclear plasmodium (schizogony), large outline represents plasmodium during subsequent development; 3-6, 17-19, multiplication of diploid nuclei; 7, 20, differentiation of nuclei into large and small; 11-12, 21-23, reduction division in macro- and microgametocytes; 14-15, 25-26, equational division; 16, macrogamete; 29, microgamete; 30, plastogamy; 31, pansporoblast; 32-37, mitoses of haploid nuclei to form fourteen; 38, two sporoblasts formed; 39, two pronuclei remaining in each spore; 40, fertilzation within spore. (After Naville, 1930b.)

isms are too widely dispersed or too effectively isolated by barriers, there will be no progeny. This difficulty simply does not exist for autogamous organisms, as the isolated individual can reproduce itself sexually without recourse to others of its kind.

Several cases of self-fertilization have already been pointed out (*Actinosphaerium, Actinophrys, Sappinia?*). In these cases fertilization is accomplished by the fusion of sister cells, which have undergone nuclear reorganization (perhaps reduction). This type of autogamy has been called pedogamy.

Another type of autogamy is common in the Cnidosporidia and is known best in the Myxosporidia. Here the nucleus divides to form several nuclei, without division of the cytoplasm. The nuclei then reunite in pairs, to form amphinuclei.

In the myxosporidian *Sphaeromyxa sabrazesi,* according to Schröder (1907, 1910), the plasmodial body contains two kinds of nuclei. Small areas become differentiated from the surrounding protoplasm. Each area, or pansporoblast, contains two nuclei, one large and one small. Both of these nuclei divide to form seven, so that there are fourteen nuclei produced in each pansporoblast. The pansporoblast divides into two halves, the sporoblasts, which are destined to become the two spores. The daughter sporoblasts receive six nuclei apiece, and the other two nuclei are expelled at the fission of the pansporoblast and degenerate as "reduction nuclei." Of the remaining six in each sporoblast, two form the capsule and shell, two form the polar capsules, and two presumably one of each original kind, remain as the pronuclei and later fuse with each other in autogamous fertilization. More recently Kudo (1926) has described a somewhat similar case of autogamy in *Myxosoma catostomi.*

Debaisieux (1924) found six chromosomes in the early mitoses of the plasmodium of *Sphaeromyxa sabrazesi.* In later stages he found the number reduced to the haploid three.

Naville (1930b) is very specific in his account of this species (Fig. 145) and of *S. balbianii.* In both forms four chromosomes are reduced to the haploid two in the plasmodium, just before the formation of the pansporoblast and after the differentiation of the nuclei into large and small types. The two types he calls macrogametocytes and microgametocytes, because they are surrounded by a zone of condensed cytoplasm. The union of two of these zones of cytoplasm in plastogamy brings a large and a small nucleus together in the formation of the pansporoblast. A considerable portion of the life cycle of these organisms is therefore passed in the haploid state. Naville also describes a similar diploid-haploid cycle of four and two chromosomes for *Myxidium incurvatum.* In this case there are more variations in the method of spore formation, but reduction from four to two chromosomes occurs in the plasmodium, as in *Sphaeromyxa,* and fertilization occurs between the two remaining nuclei of the spore.

Many variations of this process occur in other forms, but *Chloromyxum leydigi* (Naville, 1931) is of particular interest, because of its two haploid-diploid cycles. In this multinucleated plasmodium, the nuclei divide by mitosis, showing a diploid chromosome number of four. Then a heteropolar reduction division occurs, producing large and small nuclei with two chromosomes each. Internal buds are formed, wherein large and small nuclei fuse in pairs (first union). The difference in size of these fusing nuclei makes this an anisogamous fertilization. Several divisions of the fusion nuclei follow, each showing the diploid four chromosomes. The young plasmodium grows until the advent of spore formation, which is marked by the appearance of groups of four nuclei, two large and two small, each with chromosomes again reduced to the haploid two. The two small nuclei degenerate, while the larger two divide twice, to form a group of eight which become enclosed in a wall. Of the eight nuclei thus formed, six function in the formation of the spore complex and the remaining two fuse in the second fertilization of the cycle. This second fusion is comparable to fertilization in other forms, but the first fusion of nuclei in the plasmodium is a secondary development interpolated in the life cycle. Its significance is a matter of speculation. The phenomenon is actually a double autogamy and is difficult to harmonize with meiotic processes of either Metazoa or other Protozoa. Little is known of chromosome behavior in other Cnidosporidia.

Among the Actinomyxida and Microsporidia, autogamous fertilization is said to occur in a manner broadly similar to that of the Myxosporidia. One member of the Actinomyxida, *Guyenotia sphaerulosa,* has been shown by Naville (1930a) to undergo chromosome reduction in the second of three gametogenic divisions. In this case the development of the pansporoblast occurs as described in the case of other Actinomyxida. The two nuclei of the sporozoite divide mitotically, forming two large central germinal cells which ultimately give rise to eight male and eight female gametes respectively, and two smaller peripheral cells which divide again to form the four enveloping cells of the cyst wall. Of the three gametogenic divisions, the second reduces the chromosome number from the diploid four to the haploid two. The gametes are differentiated on the basis of size, the male gametes being smaller than the female. The eight microgametes unite with the eight macrogametes to form eight zygotes and reëstablish the diploid condition.

The origin of the two nuclei in the sporozoites is not known, but it is presumably by division of an original single nucleus. As the male and female gametes are produced by a single original sporozoite, it may be regarded as a hermaphroditic animal. This condition is rare in the Protozoa except in ciliate conjugation, in which the two pronuclei produced by one conjugant behave differently and in a few cases are morphologically differentiated (see p. 622 below).

While a number of other life cycles of Cnidosporidia have been worked out, knowledge of the fertilization process is fragmentary and data on chromosome behavior in meiosis are almost entirely lacking. Descriptions of "reduction" generally refer to the loss of chromatin by the degeneration of some of the nuclei which are sisters of the functional pronuclei. In other cases the extrusion of the karyosome is interpreted as reduction. Such loss of nuclear elements is not to be confused with reduction in the chromosome number from the diploid condition to the haploid. It is possible that the two processes are similar in function, but until such time as that is demonstrated, we are not justified in assuming that one is the equivalent of the other.

Autogamous fertilization has also been described for a few ciliates. According to Buschkiel (1911), the parasitic form *Ichthyophthirius multifiliis* becomes encysted and the micronucleus divides twice producing four, of which two degenerate while the remaining two fuse autogamously. Fermor (1913) described a reorganization process within the cyst of *Stylonychia pustulata,* wherein the old macronuclei degenerate and the micronuclei fuse and produce a new nuclear complex. The evidence in support of these two cases is not conclusive.

Diller (1936) has recently given a detailed account of autogamy in *Paramecium aurelia.* The process is similar to conjugation except that there is no pairing nor cross-fertilization. The macronucleus disintegrates during the process, as in conjugation and endomixis, and the micronucleus produces two pronuclei, as in conjugation. The pronuclei fuse autogamously and the synkaryon divides twice to form four nuclei; two are macronuclear *Anlagen* which separate at the first fission, and two become micronuclei. Diller challenges the very existence of endomixis in this species, on the grounds that stages of autogamy and irregular reorganization processes called "hemixis" have been mistaken for endomixis. Certainly this challenge must be met by careful reëxamination of the studies already made on endomixis (see Woodruff, Chapter XIII; also Sonneborn, Chapter XIV).

Zygotic Meiosis

In nearly all animals that have two parents, the two sets of chromosomes that are contributed to the offspring remain in the nuclei of all cells derived by mitosis from the zygote. This diploid number is characteristic of all cells except the gametes, in which case the chromosomes are separated out again by one or two meiotic divisions, giving to each gamete one half the diploid number. This haploid number is found only in the gamete, because the union of gametes reëstablishes the diploid condition.

There is evidence to show that in a few Protozoa, notably among the Telosporidia, reduction in chromosome number occurs in the division of the zygote. This means that the organism lives a haploid existence and only the zygote is diploid. Bělař (1926) argued that since the complete reduction process is known only in *Aggregata* and *Karyolysus* among Coccidia and in *Diplocystis* among gregarines, and that in all these forms reduction is zygotic, therefore reduction in all members of the two groups is probably zygotic. He suggested that Mulsow (1911) confused two species in obtaining his results, but the later work of Calkins and Bowling (1926) and of Naville (1927a) on *Monocystis* has made that conclusion untenable. Bělař further points to the frequent occurrence of odd numbers of chromosomes in these groups as evidence of the haploid condition, which implies zygotic reduction. Odd chromosome numbers could be explained by postulating a supernumerary or a sex chromosome, or by interpreting each chromosome as in reality a pair in close and prolonged synapsis; but these assumptions are unsatisfying, in the absence of more adequate evidence.

In their preliminary report, Dobell and Jameson (1915) gave the main features of their later detailed descriptions of the life cycles of the gregarine *Diplocystis schneideri* (Jameson, 1920) and the coccidian *Aggregata eberthi* (Dobell, 1925). In both cases meiotic reduction occurs in the first division of the zygote, and the organism lives all the rest of its life as a haploid animal.

In *Diplocystis* the nucleus of each pseudoconjugant divides many times, showing the haploid three chromosomes at each division. The dividing nuclei migrate to the periphery and eventually form club-shaped gametes, which pinch off from the gametocyte and fuse with those of the other pseudoconjugant. After a synaptic clumping of bead-like chromatin threads, six chromosomes appear in the prophase of the zygotic division and three go to each pole. The diploid number of chromosomes in the zygote is therefore reduced at the first amphinuclear division to the haploid three, a number which is also observed in the later divisions of the sporoblast.

Aggregata eberthi, like other Coccidia, has well differentiated male and female gametes. The female gametocyte is transformed bodily into a macrogamete after a complicated nuclear reorganization, which does not, however, involve reduction, though a spindle is formed and the six haploid chromosomes are seen. The nucleus of the male gametocyte divides by a complicated method, showing six chromosomes at each division. When the small flagellated microgamete fertilizes the macrogamete, a fertilization membrane appears. The diploid number of twelve chromosomes appears on the spindle of the first zygotic division. Pairing of homologous chromosomes follows, and the bivalent chromosomes become closely applied to each other. They later disjoin and six go to each pole, thereby reducing the number to the haploid condition. All other divisions of the nuclei are mitotic and six chromosomes appear and are divided at each mitosis.

It is possible that in both *Aggregata* and *Diplocystis* disjunction does not occur in the metaphase of the first zygotic division, and that, instead of this the bivalent chromosomes divide equationally, with six bivalents going to each pole. If at each subsequent mitosis the bivalent chromosomes divided until just before gamete formation and then disjoined, then reduction would be gametic instead of zygotic. However, the evidence seems conclusive enough to convince most biologists that in these two cases meiosis is truly zygotic.

In 1898 Dangeard found no reduction taking place during gamete formation in *Chlamydomonas* and suggested that it occurs during the germination of the egg. Pascher (1916) made no chromosome counts, but presented genetic evidence for zygotic meiosis in *Chlamydomonas.*

Hartmann and Nägler (1908) indicated that reduction is zygotic in *Sappinia (Amoeba) diploidea,* because three nuclei disintegrate, out of the four that are formed by two zygotic divisions. Diwald (1938) very recently stated that, because he could see no meiosis in the formation of the four gametes of *Glenodinium lubiniensiforme* and because a tetrad of four potential individuals are produced by two divisions of the zygote, reduction occurs in the two zygotic divisions.

It does not seem justifiable to base an assumption of zygotic meiosis on such indirect and questionable evidence. Genetic evidence must be considered, in the absence of cytological data; but only positive determination of chromosome number and identification of the stage in the cycle in which the number is reduced from diploid to haploid can be accepted as conclusive evidence of this phenomenon.

Valkanov (1935) has presented fragmentary cytological evidence of zygotic reduction in *Monocystella arndti.* He shows eleven long chromosomes in the early divisions of the pseudoconjugants. In the first zygotic division, eleven synaptic pairs condense into short, fat Ys and Xs. He concludes that reduction is zygotic, but his evidence is admittedly incomplete, as he was unable to follow the subsequent behavior of the chromosomes. Whether this is truly zygotic meiosis or whether the zygotic pairing of chromosomes is a phenomenon similar to that found in *Urospora lagidis* (see p. 602 above) remains an open question. The odd number of chromosomes lends some support to Valkanov's belief.

Weschenfelder (1938) has just published what appears to be a clear-cut case of zygotic meiosis in the gregarine *Actinocephalus parvus.* In the early nuclear divisions of the pseudoconjugants, four long, rod-shaped, haploid chromosomes are repeatedly observed. Isogametes bud off the mother cell and fertilization occurs as in other gregarines. At the first division of the zygote, eight chromosomes develop from the synkaryon as four synaptic pairs. These pairs disjoin in the anaphase and four go to each pole, reducing the number of chromosomes to the haploid condition again. Subsequent mitoses in the sporoblast reveal the haploid four chromosomes, now globular in shape, appearing in the prophase and passing to each pole in the anaphase.

Weschenfelder's observations have confirmed the suspicions of many protozoölogists that there exist other gregarines besides *Diplocystis schneideri* which undergo zygotic meiosis. The problem of inheritance in these forms presents some interesting possibilities to the geneticist. All genes possessed by the two parent organisms are passed to the zygote; therefore, if oöcyst (sporoblast) characters can be differentiated and mated, the immediate and direct effect of those genes may be observed in the resulting zygote. Furthermore, a haploid organism whose characteristics are controlled by a single set of chromosomes presents a rare opportunity for unusual genetic and cytological studies.

SIGNIFICANCE OF FERTILIZATION

The causes and effects of fertilization in Protozoa are subjects upon which a great deal has been written and some significant data obtained. The three conditions cited by Maupas (1889) as necessary for conjugation in ciliates are sexual maturity, diverse ancestry, and hunger. All three of these contributing factors have been supported by evidence from some later investigations and all three have been discounted by other investigations. In many cases the investigators have been dealing with different species of Protozoa. This in itself is probably responsible for many of the conflicting conclusions that have been reached. As the evidence accumulates, it becomes increasingly clear that different Protozoa require different conditions for conjugation and copulation, and that we are not justified in applying to all Protozoa conclusions derived from one or two or even several species.

Among the flagellates and rhizopods there are many organisms in which sexual phenomena have never been reported and in which probably none exists. These forms, then, are able to reproduce indefinitely by asexual means. Inherently, therefore, protoplasm does not seem to require sexual union.

At the other end of the sexual scale are found those Sporozoa and Foraminifera in which the life cycle is an obvious fact, and in which a sexual stage develops as one sector of that cycle, without which they could not continue their existence. If generalizations were made from these two kinds of organisms, there would be contradictions too obvious to relate.

For similar but less obvious reasons, we may partially account for the different schools of thought regarding the conditions necessary for conjugation in ciliates, upon which most of this work has been done.

In regard to ancestry, Calkins (1904) found that in *Paramecium caudatum* there are fully as many conjugations between closely related individuals as between individuals of diverse ancestry. He further indicates (Calkins 1933) that similar results have been obtained through isolation cultures of *Didinium nasutum, P. aurelia, P. bursaria, Stylnychia* sp., *Blepharisma undulans, Spathidium spathula, Oxytricha fallax, Chilodonella cucullus,* and *Uroleptus mobilis.* Sonneborn and Cohen (1936) found that under identical conditions, "The Johns Hopkins stock R of *Paramecium aurelia* can invariably be induced to conjugate," while "the Yale stock of the same species cannot." This difference appears to be clearly racial. Sonneborn's (1937) discovery of two "sex reaction types" in a race of *P. aurelia* may throw considerable new light on this question. Members of one type readily conjugate with those of the other type, but do not conjugate among themselves.

At first this looked as though something resembling the two sexes of other organisms had been found in the reaction of one ciliate to another. However, the discovery by Jennings (1938) of as many as nine sex reaction types in *P. bursaria* seems to remove these types from the category of sexes and indicates that they are simply strains which will not inbreed. The significance of these discoveries is not yet clear, but they do show that in some cases, at least, diverse ancestry is a potent factor in conjugation.

In regard to the relative importance of external conditions and internal conditions in ciliate conjugation, we again find contradictory evidence if we generalize from specific instances. The inductive method of reasoning is certainly stimulating and productive, but its misuse has led to some unjustifiably broad propositions. There is a rapidly accumulating array of evidence that external conditions, such as food, temperature, pH, population concentration, light, seasons, chemicals, condition of host in some parasitic forms, and so forth, do play an important rôle in inducing conjugation in some ciliates. However, there is valid evidence to indicate that in some forms, at least, conjugation can be induced only at certain times in the life cycle of the organism—in other words, only when the protoplasm is sexually "mature" for conjugation. Calkins (1933, p. 286) states that "One unmistakable conclusion can be drawn from the many diverse observations and interpretations of the conditions under which fertilization occurs in ciliates, viz., the protoplasmic state with which conjugation is possible is induced in large part, but not wholly, by environmental conditions."

It is a matter of common observation that when conjugation occurs in mass cultures, all the ciliates do not conjugate, but only a certain proportion of them. The proportion may vary with the culture and the species, but in any case if the conditions in the mass culture are favorable for inducing conjugation in some individuals, why do they not all conjugate? The fact that some do and some do not conjugate under conditions that appear to be identical, would indicate the existence of internal differences.

Calkins (1933, p. 290) summarizes the evidence and concludes "that environmental stimuli are without effect in producing conjugations unless the protoplasm is in a condition where such conjugations are possible." Two examples illustrate different phases of this proposition: *Uroleptus mobilis* will conjugate only after a period of from five to ten days after fertilization, and stock R of *Paramecium aurelia* (Sonneborn, 1936) will conjugate only in descendants of animals which have recently undergone conjugation or endomixis. The time factor is obviously different in these two animals, but both clearly indicate a strong cyclical differentiation which affects conjugation.

For more detailed analyses of this subject, reference should be made to Calkins's *Biology of the Protozoa* (1933) and to Chapter XIV below, by Sonneborn.

There is, perhaps, even less agreement concerning the effects of conjugation than concerning the causes. In some ciliates, e.g., *Uroleptus mobilis* (Calkins, 1919), conjugation results in a definite renewal of vitality, as indicated by an increase in the fission rate. Calkins interprets this as a fundamental process, which is an integral and normal part of the life cyle. Woodruff and Spencer (1924) found a similar renewal of vitality following conjugation in *Spathidium spathula,* but Woodruff (1925) interprets this as a rescue process to "meet the emergency of physiological degeneration *induced by environmental conditions* which are not ideal." Beers (1931) shows that conjugation increases vitality in *Didinium nasutum* which has been depressed by inadequate feeding, but that no depression occurs in well-fed animals.

In other ciliates, however, conjugation apparently reduces vitality. Thus in *Blepharisma undulans* Calkins (1912) found that all exconjugants died, although Woodruff (1927) concluded from his investigations that conjugation in this species accelerates the division rate. Jennings (1913) concluded that conjugation reduced vitality in *Paramecium* as indicated by a reduction in the average rate of fission in exconjugants.

At the present time the problem as regards ciliates seems to be: does increased vitality following conjugation mean that conjugation is a normal and essential part of the life cycle, or is it merely an emergency measure called into play when unfavorable environmental conditions have resulted in physiological degeneration? This problem is not easy to solve, because it is difficult to know what optimum or even "normal" environmental conditions are, and the two are probably not identical. Another complication is that in some species endomixis may be substituted for conjugation as a revitalizing process.

Another angle from which this problem may be approached is that of comparison with the plant kingdom. Many plants are able to reproduce themselves indefinitely by asexual methods, but at the same time sexual stages occur which, though not indispensable to their continued existence, are nevertheless certainly an integral part of their normal life cycle and valuable to the organism in other ways. In other plants, sexual processes must occur at regular intervals under "normal" conditions, or the species will die out.

Further investigation may disclose a similar situation in the ciliates, wherein some ciliates cannot continue to exist without periods of conjugation, while in others endomixis may be substituted for conjugation, and in still others asexual reproduction will carry on the line indefinitely. The final answer to this problem will come only through continued investigation.

CONJUGATION

Conjugation has been defined as the temporary union of two protozoan cells for the exchange of nuclear elements. It is a sexual process, differing from ordinary sexual union in that it is not directly related to reproduction. Two organisms enter into the relationship and the same two functional units leave the relationship; no third party—no progeny —has come into being. The two conjugants have been genetically changed by conjugation into new genetic entities, but this is actually genetic transformation rather than reproduction. Before the exconjugants return to their normal vegetative condition, they undergo one or more divisions in most cases, but these divisions are reproduction by binary fission, an asexual phenomenon. Although these divisions are modified by the previous sexual union, they are none the less asexual reproduction.

Conjugation is peculiar to the Ciliata and the process is strikingly uniform, with but few exceptions, in all ciliates which have been studied. The general course of the maturation phenomena in conjugation was first described by Maupas (1889), who studied the process in a number of ciliates and divided it for convenience into eight stages. Calkins (1933) states that "With one or two exceptions (*Trachelacerca phoenecopterus, Spirostomum ambiguum,* etc.) all of the free living ciliates thus far described agree in the general course of their maturation phenomena." Several parasitic species, however, exhibit some important differences from the usual course, and recent investigations have revealed a few interesting deviations among free-living forms.

With a few noteworthy exceptions, the union of two ciliates in conjugation takes place longitudinally and symmetrically (Fig. 146). The first sign of approaching conjugation in a mass culture is frequently a tendency to agglomerate in dense masses. Individuals appear to stick together on contact, even though they may separate soon after. Eventually, two individuals will adhere side by side, or with ventral surfaces together, and become more intimately connected in the anterior region. The extent of union varies from a thin protoplasmic bridge at the time of cross-fertilization to an intimate fusion of more than half the body length in other species.

Two individuals of *Euplotes patella* will come into contact, spiral about each other for a few moments, and then apply themselves together at their left peristomal margins, so that the appearance is similar to two turtles stuck together by their left ventral halves (Fig. 146). They swim forward together in a well coördinated manner, rotating on an axis which, owing to the symmetry of the pair, is straighter than the spiraling axis of a single individual. At this stage the pairs are joined only by their cirri. After remaining together a short time, they may separate and repeat the process with the same or with other individuals, until finally a union is made which involves an insecure adhesion of the bodies in the anterior left peristomal region. The peristome is distorted by the fusion, but the mouth continues to feed until it degenerates in the reorganization process.

While the majority of ciliates become attached along their ventral or ventro-lateral margins and fuse anteriorly, several exceptions are note-

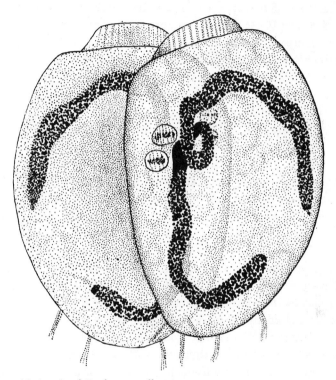

Figure 146. A pair of *Euplotes patella* in conjugation. The micronuclei have undergone preliminary division and are now in the first meiotic division; the C-shaped macronuclei are beginning to degenerate. (Turner, 1930.)

worthy. *Didinium nasutum* (Prandtl, 1906; Mast, 1917) and members of the Ophryoscolecidae (Dogiel, 1925) join end to end anteriorly, the latter forming an oral chamber by the juxtaposition of the two deep peristomal pockets. Dogiel states that the conjugants are smaller than the ordinary forms, owing to special progamic fissions. In *Parachaenia myae,* Kofoid and Bush (1936) found conjugants attached by their posterior ends, the anterior ends pointing in opposite directions. Kidder (1933b) describes the anterior tip of one member of a pair of *Ancistruma isseli* uniting asymmetrically with the peristomal groove part way back on its mate, though the two are of equal size. Miyashita (1927) shows that radically asymmetrical union occurs in *Lada tanishi,* in which the micro-conjugant, smaller than its mate, attaches its anterior end to the posterior ventral surface of the macroconjugant. In *Kidderia* (*Conchophthirius*) *mytili,* Kidder (1933a) shows an almost tandum association, with the anterior peristomal region of the slightly smaller member joined by a wide protoplasmic bridge to the aboral surface of the larger member of the pair. In *Dileptus gigas* (Visscher, 1927) fusion takes place along the ventral surfaces of the proboscides, and the mouth remains in evidence during the entire period of conjugation.

The varied methods of joining of the conjugants suggest that the location of the fusion bridge is not significant. If ciliate conjugation evolved from a process similar to pseudoconjugation as seen in present-day gregarines, as many protoölogists believe, it seems reasonable that one location would serve as well as another, provided the cortex were not too firm for a protoplasmic bridge to be formed. It is interesting in this connection to observe that in *Euplotes patella* (Turner, 1930), which has a rigid cuirass, no true cytoplasmic bridge is formed. The wandering pronucleus of one conjugant breaks out of the left anterio-ventral margin of the one conjugant and passes backwards through a tube formed by the local separation of the applied surfaces of the two conjugants, and finally enters the cytostome of the other conjugant. This method of entering the apposed conjugant probably developed simply because it was easier to penetrate the soft cytostomal membrane than the rigid cuirass. The mouth-to-mouth migration of the male pronucleus in *Cycloposthium bipalmatum* (Dogiel, 1925) is a simpler example of the same process.

In *Polyspira* and other members of the Foettingereidae, there occurs a remarkable combination of conjugation, fission, and chain formation called "syndesmogamie" by Minkiewicz (1912), and recently renamed "zygopalintomie" by Chatton and Lwoff (1935) in their comprehensive work on the Apostomea. The two conjugants unite by their lateral surfaces rather than by the usual ventral method, then proceed to undergo a series of synchronous, partial, transverse divisions, until a chain of conjugating zoöids is formed, resembling superficially a double tapeworm. After a time the fissions cease, and conjugation proceeds between members of each pair of zoöids according to the "classical scheme," although the nuclear details have not been worked out. Eventually fission is completed, and the exconjugants soon separate, reorganizing themselves in the usual way. These fissions of the paired conjugants appear to be related in kind to the special preconjugation fissions, observed in several other ciliates, which result in conjugants that may be distinguished from the vegetative forms chiefly by their smaller size. Specialized conjugants are observed in *Nicollella cteriodactyli* and *Collinella gundii* (Chatton and Pénard, 1921); the Ophryoscolecidae (Dogiel, 1925); *Dileptus gigas* (Visscher, 1927); *Balantidium* sp. (Nelson, 1934); *Nyctotherus cordiformis* (Wichterman, 1937), and in the microgametes of peritrichs.

The preliminary division of the micronucleus in *Euplotes charon* and *E. patella* (Maupas, 1889; Turner, 1930), without fission of the body, is a modification of this same tendency of the conjugants to differ from the vegetative forms.

Conjugants of many other ciliates are in some degree smaller than vegetative individuals, and this may be the result of reduced feeding or of other factors as yet unknown. It is among the copulating ciliates that the greatest difference occurs between the vegetative forms and the mature gametes (see p. 610 above).

Sexual differences are difficult to elucidate in the ciliates because the picture is confused by two kinds of possible differences. The two conjugants may show differences in size, shape, or other characteristics. These differences between the two conjugants entering the union may be interpreted as indicating maleness and femaleness. In a number of forms the differences are slight but fairly constant, as in Miyashita's (1927) "macroconjugants" and "microconjugants" of *Lada tanishi*. In the Vorticellidae, on the other hand, the difference in size and behavior of the microconjugant and the macroconjugant is very striking, far greater, in fact, than could be explained on the grounds of fluctuating variation. The small free-swimming form that seeks out and fertilizes the large sessile form could reasonably be called the male conjugant, and the large form may be considered a female conjugant. In the Vorticellidae and in *Metopus sigmoides* (Noland, 1927), mutual fertilization is not accomplished, because both pronuclei of one member—the microconjugant in the Vorticellidae—pass over with the cytoplasm into the other conjugant. One pronucleus of the donor and one pronucleus of the recipient fuse, to form the functional synkaryon. The other two pronuclei may or may not fuse, but in either case they eventually disintegrate. In *Metopus* the conjugants separate and the remnant of the donor dies; in the Vorticellidae the microconjugant fuses completely with the macroconjugant. These are obviously fertilization types intermediate between copulation and conjugation.

The other category of differences is that exhibited between the wandering and the stationary pronuclei which are produced in the same conjugant. They are usually considered to be male and female pronuclei respectively. Here the only apparent difference may be in their behavior, as is the case in the majority of ciliates studied. In *Euplotes patella* there is a slight difference in size between the wandering and the stationary pronuclei, and there is a special zone of cytoplasm which accompanies the wandering pronucleus in its migration. In *Cycloposthium bipalmatum,* however, Dogiel (1925) has described a spermatozoön-like wandering pronucleus, which is in striking contrast to the rounded stationary pronucleus. These illustrations may be considered as representing stages in the evolution of distinct sexual differences between pronuclei of ciliates.

If we assume that differences between members of a conjugating pair indicate sexual differentiation, then we would have male and female individuals both producing structurally isogamous but functionally anisogamous pronuclei, as in *Chilodonella (Chilodon) uncinatus* (Enriques, 1908; MacDougall, 1925). In other cases we would see male and female conjugants both producing pronuclei which are functionally and structurally differentiated as male and female, as in *Cycloposthium* (Dogiel, 1925).

If we consider the differences in behavior and structure between the wandering and the stationary pronuclei as indicating sexual differences, then we must consider the parent conjugants as hermaphrodites, and any differences between conjugants would then be a leaning toward maleness or femaleness on the part of an hermaphroditic organism. Viewed in this light, members of the Vorticellidae have lost their double nature, and the microconjugant has come to produce only male functional pronuclei and the macroconjugant only female functional pronuclei. The other pronuclei are produced as usual, but fail to develop. Similarly, in *Metopus sigmoides* (Noland, 1927) one conjugant contributes all of its potentialities to the other, in what may be interpreted as a male animal contributing its life and all its potential gametes to the female.

THE MACRONUCLEUS DURING CONJUGATION

In ciliates, the micronucleus is concerned with sexual activity and reproduction and is therefore frequently referred to as the generative or reproductive nucleus and represents the "germ plasm" of the Metazoa. The macronucleus, on the other hand is concerned with metabolism or vegetative activity and is considered the trophic or vegetative nucleus and represents in part the somatoplasm of the Metazoa. The chromatin of these two types of nuclei is combined in the single nucleus of other cells and in the one kind of nucleus found in the multinucleate *Opalina*.

The disintegration of the macronucleus at the time of conjugation, in all ciliates with dimorphic nuclei, represents the death of the soma and the end of the genetic unit. Differentiation of the new macronucleus in exconjugants similarly represents the development of the new somatic individual from the zygote.

In ciliates generally, the old macronucleus shows signs of disintegration during the maturation divisions, and, by the time of crossing of the pronuclei, fragmentation or other evidences of disintegration are well under way. It is during the differentiation of the new macronucleus from the synkaryon that the most rapid breakdown and the final absorption of the old macronucleus occur. This is probably due to the withdrawal from the cytoplasm of all chromatin-building elements by the developing macronuclear *Anlage* or "placenta." The old macronuclear remnants are possibly used as "fertilizer," or reserve of chemical elements in about the right proportion, for replenishing the cytoplasm and maintaining the equilibrium.

Before disintegration, the old macronucleus exhibits strange activity in several species of *Anoplophrya* (Schneider, 1886; Collin, 1909; Brumpt, 1913; Summers and Kidder, 1936), and in two species of *Chilodonella* (MacDougall, 1936). At about the time of crossing of the pronuclei, the macronucleus in each conjugant elongates and constricts in the middle, as one half pushes across the protoplasmic bridge into the apposed conjugant. This macronuclear exchange results in each exconjugant possessing half of both macronuclei. Their eventual decomposition makes the exchange difficult to explain on functional grounds. Summers and Kidder suggest that it may represent a "reminiscence of a more primitive protozoan condition before the separation of trophic nuclear materials from the germinal materials."

The odd elongations of degenerating macronuclear chromatin into ribbon or rod-like fragments in *Paramecium* may be an abortive attempt at a similar process.

CONJUGANT MEIOSIS

Because the general course of conjugation, as outlined by Maupas (1889) is followed by the vast majority of ciliates so far studied, it is convenient to use this outline in reviewing the process. His eight stages are as follows:

Stage A, in which the micronucleus swells and prepares for division;

Stage B, the first meiotic or maturation division;

Stage C, the second meiotic division;

Stage D, the third nuclear division, which produces the pronuclei;

Stage E, that of mutual exchange and the union of pronuclei;

Stage F, the first metagamic (amphinuclear) division;

Stage G, the second metagamic division;

Stage H, subsequent reorganization.

Stages A, B, C, and D are concerned with preparation for syngamy. This preparation includes meiosis and the formation of pronuclei (see Fig. 147). Stage E is the climax of the entire process, wherein the act of fertilization is consummated. Stages F, G, and H are concerned with the reorganization of the body and the reëstablishment of the usual vegetative form.

Among ciliates that normally possess more than one micronucleus there is little uniformity in the number of nuclei that undergo the two meiotic divisions. In many forms all micronuclei enter the first meiotic division. Then all products of this division may divide again, or various numbers of them may be resorbed (see Calkins, 1933, p. 295). In *Dileptus gigas*, however, Visscher (1927) has shown that only one of the large number of micronuclei undergoes maturation.

Two to eighteen micronuclei have been described as entering the first

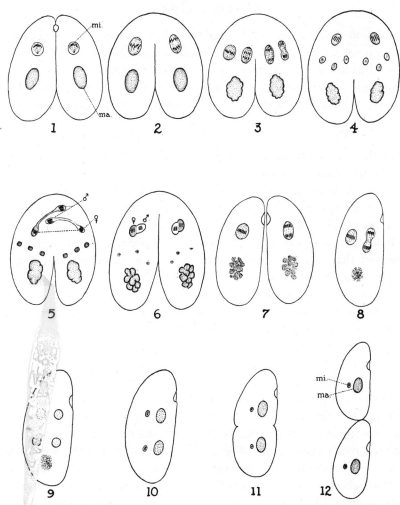

Figure 147. Diagram of ciliate conjugation. 1, two ciliates joined ventrally, micronuclei in prophase parachute stage; 2, first meiotic (equational) division; 3, second
meiotic (reduction) division in which the chromosome number is reduced from diploid
to haploid, and the macronucleus begins to degenerate; 4, third maturation division,
involving only one of the four nuclei in each animal, the other three degenerate; 5,
migration of the wandering (♂) pronuclei into the apposed animals, ♂ and ♀ pronuclei of left conjugant stippled to indicate common origin; 6, fusion of wandering
and stationary pronuclei to form synkaryon and restore diploid condition; 7, conjugants
separate, first division of amphinucleus in exconjugants; 8, second amphinuclear division;
9, four nuclei produced by the two amphinuclear divisions, the old macronucleus disintegrates; 10, two of the four new nuclei develop into new micronuclei, two into new
macronuclei; 11 and 12, first fission of the exconjugant separates out one micronucleus
and one macronucleus to each daughter cell, reëstablishing the vegetative condition.

maturation division of different ciliates. Variation in the number of nuclei involved in the maturation divisions occurs within the same species, as well as among different species. As only two pronuclei function in fertilization, and as these two are known to be sister nuclei in many ciliates, only one micronucleus really needs to undergo maturation.

STAGES A AND B, THE FIRST MEIOTIC DIVISION

In the earlier accounts of conjugation, the first maturation division in several ciliates was said to be not greatly different from ordinary vegetative mitosis. Recent accounts, based on careful cytological studies, show marked peculiarities in the prophase of the first maturation division. It seems possible, therefore, that more detailed studies, with improved techniques, may reveal these distinguishing prophase stages in all ciliates. The fact that such a careful worker as Maupas failed to observe the highly characteristic changes that occur in *Euplotes patella* (Turner, 1930) lends weight to this possibility.

In the vast majority of ciliates, the prophase of the first maturation division is highly characteristic and presages the coming reduction. In some ciliates the micronucleus takes on the form of a crescent, or comma, during the prophase, and this appearance is sufficiently characteristic to be recognized as a general type. Among ciliates that exhibit the crescent formation are various species of *Paramecium* and the Vorticellidae.

There is little agreement as to the number of chromosomes in *Paramecium* or even as to the method by which the crescent is transformed into the metaphase spindle. The chromosome number in all species is surely larger than the 8 or 9 given by Hertwig (1889) for *P. aurelia,* and probably less than 150, which has been attributed to *P. caudatum.* Calkins and Cull (1907) suggested that the 165 or more small chromatin rods or fibers seen in *P. caudatum* are comparable to the physical counterpart of the individual genes of higher animals. Aggregates of these would represent a chromosome in cases where chromosomes are formed. Perhaps the 32 chromomeres of *Euplotes patella* (Turner, 1930) also represent 32 genes, although one would expect this highly specialized hypotrich to have more genes than the more primitive holotrich.

Whatever the nature of the chromatin elements in *Paramecium* may be, these investigators, as well as Dehorne (1920), show that the first maturation spindle is formed at right angles to the prophase crescent by the migration of the division center from the apex of the crescent to the middle of the crescent, and by the pushing out of the other pole across the crescent. Earlier workers believed the spindle was formed by a shortening of the crescent. Dehorne finds no chromosomes at all, but, instead, a simple convoluted thread.

In a wide variety of other ciliates, the prophase develops a "candelabra" (Collin, 1909) or "parachute" (Calkins, 1919) stage. It is noteworthy that the parachute prophase occurs in most of the ciliates in which reasonably complete chromosome studies have been made and reduction definitely located. *Kidderia* (*Conchophthirius*) *mytili* may be an exception to this, but Kidder (1933a) admits he might have missed finding it. Tannreuther (1926) describes a simple type of chromosome formation in *Prorodon griseus,* in which chromosomes arise directly out of a central chromatin mass upon the equator of the spindle.

In *Euplotes patella* a typical parachute is formed, which is seen as a stage in the transformation of resting chromatin into the chromosomes of the metaphase spindle. The events transpire synchronously in both nuclei produced by the preliminary, or pre-maturation division of the micronucleus occurring in this species. Each nucleus swells to several times its original size, as the faintly granular chromatin becomes more basophilic, and is arranged in a reticulum filling the nuclear space (Fig. 148). The reticulum condenses in the center and becomes polarized, with most of the chromatin at one pole. Further condensation forms a dense club-like structure, which presently loosens up and is transformed into a parachute, with most of the chromatin forming the "cloth" at one pole, the spindle fibers forming the "rope," and an endosome at the other pole forming the "weight." The chromatin then forms thirty-two discrete chromatin granules, the chromomeres, which soon migrate to the equatorial plate in groups of four. These eight groups of four chromomeres apparently correspond to the eight diploid chromosomes found in other stages of the life cycle. In the anaphase of this division, sixteen chromomeres pass to each pole, and one may frequently observe them associated in pairs as loosely connected dumb-bells. The sixteen chromomeres, or eight dumb-bells, which pass to each pole represent the eight diploid chromosomes and identify this as an equational division.

In *Pleurotricha lanceolata,* Manwell (1928) found chromomeres that fuse to form about eighty dumb-bells. Since the diploid chromosome number is forty in this species, four granules (two dumb-bells) evidently represent a chromosome, just as they do in *E. patella,* and forty dumb-bells pass to each pole in the anaphase.

In *Kidderia (Conchophthirius) mytili,* Kidder (1933a) found thirty-

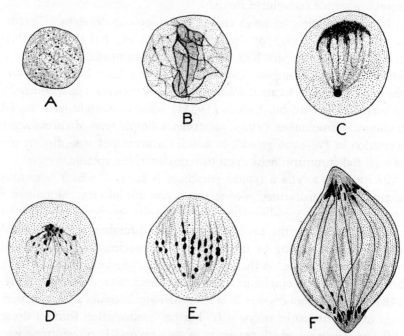

Figure 148. Stages in the first maturation division of *Euplotes patella.* A, early nucleus with finely granular chromatin; B, chromatin reticulum condensing in center; C, parachute stage; D, later parachute showing chromatin granules migrating from upper pole, endobasal bodies and intradesmose visible; E, metaphase stage with thirty-two chromomeres arranged in eight chromosome groups; F, anaphase stage with sixteen chromomeres (eight chromosomes) passing to each pole. (Turner, 1930.)

two granules forming on the spindle and sixteen passing to each pole, as in *E. patella.* Sixteen is the diploid number in this species, so the thirty-two granules represent half a chromosome each, although there is no visible association between the halves.

Gregory (1923) described forty-eight chromomeres appearing in the prophase and fusing to form twenty-four dumb-bell chromosomes in *Oxytricha fallax.* In this case twelve dumb-bells pass to each pole. If twelve were the diploid number, this would correspond exactly to the condition existing in *Euplotes patella* and in *Pleurotricha lanceolata.* However, Gregory believes that twelve is the haploid number and that the separation of the twenty-four dumb-bells into two groups of twelve each, in the first maturation division, means that this is the reduction division, a condition unusual in ciliates. This interpretation is weakened somewhat by the fact that twenty-four dumb-bells are formed in the prophase of the second maturation division, twelve passing to each pole in the anaphase. It is possible that the twenty-four dumb-bells which separate into two groups of twelve each in the first maturation division are actually tetrads, and that the twelve going to each pole are diads. This would mean that the joining of the original granules is synaptic in character and that their passing to the same pole indicates splitting, or equational division. This explanation is not completely satisfying, in view of the events of the second maturation division. Each of the twenty-four dumb-bells which are formed in the prophase of the second division would have to be derived from one granule of the first anaphase dumb-bells. If that occurred, then the second maturation division would be reductional; but if they are formed by the splitting of entire dumb-bells, as believed by Gregory, then the second division would be equational and the first would be reductional.

Whichever interpretation is correct, one thing seems clear: in all these forms the chromosomes of the first maturation division are composed of a definite number of loosely associated chromomeres. It is in their method of distribution that interpretations differ, and further investigation in this field will be welcomed by those interested in meiotic phenomena in the Protozoa.

No parachute stage was found by Noland (1927) in *Metopus sigmoides.* Instead, the chromatin forms a spireme, which condenses into a single large sausage-shaped "chromosome" on the spindle. This divides, and one part goes to each pole. The interpretation of this condition is difficult, because of the obscurity of later stages. From the appearance of the synkaria in Noland's drawings, one would judge that there are two large chromosomes in some, and four in others. If two were the diploid number, the four would represent splitting for fission. Then the single chromosome of the first maturation division could be interpreted as a synaptic pair in close union. This speculation may not be justified by further investigation, but it seems reasonable on the basis of the available facts and is in line with current theory.

Calkins (1919) describes a peculiar situation in *Uroleptus mobilis,* in which two types of metaphase stages are found (Fig. 149), one in which about twenty-four chromosomes appear and twelve go to each pole, and another in which eight chromosomes appear and eight go to each pole in an obviously equational division. Although the number is not strictly homologous, the first type is similar to that which occurs in *Euplotes patella* and other species, and the second type is what would be expected if all chromosomes were compact. Intermediate forms are conceivable, in which some chromosomes are compact and others are

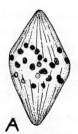

Figure 149. First maturation spindles of *Uroleptus mobilis.* A and B, two types of metaphase stages; C and D, two types of anaphase stages found in this form. Both types are equational divisions, since the diploid eight chromosomes appear in the subsequent division, in which they are reduced to four. (After Calkins, 1919.)

dispersed as several loosely associated chromomeres. This would explain many irregular counts, which otherwise seem chaotic.

In *Chilodonella* (*Chilodon*) *uncinatus* a parachute is formed after the division of an endobasal body, according to MacDougall (1925). Enriques (1908) failed to see the parachute in the same species, but described a peculiar rod formation. In MacDougall's material, four strands of chromomeres are formed from the spireme of the late parachute stage and condense into four dumb-bell chromosomes on the spindle. This author states that the exact number of granules in each strand was not determined, but her Figure 23 shows four on each strand. This is interesting, since four chromomeres to a chromosome has been found in *E. patella* (see Fig. 148 E) and other ciliates. The four chromosomes then split longitudinally and four halves migrate to each pole, at which point they fuse in pairs, forming diads before entering upon the resting stage.

A tetraploid strain arose spontaneously in a pure culture of Mac-Dougall's *Chilodonella uncinatus*. Maturation phenomena in the tetraploid form were similar to those of the usual diploid form, except that there were twice the number of chromosomes in every stage. Investigations, presented and reviewed in a later article by MacDougall (1936), show meiotic processes which are similar in six species of *Chilodonella*. In all species the diploid number of chromosomes is four. A parachute stage is followed by the formation and the synaptic pairing of chromatin threads, as in "classic leptotine and zygotine" stages, which condense to form the pachytene chromosomes. MacDougall's descriptions reveal the striking similarity of meiosis in *Chilodonella* to the general scheme of meiosis in the Metazoa. Messiatzev (1924) reported synapsis occurring in the first maturation division and again in a fifth amphinuclear division of *Lionotus lamella*, but Poljansky (1926) believes that Messiatzev confused his stages in the latter case.

The small number and the large size of the chromosomes in *Chilodonella* make this a very favorable form for study of meiotic phenomena. It seems unfortunate that more of the recent studies that have genetic significance were not made on this animal, in which chromosome behavior is clear-cut and well known, instead of on *Paramecium,* in which it is practically impossible to determine any of the significant stages in meiosis.

STAGE C, THE SECOND MEIOTIC DIVISION

The second meiotic division is the reduction division in all ciliates thus far studied, except in *Oxytricha fallax,* according to Gregory (1923). Prandtl (1906), in his work on *Didinium nasutum,* was the first to present conclusive evidence on chromosome reduction in ciliates when he described reduction from sixteen to eight chromosomes in the second meiotic division. There seems to be no general rule for the number of nuclei that enter this division. In the species of *Chilodonella* studied by MacDougall (1936), only one nucleus is involved in any of the three progamous divisions, the other products degenerate.

In perhaps the majority of ciliates both products of the first division enter the second division. In *O. fallax* (Gregory, 1923) and in forms with multiple micronuclei such as *Uroleptus mobilis* (Calkins, 1919), a variable number of nuclei may divide a second time. In *Euplotes patella* all micronuclear products undergo a second meiotic division. Because of the preliminary division, there are four in each conjugant. No resting stage occurs between divisions here, in contrast to MacDougall's (1936) account of *Chilodonella*. The daughter nuclei of the previous division are still connected by their respective drawn-out nuclear membranes when the chromatin begins to resolve itself into a reticulum in each nucleus and the granules on the reticulum condense into eight discrete, ovoid

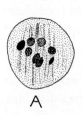  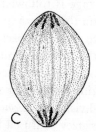

Figure 150. Second meiotic (reduction) division in *Euplotes patella*. A, eight ovoid chromosomes appearing on the spindle; B, synaptic pairing and lengthening of chromosomes; C, disjunction and separation of homologous chromosomes in the anaphase, four passing to each pole.

chromosomes (Fig. 150 A). The chromosomes now conjugate in four pairs, in what is probably a delayed synapsis, elongate somewhat, and disjoin longitudinally, four haploid chromosomes passing to each pole.

Calkins (1919) described a similar pairing and separation of the eight chromosomes in *Uroleptus mobilis*. Tannreuther (1926) presents evidence of chromosome pairing in the reduction division of *Prorodon griseus,* but in most cases where synapsis has been observed, it occurs in the first meiotic division.

STAGE D, THE THIRD DIVISION, AND THE FORMATION OF PRONUCLEI

In all ciliates thus far studied, a third division occurs. This division is equational in character and usually involves only one nucleus, while the rest degenerate. The two products of this division are the pronuclei which take part in fertilization.

In a few ciliates, two, three, and four micronuclei have been reported to divide at this stage, but in no case has it been demonstrated that the two functional pronuclei are derived from different spindles. In all cases in which only one nucleus is involved, and possibly also in those where two or more are involved, the two pronuclei must be genetically identical if the third division is equational. In *Uroleptus* (Calkins, 1919), two or three nuclei divide, but the two pronuclei are always sister nuclei. As this occurs in both members of the pair, the exconjugants should theoretically be genetically identical. This appears to be the significant feature in the third maturation division.

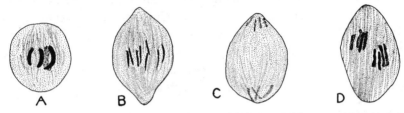

Figure 151. *Euplotes patella.* A, B, C, third maturation division, in which the four haploid chromosomes in A split longitudinally and the halves slip past each other in B, and four go to each pole in C; D, fertilization nucleus in which the ♂ and ♀ pronuclei have just joined, but their chromosome groups have not yet mingled.

In *Euplotes patella* (Turner, 1930), two nuclei enter the third division. In each nucleus (Fig. 151) the chromatin reticulum condenses into four strands of chromatin granules, which condense into four compact sausage-shaped chromosomes lying lengthwise of the spindle. The chromosomes split longitudinally, and the halves slip past each other as they migrate to separate poles in this equational division. The chromosomes are all lying in the same axis, so that as the chromosomes slip past each other in the early anaphase, they appear end to end, and the figure might easily be misinterpreted as a transverse division of chromosomes. It is possible that the descriptions of the transverse division of chromosomes in the third division, given by Enriques (1908), Calkins (1919, 1930), MacDougall (1925), and others were based on some such artifact. Calkins points out that if each chromosome represents one gene, the method of division is of no consequence. This interpretation would doubtless serve for *Paramecium,* which has a large number of chromosomes, but it is less likely . . . . genes. It seems more probable that the apparent transverse division of some chromosomes in mitosis is due to our inability to demonstrate by present techniques some of the finer structural changes which occur within the chromatin mass.

Since Prandtl (1906) first noted a difference in size between the wandering and the stationary pronuclei of *Didinium nasutum,* slight differences have been reported in a number of other cases. Calkins and Cull (1907) showed that this is due to a heteropolar third division, in *Paramecium caudatum.* Maupas (1889) was the first to record a structural difference between pronuclei, when he observed the area of dense cytoplasm in front of the migrating pronucleus of *Euplotes patella.* The most striking dimorphism appears in *Cycloposthium,* according to Dogiel (1925). The wandering pronucleus is spermatozoön-like in having an elongated tail. All these differences between pronuclei must be cytoplasmic in origin, for the nuclei, as has been pointed out, are genetically identical, if our present concepts are correct.

STAGE E, MIGRATION OF PRONUCLEI AND FERTILIZATION

Migration of the wandering nucleus occurs synchronously in the two conjugants, so that they generally pass each other in the cytoplasmic bridge which joins the two conjugants. In *Cycloposthium* (Dogiel, 1925), the spermatozoön-like male pronucleus passes out of the mouth of the parent body and into the mouth of the recipient, by way of the juxtaposed peristomal cavities (Fig. 167).

In *Euplotes patella* (Turner, 1930), the wandering pronucleus breaks out of the left anterior tip of the parent body, which is pressed into the peristomal field of its mate (see p. 620, above), passes backward between the appressed conjugants, and finally enters the cytostomal area of the recipient. Both pronuclei form spindles as the male approaches the female, and four chromosomes can be seen in each. As the pronuclei touch, their membranes dissolve and the two groups of four chromosomes mingle, as fertilization is completed and the diploid condition restored (Fig. 151 D). In *Chilodonella,* MacDougall (1925) shows that the two haploid chromosomes are visible throughout the migration period (Fig. 152), but lose their identity soon after fertilization.

The appearance of the pronuclei at the time of union varies with the species. In a number of ciliates, they are in the form of a spindle similar to those of *Euplotes,* although few show chromosomes. In other ciliates, the pronuclei are spherical and vesicular at the time of union. In still others, intermediate conditions have been reported.

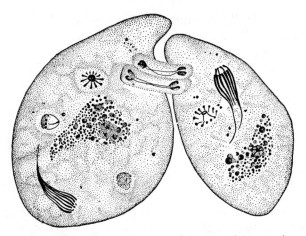

Figure 152. *Chilodonella uncinatus.* Migration of the pronuclei across the protoplasmic bridge. Each pronucleus contains two haploid chromosomes still attached by strands to their sister halves. Also visible in each conjugant are old and new oral baskets and the granular remnants of the old macronucleus. (After MacDougall, 1925.)

STAGES F, G, AND H, THE EXCONJUGANTS

The subject of reorganization is dealt with elsewhere in this volume, but we may consider briefly some of the cytological aspects of the re-organizing exconjugant.

After fusion of the pronuclei, the fertilization nucleus divides one or more times, and from the products of division the new micronuclear and macronuclear elements are formed, while extra products disintegrate. The number of divisions the synkaryon undergoes before differentiation of the macronuclei and micronuclei in various ciliates is reviewed by Kidder (1933b) in his work on *Ancistruma*. Kidder lists eight species in "group A," in which the micronucleus and the macronucleus are differentiated after the first amphinuclear division. To these we may add *Chilodonella cucullulus* (Ivanić, 1933); *C. chattoni, C. labiata, C. caudata, C. faurii* (MacDougall, 1936); and *Nyctotherus cordiformis* (Wichterman, 1937). In other ciliates, differentiation occurs after the second amphinuclear division. Kidder lists twenty-one species in this "group B," which includes a majority of the best-known ciliates. To this list may be added *Balantidium* (Nelson, 1934), from the chim-panzee.

In about half of these species, all four products remain functional,

and in the others two or three develop into micronuclei and macronuclei, while the remaining one or two degenerate.

In *Euplotes patella,* the macronucleus and the micronucleus are never
sister nuclei, so that the first amphinuclear division evidently separates
the macronuclear line from the micronuclear line. However, there is no
apparent difference, except one of position, between the two products
of the first division.

In "group C" Kidder lists twelve ciliates, in which differentiation
occurs after the third amphinuclear division. To this list, which includes
the familiar *Paramecium caudatum,* may be added *Parachaenia myae*
(Kofoid and Bush, 1936). The number of products, if any, which degenerate appears to vary with the author, as well as with the species.
In about half of these species, all nuclei remain functional.

In *Paramecium caudatum,* four of the eight amphinuclear products
become macronuclei and are distributed by fission to the four grand-
daughters, while four become micronuclei. According to Calkins and
Cull (1907), all of these micronuclei remain functional and are distributed by fission; but Maupas (1889), Jennings (1920), and Doflein-
Reichenow (1928) indicate that three degenerate, while the fourth
remains as the functional micronucleus and divides at each subsequent
fission, just as it does in *P. putrinum* (Doflein, 1916).

In some material, both of these schemes are represented in *P. caudatum;* but, perhaps because the dividing micronucleus is more easily
identified, the latter type is more clearly and unquestionably demonstrable.

In "group D," in which differentiation occurs after the fourth amphinuclear division, Kidder lists only *P. multimicronucleata* (Landis, 1925),
Kidderia (*Conchophthirius*) *mytili* (Kidder, 1933a), and *Bursaria
truncatella,* according to Prowazek (1899). But *Bursaria* belongs in
"group C," according to Poljansky (1928, 1934).

After the final division of the synkaryon, which separates macronuclear elements from micronuclear elements, the micronuclei return to
the normal condition. This involves shrinkage in size and the restoration of the chromatin to the homogeneous condition. If more than the
normal number are formed, they are separated by the subsequent body
fissions until the normal number is established, as in *P. caudatum,*
according to Calkins and Cull (1907). If the normal number were

differentiated, they divide at every subsequent body division, as in *Euplotes patella.*

The macronuclei grow to their normal size and are distributed by body fissions to the daughter and granddaughter cells, if more than the vegetative number are formed, as in *Paramecium.* Thus the ordinary vegetative form is reëstablished.

During their development, the young macronuclei undergo some striking and significant changes. Several investigators have reported marked enlargement of the macronuclear *Anlagen* in ciliates, since the early description of the "ball-of-yarn" stage in *Nyctotherus cordiformis* by Stein (1867).

Calkins (1930) showed that the macronuclear *Anlage* of *Uroleptus halseyi* at first contains no chromatin, if chromatin be defined as a nucleic acid containing substance. As the young macronucleus grows, minute chromatin granules are formed within the matrix and grow in size and number until the nucleus is filled with large, intensely staining chromatin granules. Subsequent divisions of this "placenta" and of the cell body restore the normal vegetative nuclear complex.

A number of other ciliates show enlargement of the placenta to a size greater than that of the ordinary condition, after which shrinkage, or condensation, reduces it to its vegetative size. According to MacDougall (1925), the young *Anlage* of *Chilodonella uncinata* stains very faintly at first, but more intensely later. The enlargement continues until the young macronucleus nearly fills the cell.

In *Euplotes patella* (Turner, 1930), as in *Uroleptus,* the early macronuclear *Anlage* contains little, if any, demonstrable chromatin. After several hours of growth, a fine chromatin reticulum develops and is transformed into a broken spireme. The chromatin spireme enlarges with the nucleus and resembles the "ball of yarn" in *Nyctotherus cordiformis,* as pictures by Stein (1867) and Wichterman (1937), except that the spireme is more tortuous. Earlier stages resemble those of *Metopus sigmoides* as drawn by Noland (1927), and of *Paraclevelandia simplex* (Kidder, 1938) in cystic reorganization. The spireme finally becomes shorter and very much thicker, as chromatin granules are formed along its periphery. At this stage a fine thread is discernible running through the center of the spireme. The granules become scattered throughout the macronuclear *Anlage* as the structure of the spireme disappears. The large granular ball thus formed condenses somewhat and then elongates, starts bending into its normal C-shape, and combines with one or two sizeable remnants of the old macronucleus which have persisted through the process. When the elongating macronuclear *Anlage* approaches the remnants of the old macronucleus, the latter lose their pycnotic appearance and the chromatin reorganizes itself into discrete, dispersed granules, again resembling the normal condition. The new and the old portions then unite, end to end, to produce the C-shaped nucleus of the trophic form, the reconstituted portion forming the posterior portion. Examination of hundreds of exconjugants at this stage convinces one that the proximity of the elongating *Anlage* is the influence which brings about the reorganization of the old remnants. The old chromatin is so thoroughly reorganized before joining the new *Anlage* that it possibly has little more effect on the nature of the new nucleus than if it had been dissolved and re-formed within the new nuclear membrane.

Ikeda and Ozaki (1918) first reported fragments of the old macronucleus being incorporated into the developing macronuclear *Anlage* in *Boveria labialis*.

Kidder (1933a, 1933b) has described an interesting phenomenon occurring in the macronuclear *Anlagen* of *Kidderia* (*Conchophthirius*) *mytili* and in *Ancistruma isseli*. At each of the two or three exconjugant fissions which separate the seven or eight new macronuclei, all unseparated macronuclei cast out in an orderly manner a sphere of chromatin. According to Kidder, this "may represent the sloughing off of the germinal chromatin contained in the amphinucleus, a substance that is superfluous for the further activity of a purely trophic cell element (Reichenow, 1927)." Diller (1928) suggested this as an explanation of a similar occurrence during endomixis in *Trichodina*. Chromatin extrusions from developing macronuclear *Anlagen* would appear to be even more closely analogous to the *Ascaris* type of chromatin diminution, as brought out by Boveri, than is the chromatin diminution described by MacDougall in the division of the macronucleus in *Chilodonella*.

Until recently the behavior of the macronucleus in conjugation has been given scant attention. The disintegration of the old macronucleus and the differentiation, number, and distribution of new macronuclear *Anlagen* have been noted, but no great significance has been attached to these processes. It now seems probable that more intensive investigation of the rôle of the macronucleus in conjugation will be extremely profitable.

Literature Cited

Beers, C. D. 1931. Some effects of conjugation in the ciliate *Didinium nasutum*. J. exp. Zool., 58: 455-70.

Bělař, K. 1923. Untersuchungen an *Actinophrys sol* Ehrenberg. I. Die Morphologie des Formwechsels. Arch. Protistenk., 46: 1-96.

—— 1926. Der Formwechsel der Protistenkerne. Ergebn. Zool., 6: 235-654.

Bott, K. 1907. Über die Fortpflanzung von *Pelomyxa palustris*. Arch. Protistenk., 8: 120-58.

Brumpt, E. 1909. Demonstration du rôle pathogène du *Balantidium coli*. Enkystment et conjugaison de cet infusore. C. R. Soc. Biol. Paris, 67: 103-5.

—— 1913. Études sur les Infusoires parasites. I. La Conjugaison d'*Anoplophrya circulans* Balbiani, 1885. Arch. parasit., 16: 187-210.

Buschkiel, A. L. 1911. Beiträge zur Kenntnis des *Ichthyophthirius multifiliis* Fouquet. Arch. Protistenk., 21: 61-102.

Calkins, G. N. 1904. Studies on the life history of Protozoa. IV. Death of the A series. J. exp. Zool., 1: 423-61.

—— 1907. The fertilization of *Amoeba proteus*. Biol. Bull., 13: 219-30.

—— 1912. The paedogamous conjugation of *Blepharisma undulans*. J. Morph., 23: 667-88.

—— 1919. *Uroleptus mobilis*, Engelm. I. History of the nuclei during division and conjugation. J. exp. Zool., 27: 293-357.

—— 1930. *Uroleptus Halseyi* Calkins. II. The origin and fate of the macronuclear chromatin. Arch. Protistenk., 69: 151-74.

—— 1933. The Biology of the Protozoa. Philadelphia.

Calkins, G. N., and R. Bowling. 1926. Gametic meiosis in *Monocystis*. Biol. Bull., 51: 385-99.

—— 1929. Studies on *Dallasia frontata* Stokes. II. Cytology, gametogamy and conjugation. Arch. Protistenk., 66: 11-32.

Calkins, G. N., and S. W. Cull. 1907. The conjugation of *Paramecium aurelia (caudatum)*. Arch. Protistenk., 10: 375-415.

Chatton, E. 1927. La Gamétogénèse méiotique du flagellé *Paradinium poucheti*. C. R. Acad. Sci., Paris, 185: 553-55.

Chatton, E., and B. Biechler. 1936. Documents nouveaux relatifs aux Coccidinides (Dino flagellés parasites). La Sexualité du *Coccidinium mesnili* n. sp. C. R. Acad. Sci., Paris, 203: 573-76.

Chatton, E., and A. Brodsky. 1909. Le Parasitisme d'une Cytridinee du genre *Sphaerita* Dangeard chez *Amoeba limax* Dujard. Étude comparative. Arch. Protistenk., 17: 1-18,

Chatton, E., and A. Lwoff. 1935. Les Ciliés Apostomes. I. Aperçu historique et général; étude monographique des genres et des especes. Arch. zool. exp. gén., 77, Protistologica L: 1-453.

Chatton, E., and C. Pénard. 1921. Les Nicollellidae. Infusoires intestinaux des Gondis et des Damans, et le "cycle évolutif" des ciliés. Bull. biol., 55: 87-153.

Collin, B. 1909. La conjugaison d'*Anoplophrya branchiarum* (Stein) (*A. circulans* Balbiani.). Arch. zool. exp. gén., Series 5, 1: 345-88.

Crawley, H. 1916. The sexual evolution of *Sarcocystis muris*. Proc. Acad. nat. Sci. Philad., 68: 2-43.

Dangeard, P. A. 1898. Sur les Chlamydomonadinées. C. R. Acad. Scie., Paris, 127: 736-38.

Debaisieux, P. 1924. *Sphaeromyxa sabrazesi* Laveran et Mesnil. Cellule, 35: 267-301.

Dehorne, A. 1920. Contribution à l'étude comparée de l'appareil nucleaire des Infusoires ciliés (*Paramecium caudatum* et *Colpidium truncatum*), des Euglenes et des Cyanophycees. Arch. zool. exp. gén., 60: 47-176.

Diller, W. F. 1928. Binary fission and endomixis in the *Trichodina* from tadpoles (Protozoa, Ciliata). J. Morph., 46: 521-61.

—— 1936. Nuclear reorganization processes in *Paramecium aurelia*, with descriptions of autogamy and "hemixis." J. Morph., 59: 11-67.

Diwald, K. 1938. Die ungeschlechtliche und geschlechtliche Fortpflanzung von *Glenodinium lubiniensiforme* spec. nov. Flora, Jena, 32: 174-92.

Dobell, C. C. 1908. The structure and life-history of *Copromonas subtilis*. Quart. J. micr. Sci., 52: 75-120.

—— 1925. The life-history and chromosome cycle of *Aggregata eberthi*. Parasitology, 17: 1-136.

Dobell, C. C., and A. P. Jameson. 1915. The chromosome cycle in Coccidia and gregarines. Proc. roy. Soc., B., 89: 83-94.

Doflein, F. 1916. Lehrbuch der Protozoenkunde. 4th ed., Jena.

Doflein, F., and E. Reichenow, 1928. Doflein, Lehrbuch der Protozoenkunde. 5th ed., Jena.

Dogiel, V. 1925. Die Geschlechtsprozesse bei Infusorien (speziell bei den Ophryoscoleciden), neue Tatsachen und theoretische Erwägungen. Arch. Protistenk., 50: 283-442.

Elpatiewsky, W. 1907. Zur Fortpflanzung von *Arcella vulgaris* Ehrbg. Arch. Protistenk., 10: 441-66.

Enriques, P. 1908. Die Conjugation and sexuelle Differenzierung der Infusorien. Arch. Protistenk., 12: 213-76.

Fermor, X. 1913. Die Bedeutung der Encystierung bei *Stylonychia pustulata*. Zool. Anz., 42: 380-84.

Goldschmidt, R. 1907. Lebensgeschichte der Mastigamöben *Mastigella vitrea* n. sp. u. *Mastigina setosa* n. sp. Arch. Protistenk., Suppl. 1: 83-168.

Goroschankin, J. N. 1890. Beiträge zur Kenntnis der Morphologie und Systematik der Chlamydomonaden. I. *Chlamydomonas Braunii* (Mihi). Bull. Soc. imp. Natur., Moscou, pp. 498-520.

—— 1891. Beiträge zur Kenntnis der Morphologie und Systematik der Chlamydomonaden. II. *Chlamydomonas Reinhardii* (Dang.). Bull. Soc. imp. Natur., Moscou, pp. 101-42.

—— 1905. Beiträge zur Kenntnis der Morphologie und Systematik der Chlamydomonaden. III. *Chlamydomonas cocifera* (Mihi). Flora, Jena, 94: 420-23.

Gregory, L. H. 1923. The conjugation of *Oxytricha fallax*. J. Morph., 37: 555-81.

Gross, F. 1934. Zur Biologie und Entwicklungsgeschichte von *Noctiluca miliaris*. Arch. Protistenk., 83: 178-96.

Hartmann, M. 1911. Die Konstitution der Protistenkerne und ihre Bedeutung für die Zellenlehre. Jena.

Hartmann, M., and K. Nägler. 1908. Copulation bei *Amoeba diploidea* n. sp. mit. Selbständigbleiben der Gametenkerne während des ganzen Lebenszyklus. S. B. Ges. naturf. Fr. Berl., Nr. 5, 112-25.

Hertwig, R. 1889. Über die Konjugation der Infusorien. Abh. bayer. Akad. Wiss. München, II. Kl. 17: 1-83.

—— 1898. Über Kernteilung, Richtungskörperbildung und Befruchtung von *Actinosphaerium eichhornii*. Abh. math.-phys. bayer. Akad. Wiss. München, 19: 631-734.

Ikeda, I., and Y. Ozaki. 1918. Notes on a new *Boveria* species, *Boveria labialis* n. sp. J. Coll. Sci. Tokyo, 40: 1-25.

Ivanić, M. 1933. Die conjugation von *Chilodon cucullulus* Ehrbg. Arch. Protistenk., 79: 313-48.

Jameson, A. P. 1920. The chromosome cycle of gregarines, with special reference to *Diplocystis schneideri* Kunstler. Quart. J. micr. Sci., 64: 207-66.

—— 1927. The behaviour of *Balantidium coli* Malm. in cultures. Parasitology, 19: 411-19.

Jennings, H. S. 1913. The effect of conjugation in *Paramecium*. J. exp. Zool., 14: 279-391.

—— 1920. Life and death; heredity and evolution in unicellular organisms. Boston.

—— 1938. Sex reaction types and their interrelations in *Paramecium bursaria*. I, II. Clones collected from natural habitat. Proc. Nat. Acad. Sci. Wash., 24: 112-20.

Johnson, P. L. 1930. Reproduction in *Amoeba proteus*. Arch. Protistenk., 71: 463-98.

Jones, P. M. 1928. Life cycle of *Amoeba proteus (Chaos diffluens)* with special reference to the sexual stage. Arch. Protistenk., 63: 322-32.

Kidder, G. W. 1933a. Studies on *Conchophthirius mytili* De Morgan. II. Conjugation and nuclear reorganization. Arch. Protistenk., 79: 25-49.

—— 1933b. On the genus *Ancistruma* Strand (= *Ancistrum* Maupas). II. The conjugation and nuclear reorganization of *A. isseli* Kahl. Arch. Protistenk., 81: 1-18.

—— 1938. Nuclear reorganization without cell division in *Paraclevelandia simplex* (Family Clevelandellidae), an endocommensal ciliate of the wood-feeding roach, *Panesthia.* Arch. Protistenk., 91: 69-77.

Kofoid, C. A. 1921. Symposium on fertilization. Anat. Rec., 20: 223-325.

Kofoid, C. A., and M. Bush, 1936. The life cycle of *Parachaenia myae* gen. nov., sp. nov., a ciliate parasitic in *Mya arenaria* Linn. from San Francisco Bay, California. Bull. Mus. Hist. nat. Belg., 12: 1-15.

Kudo, R. 1926. On *Myxosoma catostomi* Kudo 1923, a myxosporidian parasite of the sucker, *Catostomus cammersonii.* Arch Protistenk., 56: 90-115.

Landis, E. M. 1925. Conjugation of *Paramecium multimicronucleata,* Powers and Mitchell. J. Morph., 40: 111-67.

Lebedew, W. 1909. Über *Trachelocerca phoenecopterus* Cohn. Ein marines Infusor. Arch. Protistenk., 13: 70-114.

Le Calvez, J. 1938. Recherches sur les Foraminifères. I. Développement et reproduction. Arch. zool. exp. gén., 80: 163-333.

Léger, L. 1903. La Reproduction sexuée chez les *Stylorynchus.* Arch. Protistenk., 3: 303-57.

—— 1907. Les Schizogregarines des Trachéates. I. Le genre *Ophryocystis.* Arch. Protistenk., 8: 159-202.

Léger, L., and O. Duboscq. 1909. Études sur la sexualité chez les grégarines. Arch. Protistenk., 17: 19-134.

Liesche, W. 1938. Kie Kern- und Fortpflanzungsverhältnisse von *Amoeba proteus* (Pall.). Arch. Protistenk., 91: 135-86.

Lister, J. J. 1895. The life-history of the Foraminifera. Philos. Trans., B. 186: 401-53.

MacDougall, M. S. 1925. Cytological observations on gymnostomatous ciliates, with a description of the maturation phenomena in diploid and tetraploid forms of *Chilodon uncinatus.* Quart. J. micr. Sci., 69: 361-84.

—— 1935. Cytological studies of the genus *Chilodonella* Strand, 1926 (*Chilodon* Ehrbg., 1838). I. The conjugation of *Chilodonella* sp. Arch. Protistenk., 84: 199-206.

—— 1936. Étude cytologique de trois espèces du genre *Chilodonella* Strand. Morphologie, conjugaison, réorganisation. Bull. biol., 70: 308-31.

Manwell, R. D. 1928. Conjugation, division, and encystment in *Pleurotricha lanceolata.* Biol. Bull., 54: 417-63.

Mast, S. O. 1917. Conjugation and encystment in *Didinium nasutum* with especial reference to their significance. J. exp. Zool., 23: 335-59.

Maupas, E. 1889. Le Rejeunissement karyogamique chez les ciliés. Arch. zool. exp. gén., Ser. 2, 7: 149-517.

Messiatzev, J. 1924. The conjugation of *Lionotus lamella* (in Russian). Arch. russ. protist., 3, no. 1-2.

Metcalf, M. M. 1923. The opalinid ciliate infusorians. Bull. U. S. nat. Mus., 120.

Milojevic, B. D. 1925. Zur Entwicklungsgeschichte der *Gregarina cuneata* (F. St.), mit besonderer Berücksichtigung der Entstehung des Geschlechtskerns. Arch. Protistenk., 50: 1-26.

Minchin, E. A. 1912. An Introduction to the Study of the Protozoa. London.

Minkiewicz, R. 1912. Un cas de reproduction extraordinaire chez *Polyspira Delagei*. C. R. Acad. Sci., Paris, 155: 733.

Miyashita, Y. 1927. On a new parasitic ciliate *Lada tanishi* n. sp., with preliminary notes on its heterogamic conjugation. Jap. J. Zool., 1: 205-18.

Mulsow, K. 1911. Ueber Fortpflanzungserscheinungen bei *Monocystis rostrata* n. sp. Arch. Protistenk., 22: 20-55.

Myers, E. H. 1935. The life history of *Patellina corrugata* Williamson, a foraminifer. Univ. Cal. Publ. Zool. tech., series, 3: 355-92.

—— 1936. The life cycle of *Spirillina vivipara*, with notes on morphogenesis, systematics and distribution in the Foraminifera. J. R. micr. Soc., 56: 120-46.

—— 1938. The present state of our knowledge concerning the life cycle of the Foraminifera. Proc. nat. Acad. Sci. Wash., 24: 10-17.

Naville, A. 1927a. Le Cycle chromosomique et la meiose chez les *Monocystis*. Z. Zellforsch., 6: 257-84.

—— 1927b. Le Cycle chromosomique d'*Urospora lagidis* (de Saint Joseph). Parasitology, 19: 100-38.

—— 1927c. Le Cycle chromosomique, la fécondation et la réduction chromatique de *Chloromyxum leydigi* Mingazz. Ann. Inst. océanogr. monaco, 4: 177-208.

—— 1930a. Le Cycle chromosomique d'une nouvelle Actino myxidie: *Guyenotia sphaerulosa* n. gen.; n. sp. Quart. J. micr. Sci., 73: 547-75.

—— 1930b. Recherches sur la sexualité les myxosporidies. Arch. Protistenk., 69: 327-400.

—— 1931. Les Sporozoaires. Mém. Soc. Phys. Genève, 41, No. 1.

Nelson, E. C. 1934. Observations and experiments on conjugation of the *Balantidium* from the chimpanzee. Amer. J. Hyg., 20: 106-34.

Neresheimer, E. R. 1907. Die Fortpflanzung der Opalinen. Arch. Protistenk., (Suppl.) 1: 1-42.

—— 1908. Fortpflanzung eines parasitischen Infusors *(Ichthyophthirius)*. S. B. Ges. Morph. Physiol., München, 23.

Noland, L. E. 1927. Conjugation in the ciliate *Metopus sigmoides* C. and L. J. Morph., 44: 341-61.

Pascher, A. 1916. Über die Kreuzung einzelliger, haploider Organismen: *Chlamydomonas*. Ber. dtsch. bot. Ges., 34: 228-42.

Poljansky, G. 1926. Die Conjugation von *Dogielella sphaerii* (Infusoria Holotricha, Astomata). Arch. Protistenk., 53: 407-34.

—— 1928. Über die Konjugation von *Bursaria truncatella*. Zool. Anz., 79: 51-58.

—— 1934. Geschlechtsprozesse bei *Bursaria truncatella* O. F. Müll. Arch. Protistenk., 81: 420-546.

Prandtl, H. 1906. Die Konjugation von *Didinium nasutum*. Arch. Protistenk., 7: 229-58.

Pratje, A. 1921. *Noctiluca miliaris* Suriray. Beiträge zur Morphologie, Physiologie und Cytologie. I. Morphlogie und Physiologie. (Beobachtungen an der lebenden Zelle.) Arch. Protistenk., 42: 1-98.

Pringsheim, E. 1869. Über Paarung von Schwarmsporen, die morphologische Grundform der zeugung im Pflanzenreiche. Monatsber. preuss. Akad. Wissensch., 721-38.

Prowazek, S. 1899. Protozoenstudien. I. *Bursaria truncatella* und ihre Conjugation. Arb. zool. Inst. Univ. Wien. 11: 195-286.

Schaudinn, F. 1896. Über die Copulation von *Actinophrys sol* Ehrbg. S. B. preuss. Akad. Wiss., 83-89.

—— 1903. Untersuchungen über die Fortpflanzung. Arb. GesundhAmt., 19: 547-76.

Schellack, C. 1907. Über die Entwicklung und Fortpflanzung von *Echinomera hispida* A. Schn., Arch. Protistenk., 9: 297-345.

—— 1913. Coccidien-Untersuchungen II. Die Entwicklung von *Adelina dimidiata* A. Schn., einem *Coccidium* aus *Scolopendra cingulata* Latr. Arb. GesundhAmt. Berl., 45: 269-316.

Schneider, A. 1886. *Anoplophrya circulans* (Balb.) Tabl. Zool., 1:31-88.

Schröder, O. 1907. Beiträge zur Entwicklungsgeschichte der Myxosporidien. *Sphaeromyxa sabrazesi* (Laveran et Mesnil). Arch. Protistenk., 9: 359-81.

—— 1910. Über die Anlage der Sporocyste (Pansporoblast) bei *Sphaeromyxa sabrazesi* Laveran et Mesnil. Arch. Protistenk., 19: 1-5.

Scott, M. J. 1927. Studies on the *Balantidium* from the guinea-pig. J. Morph., 44: 417-65.

Sharp, R. W. 1934. Introduction to Cytology. New York.

Sonneborn, T. M. 1936. Factors determining conjugation in *Paramecium aurelia*. I. The cyclical factor: the recency of nuclear reorganization. Genetics, 21: 503-14.

—— 1937. Sex, sex inheritance and sex determination in *Paramecium aurelia*. Proc. nat. Acad. Sci. Wash., 23: 378-85.

Sonneborn, T. M., and B. M. Cohen. 1936. Factors determining conjugation in *Paramecium aurelia*. II. Genetic diversities between stocks or races. Genetics, 21: 515-18.

Stein, F. 1867. Der Organismus der Infusionsthiere. II. Abt. Leipzig.

Summers, F. M., and G. W. Kidder. 1936. Taxonomic and cytological studies on the ciliates associated with the amphipod family Orchestiidae from the Woods Hole district. II. The coelozoic astomatous parasites. Arch. Protistenk., 86: 379-403.

Swarczewsky, B. 1908. Über die Fortpflanzungserscheinungen bei *Arcella vulgaris* Ehrbg. Arch. Protistenk., 12: 173-212.

Tannreuther, G. W. 1926. The life history of *Prorodon griseus*. Biol. Bull., 51: 303-20.

Turner, J. P. 1930. Division and conjugation in *Euplotes patella* Ehrenberg with special reference to the nuclear phenomena. Univ. Cal. Publ. Zoöl., 33: 193-258.

Valkanov, A. 1935. Untersuchungen über den Entwicklungskreis eines Turbellarienparasiten (*Monocystella Arndti*). Z. Parasitenk., 7: 517-38.

Visscher, J. P. 1927. Conjugation in the ciliated protozoön, *Dileptus gigas,* with special reference to the nuclear phenomena. J. Morph., 44: 383-414.

Weschenfelder, R. 1938. Die Entwicklung von *Actinocephalus parvus* Wellmer. Arch. Protistenk., 91: 1-60.

Wichterman, R. 1937. Division and conjugation in *Nyctotherus cordiformis* (Ehr.) Stein (Protozoa, Ciliata) with special reference to the nuclear phenomena. J. Morph., 60: 563-611.

Woodruff, L. L. 1925. The physiological significance of conjugation and endomixis in the Infusoria. Amer. nat., 59: 225-49.

—— 1927. Studies on the life history of *Blepharisma undulans*. Proc. Soc. exp. Biol. N.Y., 24: 769-770.

Woodruff, L. L., and H. Spencer. 1924. Studies on *Spathidium spathula*. II. The significance of conjugation. J. exp. Zool. 39: 133-96.

Zederbauer, E. 1904. Geschlechtliche und ungeschlechtliche Fortpflanzung von *Ceratium hirundinella*. Ber. dtsch. bot. Ges., 22: 1-8.

Zuelzer, M. 1904. Beiträge zur Kenntnis der *Difflugia urceolata* Carter. Arch. Protistenk., 4: 240-95.

CHAPTER XIII

ENDOMIXIS

Lorande Loss Woodruff

During recent years the attention of students of the Ciliophora has been focused increasingly on macronuclear changes during the life of the individual cell and the life history of the species. This has revealed that the macronucleus is by no means a relatively passive agent in the nuclear complex, but rather a product of micronuclear activity which undergoes various radical transformations in contributing to the "somatic" functions of the cell until another is provided from the same source.

Macronuclear Reorganization

As early as 1859, Stein noted a clear band, or *Kernspalt,* in the macronuclei of hypotrichous ciliates, but chiefly within the past decade this has

Figure 153. Stages in the progress of the reorganization bands in *Aspidisca lynceus* from the center of the horseshoe-shaped macronucleus to its tips. (From Summers, 1935.)

been intensively studied by several investigators who have shown that *Kernspalten* are but a part of the regions now called reorganization bands, reconstruction bands, and so forth, which have an important function in the transformation of macronuclear substance at the time of division (see p. 21). Thus in certain cases material visibly passes from this region into the surrounding cytoplasm, as, for example, in

Uroleptus (Calkins, 1919, 1930), *Euplotes patella* (Turner, 1930), and *Aspidisca* (Summers, 1935), while in others, such as *Euplotes worcesteri* (Griffin, 1910), "extrusion bodies" are either absent or not described. Again, in some ciliates reorganization bands may not appear so clear-cut as in hypotrichous forms. Nevertheless the macronuclear material is eliminated as a residual mass, left between the two parts of a dividing macronucleus, as in *Ancistruma* and *Conchophthirius* (Kidder, 1933a, 1933b), *Colpidium, Glaucoma,* and *Urocentrum* (Kidder and Diller, 1934), and *Blepharisma* (D. Young, 1939); or beside the macronucleus

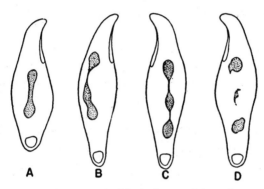

Fig. 154. Macronuclear dissolution in *Blepharisma undulans*. A, vegetative individual with dumb-bell-shaped macronucleus; B, early stage in the formation of the central bulb; C, central bulb fully formed; D, dissipation of the central bulb almost accomplished. (From D. Young, 1939.)

during or after cell division, as in *Chilodonella* (MacDougall, 1936) and *Colpoda* (Kidder and Claff, 1938) (Figs. 153, 154).

Variations in the details of such intrinsic macronuclear reorganization processes are indeed legion—reaching their climax, perhaps, in one or more of the types of "hemixis" described by Diller (1936)—but the accumulating data seem to justify the belief that they may be universal in the ciliates and perhaps represent at once a "purification" (Calkins) of the macronucleus, a regulation of the nucleo-cytoplasmic ratio, and a contribution of significance to the metabolic activities of the cytoplasm. For summaries, reference may be made to Summers (1935) and Kidder and Claff (1938).

Apparently, intrinsic macronuclear reorganization phenomena during the division of the cell suffice, in most species at least, for the continued well-being of the race, but provisions are also made for the periodic destruction of the macronucleus and its replacement from the micronuclear reserve: in some species by endomixis and autogamy, and probably in all by conjugation.

ENDOMICTIC PHENOMENA

A periodic replacement of the macronuclear apparatus, without synkaryon formation, was described in *Paramecium aurelia*, and named en-

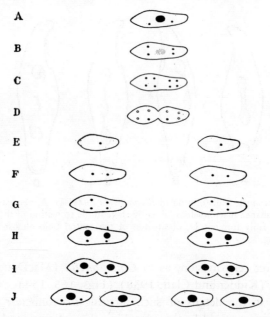

Figure 155. General plan of the usual nuclear changes during endomixis in *Paramecium aurelia*. A, typical nuclear condition; B, degeneration of macronucleus (chromatin bodies not shown) and first division of micronuclei; C, "climax": second division of micronuclei; D, degeneration of six of the eight micronuclei; E, division of the cell; F, first reconstruction micronuclear division; G, second reconstruction micronuclear division; H, transformation of two micronuclei into macronuclei; I, micronuclear and cell division; J, typical nuclear condition restored. (Constructed from the description and figures of Woodruff and Erdmann, 1914.)

domixis, by Woodruff and Erdmann (1914). Following their account, in summary, endomixis in this species involves the resolution of the old macronucleus into chromatin bodies, which disintegrate in the cytoplasm, and the transformation of one or two of the products of the micronuclear divisions into new macronuclei, to reconstitute the normal vegetative apparatus when distributed by cell division (Figs. 155, 156).

Immediately after this announcement, Hertwig (1914) described similar phenomena in *P. aurelia* as parthenogenesis, induced, he believed, by degenerative changes, and emphasized the fact that in his study of conjugation in this species (1889), he had noted stages in certain non-conjugants that were open to a similar interpretation. Thereafter en-

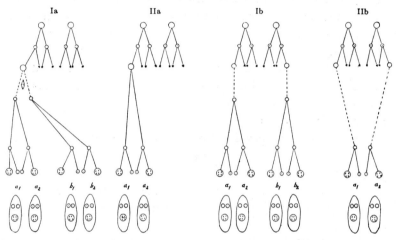

Figure 156. Possible methods of micronuclear and cell division at the climax of endomixis in *Paramecium aurelia*. Ib is typical. (From Woodruff and Erdmann, 1914, p. 448.)

domixis, or, if one prefers, diploid parthenogenesis, was reported by many investigators, including Erdmann and Woodruff (1916) in *P. caudatum*, Calkins (1915, 1919) in *Didinium nasutum* and *Uroleptus mobilis*, Moore (1924) in *Spathidium spathula*, Erdmann (1925) in *P. bursaria*, Woodruff and Spencer (1923) in *P. polycaryum*, Klee (1925) in *Euplotes longipes*, Ivanić (1928, 1929) in *Chilodonella uncinatus*, *Vorticella nebulifera*, *Euplotes charon*, and *E. patella*, Manwell (1928) in *Pleurotricha lanceolata*, Diller (1928) in *Trichodina* sp., Chejfec (1928, 1930) in *P. caudatum*, Fauré-Fremiet (1930) in *Zoothamnium alternans*, Stranghöner (1932) in *P. multimicronucleatum*, Tittler (1935) in *Urostyla grandis*, Kidder (1938) in *Paraclevelandia simplex*, and Gelei (1938) in *Paramecium nephridiatum*. In most of these studies, it must be admitted, the authors failed to follow the sequence of nuclear events in series of pedigreed animals, but fitted their findings in isolated animals into the picture of endomixis as originally portrayed. Three of these investigations are of particular significance at the moment.

Diller (1928), in a study of "binary fission and endomixis in *Trichodina* from tadpoles," gives a categorical account of the reorganization process. He shows the resolution of the macronucleus into chromatin bodies: "In most cases the macronucleus breaks up completely by forming numerous spherical bodies of varying sizes." And the origin of the primordium of the new macronucleus is from a residual micronucleus: "The final eight products of the micronuclear divisions are originally all apparently similar. Seven of them, however, rapidly differentiate into macronuclear *Anlagen,* while one remains the functional micronucleus." The process is "characterized by the absence of maturation spindles and synkaryon formation" (Fig. 157).

Stranghöner (1932), in a detailed description of endomixis in *Paramecium multimicronucleatum,* emphasizes the fact that *"Im Gegensatz zur Conjugation bildet der Macronucleus bei der Auflösung keine wurstformigen Schlingen,"* and describes and figures the incorporation of chromatin spheres from the old macronucleus into the new one (Fig 158).

Kidder (1938) observes a "nuclear reorganization without cell division in *Paraclevelandia simplex,"* in which the details of the process are unique but the end result is the same. The old macronucleus eliminates a large part of its chromatin and the remainder then becomes incorporated with one product of a single micronuclear division, to constitute the primordium of the new macronuclear apparatus (Fig. 159).

In this coöperation between macronucleus and micronucleus, described by Stranghöner and Kidder as endomixis, we seem to have, as it were, stages in the evolution of macronuclear metamorphosis intermediate between the intrinsic changes evidenced by reorganization bands, direct elimination of material, and so forth, and the complete competence of the micronucleus alone to form a new macronucleus during endomixis as it has been described in other species.

That intrinsic reorganization is adequate for the continued life of the race appears to be evident from the study of Dawson (1919) on an

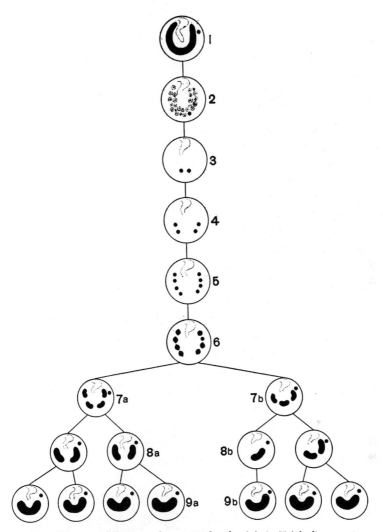

Figure 157. Diagram of the normal process of endomixis in *Trichodina* sp. 1, normal vegetative animal; 2, macronucleus fragmented and disintegrating, and not shown in subsequent diagrams (the micronucleus has migrated to the other end of the body); 3, micronucleus divided the first time; 4, micronucleus divided the second time; 5, micronucleus divided the third time; 6, seven of the nuclei differentiating into macronuclear *Anlagen*, while the eighth remains the functional micronucleus which divides before each cell division; 7a and 7b, daughters resulting from the first cell division and having four and three macronuclear *Anlagen* respectively; 8a and 8b, daughters resulting from the second cell division (the 8b monomacronucleate individual is completely reorganized); 9a and 9b, daughters resulting from the third cell division (growth of the macronucleus will reconstruct them into normal vegetative individuals). (From Diller, 1928.)

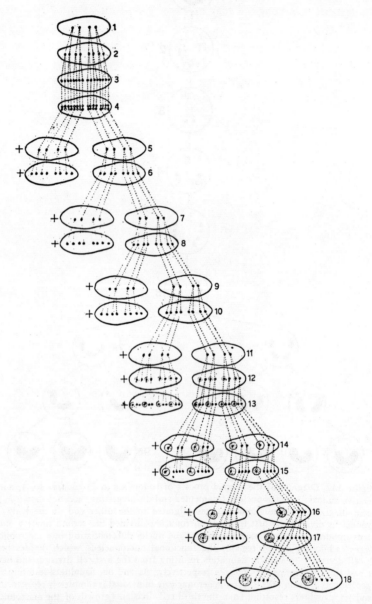

Fig. 158. Endomixis in *Paramecium multimicronucleatum*. (From Stranghöner, 1932.)

amicronucleate race of *Oxytricha hymenostoma,* in which, obviously, a micronuclear reserve could play no part. Woodruff's work (1935) on a race of *Blepharisma undulans,* without endomixis or autogamy, and D. Young's findings (1939) on the same race, showing the elimination of material from the macronucleus during division, support this thesis.

However, these results might be anticipated because several species have been cultured for long periods without showing any evidence of a

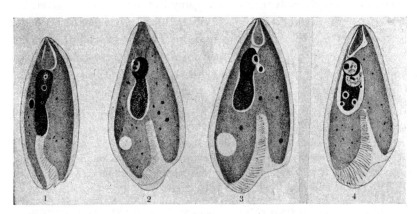

Figure 159. Endomixis in *Paraclevelandia simplex.* 1, pre-cystic form (the macronucleus differentiating into degenerating posterior, and reorganizing anterior chromatin; the micronucleus in prophase); 2, later stage (the micronucleus in anaphase; further differentiation of macronucleus); 3, telophase of the micronucleus (note the smoothly granular anterior half of the macronucleus); 4, fusion of daughter micronucleus with reorganized portion of macronucleus (posterior portion of macronucleus shrinking away from the old nuclear membrane; enlarged condition of the daughter micronuclei quite characteristic). (From Kidder, 1938.)

reorganization process, either intrinsic or endomictic, in the *free-living* animals, and probably none occurs. As examples, reference may be made to the work on *Spathidium spathula* by Woodruff and Moore (1924), on *Paramecium calkinsi* by Spencer (1924), and on *Didinium nasutum* by Beers (1929). In regard to reorganization by endomixis, the culture of *P. caudatum* studied by Metalnikov (1937), and the (to date) thirty-three-year-old culture of *P. aurelia* at Yale University may be cited, unless the future should prove that autogamy, to the exclusion of endomixis, occurs in these species (Woodruff, 1932).

Thus in the species in which reorganizational phenomena occur, it appears that intrinsic reorganization, as well as endomixis, meet the normal exigencies of existence and keep the race on the even tenor of its way. For significant possibilities of genetic change in heterozygous individuals, however, reorganization is accompanied by synkaryon formation, either in autogamy or conjugation.

AUTOGAMY

The first definite statement of autogamy in the ciliates was given in a brief article by Fermor (1913), who described the degeneration of the macronucleus and the origin of a new one from a synkaryon of micronuclear origin, during the encystment of *Stylonychia pustulata*. But the problem was not emphasized until Diller (1936) described autogamy in *P. aurelia* and stated:

I have not been able to confirm the micronuclear behavior which Woodruff and Erdmann have described for endomixis in *P. aurelia*. In the failure of such verification I am inclined to deny the existence of endomixis as a valid reorganization process. I feel that Woodruff and Erdmann have combined stages of hemixis and autogamy into one scheme, "endomixis," overlooking the maturation and syncaryon stages in autogamy.

And in regard to his own earlier description of endomixis in *Trichodina* sp., Diller remarks, "It may be that hemixis and exconjugant stages were lumped together as 'endomixis' in this account" (Figs. 160, 161).

In enthusiasm for the concept of autogamy, it may be well not to exclude endomixis in *P. aurelia*—or *Trichodina* sp.—without careful consideration, although there is no inclination to deny that autogamy occurs in *P. aurelia,* in view of the combined data presented in the cytological study by Diller and the genetical studies of Sonneborn (1939a, 1939b, 1939c). However, Sonneborn's observations were not made on the Yale race of *P. aurelia* nor on the mating-type variety which it represents, because there are as yet no known genes in this variety, and such tests therefore cannot be made.

In the opinion of the writer, the crucial cytological stages are not absolutely demonstrated, in part because most of the animals were taken from mass cultures and relatively few from isolated lines or from pedigreed lines, and therefore the sequence of events was not determined from pedigreed *series.* To demonstrate satisfactorily the exact sequence, it is necessary to follow critically the nuclear behavior in series of pedi-

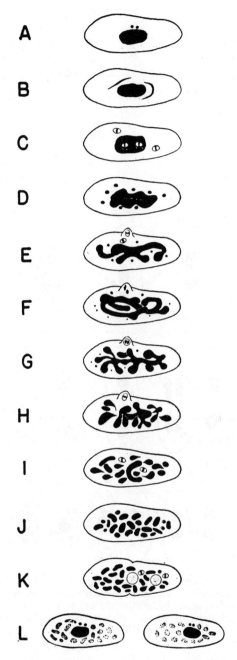

Figure 160. Nuclear changes during autogamy in *Paramecium aurelia.* (From Diller, 1936.)

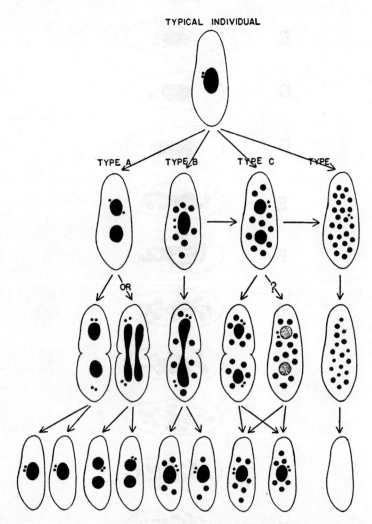

Figure 161. Hemixis. Diagram of the macronuclear behavior (exclusive of conjugation and autogamy) in *Paramecium aurelia*. The macronuclei are represented by large solid ovals; macronuclear fragments by smaller circles; micronuclei by small round dots; "*Anlagen*-like" macronuclei by stippled circles. The interrelationships of the various forms are indicated by arrows. (From Diller, 1936.)

greed animals from day to day, as emphasized by Woodruff and Erdmann (1914, pp. 457-72) and Beers (1935).

The study of pedigreed series of animals, for example, precluded, it
is believed, the possibility of "combining stages of hemixis and autogamy
in one scheme." Indeed, Erdmann and Woodruff (1916), contrasting
endomixis in *P. caudatum* and *P. aurelia,* stated that they had "some
data which suggest that under certain conditions merely a partial reorganization, not involving the formation of macronuclear *Anlagen,*

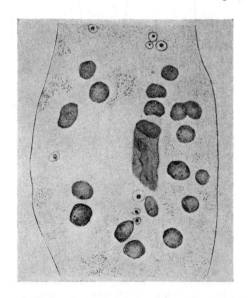

Figure 162. Climax of
endomixis in *Paramecium
aurelia.* The old macronucleus is merely in the form
of a membrane from which
the numerous chromatin
bodies have been ejected
and are free in the cyto-
plasm. Eight so-called reduction micronuclei. (From
Woodruff and Erdmann,
1914, plate 2.)

may lead, at least temporarily, to the continuance of the life of the line."
This would appear to be Diller's "hemixis."

On the other hand, "overlooking the maturation and synkaryon
stages" is a different matter, as will be appreciated by anyone who has
worked on the cytology of *Paramecium.* This may have occurred, even
though Woodruff and Erdmann naturally "expected" to find autogamy
when they observed the primordia of macronuclei in non-conjugants.
Their inability to find maturation spindles and synkaryon of course led
them to coin the name endomixis for the process. But an equally plausible
explanation, at least in the mind of the writer, as to why these investigators did not find such stages, nor even the paroral lobe, in which the
synkaryon is characteristically located, according to Diller, is that these

did *not* occur in their material. None have been observed by other investigators studying what they interpret as endomixis in various species. As already stated, none were found by Diller in endomixis in *Trichodina*. Certainly none occur in *Paraclevelandia simplex,* according to the clear-

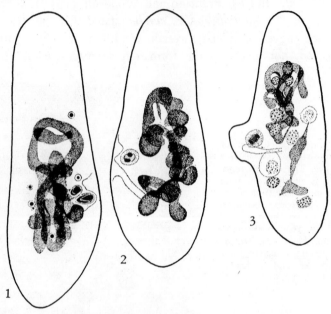

Figure 163. 1, Autogamy, Woodruff race (gamete nuclei in contact in the paroral cone at the right; five or six degenerating nuclei are visible; macronucleus in skein); 2, autogamy, Woodruff race (synkaryon formation; gamete nuclei enclosed within a common membrane; paroral cone; no degenerating micronuclei visible; macronucleus in skein); 3, autogamy, isolation, Philadelphia race (synkaryon, in paroral cone, in metaphase of first division; no degenerating micronuclei seen; macronuclear skein fragmenting; a number of macronuclear bodies of various stages of degeneration present in the cell). Animals 1 and 2 are from mass cultures. (From Diller, 1936.)

cut description of Kidder (1938). So from the latter account alone it is evident that endomictic phenomena actually do occur. Synkaryon formation is not a necessary antecedent to the formation of a macronuclear primordium (Figs. 162, 163).

But confining attention to *Paramecium aurelia,* the fact must be emphasized that almost the entire picture, and not merely the crucial detail of the presence or absence of a synkaryon, differs in the endomixis of Woodruff and Erdmann and the autogamy of Diller. In endomixis, maturation "crescents" were not observed, and the elimination of chromatin bodies was found to be the typical method of macronuclear destruction. Only one among the many hundreds of endomictic animals studied by Woodruff and Erdmann showed even a slight simulation of the macronuclear ribbon-formation so characteristic of conjugating animals, and also of autogamy according to Diller. This single animal, of the four-thousand-and-eighty-seventh generation, was figured as atypical. However, eight years later Woodruff and Spencer (1922) found, on one single day in a subculture from this same pedigreed race at about the eight-thousand-nine-hundredth generation, several animals with ribbon-like degenerating macronuclei. The publication of this exception brought a protest from Erdmann, who was convinced that conjugation must have occurred in the subculture.

Now much of Diller's work has been done on this same Yale race, and therefore it is clear that ribbon-formation does occur, other than at conjugation, in this race, under certain conditions. It is not clear how Diller's culture conditions differ from those in the Yale Laboratory, where ribbon-formation has not been observed since the instance in 1922, referred to above. A clue may be afforded by De Lamater (1939), who found that different kinds of bacteria in the culture medium of this same race of *Paramecium* had marked effects on the macronuclear changes. It is possible that other types of bacteria or other environmental changes may underly the differences between endomixis and autogamy.

PERIODICITY OF ENDOMIXIS

Another important point is the rhythmic periodicity of endomixis observed by Woodruff and Erdmann (1914, 1916) and Woodruff (1917a, 1917b), which, according to Diller, is absent in autogamy. He says: "Under the conditions of my experiments, no regular periodicity in the incidence of autogamy was evident."

Woodruff and Erdmann (1914) and Woodruff (1917) definitely stated that the interendomictic periods in both *P. aurelia* and *P. caudatum* showed some variation in length and furthermore were somewhat modified by environmental factors, but nevertheless were strikingly periodic—endomictic periods and interendomictic periods affording the rhythms in the division rate of pedigreed cultures. And this rhythmicality has appeared throughout the years in the culture of *P. aurelia* in the Yale

Laboratory, whenever tests have been made; but now, in the thirty-third year of its life, with unimpaired vitality, the interendomictic periods seem to be slightly more variable in length (Fig. 164).

A number of other investigators have studied the question of periodicity in this and other races of *P. aurelia* and *P. caudatum,* among them R. T. Young (1918), Jollos (1916, 1920), Erdmann (1920), Chejfec (1930), Galadjieff (1932), and, in particular, Sonneborn (1937a).

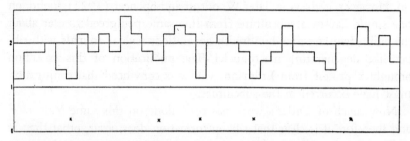

Figure 164. Graph of the division rate of *Paramecium aurelia,* line III, subculture IE, averaged for five-day periods. Endomixis occurred during the periods indicated by an X. Note that the interendomictic periods exhibit some variation in length, and the final endomixis shown is deferred. (From Woodruff and Erdmann, 1914.)

The latter compared the endomictic period in the Yale race of *P. aurelia* with that of another race under identical environmental conditions. Sonneborn shows that great variations may exist in the interendomictic interval, not only in different races, but even in the same race under carefully standardized conditions of daily isolation culture.

So in regard to the periodicity of endomixis, it now appears that the limits of approximately 25 to 30 days and 40 to 50 fissions for *P. aurelia,* and 50 to 60 days and 80 to 100 fissions for *P. caudatum,* as originally announced, are somewhat too narrow and stereotyped. Environmental and racial factors play a still greater part than these investigators believed. But withal, the endomictic process does recur with considerable regularity when the environmental and genetic factors are uniform, and so must still be regarded as periodic.

GENETICAL STUDIES ON ENDOMIXIS

Genetical studies on endomixis include those by Erdmann (1920), Jollos (1921), Parker (1927), Caldwell (1933), Kimball (1937, 1939), and Sonneborn (1937b, 1939a, 1939b, 1939c). The results obtained by Sonneborn (1939), in particular, afford evidence that the preciseness of the ratios and the segregation of sex or mating types, following the reorganization process, is quite as regular and exact as after conjugation. Thus it would appear that autogamy and not endomixis is involved. Indeed, Sonneborn states that in one race of *P. aurelia* —not the Yale race nor the mating-type variety which it represents— the alternative between endomixis and autogamy was tested genetically by determining the genotypes following reorganization in clones of type Aa (genes determining mating types). Genetic analysis of the reorganizations showed that all the resulting lines are homozygous, half of them dominant and half recessive. From any one reorganizing individual both caryonides are of the same genotype. Thus, under these conditions (mass cultures at 31° C. and isolation lines at 27° C.), autogamy, not endomixis, takes place.

Accordingly the combination of genetical and cytological data at present available justifies the conclusion that autogamy occurs, under certain circumstances at least, in some races of *P. aurelia*. Granting this raises the question whether all the investigations reported on the physiology and genetics of endomixis actually are on autogamy, because *Paramecium* is the form that has been almost universally employed in such studies. If the accumulated data are really all in regard to autogamy, then the question is essentially one of name. On the other hand, if both endomixis and autogamy occur in *Paramecium,* then, for a time, confusion is worse confounded.

Obviously, at present it is useless to attempt to generalize in regard to reorganization in *Paramecium*—that must await far more extended investigation. However, the personal judgment of the writer, at the moment, is that *both* endomixis and autogamy do occur in *Paramecium*— an opinion reached, it is believed impartially, from a consideration of the picture of the micronuclear divisions and the macronuclear destruction, as he saw them in the original work on the reorganization process and as demonstrated in the cytological preparations of Diller. Certainly the two pictures presented are quite different; indeed, in many ways as different as the stages in endomixis and conjugation appeared in the original study. And Sonneborn (1939a) remarks: "Probably it will be found that autogamy and endomixis take place in different races or under different conditions." The occurrence of these two processes, either in different races or in the same race, synchronously or otherwise, thus adds to the known repertoire of the versatile *Paramecium*.

CONCLUSIONS

A synoptic view of the rapidly accumulating data on macronuclear reorganization phenomena in the Ciliophora justifies, it is believed, the statement that these processes include intrinsic reorganization (reorganization bands, and so forth), coöperation of macronucleus and micronucleus (endomixis), a new macronucleus of micronuclear origin (endomixis), and a new macronucleus of synkaryon origin (autogamy and conjugation). These constitute a series of macronuclear metamorphoses of increasing complexity, affording progressively greater possibilities for the organism.

LITERATURE CITED

Beers, C. D. 1929. On the possibility of indefinite reproduction in the ciliate Didinium without conjugation or endomixis. Amer. Nat., 63: 125-29.

—— 1935. Structural changes during encystment and excystment in the ciliate *Didinium nasutum*. Arch. Protistenk., 84: 133-55.

Caldwell, L. 1933. The production of inherited diversities at endomixis in *Paramecium aurelia*. J. exp. Zool., 66: 371-407.

—— 1915. *Didinium nasutum:* I. The life history. J. exp. Zool., 19: 225-39.

—— 1916. General biology of the protozoan life cycle. Amer. Nat., 50: 257-70.

Calkins, G. N. 1919. *Uroleptus mobilis* Engelm. II. Renewal of vitality through conjugation. J. exp. Zool., 29: 121-56.

—— 1930. *Uroleptus halseyi* Calkins. II. The origin and fate of the macronuclear chromatin. Arch. Protistenk., 69: 151-74.

Chejfec, M. 1928. On the nuclear reorganization of *Paramecium caudatum*. Acta Biol. exp., 2: 89-121.

—— 1930. Zur Kenntniss der Kernreorganizationsprozesse bei *Paramecium caudatum*. Arch. Protistenk., 70: 87-118.

Dawson, J. A. 1919. An experimental study of an amicronucleate Oxytricha. I. A study of the normal animal with an account of cannibalism. J. exp. Zool., 29: 473-511.

De Lamater, A. J. 1939. Effect of certain bacteria on the occurrence of endomixis in *Paramecium aurelia*. Biol. Bull. 76: 217-225.

Diller, W. F. 1928. Binary fission and endomixis in the Trichodina from tadpoles. J. Morph., 46: 521-52.

—— 1936. Nuclear reorganization processes in *Paramecium aurelia*, with descriptions of autogamy and "hemixis." J. Morph., 59: 11-67.

Enriques, P. 1916. Duemila Cinquecento generazioni in un infusorio, senza coiugazione nè partenogenesi, nè depressioni. R. Accad. Sci. Bologna, Ser. VII.

Erdman, R. 1920. Endomixis and size variation in pure bred lines of *Paramecium aurelia.* Roux Arch. EntwMech. Organ., 46: 85-148.

—— 1925. Endomixis bei *Paramecium bursaria.* S. B. Ges. natur. Fr. Berl., Jahrgang, 1925: 24-27.

Erdmann, R., and L. L. Woodruff. 1916. The periodic reorganization process in *Paramecium caudatum.* J. exp. Zool., 20: 59-97.

Fauré-Fremiet, E. 1930. Growth and differentiation of the colonies of *Zoothamnium alternans.* Biol. Bull., 58: 28-51.

Fermor, X. 1913. Die Bedeutung der Enzystierung bei *Stylonychia pustulata.* Zool. Anz., 42: 380-84.

Galadjieff, M. 1932. Sur le problème de l'immortalité des Protozoaires (Vingt ans de culture de l'infusoire *Paramecium caudatum* sans conjugation). Bull. Acad. Sci. U.R.S.S., VII série, Classe Sci. Math. Nat.

Gelei, J. von. 1938. Beiträge zur Ciliatenfauna der Umgebung von Szeged. VII. *Paramecium nephridiatum.* Arch. Protistenk., 91: 343-56.

Griffin, L. E. 1910. *Euplotes worcesteri* sp. nov. II. Division. Philipp. J. Sci., 5: 315-36.

Hertwig, R. 1889. Über die conjugation der Infusorien. Abh. Bayer. Akad. Wiss., Cl. 2, 17:

—— 1914. Über Parthenogenesis der Infusorien und die Depression-szustände der Protozoen. Biol. Zbl., 34: 557-81.

Ivanić, M. 1928. Über die mit den Parthenogenetischen Reorganisations-prozessen des Kernapparates verb. Vermehrungscysten v. *Chilodon uncinatus.* Arch. Protistenk., 61: 293-348.

—— 1929. Zur Auffassung der sog. bandförmigen Grosskerne bei Infusorien; zugleich ein Beitrage zur Kenntnis der sog. parthenogenetischen und ihnen ähnlichen Reorganisationsprozesse des Kernapparates bei Protozoen. Arch. Protistenk., 66: 133-59.

Jennings, H. S. 1929. Genetics of the Protozoa. Bibliog. Genet., 5: 105-330.

Jollos, V. 1916. Die Fortpflanzung der Infusorien und die potentielle Unsterblichkeit der Einzellen. Biol. Zbl., 36: 497-514.

—— 1920. Experimentelle Vererbungsstudien an Infusorien. Z. Indukt. Abstamm.- u. VererbLehre, 24: 77-97.

Kidder, G. W. 1933a. Studies on *Concophthirius mytili.* II. Conjugation and nuclear reorganization. Arch. Protistenk., 79: 25-49.

—— 1933b. On the genus *Ancistruma* Strand. II. The conjugation and nuclear reorganization of *A. isseli.* Arch. Protistenk., 81: 1-18.

—— 1938. Nuclear reorganization without cell division in *Paraclevelandia simplex,* an endocommensal ciliate of the wood-feeding roach, Panesthia. Arch. Protistenk., 91: 69-77.

Kidder, G. W., and C. L. Claff. 1938. Cytological investigations of *Colpoda cucullus*. Biol. Bull., 74: 178-97.

Kidder, G. W., and W. F. Diller. 1934. Observations on the binary fission of four species of common free-living ciliates, with special reference to the macronuclear chromatin. Biol. Bull., 67: 201-19.

Kimball, R. F. 1937. The inheritance of sex at endomixis in *Paramecium aurelia*. Proc. Nat. Acad. Sci., Wash., 23: 469-74.

—— 1939. Change of mating type during vegetative reproduction in *Paramecium aurelia*. J. exp. Zool., 81: 165-79.

Klee, E. E. 1925. Der Formwechsel im Lebenskreis reiner Linien von *Euplotes longipes*. Zool. Jarhb., 42: 307-66.

MacDougal, M. S. 1936. Étude cytologique de trois espece du genre *Chilodonella* strand. Bull. biol., 70: 308-331.

Manwell, R. D. 1928. Conjugation, division, and encystment in *Pleurotricha lanceolata*. Biol. Bull. 54: 417-63.

Metalnikov, S. 1937. Le Rôle et la signification de la fécondation. Scientia, Milano, Mars, 1937: 167-76.

Moore, E. L. 1924. Endomixis and encystment in *Spathidium spathula*. J. exp. Zool., 39: 317-37.

Parker, R. C. 1927. The effect of selection in pedigree lines of Infusoria. J. exp. Zool., 49: 401-39.

Sonneborn, T. M. 1937a. The extent of the interendomictic interval in *Paramecium aurelia* and some factors determining its variability. J. exp. Zool., 75: 471-502.

—— 1937b. Sex, sex inheritance and sex determination in *Paramecium aurelia*. Proc. Nat. Acad. Sci., 23: 378-385.

—— 1939a. Sexuality and related problems in Paramecium. Coll. Net, 14: 77-84.

—— 1939b. Genetic evidence of autogamy in *Paramecium aurelia*. Anat. Rec., 75: 85. (Suppl.)

—— 1939c. *Paramecium aurelia:* mating types and groups; lethal interactions; determination and inheritance. Amer. Nat., 73: 390-413.

Spencer, H. 1924. Studies on a pedigree culture of *Paramecium calkinsi*. J. Morph., 39: 543-551.

Stein, F. R. 1859. Der Organismus der Infusionsthiere. I. Abth. Leipzig.

Stranghöner, E. 1932. Teilungsrate und Kernreorganisationsprozess bei *Paramecium multimicronucleatum*. Arch. Protistenk., 78: 302-60.

Summers, F. M. 1935. The division and reorganization of *Aspidisca lynceus, Diophrys appendiculata*, and *Stylonychia pustulata*. Arch. Protistenk., 85: 173-208.

Tittler, J. A. 1935. Division, encystment and endomixis in *Urostyla grandis*, with an account of an amicronucleate race. Cellule, 44: 189-218.

Turner, J. P. 1930. Division and conjugation in *Euplotes patella*, with special reference to the nuclear phenomena. Univ. Cal. Publ. Zool., 33: 193-258.

Woodruff, L. L. 1917a. Rhythms and endomixis in various races of *Paramecium aurelia*. Biol. Bull., 33: 51-56.

—— 1917b. The influence of general environmental conditions on the periodicity of endomixis in *Paramecium aurelia*. Biol. Bull., 33: 437-62.

—— 1932. *Paramecium aurelia* in pedigree culture for twenty-five years. Trans. Amer. micr. Soc., 51: 196-98.

—— 1935. Physiological significance of conjugation in *Blepharisma undulans*. J. exp. Zool., 70: 287-300.

Woodruff, L. L., and R. Erdmann. 1914. A normal periodic reorganization process without cell fusion in Paramecium. J. exp. Zool., 17: 425-517.

Woodruff, L. L., and E. L. Moore. 1924. On the longevity of *Spathidium spathula* without endomixis or conjugation. Proc. Nat. Acad. Sci. Wash., 10: 183-86.

Woodruff, L. L., and H. Spencer. 1922. On the method of macronuclear disintegration during endomixis in *Paramecium aurelia*. Proc. Soc. exp. Biol. N.Y., 19: 290-91.

—— 1923. *Paramecium polycaryum*, sp. nov. Proc. Soc. exp. Biol. N.Y., 20: 338-39.

—— 1924. Studies on *Spathidium spathula*. II. The significance of conjugation. Jour. exp. Zool., 39: 133-96.

Young, D. 1939. Macronuclear reorganization in *Blepharisma undulans*. J. Morph., 64: 297-353.

Young, R. T. 1918. The relation of rhythms and endomixis, their periodicity and synchronism in *Paramecium aurelia*. Biol. Bull., 35: 38-47.

CHAPTER XIV

SEXUALITY IN UNICELLULAR ORGANISMS

T. M. SONNEBORN

AMONG UNICELLULAR ORGANISMS, many different sexual conditions have long been known; descriptions of these are readily accessible (e.g., Calkins, 1926). In recent years, surprising discoveries concerning sexuality have been reported in some of the commonest and most studied unicellular organisms, such as *Polytoma* and *Chlamydomonas* among the flagellates, and *Paramecium* and *Euplotes* among the ciliates. In this chapter an attempt will be made to set forth this recent work and to examine critically the current interpretations of it.

The work on the flagellates began earlier and has been carried further than the work on the ciliates; it will therefore be presented first. Although in both classes of organisms investigations of similar nature have been pursued on a number of forms (see especially Moewus, 1935a, 1935b, 1935c, 1937a), they have been carried further on *Chlamydomonas* among the flagellates and on *Paramecium* among the ciliates. As these show essentially the same general relations as do other species of the same classes, the following account will be confined in the main to these two genera.

SEXUALITY IN *Chlamydomonas*

The recent work on *Chlamydomonas* has appeared in a series of extensive and detailed studies by Moewus since 1932 (Moewus, 1933, 1934, 1936, 1937b, 1938a, 1938b, 1939a, 1939b, 1939c; Hartmann, 1932, 1934). From the beginning, its great importance with relation to problems of sexuality has been apparent and, as the work progressed, these relations have been repeatedly emphasized by Hartmann, Moewus, and others. However, in order to approach the facts without theoretical bias, they will be restated here in a purely descriptive way.

Six species of *Chlamydomonas* have been most fully investigated: *C. braunii, C. dresdensis, C. eugametos, C. paupera, C. paradoxa,* and *C.*

pseudoparadoxa. In all species the vegetative cells and gametes have a haploid set of ten small dot-like chromosomes. Under appropriate conditions, differing in different species and races, the vegetative cells produce or become gametes that copulate and form a diploid zygote cyst. (In some species, vegetative cells function as gametes; in others, gametes differ from vegetative cells.) Under certain conditions, maturation divisions, restoring the haploid condition, take place in the cyst. The reduced cells emerge from the cyst and each gives rise by vegetative multiplication to a clone.

THE KINDS OF GAMETIC DIFFERENCES OBSERVED IN *Chlamydomonas*

The basic problem of sexuality in unicellular organisms is whether the copulating or conjugating cells regularly differ from each other. As will appear at once, certain differences are found only in some species or races, not in others; while other differences seem to be of general occurrence.

Morphological differences between copulating cells or gametes.—In *C. coccifera* (Moewus, 1937b), the copulating pairs invariably consist of a large, nonflagellated gamete and a small, flagellated one. In *C. braunii,* both copulants are flagellated, but one is always much smaller than the other. In the remaining species, there is no regular morphological difference between copulating gametes. Nevertheless, in particular pairs of at least certain species (e.g., *C. eugametos;* Moewus, 1933), one gamete may be as much as twice as large as the other, while in other pairs of the same species no size difference appears. All possible kinds of gamete combinations are found: large with large, large with small, and small with small. Finally, in species like *C. pseudoparadoxa* (Hartmann, 1934), the gametes are regularly smaller than vegetative cells, though the two gametes do not ordinarily differ from each other.

Functional differences between gametes.—In *C. coccifera* (Moewus, 1937b), the large gametes lack flagella, and the small ones retain them. Consequently, the small gametes are more active and must move toward the larger ones to accomplish copulation. Further evidence of the greater activity of the smaller gamete appears during copulation, for its contents regularly pass into the larger gamete. Less functional differentiation appears in *C. braunii;* here both gametes are flagellated and active, but during copulation the smaller gamete regularly empties into the larger one. This same functional difference occurs also in those pairs that differ markedly in size in *C. eugametos;* but not in the usual pairs, in which the gametes are alike in size (Moewus, 1933). Nor does it appear in other species in which the gametes are morphologically isogamous. Functional differentiation thus appears to be strictly correlated with morphological differentiation, occurring only when one gamete is flagellated and the other not, or when one is much larger than the other. Further, both morphological and functional differentiation may exist between gametes in some copulating couples and not in others of the same species (*C. eugametos,* Moewus, 1933).

Physiological differences between gametes.—In this section will be given only the general evidence of physiological differentiation between gametes, reserving for later consideration the question of the nature of such differentiation. If copulating gametes are not diverse physiologically, then any two gametes can copulate with each other; but if they are regularly diverse, there must be at least two kinds of gametes, with copulation taking place between gametes of different types. The basic question, therefore, is simply whether or not any two gametes of a species can copulate with each other.

In some species and varieties (*C. coccifera, C. braunii, C. paupera, C. eugametos* f. typica and simplex, *C. paradoxa,* and *C. pseudoparadoxa* from Coimbra, Portugal; Moewus, 1933; Hartmann, 1934; Moewus, 1936, 1937b, 1938a), the answer to this question is simple and definite. Gametes produced within a clone do not copulate with each other, but they do copulate with gametes produced by certain other clones. In these races, therefore, the copulating gametes must always be physiologically diverse. Moreover, this diversity is not invariably associated with morphological or functional diversity, for five of the seven races showing this phenomenon have morphologically and functionally isogamous gametes.

To the same category belongs a race of *C. pseudoparadoxa,* from Giessen; but the physiological difference is less apparent, requiring special methods to bring it to light. In this race, copulation does not normally occur either between gametes of the same clone or between gametes of different clones. Gametes of *C. pseudoparadoxa* are recognizable by their small size. Moewus (in Hartmann's article, 1934) found that the noncopulating gametes of the Giessen race could be rendered capable of copulating by subjecting them to filtrates from gamete cultures of the Coimbra race of the same species. (See 675 *et seq.* for a further account of these filtrates.) However, filtrates from any one Coimbra clone would activate some of the Giessen clones, but not all. The remaining Giessen clones required for activation treatment with a filtrate from a different Coimbra clone, one which would copulate with the first Coimbra clone. The Giessen gametes would now copulate only if a clone activated by the one filtrate was mixed with a clone activated by the other filtrate. Thus the Giessen clones are of two diverse physiological types, and copulation occurs only between the two types, not between gametes of the same clone.

In all the races thus far considered, there are regularly physiological differences between copulating gametes, which are invariably members of different clones. In the remaining three races investigated by Moewus (1934, 1938a), *C. eugametos* f. subheteroica and f. synoica and *C. dresdenis,* copulation occurred regularly among gametes of any one clone. Among these three races, the situation in *C. eugametos* f. subheteroica is unique. In any culture relatively few cells copulate. If the cells left over after copulation has ceased in a culture are mixed with the left overs from cultures of other clones, some of the mixtures will exhibit typical copulation. Exhaustive analysis of many clones shows that the same results hold here for the left overs as for entire clones of the species previously discussed. The left overs of any one clone are always of the same physiological type, but other clones yield left overs of a different type. There are just two kinds of clones, differing in the type of left overs they produce. Left overs of one type copulate only with left overs of the other type.

From these observations, Moewus (1934) concludes that each clone produces both types of gametes, but always one in much greater frequency than the other. Some clones regularly produce mostly one kind of gamete; other clones regularly produce mostly the other kind of gamete. Copulation then takes place within a clone until all the rarer type of gametes have found partners, so that all the left-over cells are of the prevailing type. Moewus (1934) reports that the behavior of this race can be made to simulate that of those previously discussed by subjecting the cultures to very dilute formalin or acetaldehyde. With this treatment, the cultures no longer yield copulation within a clone, but the clones are divisible into two physiological types, with copulation taking place only between the two types of clones. Under these conditions, the type of each clone is identical with the type of its left-over cells under normal conditions.

These results of Moewus show that different clones of *C. eugametos* f. subheteroica produce different kinds of gametes and that in mixtures of left-over cells and in mixtures of chemically treated cultures copulation is between physiologically diverse types of gametes. His further conclusion that the copulation occurring under normal conditions within each clone is also between the same two types of gametes has not been directly demonstrated, though it appears a reasonable inference. The possibility that copulation is here taking place between gametes of the same type has not been excluded. Convincing evidence on this important question calls for direct tests with "split pairs," as performed by Kimball (1939a) on *Paramecium* (see p. 697).

The other two races showing copulation within a clone, *C. eugametos* f. synoica and *C. dresdensis*, differ from *C. eugametos* f. subheteroica in three respects: (1) copulation occurs presumably on a much larger scale within a clone, the proportion of left overs being relatively small; (2) in different cultures of the same clone the left overs may be of different types; (3) no environmental means of suppressing or decreasing copulation within the clone has been reported. Otherwise the observations reported for these two races agree with those reported for *C. eugametos* f. subheteroica. Mixture of the left overs from different cultures in all possible combinations of two shows that in each race there are two kinds of left-over gametes, with copulation occurring only between the two kinds. The same uncertainty attaches here to Moewus's conclusion that the copulations within a clone are likewise between gametes of different type. However, the interpretation is here rendered more probable, for it has been shown that both types are producible within a single clone, different cultures of the same clone yielding left overs of different type.

The question at issue here is of theoretical importance. Does copulation ever take place between cells that are physiologically as well as morphologically and functionally identical? The preceding survey of the conditions in various species and races of *Chlamydomonas* shows that morphological and functional differences are frequently lacking, but that at least in some of these cases physiological differences do exist when no others are apparent. The only cases about which a reasonable doubt may still be entertained are those in which copulation occurs within a clone. This matter has been intensively studied by Pascher (1931) in *C. paupera* and by Pringsheim and Ondraček (1939) mainly in *Polytoma*. Their observations are in fundamental disagreement with those of Moewus, leading them to conclude that these forms show copulation without any physiological sex differentiation. Further, Pringsheim and Ondraček could not confirm Moewus's observation that the cells left over after copulation were unable to copulate with each other because they were all of one physiological type. They attribute the cessation of copulation to a change in the chemical conditions in the culture, rendering it unsuitable for copulation. Appropriate modification of the conditions leads to resumed copulation. They therefore deny the validity of the left-over technique for the analysis of the question at issue. The reader is referred to their article for a detailed criticism of numerous points in Moewus's work. Moewus (1940) has replied to these criticisms in an article that appeared too late for inclusion in this review.

THE NATURE OF THE PHYSIOLOGICAL DIFFERENCES BETWEEN GAMETES IN *Chlamydomonas*

As the union of gametes in *Chlamydomonas* is obviously a sex act, the physiological differences that usually, if not always, characterize the gametes may be considered sex differences. This section will set forth the number and interrelations of these sexes, their chemical characteristics, and their possible relation to male and female.

The number of sexes and their interrelations.—The system of breeding relations in *Chlamydomonas* was discovered by mixing together, in combinations of two, cultures of the sexes isolated from the species and varieties of *Chlamydomonas* examined. The two species *C. paradoxa* and *C. pseudoparadoxa* constitute one interbreeding system, and the four species *C. eugametos*, *C. paupera*, *C. braunii*, and *C. dresdensis* constitute another interbreeding system; but these two systems of species will not breed with each other.

Among the first group of species, Moewus (Hartmann, 1934) found two sexes in each of two races (from Giessen and from Coimbra) of *C. pseudoparadoxa* and in *C. paradoxa*. In order to discover whether the two sexes were alike in the three races, they were matched up in all possible combinations. As appears in Table 7, no two of the sexes are exactly alike. For example, the sexes I have designated A and B differ in that A will copulate with C while B will not; A and C differ in that they copulate with each other, although they are alike in their reactions to the other four sexes; and so on. The direct inference naturally drawn from these observations is that there are six diverse sexes in this group of races, and I have therefore designated them by six different letters. The question of whether such multiple sex systems can be reduced

TABLE 7: BREEDING RELATIONS IN *Chlamydomonas paradoxa* AND *C. pseudoparadoxa**

Species			C. para-doxa	C. pseudoparadoxa				C. para-doxa
	Source			Coimbra	Giessen	Giessen	Coimbra	
		Sex	A	B	C	D	E	F
C. para-doxa		A	−	−	+	+	+·	+
C. pseudo-paradoxa	Coimbra	B	−	−	−	+	+	+
	Giessen	C	+	−	−	+	+	+
	Giessen	D	+	+	+	−	−	+
	Coimbra	E	+	+	+	−	−	−
C. paradoxa		F	+	+	+	+	−	−

* + = copulation; − = no copulation. Data by Moewus (Hartmann, 1934). The designations of the sexes differ from those used by Moewus.

to two sexes, male and female, will be taken up later. Moewus holds that they can and designates them otherwise than I have done in Tables 7, 8, and 9.

A similar system of multiple sexes is indicated by the breeding relations in the second group of species, as shown in Table 8, constructed from the data of Moewus (Hartmann, 1934; Moewus, 1936, 1937b, 1938a). Two sexes have been isolated in *C. sp. (coccifera?)*, *C. braunii*, and *dresdensis*, six in *C. paupera*, and eight in *C. eugametos*. Not all of these are diverse, however. The two in *C. dresdensis* are the same as two in *C. paupera* and two in *C. eugametos;* four in *C. eugametos* are reducible to two diverse sexes identical with two others in *C. paupera;* and the remaining two in *C. paupera* are identical with two others in *C.*

eugametos. Altogether, there are eight diverse sexes designated in the table by the letters G to O. Two of these occur in *C. braunii,* two in *C. dresdensis,* six in *C. paupera,* and six in *C. eugametos.* Their interrelations are shown in condensed form in Table 8. G copulates with all others but H, H with all others except G and J, and so on. The breeding relations in this system of eight sexes is in general similar to the relations shown by the six sexes of the first group of races (Table 7): any

TABLE 8: BREEDING RELATIONS IN *Chlamydomonas* SP. (*coccifera?*), *C. braunii, C. dresdensis, C. eugametos* AND *C. paupera**

Sex	G	H	J	K	L	M	N	O
G	−	−	+	+	+	+	+	+
H	−	−	−	+	+	+	+	+
J	+	−	−	−	+	+	+	+
K	+	+	−	−	+	+	+	+
L	+	+	+	+	−	−	+	+
M	+	+	+	+	−	−	−	+
N	+	+	+	+	+	−	−	−
O	+	+	+	+	+	+	−	−

* + = copulation; − = no copulation. In each species the sexes found are as follows: in both *C.* sp. and *C. braunii,* sexes G and O; in *C. dresdensis,* sexes H and N; in both *C. eugametos* and *C. paupera,* sexes H, J, K, L, M, and N. In *C. eugametos* sexes H and N occur in form typica, J and M in form simplex, K and L in both forms subheteroica and synoica. Data from Moewus, 1936, 1937b, 1938a; and Hartmann, 1934. The designations of the sexes in this table differ from those used by Moewus and Hartmann. Not all combinations between the sexes in different varieties and species have been reported (e.g., the sexes in *C.* sp. were tested only with the sexes of *C. braunii;* and *C. braunii* was tested with all others except *C. paupera*), but every possible combination of sexes was made with at least one representative of the sexes.

sex copulates with any other sex in the group except that sometimes copulation will not take place with the sex next above or below it in the table.

Nature of the sex differences.—In the preceding section the sexes were defined in terms of the sexes with which they copulate. Two cultures are of the same sex if they do not copulate with each other and if each does copulate with all the sexes with which the other copulates. Two cultures are of different sex if they copulate with each other, or if one copulates with one or more sexes that the other does not copulate with. Thus the primary differences among the sexes lie in these breeding relations. There are at least two further kinds of sex differences that throw much light on the nature of the sexes in *Chlamydomonas*.

The first of these involves the intensity of the mating reaction. It is known that algal gametes of certain species (including *Chlamydomonas*) form groups as a preliminary to copulation (Fig. 165). When ripe cultures of gametes that can copulate with each other are mixed together, the gametes at once form clusters of as many as 100 or more

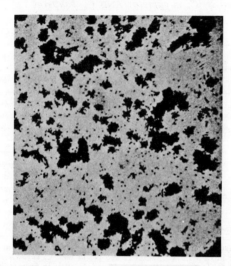

Figure 165. Group formation in *Chlamydomonas*, showing the groups formed in a mixture of cells differing in sex. (From Moewus, 1933.)

gametes. Within the clusters the gametes pair off, with the result that the cluster disintegrates into copulating pairs. The size of the initial clusters is partly determined by the number of gametes per unit of volume. When this concentration is uniform (e.g., 2×10^6 gametes per cc.), the size of the clusters depends upon which two sexes are present in the mixture. Certain combinations of sexes yield groups of 100 or more gametes, others give groups of but 10 to 20 gametes, others give only pairs, and of course some do not even give pairs. These four grades of reaction have been designated as 3, 2, 1, and 0 respectively, and the intensities of the reaction of the sex mixtures shown in Table 8 are given in Table 9. As shown in Table 9, the same sex may give different grades of reaction in mixtures with different sexes: thus G reacts to L, M, N, and O with intensity 3, to K with intensity 2, to J with intensity 1, and not at all to H. Moreover, the weak reaction of G with J is not due to the weakness of J, for J reacts with intensity 3 in mixture with L, while the latter gives but a weak reaction with N. Consequently the strength of the reaction is not a general characteristic of a given sex, but depends in some way on the particular *combination* of sexes. When the sexes are arranged as in Tables 8 and 9, the differing intensities of reaction fall into a definite system: the strength of reaction between sexes increases with their

TABLE 9: GRADES OF SEX REACTION IN MIXTURES OF SEXES G TO O FROM
THE *Chlamydomonas* SPECIES *C. braunii, C. dresdensis,* AND *C.*
eugametos (FORMS TYPICA, SIMPLEX, SUBHETEROICA,
AND SYNOICA)

Sex	G	H	J	K	L	M	N	O
G	0	0	1	2	3	3	3	3
H	0	0	0	1	3	3	3	3
J	1	0	0	0	3	3	3	3
K	2	1	0	0	2	3	3	3
L	3	3	3	2	0	0	1	2
M	3	3	3	3	0	0	0	1
N	3	3	3	3	1	0	0	0
O	3	3	3	3	2	1	0	0

* Mixtures made from cultures with cells in concentration of 2 x 10^6 per cc. O = no reaction; 1 = pairs only; 2 = clumps of 10 to 20 cells; 3 = clumps of 100 or more cells. Data from Moewus, 1938a. The designations of the sexes differ from those used by Moewus.

distance apart in the table until the maximum reaction (grade 3) is reached. These quantitative differences in intensity of sex reaction suggest that the fundamental differences among the sexes are also quantitative, a suggestion strikingly confirmed by studies of Moewus on the chemical basis of the sex reaction, as will now be set forth.

The culture fluid in which ripe gametes are living has been shown, in a number of algae, to contain material ("sex stuffs") capable of affecting the sexual behavior of other gametes. In *Chlamydomonas,* Moewus (1933, and later) obtained this material free from the organisms that produce it, by means of filtration and centrifugation, and found it to have two striking effects. Gametes grown in the dark are incapable of copulating, but treatment with the sex stuff from a ripe culture of gametes of the same sex rendered them capable of copulation. It will be recalled that activation of non-reactive gametes by filtrates from cultures of reactive gametes has previously been referred to (pp. 668-669) as the method employed in activating the peculiar gametes of the Giessen race of *C. pseudoparadoxa,* which are always normally nonreactive. This situation differs from most of the others described by Moewus in that activation is here brought about by the sex stuff from gametes of a different sex. Reference to page 672 and Table 7 will show that the sex stuff from sexes B and E were used to activate gametes of sexes C and D respectively. Similarly Moewus (1934) states that filtrates from sex K can activate gametes of sex H. This raises the question of how wide a range of sexes can be activated by sex stuffs from any one sex. Moewus (1939a), in a discussion of those sexes which I have designated G to O, states that each sex can be activated by filtrates only when they are derived from active gametes of the same sex. The earlier results with *C. pseudoparadoxa* and *C. eugametos* (Hartmann, 1934; Moewus, 1934) do not agree with this generalization.

The second effect is observed when reactive gametes of one sex are added to sex stuffs obtained from gametes of certain other sexes. The introduced gametes form groups or clusters as if they were about to copulate, but eventually the clusters disintegrate without copulation taking place. This happens when reactive gametes of one sex are added to sex stuffs from filtrates of reactive gametes of the other sex in the same race, as, for example, when gametes of sex J are added to filtrates from gametes of sex M, for sexes J and M are the two found in *C. eugametos* f. simplex. Whether similar effects of one sex stuff are producible on more than one other sex is not stated. The important point here is that sex stuffs can induce a sex reaction between cells alike in sex, but cannot induce them to copulate with each other. This indicates that there are two distinct processes in the sex act: the agglutinative sex reaction, and the actual fusion of cells and nuclei. The sex stuffs function in the former but not in the latter process. The existence of sex differences without sex stuffs (or with sex stuffs in ineffective concentrations) is also shown by the gametes of *C. pseudoparadoxa* from Giessen (see p. 668

above). These observations by Moewus are perhaps subject to a very different interpretation. The agglutinative reactions observed between gametes of the same sex are weak and transient. Failure to copulate might well be due to this, rather than to the absence of an additional factor such as a sex difference. Similar weak mating reactions between cells of the same mating type were observed by Sonneborn (1937) in *Paramecium aurelia* after the cells had been in contact with animals of another mating type. As in *Chlamydomonas,* the mating reaction was transient and did not lead to copulation. Similar behavior was also observed by Sonneborn (1938a) when cultures of mating types II and V, belonging to non-interbreeding varieties, were mixed together. Here the mating reaction occurs between animals of different mating types, and yet they fail to conjugate. Further, cultures known to belong to two mating types that will interbreed under favorable conditions will, under other conditions, give a weak and brief mating reaction without proceeding to conjugate. In view of these observations, it appears to be still an open question whether the failure of copulation to take place between cells of the same sex in *Chlamydomonas* that have given a weak sex reaction with each other is due to the weakness of the reaction or to some other aspect of sex, different from the production of diverse sex stuffs.

In later publications, Moewus (1938b, 1939a) reported the discovery of the chemical nature of the sex stuffs in the group of sexes G to O. The active stuffs for these eight sexes are all diverse percentage combinations of the cis and trans forms of the dimethyl ester of crocetin. The proportions are as follows:

Sex	G	H	J	K	L	M	N	O
Percentage cis	95	85	75	65	35	25	15	5
Percentage trans	5	15	25	35	65	75	85	95

The chemical nature of the sex stuffs aids greatly in understanding the breeding relations summarized in Tables 8 and 9. The order of sexes from G to O in the table is in the order of decreasing percentages of cis and increasing percentages of trans dimethyl crocetin. The difference in percentage of either cis or trans between any two successive sexes in the table is always 10 percent, except between sexes K and L which show a difference of 30 percent. Copulation occurs between any two sexes differing by 20 percent or more in the production of either cis or trans dimethyl crocetin, but not if they differ less than this. Further, the intensity of the sex reaction (Table 9), as measured by the size of clusters, also depends upon the difference in proportions of cis and trans dimethyl crocetin produced by the two sexes under examination: a difference of 20 percent results in the formation of pairs only (grade one reaction); a difference of 30 percent yields clusters of 10 to 20 cells (grade 2); a difference of 40 percent or more yields clusters of 100 or more cells (grade 3).

By introducing capillary tubes filled with known mixtures of cis and trans dimethyl crocetin into one edge of a drop of culture fluid and adding gametes of a known sex to the opposite edge of the drop, Moewus (1939b, 1939c) observed that the gametes aggregated at the open end of the capillary tube whenever it contained cis and trans dimethyl crocetin in proportions differing from those produced by the gametes by 19 percent or more, but not when the difference was less than this. Moreover, the time required to obtain at the mouth of the tube an aggregation of from 18 to 22 cells was from 200 to 254 seconds when the difference in proportions was 20 percent, 140 to 180 seconds when the difference was 30 percent, and 80 to 109 seconds when the difference was 40 percent. The speed of aggregation increased with increasing cis/trans difference to 22 to 37 seconds with a difference of 90 percent; hence the sex stuffs are chemotactic substances, and the grades of sex reaction are indices of the speed of chemotaxis. Moreover, in any combination of gametes that will copulate, each sex secretes chemicals that attract the other and each reacts to the chemicals secreted by the other: both gametes thus attract and both respond.

INTERPRETATION OF THE SEXUAL PHENOMENA IN *Chlamydomonas*

The sexual phenomena in *Chlamydomonas* have been interpreted by Moewus and by Hartmann in accordance with Hartmann's (1929) theory of sexuality. This theory may be formulated in the following series of propositions:

1. Sex is a universal biological phenomenon.
2. There are always two and only two sexes.
3. These two sexes are always male and female.
4. Male and female are qualitatively diverse.

5. Every cell has the full *Anlagen,* or potencies, of both male and female.

6. These potencies are not localized in any one cell component, but are general properties of all the living material.

7. The sex manifested by a cell is the result of a weakening or strengthening of the expression of either the male or female potency.

8. This weakening or strengthening may be determined by outer conditions, or by developmental conditions, or by genetic factors.

9. The degree of weakening or strengthening depends upon the effectiveness of the determiners listed in proposition 8.

10. This quantitative variation results in the appearance of each sex in a series of strengths called valences.

11. Sexual union takes place only under one or the other of two conditions: (a) when the gametes differ in sex; i.e., when one manifests a stronger male than female potency, the other a stronger female than male potency; (b) when the gametes are alike in sex, but very different in sex valence; e.g., when one is strong female, the other weak female; or when one is strong male, the other weak male.

12. Sexual union equalizes or reduces the tension resulting from difference in sex or sex valence.

The work on *Chlamydomonas* shows that physiological sex differences may exist in cases in which morphological sex differences are lacking. This is most clearly evident in those species and races in which each clone consists exclusively of one sex type. Here sexual union takes place only between gametes from different clones, the physiological sex difference of which has been demonstrated. Moewus and Hartmann further hold that similar physiological sex differences distinguish the uniting gametes in species and races manifesting copulation among the members of a single clone. The evidence for this, drawn from experiments employing the "left-over" technique, has been set forth on pages 670-671, along with the contrary evidence of Pascher and of Pringsheim and Ondraček. There thus remains some doubt, even within the genus *Chlamydomonas,* as to whether sex union is invariably accompanied by sex differences.

Hartmann's contention that sex differences are always male and female could not at first be applied to *Chlamydomonas.* Moewus, therefore, simply classified the sexes as plus ($+$) and minus ($-$). In the group of species shown in Table 7, sexes A, B, and C were called $+$, sexes D, E, and F, $—$. The three sexes of each type were assigned arbitrary strengths or valences: A and F were assigned a valence of 3; B and E, 2; and C and D, 1. Copulation was thus held to take place either

TABLE 10: SYSTEM OF MATING RELATIONS IN *Chlamydomonas braunii, C. dresdensis,* AND *C. eugametos,* SUMMARIZING THE OBSERVATIONS AND INTERPRETATIONS OF MOEWUS, 1937B, 1938A, 1938B, 1939A*

Sex (Later View)				Female				Male			
Sex (Earlier View)				$+$				$—$			
Valence				4	3	2	1	1	2	3	4
Sex as Designated in This Review				G	H	J	K	L	M	N	O
Percentage cis / Percentage trans				95/5	85/15	75/25	65/35	35/65	25/75	15/85	5/95
Female $+$	4	G	95/5	o	o	1	2	3	3	3	3
	3	H	85/15	o	o	o	1	3	3	3	3
	2	J	75/25	1	o	o	o	3	3	3	3
	1	K	65/35	2	1	o	o	2	3	3	3
Male $—$	1	L	35/65	3	3	3	2	o	o.	1	2
	2	M	25/75	3	3	3	3	o	o	o	1
	3	N	15/85	3	3	3	3	1	o	o	o
	4	O	5/95	3	3	3	3	2	1	o	o

* The numbers 0, 1, 2, 3 in the body of the table give the intensity of the sex reaction; 0 = no copulation; 1 = pairs which form directly; 2 = preliminary clusters of 10 to 20 cells; 3 = preliminary clusters of 100 or more cells. Percentage cis/Percentage trans = the proportions of the sex stuffs, cis and trans dimethyl crocetin, produced by the gamets.

between gametes differing in sex (i.e., between any $+$ and any $—$), or between two gametes of the same sex differing by as much as 2 in valence (for example, between gametes of A and C, both of which are held to be $+$, because A is $+$ 3 and C is $+$ 1). Similar interpretations were put forth for the mating relations summarized in Tables 8 and 9. The sexes G, H, J, and K were denominated $+$ in sex, with valences of 4,

3, 2, and 1, respectively; and the sexes L, M, N, and O were said to be — in sex, with valences of 1, 2, 3, and 4 respectively. In this group, Moewus (1937b, 1938a) later held that the sex previously called + was female, the one called — male. These designations, together with other pertinent information on the strength of sex reaction and sex stuffs, are shown in Table 10. (Identification of male and female in the first group of races has not yet been reported.)

The remainder of the interpretation is largely genetic and will be discussed here only insofar as appears necessary for a satisfactory understanding of the general phenomena of sexuality and their relation to the theory of Hartmann. For further details the reader should consult Chapter XV, "Inheritance in Protozoa," by H. S. Jennings. Moewus gives evidence for two series of multiple alleles affecting sex in the *braunii, dresdensis, paupera, eugametos* group of species: at one locus is a series of genes M1, M2, M3, and M4, determining the four valences of male gametes; at another linked locus is a series of genes F1, F2, F3, and F4, determining the corresponding four valences of female gametes. When crossing over takes place between these two loci, nuclei with a chromosome lacking both an M and F gene die, while those with both M and F genes survive. In the latter, if the valences are equal, sex is determined by nongenetic factors, the valence is unchanged, and both male and female gametes arise within a single clone; but if the valences are unequal, sex is determined by the gene of stronger valence and the resulting valence is the arithmetic difference between the valences of the two genes. In races such as *C. eugametos* f. subheteroica, in which each clone is always prevailing of one sex, another pair of genes determines which sex shall prevail. The genetic relations have not been worked out so fully in the *paradoxa-pseudoparadoxa* group of races, but there also multiple alleles are held to operate. Although evidence as to whether the + and the — genes are alleles has not been reported, observations on regular non-disjunction showed that the sex and valence resulting from the presence of two or more alleles was their algebraic sum.

The various genes affecting sex are considered to be the sex realisators, in agreement with Hartmann. They are held to operate by acting on the underlying sexual *Anlagen,* or potencies, A and G, the genes of differing valence acting on A and G to different extents. Sexual union then results when gametes differ in sex or in sex valence by as much as 2.

Further, the grades of reaction shown in Table 10 are presumably indices of the magnitude of sex tension between the gametes. Difference of sex always results in a grade 3 reaction, except between gametes of the lowest valence. When alike in sex, a difference of 2 in valence is required for a grade 1 reaction and a difference of 3 for a grade 2 reaction.

Certain features of Moewus's interpretation are of special interest: (1) his reduction of what appeared superficially to be many interbreeding sexes to but two, assumed to be qualitatively diverse; (2) his identification of these two sexes with male and female; (3) his distinction between unions resulting from (qualitative) difference in sex and those resulting from (quantitative) difference in sex valence. The evidence and reasoning involved in these views is set forth in the following.

The original basis for holding that only two sexes are present in each interbreeding system appears to be partly that the sexes were discovered in pairs. For example, in the *paradoxa-pseudoparadoxa* group of species (Table 7), Moewus found the two sexes here called A and F in *C. paradoxa,* B and E in the race of *C. pseudoparadoxa* from Coimbra, and C and D in the race from Giessen. Similar pairs of sexes were found in the other group of species: in *C. braunii,* G and O; in *C. dresdensis,* H and N; in *C. eugametos* f. typica, H and N; in *C. eugametos* f. simplex, J and M; in *C. eugametos* f. subheteroica and f. synoica, K and L in each. Only in *C. paupera* did an exception appear; the six types H, J, K, L, M, and N were all found together in a single natural source.

From this point on, it appears to be simply *assumed* that the two sexes in one race are qualitatively the same as the two in any other race with which it can interbreed. If this assumption be accepted, then the remaining interpretation follows naturally. For example, if in *C. paradoxa* (Table 7) the two sexes A and F are designated + and — respectively, then in the Coimbra race of *C. pseudoparadoxa* B must be + and E —, for B copulates with F (—), not with A (+); and E copulates with A (+) not with F (—). Similarly, C and D in the Giessen race are + and — respectively. This is clear from their mating relations with B and E. The exceptional copulations between like-sexed gametes (A+ with C+ and D— with F—) are interpreted as follows: A and C must both be the same sex (+) because of the mating relations set forth above; yet they must also be unlike in some sexual way, for they copulate with each other, though neither will copulate with others like itself; hence they must differ in *degree* of sex, or valence. In a similar way, the eight types G to O (Table 10) are reduced to two sexes, $+$ and $-$, each appearing in four valences. Here K and L are recognized as $+$ and $-$ of the lowest valence because they give a weaker sex reaction (grade 2) with each other than do J and M (in *C. eugametos* f. simplex), or H and N (in *C. dresdensis*), or G and O (in *C. braunii*). Of the four grades of $+$ gametes, G is most diverse from K because it gives the strongest reaction with it; hence G has the highest valence among the $+$ gametes. Similarly, O is the $-$ gamete of highest valence, and H and N are the next strongest $+$ and $-$ types (for they react less strongly with K and L than do G and O, while the others do not react at all with them). This leaves J and M intermediate between H and K and between L and N; and this is confirmed by their grades of reaction with G.

The identification of $+$ and $-$ with female and male (in the *eugametos—paupera* group of species) is based on differences in morphology, activity, and function between the gametes in certain species, and on the assumed identity of the sex differences in all the species. In *C. coccifera* and *C. braunii,* as set forth on pages 667, 668, the two kinds of gametes differ markedly in size and behavior during copulation: the smaller gamete empties into the larger one. Further, in *C. coccifera* the large gametes lack flagella and are nonmotile, while the small gametes have flagella and are motile. Moewus therefore holds that the large, nonmotile gametes of *C. coccifera* are eggs and so female, while the small, motile gametes are comparable to sperm and so are male. If this be admitted, then the large and the small gametes of *C. braunii* are also female and male, even though both are flagellated, because in combinations between the two species copulation occurs only between large and small gametes. On the same grounds, the gametes of isogamous species are female and male, because of the two physiological kinds of gametes in *C. eugametos* f. typica (types H and N), H will copulate only with the small gametes of *C. braunii* while N will copulate only with the large ones. Thus the $+$ sex has been identified with female and the $-$ with male in *C. eugametos* f. typica. And, since $+$ and $-$ were assumed to be the same in all races and species, female and male must be the same in all races and species. The copulations between female gametes (or between male gametes) of different races must then be consequences of difference of sex valence.

Critique of the works of Moewus on Chlamydomonas.—Attention should be called to certain difficulties in some of the important features of Moewus's interpretations and observations.

1. Identification of + and — with female and male. Moewus's identification of + and — with female and male is based, as set forth above, on two points: the two sexes in anisogamous species, especially in *C. coccifera*, are male and female; the two sexes are the same in all races and species. The point has already been emphasized that the latter is an assumption, not a fact of observation. The interpretation of the two sexes in *C. coccifera* as male and female is based on the proposition that female gametes are distinguishable from male gametes by their passive rôle in copulation, their larger size, and their nonmotility. Though these criteria are widely accepted as valid, one may question whether the evidence warrants this. The passive rôle of the "female" gamete in copulation is shown by the fact that the "male" gamete empties its contents into the "female" gamete. Nevertheless, the same behavior takes place in a certain race of *C. eugametos,* in which Moewus (1933) showed that it is of no sexual significance for both the + and the — gametes may play either rôle in copulation. The same holds for difference in size: either the + or the — gamete of this race of *C. eugametos* may be twice as large as its mate. The difference in behavior is correlated with the difference in size, but neither is correlated with sex. One may doubt, then, whether these two criteria are of sexual significance in *C. coccifera,* since they are clearly not significant in *C. eugametos*. The difference in motility is perhaps stronger evidence, for only the + gametes of *C. coccifera* are non-flagellated and these are generally considered to be comparable to eggs. It is important to keep clearly in mind that the use of the terms male and female for the gametes of all the races and species of *Chlamydomonas* rests finally on the single fact that the + gametes of *C. coccifera* lack flagella. Whether this is sufficient ground for holding they are female in the same sense as the eggs of higher organisms and for extending the terms male and female to the gametes in all other species of *Chlamydomonas* that interbreed with *C. coccifera* must be left to the judgment of the critical reader. The present author, in agreement with Kniep (1928) Mainx (1933) and others, holds that such facts constitute too slender a basis to justify an interpretation of such general theoretical significance.

2. Reduction of systems of multiple gamete types to two sexes. As earlier set forth, the reduction of the multiple gamete types in an interbreeding system to two sexes is based on the assumption that the two sexes in any one race or species are fundamentally the same as the two in any other race. In the case of *C. paupera,* in which six types of gametes were found in the same natural source, it is presumably assumed that three races, each with the same two sexes, were here living together. It is important to recognize clearly that this view is based on Hartmann's *theory;* it is not an observation or an induction from observation. Chemical analysis of the sex stuffs shows that reduction of the eight gamete types in the *eugametos-paupera* group of species to two qualitatively diverse sexes cannot be made on this basis, for the differences among the eight sex stuffs are exclusively quantitative. The "tension" assumed to bring the gametes together is held to be of two kinds. One kind is purely chemotactic and due to the sex stuffs; this brings the gametes into contact. It is clearly a quantitative phenomenon, dependent upon differences in relative proportions of cis and trans dimethyl crocetin. The other kind of tension determines whether gametes that have been brought into contact will unite in copulation. The evidence for this, together with considerations that render the conclusion less certain, was set forth on page 682. However, if an unknown factor determining union in copulation exists, it appears to act in the same quantitative way as the sex stuffs, for copulation takes place between any two gamete types that produce sex stuffs sufficiently diverse to attract each other. Consequently, there are no observations justifying or even suggesting the introduction of the concept of two qualitatively diverse sexes; all the observations point directly to a system of multiple, quantitatively diverse sexes.

In one respect the preceding account may not fairly represent Moewus's views. The two sex stuffs may be taken as indices of two qualitatively diverse sex tendencies or potencies, cis demethyl crocetin being the manifestation of the + sex potency, and trans dimethyl crocetin of the — sex potency. In four of the eight types of gametes, the + sex potency prevails, for these types produce more cis than trans; and this prevails to different degrees in each type. In this sense these four types of gametes may be considered as different strengths or valences of the + sex. Correspondingly, the remaining four types could be considered four diverse valences of the — sex. This view is in accord with that part of Hartmann's theory which holds that both sex potencies reside in all kinds of gametes and that the sex of the gamete is simply the potency that prevails. Thus the qualitative sex difference is not segregated into different gametes and has nothing to do with copulation; all gametes have both qualitative sex characters and differ only in the quantitative manifestation of one or the other. These quantitative differences alone determine copulation and sex reactivity. Conceivably two qualitatively diverse sexes might exist, one producing only cis, the other only trans dimethyl crocetin. But these have not been found. The observed gamete types are all quantitatively diverse grades of intersexes, some prevailingly +, others prevailingly —. Viewed in this way, the observations are in accord with part of Hartmann's theory.

3. Difficulties in Moewus's observations. There are certain difficulties in Moewus's observations that raise serious questions concerning the reliablity and accuracy of his reports. Two of these must be mentioned. The first involves the apparently irreconcilable conflict between observations of the consequences of non-disjunction of the sex chromosomes in crosses between C. paupera and C. eugametos and the later discoveries of the sex stuffs. Moewus (1939a) reports that copulation takes place between gametes of the same sex when there is at least a difference of 2 in valence. By definition, gametes of valence 5 would copulate with gametes of valence 3, but not with gametes of valence 4; and gametes of valence 6 would copulate with those of valence 4, but not with those of valence 5. In a series of crosses and back crosses involving C. eugametos f. subheteroica (valence 1) in C. paupera (valence 3), Moewus (as reported by Hartmann, 1934) obtained through nondisjunction of the sex chromosomes clones that yielded gametes of valences 5 and 6, presumably recognized as such through the breeding tests mentioned above. Moewus (1939a) shows that copulation will take place only when there is a difference of at least 20 percent in the cis or trans dimethyl croetin produced. Valence 5, by definition, copulates with valence 3; but valence 3 produces 85 percent cis or trans dimethyl crocetin. This leads to the impossible conclusion that the valence 5 gametes produced 105 percent cis or trans dimethyl crocetin. Similarly, the valence 6 gametes would be required to produce 115 percent cis or trans dimethyl crocetin. This apparently irreconcilable contradiction in the reports raises the serious question of whether the reporting is accurate and reliable.

The same question has been raised by Philip and Haldane (1939) from an analysis of data in many experiments by Moewus on crossing over and segregation in both *Chlamydomonas* and *Protosiphon*. These authors calculated that the chance of getting such close numerical agreement among the 22 experiments analyzed was once in 3.5×10^{22} trials. According to them "if every member of the human race conducted a set of experiments of this type daily, they might reasonably hope for such a success once in 50,000 million years." They suggest that this implies a conscious or unconscious adjustment of observations to fit a theory and they call for repetition of the experiments by an independent worker. The failure of Pringsheim and Ondraček (1939) in their attempts to confirm certain parts of Moewus's work, their numerous criticisms, the criticisms of Philip and Haldane, the internal inconsistencies in Moewus's data, and the great theoretical importance of the work, all make independent repetition of the work an urgent need.

SEXUALITY IN *Paramecium* AND OTHER CILIATE PROTOZOA

The ciliate Protozoa differ from *Chlamydomonas* and the flagellates in their nuclear condition and in some features of the sexual phenomena. There are two kinds of nuclei: macronuclei and micronuclei. Ordinarily the macronucleus disappears during the sexual processes and a new one is formed from a product of the micronucleus. The micronuclei alone contain recognizable chromosomes and play the leading rôle in the nuclear changes involved in sexual processes. The vegetative individuals contain diploid micronuclei that undergo maturation with reduction of the chromosomes to the haploid condition during mating. In each conjugant two reduced nuclei are formed. In most ciliates, both of these are functional: one remains within the animal in which it is formed and is known as the stationary pronucleus, or gamete nucleus; the other goes into the mate of this animal and is known as the migratory gamete nucleus, or pronucleus. The two nuclei present in each conjugant after exchange of pronuclei unite to form a synkaryon. Conjugation thus involves a reciprocal fertilization, both conjugants being fertilized, each by the other. The conjugants then separate and each reconstitutes a new nuclear apparatus and gives rise to progeny by repeated fissions. (See Chapter XII.)

The mating process is somewhat different in the peritrichous ciliates.

Unlike most other ciliates, in the peritrichs the two mates differ greatly: one is sessile and large, the other is motile and much smaller. Of the two reduced nuclei formed in each mate, only one is functional: one of those formed in the microconjugant wanders into the macroconjugant and unites with one of its nuclei. The other nuclei degenerate, as does the remainder of the microconjugant. Thus only one individual results from the mating act and this one then reproduces by repeated fissions.

Obviously the phenomena of sexuality are different in the Peritrichida from what they are in other ciliates. In the following, attention will be directed chiefly toward these other ciliates, of which *Paramecium* is an example. For both kinds of ciliates, however, the problems of sexuality are essentially the same: (1) Are the conjugant individuals sexually diverse? That is, can any two individuals conjugate with each other, or do the individuals differ morphologically or physiologically so that conjugation can occur only between individuals of these different types? (2) Are the two gamete nuclei, formed in each conjugant, sexually diverse? (3) Do conjugants differ from non-conjugants? This question involves the problem of the ciliate life cycle, with possible periods of immaturity and maturity.

SEXUAL DIFFERENCES BETWEEN CONJUGANT INDIVIDUALS

As already indicated, in one order of ciliates, the Peritrichida, the conjugants show a clear-cut differentiation into two sex types. One type, the macroconjugant, is sessile and large; the other, the microconjugant, is small and free-swimming. Conjugation takes place only between these two types, never between two individuals of the same type. In these respects the Peritrichida and a few Holotrichida (e.g. *Opalina, Trachelocerca, Ichthyophthirius*) differ from all other ciliates.

In *Metopus*, Noland (1927) observed that although the conjugants are at first morphologically indistinguishable, only one mate is fertilized and the other one degenerates. Whether this difference in behavior and fate of the two conjugants of a pair is determined by preëxisting physiological differences between them, or whether it arises first in the process of conjugation is not known.

In another order of ciliates, the Oligotrichida, a few species have been reported by Dogiel (1925) and others to show an equally clear-cut dimorphism, which is not, however, so clearly or simply viewed as a sex difference. In *Opisthotrichum,* as in the peritrichs, there are large and small individuals that differ considerably in structure, though both are motile. About 85 percent of the conjugant pairs include one large and one small member, 15 percent include two large members, and none include two small members. The small conjugants are thus sexually specialized for conjugation with large animals only; but the large type is only to a slight degree sexually specialized: it conjugates more readily with the small than with the large type, though it can conjugate with either type.

Indications of differences between the conjugants in some pairs are often observed. Doflein (1907) observed differences in size between the two conjugants in many pairs of *Paramecium putrinum,* and Mulsow (1913) observed the same thing in about 70 percent of the pairs of *Stentor.* Calkins and Cull (1907) reported frequent differences in viability between the two members of pairs of *P. caudatum.* Zweibaum (1922) found that in about 70 percent of the conjugant pairs the two members differed in the amount of glycogen they contained. These observers suggested that the larger size, greater viability, and higher glycogen content were female characters, and the reverse characters male. On the other hand, Jennings (1911) showed by thorough statistical analysis that while the two members of a pair did sometimes differ in their characters, on the whole there was a high degree of assortative mating, or tendency for like to mate with like; and, further (Jennings and Lashley, 1913a, 1913b), that after conjugation there was remarkable agreement in character between the two members of a pair (biparental inheritance), even with respect to vigor and viability. It was generally held, therefore, that in most ciliates regular or frequent differences between the two members of conjugant pairs were lacking.

Two observations made long ago raise the question of whether after all there might not be, beneath the usual superficial morphological similarity of the conjugants, a deeper-lying physiological difference. In *Chilodonella,* Enriques (1908) found that although the two conjugating individuals are indistinguishable at the start of mating, they become diverse as mating progresses: the left conjugant changes form so as to appear shorter, and its mouth migrates to the opposite side of the body. However, it is not clear whether this is an indication of a prior physiological difference between the mates, or whether it is a direct consequence of their method of union. The second observation is one made by Maupas (1889). He observed that in certain species conjugation never occurred in cultures containing animals all from a single natural source; it was necessary, in order to get conjugation in a single culture, to have animals from different natural sources. He concluded that diversity of ancestry was a necessary condition for conjugation. Many later observers found that conjugation occurred abundantly among the progeny of a single individual and so turned attention away from Maupas's contention, with its implication of physiological difference between conjugants. Until recently it was generally supposed that, in most species of ciliates, any two individuals of the same species could conjugate with each other if they were capable of conjugation at all.

The interpretations given the various observations just set forth will be deferred until the newer knowledge of sexuality in *Paramecium* has been outlined. In describing this newer work, there will be mentioned first the usual typical relations, and later, certain instructive exceptions. The facts on which the following account is based are to be found in recent articles by Sonneborn (1937, 1938a, 1938b, 1939a, 1939b, and 1939c), Kimball (1937, 1939a, 1939b, 1939c), Jennings (1938a, 1938b, 1939a, 1939b), Gilman (1939), Giese (1938, 1939), and Giese and Arkoosh (1939).

In *P. aurelia,* individuals containing macronuclei descended from one original macronucleus do not as a rule conjugate with each other. Such a group of individuals is called a caryonide. Caryonides terminate and new ones are formed when the macronuclei are destroyed and replaced by products of the micronucleus, during the reorganization following conjugation and during endomixis or autogamy. At such times usually two new macronuclei arise in each reorganizing individual, and these go into different cells at the first fission. The fact that individuals of the same caryonide do not conjugate with each other agrees with Maupas's observation that closely related individuals do not interbreed. But if several caryonides are present in the same culture, even though all come from a single original individual, they may conjugate. This agrees with the observation of the opponents of Maupas, who found conjugation within a clone.

When several caryonides are cultivated in different dishes and samples of each are mixed with samples of each of the others, in some of the

combinations nothing happens—each individual moves about independently of the others; but in other mixtures the animals quickly unite
in large clusters. The animals stick together as they collide in their ran-

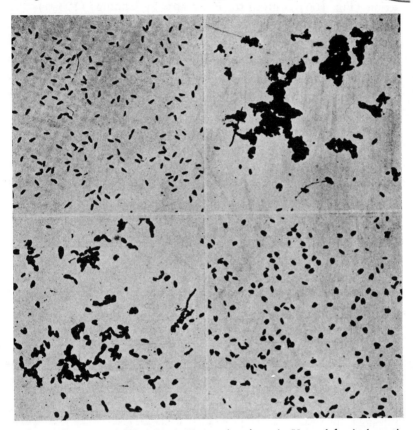

Figure 166. The mating reaction in *Paramecium bursaria*. Upper left, single mating
type with individuals scattered singly. Upper right, the clusters formed six minutes after
mixture of cultures of two different mating types. Lower left, a later stage of the mating
reaction (after five hours). Lower right, the final conjugating pairs as they appeared
twenty-four hours later. (From Jennings, 1939.)

dom movements. Animals not in contact do not attract each other; nor
are they in a specially sticky condition, as has been so often maintained,
for neither caryonide shows the least trace of stickiness until the animals
are mixed, and then only when animals of different caryonides collide.
The clusters begin with just two individuals and build up into larger

aggregations by the repeated addition of other individuals, as these collide with those already united. In the course of an hour or so, the clusters break down into conjugating pairs. A detailed account of this mating reaction (Fig. 166) is given for *P. bursaria* by Jennings (1939a).

The final pairs always consist of one individual from each of the two caryonides. When animals of the two caryonides differ in size, each pair consists of one large and one small animal. In *P. bursaria* (Jennings, 1938a) the two members of each pair differ in color when a normal green culture is mixed with one made pale as a result of recent rapid multiplication.

When all possible combinations are made among a group of caryonides, they are classifiable, on the basis of their reactions, into two groups (Table 11); no two members of the same group will conjugate with each other, but any two caryonides from different groups will. These

TABLE 11: RESULTS OF MIXING TOGETHER ANIMALS FROM DIFFERENT
CARYONIDES OF STOCK F, *Paramecium aurelia**

		Caryonides										
		2b1	2b2	3a2	4b1	4b2	1b2	2a1	2a2	3a1	5b1	5b2
Caryonides	2b1	−	−	−	−	−	+	+	+	+	+	+
	2b2	−	−	−	−	−	+	+	+	+	+	+
	3a2	−	−	−	−	−	+	+	+	+	+	+
	4b1	−	−	−	−	−	+	+	+	+	+	+
	4b2	−	−	−	−	−	+	+	+	+	+	+
	1b2	+	+	+	+	+	−	−	−	−	−	−
	2a1	+	+	+	+	+	−	−	−	−	−	−
	2a2	+	+	+	+	+	−	−	−	−	−	−
	3a1	+	+	+	+	+	−	−	−	−	−	−
	5b1	+	+	+	+	+	−	−	−	−	−	−
	5b2	+	+	+	+	+	−	−	−	−	−	−

* + = conjugation; − = no conjugation. Data from Sonneborn, 1938a. The caryonides 2b1, 2b2, 3a2, 4b1, and 4b2 are of mating type I; caryonides 1b2, 2a1, 2a2, 3a1, 5b1, and 5b2 are of mating type II.

two groups are said to be of different mating types, and in one group of races are designated as I and II. Conjugation occurs between types I and II, never between individuals of the same type, whether they be members of the same or different caryonides. In order to ascertain the type of any unknown caryonide, some of its animals are mixed with type I and some with type II; conjugation occurs in one of the mixtures, not in the other. The type of the new caryonide must then be the same as the type

TABLE 12: THE SYSTEM OF BREEDING RELATIONS IN *Paramecium aurelia,*
DATA FROM SONNEBORN, 1938A*

Variety		1		2		3	
	Mating Type	I	II	III	IV	V	VI
1	I	−	+	−	−	−	−
1	II	+	−	−	−	−	−
2	III	−	−	−	+	−	−
2	IV	−	−	+	−	−	−
3	V	−	−	−	−	−	+
3	VI	−	−	−	−	+	−

* The three varieties (1, 2, and 3) do not interbreed; conjugation occurs only between the two mating types within each variety. + = conjugation; − = no conjugation.

with which it did not conjugate, different from the one with which it did. For example, if a culture fails to conjugate with type I, but does with type II, it is type I.

Sonneborn (1938a, 1939a, 1939b) has analyzed some fifty stocks of *P. aurelia,* collected from various regions between Canada and Florida and from the Atlantic to the Pacific Coast. Nearly all showed a similar system: in each stock all caryonides were classifiable into one or the other of two mating types. The few remaining stocks consisted exclusively of but one mating type: e.g., in stock B all caryonides conjugated with type II from another stock, none with type I; so stock B consists exclusively of type I. Studied alone, stock B would be considered non-conjugating, because it never conjugates among its own members. All so-called non-conjugating stocks behave like this; they consist of only one mating type and conjugate readily when mixed with the proper type from another stock. Mating types appear to be of universal occurrence in *P. aurelia.*

Although not more than two mating types occur in any one stock, more than two must exist in the species, for both mating types in some stocks fail to conjugate with either of the types in certain other stocks. Altogether, six different mating types have been found (Table 12). One group of stocks contains types I and II; a second group contains types III and IV; a third group, types V and VI. Conjugation takes place only between the two types in the same group, never between types in different groups. This sexual isolation of the three groups of stocks makes them distinct genetical species or varieties; but they appear to be morphologically alike, all conforming to the description of the taxonomic species *P. aurelia.* However, they are physiologically diverse in a number of ways.

Each mating type is uniquely defined by the type with which it mates. The mating type of a culture can be ascertained by mixing some of its animals with standard cultures of each of the six mating types. With one, and only one, of these it will conjugate. Its mating type is the other one in the variety with which it mates. For example, if it mates with type V, it belongs to variety 3 and is of mating type VI.

In *P. bursaria,* Jennings (1938a, 1938b; 1939a, 1939b) reports somewhat different mating-type relations. Each stock of this species shows as a rule only one mating type. As nuclear reorganization is extremely rare, a stock is practically equivalent to a caryonide of *P. aurelia.* The mating types fall into three different groups, or genetical species (a fact first found in *P. bursaria*), with no conjugation between types in different groups (Table 13). In group I occur the four mating types A to D; in group II, the eight types E to M; in group III, the the four types N to Q. In each group each mating type conjugates with all the other types in that group. This system of multiple interbreeding types is in marked contrast to the system of paired types in *P. aurelia.* To discover the group to which a new stock belongs, it must be mixed with at least two types from each of the three groups. It will conjugate with one or both of the types from one group, not with any of the others. It belongs to the group with which it conjugates. To discover its mating type, it must now be mixed with all the types of this group until one is found with which it will not conjugate. It is then of the same type as this one. For example,

TABLE 13: THE SYSTEM OF BREEDING RELATIONS IN *Paramecium bursaria*, DATA FROM JENNINGS, 1939A*

Variety	Mating Type	I				II								III			
		A	B	C	D	E	F	G	H	J	K	L	M	N	O	P	Q
I	A	−	+	+	+	−	−	−	−	−	−	−	−	−	−	−	−
	B	+	−	+	+	−	−	−	−	−	−	−	−	−	−	−	−
	C	+	+	−	+	−	−	−	−	−	−	−	−	−	−	−	−
	D	+	+	+	−	−	−	−	−	−	−	−	−	−	−	−	−
II	E	−	−	−	−	−	+	+	+	+	+	+	+	−	−	−	−
	F	−	−	−	−	+	−	+	+	+	+	+	+	−	−	−	−
	G	−	−	−	−	+	+	−	+	+	+	+	+	−	−	−	−
	H	−	−	−	−	+	+	+	−	+	+	+	+	−	−	−	−
	J	−	−	−	−	+	+	+	+	−	+	+	+	−	−	−	−
	K	−	−	−	−	+	+	+	+	+	−	+	+	−	−	−	−
	L	−	−	−	−	+	+	+	+	+	+	−	+	−	−	−	−
	M	−	−	−	−	+	+	+	+	+	+	+	−	−	−	−	−
III	N	−	−	−	−	−	−	−	−	−	−	−	−	−	+	+	+
	O	−	−	−	−	−	−	−	−	−	−	−	−	+	−	+	+
	P	−	−	−	−	−	−	−	−	−	−	−	−	+	+	−	+
	Q	−	−	−	−	−	−	−	−	−	−	−	−	+	+	+	−

* The three groups (or varieties) 1, 2, and 3 do not interbreed; conjugation occurs only among the four or eight mating types within each variety. + = conjugation; − = no conjugation.

if it mates with A, B, and C, but not with D, it is of mating type D.

Five of the seven species of *Paramecium* found in the United States have been examined for mating types and all have shown them. *P. aurelia* and *P. bursaria* have already been discussed. In *P. caudatum*, Gilman (1939) finds a system of the same kind as found in *P. aurelia*: six mating types occurring in three groups, with only two interbreeding mating types in each group, and no conjugation between types in different groups. Sonneborn (1938a, 1939a) found two mating types in *P. calkinsi* and three interbreeding types in *P. trichium,* indicating a system of multiple types such as Jennings found in *P. bursaria.* Giese (1938, 1939) and Giese and Arkoosh (1939) have found mating types in *P. multimicronucleatum* and *P. caudatum.*

Kimball (1939c) found in *Euplotes,* one of the hypotrichous ciliates, a system of mating types like the one in *P. bursaria.* There are five groups of non-interbreeding types, with morphological differences between some of the groups, indicating that these may be taxonomically as well as genetically different species. In each group occur multiple interbreeding mating types, six in the group most fully studied, any one of these conjugating with any of the other five. The striking agglutinative mating reaction so characteristic of *Paramecium* appears to be lacking: conjugation first occurs several hours after mixture of the different types, and then pairs form directly without the prior formation of clusters.

In view of our present knowledge, it seems allowable to include Maupas's (1889) old evidence for the necessity of diverse ancestry as evidence for diversity of mating type. If so, at least four more species must be added to the list of those in which mating types are known: *Stylonychia pustulata, Leucophyrs patula, Onychodomus grandis,* and *Loxophyllum fasciola.*

This brings the number of species now known to have mating types to about a dozen. These belong to six different genera and two different orders of ciliate Protozoa. It appears, therefore, that mating types will be found to be widely distributed among the ciliates. The view that any two individuals of the same species can conjugate with each other, if capable of conjugating at all, is demonstrably false; on the contrary, in general, conjugation can take place only between individuals of diverse mating types.

Are there ever exceptions to this general rule? Does conjugation ever take place between animals of the same mating type? In nearly all the species examined in detail, conjugation has been observed in cultures containing only one caryonide, and, as members of the same caryonide are presumably of the same mating type, this appears to be conjugation between animals of the same mating type. Can individuals of the same caryonide ever differ in mating type? And is this the explanation of these exceptional conjugations within a caryonide? There is only one method

(Kimball, 1939a) of answering these questions directly. The two animals that come together for conjugation must be separated before they become too tightly united, and the mating types of the two members of such a split pair must be directly ascertained by placing each of them separately in standard cultures of the different types, to discover with which ones they will react sexually. If one reacts only with type I and the other only with type II, they must be of different types; but if both react with the same type, then they are alike in mating type.

This problem has been most fully studied by Kimball (1939a, 1939b). In *P. aurelia,* he found that conjugation within a caryonide occurred under two very different kinds of conditions. One kind is very common; it occurs in caryonides genotypically of type I, when the last preceding caryonide in the direct line of ancestry was of type II. Under these conditions conjugation may occur in the caryonide during the first few days of its existence. Kimball split some of these conjugant pairs, tested them directly for mating type, and showed that in each pair one animal was of type I, the other of type II. Thus both mating types can be present in a single caryonide, and the mating is between the two types only. Kimball now obtained clone cultures from the two members of such split pairs and found in every pair that both cultures were of type I and showed no further conjugation among their own members. Hence the type II animals originally present in the caryonide changed to type I. The early occurrence of type II was due to the type II character of the immediate ancestors. This phenotypic or cytoplasmic "hang-over" fades out, as the new genotype comes into action. Not all individuals accomplish this at the same speed, so for a short time some are still type II while others have completed the change to type I; at this moment conjugation may occur. A little later all have changed to type I, and conjugation is no longer possible. Similar "cytoplasmic lag" in the inheritance of other characters in *Paramecium* had been reported by both De Garis (1935) and by Sonneborn and Lynch (1934).

The other type of conjugation within a caryonide is of much rarer occurrence. In the race of *P. aurelia* examined by Kimball (1939b), less than 3 percent of the caryonides showed it. In these, conjugation occurred not only when the caryonide was young, but probably throughout its whole history. Moreover, any individual in the caryonide gave rise to progeny that conjugated with each other. Even the members of a split pair both gave rise to cultures in which conjugation took place. Nevertheless, Kimball found that the two members of a split pair were always of diverse mating types at the time they conjugated: one was type I, the other type II. Hence such caryonides are unstable in mating type. The type changes back and forth repeatedly; but when conjugation occurs, it is always between animals of different mating type.

In the Vorticellidae, the invariable morphological and functional difference between conjugants has already been mentioned. Finley (1939a, 1939b) shows clearly that both types of conjugants not only arise within a caryonide, but at a single unequal cell division. The macroconjugant and the microconjugant produced at this fission can then copulate with each other or with other similarly differentiated conjugants of the same caryonide. Here it is obvious that conjugation within a caryonide is nevertheless invariably between different mating types or sexes.

There are, however, a number of known instances of conjugation within a caryonide which require further investigation. Foremost among these are species, without morphological difference between the conjugants, in which conjugation regularly occurs within a clone or a caryonide. This has been reported as common in *P. multimicronucleatum* by Giese (1938, 1939), less common in *P. caudatum* by Gilman (1939), and very rare in *P. bursaria* by Jennings (1938a, 1938b, 1939a, 1939b). An especially interesting situation is reported for *Euplotes* by Kimball (1939c). Fluid from a culture of one mating type, added to a culture of a different mating type, induces the latter to conjugate among themselves. Likewise, in mixtures between normal animals of one mating type and double animals of another type, some of the resulting conjugant pairs are unions of singles with singles, a few are doubles with doubles, though most are, as would be expected, singles with doubles. The relations here raise the question of whether subjection to fluid from another mating type makes animals acquire a type corresponding to the fluid, as Jollos (1926) showed happens in the alga *Dasycladus*. If so, it may be difficult or impossible to analyze it satisfactorily, because in ascertaining the types of members of split pairs they have to be subjected to the very fluid that would change their type. This may be one of those exasperating problems, like attempting to determine the position and the velocity of an electron at the same time, in which the methods of investigation essentially alter the things being investigated.

Are any of these observations of conjugation within a caryonide evidence of conjugation between animals identical in mating type? The direct test has not been made in most cases; but, in the few examples in which it was made, it was demonstrated that conjugants were always of different types, in spite of the fact that they were members of the same caryonide. The evidence is therefore strongly against the occurrence of conjugation between animals of the same mating type, though final judgment must await further analysis.

MATING TYPES IN RELATION TO THE MAUPASIAN LIFE CYCLE

According to the well-known theory of Maupas (1889), the ciliate exconjugant is conceived as being a young individual producing by repeated fissions immature cells unable to mate; the cells produced after many youthful fissions become sexually mature and are capable of conjugating; after many more fissions, the cells grow old, losing their power of conjugating and showing other signs of senescence; and they finally die. If conjugation occurs during the period of maturity, the conjugants are rejuvenated and the cycle is renewed. In some ciliates, such as *Uroleptus mobilis,* investigated by Calkins (1920), this Maupasian life cycle is clearly shown. In *Paramecium,* however, there are striking specific and racial differences in presumably so fundamental a matter as the life cycle.

Many races of *P. aurelia* (Sonneborn 1937, 1938a) show a definite period of immaturity: during the first week or two after conjugation, cultures do not give the mating reaction and cannot conjugate. In a few more days, the power of conjugating rapidly develops to full strength, inaugurating a period of maturity. But the organisms remain mature indefinitely; no period of senescence appears. Only for a day or so during the periodically recurring processes of nuclear reorganization, are they unable to conjugate. As soon as reorganization is completed, the mating reaction reappears in full strength. Why the reorganized cells fail to begin again with a period of immaturity, as they do after conjugation, is at present a puzzling and probably a significant fact.

Other races of *P. aurelia* (Sonneborn, 1938a) not only lack a period of senescence, but also a period of immaturity: they are able to conjugate immediately after conjugation. Eight successive conjugations have been obtained in a period of seventeen days (Sonneborn, 1936). As the process of conjugation and nuclear reconstitution require one day, there could have been only about a day between successive conjugations, a period in which at most only three or four fissions could take place.

In *P. bursaria*, Jennings (1939a, 1939b) reports a regularly occurring period of immaturity. In group I it lasts for from two weeks to several months; in group II all clones under investigation were still immature at last reports, eight months after their origin at conjugation. Periods of immaturity have also been found regularly in *P. caudatum* by Gilman (1939) and in *Euplotes* by Kimball (1939c). In none of these species has there as yet been any report that maturity is followed by a period of senescence, with loss of ability to conjugate. Many of Jennings's clones of *P. bursaria* have been mature for over two years, without loss of sexual vigor; and in this species endomixis is so rare as scarcely to account for the results.

Thus age sometimes is and sometimes is not a factor in determining conjugation; the Maupasian life cycle is not an invariable feature of ciliate life. Immaturity may be absent, short, or long; maturity may be coextensive with life, or it may be simply preceded by a period of immaturity; or it may be delimited on either side by periods of immaturity and senescence.

THE RÔLE OF ENVIRONMENTAL CONDITIONS IN DETERMINING CONJUGATION

Maupas (1889) recognized the importance of environmental conditions in determining conjugation, and most subsequent workers have been in more or less agreement on this point; but some have carried this view to the extreme of ascribing to environmental conditions alone the determination of conjugation. The preceding account has shown that this cannot always be true, for hereditary and developmental internal factors have been demonstrated as playing a decisive rôle in many of the races and species. Nevertheless, environmental conditions, such as nutrition, temperature, and light, do have marked limiting effects on the occurrence of conjugation.

In *P. aurelia* (Sonneborn, 1938a), the mating reaction does not take place in cultures that are either overfed or completely starved. Intermediate nutritive conditions are most favorable for its occurrence. More-

over, the cultural conditions must be good in other respects: when deleterious bacteria or other unfavorable conditions injure the paramecia,
the mating reaction is weak or lacking.

In variety 1, mating types I and II will react sexually at any temperature within the range examined, 9° C. to 32° C.; but mating types III
and IV of variety 2 will not react above 24° C. and types V and VI
of variety 3 not above 27° C.

Similar differences appear in the time of day in which reactions will
occur: variety 1 will react at any time; but variety 2 reacts only between
6 P.M. and 7 A.M., while variety 3 reacts only between 1 A.M. and 1 P.M.
As might be supposed, this periodicity is an effect of the daily alternation
of light and dark. In variety 3, sexual reactivity has been completely
suppressed by exposing the organisms to continuous illumination, and
they have been made to react at all hours by keeping them in continuous
darkness. These effects have been shown (Sonneborn, 1938a) to be due
to the suppression of reactivity by light, not to its stimulation by darkness. Similar diurnal periodicities in mating occur in *P. bursaria* (Jennings, 1938a, 1939a, 1939b).

The environmental conditions thus determine whether conjugation
will occur when the proper mating types are brought together. Ordinarily
the mating types themselves are hereditary characters (see Chapter XV,
"Inheritance in Protozoa," Jennings); but in the exceptional unstable
caryonides studied by Kimball (1939b), genetic determination seems
excluded, for the mating types change repeatedly during vegetative reproduction. Here environmental conditions probably determine even the
mating types themselves, and similar relations may be the rule, instead
of the exception, in species in which conjugation within a caryonide occurs regularly. Thus investigations of possible genetic, developmental,
and environmental factors determining conjugation show all to be involved, as might have been expected.

SEX DIFFERENCES BETWEEN GAMETE NUCLEI

Careful observations on the form and behavior of the gamete nuclei
during conjugation were made by Maupas (1889) and by R. Hertwig
(1889). These and nearly all subsequent investigators have agreed that
in most ciliates the two gamete nuclei formed in each conjugant differ in

behavior: one, the stationary gamete nucleus, remains in the conjugant that produces it; the other, the migratory gamete nucleus, passes into the mate and unites with the stationary gamete nucleus located in that animal. As a rule, the gamete nuclei are morphologically indistinguishable; but in some species differences in size and form have been reported. The most extreme example of morphologically different gamete nuclei is in *Cycloposthium* (Dogiel, 1925). The spindle resulting in the formation of the gamete nuclei is heteropolar: one pole, destined to produce the

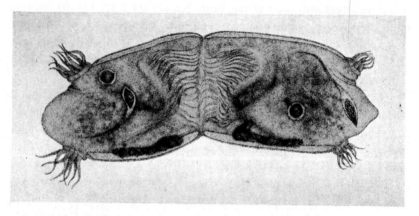

Figure 167. Conjugation in *Cycloposthium bipalmatum,* showing the sperm-like migratory pronuclei differing from the spherical stationary pronuclei. (After Dogiel.)

stationary gamete nucleus, is larger and rounder than the other smaller and more pointed pole, destined to yield the migratory gamete nucleus. The latter arises from the anterior pole of the spindle and develops a long tail-like appendage at the proximal end, and a small, pointed distal end, functional in piercing the cuticle in its passage from one mate into the other (Fig. 167). In other ciliates, lesser differences between the gamete nuclei have been observed: slight differences in size in *Didinium* (Prandtl, 1906), in *Paramecium caudatum* (Calkins and Cull, 1907), and in *P. multimicronucleatum* (Landis, 1925). Calkins and Cull (1907) concluded that the two gamete nuclei in *P. caudatum* differ in their chromatin content, as a consequence of transverse chromosomal division at the nuclear division which gives rise to them. In most ciliates, however, no morphological differences between the two gamete nuclei have been observed.

SIGNIFICANCE OF THE DIVERSITIES BETWEEN CONJUGANTS AND BE-
TWEEN GAMETE NUCLEI

There is great diversity of opinion regarding the significance of the observed differences between conjugants and between gamete nuclei. This is due partly to the variety and complexity of the observed phenomena, and partly to confusion as to the meaning of the concepts employed, particularly concepts developed primarily with relation to phenomena in higher organisms. An attempt will be made to summarize the more prominent views concerning the main types of observed relations and to set forth some general considerations concerning them.

In some ciliates, the Vorticellidae and a few others, in which the conjugants are always morphologically diverse and the gamete nuclei morphologically alike, with fertilization of only the larger conjugant, it is usually agreed that the conjugants differ sexually. Further, the gamete nuclei in each conjugant are sometimes said to be of the same sex as the conjugant. Some authors hold that the sexes here are female and male.

In *Chilodonella* (see p. 689), in which the conjugants become morphologically diverse during conjugation and fertilization is reciprocal, Enriques (1908) concluded that although both conjugants were functionally hermaphroditic (producing two sexually diverse gamete nuclei), the conjugants also showed a partial, incompletely developed sex diversity, for which he devised the term "hemisexes."

In *Opisthotrichum* (see p. 689) fertilization is also reciprocal, involving gamete nuclei with strongly marked morphological sex differences. Nevertheless, in a majority of the conjugant pairs, the two members differ in size. When the two are alike in size, both are large. Dogiel (1925) interprets these facts as follows: the conjugants are all functionally hermaphroditic, each producing male and female gamete nuclei; but the conjugants also show the beginnings of sexual differentiation, the small ones being more differentiated, for they can mate only with large individuals, while the large ones can mate either with small or large, though more commonly with the former. The small conjugants are viewed as considerably differentiated toward the male condition, in spite of their functional hermaphroditism.

In *Paramecium* and *Euplotes,* fertilization is reciprocal, the gamete nuclei show little or no morphological difference, and the conjugants

show no significant morphological differences. Yet the conjugants are
regularly differentiated physiologically into diverse mating types. The
gamete nuclei in these and similar forms are often considered to be
sexually diverse; frequently the migratory gamete nucleus is viewed as
male, the stationary one as female. This introduces the same difficulty
as in *Chilodonella* and *Opisthotrichum*. How reconcile sex differences
between the gamete nuclei with the differences between the "hermaphroditic" conjugants? Jennings (1939a) inclines toward interpreting
the mating types as manifesting phenomena of self-sterility, or incompatibility, of the kind found in certain higher plants (Stout, 1938) and
animals (Morgan, 1938), in the sense that the single clone or caryonide,
like the single self-sterile plant, ordinarily does not fertilize itself. Jennings points out the features in which the two sets of phenomena are
different, as well as those in which they are alike. More recently, Sonneborn (1939c) has shown that the periodic nuclear reorganization in variety
1 of *P. aurelia* is regularly a self-fertilization, as maintained by Diller
(1936). Consequently, *P. aurelia* is not self-sterile, but regularly self-
fertile. The failure of individuals of the same mating type to conjugate
with each other is thus not related to any incompatibility between their
gametes, for such appears not to exist. It seems, therefore, more comparable to the failure of two individuals of the same sex to unite in copulation. In higher organisms, self-sterility serves to prevent self-fertilization;
in *P. aurelia* the mating types serve to bring together for cross-fertilization
diverse sex types, each of which regularly undergoes self-fertilization.
The present author, therefore, concludes that the mating-type phenomena
are not properly to be viewed as self-sterility or incompatibility. If by
sexual differentiation is meant the differentiation of the individuals of a
species into diverse kinds, so that mating occurs regularly between different kinds, not between two of the same kind, then the mating types
of *Paramecium* are diverse sexes. As in *Opisthotrichum*, the sex differences between the conjugants are of a different kind from those existing
between the gamete nuclei: one serves to bring together the mates, the
other to bring together their gamete nuclei.

Multiple sex systems, such as those in *P. bursaria* and *Euplotes*, offer
serious difficulties to those who, like Hartmann (1929), hold there can
be but two sexes. Whether or not one agrees with this contention, the

work of Moewus (1939a) on *Chlamydomonas* shows how what appears, through biological analysis, to be a multiple sex system may be reduced, through chemical analysis, to a fundamentally dual system. Further investigation is, of course, required to ascertain whether the multiple sex systems in ciliates are, in fact, similar in this respect to the system in *Chlamydomonas*.

A number of the interpretations of sex relations in ciliates employ the concepts male and female, as set forth above. Many authors follow Hartmann (1929), who holds, as has been pointed out on page 678, that sex differences, wherever found, are always male and female. The characters by which the female is ordinarily recognized are larger size, lesser activity, greater storage of nutritive reserves, and egg-like form; and the male by the corresponding opposed characters. In attempting to apply these views, however, numerous difficulties are encountered. In the Vorticellidae, both gamete nuclei in the microconjugant are held to be male; yet only one of them shows the "male" character of activity by migrating into the macroconjugant. In *Opisthotrichum,* the migratory gamete nucleus has the form of a sperm, but it has the "female" character of much greater size than the stationary gamete nucleus. The difficulty of using size as an index of femaleness is clearly shown in the work of Satina and Blakeslee (1930) on certain bread molds. In a number of strains, two sexes were observed and found to be the same in all strains. One sex was distinctly larger than the other in each strain, yet the larger sex in one strain was shown to be identical with the small sex in others. Geitler (1932) found similar difficulties in identifying the sexes in diatoms by their activity. These and other difficulties have led Kniep (1928), Mainx (1933), and others to abandon the concepts of male and female in unicellular organisms and to view sexual union as brought about by copulation-conditioning factors, some of which operate to bring together the cells, others the nuclei. In the present state of knowledge, this point of view appears to be preferable to one that appeals to such abstract, ill-defined, and confusing concepts as fundamental maleness and femaleness.

From this point of view, the conflicts between sexual differentiation in the gamete nuclei and sexual differentiation in the conjugant individuals present far less difficulty than from the point of view which requires identification of all sex differences with male and female. There may be two kinds of copulation-conditioning factors: one functioning in bringing together the cells, the other in bringing together their nuclei. In *Chlamydomonas* and the Vorticellidae, the two kinds of nuclear factors operate in different kinds of cells; in *Paramecium* and most other ciliates, both kinds of nuclear factors operate in each of the kinds of cells. Thus by abandoning the pure assumption that sex differences, wherever found, must always be fundamentally the same (male and female), the conflict between sex differences in nuclei and sex differences in cells disappears.

LITERATURE CITED

Calkins, G. N. 1920. *Uroleptus mobilis* Engelm. III. A study in vitality. J. exp. Zool., 34: 449-70.

—— 1926. The Biology of the Protozoa. Philadelphia and New York.

Calkins, G. N., and S. W. Cull. 1907. The conjugation of *Paramecium aurelia* (*caudatum*). Arch. Protistenk., 10: 375-415.

De Garis, C. F. 1935. Heritable effects of conjugation between free individuals and double monsters in diverse races of *Paramecium caudatum*. J. exp. Zool., 71: 209-56.

Diller, W. F. 1936. Nuclear reorganization processes in *Paramecium aurelia*, with descriptions of autogamy and "hemixis." J. Morph., 59: 11-68.

Doflein, F. 1907. Beobachtungen und Ideen über die Konjugation der Infusorien. S. B. Ges. Morph. München, 23: 107-14.

Dogiel, V. 1925. Die Geschlechtsprozesse bei Infusorien (speziell bei den Ophryoscoleciden), neue Tatsachen und theoretische Erwägungen. Arch. Protistenk., 50: 283-442.

Enriques, P. 1908. Die Conjugation und sexuelle Differenzierung der Infusorien. Zweite Abhandlung. Wiederconjugante und Hemisexe bei *Chilodon*. Arch. Protistenk., 12: 213-76.

Finley, H. E. 1939. Sexual differentiation in *Vorticella microstoma*. J. exp. Zool., 81: 209-29.

—— 1939b. Further observations upon sexual differentiation in *Vorticella microstoma*. Anat. Rec., 75, (suppl.), p. 85.

Geitler, L. 1932. Der Formwechsel der pennaten Diatomeen (Kieselalgen). Arch. Protistenk., 78: 1-226.

Giese, A. C. 1938. Race and conjugation of *Paramecium*. Physiol. Zoöl., 11: 326-32.

—— 1939. Studies on conjugation in *Paramecium multimicronucleatum*. Amer. Nat., 73: 432-44.

Giese, A. C., and M. A. Arkoosh. 1939. Tests for sexual differentiation in

SEXUALITY 707

Paramecium multimicronucleatum and *Paramecium caudatum.* Physiol. Zoöl., 12: 70-75.

Gilman, L. C. 1939. Mating types in *Paramecium caudatum.* Amer. Nat., 73: 445-50.

Hartmann, M. 1929. Verteilung, Bestimmung und Vererbung des Geschlechtes bei den Protisten und Thallophyten. Handb. d. Vererbungswiss. II.

—— 1932. Neue Ergebnisse zum Befruchtungs-und Sexualitätsproblem. (Nach Untersuchungen von M. Hartmann, J. Hämmerling und F. Moewus.) Naturwissenschaften, 20: 567-73.

—— 1934. Beiträge zur Sexualitätstheorie. Mit besonderer Berücksichtigung neuer Ergebnisse von Fr. Moewus. S. B. preuss. Akad. Wiss., Phys. Math. Kl., 379-400.

Hertwig, R. 1889. Über die Conjugationen der Infusorien. Abh. bayer. Akad. Wiss., II Kl., 17: 151-233.

Jennings, H. S. 1911. Assortative mating, variability and inheritance of size in the conjugation of *Paramecium.* J. exp. Zool., 11: 1-134.

—— 1938a. Sex reaction types and their interrelations in *Paramecium bursaria.* I. Proc. Nat. Acad. Sci. Wash., 24: 112-17.

—— 1938b. Sex reaction types and their interrelations in *Paramecium bursaria.* II. Clones collected from natural habitats. Proc. Nat. Acad. Sci., 24: 117-20.

—— 1939a. Genetics of *Paramecium bursaria.* I. Mating types and groups, their interrelations and distribution; mating behavior and self sterility. Genetics, 24: 202-33.

—— 1939b. *Paramecium bursaria:* mating types and groups, mating behavior, self sterility; their development and inheritance. Amer. Nat., 73: 414-31.

Jennings, H. S., and K. S. Lashley. 1913a. Biparental inheritance and the question of sexuality in *Paramecium.* J. exp. Zool., 14: 393-466.

—— 1913b. Biparental inheritance of size in *Paramecium.* J. exp. Zool., 15: 193-99.

Jollos, V. 1926. Untersuchungen über die Sexualitätsverhältnisse von *Dasycladus clavaeformis.* Biol. Zbl., 46: 279-95.

Kimball, R. F. 1937. The inheritance of sex at endomixis in *Paramecium aurelia.* Proc. Nat. Acad. Sci. Wash., 23: 469-74.

—— 1939a. A delayed change of phenotype following a change of genotype in *Paramecium aurelia.* Genetics, 24: 49-58.

—— 1939b. Change of mating type during vegetative reproduction in *Paramecium aurelia.* J. exp. Zool., 81: 165-79.

—— 1939c. Mating types in *Euplotes.* Amer. Nat., 73: 451-56.

Kniep, H. 1928. Die Sexualität der niederen Pflanzen. Jena.

Landis, E. M. 1925. Conjugation of *Paramecium multimicronucleata,* Powers and Mitchell. J. Morph., 40: 111-67.

Mainx, F. 1933. Die Sexualität als Problem der Genetik. Jena.

Maupas, E. 1889. La Rajeunissement karyogamique chez les ciliés. Arch. zool. exp. gén. (2), 7: 149-517.

Moewus, F. 1933. Untersuchungen über die Sexualität und Entwicklung von Chlorophyceen. Arch. Protistenk., 80: 469-526.

—— 1934. Über Subheterözie bei Chlamydomonas eugametos. Arch. Protistenk., 83: 98-109.

—— 1935a. Über den Einfluss äusserer Faktoren auf die Bestimmung des Geschlechts bei Protosiphon. Biol. Zbl., 55: 293-309.

—— 1935b. Die Vererbung des Geschlechts bei verschiedenen Rassen von Protosiphon botryoides. Arch. Protistenk., 86: 1-57.

—— 1935c. Über die Vererbung des Geschlechtes bei Polytoma Pascheri und bei Polytoma uvella. Z. indukt. Abstamm. u. VererbLehre, 69: 376-417.

—— 1936. Faktorenaustausch, insbesondere der Realisatoren bei Chlamydomonas-Kreuzungen. Ber. dtsh. bot. Ges., 54: 45-57.

—— 1937a. Methodik und Nachträge zu den Kreuzungen zwischen Polytoma-Arten und zwischen Protosiphon-Rassen. Z. f. indukt. Abstamm.- u. VererbLehre, 73: 63-107.

—— 1937b. Die allgemeinen Grundlagen der Sexualität. Biologe, 6: 145-51.

—— 1938a. Vererbung des Geschlechts bei Chlamydomonas eugametos und verwandten Arten. Biol. Zbl., 58: 516-36.

—— 1938b. Carotinoide als Sexualstoffe von Algen. Jb. wiss. Bot., 86: 753-83.

—— 1939a. Untersuchungen über die relative Sexualität von Algen. Biol. Zbl., 59: 40-58.

—— 1939b. Carotinoide als Sexualstoffe von Algen. Naturwissenschaften, 27: 97-104.

—— 1939c. Über die Chemotaxis von Algengameten. Arch. Protistenk., 92: 485-526.

—— 1940. Carotinoid. Derivate als beschlecktsbestimmende Stoffe von Algen Biol. Zbl., 60: 143-66.

Morgan, T. H. 1938. The genetic and physiological problems of self-sterility in Ciona. I and II. J. exp. Zool., 78: 271-334.

Mulsow, W. 1913. Die Conjugation von Stentor coeruleus und Stentor polymorphus. Arch. Protistenk., 28: 363-88.

Noland, L. E. 1927. Conjugation in the ciliate Metopus sigmoides. J. Morph., 44: 341-61.

Pascher, A. 1931. Über Gruppenbildung und "Geschlechtswechsel" bei den Gameten einer Chlamydomonadine (Chlamydomonas paupera). Studien und Beobachtungen über die geschlechtliche Fortpflanzung und den Generationswechsel der Grünalgen. I. Jb. wiss. Bot., 75: 551-80.

Philip, V., and J. B. S. Haldane. 1939. Relative sexuality in unicellular Algae. Nature, 143: 334.

Prandtl, H. 1906. Die Konjugation von Didinium nasutum, O. F. M. Arch. Protistenk., 7: 229-58.

Pringsheim, E. G., and K. Ondraček. 1939. Untersuchungen über die Geschlechtsvorgänge bei *Polytoma*. Beih. bot. Zbl., 59A: 117-72.

Satina, S., and A. F. Blakeslee. 1930. Imperfect sexual reactions in homothallic and heterothallic Mucors. Bot. Gaz., 90 (3): 299-311.

Sonneborn, T. M. 1936. Factors determining conjugation in *Paramecium aurelia*. I. The cyclical factor: the recency of nuclear reorganization. Genetics, 21: 503-14.

—— 1937. Sex, sex inheritance and sex determination in *Paramecium aurelia*. Proc. Nat. Acad. Sci. Wash., 23: 378-85.

—— 1938a. Mating types in *Paramecium aurelia:* diverse conditions for mating in different stocks; occurrence, number and interrelations of the types. Proc. Amer. Phil. Soc., 79: 411-34.

—— 1938b. Mating types, toxic interactions and heredity in *Paramecium aurelia*. Science, 88: 503.

—— 1939a. Sexuality and related problems in *Paramecium*. Coll. Net, 14: 77-84.

—— 1939b. *Paramecium aurelia:* mating types and groups; lethal interactions; determination and inheritance. Amer. Nat. 73: 390-413.

—— 1939c. Genetic evidence of autogamy in *Paramecium aurelia*. Anat. Rec., 75, (suppl.), p. 85.

Sonneborn, T. M., and R. S. Lynch. 1934. Hybridization and segregation in *Paramecium aurelia*. J. exp. Zool., 67: 1-72.

Stout, A. B. 1938. The genetics of incompatibilities in homomorphic flowering plants. Bot. Rev., 4: 275-369.

Woodruff, L. L., and R. Erdmann. 1914. A normal periodic nuclear reorganization process without cell fusion in *Paramecium*. J. exp. Zool., 17: 425-518.

Zweibaum, J. 1921. Richerche sperimentali sulla conjugazione degli Infusori. II. Influenza della conjugazione sulla produzione dei materiali di riserva nel *Paramecium caudatum*. Arch. Protistenk., 44: 375-96.

CHAPTER XV

INHERITANCE IN PROTOZOA

H. S. Jennings

In his *Genetics of the Protozoa* (1929), the author has reviewed some-what fully the investigations and literature on inheritance in Protozoa, up to 1929. No attempt is made to repeat here these detailed reviews; the plan is rather to summarize the present state of knowledge on the subject. Very great advances have been made since 1929, particularly in the knowledge of biparental inheritance, largely through the work of Moewus (1932-38).

The question dealt with in the study of inheritance is: To what extent and how are the constitutions and characteristics of later genera-tions affected by the constitutions of their ancestors, particularly by the constitutions of the immediate parents? Certain subordinate questions arise in connection with this: To what extent and how are characteristics affected by environmental conditions? What are the relations between environmental modifications and genetic constitution?

By genetic constitution is meant the constitution insofar as it affects descendants. The genetic constitution is known from studies of multi-cellular organisms to be embodied in certain genetic materials. These are, mainly or entirely, found in the chromosomes. General genetics has shown (1) that in the chromosomes there are great numbers of diverse genetic materials (known commonly as genes or factors), having dif-ferent effects on development and characteristics; and (2) that genetic materials are transferred bodily from parents to offspring.

In the present account the term "factors" will usually be employed in place of the term "genes," since the latter has acquired, of late, certain doubtful theoretical implications.

Genetic materials have two essential properties: (1) the genetic ma-terials received from parents affect the development and characteristics of the descendants; (2) the many different kinds of genetic materials (genes or factors) reproduce themselves true to type, in development and reproduction. Each kind of genetic material assimilates, producing more material of its own type, and each unit of material, or gene, pro- duces at division new units like itself.

The Protozoa have chromosomes that are similar to those in other organisms (see Chapter XII). One question that arises in protozoan genetics is this: Are there in the Protozoa other genetic materials in ad- dition to those in the chromosomes, having the two essential properties just mentioned?

Types of Reproduction and Inheritance

In the Protozoa, as in some other organisms, there are two main types of relation of offspring to parents:

1. Uniparental reproduction; offspring arise from a single parent, as in the various types of vegetative reproduction.

2. Biparental reproduction; offspring are formed from the combined parts of two parents, as in sexual reproduction in the Protozoa by copu- lation, conjugation, and the like—followed by division.

The two kinds of reproduction differ fundamentally in their relation to the genetic constitution, or genetic materials (chromosomes and their genes). In uniparental reproduction, typically each of the genetic mate- rials of the parent is divided and duplicated, so that the genetic consti- tutions of offspring are like those of their parents. In biparental repro- duction, the complex of genetic materials present in each of the two parents is taken apart, and a new combination is made from parts of these. The genetic materials of the offspring are a new combination of those of the two parents.

In consequence of these differences, uniparental and biparental repro- duction give very different consequences in inheritance. The two will therefore be dealt with separately.

Inheritance in Uniparental Reproduction

MATERIAL PROCESSES

Details as to the material processes in uniparental reproduction are dealt with in other chapters (see Chapters XIII and XIV). The essential features, for genetics, are that the nuclei divide, each chromosome divides, each gene divides—one product from each going to each of the two offspring. In the ciliate Infusoria, the macronucleus not only divides but is in many cases reorganized (see Chapter XIII). The cytoplasmic body divides and is to a great extent (or entirely) reorganized. The general upshot is that the constitution of nucleus and cytoplasm is typically the same in the offspring as in the parent (exceptional conditions are dealt with in later pages).

INHERITANCE OF CHARACTERISTICS

Clones.—All the individuals produced by uniparental reproduction from a single individual are known collectively as a *clone*. The general rule for inheritance in uniparental reproduction is that all members of the clone are alike in genetic constitution and in inherited characteristics. That is, the new individuals (clone) produced from a single parent are like the parent and like one another in their characteristics, structural and physiological. Taken together, they form the equivalent of a set of identical twins.

There are numerous exceptions to this rule of the genetic identity of parent and offspring in uniparental reproduction, and these are among the most important and interesting phenomena of genetics. They are dealt with fully on later pages. But the relation of identity of genetic constitution in parent and offspring holds for perhaps 99.9 percent of all cases; it is the most striking feature of uniparental reproduction.

Certain manifestations of this principle of identity in genetic constitution between parent and offspring require special consideration:

1. *Biotypes.* In all Protozoa fully studied, any species consists of a great number of diverse biotypes—races differing in inherited characteristics. The different biotypes may differ in size, form, structure, and physiology (rate of multiplication and the like). Such diverse biotypes in *Paramecium, Difflugia, Arcella,* and other Protozoa are described and illustrated in the present author's *Genetics of the Protozoa* (1929).

When individuals of diverse biotypes reproduce uniparentally, as by fission, the general rule is that each biotype retains its characteristics. The offspring are like the parents in all conspicuous respects. Thus all members of a single clone belong to the same biotype and have the same inherited characteristics. In biotypes of large individuals, each individual produces a clone of large individuals; biotypes of small individuals give clones with small individuals; rapidly multiplying biotypes produce rapidly multiplying descendants; and so on. Such inheritance is shown with respect to vigor or weakness, to resistance and lack of resistance, and to structural and physiological characteristics of all sorts.

Members of a given biotype, having the same genetic constitution, may differ in ways induced by different environments, or resulting from different periods in the life of the individual. Such differences are, as a rule, not inherited in uniparental reproduction (exceptions are dealt with later). The main classes of non-heritable differences among the individuals of a single biotype are: age differences; nutritional differences, and environmental diversities resulting from differences in temperature, chemical conditions, and the like.

In addition to these, there are in some species non-heritable diversities of unknown origin between members of the same biotype, the same clone. Thus in *Difflugia corona,* which has a silicious shell bearing spines, there are within the same clone differences as to the number and size of the spines borne by the shell. In this case the differences arise at reproduction, presumably under the influence of environmental diversities. They follow the same rule as known environmental differences; they are not as a rule inherited. If parents with many spines produce descendants, the mean number of spines in these descendants is the same as in the descendants of individuals of the same clone that have few spines (exceptions noted in later pages).

Thus, as a rule, racial or inherited characters are not altered in uniparental reproduction. This is the most striking and obvious feature of such reproduction. Yet it does not hold absolutely; there are important limitations and exceptions to this rule. A large proportion of our discussion of uniparental inheritance will deal with these exceptions. They are taken up next.

Changes in Inherited Characters in Uniparental Reproduction

In a number of different categories of cases, inherited differences arise during uniparental reproduction, so that the members of a single clone are not all alike in characteristics that are inherited in vegetative reproduction. Some of these phenomena are of great interest for general genetics. They may be classified in various ways.

Beginning with an individual that has recently conjugated, if the lines of descent by vegetative reproduction are followed for great numbers of generations, certain characteristics of the individuals are found gradually to alter. The offspring produced at different periods differ. In *Uroleptus mobilis* (Calkins, 1919) or in *Paramecium bursaria* (Jennings, 1939), the individuals are at first sexually immature; they do not conjugate under any conditions. This continues for many generations of vegetative multiplication. The offspring during this period are like the parents in this respect.

But after many generations have passed, the descendants gradually become sexually mature. They now conjugate when mixed with individuals of different mating type. These descendants are thus different in this respect from their earlier ancestors. In this period their own offspring inherit from them the mature condition.

In *Paramecium bursaria,* and presumably in other species, the mature condition comes on slowly and gradually. There is for many generations partial maturity, in which the tendency to conjugate is but slight. The tendency becomes stronger as generations pass, until full maturity is reached. The period of full maturity lasts for a great number of generations, during which the mature condition is inherited in vegetative reproduction. Such periods of immaturity and maturity were described fifty years ago by Maupas (1889) for a number of species of ciliates. In some species, however, they hardly exist, or the period of immaturity if it occurs at all is very short. Such is the situation in *P. aurelia* (Sonneborn, 1936).

At a late period in the life history, in some species the individuals are found to become less vigorous as generations pass. They multiply less rapidly, become "depressed," degenerate. Whether this is an additional period in the life history, beyond the periods of immaturity and maturity; whether, in other words, it is an age change, constituting a period of senescence and final senility, or whether it is only a degenerate condition arising in consequence of living long under unfavorable conditions, appears as yet unsettled. This period of decline will therefore be considered in the next section.

In some of the Protozoa, particularly among parasitic forms, in different periods of the life history there are very great differences in form, structure, and physiology, constituting an "alternation of generations." Each condition is transmitted from parent to offspring for many generations, yet each in time transforms into a later condition.

All these phenomena are commonly thought of as matters of "life history," rather than of inheritance. Yet they represent fundamental features in those relations of successive generations that are called inheritance. The single cell, reproducing vegetatively, produces a great number of other free cells that are like itself in their special peculiarities. Later the character of the cells changes; and again the resulting condition is for a long period inherited in vegetative reproduction. In these respects the phenomena are like modifications resulting from environmental action, as shown in later paragraphs.

Are the diverse conditions that are vegetatively inherited in different periods—such as sexual immaturity and maturity—the result of changes in chromosomal materials, or changes in the cytoplasm? Dobell (1924) shows that the chromosomes do not visibly change in the series of diverse forms passed through in the life history of certain haploid Sporozoa; throughout all the changes the same set of chromosomes in the same number are present. Tartar and Chen (1940) have found that in the period of sexual maturity, in *P. bursaria,* parts of the individual consisting only of cytoplasm react sexually. Neither of these observations proves conclusively that the chromosomal materials are not altered in the different periods, but they perhaps make it probable that the different periods in the life history result rather from such interactions between chromosomes and cytoplasm as must occur in producing the bodily differentiations of a developing multicellular organism.

Whatever the seat of the different inherited conditions in different periods of the life history, it is clear that the material on which the different conditions depend must multiply itself, for long periods remaining true to type. An immature individual contains a certain small amount of the material on which immaturity depends. In ten generations this material has multiplied to more than a thousand times its original quantity, still remaining immature. Later, having attained the mature condition, it again multiplies in that condition to thousands of times its original quantity.

INHERITED DEGENERATIVE CHANGES RESULTING FROM UNFAVORABLE
CONDITIONS

In many cases, when ciliate Infusoria are cultivated for long periods in isolation cultures, in which great numbers of successive generations are produced, the organisms are found in the later generations to decline in vigor and vitality. This change is progressive; it becomes greater in later generations. The vital processes become "depressed," slow, inefficient; in particular the rate of multiplication decreases. In time the animals become degenerate—abnormal in form and structure, reduced in size. As an index of this decline in vigor and vitality, the changes in the rate of multiplication are commonly employed. Graphs of the daily number of fissions show a curve gradually descending from a high point at the beginning of the isolation culture, to nearly zero at a later period. A large number of such graphs, based on the work of many different investigators on many species, are published in the author's *Genetics of the Protozoa* (1929).

It has been held by many investigators that this decline is a matter of age; that these graphs are curves of senescence. The earlier periods of immaturity and maturity were believed to be followed inevitably by a period of senescence. Whether this is, indeed, true for some species is still uncertain. But for a number of species it has been shown that the decline need not and does not occur if the conditions are kept entirely favorable (a summary of investigations on this matter is found in the author's *Genetics of the Protozoa,* 1929). In these latter species, therefore, the decline and degeneracy are consequences of life under unfavorable conditions.

Thus unfavorable environmental conditions, acting for many successive generations, cause changes in the characteristics of the individuals, and cause them to produce in the later periods offspring that differ from those produced in the earlier periods. In the earlier periods, parents and offspring are vigorous, multiplying rapidly. Later, under the same environmental conditions as before, parents and offspring are weak, nonresistant, multiplying slowly. The effects of the unfavorable environment become cumulative as generations pass, and in vegetative reproduction they are transmitted to the offspring.

The inheritance of the depressed condition is demonstrated in the following way. Individuals from the depressed later generations are cultivated side by side, under the same conditions, with individuals from the same clone that have not lived under unfavorable conditions. One of the two sets—the latter—multiplies rapidly and at a high level of vitality. But the former set, that has lived under unfavorable conditions, multiplies slowly, at a low level of vitality, in a degenerate condition.

Thus in these Protozoa we find realized what some have held must occur in mankind: the production of inherited degeneration, by long-continued bad living conditions.

Discussion of the nature of these changes will be reserved until other inherited environmental modifications have been considered.

INHERITED ACCLIMATIZATION AND IMMUNITY

The changes in inherited characters induced by unfavorable environmental conditions are not always degenerative in character. In the unicellular organisms, as in multicellular organisms, long exposure to unfavorable conditions may result in the production of acclimatization or immunity. In the Protozoa, after removal from the unfavorable conditions, the acquired acclimatization or immunity is inherited in vegetative reproduction for many generations. Cases are on record in which such inheritance continued for many months, including hundreds of vegetative generations.

But in the course of many generations under the favorable conditions, with the injurious agent no longer present, the acquired immunity becomes gradually less marked; it slowly decreases, and, finally, in a sufficiently long period it is lost. But this may not occur until months after removal from the immunizing agent, during which time the acquired immunity is inherited.

Such acquisition of inherited immunity is most extensively known in parasitic Protozoa and in pathogenic bacteria, since in these organisms it is of medical importance. Detailed accounts of the knowledge in these fields will be found in the treatise of Taliaferro (1929).

But acquisition and inheritance of acclimatization or immunity occurs also in free-living Protozoa. A somewhat detailed review and discussion of investigations in this field will be found in the author's *Genetics of the Protozoa* (1929). Here only brief summaries of some of the more important investigations in this field can be presented.

In a famous investigation by Dallinger (1887), published fifty-two years ago, three species of flagellates were acclimatized, in seven years, to a high degree of heat. At first the organisims could not tolerate a temperature higher than 26° C. If subjected to this temperature for some time, death occurred. By raising the temperature a half degree at a time, at the average rate of two degrees a month, the tolerance was in the seven years raised to 70° C.

When the temperature was raised at any period, usually many of the individuals died. Others lived and multiplied, replacing those that died. There was thus a selective action of the heat; the individuals that did not become acclimatized died.

At certain periods, the cytoplasm of the organisms became filled with small vacuoles, which lasted for a month or more, then disappeared. The temperature could not be raised further until the vacuoles had disappeared. It has been suggested that this vacuolization is a process of getting rid of water, since it is known that protoplasm containing little water can tolerate higher temperatures than protoplasms containing much water.

In cases of acclimatization to high temperatures, it is obvious that all parts of the organism must be altered. If any part—nucleus, chromosomes, cytoplasm—failed to acquire the increased resistance to heat, the organisms would die.

Extensive experimental studies in this field were carried on for many years by Jollos (1913, 1920, 1921). He investigated the acclimatization of *Paramecium caudatum* to arsenic (solutions of arsenious acid), to certain other chemicals, and to high temperatures.

The most extensive work was on acclimatization to solutions of arsenic. It was found that *P. aurelia* could not be acclimatized to arsenic, and that some biotypes of *P. caudatum* were likewise refractory. But in other biotypes of *P. caudatum* acclimatization was successfully produced.

In the successful experiments, a method combining selective action with subjection to gradually increasing concentrations of arsenic was employed. Using as a standard concentration a one-tenth normal solution of arsenious acid, the animals were first placed in a very weak concentration, the maximum tolerance at the beginning being about one percent of the standard solution. The organisms were left for a long time (days or months) in a solution too weak to destroy them. Then the concentration was slightly increased, until many were killed. The few that survived were restored to a weak solution and allowed to multiply. Again there was an increase in concentration until most died; then a restoration to a weak solution. As this continued, it was found that the animals became able to tolerate higher concentrations. In three or four months, the toleration was thus increased from about one percent to 2.5 percent of the standard solution. In other cases the tolerance was raised in several months to 5 or 6 percent.

When the organisms that had thus acquired a higher resistance to arsenic were restored to water containing no arsenic, the increased resistance was for long periods not lost. Tests at intervals showed that they still retained the higher tolerance to arsenic. The acquired higher resistance lasted in some cases for eight months or more. As the animals were reproducing at about the rate of one fission daily, the acquired resistance was inherited for about 250 generations.

But during this time in water the acquired resistance to arsenic gradually decreased. The rate of decrease was very slow, so that in such a case as that mentioned above, the tolerance had not returned to its original low level until after a period of eight months.

The course of events may be illustrated from the history of Jollos's clone A of *P. caudatum*. In this clone the original maximum tolerance to arsenic was to 1.1 percent of the standard solution. By cultivation in gradually increasing concentrations for four months, the maximum tolerance was raised to 5 percent. Upon restoration to water containing no arsenic, the tolerance was, in tests for successive periods after the restoration, as follows: 6 days, 5 percent; 22 days, 5 percent; 46 days, 4.5 percent; 53 days, 4.7 percent; 60 days, 4.7 percent; 75 days, 4.7 percent; 130 days, 4.5 percent; 144 days, 4.2 percent; 151 days, 4.0 percent; 166 days, 3.0 percent; 183 days, 2.5 percent; 198 days, 1.25 percent; 255 days, 1.0 percent.

Thus the acquired resistance persisted in a very marked degree for more than five months, but had been entirely lost at the end of eight and a half months.

Jollos acclimatized *Paramecium* also to high temperatures, continuing the experimental cultures in some cases as long as two and a half years. The tolerance to high temperatures was increased, and the increased tolerance lasted in some cases to six months after removal from the high

temperatures. As in the case of resistance to arsenic, the acquired tolerance slowly decreased and finally disappeared when the organisms were
cultivated at moderate temperatures.

To such long-lasting modifications, inherited for many generations
but finally disappearing, Jollos gave the name *Dauermodifikationen*. This
designation is much employed, even in languages other than German.
Jollos interprets these modifications as affecting only the cytoplasm, not
the chromosomes; a matter to which we shall turn later. Phenomena of a
similar character have been described by Jollos and others in multicellular organisms. Hämmerling (1929) has published an extensive summary of what are believed to be *Dauermodifikationen* in many organisms.

Certain additional features of acclimatization in free-living Protozoa
are brought out in the work of Neuschloss (1919, 1920). He investigated the acclimatization of *P. caudatum* to certain chemicals, particularly
to quinine, methylene blue, trypan blue, fuchsin; also to arsenic and antimony. He found that he could induce increased resistance in about a
month by gradually subjecting the animals to increasing concentrations
of the substances.

Neuschloss investigated the nature of the changes in the organisms
in the following way. After acclimatizing the organisms to an injurious
substance, two equal samples of that substance were taken, in concentration somewhat greater than that to which the animals were resistant.
To one of these was added a large number of the acclimatized paramecia,
to the other an equal number of the unacclimatized animals. The animals
were in each case left in the samples until death occurred. After death,
the amount of injurious substance that had been removed by the organisms from the solution was quantitatively determined.

In all cases it was found that the acclimatized animals removed from
solution a much greater proportion of the injurious substance than did
the unacclimatized. In the case of the four organic compounds, the percentage removed from the solutions by the acclimatized and the unacclimatized animals were as follows:

Injurious Substance	Acclimatized Animals	Unacclimatized Animals
Quinine	80.0	4.5
Methylene blue	52.5	0.5
Trypan blue	54.0	0.75
Fuchsin	60.0	1.00

The substances removed from solution were found not to exist in the bodies of the animals. It appears, therefore, that the acclimatized animals in some way destroy the injurious organic compounds. Presumably they produce some secretion that has this effect.

In the case of the inorganic poisons, arsenic and antimony, the acclimatized animals were found to have acquired the power to transform the highly injurious trivalent compounds of the two elements into the relatively harmless pentavalent compounds.

The acclimatization was found to be, as a rule, specific for the substance to which the animals had been exposed. Acclimatization to one of the organic substances did not increase resistance to the other three. But in the case of the related inorganic materials arsenic and antimony, acquirement of increased resistance to one induced acclimatization to the other.

INHERITED ENVIRONMENTAL MODIFICATIONS IN FORM AND STRUCTURE

Reynolds (1923) and Jollos (1924) studied extensively the inheritance of certain abnormal forms of the shell in *Arcella;* an extended review of this work is given in the author's *Genetics of the Protozoa* (1929).

The work of Jollos shows that these abnormalities are favored by certain environmental conditions, and perhaps makes it probable that they originate as environmental modifications. Inheritance of the abnormal conditions continued for many generations.

By selection, the grades or degrees of abnormality could be changed. By long selection of the most nearly normal individuals, a stock that was nearly or quite normal could be produced. According to Jollos, the length of time required to bring the animals by selection back to a normal condition is proportional to the length of time that the ancestors have lived under the environmental conditions that favor the abnormality.

When the abnormal stock is allowed to multiply without selection, the proportion of individuals with but slight degrees of abnormality increases. Jollos holds that this indicates that the abnormal condition is essentially transient, a *Dauermodifikation,* in which the cytoplasmic tendency to abnormality is in time overcome by the nuclear tendency toward normality. On the other hand, the higher degrees of abnormality are harmful to the organism, so that natural selection tends to weed these out, causing the stock to become less abnormal in later generations. See the extended presentation of the evidence and the discussion on this point in the author's *Genetics of the Protozoa* (1929, pp. 261-70).

Moewus (1934b) has made extensive studies of inherited environmental modifications in structural characters in *Chlamydomonas debaryana*. This species is known to occur in many varieties differing slightly in form, size, and structure; Moewus describes and figures twelve of these. The different varieties are found in nature; also they occur under different cultural conditions in the laboratory.

Moewus found that he could transform the different varieties one into another by the use of different culture media. But the transformation did

TABLE 14: ENVIRONMENTAL MODIFICATIONS, *Chlamydomonas debaryana;* RELATION OF THE NUMBER OF DAYS CULTIVATED IN PEPTONE MEDIUM TO THE NUMBER OF DAYS IN SALT-SUGAR MEDIUM REQUIRED TO TRANSFORM THE ORGANISMS FROM TYPE 1 TO TYPE 5 (FROM MOEWUS, 1934b)

Days in Peptone as Type 1	Days in Salt-Sugar Medium Required for Transformation to Type 5
28	28
140	49
273	133
441	175
567	231
609	370
644	459
672	531
690	534

not occur at once, on transfer to the different culture medium; on the contrary, many generations in the new culture medium were necessary to induce the transformation.

An extremely important relation was observed to hold as to the time required for the transformation. The longer a stock has remained under conditions producing a given type, the greater the time and the number of generations required to transform it to a new type, when placed under new conditions. An example will illustrate the operation of this principle.

Individuals in peptone culture are of type 1. If transferred to a "salt-sugar" infusion, they transform to type 5. If they have been only a few generations in the peptone culture, the transformation in the salt-sugar medium occurs very quickly. But if they have lived for a long time in the peptone medium, a long time is required for transformation to type 5.

Table 14 from Moewus shows the number of days required for the transformation to type 5, in relation to the number of days the organisms had lived in peptone medium. In interpreting this table, it is important to keep in mind the fact that reproduction occurs at about the rate of one generation a day, or more rapidly.

In the last examples of Table 14, it required a year and five months in their new conditions to induce in the organisms the transformation to type 5. During this period hundreds of generations were produced. The longer the organisms remain in peptone culture as type 1, the longer it requires for the new conditions to transform them to type 5.

Moewus presents many other cases illustrating the same principle.

The different types or varieties found in nature usually require a very long period of culture in a given medium to induce them to transform to the type characteristic for that medium. Thus a certain stock 4 was found in nature as type 10, having no papilla. Placed in peptone culture, it remains as type 10 for 450 generations, then the organisms begin to transform to type 1, having a papilla; in time all transform to type 1. The type 10 was seemingly a *Dauermodifikation* in nature; it transforms to type 1 only after many generations under artificial conditions. Moewus describes many other cases in which types found in nature resist, for many generations, changes due to new conditions, but finally transform into types characteristic for the medium to which they are transferred. Thus the indications are, as Moewus points out, that many of the diverse types ("varieties") found in nature are in fact long-lasting modifications resulting from the long-continued action of certain environmental conditions.

VARIATION AND ITS INHERITANCE OCCURRING WITHOUT OBVIOUS
ACTION OF DIVERSE ENVIRONMENTS

In addition to inherited changes produced by diverse environments, already described, there occur in some cases during vegetative reproduction inherited changes, without apparent action of diverse environments. Such cases possibly differ in principle from those described above, illustrating the occurrence of inherited variations that are not brought about by environmental action. If so, they are of much theoretical interest. On the other hand, they may involve concealed action of diverse environmental conditions, as urged by Jollos (1934). The phenomena are in themselves of interest, illustrating certain methods of action not thus far described. The author's studies on *Difflugia corona* may serve as an example (Jennings, 1916, 1929).

In *D. corona* (Fig. 168) the individuals differ greatly in a number of strongly marked structural features. They have a silicious cell, produced

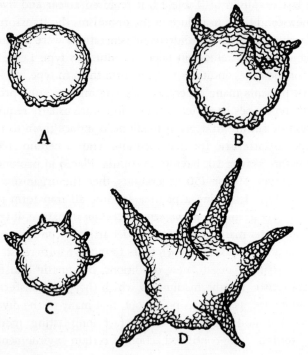

Figure 168. *Difflugia corona*. Members of four different clones, showing diversities in characteristics. (After Jennings, 1916.)

at the time of reproduction. The shell bears conspicuous spines, which differ greatly in number and in length in the different individuals. The number of teeth surrounding the mouth differs in different cases. The size of the shell shows wide variation.

Many diverse biotypes occur in nature, each characterized by a definite combination of characteristics. Some have few spines, some many. Some have long spines, others short spines. Some have few teeth, others many teeth. Some are large, others small.

Even within a single clone there is variation in characters, though much less than when different clones are taken into consideration. Different individuals of the same clone may differ greatly in number and length of spines, slightly in size of the body, very slightly in the number of teeth.

In vegetative reproduction, the offspring resemble the parent, but not completely. The general rule is that parent and offspring are alike in respect to characters that distinguish different biotypes or races, but they need not be alike with respect to characters in which different individuals of a clone are diverse. The type of inheritance will be seen from examples of the numbers of spines in successive generations in certain clones:

        Clone 1: 2—2—5—4—6
        Clone 2: 1—2—1—4—2—3
        Clone 3: 6—6—6—6—4—5—7—6
        Clone 4: 0—0—0—0—1—0—0—1—0

Thus clones differ from each other in the fact that some usually have few spines or none, others a slightly larger number, others a still larger number—though within each clone there is variation. A given biotype, or clone, is characterized by a certain average number of spines; another by a different average number. In respect to the average numbers of spines, many different biotypes may be distinguished.

When from a single clone individuals differing in number of spines are allowed to multiply until each has produced many descendants, the usual result is that the average number of spines in the two sets of descendants is the same. Thus in clone 1 above, the individual with 2 spines and that with 6 spines would, as a rule, produce descendants having the same average number of spines. The individual differences within the clone are usually not inherited; the mean differences between biotypes are inherited.

To what are due the differences in numbers of spines or in other characteristics, between the different members of the same clone, illustrated above? The shell is produced complete at the time that reproduction occurs. The individual about to reproduce buries itself in a mass of soft debris. The protoplasm swells, projects from the mouth of the parental shell, and takes on the form of the protoplasmic body. A mass of silicious particles, which have earlier been collected within the parent, passes to the surface of the new protoplasmic body and is there molded into the final form of the shell of the offspring. Meantime the nucleus has divided and one of the products passes into the young individual. Parent and offspring now separate.

During this process are formed the spines in their final form, length, and number. The parent shell may have four spines, the shell of the new individual five spines or some other number; there may be differences in the length of the spines. The differences between the characteristics of parent and offspring might conceivably be due to slight and obscure differences in the environmental conditions at the time of reproduction. Or they might be due to diversities in the internal condition or constitution of the individuals at the time of reproduction, however such diversities had been produced.

Results of long-continued selection.—From the results thus far set forth, it appears that the individual diversities within a clone of *Difflugia* are not usually or strongly inherited. Is there any tendency whatever for such inheritance? In other words, does the difference of condition or constitution that results in diversity of characters ever or in any degree persist through vegetative reproduction?

This may be tested by long-continued selective breeding within a clone. Suppose that the basis of selection is the number of spines. From a single clone two groups are segregated, one containing only individuals with high numbers of spines, the other only individuals with low numbers of spines. In the former group are retained only descendants with high numbers of spines; from the second group only descendants with low numbers of spines. In this way there are obtained in later generations from the first group individuals descended for many generations from ancestors all of which had high numbers of spines, and from the second group individuals descended for many generations from ancestors with low numbers of spines. Selective breeding of this kind may be carried on with any of the varying characteristics.

When such selection is practiced for many generations, usually for considerable periods no difference in the offspring of the two groups is to be discovered. But as the number of generations becomes greater, an effect in the direction of selection appears. The offspring of the group whose ancestors have had high numbers of spines have a greater average number of spines than the offspring of the group whose ancestors have had low numbers of spines.

Such results may be illustrated from the clone 326, in which selective breeding for high and low numbers of spines was carried on for a long period (see Table 15).

TABLE 15: EARLY RESULTS OF SELECTION FOR LOW AND HIGH NUMBERS OF SPINES, *Difflugia corona* (FROM JENNINGS, 1916)

Period	Days	Low Group		High Group	
		Parental Spines	Mean Spines of Descendants	Parental Spines	Mean Spines of Descendants
1	36	1–4	5.56	5–9	5.46
2	60	1–4	5.13	6–11	5.19
3	60	1–4	5.56	6–11	5.46
4	120	1–5	5.29	7–11	5.79

In the fourth period there is a distinct difference in the two groups. To test whether this is significant, the fourth period (four months) was divided into six successive periods, and the numbers of spines in the offspring of the two groups determined for each. They were as shown in Table 16.

TABLE 16: LATER RESULTS OF SELECTION FOR LOW AND HIGH NUMBERS OF SPINES, *Difflugia corona;* NUMBERS OF SPINES IN THE PARENTS (1-5 IN THE LOW GROUP, 7-11 IN THE HIGH GROUP) WITH THE MEAN NUMBERS OF SPINES ON THEIR OFFSPRING IN SUCCESSIVE PERIODS (FROM JENNINGS, 1916)

	PARENTS	
	1–5	7–11
Period	Offspring	
1	5.59	6.11
2	5.22	5.71
3	5.49	6.53
4	5.23	5.51
5	5.15	5.38
6	4.63	5.59

Thus in each of the six periods of the last four months, the parents with high numbers of spines produced offspring with higher mean numbers of spines than did parents with low numbers of spines, though the differences were not very great.

Selective breeding of this type was practiced also with relation to the length of the longest spines. The results here show clearly a phenomenon of much interest, namely inherited variation with "regression toward the mean" of the biotype as a whole. This appeared also in the results of

TABLE 17: INHERITANCE OF SPINE LENGTH, WITH REGRESSION TOWARD THE MEAN, *Difflugia corona,* CLONE 326; MEASUREMENTS OF SPINE LENGTHS, IN UNITS OF 4 2/3 MICRONS EACH (FROM JENNINGS, 1916)

Parents Length of Longest Spines, in Units	Number of Offspring	Mean Lengths of Longest Spine in the Offspring, in Units
4–6	21	10.38
7–9	162	11.01
10–12	451	11.85
13–15	367	12.90
16–18	129	14.39
19–21	26	14.34
22–24	15	16.34
25 and above	18	17.06

Mean length for all, 12.54

selective breeding with respect to other characters. The results of selection for length of spines are given in Table 17. (The spines were measured in units, each of which was 42/3 microns.)

Table 17 shows that parents with longer spines produce, on the average, offspring with longer spines, so that there is a distinct tendency to inheritance in spine length, even within the single clone.

But another relation is equally evident. Parents that are above the mean of the biotype (12.54) produce offspring that are above the mean, *but not so much above the mean* as are the selected parents. Parents that are below the mean produce offspring that are below the mean, but not so much below the mean as are the selected parents. That is, inheritance of the parental peculiarities occurs, but always with regression toward the mean of the biotype. On the average, the offspring diverge from the racial mean in the same direction as do the selected parents, but not so far. Only a small part of the selected parents' divergence from the mean is inherited by the offspring.

To what is due such inheritance, with regression toward the racial mean? The interpretation is not clear, but the most natural conception appears to be the following. The divergence from the racial mean in the case of the selected parents is partly a matter of environment, partly a matter of genetic constitution. The latter is inherited by offspring, the former not. The fact that a part of the parent's peculiarity is inherited shows that variation in the genetic constitution itself occurs in vegetative reproduction and is inherited. It is, however, in large degree masked by variations due to environmental conditions, and not inherited.

As a consequence of the inheritance of some portion of the parents' peculiarities, in time the single clone or biotype may by selective breeding become differentiated into a number of biotypes, differing slightly in inherited characters. Five such hereditarily different biotypes of *Difflugia corona,* produced through selective breeding from a single one, are described and figured in the paper of Jennings (1916).

For such heritable change in the genetic constitution two types of interpretation may be suggested. On the one hand, the heritable changes may be due to obscure alterations in environmental conditions, so that these changes are comparable to those in acclimatization or other *Dauermodifikationen.* On the other hand, the changes may be attributed to irregularities occurring in the genetic materials—either "mutations" or slight irregularities of distribution—when reproduction occurs. At present it appears not possible to decide between these two interpretations.

SUMMARY AND INTERPRETATION

To recapitulate, in Protozoa, a number of different types of inherited changes may occur during vegetative reproduction, giving rise to individuals or lines of descent that have different characteristics which are transmitted to descendants in further vegetative reproduction.

First, there are changes that occur in the course of the normal life history—changes as to sexual maturity, and, in some species, profound alterations in structural and physiological characters ("alternation of generations"). Each stage in such a life history includes many successive generations, during which the distinctive features of that stage (immaturity, maturity, or the like) are transmitted to descendants.

Second, there are inherited degenerative changes, resulting from life under unfavorable conditions.

Third, there are adaptive changes, fitting the organism to a changed environment—inherited acclimatization, or immaturity.

Fourth, there are inherited changes in form and structure, apparently neither adaptive nor degenerative, occurring under the influence of specific environmental conditions.

Fifth, forming perhaps an additional type, there are inherited variations in form, size, and other characters, that are not obviously due to environmental conditions.

Some of these five types fall under Jollo's concept of *Dauermodifikationen:* changes produced under the influence of environmental conditions, inherited for many generations after removal from those conditions, but gradually fading out to final disappearance. This includes the third and fourth types above, and, according to Jollos's view, also the fifth. The first type obviously appears not to fall under this concept, and the second type is not known to disappear upon restoration to a more favorable environment.

Most of these types of modification may disappear or be altered upon the occurrence of sexual reproduction. In type 1, the condition of maturity changes at sexual reproduction into one of immaturity. In type 2, the inherited degenerative changes in many cases disappear at sexual reproduction ("rejuvenescence through conjugation"), but it is a notable fact that in some cases they do not. In type 3, the inherited acclimatization and immunity likewise often disappear at sexual reproduction, though again in some cases they do not. In types 4 and 5, the effect of sexual reproduction is not known.

These inherited modifications, so far as they can be brought under the concept of *Dauermodifikationen,* are held by Jollos to have their seat in the cytoplasm only, the genetic materials of the chromosomes being unchanged. This is based in the main on the fact that the inherited modifications grow less and finally disappear, when the organisms are restored to environments that do not produce the modifications. A change in the genetic materials of the chromosomes ("mutation"), it is held, would be permanent; it would not disappear after many generations in an altered environment. But a change in the cytoplasm, it is urged, would in the course of time be overcome and dominated by the unchanged constitution of the nucleus, bringing about a return to the original characteristics—as actually occurs. These inherited modifications are, on this view, essentially transitory conditions, forming no part of the system of genuinely inherited characters.

The fact that these modifications usually (though not always) disappear at sexual reproduction is likewise held to be due to the fact that they have their seat in the cytoplasm, though the logic of this is not entirely clear. As will be shown later, it is the nucleus, not the cytoplasm, that is directly changed in constitution at conjugation. The disappearance at conjugation is seemingly attributed to the profound making over of the cytoplasm that has been believed to occur at conjugation, producing rejuvenescence. The long-continued inheritance of the modifications during vegetative reproduction is held to be an example of cytoplasmic inheritance.

If we are here indeed dealing with cytoplasmic inheritance, this appears to demonstrate that in the Protozoa the cytoplasm partakes of the essential features of genetic material. These essential features, as before set forth are: (1) the fact that the material in question affects the inherited characteristics, and (2) the fact that this material multiplies true to type. The environmental modifications are, as we have seen, at times inherited for more than 200 generations after removal from the conditions that induced them. In 10 generations, the originally modified cytoplasm has been diluted to one-thousandth of its volume, in any individual; in 20 generations, to less than one-millionth of its volume. The rest of the cytoplasm of the individual is a product of cytoplasmic growth. Yet the environmental modification still persists. If the cytoplasm is the seat of the modification, the modified cytoplasm must reproduce itself in its modified condition at fission; otherwise its effect would have disappeared under the great dilution that it has undergone.

On the question as to whether cytoplasmic inheritance indeed occurs in these organisms, evidence will be presented in the section on biparental reproduction. There also a method will be indicated by which it may be determined whether the seat of these modifications is in the cytoplasm or in the nucleus.

Inheritance in Biparental Reproduction

Biparental or sexual reproduction includes the processes that lead to the formation of new individuals from the united parts of two earlier individuals. Biparental reproduction in the Protozoa occurs in two main types, known respectively as copulation and conjugation. Copulation is characteristic of haploid species; in it two haploid individuals unite completely to form a diploid zygote, which later divides with reduction, to form haploid individuals again. This method occurs in the Flagellata. In conjugation two diploid individuals exchange halves of their nuclei, including a haploid set of chromosomes; then the two separate and each continues thereafter to multiply by fission. This is the characteristic method in the Ciliata.

BIPARENTAL INHERITANCE IN HAPLOIDS: FLAGELLATA

Knowledge of the genetics of flagellates is largely due to the recent work of Moewus (1932-38).

In flagellates the single motile individual is haploid. In copulation two such haploid cells unite completely, to form a diploid cell. The two hapoid cells thus correspond functionally to two gametes, while the diploid cell is the zygote. The zygote is inactive; it secretes a wall about itself and becomes a cyst. Later, under favorable conditions, the diploid cyst divides twice by the two "maturation divisions." At one of these divisions chromosome reduction occurs, so that the four cells formed are haploid. At times additional cell divisions occur before the cells emerge from the cyst. The cyst wall dissolves and the haploid cells are freed. Each develops flagella and swims about as a free individual. On emerging from the cyst, the individuals are often called swarmers or swarm cells; each is potentially a gamete. These free cells commonly multiply vegetatively for many generations, the descendants of each original swarm cell forming a clone.

In any species or variety the haploid individuals or gametes are differentiated into two sexes; details as to this will be found in the chapter on "Sexuality in Unicellular Organisms" (Chapter XIV). In some species or varieties all members of the same clone are of the same sex, the different sexes being in different clones (dioecious species or races). In others both sexes are found among the individuals of a single clone

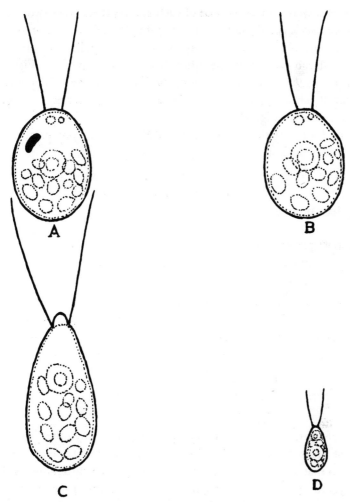

Figure 169. *Polytoma uvella* (A, B) and *P. pascheri* (C, D) ; the four races used in the breeding experiments of Moewus. (After Moewus, 1935b.)

(monoecious species or races). Details as to sex determination will be found in Chapter XIV; here sex will be dealt with only insofar as it plays a rôle in the general system of inheritance.

The flagellates exemplify in admirably simple form the system of inheritance in haploid organisms. The system is well exhibited in the studies of Moewus (1935b) on inheritance in *Polytoma*. In this work two species were employed, and of each species there were two varieties. *Polytoma uvella* is ellipsoidal in form, with no papilla at the anterior end (between the places of attachment of the two flagella). In one variety there is a large eyespot, in the other there is no eyespot (see Fig. 169A, B). In *P. pascheri* the form is a long oval, there is a papilla between the flagellar attachments, and there is no eyespot. One variety is large, the other very small (Fig. 169 C, D).

In the four varieties there are thus differences in four pairs of characters. Moewus designates these characters as follows:

Form of body: ellipsoidal, F (as in *P. uvella*)
long oval, f (as in *P. pascheri*)
Papilla: present, P
absent, p
Eyespot: present, S
absent, s
Size: large, D
small, d

Using these designations, the four varieties may be represented by formulae as follows:

P. uvella, with eyespot, FpSD
without eyespot, FpsD
P. pascheri, large, fPsD
small, fPsd

Any of these four varieties may be crossed with any other, so that six different crosses are possible. Each of the varieties has eight chromosomes, so that any cross forms a zygote with sixteen chromosomes. From each zygote arise, by the two maturation divisions, four swarm cells. These four, after becoming free, multiply vegetatively to form clones. All the individuals of each clone show the same characteristics as does the swarm cell from which the clone is derived, so that we may speak indifferently of the characteristics of each of the four swarm cells, or of each of the four clones derived from them.

When two varieties that differ in a single pair of characters are crossed, two of the four swarm cells produced show one of these characters, the other two the other character. Thus *P. uvella* with eyespot, S, is crossed with *P. uvella* without eyespot, s. Of the four swarmers arising from the zygote, two have the eyespot (S), two are without it (s). The cross may be represented as follows:

$$S \times s = S + s$$

It is obvious that the segregation of the two characters must have occurred at the reduction division. One of the sets of eight chromosomes includes the conditions for producing S, the other the conditions for producing s.

The results of the crosses in which the parents differ in two pairs of characters may be exemplified in the mating of *P. uvella* without an eyespot (FpsD) with the large variety of *P. pascheri* (fPsD) (Fig. 170 A, B). The two differ in form (F and f), and in presence or absence of the papilla (P and p). The zygote thus carries both these sets of factors; it is

FpsD
fPsD

The zygote divides into four swarm cells, with reduction. The results are as follows:

1. From any single zygote two and only two types or combinations are produced. Two of the four cells are of one combination, two of the other.

2. Different pairs of types are produced in different cases (Fig. 170). About half the zygotes yield again the two parental types, FpsD and fPsD (Fig. 170 C, D). The other half yield FPsD (ellipsoidal with papilla, Fig. 170 E) and fpsD (oval without papilla, Fig. 170 F).

Thus the characters, form, and papilla are inherited independently. The factors F and f are in the two members of one pair of chromosomes, the factors P and p in another pair of chromosomes.

When crosses are made in which the individuals differ in three pairs of characters (as FpsD $\times$ fPsd) or in all the four pairs of characters (FpSD $\times$ fPsd), all the four pairs are found to be independent in their distribution. From any single zygote only two types of offspring are produced. But from different zygotes of the three-pair cross, eight different combinations are produced; from those of the four-pair cross, sixteen different combinations. The different combinations occur in approximately equal numbers. Thus the factors for the four pairs of characters are distributed in four chromosome pairs, which at reduction are assorted independently.

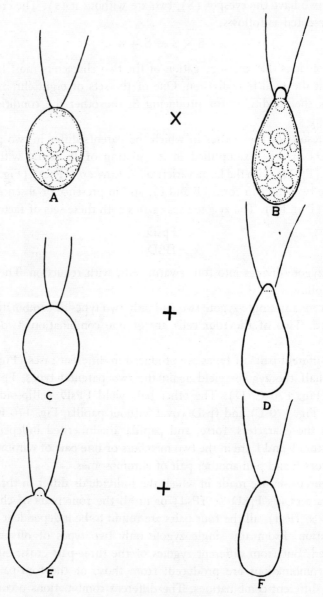

Figure 170. Results of a cross between two species of *Polytoma*. (Combined from figures of Moewus, 1936) A and B, the two parents; C and D, types produced by half the zygotes; E and F, types produced by the remainder of the zygotes.

In consequence of the independent inheritance of these four pairs of characteristics, there result from crosses certain curious combinations in which there is lack of harmony in the parts of the individuals. Thus the eyespot, S, is inherited independently of the size of the cell (D or d). The eyespot is originally in the large race, D, where its length is about one-sixth the length of the cell. By crosses it may be transferred to the small race, d. Here it retains its large size, so that there are produced small individuals with eyespots about half the length of the individual.

The fact that from any single zygote but two types or combinations appear among the offspring shows that the reduction of the chromosomes must occur at the first division of the zygote. If it occurred at the second division, there would be in some cases four different types or combinations from a single zygote.

For suppose that we have a two-factor cross, such as Fp $\times$ fP. Then the zygote has the combination FpfP. If reduction occurs at the first division, in some zygotes the two cells produced are Fp and fP. Each now divides equationally, giving two cells that are Fp, two that are fP. In other zygotes the reduction division yields FP and fp; again the second division yields two cells of each type. In either case but two types are produced from any one zygote.

But if reduction occurs at the second division, then after the first division there are two cells present, both with the combination FpfP. Now by reduction at the second division, one of these may yield Fp and fP, the other FP and fp, so that four different combinations would be produced from a single zygote.

Linkage and crossing over.—Besides the four independent pairs of characters just described, there are in *Polytoma* others that are linked with some of the four. Such a character is length of the flagella. In the original types, the length of the flagella is proportional to the length of the body; large cells (A, B, C) have long flagella, small cells (D) have short flagella (Fig. 169). This proportionality usually holds among the crosses; those that have large bodies have long flagella, those with small bodies have short flagella. This indicates either that the factors for size (D and d) and those for length of flagella are close together in the same chromosome, and so linked, or that the relation is merely a physiological one, cells of a given size always producing flagella of length proportional to the size.

To determine which of these alternatives is correct, Moewus made 1,400 crosses between the large and the small types (D and d). Of these, 1,357 yielded as usual zygotes which gave 50 percent large cells with long flagella, 50 percent small cells with short flagella. The remaining 73, or 5.2 percent of all, yielded zygotes that gave 50 percent large cells with short flagella, and 50 percent small cells with long flagella (Fig. 171). In other words, these 73 zygotes gave only cells carrying the new combinations. It appeared, therefore, that the factors for size of body and length of flagella are merely linked, through the fact that the two

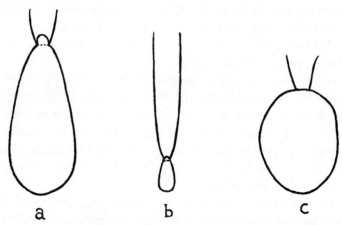

Figure 171. New combinations resulting from crossing over. a and c, large cells with short flagella, b, small cell with long flagella. (After Moewus.)

are close together in the same chromosome. Their distance apart is such as to yield crossing over in about 5.2 percent of the zygotes. This was confirmed by making 600 crosses between the new combinations, one parent having large body cells with short flagella, the other small body cells with long flagella (Fig. 171). Of the zygotes, 570 gave equal numbers that were like the parents, while 30 (or 5 percent) gave equal numbers of the original combinations—large body with long flagella, and small body with short flagella. These 30 therefore gave 100 percent crossovers.

In these cases an extraordinary situation appears, seemingly unique in crossing over. In all these cases, any zygote that yields any crossover combination yields exclusively crossovers. The 73 zygotes of the first 1,400 mentioned in the preceding paragraph yielded 50 percent of one of the crossover combinations (large cells with short flagella), 50 percent of the other crossover combination (small cells with long flagella). The same relations hold for the 30 crossover zygotes out of 600 in the recrossing of the new combinations; they yield 100 percent crossovers.

Such results can occur only if crossing over takes place between the two entire chromosomes that are in synapsis, that is, only if crossing over occurs in the so-called two-strand stage of the synapsed chromosome pair. In other organisms, so far as the matter has been analyzed, crossing over occurs only after the two synapsed chromosomes have split, so that there are four strands instead of two. Crossing over, then, occurs between but two of the four strands, with the result that two strands remain without crossovers. If this were the case in *Polytoma,* the zygotes that yield crossovers would yield but two crossover cells and two that were non-crossovers, in place of yielding only crossovers. The same situation is found in all the accounts of crossing over in Flagellata given by Moewus, up to his article of 1938. In this publication he states that he has observed in *Chlamydomonas* the occurrence of crossing over in accordance with the four-strand schema (details to be given later).

None of the four pairs of characters thus far considered was found to be linked with sex. That is, any of the alternative characters occurs equally frequently with either sex. If, therefore, sex depends on a chromosome pair, it is a fifth pair, not one of the four that carry the characters above discussed.

In an article of 1936, Moewus presents the results of extensive studies of inheritance in crosses of *Chlamydomonas eugametos* with *C. paupera.* Here eleven pairs of characters were distinguishable, some morphological, others physiological. In addition, there were sex differences, making twelve pairs of characters in all.

In the species of *Chlamydomonas* there are ten chromosomes, so that some of the twelve pairs of characters must have their factors in the same chromosome, and in fact Moewus discovered that some of the characteristics are linked. He reports that in the many crosses made, he analyzed the 8,000 haploid individuals derived from 2,000 zygotes resulting from crosses, and that he obtained 1,024 diverse types in such proportions as to show that each of the ten chromosomes bears the factors for one or more pairs of characters. Of most of these combinations no detailed accounts are given.

Linkage was found to exist between two physiological characters, (1) adaptation to acid or alkaline medium, with (2) differences in the number of cells into which the zygote divides before the cells are set free. Here, as in the former case, the results are such as to indicate crossing over in the two-strand stage. Any zygote that yields crossovers yields nothing but crossovers. Other cases of linkage involved factors for sex; these will be mentioned in the account of sex inheritance.

Certain general relations may be pointed out in the method of inheritance of non-sexual characters thus far presented, particularly as illustrated by *Polytoma:*

1. The inheritance is the typical Mendelian inheritance for haploid organisms. The characters manifested in the haploid descendants are combinations of characters that were manifested in the two parents. No characters appear in the offspring that were not manifested in the two parents, that is, no recessive characters occur in haploid organisms. In them all characters for which factors exist are manifested.

This is, of course, a consequence of the fact that in haploids the chromosomes are not in pairs, but single. Hence heterozygotes cannot occur (except in the diploid zygotes). In haploid inheritance, the combinations of characteristics that occur in the offspring depend wholely on the characteristics manifested in the immediate parents.

2. The inheritance, like that in multicellular organisms, follows the course that is to be expected if the different pairs of characters depend on factors present in the different pairs of chromosomes. Independent segregation, linkage, and crossing over are demonstrated in Protozoa. Crossing over is, however, as before mentioned, of an exceptional type.

SEX INHERITANCE AND SEX-LINKED INHERITANCE

In any species or race of the flagellates examined by Moewus, there are two sexes, or mating types. An account of these will be found in Chapter XIV of the present volume, "Sexuality in Unicellular Organisms." Moewus designates the two sexes on his earlier reports as plus and minus, in later publications as male and female, those earlier called minus being male, while the plus types are female (Moewus, 1937a). Here the different types of sex inheritance will be summarized, the account being based upon the work of Moewus.

In the flagellates investigated, some stocks are dioecious, others monoecious. Both types may occur in different races of the same species.

Dioecious races.—In these the sexes are in separate clones, all members of any one clone being of the same sex. The diploid zygote is formed by the union of two haploids of opposite sex and from different clones. By the two maturation divisions, the zygote divides into four cells. Two of these are always of one sex, two of the other. Thus the sexes are segregated at the reduction division, as if sex were determined by a single chromosome pair, one member of which produces one sex, the other the other sex. Sex is determined by the genetic constitution of the clone; sex determination is genotypic. Such dioecious stocks occur in certain races of *Chlamydomonas eugametos,* of *Polytoma pascheri,* and of *Protosiphon botryoides.*

Subdioecious races.—In certain races of *Chlamydomonas eugametos* there is found a modification of the dioecious condition. In any single clone, most of the individuals belong to one sex, a few to the opposite sex. Some clones are prevailingly plus, others prevailingly minus. A number of different matings are possible:

1. Plus gamete from a prevailingly plus race, minus gamete from a prevailingly minus race. Two of the cells from the zygote yield prevailingly plus clones, two yield prevailingly minus clones. The segregation of "prevailingly plus" from "prevailingly minus" therefore occurs at the reduction division; the difference is genotypic.

2. Plus and minus gametes from a prevailingly plus race. All clones from the zygote are prevailingly plus.

3. Plus and minus gametes from a prevailingly minus race. All clones from the zygote are prevailingly minus.

In other crosses, clones from a dioecious race, in which the clones are exclusively of one sex, are crossed with prevailingly plus or prevailingly minus clones from subdioecious races. The former will be spoken of as "pure" for sex, the latter as "mixed" for sex.

4. Plus gamete from a race pure for sex, minus gamete from a prevailingly minus clone of mixed race. Offspring: two pure plus, two prevailingly minus mixed.

In other crosses of this type, the results were similar; the "pure" condition segregates from the "mixed" condition at the reduction division. In general, the two types that unite to produce the zygote separate again at the reduction division; the differences are genotypic.

The fact that in single clones of the subdioecious races some individuals are plus and others minus is held by Moewus to be due to something in the surrounding conditions; the sex determination within the clone is phenotypic instead of genotypic, as in the dioecious races. But in the "prevailingly plus" clones the constitution is such that minus gametes are not so readily or numerously produced by the conditions as is the case in "prevailingly minus" clones. Moewus found that subjection to dilute formaldehyde or acetone causes the zygotes of the subdioecious races to produce gametes of only one sex, that sex which would have been in the majority if these substances had not been used. The precise constitution of subdioecious races, together with that of other types, is considered in a later section.

Monoecious races.—In monoecious races both sexes occur in a single clone, so that such clones may be spoken of as "mixed" as to sex. This is the situation in certain races of various species of *Chlamydomonas;* in *Polytoma uvella,* and in somes races of *P. pascheri* and of *Protosiphon botryoides.*

In monoecious stocks obviously the fission of a single individual gives rise to both sexes. The determination of sex in such stocks is largely or entirely phenotypic; that is, through external conditions. Such phenotypic sex determination is not dealt with in the present chapter.

When from monoecious stocks plus and minus gametes are mated, the descendant clones are all monoecious, that is, mixed as to sex.

Crosses between dioecious and monoecious stocks.—A number of different types of crosses may be made between dioecious and monoecious stocks, as follows:

1. Plus gamete from a "mixed" (monoecious) clone; minus gamete from a "pure" (dioecious) clone (cross of two diverse races of *Polytoma pascheri*). Result, two of the four descendant clones are mixed, two pure minus.

2. Plus gamete from pure clone, minus gamete from mixed clone (cross of two races of *P. pascheri*). Result, two clones mixed, two pure plus.

In these two crosses, two results are notable. (1) The pure condition segregates from the mixed condition at the reduction division, the two depending apparently on the two different chromosomes of a pair. The difference is genotypic. (2) The "pure" condition emerges with the same sex (plus or minus) as that with which it enters the cross. If the plus parent is pure for sex, it is the plus offspring that are pure; if the minus parent is pure, the minus offspring are pure.

Linkage and crossing over.—Certain crosses yield a small proportion of exceptional results, which are held to be due to crossing over. Such are the following:

Polytoma pascheri: pure plus clone by pure minus clone. Result, out of 2,000 zygotes, 1,843 gave four cells each (as usual), two of which were pure plus, two pure minus.

The other 157 (7.9 percent) gave but two cells each, and these all produced clones mixed as to sex.

The results given by the 1,843 zygotes are those to be expected from

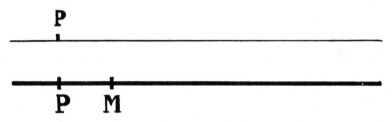

Figure 172. Diagram of the sex chromosome of *Polytoma pascheri* plus (upper line, with factor P), and of *P. uvella* minus (lower line, with factors P and M). The two together show the sex pair in the zygote of a cross between the two species.

the principles thus far set forth. How are the 157 exceptional zygotes to be accounted for?

The exceptions might be produced either by non-disjunction of two sex chromosomes, or by crossing over between them. If they were the result of non-disjunction, the 157 individuals would have received both the plus-producing and the minus-producing chromosomes of the sex pair; this would account for their mixed sex condition. There would be a pair of sex chromosomes, in place of the usual single chromosome. The exceptional individuals would therefore contain nine chromosomes instead of the eight usual for *Polytoma*. But cytological observations showed that only eight were present. The exceptional cases are therefore not the result of non-disjunction.

The alternative explanation is that they are due to crossing over. But how could crossing over between plus-producing and minus-producing chromosomes yield clones mixed as to sex?

Moewus concludes that the plus-producing chromosomes must contain a factor ("realisator") for the plus sex condition, the minus-producing chromosomes a factor for the minus sex condition, and that by crossing over, both must come into the same chromosome, which therefore produces the mixed sex condition. The cells containing such a chromosome may become either plus or minus, depending on external conditions.

But if the plus and minus factors may by crossing over come into the same chromosome, these factors are not alleles; they are not at the same locus. The condition of the chromosomes may then be represented as in Figures 172 and 173, in which P represents the plus-producing factor, M the minus-producing factor.

If crossing over occurs between the P and the M chromosomes, half of

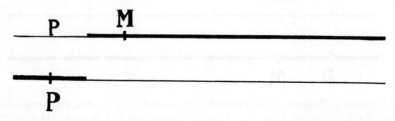

Figure 173. Diagram of the sex chromosomes produced by crossing over between the chromosomes of *Polytoma pascheri* plus (light line) and *P. uvella* minus (heavy line). The upper chromosome has the factor P from *P. pascheri*, the factor M from *P. uvella*.

the resulting chromosomes in this case contain both P and M, while the other half contains neither. Moewus holds that the cells carrying neither P nor M die. Thus is accounted for the fact that each of the 157 exceptional zygotes produces but two cells in place of the usual four. It is to be noted that here, as in former cases, the results are such as would be given by two-strand crossing over; a zygote produces either no crossover cells or exclusively crossover cells.

From these considerations and others of similar character, Moewus concludes that in clones mixed as to sex (monoecious races), the sex chromosome contains both the sex factors—the factor P for producing the plus sex, and the factor M for producing the minus sex.

Moewus (1936) reached similar results in a cross of *Chlamydomonas paradoxa* and *C. pseudoparadoxa*. Both species are dioecious, any clone being pure for one sex or the other. When plus clones of one species are mated with minus clones of the other, the majority of the zygotes yield four clones, of which two are pure plus, two pure minus. But in a particular case, out of 1,000 such zygotes, 117 gave exceptional results, all the four clones from each zygote being mixed as to sex. These exceptions are held to be due to crossing over. One of the haploid parents had the factor P, the other the factor M in the sex chromosome. By crossing over, the two factors are brought into one chromosome; the cells that receive this chromosome yield clones that are mixed as to sex. Half of the sex chromosomes in which crossing over occurred would lack both the factors P and M. Moewus holds that the cells that receive such chromosomes die, while the cells that receive both P and M divide twice before escaping from the zygote. Thus the zygote produces four cells as usual, all having the crossover combination P and M. Here, as in former cases, the results are those characteristic for two-strand crossing over.

Many other cases of crossing over of the sex factors (with results that require two-strand crossing over) are described by Moewus (1936) in crosses of *C. eugametos* and *C. paupera*—these all indicating that clones which are mixed as to sex carry a chromosome which has both sex factors, P and M. Other cases will be mentioned in later paragraphs.

Sex-linked inheritance.—In crosses of the two species of *Chlamydomonas* just mentioned, Moewus observed sex-linked inheritance. The species *C. eugametos* has an eyespot, while *C. paupera* has none. When *C. eugametos* of one sex is crossed with *C. paupera* of the other sex, in the descendants the gametes that are of the same sex as the *C. eugametos* parent have the eyespot, while those that are of the same sex as the *C. paupera* parent have none. The eyespot is thus linked with sex.

There are, however, a few exceptional cases, due to crossing over, in which the eyespot no longer goes with the parental sex. The results are complex and will not here be presented in detail. In all cases the results, as given by Moewus, are those that would be characteristic for two-strand crossing over. By analysis of the results, Moewus believes that he is able to establish the order in the chromosome of the two sex factors (P, M) and that for eyespot (S), as P-M-S.

Relative sexuality, in crosses between different species.—In crosses between different species, in some cases gametes of like sex may copulate and yield descendants (see Chapter XIV). Plus gametes from one species may unite with either plus or minus gametes from the other species, and minus gametes from one with plus or minus from the other. This phenomenon, known as relative sexuality, is accounted for by Moewus (following Hartmann) by assuming that the gametes of the different species differ in the strength, or "valence," of their sex tendency and that if two gametes differ sufficiently in the strength of the sex tendency they copulate, whether of the same or of different sex.

Exemplifying this situation is the fact that any gamete of *Polytoma uvella* may copulate with any gamete of *P. pascheri,* irrespective of the sex of the gametes. Hence it is held that in one of these species the sex tendency of the gametes is stronger than in the other species.

The relations as to crossing over described in the preceding section offer an opportunity for determining whether the sex factors present in the chromosomes of the two species differ in the strength of the sex tendency; also for determining in which species the sex factors are stronger. By crossing over, the plus factor of one species may be brought into the same chromosome with the minus factor of the other species. It is then possible to determine which of these prevails, and thus to discover in which species the sex factor is stronger.

Moewus crossed plus *P. pascheri* gametes from a clone pure for sex with minus *P. uvella* gametes from a race mixed for sex. According to the hypothesis, the *P. uvella* gametes, being from a race mixed for sex, contain a sex chromosome which carries both the plus and the minus factors, P and M. The *P. pascheri* gametes have a sex chromosome containing only the P factor. The situation as to sex chromosomes in the zygote may therefore be represented as in Figure 172. From this cross 625 zygotes were obtained. Of these, 582 gave the results that are usual without crossing over. Each zygote gave two clones that are pure plus, as in the *P. pascheri* parent, two that are mixed, as in the *P. uvella* parent.

The remaining 43, or 6.9 percent, gave exceptional results, presumably due to crossing over. In these, two of the four cells from each zygote yield pure plus clones, like the *P. pascheri* parent, while the other two yield pure minus, unlike either parent. No mixed clones are produced. How are these results accounted for?

The original condition of the chromosomes is that shown in Figure 172. By crossing over (the break occurring between P and M), the condition shown in Figure 173 is produced. One chromosome contains only the plus *P. uvella* factor (P); it is obviously the gametes containing this that yield the pure plus clones. The other chromosome contains both the P from *P. pascheri* and the M from *P. uvella*. The gametes that contain such chromosomes act, as above mentioned, like pure minus gametes. Therefore the effect of the M *P. uvella* factor completely overcomes the effect of the P factor from *P. pascheri*. The sex factors of *P. uvella* are thus stronger in sex tendency than those of *P. pascheri*.

From this the further conclusion is drawn that the gametes of *P. uvella* are stronger in sex tendency than the gametes of *P. pascheri*. Since *P. uvella* has only monoecious races (mixed as to sex), these gametes, according to the assumption, contain both a strong P and a strong M factor. It might be anticipated that the two would partly or entirely counteract each other, leaving the gametes weak or neutral in sex tendency. This, however, does not occur; if Moewus's assumption is correct, one of the two fully prevails, the other having no effect, so that the gametes are strong.

Moewus carried out numerous other crosses of these two species, which gave results that are in accord with those just set forth. They all indicate that the *P. uvella* factors have a stronger sex tendency than the *P. pascheri* factors. They agree further with the idea that the P and M factors are in different loci, and that by crossing over they may be brought into the same chromosome.

One further result of these experiments is of special interest. In certain cases there were crosses of two dioecious races (produced by hybridization), in which the gametes were of the same sex (both plus or both minus). In such crosses between plus gametes, according to the theory, each gamete contains only the P factor; no M factor is present. All the offspring in such cases are then of the plus sex, as the theory would lead one to expect. Similarly, if the two gametes each contain only the minus factor, all the offspring are of the minus sex.

An elaborate investigation of these matters was carried out by Moewus (1935a, 1935c) on diverse races of the unicellular alga *Protosiphon botryoides*. The relations here, while differing much in details, are concordant with those above set forth for *Polytoma* in the matters of crossing over of sex factors, different strengths of sex tendencies in different races, segregation at the reduction division, and the like. The results will therefore not be taken up in detail here. It is to be remarked that in this species, as in the others, crossing over (as the data are reported by

Moewus) took place in accordance with the two-strand scheme; any zygote that gave crossover combinations at all gave only crossover combinations.

In *Protosiphon* an investigation was made on the determination of sex by external conditions. *Protosiphon* includes dioecious races, in which the determination of sex is genotypic, as described in earlier pages; monoecious races, in which a single clone contains individuals of both sexes, the determination of sex being mainly phenotypic; and certain other races that agree in some respects with one type, in some respects with the other. The phenotypic determination of sex is not here dealt with.

Of interest for the nature of inheritance are the races which partake of the features of both types. Such is the race d, described by Moewus (1935c). At a certain stage in its life history, *Protosiphon* is a small club-shaped "haplont," containing many haploid nuclei imbedded in a mass of cytoplasm without cell walls. Such a haplont may be produced by a zygote resulting from the union of two gametes. In the diploid zygote, the reduction division occurs and the haploid nuclei multiply, giving rise to the haplont. Or haplonts may be produced each from a single haploid swarm cell, which comes to rest on the surface of a solid (as agar) and produces many nuclei by division of its single nucleus. Haplonts produced in either of these ways give rise later to swarm cells (gametes). The nuclei separate, each is surrounded by a small mass of cytoplasm and each such cell transforms into a flagellate swarm cell. In the race d, as in dioecious races, all the swarm cells produced by a single haplont are of the same sex.

But, in the race d, the sex of the swarm cells produced by the haplont depends on the conditions under which the haplont was produced. Haplonts grown on acid agar give rise only to plus swarm cells. Those grown on alkaline agar yield only minus swarm cells. Of those grown on neutral agar, some yield only plus swarm cells, others only minus swarm cells. The sex of the swarm cells and of all their descendants by vegetative fission is determined once for all by the conditions prevailing in the development of the haplont from which they arise. This is true not only for haplonts produced from single swarm cells, but also for haplonts produced from zygotes. Since in the zygote the reduction division occurs, yet all the four cells resulting from the maturation divisions are of the same sex, it is clear that sex segregation does not occur at the reduction division.

Moewus determined by careful experimentation at exactly what stage in the development of the haplont sex determination occurs. It is at the first division of the single nucleus of the swarmer that is dividing to produce a haplont. If at this time the conditions are acid, the swarm cells later produced from the haplont are plus; if the conditions at that time are alkaline, the later swarmers are minus.

In this case, therefore, the sex is determined by an external condition, and this sex is inherited by all the descendants of the cell in which it is thus determined. The consequences induced by an environmental condition are inherited.

In his article of 1938, Moewus sums up the conclusions to which his investigations lead as to inheritance and determination of sex, and particularly as to the constitution of the sex chromosomes in the different sex conditions. In pure dioecious races, each sex chromosome carries but one sex factor, P or M, and sex is segregated at the reduction division. In different dioecious races, the strength or valence of the sex factors may be diverse. In monoecious races, the sex chromosome of each individual carries both sex factors, P and M, in equal strength; which of these shall prevail in determining the sex of the gamete is decided by external conditions. In the subdioecious cases, in which any clone is prevailingly of one sex but produces also a few cells of the other sex, the sex chromosomes are held each to contain both sex factors, P and M, but every cell contains an additional factor which inclines it toward one sex or the other. The prevailingly plus clones contain a factor which inclines them toward the plus sex, the prevailingly minus cells a factor which inclines them toward the minus sex. But the tendencies of these additional factors are in some cases overcome by special outer conditions, so that a few cells of the opposite sex are produced.

In addition to these three types, there are other types in which the sex chromosome carries both the sex factors, P and M, but one of these is stronger than the other, so that it prevails and determines alone the sex of the clone. If P is the stronger, the sex of the clone is plus; if M is stronger, the sex of the clones is minus. Each clone is therefore pure for sex, and the race is dioecious. Such are called by Moewus complex dioecious races. They are not found in nature, but occur as a result of crossing over between dioecious races having sex factors of different strengths.

In this article of 1938, Moewus discusses the difficulties arising in his data as to crossing over. In all the data presented by him up to this time, crossing over was reported as occurring in accordance with the two-strand schema. Any zygote that gave any crossover combination gave such combinations exclusively. Moewus now concludes that this is not the normal state of affairs; he characterizes it, indeed, as pathological. Under certain conditions, he reports, crossing over occurs in the normal manner, according to the four-strand schema. He promises a future account of detailed investigations showing this.

Certainly there is great need for clearing up the confused situation as to crossing over in these organisms. In hundreds of detailed earlier reports, Moewus has given data that are consistent only with two-strand crossing over. A further serious criticism, based on other grounds, has been made as to the accuracy of Moewus's data on crossing over, by Philip and Haldane (1939).

Aside from this difficulty, the work of Moewus has placed the genetics of the Protozoa on a new footing. It has brought the phenomena of inheritance in these organisms into the same system that is manifested in the Mendelian inheritance of higher organisms. It has brought to light in the flagellate Protozoa instances of most of the phenomena in such inheritance, as before known in multicellular organisms.

BIPARENTAL INHERITANCE IN DIPLOIDS: CILIATA

The genetics of diploids is necessarily more complex than that of haploids. The individuals have during active life two sets of chromosomes instead of one. In consequence, in the sexual process two individuals do not normally unite completely, to form a diploid zygote, as in the copulation of the flagellates. In the diploid ciliates, two individuals merely come into intimate contact and exchange pronuclei that contain each a haploid set of chromosomes, a process known as conjugation. The two then separate, each carrying two haploid sets of chromosomes, one from each of the conjugant individuals. Each then multiplies vegetatively, forming clones, all members of any clone having the same diploid combination of chromosomes. Conjugation thus is like fertilization in higher organisms, in that it produces new diploid combinations of chromosomes.

Important for the understanding of inheritance in conjugation are the following:

1. The macronucleus or macronuclei of each individual disappear, being absorbed into the cytoplasm.

2. If more than one micronucleus are present, usually all but one are absorbed and disappear.

3. The single micronucleus divides three times in succession (the "maturation divisions"). By the first two divisions (Fig. 174, 1 and 2) four nuclei are produced; of these, three are absorbed and disappear. The remaining one divides once more into two (Fig. 174, at 3).

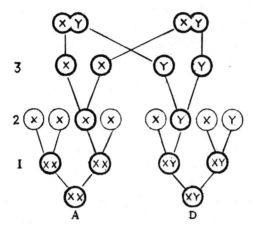

Figure 174. The three maturation divisions (1, 2, 3) and the exchange of pronuclei in the micronuclei during conjugation. A and D are the micronuclei of the two conjugants respectively. Reduction at the second division, represented by the separation of XX and XY. The diagram illustrates the fact that the two diploid nuclei produced (above) are alike in constitution.

4. The two nuclei produced by the third division are haploid, so that one of the three maturation divisions is a reduction division.

5. One of the two haploid nuclei (the "migrating pronucleus") in each individual passes over into the other individual, where it unites with the remaining one ("stationary pronucleus") of that individual. Thus there is produced again a diploid nucleus in each individual.

6. This diploid nucleus divides, some of its products becoming large as macronuclei, others remaining small as micronuclei. These macronuclei and micronuclei are distributed by fission to different individuals, the exact processes differing in different species.

7. In conjugation, only nuclei are exchanged; each individual retains its cytoplasm complete. Hence after conjugation each individual has a

new combination of chromosomes, half derived from each conjugant, but has the same cytoplasm as before. This gives an opportunity to compare the relative rôles of chromosomes and cytoplasm in inheritance, since in the ex-conjugants the chromosomal combination is changed, but the cytoplasm is not.

The only qualification required by the statement that cytoplasm is not changed is the fact that a minute bit of cytoplasm carrying an aster precedes the migratory pronucleus into the opposite conjugant. The results show, as will be seen, that this minute bit of cytoplasm is not effective in determining inheritance.

The two individuals that conjugate are, in some species at least, physiologically differentiated into diverse "mating types," which play the same physiological rôle in bringing about conjugation as do diverse sexes. For an account of these, Chapter XIV is to be consulted.

Knowledge of biparental inheritance is much less exact and extensive in the ciliates than in the flagellates. The recent discovery of diverse mating types furnishes an opportunity for exact analysis; but the novelty of this discovery has not yet given opportunity for its full investigation.

The earlier work on inheritance in conjugation before the discovery of mating types, is summarized in the author's *Genetics of the Protozoa* (1929). In the early work numerical ratios were not obtained, but qualitative relations of importance were demonstrated. Most generally expressed, the work showed that conjugation results in the production of many hereditarily diverse biotypes from the two involved in the conjugation. Production of hereditary diversities at conjugation was demonstrated by the work of Jennings in respect to the following types of characteristics: rate of fission, rate of mortality, presence of abnormalities. This demonstration of the production of inherited differences at conjugation was extended in 1930 by the work of Raffel, and in 1932 by Jennings, Raffel, Lynch, and Sonneborn to various other characteristics, including size and form, vigor, resistance, and degeneration. It was further shown that conjugation causes the descendants of the two members of a pair to become similar in fission rate, in mortality, in the occurrence of abnormalities, and probably in size (see Jennings, 1929, pp. 181-85).

Later, more exact work on inheritance in ciliates deals with the inheritance of mating type (Sonneborn, 1937-39; and Jennings, 1938-39) and with the inheritance of size (De Garis, 1935). That on mating type will be presented first.

INHERITANCE OF MATING TYPE IN *Paramecium aurelia*

In *Paramecium aurelia,* according to the work of Sonneborn (1937, 1938, 1939), there are in any variety but two mating types, members of which unite in conjugation as the two sexes unite in multicellular ani-

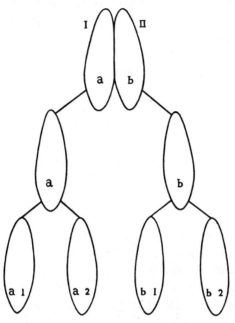

Figure 175. The four clones (a1, a2, b1, b2) produced from the two ex-conjugants (a and b) of a pair, in the experiments of Sonneborn and of Jennings.

mals. Three varieties or "groups" are known; individuals of one variety do not conjugate with individuals of the other varieties.

In Variety or Group 1, the two mating types are known as types I and II. These unite in conjugation. In studying the types produced by them, each ex-conjugant after separation was allowed to divide once, and from each of the products a clone was produced by vegetative reproduction. Thus four clones were derived from each pair, two from each ex-conjugant, as shown in Figure 175.

All members of any one of the four clones belong to the same mating type, and that type is inherited within the clone until endomixis or a new conjugation occurs. In a certain experiment described by Sonneborn (1937), there were fifty-six pairs formed by union of mating types I and II. The four clones from each pair were constituted as shown in the accompanying table.

In this table the following relations are seen:

1. In some cases (4 out of 56) all the descendants of the two parents (which were of diverse type) are of the same type. In these cases the type of one of the conjugants and its descendants was changed by con-

PARENTS, I × II

Number of Pairs	Number of Clones		Mating Types of the Four Descendant Clones
	Type I	Type II	
1	4		I, I, I, I
18	3	1	I, I, I, II
21	2	2	I, I, II, II
13	1	3	I, II, II, II
3		4	II, II, II, II

jugation. That is, an individual of type I, receiving a pronucleus from type II, becomes changed to type II, and vice versa.

2. In some cases the two ex-conjugants of a pair give clones of different mating types.

3. In most cases a single ex-conjugant gives rise to two clones of the same type.

4. But in some cases a single ex-conjugant gives clones of different mating type.

Do ex-conjugants that were of a given type before conjugation tend after conjugation to produce descendant clones of that type, or do they produce both types with the same frequency? Sonneborn tested this by mating two clones that were different in appearance, one of each type, then determining the mating type of the descendants. The results were as follows:

Of 22 ex-conjugants originally of type I, 5 gave descendants of type I only, 7 of type II only, and 10 half of type I, half of type II. Of 25 ex-conjugants originally of type II, 4 gave descendants of type II only, 3 of type I only, and 18 half of type I, half of type II.

In summary, there were 47 ex-conjugants, which give 94 descendant clones. Of these 94, 46 were of the same type as their parent's before conjugation, 48 of different type. Thus it is clear that ex-conjugants originally of a given type produce descendants of both mating types with equal frequency. The reception in conjugation of a pronucleus from an individual of different type changes the type as frequently as it leaves it unchanged. There is no tendency for the descendants to be of the same type as was the cytoplasmic body from which they are derived (certain exceptions will be described later).

Of the 47 ex-conjugants just considered, 28 gave two clones of different types, while 19 gave two clones of like type. There is no indication of a tendency for the two clones descended from a single ex-conjugant to be alike in type.

Thus the first fission after conjugation separates the ex-conjugant into two individuals which may be of different mating type, giving rise to two clones of different types; or they may be of the same type. Segregation of the diverse mating types (in cases in which it occurs) takes place at the first fission after conjugation.

What is it that decides the mating type of each clone? Sonneborn is disposed to believe that the segregation is the result of the separation of the two macronuclei, one macronucleus tending to produce type I, the other type II. How the two became diverse (if this is the case) is not known. There appears to be no evidence of a reduction division at this point, such as might give rise to nuclei differing in chromosomes. It appears equally difficult to suppose that the two nuclei of the same cell are subjected to differing conditions, such as to cause one to be of type I, the other of type II.

Sonneborn (1938) has discovered, however, that the temperature during conjugation affects the proportion of the types produced in a group of ex-conjugants. In variety or group 1, higher temperatures cause the appearance of a greater proportion of type II; in group 3, higher temperatures favor the production of type VI (rather than type V). But it is not evident how the difference of type in the two products of fission of a single individual could be induced in any such manner. The segregation of types at the first fission after conjugation remains a riddle.

Segregation of mating types at endomixis.—It is of great interest that at the first fission after endomixis segregation of the mating types may occur, just as it does after conjugation. The single individual, of a definite mating type, divides after endomixis into two that may be diverse in mating type. The nuclear processes in endomixis are not yet fully cleared up, but are known to be in many respects similar to those which occur at conjugation. According to the recent work of Diller (1936), the similarity goes so far that there is in endomixis a union of two micronuclei (presumably haploid), just as occurs in conjugation. In this autogamy the two micronuclei that unite are of course both from the same individual. After their union, the single (diploid) nucleus divides into two, then into four. Two of these become macronucei and two remain micronuclei, just as in conjugation. The two macronuclei are separated at fission, just as occurs after conjugation, and the same is true of the micronuclei. As will be seen later, there is genetic evidence that autogamy does indeed occur at endomixis.

In variety or group 1 of *Paramecium aurelia,* Kimball (1937) determined the mating type of the two clones produced by each of 181 individuals that had undergone endomixis. Of the individuals, 96 gave 2 clones of the same type (both I or both II), while 85 gave two clones of different types (one type I, one type II). In other respects also the results of endomixis in relation to the mating types are like those of conjugation (see Kimball, 1937).

As a rule, in both endomixis and conjugation the segregation of the mating types occurs at the first fission after completion of the process. But in a certain stock there was a small proportion of cases in which segregation into two mating types occurred at the second fission after conjugation. In this same stock, cytological examination showed that in about the same small proportion of cases the ex-conjugants had three or four macronuclear *Anlagen* in place of the usual two. It of course requires two fissions to separate these into different individuals. Sonneborn is disposed to consider this the cause of the occasional segregation of the mating types at the second fission, in place of the first.

Single-type clones.—There exist clones in which there is no segregation into different mating types at endomixis. In such clones endomixis makes no change in the mating type; the clone remains throughout of the same type (Sonneborn, 1938c). Of twenty-six clones examined by

Sonneborn, six showed no change of type at endomixis. All these six were of mating type I.

Such clones may be designated as single-type clones, as compared with the more common double-type clones, in which from a single clone both sexes are produced at endomixis.

Crosses between single-type and double-type clones.—Crosses between the two types (Sonneborn, 1939) show that they are inherited in typical Mendelian fashion, double type being dominant over single type. The factor for double type may therefore be designated A, that for single type a. The two original diploid clones are AA and aa. Crosses of the two (AA $\times$ aa) gave in the 149 pairs examined, all double-type (Aa) offspring. Mating together these heterozygotes yielded, in 120 pairs, 88 of the dominant double type, 32 of the recessive single type, so that the results approximate the typical three-to-one ratio (Aa $\times$ Aa $=$ AA $+$ 2A $+$ aa). The hybrids Aa were back crossed to the single-type parents (aa) in 165 pairs; these yielded 88 double-type and 77 single-type descendant clones, an approximation to the expected one-to-one ratio (Aa $\times$ aa $=$ Aa $+$ aa). Here we have diploid Mendelian inheritance, one of the characters being recessive.

Genetic evidence of autogamy.—In the further study of the heterozygotes Aa, a discovery of great interest was made. At endomixis, some of these heterozygotic individuals produce (at the first fission after endomixis) double-type clones, others single-type clones. Is the double-type clone the heterozygote Aa or the homozygote AA? This was tested by mating them with the normal single-type individuals aa. All the descendant clones are the heterozygotic double-type Aa. This shows that the double-type clones produced at endomixis are the homozygotes AA; correspondingly, the single-type clones produced at endomixis are necessarily aa. From the heterozygotes Aa there are produced at endomixis two types, both homozygotic: AA and aa.

How is this result brought about? It is the natural consequence of the occurrence at endomixis of a reduction division with subsequent union of two of the reduced nuclei (autogamy), as described cytologically by Diller (1936). The heterozygote nucleus before reduction is Aa. By reduction are produced haploid nuclei A and a: by a second division these give rise to four nuclei A, A, a, a. Three of these haploid nuclei degenerate, leaving but one, A or a. This remaining nucleus now undergoes the third division, yielding in one case two nuclei A and A, in the second case a and a. These two nuclei now reunite (autogamy), yielding in the one case the homozygote dominant AA, in the other case the homozygote recessive aa. Thus after endomixis has occurred in any stock, the clones are all homozygotes, as Sonneborn points out.

These results are obviously strong genetic evidence for the occurrence of autogamy at endomixis in *Paramecium aurelia.*

Certain rare and exceptional conditions in the genetics of the mating types in *P. aurelia* are discussed in Chapter XIV, "Sexuality in Unicellular Organisms."

Inheritance of Mating Type in *Paramecium bursaria*

In *Paramecium bursaria,* according to the work of Jennings (1939a, 1939b), there are in one of the varieties four mating types, A, B, C, and D. Another variety has eight mating types (E, F, G, H, J, K, L, and M), a third variety four types (N, O, P, and Q) distinct from the four of the first variety. Inheritance of mating type has been examined only in the first variety (some of the relations are here published for the first time).

In variety 1 (as in the other varieties), clones of any one of the types may conjugate with clones of any of the other types, but not with clones of their own type Thus in variety 1 six different matings occur, yielding pairs AB, AC, AD, BC, BD, CD.

Very rarely in this species a single clone, in which all the individuals belong originally to the same type, may differentiate into two types, which thereupon conjugate. This phenomenon is parallel to the segregation of different types from one type at endomixis in *P. aurelia.* It is presumably the result of endomixis in *P. bursaria;* in this species endomixis is known to occur very rarely (Erdmann, 1927). Thus the occasional "self-fertilization" of a clone is in fact the conjugation of two diverse mating types. A clone of the mating type D has been observed to differentiate part of its individuals into the mating type A; these then conjugate with the D individuals, giving the cross AD. The results of "self-fertilization" may therefore be considered with those of other matings between two different types.

In studies of mating-type inheritance, four descendant clones are obtained from each pair, two from the first fission of each ex-conjugant, as indicated in the diagram of Figure 175. After the attainment of sexual maturity, the mating type of each of the four clones is discovered by testing them with standard clones representing the four known mating types.

The fullest data as yet available are for the cross A × D. Of this cross the mating types have been determined for the clones descended from 61 pairs, including 175 clones. The 4 original clones did not survive in all the pairs. They did all survive in 26 pairs (104 clones). The remaining pairs had each but 1, 2, or 3 surviving clones. But in all cases all

TABLE 18: INHERITANCE OF MATING TYPES (THE FOUR TYPES ARE A, B, C, D), *Paramecium bursaria*

PARENTS, A × D

	Typical Constitution of the Four Descendant Clones of Each Pair
30 Pairs give type A only	A+A+A+A
4 Pairs give type B only	B+B+B+B
4 Pairs give type C only	C+C+C+C
23 Pairs give type D only	D+D+D+D

clones descended from any pair were of the same mating type. The results have been summarized as in Table 18.

In Table 18 the following general relations appear:

1. All the descendants of any one pair are of the same mating type, though the parents were of two different types.

2. Among the descendants of the cross of the two types A and D occur all four types A, B, C, and D.

3. The majority of the descendant clones are like one or the other of the two parents in type; only a few differ. In this case 53 of the 61 clones are either A or D (the parental types); only 8 are of types different from those of the parents.

4. By conjugation the mating type has become changed in one or both of the ex-conjugants and their descendants. In the above table, the type is changed in both ex-conjugants in 8 pairs out of 61; it is changed in but one of the ex-conjugants in the remaining 53 pairs.

The change in type is due to the exchange of migratory pronuclei, since the cytoplasm remains unmixed. Individuals of type A, receiving a micronucleus from type D, are changed to type D in nearly half of the cases; to B or C in few cases. In about half the cases they remain unchanged. Parallel statements may be made for the individuals originally of type D.

It is clear that although all four mating types are present in the descendants of a cross of but two types, the statistical make-up of the descendant population is influenced by the constitution of the two parents.

TABLE 19: MATING TYPES OF DESCENDANT CLONES FROM 131 PAIRS (INCLUDING 279 DESCENDANT CLONES), IN THE SIX POSSIBLE CROSSES OF THE FOUR MATING TYPES OF VARIETY 1, *Paramecium bursaria*

Cross	Number of Pairs	Mating Types of the Descendant Clones: Numbers of Pairs Yielding Each Type					
		A	B	C	D	Like Parents	Unlike Parents
A × B	25	13	12	0	0	25	0
A × C	6	3	1	1	1	4	2
A × D	61	30	4	4	23	53	8
B × C	10	0	10	0	0	10	0
B × D	15	5	8	0	2	10	5
C × D	14	0	0	6	8	14	0
Totals	131	51	35	11	34	116	15

The majority of the descendant clones belong to one or the other of the two parental types. This is evident in the results of all the crosses that have been made, as shown in Table 19.

As Table 19 shows, of 131 pairs in which the original type of the conjugants was known, 116, or 88.5 percent, gave descendant clones that were of the same type as one or the other of the two parent individuals, while but 15, or 11.5 percent, gave descendants that were of different mating type from either of the two parents.

What determines the type to which a particular clone belongs? In all the 131 pairs of Table 19, all surviving clones from any pair (1 to 4 clones per pair, in different cases) were of the same mating type. The four clones descended from the two members of any one pair were cultivated separately and tested separately. The fact that all four turn out to be of the same mating type shows that the type is determined at the time that the two ex-conjugants separate and before they divide. It might be determined at that time either by internal or external conditions. The fact that in the great majority of cases the descendant clones are of the same type as one or the other of the two parents shows that the constitution of the two parents is one important factor in determining the type of the descendants. But why in the same cross the descendants from some pairs should be of the same type as one parent, those from other pairs of the same type as the other parent, while a few are of different type from either parent—this is as yet quite unknown.

In extremely rare cases the four clones produced by a particular pair are not all of the same type. Thus in a certain case a pair composed of the types A $\times$ C yielded four clones of the types A, B, B, B. In another case A $\times$ C gave B, D, D, D. The great rarity of such cases indicates that such results are due to irregularities in the cytological processes, comparable to cases of non-disjunction of chromosomes.

Immaturity and partial maturity.—The relations thus far described are those that exist after the descendant clones have reached sexual or reproductive maturity. But for a long period after conjugation, varying in different clones from a few weeks to more than a year, the descendant clones are in *Paramecium bursaria* sexually immature. During this period the descendant clones do not mate at all, with any of the mating types. In group 1, if they are mixed with mature individuals of any of the four mating types A, B, C, D, there is no mating reaction, no formation of pairs. In this period descendant clones show the characters of no mating type. Later begins a period of partial maturity, in which a few members of the clone pair with mature members of certain of the mating types. During the early part of this period of partial maturity there are in variety 1 only two mating types instead of four. Some of the descendant clones form pairs with the types A or B, but not with C or D; others with C or D, but not with A or B. The former may now be said to constitute the type CD, the latter the type AB. Later these two young types become further differentiated. Some of the clones that thus far do not react with A or B acquire the power to react with A, but still do not react with B; these now belong to the definitive mating type B. Others acquire the power to react with B but not with A; these belong to the definitive type A. A similar differentiation occurs among the clones of the young type CD; some of these become type C, others type D. Thus the types A and B are closely akin, being for a time one type AB; similarly, types C and D at first constitute one type, CD. The interest of these gradual changes in maturity and type during vegetative reproduction has been emphasized in an earlier section (p. 714 above).

EFFECT OF THE CYTOPLASM AND ITS RELATION TO NUCLEAR CONSTITUTION

Certain phenomena in the biparental reproduction of ciliates throw light on the relative rôle of the nucleus and the cytoplasm in inheritance. Such phenomena are shown in the inheritance of a number of different types of characteristics, but are best seen in crosses of individuals of different inherited sizes, as set forth in the work of De Garis (1935).

By ingenious methods De Garis obtained in *Paramecium caudatum*

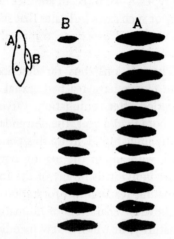

Figure 176. Change of size resulting from conjugation of individuals of large and small races, in the experiments of De Garis. Upper left, A and B, the large and small individuals of the pair, showing their relative sizes. The columns headed A and B respectively show (reading from above downward) the average sizes of the descendants of the two at successive intervals of two days each. (Diagram based on the measurements of De Garis, 1935.)

conjugation between races the members of which differed greatly in size. The nature of the consequences will best be seen from a typical example.

In a certain case one of the members of a pair (A, Fig. 176) belonged to a race in which the average length of the individuals was 198 microns, while the other belonged to a small race with an average length of but 73 microns (B, Fig. 176). The large individual of the pair had thus about twenty times the bulk of the smaller one.

Before conjugation the large and the small sizes are inherited in the two races; all individuals of race A are large, all those of race B are small. The two individuals of the different races conjugate and exchange a haploid set of chromosomes, then separate. The two are now alike in their nuclei, but are diverse in their cytoplasm—one having the cytoplasm of race A, the other that of race B. The ex-conjugant A still has the large size of race A, the ex-conjugant B the small size of race B.

The large ex-conjugant A now divides by fission. Its two offspring grow to the large size usual for race A. The small ex-conjugant B divides; its two offspring grow only to the small size usual for race B. The offspring divide again, and fission continues at the rate of once or twice a day, each ex-conjugant producing a clone.

As fission continues day after day, it is found that the adult sizes are changing in each clone. In the clone descended from A, the individuals of the successive generations grow smaller; in the clone descended from

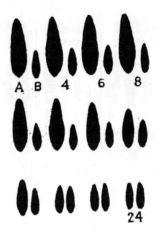

Figure 177. Changes in mean size of the descendants of the two members, A and B, of an unequal pair, in another cross. At the upper left are shown the relative sizes immediately after conjugation. Reading from left to right are shown in the three rows (from above downward) the successive sizes of the descendants of the two at intervals of two days, till at the end of twenty-four days (lower right) the two have reached a common small size that is not greatly different from the original size of B. (Based on the measurements of De Garis, 1935.)

B, they grow successively larger (Fig. 176). The average sizes of the two races approach one another. This continues for twenty-two days, including about the same number of generations. By that time the individuals in both clones have reached a size that is approximately midway between the original size of race A and that of race B. At that point, the changes in size cease; this intermediate size remains constant in the two clones until there is another conjugation. The two clones having come to a common size, now form a single race of uniform adult size.

Thus for a long period, twenty-two days, in this case, the size in the descendants is affected both by the cytoplasm and by the nucleus, but finally the size is controlled entirely by the nuclear constitution. During the intervening period the two clones differ in size, and this can be due only to the difference in their cytoplasm, since they are alike in their nuclei. The large cytoplasmic body of A is reduced only slowly and gradually to the new size, and while this is occurring, potentially mil-

764 INHERITANCE

lions of new individuals are produced, all with body size partly dependent on the cytoplasmic constitution. Similarly, the small body of B is raised to the new size only slowly and gradually through many generations. The gradual change in size to an intermediate condition, which is finally the same in both the clones, can be due only to the fact that after conjugation the two ex-conjugants and their descendant clones are alike in their nuclei. The nucleus gradually alters the cytoplasm, bringing the

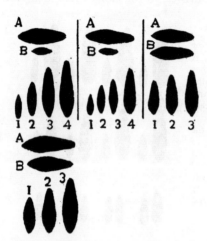

Figure 178. Different ultimate mean sizes (numbered 1, 2, 3, 4 in each case) reached by descendants of different pairs from crosses of the same two races (A and B, in each case). The results are shown for four different crosses. To be read as follows: at the upper left A and B shows the mean sizes of two races that were crossed. Four pairs of this cross gave four stocks of the ultimate mean sizes shown at 1, 2, 3, 4. Of the next cross to the right four pairs were similarly obtained, of the two others but three. (Based on the measurements of De Garis, 1935.)

size in both clones ultimately to that which is characteristic for the nuclear constitution.

Many such crosses between large and small races were made by De Garis. In all cases the cytoplasm affects the size for many generations, but finally the descendants of the two ex-conjugants come to a common size, corresponding to the common nuclear constitution. The final size is, however, not always midway between the sizes of the two conjugants of the pair. In some cases it is much nearer to that of one of the two parental clones than to that of the other. Such a case is illustrated in Figure 177. In this, two clones were crossed, in which the mean lengths were respectively 203 microns (clone A) and 81 microns (clone B). After the gradual change in size through some 24 generations, the ultimate size reached was much nearer to that of the smaller parent, B. Such results were seen in many of De Garis's crosses (see Table 20). The ultimate size obviously depends on what nuclear (chromosomal) combination is present in the two parents after conjugation.

Repeated crosses between the same two clones give different final results in different cases, just as the same pair of parents in multicellular organisms produce in different cases offspring that differ in hereditary constitution. Table 20 shows the different final sizes reached by the descendants in repeated examples of a number of different crosses. Figure

TABLE 20: LENGTHS IN MICRONS OF THE TWO RACES CROSSED (DESIGNATED A AND B IN EACH CASE) WITH THE RESULTING FINAL LENGTHS OF THE OFFSPRING, *Paramecium caudatum;* IN CASES IN WHICH SEVERAL DIFFERENT PAIRS WERE OBTAINED FROM THE SAME CROSS, THEIR DESCENDANTS ARE NUMBERED 1, 2, 3, 4 (DATA FROM DE GARIS, 1935)

Parents A B	Mean Lengths of the Descendants of the Different Pairs			
	I	2	3	4
203 × 81	143	191	91	204
191 × 75	64	139	111	192
217 × 75	170	133	170	215
172 × 172	165	132	183	137
152 × 81	114	119		
201 × 73	69	71		
217 × 75	133	200		
201 × 198	165			
198 × 73	142			
152 × 152	145			
198 × 198	204			
75 × 75	92			

178, drawn to scale, shows graphically the relative sizes of descendants of different pairs in certain crosses.

The fact that different pairs of conjugants from the crossing of the same two clones give descendants of different sizes indicates that in these Protozoa, as in Metazoa, recombination of the chromosomal materials occurs, giving different combinations in different cases. It shows also that clones of *P. caudatum* must in many cases be heterozygotic for size factors; otherwise different results would not be produced from different pairs of the same cross.

The fact that the descendants of the two ex-conjugants of any pair yield descendants that are finally of sensibly the same size shows that reduction of the chromosomes, so far as factors for size are concerned, must occur at the first or the second of the maturation divisions, not at the third. The fact that they are of the same final size shows that the two descendant clones have the same factors for size. This can occur only if the third maturation division, producing the two pronuclei, is non-reductional. In that case the two pronuclei of an individual are necessarily alike, and only when this is so will the nuclear combinations be the same in the two ex-conjugants. If reduction occurred at the third maturation division, the two pronuclei would often be diverse, leading to different final sizes.

In summary, it may be said that the final size of the descendants of the ex-conjugants is determined by the nuclear constitution, as is shown by the fact that the final sizes are the same for the clones derived from the two conjugants of a pair. But for a long time, for many vegetative generations, the nature of the cytoplasm affects the size of the descendants; descendants with cytoplasm of different constitution remain for long periods diverse. The longest period observed by De Garis during which the diversity of cytoplasm persisted was thirty-six days, which would mean about the same number of vegetative generations.

A similar differential effect of the cytoplasm is seen in the inheritance of other characteristics. Sonneborn and Lynch (1934), before the work of De Garis on size, observed such a cytoplasmic effect in crosses of clones that differed much in the rate of multiplication. If before conjugation one of the clones multiplies rapidly, the other slowly, after conjugation the clone that receives the cytoplasm from the rapid race continues for a time to multiply rapidly, while the other, receiving its cytoplasm from the slow race, continues to multiply slowly. This effect of the diverse cytoplasms continues for about ten vegetative generations. But during that time the difference in fission rate for the two ex-conjugant clones gradually disappears, till at the end of the period the rate of fission is the same in the two. For example, the two clones N 21 and B were crossed by Sonneborn and Lynch. The mean daily fission rate in N 21 was 0.67, while in B the fission rate was 1.97, three times as great as in N 21. During the first five days after conjugation, the daily rate for the clones that had received cytoplasm from the slow race N 21 was 1.10; for those that had received cytoplasm from the fast race, B, the daily rate was 2.00. In the second five-day period the rates were respectively

1.6 and 1.8. After an interval of twelve days the rates were taken again. They were now the same in the two ex-conjugant clones; during seventeen days they stood for both at 2.0 to 2.5 fissions daily. Other crosses between fast and slow clones showed similar conditions.

The effect of the cytoplasmic constitution in thus delaying the assumption of the final characteristics resulting from the nuclear constitution is commonly known as the "cytoplasmic lag."

A cytoplasmic lag of a similar sort is at times seen in the inheritance of mating type at endomixis in exceptional clones of *P. aurelia,* as described by Kimball (1939). Normally in both *P. aurelia* and *P. bursaria* no such lag is evident, unless it be in the fact that there is a period of immaturity during which no mating reaction occurs. The cases described by Kimball are in certain clones of variety 1 of *P. aurelia.* These clones were of the mating type II. At endomixis some of these are transformed to type I, as before set forth. But for a few generations after the formation of the new nucleus, they remain of type II, and will still conjugate with individuals of type I. After these few fissions, however, the new nucleus asserts itself and the members of the clone transform to type I; they now conjugate only with type II. The animals thus, after the formation of the new nucleus, remain for a time of the mating type appropriate to the cytoplasmic constitution only; then transform to correspond to the new nuclear constitution.

These phenomena illuminate the rôle of the cytoplasm in inheritance. At every fission the volume of the cytoplasm present in any individual is reduced one half; then the original volume is restored by new growth. Thus, in the case of the ex-conjugant of the large race A of Figure 176, the descendants of which are gradually diminished in size after conjugation, the size of the individual is reduced at the first fission after conjugation to one half the original size, so that if growth did not occur the body would in two or three generations be reduced to the final size. But owing to the properties of the cytoplasm of that race A, growth occurs after the first fission to practically the original racial size. At every succeeding fission the original cytoplasm is again diluted one half, so that after ten generations, it has been diluted to less than one-thousandth part, the remainder being new cytoplasm produced by growth. Yet after ten generations the nature of the original cytoplasm still has a marked effect on size. The original cytoplasm must therefore have to some extent the power of reproducing itself in its distinctive nature, at the time that growth occurs. In this respect it partakes of the character of genetic material, since it shows the two distinctive features of that material: it affects the characteristics of the individuals, and it reproduces itself in some degree true to type. But in time it is made over by the new nucleus.

These facts as to the differential effect of the cytoplasm on inherited characteristics in crosses furnish a basis in normal genetics for the idea of Jollos, set forth in a previous section, that inherited environmental modifications may have their seat in the cytoplasm. These, like the size due to the nature of the cytoplasm in the crosses made by De Garis, are inherited for a number of generations, but finally fade away. But there is so great a difference in the time, the number of generations, during which the inheritance continues, in the two cases, that it raises doubts as to the fundamental similarity of the two. In the crosses, the inherited cytoplasmic effect continues in different cases for ten, twenty, thirty generations, the extreme limit observed being thirty-six generations. By the end of such a period, the cytoplasm has been made over by the nucleus, so that it is thereafter the constitution of the nucleus that determines the characteristics. But such experimental modifications as acclimatization are inherited for hundreds of generations. If they are merely modifications of the cytoplasm, it might be anticipated that long before so many generations had passed the cytoplasm would have been made over by the nucleus, so that its modifications would have disappeared. Yet of course it is not certain that the time required for the nucleus to dominate the cytoplasm would be subject to similar limits in all cases. Here much remains to be discovered. The question whether inherited environmental modifications are exclusively cytoplasmic, or whether they affect the chromosomal materials of the nucleus, must be left open for the present. Decision of this question appears practicable by experimental breeding. What is required is to induce environmental modifications in a clone, then to cross this clone with another which lacks the modification. In conjugation only nuclei with their chromosomal materials are transferred from one clone to the other. If the modifications have affected the nuclei, they should be transferred by conjugation from the modified clone to the unmodified one. But if they affect only the cytoplasm, they will not thus be transferred. The prospects for successfully carrying through such experiments have been greatly increased by the recent discovery of diverse mating types in these organisms. This discovery renders it possible to make any desired crosses as readily in these organisms as in fruit flies or in rats.

In general terms, the relations of nucleus and cytoplasm to inheritance, revealed in crosses in the ciliates, may be expressed as follows. The primary source of diversities in inherited characters lies in the nucleus. But the nucleus by known material interchanges impresses its constitution on the cytoplasm. The cytoplasm retains the constitution so impressed for a considerable time, during which it assimilates and reproduces true to its impressed character. It may do this after removal from contact with the nucleus to which its present constitution is due, and even for a time in the presence of another nucleus of different constitution. During this period, cytoplasmic inheritance may occur in vegetative reproduction. The new cells produced show the characteristics due to this cytoplasmic constitution impressed earlier by a nucleus that is no longer present. But in time the new nucleus asserts itself, impressing its own constitution on the cytoplasm. Such cycles are repeated as often as the nucleus is changed by conjugation.

LITERATURE CITED

An extensive bibliography of contributions on the genetics of Protozoa prior to 1929 will be found in the author's *Genetics of the Protozoa* (1929). The present list includes only articles referred to in the foregoing chapter.

Calkins, G. N. 1919. *Uroleptus mobilis* Engelm. II. Renewal of vitality through conjugation. J. exp. Zool., 29: 121-56.

Dallinger, W. H. 1887. The president's address. J. R. micr. Soc., 1: 185-99.

De Garis, C. F. 1935. Heritable effects of conjugation between free individuals and double monsters in diverse races of *Paramecium*. J. exp. Zool., 71: 209-56.

Diller, W. F. 1936. Nuclear reorganization processes in *Paramecium aurelia*, with descriptions of autogamy and "hemixis." J. Morph., 59: 11-67.

Dobell, C. 1924. The chromosome cycle of the sporozoa considered in relation to the chromosome theory of heredity. Cellule, 35: 169-92.

Erdmann, R. 1927. Endomixis bei *Paramecium bursaria*. S. B. Ges. Naturf. Fr. Berl., 1925: 24-25.

Hämmerling, J. 1929. Dauermodifikationen. Handbuch der Vererbungswissenschaft (Baur u. Hartmann), Bd. 1, Liefrg. II: 1-69.

Jennings, H. S. 1916. Heredity, variation and the results of selection in the uniparental reproduction of *Difflugia corona*. Genetics, 1: 407-534.

—— 1929. Genetics of the Protozoa. Bibliogr. genet., 5: 105-330.

—— 1938. Sex reaction types of their interrelations in *Paramecium bursaria*. Proc. nat. Acad. Sci. Wash., 24: 112-20.

—— 1939a. Genetics of *Paramecium bursaria*. I. Mating types and groups, their interrelations and distribution; mating behavior and self-sterility. Genetics, 24: 202-33.

—— 1939b. *Paramecium bursaria:* Mating types and groups, mating behavior, self-sterility; their development and inheritance. Amer. Nat., 73: 414-31.

Jennings, H. S., D. Raffel, R. S. Lynch, and T. M. Sonneborn. 1932. The diverse biotypes produced by conjugation within a clone of *Paramecium*. J. exp. Zool., 63: 363-408.

Jollos, V. 1913. Experimentelle Untersuchungen an Infusorien. Biol. Zbl., 33: 222-36.

—— 1920. Experimentelle Vererbungsstudien an Infusorien. Z. indukt. Abstamm.- u. VererbLehre., 25: 77-97.

—— 1921. Experimentelle Protistenstudien. I. Untersuchungen über Variabilität und Vererbung bei Infusorien. Arch. Protistenk., 43: 1-222.

—— 1924. Untersuchungen über Variabilität und Vererbung bei Arcellen. Biol. Zbl., 44: 194-208.

—— 1934. Dauermodifikationen und Mutationen bei Protozoen. Arch. Protistenk., 83: 197-219.

Kimball, R. F. 1937. The inheritance of sex at endomixis in *Paramecium aurelia*. Proc. nat. Acad. Soc. Wash., 23: 469-74.

—— 1939. A delayed change of phenotype following a change of genotype in *Paramecium aurelia*. Genetics, 24: 49-58.

Maupas, E. 1889. La Rajeunissement karyogamique chez les ciliés. Arch. zool. exp. gén., (2), 7: 149-517.

Moewus, F. 1932. Untersuchungen über die Sexualität und Entwicklung von Chlorophyceen. Arch. Protistenk., 80: 467-526.

—— 1934a. Über Subheterözie bei *Chlamydomonas eugametos*. Arch. Protistenk., 83: 98-109.

—— 1934b. Über Dauermodifikationen bei Chlamydomonaden. Arch. Protistenk., 83: 220-40.

—— 1935a. Über den Einfluss äusserer Faktoren auf die Geschlechtsbestimmung bei *Protosiphon*. Biol. Zbl., 55: 293-309.

—— 1935b. Über die Vererbung des Geschlechts bei *Polytoma pascheri* und bei *Polytoma uvella*. Z. indukt. Abstamm.- u. VererbLehre., 69: 374-417.

—— 1935c. Die Vererbung des Geschlects bei verschiedenen Rassen von *Protosiphon botryoides*. Arch. Protistenk., 86: 1-157.

—— 1936. Faktorenaustausch, insbesondere der Realisatoren bei *Chlamyomonas*-Kreuzungen. Ber. dtsch. Bot. Ges., 54: (45)-(57).

—— 1937a. Die allgemeinen Grundlagen der Sexualität. Biologe, 6: 145-51.

—— 1937b. Methodik und Nachträge zu den Kreuzungen zwischen Polytoma-Arten und Zwischen Protosiphon-Rassen. Z. indukt. Abstamm.-u. VererbLehre., 73: 63-107.

—— 1938. Vererbung des Geschlechts bei *Chlamydomonas eugametos* und verwandten Arten. Biol. Zbl., 58: 516-36.

Neuschloss, S. 1919, 1920. Untersuchungen über die Gewöhnung an Gifte. I, II, III. Pflüg. Arch. ges. Physiol., 176: 223-35; 178: 61-79.

Philip, N., and J. B. S. Haldane. 1939. Relative sexuality in unicellular algae. Nature, 143: 334.

Raffel, D. 1930. The effects of conjugation within a clone of *Paramecium aurelia*. Biol. Bull., 58: 293-312.

Reynolds, B. D. 1923. Inheritance of double characteristics in *Arcella polypora* Penard. Genetics, 8: 477-93.

Sonneborn, T. M. 1936. Factors determining conjugation in *Paramecium aurelia*. I. The cyclical factor: the recency of nuclear reorganization. Genetics, 21: 503-14.

—— 1937. Sex inheritance and sex determination in *Paramecium aurelia*. Proc. nat. Acad. Sci. Wash., 23: 378-85.

—— 1938a. Sex behavior, sex determination and the inheritance of sex in fission and conjugation in *Paramecium aurelia*. Genetics, 23: 168-69.

—— 1938b. Mating types, toxic interactions and heredity in *Paramecium aurelia*. Science, 88: 503.

—— 1938c. Mating types in *Paramecium aurelia;* diverse conditions for mating in different stocks; occurrence, number and interrelations of the types. Proc. Amer. phil. Soc., 79: 411-34.

—— 1939. *Paramecium aurelia:* Mating types and groups, lethal interactions, determination, inheritance, relation to natural selection. Amer. Nat., 73: 390-413.

Sonneborn, T. M., and R. S. Lynch. 1934. Hybridization and segregation in *Paramecium aurelia*. J. exp. Zool., 67: 1-72.

Taliaferro, W. H. 1929. The immunology of parasitic infections. New York and London.

Tartar, V., and T. T. Chen. 1940. Preliminary studies on mating reactions of enucleate fragments of *Paramecium bursaria*. Science, 91: 246-47.

CHAPTER XVI

THE PROTOZOA IN CONNECTION WITH MORPHOGENETIC PROBLEMS

FRANCIS M. SUMMERS

THE QUEST for underlying causes of organic unity or individuality has been for many years one of the most intriguing and, at the same time, one of the most evasive problems in experimental biology. That an immense amount of work, involving a large variety of organisms, has produced little more than a background for what may be the ultimate solution is neither surprising nor discouraging. If analytical research cannot now detail all of the vital processes, then what progress should be expected in the full appreciation of the total relations between them?

One of the principal approaches to the evaluation of factors which condition growth and development is through studies of regeneration. This method consists essentially in disturbing the normal trend of development in order to obtain unusual or exaggerated expressions of one or more of the growth-conditioning or integrating factors. The fact that a fragment of an organism is frequently capable of regenerating a complete organism the total characteristics of which are homologous with those of the original is convincing evidence that some kind of fundamental organization characterizes the fragment and is responsible for the individuality of the new organism derived from it. Conversely, regional specialization or localization in the regenerating fragment presupposes the existence of overlying control mechanisms, i.e., the operation of factors or a set of energy transformations which restrict, direct, or determine the complete developmental freedom of its various parts or regions. If some such hypothesis of energy changes and translocations serves to fix our experimental point of view, then the many fragmentary accounts and dangling issues herein reported may serve as a challenge to those who are acquiring the newer instruments and techniques provided by the sciences.

In common with other organisms, the Protozoa exhibit such features as polarity and symmetry, which, according to Child (1920), appear to be largely independent of specific differences in the protoplasmic constitution. Should not the Protozoa, either as cells or organisms, present very suitable material for studying the potentialities of the protoplasmic constitution in a somewhat less complex setting than that which characterizes the higher invertebrates? Furthermore, does the disadvantage of the operative techniques involved with Protozoa offset the advantage of being able to disregard some of the physiological and mechanical relationships between the cells of compact tissues? The answers to these and other questions relative to the significance of Protozoa in connection with problems in regeneration, or their merits as material for investigating these problems, are to be attempted in this section.

It is the purpose of this section to organize under special headings the unassembled work having to do with regeneration phenomena in Protozoa. Microdissection or micrurgical studies which have as an objective the elucidation of the physical properties of protoplasm, the functions of fibrillae, membranes, and so forth, are to be treated only insofar as they have a bearing upon the topics under consideration.

PHYSIOLOGICAL REGENERATION

One aspect of regeneration pertains to those functions by which cells or organisms are able to maintain a certain structural and functional integrity, in spite of wear-and-tear processes of the normal life cycle. The maintenance functions are commonly referred to as physiological regeneration and they appear to have much in common with reparative regeneration.

As applied to Protozoa, particularly the Infusoria, the term has an added significance because it should be taken to include not only the gradual and continuous energy changes within the protoplasm, but also the grosser reorganizational changes encountered during periods of division, conjugation, endomixis, and encystment. In ciliates and flagellates generally, the structures observed to be affected include particularly the nuclei, the external motor organellae, and their associated fibrils.

It is beyond the scope of this section to consider those repair phenomena which follow the normal vegetative or cyclical variations in Protozoa. Of especial interest in this connection, however, is the "spontaneous" dedifferentiation and redifferentiation described for *Bursaria* by Lund (1917). Slightly starved *Bursaria* frequently undergo extensive reorganizational changes distinct from those which accompany division, conjugation, and so forth. According to Lund, this distinctive type of physiological regeneration has no apparent extrinsic cause. That such is the case seems improbable. Hetherington (1932) cultivated *Stentor coeruleus* under more carefully controlled conditions and was unable to find cases of spontaneous regeneration except when the environment was unfavorable. Starvation, wide variations in oxygen tension and organic content of the medium are apt to initiate these reorganizational changes. Hetherington maintains that physiological regeneration occurs very infrequently, if at all, and that it has never been demonstrated in a known medium. Both Lund and Hetherington are inclined to employ the term physiological regeneration in a special sense, not including the cyclical changes already mentioned.

Periodic reorganizations during periods of partial or complete starvation in *Stylonychia mytilis* were recently reported by Dembowska (1938). In conductivity water *Stylonychia* lives for fourteen to nineteen days. During this time the organisms undergo repeated processes of complete body reorganization by renewing the entire ciliature and by nuclear reorganization. The reorganizations are increasingly frequent up to a certain point and are unaccompanied by manifestations of division.

SOME OF THE FACTORS IN REGENERATION

1. EXTERNAL ENVIRONMENT

Calkins (1911b) and especially Peebles (1912) noted a definite correlation between the regenerative behavior of *Paramecium caudatum* and the periods of depression to which their cultures were subject. During such periods pieces cut from organisms, which under more favorable conditions exhibited a greater capacity to regenerate, were unable to regenerate or to divide. These authors concluded that when paramecia are starved or are undergoing periods of depression from other causes, the division rate is greatly diminished and the reparative activities greatly reduced or altogether lost.

Diminutive but perfectly formed regenerates were obtained when Sokoloff (1923) cut slightly starved *Bursaria* into several parts. These regenerates were proportionate in size to the pieces from which they were derived. In other words, within reasonable limits, inanition inhibits growth but does not impair the ability to regenerate; the latter remains the same as for normally fed specimens. Extreme starvation leads to a state of depression in which fragments attempt but never achieve full structural restitution. The cases of physiological reorganization, as reported by Lund (1917) and Hetherington (1932) for *Bursaria* and by Dembowska (1938) for *Stylonychia,* confirm the validity of Sokoloff's conclusion.

Chejfec (1932), following Balbiani (1893), Calkins (1911b), and Peebles (1912), accepted the possibility of a low potential for regeneration in *Paramecium caudatum.* He attempted to increase the regenerative power in this species by altering the environment in several ways. The experimental organisms, placed under conditions of starvation or in acidified media (initial pH 4.5-6.0), regenerated more readily than the control group grown in ordinary hay infusion.

A variety of alkaloids such as morphine, strychnine sulphate, and so forth, when administered in sublethal concentrations to *Blepharisma undulans,* cause the pink-colored pellicle to be discarded (Nadler, 1929). Animals so treated readily regenerate new pellicles if returned to the customary medium. If they are left in the alkaloid-containing medium, the pellicle does not reappear. Naked but otherwise normal organisms were maintained in this fashion for 110 days by Nadler.

Since reparative processes are set in motion by unfavorable alterations of an organism or the medium in which it lives, almost the entire subject of regeneration could be exhaustively but not effectively treated under this heading. The many other pertinent publications are discussed throughout the chapter.

2. CYCLICAL VARIATIONS

Division cycle.—The experiments of Calkins (1911a) on *Uronychia transfuga* were designed to test whether or not Protozoa are capable of regenerating with equal facility during the various phases of cell division. It appears that prior to that time regeneration was generally assumed to be independent of cyclical phenomena. Operations performed on *Uronychia* shortly after cell division showed the regenerative capacity of the individual to be feebly developed. The presence of both macronucleus and micronucleus was essential for such cases of regeneration as did occur at this period. At the mid-interphase, eight to sixteen hours after division, the tendency to regenerate was slightly greater than before. These fragments sometimes regenerated in the absence of micronuclear material. The highest percentage of successful regenerates was obtained as the cells entered upon the division period. When *Uronychia* was cut in the early prophase, either transversely or obliquely in a region anterior to the presumptive division plane, three complete individuals were produced (Fig. 179). Three individuals instead of two resulted in such cases, because the predetermined division plane was not appreciably

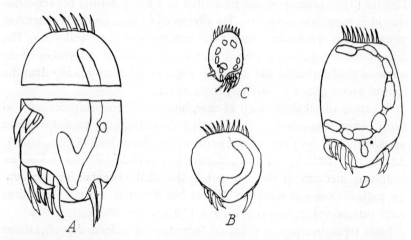

Figure 179. Regeneration in *Uronychia transfuga*. A, transection made anterior to the division plane in mid-division phase (B, C, and D show the results after twenty-four hours) ; B, small amicronucleate individual derived from the anterior fragment; C and D, anterior and posterior daughters produced when the large posterior fragment divided. (From Calkins, 1911a.)

altered by the type of cut just described. The small amicronucleate anterior piece regenerated a diminutive organism whose morphological features were normal. Fission occurred in the large posterior fragment soon after the healing of the cut surface. A small anterior regenerated daughter and a normal posterior daughter were the products, both of which were normal as regards nuclear apparatus. The regenerative power diminished in the mid-division phases, inasmuch as one or another of the three cells frequently failed to regulate. As the division process drew to a close, reparative regeneration reached its lowest ebb; those which did regenerate were low in vitality and short-lived. In regard to the progressive development of regenerative power from one division to the next, Calkins's experiences with *Uronychia* have been repeated and reaffirmed by D. B. Young (1922) and Dembowska (1926). Similar results were also reported by M. E. Reynolds (1932) for an amicronucleate race of *Oxytricha fallax*.

To the contrary, in *Spathidium* there appears to be no correlation between the degree of regeneration and the division cycle. Before, during, and after fission, small pieces excised from either extremity disintegrate immediately or regenerate without dividing. Large nucleated fragments nearly always regenerate, irrespective of the stage at which they are taken (Moore, 1924).

Peebles (1912) cut *Paramecium caudatum* at two different division stages without discovering any marked differences in the regenerative ability. Both halves of those which were transected in the division plane, at a time when the macronucleus was elongated and the body slightly constricted, usually survived to produce normal descendants. Traumatic effects were reduced in operations made during the later stages. Peebles also made note of the fact that while the power of regeneration is present in cells obtained from two to five hours after separation, approximately 90 percent of the operated individuals died as a result of injury. This was attributed to lowered viscosity in the cytoplasm of growing paramecia, such that an injury to the ectosarc allows the endoplasm to escape. The surviving organisms regenerated as readily as those cut in the late interphase.

Taking exception to Calkins's statement that the "power of regeneration" varies in different stages of the division cycle, Tartar (1939) restates the problem in terms of the observed data, without reference to the "power" of regeneration. Calkins's experiments

reveal that specimens of *Uronychia transfuga* are able to regenerate morphologically without the presence of the micronucleus when the transection removes this structure before or during division, and not when it is removed after division. The fate of the amicronucleate fragments was apparently not followed long enough to determine to what extent subsequent division is possible in the absence of the micronucleus, but it is probable from the work of others (e.g. Moore, 1924) that division would not have taken place without the micronucleus. It is for this reason that my restatement of Calkins' results restricts the restoration to morphological regeneration. A corollary of the statement is that when a fragment contains both nuclei, it regenerates regardless of the time of cutting during the division cycle. Calkins presented exceptions to this corollary: fragments from cells cut immediately after division which contained both types of nuclei did not regenerate. A more complete history of fragments than he presented would, however, be necessary to ascertain whether their failure was due to a decrease in the "power of regeneration" or merely to unsuccessful recovery from the operation. These cases seem to indicate that just after division animals may fail to regenerate even though they contain the full nuclear complement; but such failures were certainly the exception and not the rule (four cases out of twenty-two), so that demonstration of a fundamental decrease in the "power of regeneration" during this period, exclusive of the subtraction of one of the nuclei, was not intended [p. 199].

From data including only fifty operations made during and after division in one race of *Paramecium caudatum*, of which twenty-four successfully regenerated and subsequently divided, Tartar concluded that the division cycle in this species does not influence regeneration, even when both nuclei are present.

Evidence of progressive physiological differentiation between divisional periods, other than the increasing ability to regenerate, is the determination of the division plane. Lewin (1910), Calkins (1911b), and Peebles (1912) have repeatedly demonstrated its occurrence in *P. caudatum;* other amply described cases are those of *Uronychia* (Calkins, 1911a; D. B. Young, 1922; Dembowska, 1926) and Oxytricha (M. E. Reynolds, 1932). All of these species may be transected above or below the mid-region sometime prior to cell division, without injury to this definitive region. As already described for *Uronychia,* the larger piece divides unequally through what was originally the mid-region of the intact cell. Peebles identified the division plane as early as 2.5 hours after fission. She thought that several division planes develop in vegetative cells when fission has been delayed for a time. This would account for the fact that several divisions follow in rapid succession when one of the extremities is cut away from such an individual.

Conjugation.—Four giant races of *Paramecium caudatum* were reported by Calkins (1911b) as having different regeneration potentials. The incidence of regeneration in three of the races, those having very restricted regenerative powers, was found to be greater in ex-conjugants, or in cells that were operated during conjugation, than in the ordinary vegetative individuals.

The merotomy studies on *Uroleptus mobilis* (Calkins, 1921) proved that conjugating pairs may be cut apart without perceptibly altering the trend of events, which, once started by conjugation stimuli, continue to completion. The regeneration requirements, when superimposed upon the internal readjustments already in progress as a consequence of the sexual processes, appear only to prolong redifferentiation in the ex-conjugants. Calkins dissected away the apical protoplasmic junction containing the migratory pronuclei and thereby eliminated the amphinucleus in both conjugants; nevertheless the resulting cells regenerated completely and with full restoration of their vegetative and reproductive powers. Although cytological details are not supplied in this article, as seen in the living material the reorganization processes without exception followed the same general sequences as outlined for the normal ex-conjugants (Calkins, 1919).

In *P. caudatum,* anterior or posterior cut-offs made during conjugation result in a large number of fatalities. Those which survive regenerate slowly and eventually divide (Peebles, 1912).

Cases of autogamy following the separation of conjugants were recently described by Poljansky (1938) for *Bursaria.* The processes of sexual differentiation continue in the majority of individuals derived from pairs split apart four to six hours after the onset of conjugation. In these instances autogamy supplants heterogamy; the two native pronuclei of each cell fuse to produce the amphinucleus, from which the nuclear apparatus of the reorganized cell originates.

A somewhat different situation prevails in *Spathidium.* Tests for regenerative ability were made at three different phases of the conjugation process by Moore (1924). The fragments obtained while maturation is in progress, but before the exchange of pronuclei, do not regenerate. Those obtained immediately after fertilization regenerate fully, provided the amphinucleus is included. Ex-conjugant pieces containing only the degenerating macronuclei and the maturation by-products achieve a partial restoration of form, but they subsequently dedifferentiate and never divide.

Encystment.—Slightly starved pre-cystic individuals of *Spathidium spathula* are capable of regulation in a fair percentage of cases. The usual consequence of cutting at this stage is immediate dedifferentiation and encystment. Interestingly enough, the external form of a regenerator may be restored before encystment, but nevertheless this cell dedifferentiates and then encysts in order to complete the nuclear reorganization already in progress. Young excysted specimens have a lowered regenerative capacity because physiological restoration is incomplete; with increasing age, the normal vegetative existence is resumed and the regenerative capacity is restored to the normal level (Moore, 1924).

Recently Garnjobst (1937) described an interesting case of what she called regeneration cysts in *Stylonethes*. Anterior and posterior halves of bisected organisms round up and secrete a cyst wall about themselves. Regeneration occurs within the cysts. Excystment is spontaneous, releasing minute but perfectly formed individuals. Garnjobst maintained these excysted regenerates until they formed reproductive cysts, the normal condition for binary fission in the species.

3. RACIAL DIFFERENCES

The amount of literature on genetics and regeneration is pitifully small. Perhaps this is due, in part, to the difficulty of making clear-cut distinctions between the total inherent capacities and the strictly extrinsic factors which are responsible for variations in the expression of these capacities.

In contrast with Balbiani's (1893) conclusion that *Paramecium caudatum* does not regenerate as do other Protozoa, Calkins (1911b) found that the power of regeneration varied in different giant races of *P. caudatum*. Of these races for which data are given, one produced regenerates in approximately one percent of the cases; another race produced 10 percent; and a third produced 30 percent. Mention is made of a fourth race, which showed 100 percent regeneration (data not given). Peebles (1912) likewise reported four races of *P. caudatum* which, according to her conclusions, showed wide variations in regenerative power. Regeneration in the different races varied from 23 percent to 67 percent for anterior cut-offs and from 25 percent to 100 percent for posterior sections (her Table 3, p. 164). In view of her generalizations that *"Paramecium* taken from a pure line will regenerate in ninety cases out of a hundred if the cytoplasm is in a viscid state and the animals are well-fed" (p. 165) and that "The power to regenerate is not so much a characteristic of the race as it is an indication of the vitality of the individual cell" (p. 165), the data tabulated in percentages only do not distinguish the racial from the individual differences.

M. E. Reynolds (1932) described experiments with a non-regenerating species (undescribed) of *Euplotes*. The fact that *E. patella* regenerated very well under the same conditions led her to the conclusion that "under identical conditions there is a difference in the regenerative power of these two races of Euplotes" (p. 353). It should be noted that none of the undescribed species survived the operations; all distintegrated during or immediately after the operation.

Of particular merit is the work of Tartar (1939), in which rigid criteria were set up for distinguishing between recovery and regeneration and between form restoration and complete regeneration. Twenty-five races, representing seven species, were used in the experiments to determine whether there are racial or specific differences in the regenerative ability of *Paramecium*. A total of 865 anterior transections were performed, in 509 of which the organisms survived. Morphological and complete physiological regeneration occurred in 98 percent of the survivals. Having established from minor operations the working hypothesis that "any *Paramecium* able to recover from the injury of cutting is able to regenerate completely" (p. 196), Tartar performed operations of a more serious character—excisions involving the major portion of the posterior end—upon five races of *P. caudatum*. Complete regeneration occurred in 93 percent of the 121 survivals. According to these results, the ability of *Paramecium* to regenerate is much greater than previous investigations have shown and, furthermore, the regional differences in individuals of the same species are relatively slight. The final conclusion that there is no racial nor species variation in regeneration of cells, even after quite severe cuts, may be subject to further qualification. Although the compiled data show no large order variations in the incidence of regeneration among the different species or races of *Paramecium*, there remains the possibility of genetic differences in the capacity for regeneration after repeated injuries or in the rate of regeneration.

4. DEGREE OF INJURY AND REORGANIZATION

In many Protozoa the extent of cytoplasmic dedifferentiation varies with the degree of injury; but is does not necessarily follow that a small or even a large operation will be followed by complete resorption and reintegration of the cell structures. In holotrichs, *Loxophyllum* for example (Holmes, 1907), the mode of restoration may be simple and direct, with the alteration of uninjured, preëxisting parts reduced to a minimum.

Differences in the degree of reorganization in anterior and posterior halves of *Spathidium spathula* are clearly described by Moore (1924). In this species the alterations prior to restoration do not follow the same course, but vary with the size of the fragment and with the condition of the organism at the time of cutting. She observed no instance in which the cytostome remained entirely unchanged throughout the course of regeneration in anterior halves. Shortly after the cut surface heals, anterior pieces begin to round out at the apex. The assumption of spherical form extends to the oral region, the oral parts become less distinct, the neck disappears completely, and finally only an indication of the oral lips is apparent. Occasionally dedifferentiation proceeds further, the fragment forming a complete sphere with all traces of the oral apparatus totally obliterated before redifferentiation sets in. The posterior halves require a longer time for the restoration of form, although they contain none of the original oral parts to be resorbed and remodeled.

Heterotrichs are not strikingly different from holotrichs in regard to the extent of dedifferentiation following injury. The greater the relative size of the regenerator in such forms as *Spirostomum* (Sokoloff, 1923), the greater its regenerative capacity, and the sooner are such parts restored to complete, full-size individuals. According to Moore (1924), the oral structures in anterior pieces of *Blepharisma* do not dedifferentiate unless they are injured when the cell is operated upon. A somewhat similar condition obtains in Bursaria (Lund, 1917). In this heterotrich the *Anlage* of the gullet in reorganizing individuals may be torn slightly without provoking complete dedifferentiation. The injured region persists as an abnormal part of the gullet, which is normal in all other respects. This appears to be a clear-cut demonstration of embryonic localization as found in metazoan embryogeny, a definite spatial correspondence between the undifferentiated and the fully formed parts. In *Bursaria,* however, there is considerable variation in the degree of injury required to produce more drastic regenerative measures. If sufficiently great, an injury to the gullet rudiment causes complete dedifferentiation, such that all visible traces of the rudimentary structure disappear before redifferentiation begins. The regenerative processes in hypotrichs tend to obey an all-or-none rule for, once initiated by a requisite degree of injury, the dedifferentiation proceeds to completion. It is noteworthy, moreover, that minimal injuries to certain of the motor organelles will provoke reorganizational activities as far-reaching as those which occur at the time of division.

Taylor and Farber (1924) removed small drops of endoplasm from

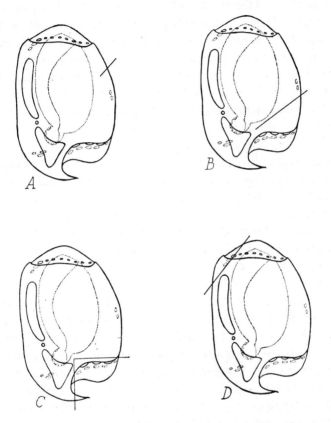

Figure 180. Regeneration in *Uronychia uncinata*. A, shallow marginal incision which did not injure any of the motor organelles, the healing of the wound was not accompanied by cell reorganization; B, deep incision not involving motor organelles, followed by complete reorganization; C, excision involving the large posterior cirri, followed by complete reorganization; D, excision of anterior portion, including several membranelles, followed by complete reorganization. (From Taylor, 1928.)

Euplotes patella without producing serious structural changes through dedifferentiation, although death invariably occurred within two to three days, if the micronucleus was taken out. In this and a good many other species, shallow marginal incisions, or wounds from small excisions, heal

almost immediately, whereas large-scale operations involving some of
the kinetic organelles are apt to incite breakdown changes that lead to
the resorption of the entire ciliary apparatus (Fig. 180). Extreme cases
were reported by Dembowska (*Stylonychia,* 1925; *Uronychia* and *Stylonychia,* 1926) and Taylor (*Uronychia,* 1928) in which the removal
of a single cirrus or even severe injury to its basal plate is sufficient to set
in motion the entire regenerative functions. External portions of the
giant cirri on *Uronychia* may be excised with none of these consequences
(Taylor).

For those hypotrichs studied by Dembowska (1926) and M. E. Reynolds (1932), the duration of the regenerative process is independent
of the degree of injury, requiring from three to five hours for complete
restitution in all cases. The time interval between cutting and the initiation of reorganization varies for different genera. Generally, the greater
the injury, the shorter the interval.

Tittler's (1938) experiments on *Uroleptus mobilis* were designed to
test whether or not ciliates injured or mutilated by high-tension currents
undergo reorganization and regeneration comparable to that which follows other types of mutilation. The induction current caused most of
the organisms to migrate toward the cathode of the break shocks. Approximately 75 percent of the exposed organisms were vacuolated or
deformed at the posterior end, i.e., the extremity nearest the anode. One-
minute exposures sufficed to produce many fragments with posterior de-
ficiencies and a few with anterior injuries. Regeneration, accompanied by
a complete de- and redifferentiation of cortical organelles and macronuclear reorganization, was the rule. Most of the reconstituted individuals
were ready to divide within thirty-four hours after treatment. The destroyed micronuclei were replaced from those remaining. Nuclear clefts
appeared in the intact macronuclei; the latter fused into a single mass
before constricting into the eight parts which characterize the normal
individual.

5. THE SIZE FACTOR

Surprisingly large portions of the cytoplasm may be removed from
most Protozoa without permanently impairing the vital processes; under
favorable growth conditions, complete regeneration occurs within a relatively short time. Moreover, the regenerative capacity can be tested further by repeated excisions, so that the reparative processes are made to

function almost continuously (cf. Hartmann, 1928). Is there a minimal protoplasmic mass of definite size within which the organization of the species can find latent expression—i.e., what is the smallest fragment capable of regeneration?

After shaking *Stentor* into fragments, Lillie (1896) learned that the smallest nucleated piece of *S. polymorphus* to become a perfect form was equal to a sphere of approximately 80 µ in diameter. Since the average diameter of the normal forms was 230 µ, the smallest piece was about one twenty-seventh of the original. The same approximate proportions were obtained for *S. coeruleus*. For the latter species, Morgan (1901) obtained minimum values which were even less than those of Lillie. The smallest regenerates produced were one sixty-fourth of the volume of large normal stentors. According to Morgan, the conclusion that a piece one twenty-seventh or even one sixty-fourth of the entire animal can produce a new individual gives only a general idea of the relative size of the minimum reorganization volume. More depends upon the size of the normal individual than on that of the smallest piece, for normal size limits vary considerably; a large normal individual may contain eight times the volume of a small normal cell. The absolute size of the smallest fragment is of greater significance than its relative size.

Sokoloff (1923, 1924) was mainly interested in the minimum volume necessary for regeneration and the rate of regeneration in pieces cut from different regions. For *Spirostomum* the minimum volume was between one fifty-third and one sixty-ninth of the initial volume, and was not the same for pieces taken from different regions. In general, he found different rates in pieces of different sizes; within certain limits, the larger the piece the greater the rate. A comparison of his results with four genera is given in the accompanying table (Table 21). In a type like *Spirostomum*, the substance of which is qualitatively different as regards ability to regenerate, the minimum volume and rate of regeneration varied according to the region from which it was taken. In other Infusoria (*Dileptus, Bursaria,* and *Frontonia*) the substance of the body appeared to be qualitatively similar throughout (Sokoloff, 1923).

Moore (1924) obtained successful regenerates from *Spathidium* fragments as small as 1.3 percent of the original volume. The minimum reorganization mass in *Chaos diffluens* was determined to be one eightieth of the normal volume (Phelps, 1926).

The proportionality of the regenerated structures to the size of the parts from which they regenerate has been emphasized by Morgan (1901), Sokoloff (1923), and Dembowska (1926). The latter noted that even the most minute parts of a small *Uronychia* regenerate were of proportionate size. The small regenerators retained none of the large, original cirri. The new cirri developed from corresponding *Anlagen* and grew only until proportionate size was attained.

The quest for minimum reorganization volume in Protozoa originated with those embryologists who were interested in the totipotency of early blastomeres and gastrula fragments. Perhaps it has been partly successful in demonstrating that a specific organization can be contained

TABLE 21: TABLE OF MINIMUM VOLUMES NECESSARY FOR REGENERATION
(Sokoloff, 1923)

TYPE	AVERAGE LIMIT OF REGENERATIVE CAPACITY	LIMITS OF REGENERATIVE CAPACITY OF DIFFERENT PARTS		
		Front	Middle	Hind Part
Spirostomum ambiguum Ehrbg.	1/61 of volume (1/53 to 1/69)	1/53	1/68	1/67
Dileptus anser O.F.M.	1/72 of volume (1/70 to 1/75)	1/70	1/72	1/73
Bursaria truncatella O.F.M.	1/35 of volume (1/30 to 1/45)	1/36	—	1/34
Frontonia leucas Ehrbg.	1/5 of volume (1/4 to 1/5)			

within bits of protoplasm much smaller than previously shown for separated blastomeres, and that the disadvantages of small volumes for the mechanical processes involved in cleavage, gastrulation, and so forth, are not necessarily the reasons for developmental failure in small blastomeres. The knowledge that minute fractional parts of a protozoan are capable of regenerating also has a practical value in experimental work. The assignment of a certain experimentally determined value as either the absolute or the relative minimal mass does not greatly enhance the theoretical implications of these works. We learn that one fifty-third or one eightieth of the protoplasm of a cell of a given species regenerates under favorable conditions, but, as Lillie (1896) aptly remarked, this does not imply that fifty or eighty diminutive regenerators can be obtained from the original cell. In uninucleate types, only one such fragment may be obtained. It is also apparent that parts of the original derived organization, cortical organelles, oral structures, and so forth are not requisite for the survival and regeneration of the smallest fragments. The requirements are primarily qualitative: a representative bit of the basic protoplasmic organization. The matter of obtaining a viable nucleus or fragment of the macronucleus has had much to do with the values so far determined.

6. THE NUCLEI IN REGENERATION

More than a century after Rösel (1755) distinguished ecto- and endoplasm in amoebae by surgical methods, the works of Brandt (1877), Nussbaum (1884), and Gruber (1886) inaugurated an era of active interest in protozoan physiology. Although not in general agreement as to the sensory and reactive capacities of enucleated fragments, these investigators, together with Balbiani (1888), Verworn (1888, 1892), Hofer (1890), Lillie (1896), Prowazek (1904), Popoff (1907), and many others, showed that both nucleus and a certain amount of cytoplasm are essential for the continuation of the vegetative, reproductive, and reparative activities of the cell. The consensus of these early publications is aptly stated by Minchin (1912): "Non-nucleated fragments may continue to live for a certain time; in the case of amoeba such fragments may emit pseudopodia, the contractile vacuole continues to pulsate, and acts of ingestion and digestion that have begun may continue; but the power of initiating the capture and digestion of food ceases, consequently all growth is at an end, and sooner or later all non-nucleated bodies die off" (p. 210).

There still exists a diversity of opinion regarding the essential nature of the nucleus or dimorphic nuclei in regenerative or regulative functions. Apropos of the latter, Moore (1924) places a greater premium than most upon ultimate physiological recovery:

In general previous investigators have centered their attention upon morphological regeneration alone, and have considered the restoration of external organelles as sufficient evidence of a completion of the process. Since form regulation in the Protozoa is of little value if not accompanied by the ability to continue normal existence, it would appear that a more valuable definition of regeneration is one which includes restoration of function as well as of structure [p. 250].

Notwithstanding the evident truth that maintenance and perpetuation are ends to which all of the vital functions are directed, our criteria of recovery need not be so inclusive, if the practical objectives of the experimental laboratory are to be realized. To those interested in the mechanisms of form determination, the ultimate fate of a regenerate is of secondary importance. Pieces from starved Protozoa do not grow, but they are frequently capable of regenerating new individuals of proportionate size. Amicronucleate fragments of some ciliates are sometimes capable of regenerating without being able to maintain the redifferentiated state or to divide. The success of regenerative processes in such instances is limited only to the extent to which they are correlated with the general maintenance functions.

According to Stolĉ (1910), cytoplasmic bits of *Amoeba* may live for as much as thirty days and are able to prehend, ingest, digest, and assimilate food. In agreement with Verworn, Lynch (1919) found evidences of all the usual catabolic activities, but his enucleated amoebulae showed no symptoms of growth, regeneration, or division. A few years later, Phelps (1926) concluded that merozoa from *Amoeba* do not carry on any of the fundamental body processes except locomotion. Nevertheless, she did note an increase in the number of crystals within the fragments which, according to her own criteria, denotes metabolic change. Others have reported only dissociated movements in cytoplasmic fragments of *Amoeba* (Willis, 1916; Mast and Root, 1916). Quite a number of merotomists have recorded instances of fusion between enucleated bits of cytoplasm with the parent cell in various Foraminifera, Heliozoa, and testate rhizopods (see p. 793). In *Actinophrys sol* the fragments fuse with one another (Looper, 1928).

The survival of enucleated ciliate fragments, even for short periods, is not the general rule. There appear to be only few instances recorded in the literature in which such fragments show any tendency to redifferentiate. Gruber (1886) reported that redifferentiation occurs in enucleated fragments derived from dividing stentors, provided a developing peristome is included in each piece. Cytoplasmic portions of *Blepharisma* are sometimes capable of form regeneration, but since they are unable to feed involution and cystolysis soon occur (Moore, 1924).

Garnjobst (1937) found evidence of secretory activity on the part of enucleate fragments of *Stylonethes*. Cyst formation after injury is the rule in this species. In one case a cyst wall was produced by a fragment which contained no nuclei.

The possibility of regeneration in Infusoria is not precluded by the absence of either micronucleus or macronucleus. It is difficult, however, to formulate any satisfactory conclusion at this time, inasmuch as observations on different species often present contrasting results. In spite of the accumulated literature, the respective rôles of each of the dimorphic nuclei remain uncertain.

Gruber (1886) regarded the micronuclei in *Stentor* as of secondary importance, since no regeneration occurred until a macronucleus differentiated from one of the amphinuclear products. Lewin (1910) agreed with Gruber and did not believe the micronucleus necessary for growth or regeneration in *Paramecium caudatum*. On the contrary, Calkins (1911b), Peebles (1912), and Schwartz (1934) obtained regenerates in this species only when both types of nuclei were present.

In Stevens's (1903) experiments on *Lichnophora auerbachii*, a species with a beaded macronucleus and one micronucleus, regeneration was limited to the production of a few oral cilia (*in situ*), new peristomes, and relatively small segments of the attachment disc. She showed that only pieces including the attachment disc, the neck, a quarter section of the oral disc, and representatives of both nuclei were capable of regenerating "fairly normal" individuals. Isolated oral discs (amicronucleate) and basal discs (with micronuclei) never regenerated, although they survived for several days. More recently, Balamuth (MSS, 1939) studied regeneration in *L. macfarlandi* and found its regenerative capacity to be greater than Stevens showed, since isolated amicronucleate oral discs definitely replaced injured adoral zones. Missing basal discs were not regenerated in *L. macfarlandi,* but it was pointed out that the two daughter basal discs form in this genus only by division of the corresponding parent structure.

Calkins (1911a) and D. B. Young (1922) described regeneration in amicronucleate fragments of *Uronychia* when the operations were made very late in the interdivisional period. The regenerates usually became abnormal after three to four days; they apparently starved to death for, according to Young, no food was ingested or assimilated. The fragments were also incapable of dividing. In the case of *U. setigera*, normally with one micronucleus, only one of the two halves of a bisected cell was able to divide, although form regeneration occurred in both. In *U. binucleata*, with two micronuclei, both halves were successful in regeneration and division. It appears from these observations that while form regeneration is possible in amicronucleate *Uronychia* fragments under special conditions, the micronucleus is indispensable for ultimate physiological regeneration. Dembowska (1926) was unable to find evidences of regeneration in amicronucleate pieces of either *U. setigera* or *U. transfuga*.

Euplotes patella does not live longer than a few days or divide more than twice, in the absence of the micronucleus. Using a mercury micropipette, Taylor and Farber (1924) demonstrated that portions of the cytoplasm or macronucleus could be removed without serious consequences, whereas the removal of the micronucleus with only a small volume of the surrounding cytoplasm ultimately proved fatal. In several of their experiments on *E. patella*, the micronuclei were completely withdrawn from the organism and then immediately replaced. The animals so treated subsequently gave rise to vigorous cultures. They concluded that the micronucleus plays more than a germinal rôle in the life history of *Euplotes*.

A majority of the publications bearing on regeneration in ciliate fragments contain casual references to the nonviability of amacronucleate pieces, but very few have treated the matter at length.

The vegetative individuals of *Blepharisma undulans* (Moore, 1924) are incapable of regenerating in the absence of macronuclear material. However, if cuts are made during the early phases of division, such pieces sometimes regenerate. Moore found that regeneration was followed by immediate dedifferentiation in all of these cases; the dedifferentiated fragments ultimately disappeared without dividing, in spite of the fact that some of them were known to contain as many as six micronuclei. According to the evidence presented, there appears to be no doubt that the restoration of external organelles may proceed in the absence of the macronucleus *if* the initial steps in the division process are under way at the time of cutting. From a posterior fragment possessing only a part of the developing membranelles, an organism of nearly perfect form may arise. The peristome, adoral zone, and undulating membranelles differentiate, but a new mouth fails to develop and the degenerative changes immediately appear.

Fortner (1933) expressed the single, large macronucleus from the body of gastrostyla-like hypotrichs (species?) by gently compressing the organism between cover glass and slide. Such operations were followed by a diminution in general mobility and contractile-vacuole pulse rate and by the persistence of food vacuoles. Schwartz (1935) concluded that amacronucleate pieces of *Stentor coeruleus* containing only micronuclei are like totally anucleate specimens as to regenerative power, digestion, and length of life. The amacronucleate pieces merely regulate external form, without undergoing the sequence of changes involved in physiological reorganization.

In physiological regeneration following conjugation, autogamy, endomixis, or analogous processes, the dimorphic nuclei have a common origin from the synkaryon or from some micronucleus-like body, but there is no convincing evidence that the micronuclei of strictly trophic individuals are able to replace artificially removed macronuclei. It is equally true that macronuclei do not give rise to new micronuclei. Numerous attempts to create amicronucleate races operatively have not met with success. The operated individuals survive no more than a few days and do not regenerate new micronuclei. The origin of amicronucleate races of *Spathidium spathula* (Moody, 1912), *Oxytricha hymenostoma* (Dawson, 1919), *O. fallax* (Woodruff, 1921; M. E. Reynolds, 1932), *Paramecium caudatum* (Lewin, 1910; Landis, 1920; Woodruff, 1921; Schwartz, 1934), *Didinium nasutum* (Thon, 1905; Patten, 1921), *Urostyla grandis* (Woodruff, 1921; Tittler, 1935), and possibly others, appears to be the result of anomalous metagamic or post-endomictic differentiation.

Small fragments of the macronucleus of ciliates usually are able to reconstitute the entire macronuclear system. In *Stentor,* for example, one or two of the macronuclear nodules ultimately regenerate the long moniliform chain. Schwartz (1935) thought that the process of reorganization in this species serves to regulate the size relationships of the different organelles. The surface relations of nucleus and cytoplasm are adjusted by fusion, division, or deformation of the macronuclear segments.

As far as regeneration is concerned, the macronucleus appears to be qualitatively homogeneous. Full regeneration is frequently possible when only a small piece of the macronucleus is present. The rate of regeneration in *Bursaria* is independent of the size of the macronuclear fragment included. The two halves of an individual, one containing a small and the other a very large portion of the macronucleus, usually regenerate in about the same length of time (Lund, 1917).

It is well established that micronuclei are capable of regenerating other micronuclei. A unimicronucleate section from a multimicronucleate species eventually regains the characteristic number, provided other conditions favor regeneration. There has been no confirmation of Lewin's (1912) experiments on *Stylonychia* in which, after merotomy, the micronuclei increased in number beyond that characteristic of the normal races. The effect of merotomy upon micronuclear number in *Pleurotricha* was investigated by Hewitt (1914), who concluded that merotomy does not permanently alter the numerical relationship between micronuclei and macronuclei. If at least a representative of each is present in the fragment, the normal nuclear complement occurs in the immediate regenerate and in its descendants.

The question arises as to whether or not the inability of an experimentally amicronucleated fragment to regenerate physiologically is likewise characteristic of the fragments derived from viable amicronucleate races. The contribution of M. E. Reynolds (1932) definitely answers the question: "The regeneration process in the fragments from the amicronucleate *Oxytricha* is similar to that which occurs in pieces from micronucleate *Oxytricha* which contain both types of nuclei" (p. 357). The absence of the micronucleus as a structural entity does not alter the course of regeneration. As in the normally constituted races, regeneration is accompanied by a protoplasmic reorganization involving the dedifferentiation of old and the redifferentiation of new ciliary fields from migrating *Anlagen*. The extent and time of regeneration, as well as the total number of regenerates obtained from the peculiar race of *O. fallax,* compare favorably with the normal strains. In view of the fact that experimentally amicronucleate fragments do not regenerate in the complete sense, we are obliged to assume that whatever the constitution of the usual micronucleus, it is somehow represented in the amicronucleate races. Following Woodruff's (1921) interpretation, Reynolds regarded the nuclear mass of the amicronucleate race as an undifferentiated product of the amphinucleus.

BEHAVIOR OF FRAGMENTS: GRAFTING AND REINCORPORATION

Balbiani (1888) and Verworn (1889) were first to remark that parts of ciliates move in the general manner of the entire animals. In his work on *Stylonychia, Oxytricha,* and *Paramecium,* Jennings (1901) emphasized body shape and the activity of heavy oral cilia as determining factors in the characteristic swimming movements, whereas Ludwig (1929) ascribed the spiral progression of ciliates to a more vigorous activity of the aboral rather than the oral cilia. Bullington (1925) and Horton (1935) discovered that the spiral movements do not necessarily relate to the configuration of the oral groove. It appears that portions of Infusoria in which oral cilia are lacking react in a fashion much the same as that of the whole organisms; and, further, that the spiral motions are not entirely dependent upon body shape or the activity of oral cilia, but are produced by the coördinated activity of the body cilia in general (Horton).

The responsiveness of ciliate fragments to mechanical and chemical stimuli was investigated in some detail by Jennings and Jamieson (1902). They found that if the fragments were not too small or irregular in form, the motor and sensory capacities compared favorably with those of the intact individuals. On the contrary, Alverdes (1922) reported some degree of sensory localization, wherein only the anterior halves of *Paramecium* and *Stentor* were responsive to chemical stimulation. More recently, Horton (1935) ascertained the sensitivity of the anterior and posterior halves of *P. caudatum* to weak acid stimulation. The results are in general agreement with the earlier work of Jennings (1901). Unlike his predecessors, Horton noted that the posterior halves of this species are somewhat more sensitive to weak acid stimuli than are the anterior halves.

The anastomosis of protoplasmic streamers in myxopods and the recombination phenomena in Foraminifera (Verworn, 1892; Jensen, 1896) was known to pioneer students of the Protozoa. In recent years numerous other workers have discovered that excised, anucleate fragments of many types of Sarcodina will recombine with the parental cell mass. This fact is even more striking in virtue of the active part taken by healthy fragments to effect the union. Kepner and Reynolds (1923) reported more than one hundred experiments with several species of *Difflugia,* in which isolated pseudopodial fragments would again enter into the protoplasmic structure of the cell (Fig. 181). Fusion in *Difflugia* is species specific, occurs along the mid-region rather than at the ends of extended pseudopods, and is limited to the fusion between fragments and nucleate cells. The fusion of anucleate fragments with each other

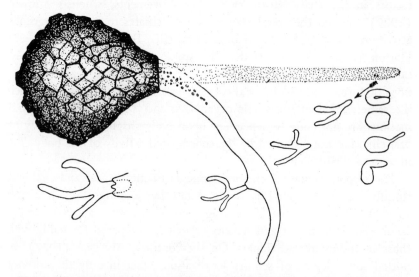

Figure 181. Reincorporation in *Difflugia pyriformis.* The small figures show form changes and directional movements of a fragment left by the sudden retraction of a pseudopod (stippled). The mode of fusion between the fragment and a second pseudopod is shown below. (From Kepner and B. D. Reynolds, 1923.)

was not observed, even when they were placed in contact. Healthy pieces were reappropriated after separation by distances as great as 1.5 mm.; even those of ectoplasmic composition appeared to move and orient with respect to the parent cell. Autoplastic and homoplastic fusions between individuals of *Actinosphaerium eichornii,* under conditions of slight compression, were made by Howland (1928). Permanent fusions were made between two medullary surfaces and between cortical and medullary surfaces. Axopodial fragments united with the individual at any point along an unsevered axopodium or with the cortical surface of the cell. Okada (1930) published fusion experiments on *Actinosphaerium* and *Arcella*, which gave results comparable with those described by Howland.

The reincorporation phenomena in some of the Sarcodina appear to offer a fertile field for studying the physical and chemical factors involved in the union of cells or fragments. As yet the attention of very few investigators employing Protozoa has been directed toward this end. Miller (1932) studied cytoplasmic reappropriation in *Arcella discoides* under the influence of different hydrogen-ion concentrations, as well as under the influence of a low-voltage *dc* current. Within the limits of pH 5.0 to pH 7.6, the rates of contact and fusion were not perceptibly altered, although in media having a pH lower than 5.0 or greater than 7.6 the reappropriation reactions were retarded. Likewise the low-voltage currents (0.3 to 2.1 *microamps*) failed to accelerate either contact or fusion. Miller regarded reappropriation as wholly beyond the control of pH or direct currents within the ranges tested. It occurred to Brehme (1933) that the various Sarcodina in which reappropriation has been reported are forms with relatively viscous protoplasm. She therefore attempted fusion experiments with *Amoeba proteus,* a type with less viscous protoplasm than those previously examined. Since *A. proteus* did not reappropriate its own or foreign fragments, she was able to conclude that increased viscosity may be an important factor in the fusion process.

Richard Hertwig's (1903, 1908) karyoplasmic relationship hypothesis has strongly influenced the more recent investigations of nuclear and cytoplasmic phenomena. Taylor and Farber (1924) excised varying quantities of the cytosome of *Euplotes patella* with a micropipette. Contrary to the results expected in accordance with Hertwig's rule, the operated individuals continued to divide in the vigorous, normal fashion. In *Chaos diffluens* the initiation of division was believed by Phelps (1926) to be a direct result of increasing the volume of cytoplasm, whereas the removal of cytoplasm retards growth, and presumably division, owing to the intervention of reconstruction processes. Similarly, Causin (1931) maintained that the excision of fragments in *Stentor* disturbs the K/P equilibrium and that this, in turn, initiates reorganizational activities which culminate in division or, more frequently, in physiological regeneration. During periods of physiological regeneration in *S. coeruleus,* the number of macronuclear segments is regulated to correspond with the size of the resulting cell, instead of showing any constant relation with the number of segments prior to the onset of reorganization. If all but one or two of the macronuclear segments are surgically removed, those remaining increase their surface values by elongation or deformation. In other instances the requisite equilibrium of surface relations between cytoplasm and nucleus is attained by nuclear coalescence or fragmentation (Schwartz, 1935). Previously Burnside (1929) made an unsuccessful attempt to set up divers biotypes by micro-dissection methods. In some of his experiments the quantity of nuclear material in one individual was two to three times greater than in another. Regulatory processes ensued, such that individuals with little nuclear material increased that amount; and individuals with large proportions of nucleus to cytoplasm decreased that proportion within a few generations.

Looper (1928) took advantage of the reincorporation phenomenon in *Actinophrys sol* as an approach to the K/P problem. By mechanical means, several individuals may be fused into temporary syncytia, and non-nucleated individuals may be fused with each other or with intact individuals. In this way he altered K/P either by increasing or decreasing the cytoplasmic volume. The division rate in this species was accelerated and the amount of nuclear material ultimately increased when fragments of cytoplasm were added to the cytosome. Removal of cytoplasm retarded the division rate. The same method was employed by Burch (1930) in studying the possible rôle of the karyoplasmic relationship as an inciting cause of cell division in pedigreed races of *Arcella vulgaris* and *A. rotunda*. Following Hegner's (1920) hypothesis that additional cytoplasm outside the sphere of influence of a single nucleus in rhizopods may stimulate that nucleus to divide, Burch made daily additions to or reductions in the volume of cytoplasm for different lines of his pedigreed strains. Trauma alone was found to have little effect upon the daily division rate. In general, the division rate varied directly with volumetric alterations of the cytoplasm, but these variations were not proportional to the volumes of cytoplasm gained or lost. As Woodruff (1905), Gregory (1909), Conklin (1912), Moody (1912), and others have suggested, Burch assumed that other intrinsic factors affect the division rate to a great degree. Our inability to distinguish between cause and effect further clouds the dying issue of the *Kernplasmaverhältnis* theory.

REGENERATION AND DIVISION

The lines of investigation summarized by Calkins (1934) and in part by Dixie Young (1939) predicate an increase of vitality through reorganizational processes in division, encystment, endomixis, and conjugation. The de- and redifferentiation incident to the physiological ("spontaneous") and reparative types of regeneration provide other opportunities for the renovation of derived structures. If we are correct in the

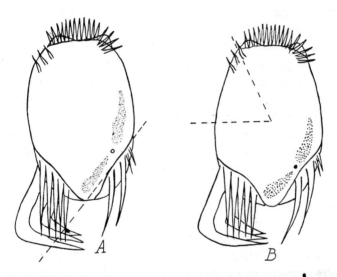

Figure 182. Divisional and physiological reorganization in *Uronychia*. Sister cells operated on four hours after division. Five hours after injury, the cell cut as shown in A regenerated by dividing. Physiological reorganization without division occurred in B after the same time interval. (Modified from Dembowska, 1926.)

belief that nuclear and cytoplasmic reorganization stimulates vitality, then regeneration in the strict sense includes not only the replacement of a part but also some degree of rejuvenescence of the whole cell. Is it a priori a beneficial process? At least it is one which can be provoked artificially.

To what extent are reproduction and regeneration dependent upon the same fundamental activities of the cell? Gruber (1886), Morgan (1901), Dembowska (1926), Causin (1931), and M. E. Reynolds (1932) have remarked that the initial steps in division and regeneration are comparable. Since the ends achieved by these processes are dissimilar, it would be of interest to discover the conditions under which differentiation leads only to restoration or to division.

Dembowska (1926) made an interesting discovery while working with *Uronychia*. Four hours after division, two sister cells were operated on in different regions. The individual cut as shown in Fig. 182 A divided five hours later, giving rise to two normal individuals. After the same interval of time the individual cut anteriorly (Fig. 182 B) began to regenerate in the usual fashion; it produced a single new individual. Dembowska suggested that the type of operation determines the mode of reorganization.

A similar condition obtained in M. E. Reynold's ·(1932) experiments on *Oxytricha fallax*. Amicronucleate individuals were able to regenerate when cut late in the interphase, from approximately five hours after one division until the beginning of the next. The majority of the variously operated organisms reorganized by division, rather than by the ordinary mode of restoration. In this species injury hastened the divisional process. Reynolds thinks that once the division processes are under way, they are completed in spite of moderate surgical disturbances.

Injuries sustained shortly before the onset of division are not repaired by divisional reorganization in some species. In *Paramecium caudatum* injured cells need not regenerate to divide, and may or may not regenerate before dividing (Calkins, 1911b). A truncated *Paramecium* frequently gives rise to a truncated and a perfectly formed daughter. The former occasionally divides again before regenerating (Fig. 183).

It is also true that regeneration may suppress or supplant division in certain species. Hartmann (1924), for example, maintained *Amoeba polypoda* in an undivided state for long periods by repeatedly cutting away portions of the cytoplasm. Each time that growth reached the point of an impending division, he removed cytoplasmic fragments amounting to as much as one-third of the total cell volume. In one experiment *A. polypoda* was operated on 21 times within 25 days. In this manner division was held in abeyance while the control strains divided 11 times. Another experiment continued for 42 days, during which the operated individual regenerated 32 times, as against 15 divisions for the controls. Phelps's (1926) experiences with *Chaos diffluens* were not in agreement with those of Hartmann. Successive excisions of one-fourth of the cytoplasm for at least 32 days did not prevent division in *C. diffluens*. Within 24 hours after each cut, the regenerative processes were complete and division occurred. Hartmann's initial attempts to substitute regeneration for division in *Amoeba proteus* met with little success. The animals lived for not more than 15 days. Later, however, he was able to contest

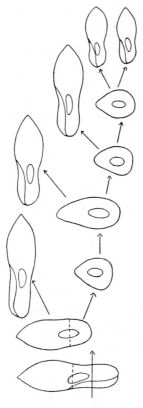

Figure 183. Diagram showing delayed regeneration in *Paramecium caudatum*. The nucleated posterior part of a *Paramecium* transected anterior to the division plane may divide several times before form restoration occurs in all of its descendants. (Modified from Peebles, 1912.)

Phelps's results by successfully culturing *A. proteus* through a long series of regenerations (Hartmann, 1928). One such series (protocol 3) was discontinued after a four-month period, during which 130 regenerations replaced the normal division process. In an equivalent period the control series produced approximately 65 generations. It is Hartmann's conviction that *Amoeba* can be maintained in a healthy undivided state for infinite periods by repeated operations.

The question of "artificial immortality" through regeneration was further investigated by Bauer and Granowskaja (1934b) in *Oxytricha*. Operations made subsequent to division, but before the end of the growth period, hastened the next division. When operations were made a short time before the appearance of division symptoms, regeneration supplanted division. Repeated operations of this kind had the effect of shortening the interdivisional period and, as a consequence, the successively operated oxytrichas progressively diminished in size until death occurred. The cytological and some of the physiological aspects of reorganization are given in a preceding publication (Bauer and Granowskaja, 1934a). Luntz (1936) suppressed division in *Stylonychia* sp. by subjecting these organisms to weak electric currents (0.95-1.0 *ma* and 2.0 *a*, voltage not stated) for approximately one hour each day. Division was averted for a period equal to twenty-seven to twenty-eight control generations, but each application of the current was followed by a transitory diminution in cell size.

The substitution of regeneration for division is suggestive, but does not prove that the vitality or longevity of the undivided cell is thereby increased. It is known that Protozoa which, for some reason, fail to divide for relatively long periods become morbid. Agonal symptoms and death are the usual accompaniments of the depression. But we have not yet succeeded in experimentally inhibiting division in a cell by means other than surgery without inducing adverse changes in its physiological behavior. It is significant, nevertheless, that physiological reorganization has been detected in a variety of species after mechanical injury, exposure to irritants, or during periods of starvation. The internal changes induced by these conditions appear to be restorative rather than adaptive.

More information is necessary before we can be confident of the conditions governing the mode of injury repair, whether by divisional or physiological reorganization. The nuclear and cytoplasmic alterations accompanying division have been studied extensively. On the cytological side, the reorganization processes in ciliates are most profound in the hypotrichs; and, as has been mentioned already, relatively small injuries often provoke a complete cycle of de- and redifferentiation. In the less specialized types, the internal changes at division are less striking. According to many observations, some of the parental organelles are passed unchanged to one or the other of the two daughters. It is in the hypotrichs that injuries are apt to be repaired during division. In the types in which dedifferentiation, preparatory to division, is less extensive, there appears to be a greater independence between regeneration and division. Whether or not the imminence of division reduces the possibility of independent physiological reorganization has yet to be determined.

POLARITY CHANGES AND PROTOPLASMIC STREAMING

The development of temporary heteromorphic individuals in *Bursaria* during physiological regeneration or in regenerating halves was first described by Lund (1917). The suppression of secondary axes and the reversal of polarity in the weaker member of the heteromorph occurred as a consequence of the dedifferentiation and reorganization of that region. In some instances the secondary axes were so feebly developed that a local reversed ciliary action gave the only clue to their existence.

The operations on *Mastigina hyale* by Becker (1928) have a more direct bearing on the mechanisms of polar organization. In normally creeping animals, the nucleus occupies a vacuolar space at the extreme anterior tip, and the endoplasm shows a type of fountain streaming, toward the anterior end in the center and posteriorly at the cell periphery. When individuals are divided into anterior and posterior halves, only the anterior half continues to move as before; motility and streaming are upset in the non-nucleated half, pseudopod formation is interrupted, and death soon occurs. The importance of the anterior extremity in the determination of polarity is demonstrated by the fact that decapitated cells behave just as posterior halves. If the anterior tip, together with the affixed nucleus, is pulled posteriorly with the surface gel, streaming and locomotion cease momentarily and then resume with reference to the new position of the shifted polar cap. But if only the nucleus is dislodged from its vacuole and expressed into the flowing endoplasm without severe injury to neighboring parts, the original polarity is undisturbed, which suggests that the nucleus alone is not responsible for the axiate pattern. Becker assumed that the resistance of the anterior peripheral layer to internal pressure is reduced by the physical presence of the "foreign body" in the gel. The kinetics of movement in *Mastigina* are interpreted in the following way: "it is the entire anterior tip of permanently gelled protoplasm which prevents gelation of the 'endoplasm' immediately behind it, and which by imperfect continuity with the outer gelled layer of protoplasm of the intermediate zone creates a circular zone of weakened elasticity or lowered resistance to the internal pressure of the plasmasol" (p. 113). The differentiated anterior region is therefore a prerequisite for normal streaming.

Fragments of *Paramecium caudatum* were examined by Hosoi (1937) in an effort to identify some of the forces involved in cyclosis. The organisms were narcotized with iso-propyl alcohol (Bills, 1922) and transversely sectioned at intervals along the primary axis. The type of protoplasmic streaming concerned with the formation and release of food vacuoles, the *Schlundfadenströmung* (Bozler, 1924), was evident only in pieces possessing a large portion of the gullet, whereas cyclosis in the strict sense occurred in all of the fragments within a few moments after operation. Hosoi found that the *Schlundfadenströmung* and the cyclical currents were largely independent phenomena, and that the nuclei play no direct part in effecting either of these movements. His suggestion that some special substances are attracted on the ecto-endo-plasmic interface, which serve as the generating force of the streaming movements, awaits further amplification.

PHYSIOLOGICAL GRADIENTS

Outstandingly important in morphogenetic studies was Child's dis-covery of physiological gradients in various animal types. After consider-able experience with metazoan forms, Child (1914) directed his atten-tion to Protozoa (*Paramecium, Stentor, Stylonychia, Vorticella,* and *Carchesium*), in which polarity is well defined. In these first experi-ments the direct susceptibility or resistance method was used, with several dilutions of KCN as the reagent in most cases. All of the forms showed the greatest susceptibility in the apical region, although local regions of still higher metabolic rate, such as the vacuolar regions in *Paramecium,* were sometimes found. These experiments brought out the fact that a close parallelism exists between the magnitude of the gradient and the general morphological and physiological differentiation of the cell. Peebles (1912) had previously noticed that anterior cuts cause greater physiological disturbances than posterior cuts in *Paramecium.*

The KCN susceptibility experiments were extended to amoeboid forms by Hyman (1917). Unlike the clearly polarized types, *Amoeba* was shown to have no permanent axial organization. A susceptibility gradient arises in *Amoeba* before a pseudopodium appears; it is greatest at the tip of the developing pseudopodia, and greater in the more recently produced than in the older ones. The bearing of axial gradients upon the physiology of amoeboid movement is discussed at length in Hyman's report.

Apparently the differential susceptibility of various regions in *P. caudatum* is not primarily dependent upon qualitative differences in the protoplasmic constitution nor upon the precise mode of action of the physical or chemical agent used to demonstrate it (Child and Deviney,

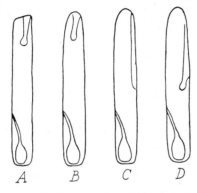

Figure 184. Successive stages in the regulation of the mouth in a piece of *Spirostomum* with the mouth at the anterior end. A, the operated individual; B, changes after one day— the rounding off of the cut surface; C, twenty-four hours later—disappearance of the original mouth and the development of a shallow peristomial depression on the side; D, appearance of the new mouth four days after the injury. (From Seyd, 1936.)

1926). For example ultra-violet radiation has a differential cytolytic action, which is identical with that of chemical poisons (Child and Deviney, 1926; Monod, 1933). It is believed that the environmental agent alters in degree the physiological activities, with the differential depending upon quantitative differences in the physiological condition. Permeability and cytolytic gradients are therefore to be regarded as manifestations, rather than causes, of an underlying physiological gradient.

The significance of physiological gradients in connection with the regulation of specific organelles in Protozoa has yet to be made clear. A more recent article devoted to the determinative action of the physiological gradients is that of Seyd (1936) on *Spirostomum ambiguum*. A

new contractile vacuole develops from the long vacuolar feeding canal
near the posterior end of an anterior body section, but never at the forward end of a posterior section. The new vacuole appears only at the
posterior end of a piece cut from the mid-body region. After a deep
transverse cut in the mid-region, a vacuole appears near the cut end of
the anterior half; but when the wound closes the new vacuole disappears, as the posterior part of the feeding canal fuses with it. In long
anterior pieces, in short anterior and posterior pieces, and in cut-out
sections, a new mouth develops at the appropriate position; those cut
in such a manner that the old mouth occupies an odd position will regenerate a new mouth in the correct position, and the old one is resorbed (Fig. 184). The specificity and location of these regenerated
structures in *Spirostomum* are attributed to the determining action of
the physiological gradients.

REGENERATION IN COLONIAL FORMS

The foregoing pages have dealt with different aspects of regeneration
as it occurs in wounded cells or in their dismembered parts. Whereas this
phase of morphogenetic investigation revolves about the cell, its fundamental organization, or the relations between different regions of the
same cell, an equally fruitful line of inquiry concerns the relationships
of one cell to another in true or temporarily colonial Protozoa.

Twenty-five years ago Runyan and Torrey (1914) became interested
in the determination problem in *Vorticella* sp. when they discovered
that, after division, the migratory cell always forms from the lateral
daughter. The cleavage plane in this peritrich coincides with the longitudinal cell axis. But instead of bisecting the aboral or attached end of
the constricting cell, the plane deviates from the mid-line enough to
disrupt the stalk connections of one (the lateral) daughter. This is the
presumptive migrant or "ciliospore." The other cell retains its continuity
with the stalk and remains behind to repeat the division process at a
later time. Metamorphosis of the lateral cell does not begin until the
protoplasmic junction between the two cells is reduced to a small thread.
These observations led Runyan and Torrey to suppose that the posterior
girdlet of cilia, the scopula, and other features of the migrant, appear
only after the cell is physiologically isolated from the stalk; and, furthermore, that such isolation does not exist until the organic connection

between the sister cells becomes very slender. They attributed this behavior to dominant influence of the stalk upon the expression of ciliospore potentialities. Their investigations also included experiments wherein solitary cells were dislodged from their contractile stalks. Thus freed, the cells metamorphosed into typical migrants within a two-hour period. Subsequent reattachment to the substrate was followed by immediate resorption of the locomotor organelles.

Autonomy cannot be ascribed to the stalk portion of peritrichs, despite the fact that the contractile core, or spasmoneme, is composed of living protoplasm. When the cell or cells are stripped from the stalk, either by natural or artificial means, no further activity is manifested; it remains inert and lifeless. Consequently, it seems unsafe to assume that the stalk, or peduncle, plays more than a passive part in the differentiation of the cell from which it develops. Changes in the external environment alone are capable of inaugurating the metamorphic processes in recently affixed ex-migrants. A migrant of *Zoothamnium* or *Carchesium*, for example, may settle upon the substrate, metamorphose into a typical trumpet-shaped cell, and secrete a section of the peduncle, then suddenly reacquire migrant characteristics and relinquish its peduncle in order to establish itself in a more favorable location.

Peritrichs of the genus *Zoothamnium* present a branching type of colonial organization in which the spatially separated cells show a relatively high degree of integration. One of the large marine species, *Z. alternans*, has proved to be especially suitable for morphogenetic studies, in virtue of its development according to a definitely determined pattern, as first described by Claparède and Lachmann (1858). Various aspects of growth and differentiation in this form have been investigated recently by Fauré-Fremiet (1930) and Summers (1938a, 1938b).

Asexual reproduction is the general rule in *Z. alternans*. Cells at predetermined nodes along the axis of a colony differentiate as ciliospores. One by one, these mature migrants break away from the parent colony and affix themselves elsewhere, to start new colonies. In this respect *Zoothamnium* resembles *Vorticella*. Unlike the latter, however, its asexual migrants are endowed with far greater developmental potentialities. An affixed ciliospore loses its aboral cilia, acquires the conical vorticellid form, and begins to elaborate the primary stalk, a process interrupted periodically by unequal division. The successive divisions of the transformed ciliospore produce, each time, an apical cell of the next generation and a smaller lateral daughter, the initial cell of the presumptive branch at that node. The initial branch cells are strictly alternate

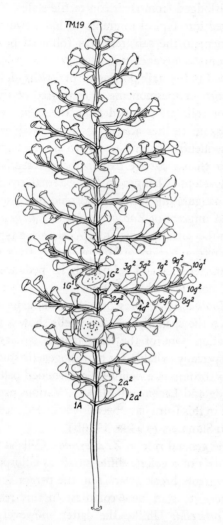

Figure 185. A relatively mature colony of *Zoothamnium alternans* showing the alternate arrangement of branches and cells. The apical cell of the nineteenth generation ($TM.19$) represents the growing point of the primary axis. The numerous common nutritive cells of each branch are division products of the terminal cell of that branch. On the seventh branch, $10g^1$ designates the terminal branch cell of the tenth generation. The lateral cell of the first branch generation (e.g., $1A$) or its two immediate descendants (e.g., $1G^1$ and $1G^2$) represent the potential ciliospores; several stages in the differentiation of these cells on several of the branches are shown. (From Summers, 1938a.)

in position along the primary axis, lying alternately on the right and left sides of the axis at successive nodes. Cells of the branch strain descend from the initial branch cell by a series of equal divisions. Here again each division results in a median and a lateral individual, with the former remaining in the terminal position on the laterally growing branch axis. Thus the primary developmental functions of a growing colony are almost exclusively limited to the terminal cells of the primary and branch axes (Fig. 185).

A painstaking cytological analysis of normal development in this species led Fauré-Fremiet (1930) to postulate that the two daughters resulting from the division of one initial cell are never equivalent as to their developmental potentialities. He assumed that quantitatively differential divisions of the apical cell series restrict the subsequent power of division in the branch strains; and, similarly, the qualitatively differential nature of the first division of the initial branch cells effects a segregation of potencies for ciliospore formation. Such a hypothesis of embryonic segregation by division apparently covers the facts of normal development. A terminal branch cell, for example, produces fewer generations than the apical cell (see Fig. 185). And, for the most part, a ciliospore differentiates only from the lateral cell of the first branch generation or from its two immediate descendants on certain of the branches.

The facts derived from regeneration studies are not in accord with Fauré-Fremiet's assumptions. Common branch cells above or lateral to the supposedly differential divisions retain, for a time at least, potentialities for regenerating large portions of the colony. An apical cell, terminal branch cells, common nutritive cells, ciliospores, and sometimes gamonts differentiate at appropriate positions on the regenerate. Furthermore, the ciliospore-forming cells can be induced to differentiate as new apical cells, which continue axial development according to the normal pattern (Fig. 186).

If the well-defined apical cell is removed, the terminal cell of the topmost branch usually differentiates as a new apical cell, the subsequent development of which is identical with that of the original. But if the apical cell *and* the first terminal branch cell are destroyed, the functions of the former are assumed by either the subterminal cell of the topmost branch or the terminal cell of the penultimate branch, more frequently the latter. A variety of operations performed by the writer (1938b) upon large and small colonies show that subordinate cells—terminal branch cells or merely the common nutritive cells, the complete developmental potentialities of which are never otherwise expressed—can be induced to assume the dominant generative functions. The regenerative behavior following simple, compound, or successive operations is another illustration of what Child (1929) referred to as physiological correlation: the relations of dominance, or control and subordination between parts. The single apical cell of *Zoothamnium* colonies exercises the controlling influence over growth and differentiation in subadjacent cells.

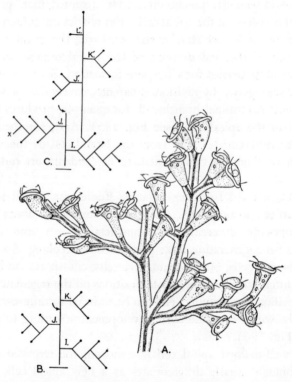

Figure 186. A, a seventy-two-hour regenerate produced from a lateral cell of the first branch generation (a cell which ordinarily represents the presumptive ciliospore); B, schematic diagram of the apical portion of the colony at the time of cutting (see arrow); C, a similar diagram of the regenerating colony seventy-two hours later, or as shown in A. (From Summers, 1938b.)

Regional coördination, according to Child, depends primarily upon quantitative rather than specific differences in the protoplasmic condition of the dominant region. Evidence that this is not necessarily the case comes from another phase of development in *Zoothamnium*. In the sexual process, the apical cell becomes the sessile macrogamont. The fusion of a free-swimming microgamont with the sexually differentiated apical cell arrests axial development for several days, pending the origin of a new apical cell from one of the ex-conjugants. In the meantime, all of the cells on three or four of the youngest (uppermost) branches begin to divide precociously. The terminal branch cells are aroused to unusual mitotic activity, producing twice as many generations as when they comprise a part of the vegetative colony. The common branch cells are likewise activated to produce secondary and even tertiary branches. This precocious development never occurs when the apical cell presides over a vegetative colony. Neither does it occur as a result of decapitation—when the apical cell is destroyed. The phenomenon appears to be initiated by qualitative changes in the coördinating mechanism, which arise in consequence of reorganizational activities in the single apical cell.

The growth relations are likewise altered by conjugation in *Z. arbuscula* (Furssenko, 1929). Each of the several primary axes in this species bears an apical cell which becomes the macrogamont during the sexual period. Conjugation on one axis stops further apical extension of that branch until two new vegetative axes spring from the two "stem cells" of the fourth ex-conjugant generation. One daughter cell from each of the first four generations differentiates into a very large macrozoöid (immature ciliospore). A single conjugant therefore produces two growing points and a cluster of from four to six bulbous macrozoöids. As is the case in *Z. alternans*, several of the small secondary branches below the conjugant-bearing node develop to the proportions of primary axes. Under the influence of a vegetative apical cell, these branches do not hypertrophy.

Furssenko accounts for the changed relations between apical and subordinate regions in terms of local variations in the food-energy requirements. In the light of the above observations, he supposed that the cluster of huge non-feeding macrozoöids at the tip of the stalk, together with the two developing apical cells, have energy needs in excess of the apical requirements in non-conjugating colonies. Multiplication of the actively feeding cells on neighboring branches presumably occurs, in order to compensate for the unusual metabolic needs at the apex.

In *Z. alternans,* however, hypertrophy of the inferior branch cells begins before the conjugant undergoes its first division. It is therefore doubtful whether any increase in the energy requirements, coincidental with the conversion of an apical cell into an exconjugant, is adequate to account for the far-reaching alterations of the normal growth pattern. It is more than likely that the combined energy demands of the actively dividing cells on subordinate branches exceed those of the single conjugant or its first few non-feeding descendants. It is probable, then, that the flux would be directed away from the cell or cells in the apical position.

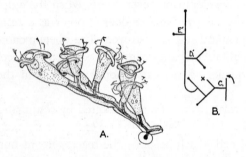

Figure 187. A, branch *C* of a colony fifty-six hours after injury to the neuromuscular cord (drawn from above) (the original colony of six branches was pinched in the midregion, isolating *ABC* from *DEF* and the apical cell; the terminal cell on branch *C* differentiated into a new apical cell, which produced two new branches as illustrated); B, schematic representation of branch *C,* as drawn in A.

We are thus confronted with two divergent interpretations relative to the specificity of form-regulating factors in *Zoothamnium.* The evidence presented by Summers suggests that qualitative physiological changes in one cell play a greater part in the development of neighboring cells than heretofore suspected. Furssenko's hypothesis, on the other hand, emphasizes the directive influence of metabolic fluctuations. It is therefore consistent with the metabolic-gradient theory, inasmuch as the nutritive factors may be quantitative and continuous.

The stalk structure of *Z. alternans,* when viewed in section, somewhat resembles a sheathed nerve fiber. There is an elastic surface membrane, a thick cortical region of hyaline, gelatinous material, and a core of protoplasm, the neuromuscular cord. The latter is continuous from branch to branch and from cell to cell. Unlike the axis cylinder of a nerve fiber, a break in the neuromuscular cord does not cause degeneration in either of its separated parts. The motor reactions of the cells joined by the distal segment of the cord are well integrated, but independent of those connected by the proximal portion. The cord alone, rather than the entire stalk structure, appears to be the medium through which regenerative functions are coördinated. Preliminary experiments have shown that a local injury to the protoplasmic portion of the stalk physiologically divides a colony. The distal part, with its original apical cell, continues to develop as before, whereas one of the cells on the proximal (basal) portion differentiates into an apical cell the activity of which produces another dominant growth axis (Fig. 187). How the single apical cell regulates developmental functions in the distantly separated cells, through the agency of the slender protoplasmic thread, remains as one of the more important problems that invite attention to these colonial Protozoa.

We are far from having more than an elementary notion of what goes on within differentiating cells, but it is probable that the recent developments in other more specific phases of cell physiology presage a period of keener interest in the age-old problems of growth and form determination. Most of the publications on regeneration now extant are important in substance, but lack the specificity of detail which characterizes the newer contributions on nutrition, respiration, and so forth. They also indicate (1) the advisability of more carefully planned pre- and post-operative culture techniques, wherein a better evaluation of extrinsic factors is possible; and (2) the value of critical distinction between the failure of cells to survive an experimental procedure and their failure to regenerate.

LITERATURE CITED

Alverdes, Friederick von. 1922. Zur Lokalisation des chemischen und thermischen Sinnes bei *Paramecium* und *Stentor,* Zool. Anz., 55: 19-21.
Balamuth, W. 1939. Studies on regeneration in Protozoa. I. Cytology and regeneration of *Lichnophora macfarlandi.* Univ. Cal. Library, MS, 1939.
Balbiani, E. G. 1888. Recherches expérimentales sur la mérotomie des Infusoires ciliés. Rec. zool. suisse, 5: 1-72.
—— 1891. Sur la régénérations successives du peristome comme characterè d'âge chez les Stentors et sur la rôle du noyau dans ce phénomène. Zool. Anz., 14: 312-16.

—— 1892. Nouvelle Recherches expérimentales sur la mérotomie des Infusoires ciliés. Ann. Microg., 4: 369-449.

—— 1893. Mérotomie des infusoires ciliés. Ann. Microg., 5: 49-84.

Bauer, E. S., and A. M. Granowskaja. 1934a. Die Rekonstruktion des Kerns und die Atmungsprozesse bei Hypotricha im Ergebnis operativer Einwirkungen auf das Protoplasma und ihre Abhangigkeit vom Alter. Biol. Zbl., 3, 457 (Berichte ü. wiss. Biol., 33: 282).

—— 1934b. Abhangigkeit der experimentellen "individuellen Unsterblichkeit" vom Alter. Biol. Zbl., 3: 609 (Berichte ü. wiss. Biol., 33, 282).

Becker, E. 1928. Streaming and polarity in Mastigina hyale. Biol. Bull., 54: 109-16.

Bills, C. E. 1922. Inhibition of locomotion in Paramecium and observations on certain structures and internal activities. Biol. Bull., 42: 7-13.

Bozler, E. 1924. Über die Morphologie der Ernährungsorganelle und die Physiologie der Nahrungsaufnahme bei Paramecium caudatum Ehrbg. Arch. Protistenk., 49: 163-215.

Brandt, K. 1877. Über Actinosphaerium eichornii. Dissertation, Halle.

Brehme, K. S. 1933. An investigation of the question of cytoplasmic fusion in Amoeba proteus. Arch. Protistenk., 79: 303-10.

Bullington, W. E. 1925. A study of spiral movement in the ciliate infusoria. Arch. Protistenk., 50: 219-74.

Burch, P. R. 1930. The effect on the division rates of Arcella vulgaris and A. rotunda 1. of the injury caused by the excision of cytoplasm; 2. of the loss of cytoplasm; 3. of the gain of cytoplasm. Arch. Protistenk., 71: 307-22.

Burnside, L. H. 1929. Relation of body size to nuclear size in Stentor coeruleus. J. exp. Zool., 54: 473-83.

Calkins, G. N. 1911a. Regeneration and cell division in Uronychia. J. exp. Zool., 10: 95-116.

—— 1911b. Effects produced by cutting Paramecium cells. Biol. Bull., 21: 36-72.

—— 1919. Uroleptus mobilis, Engelm. I. History of the nuclei during division and conjugation. J. exp. Zool., 27: 293-357.

—— 1921. Uroleptus mobilis, Engelm. IV. Effect of cutting during conjugation. J. exp. Zool., 34: 449-70.

—— 1934. Factors controlling longevity in protozoan protoplasm. Biol. Bull., 67: 410-31.

Causin, M. 1931. Régénération du Stentor coeruleus. Arch. Anat. micr., 27: 107-25.

Chejfec, M. 1932. Regulacja i regeneracja u Paramecium caudatum. Acta Biol. exp., 7: 115-34.

Child, C. M. 1914. The axial gradient in ciliate infusoria. Biol. Bull., 26: 36-54.

—— 1919. Demonstration of the axial gradients by means of potassium permanganate. Biol. Bull., 36: 133-47.

—— 1920. Some considerations concerning the nature and origin of physiological gradients. Biol. Bull., 39: 147-87.

—— 1929. Physiological dominance and physiological isolation in development and reconstitution. Roux Arch. EntwMech. Organ., 117: 21-66.

Child, C. M., and E. Deviney. 1926. Contributions to the physiology of *Paramecium caudatum*. J. exp. Zool., 43: 257-312.

Claparède, E., and J. Lachmann. 1858-60. Études sur les Infusoires et les Rhizopods. Mém. Inst. nat. genev., 5-7.

Conklin, E. G. 1912. Cell size nad nuclear size. J. exp. Zool., 12: 1-98.

Dawson, J. A. 1919. An experimental study of an amicronucleate *Oxytricha*. J. exp. Zool., 29: 473-508.

Dembowska, W. S. 1925. Studien über die Regeneration von *Stylonychia mytilus*. Arch. mikr. Anat., 104: 185-209.

—— 1926. Studies on the regeneration of Protozoa. J. exp. Zool., 43: 485-504.

—— 1938. Körperreorganisation von *Stylonychia mytilus* beim Hungern. Arch. Protistenk., 91: 89-105.

Fauré-Fremiet, E. 1930. Growth and differentiation of the colonies of *Zoothamnium alternans* (Clap. and Lachm.). Biol. Bull., 58: 28-51.

Fortner, H. 1933. Über Kernresektion bei einem Hypotrichen (nov. spec.?). Arch. Protistenk., 81: 284-307.

Furssenko, A. 1929. Lebenscyclus und Morphologie von *Zoothamnium arbuscula* Ehrenberg. Arch. Protistenk., 67: 376-500.

Garnjobst, L. 1937. A comparative study of protoplasmic reorganization in two hypotrichous ciliates, *Stylonethes sterkii* and *Euplotes taylori*, with special reference to cystment. Arch. Protistenk., 89: 317-80.

Gregory, L. H. 1909. Observations on the life history of *Tillina magna*. J. exp. Zool., 6: 383-431.

Gruber, A. 1885. Über kunstliche Teilung bei Infusorien. Biol. Zbl., 4: 717-22.

—— 1886. Beiträge zur Kenntniss der Physiologie und Biologie der Protozöen. Ber. naturf. Ges. Freiburg i. B., 1: 1-33.

Hartmann, M. 1924. Der Ersatz der Fortpflanzung von Amöben durch fortgesetzte Regenerationen. Weitere Versuche zum Todproblem. Arch. Protistenk., 49: 447-64.

—— 1928. Über experimentelle Unsterblichkeit von Protozoen-Individuen. Ersatz der Fortpflanzung von *Amoeba proteus* durch fortgesetzte Regenerationen. Zool. Jahrb., Abt. allg. Zool. Physiol. Tiere, 45: 973-87.

Hegner, R. W. 1920. The relations between nuclear number, chromatin mass, and shell characteristics in four species of the genus *Arcella*. J. exp. Zool., 30: 1-95.

MORPHOGENESIS

Hertwig, R. 1903. Über Korrelation von Zell-und Kerngrösse und ihre Bedeutung für die geschlechtliche Differenzierung und die Teilung der Zelle. Biol. Zbl., 23: 49-62; 108-19.

——— 1908. Über neue Probleme der Zellenlehre. Arch. Zellforsch., 1: 1-33.

Hetherington, A. 1932. On the absence of physiological regeneration in *Stentor coeruleus*. Arch. Protistenk., 77: 58-63.

Hewitt, J. H. 1914. Regeneration of *Pleurotricha* after merotomy with reference especially to the number of micronuclei and the occurrence of uninucleated cells. Biol. Bull., 27: 169-76.

Hofer, B. 1890. Experimentelle Untersuchungen über den Einfluss des Kerns auf das Protoplasma. Jena. Z. Naturw., 24: 105-76.

Holmes, S. J. 1907. The behavior of *Loxophyllum* and its relation to regeneration. J. exp. Zool., 4: 399-430.

Horton, F. M. 1935. On the reactions of isolated parts of *Paramecium caudatum*. J. exp. Biol., 12: 13-16.

Hosoi, T. 1937. Protoplasmic streaming in isolated pieces of *Paramecium*. J. Fac. Sci. Tokyo Univ., Zool., 4: 299-305.

Howland, R. B. 1928. Grafting and reincorporation in *Actinosphaerium eichornii* Ehr. Biol. Bull., 54: 279-88.

Hyman, L. H. 1917. Metabolic gradients in *Amoeba* and their relation to the mechanism of amoeboid movement. J. exp. Zool., 24: 55-99.

Ishikawa, H. 1912. Wundheilungs- und Regenerationsvorgänge bei Infusorien. Roux Arch. EntwMech. Organ. 35: 1-29.

Jennings, H. S. 1901. On the significance of the spiral swimming of organisms. Amer. Nat., 35: 369-78.

Jennings, H. S., and C. Jamieson. 1902. Studies on reactions to stimuli in unicellular organisms. Biol. Bull., 3: 225-34.

Jensen, P. 1896. Über individuelle physiologische Unterschiede zwischen Zellen der gleichen Art. Pflüg. Arch. ges. Physiol., 62: 172-200.

Kepner, W. A., and B. D. Reynolds. 1923. Reactions of cell bodies and pseudopodial fragments in *Difflugia*. Biol. Bull., 44: 22-46.

Landis, E. M. 1920. An amicronucleate race of *Paramecium caudatum*. Amer. Nat., 54: 453-57.

LeDantec, F. 1897. La Régénération du micronucleus chez quelques infusoires ciliés. C. R. Acad. Sci. Paris, 125: 51-52.

Lewin, K. R. 1910. Nuclear relations of *Paramecium caudatum* during the asexual period. Camb. Phil. Soc. Proc., 16: 39-41.

——— 1912. The behavior of the infusorian micronucleus in regeneration. Proc. roy. Soc., B, 84: 332-44.

Lillie, F. R. 1896. On the smallest parts of *Stentor* capable of regeneration. J. Morph., 12: 239-49.

Looper, J. B. 1928. Cytoplasmic fusion in *Actinophrys sol*, with special reference to the karyoplasmic ratio. J. exp. Zool., 50: 31-49.

Ludwig, W. 1929. Untersuchungen über die Schraubenbahnen neiderer Organismen. Z. f. vergl. Physiol., 9: 734-801.

Lund, E. J. 1917. Reversibility of morphogenetic processes in *Bursaria*. J. exp. Zool., 24: 1-19.

Luntz, A. 1936. Unsterblichkeit von Protozoenindividuen, erhalten durch periodische Reizungen. Arch. Protistenk., 88: 23-26.

Lynch, V. 1919. The function of the nucleus of the living cell. Amer. J. Physiol., 48: 258-83.

Mast, S. O., and F. M. Root. 1916. Observations on *Amoeba* feeding on rotifers, nematodes and ciliates, and their bearing on the surface-tension theory. J. exp. Zool., 21: 33-49.

Maupas, M. E. 1888. Recherches expérimentelles sur la multiplication des infusoires ciliés. Arch. zool. exp. gén., II, 6: 165-277.

Miller, E. De Witt. 1932. Reappropriation of cystoplasmic fragments. Arch. Protistenk., 78: 635-45.

Minchin, E. A. 1912. An introduction to the study of the Protozoa. London.

Monod, J. 1933. Mise en évidence du gradient axial chez les Infusoires ciliés par photolyse à l'aide des rayons ultraviolets. C. R. Acad. Sci. Paris, 196: 212-14.

Moody, J. E. 1912. Observations on the life history of two rare ciliates, *Spathidium spathula* and *Actinobolus radians*. J. Morph., 23: 349-99.

Moore, E. L. 1924. Regeneration at various phases in the life history of *Spathidium spathula* and *Blepharisma undulans*. J. exp. Zool., 39: 249-316.

Morea, L. 1935. Régénération chez *Spirostomum ambiguum*. C. R. Soc. Biol. Paris, 119: 235-37.

Morgan, T. H. 1901. Regeneration of proportionate structures in *Stentor*. Biol. Bull. 2: 311-28.

Nadler, J. E. 1929. Notes on the loss and regeneration of the pellicle in *Blepharisma undulans*. Biol. Bull. 56: 327-30.

Nussbaum, M. 1884. Über spontane und künstliche Zelltheilung. Verh. Naturh. Ver., Bonn, 41: 259.

Okada, Yô K. 1930. Transplantationsversuche an Protozoen. Arch. Protistenk., 69: 39-94.

Patten, M. 1921. The life history of an amicronucleate race of *Didinium nasutum*. Proc. Soc. Exp. Biol. N.Y., 18: 188-89.

Peebles, F. 1912. Regeneration and regulation in *Paramecium caudatum*. Biol. Bull., 23: 154-70.

Phelps, L. A. 1926. Experimental analysis of factors concerned in division in Ameba. Trans. Amer. micr. Soc., 45: 133-45.

Poljansky, G. 1938. Die Rekonstruktion des Kernapparates der *Bursaria truncatella* bei experimentellen Trennung der konjugierenden Paare. Biol. Zbl., 7: 123 (Berichte ü. d. wiss. Biol., 48: 608).

Popoff, M. 1907. Depression der Protozoenzelle und der Geschlechtszelle der Metazoen. Arch. Protistenk., (Suppl.) 1: 43-82.

Prowazek, S. 1904. Beiträge zur Kenntnis der Regeneration und Biologie der Protozoen. Arch. Protistenk., 3: 44-59.

Reynolds, B. D. 1924. Interaction of protoplasmic masses in relation to the study of heredity and environment in *Arcella polypora*. Biol. Bull., 46: 106-40.

Reynolds, M. E. 1932. Regeneration in an amicronucleate infusorian. J. exp. Zool., 62: 327-61.

Rösel von Rosenhof, A. S. 1755. Insekten Belustigung. 3 vols., Nürnberg.

Runyan, E. M., and H. B. Torrey. 1914. Regulation in *Vorticella*. Biol. Bull., 27: 343-45.

Schwartz, V. 1934. Versuche über Regeneration und Kerndimorphismus der Ciliaten. Nachr. Ges. wiss. Göttingen, math.-physik., N.F. 1: 143-55.

—— 1935. Versuche über Regeneration und Kerndimorphismus bei *Stentor coeruleus* Ehrbg. Arch. Protistenk., 85: 100-39.

Seyd, E. L. 1936. Studies on the regulation of *Spirostomum ambiguum*. Arch. Protistenk., 86: 454-70.

Sokoloff, B. 1922. Le Noyau est-il indispensable a la régénération des Protozoaires? C. R. Soc. Biol. Paris, 87: 1144-47.

—— 1923. Hunger and regeneration. J. R. micr. Soc., London, 1923, 183-89.

—— 1924. Das Regenerationsproblem bei Protozoen. Arch. Protistenk., 47: 143-252.

Stevens, N. M. 1903. Further studies on the ciliate infusoria, *Lichnophora* and *Boveria*. Arch. Protistenk., 3: 1-43.

Stolĉ, A. 1910. Über kernlosen Individuen und kernlose Teile von *Amoeba proteus*. Roux Arch. EntwMech. Organ., 29: 152-68.

Summers, F. M. 1938a. Some aspects of normal development in the colonial ciliate *Zoothamnium alternans*. Biol. Bull., 74: 117-29.

—— 1938b. Form regulation in *Zoothamnium alternans*. Biol. Bull., 74: 130-54.

Tartar, V. 1939. The so-called racial variation in the power of regeneration in *Paramecium*. J. exp. Zool., 81: 181-208.

Taylor, C. V. 1928. Protoplasmic reorganization in *Uronychia uncinata*, sp. nov., during binary fission and regeneration. Physiol. Zoöl., 1: 1-25.

Taylor, C. V., and W. P. Farber. 1924. Fatal effects of the removal of the micronucleus in *Euplotes*. Univ. Cal. Publ. Zool., 26: 131-43.

Thon, K. 1905. Über den feineren Bau von *Didinium nasutum* O. F. M. Arch. Protistenk., 5: 281-321.

Tittler, I. A. 1935. Division, encystment, and conjugation in *Urostyla grandis*. Cellule, 44: 189-218.

—— 1938. Regeneration and reorganization in *Uroleptus mobilis* following injury by induced electric currents. Biol. Bull., 75: 533-41.

Verworn, Max. 1889. General Physiology. 2d ed., London.

―― 1892. Die physiologische Bedeutung des Zellkerns. Pflüg. Arch. ges. Physiol., 51: 1-118.

Willis, H. S. 1916. The influence of the nucleus on the behavior of *Amoeba.* Biol. Bull., 30: 253-70.

Woodruff, L. L. 1905. An experimental study of the life history of hypotrichous infusoria. J. exp. Zool., 2: 585-632.

―― 1913. Cell size, nuclear size and the nucleocytoplasmic relation during the life of a pedigreed race of *Oxytricha fallax.* J. exp. Zool., 15: 1-22.

―― 1921. Micronucleate and amicronucleate races of infusoria. J. exp. Zool., 34: 329-37.

Young, D. B. 1922. A contribution to the morphology and physiology of the genus *Uronychia.* J. exp. Zool., 36: 353-90.

―― 1926. Nuclear regeneration in *Stylonychia mytilus.* Biol. Bull., 51: 163-65.

Young, Dixie. 1939. Macronuclear reorganization in *Blepharisma undulans.* J. Morph., 64: 297-353.

CHAPTER XVII

CERTAIN ASPECTS OF PATHOGENICITY OF PROTOZOA

Elery R. Becker

It is customary to recognize three functional categories of parasitic Protozoa: (1) commensals, which neither harm nor abet the host; (2) symbionts (= symbiotes), which aid the host; and (3) true parasites or pathogenes, which disarrange the host organism to a greater or less degree. This practice may be defended on academic grounds, since it serves to clarify concepts and to attract students' interest to animal microörganisms and the rôles they play in the lives of other animals and plants, but it is in reality highly artificial. The ensuing discussion will be developed principally about this point, with the deliberate intention of provoking wide consideration of the subject, particularly as regards the "pathogenic" aspects of parasite activity, as was done with the subject of host-specificity of parasites a number of years ago (Becker, 1933). Such terms as commensalism, symbiosis (symbiotism), and pathogenicity can represent no more than an expression of the state of adjustment between two separately functioning entities, the host and the parasite, coëxisting in one of the most intimate relationships, and as such are subject to analysis.

Problems of Virulence and Pathogenicity

The functional categories have no counterparts in the zoölogical scheme: that is to say, there are no classes, orders, or families which have as their distinguishing character that they are pathogenic or otherwise. The statement applies also to genera, for, as a matter of fact, we recognize "pathogenic" and "non-pathogenic" members of *Trypanosoma, Trichomonas, Entamoeba,* and other genera. The situation is seen, at the outstart, to limit itself almost entirely to a consideration of "pathogenic species," but it is actually still more complicated than that. There is indisputable evidence that many species of pathogenic Protozoa are made up of a number of strains. *Entamoeba histolytica,* for example, is believed to be a composite of many races differing both in cyst size (see Dobell and Jepps, 1918) and virulence (see Meleney and Frye, 1935, pp. 431-32). The evidence for the latter is indisputable, especially since the appearance of the work of Meleney and Frye (1933, 1935), although Craig (1936) is still skeptical regarding the existence of avirulent strains. The latter point can be conceded for the present (though it is still a live issue), without impugning the significance of observations on human cases and experimental infection in kittens and puppies pointing to the existence of strains of low virulence, medium virulence, and high virulence.

Meleney and Frye adopted a standardized procedure. Recognizing the doubtful validity of experiments performed on too few animals and conducted without due allowance for variability of individual response, they made it a practice to test each strain in a large series of kittens of standard size. Each strain was isolated in culture, young transplants were used for inoculations, and inoculations were made directly into the caecum after laparotomy incision. Furthermore, the history of the human patient was known, and there were records regarding the character of the community and the prevalence of amoebiasis in the community in which each patient resided. The criteria of pathogenicity were success or failure in infecting, extent and intensity of lesions produced, and duration of the infection. The results of the experiment showed conclusively that certain strains of *Entamoeba histolytica* of human origin exhibited more "pathogenic activity" in kittens than other strains.

Furthermore, by correlating the experimental data with field observations, they were led to the following conclusion regarding the relative pathogenicity of the strains for man: "The more pathogenic strains (i.e., in kittens), whether they were obtained from acute cases of amoebic dysentery or from so-called 'healthy carriers' were associated geographically and epidemiologically with acute dysentery, whereas the less pathogenic strains were associated both individually and epidemiologically with very little evidence of acute dysentery."

AMOEBIC DYSENTERY AND BACTERIAL COMPLICATIONS

The problem of virulence has had its reflection in matters of specificity. Brumpt (1925) described *Entamoeba dispar* as an amoeba of the *histolytica*-type dwelling in man, incapable of producing symptoms of dysentery in its human host, and producing no definite macroscopic ulcerations in the cat, but capable of penetrating the intestinal wall of the latter animal so far as the *muscularis mucosae*. Simić, in several papers, has corroborated Brumpts's claim for the validity of *E. dispar,* but his last (1935) paper presents the strong argument that *E. dispar* infection in dogs lasts only from 6 to 8 days, while *E. histolytica* infection lasts from 60 to 120 days. *E. dispar* infection in dogs is quite benign, while *E. histolytica* produces characteristic amoebic ulcerations, and amoebae with ingested red cells may be found in the stools of the infected dog. Wenyon (1936), however, in commenting on the strong case built up by Simić (1935) states:

> It still seems futile to attempt to separate *E. dispar* from *E. histolytica* on the grounds of pathogenicity. It seems that all the differences described can be more reasonably accounted for by the supposition that races of *E. histolytica* of varying virulence occur and that hosts vary in their susceptibility to the one species.

Meleney and Frye (1935) likewise prefer to consider *E. dispar* as a strain of *E. histolytica,* possessing a low degree of pathogenicity. Even the least virulent strains encountered by the latter authors, though not producing clinical symptoms in the persons in whom they had their origin, were capable of producing lesions in some kittens. Hence the skepticism of these authors regarding totally avirulent strains of *E. histolytica.*

Is virulence-level retained by a strain of *E. histolytica,* or is it subject to modifying factors, such as attenuating effects of artificial culture media or exaltation by animal passage? Meleney and Frye (1933) first noted the contrast in pathogenic activity in kittens between "A" strains of low virulence from the hill country, where symptomless carriers and persons with mild symptoms were the rule, and "B" strains of high virulence from severe cases of amoebic dysentery in the bottom lands, where acute cases were much more common than in the hills. Later (1935), they were able to report, after adequate testing in kittens, that two "A" strains and two "B" strains had retained their respective pathogenic indices after a period of three years of artificial cultivation. The highly virulent strains did not decline in pathogenicity. But what about the effect of animal passage? Meleney and Frye (1936) state that their efforts to step up the virulence of less pathogenic strains by serial passage through kittens and dogs have always failed at the first transfer, but Cleveland and Sanders (1930) have made some experiments bearing on this point, except that they ascribe the effects observed to bacteria rather than to amoebae. Using for the first passage in kittens a strain that had been carried on in culture for from 460 to 540 days, they found that only 2 out of 26 animals became infected in the first passage, 5 out of 5 in the second, 3 out of 7 in the third, and 2 out of 2 in the fourth. They conclude: "An increase in the percentage of animals that became infected with passage is demonstrated in these experiments, but this may be due, as in the liver passages, to an increase in virulence by the bacteria rather than the amoebae." The need for further work on the possibility of exalting the virulence of less pathogenic strains by animal passage is apparent, but efforts along this line, in order to obtain results of significance, will first have to eliminate the effects of bacteria accompanying the Protozoa.

The comments of Cleveland and Sanders regarding bacteria suggest the next point, the effect of bacteria on pathogenicity of *E. histolytica*. Their criterion for virulence was principally infectivity for the liver in kittens, when inoculated with a hypodermic needle directly into this organ. Pure cultures in liver-infusion, agar-horse serum saline medium lost most of their ability to establish infection in the liver after a year or more. Such a strain was reduced to an infectivity of 20 percent in the first passage. The infectivity increased, however, with succeeding passages, until by the sixth passage it amounted to 73 percent. Was the apparent increase in virulence to be attributed to the amoebae or to the bacteria accompanying them? Which had lost virulence during the year of life in the artificial media?

An attempt to settle the issue was made in a crisscross experiment. Bacteria from the fifth passage were inoculated with the culture amoeba that had not been passed; and, conversely, the passed amoebae were inoculated with bacteria that had not been passed, the latter being the nonpathogenic *Bacillus brevis*. The experiments showed that fifth-passage bacteria increased the virulence of amoebae in culture for a year, and that amoebae were not able to maintain themselves in the liver of the cat unless accompanied by bacteria capable of damaging the liver. Thus Cleveland and Sanders concluded that it was the bacteria accompanying

E. histolytica in culture, and not the amoeba, which lost virulence during the year in the artificial medium and regained virulence after repeated liver passage. They add the precautionary remark that failure to infect livers with cultures containing only nonpathogenic bacteria might not be repeated with more virulent strains of *E. histolytica*.

Frye and Meleney (1933) attacked the same problem, using severity of intestinal infection as the criterion. They, too, tried a crisscross technique, interchanging the bacteria in a culture of proved high virulence with those in a culture of proved low virulence. The interchange did not materially alter the incidence or severity of infection of the two strains of amoebae; hence their conclusion that the difference in pathogenicity of the two cultures was really due to the amoebae themselves. Thus differences in pathogenicity of strains of *E. histolytica* claimed by Meleney and Frye was shown to be due to inherent qualities of the protozoön, and not to accompanying microörganisms. Since they had previously not been able to detect any alteration of pathogenicity in artificial medium, the question of alteration of virulence of bacteria in such a medium does not enter in.

MALARIA: *Plasmodium vivax*

It is inescapable that there are strains of intestinal Protozoa differing in virulence, but what is the situation regarding the pathogenic blood Protozoa? Since the behavior of the trypanosomes in animal passage is complicated by differences in behavior of "passage" and "relapse" strains, the author prefers to evade discussing this subject. Human malaria, however, lends itself more readily to discussion, as becomes evident after reading the chapter entitled "The Complexity of the Malaria Parasite" in Hackett (1937). Malaria therapy in general paralysis (paresis) has made it possible to determine definitely whether there are strains of the human malarias differing in morphology, pathogenicity, or other behavior, and the facts learned have been rather surprising.

Plasmodium vivax is the species commonly employed in malaria therapy. Using infected *Anopheles* for inoculation, Boyd and Stratman-Thomas (1933a) showed that during an attack of malaria induced by a particular strain of this species, a patient acquires a "tolerance" which makes him refractory to reinoculation with that strain, but not with a different strain of the same species. They concluded that a person infected with benign tertian malaria acquires a homologous but not a heterologous tolerance to *P. vivax*. Later, Boyd, Stratman-Thomas, and Muench (1934) discovered that superinfections with heterologous strains appear to result in clinical attacks of milder intensity than the original attacks. Manwell and Goldstein (1939) have discovered a similar situation in *P. circumflexum* infection in birds. Using six strains, they concluded that immunity was strain specific rather than species specific, although all strains conferred at least partial protection against the others. It should be added that certain strains of *P. vivax* do have the ability to immunize (or premunize) the patient toward certain other strains.

Variability in Strains and in Host Response

Morphological differences between strains of *P. vivax* have been observed. Two strains of this species widely used in Europe for malaria therapy are the so-called Dutch and Madagascar strains. Buck (1935) has found that the Dutch strain consistently exhibits between twelve and thirteen merozoites in both mosquito-inoculated and blood-inoculated malaria, while the Madagascar strain exhibits between seventeen and eighteen. The incubation period of the former is twenty-one days, while that of the latter is but twelve days. Whether there is a relationship between merozoite number and incubation period in these cases is somewhat of a problem, especially since the discovery of extracellular schizogony of malaria organisms in the internal organs.

Strains of *P. vivax* likewise differ exceedingly in pathogenicity. Some strains are too low in virulence to be useful in malaria therapy of general paralysis. The Dutch strain referred to above is said by Hackett (1937) to give higher fever, to be less susceptible to treatment with salvarsan, and to be less virulent than the Madagascar strain. Furthermore, it often produced no immediate attack, but in 40 percent of the cases went into a long latency of several months, a phenomenon that occurred with the Madagascar strain in only 6 percent of the cases.

There appear to be likewise multiple strains of the other human malarias, viz., *P. falciparum* and *P. malariae* (see Hackett, 1937; Boyd and Kitchen, 1937).

Every case of parasitism exhibits three aspects—the parasite, the host, and the effect of the impinging of the one on the other. It has been shown that parasitic Protozoa differ not only in the response they evoke from the host according to their standing as species, but also according to strain properties. There is likewise abundant evidence that individual hosts differ in their response to the same strain of parasitic protozoön. The latter is in reality a statistical concept. It has been a general experience that when an attribute of an unselected group of individuals was measured, the plotted measurements fell into the well-known frequency distribution curve, either normal or skewed. The writer knows of no data which have been plotted to demonstrate that quantitative data on either individual resistance or susceptibility to adverse effects of parasitism could be presented in a similar sort of graph, but there are many facts, to support such a supposition.

COCCIDIOSIS IN POULTRY

For several years the writer (see Becker and Waters, 1938, 1939b) has been testing the effect of the ration on the course of caecal coccidiosis in chicks. While, in general, fatality was used as the criterion for comparing the effects of two rations, it has been possible to make a number of additional hitherto-unpublished observations bearing on the variability of host response to the disease. In one lot of thirty-three White Leghorn chicks experimentally infected with the same dosage, there were three deaths by eight o'clock in the morning of the fifth day, and five more during the remainder of that day. The next day nine succumbed, making a total of seventeen. Nine others were noted to be in an extremely precarious condition, missed succumbing only by a narrow margin, but recovered to a considerable degree. Five others were observed to be severely affected, but continued to move about and eat some feed during the entire ordeal. One was quite active throughout, though its comb paled significantly. One, a cockerel, continued to eat and move about with undiminished vigor, and its comb did not pale perceptibly, though the droppings were streaked slightly with blood. Similar observations have been common, and justify the assertion that fowls differ significantly in the morbidity they exhibit in response to uniform dosage with the same strain of *Coccidium*.

The literature is replete with evidence that similar variability of host response exists in the case of other protozoan infections. Walker and Sellards (1913) early distinguished between "contact" carriers of *En-*

tamoeba histolytica (who did not develop dysentery) and "convalescent" carriers (who have suffered with dysentery, but have become convalescent). In fact, they passed a strain of the amoeba from a convalescent carrier serially through three other men, two of whom became contact carriers, i.e., did not develop dysentery, and one of whom became a victim of an acute attack of amoebic dysentery. Meleney and Frye, in the experiments previously mentioned, found that kittens inoculated with the same strain differed as to whether or not they became infected, as to the extent and severity of the lesions in the colon, and as to the period of survival of the diseased kittens.

Individuals differ also in the degree of resistance offered to the multiplication of the malaria parasite in their blood and tissues, and in their reaction to parasite density. The existence of racial tolerance or resistance of Negroes to inoculation with *Plasmodium vivax* was pointed out by Boyd and Stratman-Thomas (1933b), though it was by no means absolute. The same authors later (1934) reported their finding that Caucasians appear to be universally susceptible. Wilson (1936, quoted by Hackett) made the observation that Bantu babies in Tanganyika Territory were all infected with the three species of human malaria by the fifth month of life, and commented as follows:

One of the striking features of this period of acute infestation, lasting about eighteen months, is the difference in degree of infestation in different individuals. These babies were constantly being reinfected by fresh invasions of sporozoites. The difference cannot therefore be due to variations in the parasites, but rather to a variation in individual resistance.

Hackett (1937) discusses the variability in the incubation period exhibited by different individuals, and states that in some cases there were as few as one parasite per cubic millimeter at the onset of symptoms, while in others there were 900.

Thus it is evident that the clinical aspects of protozoan infections may differ, owing to inherent basic characters of both the parasite and the host. The reaction of the host may be governed further by another factor that we shall designate the physiological state. Admittedly very little is known concerning the relationship between the physiological state and pathogenicity, but one would conclude a priori that a far-reaching relationship should prevail here. As a striking concrete example, nursling rats usually succumb to the long-supposed "non-pathogenic"

Trypanosoma lewisi, while older rats undergo a response to this micro-organism, in behavior of leucocytes and monocytes, that confers on them sufficient resistance for survival.

NUTRITION AND RESISTANCE

Nutrition may have a far-reaching effect on physiological state, and indirectly on resistance. The following hitherto-unpublished experiment is useful in illustrating the point. Forty young rats of about fifty grams' average weight were divided into two equal groups. One group was fed the following mixture (parts by weight): Beet sugar, 67; casein, unextracted, 10; normal salt mixture, 3; lard, 3; cod liver oil, 2; bright green alfalfa meal, 15. The other group was fed the same mixture, except that alfalfa meal was replaced with whole oats ground to a fine flour. After two weeks on these rations, the lot receiving the ground oats had made slightly greater weight gains than the other. On the fifteenth to the eighteenth days each rat was fed 10,000 recently sporulated oöcysts of *Eimeria nieschulzi,* a coccidium that develops in enormous numbers in the mucosa of the small intestine. On the sixth day the alfalfa-fed rats were obviously affected with diarrhoea, while the oat-fed animals were not showing distress. Strangely enough, on the seventh day the alfalfa-fed lot appeared to be recovering, with formed stools and return of appetite, but the oat-feds were off their feed and passing liquid stools. On the eighth and ninth days, 16 out of 20 of the latter died, a marked contrast to what happened in the alfalfa-fed lot all of which recovered. The result was rather surprising, in view of the biological assays of Becker and Derbyshire (1937, 1938) and Becker and Waters (1939a), which showed that alfalfa meal in the ration in some manner or other stimulated the development of several times as many oöcysts of *Eimeria nieschulzi* in its rat host as either oat hulls or hulled oats.

What is the explanation of the observed effects? The early development of diarrhoea in the alfalfa-feds appears to have been due to the preponderance of the parasite population, but there is a possibility that it lies in the superior accessory food factors of alfalfa meal. The following experiment suggests that vitamin B may have had something to do with it. Twenty young rats were fed the following ration: beet sugar, 71; soy-bean oil meal, expeller process, 10; casein, commercial medium fineness, 10; normal salt mixture, 4; lard, 3; cod-liver oil, 2. Another lot of 20 was fed the same mixture with 10 micrograms of thiamin chloride (vitamin B) per rat daily. The second lot made much greater weight gain during the next ten days than the first. On the tenth to the fifteenth days each rat was inoculated with daily doses of 6,000 sporulated oöcysts of *E. nieschulzi*. Twelve rats out of 20 in the first lot succumbed to the infection, and the remaining 8 all lost weight upon recovery. The recipients of vitamin B all lived and, by the time the infection had cleared up, all had gained in weight.

Thus it is evident that the physiological state may be of prime importance in determining whether or not an animal survives an infection. In one state it may show few or no outwardly visible symptoms, while in another it may be seriously affected, or even succumb.

CONCLUSIONS

Such terms as commensalism and true parasitism lose their significance when a comprehensive analysis is made of the circumstances surrounding an infection with any particular protozoan species. Pathogenicity in the generally accepted sense is a matter of degree, subject in the first place not only to the species, but also to the strain, of the microörganism concerned in the infection. The degree of pathogenicity exhibited by a particular strain in its host may vary from nil to fatal termination, depending upon the inherent defense mechanisms and the other conditions affecting the resistance of the host. The effectiveness of this resistance, in turn, may vary according to changes in the physiological state of the host. These considerations are of fundamental importance to the investigator who conducts researches on the reaction of any host to the invasion of a protozoan parasite.

LITERATURE CITED

Becker, E. R. 1933. Host-specificity and specificity of animal parasites. Amer. J. Trop. Med., 13: 505-23.

Becker, E. R., and R. C. Derbyshire. 1937. Biological assay of feeding stuffs in a basal ration for *coccidium*-growth-promoting substance. I. Procedure, yellow corn meal, oats, oat hulls, wheat, linseed meal, meat scraps. Iowa St. Coll. J. Sci., 11 (1938): 311-22. II. Barley, rye, wheat bran, wheat flour middlings, soy bean meal. Ibid., 12: 211-15.

Becker, E. R., and P. C. Waters. 1938. The influence of the ration on mortality from caecal coccidiosis in chicks. Iowa St. Coll. J. Sci., 12: 405-14.

—— 1939a. Biological assay of feeding stuffs in a basal ration for *coccidium*-

growth-promoting substance. III. Dried fish meal, alfalfa meal, white wheat flour. Iowa St. Coll. J. Sci., 13: 243.

—— 1939b. Dried skim milk and other supplements in the ration during caecal coccidiosis of chicks. Soc. Exp. Biol. N.Y., 40:439.

Boyd, M. F., and S. F. Kitchen. 1937. The duration of the intrinsic incubation period in *Falciparum malariae* in relation to certain factors affecting the parasites. Amer. J. Trop. Med., 17: 845-48.

Boyd, M. F., and W. K. Stratman-Thomas. 1933a. Studies on benign tertian malaria. I. On the occurrence of acquired tolerance to *Plasmodium vivax*. Amer. J. Hyg., 17: 55-59.

—— 1933b. Studies on benign tertian malaria. IV. On the refractoriness of Negroes to inoculation with *Plasmodium vivax*. Amer. J. Hyg., 18: 485-89.

—— 1934. Studies on benign tertian malaria. V. On the susceptibility of Caucasians. Amer. J. Hyg., 19: 541-44.

Boyd, M. F., W. K. Stratman-Thomas, and H. Muench. 1934. Studies on benign tertian malaria. VI. On heterologous tolerance. Amer. J. Hyg., 20: 482-87.

Brumpt, E. 1925. Étude sommaire de l'*"Entamoeba dispar"* n.sp. Amibe à kystes quadrinucleés, parasite de l'homme. Bull. Acad. Med., Paris, 94: 943.

Buck, A. de. 1935. Ein morphologischer Unterschied zwischen zwei *Plasmodium vivax-Stämmen*. Arch. Schiffs- u. Tropenhyg., 39: 342-45.

Cleveland, L. R. and E. P. Sanders. 1930. The virulence of a pure line and several strains of *Entamoeba histolytica* for the liver of cats and the relation of bacteria, cultivation, and liver passage to virulence. Amer. J. Hyg., 12: 569-605.

Craig, C. F. 1936. Some unsolved problems in the parasitology of amebiasis. Parasitology, 22: 1.

Dobell, C., and M. W. Jepps. 1918. A study of the diverse races of *Entamoeba histolytica*. Parasitology, 10: 320-51.

Frye, W. W., and H. E. Meleney. 1933. Studies of *Endamoeba histolytica* and other intestinal Protozoa in Tennessee. VI. The influence of the bacterial flora in cultures of *E. histolytica* on the pathogenicity of the Amoebae. Amer. J. Hyg., 18: 543-54.

Hackett, L. W. 1937. Malaria in Europe. London.

Manwell, R. D., and F. Goldstein. 1939. Strain immunity in avian malaria. Amer. J. Hyg., sec. C, 30: 115-22.

Meleney, H. E., and W. W. Frye. 1933. Studies of *Endamoeba histolytica* and other intestinal Protozoa in Tennessee. V. A comparison of five strains of *E. histolytica* with reference to their pathogenicity for kittens. Amer. J. Hyg., 17: 637-55.

—— 1935. Studies of *Endamoeba histolytica* and other intestinal Protozoa in Tennessee. IX. Further observations on the pathogenicity of certain strains of *E. histolytica* for kittens. Amer. J. Hyg., 21: 422-37.

—— 1936. The pathogenicity of *Endamoeba histolytica.* Trans. R. Soc. trop. Med. Hyg. 29: 369.

Simić, T. 1933. L'Infection du chien par l'*Entamoeba dispar* Brumpt. Ann. Parasit. hum. corp., 11: 117-28.

—— 1935. Infection expérimentale du chat et du chien par *Entamoeba dispar* et *Entamoeba dysenteriae.* Ann. Parasit. hum. corp., 13: 345-50.

Walker, E. L., and A. W. Sellards. 1913. Experimental entamoebic dysentery. Philipp. J. Sci. (B., Trop. Med.), 6: 259.

Wenyon, C. M. ("C. M. W.") 1936. Trop. Dis. Bull., 33:534.

CHAPTER XVIII

THE IMMUNOLOGY OF THE PARASITIC PROTOZOA

William H. Taliaferro

The central theme of the science of immunology is the study of the defense mechanisms of the host against the invasion of parasitic organisms or against the introduction of their products or of other inanimate materials. In the present chapter emphasis is placed almost entirely on the defense mechanisms against living parasites. A complete analysis of these mechanisms involves such widely diverse subjects as the origin, nature, and developmental potencies of the cells and tissues of the host, the physiological action and chemical nature of the humoral forces marshaled by the host in defense, the activity of the invading parasite, the chemical nature of the products of the parasite which stimulate the immune processes in the host, and the effects of the various immune processes on the parasite. As protozoan immunity is just one aspect of the general field of immunology, most of the general principles of immunity can be applied directly to the protozoan parasites. Work on protozoan immunity itself, however, has been restricted more or less to the biological aspects, such as the study of the cellular and the serological mechanisms of the host and the effects of resistance on the parasite, with very little emphasis on chemical phases.

The Physical Bases of Immunity

Immunity or resistance, in the broad sense, denotes various mechanisms of the host which counteract the invasion and the activities of a parasite. It may be manifested as hindrances to the action of invasion, as conditions arising in the body of the host adverse to the parasite, as efforts on the part of the host to make good the deleterious effects of the parasite (as evidenced by the hyperactivity of hematopoietic organs after the destruction of red cells in malaria), or the production of antitoxins in those infections in which toxins are formed. It may be natural (innate) or acquired. Natural immunity is generally correlated with nonspecific factors which are incompatible with or unfitted to the life of the parasite in the unimmunized host. The specificity of parasites for various hosts (Chapter XVII) is largely an expression of natural immunity. Acquired immunity, on the other hand, denotes the various conditions arising in a host as a result of infection or other immunizing procedure and is generally thought of as resulting in large measure from the production of antibodies in the host.

Immunity is the reciprocal of virulence, which in this sense is an expression of the ability of the parasite to invade and parasitize the host. Both immunity and virulence are relative and represent the resultant of the invasive activities of the parasite and the defense activities of the host; they may, therefore, vary in degree from zero to 100 percent.

THE CELLS INVOLVED IN IMMUNITY

The defense of the vertebrate body against invading parasites, or even against inanimate foreign material introduced parenterally, appears to be taken care of predominantly by some of the cells of the connective tissue and is a specialized or accentuated aspect of their normal functions. The connective tissue has manifold normal functions, such as respiration, intermediate metabolism, storage, and mechanical support and in its widespread distribution throughout the body consists of the blood and lymph, cartilage, bone, the reticular (blood-forming) tissue of the myeloid and lymphatic organs, and loose and dense connective (including adipose) tissues associated with the skin, omentum, liver, lung, and so forth. The cells of this tissue arise embryonically from the mesenchyme and may be either fixed or free. Those of the blood and lymph and of the reticular and loose connective tissues are chiefly concerned in defense.

The terminology of the connective tissue cells is complicated by the frequent use of several names for the same cell. This condition has arisen (1) because connective tissue is so widespread and involves so many organs that it has been studied by histologists, hematologists, pathologists, and so forth, some of whom have not correlated the knowledge in fields other than their own; and (2) because investigators have disagreed as to the nature and developmental potencies of various cells. In the following brief review we have defined only those cells of the connective tissue which are known to be involved in defense against the infections to be described herein and have followed in the main Maximow's views with regard to the origin and potencies of the various cells. We have simplified and used uniform terms wherever possible.

The rôle of the various connective tissue cells in immunity is shown by direct histological studies and by other experimental work. Thus histological studies of defense reactions have demonstrated directly that some cells remove parasites and various types of debris by phagocytosis; that others wall off nonremovable objects and repair damage by filling in cavities, regenerating certain tissues, and so forth; and that still others, such as the eosinophils, show a definite pattern of behavior and seem correlated with certain phases of immunity, although their exact function is still uncertain. On the other hand, removal or impairment by splenectomy, blockading procedures, and the like, of an appreciable portion of the connective tissue cells have furnished evidence of the rôle of phagocytes in the immunity of certain infections and of the rôle of the macrophages in the production of antibodies.

In the successful carrying out of these studies, certain technical difficulties have to be recognized and overcome. To study cellular details and especially to see transitional forms, migrating cells, and so forth, early and closely spaced stages in an infection should be studied, fresh material should be used, and this should be adequately fixed and stained by a satisfactory technique. One of these techniques involves fixing in Helly-Maximow's Zenker formol, preferably embedding in celloidin, staining with dilute Delafield's hematoxylin, and counterstaining with eosin azure II. In impairing the macrophage system, the time when splenectomy and blockade are performed is important, inasmuch as impairment is partially made good by the host in time. Furthermore, splenectomy is more effective in impairing the macrophage system in certain infections in which the spleen is especially active and in certain laboratory animals having a high spleen weight-body weight ratio. Thus the most conclusive results may be expected when certain blood infections are studied in dogs, rats, and mice splenectomized and blockaded as rapidly and thoroughly as possible; whereas inconclusive or negative results may be expected from inadequate blockade, splenectomy a week or more before infection. In fact, if impairment is slight, the system may even be stimulated to greater activity.

A. *Predominantly Fixed Connective Tissue cells.*—From the strictly functional aspect of immunity, the predominantly fixed cells of the reticular and loose connective tissues may be divided into two great groups: (1) fixed and free macrophages (including the reticular cells), and (2) the fibroblasts of connective tissue and the endothelial cells lining the ordinary blood vessels.

The term macrophage is essentially a physiological designation for almost any large mononuclear connective-tissue cell which is or may become phagocytic. Under macrophages are classified a group of fixed mesenchymal cells, which retain many embryonic characters and a wide range of potencies for development. The concept that the connective tissue of the adult body possesses fixed cells retaining mesenchymal or embryonic potencies for development is largely due to Marchand (1924, review) and Maximow (1927a, review). There are three chief categories: (1) Pericytes (Maximow) which are fixed, undifferentiated, outstretched cells in the adventitia of all of the small blood vessels of loose connective tissue throughout the body; (2) reticular cells, which, together with fibers, form the stroma of all reticular (myeloid and lymphatic) tissues; (3) littoral cells (Siegmund), which line the sinuses or sinusoids of the reticular tissues, the liver, hypophysis, and adrenal (Pls. 1 and 2). Where phagocytic in the liver, they are generally designated Kupffer cells (Pl. 1, Fig. 1; Pl. 2, Fig. 1). The cells lining the sinuses of the reticular tissues are actually reticular cells. The littoral cells are often called endothelial cells or cells of the special endothelium, but this is unfortunate because the littoral cells have wide developmental potencies, whereas the ordinary endothelial cells lining the blood vessels have restricted developmental potencies.

There is general agreement that under proper stimuli the cells of these three categories can divide by mitosis, can become phagocytic, can develop into fibroblasts, or can develop into practically any other type of cell of the blood or connective tissue. From the standpoint of the present discussion, it is important that they can become phagocytic either in their fixed position (fixed macrophages) or after rounding up and becoming free (free macrophages). It is not definitely known, however, whether, while engorged, they temporarily or permanently lose their mesenchymal potencies. There may be a difference, for example, between the primitive outstretched reticular cell and the same cell after it has become free and phagocytic.

In addition to the cells which are generally admitted to retain mesenchymal potencies, free cells occur in the loose connective tissue, which we have called macrophages and which are variously known as histiocytes, clasmatocytes, rhagiocrine cells, or resting wandering cells. Just as in the case of the phagocytic mesenchymal cells, there is no unanimity of opinion as to whether these free cells retain all hematopoietic functions, but in any case they can become phagocytic without morphological change, can reproduce by mitotic division, and can transform into fibroblasts.

Many other macrophages occur throughout the body, the developmental capacities of which have not been adequately studied. Thus the stroma cells of the *lamina propria* of the intestine probably have developmental potencies identical with those of the reticular cells.

As would be expected, macrophages in different locations and before and after becoming phagocytic vary somewhat in structure with regard to the amount of their cytoplasm, the size and shape of their nucleus, and the amount and size of the chromatin granules and nucleoli in their nucleus. They generally possess, however, well defined cytoplasm and a large, vesicular, often indented nucleus, in which are found fine chromatin granules and a few small nucleoli (see reticular cell and macrophage in Pl. 3, Fig. 1).

Fibroblasts of loose connective tissue have outstretched, ill-defined cytoplasm and a large, regularly oval, vesicular nucleus containing dust-like chromatin granules and small nucleoli. They can divide by mitosis, are instrumental in repair and in walling off foreign material, but are rarely phagocytic and do not generally develop into other cells (except in bone and cartilage).

Endothelial cells line the larger blood vessels and capillaries. (The term as herein used, does not include the littoral cells lining the sinuses and sinusoids of the reticular tissues and elsewhere, which have wide developmental potencies). The endothelial cells can divide by mitosis, can form endothelium of new blood vessels, and can develop into fibroblasts, but are rarely phagocytic and do not generally develop into other cells.

B. *Free Blood and Connective Tissue Cells.*—In accordance with common usage, cells of the blood and lymph are classified according to whether they are of myeloid or lymphoid origin. The lymphoid cells of the blood and the cells of the lymph consist of various-sized lymphocytes, which together with monocytes are termed agranulocytes. The myeloid cells of the blood are the various granulocytes (heterophils or polymorphonuclears, eosinophils, and basophils), the erythrocytes, and the platelets. Some authors classify monocytes as lymphoid and others as myeloid cells, but they are classified in this chapter as both, since we believe that they arise from lymphocytes of the lymphatic tissue and from hemocytoblasts (equivalent to lymphocytes) of the bone marrow.

The heterophils are functional in immunity by virtue of their obvious phagocytic activities and probably because of their secretion of enzymes. They are end cells, however, which do not reproduce or develop into other cells. Lymphocytes and presumably monocytes, on the other hand, can divide mitotically (Pl. 4, Fig. 3) and both lymphocytes and monocytes can develop into macrophages, with all of their developmental potencies. Lymphocytes possess basophil cytoplasm and a relatively large, deeply staining, often indented nucleus, with large acidophil nucleoli Pl. 3, Fig. 1. The monocytes may be the same size as, but in most cases are larger than the medium lymphocytes, their cystoplasm is less basophil and is increased in amount, and their nucleus is more vesicular, more deeply indented with smaller chromatin granules and smaller and more numerous nucleoli (cf. monocytoid lymphocyte in Pl. 3, Fig. 1). As the lymphocytes and monocytes transform into macrophages, they show increased amounts of cytoplasm, their nuclei gradually take on macrophage characteristics, and they become phagocytic (polyblasts 1-5, in Pl. 3, Fig. 2). These intermediate forms, together with lymphocytes, monocytes, and macrophages, are grouped under the term lymphoid-macrophage system.

There is general agreement that free lymphoid cells, more or less similar to lymphocytes, occur in varying numbers under various physiological and pathological conditions in the reticular tissues and the loose connective tissue, and that in such sites they act as "stem" cells of lymphoid and myeloid cells. The nature, classification, and even exact morphology of these different stem cells are subject to such controversy that they are termed by various authors lymphocytes, hemocytoblasts, lymphoblasts, myeloblasts, monoblasts, and so forth, according to the particular theory of blood formation held by the author (see Bloom, 1938). We have adopted essentially the unitarian viewpoint of Maximow and have called all free mesenchymal stem cells, with wide potencies for development, hemocytoblasts in the bone marrow and lymphocytes in all other locations. Lymphocytes and hemocytoblasts are identical morphologically and probably in their developmental potencies. Under physiological conditions, lymphocytes in lymphatic tissue give rise only to lymphocytes (Pl. 4), and hemocytoblasts in bone marrow give rise only to myeloid cells (erythroblasts, myelocytes, and so forth), but under abnormal stimuli they may exhibit their full potencies for development. In general, these free stem cells are self-perpetuating, but they may arise from the fixed mesenchymal cells of the preceding section.

C. *So-called Systems of Cells.*—The foregoing classification of cells should be brought into line with the so-called systems of cells frequently used by various authors. Modern concepts of the cellular basis of immunity have been largely based on studies of inflammation. Credit should be given to Metschnikoff (1892) for insisting upon the essential rôle of the mesenchymal cells in inflammation and to Cohnheim, Ziegler, Marchand (1924, review), and Maximow (1927a, 1927b, review), among others, for studying the histogenesis of the local inflammatory reactions. Metschnikoff (1892 and 1905, among other studies) laid the whole foundation for the modern concept of the defense function of fixed and mobile cells of the connective tissue by phagocytosis. His concept was essentially physiological. He distinguished (1) microphages, herein designated heterophils; and (2) macrophages, which are identical with macrophages as herein defined, except that he included the phagocytic microglial cells of the brain, which are possibly of mesenchymal origin. The modern understanding of macrophages is based largely upon the studies of vital staining and the storage of colloidal dyes, chiefly by Renaut, Maximow, Goldman, Tschaschin, Kiyono, and Aschoff. The *Gefässwandzellen* of the Marchand-Herzog school (see Marchand, 1924) include pericytes and perivascular macrophages (adventitial cells) which are supposed to arise from the endothelium of developing vessels. Aschoff's (1924, review) reticulo-endothelial system, broadly defined, consists of the macrophages as we have outlined them.

It has unfortunately been assumed by most writers that the increase of macrophages associated with immunity, i.e., "hyperplasia of the reticulo-endothelial system," is due to the proliferation of macrophages or cells of the reticulo-endothelial system. This is an admitted source, but detailed studies of a wide variety indicate that most of the new macrophages arise from lymphocytes, with or without the intervention of a monocyte stage (see Pl. 4). In order to include both macrophages and all of their precursors under one term, which would indicate the cytogenesis of macrophages from agranulocytes (lymphocytes and monocytes) as well as from reticulo-endothelial cells, W. H. Taliaferro and Mulligan (1937) proposed the term, lymphoid-macrophage system. This term includes the mononuclear exudate cells, or Maximow's polyblasts, which form the cellular exudate in inflammation.

ANTIBODIES AND ANTIGENS INVOLVED IN IMMUNITY

Infective organisms, derivatives of them, or other foreign, colloidal, protein materials can generally act as antigens. When an antigen is introduced parenterally into an animal, it calls forth a substance in the blood of the animal, known as an antibody, which will react with the antigen specifically *in vivo* and generally *in vitro* and is passively transferable. Such an antibody is often termed an immune antibody, to differentiate it from natural antibodies, which sometimes exist in blood without immunization. Serum from the blood of an animal containing an antibody is known as antiserum. Some antibodies or antiserums, in addition to reacting with their specific complete antigens, may also react with isolated carbohydrate or lipoid parts of the antigen *in vitro*. These substances have been differentiated from true antigens by the terms haptenes or partial antigens, since they generally do not stimulate the production of antibodies *in vivo*. Both complete antigens and haptenes have been isolated in high states of purity. Antibodies result from antigenic stimulation and are metabolic products of cells. Thus the amount of circulating antibody is often decreased by removing the spleen, which is rich in cells of the lymphoid-macrophage system, or by filling the macrophages along the blood stream with colloidal or particulate matter. Furthermore, antibodies undoubtedly represent definite substances which are closely associated with the globulin fraction of the serum. It is, however, impossible to say at present whether they are actually globulins and, if they are globulins, whether they are new globulins or the regular serum globulins slightly modified (see Wells, 1929; Marrack, 1938).

Antibodies are variously named according to the effect produced when mixed with antigen. The antibody is a precipitin if it produces a precipitate on mixing with a soluble antigen (precipitinogen). It is an agglutinin if it induces clumping or agglutination of cellular antigens (agglutinogens), such as Bacteria, Protozoa or blood cells. It is an opsonin if it sensitizes the antigen and makes it more readily ingested by phagocytes. It is a lysin if it sensitizes cellular antigens so that, on the addition of a thermolabile component of normal serum known as complement or alexin, the cell undergoes death and lysis, during which many of its internal substances diffuse through the cell membrane. In both of the foregoing processes, the antigen is first sensitized by antibody. After such a preparation, it is then lysed by intracellular enzymes (phagocytosis) or by extracellular enzymes (lysis) (see Wells, 1929). The antibody is known as an antitoxin if it neutralizes the biological action of a toxic antigen (exotoxin). Definite antitoxins and exotoxins have not been demonstrated in protozoan infections (see W. H. Taliaferro, 1929).

An increasing number of immunologists accept the unitarian viewpoint that the introduction into the body of a single antigen results in the formation of a single antibody, which is an agglutinin, precipitin, and so forth, according to the nature of the antigen or the particular method of testing. This does not mean that a complex cell will not contain many different antigens. Furthermore, a given antibody in a specific infection may act as one type of antibody, and not as another, because of the position of various antigens on or within the cell (see Topley, 1935).

Finally, an antibody-like substance is known as ablastin if it inhibits the reproduction of organisms when mixed *in vivo*. So far, it has been demonstrated only for certain nonpathogenic trypanosomes. Like other antibodies, it is associated with the globulin fraction of serum and is passively transferable, but differs from them in that it has no *in vitro* affinity for its specific antigen. In the latter respect, it appears to resemble certain nonabsorbable antibodies reported in bacterial, virus, and worm infections.

The rôle of antibodies is studied by *in vivo* protective (passive transfer) and curative tests and by *in vitro* studies involving various serological tests. Protective and curative tests differ only as regards the time of injecting the serum and organisms. In protective tests the serum is injected at the same time (or not more than a day before or after) the organisms are injected, whereas, in curative tests the serum is injected some time after the organisms have been injected and generally when they can be found in some particular part of the body. The effect of the latter may obviously be more variable, since the organisms already have a start in the body and may be more difficult to check.

THE CELLULAR AND HUMORAL ASPECTS OF IMMUNITY

The reader is referred to Maximow (1927b) for a general description of the histogenesis of the inflammatory and defense reactions, to Aschoff (1924), Jungeblut (1930), Gay (1931), and Jaffé (1931, and 1938) for a general consideration of the function of cells and, in particular, of macrophages; and to Linton (1929), W. H. Taliaferro (1929 and 1934), and W. H. Taliaferro and Mulligan (1937) for a specific consideration of the rôle of cells in protozoan immunity.

The way in which the cells of the connective tissue, in particular the granulocytes and the cells of the lymphoid-macrophage system, are involved in local defense can be seen during the inflammation which follows the introduction of foreign material into connective tissue of the skin. The heterophils migrate early from the blood vessels. Their number and activity depend upon the nature of the inflammatory stimulus and whether it is sterile or septic. They are generally not numerous in protozoan infections and soon disappear when the inflammatory material is bacteriologically sterile. Under sepsis, however, they continue to migrate from the blood vessels and to combat the invading organisms in many visible ways—by active phagocytosis and digestion, by the secretion of bactericidal and proteolytic ferments, and the like. They represent an important first line of defense since they are the most easily mobilized cells, but their functions are limited since they generally disintegrate within a few days, ordinarily are recruited only from the blood stream, i.e., do not multiply *in situ,* and cannot develop into other cells of the area.

The cells of the lymphoid-macrophage system are the most important cells in local defense. The lymphocytes and the monocytes migrate from the blood vessels, as do the heterophils; but unlike the heterophils they are long-lived, may multiply in the tissues, may develop into macrophages with phagocytic potencies, and from macrophages may progressively develop into fibroblasts with reparative functions. As macrophages, they, together with the macrophages previously present in the area, actively phagocytose and digest certain invading organisms, remove cellular and other debris, and after the acquisition of immunity probably elaborate antibodies which aid in phagocytosis. When large bodies are present, the macrophages may fuse to form foreign body giant cells; when microörganisms are indigestible, they may form giant cells around them, such as the epithelioid cells of the tubercle; or, when large areas are necrotic, they may surround the area, become transformed into fibroblasts, and effectively wall it off. The fibroblasts, both those of the local area and those arising from macrophages, react slowly and probably play an active part only in the later stages of local inflammation during regenerative and reparative processes, the formation of scar tissue, and the walling off of foreign bodies.

Several other cells may come into play, generally during late stages in the defense reaction. Of these, the eosinophils seem to play a part in the detoxification of foreign proteins and their disintegration products and are particularly prominent after the body has become sensitized to the proteins. Like the heterophils, they do not multiply and cannot develop into other cells of the area. Some investigators believe that the plasma cells are also associated with the detoxification of foreign materials. They are not phagocytic, do not seem to have any developmental potencies, and proliferate rarely, if at all. The exact function of the basophils is unknown.

Ordinarily, when the stimulant is distributed over a large part of the body, the reaction is designated as a general defense reaction, in contradistinction to the local defense reaction just described, but as a matter of fact such distribution usually signifies that the stimulant is in the blood stream and is combated by macrophages of organs most closely associated with the blood, such as the spleen, liver, and bone marrow. In some cases, as in malaria, these general reactions can actually be considered local ones in strategically placed organs (see W. H. Taliaferro,

1934). The same types of cells are involved, and the extent to which they are involved depends, as in other sites, upon the nature of the foreign material or infectious agent. The heterophils are often mobilized first, and the lymphoid-macrophage series shows the most pronounced histological changes, with the macrophages seeming to bear the brunt of the activity. Fibroblasts rarely come into contact with foreign material in the blood and are rarely active. Endothelial cells, although they come into contact with hematogenous material, show extremely little histological change or phagocytic activity.

The foregoing account of defense reactions, involving the disposal of foreign material and tissue debris with eventual repair, is characteristic when either antigenic or nonantigenic materials are introduced into a normal animal, but certain quantitative differences are noted when antigenic materials, including parasites, are introduced into an immune animal. These differences are chiefly due to antibodies and are specific. When antigens are introduced into the immune body, they are generally localized by agglutination if they are cells, or by precipitation if they are in solution, and are made more readily phagocytable by opsonification. Such localization and opsonification are particularly well seen in local reactions in the loose connective tissue. They are often limited to organs such as the spleen, liver, and bone marrow in general reactions, as is well-illustrated in malaria. In trypanosomiasis, these phases have not been completely studied, but ablastin (the reproduction-inhibiting antibody) at least does not involve either localization or phagocytosis. When antigen and antibody meet in the tissues of an immune animal, not only do localization and opsonification of the antigen occur, but there is generally a much heightened inflammatory reaction (hypersensitivity, sometimes evidenced by a local reaction—the so-called skin test—when suitable amounts of antigen are injected intradermally). Provided this heightened inflammation is not so intense as to overwhelm the body, it represents a speeding up of the whole cellular response in the immune animal. In addition to the specific action of antibodies after the acquisition of immunity, a residual increase in the number of cells of the lymphoid-macrophage system sometimes is seen at strategic sites, which results in a much more rapid mobilization of macrophages during immunity. This is well illustrated in the spleen in malaria.

Rôle of Immune Processes in the Development of Protozoan Infections

GENERAL METHODS

The detailed considerations in the succeeding sections deal largely with the rôle of antibodies and cells in modifying the course of infection, together with such allied subjects as recovery, relapse, and immunity to super- and reinfection. The Protozoa offer the advantage of being large enough so that one can ascertain the effect of these processes in a way that is impossible with the smaller bacterial and virus invaders. This analysis has been further facilitated by selecting certain plasmodia and trypanosome infections in which practically all stages in the life cycle of the parasite are accessible for study (i.e., are more or less evenly distributed in the peripheral blood and are not localized in the deeper tissues).

The course of these infections can be roughly indicated by changes in the number of organisms. Since, however, the number curve is the resultant of the number of parasites produced by reproduction and the number of organisms which die or are actually destroyed, the only deductions that can be drawn from it are that reproduction is going on if the numbers increase, although the rate may be actually decreasing; and that the rate of reproduction is being inhibited, or that the parasites are being killed, or that both activities may be operating, if the numbers remain constant or decrease. The rate of reproduction, if ascertained, however, in conjunction with number counts throughout an infection, will adequately indicate whether the host acquires a defensive mechanism, and, if so, whether it is directed toward inhibiting reproduction of the parasites, or destroying the parasites after they are formed, or both.

All valid measures of the rate of reproduction so far devised, which are independent of the number of organisms killed, depend upon some measurement of size (since organisms usually grow before they reproduce) or upon some determination of division forms. A direct measure can be devised for the plasmodia, which divide more or less synchronously; but an indirect measure has to be resorted to for the trypanosomes, since they neither divide synchronously nor can their fission rate be ascertained directly as can be done among the free-living forms. The particular criteria used are of course more or less arbitrarily selected and, to be satisfactory, necessitate a nice adjustment between the validity of the criteria selected and the time required to make measurements.

The most convenient measure of the rate of reproduction of the plasmodia so far devised consists in ascertaining the length of the asexual cycle directly (i.e., the time it takes for a young merozoite to become a mature schizont and divide into the next generation of young merozoites), in conjunction with the number of merozoites produced. Thus the percentage of segmenters are computed in samples of 50 to 100 parasites from stained blood smears made every 4 to 12 hours whenever parasites can be found. A regularly recurring percentage of segmenters, considered arbitrarily, for example in *P. brasilianum,* to have 5 or more nuclei by W. H. Taliaferro and L. G. Taliaferro (1934a), indicates a constant rate of reproduction, provided the number of merozoites produced by each segmenter remains approximately constant. The most satisfactory measure of the rate of reproduction among the trypanosomes consists in comparing the percentage of division forms in samples of 50 to 100 forms from stained blood smears made every 6 to 24 hours throughout an infection. Among the pathogenic trypanosomes, in which dividing forms are numerous, division forms may simply be considered as those with some duplication of parts (see Krijgsman, 1933), but among the nonpathogenic *Trypanosoma lewisi* and *T. duttoni,* in which actual dividing forms are rare, division forms are considered to be dividing forms plus short young forms 25 μ or less in length (see W. H. Taliaferro and Pavlinova, 1936). The higher the percentage of division forms among the trypanosomes is, the higher the rate of reproduction. Valid measures have also been devised for malaria by L. G. Taliaferro (1925), G. H. Boyd (1929a), Lourie (1934), and Mulligan (1935); and for trypanosomes by Robertson (1912), Krijgsman (1933), and W. H. Taliaferro and L. G. Taliaferro (1922). The last authors' coefficient of variation method depends upon the fact that, within certain limits, the variability in total length increases proportionately as the young and growing forms resulting from reproduction increase.

MALARIA

This analysis applies only to those malarial species which parasitize erythrocytes and not to such species as *Plasmodium elongatum* (Huff and Bloom, 1935), which undoubtedly infect other blood and connective tissue cells. We have omitted from discussion the so-called exo-erythro-cytic and fixed tissue cell stages described in the life cycles of some plasmodia because of the lack of agreement which now exists among malariologists as to their nature (Boyd and Coggeshall, 1938, review)`.

The course of untreated malarial infections has been most thoroughly studied in avian malaria. This work commenced with the careful statistical studies of the Sergents (1918) and was extended by Ben Harel 1923), L. G. Taliaferro (1925), G. H. Boyd (1929a), Hartman (1927), Gingrich (1932), Lourie (1934), and others. Treated cases will not be considered in this analysis because treatment itself has been shown greatly to affect the length of the asexual cycle and the number of merozoites produced (G. H. Boyd, 1933; G. H. Boyd and Allen, 1934; Lourie, 1934; and Boyd and Dunn, 1939).

Infections with *P. cathemerium* in canaries are extremely stereotyped and therefore afford an excellent base line for considering the so-called benign infections, which tend to recover and which constitute the majority of malarial infections. When a few parasites are injected into a bird, an incubation period follows during which no parasites can be detected in the peripheral blood. As soon as parasites appear, they increase from day to day at a constant rate (the intersporulation death of parasites will be taken up later), according to a geometrical progression, until sometimes as many as half of the red blood cells are infected (acute rise of infection). At this point, if the bird does not succumb to the infection, recovery is initiated and is manifested by the rapid disappearance of many of the parasites from the peripheral blood (crisis). Following the crisis, parasites remain few in number, but may fluctuate to some extent (developed infection). Sooner or later, the number of parasites is reduced to a level at which they can no longer be detected in peripheral blood films; but a few persist, since transfers of large amounts of blood will infect other birds. This latent period may last for several years, but it may be interrupted periodically by spontaneous or induced relapses (much rarer in *P. cathemerium* than in many other species), which are similar to, though generally quantitatively less than the acute rise of the initial infection, and which are terminated by a crisis. Occasionally, such a relapse may be fatal. By Hegner's (1926) terminology such an infection is divided into prepatent (incubation), patent (acute rise, crisis, and developed infection), and subpatent (latent) periods, with second and third patent periods representing first and second relapses and crises.

When this infection was analyzed, it was found that the basic rate of reproduction remains comparatively constant, whenever parasites are

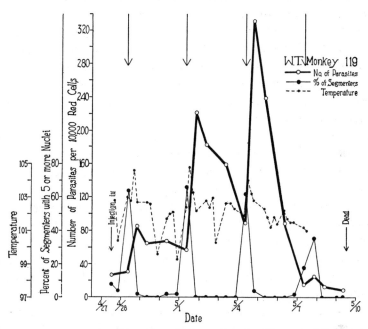

FIGURE 188. The changes in number of *Plasmodium brasilianum* and the percentage of segmenters during the acute rise and crisis of the infection in Central American monkey 119. A natural parasiticidal immunity is operative, as evidenced by the inter- and intrasporulation death of parasites; and as acquired immunity is developed at the crisis, further parasiticidal effects are operative, as evidenced by the tremendous death of parasites. The rate of reproduction is temporarily affected, as evidenced by the irregular percentage of segmenters. Had the animal lived, reproduction would have resumed its normal rate, as ascertained from the study of other monkeys similarly infected. (From W. H. Taliaferro, 1932.)

demonstrable in the blood, since the schizonts produce between ten and fifteen and a half progeny (merozoites) continuously and produce them every twenty-four hours. (This statement is relatively true, since there is no prolonged inhibition of reproduction. Temporary deviations and fluctuations do occur, however. Thus, Boyd and Allen [1934] and Boyd [1939] found that the number of merozoites produced decreases as the crisis approaches and then rises thereafter.) Furthermore, since parasites reproduce at a high rate whenever they are found, it seems reasonable to assume that they reproduce at a high rate during latency, when they cannot be found in sufficient numbers to study. This assumption is in accord with the view held by Ross (1910, review), Bignami (1910), James (1913), and others.

Essentially the same results were obtained from the studies of *P. brasilianum* in Panamanian monkeys, except that during the crisis of the initial infection (W. H. Taliaferro and L. G. Taliaferro, 1934a) one asexual cycle may deviate or be retarded for a day or two. This parasite shows a quartan periodicity. Thus in Figure 188 the percentage of segmenters sharply increases and decreases every third day (4/28, 5/1, 5/4) until at the time the number crises is reached (5/7) the percentage of segmenters does not rise as high as before and does not decrease as precipitously. The other so-called benign malarias seem to be similar, as far as the data on the following species go: *P. cynomolgi*, in both *Silenus rhesus* and *S. irus; P. knowlesi*, in *S. irus* (Sinton and Mulligan, 1933a, 1933b; Mulligan, 1935); and *P. vivax* and *P. malariae* in man. The rapidly fatal infection of *P. knowlesi* in *S. rhesus* is similar to the acute rise of benign infections without a crisis. These statements do not mean that temporary derangements of the cycle may not occur during the crisis, as in *P. brasilianum;* or after treatment with quinine, as in *P. cathemerium* (previously cited); or after changes in host habits, as has been shown to occur in both the latter species (L. G. Taliaferro, 1928; G. H. Boyd, 1929b; W. H. Taliaferro and L. G. Taliaferro, 1934b; and Stauber, 1939).

Since there is no prolonged inhibition of reproduction, the number curve can be interpreted chiefly in terms of parasiticidal effects. In other words, the number of parasites after each asexual cycle should increase by the number of progeny in each mature schizont (minus the number of merozoites which develop into sexual forms), provided no death of parasites occurs. The geometrical rate at which *P. cathemerium* increases does not, however, account for all the progeny produced. They evidently die at all stages of growth and segmentation (L. G. Taliaferro, 1925; Hartman, 1927; Hegner and Eskridge, 1938) and die at a greater rate during the latter part of the acute infection than during the early part (Boyd, 1939). Hegner and Hewitt (1938) and Hewitt (1938) suggested that their death is due to the destruction of multiple infected red cells.

It is interesting that in 1888 Golgi noted that malaria would always be progressive until pernicious symptoms were evident, if the parasites arriving at maturity every two days in tertian or every three days in quartan malaria should complete their life cycle. W. H. Taliaferro and L. G. Taliaferro (1934a) found that out of an average of 9 progeny produced by P. brasilianum, never more than 1.5 complete their development and of the 7.5 which fail to complete their development, about 6 fail to get into new cells and 1.5 die or are killed during intracorpuscular growth. Thus, in Figure 188 on the morning of May 1 there were 60 parasites per 10,000 red blood cells. That evening after sporulation, there were 221, an increase of 3.66 times instead of 9 times. During the subsequent 2.5-day period of growth and division, about two-thirds of these were destroyed, so that on the morning of May 4 there were only 83 per 10,000 red cells. Similar data are furnished by Brug (1934) and W. H. Taliaferro and Mulligan (1937), who worked on infections with P. knowlesi and P. cynomolgi; and by Pijper and Russell (1925, quoted by Sinton et al., 1931), by Rudolf and Ramsay (1927), by Sinton et al. (1931), and by Lowe (1934), all of whom worked on one or both of the tertian and quartan malarias of man. Knowles and Das Gupta (1930) believed that the destruction of parasites takes place only during the free merozoite stage. Examination of their tables, however, shows that the infection they studied was made up of several broods of parasites, the sporulation of one of which would obscure a decrease of parasites during the intrasporulation period of another brood.

The constant rate of death of large numbers of parasites during the initial part of the infection is a manifestation of natural immunity and represents the suitability of the hosts' blood to the malarial organism. The crisis, on the other hand, represents the beginning of the immune reaction, when more progeny die than survive and the infection therefore declines. Thus the crisis in Figure 188 takes place after the sporulation of May 4 and between May 5 and May 7, as shown not only by the conspicuous death of parasites during the intersporulation period of May 5 and May 6, but by the relatively slight increase of parasites during the sporulation on May 7. This infection, which was the most acute encountered, caused the death of the monkey on May 9. The temperature curve is extraneous to the present discussion, but is interesting because of the stiletto-like peaks it shows at each sporulation. Some work on the infection after the crisis has been done on human malaria by Böhm (1918), Knowles and Das Gupta (1930), and Sinton *et al.* (1931).

Throughout the developed infection, an equilibrium is established between the number of parasites killed and the number produced. Later, the defensive factors are usually successful in suppressing the parasites arising by reproduction, so that latency ensues. Latency may last for years, but may be interrupted by relapses. Although there is no unanimity of opinion on the mechanism of relapse, the best evidence indicates that it represents simply the removal of the defensive factors, so that the parasites, which are continuously reproducing at a constant rate, reaccumulate in the blood. The relapse may be fatal, or the defensive factors may again materialize and successfully suppress it. Accordingly, the severity of the relapse depends upon its extent and upon the length of time the defensive factors are removed. These statements are substantiated by work on *P. cathemerium* and *P. brasilianum,* references for which have already been given.

Koch (1899) believed that immunity persisted after complete recovery, but Wasielewski (1901) suggested, and subsequent workers have supported the conclusion, that parasites remain in small numbers for longer periods than was at first supposed and that this latent infection accounts for the long continued immunity (see Thomson, 1933, review; Chopra and Mukherjee, 1936; Sergent, 1936). More recently, Nauck and Malamos (1935) and Coggeshall (1938) have shown that a certain amount of immunity is retained after complete cure of *P. knowlesi.*

Acquired immunity to malaria can be demonstrated not only by the crisis, recovery from initial infection, and recovery from relapse, but by superinfection. Such immunity is usually species-specific or even strain-specific in human, simian, and avian infections. A series of investigators, beginning with Koch (1899), have worked on this aspect of the subject (for literature and for especial work, see W. H. Taliaferro and L. G. Taliaferro, 1929a, 1934c; Gingrich, 1932; Mulligan and Sinton, 1935; M. F. Boyd *et al.*, 1936; Manwell, 1938; Redmond, 1939; and Manwell and Goldstein, 1939). The work on human malaria has been recorded chiefly as a result of the use of therapeutic infections in the treatment of paresis (see Mulligan and Sinton, 1933a). W. H. Taliaferro and

L. G. Taliaferro (1929a, 1934c) intravenously superinfected birds and monkeys with such large numbers of parasites that they could study the superinfection quantitatively. They found that in immune animals the parasites, after reinfection, begin to be removed at once, with the same effectiveness that they are removed at the time of the crisis in initially infected animals, and that in monkeys the asexual cycle occasionally exhibits a transitory delay, accompanied by the production of fewer merozoites. In other words, superinfection, since it occurs in an effectively immune animal instead of in an uninfected animal, has no incubation period or acute rise, but begins at once with a crisis and proceeds immediately to latency.

The problem now arises: what is the mechanism whereby the plasmodia are removed throughout the course of the infection? The death of all of the parasites, whether it takes place before, during, or after the crisis (i.e., during natural or acquired immunity), is associated with phagocytosis by macrophages, chiefly of the spleen, liver, and bone marrow. Other cells play insignificant rôles. Furthermore, the macrophages phagocytose free parasites, parasitized red blood cells, and residues of parasites such as malarial pigment and uninfected red cells, which are probably injured by the infection. Malarial pigment is the most indigestible part of the parasite red-cell complex, since it is often found in macrophages months after all other vestiges of the parasites have disappeared. Large monocytes containing malarial pigment may be found in the peripheral blood, especially during the crisis and thereafter. A review of the literature covering the considerable amount of work done on this aspect of the subject will be found in W. H. Taliaferro and Mulligan (1937) and involves direct evidence from necropsy findings and indirect evidence from splenomegaly (see Stratman-Thomas, 1935; Coggeshall, 1937; Afridi, 1938) and splenectomy and blockading procedures.

Although phagocytosis has been known for years to occur in malarial infections, Cannon and W. H. Taliaferro (1931) and W. H. Taliaferro and Cannon (1936) were the first to study infections and superinfections at closely spaced intervals and to correlate phagocytosis in detail with the course of the infection and with immunity. The rate of phagocytosis during natural immunity, i.e., before the crisis, increases as the number of parasites increases, but is always comparatively sluggish (Pl.

1, Figs. 1 and 2). When the parasites disappear from the peripheral blood at the time of the crisis, they are not at first phagocytosed, but are filtered out and concentrated in the Billroth cords of the spleen (Pl. 1, Fig. 3). This initial manifestation of acquired immunity probably represents a localization of the parasites due to an antibody, which may be an agglutination, as demonstrated *in vitro* in P. *knowlesi* by Eaton (1938), or an attachment of parasites to the macrophages, as found in cultures of P. *falciparum* by McLay (1922). In a few hours or days the concentrated parasites are ingested in great numbers by the macrophages (Pl. 2, Figs. 1 and 2), owing undoubtedly to an opsonification of the parasites and parasitized red cells by antibody. This antibody probably accounts for the protective property of serum taken from latent infections in avian malaria (Findlay and Brown, 1934), in human malaria (Sotiriadés, 1917; Kauders, 1927; Neumann, 1933; and Lorando and Sotiriadés, 1937), and in simian malaria (Coggeshall and Kumm, 1938; Coggeshall and Eaton, 1938). The fact that protective antibodies have not always been found (W. H. Taliaferro and L. G. Taliaferro, 1929b and 1934b) led W. H. Taliaferro and Cannon (1936) to suggest that antibodies are produced locally in sufficient quantities to be operative in the specific organs *in situ,* but not in sufficient quantities to be easily demonstrated in the peripheral blood. A day or two after concentration in the Billroth cords, the parasites are quickly digested except for the ma-

CAPTION FOR PLATE ON FACING PAGE

Plate 1. Sluggish phagocytosis of *Plasmodium brasilianum* by macrophages in control of the American monkeys during the acute rise of the malarial infection (Figs. 1 and 2) and the concentration of P. *brasilianum* in Billroth cords of the spleen at the initiation of the crisis (Fig. 3). × 1450. (From W. H. Taliaferro and P. R. Cannon, "Cellular Reactions during Primary Infections and Superinfections of *Plasmodium Brasilianum* and Panamanian Monkeys," *Journal of Infectious Diseases,* LIX [July-August, 1936], 72-125.)

Figure 1. A Kupffer cell in the liver containing a parasite, an unparasitized erythrocyte, and four small masses of malarial pigment. Acute rise of infection.

Figure 2. Two macrophages in a Billroth cord in the red pulp of the spleen, containing a small number of residues of parasites, red cells, and malarial pigment. Acute rise of infection.

Figure 3. Large numbers of parasitized erythrocytes concentrated in Billroth cords of the spleen. Note that parasites are absent from the venous sinus. Also note that the macrophages in the Billroth cords show the sluggish phagocytosis characteristic of the acute rise and that the littoral cells lining the sinus are not phagocytic. Initiation of crisis of infection.

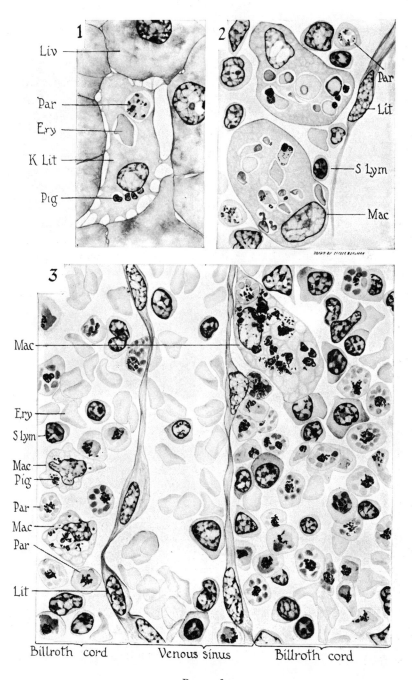

PLATE I

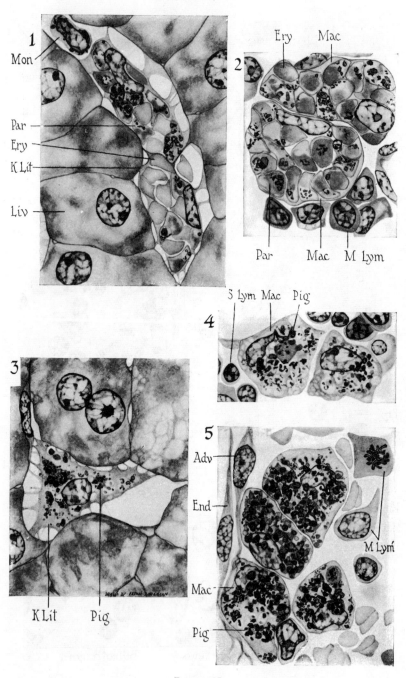

PLATE II

larial pigment (Pl. 2, Figs. 3-5, which are from monkey 119; see also Fig. 188). The pigment is digested within a few months, as may be seen by a study of animals in which relapses or superinfections do not intervene. The various tissues gradually return to their normal histological appearance, but as long as immunity lasts they retain their ability to react more quickly than the tissues of nonimmune animals.

These data raise the interesting immunological question as to whether the death of the parasites before the crisis and that during or after the crisis are due to the same factors. In other words, is acquired immunity simply an enhancement of the high-grade natural immunity present from the beginning, or is it due to an entirely new set of factors superimposed on the natural condition? Certain facts tend to indicate that the two mechanisms are essentially different. In the first place, during natural immunity, all evidence indicates that phagocytosis is nonspecific, and there is some evidence that only those parasites are phagocytosed which are moribund or otherwise abnormal (Hartman, 1927; Gingrich, 1934; Hegner and Hewitt, 1938; Hewitt, 1938). In the second place, Gingrich found that the injection of large numbers of red cells breaks down

---

CAPTION FOR PLATE ON FACING PAGE

Plate 2. Intense phagocytosis of *Plasmodium brasilianum* by macrophages in Central American monkeys at the height of the crisis of the malarial infection (Figs. 1 and 2) and malarial pigment, the residue of parasites after the intense phagocytosis, in the macrophages shortly thereafter (Figs. 3-5). × 1450. (From W. H. Taliaferro and P. R. Cannon, "Cellular Reactions during Primary Infections and Superinfections of Plasmodium Brasilianum and Panamanian Monkeys," *Journal of Infectious Diseases*, LIX [July-August, 1936], 72-125.)

Figure 1. A Kupffer cell in the liver containing many recognizable parasitized and unparasitized red cells, and a monocyte (Mon.) containing malarial pigment. Crisis of infection.

Figure 2. Two macrophages in the Billroth cord of the spleen containing many recognizable parasitized and unparasitized red cells. Crisis of infection.

Figure 3. A Kupffer cell in the liver containing malarial pigment. Two days after the crisis.

Figure 4. Two macrophages in the bone marrow containing malarial pigment. Two days after the crisis.

Figure 5. Four macrophages in the Billroth cord of the spleen containing malarial pigment. Two days after the crisis. Figures 3, 4 and 5 were all taken from monkey 119 (see text Figure 188). Note that the Kupffer cell of the liver contains intermediate amounts of malarial pigment between that found in the macrophages of the spleen and of the bone marrow.

acquired resistance to *P. cathemerium,* but, even when pushed to the maximum, does not affect natural resistance. This work, however, is open to the possible criticism that the large amounts of pigment in the macrophages after the crisis, which are not there before the development of immunity, may simply augment the blockading doses to produce the observed effect. In the third place, as pointed out previously, accumulating evidence indicates that the greatly superior mechanism for disposing of parasites associated with acquired immunity is highly specific and is probably associated with an antibody.

The limitation of phagocytosis to the macrophages of the spleen, liver, and bone marrow is probably a question of opportunity, as such macrophages are advantageously placed where they can remove material from the blood stream. The so-called general immunity in malaria, therefore, is actually a local reaction in strategically placed organs (W. H. Taliaferro, 1934). It should be noted, however, that in overwhelming infections the parasites may be so numerous that secondary complications, such as stasis of the blood, clogging of the capillaries, and hemorrhage may occur in various organs. In such an event the macrophages of the brain, lungs, suprarenals, and kidneys, as well as those of the bone marrow, liver, and spleen, may actively phagocytose malarial material.

CAPTION FOR PLATE ON FACING PAGE

Plate 3. Portions of a venous sinus and Billroth cords in the spleen of an uninfected rhesus monkey (Fig. 1) and comparable portions from a rhesus monkey during the late acute rise of an infection with *Plasmodium cynomolgi* (Fig. 2). × 1400. (From W. H. Taliaferro and H. W. Mulligan, "Histopathology of Malaria with Special Reference to the Function and Origin of the Macrophages in Defence," *Indian Medical Research Memoirs,* No. 29 [May, 1937], pp. 1-138.)

Figure 1. The normal constituents of the venous sinus are chiefly the cells of the circulating blood (lymphocytes, granulocytes, and red cells), and those of the Billroth cords are also cells of the circulating blood with a greater proportion of large lymphocytes and, in addition, reticular cells with indeterminate cytoplasm and slightly phagocytic macrophages.

Figure 2. The additional constituents of the venous sinus and of the Billroth cords in the spleen of a monkey during the late acute rise of an infection of *P. cynomolgi* are parasitized red cells and many transitional cells (polyblasts) between lymphocytes and macrophages, containing malarial pigment. The progressive hypertrophy of lymphocytes into macrophages is shown by nuclear changes and by increased amounts of cytoplasm in the lymphocytes (see especially Med Lym 1, which also contains two small granules of malarial pigment) and by further nuclear changes, increased size, and increased phagocytic activity in the polyblasts (see especially polyblasts 1 through 5). Throughout this transitional series, phagocytosis increases approximately with the size of the cell.

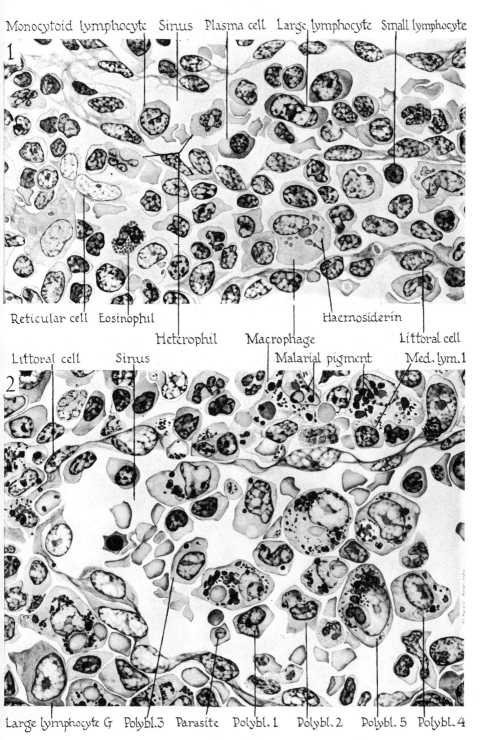

Monocytoid lymphocyte   Sinus   Plasma cell   Large lymphocyte   Small lymphocyte

1

Reticular cell   Eosinophil
                 Heterophil   Macrophage   Haemosiderin
                                                              Littoral cell
Littoral cell   Sinus   Malarial pigment   Med. lym.1

2

Large lymphocyte G   Polybl.3   Parasite   Polybl.1   Polybl.2   Polybl.5   Polybl.4

PLATE III

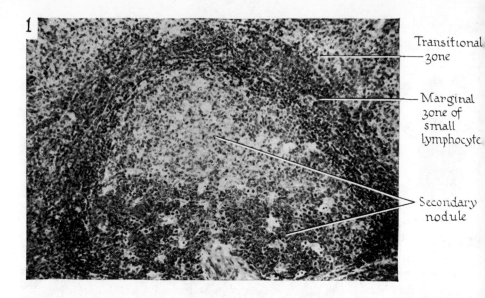

1   Transitional zone

Marginal zone of small lymphocyte

Secondary nodule

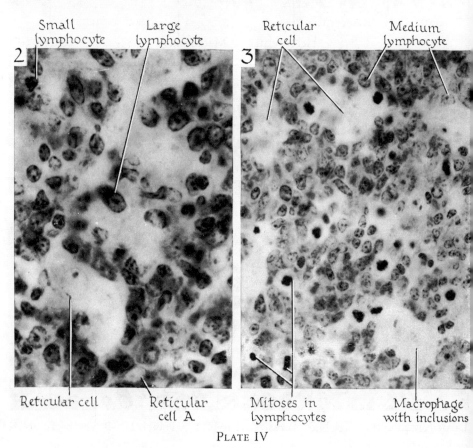

2   Small lymphocyte   Large lymphocyte

Reticular cell   Reticular cell A

3   Reticular cell   Medium lymphocyte

Mitoses in lymphocytes   Macrophage with inclusions

PLATE IV

The fact has been stressed that the greatly enhanced phagocytic activity of individual macrophages during acquired immunity is specific and is probably related to antibodies. In addition, there is a nonspecific increase of macrophages at strategic points, particularly in the spleen and the bone marrow, due to their cytogenesis from lymphocytes through polyblast stages and from histogenous macrophages. The latter source is acknowledged by Bruetsch (1927, 1932a, 1932b) and others, and both sources have been demonstrated by W. H. Taliaferro and Cannon (1936) and W. H. Taliaferro and Mulligan (1937). In fact, the chief source of new macrophages is from lymphocytes (Pl. 3). The lymphocytes themselves arise by mitotic proliferation (Pl. 4). This is a prominent part of the so-called lymphoid hyperplasia of the spleen and occasionally of other organs, if the malarial infection is long drawn out.

In passing it may be mentioned briefly that the parasites destroy red blood cells in large quantities, flood the blood plasma with foreign matter such as corpuscular debris, free malarial parasites and malarial pigment, sometimes block the capillaries and damage various tissues, especially the spleen and liver. All these losses and destructions are made good by various nonspecific hyperplastic and reparative activities of the host. The pathological and regenerative changes have been extensively studied, especially in human malaria (see W. H. Taliaferro and Mulligan, 1937, for a review of the literature).

The foregoing results indicate that acquired immunity against malaria largely involves parasiticidal effects, with no pronounced inhibition of the rate of reproduction for extended lengths of time. The parasiticidal effects can be correlated with phagocytosis. Phagocytosis is sluggish dur-

---

CAPTION FOR PLATE ON FACING PAGE

Plate 4. A nodule in the white pulp of the spleen during lymphoid hyperplasia associated with the late acute rise of *Plasmodium cynomolgi* in a rhesus monkey. (From W. H. Taliaferro and H. W. Mulligan, "Histopathology of Malaria with Special Reference to the Function and Origin of the Macrophages in Defence," *Indian Medical Research Memoirs*, No. 29 [May, 1937], pp. 1-138.)

Figure 1. This activated splenic nodule due to malaria is slightly enlarged, shows a pronounced transitional zone and a markedly active secondary nodule, in which occur swollen phagocytic reticular cells and mitoses, among many lymphocytes and a few reticular cells. × 180.

Figure 2. A detail of the upper portion of the secondary nodule shown in Figure 1 which consists mainly of medium lymphocytes and swollen, slightly phagocytic reticular cells. Reticular cell A is beginning to mobilize. × 1015.

Figure 3. A detail of the lower portion of the secondary nodule shown in Figure 1, which consists mainly of lymphocytes many of which are dividing. × 655.

ing the acute rise of the infection and is greatly enhanced at the time of the crisis, when immunity, which is probably associated with the elaboration of antibodies, is developed. Thereafter the infection progressively subsides, unless the immunity is lowered. If the immunity is lowered, the parasites, because their reproduction has not been inhibited, reaccumulate in the blood until immunity again develops and becomes operative.

LEISHMANIASIS

The course of kala azar cannot be studied and analyzed as was done in the case of malaria because its causative agent, *Leishmania donovani*, is not accessible for study; but it is of interest here because *L. donovani* lives in the macrophages themselves, as has been shown by Christophers (1904), Meleney (1925), Hu and Cash (1927), and others. It not only lives in the macrophages of the spleen, liver, bone marrow, and intestinal wall, and, in extreme cases, the macrophages of almost all organs and tissues, but proliferation of the macrophages constitutes the chief characteristic of the disease. For a review of the literature, see C. J. Watson, 1928; Linton, 1929. See the former reference also for a seemingly similar condition in the little-known histoplasmosis. The parasites, therefore, instead of being digested, find the cytoplasm of the phagocytes a suitable medium in which to grow and multiply. Splenectomy should be particularly illuminating in trying to decide whether the macrophage system is valuable, imperfect as it is, as the only defense the body has; or is deleterious, as being the most suitable location for the parasites. Some work on kala azar has been done (see Laveran, 1917), but further systematic experimental work on animals should prove valuable. The fact that the disease is so often fatal indicates that reproduction of the parasites is continuous, as in malaria. Immunity, nevertheless, is developed in approximately 10 percent of the infections, but it is not apparent whether the suppression of the infection is predominantly due to an increase in the ability of the macrophages to digest the parasites or to an inhibition of reproduction of the parasites.

Oriental sore, a cutaneous leishmaniasis caused by *L. tropica,* on the other hand, usually spontaneously heals and confers an immunity to reinfection. Sections of the skin at the site of the sores often show pronounced local accumulations of macrophages. As in the case of the small percentage of human beings recovering from kala azar, it is impossible to decide from the available data whether suppression of the infection is predominantly due to a destruction of the parasites or to an inhibition of their reproduction. The fact that immunity is more or less generalized indicates, however, that some humoral principle is involved.

NONLETHAL INFECTION WITH THE *Trypanosoma lewisi* GROUP OF TRYPANOSOMES

The trypanosomes differ from the plasmodia in that they live in the blood stream and do not infect the red blood cells. Some are pathogenic. Others are nonpathogenic. Among the latter is a large group of trypanosomes which produce nonlethal infections in rodents, are morphologically identical or similar to *T. lewisi* of the rat, and are differentiated almost entirely by their specificity for their rodent hosts. Of these, *T. lewisi* of the rat and *T. duttoni* of the mouse have been extensively studied and furnish the basis for the conclusions in the following discussion.

The number curve of *T. lewisi* in the rat when a few parasites are injected, as shown by Steffan (1921), W. H. Taliaferro and L. G. Taliaferro (1922), W. H. Taliaferro (1924), and Coventry (1925), starts, as does a malarial infection, with an incubation period and an acute rise, until the trypanosomes may reach 300,000 or more per cu. mm. Then there is a crisis between the eighth and the fourteenth days, during which most of the parasites are destroyed. Those that remain continue to live in the blood for some time (varying from several weeks to several months), until they are removed either gradually or suddenly. Thereafter they are not found in the blood, and relapses seldom occur, but a few may persist, as ascertained by relapses which sometimes ensue after splenectomy and blockade with India ink, or after other conditions which lower the immunity of the host. Whether a few always persist cannot be determined from the available data. The rat, however, is immune to reinfection for long periods, as was first shown by Kanthack et al. (1898). Fatal infections of *T. lewisi* in young rats were first reported by Jürgens in 1902 (see also W. H. Brown, 1914; Herrick and Cross, 1936; Duca, 1939; Culbertson and Wotton, 1939). They are often complicated by a concomitant occurrence of either or both of the following: a subnormal condition of the rats or infections, such as Bartonella or paratyphoid. In any case Culbertson and Wotton (1939) have found that fatal infections develop in rats in which the content of ablastin is low.

Early investigators (Rabinowitsch and Kempner, 1899; von Wasielewski and Senn, 1900; especially Laveran and Mesnil, 1901; MacNeal,

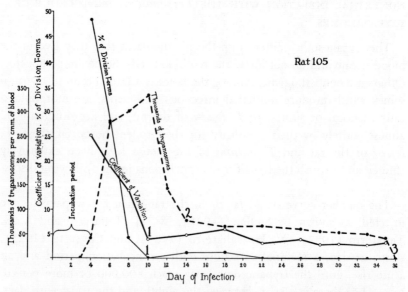

FIGURE 189. The changes in number of *Trypanosoma lewisi* and the coefficient of variation and percentage of division forms during the course of infection in rat 105. As acquired immunity develops, the rate of reproduction is inhibited by ablastin, as evidenced by the low coefficient of variation and low percentage of division forms beginning at location 1, and the parasites are killed by trypanolysins operative at locations 2 and 3. Whether, in addition, natural immunity operates has not been ascertained. (From W. H. Taliaferro, 1924; division forms added.)

1904; W. H. Brown, 1915) were convinced by their microscopical studies that *T. lewisi* reproduces only during the first few days in the rat, after which the trypanosomes live in the blood as nonreproducing adults. This conclusion has been substantiated by W. H. Taliaferro (1924), Coventry (1925), and Regendanz and Kikuth (1927). Thus in Figure 189 the coefficient of variation of the total length of the trypanosomes and the percentage of division forms, each of which, as will be recalled, measure the rate of reproduction, are high when trypanosomes appear in the blood on the fourth day after infection and then drop precipitously until on the tenth day of the infection they reach a low level (at location 1, in Figure 189), from which they do not thereafter deviate. Provided such a large number of adult trypanosomes are injected intravenously that they appear in the blood and can be studied immediately, the coefficient is low on the first day and rises precipitously, as may be seen in control rat 980 in Figure 190. The inhibition of reproduction, as will be shown later, is due to the development of an acquired immunity involving an antibody which has been called ablastin (W. H. Taliaferro, 1924, 1932). The rate of reproduction of *T. lewisi* is similarly retarded and inhibited when grown in an abnormal host, the guinea pig, as ascertained by Coventry (1929). Essentially the same results were found for *T. duttoni* in the mouse, except that the rate of reproduction is never as high and is not as completely inhibited, according to W. H. Taliaferro and Pavlinova (1936) and W. H. Taliaferro (1938). Hence the trypanosomes are never as numerous during the acute rise and may increase slightly in numbers during the first part of the developed infection. *T. iowensis* in the striped ground squirrel, as described by Roudabush and Becker (1934), closely parallels the development of *T. duttoni*. Since *T. nabiasi* in its natural host, the rabbit, increases in numbers only during the first few days of the infection and thereafter does not show division forms, as reported by Króo (1936), the rate of reproduction of this trypanosome may also be inhibited.

The question arises: Is there a natural immunity during the acute rise of these infections? For it must be realized that in spite of the rise in numbers a constant percentage of the parasites formed may be being killed as was demonstrated in malaria. There are two ways of demonstrating natural immunity. The first applies to the death of the organisms and can be used only in such infections as malaria, in which it can be demonstrated that all of the progeny formed do not survive. This is impossible in the trypanosome infections, in which reproduction is not synchronous and in which no method of ascertaining the total number of progeny produced has so far been devised. The second method applies not only to the death of the organisms, but also to the rate of reproduction, and involves various procedures such as comparisons of the same species in various hosts and splenectomy combined with blockade. Positive experiments of this kind will indicate the existence of a natural immunity, but, since natural immunity may be due to many factors, the only conclusion that is warranted from negative experiments is that the experimental method used did not disclose any natural immunity. With these facts in mind, we may conclude the following: Guinea pigs seem to possess a high natural parasiticidal immunity to *T. lewisi,* since *T. lewisi* hardly increases in numbers at all in spite of the fact that it goes through the same reproductive cycle as it does in the rat. Mice seem to possess a high natural ablastic immunity to *T. duttoni,* since *T. duttoni* in splenectomized and blockaded mice reproduce at a higher rate than in normal mice. Rats have not been shown to possess any natural immunity against *T. lewisi.*

The drop in numbers at the crisis (at location 2, in Fig. 189) represents the acquisition of a trypanocidal response on the part of the host, in addition to the ablastic effect, since if only the latter were present the numbers would remain constant. Also, the disappearance of the trypanosomes at the end of the infection (at location 3, in Fig. 189) represents another acquisition of the same type of immune response. The complementary action of the ablastic and trypanocidal effects not only effectively suppresses the infection, but also prevents relapses and reinfections for long periods (at least 325 days). Thus a few trypanosomes left from the initial infection or introduced by reinfection may be killed at once, or, if they are not killed at once, their reproduction is inhibited until they are killed. These statements hold for *T. duttoni* and, as far as they have been tested, for *T. nabiasi* (previous citations).

The three effects of immunity which operate at points 1, 2 and 3 in Figure 189 are all due to humoral antibodies, as tested by passive transfer experiments. They are associated with the globulin fraction of serum, are acquired as a result of specific infection or specific immunization, and are decreased in amount or delayed in time of appearance by splenectomy and blockade. They differ in the following ways: The titer of the three varies independently, as far as can be tested. The trypanocidal effects are due to typical lysins which may, however, act as opsonins *in vivo.* The trypanolysin which terminates the infection kills either adult or dividing trypanosomes taken at any time during the course of infection, whereas that which causes the first number crisis kills only those trypanosomes which have just appeared in a rat's blood. The parasites that survive the first number crisis are either basically nonsusceptible to this antibody or acquire a resistance to it. So far, the reproduction-inhibiting properties of ablastin have not been demonstrated *in vitro*, but suitable amounts of serum, containing ablastin together with adult trypanosomes, when tested *in vivo* in a rat allow the trypanosomes to live, but prevent them from reproducing. (Adult trypanosomes have to be used for this

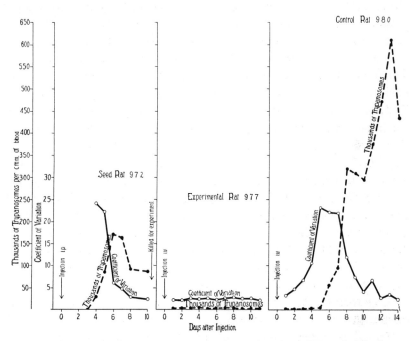

FIGURE 190. The demonstration of ablastin against *Trypanosoma lewisi* by passive transfer. The rate of reproduction of the trypanosomes, as shown by the constant coefficient of variation, is completely inhibited in experimental rat 977, which received 2 cc. of serum and a large number of adult trypanosomes from seed rat 972; whereas the rate of reproduction goes through the normal cycle in control rat 977, which received a similar number of adult trypanosomes, but no serum, from seed rat 972. (From W. H. Taliaferro, 1924.)

test since a curtailment of the reproductive activity of dividing trypanosomes, which is all that could be expected, is difficult to demonstrate with certainty.) Thus the difference between the coefficient of variation curve in experimental rat 977, which was given ablastic serum plus adult trypanosomes, and in control rat 980, which was given normal serum plus adult trypanosomes, is seen in Figure 190. The number curve remains constant in experimental rat 977 because the trypanosomes are not reproducing. As was indicated above, ablastin is not absorbed by the trypanosomes *in vitro,* whereas the trypanocidal antibodies are. There is thus no lasting union of trypanosomes and ablastin, nor is there a sensitization of trypanosomes by ablastin as there is with the trypanolysins. Moreover, if smaller and smaller doses of trypanocidal antibody are given, a point is reached at which the trypanosomes are not killed, but their reproduction remains unaltered. A group reaction can be demonstrated between *T. duttoni* and *T. lewisi* and their ablastins *in vivo,* and between *T. lewisi* and anti*duttoni* trypanocidal antibody *in vivo* and *in vitro;* but whether the reaction of anti*lewisi* trypanocidal antibody against *T. duttoni in vivo* and *in vitro* is a true group reaction of an immune anti*lewisi* antibody is not evident because normal rat serum is also trypanocidal to *T. duttoni.* These statements are based on a series of investigations involving either or both of the following: *in vitro* work for the trypanocidal effects and *in vivo* passive transfer experiments for all three effects by W. H. Taliaferro and coworkers (*vide infra*), Regendanz and Kikuth (1927), Perla and Marmorston-Gottesman (1930) and coworkers, W. H. Taliaferro, Cannon, and Goodloe (1931), and W. H. Taliaferro (1932). Culbertson (1938) has shown that the immunity to *T. lewisi* is passed through the mother's placenta and milk to young rats where it persists for several weeks. Later, Culbertson and Wotton (1939) found that the young rats do not appear to produce ablastin as promptly or as well as older rats.

Various procedures designed to lower the macrophage function, such as splenectomy, especially if combined with India-ink blockade or some infection such as *Bartonella* which affects the macrophage system, decrease the strength and delay the appearance of ablastin and the terminal trypanolytic antibody (Regendanz and Kikuth, 1927; Perla and Marmorston-Gottesman, 1930; Regendanz, 1932; and W. H. Taliaferro, Cannon, and Goodloe, 1931). This is an effect on acquired immunity.

The next question which arises is whether the macrophages or other phagocytic cells assist in passive immunity or in the action of ablastin or the trypanolysins in the body. The work of W. H. Taliaferro (1938) indicates the following: Splenectomy and blockade have no effect on the passive transfer of ablastin, but the following interesting secondary effect results: The passive transfer of ablastin lasts only for a few days.

Thereafter it is not adequately augmented and supplemented by an active ablastic immunity in splenectomized and blockaded animals, as it is in normal rats, because the active immunity is slow in developing and decreased in amount. Splenectomy and blockade definitely decrease the effectiveness of the trypanolysins. It would seem that such an effect could be adequately explained by a decrease in the amount of complement which would prevent the lysis of sensitized trypanosomes, or by a decrease of macrophages which would prevent the removal of opsonized parasites. The fact that previously sensitized trypanosomes are as readily removed in splenectomized and blockaded animals as in normal rats seems to negate both of these suppositions, unless the sensitized trypanosomes are agglutinated and removed mechanically—a possibility which because of technical difficulties has not yet been ruled out. A more likely explanation is that there is an interference with the union of antigen and antibody.

It has already been indicated that a lysin differs from an opsonin only in that the terminal lysis and death of sensitized organisms may be effected by extra- rather than intracellular enzymes. Just as in the indirect studies discussed in the preceding paragraph, however, direct studies on phagocytosis have failed to indicate whether phagocytosis or lysis is more important in acquired immunity. Laveran and Mesnil (1901) considered that the parasites are actively phagocytosed, and Roudsky (1911) and Delanoë (1912), studying the acquired immunity of mice to *T. lewisi,* came to the same conclusion. Regendanz and Kikuth (1927) believed that the parasites are phagocytosed in a nonspecific way. MacNeal (1904), Manteufel (1909), W. H. Taliaferro (1924), and Coventry (1929), on the other hand, considered that they are lysed. In studying the tissues for evidence of phagocytosis, we are handicapped by the fact that no easily recognizable vestiges of trypanosomes, such as malarial pigment, remain in macrophages for any appreciable length of time. The fact that pigment by itself may be phagocytosed does not invalidate this statement, since the whole complex, consisting of red cell, parasite, and pigment, is often recognized intact in macrophages.

To sum up: acquired immunity against nonpathogenic trypanosomes primarily involves ablastin and trypanolysins, the first of which prevents the trypanosomes from undergoing growth and cell division, and the second of which kill the trypanosomes. They are both humoral antibodies which are associated with the globulin fraction of immune serum, are passively transferable, and are probably a product of the lymphoid-macrophage system; but whereas ablastin possesses no *in vitro* affinity for the trypanosomes, the trypanolysins are typical antibodies (amboceptor) and can sensitize the trypanosomes *in vitro*. Furthermore, the macrophage system does not appear to intervene in the passive transfer of ablastin, but in some way functions in the union of antigen and antibody during the passive transfer of the trypanolysins.

CONTINUOUS FATAL TRYPANOSOMIASIS IN THE MOUSE AND SOMETIMES IN THE RAT

Most species of trypanosomes appear to be nonpathogenic, and there is a growing mass of evidence that even many of the pathogenic forms

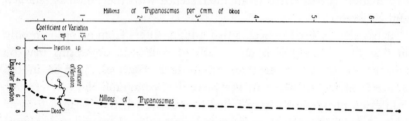

FIGURE 191. The changes in number of *Trypanosoma rhodesiense* and the coefficient of variation during the course of infection in a mouse. No acquired immunity is developed in the mouse, since the number of trypanosomes more or less steadily increase and their rate of reproduction, as evidenced by the high coefficient of variation, is not inhibited. (Redrawn from W. H. Taliaferro and L. G. Taliaferro, 1922.)

in man and domesticated animals may be nonpathogenic in their natural hosts (cf. Duke, 1936). Little is known of the course of their infection in their natural hosts, but they have been extensively studied in laboratory animals, in which they are all pathogenic. The well-known pathogenic trypanosomes, which produce disease in man and domestic animals, are *T. gambiense, T. rhodesiense, T. brucei, T. congolense, T. vivax, T. evansi, T. equinum,* and *T. equiperdum.* When injected into mice, the parasites almost invariably appear in the blood after a short incubation period, and increase in number more or less steadily until the death of the host. This type of infection is composed of an incubation period and acute rise, with no crisis or developed infection. It sometimes occurs in the rat. Its continuous nature was clearly pointed out by Massaglia (1907) and is illustrated in Figure 191 by *T. rhodesiense* in the mouse.

Whether there is a natural immunity against the trypanosomes in the mouse cannot be answered at present, because of insufficient data. On the one hand, W. H. Taliaferro, Johnson and Cannon (unpublished work, see W. H. Taliaferro, 1929) reported no effect of splenectomy on mice infected with *T. equinum,* and on the other hand Schwetz (1934) and Russeff (1935) found a slight effect of splenectomy in mice infected with *T. congolense* and *T. equiperdum* respectively.

With regard to acquired immunity, most data are in accord in showing that the mouse does not develop any appreciable amount. As may be seen in Figure 191, the rate of reproduction of the parasites remains constant and fairly high (the C.V. varies between 8.9 and 10.5 percent) and the parasites progressively increase in number until the death of the host (see also control mouse, infected with *T. equinum,* in Figure 193, which is drawn on a semilogarithmic scale).

In 1933 Krijgsman showed that the rate of increase of *T. evansi* in the mouse and rat is not uniform during the acute rise, but that two periods of high rates of increase, approximately between eighteen and thirty-two hours and sixty and sixty-six hours, alternate with three periods of lower rates of increase at the beginning, middle, and end of the infection. He believed that the terminal low rate of increase is due to a destruction of parasites, as evidenced by the occurrence of more degenerating forms, but that the earlier low rates are due to a partial inhibition of reproduction. He reached this conclusion because he found no increase in degenerating trypanosomes in the blood and in spite of the fact that he found no diminution in the percentage of division forms. He visualized the mechanism of this partial inhibition of reproduction as a uniform retardation of all stages of the cycle of growth and cell division. There is no doubt that theoretically such a uniform lengthening of all stages would give a partial inhibition of reproduction, without affecting either the percentage of division forms or the coefficient of variation. Nevertheless, the existence of such a mechanism is doubtful, in view of the fact that in *T. lewisi,* as well as in cells in general, a retardation of cell division is characterized by an increased length of the resting stage (the so-called adult stage of *T. lewisi*), and not by a gradual slowing down of the whole process with

uniform increases in the length of each stage. Furthermore, Krijgsman's inability to find degenerating stages in the blood early in the
infection and his finding them during the terminal phases is not conclusive evidence that trypanosomes are not dying during the early low
rates of increase. It is very likely that the macrophages remove such
forms less quickly during the latter part than during the early part of
the infection, because they have become partially blockaded. The interpretation of the varying rates of increase of *T. evansi* in the mouse, as
found by Krijgsman, will therefore have to await further analysis. From
an enormous mass of work on the mouse as a carrier of so-called passage strains, however, it appears that no trypanolysin usually develops
in the mouse which kills most of them and toward which the residue
become antigenically resistant, as has been demonstrated in the infections to be described in the following section.

We may accordingly conclude that the course of the infection in the
mouse and sometimes in the rat most closely approximates the simplest
type of infection, which increases as a geometrical progression and in
which little, if any, immunity is acquired of either an ablastic or trypanocidal type.

Intermittent Fatal Trypanosomiasis in Various Laboratory Animals

When the same pathogenic trypanosomes considered in the preceding section are grown in the guinea pig, the infection is typically characterized by an incubation period, followed by alternate increases (the
first is an acute rise and the succeeding ones are relapses) and decreases
(crises) in the parasite population until the animal dies. Besides the
guinea pig, this kind of infection is observed in rabbits, dogs, cats, and
occasionally in rats infected with these same trypanosomes, in man
infected with *T. rhodesiense* and sometimes in mice infected with *T.
congolense*. Sometimes the initial acute rise and crisis do not develop,
as is shown in Figure 192. In other animals, such as sheep, the entire
infection is of such low grade that trypanosomes are rarely found and
then only in thick film. These statements are based on work by Ross
and D. Thompson (1910, 1911), J. G. Thomson (1912), W. H.
Taliaferro and L. G. Taliaferro (1922), Knowles and Das Gupta
(1928), Davis (1931), Krijgsman (1933), Browning *et al.* (1934),

and others. The fact that animals often die when trypanosomes are scarce in their blood has been explained as being due to sugar depletion, asphyxiation, toxins, and so forth (see von Brand, 1938, review). A possibility that has not been adequately discussed in the literature is that death shortly after a crisis may be due to the severity of the immune reaction, comparable to that seen in overwhelming hypersensitivity (cf. graph of malaria, Figure 188).

In analyzing such infections, W. H. Taliaferro and L. G. Taliaferro (1922) found that the basic rate of reproduction remains relatively

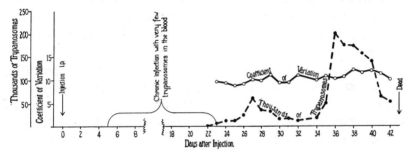

FIGURE 192. The changes in number of *Trypanosoma rhodesiense* and the coefficient of variation during the course of infection in a guinea pig. As acquired immunity develops, the parasites are killed by trypanolysins operative during the long chronic period and at the two crises thereafter, but the rate of reproduction is not inhibited, as evidenced by the high coefficient of variation. (Redrawn from W. H. Taliaferro and L. G. Taliaferro, 1922.)

constant whenever parasites can be found in the blood in sufficient numbers for study. Thus in Figure 192 the coefficient of variation remains between 8.5 and 12.0 percent from twenty-three through forty-three days after infection. Knowles and Das Gupta (1928) corroborated this finding for *T. evansi* in the rat, and Davis (1931) for *T. rhodesiense* in the cat. There is some evidence that these trypanosomes, when originally isolated in Africa, exhibit infections essentially similar, except that superimposed on the reproductive activity seen in Figure 192 are intermittent periods of heightened reproductive activity. Thus Robertson (1912) reported that periods of active reproduction of *T. gambiense* in the monkey alternate with periods of less active reproduction. Since reproduction continues at a relatively high rate, even during the periods of less active reproduction, and is never completely inhibited, these infections may be considered with the constantly repro- ducing experimental infections discussed in this section.

Little is known about natural immunity because, as pointed out in the consideration of the *T. lewisi* group of trypanosomes, the rate of reproduction of the trypanosomes cannot be measured directly and few attempts have been made to raise the rate of reproduction (see *T. duttoni* in the mouse) or to increase the percentage of surviving trypanosomes. Nieschulz and Bos (1931), however, reported a slightly shorter incuba- tion period in dogs infected with *T. evansi* as the result of splenectomy.

Acquired immunity, which develops later and which is superim- posed on any natural immunity which may be present, is entirely para- siticidal, with no evidence of an inhibition of reproduction. When the infection shows a typical acute rise in numbers, the first manifesta- tion of acquired immunity is a crisis (see guinea pig, in Figure 193). Thereafter, one or several relapses and crises follow. When a pro- longed chronic low-grade infection ensues, with no typical acute rise as in Figure 192, it is probable that parasiticidal effects of acquired immunity, similar to those produced at the two typical crises later, hold the numbers down. Since all the animals die, the acquired immunity is obviously ineffective. In other words, once the parasites are intro- duced into the host, they reproduce during the entire infection, and although at intervals most of those that have accumulated in the blood are destroyed, the few that escape destruction repopulate the blood again and again until the host dies.

An extensive series of *in vivo* and *in vitro* investigations (see Laveran and Mesnil, 1912; W. H. Taliaferro, 1929, for reviews of the pioneer work) indicate (1) that the periodic destructions of the trypanosomes are due to typical trypanolysins which can be demonstrated *in vivo* and *in vitro* and which are acquired by the host as a result of specific infec- tion or specific immunization; (2) that the trypanosomes reaccumulate during the relapses not because the trypanolysins disappear, but be- cause the parasites have become resistant to them; and (3) that the relapse trypanosomes differ antigenically and that the majority of them are subsequently killed by a new trypanolysin. In all of this work, the strains of trypanosome used and their continuous maintenance are of paramount importance. The experimental infection is started with what is known as the passage strain and is usually maintained in mice by serial transfer, since it has been found that strains remain immunologically unchanged for long periods in mice. The trypanosomes which re-populate the blood after each trypanolytic crisis are immunologically different from the passage strain and are resistant to the lysin producing the crisis. They are known as relapse strains and also have to be maintained in mice by serial transfer. Thereafter, as many relapse strains as are studied have to be extracted and maintained separately and con-

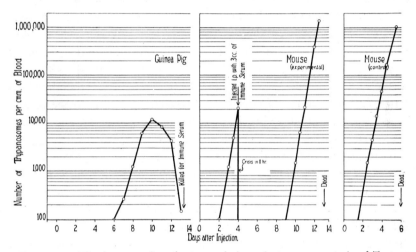

FIGURE 193. The demonstration of a trypanolysin against a passage strain of *Trypanosoma equinum* by passive transfer. An artificial crisis is produced in the trypanosome infection in the experimental mouse, which received 0.3 cc. of serum taken from the seed guinea pig after its infection had undergone a crisis, whereas the infection in the control mouse which was not given serum proceeded uniformly until the mouse died. (From W. H. Taliaferro and T. L. Johnson, 1926.)

tinuously until the investigation is terminated. Reference to Figure 193 will make this clearer. Both the guinea pig and the two mice were infected at appropriate intervals with a passage strain. Serum taken from the guinea pig at the time of the naturally occurring crisis produced an artificial crisis when injected into the experimental mouse, but would have been ineffective had the experimental mouse been infected with the relapse strain, which would have repopulated the blood of the guinea pig after the crisis. (The artificial crisis produced by the immune serum in this so-called curative test is very temporary.) A brief résumé of the investigations follows.

Rouget (1896) first found that the serum of rabbits and dogs, which had been infected with *T. equiperdum* and had become cachetic, exerted a protective action, in a dose of 0.3 cc., on mice infected with the passage strain as measured by the survival time of the mice. The fact that immune serum is protective led Laveran and Mesnil (1901) to hope that serotherapy might possibly be developed against pathogenic trypanosomiasis, but so far this hope has not materialized. Schilling (1902) was the first to recognize the phenomenon of trypanolysis *in vitro*. Rodet and Vallet (1906) studied the lysins systematically, and Massaglia (1907) showed that the trypanosomes which repopulate the blood after each trypanolytic crisis are immunologically different from the original strain and are resistant to the lysin producing the crisis. Thus serum from an infected guinea pig before a crisis is only slightly lytic, whereas that during and after the crisis is strongly lytic to the original strain of trypanosomes, but has no deleterious effect on the trypanosomes reappearing after the crisis. Levaditi and Mutermilch (1909) reported that the lysis is a complement-amboceptor reaction (i.e., involves a heat labile component of serum and the heat stable antibody), and Leger and Ringenbach (1911 and 1912) found a group specificity between trypanolytic immune serums and different species of pathogenic trypanosomes.

W. H. Taliaferro and T. L. Johnson (1926), in a study of the production of artificial crises (Figure 193) by immune serum against *T. equinum* in mice, found that zones of inhibition may occur. T. L. Johnson (1929), in a continuation of this work, found that the production of the artificial crisis, with resulting lengthening of life in the mouse, is dependent not only upon the amount of immune serum, but upon the absolute number of parasites present and upon the strain of parasite used. For example, when a given serum was injected into mice whose blood showed one to five parasites of a particular strain per microscopic field, it caused lysis of the trypanosomes uniformly in all doses greater than the minimal effective dose; when injected into mice the blood of which contained from ten to twenty-eight parasites of the same strain per microscopic field, it caused alternate zones of lysis and non-lysis (zone phenomenon); whereas when injected into mice whose blood showed fifty parasites per microscopic field, no lysis occurred, no matter what dose of serum was given. Moreover, Johnson was able to subject this strain to immune serum and to secure a relapse strain which showed the zone phenomenon with one to five parasites per field. Such data give a basis for the interpretation of the variable and often contradictory results obtained by investigators doing only a few tests.

The relapse strains can be differentiated from the passage strain not only by their resistance to lysins, but by their behavior in other serological tests, such as the Rieckenberg blood platelet test (Rieckenberg, 1917; see also Brussin and Kalajev, 1931). They also differ antigenically and therefore stimulate different immune mechanisms, as is shown, for example, by cross-immunity tests.

The difference in antigenic constitution of various strains was originally studied by Ehrlich and his coworkers in infections in mice in which artificial crises were produced by incomplete cures with drugs. Of the earlier papers, that of Ritz (1914) is particularly interesting. He incompletely cured a mouse twenty times, during which seventeen immunologically different relapse strains were produced, as tested by cross-immunity in mice after cure. Some of these strains were identical with those of another mouse which had been incompletely cured nineteen times, during which nine immunologically different strains had been produced. The immunological variations may be inherited, but in time may be lost. Ritz (1916) also showed that the strains arising naturally in the rabbit could be differentiated by the same methods. In the succeeding years, more or less similar studies have been made with both antibody and drug-induced relapse strains. Recently, Lourie and O'Connor (1937), in an *in vitro* study of relapse strains after drug treatment, obtained twenty-two relapse strains of which thirteen were immunologically distinct. In addition, they ascertained that certain strains tended to occur more frequently than others, that a strain may be a combination of two or several strains, and that individual strains may disappear from such compound strains.

The acquisition of this antibody resistance by the trypanosomes, with a concomitant antigenic change, is an interesting case of an environmentally induced persistent modification which is inherited for many asexual generations, sometimes through 400 mouse passages. It seems to be similar to the acquisition of drug resistance by free-living Protozoa. It can be produced not only *in vivo,* but also *in vitro.* It is always associated with the destruction of many organisms, and hence involves a selection, but the selection is effective within a clone, i.e., within the progeny of a single trypanosome. At the present time, however, it is impossible to decide whether such persistent modifications are due to changes in gene constitution, or, if they are not, whether they may eventually lead to such changes (see Robertson, 1929; W. H. Taliaferro and Huff, 1940; and, in part, Dobell, 1912). Nevertheless, they are of extreme importance in allowing the parasite to overcome the defensive processes of the host and are probably largely responsible for the continued survival of the parasite.

The lymphoid-macrophage system and particularly the macrophages along the blood stream, appear to be involved in immunity, as indicated by enlargement and histological changes in the spleen (Laveran, 1908; Van den Branden, 1935; and others) and by the decreased length of life of splenectomized animals infected with various trypanosomes (see Davis, 1931, for most of the work prior to 1931; Nieschulz and Wawo-Roentoe, 1930; Nieschulz and Bos, 1931; Russeff, 1935). Negative results, as might be expected from the complexity of the problem as explained previously, have also been reported by some of the earlier workers (see Davis, 1931), and also by Davis (1931) and Browning et al. (1934), whereas increased length of life was noted in partially blockaded rats infected with T. equiperdum by Kolmer et al. (1933). Data on splenectomized and blockaded animals which were treated are omitted from consideration because the treatment itself may affect the course of infection.

Whether the trypanocidal antibody acts within the body as a trypanolysin, or as an opsonin with resulting phagocytosis, or both, has been variously answered. Some authors have maintained that one or the other is the sole method of defense; some that they share equal honors; and some that, although lysis is the fundamental mechanism, phagocytosis is responsible for clearing up the debris, and so forth. No one can doubt the occurrence of phagocytosis after its careful description by so many competent observers (Neporojny and Yakimoff, 1904; Sauerbeck, 1905; Yakimoff, 1908; Mesnil and Brimont, 1909; Levaditi and Mutermilch, 1910). On the other hand, W. H. Taliaferro and T. L. Johnson (1926) reported the finding of disintegrating trypanosomes in the blood during experimental crises, which they interpreted as stages in lysis. This question can probably be answered as was the similar question with regard to the trypanolysins in *T. lewisi* infections. The trypanosomes become sensitized by antibody, and the process may be completed by lysis (extracellular enzymes) on one hand, or by digestion within phagocytes (intracellular enzymes) on the other hand. Which occurs may depend to a certain extent on the strength of the antibody.

Besides the cellular basis for the production of the trypanocidal antibody and the coöperation of phagocytes in removing sensitized parasites, Kuhn (1938) has shown a peculiar rôle of the lymphoid-macrophage system in passive transfer of anti *T. equiperdum* protective serum to mice. Thus immune serum, which is effective in protecting normal mice in doses of 0.4 cc. per 20 gm. body weight, gives only partial protection to splenectomized mice, blockaded with India ink, in doses as high as 1.7 cc. per 20 gm. body weight. Suitable experiments indicate (as in previous work with *T. lewisi*) that this finding is due neither to the lowering of complement nor to the removal of phagocytic cells which might be necessary in removing opsonized parasites, but rather to the prevention of antibody uniting with trypanosomes. An interesting, but confusing element in these experiments was that unilateral nephrectomy or ureterotomy was accompanied by a slight reduction in the protective titer of the serum.

A comparison of the resistance acquired by hosts against pathogenic and against nonpathogenic trypanosomes is very illuminating. In the first case, the host acquires practically no resistance (mouse) or it periodically forms trypanolysins (guinea pig, dog, etc.) which hardly ever effectively rid the animal of infection because a few of the pathogens generally become resistant and repopulate the blood again and again until the host dies. In the second case, the host first produces an antibody which inhibits cell division of the parasites and then periodically forms trypanolysins which get rid of the nonreproducing parasites.

PRACTICAL APPLICATIONS OF IMMUNE REACTIONS

By far the most extensive literature on the immunology of the parasitic Protozoa deals with experiments fundamentally planned in the hope of achieving some practical method of preventing, curing, or diagnosing infections. This work has yielded many facts of great interest, but actual practical applications have been limited.

ARTIFICIAL IMMUNIZATION

The earlier literature on this subject has been critically reviewed by W. H. Taliaferro (1929), to which publication the reader is referred for details. Only a few of the more successful examples are cited.

The greatest success with artificial immunization has been attained in *Babesia* infections of cattle and consists of inducing in young healthy animals a low-grade or latent infection which is frequently controlled with drugs. During this latent infection, the animal possesses a solid immunity to superinfection, similar to the condition in malaria. Like malaria, however, the host's defenses may weaken and permit severe and even fatal relapses.

Mention has been made of the fact that one attack of oriental sore in man generally confers a lasting immunity. As the natural sores occur on the face or other exposed portions of the body and leave disfiguring scars, it has been the practice in many endemic centers for centuries to inoculate children on unexposed portions of the body. In a sense this is the crudest type of immunization, in that the highly virulent virus is employed to induce the ordinary disease. The use of attenuated organisms has not met with particular success.

Several investigators have been able to immunize laboratory animals with dead trypanosome vaccines. So far, however, such vaccines have not been extensively applied in a practical way and the outlook is not favorable. Among other difficulties, the attainment of an adequately polyvalent vaccine can hardly be hoped for, owing to the existence of so many immunologically different strains of trypanosomes.

IMMUNOLOGICAL REACTIONS USED IN DIAGNOSIS

Considerably more success has followed the practical application of immunological reactions in diagnosis than in immunization, but even here the success has been limited. This is due in part to the technical difficulty of perfecting the tests, especially when only weak reactions ensue, and in part to the fact that they have to be as satisfactory as or better than the demonstration of the parasites, which has been rendered remarkably delicate in certain blood infections, notably malaria and trypanosomiasis, by the use of stained thick-blood films (see Barber, 1936). From the great mass of literature only the tests which have been perfected or show considerable promise will be mentioned. Detailed protocols and specific methods of procedure can be found in W. H. Taliaferro (1929) or in some of the more recent articles.

A. *Specific Immunological Reactions.*—The reactions between antigen (either complete or haptene) and antibody are so specific that, within certain limits, the presence of a suspected antigen can be ascertained with a known antibody, or, vice versa, a suspected antibody can be verified with a known antigen. Both have been used in diagnosis. When the invading organism liberates some antigen, either whole or partial (haptene) into the blood, sputum, urine, and so forth, the antigen may be detected and identified by its reaction with a high titer immune serum, generally prepared in the laboratory. Sometimes, if the organism isolated from an infected host cannot be fully identified by morphological criteria, it may be further classified in this way (see section on immunological methods of classification). Or, if the invading organism during infection stimulates the formation of a specific antibody in the blood, it may be identified by its reaction with a known antigen which is prepared from the organism in the laboratory. In the Protozoa only the last type of reaction has been extensively used.

The specific complement fixation test is one of the most highly standardized laboratory tests. It is based on the fact that antibody will combine with antigen, and the resultant sensitized antigen will then combine further with complement (a heat labile component of serum), but neither antigen nor antibody will combine with complement alone. In practice, serum suspected of containing an antibody is first heated at 56° C. for twenty minutes to inactivate the complement which it also contains and then is added in varying proportions to a known antigen. To such mixtures, known quantities of complement (generally fresh guinea-pig serum) are subsequently added. The actual fixation of complement gives no visible sign, but is tested by adding to the system at this point a suspension of red blood cells which have been previously sensitized with their specific lysin (sheep cells and antisheep lysin are generally used). Obviously, if the complement was previously fixed, there will not be enough left to lyse the sensitized cells. In terms of the original test, if the red blood cells undergo lysis and their hemoglobin colors the solution, the suspected antibody was not present (test negative); if the cells remain entire and unlysed and the supernatant clear, the suspected antibody was present in the serum (test positive). From the foregoing brief résumé, it is obvious that this test demands careful preparation and standardization of the component parts for its successful execution. It differs from the nonspecific complement fixation or Wassermann test widely used in syphilis only in that the test antigen is derived from the immunizing organism or antigen. The test antigen for the Wassermann test, on the other hand, involves the use of a lipoid extracted from normal tissue, such as beef heart.

The specific complement fixation test has been most satisfactorily standardized in amoebiasis and in dourine of horses. The successful cultivation of *Endamoeba histolytica,* the causative agent of amoebic dysentery, made amoebae available in sufficient quantities to provide a suitable antigen, and since the work of Craig in 1927 the complement fixation test for amoebiasis has been intensively studied (see Craig, 1937; Meleney and Frye, 1937; Paulson and Andrews, 1938). The consensus of opinion seems to be that the test has to be carefully carried out to be dependable and that at best it can be used only as an adjunct to fecal diagnosis.

The sum total of published work through 1910 indicated that complement-fixing antibodies could be demonstrated in various trypanosomiases under controlled conditions, but there was little to indicate that they could be used for diagnosis. From 1911 onward, however, the test was perfected and used extensively for the diagnosis of dourine caused by *Trypanosoma equiperdum* in horses and mules. It was standardized mainly through the efforts of Mohler, Eichhorn, and Buck (1913), E. A. Watson (1920), who used an aqueous antigen; and Dahmen (1922), who used both aqueous and alcoholic extracts. According to Watson, the test is often positive before symptoms are apparent and during latent stages, and in practice no animal should be considered free of the disease unless negative two months after a last exposure. C. M. Johnson and Kelser (1937) concluded that the test is distinctly valuable in revealing active cases of Chagas's disease.

Little success has attended workers using specific complement fixation in malaria and the leishmaniases, especially kala azar, owing perhaps to the low titer of serums from infected persons and the difficulty of obtaining antigens. Recently Coggeshall and Eaton (1938) have reported good results in simian malaria with an aqueous antigen obtained from heavily infected blood or spleen.

The red-cell adhesion test grew out of Rieckenberg's (1917) blood platelet test. As used by Duke and Wallace (1930), it involves the addition of one drop of a citrated trypanosome suspension to one drop of equal parts of blood from the suspected animal and 2-percent sodium citrate. If the blood comes from an infected animal, red blood cells (occasionally also blood platelets) adhere to the trypanosomes within ten to fifteen minutes. In 1931 Wallace and Wormall concluded that complement is necessary, and H. C. Brown and Broom (1938) found that the concentration of trypanosomes should be between 3,000 and 100,000 per cu. mm. and the red cells between 300,000 and 1,250,000. This test compared favorably with specific complement fixation, when untreated horses infected with *T. hippicum* were tested by W. H. and L. G. Taliaferro (1934d).

B. *Nonspecific Serological Reactions.*—Infection often results in definite changes in serum which can be detected by various physical and chemical means and which, although not specific in the immunological sense, are characteristic enough to be useful in diagnosis. Even when the same changes occur in several infections, they may still be used in conjunction with other criteria or if the infections have different geographical distributions.

Several miscellaneous tests have been devised for kala azar which are associated with an increase in the euglobulins of the serum. They include the serum-globulin test of Brahmachari, the aldehyde test of Napier, and the urea-stibamine test of Chopra, Gupta, and David. These tests have been modified and combined by these same and other workers. In general, upon the addition of distilled water, formaldehyde, or urea-stibamine in proper proportions to serum from a person infected with kala azar, the mixture becomes characteristically opaque, owing to the formation of a precipitate within a comparatively short time. These tests appear to be extremely useful and Menon *et al.* (1936) advocate testing a serum by both the aldehyde and the urea-stibamine test (see Menon *et al.*, 1936, and Chorine, 1937, for the literature on this subject).

Some of these tests may also be of value in trypanosomiasis (see

Hope-Gill, 1938), especially in areas in which kala azar is absent.

In 1927 Henry described certain serological tests for the diagnosis of malaria, based on the observation that the serums of malarial subjects flocculate in solutions of metharsenate of iron (ferroflocculation test) and of melanin pigment (Henry's test, or the melanoflocculation test). As a reagent for the Henry test, which was shown by later work to be more sensitive than the ferroflocculation test, Henry (1934) used the filtered supernatant from a suspension of finely ground choroid tissue of ox eye in distilled water. This material after formalin had been added and it had been kept on ice for at least several hours, is added in proper proportions to the serum to be tested, and flocculation is looked for after a half hour or more, preferably by means of the photometer of Vernes, Bricq, and Yvonne. Many subsequent papers on this test have been ably reviewed by Greig, Von Rooyen, and Hendry (1934), Trensz (1936), Villain and Dupoux (1936), de Alda Calleja (1936), Vaucel and Hoang-Tich-Try (1936), and Proske and Watson (1939). The upshot of this work seems to indicate that the test may serve as an adjunct to the search for malarial parasites in diagnosing malaria, but that its use is restricted to laboratories equipped with a photometer and to areas in which kala azar, certain types of leprosy, and certain other diseases are not common. Since it has been shown to be due to an increase of the euglobulin fraction, which flocculates upon dilution with distilled water or weak salt solutions, Proske and Watson (1939) have developed the protein-tyrosin reaction, which is a quantitative chemical estimation of the euglobins of the serum.

IMMUNOLOGICAL REACTIONS IN RELATION TO CLASSIFICATION

Various immunological reactions, since they are frequently species-specific, have been used to check and extend other biological classifications. In other words, the more closely two species are related, the stronger, in general, is the group reaction between them. This specificity seems to depend on the basic structure of the antigens and haptenes, which react specifically with immune serum *in vitro*. It also probably depends upon the quantitative proportions of the various antigens contained in a particular organism (see Wells, 1929). Immunological reactions can therefore be employed to compare chemical structure with anatomical structure. The reactions have to be studied, however, to see if, on the one hand, they vary too much within what is a generally recognized species, or if, on the other hand, they do not differentiate sufficiently among large groups. The extreme specificity within a species may be exemplified by the diversification of a single cell strain of trypanosomes, through the mediation of immune serums or drugs, into a large number of strains which will remain immunologically distinct for long periods.

In a sense most of the work on the serology of parasites can be used in classification. For example, an investigator, in attempting to discover a serological test for a given infection, generally considers at once the specificity of the reaction by ascertaining to what extent group reactions with other species exist. On the whole, however, the study of the immunological relationships of organisms can best be attained by using antiserums from artificially immunized laboratory animals. By this method animals such as rabbits, in which antibodies are readily produced, can be immunized until high titer antiserums are obtained.

Immunological methods have been employed extensively to establish the identity or nonidentity of various proposed species of *Leishmania,* which are morphologically identical, and their relationship to certain insect and plant herpetomonads which resemble the cultural forms of *Leishmania.* This work is fairly consistent in showing that the members of the genus *Leishmania* are a closely related group and are entirely distinct from the genus *Herpetomonas* (Noguchi, 1926; Wagener and Koch, 1926; Zdrodowski, 1931).

Since trypanosomes, like the Leishmanias are frequently morphologically indistinguishable from one another, various immunological tests, as well as biological criteria, have been employed to distinguish them. The *in vivo* cross-immunity test has been most extensively used (Braun and Teichmann, 1912; Laveran, 1917; Kroó, 1925, 1926; Schilling and Neumann, 1932), but *in vitro* tests have also been used, such as complement fixation by Robinson (1926), the phenomenon of "attachment" by several authors (see Levaditi and Mutermilch, 1911) and *in vitro* trypanolysis by Leger and Ringenbach (1912) and others. In evaluating the results of these methods, it appears that they need to be reworked, because of the advance in modern technique and because authorities such as Wenyon (1926) believe that many of the species formerly recognized as distinct should be combined (cf. Becker, Chapter XVII).

The work on piroplasms, although lacking in conclusiveness, has at least served to direct the attention of systematists to the problems of classification (Theiler, 1912; Stockman and Wragg, 1914; du Toit, 1919; Brumpt, 1920).

In malaria cross-immunity tests have been extensively employed (Manwell, 1938), whereas in amoebiasis complement fixation tests have occasionally been used (Menendez, 1932).

LITERATURE CITED

Afridi, M. K. 1938. Observations on extra-abdominal spleen in monkeys infected with *P. cynomolgi* and *P. knowlesi.* J. Malar. Inst. India., 1: 355-90.

Alda Calleja, M. de. 1936. Estado actual de los estudios sobre suerofloculacion en el paludismo. Med. Países Câlidos, 9: 203-36.

Aschoff, L. 1924. Das reticulo-endotheliale system. Ergebn. inn. Med. Kinderheilk., 26: 1-118.

Barber, M. A. 1936. The time required for the examination of thick blood films in malaria studies, and the use of polychromatophilia as an index of anemia. Amer. J. Hyg., 24: 25-31.

Ben Harel, S. 1923. Studies of bird malaria in relation to the mechanism of relapse. Amer. J. Hyg., 3: 652-85.

Bignami, A. 1910. Sulla patogenesi delle recidive nelle febbri malariche. Atti Soc. Studi Malar. 11: 731-45. Translated by W. M. James: Sth. med. J. Nashville, 1913 (Feb.).

Bloom, W. 1938. Lymphocytes and monocytes: theories of hematopoiesis. Downey: Handbook of Hematology, 1: 373-436.

Böhm. 1918. Hämatologische Studien bei Malaria. Arch. Schiffs- u. Tropenhyg., 22: 49-55.

Boyd, G. H. 1929a. Induced variations in the asexual cycle of *Plasmodium cathemerium.* Amer. J. Hyg., 9: 181-87.

—— 1929b. Experimental modification of the reproductive activity of *Plasmodium cathemerium.* J. exp. Zool., 54: 111-26.

—— 1933. Effect of quinine upon reproduction of the avian malaria parasite, *Plasmodium cathemerium.* J. Parasit., 20: 139-40.

—— 1939. A study of the rate of reproduction of the avian malaria parasite, *Plasmodium cathemerium.* Amer. J. Hyg., 29 (Sect. C): 119-29.

Boyd, G. H., and L. H. Allen. 1934. Adult size in relation to reproduction of the avian malaria parasite, *Plasmodium cathemerium.* Amer. J. Hyg., 20: 73-83.

Boyd, G. H., and M. Dunn. 1939. Effects of quinine and plasmochin administration upon parasite reproduction and destruction in avian malaria.
Amer. J. Hyg., 30 (Sect. C) : 1-17.

Boyd, M. F., and L. T. Coggeshall. 1938. A résumé of studies on the host-
parasite relation in malaria. Trans. Third Inter. Congress Trop. Med.
and Malaria, 2:292-311.

Boyd, M. F., W. K. Stratman-Thomas, and S. F. Kitchen. 1936. On the
duration of acquired homologous immunity to *Plasmodium vivax*. Amer.
J. trop. Med., 16: 311-15.

Brand, T. von. 1938. The metabolism of pathogenic trypanosomes and the
carbohydrate metabolism of their hosts. Quart. Rev. Biol., 13:41-50.

Braun, N., and E. Teichmann. 1912. Versuche zur Immunisierung gegen
Trypanosoma. Jena.

Brown, H. C., and J. C. Broom. 1938. Studies in trypanosomiasis. II. Observations on the red cell adhesion test. Trans. R. Soc. trop. Med. Hyg.,
32:209-22.

Brown, W. H. 1914. A note on the pathogenicity of *T. lewisi*. J. exp. Med.,
19: 406-10.

—— 1915. Concerning changes in the biological properties of *Trypanosoma lewisi* produced by experimental means, with especial reference to
virulence. J. exp. Med., 21: 345-64.

Browning, C. H., D. F. Cappell, and R. Gulbransen. 1934. Experimental
infection with *Trypanosoma congolense* in mice: the effect of splenectomy. J. Path. Bact., 39: 65-74.

Bruetsch, W. L. 1927. Ein Beitrag zur Wirkungsweise der Impfmalaria auf
den histopathologischen Prozess bei progressiver Paralyse. Z. ges. Neurol.
Psychiat., 110: 713-28.

—— 1932a. The histopathology of therapeutic (tertian) malaria. Amer.
J. Psychiat., 12: 19-65.

—— 1932b. Activation of the mesenchyme with therapeutic malaria. J.
nerv. ment., Dis., 78: 209-19.

Brug, S. L. 1934. Observations on monkey malaria. Riv. di Malariol., 13: 3-
23.

Brumpt, E. 1920. Les Piroplasmes des bovidés et leurs hôtes vecteurs. Bull.
Soc. Path. exot., 13: 416-60.

Brussin, A. M., and A. W. Kalajev. 1931. Die Bedeutung des Komplements
and der Blut-plättchen für die Feststellung der Thrombozotobarine. Z.
ImmunForsch., 70: 497-521.

Cannon, P. R., and W. H. Taliaferro. 1931. Acquired immunity in avian
malaria. III. Cellular reactions in infection and superinfection. J. prev.
Med. Lond., 5: 37-64.

Chopra, R. N., and S. N. Mukherjee. 1936. The trend of immunity studies
in malaria. Indian med. Gaz., 71: 34-39.

Chorine, V. 1937. Les Réactions sérologiques dues aux euglobulines. Ann. Inst. Pasteur, 58: 78-124.

Christophers, S. R. 1904. A preliminary report on a parasite found in persons suffering from enlargement of the spleen in India. Sci. Mem. Offrs. Med. san. Dept. Gov. India, N.S., 8: 1-17.

Coggeshall, L. T. 1937. Splenomegaly in experimental monkey malaria. Amer. J. trop. Med., 17: 605-17.

—— 1938. The cure of *Plasmodium knowlesi* malaria in Rhesus monkeys with sulphanilamide and their susceptibility to reinfection. Amer. J. trop. Med., 18: 715-21.

Coggeshall, L. T., and M. D. Eaton. 1938. The complement fixation reaction in monkey malaria. J. exp. Med., 67: 871-81.

—— 1938. The quantitative relationship between immune serum and infective dose of parasites as demonstrated by the protection test in monkey malaria. J. exp. Med., 68: 29-38.

Coggeshall, L. T., and H. K. Kumm. 1938. Effect of repeated superinfection upon the potency of immune serum of monkeys harboring chronic infections of *Plasmodium knowlesi*. J. exp. Med., 68: 17-27.

Coventry, F. A. 1925. The reaction product which inhibits reproduction of the trypanosomes in infections with *Trypanosoma lewisi*, with reference to its change in titer throughout the course of infection. Amer. J. Hyg., 5: 127-44.

—— 1929. Experimental infections with *Trypanosoma lewisi* in the guinea-pig. Amer. J. Hyg., 9: 247-59.

Craig, C. F. 1937. Observations upon the practical value of the complement-fixation test in the diagnosis of amebiasis. Amer. J. publ. Hlth., 27: 689-93.

Culbertson, J. T. 1938. Natural transmission of immunity against *Trypanosoma lewisi* from mother rats to their offspring. J. Parasit., 24: 65-82.

Culbertson, J. T., and R. M. Wotton. 1939. Production of ablastin in rats of different age groups after infection with *Trypanosoma lewisi*. Amer. J. Hyg., 30 (Sec. C): 101-13.

Dahmen, H. 1922. Die Serodiagnostik der Beschälseuche. Arch. wiss. prakt. Tierheilk., 47: 319-53.

Davis, L. J. 1931. Experimental feline trypanosomiasis with especial reference to the effect of splenectomy. Ann. trop. Med. Parasit., 25: 79-90.

Delanoë, P. 1912. L'Importance de la phagocytose dans l'immunité de la souris à l'égard de quelques flagelles. Ann. Inst. Pasteur, 26: 172-203.

Dobell, C. 1912. Some recent work on mutation in microorganisms. J. Genet., 2: 201-20.

Duca, C. J. 1939. Studies on age resistance against trypanosome infections: II. The resistance of rats of different age groups to *Trypanosoma lewisi*, and the blood response of rats infected with this parasite. Amer. J. Hyg., 29: (Sect. C) 25-32.

Duke, H. L. 1936. Recent observations on the biology of the trypanosomes of man in Africa. Trans. R. Soc. trop. Med. Hyg., 30: 275-96.

Duke, H. L., and J. M. Wallace. 1930. "Red cell adhesion" in trypanosomiasis of man and animals. Parasitology, 22: 414-56.

Du Toit, P. J. 1919. Experimentelle Studien über die Pferdepiroplasmose. Arch. Schiffs- u. Tropenhyg., 23: 121-35.

Eaton, M. D. 1938. The agglutination of *Plasmodium knowlesi* by immune serum. J. exp. Med., 67:857-70.

Ehrlich, P., and K. Shiga. 1904. Farbentherapeutische Versuche bei Trypanosomenerkrankung. Berl. klin. Wschr., 41: 329-32, 362-65.

Findlay, G. M., and H. C. Brown, 1934. The relation of the electric charge of the red cells to phagocytosis in avian malaria. Brit. J. exp. Path., 15: 148-53.

Gay, F. P. 1931. Tissue resistance and immunity. J. Amer. med. Ass., 97: 1193-99.

Gingrich, W. 1932. Immunity to superinfection and cross-immunity in malarial infections of birds. J. prev. Med. Lond., 6: 197-246.

—— 1934. The effect of an increased burden of phagocytosis upon natural and acquired immunity to bird malaria. J. Parasit., 20: 332-33.

Golgi, C. 1888. Il fagocitismo nell'infezione malarica. Rif. med., 4.

Greig, E. D. W., C. E. van Rooyen, and E. B. Hendry. 1934. Observations on the melano-precipitation serological reaction in malaria. Trans. R. Soc. trop. Med. Hyg., 28: 175-92.

Hartman, E. 1927. Certain interrelations between *Plasmodium praecox* and its host. Amer. J. Hyg., 7: 407-32.

Hegner, R. W. 1926. The biology of host-parasite relationships among Protozoa living in man. Quart. Rev. Biol., 1: 393-418.

Hegner, R., and L. Eskridge. 1938. Mortality of merozoites in infections with *Plasmodium cathemerium* in canaries. Amer. J. Hyg., 28: 299-316.

Hegner, R., and R. Hewitt. 1938. The influence of young red cells on infections of *Plasmodium cathemerium* in birds. Amer. J. Hyg., 27: 417-36.

Henry, X. 1934. Seroflokkulation bei Malaria. Technik und anwendung in der Praxis. Arch. Schiffs- u. Tropenhyg., 38: 93-100.

Herrick, C. A., and S. X. Cross. 1936. The development of natural and artificial resistance of young rats to the pathogenic effects of the parasite *Trypanosoma lewisi*. J. Parasit., 22: 126-29.

Hewitt, R. 1938. Multiple-infected red cells in avian malaria. Amer. J. Hyg., 28: 321-44.

Hope-Gill, C. W. 1938. A study of the reaction rate of the serum-formalin reaction in *Trypanosoma gambiense* sleeping sickness. Trans. R. Soc. trop. Med. Hyg., 31: 507-16.

Hu, C. H., and J. R. Cash. 1927. Considerations of the relationship of the reticulo-endothelial system to kala-azar. Proc. Soc. Exp. Biol. N.Y., 24: 469-72.

Huff, C. G., and W. Bloom. 1935. A malarial parasite infecting all blood and blood-forming cells of birds. J. infect. Dis., 57: 315-36.

Jaffé, R. 1931. The recticulo-endothelial system in immunity. Physiol. Rev., 11: 277-327.

—— 1938. The reticulo-endothelial system. In Downey: Handbook of Hematology, 2: 973-1272.

James, W. M. 1913. Notes on the etiology of relapse in malarial infections. J. infect. Dis., 12: 277-325.

Johnson, C. M., and R. A. Kelser. 1937. The incidence of Chagas' disease in Panama as determined by the complement-fixation test. Amer. J. trop. Med., 17: 385-92.

Johnson, T. L. 1929. In vivo trypanolysis with especial reference to "zones of inhibition," relapse phenomena and immunological specificity. Amer. J. Hyg., 9: 260-82.

Jungeblut, C. W. 1930. Die Bedeutung des retikulo-endothelialen Systems für die Infektion und Immunität. Ergebn. Hyg. Bakt., 11: 1-67.

Jürgens, R. J. 1902. Beitrag zur Biologie der Rattentrypanosomen. Arch. Hyg. Berl., 42: 265-88.

Kanthack, A. A., H. E. Durham, and W. F. H. Blandford. 1898. On nagana, or tsetse-fly disease. Proc. roy. Soc., 64: 100-18.

Kauders, O. 1927. Immunitätsstudien bei Impfmalaria. Zbl. Bakt., I, Orig., 104: 158-60.

Knowles, R., and B. M. Das Gupta. 1928. Laboratory studies in surra. Indian J. med. Res., 15: 997-1058.

—— 1930. Studies in untreated malaria. Indian med. Gaz., 65: 301-10.

Koch, R. 1899. Über die Entwickelung der Malariaparasiten. Z. Hyg. InfektKr., 32: 1-24.

Kolmer, J. A., and J. F. Schamberg, with the assistance of A. Rule and B. Madden. 1933. The influence of reticuloendothelial "blockade" and splenectomy upon experimental trypanosomiasis and syphilis and the chemotherapeutic properties of arsphenamine and neoarsphenamine. Amer. J. Syph., 17: 176-87.

Krijgsman, B. S. 1933. Biologische Untersuchungen über das System: Wirtstier-Parasit. 1 & 2 Teil: die Entwicklung von Trypanosoma evansi in Maus und Ratte. Z. Parasitenk. 5: 592-678.

Kroó, H. 1925. Beitrag zur Immunbiologie der Trypanosomen. Ueber Stammeinheit und Arteinheit des Trypanosomo brucei. Z. Hyg. InfektKr., 105: 247-53.

—— 1926. Weiterer Beitrag zur Immunbiologie der Trypanosomen. Zur Kritik des Kreuzinokulationsverfahrens als immunbiologische Methode der Artabgrenzung. Z. Hyg. InfektKr., 106: 77-82.

—— 1936. Die spontane, apathogene Tryanosomeninfektion der Kaninchen. Z. ImmunForsch., 88: 117-28.

Kuhn, L. R. 1938. The effect of splenectomy and blockade on the protective titer of antiserum against *Trypanosoma equiperdum*. J. infect. Dis., 63: 217-24.

Laveran, A. 1908. Sur quelques altérations de la rate chez les cobayes infectés de trypanosomes. Bul. Soc. Path. exot., 1: 393-98.

—— 1917. Leishmanioses. Paris.

Laveran, A., and F. Mesnil. 1901. Recherches morphologiques et expérimentales sur le trypanosome des rats (*Tr. lewisi* Kent.). Ann. Inst. Pasteur, 15: 673-715.

—— 1912. Trypanosomes et trypanosomiases. Paris.

Leger, A., and J. Ringenbach. 1911. Sur la spécificité de la propriété trypanolytique des sérums des animaux trypanosomiés. C. R. Soc. Biol. Paris, 70: 343-45.

—— 1912. Sur la spécificité de la propriété trypanolytique des sérums des animaux trypanosomiés. C. R. Soc. Biol. Paris, 72: 267-69.

Levaditi, C., and S. Mutermilch. 1909. Recherches sur la méthode de Bordet et Gengou appliquée à l'étude des trypanosomiases. Z. ImmunForsch., Orig., 2: 702-22.

—— 1910. I. Mécanisme de la phagocytose. C. R. Soc. Biol. Paris, 68: 1079-81.

—— 1911. Le Diagnostic de la maladie du sommeil par l'examen des propriétés attachantes du sérum. C. R. Acad. Sci. Paris, 153: 166.

Linton, R. W. 1929. The reticulo-endothelial system in protozoan infections. Arch. Pathol., 8: 488-501.

Lorando, N., and D. Sotiriades. 1937. Treatment of malaria with immune Blood. Trans. R. Soc. trop. Med. Hyg., 31: 227-34.

Lourie, E. M. 1934. Studies on chemotherapy in bird malaria. Ann. trop. Med. Parasit., 28: 151-69; 255-77; 513-23.

Lourie, E. M., and R. J. O'Connor. 1937. A study of *Trypanosoma rhodesiense relapse strains in vitro*. Ann. trop. Med. Parasit., 31: 319-40.

Lowe, J. 1934. Studies in untreated malaria. Numerical studies of the parasites in relation to the fever. Rec. Malar. Surv. India, 4: 223-41.

McLay, K. 1922. Malaria in Macedonia, 1915-1919. Part III. Haematological investigations on malaria in Macedonia. Jour. R. Army med. Cps., 38: 93-105.

MacNeal, W. J. 1904. The life history of *Trypanosoma lewisi* and *Trypanosoma brucei*. J. infect. Dis., 1: 517-43.

Manteufel, P. 1909. Studien über die Trypanosomiasis der Ratten mit Berücksichtigung der Übertragung unternatürlichen Verhältnissen und der Immunität. Arb. GesundhAmt. Berl., 33: 46-83.

Manwell, R. D. 1938. Reciprocal immunity in the avian malarias. Amer. J. Hyg., 27: 196-211.

Manwell, R. D., and F. Goldstein. 1939. Strain immunity in avian malaria. Amer. J. Hyg., 30, Sec. C.: 115-22.

Marchand, F. 1924. Die örtlichen reacktiven Vorgänge in Krehl und Marchand: Handb. d. Allgemein. Pathol., 4: 78-649.

Marrack, J. R. 1938. The chemistry of antigens and antibodies. Special Rep. Ser. Med. Res. Counc., No. 230.

Massaglia, M. A. 1907. Des causes des crises trypanolytiques et des rechutes qui les suivent. C. R. Acad. Sci. Paris, 145: 687-89.

Maximow, A. 1927a. Bindegewebe und blutbildende Gewebe. In Handb. Mikrosk. Anatomie d. Menschen, 2:232-583. Berlin.

—— 1927b. Morphology of the mesenchymal reactions. Arch. Pathol. Lab. Med., 4: 557-606.

Meleney, H. E. 1925. The histopathology of kala-azar in the hamster, monkey and man. Amer. J. Path., 1: 147-67.

Meleney, H. E., and W. W. Frye. 1937. Practical value and significance of the complement-fixation reaction in amebiasis. Amer. J. publ. Hlth., 27: 505-10.

Menendez, P. E. 1932. Serological relationships of Entamoeba histolytica. Amer. J. Hyg., 15: 785-808.

Menon, T. B., D. R. Annamalai, and T. K. Krishnaswami. 1936. The value of the aldehyde and stiburea tests in the diagnosis of kala-azar. J. trop. Med. Hyg., 36: 92-95.

Mesnil, F., and E. Brimont. 1909. Sur les propriétés protectrices du sérum des animaux trypanosomiés: Races résistantes à ces sérums. Ann. Inst. Pasteur, 23: 129-54.

Metschnikoff, E. 1892. Leçons sur la pathologie comparée de l'inflammation. Paris.

—— 1905. Immunity in infective diseases (Trans. of French book of 1901). Cambridge.

Mohler, J., A. Eichhorn, and J. Buck. 1913. The diagnosis of dourine by complement fixation. J. agric., Res., 1: 99-107.

Mulligan, H. W. 1935. Descriptions of two species of monkey Plasmodium isolated from Silenus irus. Arch. Protistenk., 84: 285-314.

Mulligan, H. W., and S. A. Sinton. 1933. Studies in immunity in malaria. Rec. Malar. Surv. India, 3:529-68; 809-39.

Nauck, E. G., and B. Malamos. 1935. Über Immunität bei Affenmalaria. Z. ImmunForsch, 84: 337-58.

Neporojny, S. D., and W. L. Yakimoff. 1904. Über einige pathologischanatomische Veränderungen bei experimentellen Trypanosomosen. Zbl. Bakt., Ref. 35: 467-68.

Neumann, H. 1933. Der Nachweis des parasitiziden Antikörpers bei der Malaria des Menschen. Riv. Malariol., 12: 319-34.

Nieschulz, O., and A. Bos. 1931. Über den Einfluss der Milz auf den Infektionsverlauf von Surra bei Hunden. Dtsch, tierärztl. Wschr. 39:488-89.

Nieschulz, O., and F. K. Wawo-Roentoe. 1930. Über den Einfluss der Milzexstirpation bei Infektionen mit *Trypanosoma gambiense* und *Schizotrypanum cruzi*. Z. ImmunForsch., 65: 312-17.

Noguchi, H. 1926. Comparative studies of herpetomonads and leishmanias. J. exp. Med., 44: 327-37.

Paulson, M., and J. Andrews. 1938. Complement fixation in amebiasis. A comparative evaluation in clinical practice. Arch. intern. Med., 61:562-78.

Perla, D., and J. Marmorston-Gottesman. 1930. Further studies on *T. lewisi* infection in albino rats. J. exp. Med., 52: 601-16.

Proske, H. O., and R. B. Watson. 1939. The protein tyrosin reaction. Publ. Hlth. Rep. Wash., 54: 158-72.

Rabinowitsch, L., and W. Kempner. 1899. Beitrag zur Kenntniss der Blutparasiten, speciell der Rattentrypanosomen. Z. Hyg. InfaktKr., 30: 251-94.

Redmond, W. B. 1939. The cross-immune relationship of various strains of *Plasmodium cathemerium* and *P. relictum*. J. infect. Dis. 64: 273-87.

Regendanz, P. 1932. Über die Immunitätsvorgänge bei der Infektion der Ratten mit *Trypanosoma lewisi*. Z. ImmunForsch., 76: 437-45.

Regendanz, P., and W. Kikuth. 1927. Über die Bedeutung der Milz für die Bildung des vermehrungshindernden Reaktionsproduktes (Taliaferro) und dessen Wirkung auf den Infektionsverlauf der Ratten-Trypanosomiasis (*Tryp. lewisi*). Zbl. Bakt., Abt. 1. Orig. 103: 271-79.

Rieckenberg, P. 1917. Eine neue Immunitätsreaktion bei experimenteller Trypanosomen-Infektion: die Blutplättchenprobe. Z. ImmunForsch., 26: 53-64.

Ritz, H. 1914. Über Rezidive bei experimenteller Trypanosomiasis. Dtsch. med. Wschr., 40: 1355-58.

—— 1916. Über Rezidive bei experimenteller Trypanosomiasis. Arch. Schiffsu. Tropenhyg., 20: 397-420.

Robertson, M. 1912. Notes on the polymorphism of *Trypanosoma gambiense* in the blood and its relation to the exogenous cycle in *Glossina palpalis*. Rept. Sleep. Sickn. Comm. roy. Soc., 13: 94-110.

—— 1929. The action of acriflavine upon *Bodo caudatus*. Parasitology, 21: 375-416.

Robinson, E. M. 1926. Serological investigations into some diseases of domesticated animals in South Africa caused by trypanosomes. Eleventh and Twelfth Rept. vet. Res. S. Afr. (Sept.), pp. 9-25.

Rodet, A., and G. Vallet. 1906. Contribution à l'étude des trypanosomiasis. Arch. Méd. exp., 18: 450-94.

Ross, R. 1910. The prevention of malaria. New York.

Ross, R., and D. Thomson. 1910. A case of sleeping sickness studied by precise enumerative methods: regular periodical increase of the parasites disclosed. Proc. roy. Soc., B, 82: 411-15.

—— 1911. A case of sleeping-sickness studied by precise enumerative methods. Further observations. Proc. roy. Soc., B, 83: 187-205.

Roudabush, R. L., and E. R. Becker. 1934. The development of *Trypanosoma iowensis* in the blood of the striped ground squirrel, *Citellus tridecemlineatus*. Iowa St. Coll. J. Sci., 8: 533-35.

Roudsky, D. 1911. Mécanisme de l'Immunité naturelle de la souris vis-a-vis du *Trypanosoma lewisi* Kent. C. R. Soc. Biol. Paris, 70: 693-94.

Rouget, J. 1896. Contribution à l'étude du trypanosome des mammifères. Ann. Inst. Pasteur, 10: 716-28.

Rudolph, G. de M., and J. C. Ramsay. 1927. Enumeration of parasites in therapeutic malaria. J. trop. Med. Hyg., 30: 1-8.

Russeff, C. 1935. Der Einfluss der Milzexstirpation auf den Verlauf der Dourineinfektion bei verschiedenen Versuchstieren. Z. ImmunForsch., 84: 295-99.

Sauerbeck, E. 1905. Beitrag zur pathologischen Histologie der experimentellen Trypanosomen-Infection (mit *Trypanosoma brucei*). Z. Hyg. InfektKr., 52: 31-86.

Schilling, C. 1902. Bericht über die Surra-Krankheit der Pferde und Rinder im Schutzgebiete Togo. Zbl. Bakt. I. Orig. 31: 452-59.

Schilling, C., and H. Neumann. 1932. Zur Methodik der immunologischen Differenzierungsmethoden von Trypanosomenstämmen. Arch. Schiffs- u. Tropenhyg., 36: 214-29.

Schwetz, J. 1934. L'Influence de la splénectomie sur l'evolution des trypanosomes pathogènes (*Tr. gambiense* et *Tr. congolense*) chez les rats et les souris. Bull. Soc. Path. exot., 27: 253-60.

Sergent, Ed. 1936. Immunité ou prémunition dans les maladies à hémocytozoaires (paludismes, piroplasmoses). Arch. Inst. Pasteur Algé., 14: 413-17.

Sergent, Ed., and Et. Sergent. 1918. Sur le paludisme des oiseaux du au *Plasmodium relictum* (*vel Proteosoma*). Ann. Inst. Pasteur, 32: 382-88.

Sinton, J. A., Harbhagwan, and J. Singh. 1931. The numerical prevalence of parasites in relation to fever in chronic benign tertian malaria. Indian J. med. Res., 18: 871-79.

Sinton, J. A., and H. W. Mulligan. 1933a. Mixed infections in the malaria of the lower monkeys. I. Mixed infections as the cause of apparent variations in the morphology and pathogenicity of simian Plasmodia. Rec. Malar. Surv. India, 3:719-67.

—— 1933b. Mixed infections in the malaria of the lower monkeys. II. The probable occurrence of mixed infections in some of the older records of monkey malaria. Rec. Malar. Surv. India, 3: 769-808.

Sotiriadés, D. 1917. Essais de sérothérapie dans la malaria. Grèce méd., 19: 27-28.

Stauber, L. A. 1939. Factors influencing the asexual periodicity of avian malarias. J. Parasit., 25: 95-116.

Steffan, P. 1921. Beobachtungen über den Verlauf der künstlichen Infektion der Ratte mit *Trypanosoma lewisi*. Arch. Schiffs- u. Tropenhyg., 25: 241-47.

Stockman, S., and W. G. Wragg. 1914. Cross immunisation with *Piroplasma bigeminum* and *Piroplasma divergens*. J. comp. Path., 27: 151-55.

Stratman-Thomas, W. K. 1935. Studies on benign tertian malaria. Amer. J. Hyg., 21: 361-63.

Taliaferro, L. G. 1925. Infection and resistance in bird malaria, with special reference to periodicity and rate of reproduction of the parasite. Amer. J. Hyg., 5: 742-89.

—— 1928. Return to normal of the asexual cycle in bird malaria after retardation by low temperatures *in vitro*. J. prev. Med. 2: 525-40.

Taliaferro, W. H. 1924. A reaction product in infections with *Trypanosoma lewisi* which inhibits the reproduction of the trypanosomes. J. exp. Med., 39: 171-90.

—— 1929. The immunology of parasitic infections. New York.

—— 1932. Trypanocidal and reproduction-inhibiting antibodies to *Trypanosoma lewisi* in rats and rabbits. Amer. J. Hyg., 16: 32-84.

—— 1934. Some cellular bases for immune reactions in parasitic infections. J. Parasit. 20: 149-61.

—— 1938. The effects of splenectomy and blockade on the passive transfer of antibodies against *Trypanosoma lewisi*. J. infect. Dis., 62: 98-111.

Taliaferro, W. H., and P. R. Cannon. 1936. The cellular reactions during primary infections and superinfections of *Plasmodium brasilianum* in Panamanian monkeys. J. infect. Dis., 59: 72-125.

Taliaferro, W. H., P. R. Cannon, and S. Goodloe. 1931. The resistance of rats to infection with *Trypanosoma lewisi* as affected by splenectomy. Amer. J. Hyg., 14: 1-37.

Taliaferro, W. H., and C. G. Huff. 1939. The genetics of the parasitic protozoa. (In press.)

Taliaferro, W. H., and T. L. Johnson. 1926. Zone phenomena in in vivo trypanolysis and the therapeutic value of trypanolytic sera. J. prev. Med. 1: 85-123.

Taliaferro, W. H., and H. W. Mulligan. 1937. The histopathology of malaria with special reference to the function and origin of the macrophages in defence. Indian med. Res. Mem., No. 29. Pp. 138.

Taliaferro, W. H., and Y. Pavlinova. 1936. The course of infection of *Trypanosoma duttoni* in normal and in splenectomized and blockaded mice. J. Parasit., 22: 20-41.

Taliaferro, W. H., and L. G. Taliaferro. 1922. The resistance of different hosts to experimental trypanosome infections, with especial reference to a new method of measuring this resistance. Amer. J. Hyg., 2: 264-319.

—— 1929a. Acquired immunity in avian malaria. I. Immunity to superinfection. J. prev. Med. 3: 197-208.

—— 1929b. Acquired immunity in avian malaria. II. The absence of protective antibodies in immunity to superinfection. J. prev. Med. 3: 209-23.

—— 1934a. Morphology, periodicity and course of infection of Plasmodium brasilianum in Panamanian monkeys. Amer. J. Hyg., 20: 1-49.

—— 1934b. Alteration in the time of sporulation of Plasmodium brasilianum in monkeys by reversal of light and dark. Amer. J. Hyg., 20: 50-59.

—— 1934c. Superinfection and protective experiments with Plasmodium brasilianum in monkeys. Amer. J. Hyg., 20: 60-72.

—— 1934d. Complement fixation, precipitin, adhesion, mercuric chloride and Wassermann tests in equine trypanosomiasis of Panama (murrina). J. Immunol., 26: 193-213.

Theiler, A. 1912. Weitere Unterschungen über die Anaplasmosis der Rinder und deren Schutzimpfung. Z. InfektKr. Haustiere, 11: 193-207.

Thomson, J. G. 1912. Enumerative studies on T. brucei in rats and guinea-pigs, and a comparison with T. rhodesiense and T. gambiense. Ann. trop. Med. Parasit., 5: 531-36.

—— 1933. Immunity in malaria. Trans. R. Soc. trop. Med. Hyg., 26: 483-514.

Topley, W. W. C. 1935. An outline of immunity. Baltimore.

Trensz, F. 1936. La Valeur pratique de la mélanofloculation de Henry. Arch. Inst. Pasteur Algér., 14: 353-90.

Van den Branden, F. 1935. Sur le rapport du poids de la rate ou du foie au poids du corps, chez des rats blancs (variété albinos de Mus decumanus) non infectés, ainsi que chez les animaux de même espèce, préalablement infectés de Trypanosoma congolense ou de Trypanosoma brucei, puis guéris ou non guéris par traitement. C. R. Soc. Biol. Paris, 119: 529-30.

Vaucel, M., and Hoang-Tich-Try. 1936. Reactions de malaria-floculation au Tonkin. Bull. Soc. méd.- chir. Indochine, 14: 1101-14.

Villain, G., and R. Dupoux. 1936. Contribution à l'étude serologique du paludisme. Utilisation d'une mélanine artificielle: la M. A. floculation. Arch. Inst. Pasteur Afr. N., 25: 469-551.

Wagener, E. H., and D. A. Koch. 1926. The biological relationships of Leishmania and certain herpetomonads. Univ. Cal. Publ. Zool., 28: 365-88.

Wallace, J. M., and A. Wormall. 1931. Red-cell adhesion in trypanosomiasis of man and other animals. II. Some experiments on the mechanism of the reaction. Parasitology, 23: 346-59.

Wasielewski, T. K. W. N. von. 1901. Über die Verbreitung und künstliche Ubertragung der Vogelmalaria. Arch. Hyg. Berl., 41: 68-84.

Wasielewski, T. K. W. N. von, and G. Senn. 1900. Beiträge zur Kenntniss der Flagellaten des Rattenblutes. Z. Hyg. InfektKr. 33: 44-72.

Watson, C. J. 1928. The pathology of histoplasmosis (Darling) with special reference to the origin of the phagocytic cells. Folia haemat., 37: 70-93.

Watson, E. A. 1920. Dourine in Canada, 1904-1920: History, research and suppression. Canada, Dept. Agric. Health of Animals Branch.

Wells, H. G. 1929. The chemical aspects of immunity. 2d ed., New York.

Wenyon, C. M. 1926. Protozoology. 2 vols., New York.

Yakimoff, W. L. 1908. Contribution aux altérations du sang des animaux atteints de trypanosomiases expérimentales. Arch. Sci. biol. St. Pétersb., 13: 243-76.

Zdrodowski, P. 1931. Sur la sérologie comparée du groupe de Leishmanies d'origine humaine et canine. Bull. Soc. Path. exot., 24: 37-41.

CHAPTER XIX

RELATIONSHIPS BETWEEN CERTAIN PROTOZOA AND OTHER ANIMALS

HAROLD KIRBY, JR.[1]

IN THE LITERATURE in which consideration is given to close relationships between organisms of different species the effort is often made to group the associations discussed under definite categories. The categories are defined, and it is shown in what manner and to what extent each separate association can be referred to its proper position. A reader of this literature soon becomes sensible of the lack of agreement in almost every major particular. Unlike names are given to the categories, definitions are dissimilar, there is difference of opinion or lack of exact information on the nature of the relationship itself, and the impossibility of making unequivocal distinctions is apparent in many instances. In order to make an advance toward harmony of opinion, the bionomics of many groups of living things must be taken into consideration. The author must here confine his discussion to a statement of what he regards as a satisfactory denomination of the types of relationship between Protozoa and other organisms, including other Protozoa, with which they are intimately associated.

At the outset, it is apparent that a comprehensive term is necessary to designate all types of relationship between Protozoa and their hosts, whether the Protozoa are epibiotic or endobiotic, whether they live at the expense of their hosts or aid them in some way. Such a term would be applicable also in the instances in which it is uncertain in what subdivision an association belongs. The choice should be between two existing terms, parasitism and symbiosis, the latter of which has etymologically exactly the meaning desired. Both words have been used in the general sense. The word parasite, however, has by universal agreement been used to designate an organism that lives at its host's expense,

[1] Assistance rendered by personnel of Work Projects Administration, Official Project number 65-1-08-113, Unit C1, is acknowledged.

obtaining nutriment from the living substance of the latter, depriving it of useful substance, or exerting other harmful influence upon it. In the interests of exactitude, a word should, if possible, express a single definite idea; and therefore it seems undesirable to use parasitism also in a general sense if it can be avoided.

There is much justification for applying the term symbiosis to the general relationship under consideration, although many authors have given it a restricted applicability to mutually advantageous associations only. The word in this restricted meaning has, in fact, acquired whatever sanction general usage confers. Most textbooks in biology and zoölogy, as well as protozoölogy and parasitology, so define it, the oldest one noted being T. J. Parker's (1893); and Hegner (1926b) restricted the meaning further, stating that in symbiosis life apart is impossible. J. A. Thomson (1934; with Geddes, 1931; also in the *Encyclopaedia Britannica* fourteenth ed.) contended that symbiosis is a mutually beneficial internal relationship, and that externally mutualistic relationships are commensalism. To Haupt (1932) symbiosis includes what others consider commensalism, but does not include parasitism; the same sense is implicit in some dictionary definitions. The extended meaning of the word has been found in only two general biology texts (McFarland, 1913; Eikenberry and Waldron, 1930); it has that meaning also in the article on symbiosis in the *New International Encyclopaedia* (1925). Most important, however, is the report of Hertig, Taliaferro, and B. Schwartz (1937).

The sense in which the word was employed by its originator, A. de Bary (1879), is of decisive importance. The three members of the Committee on Terminology of the American Society of Parasitologists, as well as W. Schwartz (1935), appear to have been the only ones among recent authors to understand de Bary. It has been widely stated that he meant symbiosis to designate mutually beneficial relationships (by Caullery, 1922; Hegner, 1926b; as well as by the others cited in the Committee's report). The Committee gave quotations from de Bary showing clearly that he used symbiosis as a collective term, the subdivisions of which include parasitism and mutualism; he recognized two main categories, antagonistic and mutualistic symbiosis. The results of the writer's examination of de Bary's paper are in complete agreement with their interpretation; there is no ambiguity in de Bary's usage. Hertwig (1883) had a similar understanding of the meaning of the word; parasitism and mutualism, he stated, are types of symbiosis. As the Committee pointed out, he changed his usage later.

In recent literature other opinions that the words should be used in this original sense have appeared. W. Schwartz (1935) took that attitude, although he would restrict symbiosis to the relationship in which there is physiological dependence of one partner on another. Cleveland (1926) remarked that it would be much better to use it in the general sense, if the change could be made.

Van Beneden (1876) referred to certain associated animals as mutualists, before the term symbiosis had been coined. The word conveys the idea of reciprocal benefit, although the examples he described indicate a vague concept on his part of the relationships concerned, and none of them would now be regarded as mutualistic. (He discussed among mutualists parasitic copepods, opalinids, endozoic rotifers, and even *Vaginicola* on *Gammarus*). He recognized the three types of association: commensalism, mutuality, and parasitism. If symbiosis is used in the broad sense, reciprocal relationships should be termed mutualism or mutualistic symbiosis.

Protozoa that live in natural cavities of the body, such as the mantle cavities of molluscs and the lumen of the alimentary canal, but do not nourish themselves at the expense of the host, have been termed inquilines (Caullery, 1922; Grassé, 1935). All inquilines are commensals, but not all commensals are inquilines. There are also ectozoic commensals, or ectocommensals; endocommensalism is equivalent to inquilinism. In a sense, inquilines, like ectozoic symbionts, have not invaded the body itself. They occupy cavities open to the outside.

Endozoic Protozoa which invade the interior of the body proper, living intracellularly, among tissue cells, or in blood or coelomic cavities, are all parasitic. The conditions of their nutrition necessarily involve strict dependence on the host. Ectozoa and inquilines, which usually are commensals, may become parasites when in their nutritive processes they develop one or another means of using the substance of the host, generally by attack upon, or extraction of substance directly from, the epithelial cells. They are also called parasites when they consume enough material that would otherwise be used by the host to make a difference to it; or when in some way not connected with nutrition they injure the host without invading its substance. There are few instances among Protozoa in which injurious effects of this type have been proved.

There are only two ways in which a protozoön can directly benefit a larger animal. It may contribute its own body to be used in the nutrition of the host; but such a relationship would not constitute mutualism unless some essential substance not otherwise obtained is thus supplied. Another way is action upon the food materials such as to make a substance usable that otherwise would not be. The latter is the situation in the only proved instance of significant mutualism between Protozoa and their hosts—that between flagellates and certain termites, as well as *Cryptocercus punctulatus.* Another, indirect, benefit might be conferred by aid in controlling an injurious organism or substance, but no instance of that is known with certainty.

In all its ramifications, the problem of the symbiotic relationships between Protozoa and other animals is far too large for concise treatment. It has been considered in various textbooks of protozoölogy, particularly in the chapter on "Ecology, Commensalism, and Parasitism" in Calkins (1933) and in Doflein-Reichenow ((1927-29). Protozoan relationships are discussed, together with associations in other groups of animals, in Caullery (1922) and Grassé (1935). An important general article on the subject is that by Wenrich (1935); and a general discussion was published by Fantham (1936). Various aspects have been discussed by Hegner (1924, 1926a, 1926b, 1926c, 1928, 1937), by Metcalf (1923, 1929), by Cleveland (1926, 1934), by contributors to Hegner and Andrews (1930), by Becker (1932, 1933), and by Kirby (1937).

In undertaking to make a contribution to the subject, the author recognized two possible paths of approach. Either he could attempt to make a comprehensive scrutiny of the entire expanse of pertinent information, or he could explore in as much detail as possible certain chosen fields of inquiry. The former approach would lead to a generalized account, with selected and perhaps original illustrations; and it would in large part reiterate existing, readily accessible, sometimes commonplace concepts. The latter course, although less exhaustive, permits selection for more detailed consideration of certain representative topics; that is the course which has in the main been followed here.

In the ecology of symbiotic relationships, an introductory chapter is provided by accidental and facultative parasitism. Among Protozoa this is best exemplified by certain holotrichous ciliates. As the subject develops, instances in which certain genera contain both free-living and symbiotic species come under consideration, together with examples of closely related genera in the two types of habitat. The situation may be illustrated by certain euglenid and some polymastigote flagellates, and again by certain holotrichous ciliates. In certain groups of Protozoa, instructive series in degree of adaptation and types of relationship appear. One of the most fruitful of these series, in which the range is from incidental commensalism to strict parasitism, is found in the Thigmotricha. Host relationships of a variety of types, which bring out also the adaptation of life cycles to the conditions of symbiotic existence, are found among certain holotrichs of the marine Crustacea, *Conidiophrys pilisuctor,* and the Apostomea. There is more or less morphological alteration and adaptation in the symbiotic holotrichs considered above, particularly in the Thigmotricha; and this is also well brought out in the development of attachment structures in Ptychostomidae, Astomata, peritrichs, and certain flagellates of termites.

Animals of certain specific groups are characterized by protozoan faunules of particular types. That does not express only the commonplace fact that examination of particular hosts will reveal particular symbionts, but makes the point that related hosts often have faunules of similar composition. In this matter of distribution of symbionts among species of hosts is incorporated the problem of specificity in symbiosis, the quality of a symbiont of being restricted to certain hosts. This quality is known as host-specificity. The incorrectness of the term "host-parasite specificity" has been commented on by Hertig, Taliaferro, and Schwartz (1937). Faunules of two groups of animals are discussed here in their composition, host-specificity, and distributional characteristics: the ciliates of sea urchins, in which there is in part relationship to free-living species; and the Protozoa of termites and *Cryptocercus,* in which the symbiotic relationship has reached a maximum development.

Finally, physiological host relationships are discussed in detail in two groups of Protozoa: Ophryoscolecidae in ruminants, and flagellates in termites and *Cryptocercus.* The former relationship, though long suspected of being mutualism, is probably simple commensalism; there is general agreement on the mutualistic character of the latter. To complete this series, relationships of strict parasitism should be discussed from the standpoint of the physiology of the parasite and the effect on the host. The absence of this is readily compensated for, however, in the abundant literature of parasitology; and examples of parasitism are described in almost all groups considered in this chapter.

As is apparent, the material is grouped under certain headings; but in every section data bearing on various topics in symbiotic relationships will be found. To group all facts under specific topics would involve much dislocation in other respects; and it has seemed preferable to preserve systematic continuity to a considerable extent.

ACCIDENTAL AND FACULTATIVE PARASITISM

In connection with the origin of host relationships, it is of interest to consider instances in which organisms can develop in both free-living and symbiotic habitats. Accidental and facultative parasitism, therefore, come up for primary consideration. Facultative parasites, as opposed to obligate parasites, are able to live either associated with hosts or not, but parasitism is a natural occurrence in the bionomics of the species. Accidental parasitism is that of a naturally free-living species, which happens through some accident to become parasitic. Although the two types are not identical, they are obviously closely related and categorical separations are not attempted in the discussion.

Mercier and Poisson (1923) pointed out that those forms that are most ubiquitous and are preadapted to varied modes of nutrition have the best chance of surviving in the new medium into which they are introduced accidentally. For such forms Giard (1880) used the term "inchoate parasitism"; Giard had reference to the incomplete and temporary parasitism illustrated by the occurrence of geophilids in the nasal cavities of man.

These conditions are met by the ciliate *Glaucoma pyriformis*. Under natural conditions *G. pyriformis* ingests bacteria. Hetherington (1933) stated that it is one of the commonest fresh-water Protozoa, appearing in the early stages of the usual infusions of hay, wheat, or lettuce if they are inoculated with pond water. (At that time he named the ciliate *Colpidium campylum*, but later [1936] reported that it is *G. pyriformis*.)

In 1923 Lwoff reported that he had succeeded in growing *"Colpidium colpoda"* in pure culture, with neither living nor dead microörganisms, in a medium of peptone broth. He later revised the identification to *Glaucoma piriformis*, and in 1932 considered it to be exceptional and unique in its utilization of dissolved nutrient material only. Since then some other ciliates have been maintained in sterile, non-particulate culture; but the number is limited, and some have turned out to be actually *G. pyriformis* (Hetherington, 1936).

It has been shown that *G. pyriformis*, when introduced into the hemocoele of certain insects, multiplies rapidly and exhibits marked pathogenic potentialities. Lwoff (1924) inoculated the ciliate from pure culture into about thirty caterpillars of *Galleria mellonella*, all of which succumbed to the infection in from eight to fifteen days. Shortly before death, the blood contained no more leucocytes, but only great numbers of ciliates invading all parts of the body. The ciliates nourished themselves phagocytically at least in part, and contained many globules of fat from the fat bodies of the caterpillar.

Janda and Jírovec (1937) injected bacteria-free cultures of *G. pyriformis* into the body cavities of various invertebrates and vertebrates, and also brought them into contact with artificially produced wounds. Attempts to infect annelids, molluscs, crustacea, fish, and amphibia failed, but many insects were successfully inoculated. The ciliates multiplied so rapidly as almost completely to fill the hemolymph in a few days. The fatty tissues especially were destroyed, the ciliates became larger than normal, and the infected insects usually died in a few days. Infection through wounds was achieved only in the aquatic larvae of *Aeschna cyanea*. Infection by mouth did not occur. *Glaucoma* that had been parasitic for some time when returned to the water survived and multiplied normally.

It appears, then, that insects' blood is a favorable medium in which *G. pyriformis* may grow, and that the tissues often provide no protection against the organism once it has entered. One would expect that occasionally so common a ciliate might enter an aquatic insect through an external wound or a damaged gut wall, and multiply in the same way with disastrous consequences to the host. That has, indeed, been found to take place.

It is possible, as Wenyon (1926) suggested, that *Lambornella stegomyiae* (Fig. 194), found by Lamborn (1921) in mosquito larvae (*Stegomyia scutellaris*) in an earthenware pot in the Malay States, may actually be this species. All the infected larvae died in a few days; the ciliates escaped while the host was still living or soon after death. Keilin (1921), who described the species from formalin-preserved ma-

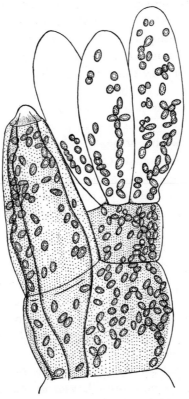

Figure 194. Posterior end of larva of *Aëdes* (*Stegomyia*) *scutellaris* parasitized by ciliates, *Lambornella stegomyiae* Keilin (=*Glaucoma pyriformis ?*). (After Keilin, 1921.)

terial, regarded the ciliate as a true parasite, and others have agreed with him, apparently largely because of the epizoic character of the supposed cysts which Keilin found studding the external surface of one mosquito larva. There is no proof that these were cysts of the ciliate—one may, in fact, be justified in thinking it improbable that they were. If they were not, there seems to be no reason why Codreanu

(1930), Lwoff (1932), and others should except it from the list of accidental parasites. The ciliates found by MacArthur (1922) in larvae of *Theobaldia annulata* and studied also by Wenyon (1926) showed but little difference in habit or appearance from *Lambornella;* and, considering the inadequacy of Keilin's material, may well have been the same. Wenyon concluded they were *Glaucoma pyriformis.*

Next came the report by Treillard and Lwoff (1924) of the finding of ciliates corresponding to *G. pyriformis* in larvae of *Chironomus plumosus* bought at a market and probably obtained in the vicinity of Paris. Of 300 larvae, 13 were parasitized. The ciliates multipled actively, causing death of the host in about eight days. In the cytoplasm were granules of yellow pigment, probably derived from hemoglobin. In 5 of the hosts conjugation was in progress, with all ciliates in any one larva at about the same stage.

From another chironomid, *Culicoides peregrinus* in India, Ghosh (1925) reported *Balantidium knowlesii.* The ciliates were numerous in the "coelomic cavity"; there is no statement as to whether the host was a larva or adult, or how many hosts there were. Though Grassé and Boissezon (1929) proposed the new genus *Leptoglena* for this very inadequately described ciliate, and it seemed to Lwoff (1932) and Codreanu (1930) to be a *Glaucoma,* it is impossible to recognize it from the description as any one of a considerable number of ciliates. No doubt it belongs in the list of accidental or facultative parasites.

The same is true of *Turchiniella culicis,* a new genus and species, described, from sections only, by Grassé and Boissezon (1929). The ciliates occurred in the hemocoele of an adult female *Culex.* Boissezon (1930) suggested that adults may die on the surface of the water and the ciliates may escape and infect larvae; in the original paper it was considered that the parasites lived in larvae, and occurrence in the adult was an impasse. It must, in fact, be rare in adults if the ciliates are as pathogenic as *G. pyriformis* in other hosts is known to be, for an infected larva would then seldom transform into an adult. Codreanu (1930) and Lwoff (1932) considered this ciliate to be *Glaucoma.*

Glaucoma or *Glaucoma*-like ciliates have been found also in other endozoic habitats than the hemocoele of aquatic larvae of Nemocera. *G. parasiticum* was observed by Penard (1922) in the gills of *Gammarus pulex,* not only on the surface but also in the interior, where it consumes the soft parts of the parenchyma and blood cells. Penard considered that it may be a temporary parasite only, and is closely related to *G. "pyriforme."*

There is one record of a similar ciliate in the tissues of a vertebrate. Epstein (1926) studied an infection of very young fish, *Abramis brama* L., with *Glaucoma*, probably *G. pyriformis* according to Lwoff (1932). Two to three percent were naturally infected in an aquarium at a lake near Moscow. The infection began with the yolk sac, which the ciliates reached through the gut. They then entered the heart and spread throughout the vascular system. In two or three days the hosts succumbed, with all except the resistant parts consumed. The ciliates occurred in great abundance in the canal of the spinal cord.

Related ethologically to the invasion of the bodies of aquatic larvae of Nemocera by *G. pyriformis* or related ciliates is the occurrence of the common marine ciliate, *Uronema marinum* Duj., in the coelom of a sipunculid. Madsen (1931) mentioned the observation by Mrs. E. Wesenberg-Lund of masses of ciliates in several *Halicryptus spinulosus* that had been kept in Copenhagen for several months without food. After some days the sipunculids died, and the ciliates lived longer in the cadavers. He regarded this invasion as following a bacterial infection, *Uronema* feeding on the bacteria, but did not suggest how the ciliates may have entered *Halicryptus*.

Accidental parasitism similar to that of *G. pyriformis* is the relationship of *Anophrys sarcophaga* to crabs, noted by Cattaneo (1888) and studied exhaustively by Poisson (1929, 1930). This marine ciliate normally lives in decomposing animal matter. Under certain circumstances it invades the hemocoele of *Carcinus maenas*, but natural infection is rare. Cattaneo found it in one of 300; Poisson in 7 of more than 3,000 at the biological station of Roscoff. The ciliates multiply actively in the blood, consuming the amoebocytes, and when these are exhausted feeding on plasma. When the host dies, the ciliates devour bacteria and fragments of tissue, surviving for some hours until decomposition is advanced, when they encyst or die.

Artificial transmission was easily accomplished. Of 25 *Carcinus maenas* inoculated, 20 died within 7 days, usually with a massive infection. Five crabs survived and soon lost the ciliates. Attempts were made to inoculate 7 other crabs of the genera *Cancer, Portunus, Maia,* and *Eupagurus*.

Of these only *Portunus depurator* developed a heavy infection and died. Some were naturally immune, the serum agglutinating and destroying the ciliates. In others the serum was not toxic *in vitro,* but the ciliates were arrested in certain lymphatic spaces, killed, and phagocytized.

Accidental parasitism of a nymph of the hemipteran *Nepa cinerea* was reported by Mercier and Poisson (1923). A species of *Colpoda* had invaded the body, probably through a wound in the integument, and produced a tumor the size of a pinhead on the lateroventral surface of the metathorax. The tumor extended part inside and part outside of the body, and in it ciliates were numerous. Large ones contained numerous inclusions, especially phagocytized amoebocytes. There were also very small ones with no inclusions; these were believed to be nourishing themselves by absorption of dissolved substances. Though locally destructive, the parasite did not prevent growth of the nymph up to the imaginal molt, when it was killed by the observers.

Instances of accidental parasitism by Protozoa have been noted in sea urchins. Lucas (1934), in examining Bermuda sea urchins, encountered transient Protozoa in the body fluids. She stated that these "were normally free-living forms, which probably gained entrance through the water-vascular system, and gave no evidence of colonization." André (1910) reported *Euplotes charon* in certain abundance in the perivisceral fluid of the sea urchin *Echinus esculentis,* as well as on the surface of the host. Accidental invasion of the body cavity of these marine echinoderms is apparently not infrequent.

Warren (1932) studied at Pietermaritzburg, Natal, a ciliate which possibly, according to his account, was a facultative parasite, in the common garden slug *Agriolimax agrestis.* He considered it to belong to a new genus and species, *Paraglaucoma limacis.* Kahl (1926) had already established a genus *Paraglaucoma* for *P. rostrata,* found in moss in Germany; later he found the species in moss from California and Wisconsin. Apparently Warren knew nothing of Kahl's work, but the two ciliates appear similar. The length (60-80 μ) as reported by Kahl (1931) is greater than that usual in Warren's form (40 μ in the free-living form; mean lengths 41-63 μ in the parasitic form); but in 1926 Kahl had reported the length as 45-55 μ. Warren did not report the posterior bristle which Kahl observed. The species also resembles *Glaucoma maupasi* Kahl, 1926, the ciliate Maupas (1883) described as *G. pyriformis.*

Warren found the ciliate swarming in certain fecal deposits, and then determined that they live in the lumen of the liver tubules, some at times passing into the stomach and being discharged in "fecal chambers of mucus." The incidence of infection varied from 50 to 87 percent at different times of the year, and in one slug 18,000 ciliates were present. What seemed to be the same ciliate was found in the "greenish incrustation of earthy matter underneath bricks and flower pots." This ciliate seemed to have no injurious effect on the slugs, even when present in large numbers.

Reynolds (1936) observed ciliates in freshly passed feces of the same species of slug in Virginia. He determined these as *Colpoda steini,* but as he gave no illustration or description, and even made the statement that the parasitic stage of this (holotrich) resembles (the heterotrich) *Balantidium* more closely than it does its own free-living stage, we may not unreasonably consider the systematic status to be unsettled. He determined in sections that the ciliates may be widely distributed in the tissues of the body, and were most abundant in the respiratory chamber and the anterior and posterior ends of the alimentary tract. In one region more than 94 percent of the slugs were infected, in another 25 percent. Infection occurred by ingestion, presumably, of the free-living ciliates in the soil, where *C. steini* was also found. Unlike Warren, Reynolds considered that many slugs are killed by the ciliate, and even suggested that the ciliates may be useful in combating molluscan pests. Warren had also examined sections, but did not find invasion of the tissues other than the liver tubules. It is likely that the extensive invasion noted by Reynolds would be more harmful to the slugs.

Probably the ciliate described by van den Berghe (1934) as *Glaucoma paedophthora* n. sp. belongs in this group of facultative parasites. At any rate, it seems to be a form that has been directly adapted from a free-living habitat to parasitism in the egg masses of *Planorbis* and *Physopsis.* At Elizabethville, Belgian Congo, van den Berghe found the ciliates in certain eggs, generally two or three in an egg mass, numbering from four or five to a great many. They were not found in the genital organs of the snails, and were abundant in the water of the aquarium. Infection of all eggs in a dish took place quickly if ciliates from an infected egg were introduced into the water. In the egg, multiplication from a few ciliates to an intense infection occurred within twelve hours. The embryo was killed by the parasites and within twenty-four hours had disappeared, the eggshell bursting and the ciliates escaping. Though the author stated decisively that the ciliate belongs to the genus *Glaucoma,* the description and figure do not prove that systematic position.

Along with the adaptation of free-living Protozoa to a symbiotic environment, there should be considered a number of instances of a secondary type of infection of associates in the same hosts. These have been referred to as facultative parasites, from the standpoint of the secondary hosts. Facultative parasitism of *Heterocineta janickii* on the oligochaete *Chaetogaster limnaei,* which occurs with the hypocomid ciliate in the mantle cavity of snails, is described below (p. 940). Theiler and Farber (1932, 1936) found *Trichomonas muris* present with considerable frequency in oxyurid nematodes in white mice, and division took place in the intestine of the worms. They even found trichomonads in nematodes when the flagellates could not be demonstrated elsewhere in the mice. J. G. Thomson (1925) found *Giardia* present in abundance in all of hundreds of nematode worms, *Vianella* sp., from a specimen of the South American rodent *Viscacia viscacia.* Although he found no trophozoites or cysts of *Giardia* elsewhere in the intestine of the rodent, he observed the flagellate from the nematode to be morphologically identical with *G. viscaciae* Lavier. Graham (1935) found *Giardia* in nematodes, probably *Cooperia oncophora,* from a bull; but was unable to find the flagellates in the intestine of the bull. A comparison with *G. bovis* Fantham would be of interest. As species of *Giardia* are otherwise exclusively parasites of vertebrates, it is likely that the nematodes with *Giardia* had, like those with *Trichomonas,* been secondarily infected with the mammalian flagellates. Flagellates can evidently survive for long and even multiply in the worms, so that their presence in them without simultaneous occurrence in the lumen of the vertebrate intestine is not significant.

SYSTEMATICALLY RELATED FREE-LIVING AND SYMBIOTIC PROTOZOA

MASTIGOPHORA

In addition to the existence of accidental and facultative parasitism, it is significant in connection with the origin of symbiotic relationships that certain genera contain both free-living and symbiotic species, or that the two types of habitat are occupied by members of closely related genera. The organisms have become closely adapted to their biotic environment, but have not undergone extensive modification. That does not necessarily imply recent adaptation, since stability of characteristics would equally well explain it; but it does indicate a direct origin from free-living types.

There are some epibiotic euglenids, including species of *Ascoglena* and *Colacium, Euglena cyclopicola* described by Gicklhorn (1925), and *Euglena parasitica* described by Sokoloff (1933). The last species adhered by the anterior end to all of numerous colonies of *Volvox* in a tank in Mexico City, and was not found free in the water. In the green color, stigma, and other structures, except for lack of a flagellum, this is a typical *Euglena*. It is not certain, however, whether the relationship is more than occasional phoresy. *E. cyclopicola* is normally epibiotic, occurring on *Cyclops strennuus* and species of *Daphnia*. Epibiotic euglenids have been observed on plankton Crustacea in reservoirs in the vicinity of Berkeley, California.

The euglenids, *Euglenamorpha hegneri, E. pellucida,* and *Hegneria leptodactyli,* are obligate inquilines of amphibia. They have never been found free-living. Species of *Euglena* and *Phacus,* with normal green color and activity, have, however, been found living in frog tadpoles (Alexeieff, 1912; Hegner, 1923; Wenrich, 1924a). This is merely a survival of free-living forms in the intestine, and Alexeieff may be unjustified in terming it facultative parasitism. A colorless euglenid of the genus *Menoidium* was found living in the intestine of one specimen of *Spirobolus marginatus* by Wenrich (1935); and it occurred free-living in damp *Sphagnum* in the aquarium jar. He reported no observations on how long this flagellate might survive in the host. *Euglena gracilis* fed to the millipeds could be in part recovered alive in one or two days.

Euglenamorpha hegneri Wenrich was observed by Hegner (1922) and described by Wenrich (1923, 1924a) and Hegner (1923) from tadpoles of frogs and toads and from *Hyla* in the North Atlantic states. The typical form has green chloroplasts, a red stigma, and three flagella. In 0.6 percent salt solution it survived for weeks in a hanging drop, and multiplied at first, but continued cultivation was not achieved. A colorless form, distinguished as the variety *pellucida* by Wenrich, is also present in tadpoles. This differs from the type in several respects, the most important being the lack of a stigma and the presence of from two to six flagella. Most frequently there are from four to six flagella. According to Wenrich, six is the doubled number, three new ones growing out very early in preparation for division. In other numbers above three, there are various stages of outgrowth. Division of a flagellate with four flagella results in daughter flagellates with two.

Brumpt and Lavier (1924) considered Wenrich's colorless variety to be a separate species, and Wenrich (1935) seemed inclined to the same opinion. Brumpt and Lavier described a similar colorless form with no stigma, from tadpoles of *Leptodactylus ocellatus* at São Paulo, Brazil, as *Hegneria leptodactyli*. That flagellate has seven flagella ordinarily, but may have only six. The authors did not mention the presence of an accompanying green form with fewer flagella, and Wenrich (1935) stated that he found the colorless flagellate in some hosts, unaccompanied by the green one. The six-flagellated forms of *Hegneria* seem to resemble very closely the six-flagellated forms of *Euglenamorpha hegneri* var. *pellucida*, so that it may be necessary to revise the taxonomy of the flagellates.

One is tempted to find, in this interesting series of forms, as Wenrich has brought out, adaptation to the conditions of an endobiotic habitat in loss of chloroplasts and increase of the number of flagella.

Endozoic colorless euglenid flagellates of the *Astasia* type have often been found, especially in Turbellaria, but also in rotifers, Gastrotricha, fresh-water nematodes, fresh-water oligochaetes, nudibranch eggs, and copepods. They usually are in vigorous metabolic movement, and generally lack a flagellum when in the host.

Haswell (1892) found them abundant in parenchymal cells in all specimens examined of a rhabdocoele turbellarian in Sydney. A flagellum was present in many but not in most cases. No stigma is mentioned. In 1907 Haswell described a similar euglenid in many specimens of another rhabdocoele, within cells of the digestive epithelium and in the spaces between the gut and the body wall. In the host, it was motionless or executed slow movements, but was more active when freed. No flagellum was present until two hours or more after the organisms were freed from the host. They were kept alive outside of the host for several days, but no euglenids were found normally free in the water.

Playfair (1921), who made his studies in the vicinity of Sydney also, stated that on one occasion he found half a dozen specimens of *Astasia margaritifera* Schmarda within the tissues of a turbellarian. This species he also found in the water of ponds, not in a free-swimming form and very often lacking a flagellum. This is the only species of *Astasia* that he reported in the survey of Australian fresh-water flagellates, and his identification is not convincing proof that the form in Turbellaria is the fresh-water species named.

Astasia captiva was described by Beauchamp (1911) from the rhabdocoele *Catenula lemnae* in France. In one pond almost all individuals were infected, while in another a mile away the flagellates occurred in a small percentage only. In some there were only one or two to a chain of zoöids, whereas in others the flagellates were very abundant. They were in continual movement in the "pseudocoele," between the parietal cells. A flagellum was present sometimes even on flagellates in the tissue, but most of the organisms lacked that structure. A colorless rudiment of a stigma, which was invisible in life, was seen frequently in stained preparations. Beauchamp stated that no euglenid was seen in other species of rhabdocoeles, including the common *Stenostomum leucops*. Howland (1928) identified as *Astasia captiva* an actively metabolic euglenoid flagellate, without flagellum or stigma, which she observed in *Stentor coeruleus* and *Spirostomum ambiguum*.

S. R. Hall (1931) found euglenids rarely in the mesenchyme of another species of *Stenostomum* and in *S. predatorium* in Virginia, where Kepner and Carter (1931) doubted the existence of *S. leucops*. While the flagellate was in the host, the flagellum did not extend beyond the edge of the body; but when it was liberated into water the flagellum soon grew out, metabolic movement ceased, and the organism swam rapidly. The euglenids could be kept alive in spring water for three or four days, but attempts to cultivate them failed. When infected hosts were added to a culture of the rhabdocoeles, practically all became infected within a week. In one instance, when an infected worm was devoured by another, several flagellates were observed to pass through the wall of the enteron into the mesenchyme, where they multiplied. There was no apparent effect on the host except in instances in which two or three hundred were present; then the rhabdocoeles became sluggish and bloated, ruptured with liberation of the flagellates, and died.

Because of the presence of a red stigma and bifurcation of the root of the flagellum, Hall assigned this flagellate to the genus *Euglena*, although it is colorless, naming it *E. leucops*.

Nieschulz (1922) examined large numbers of the fresh-water nematode *Trilobus gracilis*, in the hope of finding *Herpetomonas* (= *Leptomonas*) *bütschlii*. This was not found, but he reported *Astasia* from some specimens, usually only one or two in a host. There was no stigma and no flagellum. He did not state in what part of the body the parasites occurred.

In the rotifer *Hydatina senta*, *Astasia* has been reported on three occasions. Leydig (1857) observed it in the alimentary tract of almost all of the hundreds of rotifers that were examined. Metabolic movements were very active, a red stigma was present, and no flagellum was mentioned. Hudson and Gosse (1889) wrote: "*H. senta*, too, suffers from an internal parasite. It . . . swims up and down its host's stomach by jerking the contents of its body constantly backwards and forwards." Their figures show no flagella, and one, in color, shows a red stigma. Valkanov (1928), without reference to other observers, named the organism he found parasitic in the intestine of the same species of rotifer, *A. hydatinae*.

In the intestine of gastrotrichs, *Astasia*-like inquilines were reported by Voigt (1904). He found them in some specimens of a gastrotrich that he later (1909) named *Chaetonotus ploenensis*, and was unable, despite careful search, to find free-living examples of *Astasia* in the material. Remane (1936, p. 231) stated that he found the same species in the intestine of another species of *Chaetonotus*.

Astasia doridis was found by Zerling (1933) to be rather abundant in some eggs of the egg masses of the nudibranch *Doris tuberculatus* at Wimeraux. When heavily parasitized, larvae were destroyed. The parasites lacked flagella and stigmas and showed intense euglenoid activity. Freed from the eggs, they lived many days with no change in morphology and behavior. The flagellate was not found in the genital tract of the adult. Zerling believed it probable, nevertheless, that adult molluscs are infected by the parasites liberated into sea water at the hatching of infected larvae, and that they transmit the parasites to their egg masses. This is the only published record of a euglenid parasite in a marine host.

Codreanu and Codreanu (1928) found a considerable percentage of the fresh-water oligochaete *Chaetogaster diastrophus* Gruith in the vicinity of Bucharest infected by a euglenid parasite that they named *Astasia chaetogastris*. The flagellates multiplied rapidly in the coelom, and the infection was always fatal in from eight to thirteen days. When freed into water, metabolic movement lessened and a flagellum developed. Both forms had a stigma, and the free form as well as the parasitic one was capable of division. This euglenid is more pathogenic than any other described. One is reminded of the invasion of the hemocoele of insect larvae by *Glaucoma pyriformis*.

Foulke (1884) wrote concerning the fresh-water sabellid *Manayunkia speciosa:* "Several individuals of *Manayunkia* were observed to be preyed upon, while still living, by large monads, embedded in one or more of the segments, which were sometimes excavated to a considerable degree." It is possible that in this statement there is reference to a situation analogous to that of *Astasia chaetogastris.*

Finally, in copepods, occurs *Astasia mobilis,* which was the first endozoic euglenid to be observed (Rehberg, 1882). Alexeieff (1912) studied it in *Cyclops,* finding it not only in the lumen of the intestine but also twice in the eggs. It sometimes had a flagellum, and a stigma was described. The metabolic activity and some features of the structure of this organism have suggested to some sporozoan affinities. By Labbé (1899), for example, it was included in the genus *Monocystis.* Alexeieff discussed the possible euglenid origin of Sporozoa, and Stein (1848) had long before remarked upon the apparent relationship between euglenids and *Monocystis.*

Jahn and McKibben (1937) studied a colorless, stigma-bearing euglenid flagellate whose habitat is given as putrid leaf infusion. They found the root of the flegellum to be bifurcated, as in Euglenidae; whereas in Astasiidae, according to Hall and Jahn (1929), it is not bifurcated. The new genus *Khawkinea* was established by Jahn and McKibben for flagellates whose characteristics agree with those of *Euglena* except that they are permanently colorless; and they assigned to this genus not only their new species, *K. halli,* but also the free-living form that had been known as *Astasia ocellata* Khawkine, *A. captiva* Beauchamp, *A. mobilis* Alexeieff, *A. chaetogastris* Codreanu and Codreanu, and *E. leucops* Hall.

In the question of the relationship of free-living and endozoic Protozoa, the flagellate recently discovered in pond water by Bishop (1935, 1936) in England, and by Lavier (1936c) in France is of much interest. Bishop found it on four different occasions in the course of thirteen months, in a small pond with thick, black mud and much decaying organic matter; and Lavier found it in samples from two separate places. Many of its characteristics are those of a trichomonad and, as Lavier pointed out, it seems to be the only free-living member of the Trichomonadidae. There is a slender axostyle, which often is extended into a pointed, posterior projection of the cytosome, or itself projects from the body. Sometimes the flagellate anchors itself to an object by the end of the axostyle. There are three anterior flagella and a trailing flagellum that usually adheres to the body, forming an undulating membrane, but that sometimes, according to Lavier, is free. The nucleus is trichomonad in position, structure, and division, and there is a well-defined paradesmose. Its manner of progression, which differs from that of other free-living forms (Lavier), impressed Bishop with its similarity to the movement of *Trichomonas*. Bishop (1935) wrote of it under the name *"Thichomonas" Keilini* n.sp.

The flagellate differs from *Tritrichomonas* in the absence of a costa. Lavier assigned it to the genus *Eutrichomastix,* which resembles *Trichomonas* in all respects except the lack of the costa and undulating membrane. Although it has been shown that the trailing flagellum of *Eutrichomastix* may adhere to the body under certain conditions, the usual presence of an undulating membrane in the pond flagellate differentiates it from that genus. Neither Bishop nor Lavier made any mention of the parabasal body or of an attempt to demonstrate it. If this structure, so characteristic of *Monocercomonas* (*Eutrichomastix*) and *Trichomonas,* is present, it would leave no doubt of the trichomonad affinities of the organism; if absent, the flagellate would not show so close a relationship to endozoic forms. Bishop (1939) proposed the new genus *Pseudotrichomonas* for the organism.

It is not possible to state that in this organism there is evidence of the origin of trichomonads, which are widespread and evidently have been adapted for a great period of time to endozoic existence. It may be a survival of an ancestral type; on the other hand, there is the possibility, which Bishop considered, that it might be a parasite of some cold-blooded host that had survived and multiplied in the water. Rosenberg

(1936) found that *Tritrichomonas augusta* sometimes survived in salt solution, on slides ringed with vaseline, for nearly a year. Cleveland (1928b) was able to cultivate indefinitely *T. fecalis* in water with feces or tissue, in hay infusion, and in other ways, at temperatures from —3° C. to 37° C.; and this, although it was supposed to have been derived from a warm-blooded host, man. Cleveland also maintained *T. augusta* in tap water with feces. He did not report on the ability of *T. batrachorum*, which Wenrich (1935) stated is morphologically indistinguishable from *T. fecalis*, to grow under the conditions supplied. It would not be surprising if flagellates that have such marked ability to survive and even to multiply outside of the host, under such simple conditions, might find natural circumstances occasionally favorable to outside maintenance of life. They might, at times, be found by collectors. This does not apply immediately to the studies of the species *P. keilini*, however, as no endozoic flagellate just like it is now known, and Bishop (1936) found that it would not live in tadpoles; but it raises a general question.

Hollande (1939) described as a free-living trichomonad the new genus and species *Coelotrichomastix convexas*. The flagellate was found in liquid manure. It has four flagella, one of them trailing and said to border an undulating membrane in a deep groove of the body; but there is no costa. There is a unique axostyle, ribbon-like in its posterior part and located superficially near the groove, anteriorly expanded to a hemispherical cupule covering a considerable part of the large nucleus. All parts of the axostyle are covered by small siderophile granules. A very small bacilliform parabasal body was reported. In considering the characteristics of *Coelotrichomastix*, Hollande failed to comment on the striking similarity in many respects that exists between it and certain flagellates that have been assigned to the genus *Tetramitus*. This cannot fail to impress the reader of the accounts by Klebs (1893), Bunting (1926), Bunting and Wenrich (1929), and Kirby (1932a). In those papers, furthermore, especially the second and third, one will find facts that suggest the possibility of a different interpretation of certain unexpected characteristics described by Hollande. In assigning *"Trichomastix" salina*, originally described by Entz (1904), to *Coelotrichomastix*, Hollande made no comment on the writer's account of what seemed to be the same flagellate under the name *Tetramitus salinus* (Entz).

Trepomonas agilis, the only species that has been described in that genus, is a common flagellate associated with decaying organic matter in fresh water, and has been found in a coprozoic habitat in human feces (Wenyon and Broughton-Alcock, 1924). The writer on several occasions found *Trepomonas* in salt-marsh pools, associated with marine flagellates and ciliates. Whether it was *T. agilis* was not determined. Flagellates of the genus have also become adapted to an endobiotic habitat in fish, amphibia, and reptiles. Alexeieff (1910) observed *Trepomonas* in *Box salpa;* and Lavier (1936b) found *T. agilis* once in that fish, where, he stated, it is doubtless an accidental saprozoite. According to Alexeieff (1909) and Lavier (1935), the endozoic *Trepomonas* common in amphibia is probably *T. agilis.* Lavier found the flagellate rather constantly in tadpoles of *Rana temporaria, R. esculenta,* and *Alytes obstetricans,* and in one adult *Triton.* He discussed it as an interesting possibility of parasitism in a flagellate normally living free, and possibly finding the endozoic habitat more favorable than the free-living. Das Gupta (1935) found a species of *Trepomonas,* usually in small numbers, in the caeca of three different species of turtles: *Terrapene major, Kinosternon hippocrepis,* and *Chelydra serpentina.* A cytological study of the flagellate has recently been made by Bishop (1937).

The genus *Hexamita* includes both free-living and endozoic species. The former are common in fresh water and infusions with decaying organic matter; they also occur in salt water; and a species resembling *H. inflatus,* and only 11 μ long, has been observed by the writer in a salt-marsh pool with decaying algae and a salinity of fifty parts per thousand. The type of habitat of the "trichomonad" named *Pseudotrichomonas keilini* by Bishop (1939) and of *Trepomonas* and *Tetramitus* is similar to that of *Hexamita;* Lavier reported them all from one sample taken in France.

Urophagus and *Octomastix* are considered by most protozoölogists to be synonymous with *Hexamita.* This was the opinion of Lavier (1936a), who also rejected *Octomitus;* but he proposed two new genera, *Spironucleus* and *Syndyomita* (the latter of which is of the original *Octomitus* type) for morphological types of *Hexamita*-like flagellates in amphibia. Lavier retained the name *Hexamita* for the common form in amphibia, the type of which, among those he considers, is most like that of free-living *Hexamita.* This, *H. intestinalis,* has undergone little modification. The others differ from the free-living type, according to him; but there seems to be a similarity in type of *Spironucleus* to *H. rostrata* (?), as figured by Wenrich (1935) from the outside of a dead fresh-water snail. At least until a systematic review of the whole group of *Hexamita*-like forms is made, it appears to be necessary, for the sake of clarity, to use only the one genus name.

The endozoic species of *Hexamita* are many and are found in a wide variety of hosts. Though given species are restricted to single or related hosts, the tendency to give different names to those in different hosts, without adequate comparison with other described species, has been manifest.

Certes (1882) found *Hexamita* frequently in the stomach of oysters from certain localities on the coast of France. Though he considered this to be *H. inflata,* he regarded it as a normal, reproducing inhabitant of the stomach; and the identification is in no way positive. In other invertebrates, *Hexamita* has been recorded from the reproductive organs of the trematode *Deropristis inflata* in marine eels, but not in the intestine of the eel (Hunninen and Wichterman, 1936); from the cockroaches *Blatta orientalis* (Bishop, 1933), *Periplaneta americana,* and *Cryptocercus punctulatus* (Cleveland, 1934); from the horse-leech *Haemopis sanguisugae* (Bishop, 1932, 1933); from the milliped *Spirobolus marginatus* (Wenrich, 1935); from the larvae of *Tipula* (Mackinnon, 1912; Geiman, 1932); and from *Tubifex* (Ryckeghem, 1928).

These all occur in the gut, except for the trematode form, as noted, and the one in *Tubifex.* The latter, furthermore, is the only one in invertebrates to which probable pathogenicity has been ascribed. *Hexamita tubifici* was encountered at intervals in the course of fifteen years in the body cavity of *Tubifex* kept in culture in the laboratory in Louvain. Worms that lost their power of activity, appeared whitish, and died were found to have a more or less intense infection with *Hexamita.* Ryckeghem considered the question as to whether the flagellate is a parasite or a free-living form invading decomposing tissue. He concluded that the former relationship exists, for he found it in living worms in apparently healthy tissue, decomposing chironomid larvae were not invaded by the flagellate, and it was encountered in different collections at long intervals. Each time it was a source of trouble.

Hexamita species occur in vertebrates of all classes. Fry and young fingerlings of trout and salmon in hatcheries in the United States were found by Moore (1922, 1923a, 1923b, 1924) and Davis (1923, 1926) to be extensively infected, especially in the anterior part of the intestine; and they believed the flagellate to be severely pathogenic and to constitute a serious menace to the success of trout culture. The flagellates have been found also in European trout (Moroff, 1903; Schmidt, 1920) and in the fan-tailed darter (*Etheostoma flabellare*) (Davis, 1926). Davis and Moore did not prove that the flagellates were not secondary in diseased fish, as Schmidt believed.

Lavier (1936b) examined 33 species of marine fish and found six species of *Hexamita*, five of them new, in seven of these. He remarked that the morphology of *Hexamita* is much varied if one does not think in general terms, and that an attentive study enables one to recognize clear and constant morphological differences.

Hexamita is commonly found in the intestine of amphibia, and has been reported from the intestine of turtles and tortoises, as well as from the bladder of *Emys orbicularis* (Grassé, 1924) and from the stomach, oesophagus, and small intestine of the snake *Natrix tigrina* (Matubayasi, 1937). It occasionally invades the blood of amphibia (Lavier and Galliard, 1925; and others) and tortoises (Plimmer, 1912) through a damaged intestinal wall.

Among birds, *Hexamita* occurs in pigeons (Nöller and Buttgereit, 1923), ducks (*Anas boschas*, Kotlán, 1923), turkeys (Hinshaw, McNeil, and Kofoid, 1938), and various wild birds in Brazil (Cunha and Muniz, 1922, 1927).

Of mammals, rodents especially have been found infected with *Hexamita*. In addition to rats, mice, ground-squirrels, and woodchucks (Crouch, 1934), the South American hystrichoid rodent *Myopotamus coipus* [=*Myocastor coypus* (Molina)] contains a species (Artigas and Pacheco, 1932).

Hexamita has also been reported in primates, including man (Cunha and Muniz, 1929; Wenrich, 1933; Chatterji, Das, and Mitra, 1928; Perekropoff and Stepanoff, 1931, 1932). Dobell (1935), discussing all these records except the third, believed that diplozoic forms of *Enteromonas*, which "are very frequently found in feces, in intestinal contents, and in cultures," were misidentified. As regards Wenrich's record, however, from *Macacus rhesus*, this is improbable when one considers his extensive knowledge of the genus as well as the exactness of his description and figures.

From this survey of the distribution of members of the genus *Hexamita*, it is apparent that the flagellates are as widespread in animals as are members of the strictly endozoic genus *Trichomonas*. In their case, however, flagellates equally or more closely related than most of the endozoic forms to the ancestral type are common free-living forms. The endozoic forms, nevertheless, are for the most part as strictly adapted to their habitat as trichomonads.

There is no evidence, except possibly in certain species in Amphibia and invertebrates, that the obligate symbionts have been recently adapted from facultatively endozoic forms; any more than that *Trichomonas* can be supposed to have recently so originated. There is little evidence of parallelism in phylogenetic development in members of these two genera and their hosts (Wenrich, 1935).

In most instances *Hexamita* has been regarded as a commensal in its hosts. A possible exception in an invertebrate host is *Hexamita tubifici*. In the body cavity of the aquatic annelid the flagellates may be fatal to the host, in a manner comparable to the effect of *Glaucoma* in dipteran larvae and *Astasia* in *Chaetogaster*. Hinshaw, McNeil, and Kofoid (1938b), on the basis of experimental data which they obtained, suggested a possible relationship between a condition of enteritis in young turkeys and a heavy infection of *Hexamita* that occurred in the affected part of the small intestine. They also reviewed reports of possible relationship in other vertebrates between pathological conditions and the occurrence of *Hexamita*.

HOLOTRICHA

Among holotrichous ciliates, all types of biotic relationship exist, so that the group is especially favorable for study of the development of symbiosis and host-specificity. In this section will be considered holotrich groups in which free-living and symbiotic species are closely related.

In some instances it seems that there has been no more than survival of ordinarily free-living forms in or on a host, where certain conditions of nutrition or protection favored the occurrence of the associate. Perhaps the occurrence of *Coleps hirtus* on the rhabdocoele *Vortex sexdentatus* as a common epizoön, as recorded by Graff (1882), is a relationship of this type.

The relationship of *Enchelys difflugiarum* Penard to *Difflugia acumi-*

nata (Penard, 1922), and that of *E. nebulosa* Entz to *Cothurnia* is apparently obligatory predatism; this relationship, of course, is comparable to parasitism, as, if the host were a metazoan and were only partially destroyed by the attacks of the ciliate, we would doubtless consider the latter a true parasite.

Haematophagus megapterae Woodcock and Lodge and *Metacystis megapterae* Kahl are commensals on the bristles of the whale *Megaptera nodosa* (Kahl, 1930).

A number of pleurostomatous and hypostomatous gymnostomes have become associated with animal hosts. In the former group there is *Amphileptus claparèdei* Stein, "parasitic on the stalks of colonial Vorticellidae" (Kahl, 1933); and *A. carchesii* Stein in a similar situation. Edmondson (1906) reported that after feeding upon a zoöid the latter species (discussed by him as *A. meleagris* Ehr.) attaches itself to a stalk. He found that many were present on *Carchesium polypinum*, clasping the stalks by a deeply cleft posterior end. In addition to these more or less predatory species, there is *Lionotus branchiarum* (Wenrich) Kahl, described by Wenrich (1924b) as *A. branchiarum*. It is a true parasite on the gills of the tadpoles of several species of *Rana,* where it lives in a capsule under the cuticular membrane and occasionally detaches and engulfs gill cells. Wenrich (1935) discussed the possibility that *A. branchiarum* is transitional between a predatory and parasitic status. There is a predaceous, free-swimming phase on the surface of the gills by which other Protozoa may be devoured. Three species of *Lionotus, L. impatiens* Penard, *L. aselli* Kahl, and *L. hirundi* Penard are commensal among the gills of *Asellus;* and one, *L. agilis* Penard, occurs on the ventral surface, among the legs, and on the egg masses of *Cyclops.* The pleurostome genus *Branchioecetes* Kahl is very closely related to *Loxophyllum,* in which Švec (1897) and Penard (1922) put the species. The two species *B. aselli* (Švec) and *B. gammari* (Penard) are commensals on their hosts, to which they adhere by thigmotactic cilia.

Commensal hypostomes belong to the genera *Trochilia* and especially *Chilodonella. Trochilia* (*Dysteropsis*) *minuta* (Roux) has been found free-living as well as commensal with *Cyclops, Gammarus,* and *Asellus* (Penard, 1922). Commensalism is widespread in *Chilodonella,* several species of which are apparently obligatory commensals on fish, others on certain rotifers, amphipods, and isopods. The species on the gills of fish have often been thought to exert direct or indirect pathogenic action, but proof that they are more than ectocommensals is lacking. Kidder and Summers (1935) distinguished several species on the carapaces of three species of Orchestiidae from beaches in the region of Woods Hole; they noted that no similar ciliates were found free in the sand or seaweeds, and the commensals lived only a short time when separated from their hosts. *C. capucinus* Penard, 1922, and *C. granulata* Penard, 1922, are commensal on *Asellus* and *Gammarus,* and *C. porcellionis* occurs in the gill cavities of the terrestrial isopod *Porcellio* sp. (Dogiel and Furssenko, 1921). In aquatic hosts, transmission would take place through the water; in *C. porcellionis* it must be through survival, at least for a short period, in moist soil.

A number of hymenostomes of the new genus *Allosphaerium* were also described by Kidder and Summers (1935) from the one species of *Talorchestia* and two of *Orchestia* that they examined at Woods Hole, Massachusetts. They remarked, concerning the ectocommensal holotrichs of amphipods and isopods, that the external characteristics are singularly well adapted to the environment.

They are all small flat forms and possess ventrally placed thigmotactic cilia (*Chilodonella, Trochilia, Allosphaerium*). When one considers the forces, mainly in the form of water currents, to which they must be subjected and which would tend to effect their removal from the carapace of their various hosts, it is seen that the flatness of their bodies and the adhesive powers of their ventral cilia are of absolute necessity. Existing under the same conditions, it is perhaps not surprising that representatives of two orders of ciliates exhibit convergence to such a degree as to render them practically indistinguishable one from the other except under extreme magnifications.

Genera with free-living Trichostomata and Hymenostomata include only a few commensal and parasitic species, but there are numerous genera all members of which are associated with animal hosts. *Frontonia branchiostomae* was found in abundance at Banyuls-sur-Mer by Codreanu (1928) in the atria of most specimens of *Branchiostoma lanceolatum* exceeding 3 cm. in length. The genus *Glaucoma* has been discussed at length under facultative parasitism. *Uronema rabaudi* was believed by Cépède (1910) to be a coelomic parasite of *Acartia clausi* and *Clausia elongata,* in the empty carapaces of which it was observed. Without free-living congeners, but similar enough to *Uronema* to have been put in that genus by Bütschli (1889) and Cuenot (1891), is *Philaster digitiformis,* which was described by Fabre-Domergue (1885) in mucus on the body of *Asterias glacialis,* multiplying abundantly on damaged and disintegrating starfish, but disappearing with death of the host.

The genus *Ophryoglena* comprises large holotrichous ciliates of which some species are free-living and others endozoic. Kahl (1931) listed eleven of the former and five of the latter. Within the genus there is a range from free-living habits, often with predatism, through commensalism to strict parasitism in close relationship to the developmental cycle of the host.

The free-living species *O. flava,* according to Penard (1922), is voracious and usually preys upon animals larger than itself, including rotifers, small worms, and small Crustacea, especially *Cyclops.* It passes under the carapace of *Cyclops* and consumes the living animal, the soft parts of which are converted into food balls in the cytoplasm.

Ophryoglena maligna, described by Penard (1922), preys upon *O. flava* as a parasite. It invades the cytoplasm, in which the number is one to four or more, and devours the host little by little until it is empty. The ciliates were also found free in the water, but Penard believed that before long they would attach themselves to *O. flava.*

The three species that have been found in the intestine of Turbellaria appear to be commensals. These are *O. parasitica,* reported by André (1909) from 11 of 234 *Dendrocoelum lacteum; O. pyriformis* found infrequently by Rossolimo (1926) in *Sorocoelis maculosa* and *Planaria nigrofasciata* at Lake Baikal; and *O. intestinalis* from a large turbellarian of the genus *Dicotylus* at Lake Baikal. It was shown that the last two species cannot survive long in the water.

Truly parasitic, however, are the species reported by Lichtenstein (1921) and Codreanu (1930, 1934) from May-fly naiads. The former found the parasites in the schizocoele and gonads of *Baetis* sp. near Montpellier; the latter in five Ephemerida from the Alps and the Carpathians. Codreanu believed that parasitism by these ciliates may occur widely in Ephemeroptera. In young *Rithrogena* the ciliates occur as cysts, division taking place within the cyst, but in *Baetis* they are not encysted at any time. When the reproductive organs develop in the females, most or all of the ciliates invade the ovaries, the contents of which they ultimately destroy. The May flies nevertheless become adults and in the act of what would normally be egg-laying, ciliates are deposited in the water instead of eggs. Only the female hosts are able to propagate the infection. Codreanu (1934) remarked that this is the only sufficiently defined case of true parasitism of the schizocoele of insects by ciliates. The species found in *Baetis* by Lichtenstein was named by him *O. collini;* that studied in *Baetis* by Codreanu (1930) was, he stated, probably the same.

Haas (1933) noted the similarity between the oral apparatus of the swarmers of *Ichthyophthirius multifiliis* Fouq. and that of *Ophryoglena;* Kahl (1935), in consequence, placed that important parasite of fish in the family Ophryoglenidae. Commensalism and parasitism being so well developed in *Ophryoglena*, although along with free-living habits, there are clear ethological relationships between it and *Ichthyophthirius.*

Pleuronema anodontae, the only commensal species of that genus, was reported by Kahl (1926) in small crushed mussels. He stated later (1931) that it is infrequent in *Anodonta,* but occurs regularly in *Sphaerium* species. (Perhaps one should investigate the possibility that this may be one of the Ancistrumidae, not *Pleuronema.*) Very close to *Pleuronema* is *Pleurocoptes hydractiniae* Wallengren, an ectocommensal on the hydromedusan *Hydractinia echinata.*

DISTRIBUTIONAL HOST RELATIONSHIPS AND HOST-SPECIFICITY IN REPRESENTATIVE SYMBIOTIC FAUNULES

GENERAL CONSIDERATIONS

There are some generic groups of Protozoa that have a rather wide distribution among animals; these groups are represented by species in hosts widely separated systematically. That is true, for instance, of *Hexamita* and *Trichomonas* among polymastigotes, of *Endolimax* among endamoebae, of *Nosema* and *Eimeria* among sporozoa, of *Nyctotherus, Balantidium,* and urceolarids among ciliates. These examples have been discussed by Wenrich (1935). The genus *Trypanosoma* is represented in a very large number of vertebrates of all classes, but is limited to them, as is also *Giardia.* (The occurrence of *Giardia* in nematodes is facultative; see p. 902.) A more or less closely restricted host distribution is, however, characteristic of many generic, familial, and even higher groups of Protozoa. Entodiniomorphina occur only in certain herbivorous mammals, chiefly ruminants and Equidae; opalinids are most likely to have anurous amphibian hosts, although a few have been found in Urodeles, fish, and reptiles; hypocomids (except the small genus *Hypocoma*) are parasites of certain groups of molluscs; Astomata are mainly inhabitants of annelids, to which most genera are limited; hypermastigote flagellates occur only in termites and roaches; and certain groups of polymastigotes are restricted to certain groups of termites (p. 923).

The problem of host-specificity is ordinarily approached from the standpoint of the individual species; that is, the degree in which it is limited to a particular host species. In strict host-specificity, the host is rigorously determined; there is only one host for a species of symbiont. As has been pointed out by Grassé (1935) and Wenrich (1935), strict host-specificity is not a general phenomenon. Surveys of lists of species and their hosts often bring out many instances in which there is only one host for a species, for example, in the genera *Giardia, Babesia, Plasmodium, Haemoproteus, Leucocytozoön,* and *Eimeria*. But such data cannot be taken at face value, because the apparent strict host-specificity may be based on insufficient search for the organism in other hosts, or on a tendency of taxonomists to differentiate species on insufficient grounds. More intense study in certain groups, as *Trypanosoma* and Devescovininae in termites, has shown less rigorous limitation than at first seemed to exist. More commonly, host-specificity is relative. The limitation is to more or less related animals; and it depends, as Becker (1933) and Wenrich (1935) have pointed out, on the characteristics of the symbiotic environment, the opportunities for transmission, and the evolutionary tendencies of the Protozoa. The phenomenon is of the same nature as that of the geographical distribution of free-living organisms, though of course it is more complex.

It is a commonplace that given animals have characteristic protozoan faunules; this phenomenon is of particular interest when there are faunules of particular types peculiar to major groups. In instances of the highest development of this tendency, it can be predicted what types of Protozoa will be found in unexamined hosts. One may be reasonably certain of finding Opalinidae in anuran species, Ophryoscolecidae in ruminants, and certain types of polymastigotes and hypermastigotes in all termites other than Termitidae .

Questions of distributional host relationships and host-specificity will now be considered in greater detail in certain representative symbiotic faunules. Two faunules have been selected for this purpose: ciliates of sea urchins, and flagellates in termites and roaches. The former is interesting also from the standpoint of the relationship of inquilines and free-living forms; the sea-urchin intestine is one of the least specialized environments of its type. The flagellates, however, occupy one of the most specialized of symbiotic habitats. Not only are the circumstances under which they live and are transmitted exceptional, but the hind-gut of the host has actually undergone structural modification to accommodate them.

CILIATES OF SEA URCHINS

Faunules of ciliates occur in the greater number of sea urchins that have been examined, but there are some without any. Uyemura (1934), giving positive reports from eight species of sea urchins of Japan, found none in *Brissus agassizi*. At Amoy, Nie (1934) found none in *Temnopleurus toreumaticus*. Of the species at Yaku Island, Japan, Yagiu (1935) found two uninfected: *Colobocentrotus mertensii* and *Cidaris* (*Goniocidaris*) *biserialis*. Powers (1935) found no faunule in *Eucidaris tribuloides* at Tortugas, nor were ciliates present in members of the genus *Arbacia* at Beaufort, Woods Hole, and Naples (1933a), in spite of association with infected species. Why a few species possess no faunules, while so many have ciliates in abundance, is an interesting question.

The intestinal faunules of sea urchins consist mostly of ciliates, which, in whatever part of the world the host occurs, are members of a number of characteristic genera. Most of them are holotrichs; outside of this group are a few species of *Metopus* and one of *Strombilidium,* heterotrichs with many free-living congeners. There are sometimes as many as twelve distinct species in eight genera in *Strongylocentrotus purpuratus* (Lynch, 1929); and Yagiu (1933, 1934) found twelve species in *Anthocidaris crassispina*. On the other hand, from *S. franciscanus* in Japan, Yagiu (1935) reported only *Conchophthirus striatus*. Four or five species is perhaps the average infection. The occurrence of amoebae, *Chilomastix echinorum* (Powers, 1935), nematodes, and rhabdocoeles (*Syndesmis*) is much less prominent than that of ciliates.

In given hosts there is variability in the occurrence of different species of ciliates. Some occur in abundance almost or quite universally; others have a lower incidence, some being of rare occurrence. *Plagiopyla minuta*

Powers (1933a) occurred in only about 10 percent of *Strongylocentrotus dröbachiensis,* and then there were not more than twelve in a host; whereas some ciliates have been found in all sea urchins of the species. Most often the incidence is not 100 percent.

Powers (1933a) pointed out the existence of two groups of ciliates in sea urchins. One group contains diverse species with many free-living congeners, which he regarded as chance or vagrant forms that were engulfed with food and survived; the other consists of obligatorily endozoic species. The members of the first group are "apparently free-living and only occasionally or accidentally associated with their host." There is, however, no evidence in the literature that many, if any, of the intestinal ciliates in sea urchins are accidentally introduced free-living forms. Though there are species belonging to genera of which most members are free-living, that in itself is no indication that they are not obligatory inquilines.

Colpidium echini Russo, found also by Powers (1933a) in all specimens of *Strongylocentrotus lividus* examined at Naples, probably, according to Kahl (1934), is not a *Colpidium. Uronema socialis,* described by Powers (1933a) from *S. dröbachiensis,* was later renamed by him (1935) *Cyclidium stercoris.* Kahl (1934) doubted the generic assignment of *Colpoda fragilis,* described by Powers (1933a) from *Toxopneustes variegatus* of Beaufort, North Carolina. These forms, which Powers mentioned in the occasional associate group, together with *Plagiopyla,* may be obligatory commensals. The *Euplotes* sp. found by Powers (1933a) in the gut and outside of *S. dröbachiensis* is possibly an accidental invader. He also reported *Trichodina* from the sea urchin and in seaweed.

Cyclidium stercoris, which occurs in great abundance in *S. dröbachiensis,* will live and reproduce in sea water (Powers, 1933a); but it is not known that it does so under natural conditions. *"Colpoda" fragilis,* on the other hand, is very sensitive to changes in its environment. Many of the ciliates can survive for more or less prolonged periods outside of the host. *Entodiscus borealis,* one of the strictly endozoic forms, was kept in sea water from fifteen to twenty-three days (Powers, 1933b).

Species of *Cyclidium* occur also in sea urchins of China (Nie, 1934) and Japan (Yagiu, 1933, 1934). Several species of *Anophrys* have been reported from various echinoids. There is only one free-living species of *Anophrys, A. sarcophaga* Cohn, which has been discussed above as a facultative parasite of crabs. Kahl (1934) suggested a relationship of certain of these ciliates to *Philaster digitiformis,* which occurs, as mentioned above, on the body of starfish. He was doubtful about the correctness of their position in the genus *Anophyrs.* There is confusion about the taxonomy of many sea-urchin ciliates.

Genera that are restricted to sea urchins, and may be supposed to have evolved in the shelter of these hosts, are *Lechriopyla* Lynch, *Entorhipidium* Lynch, *Entodiscus* Madsen, *Bigggaria* Kahl, *Madsenia* Kahl. *Cryptochilidium* Schouteden, in part included in *Biggaria,* has a species in the annelid *Phascolosoma vulgaris.*

Lechriopyla mystax Lynch, commensal in the Pacific Coast sea urchins *Strongylocentrotus franciscanus* and *S. purpuratus,* is closely related to *Plagiopyla.* It is markedly thigmotactic: "Although almost continuously in movement [it] adheres almost constantly to surfaces. The large peristomal groove seems to act as a sucker" (Lynch, 1930). *Lechriopyla* apparently has diverged from *Plagiopyla* in relation to its obligatory endocommensalism, but there are no profound alterations.

Four species of *Entorhipidium* were distinguished by Lynch (1929) in *Strongylocentrotus purpuratus* in California. None of these flattened, fan-shaped trichostomes was present in *S. franciscanus* from the same localities, so there seemed to be marked host-specificity. Uyemura (1934), however, found one of the same species in another sea urchin of Japan, and described a new species, *E. fukuii,* which occurs in five hosts of four genera.

Related to *Entorhipidium* is *Entodiscus,* represented by *E. borealis* (Hentschel) from several different hosts of the North Atlantic and Japan; and *E. sabulonis* Powers found in all individuals examined of two species of *Clypeaster* at Tortugas. *E. borealis* is present in great abundance and, with its greatly flattened form, probably in appearance and occurrence suggests *Opalina* in Amphibia. According to Powers (1933b), besides swimming about in the lumen of the intestine, it adheres by the ventral side to the intestinal mucosa. The food vacuoles, he stated, contain rods, probably bacteria, and objects resembling nuclei of epithelial cells. At that time Powers thought that the ciliate might attack the intestinal mucosa, secreting cytolytic enzymes, thus being definitely parasitic; but later (1935) he did not stress this ill-founded conclusion.

Powers (1933a) discussed the possibility that *Cryptochilidium echini* (Maupas), abundant and universal in *Strongylocentrotus lividus* at Naples, is a true parasite. In several instances the body was found partly embedded in the intestinal mucosa. As he probably recognized later, this observation does not constitute adequate proof for his conclusion. The genus *Cryptochilidium,* together with *Biggaria,* Kahl's genus for some of the forms described as *Cryptochilidium,* is well represented in sea urchins of all regions.

Metopus histophagus Powers, as the species name indicates, contains in its food vacuoles epithelial cells from the intestine of its host (Powers, 1935); but it was not observed to cause lesions, and probably simply ingests cellular debris, as does *M. circumlabens* (Lucas, 1934). The species occurs only in *Clypeaster subdepressus* of Tortugas. The species *M. circumlabens* Biggar occurs in a number of hosts at Bermuda, Tortugas, Amoy, and Japan, but several other species seem to have a limited host-specificity.

Questions of host-specificity and geographical distribution of the ciliates have been discussed by Powers (1935, 1937). He remarked that there is little evidence of rigid host-specificity. Species differ in that respect. Yagiu (1935) found *Cryptochilidium echini* and *Anophrys elongata* in all but one of the host species, examined by him at Yaku Island, which contained any ciliates; and Powers (1935) found *Cryptochilidium bermudense* (=*Biggaria bermudense*) and *Anophrys elongata* in all sea urchins at Tortugas that were infected with ciliates. Nor are those ciliates limited to those regions; they have been found in various localities. There are some ciliates that have been found in only one or a few hosts, these being sometimes in one region only but also sometimes in widely separated localities. There is nothing, however, which leads us to expect that, with the accumulation of more data, most or all of them will not be known to be in various hosts in various parts of the world. There is no limitation to certain genera or other taxonomic groups of sea urchins, as would occur in evolutionary development of associations with strict specificity. Though no experimental work has been done, it seems likely that cross infection would ordinarily be easy to accomplish; nevertheless it is noteworthy that given species have characteristic faunules, and there are a few sea urchins with no faunules, facts that call for experimental investigation of the host relationships.

Another problem that calls for further investigation is the type of faunule in the same host species in different localities, data on which are meager. *Strongylocentrotus franciscanus* in California harbors *Lechriopyla mystax,* as well as ciliates of four other genera (Lynch, 1929, 1930); from *S. franciscanus* at Yaku Island, Japan, Yagiu (1935) reported only *Conchophthirus striatus;* and *S. franciscanus* examined by Powers (1936, 1937) at Acapulco, Mexico, was found to harbor "entirely different ciliates" from those on the coast of California. As regards similarity of faunules, there is the presence of *Entodiscus borealis* and *Madsenia indomita* in *Strongylocentrotus dröbachiensis* from both Sweden and the Bay of Fundy.

PROTOZOA OF TERMITES AND THE ROACH *Cryptocercus*

Flagellates have undergone no more spectacular development than is exemplified in the faunules now existing in certain termites and in *Cryptocercus.* Elsewhere in that class of Protozoa, in fact, there is nothing that is comparable to it. Many groups of the Polymastigida and all but two species of the Hypermastigida have been found only in those insects. There are also a few Protozoa of more ordinary types. Such are among flagellates *Trichomonas* and related forms, *Retortamonas, Monocercomonas, Monocercomonoides, Hexamita,* and *Chilomastix;* flagellates of these types occur only rather sparingly in higher termites and, except for *Trichomonas,* in most roaches. There are also *Nyctotherus, Balantidium,* amoebae, gregarines, and coccidia. But, in insects ancestral to modern termites and roaches, flagellates originating in the *Monocercomonas, Monocercomonoides,* and *Trichomonas* type have undergone a remarkable evolution, giving us the main polymastigote components of the faunules that today exist in lower termites and *Cryptocercus.* Hypermastigotes doubtless developed from polymastigotes, but their origin has not been traced.

A table of the classification of termites, giving the approximate number of species and the number examined, is given by Kirby (1937). About a quarter of the 1,600 termites are in the four lower families: Mastotermitidae, Hodotermitidae, Kalotermitidae, and Rhinotermitidae; three-quarters are in Termitidae. Flagellate infections in Termitidae are sparse, and the species are small and of common types. In certain Termitidae, faunules have developed consisting mainly of amoebae, which are almost completely lacking in lower termites. In lower termites there have been recognized, in examinations of less than a third of the known species, 30 genera with 133 species of polymastigotes, and 18 genera with 63 species of hypermastigotes; and certainly thorough study will reveal many more genera and species even in that third. In *Cryptocercus punctulatus* alone, Cleveland *et al.* (1934) found 9 genera of hyper-mastigotes with 20 species (only one genus and no species of which occur in termites); and 5 polymastigotes in 3 genera, including *Hexamita* and *Monocercomonoides*.

Every termite species in the lower families, so far as has been learned, has a flagellate faunule; individual termites lack the Protozoa only in certain phases of the life history, as when they are very young, imme-diately preceding and following a molt, and in certain functional repro-ductive stages. For the most part, any termite of a species, wherever obtained, will be found to have the same group of flagellate species. Sometimes one or more flagellates are absent, but uniformity in com-position of the faunules is the rule. This fact is an aid in termite sys-tematics. Identical faunules do occur in different termite species of cer-tain groups; the fact that the faunules are identical does not necessarily indicate that the hosts belong to the same species. There are often more or less well-marked differences, and this is a strong indication for specific differentiation of the hosts. The flagellates often provide a ready means of distinguishing nymphs in regions where both the termites and their faunules are known.

Individual faunules of flagellates in termites may comprise from two to ten or, occasionally, more species. Often a genus is represented in a host by more than one species. In *Zootermopsis angusticollis* and *Z. nevadensis* there are three species of *Trichonympha* (Kirby, 1932b). Nine of sixty-seven hosts of *Devescovina* contain two species. The genus *Foaina* is represented by two species in thirty-four, and by three species in three of eighty-three hosts. In *Cryptocercus punctulatus* Cleveland *et al.* (1934) differentiated seven species of *Trichonympha,* four of *Bar-bulanympha,* three of *Leptospironympha,* and three of *Saccinobaculus.* Koidzumi (1921) distinguished six species of *Dinenympha* in *Reticu-litermes speratus.*

The degree of host-specificity varies in different genera and species. Many species are known from one host only, but as more flagellate faunules become known the tendency probably will be relatively to reduce this number. Many species are known from several or many hosts. *Trichonympha agilis* probably occurs in all species of *Reticulitermes,* but has not been found in other termites. *Staurojoenina* is widespread in *Kalotermes* sensu lato, and there are few if any differences between species of different hosts. Of twenty species of the genus *Devescovina,* only nine have been found in but one host each. On the other hand, there are species with many hosts widely separated geographically. *D. glabra* has been identified in eighteen termites from Africa, Madagascar, Java, and Sumatra; *D. lemniscata* has seventeen hosts in Central and South America, the West Indies, Australia, the Pacific Islands, Africa, Madagascar, Java, and India. A unique, elaborately organized devescovinid, when first found in a Ceylon termite, was thought to be a strictly host-specific form; but it has since been found also in a termite from Australia. The small, simply organized *Tricercomitus,* which occurs in most if not all species of *Kalotermes* sensu lato, apparently is one species, *T. divergens,* in all those in which it has been studied. Another species exists in *Zootermopsis.*

Many species of termite flagellates in all groups have a present host distribution which indicates greater stability in characteristics than existed in the same period of time in the insects. Speciation has occurred in the hosts without having taken place in certain of the symbionts. That there are other termite flagellates which have evolved into different species in single hosts is probable; but we cannot designate any one as certainly rigidly host-specific. Although there are many one-host forms, the situation is such that finding any one of them in a termite, even in another part of the world, would not be astonishing. Even although there is only a single extant host, it would in no instance be unlikely that formerly existing species not directly ancestral served as hosts of the flagellate.

But whether a flagellate species occurs in one or in several host species of termites is far from being the question of greatest interest in host distribution. It has, in fact, little significance for general considerations. More important is the fact that there is limitation of certain flagellate types to certain groups of termites. That is true mainly among polymastigotes. There are also some very widely distributed flagellate types, but that only adds to the significance of the instances of strict limitation.

Trichomonas, as is not unexpected, is one of the most widely distributed forms, occurring in termites of all families, including Termitidae. *Trichonympha,* although absent from *Mastotermes* and Termitidae, not only has a wide distribution among other termites but occurs also in *Cryptocercus punctulatus.* The genus has been found represented in forty-five termites of ten genera or subgenera in three families; and among those that have been studied for detailed characteristics fourteen species have been distinguished. Various Holomastigotidae in termites are related to hypermastigotes of this family in *Cryptocercus,* although no genus is the same. Several genera are distributed widely in termites, the situation being comparable to that of *Trichonympha.*

In the distribution of the polymastigote family Pyrsonymphidae, there is a high degree of correlation with the systematic relationships of the hosts. Cleveland *et al.* (1934) extended this family (as Dinenymphidae) to include other forms than *Pyrsonympha* and *Dinenympha,* on the basis of the type of division figure and structural similarities. There are three subfamilies: Saccinobaculinae, in which the flagella are free and there is no attachment organelle; Oxymonadinae, in which the flagella are free and there is an attachment organelle, the rostellum, developed to a high degree; and the Pyrsonymphinae, in which there is a slightly developed attachment organelle and the flagella are adherent to the surface of the body for most of its length. Saccinobaculinae have been found only in *Cryptocercus punctulatus;* Oxymonadinae are known only from *Kalotermes* sensu lato, in which group they occur in most species; Pyrsonymphinae seem to be restricted to the genus *Reticulitermes.* It seems possible that evolutionary development of the groups has taken place within the confines of the host groups concerned; although it is unsafe to state that the distribution of the flagellates may not be wider than we now know it to be.

The polymastigote subfamily Devescovininae, of which *Monocercomonas* (*Eutrichomastix*) appears to be an ancestral type, is represented in all but five or six of ninety-seven species of *Kalotermes* sensu lato that have been examined. There has been a most elaborate evolutionary development in the group; but devescovinids also occur in Mastotermitidae and Hodotermitidae. They appear to be absent, however, from Rhinotermitidae and Termitidae.

The polymastigote family Calonymphidae is of particular interest to the evolutionist, and it appears to have affinities in common with the Devescovininae (Kirby, 1939). It is restricted, except for one enigmatic form that may not belong in the group, to the genus *Kalotermes sensu lato*.

Amoebae rarely occur in lower termites, but among the Termitidae they are not infrequent. Small amoebae were present in almost all species of *Amitermes* from the United States, Africa, and Madagascar that were examined by the writer; and many larger amoebae, some with unusual nuclear characteristics, were found consistently in Central American and African species of *Mirotermes* and in African termites of the *Cubitermes* group (Kirby, 1927; Henderson, MS). It is likely that further study of these amoebae will yield results significant for problems of host-specificity.

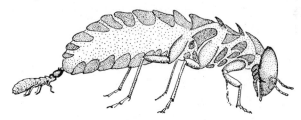

Figure 195. One-day-old nymph of *Kalotermes flavicollis*, receiving proctodaeal food from the female termite, showing the manner in which infection with flagellates takes place. (After Goetsch, 1936.)

Transmission of the flagellates of termites takes place in the active state (see Andrews, 1930). There is no evidence for true encystation, though observations by Trager (1934) and Duboscq and Grassé (1934) indicate a possibility of this in some small polymastigotes. Flagellates of most species disappear prior to each molt except the last. Infection, then, must take place not only at the beginning, but following each molt in the growth period. Refaunation takes place when termites, either naturally or experimentally defaunated, are left in contact with normally faunated individuals. Experimentally, termites can be infected by placing flagellate-containing material on the mouth parts. Under natural conditions, except for cannibalism, flagellate-containing material can ordinarily be obtained only directly from the anal opening of another termite, as the Protozoa do not survive long after deposition. Proctodaeal feeding is a common habit among termites. Goetsch (1936) has described the early infection of young nymphs of *Kalotermes flavicollis* by direct application of the mouth parts, accompanied by sucking, to the end of the abdomen of the deälates (Fig. 195).

Certain small polymastigotes are often retained through the molting period (Kirby, 1930; Child, 1934). In *Zootermopsis* this is true of the minute forms *Tricercomitus* and *Hexamastix*. At the last molt the situation differs from that in the preceding molts. Child (MS) reported that in the last molt of *Zootermopsis,* although the number of flagellates is greatly reduced, all species are carried through from the seventh instar nymph to the winged imago. Cross (MS) and May (MS) have confirmed this fact in *Kalotermes minor* and *Zootermopsis;* the shed intima of the nymph, still containing Protozoa, is retained within the gut; and the Protozoa later escape into the lumen of the imago's intestine.

In *Cryptocercus* the Protozoa are not lost at the time of molting; but then, and only then, most of them form either well-defined cysts (in *Trichonympha* and *Macrospironympha*) or resistant stages (Cleveland, et al., 1934). Flagellates are present, often in great numbers, in pellets passed in the first few days after ecdysis. Cleveland found that some pellets, passed immediately after molting, consisted mostly of Protozoa in encysted or resistant form. Reinfection of defaunated roaches took place when they were placed with molting roaches; but not, except occasionally with smaller polymastigotes, by association with other infected roaches. Proctodaeal feeding, then, does not have the same rôle in transmission in *Cryptocercus* as in termites. Cleveland could not find out the exact manner in which infection is first acquired, but thought it probable that it is by association with molting individuals. Once acquired, the faunule persists until the death of the roach.

It is probable that cross infection has not been a significant factor in determining the present distribution of flagellates in termites, below the Termitidae at least. Furthermore, the unique characteristics of almost all the flagellates, which have no close relatives except in roaches, indicate that the faunules do not to any great extent include acquisitions from other arthropods or other animals. The present distribution of the flagellates, the absence of resistant stages, and the isolated habits of termites support this opinion. If it is sustained by further studies the flagellates of termites will be shown to be easily the leading group of animals for correlative studies in phylogeny of symbionts and their hosts.

If it could be shown that there is resistance to cross infection, such that flagellates introduced experimentally from a natural host species into another one would not survive, this thesis would of course be supported. The writer (1937), however, stated that there seems to be no resistance to cross infection, basing this opinion on experiments by Light and Sanford (1927, 1928) and Cleveland *et al.* (1934). No experiments yet reported, however, have been continued long enough to warrant any definite conclusion. Unpublished experiments by Dropkin, furthermore, showed that Protozoa of *Reticulitermes flavipes, Kalotermes schwartzi,* and *K. jouteli* could not establish a physiological relationship with *Zootermopsis* sufficient to permit survival of the termite for more than fifty days in the absence of the normal faunule.

Although there has been evolutionary development of the flagellates within termites of groups that exist today, many of the types doubtless go back to ancestral insects. The genus *Trichonympha,* being found in both termites and *Cryptocercus,* may be supposed to have passed into both these insects from ancestral protoblattids (Kirby, 1937). The distribution of *Trichonympha* in termites alone would indicate its antiquity and stability (Kirby, 1932b). The existence of representatives of other hypermastigote groups in *Cryptocercus* indicates the very ancient differentiation of those flagellate types. By loss of members of the faunules here and there, together with continued but less drastic evolutionary changes, the present composition of the faunules may have originated. The flagellates were probably present in ancestors of Termitidae, but were, in the course of differentiation of those insects, dropped out. The origin of the amoebae needs to be explained; possibly they were acquired later.

ADAPTIVE HOST RELATIONSHIPS IN MORPHOLOGY AND LIFE HISTORY

GENERAL CONSIDERATIONS

Structural modifications in animals that live in association with hosts take two general forms. There are morphological changes in direct adaptation to the requirements of the habitat; and there are changes unrelated directly to that habitat, but made possible by various factors in it. In the former group, among Protozoa, is the development of organelles of fixation, though this development is not restricted to Protozoa that live in close relationship with other animals. Special adaptations may appear for nutrition. Probably also in that category is the increase of the number of flagella and the development of undulating membranes and axostyles in certain groups of flagellates. In the latter category are the reduction or loss of cilia, the reduction or loss of mouth structure, the elaborate development of the parabasal apparatus and other organelles in certain polymastigote flagellates, the complex characteristics

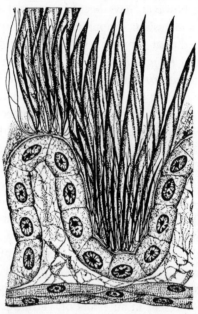

Figure 196. *Streblomastix strix* attached to the lining of the hind-gut of *Zootermopsis angusticollis*. (After Kofoid and Swezy, 1919.)

of many hypermastigotes, the elaborate morphological specialization of Ophryoscolecidae.

Organelles of fixation appear among flagellates in epibiotic dinoflagellates (Chatton, 1920; Steuer, 1928); in *Streblomastix strix,* which often is attached (Fig. 196) to the wall of the hind-gut of its termite host, by a holdfast (Kofoid and Swezy, 1919; Kidder, 1929); in *Pyrsonympha* and *Dinenympha,* which occur free in the gut lumen of *Reticulitermes* or attached by a small, simple, anterior knob (Koidzumi, 1921); and in Oxymonadinae. In the last group the holdfast, which is applied to the intima of the termite gut, is at the end of a rostellum, which may reach a relatively great length and often contains many fibrils (Kirby, 1928;

Figure 197. Fixation mechanisms in peritrichs. A, *Ellobiophrya donacis,* with ring formed by two posterior limbs applied at the ends; B, *Ellobiophrya* suspended from the bridges uniting the gill filaments of *Donax vittatus;* C, section of *Trichodina pediculus* on the ectoderm of *Hydra;* D, *Cyclochaeta* (*Urceolaria*) *korschelti* from *Chiton marginatus.* (A, B, after Chatton and Lwoff, 1929; C, D, after Zick, 1928.)

Cleveland, 1935). *Giardia* adheres by a sucking disc to the wall of the small intestine.

In some ectoparasitic dinoflagellates, the organelle of fixation is prolonged by rhizoids into the tissues of the host, and apparently nutriment is absorbed by this mechanism. In the polymastigotes in termites, fixation is only to the gut intima; there is no relationship to the epithelial cells.

In gregarines, the epimerite, which often is elaborately developed with hooks or other appendages and inserted into the cell, may serve also for absorption (see Watson, 1915).

As regards ciliates, development of fixation habits and structures in the holotrichous groups of Thigmotricha, Ptychostomidae, and Astomata is discussed below. There is among heterotrichs a well-developed fixation apparatus in *Licnophora* (Stevens, 1901; Balamuth, MS). Urceolarids have an elaborately organized scopula, a cup-like apparatus

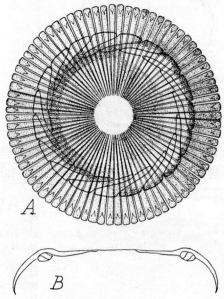

Figure 198. Fixation apparatus of *Cyclochaeta* (*Urceolaria*) *korschelti*. A, seen from above, showing radially arranged ribs bent downward in hook-like points and ring composed of overlapping, sickle-shaped individual pieces; B, cross section. (After Zick, 1928.)

supported by radially arranged ribs and a ring of denticles (Fig. 197D; Fig. 198) (see for description and illustrations, Zick, 1928, on *Cyclochaeta* (*Urceolaria*) *korschelti;* Fulton, 1923, on *Trichodina pediculus*). As has been described in *T. pediculus* on *Hydra* and *Trichodinopsis paradoxa* in *Cyclostoma elegans,* the epithelial cells of the host may be elevated into this sucker (Fig. 197C). A unique attachment mechanism is that of *Ellobiophrya donacis,* an inquiline of the gill cavity of a lamellibranch. The posterior end of the body of this peritrich is prolonged into two limbs hollowed into cups at the ends, the homologues of halves of the scopula (Fig. 197A). The two cups are sealed firmly together, so that the limbs form a closed ring by which the ciliate is suspended (Fig. 197B) from the framework of the gills (Chatton and Lwoff, 1923b, 1929).

In Protozoa that live in association with other animals as hosts, the developmental cycle must be adjusted to the requirements of the habitat. This is so arranged as to insure transmission of the organism from one host to another; production of a sufficient number of infective forms so that the likelihood of some reaching a place where development can continue is not too small; protection of the organism, if necessary, in the period when it is out of its host; and often correlation with the life cycle and habits of the host, so that escape from one host and infection of another can take place. The situation is most complex in heteroxenous forms, in which the life cycle is shared between two different species of animals, and is correlated with the bionomics of each of them. Cyclic adaptation has perhaps achieved its most perfect development when there is a regular and direct transmission to the next generation through infection of the eggs or embryos. There are not many instances in Protozoa of this last method, which is so perfectly exemplified among the cyclic endosymbionts of insects (Buchner, 1930; W. Schwartz, 1935). *Nosema bombycis* invades the eggs of silkworms, but this may be incidental and is not the only method of transmission. Other instances occur among heteroxenous Sporozoa in the invertebrate host in *Babesia* and *Karyolysis* (see also Lavier, 1925).

The developmental cycle and methods of transmission have been considered widely in textbooks and in a general way by many authors, including Hegner (1924) and Grassé (1935). The situation in certain epibiotic Protozoa is also of considerable interest; and in that connection accounts are given below of the holotrichs *Conidiophrys pilisuctor* and apostomatous ciliates.

THIGMOTRICHA

Chatton and Lwoff (1922c) proposed the name Thigmotricha for a group (suborder) of holotrichs including the families Ancistridae, Hypocomidae, and Sphenophryidae. Most of these ciliates are inquilines, commensals, or parasites on the gills or palps of molluscs, though some occur on Protozoa or other invertebrates. They are provided with thigmo-tactic cilia; and they show a series in the evolution of thigmotacticism, in the course of which there is developed a penetrative and absorptive organelle, and in the regression of body ciliature. Chatton and Lwoff (1923a) described *Thigmophyra,* which was later (1926) placed in a fourth family, Thigmophryidae. *Thigmophrya,* it was stated, closely resembles *Conchophthirus;* but, unlike *Conchophthirus,* it possesses a well-defined thigmotactic area identical with that of other Thigmotricha. Kahl (1934) summarized the characteristics of the suborder Thigmo-tricha and included in it the Conchophthiridae, which he had formerly (1931) treated in the suborder Trichostomata, and which Chatton and Lwoff (see 1937) evidently did not intend to include in their group.

Though Calkins (1933) considered adaptations to parasitism in the thigmotrichs as a group, he included most of the genera in the Tricho-stomata, including the Ancistrumidae which Kahl (1931) had put in the Hymenostomata. Calkins, on the other hand, separated *Hemispeira* from other Ancistrumidae, putting it in the Hymenostomata. Whether the Thigmotricha constitute a homogeneous group may be questionable (Fauré-Fremiet, 1924, p. 7); but for consideration of the ethological relationships and adaptations to symbiotic existence, the object of interest in this account, it is convenient to treat them together.

Most Conchophthiridae occur in the mantle cavity of Pelecypoda, both marine and fresh-water species. *Andreula antedonis* (André) Kahl (=*Conchophthirus antedonis* André) occurs abundantly in the alimen-tary canal of a crinoid echinoderm; and Uyemura (1934) described as *C. striatus* a ciliate in the intestine of several sea urchins of Japan. *Myxo-phyllum steenstrupii* (Stein) lives in the slime covering the body of a variety of land pulmonates. The species of *Morgania* Kahl and all ex-cept the one species of *Conchophthirus* Stein (the original spelling by Stein, 1861, not *"Conchophthirius"* as given by Strand, 1928) men-tioned above are restricted to bivalves.

The most detailed studies of the genus *Conchophthirus* are contained in several articles published in 1933-34 by Kidder and by Raabe. Uye-mura (1935) found three species in great abundance in a fresh-water mussel of Japan, *Anodonta lauta.* There is no doubt that in all parts of the world certain lamellibranchs will be found abundantly infected with these commensals. Only a beginning has been made in their study, as in that of all the Thigmotricha, insofar as a knowledge of geographical distribution and host-specificity is concerned.

In the mantle cavity of the hosts, some species are not localized, whereas others are. Kidder (1934a) found *C. curtus* and *C. magna* on all exposed surfaces and also swimming freely in the mantle fluids; *C. anodontae,* on the other hand, he found to be invariably localized on the nonciliated surface of the palps. The cilia of the flat left side (left if, with Kahl and Raabe, we consider the flattening to be lateral; according to De Morgan and Kidder, it is dorsoventral, and the attachment is by cilia of the ventral surface) are thigmotactic. The thigmotactic area usually covers the whole broad side, but in *C. discophorus* there is more specialized adhesive apparatus, a circular, sharply outlined area, which occupies only part of the left side, is markedly concave, and is provided with differentiated cilia (Raabe, 1934b). *C. discophorus* swims slowly, and often fastens itself firmly by the thigmotactic region. *C. mytili* (Fig. 199A) also swims about or clings firmly to surfaces (Kidder, 1933a). *C. anodontae* on *Elliptio complanatus* seems to be most markedly thigmotactic (Kidder, 1934a), remaining quiet, attached to the surface of the palp.

Kidder (1933a) found the food vacuoles of *C. mytili* (*Morgania mytili,* according to Kahl, 1934) to contain plankton organisms, including algae, and sperm cells of the host. *C. caryoclada* (*Morgania caryoclada,* according to Kahl) contained mostly algae (Kidder, 1933d). Other species contained algae, bacteria, and sloughed-off epithelial cells. The relationship appears to be simple commensalism, but Kidder (1934a), finding only well-preserved epithelial cells in the food vacuoles of *C. magna,* was "a little in doubt as to its purely commensal rôle."

Kidder (1934a) remarked that there is a fair degree of host-specificity. In nature certain species are characteristic of certain molluscs; and the faunules may differ, even though in nature the hosts are very closely associated. Rarely there are as many as three species in one host. A number of species have been found in only one or a few related hosts, but this may be a consequence of the relatively few examinations. A cosmopolitan distribution is characteristic of such species as *C. curtus,* reported from various fresh-water clams in Europe, the eastern United States, and Japan. *Morgania mytili* is a commensal of *Mytilus edulis* in various localities on both sides of the North Atlantic.

In the family Thigmophryidae, *Thigmophrya bivalviorum,* which occurs on the gills of the marine pelecypods *Mactra solida* and *Tapes pullastra,* has a thigmotactic region reduced to an elliptical area in the anterior fifth of the body (Chatton and Lwoff, 1923a). The movements of the cilia of this area are not synchronous with those of the rest of the body. The ciliate swims in the mantle cavity or fixes itself to the gills.

The family Ancistrumidae is large and diverse. In general, the ciliates are more sedentary than those previously considered in the order Thigmotricha and the thigmotactic area is still more restricted. Although the most frequent habitat is the mantle cavity of Pelecypoda, other molluscs as well as members of other phyla of invertebrates serve as hosts for species of the family. Probably, however, the original hosts were Pelecypoda.

The two principal genera are *Ancistruma* Strand, 1928 (given incorrectly as 1926 by Kahl and Kidder) (Fig. 199B, C) and *Boveria* Stevens, 1901, but there are many others: *Eupoterion* MacLennan and Connell, *Ancistrina* Cheissin, *Ancistrella* Cheissin, *Plagiospira* Issel (Fig. 199D), *Ancistrospira* Chatton and Lwoff, *Proboveria* Chatton and Lwoff, *Tiarella* Cheissin, *Hemispeira* Fabre-Domergue (Fig. 200C), *Hemispeiropsis* König (Fig. 200A, B). Kahl (1934) put into the family, though doubtfully, two ciliates parasitic in *Littorina, Protophrya ovicola* Kofoid and *Isselina intermedia* Cépède.

The Ancistrumidae possess more or less conspicuous peristomal cilia; often these rows constitute a prominent fringe. In typical forms the organisms adhere to the surfaces on which they live by the thigmotactic cilia in a tuft at the anterior end. *Ancistrella choanomphthali,* however, adheres to the gills by its entire concave, ventral surface (Cheissin, 1931). In the Ancistrumidae an evolutionary series is apparent in the shifting posteriorly of the mouth and of the peristome, which becomes spiraled. Chatton and Lwoff (1936b) suggested that the Ancistrumidae constitute so extraordinarily homogeneous a family that we may consider that there is only one genus, subdivided into subgenera, an opinion that expresses the homogeneity, though possibly the conclusion that there should be only one genus is not sound taxonomically. These authors remarked that the characteristics separating the genera or subgenera are purely quantitative, consisting of more and more accentuated retrogradation of the mouth and prostomal ciliary lines from the anterior half of the body to the posterior end (Fig. 200K-N).

Other habitats than the mantle cavity of Pelecypoda have been adopted by various species of *Ancistruma* and *Boveria,* as well as by members of other genera. Thus Issel (1903) found *A. cyclidioides* on certain chitons and gasteropods (*Natica heraea*) as well as on Pelecypoda; and he described *A. barbatum* solely from gasteropods of the genera *Fusus* and *Murex.* Adaptation to these hosts is, as was stated above, probably secondary. In a similar manner, one species of *Boveria,* the type species *B. subcylindrica* Stevens, is attached to the membrane of the respiratory tree of the holothurian *Stichopus californicus* (Stevens, 1901). So similar to this, however, that it has been classified as a variety of the same species, *B. s.* var. *concharum* Issel, is a *Boveria* that occurs on the gills of ten of fourteen Pelecypoda that harbor Ancistrumidae at Naples (Issel, 1903). *B. labialis* lives in the respiratory trees of holothurians as well as on the gills of a clam (Ikeda and Ozaki, 1918).

Eupoterion pernix, which has many characteristics of a species of *Ancistruma,* inhabits the intestine of the limpet *Acmaea persona* (MacLennan and Connell, 1931). The aberrant *Hemispeira asteriasi* Fabre-Domergue (1888) and *Hemispeiropsis antedonis* (Cuenot, 1891) occur on echinoderms, the former on the dermal branchiae of a starfish, and the latter on the pinnules of a crinoid (Cuenot, 1894; König, 1894). *Protophrya ovicola* Kofoid occurs upon the surface of the egg capsules in the brood sac of the gasteropod *Littorina rudis;* and *Isselina intermedia* is found in the mantle cavity of *Littorina obstusata.* The two latter species, at least *Protophrya,* are more truly parasitic than other Ancistrumidae, and they have undergone some retrogressive changes.

For the most part, Ancistrumidae feed on bacteria, diatoms, and other material extracted from the currents of water. Issel (1903) noted that two bivalves constantly rich in the ciliates, *Capsa fragilis* and *Tellina exigua,* live under conditions most suitable for offering their inquilines copious food. They occur in calm, muddy water, rich in organic substances. The diet of plankton organisms may be supplemented by sloughed-off epithelial cells, as noted by Stevens (1901) in *Boveria subcylindrica* and Pickard (1927) in *B. teredinidi.* The account by Ikeda and Ozaki (1918) of tissue invasion by *B. labialis* is not acceptable without corroboration. The changes said to be undergone by the encysted ciliate within the tissue are bizarre.

Protophrya ovicola in the brood sac of *Littorina* has a destructive chemical action upon the eggs (Kofoid, 1903) and a teratogenic action on the embryo (Cépède, 1910). The parasites do not act directly on the embryos, but on the medium, which exerts an injurious effect on embryos in the early stages. Abnormal embryos result, in which the shell is more or less unrolled; not only may the shell be misshapen, but the cells that secrete the shell may fail to function normally.

The Hypocomidae (Fig. 199E-I) are true parasites, and occur mostly in marine and fresh-water bivalves and snails. There is no mouth, but the anterior end is provided with a short retractile tentacle. Normally the ciliates are attached to the gills or skin of the mollusc, the tentacle being embedded in an epithelial cell. The parasites obtain nutriment by extracting the contents of the cells to which they are attached, the tentacle combining suctorial functions with those of attachment. The tentacle continues in a tubular structure, extending more or less deeply into the cytoplasm. In many Hypocomidae a fine inner canal has been observed extending from the apex of the tentacle into the deeper cytoplasm. According to Chatton and Lwoff (1922c), this adherent organelle is derived from structures of Ancistrumidae, where it is indicated in *Ancistruma mytili* and is well developed in *A. cyclidioides*.

In relation to the attached parasitic condition of hypocomid ciliates are the regression of the mouth and peristomal ciliature and the reduction of the general ciliature. The former structures have for the most part already disappeared. There is no mouth, but Chatton and Lwoff (1924) stated that in some genera there are residual segments of the adoral ciliary zone.

In reduction of the general ciliature there is in Hypocomidae a well-integrated series. The body of *Hypocomagalma dreissenae* Jarocki and Raabe, 1932, is covered with cilia except for a small ventroterminal area. *Ancistrocoma pelseneeri* Chatton and Lwoff, as figured by Raabe (1934a), has a larger cilia-free area, occupying a large part of one side of the body, and an anterolateral peristomal fringe. Perhaps the less firm fixation of these forms is also a phylogenetically primitive character. Raabe stated that *Ancistrocoma* adheres rather weakly to the gills of its host and separates readily. *Hypocomagalma* swims more rapidly than some other ciliates of the group. Reduction of the ciliature continues through *Hypocomides* Chatton and Lwoff and *Hypocoma* Gruber, in which it occupies the inner area of the ventral surface of the body. In *Heterocineta* (=*Hypocomatomorpha*) *unionidarum* Jarocki and Raabe

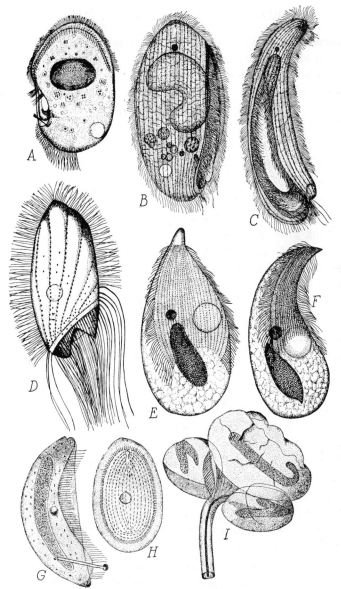

Figure 199. Thigmotricha. A, *Morgania* (*Conchophthirus*) *mytili* from *Mytilus edulis;*
B, C, *Ancistruma mytili* from *Mytilus edulis,* B, dorsal view, C, lateral view, tuft of
straight tactile cilia near anterior end; D, *Plagiospira crinita* from *Loripes lacteus;* E, F,
Hypocomina carinata from *Mytilus edulis;* G-I, *Hypocoma parasitica:* G, lateral view,
tentacle and tubular structure below, H, ventral view, I, two hypocomids attached to
Zoothamnium. (A, after Kidder, 1933a; B, C, after Kidder, 1933c D, after Issel, 1903;
E, F, after Raabe, 1934a; G-I, after Plate, 1888.)

and *Hypocomatidium sphaerii* Jarocki and Raabe, from the gills of fresh-water mussels, reduction is far advanced to a thigmotactic zone restricted almost entirely to the anterior half of the ventral side. Cilia are not lost altogether in known Hypocomidae.

Syringopharynx pterotracheae, which lives on the gills of the heteropod *Pterotrachea coronata* either swimming free or fixed to epithelial cells by the rostrum (Collin, 1914), was included by Kahl (1934) among the parasitic gymnostomes. Probably, however, this is a hypocomid ciliate, one with a general body ciliature like *Ancistrocoma* and *Hypocomagalma. Parachaenia myae,* described from *Mya arenaria* by Kofoid and Bush (1936), may also be a hypocomid, although attachment to the cells of the gills of its host was not described. Chatton and Lwoff's statement (1926, p. 351) about the prolongation of the anterior individual in a spur covering the dorsal anterior region of the posterior individual in binary fission of *Ancistrocoma pelseneeri* is in exact agreement with the division process of *Parachaenia myae.* Another point of agreement with *Ancistrocoma* is the type of conjugation. In the shape of the ciliates, the unique attachment of the conjugants by the posterior ends, and the shape and arrangement of the nuclei of *P. myae,* there is almost complete agreement with *A. pelseneeri* as figured by Raabe (1934a).

Among the many genera of Hypocomidae, not all of which can even be named here, the number of which we may expect will be markedly reduced with further study, only the species of *Hypocoma* (Fig. 199G-I) do not occur on bivalves or snails. They are parasites of Protozoa, and are discussed elsewhere in this book (p. 1083).

Hypocomidae are host-specific in marked degree. They are obligatory parasites on certain individual or closely related molluscs, and do not readily infect other molluscs. Jarocki (1935) found that *Heterocineta janickii,* placed free-swimming into an aquarium with various molluscs, would attach only for periods of from fifteen to eighteen hours in the absence of its natural host *Physa fontinalis.* When various molluscs, including *Bithynia tentaculata,* were put together in an aquarium, *Heterocineta krzysiki,* though abundant on the body of *Bithynia,* did not infect any other species. Sometimes two hypocomids are present on the same host, and Raabe (1934a) noted that there seems to be a tendency to inhibition by one parasite of the development of the other.

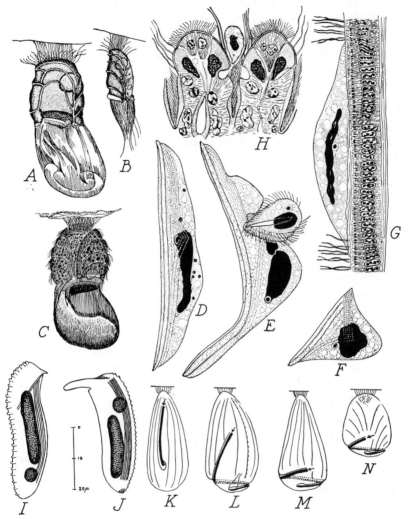

Figure 200. Thigmotricha. A, B, *Hemispeiropsis comatulae* from *Antedon* (*Comatula*) *mediterranea;* C, *Hemispeira asteriasi* from *Asterias glacialis;* D-H, *Sphenophrya dosiniae* from *Dosinia exoleta:* D, adult, E, budding individual, F, young individual, G, longitudinal section of *Sphenophrya* on branchial filament, H, transverse section showing ciliate in furrow between branchial filaments; I, J, *Gargarius gargarius* (*Rhynchophrya cristallina*) from *Mytilus edulis;* K-N, series of diagrams showing retrogradation of peristome in Ancistrumidae: K, *Ancistruma,* L, *Proboveria,* M, *Boveria,* N, *Hemispeira.* (A, B, after König, 1894; C, after Wallengren, 1895; D-H, after Chatton and Lwoff, 1921; I, J, after Raabe, 1935; K-N, after Chatton, 1936.)

In connection with host-specificity among hypocomids, Jarocki's observation (1934, 1935) that *Heterocineta janickii* is also a facultative parasite on the oligochaete *Chaetogaster limnaei* is of considerable interest. The oligochaete is usually present as an inquiline in the mantle cavity of *Physa fontinalis* and other snails. The hypocomids are almost always present on the snails; and they also infect almost all the oligochaetes, attaching to various parts of the body and inserting the suctorial tentacles into the hypodermal cells. Parasite-free worms quickly became infected if brought into contact with ciliates either in or out of the mantle cavity. Parasite-free *Physa* became infected if parasitized oligochaetes were introduced into the mantle cavity. The worms pass freely from one host to another, and thus facilitate the spread of the infection. *Chaetogaster limnaei* in other snails became facultatively parasitized by their specific hypocomids; but *Heterocineta* species could not be introduced into unnatural hosts on the oligochaetes. To two other species of *Chaetogaster, Heterocineta janickii* became attached temporarily, but soon dropped off.

The Sphenophryidae (Fig. 200D-H)) all occur on the gills of marine lamellibranchs. They are sedentary, immobile, and nonciliated in the adult phase. They are not true parasites; the relationship as defined by Chatton and Lwoff (1921) is "inquilinism complicated by phoresy." The ciliates are rather large, mostly flattened laterally, and adhere to the surface by a long ventral edge (Fig. 200G). Sometimes they adhere in a very precise and constant position, as *Sphenophrya dosiniae* in the furrows separating adjacent branchial filaments of *Dosinia exoleta* (Fig. 200H). There is no mouth opening, and the ciliates apparently feed osmotically; but Mjassnikowa (1930a) found evidence that *S. sphaerii* ingests cells of the gill epithelium. She may, however, have misinterpreted the nature of certain cytoplasmic spherules. Reproduction is by development of motile buds. Cilia develop from the infraciliature, which consists of a few rows of granules that are present in the vegetative individual.

An unusual sphenophryid is *Gargarius gargarius,* described by Chatton and Lwoff (1934a) from *Mytilus edulis* at Roscoff. *Rhynchophrya cristallina* from *M. edulis* in the Baltic Sea, of which a more complete account was later given by Raabe (1935), is evidently the same ciliate (Fig. 200I, J). Along one surface are two longitudinal, parallel, comb-like structures, producing a plaited surface suggesting *Aspidogaster* to Raabe. At the anterior end is a beak-like process, which is embedded in the cells of a filament, whereas the plaited surface adheres to another filament. Nutrition is osmotic.

PTYCHOSTOMIDAE

The holotrichous ciliates of the family Ptychostomidae are considered by some to be related to the Thigmotricha, if indeed they do not belong in that group. Beers (1938b), following Jarocki (1934), gave preference to Hysterocinetidae Diesing, 1866, as having priority over Ptychostomidae Cheissin, 1932. There is, however, no general recognition of priority in family names, but these are based on the name of the type genus. In this instance it seems that *Ptychostomum* Stein, 1860, the first described, best known, and largest genus, should not be supplanted as the type by *Hysterocineta* Diesing, 1866. Rossolimo (1925) suggested that the Thigmotricha and *Ptychostomum* represent two parallel evolutionary series, derived from the same group of free-living organisms, but adapted in somewhat different ways to attachment and the requirements of sedentary life.

The family Ptychostomidae now includes some eighteen species, of which nearly half are from Lake Baikal. The ciliates occur in the intestine of fresh-water oligochaetes, except for *Hysterocineta eiseniae* described by Beers (1938b) from a terrestrial oligochaete, and three species from the intestine of gasteropods. In oligochaetes, Ptychostomidae are associated with astomatous ciliates. Cheissin (1932) remarked that there is a tendency for Astomata to be located more anteriorly in the intestine, whereas Ptychostomidae occur in the posterior part. Beers found that 90 percent of *H. eiseniae* are localized in the third quarter of the gut of *Eisenia lönnbergi;* and an astomatous ciliate occurs more anteriorly. Cheissin (1928) stated that *"Ladopsis" (=Hysterocineta) benedictiae* is found in the mantle cavity (?) of *Benedictia baikalensis;* later (1932) he wrote that that ciliate occurs mostly in the intestine and enters the mantle cavity seemingly only accidentally.

It may be expected that study of fresh-water oligochaetes in various parts of the world will greatly increase the size of the group. There may then be a tendency to greater subdivision, but at present there are only two genera, *Ptychostomum* and *Hysterocineta, Lada* Vejdovsky being a synonym of the former and *Ladopsis* Cheissin of the latter. According to Jarocki (1934), the points of distinction are the shape of the macronucleus, the size and position of the micronucleus, and the position of the contractile vacuole. Whether such points are sufficient for distinction of genera is questionable.

The power of attachment, achieved by a sucker-like organelle, is marked (Fig. 201D). Miyashita (1927) found that *Ptychostomum tanishi,* when observed in the dissected-out gut, was in part attached to the inner surface of the intestine and in part swam freely in the fluid.

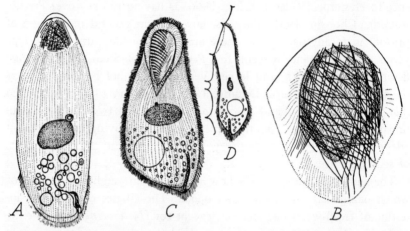

Figure 201. Ptychostomidae. *A, Ptychostomum rossolimoi* from *Limnodrilus newaensis;* B, skeletal fibrils of the sucker area of that species; C, *Ptychostomum chattoni* from *Lumbriculus variegatus;* D, *Pt. chattoni* adherent by sucker to the intestinal wall of the annelid. (A, B, after Studitsky, 1930; C, D, after Rossolimo, 1925.)

When put into water, the ciliates eventually attached themselves to surfaces. Heidenreich (1935) described strong attachment by the sucker in *P. rhynchelmis,* folds of the intestinal wall being drawn into it. Beers, however, remarked that in *H. eiseniae* the sucker appears to be only weakly functional, and most specimens were swimming freely in the lumen of the intestine. In relation to the sedentary position, there is more or less marked dorsoventral flattening of the elongated body.

According to Studitsky (1930), the first step in the development of the fixation apparatus is represented by the horseshoe-shaped, non-ciliated area in the anterior part of the ventral surface of *Ptychostomum saenuridis,* in which no skeletal structures have been described. The next step is the strengthening of this area by skeletal fibrils, consisting, first, of a set of longitudinal fibrils, and second, of a set of fibrils crossing these. In further development there is differentiation and strengthening of the two systems. The skeletal fibrils form an irregular network in the floor of the sucker. The sucker itself is a simple concavity in *P. rossolimoi* (Fig. 201A, B) and some other species; sometimes there are a few rows of cilia on an elevated area in its floor. In other species, probably more advanced in the evolutionary series, the border of the sucker is a lip-like elevation. In some forms the sucker is circular (*P. tanishi*), but often it is pointed posteriorly (*P. chattoni,* Fig. 201C) or has an opening (*P. wrzesniewskii*) or indentation (*P. elongata*). In *H. eiseniae* it is V-shaped.

Cheissin (1932) described myonemes in addition to the skeletal fibrils. Heidenreich (1935) stated that the fibrils of the sucker, as described by authors, are contractile, thus being not skeletal structures but myonemes. Beers, however, found no myonemes in the species he studied, and concluded that all the fibrils have a supporting function. *P. rhynchelmis* has, Heidenreich stated, unlike other species, a skeleton in the sucker. The sucker is bordered by two sickle-formed skeletal bows, forming a ring open posteriorly, and each sickle is prolonged posteriorly in a handle. The two handles form a canal, the neck of which is surrounded by three or four myoneme bands.

In Ptychostomidae the oral apparatus is situated at what has been regarded as the posterior end. Beers described a shallow, transverse peristomal groove, bordered by lips bearing cilia, and a small cytostome leading into a short, tubular cytopharynx. He found no food vacuoles and no ingestion of ink particles, and consequently concluded that the mouth is non-functional. In this group of ciliates the feeding apparatus is in process of reduction. It functions in some species, possibly together with saprozoic nutrition; in others it has, though still present, little rôle to play in nutrition.

ASTOMATA

The suborder Astomata is systematically heterogeneous, lacking phylogenetic unity, as Kahl (1934) remarked. Cépède (1910) himself noted that fact. The group includes many forms that lack complete descriptions. There is no systematic unity to be obtained by bringing together forms according to the negative characteristic of absence of a cytostome, as ciliates of quite divergent relationships may have suffered regression of that structure. Since a cytostome is not absent in free-living ciliates, even of the most primitive type, it seems most likely that there has been regression, rather than that lack of it is a primitive condition. Furthermore, we have to take account of the fact that oral structures may occasionally have been overlooked.

Cépède (1910) removed the opalinids from the group, and since then several other genera have been excluded. The Ptychostomidae have gone; in them the mouth structures are not absent, as at first supposed. *Protophrya* appears to have affinities with the Thigmotricha. Chatton and Lwoff (1935) stated that *Metaphrya sagittae* from the coelomic cavity of *Sagitta* sp. is an apostome; and *Kofoidella eleutheriae* may also be a foettingeriid. The description of *Kofoidella,* from the gastrovascular canals of a medusa, is too inadequate for systematic purposes; but so far as it goes relationship to *Pericaryon,* from the gastrovascular canals of *Cestus veneris,* is not excluded. The macronucleus is described as compact and central, and quite variable in size. The macronucleus of *Pericaryon cesticola* is reticular and peripheral; but it is not impossible that Cépède, who stated that the supposed macronucleus of *Kofoidella* could be demonstrated (by Maupas) only after treatment with acetic acid, was referring instead to the trophic mass.

The greater part of the Astomata inhabit the intestine of Oligochaeta. In the table of distribution given by Cheissin (1930), records are given of 69 species in the intestine of Oligochaeta, and of 41 elsewhere. Among the latter, omitting *Kofoidella* and *Chromidina,* the affinities of which are doubtful, there are 36 species, and 12 of them occur in polychaetes and in the coelom and gonads of oligochaetes. Heidenreich (1935) added 11 species in the intestine of oligochaetes, and Beers (1938a) added one. With about 75 percent of the known species in the intestine of oligochaetes and polychaetes, and the affinities of many of those found elsewhere doubtful, we may correctly consider the Astomata to have a close ecological relationship to that group of animal hosts.

Cheissin (1930), examining invertebrates of Lake Baikal for Astomata, found none in many Turbellaria, molluscs, and polychaetes, and only a few amphipods had *Anoplophrya* in the body cavity. Of 24 species of oligochaetes examined, all the Lumbriculidae and most of the others had one or more species of Astomata; three-quarters of 2,062 individuals were infected. Some hosts have several species of the ciliates; in one there are as many as 7, but all may not be present at the same time. Heidenreich (1935) examined worms, mostly oligochaetes and turbellarians, collected in the vicinity of Breslau. He noted that very few ciliates are found in worms from flowing water, presumably because of the fact that cysts are carried away.

There is a certain amount of host-specificity in the group. Hoplophryidae and Intoschellinidae appear to be restricted to annelids. Many species have been described from one host only, but some occur in many hosts. Cheissin (1930) found *Radiophrya hoplites* Rossolimo and *Mesnilella rostrata* Rossolimo in most of the Lumbriculidae; the former occurred only in worms of that family, the latter was found also in an enchytraeid. There has been a high degree of differentiation of species and genera in the Astomata, although there are comparatively few characteristics in which that differentiation can be exhibited.

There are two large groups of the Astomata from annelids, those without and those with skeletal structures. The former constitute the family Anoplophryidae; the latter were put by Heidenreich (1935) into the two families Hoplitophryidae and Intoshellinidae. The skeletal structures are differentiations of the ectoplasm or endoplasm, or of both, in the form of resistant, refractile, and stainable rods, hooks, rays, or fibrils. They are completely renewed at division. According to the scheme of development outlined by Heidenreich, the simplest form is a small, ectoplasmic, skeletal plate with a short tooth that scarcely extends from the pellicle (*Eumonodontophrya kijenskiji,* Fig. 202K). The plate elongates to the rod-like spicule characteristic of *Hoplitophrya,* in some species of which there is a point projecting from the anterior end of the body. In *H. fissispiculata* (Ch.) (=*Protoradiophrya fissispiculata* Cheissin) the spicule is divided in a narrow-angled cleft in its posterior part. This is an approach toward the V-shaped ectoplasmic skeletal element of *Radiophrya* (Fig. 202 C). The latter is usually provided with an apical point, which projects from the body surface (Fig. 202, J). Sometimes a second element, a hook or tooth attached to the central, basal part of the arrowhead and projecting free of the body, is present. This pointed tooth typically projects backward at an angle. In *R. hoplites,* Cheissin (1930) observed that the tooth is capable of movement. There

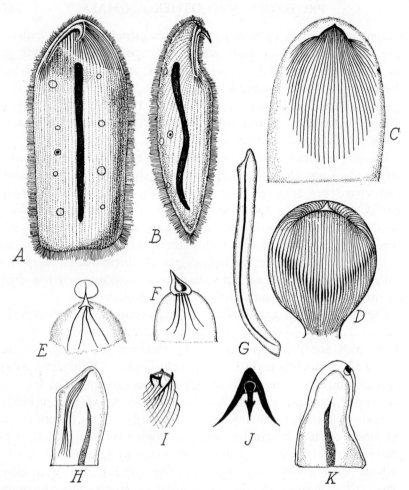

Figure 202. Skeletal structures and attachment organelles in Astomata. A, B, *Metaradiophrya asymmetrica* from the oligochaete *Eisenia lönnbergi*; C, *Radiophrya tubificis* from *Tubifex tubifex*; D, *Mrazekiella costata* (anterior end) from *Rhynacodrilus coccineus*; E, *Buchneriella criodrili* from *Griodrilus lacuum* (the sphere around the projecting spine is normally formed in the tissue cell to which the ciliate attaches); F, *Maupasella nova* from *Eisenia foetida* and species of *Lumbricus*; G, *Mesnilella fastigata* from *Enchytraeus möbii*; H, *Mesnilella maritui* from oligochaetes; I, *Intoschellina poljanskyi* from *Limnodrilus arenarius*; J, *Radiophrya lumbriculi* from *Styloscolex* sp. and *Lamprodrilus* sp., attachment apparatus; K, *Eum monodontophrya kijenskiji* from *Tubifex inflatus*. (A, B, after Beers, 1938a; C, after Rossolimo and Perzewa, 1929; D, E, F, after Heidenreich, 1935; G, after Cépède, 1910; H-K, after Cheissin, 1930.)

is a third skeletal element in *Radiophrya;* more or less numerous ecto-plasmic skeletal strands attached along the entire inside of the V and extending posteriorly on the ventral region of the body. These spread laterally and, although usually restricted to the anterior part, may reach almost to the posterior end. Skeletal elements of similar type are present in *Mrazekiella* (Fig. 202D) and *Metaradiophrya.*

The attachment organelle of *Metaradiophrya asymmetrica,* described by Beers (1938a) from the terrestrial oligochaete *Eisenia lönnbergi* in North Carolina, consists of a shaft embedded in the ectoplasm of the anterior part of the body and a stout projecting hook, which Beers found to be immovable (Fig. 202A, B). The left half of the V-shaped element, such as is present in *Radiophrya,* is lacking. From the attachment organ-elle, skeletal fibrils radiate in the ventral ectoplasm, very close to the surface, the principal group originating near the base of the hook, fol-lowed by an area of the shaft devoid of fibers, then a group of a few short fibers at the posterior end. The asymmetrical arrangement of the fibrils, which is contrasted with the bilaterally symmetrical systems of *M. falcifera* and *M. lumbrici,* is the source of the specific name.

In other forms the skeleton is completely or partly endoplasmic. In *Maupasella* (Fig. 202F) the side arms of the V-shaped element have become reduced and the point has developed into a prominent, project-ing, pointed organelle, that serves for fixation. The longitudinal rays have become endoplasmic. Related to *Maupasella* is *Buchneriella crio-drili* Heidenreich, which has a particularly well-developed movable spine (Fig. 202E). This penetrates into cells of the intestinal epithelium, anchoring the ciliate firmly. In many ciliates torn from attachment, the end of the spine was surrounded by a globule of differentiated host tissue.

In *Mesnilella* the V-formed element is lacking, and the longitudinal rays are endoplasmic and often reduced in number. A series may be ar-ranged from a many-rayed condition (Fig. 202H) to that in which there is only one spicule, reaching almost the full length of the body (Fig. 202G).

Intoschellina has a different type of skeletal apparatus (Fig. 202I). It is an open ring in the ectoplasm surrounding the apex of the body. From this ring three short spines project above the body surface an-teriorly, and three extend in the ectoplasm posteriorly. Two of the posterior spines are short; one, located at one end of the ring, is rela-tively long.

A more or less marked concavity is present on one side near the anterior end of many species of the Astomata mentioned above. This concavity, often supported by skeletal fibrils, may fit easily on the convex surface of the intestinal folds. It is not differentiated as a true sucker, however. The projecting spines and hooks of the skeletal apparatus of many forms serve definitely for attachment. These are adaptations to the requirements of the habitat, but it is a question whether the skeletal apparatus as a whole can be considered to be strictly a fixation apparatus.

In the Astomata of the family Haptophryidae, there is a true sucker. If there is a systematic unity in the family, the wide separation of the two groups of hosts, Turbellaria and Amphibia, is noteworthy. The species that have spicules, *Lachmannella* without and *Steinella* with an anterior acetabulum-like concavity, occur only in various Turbellaria, and since there are no complete and modern descriptions, comparison with other Haptophryidae is difficult to make. The several species of *Haptophrya* are better known, especially *H. michiganensis* Woodhead, 1928, as described by Bush (1933, 1934). *H. gigantea* has been found in certain European and Algerian frogs and toads; and *H. michiganensis* in several American salamanders and one frog. Rankin (1937) reported the latter species from 5 of 19 species of North Carolina salamanders, in incidence of 6.3-21.4 percent; and Hazard (1937) found it once in *Plethodon cinereus* in Ohio, which species Rankin had reported negative. Hazard also found the ciliate in 20 percent of *Rana sylvatica* in Ohio. There may be some difference in infection in the same host species in different geographic regions. Cépède (1910) noted that *R. esculenta* harbors *H. gigantea* in Algeria, but lacks it in Northern France. Rankin found what he considered to be *H. gigantea,* together with *H. michiganensis,* in a few of the many *Plethodon glutinosus* studied. Meyer (1938) reported *H. virginiensis,* a new species, in *R. palustris.*

The occurrence in a turbellarian of a species often put into the same genus, *Haptophrya,* is of interest from the standpoint of host-specificity. The species *planariarum* occurs in various marine and fresh-water Turbellaria (Cépède, 1910), principally in *Planaria torva.* Bishop (1926) found it in 70 percent of that triclad at Cambridge. Finding certain differences from the forms in vertebrates, she kept it in the genus *Sieboldiellina;* but Cheissin (1930), followed by Bush (1934), did not recognize any generic differences. Speculation on the origin of this diversity of hosts would, with our present information, be vain.

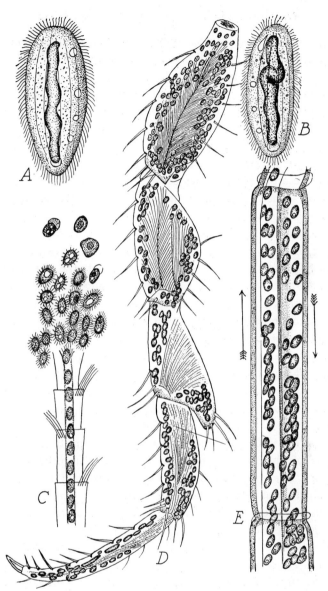

Figure 203. *Anoplophrya* (*Collinia*) *circulans* in *Asellus aquaticus*. A, B, large individuals showing nuclei and pulsating vacuoles; C, terminal portion of antenna broken at end, ciliates enclosed in blood vessel and escaping into water, on contact with which some disintegrate; D, thoracic leg containing ciliates; E, segment of the basal part of an antenna, ciliates carried in opposite directions in the currents of blood. (After Balbiani, 1885.)

Astomatous ciliates that occur in other hosts than annelids, except the Haptophryidae and Chromidinidae, were placed by Cheissin (1930) in the family Anoplophryidae. Heidenreich (1935) separated many of these from that family, without giving them other systematic assignment. So separated by him were the species of the genus *Collinia* Cépède, which occur in the hemocoele of amphipods and isopods. According to Cheissin (1930) and Summers and Kidder (1936), *Collinia* is a synonym of *Anoplophrya;* so that members of that genus occupy very diverse situations. There are several species of the ciliates which are evidently not uncommon in asellids and gammarids. Summers and Kidder believed that there is a relatively strong host-specificity.

When Balbiani (1885) described *Anoplophrya circulans* (Fig. 203), he stated that it was the first example of a ciliate living in the blood of its host (*Asellus aquaticus*) and circulating with the corpuscles. When the ciliates become too crowded to pass through orifices they constitute an obstruction that impedes the circulation. Here and there they pass out through orifices perforating the walls of the arteries, and return with the current to the heart. Only a few continue to the ends of the arteries. As the oxygen is used up in a dead isopod, the ciliates slow down and die; and they ordinarily perish quickly in fresh water. Some, however, survive and encyst on plants or on the legs and antennae of *Asellus,* later escaping from the cyst and becoming active for a time in the water.

The species of *Dogielella* are tissue parasites which occur in the parenchyma of the mollusc *Sphaerium corneum* and the rhabdocoeles *Stenostomum leucops* and *Castrada* sp. in Russia (Poljanskij, 1925). Poljanskij did not refer to Fuhrmann's statement (1894, p. 223) that numerous holotrichs occurred in the parenchyma of two individuals of *S. leucops* near Basel; but he believed that *"Holophrya virginia"* described by Kepner and Carroll (1923) from the same rhabdocoele in Virginia is *Dogielella.* The ciliates seem to have no unfavorable effect upon *Sphaerium corneum,* even in a moderately heavy infection, but with excessive multiplication the host-parasite balance is disturbed and the molluscs perish from mechanical injury. Rarely, the ciliates may infect the developing embryos in the brood chamber. The forms in rhabdocoeles are apparently harmless to the host.

Cépèdella hepatica occurs in the hepatic caecum of *Sphaerium corneum* in France. An organelle of fixation, a slightly concave plate to which a cone of myonemes is related, is developed at the anterior extremity. The ciliate may penetrate into the hepatic cells. The parasitized cell undergoes degenerative vacuolization, which extends to neighboring cells (Cépède and Poyarkoff, 1909). Cysts have been found in the liver (Poyarkoff, 1909); these may persist in the outer medium and infect a new molluscan host.

Another tissue parasite is *Orchitophrya stellarum* Cépède, a rare ciliate which was found in 3 of more than 6,000 *Asteracanthion rubens* (Cépède, 1910). The infected sea stars were all males, and the ciliates occurred in the gonads, among the reproductive cells. Cépède found that the parasites were well adapted to life in the sea water, underwent no pathological changes, and survived for a long time. In a putrefying genital gland, removed from the starfish, the ciliates lived well after a day and multiplied. In the host, the parasites bring about what Cépède termed partial castration. The ciliate absorbs material in the gonad and transforms the contents by so doing and by adding its waste products; and it also brings about mechanically detachment and degeneration of certain sexual cells. Is *Orchitophrya* an obligate parasite, or is it an accidentally invading free-living type, in which Cépède overlooked the mouth structures? Consideration of instances of accidental parasitism among holotrichs (*Glaucoma, Anophrys*), as well as of the great infrequency of the occurrence of *Orchitophrya* and its ready adaptation to sea water, suggest that the latter may be true.

Conidiophrys

One of the most complete accounts of the life history and host relationships of an epibiotic ciliate, which is probably a trichostomatous holotrich, is that of *Conidiophrys pilisuctor* Chatton and Lwoff, 1934 (Fig. 204). In its profound modification in relation to its mode of life, it is approached by no other member of its suborder, and, in fact, by few other ciliates. *C. pilisuctor* occurs on the secretory hairs, frequently on the thoracic appendages, of a number of freshwater amphipods, especially *Corophium acherusicum,* in France. A second species, *C. guttipotor* Chatton and Lwoff, 1936, is attached to the hairs of *Sphaeroma serratum.* These ciliates were placed in a new family of *Trichostomata,* named Pilisuctoridae, by Chatton and Lwoff (1934b), though the International

Rules of Zoölogical Nomenclature demand Conidiophryidae. A complete account of *Conidiophrys* was given by these authors in their second article (1936a).

In a manner suggesting the case of *Sacculina,* the determination of the systematic position of *Conidiophrys* is possible only through study of its early development.

The form attached to the hairs (Fig. 204A) is immobile, nonciliated (though an infraciliature is present), and is enclosed in a shell-like pellicle which has no opening and beyond the body proper closely encases the hair (Fig. 204B). The cucurbitoid trophont undergoes several transverse divisions within the capsule, toward its distal end, producing normally two or three (Fig. 204C), or sometimes as many as six tomites. One specimen was observed with eleven tomites and a twelfth forming, but the distal seven were degenerate (Fig. 204D). When completely formed, the tomite, the longitudinal axis of which is transverse to the longitudinal axis of the trophont, is provided with cilia, with a cytostome opening on the ventral surface, and with a relatively long, incurved, ciliated cytopharynx (Fig. 204F). Tomites are liberated periodically and have a very short period of free-swimming existence.

When the cytostome comes in contact with the end of a secretory hair, this is drawn in and the tomite becomes impaled obliquely on it (Fig. 204F). The form rapidly changes to that of a tear drop and the cilia are lost (Fig. 204G). Growth to the typical trophont proceeds. Chatton and Lwoff maintained that *Conidiophrys* is not nourished by diffusion from the surrounding water, but depends on the fluid secretion that enters it through the pores at the end of the secretory hairs. Dependence of the trophic form (trophont) upon the host is thus absolute.

In discussing the multiplicative polarity of *Conidiophrys,* fission being localized at the distal pole, Chatton and Lwoff (1936a) speculated concerning a possible trophic or humoral influence emanating from the host. Instances of inhibition of division, complete or partial, under the influence of parasitic nutrition are given among parasitic dinoflagellates, apostomatous ciliates, and other Protozoa. (The authors did not comment, however, on the absence of any indication of such inhibition in a great number of endozoic forms, a fact which is an impediment to the acceptance of their theoretical explanation.) In *Conidiophrys* inhibition is exhibited in the removal of the zone of multiplication to a distance

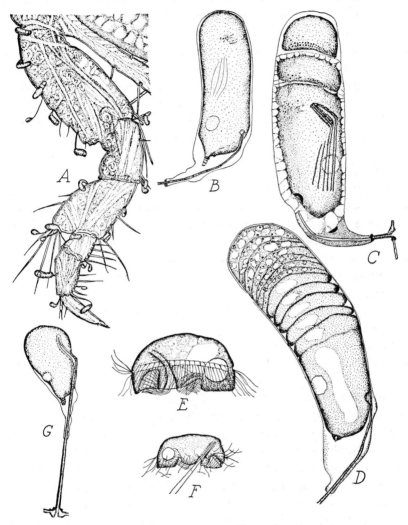

Figure 204. *Conidiophrys pilisuctor* on *Corophium acherusicum*. A, trophonts on appendage of the amphipod host, attached to hairs; B, trophont at beginning of reproductive period; C, trophont that has formed two tomites, and third forming; D, large trophont with eleven tomites, and a twelfth forming, the distal seven degenerated; E, unattached tomite; F, tomite impaled on a hair by its cytostome; G, young trophont, cilia lost. (After Chatton and Lwoff, 1936a.)

from the pole of communication with the host. When this influence is reduced on the commencement of molting, supernumerary tomites may be produced. There are, Chatton and Lwoff stated, many examples of parasites certain phases of the development of which are conditioned by the molt or sexual maturity of their host. The influence may be chemical, absorbed substances preventing a denaturation of proteins, which may be the essence of cell division. The existence, in trophoepibiotic ciliates, of a trophohumoral gradient of inhibition, susceptible to analysis and analogous to other types of biological gradients, is suggested.

APOSTOMEA

Though certain ciliates that are now included in the suborder Apostomea have been known for a long time, it is only recently that the group has become well known. Chatton and Lwoff (1935) published an outstanding memoir on the Apostomea, which, they stated, is only the first of three parts. This first part is a monographic study of the genera and species. In the suborder, according to this account, there are two families, by far the more important of which is the Foettingeriidae, with thirteen genera and twenty-six species. In the Opalinopsidae there are only two genera. Chatton and Lwoff expressed doubt that one of these, *Opalinopsis,* really is an apostome; and the other genus, *Chromidina,* was included by Cheissin (1930) in the Astomata. Kudo (1939) listed the Opalinopsidae in the Astomata.

The active, growing, vegetative phase of a foettingeriid ciliate is the trophont. The ciliature is in dextral spirals. In the process of growth the basal granules are spaced without multiplying. At the end of the period of growth the organism may encyst, the cilia are lost, and the infraciliature undergoes detorsion, the lines becoming meridional. This phase is called the protomont. It passes into the multiplicative phase, or tomont, which produces by transverse fission a variable number of tomites. The tomite is a small free-swimming ciliate. The ciliary rows are more or less meridional, with a tendency to turn in a spiral. There is a thigmotactic ciliary field, consisting of the parabuccal ciliature. Chatton and Lwoff maintained that the tomite represents the free-living, ancestral type. In twenty-two of the twenty-six species, and possibly in the others also, the tomite becomes fixed to the body surface of a crustacean, and transforms into an encysted phase, the phoront. In the phoront there is renewed multiplication of the basal granules and torsion of the ciliary lines, leading to the characteristics of the trophont. In the active phases of most species there is a more or less rudimentary, ventrally placed mouth, which is surrounded by a characteristic rosette; sometimes the mouth and rosette are lacking.

Almost all the apostomes occur on or in marine animals. Chatton and Lwoff (1935) assigned to the genus *Gymnodinioides* three species from fresh-water Crustacea, two of which were described by Penard (1922) as *Larvulina,* commensals on *Gammarus,* the third by Miyashita (1933) as *Hyalospira,* from Japanese shrimps.

Among the apostomes are the only ciliates with heteroxenous cycles, cycles that alternate as regularly as those in many Sporozoa, though there is no obligatory sexuality.

In one group of apostomes, the phoront occurs on copepods, fixed to the integument; and excystation with subsequent development occurs, normally when the host is wounded or is ingested by a predator. The ciliates, however, do not remain long enough in the predator for it to be regarded as a second host. The predators involved are mostly co-elenterates. The hydroid *Cladonema radiatum* appears to be a very special site for the trophont of *Spirophrya subparasitica,* the phoront of which is fixed to the integument of the benthonic copepod *Idya furcata.* When the copepod is ingested, *Spirophrya* excysts and grows rapidly in its remains, accumulating fluid or tissue material in a central vacuole. The trophont does not encyst within the predator, but is expelled with the residues of digestion. Encystation takes place on the carcass of the copepod, in the environment, or on the stalk of *Cladonema,* producing a tomont. This divides into a number of tomites, which may live free for a few days, and eventually degenerate or become fixed to *Idya.* The phoronts of apostomes of this group will excyst when the copepod molts, but subsequent development is not normal (see Kudo, 1939, Fig. 257).

In a second group of apostomes there are encysted phoronts on Crustacea, excystation occurs at the molt, and the trophonts develop in the exuvial fluid. Species are associated with a great variety of Crustacea, including Entomostraca, balanids, copepods, and many Malacostraca. The widely distributed genus *Gymnodinioides* belongs in this group. *Polyspira* is another genus. *P. delagei* is phoretic on the gill leaves of pagurids (*Eupagurus bernhardus*). Excystation occurs at the molt, and

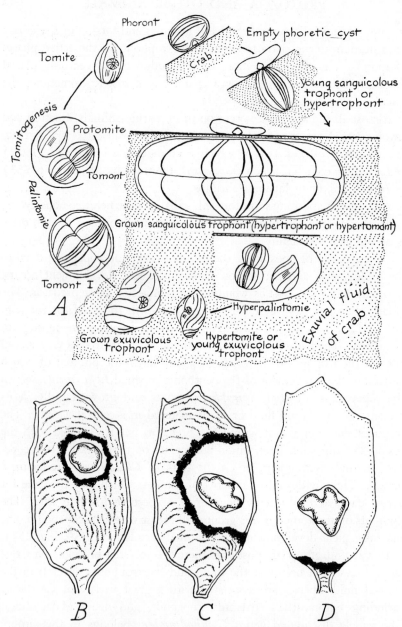

Figure 205. *Synophrya hypertrophica.* A, diagram of cycle of development as a parasite of *Portunus* or *Carcinus;* B-D, parasitized branchial lamellae of *Portunus holsatus,* showing different types of reactional cysts. (After Chatton and Lwoff, 1935.)

the young trophonts grow in the fluid contained in the discarded exo-skeleton. The proteins accumulate in a violet trophic mass, giving the color (which in other genera may be orange, red, and so forth) so characteristic of ciliates of the family Foettingeriidae. Unlike many apostomes, the trophont does not become encysted. Linear palintomy occurs in the motile stage, producing from eight to sixty-four daughter tomonts, which metamorphose into tomites. In addition to the natural host, the tomites will become fixed on the gills of *Portunus holsatus,* on which development proceeds normally.

In a third group of apostomes, in which there is only the genus *Synophrya,* the trophont is at one stage parasitic in the tissues of the crustacean to which the phoront is attached. *Synophrya hypertrophica* (Fig. 205) is phoretic on *Portunus depuratus,* and also on other species of *Portunus* and *Carcinus maenas.* The sanguicolous trophonts are in-ternal parasites in the branchial sinus of *Portunus* or the subcutaneous sinus of *Carcinus.* They are large, mouthless, immobile, irregularly lobed masses under the integument, enclosed in a double envelope. The re-sulting lesions of *Carcinus* appears as brown or black spots 1-4 mm. in diameter, found chiefly on the dorsal surface of the carapace. They oc-cur in a high percentage of crabs less than two centimeters in diameter, but not on large crabs. At ecdysis, tomites are produced which disperse in the molted exoskeleton and develop into exuvicolous trophonts. When growth is completed, these encyst as tomonts, each of which produces a number of tomites. The tomites fix themselves and become phoretic cysts on the integument of the gills or branchial cavity of the crab. The parasites then migrate from the cyst into the underlying tissue.

A fourth type of cycle is that of *Foettingeria actiniarum,* which is heteroxenous. It was first known as an inquiline in the gastrovascular cavity of sea anemones, some species of which are almost always in-fected. It has been found in various sea anemones on the coast of France, but it was not found in three species at Woods Hole. Chatton and Lwoff (1935, p. 313) listed ten host species. The ciliates are chymo-trophic. They enter the digestive mass when the coelenterate feeds and there find their sustenance. The ciliates eventually leave the host and encyst, the tomont undergoes palintomy, and the tomites become fixed to a crustacean. The host-specificity of the phoronts is almost nil; the

list of hosts given by Chatton and Lwoff (1935, p. 371) includes copepods, ostracods, amphipods, caprellids, the isopod *Sphaeroma,* and the
decapod *Carcinus.* When the crustacean is ingested by a sea anemone,
the phoronts excyst and become young trophonts.

Apostomes of the genus *Phtorophrya* are hyperparasites on other
apostomes. The phoront is fixed on the phoront of the host species, and
the parasite introduces itself into the body of the other ciliate. It grows
rapidly and soon comes to occupy a cyst otherwise empty. Tomites are
eventually produced; these leave the empty cyst of the host and swim
actively in search of another host phoront.

Rose (1933, 1934) reported two unnamed ciliates, considered by him
to be Foettingeriidae, parasitic in the oil drop in the oleocyst of the
siphonophore *Galeolaria quadrivalvis.* He thought it probable that the
cysts are attached to pelagic copepods.

Apostomatous ciliates have been found in the digestive cavity of
certain ophiurans and the ctenophore *Cestus veneris. Pericaryon cesticola*
is unusual among Foettingeriidae in adhering firmly to the walls of the
gastrovascular cavity of its host. It has an apical stylet, which seems to
be an organelle of fixation.

Sexual processes have been described in a number of Foettingeriidae.
Conjugation is contingent, as in other ciliates, and is of a common type
throughout the family. The trophonts conjugate and remain associated
during the formation of tomites. At the end of the series of fissions,
meiosis occurs, pronuclei migrate, and the tomites separate.

While phoresy on Crustacea is known or presumed to occur in all the
Foettingeriidae, except in *Phtorophrya,* the host phoront of which occurs
on Crustacea, it is unknown in the Opalinopsidae. The vermiform,
elongated (up to 1,200 μ), vegetative forms of *Chromidina elgans* are
fixed to the renal cells of cephalopods by an apparently retractile apical
papilla. There is no mouth. Multiplication is by simultaneous or successive fissions, producing chains of daughter individuals. The tomite has
a buccal ciliature and a buccal orifice, but no rosette. It is believed that
a crab may be involved in the cycle. *Opalinopsis* occurs in the liver and
intestine of cephalopods, and one species has been found in the liver
of the pelagic gasteropod *Carinaria mediterranea.*

PHYSIOLOGICAL HOST RELATIONSHIPS ILLUSTRATIVE OF
MUTUALISM AND COMMENSALISM

FLAGELLATES OF TERMITES AND *Cryptocercus*

Before discussing the relationship between the xylophagous flagellates and their wood-eating termite or roach hosts, it is desirable to give consideration to the problem of nutrition in some other invertebrates that ingest material consisting largely of cellulose.

The most abundant single constituent of wood is cellulose, which averages in general between about 54 and 64 percent (Pringsheim, 1932 after Schorger). Among other important carbohydrates are hemicelluloses, which Pringsheim stated is a poorly defined collective name for polysaccharides. A small amount of starch may be present in wood, about 3 to 4 percent, or less; and a certain amount of sugar (Schorger). Lignin is a noncarbohydrate incrustation substance in wood and makes up from about 23 to 28 percent of its bulk (Pringsheim). There are also in wood ash, less than one percent; proteins, a little under one percent, according to Pringsheim, in fir, pine, oak, and beech; fats; waxes; resins; and other substances. Straw and hay have about 30 to 35 percent cellulose, about 20 to 30 percent lignin, 3 to 10 percent protein, and 20 to 30 percent starch.

The animals that ingest these materials may use one or more of the constituents, and that is not necessarily cellulose. The larva of the goat moth *Cossus cossus,* though ingesting wood, does not affect the cellulose (Ripper, 1930). It has no cellulase and contains no symbiotic microörganisms. Ripper found that the carbohydrate used is supplied at least in part by soluble sugars, perhaps also by hemicelluloses. Mansour and Mansour-Bek (1934a) concluded that larvae of the cerambycid *Xystrocerca globosa,* with no cellulase and no microörganisms, derive their sustenance from the relatively high content of sugars and starch in the wood attacked (10.4 percent). Data bearing on the fact that some wood-eating insects seem to make no use of cellulose, but depend on the starch and sugars in the wood, being limited therefore to certain kinds of wood rich in these substances, were discussed by Mansour and Mansour-Bek (1934b). Ullmann (1932) stated that the carbohydrate requirements of invertebrates are met chiefly by sugars and hemicelluloses.

PROTOZOA AND OTHER ANIMALS

On the other hand, it has been found that certain termites can survive indefinitely on cotton cellulose or a cellulose-lignin complex (Cleveland, 1925b); and larvae of the rose beetle *Potosia cuprea* lived for more than six months on filter paper (Werner, 1926). According to Dore and Miller (1923), the wood that is ingested by *Teredo navalis* loses in the alimentary tract 80 percent of its cellulose, as well as from 15 to 56 percent of the hemicellulose, but the amount of lignin is not reduced. Digestion of cellulose undoubtedly occurs in the alimentary tract of many beetles, as, for example, the anobiid *Xestobium rufovillosum* (Campbell, 1929; Ripper, 1930) and the cerambycids *Hylotrupes bajulus* (Falck, 1930), *Stromacium fulvum,* and *Macrotoma palmata* (Mansour and Mansour-Bek, 1933, 1934a). The wood eaten by this last species was found to have very little soluble sugar and starch (0.47 percent). There are many other instances of cellulose digestion among vertebrates and invertebrates. Yonge (1925) published a review of cellulose digestion in invertebrates, but his statement that no cellulase has been found in Insecta is not true today.

Xystrocerca globosa is reported to have a strong amylase, as well as maltase and saccharase, enabling it to make use of the starches and sugars in wood (Mansour and Mansour-Bek, 1934a). Tissue-produced cellulase has been demonstrated in a number of gasteropods and insects. Among xylophagous insects, cellulase appears to be produced by the digestive epithelium of certain cerambycid and anobiid larvae.

Most cellulose decomposition in nature is brought about by bacteria, filamentous fungi, and certain Protozoa. In many animals that make ultimate use of cellulose in nutrition, the material is first acted on by microorganisms living in the alimentary tract. This is the only method of cellulose breakdown in vertebrates, and it is true also of the process in many invertebrates. Herbivorous mammals harbor bacteria capable of acting on cellulose. *Bacillus cellulosam fermentens* was isolated by Werner (1926) from larvae of *Potosia cuprea,* which feed mainly on spruce and pine needles. Bacteria in the intestine of the lamellicorn beetles *Oryctus nasicornis* and *Osmoderma eremita* are able to break down cellulose (Wiedemann, 1930). Cleveland *et al.* (1934) found evidence that symbiotic bacteria are the agents of cellulose decomposition in the xylophagous roach *Panesthia javanica.*

In the above-mentioned animals, bacteria dwell in the lumen of the gut. Very widespread in wood-eating insects, but by no means restricted to hosts of that group, are the intracellular symbionts studied intensively by Buchner and his associates, as well as by many others (see Buchner, 1930). These bacteria or yeast-like fungi live with their hosts in "cyclic endosymbiosis," being regularly transmitted to the next generation. Cyclic endosymbionts exist in the termite *Mastotermes* (Jucci, 1932; Koch, 1938a, 1938b). Buchner believed that these symbionts might play a rôle in the digestive processes of the host, but this opinion, lacking experimental proof, has not been generally adopted (Mansour and Mansour-Bek, 1934b; Schwartz, 1935).

Protozoa are present in many of these insects, and sometimes themselves derive nutriment from cellulose-rich materials. Beetle larvae frequently harbor a moderate number of small flagellates (*Polymastix, Monocercomonoides*) which feed on bacteria (Wiedemann, 1930). As stated above (p. 916), a limited number of small, non-xylophagous flagellates (mainly *Trichomonas*) and occasionally *Nyctotherus* are present in many termites of the family Termitidae (Kirby, 1932b, 1937). Some of these higher termites, especially species of *Mirotermes* and *Cubitermes,* harbor large amoebae which ingest wood or other cellulose-rich material on which the termite feeds (Kirby, 1927; Henderson, MS). The wood-feeding roach *Panesthia javanica* contains two small flagellates, *Monocercomonoides* and *Hexamita;* large xylophagous amoebae; smaller amoebae; and a number of ciliates (Kidder, 1937). Mutualistic symbiosis, however, finds its best illustration, so far as Protozoa are concerned, in the abundant and diverse xylophagous flagellates of termites other than Termitidae and of *Cryptocercus punctulatus.*

According to some investigators, Protozoa and other organisms of the gut may serve the host as a supplementary food source. Wiedemann's observation of cellulose-decomposing bacteria in lamellicorn larvae was mentioned above. He believed that the breakdown products are entirely used in the metabolism of other bacteria. The bacteria multiply rapidly in the large intestine, where they live in association with the small flagellates. The mid-gut, he found, secretes protease which is inactive in the alkaline medium there, but in the hind-gut, where bacterial acids accumulate, it digests the bacteria and flagellates. Mansour and Mansour-Bek (1934a, 1934b) and Mansour (1936) have discussed the possibility that the flagellates in termites do not benefit the hosts in nutritive processes except that, multiplying and being digested continually, they are a direct and supplementary food source for the insects. This seems unlikely, however. The flagellates in termites multiply rapidly for a few days after a molt following which there has been a new infection; then there is little division, and they are destroyed, usually, on the approach of the next molt. What use the host might make of the disintegration products at that time is entirely unknown, but certainly there is no evidence that the Protozoa could be available at any other period as a supplementary food source accounting for general nutrition (see p. 968).

Cyclic endosymbionts seem to be necessary for normal development of the host in some instances. Aschner and Ries (1933) and Aschner (1932, 1934) succeeded in freeing *Pediculus* of the symbionts that normally inhabit the mycetome, and found that without them larvae died sooner or later. The harmful effects of the absence of symbionts in *Pediculus* were reduced by rectal injection of yeast extract. Koch (1933a, 1933b) obtained symbiont-free larvae of the anobiid *Sitodrepa panicea* and found that they would not develop normally unless yeast was added to the food. (Koch, however, also reported freeing the saw-tooth grain beetle *Orzyaephilus surinamensis* of symbionts in the mycetomes by incubation [1933b, 1936]; and absence of the microörganisms seemed to be without detrimental effect.) It has been suggested that the symbionts are sources of vitamins or growth factors. It is possible, in the light of these facts, that certain symbiotic Protozoa may be necessary to the life of the host, without participating in the digestive processes or serving as a food source important in bulk.

We now come to a consideration of the demonstration—one of the outstanding advances in modern protozoölogy, though not yet complete —that wood-eating flagellates in termites and *Cryptocercus* are necessary for the survival of their hosts in making the products of decomposition of cellulose available for the nutrition of the insects. This has justly received very wide attention, so it is unnecessary to recount all details of the demonstration here (see Cleveland 1924, 1926, 1928a, 1934).

Termites feed primarily upon wood. This is especially true of the members of the families Mastotermitidae, Kalotermitidae, and Rhinotermitidae. Many Termitidae attack wood also; others bring into their nests dried grass, ingest soil and extract from it the nutrient materials it contains, or devour leaves; in fact, almost all types of vegetable matter are utilized by certain members of the group. Hodotermitidae forage for grass and herbs, even eat straw from unbaked bricks; some on the Karroo collect twigs. Kalotermitidae and Rhinotermitidae can live on paper; even, as stated above, cotton cellulose and a lignin-cellulose complex (see Cleveland, 1924, 1925b). The wood-boring roach *Cryptocercus punctulatus* eats the wood of fallen timber, well-decayed or sound.

Cleveland (1923) pointed out that, in the groups of termites that use a uniform diet of wood, all species examined had rich faunas of Protozoa; and this has been confirmed by studies by the writer of more than a hundred additional species. In Termitidae, with varied food habits, such faunas are absent, though there are some Protozoa in many species (Kirby, 1937). Cleveland *et al.* (1934) remarked that the correlation of wood feeding and intestinal flagellates is not so close as he at first supposed, since there are some Termitidae that eat wood and have no (xylophagous) flagellates.

We know very little of the actual nutrient substances among the varied materials taken in by termites as a group. Matter that has passed through the digestive tract is used extensively by higher termites in building mounds, fungus gardens, and carton nests. Cohen (1933) and Holdaway (1933) analyzed mound material of *Eutermes exitiosus,* which contains no Protozoa and feeds on wood. They found cellulose to be much reduced, though some passes out undigested, whereas lignin is unaffected. These results agree with those reported by Oshima (1919) after analyses of wood and nest material of *Coptotermes formosanus,* which does contain xylophagous flagellates. Oshima concluded that the principal food of that termite is cellulose and that there is no decrease of lignin. In termites of still another group, *Zootermopsis,* Hungate (1936, 1938) found essentially the same thing by analyses of uneaten wood and pellets.

Tissue-produced cellulase is absent from *Cryptocercus punctulatus* (Trager, 1932), *Kalotermes flavicollis* (Montalenti, 1932), and *Zootermopsis angusticollis* (Hungate, 1938). Probably none is present in any Kalotermitidae or Rhinotermitidae, though, in the light of the situation with wood-boring beetles, one should not generalize from limited data. Termitidae have not been investigated for cellulase. Mansour and Mansour-Bek (1934a) suggested that some termites may be found to have cellulase, and it is in that higher group that one might be most likely to find the enzyme.

The literature on this subject, so far as the writer has determined, contains no discussion of the hemicellulase lichenase, which Ullmann (1932) reported to occur in all the insects, including roaches, and the snails that he tested; and which Oppenheimer (1925) stated is widespread in invertebrates. Montalenti (1932) wrote that in the fore-gut of *K. flavicollis* he found a trace of amylase, which was probably present in the salivary secretion; in the mid-gut, amylase and invertase, as well as a protease that acted only in acid, though the mid-gut is basic; and in the hind-gut, amylase and invertase probably derived from the mid-gut. Hungate (1938) found amylase in an extract of the fore-gut, and protease in the mid-gut of *Zootermopsis angusticollis*. On the basis of these findings, it should be possible for the termites to hydrolyze starch; to invert sucrose; to digest the small amount of protein in wood and possibly also some of their own microörganisms, when the resistance of the latter to the enzyme has been overcome.

As remarked above, bacteria in many cellulose-utilizing animals are necessary for the preliminary breakdown of cellulose. Cleveland et al. (1934) suggested the possibility that some Termitidae may profit from the presence of bacteria in the same way. But in those termites that have been examined for cellulose-decomposing bacteria, it appears that the latter cannot account for cellulose digestion. A few positive results have been obtained. Dickman (1931) found them in one of six nitrate-cellulose tubes inoculated with gut contents of *Reticulitermes flavipes*, and Tetrault and Weis (1937) obtained some from the same termite; but Cleveland (1924) failed in many and varied attempts to isolate cellulose-decomposing bacteria or other fungi from *R. flavipes*. Beckwith and Rose (1929), using termites of six genera, including one of the Termitidae, obtained cellulose-digesting bacteria in some instances, but not at all in two species. Their results, however, are subject to criticism (Dickman, 1931; Hungate, 1936). Hungate (1936) was unsuccessful in efforts to show cellulose decomposition by bacteria from the gut of *Z. nevadensis*, and concluded that bacteria in the alimentary tract are of no importance in the digestion of cellulose. A possible explanation of the occasional positive tests is found in Cleveland's discovery that in *Cryptocercus punctulatus*, feeding on its normal diet of

wood, it is usually possible to obtain cellulose-digesting bacteria in culture from all regions of the alimentary canal, especially the fore-gut.
These disappear in time when roaches are fed on paper, and he believed
that they are forms living in the wood and accidentally ingested by
the insects.

Numerous fungi were isolated by Hendee (1933) from wood in-
habited by *Zootermopsis angusticollis, Reticulitermes hesperus,* and
Kalotermes minor. Dickman (1931) obtained cellulose-digesting organisms, both bacteria and molds, from material attacked by termites,
probably *R. flavipes* and *Zootermopsis* sp. Cellulose-decomposing molds
were found by Hungate (1936) in burrows and pellets of *Zootermopsis.* He concluded, after analyses of sawdust acted on by external organisms and material that had passed through the termites (possibly
several times), that cellulose decomposition by bacteria and molds in
the wood of the colony is negligible in comparison with that digested
in the termites. That fungous action *can* render cellulose usable by
termites is shown, however, by an observation of Cleveland's (1924).
Termites deprived of Protozoa died soon on a cellulose diet, but lived
indefinitely when a cellulose-decomposing fungus accidentally developed
in certain vials.

The flora of spirochetes and other bacteria in the gut of termites, and
this applies also to Termitidae, is considerable. They live free in the
lumen, attached to certain Protozoa, or attached to the lining of the
walls. Spirochetes do not grow on the usual laboratory media (Dickman, 1931). The possibility that they may participate in digestion of
cellulose and hemicellulose in termites was admitted by Cleveland
(1928a). In *Cryptocercus,* however, Cleveland killed the Protozoa,
leaving the spirochetes, by contrifuging; and cellulase disappeared in
twenty-four hours. The enzyme was not found after defaunated roaches
were reinfected with bacteria and spirochetes.

Excepting certain castes and brief phases of development, all termites
except Termitidae have great numbers of flagellates in the hind-gut.
The vestibule, large intestine, and caecum become voluminous organs to
accommodate these symbionts. Hungate (1939) estimated that the gut
contents containing the Protozoa amount to from a seventh to a fourth
of the total weight of *Zootermopsis angusticollis.* Katzin and Kirby
(1939) found the gut contents to be about a third of the weight of

nymphs of *Z. angusticollis* and *Z. nevadensis,* and about a sixth of the weight of soldiers. In this fluid gut contents the Protozoa are about as thick as they could possibly be. Hungate, by centrifuging, showed that about half consists of fluid, half of organisms. The organisms are in mass mostly Protozoa, but there are also a great many bacteria and spirochetes. Lund (1930) estimated the number of *Trichonympha, Streblomastix,* and *Trichomonas* in *Zootermopsis* as 54,000; but obviously this would vary greatly with the size of the termite.

In *Cryptocercus punctulates* the colon is enlarged to a relatively greater degree than in termites, becoming "an immense thin-walled bag completely filled with Protozoa" (Cleveland *et al.,* 1934). There are probably millions of flagellates in a single full-grown roach.

Most species of these flagellates ingest particles of wood. None of them possesses cytostomes. Ingestion is through the surface of the body. In *Trichonympha* wood ingestion has been described by Swezy (1923), Cleveland (1925a), and Emik (MS). Ordinarily most of the wood in the faunated portions of the hind-gut is enclosed in the cytoplasm of the flagellates. Cleveland (1924) stated that in *Reticulitermes flavipes* nearly all the particles of wood are taken into the Protozoa, whereas in *Zootermopsis* he found many particles free in the lumen of the gut.

Bacteria and other flagellates are sometimes ingested by *Trichonympha collaris* (Kirby, 1932b) and other flagellates of termites. This predatory habit is more frequent in some species than in others; and ingestion of other organisms occurs more frequently under the conditions of filter-paper feeding. Yamasaki (1937b) observed *Dinenympha* in many *T. agilis* after oxygenation. Wood is, however, the chief and usually the only material taken into holozoic forms. Lund (1930) noted that when *Zootermopsis* was fed on cornstarch, many *Trichonympha* and *Trichomonas* ingested starch grains. *Trichomonas* and *Hexamastix* in *Zootermopsis* are, according to Cleveland, able to use starch. Grains of rice starch were taken in by three of the hypermastigotes in *Cryptocercus,* and had some food value for them; and *Monocercomonoides* in the roach could make full use of starch (Cleveland *et al.* 1934).

Some flagellates in termites are saprozoic and do not take in solid particles. That is true of *Streblomastix* in *Zootermopsis,* of *Hoplo-*

nympha in *Kalotermes hubbardi,* and probably of some forms of *Dinenympha* in *Reticulitermes.* It is also true of certain very small flagellates.

That the flagellates possess an enzyme capable of acting on the cellulose of the ingested wood has been clearly demonstrated by a number of investigators. Trager (1932, 1934) proved that *Trichomonas termopsidis* produces cellulase. He maintained the flagellate in culture for several years in the presence of only one species of bacteria, which was not capable of fermenting cellulose or cellobiose. The addition of finely divided cellulose to the medium was necessary, and *Trichomonas* did not live when that was replaced by other polysaccharides. An extract of the ground bodies of the flagellates, concentrated from cultures, acted on cellulose. Emik (MS) obtained fairly pure concentrations of *Trichonympha* from *Zootermopsis* by gravity filtration. Extracts of these concentrates were able to digest certain preparations of cellulose as shown by osazone tests, demonstrating crystals of glucosazone and cellobiosazone. Emik concluded that two enzymes were present, derived from *Trichonympha:* cellulase, hydrolyzing cellulose to cellobiose; and cellobiase, hydrolyzing cellobiose to glucose.

It is not difficult to show the action of cellulase in the contents of the hind-gut, and, in view of the absence of tissue-produced cellulase and the virtual absence of cellulose-digesting bacteria or fungi, the Protozoa must be its source. Both cellulase and cellobiase were found there by Trager (1932). Cleveland *et al.* (1934) and Hungate (1938) so identified cellulase in flagellates of *Cryptocercus punctulatus* and *Zootermopsis angusticollis.*

Substance stained brown or reddish brown by iodine dissolved in potassium iodide, and assumed, as is customary, to be glycogen, has been found in many of these xylophagous flagellates. The earliest demonstration, which was discussed critically by Cleveland (1924), was made in *Trichonympha agilis* by Buscalioni and Comes (1910). Kirby (1932b) mentioned iodine-staining granules in *T. campanula.* Yamasaki (1937a, 1937b) described abundant glycogen deposits in the species of *Trichonympha, Teratonympha, Holomastigotes, Pyrsonympha, Dinenympha, Pseudotrichonympha, Holomastigotoides,* and *Spirotrichonympha* in Japanese termites, preparing the material by staining in Ehrlich's hematoxylin and Best's carmine after fixation in 90-percent alcohol. Diminution of the glycogen in *T. agilis* under conditions of incubation, starvation, and oxygenation was studied by Yamasaki (1937b). Kirby (1931) stained the axostyle of *Trichomonas termopsidis* brown in Lugol's solution; and stated that this may be taken, as Alexeieff pointed out in the case of *Tritrichomonas augusta,* to be indicative of the possible presence of glycogen in the axostyle. In the light of these results, it seems likely that carbohydrate is stored as glycogen in these Protozoa. (Cleveland *et al.* [1934], however, considered it possible that the substance colored by iodine may not be glycogen, but a breakdown product of cellulose which gives the same reaction as glycogen. See also page 981.)

Since only in the bodies of the flagellates can cellulose be digested, and termites live and develop normally when only cellulose is eaten, the rôle of the symbionts is evident. According to Hungate (1938), about one-third of the total material removed from wood, adding that acted on in the fore- and mid-gut to the soluble materials present, can be obtained without the aid of the Protozoa. It is possible that materials adequate for nutrition of the insect may be obtained in the diet without the Protozoa, as Cleveland (1924) found by feeding humus and fungus-decomposed cellulose. Presumably sufficiently rotted wood would also be adequate; Cleveland (1930) stated that defaunated *Cryptocercus,* which dies in two or three weeks on partially decayed wood or cellulose, will live two or three months on completely decayed wood. Cleveland has conclusively demonstrated that continued survival of defaunated termites and *Cryptocercus* is impossible on a natural diet of wood. Hungate's third, therefore, could not provide all necessary substances, it appears; as, if it did, the amount could be multiplied merely by the ingestion of more wood, or further use of that which ordinarily passes to the hind-gut for use of the Protozoa in faunated individuals.

Experiments in feeding various cellulose-free carbohydrates to termites have been made by Montalenti (1927) and Lund (1930); and to *Cryptocercus* by Cleveland (1930, 1934). Montalenti kept *Kalotermes flavicollis* alive for several months on soluble starch, alone or mixed with glucose, though the hypermastigotes soon disappeared and the polymastigotes greatly diminished in number. He concluded that the termite could live a long time, if not indefinitely, on soluble carbohydrates without Protozoa, but no other worker has confirmed this. Lund's studies were made to determine the effect of various diets on the

Protozoa of *Zootermopsis*, not the maximum period of survival of the termites. Cornstarch caused death of the Protozoa and the termites after twenty-three days; the starch was apparently in granular form. (In comparison with Montalenti's results, light may be thrown on the discrepancy by the statement of Ullmann (1932) that invertebrates are unable to use the starch of the plant food, but that soluble cooked starch is very well digested by all animals.) Lund used a variety of carbohydrates, on most of which the maximum survival of both Protozoa and termites did not greatly exceed the effects of starvation; and on some they died more quickly. On inulin, dextrin, and lactose *Trichomonas* and *Streblomastix* were living in the last termites reported at forty-eight, forty-four, and sixty-five days. Cleveland (1925c) found that *Trichomonas* (accompanied by *Streblomastix*) can keep the host alive from forty to fifty days longer than when no Protozoa are present, but "very few if any [termites] were able to live indefinitely." The hypermastigotes are most important in the mutualistic symbiosis in *Zootermopsis*.

Cleveland *et al.* (1934) studied the effects of various diets on *Cryptocercus punctulatus* and its Protozoa, using various cellulose-free carbohydrates, peptone, gelatin, and glycogen. On no substance did any except the smaller polymastigotes survive very long; nor did the roaches live more than a few days longer than when water alone was given. These authors found that dextrose is of more food value than the other substances, and considered it likely that a diet including dextrose might be found upon which the insects could live for a long time, if not indefinitely, without Protozoa.

Dextrose prolonged the survival period also of defaunated *Reticulitermes flavipes* (Cleveland, 1924). Trager (1932) demonstrated dextrose in the presence of extract from the hind-gut contents of *Cryptocercus*. Cleveland *et al.* (1934) suggested that dextrose, produced from cellulose by the action of cellulase and cellobiase in the cytoplasm of the flagellates, insofar as it is not used in their metabolism or stored as glycogen, diffuses from their bodies. Hungate (1939), however, identified acetic acid, carbon dioxide, and hydrogen as metabolic products of the Protozoa; and thought it likely that most of the sugar resulting from their digestive processes undergoes anaërobic dissimilation by the Protozoa. According to this view, the termites would make use of the acetic acid.

A further problem arises in the absorption of the substances released from the flagellates by the termite or roach tissues. Either absorption must take place through the chitinous layer of the hind-gut, or fluid must be passed forward into the mid-gut. The problem has been discussed by Buchner (1930) and Cleveland et al. (1934). There are differences of opinion as to whether absorption in the hind-gut is possible, and some authors are inclined to the view that it is. Buchner is one of those. Abbott (1926) found that the hind-gut of Periplaneta australasiae is permeable to dextrose. Cleveland et al. took the opposite view, as a result of osmotic experiments on the colon of Cryptocercus, which showed it to be impermeable to dextrose and water. The peritrophic membrane also seemed to be largely impermeable to dextrose. The iliac valve controls the passage of materials between the mid-gut and the hind-gut, and when the mid-gut is severed it permits no material to flow out from the hind-gut. Cleveland concluded that fluid containing dextrose passes forward at times through the iliac valve into the space between the peritrophic membrane and the wall of the mid-gut.

A problem in the metabolism of xylophagous animals is the source of nitrogen. The small amount of protein present in wood, and the larger amount in straw and hay (important in the case of Hodotermes and some higher termites), may account for the nitrogen metabolism of the Protozoa; but in the absence of action on cellulose outside of the Protozoa it might not be directly available to the termite. Bacteria and molds ingested with the wood might account for some, but that would probably be very little. Pierantoni (1937) hypothesized a fixation of nitrogen by bacteria in the gut, and Green and Breazeale (1937) reported the isolation of nitrogen-fixing bacteria from an unidentified species of Kalotermes. Wiedemann (1930) stated that bacteria in certain lamellicorn larvae can use inorganic nitrogen, and the host satisfies its nitrogen need by digesting these microörganisms. Use by termites and Cryptocercus of some dead Protozoa or pieces of cytoplasm from their bodies (Cleveland et al., 1934), while it could not account for any important part of general nutrition (p. 961), might be significant in providing nitrogen.

We have seen in the foregoing discussion that many animals ingest substances of which cellulose is an important constituent. In the digestive tract of some of them cellulose is broken down, whereas certain others make use only of other constituents of the ingested material. These two nutritional variants may be found in members of the same group, as in Cerambycidae. The decomposition of cellulose may, in some invertebrates, be accomplished by means of a tissue-produced cellulase; in other, even related, forms it may require the action of symbiotic bacteria. It is possible, though the truth of the hypothesis remains to be shown, that endobiotic bacteria and Protozoa may, in some instances, benefit their hosts as a supplementary food source. There is a hypothesis also that certain symbiotic microörganisms are a source of vitamins or growth factors, or play a rôle in nitrogen economy.

In all termites below Termitidae, except in certain functioning reproductive castes, and in the roach *Cryptocercus punctulatus*, xylophagous flagellates are exceedingly abundant in a specially enlarged part of the hind-gut. These flagellates possess cellulase and cellobiase, and reduce cellulose taken into their cytoplasm to dextrose. The insects possess no tissue-produced cellulase, and few if any cellulose-decomposing bacteria or other fungi are present. These insects cannot live for long on their usual diet or on cellulose without the flagellates, which presumably release part of the dextrose or its dissimilation products for the use of the host. This may be passed forward into the mid-gut to be absorbed, or perhaps may be absorbed in the hind-gut. To what extent the nitrogen needs of termites may be provided for by occasional digestion of Protozoa or fragments of the cytoplasm, or by symbiotic nitrogen-fixing bacteria, remains to be determined.

CILIATES OF RUMINANTS

Among the most notable endozoic faunules of Protozoa are the ciliates in ruminants and certain other herbivorous mammals. There are some holotrichs, sparsely represented among the species of ciliates in ruminants, but constituting an important and diversified part of the faunules of the caecum and colon of the horse (Hsiung, 1930). There are also a few flagellates and amoebae, but the most characteristic forms belong to the Entodiniomorphina. Of the two families of this suborder of highly organized spirotrichs, Ophryoscolecidae occur chiefly in ruminants (see Dogiel, 1927; Kofoid and MacLennan, 1930, 1932, 1933; Wertheim, 1935); and Cycloposthiidae are best known in the horse (see Hsiung, 1930). The latter family is represented also in a number of other mammals, including the tapir, rhinoceros, chimpanzee, gorilla (Reichenow, 1920), and elephant (Kofoid, 1935).

Ciliates in ruminants, except for certain species less constant in occurrence (as *Buxtonella sulcata* Jameson, in the caecum of cattle) are localized in the rumen and reticulum. Their relative abundance in the rumen and reticulum is approximately equal (Dogiel and Fedorowa, 1929; Wertheim, 1934a); and they are distributed throughout the contents. Dogiel and Fedorowa found that the ciliates are somewhat more abundant in the central part than at the periphery, but that the difference is not very great. Distribution is sufficiently uniform so that counts of a small sample from the rumen have been used for an estimate of the total population.

Poor nutrition of the ruminant can cause a rapid reduction in the number of ciliates, and this may be responsible for low counts, in slaughterhouse animals, of under 100,000 per cc. (Dogiel and Fedorowa, 1929), under 200,000 per cc. (Wertheim, 1934a), and under 400,000 per cc. (Winogradowa-Fedorowa and Winogradoff, 1929). Under conditions of normal nutrition, many counts above 500,000 per cc. have been obtained. Mangold (1929, 1933) stated that in sheep and goats the normal number remains with much constancy at about 1,000,000 per cc.; and Mowry and Becker (1930) agreed with this as regards goats. Under certain conditions, the population may be much denser than this. By experimental feeding Mowry and Becker obtained a count of nearly 7,600,000 of *Entodinium* and *Diplodinium* alone. Ferber (1928) estimated that at 900,000 per cc. a gram of rumen contents would contain about one-twentieth of a gram of ciliates.

The total number of ciliates in an individual ruminant is enormous. Calculating from a volume of material in the rumen and reticulum of goats of from 2.8 to 5.2 liters, and a ciliate count of from 121,000 to 391,000 per cc., Winogradowa-Fedorowa and Winogradoff (1929) estimated a population of from 471,000,000 to 1,548,000,000; for the normal condition these figures should probably be multiplied by about three. In an ox with from 56 to 87 liters and from 70,000 to 117,000 ciliates per cc., there would be nearly 10,000,000,000; and probably the population may be at least five times as dense as that.

The ciliates are consistently absent from suckling animals, but, as soon as a diet of plant food begins, the faunule of the rumen and reticulum develops.

The kind of food taken has a striking influence on the ciliate population. Green plant material was regarded as of fundamental importance by Trier (1926), with emphasis on the chlorophyll content. Weineck (1934) expressed agreement with this view, but Westphal (1934b) denied that there is proof of chlorophyll need. In experiments by Mowry and Becker (1930), green fodder alone maintained a low population. Hay and water alone also maintained a low population, which was more than doubled when cornstarch was added. A much greater increase occurred when a grain mixture, consisting of ground corn, ground oats, wheat bran, and linseed oil meal, was given with hay. There is a limit to this increase, however. Although the densest population of all was developed on grain alone, there was soon a very great decrease in the number of ciliates. As was pointed out by Mangold (1929), some coarse food is essential. Hay with cornstarch and either plant or animal protein, instead of with grain, also maintained a high level of population density. Apparently in the grain both the starch and the protein constituents are stimulating factors, although the Mangold school has maintained that the protein alone is determinative.

Other factors influencing the ciliate population of the rumen that have been discussed are density of the contents and pH. Mowry and Becker (1930) could not corroborate the findings of Dogiel and Fedorowa (1929) and Ferber (1929b) that thick rumen contents contain relatively more ciliates than thin fluid contents. Mowry and Becker found the average pH in the rumen of goats to be 7.7, with two-thirds from 7.6 to 7.8 and extremes of from 6.7 to 8.2. Within these limits, there seemed to be no notable changes in the ciliate population that could be attributed to the pH itself. Mangold and Usuelli (1930) found the pH of fresh rumen contents of sheep to be from 7.5 to 7.8.

When the ciliates pass into the omasum, abomasum, and intestine, they are destroyed. As nutriment passes posteriorly from the reticulum, a vast number of ciliates must go with it; it is difficult to conceive of any mechanism by which they could be kept back. The population could be maintained only by an adequate rate of multiplication, going on continuously. The rapid disappearance of the ciliates on starvation of the host probably could not be explained simply by starvation, followed by death of the ciliates in the rumen and reticulum; it is more likely that the rate of reproduction declines, and the passage of ciliates into the third stomach rapidly reduces the population.

Several attempts have been made, with varying results, to estimate the reproductive rate by counting dividing ciliates. No adequate determination has been made of the rate of reproduction in a day, a calculation which cannot be based only on the amount of fission seen on one occasion. Rate of reproduction in culture, furthermore, at least in the absence of completely satisfactory culture methods, is not necessarily the same as that in the rumen. Mowry and Becker (1930) found in goats usually less than 0.5 percent of dividing forms, and never as many as one percent. Ferber and Winogradowa-Fedorowa (1929) found in a ram on different occasions from 0.9 to 15 percent in division, the average being 7 percent. Examinations were made twice a day, but they failed to note, as Mowry and Becker pointed out, that from observation of 7 percent dividing forms in two samples per day, it does not follow that 7 percent of the ciliates are dividing in a day. The rate of multiplication would probably be much higher than that. Westphal (1934a) found that in culture of certain forms each ciliate divided an average of once in fourteen hours, and the population became 3-fold in a day. It reached in more dilute medium a rate of 5.8-fold in twenty-four hours. Dogiel and Winogradowa-Fedorowa (1930) published a report that from 50 to 90 percent of the ciliates were observed in division in goats under normal conditions of nutrition, and from 12 to 50 percent in slaughterhouse oxen. Westphal (1934a) calculated that there must be daily at least a quadrupling of the number of ciliates.

Rumen ciliates live in a chemically complex and delicately balanced environment, and *in vitro* culture has been a difficult problem. Becker and Talbott (1927) and earlier workers failed to obtain more than limited survival. Knoth (1928) obtained longer survival in a medium of rumen fluid, with controlled pH, and with partly anaërobic conditions provided by a mixture of carbon dioxide and methane, the maximum being the life of *Entodinium,* with daily change to fresh solution, for five days. Margolin (1930), in media of hay infusion with rice starch and filter paper acted on by cellulose-decomposing bacteria, the pH being kept at 6.8, reported maintenance of cultures for twenty-four days; but others have been unable to use his methods successfully (Becker, 1932; Westphal, 1934a). Westphal (1934a, 1934b) reported real success with a medium of rumen fluid kept under anaërobic conditions, with urea and starch added. There was active multiplication in the cultures, which, with daily renewal of medium, were kept several weeks, and, he stated, might be continued indefinitely. *Entodinium* lived particularly well.

The rumen ciliates, so far as is known, do not form cysts. Trophozoites have been found in the mouth fluids (Becker and Hsiung, 1929), and ruminants have been infected by giving this material with the food (Mangold and Radeff, 1930). Natural transmission is by contact, in common feeding, in which there is a certain period when the trophozoites are exposed to the external environment (Becker and Hsiung, 1929; Mangold and Radeff, 1930; Strelkow, Poljansky, and Issakowa-Keo, 1933). Their ability to withstand external conditions is therefore of crucial significance. The holotrichs and *Entodinium* are more resistant than the larger Ophryoscolecidae. Strelkow *et al.* reported that after six hours at room temperature all ciliates were still active, and many survived longer. At 0° C. all continued normal activity for an hour. On dilution of the rumen fluid, they survived for various periods of from one to thirty-two hours; and most were alive after six hours in material two-thirds evaporated. They are thus clearly able to live long enough on feed or in water to infect other animals using the same containers. The interesting report of Fantham (1922) that "species of *Entodinium* and *Diplodinium* may be found on wet grass and in aqueous washings of fresh grass and even of dried grass (fodder) from sheep runs and pasturage" leaves the reader desirous of details concerning his observations.

It has been found to be a simple matter to bring about elimination of the ciliates from the rumen and the reticulum. Modern investigators have used three defaunation treatments. Mangold and his coworkers have found starvation alone to be satisfactory. The ciliates may apparently be absent after only three or four days, but Dogiel and Winogradowa-Fedorowa (1930) found six to seven days without water necessary for complete defaunation. They found preferable, however, partial feeding with dry food and water and a liter of milk daily. Milk feeding was used by Falaschini (1935). On the basis of studies made on material kept in the thermostat, Mangold and Usuelli (1930) concluded that the increased acidity induced in the rumen contents is responsible for the incompatibility of milk and ciliates. The best method of defaunation, according to Strelkow, Poljansky, and Issakowa-Keo

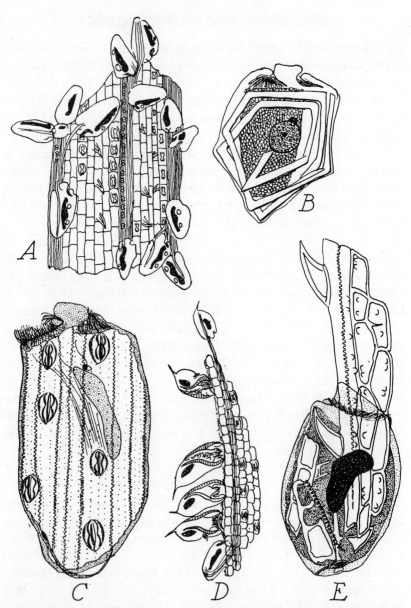

Figure 206. Ingestion of plant material by Ophryoscolecidae. A, a piece of grass with three fiber bundles, which are beset with *Diplodinium gracile:* B, long cellulose fiber rolled up in *D, gracile;* C, large fragment of grass in *Elytroplastron bubalidis;* D, a piece of grass with two *D. gracile* and four *Opisthotrichum janus.* The former have ingested the ends of fibers, the latter ingests fine detritus from the surface; E, *E. bubalidis,* which has partly ingested a large piece of grass. (After Dogiel, 1925.)

(1933), who compared it with milk feeding, is the one discovered by Becker (1929). Becker starved his animals for three days, then gave two doses, twenty-four hours apart, of fifty cc. of 2-percent copper sulphate, passed through a rubber tube into the rumen. Strelkow *et al.* shortened the starvation period to one day and gave three doses of copper sulphate, thus shortening the defaunation period to three days.

The ciliates take up plant fragments in the rumen, and Trier (1926) stated that they are apparently exclusively plant feeders. Bacteria, flagellates, amoebae, and ciliates may also be ingested, however. According to Dogiel (1925), no Ophryoscolecidae are entirely predatory, as plant debris is always to be found in the plasma. Kofoid and MacLennan (1930, 1932, 1933) and Kofoid and Christenson (1934) recorded the food contents of most of the species they studied, and showed clearly that food habits differ. Some appear to feed only on bacteria and other Protozoa, especially small flagellates. Others use various combinations of plant and animal material, some only plant material, and bacteria with plant debris are ingested by a large proportion. Some species are more inclined than others to be predatory on ciliates. *Entodinium vorax,* according to Dogiel (1925), almost always contains the remains of one or more smaller *Entodinium.*

The plant material is often in relatively small particles, but some ophryoscolecids take in large pieces that may distort the body. Dogiel (1925) described how *Diplodinium gracile* may seem actually to tear away fibers (Fig. 206A, B); *Opisthotrichum janus* may bite from the surface of a plant piece the remains of ruptured tissue (Fig. 206D); and *D. bubalidis, D. medium,* and *D. maggii* may devour large irregular or flat grass pieces (Fig. 206C), but not fibers. *Ostracodinium* sp. can take in and roll up large cellulose fibers (Weineck, 1934).

Green plant fragments are taken in preference to non-green ones, according to Usuelli (1930b), who offered a choice by feeding hay and barley. A third to a half of the green fragments were in about half of the ciliates; whereas less than 10 percent of them took in non-green fragments, all but a few of which remained free in the lumen. For this selectivity, Usuelli contended, the softer, smoother characteristics of the green plant pieces are responsible.

When available, starch grains are ingested avidly by the ciliates, both in the rumen and in the thermostat. Four hours after giving a sheep fifty grams of cornstarch, 76 percent of the ciliates had taken in grains; and in from four to six hours 87 percent took in rice starch (Usuelli, 1930a). If the amount of starch is not excessive, most of it is eventually taken into the Protozoa. The size of the grains is a factor in ingestion.

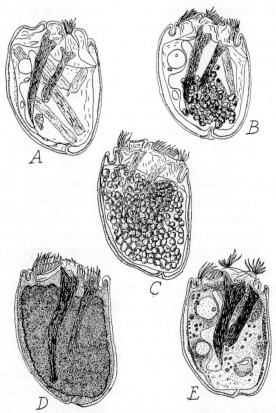

Figure 207. Ingestion and digestion of starch in *Eudiplodinium medium*. A, before feeding in culture; B, 2.5 minutes after feeding, starch grains in cytoplasm; C, 12 minutes after feeding, cytoplasm filled with starch; D, 2.5 hours after feeding, abundant deposits of glycogen; E, 16 hours after feeding, some residues of starch, glycogen deposits in certain areas. (After Westphal, 1934a.)

For example, fewer ciliates take in potato starch, the diameter of many of the grains of which exceeds 100 μ.

In material kept in the thermostat, it has been found that fat droplets in milk will be ingested (Ferber, 1928); but ingestion of accessible material is not indiscriminate (Westphal, 1934a).

That there is digestion of the starch has been well established (Trier, 1926; Westphal, 1934a, 1934b). Figures from Westphal (1934a) are reproduced here (Fig. 207) showing successive stages in the rapid ingestion by *Eudiplodinium medium* in culture of rice starch and the dissolution of this. Although there seems to be no good reason for denying starch-splitting ability to the ciliates themselves, Ullmann (1932) considered it possible that starch-digesting bacteria, taken in with the food and continuing their action in the digestive vacuoles, are responsible.

As starch is digested, glycogen (paraglycogen) accumulates in the cytoplasm in granular form. The reserve material is stored in the ectoplasm, in the region of the gullet and the rectal tube, and sometimes also in the endoplasm. It has been asserted by many that, in addition to the other storage areas, the skeletal plates contain deposits of glycogen (Schulze, 1924; Trier, 1926; Weineck, 1934; Westphal, 1934a, 1934b; MacLennan, 1934). Dogiel disagreed with this concept; and Brown (MS), by a series of chemical tests, solubility tests with substances that would have been expected to extract glycogen, and enzymatic reactions, found no evidence that the contents of the skeletal prisms is glycogen. Brown pointed out that the results of iodine reactions are insufficient in themselves to identify glycogen, as other substances stain in the same way.

Glycogen may be built up from simpler carbohydrates, appearing after feeding with dextrose (Trier, 1926; Weineck, 1934) and lactose (Trier).

When, after deposits of glycogen have accumulated, the ciliates remain without food, the reserve is used up in cell metabolism. Trier found that within forty-eight hours after ingestion of the starch most of the accumulated glycogen had disappeared.

There is disagreement as to whether the ciliates can use cellulose, though certainly a quantity is ingested. Much quoted has been the statement by Dogiel (1925) that, in the endoplasm of the Ophryoscolecidae, cellulose pieces undergo no morphological change and leave by the anus, still with sharp margins and no wrinkling or swelling. The statement was evidently based on observations on *Diplodinium maggii* and *D. medium,* which ingest large particles (Dogiel and Fedorowa, 1925). Westphal (1934b) reported that he had confirmed this account of ejection of large particles, but Weineck (1934) wrote that it has very seldom been seen, and that ejected pieces are those with excessively heavy membranes. Reichenow (in Doflein, 1927-29) did not agree that it provides evidence against the use of cellulose, stating that Protozoa, especially when in unfavorable circumstances, may give up nutriment useful to them. Usuelli (1930b) saw no microscopic indications of corrosion of ingested fibers, and commented that in any case the green plant fibers, that are the ones chiefly used, contain relatively little cellulose. Westphal (1934a, 1934b) denied cellulose digestion in ophryoscolecids, and showed that, in spite of the presence of cellulose and chlorophyll, cultures died out in the absence of starch. According to Mangold, the colorless pieces persist for a long time, even for four days, and during that time digestible substances are extracted.

Earlier opinions that there is digestion of cellulose have been summarized by Becker, J. A. Schulz, and Emmerson (1930). P. Schulze (1924, 1927) and Trier (1926) believed that cellulose particles are reduced chemically and structurally. Recently, Weineck (1934) obtained positive evidence, including observations on corrosive changes and loss of the original double reactivity in polarized light.

No cellulose-splitting enzymes have been isolated from the bodies of ophryoscolecids; that would be difficult in the presence of so many cellulose-decomposing bacteria. It was the suggestion of Trier (1926), and the opinion of Mangold (1929, 1933) and Westphal (1934a), that bacteria taken in with plant fragments are responsible for what cellulose decomposition has been observed within the ciliates. It is, of course, well known that the main rôle, at least, in cellulose-splitting in ruminants is performed by bacteria (Schieblich, 1929, 1932). If the ciliate had a cellulase, according to Mangold (1929), more intensive cellulose decomposition would be observed than has been possible.

Doflein (Reichenow ed., 1927-1929) stated that there is no instance of fat digestion in Protozoa. Ferber (1928) observed, in successive rumen samples, that milk-fat droplets ingested *in vivo* underwent deformation and eventually disappeared, but he recognized the probability that bacteria, ingested also, were responsible for this breakdown of fat. Weineck, in experimental feeding with lipoids, found that no fat was ingested, but he observed a small amount in the ciliates that probably had been taken up from plant materials.

Inasmuch as the ciliate population can be increased, up to a limit, by the addition of protein to the diet (Ferber, 1928; Mowry and Becker, 1930), and declines when protein is deficient, it is evident that, as Mangold (1929) stated, the ciliates have an important protein need. They obtain the protein ordinarily from the plant food. Mangold (1929, 1933) thought it unlikely that ingested bacteria or other Protozoa could sufficiently provide for this need. It is normally supplied by the addition of grain. Whether or not the cellulose of the plant food is fully utilized, the starch and protein of the plant cell plasma supply energy and materials for the activity and growth of the Protozoa.

Many investigators who have concerned themselves with the ciliates of ruminants have sought an answer to the question of their possible value to their hosts. The literature was reviewed by Becker, Schulz, and Emmerson (1930); and the subject has been discussed by Becker (1932) and Mangold (1933). Of the various opinions advanced, Zurn's belief that the ciliates could cause injury has been found entirely untenable. They are present in every ruminant in good condition; and Ferber (1929b) pointed out that the number of ciliates may serve as a guide to the host's well-being. Favorable conditions of nutrition and optimum physiological activity of the host at the same time favor a large ciliate population; and under the best conditions there may be approximately a doubling of the average density, to 2,000,000 per cc. Unfavorable conditions are rapidly reflected in a decline of that population.

Some, beginning with the first observers of the ruminant faunule (Gruby and Delafond, 1843), have supposed that the relationship is one of mutualistic symbiosis, the ciliates being in one way or another beneficial to their hosts. It is not disputed that great numbers of the ciliates are digested, but mutualism does not follow from that, unless some special contribution is made to the economy of the host.

There is still no general agreement as to whether or not the ciliates can break down cellulose, but opinions have been widely published that they aid their hosts in cellulose decomposition. Reichenow (1920) observed morphological changes in ingested cellulose fragments in *Troglodytella;* and suggested that the significance of this ciliate to its primate hosts, as well as that of Ophryoscolecidae in ruminants, lies in the use of cellulose in constructing their own easily digestible bodies, which serve as animal nutriment for the mammals. In Doflein (Reichenow ed., 1927-1929) he expressed the same opinion, and stated that the physiological rôle of these ciliates is evidently similar to that of termite flagellates, admitting, however, that they are not so important to the life of the host as these, because ruminants have other aids in cellulose decomposition. Usuelli (1930b) remarked that the softer green plant parts, which are those chiefly ingested, contain relatively little cellulose; so that even if intracellular digestion of this did take place, it would have little quantitative significance in the decomposition of cellulose in the rumen.

Mangold (1929, 1933) and his coworkers have emphasized the rôle of the ciliates in protein economy, reasoning that their bodies contain a significant part of the nitrogen available to the host, and that they derive this from plant protein, and presupposing that the ruminants can make better use of this animal protein than they can directly of plant protein. The last fact is certainly fundamental to their thesis, but Becker (1932) remarked that there is no proof of it. Becker, Schulz, and Emmerson (1930) found that goats digested slightly more protein when the ciliate population was present; but the difference was so small as to have little significance without further studies.

According to analyses by C. Schwarz (1925) of the rumen contents of slaughterhouse cattle, 20 percent of the nitrogen is in the ciliates and 11.7 percent in bacteria. Ferber (1928) found the ciliate nitrogen in sheep and goats, with a population of from 837 to 2079 ciliates per cubic millimeter, to be from 10.27 to 20.33 percent of the total, averaging about 15 percent. Ferber and Winogradowa-Fedorowa (1929) calculated that with a population of 900,000 ciliates per gram and with the total nitrogen 0.166 percent, there would be, in a 3-kilogram rumen content, 150 grams of ciliates. These would contain about 4.7 grams of protein, and the estimate would be nearly doubled by use of the figures of Mangold and Schmitt-Krahmer (1927) for the total nitrogen. There is ciliate protein also in the reticulum, but this amounts to only a fraction of that in the rumen.

There are no exact estimates of the amount of ciliate protoplasm digested in a day. Ferber and Winogradowa-Fedorowa (1929), on the basis of a fallacious estimate of a 7-percent daily division (see p. 973), calculated that 2 percent of a sheep's daily protein requirement of thirty grams might be met by the ciliates. In this estimate the higher figures for total nitrogen were used. Mangold (1933) accepted the estimate that the ciliates provide about one percent of the protein used daily and, under certain circumstances, from 2 to 3 percent. C. Schwarz (1925), however, thought it probable that the greater part of the protein requirements are met by the microörganisms; and Dogiel and Winogradowa-Fedorowa (1930), as well as Westphal (1934a), considered the daily reproductive rate to be much higher than 7 percent.

The conclusion that the ciliates are mutualists, participating in the protein economy of the host, was deduced by Ferber (1928, 1929a) from the observed fact that the population density increases at the time of growth, reproduction, and lactation. He himself, like Mowry and Becker (1930), attributed the increase to the increased nutrition and especially to larger protein supply; but, as the latter authors pointed out, the facts do not warrant the deduction.

Ferber pursuing the ideas advanced by C. Schwarz (1925), suggested that the rôle of the ciliates may be transformation of the protein in the plant food into easily digestible animal protein, and that in times of increased protein need by the host, this activity is enhanced by additions to the ciliate population. The result, of course, would be an increased ratio of ciliate nitrogen to total nitrogen. Becker, Schulz, and Emmerson (1930) remarked that although it is an observed fact that such protein transformations take place, and ciliates are eventually digested, it is doubtful if this substantially benefits the host.

Mangold (1933) recognized usefulness to the host, in the mechanical rôle of the ciliates in mixture and trituration of the rumen contents, a rôle which had been suggested by Bundle (1895), Scheunert (1909), and Dogiel (1925). Conclusive evidence that there is any essential aid to the digestive processes in the mechanical activities of the ciliate is, however, lacking.

It has been found that cattle will develop and reproduce normally on a diet that will produce symptoms of vitamin B (B_1) deficiency in other animals. It has therefore been believed that vitamins of the B complex are synthesized within the alimentary tract of ruminants, and the microorganisms have been investigated in this connection. Bechdel, Honeywell, Dutcher, and Knutsen (1928) found evidence of synthesis of the B complex by bacteria, as others have also reported. Manusardi (1933) investigated the possibility of synthesis of the vitamin (antiberiberic) by ciliates. Separating by filtration the ciliate, bacterial, and food fractions, he fed each to pigeons, together with polished rice. He concluded that it is extremely improbable that the ciliate fauna has the capacity of synthesizing vitamin B. The bacterial fraction proved to be the most antiberiberic.

Becker was the first to carry on the obvious experiments which should be made in investigating the significance of the ciliates to their hosts. That is defaunation, which led to such dramatic results in Cleveland's work with termites. If the ciliates perform any necessary function, the effect should become apparent in animals deprived of them. Becker, Schulz, and Emmerson (1930) defaunated four goats, and for periods of two and three weeks made detailed analyses of their use of nutrients. The goats were then reinfected, and after ten days analyses were continued for the same period as before. The presence or absence of ciliates was accompanied by little difference in the coefficients of digestibility. There were no differences of practical significance in the digestion of cellulose, and the goats with ciliates used only slightly more protein than those without. Winogradow, Winogradowa-Fedorowa, and Wereninow (1930) had found that raw fiber was 12.8 percent better digested in a normally faunated than in a ciliate-free ram.

Becker and Everett (1930) compared during nineteen weeks the growth of seven lambs with ciliates and seven without, the Protozoa having been removed by giving some lambs copper sulphate with milk. They found that the defaunated lambs actually grew a little more rapidly than the others.

Poljansky and Strelkow (1935) made observations on growth of four pairs of twin goats for ten months, beginning at the age of from one to two and a half months. The goats were isolated from the parents so early that they did not become naturally infected. One member of each pair was given a ciliate population. In this experiment, also, the ciliate-free goats of three pairs grew a little faster than the others; in one pair the goat with ciliates gained more.

Falaschini (1935) compared for a period of fourteen months the growth of four lambs. Two were defaunated by a milk diet after six months, then in five weeks reinfected with ciliates. The other two were defaunated after eight months, then continued on a normal diet without ciliates. The growth curves of the four lambs corresponded.

These experiments demonstrate conclusively that, at least in a period of a year or so, the host suffers no apparent detriment from lack of ciliates. The Protozoa perform no necessary rôle in nutrition, nor are their services necessary to aid the host mechanically or in keeping down the bacterial flora. When present, they may be a source of certain incidental benefits, but apparently the relationship can still best be defined in the words of Doflein (1916). He regarded it as ordinary commensalism, any value that the ciliates might have to the host being minimal and incidental.

LITERATURE CITED

Abbott, R. L. 1926. Contributions to the physiology of digestion in the Australian roach, *Periplaneta australasiae* Fab. J. exp. Zool., 44: 219-54.

Alexeieff, A. 1909. Les Flagellés parasites de l'intestin des batraciens indigènes. C. R. Soc. Biol. Paris, 67: 199-201.

—— 1910. Sur les Flagellés intestinaux des poissons marins. (Note préliminaire.) Arch. zool. exp. gén., (5) 6: N. et R., 5-20.

—— 1912. Le Parasitisme des Eugléniens et la phylogénie des Sporozoaires sensu stricto. Arch. zool. exp. gén., (5) 10: N. et R., 73-88.

André, E. 1909. Sur un nouvel Infusoire parasite des Dendrocoeles (*Ophryoglena parasitica* n. sp.). Rev. suisse Zool., 17: 273-80.

—— 1910. Sur quelques Infusoires marins parasites et commensaux. Rev. suisse Zool., 18: 173-87.

Andrews, B. J. 1930. Method and rate of protozoan refaunation in the termite *Termopsis angusticollis* Hagen. Univ. Cal. Publ. Zool., 33: 449-70.

Artigas, P. de T., and S. Pacheco. 1932. Sobre um flagellado parasito do *Myopotamus coipus*. *Octomitus myopotami* n. sp. Ann. Fac. Med. S. Paulo, 8: 79-81.

Aschner, M. 1932. Experimentelle Untersuchungen über die Symbiose der Kleiderlaus. Naturwissenschaften, 27: 501-5.

—— 1934. Studies on the symbiosis of the body louse. I. Elimination of the symbionts by centrifugalisation of the eggs. Parasitology, 26: 309-14.

Aschner, M., and E. Ries. 1933. Das Verhalten der Kleiderlaus bei Ausschaltung ihrer Symbionten. Eine experimentelle Symbiosestudie. Z. Morph. Ökol. Tiere, 26: 529-90.

Balamuth, W. 1939. Studies on regeneration in Protozoa. I. Cytology and regeneration of *Licnophora macfarlandi*. MS, Univ. Cal. Library.

Balbiani, E. G. 1885. Sur un Infusoire cilié parasite du sang de l'Aselle aquatique (*Anoplophrya circulans*). Rec. zool. suisse., 2: 277-303.

Bary, A. de. 1879. Die Erscheinung der Symbiose. Strassburg.

Beauchamp. P. de. 1911. *Astasia captiva* n. sp. Euglénien parasite de *Catenula lemnae* Ant. Dug. Arch. zool. exp. gén. (5) 6: N. et R., 52-58.

Bechdel, S. I., H. E. Soneywell, R A. Dutcher, and M. H. Knutsen. 1928. Synthesis of vitamin B in the rumen of the cow. J. biol. Chem., 80: 231-38.

Becker, E. R. 1929. Methods of rendering the rumen and reticulum of ruminants free from their normal infusorian fauna. Proc. nat. Acad. Sci. Wash., 15: 435-38.

—— 1932. The present status of problems relating to the ciliates of ruminants and Equidae. Quart. Rev. Biol., 7: 282-97.

—— 1933. Host-specificity and specificity of animal parasites. Amer. J. trop. Med., 13: 505-23.

Becker, E. R., and R. C. Everett. 1930. Comparative growths of normal and Infusoria-free lambs. Amer. J. Hyg. 11: 362-70.

Becker, E. R., and T. S. Hsiung. 1929. The method by which ruminants acquire their fauna of Infusoria, and remarks concerning experiments on the host-specificity of these Protozoa. Proc. nat. Acad. Sci. Wash., 15: 684-90.

Becker, E. R., J. A. Schulz, and M. A. Emmerson. 1930. Experiments on the physiological relationships between the stomach Infusoria of ruminants and their hosts, with a bibliography. Iowa St. Coll. J. Sci., 4: 215-51.

Becker, E. R., and M. Talbott. 1927. The protozoan fauna of the rumen and reticulum of American cattle. Iowa St. Coll. J. Sci., 1: 345-71.

Beckwith, T. D., and E. J. Rose. 1929. Cellulose digestion by organisms from the termite gut. Proc. Soc. exp. Biol. N.Y., 27: 4-6.

Beers, C. D. 1938a. Structure and division in the astomatous ciliate *Metaradiophrya asymmetrica* n. sp. J. Elisha Mitchell sci. Soc., 54: 111-25.

—— 1938b. *Hysterocineta eiseniae* n. sp., an endoparasitic ciliate from the earthworm *Eisenia lönnbergi*. Arch. Protistenk., 91: 516-25.

Beneden, P. J. van. 1876. Animal parasites and messmates. New York. Also German and French editions.

Berghe, L. van den. 1934. Sur un ciliate parasite de pontes de mollusques d'eau douce *Glaucoma paedophthora*, n. sp. C. R. Soc. Biol. Paris, 115: 1423-26.

Bishop, A. 1926. Notes upon *Sieboldiellina planariarum* (Siebold), a ciliate parasite of *Planaria torva*. Parasitology, 18: 187-94.

—— 1932. *Entamoeba aulastomi* Nöller. Cultivation, morphology, and method of division; and cultivation of *Hexamita* sp. Parasitology, 24: 225-32.

—— 1933. The morphology and division of *Hexamita gigas* n. sp. (Flagellata). Parasitology, 25: 163-70.

—— 1935. Observations upon a *"Trichomonas"* from pond water. Parasitology, 27: 246-56.

—— 1936. Further observations upon a *"Trichomonas"* from pond water. Parasitology, 28: 443-45.

—— 1937. The method of division of *Trepomonas agilis* in culture. Parasitology, 29: 413-18.

—— 1939. A note upon the systematic position of *"Trichomonas" keilini* (Bishop, 1935). Parasitology, 31: 469-72.

Boissezon, P. de. 1930. Contribution à l'étude de la biologie et de l'histophysiologie de *Culex pipiens* L. Arch. zool. exp. gén., 70: 281-431.

Brown, W. H. 1937. Nature and function of the skeletal platelets in the Ophryoscolecidae. MS, Univ. Cal. Library.

Brumpt, E., and G. Lavier. 1924. Un Nouvel Euglénien Polyflagellé parasite du tetard de *Leptodactylus ocellatus* au Brésil. Ann. Parasit. hum. comp., 2: 248-52.

Buchner, P. 1930. Tier und Pflanze in Symbiose. Berlin.

Bunting, M. 1926. Studiees on the life cycle of *Tetramitus rostratus* Perty. J. Morph., 42: 23-81.

Bunting, M., and D. H. Wenrich. 1929. Binary fission in the amoeboid and flagellate phases of *Tetramitus rostratus*. J. Morph., 47: 37-87.

Bundle, A. 1895. Ciliate Infusorien im Cöcum des Pferdes. Z. wiss. Zool., 60: 284-350.

Buscalioni, L., and S. Comes. 1910. La digestione delle membrane vegetali per opera dei Flagellati contenuti nell' intestino dei Termitidi e il problema della simbiosi. Atti Accad. gioenia, (5) 3: mem. 17, 1-16.

Bush, M. 1933. The morphology of the ciliate *Haptophrya michiganensis* Woodhead and its relation to the other members of the Astomatea. Trans. Amer. micr. Soc., 52: 223-32.

—— 1934. The morphology of *Haptophrya michiganensis* Woodhead, an astomatous ciliate from the intestinal tract of *Hemidactylium scutatum* (Schlegel). Univ. Cal. Publ. Zool., 39: 251-76.

Bütschli, O. 1889. Protozoa. III. Abt. Infusoria und System der Radiolaria in Bronn: Klassen und Ordnungen des Thier-Reichs. Leipzig.

Calkins, G. N. 1933. The Biology of the Protozoa. 2d ed. Philadelphia.

Campbell, W. G. 1929. The chemical aspect of the destruction of oak wood by powder-post and death watch beetles-*Lyctus* spp. and *Xestobium* sp. Bio-chem. J., 23: 1290-93.

Cattaneo, G. 1888. Su di un Infusorio ciliato, parassito del sangue del *Carcinus maenas*. Zool. Anz., 11: 456-59.

Caullery, M. 1922. Le Parasitisme et la symbiose. Paris.

Cépède, C. 1910. Récherches sur les Infusoires astomes. Anatomie, biologie, éthologie, parasitaire, systématique. Arch. zool. exp. gén (5) 3: 341-609.

Cépède, C., and E. Poyarkoff. 1909. Sur un Infusoire astome *Cepedella hepatica* Poyarkoff parasite du foie des *Cyclas* (*S. corneum* L.). Bull. sci. Fr. Belg., 43: 463-75.

Certes, A. 1882. Sur les parasites intestinaux de l'huitre. C. R. Acad. Sci. Paris, 95: 463-65.

Chatterji, G. C., K. N. Das, and A. N. Mitra. 1928. On an *Octomitus* n. sp. found in the intestinal contents of *Hylobates hoolock*. J. Dep. Sci. Calcutta Univ., 9: 21-24.

Chatton, E. 1920. Les Péridiniens parasites. Morphologie, reproduction, éthologie. Arch. zool. exp. gén., 59: 1-475.

—— 1936. Les Migrateurs horizontalement polarises de certains Péritriches. De leur signification. Mém. Mus. Hist. nat. Belg., 3: 913-40.

Chatton, E., and A. Lwoff. 1921. Sur une famille nouvelle d'Acinétiens, les Sphenophryidae adaptés aux branchies des mollusques acéphales. C. R. Acad. Sci. Paris, 173: 1495-97.

—— 1922a. Sur l'évolution des Infusoires des Lamellibranches. Relations des Hypocomidés avec les Ancistridés. Le genre *Hypocomides* n. gen. C. R. Acad. Sci. Paris, 175: 787-90.

—— 1922b. Sur l'évolution des Infusoires des Lamellibranches. Le Genre *Pelecyophrya,* intermédiaire entre les Hypocomidés et les Sphénophryidés. Bourgeonnement et Conjugasion. C. R. Acad. Sci. Paris, 175: 915-17.

—— 1922c. Sur l'évolution des Infusoires des Lamellibranches. Relations des Sphénophryidés avec les Hypocomidés. C. R. Acad. Sci. Paris, 175: 1444-47.

—— 1923a. Sur l'évolution des Infusoires des Lamellibranches. Les formes primitives du phylum des Thigmotriches; le genre *Thigmophrya*. C. R. Acad. Sci. Paris, 177: 81-83.

—— 1923b. Un Cas remarquable d'adaptation: *Ellobiophrya donacis* n. g., n. sp., Péritriche inquilin des branches de *Donax vittatus* (Lamellibranche). C. R. Soc. Biol. Paris, 88: 749-52.

—— 1924. Sur l'évolution des Infusoires des Lamellibranches: Morphologie comparée des Hypocomidés. Les nouveaux genres *Hypocomina* et *Hypocomella*. C. R. Acad. Sci. Paris, 178: 1928-30.

—— 1926. Diagnoses de Ciliés thigmotriches nouveaux. Bull. Soc. zool. Fr., 51: 345-52.

—— 1929. Contribution a l'étude de l'adaptation. *Ellobiophrya donacis* Ch. et Lw. Péritriche vivant sur les branchies de l'Acéphale *Donax vittatus* da Costa. Bull. biol., 63: 321-49.

—— 1934a. Sur un cilié thigmotriche nouveau: *Gargarius gargarius* n. gen., n. sp., de *Mytilus edulis*. Bull. Soc. zool. Fr., 59: 375-76.

—— 1934b. Sur un Infusoire parasite des poils sécréteurs des Crustacés Edriophthalmes et la famille nouvelle des Pilisuctoridae. C. R. Acad. Sci. Paris, 199: 696-99.

—— 1935. Les Ciliés apostomes. Morphologie, cytologie, éthologie, évolution, systématique. Première partie. Aperçu historique et général. Étude monographique des genres et des espèces. Arch. zool. exp. gén., 77: 1-453.

—— 1936a. Les Pilisuctoridae Ch. et Lw. Ciliés parasites des poils sécréteurs des Crustacés Edriophthalmes. Polarité, orientation, et desmodexie chez les Infusoires. Bull. biol., 70: 86-144.

—— 1936b. Les Remaniements et la continuité des cinétome au cours de la scission chez les Thigmotriches Ancistrumidés. Arch. zool. exp. gén., 78: N. et R., 84-91.

—— 1937. Sur l'enkystement d'un Thigmotriche Ancistrumidé: *Proboveria loripedis* Chatton et Lwoff, et les phénomènes qui l'accompagnent. C. R. Soc. Biol. Paris, 124: 807-10.

Cheissin, E. [E. Chejsin]. 1928. Vorlaufige Mitteilung über einige parasitische Infusorien des Bajkal-Sees. C. R. Acad. Sci. U.R.S.S., 1928 (A) : 295-99.

—— 1930. Morphologische und systematische Studien über Astomata aus dem Baikalsee. Arch. Protistenk., 70: 531-618.

—— 1931. Infusorien Ancistridae und Boveridae aus dem Baikalsee. Arch. Protistenk., 73: 280-304.

—— 1932. Sur la morphologie et la classification des Infusoires parasitiques de la famille des Ptychostomidae. Trav. Sta. limnol. Lac Bajkal, 2: 29-53.

Child, H. J. 1933. The anatomy and histology of the digestive tract and associated organs in *Zootermopsis nevadensis* (Hagen), with some observations on morphogenesis, behavior during ecdysis, and the protozoan fauna. MS, Univ. Cal. Library.

—— 1934. The internal anatomy of termites and the histology of the digestive tract *in* Kofoid et al., Termites and termite control. Berkeley, Cal.

Cleveland, L. R. 1923. Correlation between the food and morphology of termites and the presence of intestinal Protozoa. Amer. J. Hyg., 3: 444-61.

—— 1924. The physiological and symbiotic relationships between the intestinal Protozoa of termites and their host, with special reference to *Reticulitermes flavipes* Kollar. Biol. Bull., 46: 178-227.

—— 1925a. The method by which *Trichonympha campanula,* a protozoön in the intestine of termites, ingests solid particles of wood for food. Biol. Bull., 48: 282-88.

—— 1925b. The ability of termites to live perhaps indefinitely on a diet of pure cellulose. Biol. Bull., 48: 289-93.

—— 1925c. The effects of oxygenation and starvation on the symbiosis between the termite, *Termopsis,* and its intestinal flagellates. Biol. Bull., 48:309-26.

——1926. Symbiosis among animals with special reference to termites and their intestinal flagellates. Quart. Rev. Biol., 1: 51-60.

—— 1928a. Further observations and experiments on the symbiosis between termites and their intestinal Protozoa. Biol. Bull. 54: 231-37.

—— 1928b. *Tritrichomonas fecalis* nov. sp. of man; its ability to grow and multiply indefinitely in faeces diluted with tap water and in frogs and tadpoles. Amer. J. Hyg., 8: 232-55.

—— 1930. The symbiosis between the wood-feeding roach, *Cryptocercus punctulatus* Scudder, and its intestinal flagellates. Anat. Rec., 47: 293-94.

—— 1935. The intranuclear achromatic figure of *Oxymonas grandis* sp. nov. Biol. Bull., 69: 54-63.

Cleveland, L. R., S. R. Hall, E. P. Sanders, and J. Collier. 1934. The wood-feeding roach *Cryptocercus,* its Protozoa, and the symbiosis between Protozoa and roach. Mem. Amer. Acad. Arts Sci., 17: i-x, 185-342.

Codreanu, M., and R. Codreanu. 1928. Un Nouvel Euglénien (*Astastia Chaetogastris* n. sp.) parasite coelomique d'un Oligochete (*Chaetogaster diastrophus* Gruith). C. R. Soc. Biol. Paris, 9: 1368-70.

Codreanu, R. 1928. Un Infusoire nouveau (*Frontonia branchiostomae* n. sp.) commensal de l'Amphioxus (*Branchiostoma lanceolatum* Pall). C. R. Soc. Biol., Paris, 98: 1078-80.

—— 1930. Sur la phase interne du cycle évolutif de deux formes d'*Ophryoglena,* Infusoires endoparasites des larves d'Ephémères. C. R. Acad. Sci. Paris, 190: 1154-57.

—— 1934. La Présence d'*Ophryoglena,* Ciliés endoparasites chez les nymphes de l'Ephémère *Oligoneuria rhenana* Imhoff en France. Ann. Protist., 4: 181-83.

Cohen, W. E. 1933. An analysis of termite (*Eutermes exitiosus*) mound material. Jour. Counc. sci. industr. Res. Aust., 6: 166-69.

Collin, B. 1914. Notes protistologiques. Arch. zool. exp. gén., 54; N. et R., 85-97.

Cross, J. B. A study of *Oxymonas minor* Zeliff from the termite *Kalotermes minor* Hagen. Univ. Cal. Publ. Zool. (*In press.*)

Crouch, H. B. 1934. Observations on *Hexamita mormotae* n. sp., a protozoan flagellate from the woodchuck *Marmota monax* (Linn). Iowa St. Coll. J. Sci. 8: 513-17.

Cuénot, L. 1891. Protozoaries commensaux et parasites des Échinodermes. Rev. biol. nord Fr., 3: 285-300.

—— 1894. Über *Hemispeiropsis antedonis* Cuén., ein an den Comatulen lebendes Infusorium. Zool. Anz., 17: 316.

Cunha, A. M. da, and J. Muniz. 1922. *Octomitus avium* n. sp. Brazil-med., 36: 386-88.

—— 1927. Estudo sobre os Flagellados intestinaes das Aves do Brazil. Mem. Inst. Osw. Cruz, 20: 19-33.

—— 1929. Nota sobre os parasitas intestinaes do *Macacus rhesus* com a descripcão de uma nova especie de *Octomitus.* (Suppl.) Mem. Inst. Osw. Cruz, 5: 34-37.

Das Gupta, B. M. 1935. The occurrence of a *Trepomonas* sp. in the caecum of turtles. J. Parasit., 21: 125-26.

Davis, H. S. 1923. Observations on an intestinal flagellate of the trout. J. Parasit., 9: 153-60.

—— 1926. *Octomitus salmonis,* a parasitic flagellate of trout. Bull. U. S. Bur. Fish., 42: 9-26.

Dickman, A. 1931. Studies on the intestinal flora of termites with reference to their ability to digest cellulose. Biol. Bull., 61: 85-92.

Diesing, K. M. 1866. Revision der Prothelminthen. Abtheilung: Amastigen. I. Amastigen ohne Peristom. S. B. Akad. Wiss. Wien., Math.-naturw. Cl., 52: 505-79.

Dobell, C. 1935. Researches on the intestinal Protozoa of monkeys and man. VII. On the *Enteromonas* of macaques and *Embadomonas intestinalis*. Parasitology, 27: 564-92.

Doflein, F. 1916. Lehrbuch der Protozoenkunde 4th ed., Jena.

Doflein, F., and E. Reichenow. 1927-29. Lehrbuch der Protozoenkunde. 5th ed., Dofleins, Lehrbuch, Jean.

Dogiel, V. A. 1925. Über die Art der Nahrung und der Nahrungsaufnahme bei den im Darme der Huftiere parasitierenden Infusorien. Trav. Soc. Nat. St-Pétersb. (Leningr.) Sect. Zool. et Physiol., 54: 69-93.

—— 1927. Monographie der Familie Ophryoscolecidae. Teil I. Arch. Protistenk., 59: 1-288.

Dogiel, V. A., and T. Fedorowa. 1925. Über den Bau und die Funktion des inneren Skeletts der Ophyroscoleciden. Zool. Anz., 62: 97-107.

—— 1929. Über die Zahl der Infusorien im Wiederkauermagen. Zbl. Bakt., (I) 112: 135-42.

Dogiel, V. A., and A. V. Furssenko. 1921. Neue ektoparasitische Infusorien von Landisopoden. Trav. Soc. Nat. St-Pétersb. (Leningr.), 51: 147-58, 199-202.

Dogiel, V. A., and T. Winogradowa-Fedorowa. 1930. Experimentelle untersuchungen zur Biologie der Infusorien des Wiederkäuermagens. Wiss. Arch. Landwirtsch. (B) 3: 172-88.

Dore, W. H., and R. C. Miller. 1923. The digestion of wood by *Teredo navalis*. Univ. Cal. Publ. Zool., 22: 383-400

Duboscq, O., and P. Grassé. 1934. Notes sur les protistes parasites des termites de France. IX. L'Enkystement des flagellés de *Calotermes flavicollis*. Arch. zool. exp. gén., 76: N. et R., 66-72.

Edmondson, C. H. 1906. The Protozoa of Iowa. Proc. Davenport Acad. Sci., 11: 1-124.

Eikenberry, W. L., and R. A. Waldron. 1930. Educational biology. Boston.

Emik, L. O. MS$_1$. Studies on the nutrition of *Trichonympha*. (Univ. Cal. Library, 1937.) MS$_2$. Ingestion of food by *Trichonympha*. Trans. Amer. micr. Soc. (*In press*).

Entz, G. 1904. Die Fauna der kontinentalen Kochsalzwässer. Math. naturw. Ber. Ung., 19: 89-124.

Epstein, H. 1926. Infektion des Nervensystems von Fischen durch Infusorien. Arch. russ. protist., 5: 169-80.

Fabre-Domergue, P. 1885. Note sur les Infusoires ciliés de la Baie de Concarneau. J. Anat. Paris., 21: 554-68.

—— 1888. Étude sur l'organisation des Urcéolaires et sur quelques genres d'Infusoires voisins de cette famille. J. anat. Paris, 24: 214-60.

Falaschini, A. 1935. Gli infusori dell'apparato digerente degli erbivori sono indispensabili per la vita dei loro ospiti? Boll. Lab. Zool. agr. Bachic. Milano, 4 (II): 151-60.

Falck, R. 1930. Die Scheindestruktion des Koniferenholzes durch die Larven des Hausbockes (*Hytotrupes bajulus* L.). Cellulose-chem., 11: 89-91.

Fantham, H. B. 1922. Some parasitic Protozoa found in South Africa. -V S. Afr. J. Sci., 19: 332-39.

—— 1936. The evolution of parasitism among the Protozoa. Scientia, Bologna, 59: 316-24.

Fauré-Fremiet, E. 1924. Contribution à la connaissance des Infusoires planktoniques. Bull. biol., (Suppl.), 6: 1-171.

Ferber, K. E. 1928. Die Zahl und Masse der Infusorien im Pansen und ihre Bedeutung für den Eiweissaufbau beim Wiederkäuer. Z. Tierz. Zücht-Biol., 12: 31-63.

—— 1929a. Die Veränderungen der Infusorienzahl im Pansen der Wiederkäuer im Zusammenhang mit den Veränderungen des Eiweissumsatzes. Z. Tierz. ZüchtBiol., 15: 375-90.

—— 1929b. Über die optimale Beschaffenheit des Panseninhalts der Wiederkäur. Wiss. Arch. Landw., 1: 597-600.

Ferber, K. E., and T. Winogradowa-Fedorowa. 1929. Zählung und Teilungsquote der Infusorien im Pansen der Wiederkäuer. Biol. Zbl., 49: 321-28.

Foulke, S. G. 1884. Some notes on *Manayunkia speciosa*. Proc. Acad. nat. Sci. Philad., 1884: 48-9.

Fuhrmann, O. 1894. Die Turbellarien der Umgebung von Basel. Rev. suisse Zool., 2: 215-90.

Fulton, J. F. 1923. *Trichodina pediculus* and a new closely related species. Proc. Boston Soc. nat. Hist., 37: 1-29.

Geiman, Q. M. 1932. The intestinal Protozoa of the larvae of the crane fly, *Tipula abdominalis*. J. Parasit., 19: 173.

Ghosh, E. 1925. On a new ciliate, *Balantidium knowlesii* sp. nov., a coelomic parasite in *Culicoides peregrinus*. Parasitology, 17: 189.

Giard, A. 1880. Note sur l'existence temporaire de Myriapodes dans les fosses nasales de l'homme, suivie de quelques reflexions sur le parasitisme inchoatif. Bull. Sci. Fr. Belg., 12: 1-11.

Gicklhorn, J. 1925. Notiz über *Euglena cyclopicola* nov. sp. Arch. Protistenk., 51: 542-48.

Goetsch, W. 1936. Beiträge zur Biologie des Termitenstaates. Z. Morphol. Ökol. Tiere, 31: 490-560.

Graff, L. von. 1882. Monographie der Turbellarien. I. Rhabdocoelida. Leipzig.

Graham, G. L. 1935. *Giardia* infections in a nematode from cattle. J. Parasit., 21: 127-28.

Grassé, P. P. 1924. *Octomastix parvus* Alex., diplozoaire parasite de la Cistude d'Europe. C. R. Soc. Biol. Paris, 91: 439-42.

—— 1935. Parasites et parasitisme. Paris.

Grassé, P. P., and P. de Boissezon. 1929. *Turchiniella culicis* n.g., n. sp.

Infusoire parasite de l'hemocoele d'un *Culex* adulte. Bull. Soc. zool. Fr., 54: 187-91.

Green, R. A., and E. L. Breazeale. 1937. Bacteria and the nitrogen metabolism of termites. J. Bac., 33: 95-96.

Gruby, and O. Delafond. 1843. Récherches sur des animalcules se développant en grand nombre dans l'estomac et dans les intestins, pendant la digestion des animaux herbivores et carnivores. C. R. Acad. Sci. Paris, 17: 1304-08.

Haas, G. 1933. Beiträge zur Kenntnis der Cytologie von *Ichthyophthirius multifiliis* Fouq. *Arch. Protistenk.*, 81: 88-137.

Hall, R. P., and T. L. Jahn. 1929. On the comparative cytology of certain euglenoid flagellates and the systematic position of the families Euglenidae Stein and Astasiidae Bütschli. Trans. Amer. micr. Soc., 48: 388-405.

Hall, S. R. 1931. Observations on *Euglena leucops*, sp. nov., a parasite of *Stenostomum*, with special reference to nuclear division. Biol. Bull., 60: 327-44.

Haswell, W. A. 1892. Note on the occurrence of a flagellate infusorian as an intracellular parasite. Proc. Linn. Soc. N. S. W., (2) 7: 197-99.

—— 1907. Parasitic Euglenae. Zool. Anz., 31: 296-97.

Haupt, A. W. 1932. Fundamentals of biology, 2d. ed., New York.

Hazard, F. O. 1937. Two new host records for the protozoan *Haptophrya michiganensis* Woodhead. J. Parasit., 23: 315-16.

Hegner, R. W. 1922. Frog and toad tadpoles as sources of intestinal Protozoa for teaching purposes. Science, 56: 439-41.

—— 1923. Observations and experiments on Euglenoidina in the digestive tract of frog and toad tadpoles. Biol. Bull., 45: 162-80.

—— 1924. Parasitism among the Protozoa. Sci. Mon., N.Y., 19: 140-45.

—— 1926a. Host-parasite relationships among human Protozoa. Proc. R. Soc. Med., 19: 41-44.

—— 1926b. The biology of host-parasite relationships among Protozoa living in man. Quart. Rev. Biol., 1: 393-418.

—— 1926c. Homologies and analogies between free-living and parasitic Protozoa. Amer. Nat., 60: 516-25.

—— 1928. The evolutionary significance of the protozoan parasites of monkeys and man. Quart. Rev. Biol., 3: 225-44.

—— 1937. Parasite reactions to host modifications. J. Parasit., 23: 1-12.

Hegner, R., and J. Andrews (ed.). 1930. Problems and methods of research in protozoology. New York.

Heidenreich, E. 1935. Untersuchungen an parasitischen Ciliaten aus Anneliden. Teil I: Systematik. Arch. Protistenk., 84: 315-92.

Hendee, E. C. 1933. The association of the termites, *Kalotermes minor, Reticulitermes hesperus*, and *Zootermopsis angusticollis* with fungi. Univ. Cal. Publ. Zool., 39: 111-34.

Henderson, J. C. Studies of some amoebae from a termite of the genus *Cubitermes*. Univ. Cal. Publ. Zool. (In press.)

Hertig, M., W. H. Taliaferro, and B. Schwartz. 1937. Report of the Committee on Terminology. J. Parasit., 23: 325-29.

Hertwig, O. 1883. Die Symbiose oder das Genossenschaftsleben im Thierreich. Jena.

Hetherington, A. 1933. The culture of some holotrichous ciliates. Arch. Protistenk., 80: 255-80.

—— 1936. The precise control of growth in a pure culture of a ciliate, *Glaucoma pyriformis*. Biol. Bull., 70: 426-40.

Hinshaw, W. R., E. McNeil, and C. A. Kofoid. 1938a. The presence and distribution of *Hexamita* sp. in turkeys in California. J. Amer. med. Ass., 93: 160.

—— 1938b. The relationship of *Hexamita* sp. to an enteritis of turkey poults. The Cornell Veterinarian, 28: 281-93.

Holdaway, F. G. 1933. The composition of different regions of mounds of *Eutermes exitiosus* Hill. J. Counc. sci. industr. Res. Aust., 6: 160-65.

Hollande, A. 1939. Sur un genre nouveau de Trichomonadide libre: *Coelotrichomastix convexas* nov. gen. nov. sp. Bull. Soc. zool. Fr., 64: 114-19.

Howland, R. B. 1928. A note on *Astasia captive* Beauch. Science, 68: 37.

Hsiung, T. S. 1930. A monograph on the Protozoa of the large intestine of the horse. Iowa, St. Coll. J. Sci., 4: 359-423.

Hudson, C. T., and P. H. Gosse. 1889. The Rotifera: or wheel-animalcules, both British and foreign. London.

Hungate, R. E. 1936. Studies on the nutrition of *Zootermopsis*. I. The rôle of bacteria and molds in cellulose decomposition. Zbl. Bakt., (II), 94: 240-49.

—— 1938. Studies on the nutrition of *Zootermopsis*. II. The relative importance of the termite and the Protozoa in wood digestion. Ecology, 19: 1-25.

—— 1939. Experiments on the nutrition of *Zootermopsis*. III. The anaerobic carbohydrate dissimilation by the intestinal Protozoa. Ecology, 20: 230-45.

Hunninen, A. V., and R. Wichterman. 1936. Hyperparasitism: A species of *Hexamita* (Protozoa, Flagellata) found in the reproductive systems of *Deropristis inflata* (Trematoda) from marine eels. J. Parasit., 22: 540.

Ikeda, I., and Y. Ozaki. 1918. Notes on a new *Boveria* species, *Boveria labialis* n. sp. J. Coll. Sci. Tokyo, 40: art. 6, 1-25.

Issel, R. 1903. Ancistridi del Golfo di Napoli. Studio monografico sopra una nuova famiglia di cigliati commensali di molluschi marini. Mitt. Zool. Sta. Neapel, 16: 63-108.

Jahn, T. L., and W. R. McKibben. 1937. A colorless euglenoid flagellate, *Khawkinea halli* n. gen., n. sp. Trans. Amer. micr. Soc., 56: 48-54.

Janda, V., and O. Jírovec. 1937. Über künstlich hervorgerufenen Parasitismus eines freilebenden ciliaten *Glaucoma piriformis* und Infekionsversuche mit *Euglena gracilis* und *Spirochaeta biflexa*. Mém. Soc. Zool. Tchéco-slovaque de Prague, 5: 34-57.

Jarocki, J. 1934. Two new hypocomid ciliates, *Heterocineta janickii* sp. n. and *H. lwoffi* sp. n., ectoparasites of *Physa fontinalis* (L.) and *Viviparus fasciatus* Müller. Mém. Cl. Sci. Acad. polon. math. nat. (B) 1934: 167-87.

—— 1935. Studies on ciliates from fresh-water molluscs. I. General remarks on protozoan parasites of Pulmonata. Transfer experiments with species of *Heterocineta* and *Chaetogaster limnaei*, their additional host. Some new hypocomid ciliates. Bull. int. Acad. Cracovie, Cl. Sci. math. nat. (B:II) 1935: 201-30.

Jarocki, J., and Z. Raabe. 1932. Über drei neue Infusorien-Genera der Familie Hypocomidae (Ciliata Thigmotricha), Parasiten in Süsswasser-muscheln. Bull. int. Acad. Cracovie, Cl. Sci. math. nat. (B:II) 1932: 29-45.

Jucci, C. 1932. Sulla presenza di batteriociti nel tessuto adiposo dei Termitidi. Arch. zool. (ital.) Torino, 16: 1422-29.

Kahl, A. 1926. Neue und wenig bekannte Formen der holotrichen und heterotrichen Ciliaten. Arch. Protistenk., 55: 197-438.

—— 1930. Urtiere oder Protozoa I: Wimpertiere oder Ciliata (Infusoria). I. Allgemeiner Teil und Prostomata, *in* Dahl: Die Tierwelt Deutschlands, 18 Teil, Jena.

—— 1931. *Ibid.,* II. Holotricha, 21 Teil.

—— *Ibid.,* III. Spirotricha, 25 Teil.

—— 1933. Ciliata libera et ectocommensalia, *in* Grimpe, A. and E. Wagler: Die Tierwelt der Nord- und Ostsee. Lief. 23, Teil II. C3, 29-146. Leipzig.

—— 1934. Ciliata ectocommensalia et parasitica, *in* Grimpe, G. and E. Wagler: Die Tierwelt der Nord- und Ostsee, Lief. 26, Teil II C4, 147-83. Leipsig.

—— 1935. *Ibid.,* IV. Peritricha und Chonotricha, 30 Teil.

Katzin, L. I., and H. Kirby, Jr. 1939. The relative weights of termites and their Protozoa. J. Parasit., 25: 444-45.

Keilin, D. 1921. On a new ciliate: *Lambornella stegomyiae* n. g., n. sp., parasitic in the body-cavity of the larvae of *Stegomyia scutellaris* Walker (Diptera, Nematocera, Culicidae). Parasitology, 13: 216-24.

Kepner, W. A., and R. P. Carroll. 1923. A ciliate endoparasitic in *Stenostoma leucops*. J. Parasit., 10: 99-100.

Kepner, W. A., and J. S. Carter. 1931. Ten well-defined new species of *Stenostomum*. Zool. Anz., 93: 108-23.

Kidder, G. W. 1929. *Streblomastix strix,* morphology and mitosis. Univ. Cal. Publ. Zool., 33: 109-24.

—— 1933a. Studies on *Conchophthirius mytili* De Morgan. I. Morphology and division. Arch. Protistenk., 79: 1-24.

—— 1933b. Studies on *Conchophthirius mytili* De Morgan. II. Conjugation and nuclear reorganization. Arch. Protistenk., 79: 25-49.

—— 1933c. On the genus *Ancistruma* Strand (*Ancistrum Maupas*). I. The structure and division of *A. mytili* Quenn. and *A. isseli* Kahl. Biol. Bull., 64: 1-20.

—— 1933d. *Conchophthirius caryoclada* sp. nov. (Protozoa, Ciliata). Biol. Bull., 65: 175-78.

—— 1933e. On the genus *Ancistruma* Strand (*Ancistrum* Maupas). II. The conjugation and nuclear reorganization of *A. isseli* Kahl. Arch. Protistenk., 81: 1-18.

—— 1934a. Studies on the ciliates from fresh water mussels. I. The structure and neuromotor system of *Conchophthirius anodontae* Stein, *C. curtus* Engl., and *C. magna* sp. nov. Biol. Bull., 66: 69-90.

—— 1934b. Studies on the ciliates from fresh water mussels. II. The nuclei of *Conchophthirius anodontae* Stein, *C. curtus* Engl., and *C. magna* Kidder, during binary fission. Biol. Bull., 66: 286-303.

—— 1937. The intestinal Protozoa of the wood-feeding roach *Panesthia*. Parasitology, 29: 163-205.

Kidder, G. W., and F. M. Summers. 1935. Taxonomic and cytological studies on the ciliates associated with the amphipod family Orchestiidae from the Woods Hole district. I. The stomatous holotrichous ectocommensals. Biol. Bull., 68: 51-68.

Kirby, H., Jr. 1927. Studies on some amoebae from the termite *Mirotermes*, with notes on some other Protozoa from the Termitidae. Quart. Jour. micr. Sci., 71: 189-222.

—— 1928. A species of *Proboscidiella* from *Kalotermes* (*Cryptotermes*) *dudleyi* Banks, a termite of Central America, with remarks on the oxymonad flagellates. Quart. Jour. micr. Sci., 72: 355-86.

—— 1930. Trichomonad flagellates from termites. I. *Tricercomitus* gen. nov., and *Hexamastix* Alexeieff. Univ. Cal. Publ. Zool., 33: 393-444.

—— 1931. Trichomonad flagellates from termites. II. *Eutrichomastix* and the subfamily Trichomonadinae. Univ. Cal. Publ. Zool., 36: 171-262.

—— 1932a. Two Protozoa from brine. Trans. Amer. micr. Soc., 51: 8-15.

—— 1932b. Flagellates of the genus *Trichonympha* in termites. Univ. Cal. Publ. Zool., 37: 349-476.

—— 1932c. Protozoa in termites of the genus *Amitermes*. Parasitology, 24: 289-304.

—— 1937. Host-parasite relations in the distribution of Protozoa in termites. Univ. Cal. Publ. Zool., 41: 189-212.

—— 1939. The Templeton Crocker Expedition of the California Academy of Sciences, 1932, No. 39. Two new flagellates from termites in the genera *Coronympha* Kirby, and *Metacoronympha* Kirby, new genus. Proc. Cal. Acad. Sci. (4) 22: 207-20.

Klebs, G. 1893. Flagellatenstudien. Theil I. Z. wiss. Zool., 55: 265-351 (p. 328).

Knoth, M. 1928. Neue Versuche sur Züchtung der im Pansen von Wiederkäuern lebenden Ophryoscoleciden (Ciliata). Z. Parasitenk., 1: 262-82.

Koch, A. 1933a. Über das Verhalten symbiontenfreier Sitodrepalarven. Biol. Zbl., 53: 199-203.

—— 1933b. Über künstlich symbiontenfrei gemachte Insekten. Zool. Ans., (Supplb.) 6: 143-50.

—— 1936. Symbiosestudien. II. Experimentelle Untersuchungen an *Oryzaephilus surinamensis* L. (Cucujidae, Coleopt.) Z. Morph. Ökol. Tiere, 32: 137-80.

—— 1938a. Die Bakteriensymbiose der Termiten. Zool. Anz. (Supplb.) 11: 81-90.

—— 1938b. Symbiosestudien. III. Die intrazellulare Bakteriensymbiose von *Mastotermes darwiniensis* Froggatt (Isoptera). Z. Morph. Ökol. Tiere, 34: 534-609.

Kofoid, C. A. 1903. On the structure of *Protophrya ovicola,* a ciliate infusorian from the brood-sac of *Littorina rudis* Don. Mark Anniversary Volume, Art. 5, 111-20. New York.

—— 1935. On two remarkable ciliate Protozoa from the caecum of the Indian elephant. Proc. nat. Acad. Sci. Wash., 21: 501-6.

Kofoid, C. A., and M. Bush. 1936. The life cycle of *Parachaenia myae* gen. nov., sp. nov., a ciliate parasitic in *Mya arenaria* Linn. from San Francisco Bay, California. Bull. Mus. Hist. nat. Belg., 12 (22) : 1-15.

Kofoid, C. A., and J. F. Christenson. 1934. Ciliates from *Bos gaurus* H. Smith. Univ. Cal. Pub. Zool., 39: 341-92.

Kofoid, C. A., and R. F. MacLennan. 1930. Ciliates from *Bos indicus* Linn. I. The genus *Entodinium* Stein. Univ. Cal. Publ. Zool., 33: 471-544.

—— 1932. Ciliates from *Bos indicus* Linn. II. A revision of *Diplodinium* Schuberg. Univ. Cal. Publ. Zool., 37: 53-152.

—— 1933. Ciliates from *Bos indicus* Linn. III. *Epidinium* Crawley, *Epiplastron* gen. nov., and *Ophryoscolex* Stein. Univ. Cal. Publ. Zool., 39: 1-34.

Kofoid, C. A., and O. Swezy 1919. Studies on the parasites of the termites. On *Streblomastix strix,* a polymastigote flagellate with a linear plasmodial phase. Univ. Cal. Publ. Zool., 20: 1-20.

Kofoid, C. A., et al. (ed.) 1934. Termites and termite control. Berkeley, Cal.

Koidzumi, M. 1921. Studies on the intestinal Protozoa found in the termites of Japan. Parasitology, 13: 255-309.

König, A. 1894. *Hemispeiropsis comatulae,* eine neue Gattung der Urceolariden. S.B. Akad. Wiss. Wien, 103: 55-60.

Kotlán, A. 1923. Zur Kenntnis der Darmflagellaten aus der Hausente und anderen Wasservögeln. Zbl. Bakt., (I) Orig. 90: 24-28.

Kudo, R. R. 1939. Protozoology. Springfield.

Labbé, A. 1899. Sporozoa *in* Schulze, F. E.: Das Tierreich, Lief. 5, 1-180. Berlin.

Lamborn, W. A. 1921. A protozoon pathogenic to mosquito larvae. Parasitology, 13: 213-15.

Lavier, G. 1925. Infections héréditaires par les parasites animaux. Ann. Parasit. hum. comp., 3: 306-21.

—— 1935. Sur le parasitisme dans l'intestin d'amphibiens, de Flagellés du genre *Trepomonas* Duj. C. R. Soc. Biol. Paris, 118: 991-92.

——1936a. Sur la structure des Flagellés du genre *Hexamita* Duj. C. R. Soc. Biol. Paris, 121: 1177-80.

—— 1936b. Sur quelques Flagellés intestinaux de poissons marins. Ann. Parasit. hum. comp., 14: 278-89.

—— 1936c. Sur un Trichomonadidé libre des eaux stagnantes. Ann. Parasit. hum. comp., 14: 359-68.

Lavier, G., and H. Galliard. 1925. Parasitisme sanguin d'un *Hexamitus* chez un crapaud *Bufo calamita*. Ann. Parasit. hum. comp., 3: 113-15.

Leydig, F. 1857. Über *Hydatina senta*. Arch. Anat. Physiol. wiss. Med., 24: 404-16.

Lichtenstein, J. L. 1921. *Ophryoglena collini* n. sp. parasite coelomique des larves d'Ephémères. C. R. Soc. Biol. Paris, 85: 794-96.

Light, S. F., and M. F. Sanford. 1927. Are the protozoan faunae of termites specific? Proc. Soc. exp. Biol. N.Y., 25: 95-96.

—— 1928. Experimental transfaunation of termites. Univ. Cal. Publ. Zool., 31: 269-74.

Lucas, M. S. 1934. Ciliates from Bermuda sea urchins. I. *Metopus*. J. R. micr. Soc., 54: 79-93.

Lund, E. E. 1930. The effect of diet upon the intestinal fauna of *Termopsis*. Univ. Cal. Publ. Zool., 36: 81-96.

Lwoff, A. 1923. Sur la nutrition des Infusoires. C. R. Acad. Sci. Paris, 176: 928-30.

—— 1924. Infection expérimentale à *Glaucoma piriformis* (Infusoire) chez *Galleria mellonella* (Lépidoptère). C. R. Acad. Sci. Paris, 178: 1106-08.

—— 1929. Milieux de culture et d'entretien pour *Glaucoma piriformis* (Cilié) C. R. Soc. Biol. Paris, 100: 635.

—— 1932. Réchêrches biochemiques sur la nutrition des Protozoaires, le pouvoir de synthèse. Monogr. Inst. Pasteur.

Lynch, J. E. 1929. Studies on the ciliates from the intestine of *Strongylocentrotus*. I. *Entorhipidium* gen. nov. Univ. Cal. Publ. Zool., 33: 27-56.

—— 1930. Studies on the ciliates from the intestine of *Strongylocentrotus*. II. *Lechriopyla mystax* gen. nov., sp. nov. Univ. Cal. Publ. Zool., 33: 307-50.

MacArthur, W. P. 1922. A holotrichous ciliate pathogenic to *Theobaldia annulata* Schrank. J. R. Army med. Cps., 38: 83-92.

Mackinnon, D. L. 1912. Protists parasitic in the larva of the crane fly, *Tipula* sp. Parasitology, 5: 175-89.

MacLennan, R. F. 1934. The morphology of the glycogen reserves in *Polyplastron*. Arch. Protistenk., 81: 412-19.

MacLennan, R. F., and F. H. Connell. 1931. The morphology of *Eupoterion pernix* gen. nov., sp. nov., a holotrichous ciliate from the intestine of *Acmaea persona* Eschscholtz. Univ. Cal. Publ. Zool., 36: 141-56.

Madsen, H. 1931. Bemerkungen über einige entozoische und freilebenden marine Infusorien der Gattungen *Uronema, Cyclidium, Cristigera, Aspidisca* und *Entodiscus* gen. nov. Zool. Anz., 96: 99-112.

Mangold, E. 1929. Die Verdauung der Wiederkäuer *in* Handbuch der Ernährung und des Stoffwechsels der landwirtschaftlichen Nutztiere, als Grundlagen der Fütterungslehre, 2: 107-237 (Berlin, Springer).

—— 1933. Die Infusorien des Pansens und ihre Bedeutung für die Ernährung der Wiederkäuer. Biederm. Zbl., (A), n.f. 3: 161-87.

Mangold, E., and T. Radeff. 1930. Die Quelle für die Infektion des Wiederkaüermagens mit Infusorien. Wiss. Arch. Landw. (B), 4: 173-99.

Mangold, E., and C. Schmitt-Krahmer. 1927. Die Stickstoffverteilung im Pansen der Wiederkäuer bei Fütterung und Hunger und ihre Beziehung zu den Pansen-Infusorien. Biochem. Z., 191: 411-22.

Mangold, E., and F. Usuelli. 1930. Die schädliche Wirkung der Milch und der Veränderung der H.- Ionkonzentration auf die Infusorien des Wiederkäuermagens. Wiss. Arch. Landw. (B), 3: 189-201.

Mansour, K. 1936. The problem of the nutrition of wood eating insects. C. R. XIIᵉ Int. Cong. Zool., 1: 233-41.

Mansour, K., and J. J. Mansour-Bek. 1933. Zur Frage der Holzverdauung durch Insektenlarven. Proc. K. Akad. Wetensch., 36: 795-99.

—— 1934a. On the digestion of wood by insects. Jour. exp. Biol., 11: 243-56.

—— 1934b. The digestion of wood by insects and the supposed rôle of microorganisms. Biol. Rev., 9: 363-82.

Manusardi, L. 1933. Gli Infusori ciliati del rumine sintetizzano la vitamina B? Boll. Lab. Zool. agr. Bachic. Milano, 4(I): 140-48.

Margolin, S. 1930. Methods for the cultivation of cattle ciliates. Biol. Bull., 59: 301-5.

Matubayasi, H. 1937. Studies on parasitic Protozoa in Japan. I. On flagellates parasitic in snakes. Annot. zool. jap., 16: 245-52.

Maupas, E. 1883. Contribution a l'étude morphologique et anatomique des Infusoires ciliés. Arch. zool. exp. gén. (2)1: 427-664.

May, E. 1939. The behavior of the intestinal Protozoa of termites at the time of the last ecdysis. MS, Univ. Cal. Library.

McFarland, J. 1913. Biology general and medical. 2d ed., Philadelphia.

Mercier, L., and R. Poisson. 1923. Un Cas de parasitisme accidentel d'une Nèpe par un Infusoire. C. R. Acad. Sci. Paris, 176: 1838-41.

Metcalf, M. M. 1923. The opalinid ciliate infusorians. Bull. U. S. nat. Mus., 120: 1-484.

—— 1929. Parasites and the aid they give in problems of taxonomy, geographical distribution, and paleogeography. Smithson. misc. Coll., 81(8): 1-36.

Meyer, S. L. 1938. *Haptophrya virginiensis* nov. sp., a protozoan parasite of the pickerel frog, *Rana palustris* Le Conte. Anat. Rec., 72, Suppl.: 54-55.

Miyashita, Y. 1927. On a new parasitic ciliate, *Lada tanishi* n. sp. with preliminary notes on its heterogamic conjugation. Jap. J. Zool., 1: 205-18.

—— 1933. Studies on a freshwater foettingeriid ciliate *Hyalospira cardinae* n. g. n. sp. Jap. J. Zool., 4: 439-60.

Mjassnikowa, M. 1930a. *Sphenophrya sphaerii*, ein neues Infusorium aus *Sphaerium corneum* L. Arch. Protistenk., 71: 255-94.

—— 1930b. Über einen neuen Vertreter der Familie Sphenophryidae aus *Myatruncata* L. Arch. Protistenk., 72: 377-89.

Montalenti, G. 1927. Sull' allevamento dei termiti senza i protozoi dell' ampolla cecale. R.C. Accad. Lincei, (6)6: 529-32.

—— 1932. Gli enzimi digerenti e l'assorbimento delle sostanze solubili nell' intestino delle termiti. Arch. zool. (ital.) Tonino, 16: 859-64.

—— 1934. Un interessante caso di simbiosi: i flagellati dell' intestino delle termiti. Rasseg. faunist. Roma, 1: 25-35.

Moore, E. 1922. *Octomitus salmonis*, a new species of intestinal parasite in trout. Trans. Amer. Fish, Soc., 52: 74-97.

—— 1923a. Diseases of fish in state hatcheries. 12th Ann. Rep. N.Y. St. Conserv. Comm., 66-79.

—— 1923b. A report of progress on the study of trout diseases. Trans. Amer. Fish. Soc., 53: 74-94.

—— 1924. The transmission of *Octomitus salmonis* in the egg of trout. Trans. Amer. Fish. Soc., 54: 54-56.

Moroff, T. 1903. Beitrag zur Kenntnis einiger Flagellaten. Arch. Protistenk., 3: 69-106.

Mowry, H. A., and E. R. Becker. 1930. Experiments on the biology of Infusoria inhabiting the rumen of goats. Iowa St. Coll. J. Sci., 5: 35-60.

Nie, Dashu. 1934. Studies of the intestinal ciliates of sea urchin from Amoy. Rep. Mar. Biol. Ass. China, 3: 81-90.

Nieschulz, O. 1922. Über eine *Astasia*—Art aus dem Süsswassernematoden *Trilobus gracilis* Bst. Zool. Anz., 54: 136-38.

Nöller, W., and F. Buttgereit. 1923. Über ein neues parasitisches Protozoön der Haustaube (*Octomitus columbae* nov. spec.) Zbl. Bakt., Refer (I), 75. 239-40.

Oppenheimer, C. 1925. Die Fermente und ihre Wirkungen. Leipzig.

Oshima, M. 1919. Formosan termites and methods of preventing their damage. Philipp. J. Sci., 15: 319-83.

Parker, T. J. 1893. Lessons in elementary biology. 2d, ed., London.

Penard, E. 1922. Études sur les Infusoires d'eau douce. Génève.

Perekropoff, G. I., and P. I. Stepanoff. 1931. On an intestinal disturbance in man caused by *Octomitus*. (In Russian.) J. Mikrobiol., 13: 191-96.

—— 1932. Zur Frage der Darmerkrankungen des Menschen, die durch *Octomitus* bedingt sind. Zbl. Bakt., Orig. (I)123: 324-30.

Pickard, E. A. 1927. The neuromotor apparatus of *Boveria teredinidi* Nelson, a ciliate from the gills of *Teredo navalis*. Univ. Cal. Publ. Zool., 29: 405-28.

Pierantoni, O. 1934. La digestione della cellulosa e del legno negli animali e la simbiosi delle termiti. Riv. Fis. Mat. Sci. nat., 9: 57-64.

—— 1936. La simbiosi fisiologica nei termitidi xilofagi e nei loro flagellati intestinali. Arch. zool. (ital.) Torino, 22: 135-71.

—— 1937. Osservazioni sulla simbiosi nei termitidi xilofagi e nei loro flagellati intestinali.—II. Defaunazione per digiuno. Arch. zool. (ital.) Torino, 24: 193-207.

Playfair, G. I. 1921. Australian freshwater flagellates. Proc. Linn. Soc. N.S.W., 46: 99-146.

Plimmer, H. G. 1912. On the blood-parasites found in animals in the Zoological Gardens during the four years 1908-1911. Proc. zool. Soc. Lond., 1912: 406-19.

Poisson, R. 1929. Apropos de l'*Anophrys maggii* Cattaneo, Infusoire parasite du sang du *Carcinus maenas L.* (Crustacé Décapode) ; Sur son identité avec l'*Anophrys sarcophaga* Cohn. C. R. Soc. Biol. Paris, 102: 637-39.

—— 1930. Observations sur *Anophrys sarcophaga* Cohn (=*A. maggii* Cattaneo) Infusoire holotriche marin et sur son parasitisme possible chez certains Crustacés. Bull. biol., 64: 288-331.

Poljanskij, J. I. 1925. Drei neue parasitische Infusorien aus dem Parenchym einiger Mollusken und Turbellarien. Arch. Protistenk., 52: 381-93.

Poljansky, G., and A. Strelkow. 1935. Über die Wirkung der Panseninfusorien auf das Wachtum der Wiederkäuer. Trav. Inst. Biol. Peterhof., 13-14: 68-87.

Powers, P. B. A. 1933a. Studies on the ciliates from sea urchins. I. General Taxonomy. Biol. Bull., 65: 106-21.

—— 1933b. Studies on the ciliates from sea urchins. II. *Entodiscus borealis* (Hentschel), (Protozoa, Ciliata), behavior and morphology. Biol. Bull., 65: 122-36.

—— 1933c. Studies on the ciliates from Tortugas echinoids. Yearb. Carneg. Instn., 32: 276-80.

—— 1935. Studies on the ciliates of sea urchins. A general survey of the infestations occurring in Tortugas echinoids. Publ. Carneg. Instn., 452: 293-326.

—— 1936. Ciliates infesting Acapulco sea urchins. J. Parasit., 22: 541.

—— 1937. Studies of the ciliates of sea urchins. Ann. Rep. Tortugas Lab., 1936-1937: 101-3.

Poyarkoff, E. 1909. *Cepedella hepatica,* Cilié astome neuveau, parasite du foie des *Cyclas.* C. R. Soc. Biol. Paris, 66: 96-97.

Pringsheim, H. 1932. The chemistry of the monosaccharides and of the poly-saccharides. New York.

Raabe, Z. 1933. Untersuchungen an einigen Arten des Genus *Conchophthirus* Stein. Bull. int. Acad. Cracovie, Cl. Sci. math. nat., (B: II) 1932 (8-10) : 295-310.

—— 1934a. Über einige an den Kiemen von *Mytilus edulis* L. und *Macoma balthica* (L.) parasitierende Ciliaten-Arten. Ann. Mus. zool. polon., 10: 289-303.

—— 1934b. Weitere Untersuchungen an einigen Arten des Genus *Conchophthirus* Stein. Mém. Cl. Sci. Acad. polon., math. nat., (B) 6: 221-35.

—— 1935. *Rhynchophrya cristallina* g. n., sp. n. nouvelle forme d'Infusoire de la famille des Sphaenophryidae Chatton et Lwoff. Bull. Inst. océanogr. Monaco, 676: 1-6.

Rankin, J. S. 1937. An ecological study of pasasites of some North Caro-lina salamanders. Ecol. Monogr. 7: 171-269.

Rehberg, H. 1882. Eine neue Gregarine. *Lagenella mobilis* n. g. et n. sp. Abh. naturw. Ver. Bremen, 7: 68-71.

Reichenow, E. 1920. Den Wiederkäuer-Infusorien verwandte Formen aus Gorilla und Schimpanse. Arch. Protistenk., 41: 1-33.

Remane, A. 1936. Gastrotricha und Kinorhyncha, Bronn's Klassen, Bd. 4, Abt. 2, Buch 1, Teil 2, Lief 2, Leipzig.

Reynolds, B. D. 1936. *Colpoda steini,* a facultative parasite of the land slug, *Agriolimax agrestis.* J. Parasit., 22: 48-53.

Ripper, W. 1930. Zur Frage des Celluloseabbaus bei der Holzverdauung xylophager Insektenlarven. Z. vergl. Physiol., 13: 314-33.

Rose, M. 1933. Sur un Infusoire Foettingéridé parasite des Siphonophores C. R. Acad. Sci. Paris, 197: 868-69.

—— 1934. Pluralité des espèces de Foettingeridae (Infusoires Apostomes) parasites des Siphonophores de la baie d'Alger. Bull. Soc. Hist. nat. Afr. N. 25: 149-51.

Rosenberg, L. E. 1936. On the viability of *Tritrichomonas augusta.* Trans. Amer. micr. Soc., 55: 313-14.

Rossolimo, L. 1925. Infusoires parasites du tube digestif des Oligochètes-genre *Ptychostomum* St. Arch. russ. protist., 4: 217-33.

—— 1926. Parasitische Infusorien aus dem Baikalsee. Arch. Protistenk., 54: 468-509.

Rossolimo, L. L., and T. A. Perzewa. 1929. Zur Kenntnis einiger astomen Infusorien: Studien an Skelettbildung. Arch. Protistenk., 67: 237-52.

Ryckeghem, J. van 1928. *Hexamitus tubifici* nov. sp. Ann. Soc. Sci. Brux. (B), 48, pt. 2, Mémoires: 139-43.

Scheunert, A. 1909. Verdauung. IV. Besonderheiten der Verdauung bei Tieren mit mehrhöligen Mägen (Wiederkäuer), Kaltblütern und Vögeln, *in* Oppenheimer, C., Handbuch der Biochem. des Menschen und der Tiere, 3: 2: 152-70. Jena. (Also, 1924, in 2d ed., vol. 5.)

Schieblich, M. 1929. Die Mitwirkung der Bacterien bei der Verdauung *in* E. Mangold: Handbuch der Ernährung und des Stoffwechsels der landwirtschaftliche Nutztiere als Grundlagen der Fütterungslehre, 2: 310-48. Berlin.

—— 1932. Die Bedeutung der normalen Magendarmflora für den Wirtsorganismus. Biederm. Zbl. (A) n. f. 2: 483-501.

Schmidt, W. 1920. Untersuchungen über *Octomitus intestinalis truttae*. Arch. Protistenk., 40: 253-89.

Schulze, P. 1924. Der Nachweis und die Verbreitung des Chitins mit einem Anhang über das komplizierte Verdauungsystem der Ophryoscoleciden. Z. Morph. Ökol. Tiere, 2: 643-66.

—— 1927. Noch einmal die "Skelettplatten" der Ophryoscoleciden. Z. Morph. Ökol. Tiere, 7: 678-89.

Schwartz, W. 1935. Untersuchungen über die Symbiose von Tieren mit Pilzen und Bakterien. IV. Der Stand unserer Kenntnisse von den physiologischen Grundlagen der Symbiosen von Tieren mit Pilzen und Bakterien. Arch. Mikrobiol., 6: 369-460.

Schwarz, C. 1925. Die ernährungsphysiologische Bedeutung der Mikroorganismen in den Vormägen der Wiederkäuer. Biochem. Z., 156: 130-37.

Sokoloff, D. 1933. Algunas nuevas formas de flagelados del Valle de Mexico. An. Inst. Biol. Univ. Mex., 4: 197-206.

Stein, F. (Ritter) von. 1848. Über die Natur der Gregarinen. Arch. Anat. Physiol. wiss. Med., 1848: 182-223.

—— 1860. Über die Eintheilung der holotrichen Infusionsthiere und Stellte einige neue Gattungen und Arten aus dieser Ordnung. S.B. böhm. Ges. Wiss., Nat.-Math., 1860: 57-63.

—— 1861. Über ein neues parasitisches Infusionsthier (*Ptychostomum Paludinarum*) aus dem Darmkanal von Paludinen und über die mit demselben zunächst verwandten Infusorienformen. S.B. böhm. Ges. Wiss., 1861: 85-90.

Steuer, A. 1928. Über *Ellobiopsis chattoni* Caullery 1910, einen ektoparasitischen Flagellaten mariner Copepoden. Arch. Protistenk., 60: 501-10.

Stevens, N. M. 1901. Studies on ciliate Infusoria. Proc. Calif. Acad. Sci. (3), 3: 1-42.

Strand, E. 1928. Miscellanea nomenclatorica zoologica et paleontologica. Arch. Naturgesch., 92: 30-75.

Strelkow, A., G. Poljansky, and M. Issakowa-Keo. 1933. Über die Infektionswege der im Pansen und in der Haube der Wiederkäuer befindlichen Infusorien. Arch. Tierernähr. Tierz., 9: 679-97.

Studitsky, A. 1930. Eine neue Art der Gattung *Ptychostomum* Stein (*Lada*

Vejdovsky) *Pt. rossolimoi* n. sp. Eine systematische Studie. Zool. Anz., 87: 247-56.

Summers, F. M., and G. W. Kidder. 1936. Taxonomic and cytological studies on the ciliates associated with the amphipod family Orchestiidae from the Woods Hole district. II. The coelozoic astomatous parasites. Arch. Protistenk., 86: 379-403.

Švec, F. 1897. Beiträge zur Kenntnis der Infusorien Böhmens. I. Die ciliaten Infusorien des unterpočernitzer Teiches. Bull. int. Acad. Prag, Sci. Math. Nat., 4(2): 29-47.

Swezy, O. 1923. The pseudopodial method of feeding by trichonymphid flagellates parasitic in wood-eating termites. Univ. Cal. Publ. Zool., 20:391-400.

Tetrault, P. A., and W. L. Weis. 1937. Cellulose decomposition by a bacterial culture from the intestinal tract of termites. J. Bact., 33: 95.

Theiler, H., and S. M. Farber. 1932. *Trichomonas muris,* parasitic in oxyurids of the white mouse. J. Parasit., 19: 169.

—— 1936. *Trichomonas muris,* parasitic in the oxyurid nematodes, *Aspicularis tetraptera* and *Syphacia obvelata,* from white mice. Parasitology, 28: 149-60.

Thomson, J. A. 1934. Biology for Everyman. London.

Thomson, J. A., and P. Geddes. 1931. Life: Outlines of general biology. London.

Thomson, J. G. 1925. A *Giardia* parasitic in a bursate nematode living in the Viscacha. Protozoology (Suppl., J. Helminth.), No. 1, 1-6.

Trager, W. 1932. A cellulase from the symbiotic intestinal flagellates of termites and of the roach, *Cryptocercus punctulatus.* Bio-chem. J. 26: 1762-71.

—— 1934. The cultivation of a cellulose-digesting flagellate, *Trichomonas termopsidis,* and of certain other termite Protozoa. Biol. Bull., 66: 182-90.

Treillard, M., and A. Lwoff. 1924. Sur un Infusoire parasite de la cavité générale des larves de chironomes. Sa sexualité. C. R. Acad. Sci. Paris, 178: 1761-64.

Trier, H. J. 1926. Der Kohlehydratstoffwechsel der Panseninfusorien und die Bedeutung der grünen Pflanzenteile für diese Organismen. Z. vergl. Physiol., 4: 305-30.

Ullmann, T. 1932. Über die Einwirkung der Fermente einiger Wirbellosen auf polymere Kohlenhydrate. Z. vergl. Physiol., 17: 520-36.

Usuelli, F. 1930a. Stärkeaufnahme und Glycogenbildung der Panseninfusorien. Wiss. Arch. Landw., 3: 4-19.

—— 1930b. Das Verhalten der Panseninfusorien gegenüber cellulose und grünen Pflanzenteilen. Wiss. Arch. Landw., 3: 368-82.

—— 1930c. Gli infusori Ciliati che vivono nell'apparato digerente degli erbivori. Clin. vet. Milano.

Uyemura, M. 1934. Über einige neue Ciliaten aus dem Darmkanal von japanischen Echinoideen (I). Sci. Rep. Tokyo Bunrika Daig. (B) 1: 181-91.

—— 1935. Über drei in der Süsswasser-Muschel (*Anodonta lauta* v. Martens) lebende Ciliaten (*Conchophthirus*). Sci. Rep. Tokyo Bunrika Daig. (B) 2: 89-100.

Valkanov, A. 1928. Protistenstudien. II. Notizen über die Flagellaten Bulgariens. Arch. Protistenk., 63: 419-50.

Voigt, M. 1904. Die Rotatorien und Gastrotrichen der Umgebung von Plön. Foschber. biol. Sta. Plön, 11: 1-180.

—— 1909. Nachtrag zur Gastrotrichen-Fauna Plöns. Zool. Anz., 34: 717-22.

Wallengren, H. 1895. Studier öfver Ciliata infusorier. Årsskr., Lunds Univ. 31: 77 pp.

Warren, E. 1932. On a ciliate protozoan inhabiting the liver of a slug. Ann. Natal. Mus., 7: 1-53.

Watson, M. E. 1915. Studies on gregarines. Illinois biol. Monogr., 2: 215-468.

Weineck, E. 1934. Die Celluloseverdauung bei den Ciliaten des Wiederkäuermagens. Arch. Protistenk., 82: 169-202.

Wenrich, D. H. 1923. Variations in *Euglenamorpha hegneri* n. g., n. sp., from the intestine of tadpoles. Anat. Rec., 24: 370-71.

—— 1924a. Studies on *Euglenamorpha hegneri* n. g., n. sp., a euglenoid flagellate found in tadpoles. Biol. Bull., 47: 149-74.

—— 1924b. A new protozoan parasite, *Amphileptus branchiarum*, n. sp., on the gills of tadpoles. Trans. Amer. micr. Soc., 43: 191-99.

—— 1933. A species of *Hexamita* (Protozoa, Flagellata) from the intestine of a monkey (*Macacus rhesus*). J. Parasit., 19: 225-28.

—— 1935. Host-parasite relations between parasitic Protozoa and their hosts. Proc. Amer. phil. Soc., 75: 605-50.

Wenyon, C. M. 1926. Protozoology, a manual for medical men, veterinarians and zoologists. New York.

Wenyon, C. M., and W. Broughton Alcock. 1924. A *Trepomonas* coprozoic in human faeces. Trans. R. Soc. trop. Med. Hyg., 18: 9.

Werner, E. 1926. Der Erreger der Zelluloseverdauung bei der Rosenkäferlarve (*Potosia cuprea Fbr.*) *Bacillus cellulosam fermentans* n. sp. Zbl. Bakt., 67 (II): 297-330.

Wertheim, P. 1934a. Über die Infusorienfauna im Magen von *Bos taurus* L. Ann. Mus. zool. polon., 10: 251-66.

—— 1934b. Wärmeversuche mit Wiederkäuermageninfusorien. Biol. Zbl., 54: 390-402.

—— 1935. Infusorien aus dem Wiederkäuermagen vom Gebiete Jugoslawiens nebst einer Übersicht dieser Tierchen vom Balkanhalbinsel-Bereich und ein kurzer Bericht über die Pferdedarminfusorien, zugleich Revision der Familie Ophryoscolecidae. (Croat, German summary). Arhiva vet., 5: 388-526.

Westphal, A. 1934a. Studien über Ophryoscoleciden in der Kultur. Z. Parasitenk., 7: 71-117.

—— 1934b. Ein Züchtungsverfahren für die Wiederkäuerinfusorien und dessen Ergebnisse für die Frage nach der Bedeutung der Infusorien für den Wirt. Zool. Anz., (Suppl.) 7: 207-10.

Wiedemann, J. F. 1930. Die Zelluloseverdauung bei Lamellicornierlarven. Z. Morph. Ökol. Tiere, 19: 228-58.

Winogradow, M., T. Winogradowa-Fedorowa, and A. Wereninow. 1930, Zur Frage nach der Einwirkung der Panseininfusorien auf die Verdauug der Wiederkäuer. Zbl. Bakt., II, 81: 230-44.

Winogradowa-Fedorowa, T., and M. P. Winogradoff. 1929. Zählungsmethode der Gesamtzahl der im Wiederkäuermagen lebenden Infusorien. Zbl. Bakt., 78(II): 246-54.

Woodhead, A. E. 1928. *Haptophrya michiganensis* sp. nov., a *protozoan* parasite of the four-toed salamander. J. Parasit., 14: 177-82.

Woodhead, A. E., and F. Kruidenier. 1936. The probable method of infection of the four-toed salamander with the protozoön, *Haptoyhrya michiganensis*. J. Parasit., 22: 107-8.

Yagiu, R. 1933. Studies on the ciliates from the intestine of *Anthocidaris crassispina* (A. Agassiz). I. *Cyclidium ozakii* sp. nov. & *Strombilidium rapulum* sp. nov. J. Sci. Hiroshima Univ., (B:I)2: 211-22.

—— 1934. Studies on the ciliates from the intestine of *Anthocidaris crassispina* (A. Agassiz). II. *Cryptochilidium sigmoides* sp. nov. and *Cryptochilidium minor* sp. n. J. Sci. Hiroshima Univ., (B:I)3: 25-31.

—— 1935. Studies on the ciliates from sea urchins of Yaku Island, with a description of a new species, *Cryptochilidium ozakii* sp. nov. J. Sci. Hiroshima Univ. (B:I)3: 139-47.

Yamasaki, M. 1937a. Studies on the intestinal Protozoa of termites. III. The distribution of glycogen in the bodies of intestinal flagellates of termites, *Leucotermes (Reticulitermes) speratus* and *Coptotermes formosanus*. Mem. Coll. Sci. Kyoto, (B)12: 211-24.

—— 1937b. Studies on the intestinal Protozoa of termites IV. Glycogen in the body of *Trichonympha agilis* var. *japonica* under experimental conditions. Mem. Coll. Sci. Kyoto, (B)12: 225-35.

Yonge, C. M. 1925. The digestion of cellulose by invertebrates. Sci. Prog., 20: 242-48.

Zerling, Mlle. 1933. Une *Astasia* sans flagelle, *Astasia doridis* n. sp., parasite des pontes de nudibranches. C. R. Soc. Biol. Paris, 112: 643-44.

Zick, Karl. 1928. *Urceolaria korscheltii* n. sp., eine neue marine Urceolarine, nebst einem Überblick über die Urceolarinen. Z. wiss. Zool., 132: 355-403.

CHAPTER XX

ORGANISMS LIVING ON AND IN PROTOZOA

Harold Kirby, Jr.[1]

Protozoa as a group may be hosts of a great variety of other organisms. Some of these are epibiotic, and in them the relationship ranges from occasional phoresy to obligatory and constant association. True ectoparasitism exists in some epibionts, though often the distinction from predatism is disputable. Protozoa are not so constituted as to be capable of harboring inquilines; all endobionts are intracellular. Unless autotrophic, therefore, endobiotic forms are parasites in the sense that they are dependent on their hosts in nutrition. In many instances, however, the protozoan suffers no apparent detriment from the relationship; and sometimes the association of host and symbiont is constant. (Symbiosis is used as a collective term, including commensalism, mutualism, and parasitism.) It may be, even, that there is mutual advantage; but only autotrophic forms are in a position to confer the commonest type of benefit in mutualism, a nutritive one. It has been suggested, although not demonstrated, that certain intracellular microörganisms may produce enzymes that function in the nutritional metabolism of the host. Many endobionts are more or less destructive parasites, which cause injury or death; that is true, so far as is known, of all that invade the nucleus.

When the association has an obligatory and constant character, as in the occurrence of bacteria on the body surface of certain Protozoa, or of bacteria present in certain areas of the cytoplasm of all or almost all specimens, the error has often been made of interpreting the symbionts as structures of the host. When they are present only occasionally, they have sometimes been mistakenly regarded as representing occasional phases in the life history of the host, that is, reproductive phases. Increasing knowledge of the symbionts of the Protozoa has corrected most

[1] Assistance rendered by personnel of Work Projects Administration, Official Project number 65-1-08-113, Unit C1, is acknowledged.

of these errors, but there are sometimes greater difficulties in the way of correct interpretation than might be supposed.

A large number of organisms symbiotic with Protozoa are Schizomycetes or Phycomycetes. In the latter group, Chytridiales are especially widespread, as parasites of both the cytoplasm and the nucleus. There are in all the major groups of Protozoa species that are symbiotic with other Protozoa. Some groups of these, such as the Metchnikovellidae and endozoic Suctoria, are known only from hosts of this phylum. Only a few Metazoa occur as parasites in Protozoa. Although the relative size relationships are sometimes such as to make it possible, parasitization by these higher forms is not less prevalent than might be expected, in the light of the infrequency, in general, of intracellular parasitism by Metazoa.

Cyanellae, chlorellae, and xanthellae have not been included in this account of symbionts of Protozoa. The last two types, at least, are widespread in members of many free-living groups, the former in freshwater forms, the latter in marine species. Although inhabiting the cytoplasm, the nutritive processes of these organisms are autotrophic; and they are not necessarily dependent on the host. Their relationship to their hosts is often cited as mutualistic, in one way or another. The problem is in part the same as that of the relationship of similar endosymbionts to many invertebrate animals—a problem that has been reviewed by Buchner (1930). Pascher (1929) discussed the endosymbiosis of blue-green algae in *Paulinella chromatophora* and some other Protozoa, as well as in algae. Lackey (1936) described "blue chromatophores" in *Paulinella* and a number of flagellates, although he failed to recognize them as resembling Pascher's cyanellae. Goetsch and Scheuring (1926) discussed the relationship of the alga *Chlorella* to protozoan and some metazoan hosts. The work of Pringsheim (1928) and others has sustained the thesis of mutualistic symbiosis between *Chlorella* and Protozoa. Most of the xanthellae have been placed in the cryptomonad genus *Chrysidella*. Hovasse (1923a) maintained on the basis of nuclear characteristics that xanthellae are dinoflagellates.

EPIBIOTIC SCHIZOMYCETES

SCHIZOMYCETES ON MASTIGOPHORA

Bacteria attached to the surface, either by one end or applied full length, occur on many flagellates. They have been known for some time on *Mastigamoeba aspera,* small rods besetting the surface having been described by Schulze (1875) in the original description of this type species of the genus. He stated that these rods could best be compared with certain bacteria, such as *Bacterium termo.* Most of them are applied full length to the body surface. Of other observers, some recognized the similarity of the rods to bacteria and others opposed this interpretation, but Penard (1905c, 1909) definitely established their bacterial nature. The bacteria, he found, vary in number, but there are few individuals of the species without them. Lauterborn (1916) stated that a sapropelic flagellate, possibly belonging to the genus *Mastigamoeba,* possessed a yellow-green mantle of radially adherent chlorobacteria, which he named *Chlorobacterium symbioticum.*

These are the only free-living flagellates, to the writer's knowledge, on which bacteria have been reported. Many endozoic flagellates bear Schizomycetes. Grassé (1926a, 1926b, and elsewhere) has done much to increase our understanding of those microörganisms, which many earlier observers had mistaken for pellicular differentiations, cilia, or flagella. Duboscq and Grassé (1926, 1927) found short rods adherent by full length on many specimens of *"Devescovina" hilli,* showed that these are bacteria, and named them *Fusiformis hilli;* and they also found spirochetes, named *Treponema hilli,* adherent to all parts of the body surface. The former report of the rods first established the true nature of the "striations," which Foà (1905), Janicki (1915), and Kirby (1926b) had described on *Devescovina.*

In many instances the presence of certain microörganisms is characteristic of the species, and may indeed be considered, together with its morphological characteristics, as an aid in taxonomy. Simpler phoretic relationships, in which the presence of the adherent forms is only occasional, do, however, exist between microörganisms and some flagellates. Examples are the rod-like bacteria occasionally adherent, full length, to *Hexamastix claviger* (Kirby, 1930); the bacteria sometimes present on the larger forms of *Tricercomitus termopsidis* (Kirby, 1930); the occasional spirals and rods on *Eutrichomastix trichopterae, E. colubrorum,* and *Octomitus intestinalis* (Grassé, 1926b); and the lumen-dwelling types of bacteria and spirochetes often adherent sporadically to various termite flagellates. Boeck (1917) stated that rod-shaped bacteria at times, in certain preparations, covered the body and adhered to the flagella of *Giardia microti.*

Fusiformis-like Rods Adherent Full Length.—The genus *Fusiformis* has the type species *F. termitidis* Hoelling, 1910. This organism is generally free in the lumen of the hind-gut of termites. It was originally reported in *Coptotermes* sp. of Brazil, has been observed by the writer

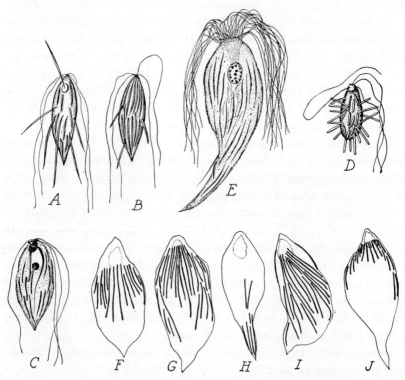

Figure 208. *Fusiformis*-like rods adherent to the surface of flagellates. A, B, *Fusiformis grandis* and *F. melolonthae* on *Polymastix melolonthae;* C, *F. melolonthae* on *P. melolonthae;* D, *F. legeri* on *P. legeri;* E, *F. lophomonadis* on *Lophomonas striata;* F-J, rod-shaped microörganisms on *Devescovina* sp. from *Neotermes dalbergiae.* (A-E, after Grassé, one 926b; F-J, original.)

in *C. niger* of Panama, and was found by Duboscq and Grassé (1926, 1927) in *Glyptotermes iridipennis,* where it occasionally adheres to *"Devescovina" hilli.* Chromatic granules number from one to sixteen, according to Hoelling's account; Hoelling identified these with nuclei. In the form studied by the writer in *C. niger,* the chromatic granules are well defined, usually extend the full width of the cell, and in most cases number from one to four. *Fusiformis hilli,* although much smaller, is similar in shape and contains one or two chromatic granules that occupy the full width. The bacteria were also observed in transverse fission, the occurrence of which, together with the structure, readily distinguishes them from pellicular striations.

Ecologically, *Fusiformis hilli* is closer to *F. termitidis* than to the "pellicular striations" on species of true *Devescovina.* It is abundant in the lumen of the intestine, as well as on the surface of *Crucinympha hilli,* which is not true of the "striations." On the surface it is arranged in a manner scarcely suggesting striations; it sometimes adheres by one end, and it is inconstant in occurrence. Kirby (1938b) noted that of 100 *Crucinympha,* 78 had no adherent rods, or very few.

Later in the same year that *Fusiformis hilli* was reported, Grassé (1926a) described *F. grandis* and *F. melolonthae* on *Polymastix melolonthae* (Fig. 208A, B, C,); *F. legeri* on *Polymastix legeri* (Fig. 208D); and *F. lophomonadis* on *Lophomonas striata* (Fig. 208E). *F. grandis* and *F. legeri* adhere by one extremity. This article, and the more extended account by Grassé (1926b), first established the true nature of the microörganisms that adhere by full length to *Polymastix* and *Lophomonas,* which had been regarded as pellicular structures.

The writer has observed *Fusiformis*-like microörganisms on many polymastigote flagellates of termites. They are present on all the twenty species of *Devescovina,* so constantly that no individual lacks them (Fig. 208A; Fig. 209F-J). On all except *D. elongata,* the rods appear almost identical with *F. lophomonadis* of *Lophomonas striata,* both in their morphology and their morphological relationship to the host flagellate. They are evenly spaced, generally occur over the whole body except the papilla, and are usually situated closer together than Grassé (1926b, Pl. 15) indicated for *F. lophomonadis.* They always adhere firmly by the full length, and are not subject to detachment in the ordinary course of technical manipulations. On smears in which the cytosome of the flagellate has been much disturbed, they may have been partly or completely detached, and can then be conveniently studied. Specimens subjected to such treatment may sometimes be ,bent considerably, as was noted by Grassé (1926b) in *F. melolonthae.* Surface microörganisms were absent from many specimens of *Devescovina lemniscata* of *Neotermes insularis* that had been fed on filter paper soaked previously in

10-percent acid fuchsin, but were never absent from normal material. Presumably the treatment had brought about detachment of many of the bacteria. Generally these slender microörganisms appear homogeneous, but chromatic granules can be demonstrated by suitable methods.

Similar microörganisms have been found by the writer on three of the five species of *Caduceia*. On *C. nova* and *C. theobromae* (Fig. 209C), they are short, slender, and are confined to a limited, sharply bounded area at the posterior end of the body (Kirby, 1936, 1938a;

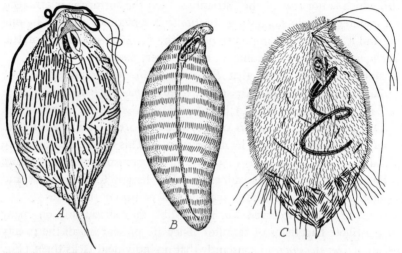

Figure 209. Adherent microörganisms on flagellates of termites. A, *Fusiformis*-like rods on *Devescovina* sp. from *Glyptotermes niger*; B, regularly arranged rods on *Caduceia* sp. from *Neotermes greeni*; C, investment of spirochaetes and posteriorly localized rods on *Caduceia theobromae*. (A, B, original; C, after Grassé, 1938.)

Grassé, 1937, 1938). On a species of *Caduceia* from *Neotermes greeni*, they are abundant on the entire surface, except the papilla; and they show a tendency to arrangement in transverse bands, between which, in many specimens, there is regularity in spacing (Fig. 209B).

On other genera of Devescovininae from termites, the striation-like bacteria are altogether absent. None have been found in *Foaina*, on most species of *Metadevescovina*, on *Pseudodevescovina*, or on *Macrotrichomonas*, except for an occasional, irregular occurrence on the surface, in a manner comparable to the situation in *Crucinympha hilli*. The absence is the more striking because of their universality on *Devescovina* and certain species of *Caduceia*.

Rods adhere also to certain hypermastigote flagellates. They were noted by Kirby (1926a) on *Staurojoenina assimilis,* but were wrongly regarded as pellicular striations; and Cleveland *et al.* (1934) found them on *Barbulanympha, Rhynchonympha,* and *Urinympha* of *Cryptocercus punctulatus.* Cleveland considered them to be cuticular striations, but noted their resemblance to the adherent bacteria described by Duboscq and Grassé (1926, 1927). Grassé (1938) identified them as bacteria.

Spirochetes and Rods Adherent by One End.—Spirochetes occur in great abundance in termites, mostly free in the lumen or attached to the lining of the hind-gut, but also adherent by one extremity to certain flagellates. It is not known whether this phoresy is obligatory or occasional from the standpoint of the spirochete, but the former condition is probable, at least in many instances. The presence of adherent spirochetes is especially characteristic of certain Pyrsonymphinae, Oxymonadinae, Devescovininae, and Calonymphidae among polymastigotes. Spirochetes are less frequent on hypermastigotes, but do occur on some genera (*Holomastigotoides,* Koidzumi, 1921; *Spirotrichonympha,* Sutherland, 1933; Dogiel, 1922a; *Spirotrichonymphella,* Sutherland, 1933; *Rostronympha,* Duboscq, Grassé, and Rose, 1937). Cleveland *et al.* (1934) did not report them on either polymastigotes or hypermastigotes of *Cryptocercus.*

On many of these flagellates, spirochetes are invariably present, either distributed widely on the surface or localized on very definite areas of the body. Localization is illustrated by the distribution of spirochetes on *Foaina nucleoflexa* and *Oxymonas grandis.* On the former flagellate, spirochetes are always present on the anterior and posterior parts of the body; many of those on the anterior part are arranged in a row along the surface just over the parabasal filament (Fig. 210C). Grassé (1938) noted in another species of *Foaina* (mistakenly named by him *Parajoenia decipiens*) that an anterior tuft of spirochetes obeyed a fixed rule in its distribution. Localization is even more marked in *Oxymonas grandis,* which bears a dense group of spirochetes on an elongated, limited area at the base of the rostellum (Fig. 210A, B). The rest of the body surface of this flagellate is covered with minute epibiotic bacilli (Fig. 210A).

Many observers have mistaken spirochetes for flagella or cilia in a wide range of flagellates of termites; and sometimes the same observer interpreted them correctly on one flagellate and wrongly on another, or even reached different conclusions concerning the filaments on different parts of the body of the same protozoan. The spirochetes have even

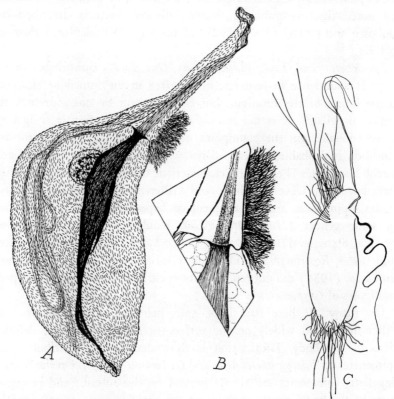

Figure 210. Adherent microorganisms on flagellates of termites. A, B, small rods on surface and localized spirochaetes at base of rostellum of *Oxymonas grandis*; C, spirochaetes on the posterior part and on a localized area of the anterior part of *Foaina* sp. from *Cryptotermes merwei*. (Original.)

been responsible for the erroneous classification of some polymastigote flagellates as hypermastigotes. Historical data on the interpretation of adherent spirochetes has been reviewed by Kirby (1924, 1926a); Duboscq and Grassé (1927); and Grassé (1938). Cleveland (1928) discussed their occurrence on flagellates.

The spirochetes range in length from only 2 μ, the minimum for

Treponema hilli of *Crucinympha hilli,* to 10 µ, which is about the average length of those on species of *Devescovina;* and up to 20 µ, as in the forms on *Metadevescovina cuspidis* of *Kalotermes minor,* or even 40 µ in the longer species on *Caduceia theobromae* (Grassé, 1938). On *Stephanonympha* sp. from *Neotermes insularis,* long spirochetes of from 40 to 60 µ are adherent (Fig. 211). Those of an investment or group are often comparatively uniform in length. On *Pseudodevescovina uni-*

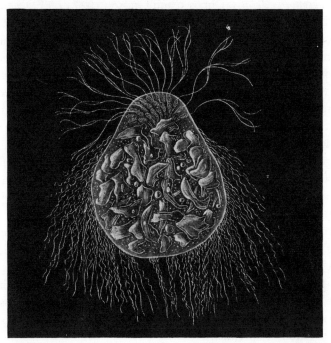

Figure 211. Spirochetes adherent to *Stephanonympha* sp. from *Neotermes insularis.* (Original.)

flagellata, which is completely covered with spirochetes, simulating a dense coat of cilia, the majority have a length of from 8 to 10 µ. *Caduceia theobromae* is similarly covered by spirochetes from 4 to 6 µ long, except for the area occupied by the above-mentioned *Fusiformis-*like rods (Fig. 209C).

The spiral of short spirochetes has only one or two turns. Two turns were counted in those of average length (9 to 13 µ) on *Metadevescovina cuspidis,* while in the longer ones, up to 20 µ, there were three or four.

The very long ones on the above-mentioned *Stephanonympha* had a much larger number of turns, but none like those have been seen on Devescovininae.

Normally the spirochetes are in continual, very active flexuous movement. They are not rigid, like Spirillaceae, although some observers have compared them with spirilla. Their activity has been described by Koidzumi (1921) on *Holomastigotoides hartmanni,* Light (1926) on *Metadevescovina debilis,* Duboscq and Grassé (1927) on *"Devescovina" hilli,* Kirby (1936) on *Pseudodevescovina uniflagellata,* and by others. They can be studied best in living material by dark-field illumination. Their movements are not synchronized, and are uncoördinated either in direction or activity. The difference between this movement and that of cilia or flagella has impressed all students who have observed it. As noted by Kirby (1936), they may move at an equally active rate under the same environmental conditions, on moving flagellates, quiet flagellates, dead flagellates, and detached balls of cytoplasm. This activity, together with their form, readily distinguishes them from flagella; but distinction is less easy in fixed material, in which the form is often less evident.

The spirochetes do not, in the writer's experience, detach readily in preparation of smears. Grassé (1938) stated that certain flagellates lacking spirochetes may have lost them in consequence of fixation, but offered no proof that this occurs. Spirochetes of *P. uniflagellata* were observed to be rubbed off by movement of the large flagellate in close contact with the cover glass, and severe manipulation might cause their loss; but that treatment is more drastic than would ordinarily occur in the preparation of specimens.

The spirochetes can be removed, however, by relatively simple methods. Light found that treatment with iodine in 70-percent alcohol freed the bodies of most *Metadevescovina debilis* of spirochetes. Cleveland (1928) discovered that all the spirochetes could be removed, both from the surface of the Protozoa and the lumen of the gut, by feeding the termites on cellulose thoroughly moistened with 5-percent acid fuchsin. That method was used by Sutherland (1933) to remove spirochetes from *Spirotrichonymphella* and *Stephanonympha;* and it has been used in the study of many Devescovininae by the writer. Feeding *Kalotermes hubbardi* for twelve days on filter paper moistened in 5-percent aqueous acid fuchsin removed the attached spirochetes, and examination by the dark-field method showed that the tertiary flagella described by Light (1926) were absent, proving that they also were spirochetes. The spirochetes of *Pseudodevescovina* of *Neotermes insularis* were not removed by this method, however, showing that it cannot be depended upon as always effective.

Rods of types other than the longitudinally adherent bacteria are less frequent than spirochetes. They occur occasionally on devescovinids, adherent by one end; have been found abundant on *Proboscidiella kofoidi* (Kirby, 1928), *Joenia annectens* (França, 1918), *Oxymonas dimorpha* (Connell, 1930), and *Microrhopalodina enflata* (Duboscq and Grassé, 1934). Occasionally there are also long filamentous organisms, which occur, for example, among the spirochetes on *Metadevescovina debilis.*

Microörganisms adherent by one end to flagellates of termites sometimes seem to be actually embedded in the ectoplasm, or to be associated with cytoplasmic differentiations. This was described by the writer (1936) in *Pseudodevescovina uniflagellata,* and it was noted that the apparent embedded part may stain more deeply than the rest and appear thicker (Fig. 212C). In *P. ramosa* (Kirby, 1938a) and *P. punctata* (Grassé, 1938), there are bacteria in the ectoplasm not directly associated with the spirochetes, but in *P. uniflagellata* the apparent granules are not of the same nature. Grassé (1938) made similar observations on adherent spirochetes of *Caduceia theobromae,* and interpreted the thickening not as part of the spirochete but as a modification of the cytoplasm in reaction to the microörganism. Rounded corpuscles associated with the point of attachment of spirochetes on *Parajoenia grassii* were described by Janicki (1915) and by Kirby (1937). These bodies seem to be cytoplasmic structures, neither part of the spirochetes nor parasites. A notable instance of a cytoplasmic differentiation, associated with the point of adherence of a microörganism, was seen by the writer in a species of *Macrotrichomonas* in *Procrytotermes* sp. from Madagascar. Rods 2.5-7 $\mu \times \frac{1}{4}$-$\frac{1}{2}$ μ adhere in large numbers to the posterior part of the body of almost all specimens. Where each of the rods meets the body is a deep-staining, cup-shaped structure (Fig. 212D, E). Rods frequently become detached from these cytoplasmic structures, not being so firmly adherent as are spirochetes.

It has been observed in several devescovinid flagellates with complete investment of short spirochetes, that pockets may be formed inward from the surface, enclosing some spirochetes. This was noted by Kirby (1936) in *Pseudodevescovina uniflagellata*, and by Grassé (1938) in *Caduceia theobromae* and *P. punctata*. Grassé noted further that the pockets in the former species may become closed, so that spirochetes are enclosed

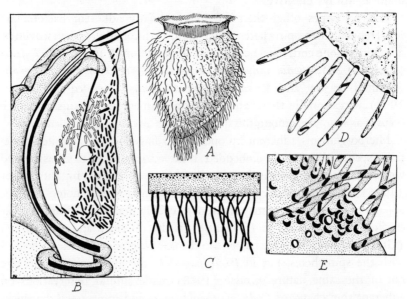

Figure 212. A, spirochetes adherent to *Trichodinopsis paradoxa;* B, microörganisms on the capitulum of the axostyle of *Macrotrichomonas pulchra* from *Glyptotermes dubius;* C, adherent spirochetes on *Pseudodevescovina uniflagellata,* with enlargement in ectoplasm at point of adherence of each; D, E, rods adherent to *Macrotrichomonas* sp. from *Procryptotermes* sp., showing the cup-shaped structure in the ectoplasm at the end of each rod. (A, after Cépède and Willem, 1912; B-E, original.)

in vacuoles in the cytoplasm. He believed also that the external, fusiform bacilli may at times enter the cytoplasm and be digested; but this opinion may have been based on the presence of an intracytoplasmic symbiont, which actually is quite different, as was noted in *Caduceia nova* and *C. theobromae* by the writer (1936, 1938a).

The possible physiological relationship between the adherent spirochetes and their flagellate hosts has been considered by Cleveland (1928) and Grassé (1938). Cleveland thought at first that they might live in some sort of mutualistic relationship; but he found that when the spirochetes were removed, no apparent detrimental effect on the Protozoa developed within three months. Sutherland (1933) found that *Stephano-nympha* died after detachment of the spirochetes, while *Spirotrichonym-phella* showed no impairment; but other factors may have been responsible for the death of the polymastigote. The writer found a marked reduction in size of *Metadevescovina debilis* after spirochete detachment, but the relationship was not proved. Grassé (1938) concluded that the relationship with the host is at least not simple phoresy. He discussed the possibility that diffusing substances attract and nourish the spirochetes, and that localization in certain regions may be related to certain areas of greatest diffusion, or to the chief phagocytic and absorptive areas.

Schizomycetes on Sarcodina.— Lauterborn (1916) found *Amoeba chlorochlamys,* a sapropelic *limax* type of rhizopod, to be characterized by a yellow-green mantle of close-set, radially adherent chlorobacteria. These bacteria, which he named *Chlorobacterium symbioticum,* were rods about 2 μ long. When the amoeba was inactive, the mantle completely surrounded it; in activity it opened more or less before advancing pseudopodia. As stated above, Lauterborn found the same microorganism on a colorless, sapropelic flagellate.

It is probable that a similar mantle of bacteria is present on *Dinamoeba mirabilis.* Leidy (1879) described the surface of the body as "bristling with minute spicules or motionless cils." In the majority of specimens he found a thick investment of hyaline jelly, at the surface of which were abundant, minute, perpendicular rods, termed by Leidy "bacteria-like cils." The rods covering the body surface were sometimes absent; and Leidy recorded instances of their disappearance from individuals that possessed them when first observed.

Schizomycetes on Ciliophora.—Certain schizomycetes occur in specifically phoretic relationship to a number of ciliates and suctorians. The instance earliest known was the adherence of spirochetes to *Trichodinopsis paradoxa,* a peritrich in the intestine of *Cyclostoma elegans* (Fig. 212A). Earlier authors described this as possessing, unlike other peritrichs, a general investment of vibratile cilia (Issel, 1906). Issel found basal granules for the supposed cilia; according to Pellissier (1936), these are mitochondria disposed under the pellicle. Observations on microörganisms adherent to Devescovininae (p. 1013) suggest another explanation. Fauré-Fremiet (1909) noted that the filaments have an undulatory movement unlike that of cilia, have an uneven distribution, and may quit the host entirely. They differ in staining from true cilia, and their movement remains unmodified, even though the ciliate be crushed. Fauré-Fremiet considered them to be spirilla, but later observers (Cépède and Willem, 1912; Bach and Quast, 1923; Pellissier, 1936) recognized them as spirochetes. Bach and Quast reported spirochetes also in the gut lumen, but found them present only when *Trichodinopsis* was also present.

Collin (1912) recorded a number of instances of the presence of adherent mircoörganisms on Suctoria. Short, rod-shaped bacteria, many of them in division, were shown adherent full length, in a close-set investment, on *Discophrya lyngbyei* (Collin, 1912, his Fig. 17). Schizophytes adherent by one end, often obviously simple phoretic microorganisms with no closer relationships, were found on the lorica of *Acineta tuberosa* and on the tentacles of *Choanophrya infundibulifera*. Bacteria were adherent in a gelatinous investment on the surface of several species of *Paracineta*.

More recently, especially through the work of Kahl (1928, 1932), the presence of characteristic types of rod-shaped bacteria on the surface of certain marine ciliates—chiefly sapropelic—has become known. The rods adhere either by one end, as on *Parablepharisma pellitum* (Fig. 213G), *P. collare* (Fig. 213H), *Metopus contortus* var. *pellitus* (Fig. 213F), and the stalk of *Epistylis barbata* (Fig. 213E); or flat, as on *Cristigera vestita* (Fig. 213C, D), *C. cirrifera* (Fig. 213B), *Blepharisma vestitum*, *Parablepharisma chlamydophorum* (Fig. 213I), and species of *Sonderia*. The presence of adherent bacteria is characteristic of all members of the genus *Sonderia*. Many of these ciliates are covered by a gelatinous layer, and it is to this that the bacteria adhere (Fig. 213A). Yagiu (1933) and Powers (1933, 1935) found bacilli constantly adherent longitudinally to the surface of three species of *Cyclidium* from sea urchins, a different type on each species (Fig. 214). Kirby (1934) believed the protuberances of *Metopus verrucosus* (Fig. 213J), which are irregular in distribution, to consist of groups of vertically adherent bacteria. Kahl questioned this statement, but in view of what we now know of bacteria on flagellates and ciliates it is not improbable. Kahl (1933) maintained that the adherent bacteria are advantageous symbionts, contributing somehow to the nutrition of their hosts; but that

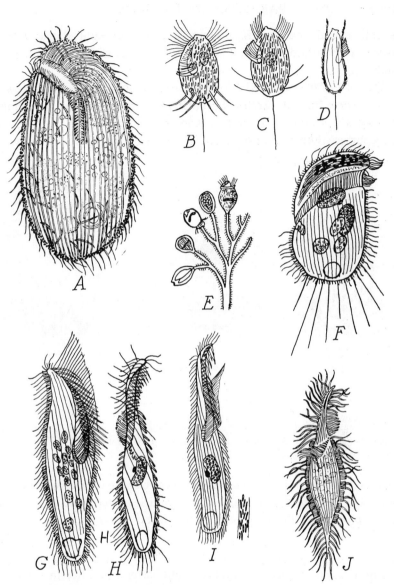

Figure 213. Bacteria adherent to ciliates. A, bacteria in the gelatinous investment of
Sonderia pharyngea Kirby (= *S. schizostoma* Kahl ?); B, surface bacteria on *Cristigera
cirrifera* Kahl; C, D, bacteria on *Cristigera vestita* Kahl; E, bacteria on stalk of *Epistylis
barbata* Gourret and Roeser; F, vertically adherent bacteria on *Metopus contortus* var.
pellitus Kahl; K, on *Parablepharisma pellitum* Kahl; H, on *Parabl. collare* Kahl; I, *Parabl.
chlamydophorum* Kahl, with Ia, longitudinally adherent bacteria; J, vertical rodlets,
possibly adherent bacteria, on *Metopus verrucosus* (da Cunha) (*Spirorhynchus verrucosus* da Cunha). (A, J, after Kirby, 1934; B-D, after Kahl, 1928; E, after Kahl, 1933
from Gourret and Roeser; F-I, after Kahl, 1932.)

is only speculation and is, in fact, improbable. Kahl stated that it seemed to him inconceivable that this symbiosis is without advantage to the host.

Dogiel (1929) observed *Sarcina*-like bacterial epiphytes on *Didesmis ovalis*. These formed a group of regularly quadrangular form, in definite number and arrangement, and were located in a preferred place on the body surface. Other epiphytes, regarded as being probably bacteria, were

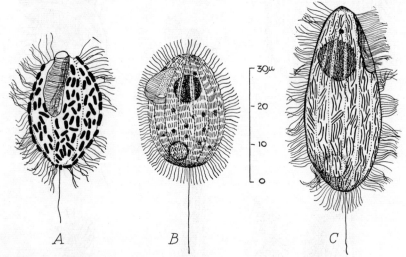

Figure 214. Characteristic bacteria adherent to the pellicle of *Cyclidium* from the intestine of sea urchins. A, *C. rhabdotectum* Powers; B, *C. ozakii* Yagiu; C, *C. stercoris* Powers. (After Powers, 1935.)

found on two species of *Diplodinium*. These had the form of an elongate oval body, attached by a stalk to the pellicle. They were found also free, ophryoscolecids being only an accidental substrate.

Endobiotic Schizomycetes

The relationship between Protozoa and bacteria that live in the cytoplasm or, less frequently, the nucleus, is closer than that of the surface forms. The bacteria must obtain all their nutriment from the host, in the body of which they multiply. The association is sometimes a constant one, the host seldom, or never, being found without the customary microörganisms. These are then probably not detrimental to the host, but have come to occupy a normal place in the metabolism of the combination. Other bacteria occur as occasional endobionts, present in a variable percentage of hosts; often there are certain types that are more or less likely to occur. Some of these are not noticeably detrimental to the maintenance by the host of normal activity. At the other extreme are some that cause fatal diseases; this is true especially of the nuclear parasites.

ASSOCIATIONS OF A CONSTANT CHARACTER

Instances in which endobiotic bacteria are always or at least usually present are known chiefly in flagellates of termites and amoebae of the genus *Pelomyxa*. Doubtless there are many other such associations among Protozoa. Miyashita (1933) found the occurrence of abundant rod-like bacteria in the cytoplasm of *Ptychostomum (Hysterocineta) bacteriophilum* to be characteristic of the ciliate; and similar rods were seen by Studitsky (1932) in the endoplasm of *P. chattoni*. Flexuous rods from 8 to 20 μ long were observed by Chatton and Lwoff (1929) in all specimens of *Ellobiophrya donacis,* but not in the mantle cavity of the lamellibranch host of the peritrich. Furthermore, the rods often showed division and were never corroded; so that these authors concluded that the microörganism is a specific symbiont.

Schizomycetes in Pelomyxa.—So characteristic are bacteria in *Pelomyxa* that Penard (1902) designated the genus as always provided with symbiotic bacteria. Greeff (1874), who observed them in *Pelomyxa palustris,* considered them to be crystals, and he had at first held them to be seminal threads. F. E. Schulze (1875), though noting their similarity to bacteria, agreed with Greeff that the rods are peculiar structures of the *Pelomyxa* body. Leidy (1879) observed rods in his *Pelomyxa villosa* (which species, according to Penard [1902], represents nothing in reality, the name having been applied to an aggregate of several species of *Pelomyxa*), and noticed, as had Greeff, that many appeared to be transversely striated. Bourne (1891) identified the rods as bacteria, and Penard (1893) expressed the same opinion as to their nature.

Penard (1902) described the bacteria as having a length of from 10 to 15 μ, or sometimes 20 μ; and in one individual there were rods of from 40 to 50 μ. Leiner (1924) found the rods varying in length from 1.5 to 22 μ (Fig. 215B). Penard found that all the rods were divided by equidistant transverse partitions, usually into two or three sections. Gould (1894) (Fig. 215A) noted their division into from two to sometimes as many as nine sections; and later (1905), under the name of Veley, she observed transverse fission of the rods. Veley stated that when the rods are set free from the cytoplasm, they are capable of independent movement, of a kind associated with the presence of flagella,

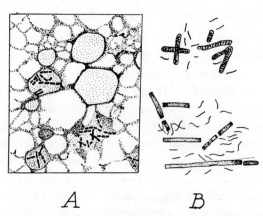

A B

Figure 215. Bacteria (*Cladothrix pelomyxae* Veley, and a small species) in *Pelomyxa palustris* Greeff. (A, after Gould, 1894; B, after Leiner, 1924.)

though none were demonstrated. This movement was at first rapid, and could readily be distinguished from Brownian movement. Leiner observed no independent movement.

Attempts to cultivate the rods have given inconclusive results. Though Veley (1905) cultured in sheep's serum, inoculated from washed *Pelomyxa*, rods which she considered to be identical with those in the cytoplasm, Leiner (1924) concluded that the amoebae cannot be certainly freed of foreign bacteria, and that the sources of error are too great.

Veley (1905) named the organism *Cladothrix pelomyxae*, noting its resemblance to the two existing species of *Sphaerotilus*, of which genus, according to Buchanan (1925), *Cladothrix* may probably be regarded as a synonym.

The rods are generally aggregated in proximity to the nuclei and refractile bodies (Fig. 215A). Penard (1902) noted that in certain individuals of *Pelomyxa vivipara,* all the nuclei are enveloped by close-set bacteria applied to the surface. Leiner stated that in *P. palustris* bacteria may thickly invest the nuclei, especially in animals in which the refractile bodies are small. Veley observed jointed rods attached to the refractile bodies, and believed that these afford them a point of attachment without which the cycle would not be completed. She thought it probable that the refractile bodies, which she considered to be protein in nature, serve the bacteria as a food supply. Leiner found evidence that the long rods extract glycogen from the refractile bodies.

Fortner (1934) studied the occurrence of bacteria in different forms of *P. palustris.* The large, club-shaped, gray-greenish forms were free of, or poor in bacteria. The yellow ones contained very numerous small refractile bodies, with large numbers of bacteria in proximity to these. The small, spherical or pyriform, milky-white type contained no refractile bodies, in place of which were the characteristic bacteria in vacuoles. The white forms he believed to be degenerate, and thought it conceivable that the whole metamorphosis of *Pelomyxa* might be conditioned by the bacterial infection. Leiner (1924) also noted the variability in the number of the rods and their abundance in yellow animals. He distinguished a second species of parasite, smaller and less numerous than the other, distributed in the cytoplasm (Fig. 215B).

Leiner found reason to believe that when the bacteria are excessively abundant, they become definitely injurious to the host. They may cause hypertrophy, structural alteration, and eventual dissolution of the nuclei; and the trophic functions of the cell appear to be disturbed. There is decreased storage of glycogen. When *Pelomyxa* dies, the bacteria multiply extraordinarily.

Schizomycetes in Flagellates of Termites.—There are numerous instances of a constant association with intracytoplasmic bacteria among flagellates of termites, and often the microörganisms are restricted to specific areas of the cell. There is not positive proof in all instances discussed below that these are bacteria; but reaction to certain fixatives, staining properties, comparison with known cytoplasmic inclusions, and, frequently, observation of fission stages make it extremely probable.

There is no evidence that the bacteria are harmful to the flagellates, though it is possible that as regards certain ones some such evidence may eventually be adduced. They may be referred to as intracellular symbionts with as much justification as may the microörganisms in the bacteriocytes and mycetocytes of certain insects. Pierantoni (1936) proposed a hypothesis concerning the function of bacteria in flagellates of termites. According to this hypothesis, with which Grassé (1938) expressed agreement, the bacteria function in the xylophagous nutrition of the flagellates. The flagellates, which are unique among Protozoa in their xylophagous habits, are, like so many wood-ingesting animals, incapable of digesting cellulose, but depend upon symbiotic bacteria. The bacteria are sometimes localized in "symbiotic organelles," sometimes diffusely distributed in the cytoplasm. The hypothesis rests on grounds similar to that of the supposed rôle of the intracellular symbionts in insects—the constant association and frequent localization suggest the likelihood of a fundamental significance in the relationship. The weakness of both hypotheses is patent; there is no physiological evidence in their support. Furthermore, the existence of "organelles" of the type mentioned is exceptional in xylophagous flagellates.

It is probable that the so-called chromidial zone of *Joenia annectens* is a bacterial aggregate. It occurs constantly in that hypermastigote, and has been shown or described by all students of the flagellate (Grassi and Foà, 1904, 1911; França, 1918; Duboscq and Grassé, 1928, 1933, 1934; Cleveland *et al.,* 1934; Pierantoni, 1936). According to Duboscq and Grassé, the bodies in this group are rods. The group surrounds the axostyle posterior to the nucleus, often forming a broad ring. Duboscq and Grassé (1934) figured an instance in which they form a spherical group, not encircling the axostyle, in a nondividing flagellate; and apparently in division stages they disperse. In the figures presented by these authors there are forms of the small rods that might be interpreted as division stages.

Grassi and Foà (1911) concluded that the area they earlier called the chromidial zone acts as a phagocytic organ, since in experimental feeding they found granules of carmine included in it. Duboscq and Grassé regarded the rods as mitochondria, but their reaction to fixatives and stains rather suggests bacteria. They are resistant to fixatives after which true mitochondria are not demonstrable. Pierantoni considered them to be bacteria, and showed differential staining of them and mitochondria in Altmann-Kull.

The writer has observed a comparable aggregate of granules or rods around the axostyle of *Devescovina glabra* a short distance posterior to the parabasal body. It is conspicuous in the species in several hosts, in some of which it has been found in all specimens from several colonies. Usually the limits of the group are well defined, and it has the form of a thick ring around the trunk of the axostyle. The bodies are generally short rods or granules; in some specimens from one host the group consisted of long rods. Their properties of fixation and staining exclude the possibility that they are mitochondria. In division stages the bacteria are dispersed. Most species of *Devescovina* have no such bacterial aggregate.

Pierantoni found no "chromidial zone" in *Mesojoenia decipiens,* but a similar area was occupied by filamentous bacteria. Kirby (1932a) described as a "proximo-nuclear parasite" a group of bacteria, usually rod-formed or filamentous and often bent or curved (Fig. 219A), located in a mass surrounding, or in proximity to the nucleus of all specimens of *Trichonympha campanula* and *T. collaris.* The suggestion was made that the organism may depend for its nutrition upon immediate proximity to the source of the nuclear influences upon metabolism.

The peripheral granules located in the outer zone of the ectoplasm of the flagella-bearing region of *T. collaris* and *T. turkestanica* (Kirby, 1932a), which stain with the Feulgen reaction and show stages of apparent division, are probably localized bacterial symbionts. They have not been found in other species of *Trichonympha.* Ectoplasmic granules, considered by the authors to be bacteria and constant in occurrence and distribution, were described by Kirby (1938a) as immediately under the surface layer of *Pseudodevescovina ramosa,* and by Grassé (1938) as in a similar position in *P. punctata.* Similar peripheral bacteria occur in *Bullanympha silvestrii* (Kirby, 1938b). Grassé found minute bacteria more deeply situated under the surface of *P. brevirostris.* Kirby (1932a) discussed the similarity between the peripheral granules of *Trichonympha* and those described in certain ciliates.

A still more remarkable localization is that of certain bacteria which occur on the capitulum of the axostyle. Not all granules that occur in that location are held to be bacteria, but some seem definitely to be. Rods on the capitulum of *Pseudodevescovina uniflagellata* were described by Kirby (1936) as probably bacteria. These were present in a large percentage of specimens, but not in all; and in only a very small percentage were they present also in the general cytoplasm, particularly the ectoplasm. Rod-shaped microörganisms were found on the capitulum of *Macrotrichomonas pulchra* from *Kalotermes contracticornis* (Kirby, 1938a) and have since been observed in that species from other hosts (Fig. 212B). Their variability in number and the fact that they do not occur on all specimens, or on any specimens in certain hosts, are in agreement with the view that they are symbionts rather than structures of the flagellate.

The greatly expanded capitulum of a large devescovinid flagellate from *Neotermes insularis* is constantly encrusted with short, stout rodlets which show evidence of fission. In this remarkable flagellate there are similar rodlets in the peripheral cytoplasm, usually separated by a narrow or a broad bacteria-free band from the capitular group.

Bacteria distributed generally in the endoplasm of termite flagellates are frequent and of many types. Some of them are practically constant in occurrence, as are the slender granule-containing rods in the endoplasm of *Caduceia nova* and *C. theobromae* (Kirby, 1936; 1938a). Devescovinids often contain many deep-staining cytoplasmic granules which may be bacteria. Jírovec (1931b) found two kinds of bacteria very abundant in the cytoplasm of every specimen of *Trichonympha serbica* he studied. They were not present in *T. agilis,* of which *T. serbica* is probably a synonym, from termites in Spain; so probably they are either present or absent in flagellates of different termite colonies. Parasites similar to one type (paired cocci) were found by Georgevitch (1929, 1932) in *T. serbica.* Pierantoni (1936) found minute bacteria present in large numbers in the cytoplasm of *Trichonympha minor* and *T. agilis,* and he reported their occurrence to be constant.

ASSOCIATIONS OF AN OCCASIONAL CHARACTER

Schizomycetes in Mastigophora.—Bacteria in the cytoplasm of flagellates of termites have been reported by Kirby (1924) in *Dinenympha fimbriata,* by Duboscq and Grassé (1925) in *Pyrsonympha vertens,* by Connell (1930) in *Oxymonas dimorpha,* by Connell (1932) in *Gigantomonas lighti,* by Powell (1928) in *Pyrsonympha major,* and by Kirby (1932a) in *Trichonympha.* Forms like the crescentic organism described by Connell have been seen by the writer in occasional specimens of a number of devescovinids. They are not stages of *Sphaerita,* as Connell suggested that they might be. Organisms with a peg form, differentiated sharply into a clear section at the broader end and a deep-staining section, are frequent in *Stephanonympha* (Fig. 216A). These have a general resemblance to the peg-formed parasite (Fig. 219D, E) in *Trichonympha campanula* (Kirby, 1932a), but are larger. A unique type is the spindle-shaped organism (Fig. 216B), up to 20 μ in length, with a deeply stainable area at one end, which occurs usually

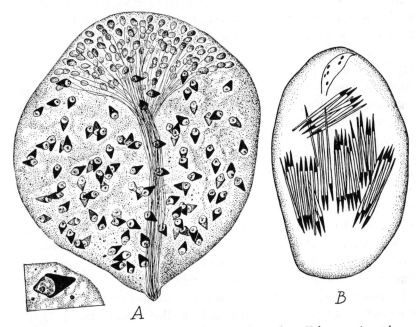

Figure 216. A, microörganisms in *Stephanonympha* sp. from *Kalotermes jeannelanus;* B, spindle-shaped organisms in *Caduceia* sp. from *Neotermes greeni*. (Original.)

in groups in a species of *Caduceia* from *Neotermes greeni*. In most termites the infection of *Caduceia* was only from 1 to 2 percent, but occasionally about half the flagellates were parasitized. With the larger forms were shorter, slender ones. The smallest forms were usually gathered into stout spindle-shaped groups, suggesting origin by splitting of a short, stout spindle. Many other bacterial parasites in termite flagellates have been studied by the writer.

Spirochetes frequently invade the cytoplasm of flagellates of termites, especially when the insects are on a filter-paper diet, and have been observed in active movement. Some of these are in vacuoles, but often they lie directly in the cytoplasm. Their condition and activity suggest that many of them are not merely ingested as food, but invade the body as facultative parasites.

Motile intracytoplasmic organisms were reported by Kirby (1932a) in *Trichonympha* from *Zootermopsis*. They were, indeed, recognized only in consequence of their relatively rapid movements, which left clear tracks in the granular prenuclear endoplasm. One was seen in a nucleus. Nothing is known of the detailed structure or the affinities of these parasites, which vary greatly in size. Abundant, vigorously vibrating bodies were present in many specimens of *Pseudodevescovina uniflagellata,* among the particles of wood. Sometimes these were sufficiently numerous to give an appearance of great activity in the cytoplasm. Like the parasite of *Trichonympha,* they have not been found in preserved material.

Few reports have been published of bacteria in flagellates other than these of termites. Yakimoff (1930) named *Micrococcus batrachorum* (*sic*) a coccus-like form which occurred, grouped in masses of irregular form or isolated, in a small percentage of *Trichomonas batrachorum.* Dangeard (1902) named *Caryococcus hypertrophicus* a bacterial parasite of the nucleus of *Euglena deses,* which caused a disease developing in great intensity. This was manifested by considerable hypertrophy of the nucleus, discoloration of plastids, and loss of the power of division. No figure of *Caryococcus* exists. As Dangeard had previously described *Nucleophaga,* it is probably distinct from that chytrid. Boeck (1917) found rod-shaped bacteria in the cytoplasm of *Giardia microti* in certain preparations. He considered these to be the same as those which occasionally adhered externally, and referred to the relationship as parasitism, leading to deleterious results.

Schizomycetes in Sarcodina.—A number of records exist of cocci and other bacteria in amoebae. Nägler (1910) found a heavy parasitic infection of amoebae, similar to *Amoeba albida* Nägler, cultivated on an agar plate. The cytoplasm was crowded with small, rounded, deep-staining granules of variable size, considered to be micrococci; and rod-formed and fusiform bacilli, likewise parasites, also occurred. Eventually all the amoebae were destroyed, with swelling and degeneration of the nuclei. The author regarded the micrococcus as a facultative parasite, which multiplied in the body after ingestion. Mackinnon (1914) reported, in *Entamoeba minchini* (*Löschia hartmanni*) in tipulid larvae, a similar organism, which occurred in the nucleus as well as in the cytoplasm; and she stated that *Polymastix melolonthae* and *Monocercomonas melolonthae* are also infected. Micrococcus was reported by Dangeard (1896) in *Sappinia* and by Wenyon (1907) in *Entamoeba muris*. Granules filling the cytoplasm of *Mastigina hylae*, resembling in size, shape, and distribution the bodies called micrococci by others, were found by Sassuchin (1928b). He mentioned their resemblance to Chytridiales, but no sporangia were shown.

Cytoplasmic granules like micrococci were described by Alexeieff (1912) in an amoeba, probably *Lecythium* sp., and in *Tetramitus rostratus*. He considered these, however, to be produced by multiplication of elementary corpuscles, existing as a corona of small granules around an initial corpuscle. The initial corpuscle was a small spherule with a central nucleus. He thought this to be a chlamydozoön, and named it *Chlamydozoön bütschlii*. Later Alexeieff (1929) described the same parasite in *Monas vulgaris*. The evidence, however, that these bodies in amoebae and flagellates are filtrable viruses and his speculations on the relationship of Chlamydozoa and Chytridiales are no more convincing to the reader than his argument that the *parasite* of cancer is a chlamydozoön, which he named (1912) *Chlamydozoön perniciosum*. The reports of Alexeieff are considered here because of the possibility that the granules described are bacterial parasites. The "initial corpuscles" are inclusions of another nature.

Mercier (1910) found bacterial filaments made up of short rods in *Endamoeba blattae*. Eventually they invaded the whole cytoplasm of a parasitized amoeba, which disintegrated. He also observed granular bodies, believed to be parasites, in the cytoplasm of cysts.

Bacilli in masses in the cytoplasm of an amoeba, probably *Vahlkampfia* sp., grown on an agar plate, were considered by Epstein (1935) to be parasites. There were no indications of digestion of the bacteria, and the masses increased in volume. In consequence of the infection, division was supposed to be delayed; the nuclei continuing to multiply, several nuclei occurred in the hypertrophied bodies. This article does not carry conviction; the possibility is not excluded that the large multinucleate bodies occur in consequence of abundant nutrition, and it is not proved that the bacteria are parasites.

Schizomycetes in Ciliophora and Sporozoa.—Bacterial parasites of the nucleus and the cytoplasm have been observed repeatedly in free-living ciliates, and some of these have been studied more intensely than any other bacteria in Protozoa. They are of considerable interest, too, in the history of protozoölogy, having frequently come to the attention of students, in the latter half of the nineteenth century, and having been variously interpreted.

J. Müller (1856) considered the question of spermatozoa in Infusoria, on the basis of observations by himself and his students, Lieberkühn, Claparède, and Lachmann. They observed undulatory filaments in the cytoplasm of *Stentor* (which were parasites, as discussed below), and found fine curved threads in hypertrophied nuclei (macronuclei) of *Paramecium "aurelia"* (=*caudatum*). Müller commented on the presence of these threads in what Ehrenberg had regarded as a seminal gland, but did not commit himself as to their real nature. Claparède and Lachmann (1857; 1860-61, p. 259) reported the observation of immobile rods in the nucleus of *Chilodon cucullulus* and in *Paramecium.* Believing that the nucleus plays an important rôle in "embryo formation" (*Sphaerophrya*), they advanced the hypothesis that it may at certain times play the rôle in some individuals of a testis, in others of an ovary (1857). Stein (1859) observed hypertrophied macronuclei of *Paramecium "aurelia"* (=*caudatum*) containing fine, straight rods. He stated that he was at first inclined to regard individuals with such nuclei as males, the nucleus functioning as a testis and producing spermatozoa, while in others it had the rôle of an ovary. Later, he concluded that the spermatozoa developed in the nucleolus (micronucleus), then penetrated the nucleus, and he pointed out the analogy with fertilization in *Volvox*.

Balbiani (1861) observed the rods in nuclei of the same species of ciliate, and noted, as had the others, that they were immobile. He concluded that they are parasites, which develop in the interior of the female reproductive organ, and pointed out fundamental differences between them and the "spermatic filaments" enclosed in the "nucleolus" of this animal. It was then Balbiani's contention, developed in 1858 (1858a, 1858b), that Infusoria are hermaphroditic sexual animals, the nucleus (macronucleus) an ovary, and the nucleolus (micronucleus) a testis (the chromosome filaments having been mistaken for spermatozoa). Engelmann (1862), finding these parasites in nuclei of *Paramecium caudatum* and *Blepharisma lateritia,* nevertheless continued to regard them as spermatozoa; but Engelmann (1876) wrote of bacteria parasitic in nuclei of *Stylonychia mytilus.* Bütschli (1876), severely criticizing the Balbiani-Stein hypothesis of sexual reproduction, reported rods in the nucleus (macronucleus) of *P. aurelia,* and agreed with Balbiani that they are parasites. He remarked that similar parasites are sometimes found in the nucleolus (micronucleus) of many Infusoria. A review of the earlier observations is given by Bütschli (1889, p. 1828).

Balbiani (1893), giving a figure and a brief discussion of parasites in the macronucleus of *Stentor polymorphus,* remarked on the fact that the parasitized animal appeared to be perfectly normal and continued its ordinary functions. Zoöchlorellae were neither more nor less abundant than usual. The alteration of the nucleus might seem to be equivalent to its artificial extirpation, as the nuclear substance had completely or almost completely disappeared. He discussed these facts in connection with his experimental results in the removal of the macronucleus, and concluded that they were not in disagreement. A minimal amount of the nuclear substance might still have been present and sufficed; or the animal observed might have been in the limited period during which normal life continues.

Studies of the nuclear parasites of *Paramecium* and their effects on the animal were made by Hafkine (1890), Metschnikoff (1892), and Fiveiskaja (1929); and certain observations on the consequences of parasitism were reported by Bozler (1924). A similar parasite, studied in detail by Petschenko (1911), was believed to be cytoplasmic. This author stated that the possibility of confusing microörganisms which live in the micronucleus with those that live in the cytoplasm cannot be denied.

Hafkine distinguished three species of *Holospora. H. undulata* invades the micronucleus and is spiraled. At the beginning the organism is a small, fusiform corpuscle. The second species, *H. elegans,* occurs also in the micronucleus, but is never associated with the first. The vegetative stage is fusiform and more elongated and slender than the others. The third species, *H. obtusa,* invades the macronucleus. It is not spiraled and the two ends are rounded, instead of both or one being pointed. According to Hafkine, multiplication of *Holospora* takes place in two ways. During development, for a time there is a rapid transverse division. The spiral form of *H. undulata* develops after division ceases; the other two species remain straight. The dimensions increase in this phase, which was supposed by Hafkine to represent a transformation into resistant spores. *H. undulata* in this phase loses its pale, transparent aspect, and becomes more refractive at one of its extremities. This modification extends until the whole organism is refractive; sometimes it is divided into three or four parts of different refractivity. Of the other reproductive process, there are only traces in *H. undulata;* but it is more frequent in the other two species. A bud forms at one of the extremities, and grows into a cell like that from which it originated. This type of reproduction, according to Hafkine, makes *Holospora* transitional between yeasts and Schizomycetes. He maintained, as did Metschnikoff and Fiveiskaja, that there are grounds for not placing the parasite among the typical bacteria. Fiveiskaja studied only *H. obtusa* in the macronucleus, and did not find any parasitized micronuclei. She found the parasites to be elongated, straight, or slightly curved rods from 12 to 30 μ in length by 0.6 to 0.8 μ in width, often showing differential refractivity or stainability. Part, up to half, or all, of a rod might be dark and the other part or other entire rods clear— a characteristic that appears in many of the early drawings (Bütschli, 1876; Balbiani, 1893).

Petschenko (1911), who studied an organism considered to be a cytoplasmic parasite, named it *Drepanospira mülleri* and assigned it to the Spirillaceae. He remarked that the external aspect of this organism is the same as that of *Holospora undulata* and *H. elegans.* The parasites develop from a group of curved rods in the cytoplasm to a large ellipsoidal mass, almost filling the body. In the vegetative period, the microorganism is a spiral with a nuclear portion near the anterior end, and it shows helicoidal movement. The karyoplasm separates into granules and bands, and endospores are said to be developed. There is a resting period in which the rod is small, curved only once, and the nuclear substance occupies from half to almost all of the cell. There is said to be no cell division, reproduction being only by the endospores. Petschenko stated that the essential difference from *Holospora* lies in the fact that Hafkine did not establish in his parasites the presence of nuclear elements; but it seems possible that the differences in refractivity noted by Hafkine may have indicated the same structure upon which Petschenko based his interpretation of nuclear organization. The latter author found micronuclei to be absent in ciliates with *Drepanospira;* so that, allowing for differences in detail of observation and interpretation, it seems not impossible that *H. undulata* and *D. mülleri* are actually the same.

Hafkine spread the infection by introducing infected paramecia into cultures. In early infection he noted no anomalies in the ciliates, but development of the parasites proceeds rapidly and by the next day all the contents of the infected nucleus are used up. When the parasite fills up a large part of the ciliate, development of the latter is arrested, and the same phenomena appear as in insufficient nutrition. Bozler (1924) and Fiveiskaja (1929) noted the vacuolization of animals in which the macronucleus was parasitized, the accumulation of fat drops and excretion granules, and the reduction and disarrangement of the trichocysts. Food-taking, digestion, and defecation continue for a long time, according to Fiveiskaja, but the number of food vacuoles formed becomes progressively fewer. In late stages food currents are absent, there are no food vacuoles, and the mouth, gullet, and cytopyge may disappear. There is partial atrophy of the ciliary coat. At first there is no change in the activity of the pulsating vacuole; later the shape of the canals changes and the pulsations become slower. Eventually, with large masses of parasites, the pulsations stop and the ciliate dies.

Petschenko found that the cytoplasmic (or micronuclear) parasite *Drepanospira mülleri* causes the cytoplasm to take on an alveolar character, and that the macronucleus undergoes degenerative changes and may fragment and dissolve. He noted that a chemical action on the cell is indicated, waste products and secretions of the parasite entering the cytoplasm of the cell. There is intoxication of the cell, acting directly on the cytoplasm, indirectly on the nucleus. Bozler stated that the possibility that degenerative changes are to be traced back to the influence of a bacterial toxin is not excluded; and Fiveiskaja thought that the changes external to the nucleus were influenced by a substance excreted from the macronuclear parasites, probably a toxin. Destruction of chromatin, with consequent disturbance of the normal macronuclear control of metabolism, together with the mechanical influence of large masses of parasites, may be sufficient, however, to account for the effect of *Holospora obtusa.*

Calkins (1904) recorded an infection of the macronuclei of 80 percent of *Paramecium caudatum,* in preparations from one culture, with a parasite that he named *Caryoryctes cytoryctoides.* The organisms appeared as scattered bodies in parts of each nucleus, and have no resemblance to *Holospora* or *Nucleophaga.*

Bacteria are especially prevalent in the cytoplasm of many ciliates that live in decaying matter, and Kahl (1930) stated that often these may be regarded as symbionts (mutualists), rather than parasites. He showed, for example, in *Epalxis antiquorum,* large symbiotic bacilli which Penard (1922) had described as rods. Recently the occurrence and significance of bacteria in sapropelic ciliates has been the subject of studies by Liebmann (1936a, 1936b, 1937).

Liebmann (1936a) found in *Colpidium colpoda* chlorobacteria that appeared to live as facultative symbionts. They were enclosed in vacuoles occurring more or less abundantly in the cytoplasm under conditions of anaërobiosis and H_2S content of the water, but not in the presence of oxygen. Their appearance was definitely correlated with the amount of H_2S. Similar chlorobacteria were present in the hay infusion from which the *Colpidium* came, and in normal oxygenated water these were ingested and digested. Under anaërobic conditions with H_2S, they remained alive in the vacuoles; and Liebmann believed that through their assimilating activity in the presence of light, H_2S is reduced in amount, and energy is contributed to the ciliate. After reserve glycogen is used up, the bacterial vacuoles may be attacked by digestive processes, and the ciliates die soon thereafter.

In many other sapropelic ciliates, Liebmann (1936b, 1937) found, together with dead bacteria, large numbers of living bacteria. These were either packed together in parallel arrangement in bundles (*Metopus*) or distributed in the cytoplasm (*Chaenia*). After a time these symbionts may be digested, and the loss is made up by taking in new saprophytes. When this is prevented, and all symbionts are used in nutrition, the ciliates perish, in spite of filled food vacuoles. Certain living bacteria are therefore necessary for the ciliates' continued life, under existing anaërobic conditions with hydrogen sulphide. In this connection Liebmann suggested that the bacteria split off oxygen, which the ciliates use.

Hetherington (1932) mentioned an extensive cytoplasmic invasion of *Stentor coeruleus* by bacilli, the ciliates losing their bright blue-green color and some of their capacity for motor response. Pale stentors from mass cultures are, he stated, often infected with a great number of bacilli. Numerous instances of physiological regeneration occurred in the recovery of these animals. The report did not indicate whether the bacilli were isolated or grouped.

In the cytoplasm of *Spirostomum ambiguum,* a motile spirillum was found in large numbers by Takagi (1938). He studied nine ciliates and found all infected, and considered it probable that all in the culture were parasitized. One hundred and six were present in one ciliate; in another, 10 of the 67 were undergoing binary fission. A flagellum was detected at one end of the parasite, which swam about actively in the cytoplasm. Takagi stressed the fact that his is the first report of a cytoplasmic parasite with active motility in a protozoan. He did not comment, however, on the observations by Müller, Claparède and Lachmann, and Stein, nor on that by Kirby. Müller (1856) mentioned observations by himself, Lieberkühn, Claparède and Lachmann of motile threads in *Stentor;* and the isolation of these by the last-named observers, when their motility soon disappeared in the water. Bütschli (1889, p. 1831) discussed these observations in his account of parasites of ciliates. The threads occurred in the vacuoles in bundles, and displayed active movement. Claparède and Lachmann (1857) thought their parasitic nature not improbable, noting their great similarity to certain vibrios. Bütschli, while admitting that the threads might be ingested food, believed it more likely that they were parasites. These forms differed from Takagi's in being in bundles instead of isolated.

Mangenot (1934) found rhodobacteria sufficiently abundant in a ciliate identified as *Spirostomum teres* to impart to it a rose color. They were distributed mostly in the peripheral cytoplasm. He regarded the relationship between them and their host as parasitic or symbiotic, and compared the "rhodelle" association to that with chlorellae, xanthellae, and cyanellae.

Irregular aggregations of minute granules (Fig. 217D) were found in the cytoplasm of many individuals of *Nyctotherus ovalis* by Sassuchin (1928a, 1934). He made various microchemical tests, excluding the possibility that these were glycogen or glycoprotein granules, chondriosomes, or volutin; and he concluded that they were bacterial parasites, which do not occur in all ciliates. Similar groups of granules were reported by Kirby (1932b) in *Nyctotherus silvestrianus*.

Bacteria were found by Hesse (1909) in monocystid gregarines from the seminal vesicles of oligochetes. Each of the species *Monocystis lumbriculi, M. agilis, M. striata, Rhynchocystis pilosa,* and *Stomatophora coronata* had its own peculiar parasite which was unlike those of the others. Their forms varied, in different species, from ovoid to filamentous. Hesse remarked that the bacterial parasites were uncommon, but when present attacked most individuals of a species, and often led to the destruction of the invaded gregarines.

Sphaerita and Nucleophaga

HISTORICAL ACCOUNT AND DISTRIBUTION

In Free-living Protozoa.—Most of the fungi of the order Chytridiales are parasitic in plants or animals (Fitzpatrick, 1930; Minden, 1915). In the lower plants they occur mainly on or in algae; and a considerable number have been found in Phytomastigophora. Though most abundant in this group of Protozoa, they attack also other free-living forms, especially Sarcodina and cysts of ciliates (Bütschli, 1889; see also p. 1059), and many have been encountered in parasitic Protozoa. The chytrids that are known to be hyperparasitic in Protozoa all belong to the genera *Sphaerita* and *Nucleophaga*. These are the chytrids, also, that have most often been found in free-living species, except for euglenid flagellates.

Carter (1856) described "irregular, botryoidal masses, dividing up into spherical cells" in *"Astasia"* (=*Peranema*). It is likely that he was observing *Sphaerita,* and that the enlarged granular nuclei described in *Amoeba radiosa* (?) were parasitized by *Nucleophaga.* The specimens of *A. verrucosa,* "partly filled with spherical ovules in the granuliferous stage of development," were probably heavily parasitized by chytrids. Wallich (1863a) found that a large subspherical, granular mass appeared in each of the specimens of *A. villosa* in a saucer; and later from five to a dozen of these masses developed in individual specimens. He observed extrusion and rupture of these, which he regarded as of the nature of nuclei. He evidently was describing an increasingly heavy infestation of a culture of amoebae by *Sphaerita.* The granulation of the nucleus described by Carter (1863) in *A. princips,* accompanied by the enlargement of the nucleus to between three and four times its normal diameter, indicates the presence of *Nucleophaga*. Greeff (1866) mistook the early plasmodial stages of *Nucleophaga* for young amoebae entering the nucleus of *A. terricola.*

Stein, who had mistaken parasitic bacteria for reproductive elements of ciliates (p. 1034), observed *Sphaerita*-like Chytridiales in a number of flagellates (1878, 1883). His plates include figures of what are probably such fungi in *Monas guttula, Chlamydomonas alboviridis, Euglena viridis, Trachelomonas volvocina, T. hispida, Phacus pleuronectes, Tropidocyphus octocostatus, Anisonema grande (A. acinus), Glenodinium pulvisculus, Heterocapsa triquetra,* and *Dinopyxis laevis.* In many of these he represented the escape of minute, flagellated organisms. He likewise concluded that these are reproductive elements, the nucleus undergoing growth and fragmentation, and giving rise thus to endogenous germs reproducing the flagellate. This theory of flagellate reproduction was accepted by Kent (1880-82), who confirmed the observations of Stein on *Euglena* and other euglenid flagellates. Ryder (1893) compared an "endoblast" figured in *E. viridis,* from which flagellate "germs" were said to escape and become amoeboid forms developing into adult euglenas, with Stein's reproductive stage. Discussion of the early errors of interpretation is given in many of the publications on chytrid parasites of Protozoa, particularly those of Dangeard (1886b, 1895), Penard (1905b), Chatton and Brodsky (1909), and Mattes (1924).

The evidence for Stein's notion of reproduction did not satisfy Klebs (1883), who pointed out that the "Keimkugel" was a sporangium of *"Chytridium* spec.," which, he stated, is one of the most frequent parasites of *Euglena.* The problem was studied independently by Dangeard, and he arrived at the same conclusion (1886a, 1886b). The name *Sphaerita endogena* was given (1886a) to cytoplasmic chytrids in the rhizopods *Nuclearia simplex* and a species of *Heterophrys,* which was later (1886b) named *H. dispersa.* The illustrated account of the parasite (1886b) included a report of its occurrence in *Euglena viridis.* Dangeard (1889a), recorded *S. endogena* in *Phacus pyrum, Trachelomonas volvocina,* and *T. hispida;* later (1889b) he described it in *Euglena sanguinea* and *P. alata;* and in 1895 he gave a fairly complete and well-illustrated description of the life history of the parasite in *Euglena (viridis?).* Serbinow (1907) studied the chytrid in *E. viridis* and *E. sanguinea.*

In more recent work, *Sphaerita*-like Chytridiales in free-living rhizopods and flagellates have been differentiated into several species, but a comparative account of the differential characteristics is lacking. Chatton and Brodsky (1909) proposed to give the parasite of euglenids, if separated from *S. endogena*, the name *S. dangeardi;* Skvortzow (1927) briefly designated as *S. trachelomonadis* a parasite of *Trachelomonas teres* var. *glabra* and *T. swirenkoi* in Manchuria; Jahn (1933) differentiated *S. phaci* from *Phacus pleuronectes* and *P. longicauda,* and Gojdics (1939) reported the same species from *Euglena sanguinea.* Puymaly (1927) failed to recognize *S. dangeardi,* describing the life history of a chytrid of *E. viridis* under the name *S. endogena.* Dangeard (1895), with no great positiveness, proposed the name *Pseudosphaerita euglenae* for a parasite of *E. viridis* in which, in the formation of the sporangium, there is fragmentation into islets, and the contour of the sporangium often becomes irregular and cord-like. Mitchell (1928) suggested assignment to *Pseudosphaerita* of a parasite, found in species of *Euglena,* which showed neither of these characteristics; Jahn (1933) considered at least those Mitchell described in *E. viridis* to be *S. dangeardi.*

The parasites found by Nägler (1911b) in *Euglena sanguinea* are *Sphaerita*-like; but the form in the cyst, with prominent protuberances, does not resemble *Sphaerita.* Mainx (1928) found *Sphaerita* often in *E. sanguinea* and *E. viridis;* Günther (1928) reported it in *E. geniculata.* Further records of *Sphaerita,* by Alexeieff (1929), are from *Monas vulgaris* and *Dimastigamoeba gruberi.*

Since Dangeard's accounts (1886a, 1886b), *Sphaerita* in free-living rhizopods has been studied by Chatton and Brodsky (1909) in *Amoeba limax,* by Penard (1912) in *A. alba,* and by Mattes (1924) in *A. sphaeronucleolus.* The last observer described two new species, *S. amoebae* and *S. plasmophaga.* The confused and improbable cycle of *Allogromia* sp. (*Cryptodifflugia* sp., according to Doflein, 1909, 1911), outlined by Prandtl (1907), probably was based on a free-living testacean, certain small free-living flagellates, ingested Testacea, and an infection of *A. proteus* with *Sphaerita.* Prandtl discussed the observations on supposed reproduction by Carter (1863), Wallich (1863a), Greeff (1866), and even those of Stein (1878), which as stated above were based on parasitization by chytrids; and he considered that they were really made on "gamete formation" by *Allogromia* or other parasitic rhizopods. A parasite described by Penard (1912) in *A. terricola* appears to be a chytrid, but it is not like typical *Sphaerita.*

There seems to be only one record of *Sphaerita* in a free-living ciliate —the brief account of Cejp (1935) of the parasite in *Paramecium,* up to eleven sporangia occurring in a cell. Bodies like the sporangia of *Sphaerita,* but with exit tubes, were shown by Collin (1912) in *Acineta tuberosa.* Chytrids in other Suctoria, found by Claperède and Lachmann and by Stein, are mentioned below (p. 1064).

Dangeard (1895) established the genus *Nucleophaga* for a parasite, *N. amoebae* (not *amoebaea* as Penard, 1905b, and Doflein, 1907, have it), which he studied in the nucleus of *Amoeba verrucosa* (*A. proteus* according to Penard, 1905b). Gruber (1904) found *Nucleophaga* in *A. viridis,* and supposed it to be different from Dangeard's species, but according to Penard (1905b) it is probably the same. Penard described *N. amoebae* in *A. terricola* and *A. sphaeronucleolus;* and Doflein recorded the parasite in *A. vespertilio* Penard. Mattes also found, in the nuclei of *A. terricola* and *A. sphaeronucleolus,* parasites which he named *Sphaerita nucleophaga.* He believed that the forms of Penard and Doflein belonged to this same species, those of Dangeard and Gruber each being a different species. Although he did not comment on the relationship of the genera *Sphaerita* and *Nucleophaga,* his treatment of the chytrids seems to indicate that he regarded the latter as synonymous with the former. Indeed, no difference exists between the two except the habitat, and the basis for their separation seems scarcely valid.

In Endozoic Protozoa—Because most of the studies of Chytridiales in endozoic Protozoa are comparatively recent, there have been few errors of interpretation. At about the time when Stein was describing germ balls in euglenid and other flagellates, Leidy (1881) observed what is clearly *Sphaerita* in *Trichonympha agilis.* He "suspected that they are masses of ova-like bodies or spores," but discussed them as inclusions in the endosarc, not as reproductive elements. Casagrandi and Barbagallo (1897) described nuclei in *E. coli* containing small, round bodies, equal in size, and sometimes so numerous as to fill the entire nucleus; and they figured several of them in certain vegetative amoebae (Pl. 2, Fig. 13). These, as suggested by Cragg (1919), were doubtless parasites, probably *Sphaerita,* and not nuclear parasites; Cragg suggested also that the account by Craig (1911) of vegetative schizogony in this amoeba was based on parasites. The statement by Craig that the nuclei were visible in life as "brightly refractile masses of granules" is in keeping with the probability that the supposed nuclei were *Sphaerita*. Dogiel (1916), finding sporangia of *Nucleophaga* in *Myxomonas polymorpha,* thought he was observing chromosomes. The same investigator noted what are probably *Sphaerita* and *Nucleophaga* in *Joenia intermedia* and recognized the former as one of the lower fungi, but he hesitated to interpret the latter as parasitic. Early students of *Trichomonas* (Wenyon, 1907; Kofoid and Swezy, 1915; Kuczynski, 1918; Mayer, 1920; Wenrich, 1921) showed chytrids without interpreting them correctly.

Sphaerita has been found in many parasitic flagellates, especially those in termites. Several species have been differentiated. Cunha and Muniz (1923) gave the name *Sphaerita minor* to a parasite of *Trichomonas muris* and *T. gallinorum;* chytrids in *Trichomonas vitali* from *Bufo marinus* (Pinto and Fonseca, 1926), *Trichomonas muris, T. caviae,* and *Eutrichomastix lacertae* (Grassé, 1926b) have been assigned to the same species. Grassé also stated that parasites, probably *Sphaerita,* invade the plasma of *Eutrichomastix colubrorum. Sphaerita trichomonadis* was described by Crouch (1933) from *Trichomonas wenrichi* of *Marmota monax,* and *S. chilomasticis* by Cunha and Muniz (1934) from *Chilomastix intestinalis.* Sassuchin (1931) found *Sphaerita*-like parasites in *Chilomastix magna* of ground squirrels. In *Mastigina hylae,* Sassuchin (1928b) noted parasites that are *Sphaerita*-like in some respects, but are not shown grouped in sporangia; perhaps these are cocci.

Among flagellates of termites, *Sphaerita,* which in no instance has been given a specific name, has been reported or figured in *Joenia intermedia* (Dogiel, 1917), *Staurojoenina assimilis* (Kirby, 1926a), *Metadevescovina debilis* (Light, 1926), *Trichonympha chattoni* (Duboscq and Grassé, 1927), *Stephanonympha dogieli* (Bernstein, 1928), *Coronympha clevelandi* (Kirby, 1929), *Oxymonas minor* (Zeliff, 1930), *Pyrsonympha* and *Dinenympha* (Jírovec, 1931b), *Pyrsonympha elongata* (Georgevitch, 1932), *Gigantomonas lighti* (Connell, 1932), and several species of *Trichonympha* (Kirby, 1932a). In undescribed polymastigote and some hypermastigote flagellates in the writer's collection, *Sphaerita* has been found to be extremely prevalent.

The presence of *Sphaerita* has been indicated in all species of intestinal amoebae of man: *Entamoeba histolytica* (Nöller, 1921; Lwoff, 1925;

Greenway, 1926; Bacigalupo, 1927, 1928), *E. coli* (Cragg, 1919; Nöller, 1921; Epstein, 1922; Lwoff, 1925; Bacigalupo, 1927, 1928), *Endolimax nana* (Dobell, 1919; Nöller, 1921; possibly Epstein, 1922; Greenway, 1926; Wenyon, 1926; Bacigalupo, 1927, 1928), *Iodamoeba bütschlii* (Nöller, 1921; Wenrich, 1937), and *Dientamoeba fragilis* (Nöller, 1921; Wenrich, 1940). Lwoff (1925) thought that the parasite of *Entamoeba dysenteriae* (*histolytica*), *E. coli,* and *Endolimax nana* is identical with *Sphaerita endogena;* but he provided the name *S. normeti* for use if it is proved to be a new species.

The chytrid is equally prevalent in other endozoic amoebae. Leger and Duboscq (1904) stated that an amoeba in *Box boops* (*E. salpae*) is often ravaged by microspheres which lead to its destruction. Wenyon (1907) noted the parasites in *Entamoeba muris* as "vacuoles containing cocci"; Kessel (1924) recorded *Sphaerita* from the same amoeba. Becker (1926) described *S. endamoebae* from *Entamoeba citelli,* and the same chytrid was found by Sassuchin, Popoff, Kudrjewzew, and Bogenko (1930) in this amoeba of ground squirrels in Russia, though without reference to Becker's account. A parasite of *Hyalolimax cercopitheci* was named *S. parvula* by Brumpt and Lavier (1935b). Other records are from *Endamoeba simulans* of termites (Kirby, 1927), *Entamoeba bobaci* of *Marmota bobaci* (Yuan-Po, 1928), *E. pitheci* from *Macacus rhesus* (Sassuchin *et al.,* 1930), and *Entamoeba* sp. of cattle (Jírovec, 1933). Third-degree parasitism is that of *Sphaerita* in entamoebae in *Zelleriella,* reported by Stabler and Chen (1936). In almost all instances, in intestinal amoebae, the parasites have been encountered only in the trophozoites.

Among endozoic ciliates, aside from Chen and Stabler's (1936) statement that *Sphaerita* has been found in *Zelleriella* as well as in its entamoeba parasites, the chytrid has been reported only in *Nyctotherus* and Ophryoscolecidae. Sassuchin (1928a, 1934) found it to be common in *N. ovalis* from *Periplaneta* (Fig. 217C, D). In ciliates of antelopes, Dogiel (1929) described *Sphaerita diplodiniorum* in *Diplodinium costatum* and *Ostracodinum gracile.* Jírovec (1933) gave the name *S. entodinii minor* to a chytrid in *Entodinium simplex* (?), and *S. entodinii major* to one in *Entodinium longinucleatum.* He observed also a *Sphaerita*-like parasite in an undetermined species of *Entodinium,* and stated that in other Ophryoscolecidae none of these chytrids were observed.

Winogradowa (1936) reported *Sphaerita,* as well as larger, distributed, probably bacterial, parasites in *Entodinium* (Fig. 217B).

Discussing *Joenia annectens* and *Mesojoenia decipiens,* Grassi and Foà (1911) mentioned an enormous enlargement of the nucleus by the presence of a parasite, and reported also a parasite in the cytoplasm. This probably is the first record of *Nucleophaga* in a flagellate. Its presence in *Joenia intermedia* and *Myxomonas polymorpha* (= *Gigantomonas herculea*), noted by Dogiel (1917, 1916), has been mentioned above. In *Hexamastix termitis,* Kirby (1930) showed some parasitized nuclei, as Duboscq and Grassé (1933, p. 392) pointed out, but failed to interpret them correctly. The large nuclei with numerous small, uniform-sized granules contained *Nucleophaga,* and the parasite has been found, on reëxamination of the material. The parasite has been observed by the writer in many Devescovininae, but not in the smaller species of *Foaina.* There seems to be a lower limit in the size of nuclei in which it can develop. Nuclear parasites of *Trichonympha* are considered below (p. 1059). *Pseudospora volvocis,* a parasite of *Volvox* with apparent affinities to the Bistadiidae, has been found infected with intranuclear chytrids by Roskin (1927) and by Robertson (1905), the latter of whom misinterpreted the parasite as representing gamete formation by *Pseudospora.*

Nucleophaga has been found in many endozoic Amoebidae, and several species have been named. Lavier (1935b) reviewed most of the accounts, with the exception of those of Kirby (1927, 1932b) and Sassuchin (1931). The earliest observations were made in *Endamoeba blattae* (Mercier, 1907, 1910; Janicki, 1909). Tyzzer (1920) found a nuclear parasite in *Pygolimax gregariniformis* of chickens and turkeys. Two amoebae of man are known to be parasitized: *Endolimax nana* and *Iodamoeba bütschlii,* in which *Nucleophaga* was first recorded by Nöller (1921). Epstein (1922) named *Nucleophaga hypertrophica* a nuclear parasite of *Endolimax nana;* in 1935 he stated that he had studied then (1922) a nuclear infection of both *E. nana* and *I. bütschlii.* Brug (1926) independently named a nuclear parasite of the latter amoeba *N. intestinalis;* according to Brumpt and Lavier (1935a), that parasite is the same as the one (Fig. 218F-J) which they also studied in *E. nana,* and Brug's name is a synonym for *N. hypertrophica.* Kirby (1927) described an unnamed *Nucleophaga* (Fig. 218A-E) in *Endamoeba disparata* of *Miro-*

termes hispaniolae, and reported it also in *E. majestas, E. simulans,* and *Endolimax termitis;* and Sassuchin (1931) found a chytrid in the nucleus of *Entamoeba citelli*. In *Entamoeba ranarum*, Lavier (1935a, 1935b) found a parasite described as *Nucleophaga ranarum*. Although not a protozoan parasite, and one that is of doubtful affinities, the organism named *Erythrocytonucleophaga ranae* by Ivanić (1934), which invades the nuclei of the red blood cells of *Rana esculenta,* is interesting to consider in this connection.

LIFE HISTORY AND STRUCTURE

Sphaerita.—Chytrids of the family Olpidiaceae, to which *Sphaerita* and *Nucleophaga* belong, have a one-celled, intramatrical thallus, enclosed from an early period by a delicate membrane, amoeboid in nature, which at maturity changes into a single sporangium or resting sporangium. The sporangium of *Sphaerita* lacks elongate discharge tubes, the spores escaping through an opening or papilla at one or both ends. The zoöspores of chytrids of this family are uniflagellate, according to Minden (1915), Fitzpatrick (1930), and Gwynne-Vaughan and Barnes (1937); but in the spores of many forms in Protozoa, either two flagella or no flagella have been observed.

Sphaerita has often been encountered in only a small percentage of the host species, but some records report a high incidence. Nöller (1921) found, in certain material, the majority of *Endolimax nana* and a very high percentage of *Entamoeba coli* infected; and Dobell (1919) saw several *E. nana* infections in which a considerable proportion of the amoebae were parasitized. *Sphaerita* was present in 80 percent of *E. coli* and *E. histolytica* studied by Lwoff (1925). Both Becker (1926) and Sassuchin *et al.* (1930) found *E. citelli* in certain ground squirrels very heavily parasitized, and Yuan-Po (1928) reported about 60 percent infection of *Entamoeba bobaci*. Almost all *Chilomastix* in a guinea pig contained *S. chilomasticis* (Cunha and Muniz, 1934). In flagellates of termites, infection varies from light to heavy. In a few instances almost every individual of certain devescovinids on some slides has been parasitized; on the other hand, the parasite may be infrequent or absent in other host faunules of the same species. Distributional factors would facilitate the presence of the chytrids in higher incidence in endozoic than in free-living Protozoa under natural conditions, but an infection in free-living amoebae in cultures may develop a high incidence in a short time. Dangeard (1886b) found *S. endogena* in great abundance in cultures of its two rhizopod hosts; and Ivanić (1925) stated that cultures often perish from severe infection. A host may be parasitized by two or more species (Mattes, 1924; Yuan-Po, 1928), but usually only one species has been distinguished.

Brumpt and Lavier (1935b) described two different sphaeritas in two amoebae on the same smears: *S. parvula* from *Hyalolimax cercopitheci,* and one with larger spores in an *Entamoeba* of the *minuta* type. This indicates host-specificity; but various amoebae of man, it appears, contain a common species (p. 1042), and Lwoff (1925) stated that the chytrids do not seem to manifest a narrow host-specificity.

The earliest stage in the cytoplasm of the host is a small, amoeboid, uninucleate thallus. Dangeard (1895) described the parasite in *Euglena* as at first smaller than the flagellate's nucleus, with a dense, homogeneous cytoplasm and a vesicular nucleus with a large nucleolus. Mitchell (1928) found the earliest stages to be from 2.5 to 3.5 μ in diameter, with a vesicular nucleus from 1.3 to 1.5 μ in diameter. Early stages of the parasites in amoebae have been found as small as 1.5 μ (Chatton and Brodsky, 1909) and 2 μ (Mattes, 1924); the latter observer failed to find a distinct nucleus. *Sphaerita endamoebae,* according to Becker (1926), is from 1.9 to 2.5 μ in its early intracytoplasmic stage, with a fine cell membrane and a relatively large, solid nucleus.

Most accounts describe increase in the size of the cytosome and the nucleus before the nucleus begins to divide. In *Euglena viridis* (Dangeard, 1895; Mitchell, 1928) the uninucleate thallus may become larger than the host nucleus, and its nucleus becomes correspondingly large. Its shape is spheroidal, ellipsoidal, or elongated. The shape in this phase, together with its size and the presence of vacuoles, is regarded as of taxonomic significance by Jahn (1933), who distinguished *E. phaci* on such grounds. Other sphaeritas appear to attain no such size before nuclear multiplication sets in. *Sphaerita* in *Vahlkampfia limax* attains only about triple its diameter before nuclear divisions begin (Chatton and Brodsky, 1909). In *S. endamoebae,* according to Becker (1926), nuclear multiplication keeps pace with growth, and there are binucleate stages no larger than uninucleate ones. Similar development has been noted by the writer in *Sphaerita* (Fig. 217A) in several species of Devescovininae. There are very few observations on actual nuclear division; Dangeard (1895) interpreted as this some figures he observed, but did not see nuclear division in the larger nuclei of the early stages. Nägler (1911b) reported dumb-bell-shaped figures, as well as granular fragmentation stages, in the parasite of *Euglena sanguinea*. The outcome, in any event, is in typical *Sphaerita* a multinucleate thallus, which is converted entire into the sporangium.

Parasites, which in some phases are much like *Sphaerita* but lack a

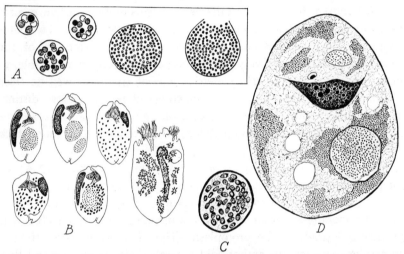

Figure 217. A, various stages in development of *Sphaerita* in *Devescovina* sp. from *Neotermes tectonae;* B, *Sphaerita* and other microorganisms in *Entodinium* sp. and *Eudiplodinium* sp. from ruminants; C, mature sporangium of *Sphaerita* from *Nyctotherus ovalis;* D, developmental stages of *Sphaerita* and aggregations of bacteria in *Nyctotherus ovalis*. (A, original; B, after Winogradowa, 1936; C, D, after Sassuchin, 1928a.)

multinucleate structure, have been described, however. Mitchell (1928) reported a parasite in *Euglena caudata,* which, after growing to a relatively large uninucleate body, underwent repeated division of both nucleus and cytoplasm to form spores. Ivanić (1925), describing in free-living amoebae parasites which appear to be *Sphaerita*-like, stated that the uninucleate forms grow and multiply by binary fission before the plasmodial period begins. When, as is often true in preparations, the cytoplasm of the parasite is not apparent, division of the nuclei may be mistaken for division of individuals within a vacuole. Individual parasites multiplying in this way alone would probably be cocci. That may account for possible errors, not made in the above examples, but probably involved in certain accounts of nuclear parasites, as discussed below.

When the parasite has reached a certain size, growth stops and sporulation sets in. In *Sphaerita amoebae* the size when spores are formed is very variable; sporangia are larger when only a few are present in an amoeba (Mattes, 1924). The number of spores produced is also variable in this species, ranging from less than a hundred to several hundred. In sporulation the protoplasm simultaneously organizes into membrane-confined bodies around each nucleus, and the spores appear as spheroidal or ellipsoidal structures. The membrane of the sporangium may remain very thin, so as to be scarcely recognizable, as in *Sphaerita* in *Amoeba alba,* where groups of spores showed no trace of an envelope (Penard, 1912). Sometimes it becomes more distinct at sporulation; and in the unique case of *Sphaerita* from *Nyctotherus ovalis,* according to Sassuchin (1928a), it becomes 1 μ or more thick (Fig. 217C).

The account by Sassuchin *et al.* (1930) and Sassuchin (1934) of the parasite of *Entamoeba citelli* is not easy to understand. The parasites are said to occur either in groups, varying considerably in size, which resemble sporangia, but around which a membrane was never observed; or arranged singly in the protoplasm. Though these parasites show a spore-like character, the authors did not call them spores, nor did they discuss multiplication. The description by Becker (1926) of *Sphaerita endamoebae* from *Entamoeba citelli,* to which Sassuchin did not refer even in his later article, is in essential agreement with the usual concept of the life cycle of *Sphaerita.*

The parasite of *Euglena caudata* (probably *Sphaerita dangeardi*) may form as many as 500 spores (Mitchell, 1928). Sporangia of *S. endogena* contain 100 or more (Dangeard, 1886a). Dogiel (1929) found only from 30 to 40 spores in the "spore balls" of *S. diplodiniorum.* In *Sphaerita* of *Monas vulgaris,* nuclear divisions preceding spore formation proceed to stages 16, 32, or sometimes 64 (Alexeieff, 1929). Pinto and Fonseca (1926) mentioned sporangia of only from 7 to 9 "individuals" in *S. minor* of *Trichomonas vitali;* Cunha and Muniz (1934) found from 20 to 30 spores in *S. chilomasticis.*

There is variability in the number of spores and the size of sporangia within a species. That sporulation can occur at different stages of growth was noted by Chatton and Brodsky (1909), in *Sphaerita* of *Vahlkampfia limax;* sporangia ranged from 20 μ in diameter down to small ones, with few spores, of 4 μ. Mattes (1924) stated that the size at which sporulation starts in *S. amoebae* is very variable; and Lwoff (1925), in *Sphaerita* of entamoebae of man, found that sporulated parasites are of different sizes. The size of the sporangium and the number of spores must not be used indiscriminately for definition of species.

The spores are spherical, ovoidal, or ellipsoidal in shape—most frequently the first. They range in size, in different species, from a diameter of 0.25 to 0.30 μ (*S. parvula,* Brumpt and Lavier, 1935b) to elongated forms of from 2.5 to 3 μ (*Sphaerita* from *Euglena;* Mitchell, 1928; Puymaly, 1927). Yuan-Po (1928) reported spores of from 2.5 to 4 μ in a parasite of *Entamoeba bobaci;* this is exceptional in *Sphaerita* of endozoic Protozoa. In the parasite of *Nyctotherus ovalis,* the spores measure from 1.5 to 2 μ (Sassuchin, 1928a). Dangeard (1886a, 1886b) stated that the spores of *S. endogena* from amoebae have a size of 1.5 μ. There appears in general to be only a limited variability in the size of mature spores; but Becker (1926) found that in *S. endamoebae,* the spores of which usually were from 1.0 to 1.6 μ, some were as small as 0.5 μ. The size and shape of spores is, used discretely, a valuable taxonomic guide.

The spore of *S. endogena* in the rhizopods *Nuclearia simplex* and *Heterophrys dispersa* has, according to Dangeard (1886a, 1886b), a long flagellum ("cil") placed anteriorly and strongly recurved. Its movements are very active and jerky, and sometimes there is simple rotation in one position. When he studied the zoöspores of *Sphaerita* of *Euglena sanguinea* (1889b), Dangeard found, in addition to the posteriorly directed flagellum, a very short one directed anteriorly. Serbinow (1907) found only one flagellum on the zoöspore of *Sphaerita* of *Euglena,* and thought it possible that Dangeard's biflagellate zoöpores belonged to some other organism, possibly to the parasite of *Sphaerita, Olpidium sphaeritae* Dang. Serbinow described their jerky, irregular movement. In *Sphaerita* of *E. viridis,* Puymaly (1927) also reported biflagellate zoöspores, the larger flagellum directed posteriorly, as Dangeard noted again in 1895; and he described the movement as rotation around an axis and rapid, oscillatory swimming. These observations agree with those of Stein (1878, 1883) on the escape of minute, flagellated organisms from the so-called germ balls. Cejp (1935) observed two flagella on the zoöspores of *Sphaerita* of *Paramecium*. Ivanić (1925) stated that he repeatedly observed release of the flagellated swarm sports of the parasite (*Sphaerita?*) of *Amoeba jollosi*.

Mattes (1924), however, though he found flagellated zoöspores of *Olpidium amoebae* of *Amoeba sphaeronucleolus*, failed to see any flagella or motility in the spores of two *Sphaerita* species of the same amoeba. The same is true of the observations of all other investigators of the parasite in Protozoa. It appears that *Sphaerita* in endozoic Protozoa lack flagellated zoöspores, and that most of those of free-living amoebae also do.

A central or eccentrically placed nucleus in the spore was reported by Dangeard (1895) in *Sphaerita* of *Euglena*, and by Penard (1912) in *Sphaerita* of *Amoeba alba*; and it was shown by Cejp (1935) in the parasite of *Paramecium*. Mitchell (1928) noted a nucleus in the spores of the chytrid of *Euglena sanguinea*, and he alone described any detail in the nuclear structure. In sphaeritas of endozoic Protozoa, the nucleus has not been found, and there appears to be a thicker spore membrane. The membrane appears in optical section as a well-defined ring, especially in spherical spores. In elongated spores there is frequently a stainable area at one end, appearing often as a crescentic thickening, as in *Sphaerita* (Fig. 217C) in *Nyctotherus ovalis* (Sassuchin, 1928a, 1934). This structure was noted also by Becker (1926) in *Sphaerita endamoebae*, by Yuan-Po (1928) in the larger species in *Entamoeba bobaci*, and by Connell (1932) in *Sphaerita* of *Gigantomonas lighti*. It has been observed by the writer in the parasites in a number of flagellates in termites.

Rupture of the sporangium takes place in the cytoplasm of the host, then the body of the host may rupture and the spores be released into the water. In most instances no previously apparent pore or papilla has been shown. Dangeard (1889b) stated that the zoöspores of *Sphaerita* in *Euglena sanguinea* escape by a papilla at one end. Serbinow's account (1907) of an elongated or fusiform sporangium in *Sphaerita* in *E. viridis* and *E. sanguinea*, with a short exit papilla at one or both ends, does not apply to most forms that have been placed in the genus. *Sphaerita* cannot, then, be diagnosed on the basis of this account, as was done by Minden (1915), without excluding many forms.

There is slight evidence of fusion of spores in *Sphaerita,* and none of this carries the conviction of cytological demonstration. Dangeard (1889b) and Puymaly (1927) reported that in sphaeritas of *Euglena* zoöspores may touch or adhere, simulating conjugation of gametes, but that they end by separating. Mattes (1924) found no fusion of spores of *Sphaerita amoebae.* Chatton and Brodsky (1909) thought copulation of spores probable, but did not see it. In *Sphaerita* of *Amoeba alba,* Penard (1912) stated that he sometimes encountered the spores in conjugation; and fusion was reported in *Sphaerita*-like parasites of *Amoeba jollosi* by Ivanić (1925), as well as in the so-called gametes of *Allogromia* by Prandtl (1907), which possibly were also *Sphaerita.*

Dangeard (1889b) reported fixation of zoöspores to the wall of *Euglena sanguinea,* and penetration into the cytoplasm. Puymaly (1927) described adherence, loss of flagella, and development of a fine surrounding membrane; following which the spore probably emits a fine papilla, which perforates the flagellate, and empties abruptly into the cytoplasm. In rhizopods, spores are ingested (Dangeard, 1886a; Chatton and Brodsky, 1909; Mattes, 1924; Lwoff, 1925); this probably is the general method of infection of holozoic Protozoa by the non-flagellated spores.

The thallus of Olpidiaceae may develop also into a resting sporangium, which is ordinarily thicker-walled, and may sometimes bear spines, but otherwise in structure and development corresponds to the ordinary sporangium. Spinous cysts or resting sporangia were described by Dangeard (1889b) in *Sphaerita* of *Euglena sanguinea;* by Serbinow (1907) in *Sphaerita* of *E. viridis* and *E. sanguinea;* by Skvortzow (1927) in *S. trachelomonadis;* and by Mattes (1924) in *S. plasmophaga* of *Amoeba sphaeronucleolus.*

Nucleophaga.—*Nucleophaga* has been found by the writer in low incidence in almost all devescovinid flagellates in termites. Sometimes the infection is greater. In some material of *Endamoeba disparata* it was from 6 to 12 percent (Kirby, 1927); Brumpt and Lavier (1935a) found it in 78 percent of trophozoites of *E. nana;* on one occasion 90 percent of a group of *Amoeba sphaeronucleolus* were parasitized (Mattes, 1924); and Gruber (1904) lost an entire culture of *A. viridis,* which had been kept ten years, on account of the fungus.

The parasite apparently occurs exclusively in the nucleus. Brumpt and Lavier found it only twice, among thousands of specimens, in the cytoplasm of *E. nana;* and in those two instances, as they stated, it probably was not developing there. A given species of *Nucleophaga* invades certain hosts, and not others. Brumpt and Lavier failed to find it in *E. dispar,* which was associated with heavily parasitized *E. nana.* Many related species of devescovinids, however, contain what is probably the same species; although there is more than one species of *Nucleophaga* in that group of flagellates.

In the recent review of studies on *Nucleophaga* by Lavier (1935b), it was noted that life-history accounts indicate two modes of development. One is in agreement with the life cycle of *Sphaerita,* as outlined above. A thallus enlarges and its nuclei multiply, it is converted into a sporangium, and a spore forms around each nucleus (Dangeard, 1895; Penard, 1905b; Mercier, 1907, 1910; Mattes, 1924; Lavier, 1935b). *Nucleophaga* of *Endamoeba disparata* (Fig. 218B-E) is considered to have this type of life history (Kirby, 1927). In the second type there is no multinucleate structure, and no sporangial membrane, the individual invading parasite (Fig. 218G-J) multiplying repeatedly within the nucleus by division (Epstein, 1922; Brug, 1926; Brumpt and Lavier, 1935a). This is the type of reproduction, evidently, in the nuclear parasite of *Entamoeba citelli,* according to Sassuchin (1931); and probably such a parasite would not seem to differ essentially from *Caryococcus* (Dangeard, 1902). It is not conceivable that there should be such fundamentally different types of development in members of the same genus. Either the latter type is nothing but a misinterpretation of the ordinary chytrid life history, because of failure to see the cytoplasm of the parasite, or the parasite is not *Nucleophaga.* The figures of *Nucleophaga* supposed to show this second type of reproduction do not differ essentially from the other accounts—it is probably a matter of differing interpretations of what is actually one form of development. The similarity to *Sphaerita* appears to be too great to separate *Nucleophaga* distantly from that genus; but the relationship of *Caryococcus* to this needs further investigation.

In its early phases, in any event, *Nucleophaga* appears as a group of granules in the interior of the nucleus, occupying a limited area, whereas the nuclear structure elsewhere is essentially unchanged. This group of granules presumably, though it cannot always be ascertained with certainty, represents the nuclei of a thallus. *Nucleophaga amoebae,*

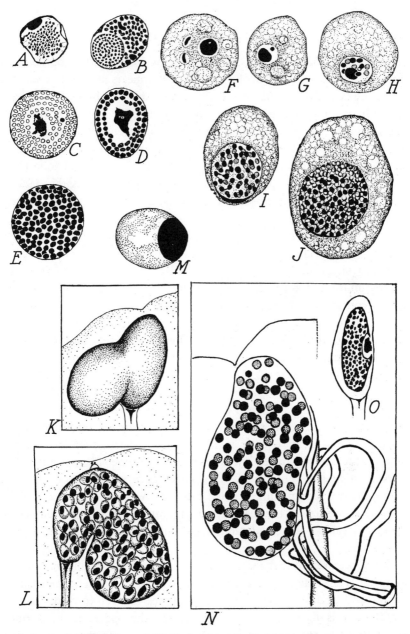

Figure 218. *Nucleophaga*. A-E, *Nucleophaga* in *Endamoeba disparata:* A, normal nucleus of *E. disparata* Kirby; B, developmental stage of *Nucleophaga;* C, D, later stages of *Nucleophaga,* with residue of chromatin in center; E, surface view of same stage. F-J, *Endolimax nana* parasitized by *Nucleophaga hypertrophica: F,* normal amoeba; G, beginning of nuclear parasitism; H, I, multiplication of spherules; J, mature spores, nuclear membrane appears on the point of rupturing. K-O, *Nucleophaga* in *Caduceia theobromae:* K, surface view of parasitized nucleus bulged out at one side; L, mature sporangium, with nucleus similarly formed; M, detail of spore; N, parasitized, greatly hypertrophied nucleus; O, normal nucleus, drawn to same scale as K, L, N. (A-E, after Kirby, 1927; F-J, after Brumpt and Lavier, 1935a; K-O, original.)

described by Dangeard (1895) in a host he considered to be *Amoeba verrucosa,* invades the nucleolus, in which it appears at first as a vacuole enclosing a granule—the former its cytoplasm, the latter its nucleus. Growth of the parasite is rapid (Mattes, 1924), and as it proceeds the chromatin is used up. The stainable nuclear material becomes restricted to the periphery in a reticulated structure (as noted by the writer in *Nucleophaga* of devescovinid flagellates), or to the central zone (Fig. 218C, D), as in hyperparasitized *Endamoeba disparata* (Kirby, 1927). Eventually the nuclear material disappears, and the interior is entirely occupied by the parasite.

The parasitized nucleus hypertrophies considerably (cf. Fig. 218F and J; O and N), up to several times its original diameter. The parasite must obtain material for its continued growth by diffusion from the cytoplasm through the nuclear membrane. Though Mercier (1910) mentioned considerable hypertrophy of the nucleus of *E. blattae,* he showed spores in nuclei in which there seems to have been little enlargement. This is unlike the usual situation. Perhaps the very thick membrane of the nucleus of *E. blattae* has an influence in restraining the growth of the parasite. Lavier (1935b) noted precocious spore formation in *Nucleophaga* of *Entamoeba ranarum,* but stated that it generally occurs when the parasite has attained a large size. In *Nucleophaga* of devescovinid flagellates, notably that in *Caduceia theobromae,* expansion of the nucleus cannot occur equally in all directions, because of its relationship to the axostyle. Instead, it is pushed out on one side, and often has a bilobed figure (Fig. 218K, L). This figure sometimes is retained in the mature sporangium; often it fills out. As has been noted also of *Sphaerita,* the size of the mature sporangium, as well as the number of spores, is subject to considerable variation in the same species of *Nucleophaga.*

It has been stated that spore formation is in certain forms continuous, and that there may be present in an individual at a given time mature spores and granules corresponding to spores not yet formed (Lavier, 1935b). It is more general, however, for the spores to be formed simultaneously, the entire thallus being converted into the group. Other granules, which have been seen by the writer among the spores, probably represent residual or discarded material; there is no evidence for maturation of later spores.

The spores are very much like those of *Sphaerita*. Their shape is spheroidal or ellipsoidal, and the wall stains more intensely than the contents. Some show a thickening at one side, in crescentic form (Epstein, 1922; Brumpt and Lavier, 1935a; Lavier, 1935b). In the interior, often no structure is discernible, or one or two granules are seen, or a central nucleus may be observed. Dangeard (1895) and Epstein (1922) reported a vesicular nucleus; Brumpt and Lavier, however, failed to observe a definite nucleus in the parasite of *Endolimax nana*, studied by Epstein. In *Nucleophaga* of *Caduceia theobromae*, a spheroidal granule of relatively large size was observed toward one end of the spore (Fig. 218M). This may be a nucleus. In size, the spores range from one to 2 μ, some being reported as only about one μ (Brumpt and Lavier, 1935a), some 2 μ (Lavier, 1935b; Penard, 1905b), others as having a variability from about one μ to 2 μ (Mercier, 1910; Mattes, 1924).

No flagella have been observed on spores of *Nucleophaga*, with the possible exception of those mentioned in the account by Robertson (1905). She described what she supposed to be gametogenesis of *Pseudospora volvocis*, in a rather complete account of what is probably the development of *Nucleophaga*. The "gametes," as figured and described, are each provided with one flagellum; and they are reported to fuse, producing a biflagellate zygote.

EFFECT ON HOST

Minden (1915) wrote (translation):

In the lower plants, mainly algae, which in the widest variety are sought by parasitic Chytridiales, the injuries are usually so striking that these fungi are designated as dangerous parasites of algae. In a short time large cultures of diatoms, flagellates, and other unicellular organisms may be completely destroyed; but also filamentous algae die cell by cell. The first indication of injury is in the discoloration and disorganization of the cell contents . . . finally there remain only granular vestiges.

Infection with *Sphaerita* may be observed in many Protozoa that appear entirely normal, but it often ends fatally. The host may sometimes rid itself of the parasite and continue normal life (Dangeard, 1895; Penard, 1912); on the other hand, many observers report death of the host at the time of sporulation.

PARASITES OF PROTOZOA

Parasitized euglenid flagellates lose their green color, chlorophyll first being affected and chromatophores degenerating (Dangeard, 1889b, 1895; Puymaly, 1927; Mitchell, 1928; Jahn, 1933). Puymaly found a decrease in flagellar activity, whereas euglenoid movement continued to the last moment and became even more energetic. There is alteration of the nucleus, according to some, though Puymaly stated that there is none; and the cytoplasm becomes vacuolated. Finally, in many cases, the flagellate ruptures and zoöspores are liberated.

Chatton and Brodsky (1909) found that parasitized amoebae tend to assume a spherical form with radial pseudopodia, instead of progressing; and Sassuchin (1928a) noted a progressive slowing of the ciliary action in *Nyctotherus*. The pulsating vacuole in these heavily infected hosts slows or loses its rhythm. Degenerative changes were observed in the nucleus of parasitized *Iodamoeba bütschlii* by Wenrich (1937); in that of *Entamoeba citelli* by Becker (1926); and in the macronucleus of *Nyctotherus* by Sassuchin (1938a). When the sporangium ruptures, or shortly before, the host may perish (Chatton and Brodsky, 1909; Mattes, 1924; Yuan-Po, 1928; Sassuchin, 1928a), especially if the infection is heavy.

The fact that in amoebae of man the parasites have been found only in trophozoites has been taken to indicate either that they hinder the amoebae from encysting or that infected cysts degenerate rapidly (Lwoff, 1925). Lwoff therefore pointed out a possible use of *Sphaerita* as a means of biological control, following Nöller's suggestion that it might be worth while to devote more attention to these natural enemies of amoebae. If one could transmit the infection to carriers of cysts, Lwoff stated, there would be a means of diminishing the number of cysts, this in addition to the inhibition of multiplication. The practicability of this, however, is doubtful.

The protozoan whose nucleus is parasitized by *Nucleophaga* continues activity until the end. Usually there is no apparent change in protoplasmic activity or in structure, aside from the nucleus, even though all stainable chromatin material has disappeared. Lavier (1935a) observed increased size and activity in parasitized *Entamoeba ranarum,* and remarked that the hyperactivity may be provoked by irritation, and may constitute a defense reaction on the part of the amoeba. By the time the parasite reaches the stage of sporulation, however, some changes may have occurred in the cytoplasm (Dangeard, 1895; Sassuchin, 1931), and there may have been some hypertrophy of the host's body (Epstein, 1922). Epstein, indeed, stated that giant amoebae reaching from 10 to 30 times normal size, may result; but that is very much more than is usually observed.

When the membrane breaks and the spores escape, the host perishes. For that reason, even in heavily parasitized groups of Protozoa, individuals with spores dispersed in the cytoplasm are seldom observed.

PARASITES OF THE NUCLEUS OF TRICHONYMPHA

Except for mention (Kirby, 1940) of the parasites described below, the only report of parasitization of the nucleus of *Trichonympha* is the description (Kirby, 1932a) of a form in *T. saepiculae* (Fig. 219B). Numerous spheroidal bodies, each apparently subdivided into compartments, filled several nuclei, in which the vestiges of chromatin were confined to the central part. Few specimens of this organism were found, and its affinities were not discussed, except for the remark that it is unlike *Nucleophaga*.

An unusually interesting parasitization of the nucleus has been studied by the writer in *Trichonympha* in certain termites of Madagascar and in one from Java. In several series of preparations from *Procryptotermes* sp. of Madagascar, a large proportion of the hypermastigotes had parasitized nuclei. Apparently, in the flagellate from this host two different parasites are involved. One of these has a life history like that of *Nucleophaga:* growth of a multinucleate parasite, using up the chromatin, which is restricted to a peripheral reticulum and finally disappears; and formation of numerous spores. The size and structural detail of the spores distinguish them from those of any described *Nucleophaga,* and suggest a possible affinity to the Haplosporidia.

The normal, interkinetic nuclei of these species of *Trichonympha* have the chromatin in the form of stout, varicose strands which extend throughout the intranuclear area. They may extend entirely to the periphery, but often in the preparations there is a clear outer zone of variable width. In this zone are minute granules. In some nuclei, which possibly show the beginning of kinetic changes, the strands tend to be peripheral, and the central part of the nucleus is occupied by a granular and reticulo-fibrillar matrix.

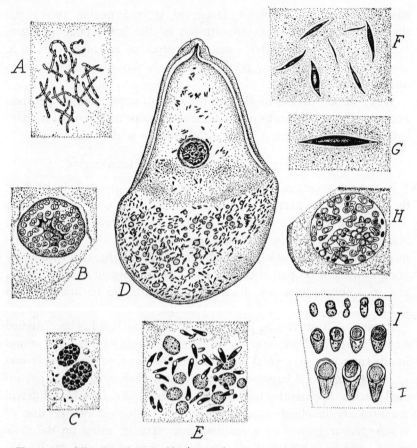

Figure 219. Microörganisms in *Trichonympha*. A, constantly present organism that forms an aggregate surrounding or near to the nucleus of *T. campanula;* B, nuclear parasite of *T. saepiculae;* C, *Sphaerita* in *T. sphaerica;* D, E, peg-formed organisms in *T. campanula;* F, G, fusiform organisms in *T. campanula;* H, group of parasites (microsporidia?) in *T. magna;* I, developmental stages of organism shown in H. (After Kirby, 1932a.)

In early stages of invasion by the *Nucleophaga*-like parasite, a body, apparently amoeboid, is observed in the process of penetration into the chromatin mass. In the earliest stages so far found, it is already multinucleate. This becomes located in the interior of the chromatin mass, and as it grows its nuclei multiply and the chromatin of the *Trichonympha* nucleus becomes restricted to a peripheral reticulum (Fig. 220E).

By the time the parasite reaches its full size, the nucleus of its host has become greatly hypertrophied and has left its normal position. A variable number of spores are produced; in one instance there were only 17, but usually there are from 150 to 200 or more, located within an ellipsoidal membrane 25 to 44 $\times$ 22 to 36 μ. The individual spore is ellipsoidal and has a size of 2.5 to 4 μ $\times$ 2 to 3 μ. The spores are larger in size when their number is smaller.

The structure of the spores (Fig. 220C-E) is the characteristic of greatest interest in this organism, as nothing like it is known in any other nuclear parasite, or indeed in any known parasite of Protozoa. The nucleus is located at one end, and is usually relatively very large, having a diameter almost equal to the width of the spore. When heavily stained, or when not well fixed, it appears homogeneous, but in good preparations it is resolved into closely packed granules. In the cytoplasm of the spore are a variable number of granules which are relatively large for cytoplasmic granules. These stain intensely with hematoxylin, and possibly are volutin, though no tests could be made to support that view. The cytoplasmic granules are often arranged in an equatorial ring (Fig. 220D), which appears solid in some preparations. Sometimes there are no granules outside of the nucleus except in this ring; at the other extreme, the ring constitutes the margin of a solid hemispherical mass of granules, which occupies all the area at its end of the spore (Fig. 220E). Between these extremes are conditions in which, in addition to the ring, there are only a few granules at the periphery of the hemispherical area, or more abundant granules in a peripheral, semicircular row at right angles to the ring (Fig. 220C).

At the periphery of the mass of mature spores of the hyperparasite from *Procryptotermes* sp., there is constantly present a single, apparently crystalloid body (Fig. 220B). In different parasites this body is relatively uniform in size and shape; and it is generally so located as to cause a protrusion of the membrane. It has the form of a conventional diamond, and stains deeply with Heidenhain's iron-hematoxylin but not with Delafield's hematoxylin. It is unlikely that it is to be regarded as residual chromatin.

What seems to be a second parasite (Fig. 220G, H) of the nucleus of *Trichonympha* occurred in from 70 to almost 100 percent of the hypermastigote in certain preparations from the Madagascar *Procrypto-*

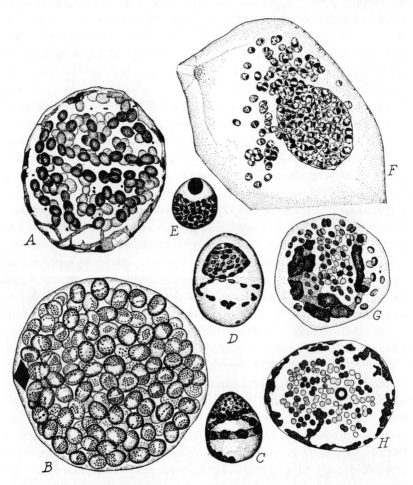

Figure 220. Nuclear parasites of *Trichonympha* sp. from *Procryptotermes* sp. of Madagascar. A-F, *Nucleophaga*-like parasite, nucleus hypertrophied; G, H, coccoid parasite, nucleus not hypertrophied. A, developmental stage, probably a multinucleate plasmodium, chromatin of nucleus restricted to periphery; B, mature sporangium, diamond-shaped crystalloidal body at periphery of nucleus; C-E, details in structure of spores; F, nucleus ruptured, with some spores in cytoplasm; G, chromatin masses present with parasites, many of latter show crescentic stainable area at one side; H, chromatin peripheral, some dividing forms of parasite. (Original.)

termes. Generally the parasite appears as a mass of spherical bodies, each about one μ or less in diameter and often with a stainable crescent at one side, located in the central part of the nucleus. The spherical bodies are a good deal smaller than are the nuclei in a plasmodium of comparable size in the other parasite. The chromatin is usually restricted to the periphery of the host nucleus; but, except for this removal of the central part of the mass, it is little altered; and there is no marked hypertrophy of the nucleus. In the nuclei of some hosts, the proportion being greater on certain slides, rounded bodies with a similar crescentic stainable area are located peripherally, the chromatin mass being concentrated in the center. The distribution of these peripheral bodies is often such that a common embedding cytoplasm appears unlikely. Further investigation is necessary to establish the nature of this parasite. It may be a bacterial, coccoid parasite of the nucleus, comparable in certain ways, possibly, to *Caryococcus*, described by Dangeard (1902).

In several instances multinucleate trichonymphas, with all nuclei parasitized, have been found. These are the only multinucleate flagellates of this genus that have ever been seen. Cytotomy generally accompanies division of the single nucleus, but binucleates occasionally occur.

PHYCOMYCETES OTHER THAN SPHAERITA AND NUCLEOPHAGA

Chytridiales of a number of genera other than *Sphaerita* and *Nucleophaga* have been found parasitic on Protozoa, especially autotrophic flagellates. Those described up to 1915, in the genera *Olpidium, Pseudolpidium, Rhizophidium, Phlyctochytrium, Rhizidiomyces, Saccomyces, Rhizophlyctis,* and *Polyphagus,* were discussed by Minden. They occur in or on *Euglena, Cryptomonas, Chloromonas, Chroococcus, Glenodinium, Haematococcus, Chlamydomonas, Pandorina,* and *Volvox. Olpidium arcellae* is considered doubtful.

Fungus parasites, which are probably Chytridiales, have been found in cysts of a number of ciliates. Stein (1854) found many cysts of *Vorticella microstoma* with up to three or four protuberances perforating the wall and extending a short distance free. Each protuberance was an extension of a rounded body (*Mutterblase*) within the cyst. From the terminal opening a thin, gelatinous, clear fluid was reported to escape, forming a globule enclosing about thirty "young," resembling certain *Monas* forms, and, when dispersed, having movements like them. Similar structures were shown in a cyst of *Vorticella nebulifera.* At that time Stein considered this to be a mode of reproduction of *Vorticella.* Cienkowsky (1855b) recorded similar bodies in cysts of *Nassula ambigua* ("*N. viridis*"), describing the appearance of clear vacuoles in the cyst contents and the development of "spores," from many of which a short process broke through the wall of the cyst and permitted the escape of the swarm spores. Lachmann (1856) mentioned these observations as showing another kind of reproduction in ciliates. Cohn (1857), however, remarked on the resemblance of these "microgonidia," with their flask-formed "mother cells," to the chytrids of many plants. In their text Claparède and Lachmann (1860-61) discussed the phenomena as forms of reproduction by embryos, adding observations of their own on similar structures in *Urnula epistylidis;* but in their footnotes they stated that these were *Chytridium.* Stein (1859) regarded them as parasites, comparing them with Saprolegniales and in particular with *Pythium entophytum* Pringsheim; but their characteristics are suggestive of *Olpidium.* Stein recorded similar bodies from cysts of *Stylonychia pustulata, Holosticha (Oxytricha) mystacea,* dead *Tokophrya (Acineta) lemnarum,* and *Metacineta (Acineta) mystacina* (observations of 1854). On the motile bodies escaping from a *Vorticella* cyst he saw a single flagellum.

Species of *Olpidium,* which differ from *Sphaerita* in the elongation of the exit tube, occur in certain rhizopods and in Suctoria, as well as in *Euglena. O. amoebae* was described by Mattes (1924) from *Amoeba sphaeronucleolus;* it is said to parasitize a rotifer also. Gönnert (1935) named *O. acinetarum* a chytrid which destroyed a culture of *Lernaeophrya capitata* and *Podophrya maupasii* within a few days. The spores are relatively large, from 2.5 to 3 μ in diameter, equaling the larger ones of *Sphaerita;* and Mattes found a relatively long posterior flagellum.

Rhizophidium and *Polyphagus* belong to the Rhizidiaceae, in which there is a restricted mycelium. *Rhizophidium beauchampi* has recently been described by Hovasse (1936) in *Eudorina illinoisensis.* A heavy infection, exceeding 90 percent, occurred homogeneously in these phytomonads in a large lake. The zoöspore, which has a single long flagellum, becomes fixed to the coenobial surface and germinates by the emission of a tube which penetrates a cell and functions as a sucker. The part of the tube that remains external to the cell swells and becomes a sporangium, in which by simultaneous partitioning from 20 to 100 zoöspores are produced. Most parasitized colonies had not more than 25 to 28 normal cells, and heavily parasitized colonies may be destroyed.

Polyphagus euglenae, whose structure and life history have been described by Nowakowski (1876), Dangeard (1900b), and Wager (1913), appeared at various times in cultures of *Euglena,* which were destroyed in a few days. Serbinow (1907) at Petersburg, and Skvortzow (1927) in East Mongolia found it parasitic on *Chlamydomonas.* The parasite germinates free in the water, and a single cell, by branched haustoria, may attack many flagellates. A haustorium perforates the cell wall, branches, and the cell contents rapidly disintegrate. A sporangium develops as an outgrowth from the protoplast, and produces a variable, usually very large, number of uniflagellate zoöspores. *Polyphagus* is one of the few Chytridiales in which sexual reproduction has been satisfactorily demonstrated. A zygote is formed by the fusion of two vegetative cells, and becomes a resting spore, with smooth or spinous membrane.

Skvortzow (1927) reported two other Chytridiales from *Eudorina elegans* in Manchuria: *Phlyctidium eudorinae* n. sp. and *Dangeardia mamillata.* The latter was originally described by Schröder from *Pandorina. Phlyctidium* is epibiotic, with a haustorium penetrating a cell. The sporangia of *Dangeardia* are located in the gelatinous sheath of the volvocid.

Lagenidium trichophryarum, which belongs near the Chytridiales in the Ancylistales, was described by Gönnert (1935) in *Trichophrya epistylidis.* The parasite, which appeared once in abundance, was fatal to the suctorian. *Lagenidium* is rare in Protozoa. Cook's revision of the genus (1935), which is in the same number of the *Archiv für Protistenkunde* as Gönnert's article, reports no species from them; the habitat is filaments of green algae, diatoms, pollen grains, and rhizoids of mosses.

Filamentous appendages on the posterior end of certain large freshwater amoebae (Fig. 221) have long been known. Leidy (1879) observed them, and was uncertain as to their nature, regarding them at first as a bundle of mycelial threads dragged behind *Amoeba proteus,* but finally concluding that they were structural elements of the amoebae. He made the presence of these appendages diagnostic of the new genus

Ouramoeba. Korotneff (1880), encountering an amoeba with similar posterior prolongations, created for it a new genus, *Longicauda.* Penard (1902), as others had already suggested, considered filaments on *Amoeba nobilis* to be parasites, and reported observation of appendages of different types on *A. proteus* and *A. vespertilio.* He noted long, fine filaments also on *Pelomyxa tertia.* He recounted these observations again later (1905c), and stated that the fungi probably belong very close to the

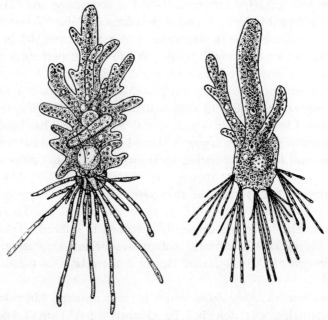

Figure 221. Filamentous fungi (*Amoebophilus*) parasitic on *Amoeba proteus* (*Ouramoeba vorax* Leidy). (After Leidy, 1879.)

Entomophthorales or Saprolegniales, resembling in the former group *Empusa,* in the latter *Leptomitus lacteus.* Dangeard (1910) studied filaments on *Pelomyxa vorax,* and named them *Amoebophilus penardi.* He gave the name *Amoebophilus caudatus* to the parasite described by Penard on *Amoeba nobilis;* and *A. korotneffi* to that of "*Longicauda amoebina.*" He thought it possible that they might belong to the Ascomycetes. Geitler (1937) studied in *A. proteus* what is apparently the same as Penard's parasite (*Amoebophilus caudatus* Dangeard), but he made no reference to Dangeard's account. Geitler stated that the fungus probably belongs in the Cladochytriaceae of the Chytridiales.

Geitler found the fungi on a narrowly defined area of the body, the protruding filaments vertical to the surface of the protoplasm and from 100 to 200 μ long. The filaments of a plant are non-septate and arise from a deep-staining, irregularly lobed vesicle, the haustorium, located in the endoplasm of the amoeba at the limits of the ectoplasm. From the vesicle, which was also noted by Penard, delicate hyphal threads extend through the ectoplasm to the surface of the body, where they broaden and continue as extracellular threads. Basal branching is common, and there may also be secondary branching. Infected amoebae show a polar organization, with the fungi at the posterior end; this polarity, Geitler concluded, is probably not called forth by the infection, but was present before.

Filaments seen by Penard (1905c) on *Amoeba proteus,* the same as those shown by Leidy on *"Ouramoeba botulicauda,"* were, when of some length, divided by constrictions into two or more equal parts. The figure of a filament on *A. vespertilio* shows constrictions marking short subdivisions. Dangeard (1910), who observed nuclei in the filaments on *Pelomyxa vorax* (*Amoebophilus penardi*), also figured constrictions demarcating long sections, which he considered to represent budding.

The incidence of these parasitic fungi on amoebae is sometimes high. At one period Geitler found 95 percent of *A. proteus* infected; later the incidence declined. Penard (1902) found the fungi on three out of five *A. nobilis,* and Dangeard (1910) on a rather large number of *Pelomyxa.*

A filamentous, cylindrical fungus, 0.75 μ in diameter, was found by the writer fairly frequently in certain material of *Devescovina hawaiensis* from *Neotermes connexus.* A large part or all of the filament was embedded in the cytoplasm, but characteristically a part projected beyond the surface. The surface was penetrated at any point.

Fungi which develop in the cytoplasm and then extend projections beyond the surface were described by Penard (1912) in *Amoeba terricola* and *A. alba.* These parasites, which he found usually fatal to the amoebae, belonged, he thought, in or near the Saprolegniaceae.

A fungus assigned to the Saprolegniales was found by Sand (1899) infesting more than half the specimens of *Acineta tuberosa* collected from the sea at Roscoff. Developing within the cytoplasm, the parasite soon destroyed the cell and formed isolated spheres in the empty lorica. These developed into long tubes, wound in the lorica or projecting free.

One of the tubes terminated in a large, spherical sporangium. Leger and Duboscq (1909c) reported parasitic fungi which developed a mycelium in cysts of the gregarine *Nina gracilis.*

Galleries excavated in the non-protoplasmic parts of calcareous tests of Foraminifera are the work of an organism behaving somewhat in the manner of the mycelium of certain fungi, according to Douvillé (1930). The relationship of this organism to calcareous shells suggests the habitat of *Didymella conchae,* an ascomycete which Bonar (1936) described from the shells of marine gasteropods and barnacles.

PROTOZOA

PHYTOMASTIGOPHORA

An unusual phoretic relationship described by Penard (1904) existed between a heliozoan and an undetermined species of *Chlamydomonas.* He found this organism fixed to the surface of *Actinosphaerium eichhornii* by its two flagella, which were applied by their full length. Often it was so abundant that the surface of the host was spotted with close-set organisms, and the heliozoan appeared covered with a green envelope. When the chlamydomonads were scattered mechanically, they later reassembled at the surface of *Actinosphaerium.* Sokoloff (1933) found a euglenid flagellate, named *Euglena parasitica,* adherent in abundance to the surface of *Volvox* coenobia in a tank. There was a conical prolongation anteriorly by which this adherence was effected; no flagella were mentioned or figured. Sokoloff did not observe the flagellate in the free state.

Endozoic, colorless flagellates that probably belong to the genus *Khawkinea* have often been found, especially in Turbellaria, but also in rotifers, Gastrotricha, fresh-water nematodes, fresh-water oligochaetes, nudibranch eggs, and copepods. In different hosts they occur in the alimentary canal, in tissues, or in the coelom. Howland (1928) identified as *Astasia captiva,* which Beauchamp had described from a rhabdocoele (p. 905), an actively metabolic euglenoid flagellate, without flagellum or stigma, found in the cortical ectoplasm of *Stentor coeruleus* and *Spirostomum ambiguum.* Jahn and McKibben (1927) assigned this species to their new genus *Khawkinea* (see p. 907).

Parasitic dinoflagellates occur in Tintinnoinea, in Radiolaria, and in other dinoflagellates. In the first two groups, as in so many Protozoa, development of the parasites has been mistaken for a phase of the cyclical development of the host. In the tintinnids these errors were first pointed out by Duboscq and Collin (1910); in the Radiolaria by Chatton (1920b).

Chatton (1920b) established the genus *Duboscquella* for the parasite of tintinnids, and recorded *D. tintinnicola* as occurring in *Codonella galea, Tintinnopsis campanula,* and *Favella* (as *Cyttarocylis*) *ehrenbergii.* In the last ciliate, Duboscq and Collin (1910) observed the parasite in abundance at Cette. It is a subspherical body which grows to a large size (100 μ) without apparent inconvenience to the host. Repeated division gives rise to a dense mass of gametocytes, each of which, after ejection from the host, undergoes two divisions inside or outside of the host, to produce biflagellate gametes. Hofker (1931) found *Duboscquella tintinnicola* in *Favella ehrenbergii* and *F. helgolandica.*

Although the enigmatic organism described by Campbell (1926) as *Karyoclastis tintinni* is apparently not a dinoflagellate, it may be mentioned here because of its occurrence in this same group of ciliates. Campbell found it to be primarily an intranuclear parasite of *Tintinnopsis nucula,* but, unlike most other described nuclear parasites, it has a cytoplasmic phase. In the macronucleus the parasites occur as numerous small bodies, each with a gray-staining mantle, a clear central area, and within a central granule which undergoes division. The parasites multiply within the nucleus, then the membrane partially disintegrates, and the parasites emerge and form a cloud-like mass in the cytoplasm. Campbell noted that the parasites are distinct in structure from *Nucleophaga* and *Sphaerita.* Further investigation is necessary to elucidate the complete life cycle and establish the systematic relationships of *Karyoclastis.* Hofker (1931) found a resemblance to *Karyoclastis* in round bodies associated in the test, in some instances, with *Tintinnopsis fimbriata;* but he recognized the possibility that their occurrence was the result of a fragmentation phenomenon.

Chatton (1920a, 1920b) pointed out that the so-called anisospores, or gametes, in *Thalassicolla, Sphaerozoum,* and *Collozoum* (Brandt), the origin of which in the first genus from intracapsular plasmodial masses was described by Hovasse (1923a), belong not to the radiolarians but to the parasitic dinoflagellates similar to *Syndinium* of the pelagic copepods. Chatton (1923) proposed the genus *Merodinium* for these organisms, establishing five species for cytoplasmic parasites of *Collozoum, Sphaerozoum,* and *Myxosphaera,* and a sixth species, in the subgenus *Solenodinium,* for the intranuclear parasite of *Thalassicolla spumida.* The dinoflagellate affinities of these organisms are shown by the nuclear structure, the mode of mitosis, and the morphological characteristics of the spores. The dinospores are reniform, constricted at the equator, and have two unequal flagella in typical dinoflagellate arrangement.

Species of *Peridinium* and related dinoflagellates may be parasitized by *Coccidinium,* which, according to Chatton and Biecheler (1934), resembles coccidia in the vegetative and multiplicative stages, whereas the spores are typical of dinoflagellates. Chatton and Biecheler (1936) reported having observed copulation and total fusion of spores of two types in *Coccidinium mesnili,* and considered this to be the first observation of an indisputable sexual process in an authentic dinoflagellate.

Keppen (1899) described from marine dinoflagellates (*Ceratium tripos, Ceratium fusus,* and *Ceratocorys horrida*) the parasite *Hyalosaccus ceratii,* which he considered to be a parasitic rhizopod. It is impossible to obtain a complete understanding of the structure, life history, and relationships of the organism from Keppen's account and illustrations; but certain similarities to *Coccidinium* are apparent in the structure and nuclear multiplication of the intracytoplasmic stages. Keppen did not describe spores. As did the French authors in *Coccidinium,* Keppen pointed out a resemblance of *Hyalosaccus* to coccidia. He considered this to be the same parasite as that observed by Bütschli (1885) in *Ceratium fusus.*

ZOOMASTIGOPHORA

Chlamydomonads may be attacked by *Colpodella pugnax,* which is more of a predator than a parasite. Cienkowsky (1865), who first described it, found it on *Chlamydomonas pulvisculus.* Dangeard (1900a) studied it mainly on *C. dilli,* but remarked that it would attack more or less all species of the genus. He never, however, observed it on other Protozoa. The free-swimming *Colpodella* is colorless, crescentic, about 12 μ in length, with a terminal flagellum. It becomes fixed to *Chlamydomonas,* perforates its membrane, and within a few minutes the cytoplasm begins to flow into *Colpodella.* The envelope of *Chlamydomonas* is soon emptied, and *Colpodella* takes on a stouter form and a green color. The substance of its prey collects in a large digestive vacuole and is absorbed. In multiplication, the organism rounds up and undergoes thrice-repeated binary fission within a membrane, from which crescentic flagellates escape. Thick-walled cysts were described by Dangeard.

Hollande (1938) gave the name *Colpodella raymondi* to a parasite found by Raymond (1901) on *Chlamydomonas*. The parasite, reported

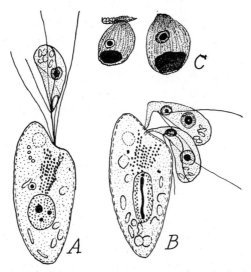

Figure 222. A, B, *Bodo perforans* Hollande, ectoparasitic on *Chilomonas paramaecium*; C, ectoparasite of *Colpoda cucullus*. (A, B, after Hollande, 1938; C, after Gonder, 1910.)

Raymond, occurs in one to several spheroidal masses on the surface of *Chlamydomonas*. Exceptionally there are more than a hundred; the usual size appears to be very much less than that of *C. pugnax*. According to Raymond, the host appears not to suffer from its presence, at least unless the infection is very heavy.

An interesting ectozoic organism on *Chilomonas paramecium* was named *Bodo perforans* by Hollande (1938). This flagellate possesses a long, slender rostrum by which it is fixed in a constant position near the anterior end of *Chilomonas,* at the base of the flagella (Fig. 222A, B). It has two unequal flagella, inserted at the base of the rostrum. Rarely two or three parasites are attached to one host. The rostrum, according to Hollande, penetrates the cytoplasm shallowly; and he found evidence that material may be extracted from the host. Many parasitized chilomonads had lost their flagella, but otherwise they were apparently not injured. *Bodo perforans* was rarely seen free from attachment.

Gregarella fabrearum, studied by Chatton and Brachon (1936) and Chatton and Villeneuve (1937), shows few characteristics that can be used for taxonomic purposes; yet it was considered by them to be a much regressed flagellate, representative of a new group of zoöflagellates, the Apomastigina. It is reported to undergo a cycle of development as follows: ingestion by *Fabrea,* fixation to the wall of a vacuole, with growth and transverse division, discharge from the body and fixation to the ectoplasm around the cytopyge, where feeding and slower multiplication take place. The fixed parasites are claviform with the large end adherent, are capable of slow changes of shape, contain large refractile inclusions, and have no flagella or other permanent differentiations.

The parasite on *Colpoda cucullus* (Fig. 222C), observed in a hay infusion by Gonder (1910), recalls in some ways certain of these ectozoic flagellates. The organism has a round or oval figure, with the narrowed end extended through the pellicle. The nucleus is single, vesicular, with a large endosome—not an ordinary ciliate type. No flagella or cilia were seen. Gonder was vague about its relationships.

Small mastigamoebae were recorded by Doflein (*Lehrbuch,* 1909, and later editions) as not infrequent parasites of *Stentor coeruleus;* and he figured an instance of heavy infection. Infected ciliates were faded and somewhat contracted, and eventually often burst.

Flagellates were found in *Craspedophrya rotunda* and other Suctoria by Rieder (1936). The incidence was high in a culture of the first species and light in four other species. At first only a few *Craspedophrya* were parasitized, but in a few days many were infected. The organisms were colorless, actively metabolic, from 6 to 11 microns in length, and had two flagella 1.5 times the body length, of which one was anterior and the other trailed. They entered the suctorian through the envelope, sometimes at the thinnest place, as over the brood chamber from which a swarmer had escaped, but also elsewhere. They were observed swimming about actively within the pellicle, taking up suctorian plasma, and undergoing binary fission. Eventually the flagellates often rounded up and lost the flagella; some left through the pellicle, probably in the way they entered. The host may die and disintegrate, according to Rieder, even when only one or two parasites are present. He considered the organism to be a strict endoparasite of *Craspedophrya*.

A number of enigmatic forms, which at least seem to show certain flagellate relationships, may be considered here.

Dangeard (1908) gave the name *Lecythodytes paradoxus* to a parasite of cysts of *Chromulina*, which decimated cultures within a few days. Within the cyst the organism is amoeboid, and grows until it occupies the whole interior. Division, he stated, results in eight, or less often four or sixteen zoöspores, which escape from the cyst and may infect another host. The zoöspores are elongated and narrowed at the extremities, each of which, according to Dangeard, terminates in a long flagellum.

Uncertain is the proper systematic position of *Sporomonas infusorium*, which Chatton and Lwoff (1924a) encountered in the marine ciliates *Folliculina elegans, Vorticella* sp., and once in *Lacrymaria lagenula*. Potts found it in *Folliculina ampulla* at Woods Hole, Massachusetts. In the cytoplasm the parasite occurs as a reniform body, provided with a long flagellum, in active rotation. It increases greatly in size, up to 70 μ, and the flagellum is lost. The parasite is then expelled, and multiplication takes place only outside. There is rapidly repeated nuclear division and binary fission without growth (palintomy), resulting in small, virgulate bodies, each provided with a lateral flagellum. Chatton and Lwoff considered this organism to be a flagellate, but discussed its resemblance to Chytridiales. They stated that it differs from that group in multiplying by palintomy, with rapidly repeated mitosis after growth, instead of by syntomy following nuclear multiplication accompanying growth. Mitchell (1928), however, described multiplication of the same type as that in *Sporomonas* in a chytrid of *Euglena caudata;* and on other grounds also it appears that the distinction is not of crucial significance for classification. The chief differences from chytrid parasites of Protozoa are the expulsion of the organism from the host before multiplication occurs and the active motility, by means of a flagellum, of the early intracytoplasmic growth stages.

Georgevitch (1936a, 1936b) assigned to the genus *Leishmania*, as *L. esocis* n. sp., a hyperparasite of *Myxidium lieberkühni* in the urinary bladder of pike. The intracellular phase is pyriform, with one nucleus and a rod-formed structure called by Georgevitch a blepharoplast. Growth, nuclear division, and plasmotomy result in rosettes, usually of eight individuals and a residual body. Longitudinal division also occurs. Flagellated stages were rarely found. The evidence that the parasite is *Leishmania* is unconvincing, and is certainly inadequate as a basis for extending the distribution of that genus to such an unusual location.

SARCODINA

The genus *Pseudospora*, which comprises parasites of true algae and of Volvocidae, has been placed in the Proteomyxa; but there appears to be good reason for accepting the suggestion of Roskin (1927) that it belongs rather in the Bistadiidae of the Amoebida. *P. volvocis*, first described by Cienkowski (1865), was later reported from *Volvox* by Robertson (1905) and Roskin (1927). It seems possible that the amoebae found by Molisch (1903) and Zacharias (1909) in *Volvox minor* (*V. aureus*) belonged to this genus, though the authors do not refer to earlier observations on parasites of *Volvox,* and their work is not cited in later accounts of *Pseudospora*. Roskin (1927) described *P. eudorini* from *Eudorina,* in which flagellate Robertson (1905) had reported *P. volvocis.*

In its free-living state both these species are small, heliozoa-shaped forms, with immobile or slow-moving pseudopodia. The heliozoan form becomes amoeboid, with lobose pseudopodia, on contact with the host, which it enters. Within the coenobium, the amoeboid form engulfs the cells and undergoes repeated division. After a period, the parasite comes to the surface of the colony, and there is rapid transition to a form with two relatively long flagella. The organism may lose the flagella and become amoeboid, may form cysts and the free heliozoan form, and the heliozoan type may transform into a flagellate, as well as into an amoeboid form.

It is generally considered that confusion with parasites is the basis of the accounts of complicated life cycles in *Arcella*. The earlier observations on this testacean were discussed by Dangeard (1910), who offered convincing evidence that the small amoeboid bodies, supposed to be produced in numbers by repeated exogenous or simultaneous endogenous budding (Awerinzew, 1906; Elpatiewsky, 1907; Swarczewsky, 1908), and variously interpreted by authors as reproductive phases of the life cycle, are really parasites. The amoebulae were reported to take on a heliozoa-like form on becoming free; Dangeard noted a similarity to *Nuclearia*. An amoeboid parasite of *Arcella* was reported by Gruber (1892), and evidently it is quite common, so many have been the accounts of it. Doflein (Reichenow ed., 1927-29) and Deflandre (1928) were inclined to accept the parasite interpretation of these supposed reproductive bodies; but Cavallini (1926a, 1926b), without reference to Dangeard's paper, reported in *Arcella vulgaris* and *Centropyxis aculeata* division of the protoplasmic body into many small amoebae, which leave the shell and develop into the mature testaceans. Nevertheless, it is probable that binary fission is, as Deflandre (1928) stated, the only mode of reproduction that has been satisfactorily demonstrated in *Arcella*.

Penard (1912) found what he considered to be small parasitic amoebae in *Amoeba terricola* and other species. They were often observed moving actively within the pellicle of *dead* amoebae, from which they eventually emerged and moved about freely, feeding on bacteria. Penard found indication that these are parasites, the development of which begins in the body of the large amoebae, but proof of this is lacking.

There is, to the writer's knowledge, only one record of amoebae parasitic in a free-living ciliate. Chatton (1910) observed a very small species living as a true parasite in *Trichodina labrorum* from the rockfish. Opalinid ciliates are, however, not uncommonly parasitized.

The hyperparasites of Opalinidae (Fig. 223A) resemble *E. ranarum* (Carini and Reichenow, 1935; Brumpt and Lavier, 1936; Stabler and Chen, 1936). Those in different opalinids have not been found to show any taxonomic distinctions; and the systematic name to be used, for some forms at least, is *Entamoeba paulista* (Carini); to be used, that is, if this amoeba is truly an independent species. Carini and Reichenow were of the opinion that the hyperparasite is either identical with *Entamoeba ranarum* or is a race or species derived from this. Stabler and Chen considered the question of the amoeba's synonymy with *E. ranarum* to be still open. Brumpt and Lavier, though recognizing a probable distinction from *E. ranarum*, discussed the relationship as *paranéoxénie*, in which an intestinal parasite of the amphibian attacks another parasite that accompanies it, which then becomes a subhost. From the standpoint of general host-parasite relationships, it seems to the writer unlikely that the same amoeba both parasitizes opalinids and lives in the intestinal lumen; certainly, to date, there is no proof of either possibility.

Published reports seem to indicate a greater prevalence of entamoebae in *Zelleriella* than in other Opalinidae, but they have been found also in *Protoopalina, Cepedea,* and *Opalina* (Chen and Stabler, 1936). The geographical distribution of the amphibian hosts of parasitized Opalinidae has been found to be very wide, but Chen and Stabler stated that "only certain species of anurans and a certain percentage of individuals of a given species harbor the parasitized opalinids." In a given host, the percentage of ciliates containing hyperparasites varies greatly from none to quite all, as noted by Carini (1933) and Brumpt and Lavier (1936) in *Zelleriella* of *Paludicola signifera,* and by Chen and Stabler (1936) in *Zelleriella* of *Bufo marinus.* The number of amoebae in a ciliate ranges from one or very few up to a condition in which the cytoplasm of the host is almost completely filled, more than a hundred being present. They have been found mainly in the trophozoites, but may occur also in precystic and encysted ciliates, according to Chen and Stabler (1936), though Brumpt and Lavier (1936) stated that they had never found them in the cysts. Chen and Stabler pointed out the importance of the parasite in cysts, in establishing infections in a new generation of the hosts.

In opalinids, entamoebae occur either as vegetative forms or as cysts. Stabler and Chen (1936) stated that they apparently feed on the endospherules, and they observed instances of mitotic division. It has been noted (Stabler, 1933; Carini and Reichenow, 1935; Stabler and Chen, 1936) that the cysts in the ciliates are all uninucleate; the same observers reported binucleate and quadrinucleate cysts outside of the hosts. It seems, in general, that all the parasites in one ciliate are in the same stage of development; but Brumpt and Lavier (1936) gave an unconvincing figure of a *Zelleriella* said to contain a uninucleate and a binucleate vegetative form, a precystic form, and a uninucleate cyst. Furthermore, in their statement that certain cysts contain two or four nuclei, though most are uninucleate, it is not clear whether or not they found multinucleate cysts in the opalinid host.

It has been noted that the parasites seem to have no injurious effect on the host. The fact that parasitized ciliates may yet undergo normal division (Brumpt and Lavier, 1936; Stabler and Chen, 1936) is the most important evidence for this benignity.

No strictly parasitic relationship has been established in any member of the group Heliozoa proper, but Wetzel (1925) discussed as temporary parasitism the association between *Raphidiocystis infestans* and ciliates (Fig. 223B). This organism is normally predatory on flagellates and small Infusoria, but when it attacks larger ciliates, such as *Paramecium, Colpidium, Glaucoma, Nassula,* and *Trachelius,* it passes,

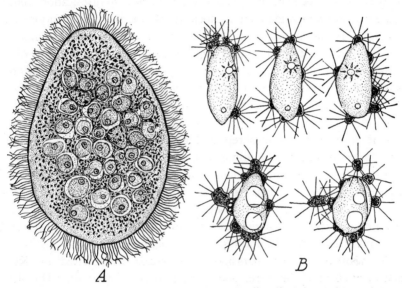

Figure 223. A, trophozoites of *Entamoeba* sp. in *Zelleriella* (after Stabler and Chen, 1936) ; B, *Raphidiocystis infestans* Wetzel on *Paramecium* (After Wetzel, 1925.)

according to Wetzel, to the parasitic manner of life. The heliozoans become attached by means of pseudopodia to various parts of the body, those pseudopodia gradually shorten, and eventually they lie very flat on the surface. The body of *Paramecium* may be completely enclosed by many fused *Raphidiocystis,* which extract dissolved nutriment until only a remnant of the ciliate is left—a process which requires from twenty or thirty minutes to two days—after which the heliozoa disperse. Wetzel discussed the significance of this type of relationship in the phylogenetic origin of parasitism as transitional between strict predatism and pure parasitism.

Ivanić (1936) stated that the parasite of *Entamoeba histolytica,* which he described as *Entamoebophaga hominis,* shows affinities to the Mycetozoa, but this is hardly apparent from his unconvincing account of the structure and life history. The earliest stages are reported to occur within the host cyst, and growth leads to an amoeboid body. At first uninucleate, this becomes multinucleate. When the cytoplasm of the host cyst has been largely consumed, the parasite breaks out and carries on for a time an active, free-living existence in the intestinal lumen, as an amoeboid organism. In this phase there is nuclear multiplication, binary fission, growth to a giant, multinucleate plasmodium, and endogenous budding. Ivanić found evidence that the organism was originally a commensal of the human intestine before it became a parasite of *E. histolytica.* This bizarre account probably contains a good deal of misinterpretation and confusion of distinct organisms.

SPOROZOA

The sporozoan parasites reported in Protozoa belong for the most part to the Microsporidia and Haplosporidia. Dogiel (1906) assigned to the coccidia a parasite, named *Hyalosphaera gregarinicola,* of a gregarine from a holothurian. Caullery and Mesnil (1919) considered this systematic determination doubtful, but were certain that the parasite is not a *Metchnikovella.* Dogiel described macrogametes, microgametocytes, and sporulation, but did not observe schizogony.

Four microsporidian parasites of Protozoa were listed by Kudo (1924). Three of these are species of *Nosema: N. marionis* (Thélohan, 1895) Stempell, 1919, in the myxosporidian *Ceratomyxa (Leptotheca) coris* from the gall bladder of *Coris julis* and *C. giofredi; N. balantidii* Lutz and Splendore, 1908, in *Balantidium.* sp. from *Bufo marinus;* and *N. frenzelinae* Leger and Duboscq, 1909, in the gregarine *Frenzelina conformis* from *Pachygrapsus marmoratus.* The last species shows a certain amount of correlation with the life cycle of the host, in that sporulation occurs at the moment of encystation of the gregarine. The gregarines develop normally up to a certain point; then the formation of gametes does not take place (Leger and Duboscq, 1909a, 1909c). The fourth species, *Perezia lankesteriae,* also parasitizes a gregarine, *Lankesteria ascidiae* from *Ciona intestinalis* (Leger and Duboscq, 1909b).

Microsporidia are probably much more widespread as parasites of

Protozoa than the published accounts indicate. Duboscq and Grassé (1927) showed certain parasites in *"Devescovina" hilli* which they considered to be possibly Microsporidia. The organism found by Kirby (1932a) in *Trichonympha magna* gives certain indications of microsporidian affinities (Fig. 219H, I). An organism with resemblances to *Nosema,* though enigmatic in relationship, has been observed by the writer in *Gigantomonas herculea* from the termite *Hodotermes mossambicus.* All the above-mentioned Microsporidia are hyperparasites, but there is probably at least one recorded instance of their occurrence in a free-living ciliate. A number of authors have reported "nematocysts" in the large vorticellid *Epistylis (Campanella) umbellaria;* these were discovered by Claparède and Lachmann (1858). They are arranged in pairs in the ciliate, but are not always present. Fauré-Fremiet (1913), although having often observed *Campanella,* found "nematocysts" only once. Chatton (1914) suggested that the structures belonged not to the vorticellid but to a cnidosporidian parasite, and in a recent note Krüger (1933) supported this view. Krüger observed in the cytoplasm of the ciliate granules that he thought might be nuclei of developmental stages of the parasite.

From the standpoint of host-specificity, the Metchnikovellidae are of particular interest, for all members of this family, and there are many, occur in gregarines. The first metchnikovellid was seen by Claparède, but he failed to interpret it correctly, mistaking the cysts for spores of the gregarine (Caullery and Mesnil, 1919, p. 232.) This group of Haplosporidia has been studied chiefly by Caullery and Mesnil (1897, 1914, 1919), but contributions have been made also by Awerinzew (1908), Dogiel (1922b), and Schereschewsky (1924). An account of the life cycles of two species of *Amphiacantha* has been prepared by Stubblefield (MS). MacKinnon and Ray (1931) reported some observations on species of *Metchnikovella* from two species of *Polyrhabdina* at Plymouth; and Ganapati and Aiyar (1937) noted the occurrence of *Metchnikovella* in *Lecudina brasili* from a species of *Lumbrinereis* at Adyar. In the absence of any description or figure, it is not certain that this is not a species of *Amphiacantha,* as found in related gregarines in *Lumbriconereis* elsewhere.

Stubblefield (MS) listed twenty species, including the two new species of *Amphiacantha* recognized by him, in four genera. The largest genus is *Metchnikovella* Caullery and Mesnil, with thirteen species; there are three species of *Amphiamblys* C. and M., three of *Amphia-cantha* C. and M., and one of *Caulleryella* Dogiel. All the grega-rines that have been found to contain these hyperparasites occur in annelids, and all but one in marine polychaetes. This one, *Metchnikovella hessei* Mesnil, 1908, is found in a monocystid gregarine of the terrestrial oligochaete *Fridericia polycheta*. The parasitized gregarines belong to various groups, and, according to Caullery and Mesnil (1919), there is no parallelism between the structure of the Metchnikovellidae and that of the gregarine hosts. The host-specificity is apparently on an ethological rather than a phylogenetic basis. The species of *Amphiacantha,* however, have been found in gregarines of the genus *Ophiodina* (*Lecudina*) or related forms in *Lumbriconereis* in France and California.

Caullery and Mesnil (1919) stated that when there is an infection, the greater part of the gregarines of a host are invaded. Stubblefield (MS) found a high frequency of *Amphiacantha* in about 20 percent of the worms collected, almost all of which contained gregarines.

Published literature gives little information about details of the life cycles of Metchnikovellidae. Caullery and Mesnil (1919) regarded the individualized, nucleated bodies enclosed by the cyst membrane as spores ("germes sporaux"). When the cysts are ingested by an annelid, these are released in the digestive tract, and penetrate into the cytoplasm of the gregarines. Growth and nuclear division lead in some instances to multinucleate plasmodia. In other instances there are numerous indi-vidual, uninucleate bodies, isolated or arranged in series. Caullery and Mesnil supposed that cysts develop by the formation of a membrane around groups of these cells or the plasmodium. The cyst contents is thus either multinucleate or in individualized uninucleate bodies from the beginning. Such a manner of cyst formation is difficult to understand.

Stubblefield (MS) prepared an account of the life cycle of *Amphia-cantha,* which is in closer agreement with that of *Haplosporidium*. He found evidence for the penetration of the gregarine by an active sporozoite; the growth of the sporozoite, followed by schizogony, to produce trophozoites; the development of the trophozoite into a cyst, which is at first uninculate; nuclear and cytoplasmic division, to pro-duce bodies in the cyst (Fig. 224), which he considered to be gametocytes; the release of these by the rupture of the cyst within the gregarine, after which they undergo reduction, producing gametes which fuse; and finally the development of sporozoites from the zygote. Stubblefield's observation that cysts rupture within the gregarine host is in

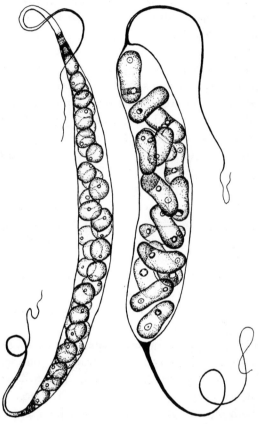

Figure 224. Cysts of two species of *Amphiacantha*, metchnikovellids parasitic in the gregarine *Ophiodina elongata* from *Lumbricoinereis*. (After Stubblefield, MS.)

agreement with the statement of Mackinnon and Ray (1931) that the "spores" of *Metchnikovella caulleryi* have been seen escaping from the cysts into the endoplasm of the gregarines.

Caullery and Mesnil stated that Metchnikovellidae seem to have little pathogenic action on the host, particularly in the vegetative stages. What injury there is, is mechanical, when infection is heavy. Ganapati and Aiyar (1937) noted that the entire cytoplasm may be packed with cysts, and the body become much misshapen, the gregarine nucleus seemingly degenerating. Stubblefield, however, found that more than six cysts rarely occur in a gregarine. Caullery and Mesnil believed that heavily parasitized gregarines are incapable of completing their sexual development; and Ganapati and Aiyar remarked that parasitized gregarines were not observed to associate.

The affinities of the Metchnikovellidae are uncertain. They have been related to fungi (Chatton, 1913, yeasts), to Microsporidia (Schereschewsky, 1924), and to Haplosporidia (Awerinzew, 1908). Caullery and Mesnil (1919), while remarking on a certain similarity in nuclear structure to Myxomycetes and Chytridiales, concluded that they are isolated among the lower Protista. Doflein (Reichenow ed., 1927-29) accepted their allocation to the Haplosporidia, and this position was supported by Stubblefield (MS).

Caullery and Mesnil (1919) provisionally designated as *Bertramia selenidicola* a parasite of a species of *Selenidium* from certain polychaetes, and reported a related parasite in *Selenidium virgula*. Other species of *Bertramia* are parasites in the body cavities of worms and rotifers. Another parasite with apparent haplosporidian affinities, but unlike *Metchnikovella,* was observed in a species of *Polyrhabdina.* It existed as isolated granules and multinucleate masses, the schizonts and sporonts; and as separate ovoid bodies, not enclosed in a cyst, which were evidently spores.

Elmassian (1909) found a hyperparasite, *Zoomyxa legeri,* in *Eimeria rouxi,* which causes fatal coccidiosis in tench. He considered this to be a haplosporidian, but also discussed its similarities to lower Mycetozoa. It is likely that his account is at least in part incorrect. The parasite is said to occur both on the surface of the epithelium and in the cells of *Eimeria,* the intracoccidian parasitism being accidental. There are said to be several types of schizogony, within coccidia or not; and in this supposed haplosporidian the author described a sexual cycle with coccidian-like development of microgametes and macrogametes, and the formation of resistant cysts containing from six to twelve sporozoites. The parasite has pathogenic effects on the nucleus and cytoplasm of *Eimeria,* causing hypertrophy and eventual dissolution of the cell. Elmassian thought that the effects are brought about by toxic secretions, which act not only on the coccidia but also on the neighboring fish cells.

CILIOPHORA

Euciliata.—There are few reports of ciliates parasitizing other Protozoa, except for *Phtorophrya* and *Hypocoma*. Penard (1904) found a ciliate, which seemed to resemble *Blepharisma,* in a large percentage of the heliozoan *Raphidiophrys viridis.* The intracytoplasmic forms showed different degrees of development and lived many days in isolated Heliozoa. An immobile organism, with a large contractile vacuole but no cilia or flagella, was found parasitic in three-fourths of a large number of *Pseudodifflugia horrida* by Penard (1905a). This, he stated, suggested the larger ciliate in the heliozoan, but its affinities are uncertain. Hertwig (1876) reported that a hypotrich bored into the body of *Podophrya gemmipara,* in the region in which the body is joined by the stalk, and destroyed the acinetid.

A number of apostomatous ciliates are parasitic in other Foettingeriidae. The most completely known of these is *Phtorophrya insidiosa* Chatton, A. Lwoff and M. Lwoff, 1930, which is parasitic on *Gymnodinioides corophii.* The phoront of *Phtorophrya* is attached to the phoront of *Gymnodinioides,* which occurs on *Corophium acutum.* The body of the parasite leaves the phoretic cyst and introduces itself into the body of its host, becoming a parasitic trophont. It grows rapidly, ingesting the cytoplasm of its host, and soon comes to occupy entirely the otherwise empty cyst of *Gymnodinioides.* By division, four to eight small ciliates, the tomites, are produced. These escape from the host cyst and swim actively in search of another phoront of *Gymnodinioides.*

Chatton and Lwoff (1930, 1935) described also the following incompletely known species of this genus of ciliate parasites: *Phtorophrya mendax* in the phoronts of *Gymnodinioides inkystans; P. fallax* in this same host species; *P. steueri* in *Vampyrophrya* (?) *steueri; P. bathypelagica* in *Vampyrophrya bathypelagica.*

The Hypocomidae, like most other Thigmotricha, occur on bivalves or snails, except for species of the genus *Hypocoma,* which are parasitic on other Protozoa. *Hypocoma parasitica* (Fig. 199G-I) was found by Gruber (1884) and Plate (1888) on marine vorticellids, especially *Zoothamnium,* on the coast of Italy. Plate recognized a second species, *H. ("Acinetoides") zoothamni.* The ciliates occur firmly fixed to the host, and suck out the contents of the zoöids. *Hypocoma acinetarum* Collin is ectoparasitic on Suctoria. Collin (1907) found it on various occasions on *Ephelota gemmipara* and *Acineta papillifera;* and Chatton and A. Lwoff (1924b) encountered it on *Trichophrya salparum.* In *Ephelota* it attacks chiefly the region in which the stalk is attached to the body. It sucks out plasma, and its presence leads to fragmentation of the nucleus and degeneration of the whole cytoplasmic mass. The parasite then detaches and swims to another suctorian. *Hypocoma ascidiarum* Collin was found on a tunicate, but probably actually is a parasite of *Trichophrya salparum,* and may not be different specifically from *H. acinetarum* (Chatton and A. Lwoff, 1924b).

Ectozoic Suctoria.—In connection with the relationship between externally attached Suctoria and their ciliophoran hosts, we must keep in mind the fact that phoresy is widespread among Suctoria. Many forms occur attached to other organisms, and often a species has been found only on a particular host species. The host is not directly concerned in the nutritive processes of the suctorian. Ciliophora may, like many Metazoa, serve as hosts for these ectocommensals. Examples of such phoretic forms are *Ophryocephalus capitatum* Wailes on species of *Ephelota; Urnula epistylidis* Claparède and Lachmann on *Epistylis* and other Suctoria (see Gönnert, 1935); *Tokophrya quadripartita* (Cl. and L.) on *Epistylis; Trichophrya epistylidis* (Cl. and L.); *Metacineta mystacina* (Ehrbg.) on *Carchesium;* and *Tokophrya carchesii* (Cl. and L.) on *Carchesium.* Ectocommensalism in such attached forms may be obligatory or facultative.

Pseudogemma Collin is more closely adapted to an ectoparasitic manner of life on other Suctoria. Reproduction is by internal embryos, and fixation to the host is by a short, stout peduncle embedded in the cytoplasm. Tentacles are absent, and Collin (1912) considered it possible that the fixation organelle has an absorptive function. Collin listed three species: *P. fraiponti* Collin, 1909, on *Acineta dirisa; P. pachystyla* Collin, 1912, on *Acineta tuberosa;* and *P. keppeni* Collin, 1912, on *Acineta papillifera.* The last species is said to have a rounded form and apparently no pedicle, and its location in its host is sometimes external and sometimes almost entirely internal. Collin (1912) believed that it furnishes a natural transition from *Pseudogemma* to *Endosphaera.*

The species of *Allantosoma* occur in the intestine of the horse, an endozoic habitat which is unique for Suctoria. According to Hsiung (1928), the species *A. dicorniger* is strictly a lumen-dweller, and is not attached to any other organism. The other two species have certain relationships to ciliates, but apparently that is only occasional. Hsiung wrote of *A. intestinalis* that some were attached to *Cycloposthium bipalmatum* and *Blepharocorys curvigula;* Gassovsky (1919) recorded the species as occurring in the colon, rarely the caecum, of horses, without mention of any attachment to ciliates. *A. brevicorniger,* Hsiung states, is "often found attached to the body of the ciliate *Paraisotricha colpoidea* by one tentacle." Apparently these Suctoria prey upon the ciliates, but are not constantly attached, as obligatory ectoparasites would be.

The only account of a suctorian clearly ectoparasitic on Euciliata is Chatton and A. Lwoff's (1927) description of *Pottsia infusoriorum.* The chief host is *Folliculina ampulla,* but it has been found also on *Cothurnia socialis.* The parasites may occur in numbers on the body of *Folliculina,* within the lorica. There are four tentacles at one end, on the surface in contact with the host, prolonged rather deeply into the body of the host. Embryos develop endogenously, swim actively, and become fixed to the body of *Folliculina.* On different occasions, from none to 75 percent of the heterotrichs have been found parasitized with as many as twenty-two parasites. When the number of *Pottsia* is large, the host undergoes degeneration. The parasites may survive for a time among the remains, but eventually themselves disintegrate.

Tachyblaston ephelotensis, as described by Martin (1909), has a curious life cycle, involving both an external phoretic existence with multiplication, and an intracytoplasmic parasitism, also with multiplication. It seems not impossible that reinvestigation will show that two organisms have been confused in this cycle, since it is so unlike the life histories of other Suctoria. The intracellular phases occur as rounded bodies in *Ephelota gemmipara.* There is equal division, followed by the formation of buds. Ciliated "spores" escape from the host and after a brief period of existence become attached to the stalk of *Ephelota,* developing a stalked lorica. The fixed form undergoes rapid budding. Each bud is provided with a single tentacle, with the aid of which, together with "euglenoid changes of shape," the bud travels up the stalk to penetrate into the body of *Ephelota.* The internal parasitic phase destroys the cytoplasm of the host.

Endozoic Suctoria.—The Suctoria that occur as internal parasites of other Ciliophora, and have a wide variety of hosts, belong to the genera

Sphaerophyra and *Endosphaera*. In connection with them, it is interesting to consider the important rôle they have played in the development of protozoölogy. The Acineta theory and the embryo theory of ciliate development held an important place in the thinking of protozoölogists in the third quarter of the nineteenth century.

Stein's Acineta theory was in the first instance not related to parasitic Suctoria. He came (1849, 1854) to the conclusion that free-living acinetids are the result of metamorphosis of vorticellids, and that they give rise to embryos from which the vorticella form is again produced. This embryo production, of course, is the result of the internal budding process characteristic of acinetids. This theory was successfully attacked by Cienkowski (1855a), Lachmann (1856), and Claparède and Lachmann (1860-61). Stein later (1859) modified the Acineta theory as it was originally stated, but still did not admit that acinetids are independent organisms. The embryos of various Infusoria, he said, have all the characteristics of acinetids; and he believed that various higher Infusoria in their development pass through Acineta-like phases; for example, that podophryids were developmental phases of *Paramecium*.

Authors credit Focke (1844) with the first observation of the so-called motile embryos. He discovered them in *Paramecium bursaria,* in which they were soon found by many other observers. They were found also in a variety of other Euciliata. As late as 1867, Stein could state that "today no one can doubt that those Infusoria whose reproductive organization consists of nucleolus and nucleus are in fact hermaphrodites, the nucleus playing the rôle of a female, the nucleolus of a male sex organ"; and could maintain that the embryonal spheres were produced from the nucleus.

Stein's thesis, however, had already been discredited. Claparède and Lachmann (1858-59) had described *Sphaerophrya pusilla* in water, associated with numerous oxytrichids; yet they were not firm in their opinion that *Sphaerophrya* might not be an embryo of *Oxytricha*. It was the view of Balbiani (1860) that the so-called embryos of ciliates were parasites belonging to the genus *Sphaerophrya;* and in support of this he adduced his observations on entry into ciliates, and on the spread of an infection among *Paramecium* by the introduction into a sound culture of a few infected ciliates. Metschnikoff (1864) observed the cycle, from separation from one *Paramecium* host through entry into another, and considered the parasite nature of the so-called embryos to be completely proved.

Sphaerophrya shows suctorian characteristics in the presence of tentacles. The so-called embryos of certain vorticellids, however, do not have this characteristic; they are simple spherical or ovoidal bodies with equatorial bands of cilia. Stein (1867), in his efforts to combat Balbiani, had only weak arguments against the parasitic nature of the *Sphaerophyra*-type "embryos"; but he was firm in his conviction that the "embryos" of vorticellids (*Epistylis plicatilis*) could not be parasitic Infusoria. Such a concept, he stated, would be ludicrous. Engelmann (1876), reviewing the whole question in support of the parasite theory (which he had vigorously opposed in 1862), reported having observed the entry of the supposed non-tentaculated embryos of *Vorticella microstoma* into that host. Thus he proved the parasitic nature of that organism also, a view also stated by Bütschli (1876), and gave it the name *Endosphaera*.

The endozoic forms of *Sphaerophrya* are but little modified in consequence of parasitism, and the majority of species of the genus are entirely free-living. *S. stentoris* Maupas is free-living or parasitic in species of *Stentor;* recently Kalmus (1928) reported it in *S. roeseli*. The parasites of other ciliates have all been placed in a second species, which also is either free-living or endozoic. Bütschli (1889) and Sand (1899) identified this second species with Claparède and Lachmann's free-living *S. pusilla.* Collin (1912), however, considered it to be *S. sol* Metchnikoff which also was originally described as a strictly free-living species. Sand regarded *S. sol* as a synonym of *S. pusilla.*

Species of the genus *Sphaerophrya* differ from those of *Podophrya* in the absence of a stalk. The body is spheroidal or ellipsoidal, and tentacles radiate from the entire surface. Reproduction is by equal or unequal fission or by external budding, except in *S. stentoris,* which is reported to show a transition to internal budding. The free-swimming forms produced by budding are provided with cilia that are localized at one extremity, in a girdle, or generally distributed; and they possess tentacles. This form, in parasitic phases, penetrates the surface of a ciliate and takes up a position in the cytoplasm, losing cilia and tentacles. There reproduction by division and budding takes place.

Endosphaera (Fig. 225) has become more closely adapted to parasitism. It does not occur as a free-living organism, except briefly in the motile phase that passes from one host to another; and it has no tentacles at any time. *Endosphaera* has been found in vorticellids of the genera *Vorticella, Zoothamnium, Epistylis, Carchesium, Trichodina,* and *Opisthonecta.* All these have been assigned to the species *E. engelmanni* Entz, the most adequate study of which was published by Lynch and Noble (1931). Gönnert (1935) described *E. multifiliis,* reporting it from the Suctoria *Lernaeophrya capitata* Perez, *Trichophrya epistylidis* Cl. and L., Tokophryidae, and *Dendrosoma;* and from vorticellids.

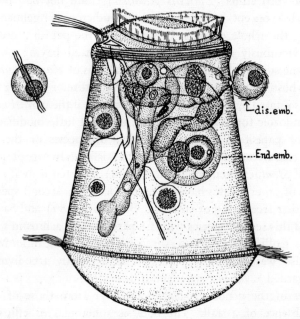

Figure 225. *Opisthonecta henneguyi,* parasitized by *Endosphaera engelmanni. End. emb., Endosphaera* containing an embryo; *dis. emb.,* embryo being discharged through birth pore. (After Lynch and Noble, 1931.)

Lynch and Noble found a high incidence of infection in *Opisthonecta henneguyi,* with as many as twelve parasites, most of which contained one or occasionally two or three internal embryos. They found each parasite to be attached to the pellicle of the host by a short stalk, perforated by a canal terminating in a birth bore. The spherical embryo, provided with three equatorial bands of cilia, was discharged through this pore. Embryos were observed to attach themselves to the host, and successive stages of penetration were studied in preparations. The authors found no evidence that the parasite pushed an extensible pellicle before it, forming an invaginated chamber in which it dwelt, such as was described by Balbiani (1860) and Bütschli (1876) in *Sphaerophrya*. *Endosphaera* was observed in cysts of *Opisthonecta*, which could account for the survival of the parasite under unfavorable conditions.

The embryos of *E. multifiliis* Gönnert have five bands of cilia. Gönnert (1935) observed penetration into *Lernaeophrya*, preceded by the resorption of cilia and the development of a long, mobile, penetrating protoplasmic process. He observed no canal connecting the internal parasite to the surface of the host. *Endosphaera* lives, he stated, four or five days, and an embryo may be produced every half hour.

Sphaerophrya and *Endosphaera* appear to be relatively benign parasites, except when present in large numbers. The effect is then evidently mechanical. Balbiani (1860) remarked that oxytrichids with more than fifty parasites were greatly swollen and deformed, but that ordinarily the host seemed to be not at all inconvenienced. Gönnert found that *Endosphaera*, when present singly, had slight effect on the host, but that the host often perished from multiple infection.

THE GENUS AMOEBOPHRYA KOEPPEN

Amoebophrya is even more of a zoölogical enigma than is *Sticholonche*, one of its hosts, which Korotneff (1891) wrote of as a "zoölogical paradox." A modern study of the structure and development of the organism, which would throw light on its affinities, is much to be desired. The evidence that it is a suctorian, accepted by Koeppen (1894), Borgert (1897), Sand (1899), and Hartog (1909, Cambridge Natural History), is not convincing. Its assignment to the Mesozoa, made by Korotneff (1891) and Neresheimer (1904, 1908, and later) and agreed to by Collin (1912), does little more than emphasize its enigmatic qualities.

Hertwig (1879) described what he regarded as a very peculiar nuclear form within the central capsule of three species of acanthometrid Radiolaria. (Fig. 226D). He stated that he found this body twice in *Acanthostaurus purpurascens* and once each in *Acanthometra serrata* and *A. claparèdei;* and he showed what is doubtless the same thing in *Amphilonche belonoides*. He described this as a large vesicle containing a very large nucleolus, around the sharpened end of which was a conical structure, the membrane of which was marked by circular striations.

Fol (1883) found structures, which he described as analogous to those seen by Hertwig, in the ectoplasm of *Sticholonche zanclea*. Some of the Radiolaria contained spherical bodies, of rather complex structure, which increased in size as the host became older, and contained an ill-defined "spiral body." At maturity, these bodies left the host and were capable of rapid movement, comparable to that of very active Infusoria. The free organism had a spiral groove turning from left to right and was completely covered with short, fine cilia. Other specimens of the radiolarian contained a mass of globules, which increased in size and number, finally becoming in volume equal to the rest of the body. Fol advanced two hypotheses in interpretation of these structures: one, that the globules are female reproductive elements, while the spiral body is a sort of spermatophore; the other, which he regarded as also reasonable, that the structures represent parasites. On the last supposition, he stated, it would be difficult to explain the fact that the two kinds of inclusions occur only in different individuals in approximately equal numbers.

Korotneff (1891), who studied the "spiral body" in *Sticholonche* obtained at Villafranca, concluded that it is a parasite and made the first suggestion as to its affinities. Believing himself to have demonstrated an endoderm of a few cellular elements and a cellular ectoderm, he considered the parasite to be closely related to the orthonectids and possibly a stage in their development.

The parasites in both *Sticholonche* and acanthometrids were studied by Koeppen (1894), who gave them the names *Amoebophrya sticholonchae* and *A. acanthometrae,* and who was convinced that the organisms are acinetids. He stated that he had studied all phases of development in the same specimen. He based his taxonomic conclusion on supposed development, in the parasitic phase, of an embryo, the spiral body, by internal budding; and on the existence of tentacles for a short period after this embryo became free and lost its cilia. The so-called tentacles soon disappeared, and the body commenced to vary in form, showing slow amoeboid movements. There is no proof that the protoplasmic processes were actually tentacles; evidence is lacking that the behavior was observed repeatedly under normal conditions; and there are no supporting illustrations.

Borgert (1897) found parasitized *Sticholonche* and acanthometrids in the Gulf of Naples and prepared the most complete existing account of the organisms. Although he disagreed with many of Koeppen's interpretations and found a large number of nuclei in the outer layer, he nevertheless agreed with him that *Amoebophrya* is a parasitic suctorian.

Amoebophrya sticholonchae (Fig. 226A-C) is a common parasite of *Sticholonche zanclea* in the Mediterranean. Borgert found parasites

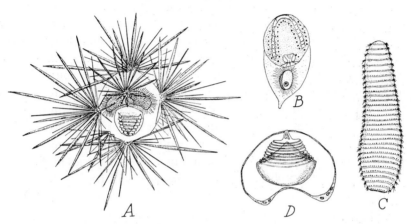

Figure 226. *Amoebophrya* in Radiolaria. A, *Sticholonche zanclea* containing *A. sticholonchae;* B, longitudinal section of *Amoebophrya in Sticholonche,* and, in lower half of figure, section of host and its central capsule; C, *A. sticholonchae* emerged from its host; D, *A. acanthometrae* in *Acanthometra serrata.* (A-C, after Borgert, 1897; D, after Hertwig, 1879.)

only in the latter part of March, 1895, though the radiolarian was abundant also before and after that period. In its host (Fig. 226A) it is an approximately spherical body, located on the concave side of the capsule, transparent, and pale yellowish in color. Within the sphere is a conical body, the point of which is directed toward the body surface of the host (Fig. 226B). The outer surface of the cone is marked by furrows in a close-set, left-wound spiral. The outer surface of the cone is continuous at its base with the inner surface of the sphere, and the spiral furrow continues on the latter. The form of the parasite in this stage has been compared to that of a half-invaginated glove finger. Borgert described a large number of very small nuclei arranged in rows between the furrows. In younger parasites there were fewer nuclei, and in an appendix he reported having found a few individuals with single large nuclei. No cell boundaries were seen, and the nuclei varied in size. The outer layer of the body does not have an epithelial structure; and perhaps the evidence for the nuclear nature of the inclusions is inconclusive.

The parasites can easily be induced to leave the host. Borgert found it sufficient to put *Sticholonche* in a small amount of water on a slide, when escape was apparently stimulated by the increase in salinity and possibly in temperature. At the beginning of the transformation to the free stage, the tip of the conical part breaks through the surrounding sphere, and cilia appear and become active. The entire body, having become everted, emerges and swims actively in the water. Its form is elongated and more or less cylindrical, and it possesses a spiral furrow in which arise abundant small cilia (Fig. 226C). In the interior is a cavity, larger in younger specimens, reduced to a tubular form in older ones, which sometimes is open at the posterior end of the body.

Amoebophrya acanthometrae was found in four acanthometrids by Hertwig, in two others by Haeckel, and in a seventh species by Borgert. Borgert stated the probability that the parasite will be found to occur in all acanthometrids the skeletal structure of which permits. He observed it only in uninucleate phases of the host. In 1895 at Naples, after *Amoebophrya* disappeared from *Sticholonche,* parasites were found repeatedly in acanthometrids.

Unlike the other species, *A. acanthometrae* occurs within the central capsule. According to Borgert, it encloses the nucleus of the radiolarian; but this fact does not discommode the latter. Nuclei are extraordinarily small (up to 1 to 2 μ), and were not observed at all in some, especially young, specimens. Emergence of the free phase, which is so easy to observe in *A. sticholonchae,* happens only occasionally. Apparently the nucleus of the host is removed in this process. The free form has a plumper figure than that from *Sticholonche,* and the cilia are better developed.

There remain to be considered the groups of small spherules which occur usually in specimens of *Sticholonche* without *Amoebophrya,* though sometimes, contrary to the opinion of Fol (1883), the two are found in a single host. Younger stages, according to Borgert, consist of a spherical protoplasmic mass with a few spherical nuclei. A great number of small nuclei result from division of these. Eventually these nuclei become the center of vesicles, which become free in the host cytoplasm by dissolution of the earlier common plasma mass. Borgert regarded these bodies as parasites of the radiolaria, unrelated to *Amoebophrya*. Though probably it is only an analogy, certain features in their development suggest the life history of *Sphaerita*.

It appears from statements by Neresheimer (1904, 1908) that *Amoebophrya* is not restricted entirely to Radiolaria. Doflein, he wrote, showed him preparations of *Noctiluca miliaris* in which the parasite was present.

METAZOA

A number of rotifers live attached to other animals as ectocommensals or ectoparasites. Ehrenberg (1838) found *Proales petromyzon* (Ehrbg.) and its eggs attached to the branched vorticellids *Epistylis digitalis, Carchesium polypinum,* and *Zoothamnium geniculatum,* and stated that it devours the vorticellid. Wesenberg-Lund (1929) showed it and its eggs on *Zoothamnium*. It is a predator rather than a parasite, but differs from ordinary predators in its attachment. Hudson and Gosse (1889), however, found it always free, though often in close association with *Epistylis* and *Carchesium*.

Approaching closer to parasitism are certain species of *Proales*, which live in certain algae and Protozoa. *P. werneckii* (Ehrbg.) occurs rather commonly in galls on *Vaucheria; P. parasita* (Ehrbg.) is parasitic in *Volvox;* and *P. latrunculus* Penard invades the heliozoan *Acanthocystis turfacea*.

Proales parasita was found by Ehrenberg (1838) and Cohn (1858) in *Volvox*. Plate (1886) described *Hertwigia volvocicola* from *Volvox globator,* considering this to be a different species from the preceding. It is listed as a synonym of *P. parasita* by Hudson and Gosse (1889), but Wesenberg-Lund (1929) considered it again as a separate species. Whether different or not, the habits of the forms are the same. They swim about within *Volvox* coenobia and feed on the cells. The males live only a day or two, remaining entirely within the host in which they are hatched. The females may be found within or outside the coenobium. Eggs are laid in the host, where they hatch and, according to Hudson and Gosse, the young rotifer either enters a daughter colony and is expelled with it or emerges to swim free. Hudson and Gosse stated that "*Volvox* appears to suffer little from the depredations of its ungrateful guest."

Penard (1904, 1908-9) has given the most complete account of *Proales latrunculus,* certain observations on which had been made by Archer (1869), Leidy (1879), and Stokes (1884). Penard studied an epidemic outbreak of the parasite, which eventually carried off most of a group of the Heliozoa. He stated, however, that it is rare, in the sense that many *Acanthocystis turfacea* in various localities may be examined without encountering it. It is widespread geographically, as indicated by the records from Switzerland, England, and the United States.

After being introduced into the body of *Acanthocystis,* probably, according to Penard, by being ingulfed as prey, the rotifer moves about actively in the cytoplasm. It devours the zoöchlorellae and the substance of the heliozoan. In two or three hours an egg may be laid, after which the rotifer may continue to feed and lay a second, smaller egg. The heliozoan occasionally frees itself of the invader, but usually it perishes before the end of the first day. After laying its eggs, the rotifer escapes by an orifice in the then empty envelope of spicules—empty, that is, except for the few small eggs. The young rotifers develop rapidly, hatching in two or three days, when they leave by the orifice through which the parent escaped.

As Penard (1908-9) remarked, these rotifers are not true parasites, as they are not adapted to continuous existence in their host. They behave rather as predaceous forms which consume the host from within. One notes a marked specificity to certain hosts or related hosts in the rotifers ectozoic on colonial vorticellids and those endozoic in *Vaucheria, Volvox,* and *Acanthocystis.*

Ehrenberg (1888) on one occasion found the usually ectozoic *P. petromyzon* within *Volvox globator;* and Wesenberg-Lund (1929) stated that *Volvox* contained also species of *Diglena,* rotifers that are naturally free-living.

Living nematode worms have occasionally been encountered in Protozoa. It is not known whether this ever represents obligatory parasitism, or is only an invasion by accidentally or facultatively parasitic forms. Wesenberg-Lund (1929) stated that free-living nematodes have been found in *Volvox;* and Schubotz (1908) wrote that Hartmann informed him of having seen nematodes in that flagellate. Schubotz found as many as three nematodes in approximately a tenth of *Pycnothrix monocystoides* from *Procavia capensis.* He stated that for entry into this large ciliate, the worms use openings in the ectosarc or, in undamaged animals, the excretion pore. They are then found wholly or partly in the canal system, whose walls they at times break through. Myers (1938) found nematode worms in the foraminiferan *Rotalia turbinata* in an incidence, in the colder months of the year, of 5 percent.

LITERATURE CITED

Alexeieff, A. 1912. Sur un Chlamydozoaire parasite des Protozoaires. Sur le Chlamydozoaire du cancer. Arch. zool. exp. gén., (5) 10: N. et R., 101-10.

—— 1929. Matériaux pour servir à l'étude des protistes coprozoites. Arch. zool. exp. gén., 68: 609-98.

Archer, W. 1869. On some freshwater Rhizopoda, new or little known. Quart. J. micr. Sci., n. s., 9: 250-71.

Awerinzew, S. 1906. Die Süsswasser-Rhizopoden. Trav. Soc. Nat. St.-Pétersb. (Leningr.), 36: 1-351. (Russian, with German résumé.)

—— 1908. Studien über parasitische Protozoen. Trav. Soc. Nat. St.-Pétersb. (Leningr.), 38: v-xii, 1-139.

Bach, F. W., and P. Quest. 1923. Über Spirochäten im Darme von *Cyclostoma elegans* Drap. und ihre Beziehungen zu *Trichodinopsis paradoxa* Clap. Zbl. Bakt., 90: 457-60.

Bacigalupo, J. 1927. *Entameba coli* parasitada con *Sphaerita*. Rev. Soc. argent. Biol., 3: 694-98.

—— 1928. *Entamoeba coli* parasitée par une *Sphaerita*. C. R. Soc. Biol., Paris, 98: 170-71.

Balbiani, E. G. 1858a. Note relativ à l'existence d'une génération sexuelle chez les Infusoires. J. Physiol. Path. Gén., 1: 347-52.

—— 1858b. Recherches sur les organes générateurs et la reproduction des Infusoires. C. R. Acad. Sci. Paris, 47: 383-87. Translation in Ann. Mag. Nat. Hist. (3) 2: 439-43.

—— 1860. Note sur un cas de parasitisme improprement pris pour un mode de reproduction des Infusoires ciliés. C. R. Acad. Sci. Paris, 51: 319-22.

—— 1861. Recherches sur les phénomènes sexuels des Infusoires. Arch. physiol. norm. path., 4: 102-30; 194-220; 465-520.

—— 1893. Nouvelles Recherches expérimentales sur la mérotomie des Infusoires ciliés. Ann. micrographie, Paris, 5: 1-25; 49-84; 113-37.

Becker, E. 1926. *Endamoeba citelli* sp. nov. from the striped ground squirrel *Citellus tridecemlineatus,* and the life history of its parasite *Sphaerita endamoebae* sp. nov. Biol. Bull., 50: 444-54.

Bernstein, T. 1928. Untersuchungen an Flagellaten aus dem Darmkanal der Termiten aus Turkestan. Arch. Protistenk., 61: 9-37.

Boeck, W. C. 1917. Mitosis in *Giardia microti.* Univ. Cal. Publ. Zool., 18: 1-26.

Bonar, L. 1936. An unsual Ascomycete in the shells of marine animals. Univ. Cal. Publ. Bot. 19: 187-94.

Borgert, A. 1897. Beiträge zur Kenntniss der in *Sticholonche zanclea* und Acanthometridenarten vorkommenden Parasiten. Z. wiss. Zool., 63: 141-86.

Bourne, A. G. 1891. On *Pelomyxa viridis,* sp. n., and on the vesicular nature of protoplasm. Quart. Jour. micr. Sci., 32: 357-74.

Bozler, E. 1924. Über die Morphologie der Ernährungsorganelle und die Physiologie der Nahrungsaufnahme bei *Paramecium caudatum* Ehrbg. Arch. Protistenk., 49: 163-215.

Brug, S. L. 1926. *Nucleophage intestinalis* n. sp., parasiet der Kern van *Endolimax williamsi* (Prow.) = *Endolimax bütschlii* (Prow.) Dutch East Indies Volksgesundheid, 1926: 466-68.

Brumpt, E., and G. Lavier. 1935a. Sur une *Nucleophaga* parasite d'*Endolimax nana.* Ann. Parasit. hum. comp., 13: 439-44.

—— 1935b. Sur un genre nouveau d'amibe parasite *Hyalolimax* n. g. Ann. Parasit. hum. comp., 13: 551-58.

—— 1936. Sur l'hyperparasitisme d'Opalines par des amibes. Ann. Parasit. hum. comp., 14: 349-58.

Buchanan, R. E. 1925. General systematic bacteriology; history, nomenclature, groups of bacteria. Baltimore.

Buchner, P. 1930. Tier und Pflanze in Symbiose. Berlin.

Bütschli, O. 1876. Studien über die ersten Entwicklungsvorgänge der Eizelle, die Zelltheilung und die Conjugation der Infusorien. Abh. senckenb. naturf. Ges., 10: 213-464.

—— 1885. Einige Bemerkungen über gewisse Organisationsverhältnisse der sog. Cilioflagellaten und der *Noctiluca.* Morph. Jb., 10: 529-77.

—— 1889. Protozoa. III Abt. Infusoria und System der Radiolaria *in* Bronn: Klassen und Ordnungen des Thier-Reichs, 1. Leipzig.

Calkins, G. N. 1904. The life-history of *Cytoryctes variolae* Guarniere. Stud. Path. Etiol. Variola (Office J. med. Res.) : 136-72.

Campbell, A. S. 1926. The cytology of *Tintinnopsis nucula* (Fol) Laachmann, with an account of its neuromotor apparatus, division, and a new intranuclear parasite. Univ. Cal. Publ. Zool., 29: 179-236.

Carini, A. 1933. Parasitisme des Zellerielles par des Microorganismes nouveau (*Brumptina* n. g.) Ann. Parasit. hum. comp., 11: 297-300.

Carini, A., and E. Reichenow. 1935. Über Amöbeninfektion in Zelleriellen. Arch. Protistenk., 84: 175-85.

Carter, H. J. 1856. Notes on the freshwater infusoria of the island of Bombay. No. 1. Organization. Ann. Mag. Nat. Hist., (2)18: 115-32; 221-49.

—— 1857. Additional notes on the freshwater infusoria in the island of Bombay. Ann. Mag. Nat. Hist., (2)20: 34-41.

—— 1863. On *Amoeba principes* and its reproductive cells, etc. Ann. Mag. Nat. Hist., (3)12: 30-54.

Casagrandi, O., and P. Barbagallo. 1897. *Entamoeba hominis* s. *Amoeba coli* (Lösch). Studio biologicoe clinico. Ann. Igiene (sper.), 7: 1-64.

Caullery, M., and F. Mesnil. 1897. Sun un type nouveau (*Metchnikovella* n. g.) d'organismes parasites des grégarines. C. R. Soc. Biol. Paris, 49: 960-62.

—— 1905. Recherches sur les Haplosporidies. Arch. zool. exp. gén., (4)4: 101-81.

—— 1914. Sur les Metchnikovellidae et autres Protistes parasites des grégarines d'Annélides. C. R. Soc. Biol. Paris, 77: 527-32.

—— 1919. Metchnikovellidae et autres parasites des Grégarines d'Annélides. Ann. Inst. Pasteur, 33: 209-40.

Cavallini, F. 1926a. The asexual cycle in *Centropyxis aculeata* and its variability in relation to heredity and environment. J. exp. Zool., 43: 225-43.

—— 1926b. The asexual cycle of development in *Arcella vulgaris*. J. exp. Zool., 43: 245-55.

Cejp, K. 1935. *Sphaerita,* parasit Paramecií. Příspěvek k poznání houbových parasitů Protozoí. Spisy Přírodovědeckou Fakultou Karlovy University, Praha (Publ. Fac. Sci. Univ. Charles), 141: 3-7.

Cépède, C., and V. Willem. 1912. Observations sur *Trichodinopsis paradoxa*. Bull. Sci. Fr. Belg., 45: 239-48.

Chatton, É. 1910. Protozoaires parasites des branchies des Labres: *Amoeba mucicola* Chatton, *Trichodina labrorum* n. sp. Appendice: parasite des Trichodines. Arch. zool. exp. gén., (5)5: 239-66.

—— 1913. *Coccidiascus* n. g., n. sp., levure ascosporée parasite des cellules intestinales de *Drosophila funebris* Fabr. C. R. Soc. Biol. Paris, 75: 117-20.

—— 1914. Les Cnidocystes du Péridinien *Polykrikos schwartzi* Bütschli. Arch. zool. exp. gén., 54: 157-94.

—— 1920a. Les Péridiniens parasites. Morphologie, reproduction, ethologie. Arch. zool. exp. gén., 59 1-475.

—— 1920b. Existence chez les Radiolaires de Péridiniens parasites considerés comme formes de reproduction de leurs hôtes. C. R. Acad. Sci. Paris, 170: 413-15.

—— 1923. Les Péridiniens parasites des Radiolaires. C. R. Acad. Sci. Paris, 177: 1246-49.

—— 1934. L'Origine péridinienne des Radiolaires et l'interprétation parasitaire de l'anisosporogénèse. C. R. Acad. Sci. Paris, 198: 309-12.

Chatton, É., and B. Biecheler. 1934. Les Coccidinidae, Dinoflagelles coccidiomorphes parasite de Dinoflagelles, et le phylum des Phytodinozoa. C. R. Acad. Sci. Paris, 199: 252-55.

—— 1936. Documents nouveau relatifs aux Coccidinides (Dinoflagelles parasites). La sexualite du *Coccidinium mesnili* n. sp. C. R. Acad. Sci. Paris, 203: 573-76.

Chatton, É., and S. Brachon. 1936. Sur un Protiste parasite du Cilié *Fabrea salina* Henneguy: *Gregarella fabrearum* n. gen., n. sp., et son evolution. C. R. Acad. Sci. Paris, 203: 525-27.

Chatton, É., and A. Brodsky. 1909. Le Parasitisme d'une Chytridinée du genre *Sphaerita* Dangeard chez *Amoeba limax* Dujard. Étude comparative. Arch. Protistenk., 17: 1-18.

Chatton, É., and A. Lwoff. 1924a. Sur un flagellé hypertrophique et palintomique parasite des Infusoires marins: *Sporomonas infusorium* (n. g., n. sp.) C. R. Soc. Biol. Paris, 91: 186-90.

—— 1924b. Sur l'évolution des Infusoires des Lamellibranches: Morphologie comparée des Hypocomidés. Les nouveaux genres *Hypocomina* et *Hypocomella*. C. R. Acad. Sci. Paris, 178: 1928-30.

—— 1927. *Pottsia infusoriorum* n. g., n. sp., Acinétien parasite des Folliculines et des Cothurnies. Bull. Inst. océanogr. Monaco, 489: 1-12.

—— 1929. Contribution a l'étude de l'adaptation. *Ellobiophrya donacis* Ch. et Lw. Péritriche vivant sur les branchies de l'Acephale *Donax vittatus* da Costa. Bull. biol., 63: 321-49.

—— 1935. Les Ciliés apostomes. Morphologie, cytologie, éthologie, évolution, systématique. Première partie: Aperçu historique et général. Étude monographique des genres et des espèces. Arch. zool. exp. gén., 77: 1-453.

Chatton, É., A. Lwoff, and M. Lwoff. 1930. Les *Phtorophrya* n. g., Ciliés Foettingeriidae, hyperparasites des Gymnodinioides, Foettingeriidae parasites des Crustaces. C. R. Acad. Sci. Paris, 190: 1152-54.

Chatton, É., and S. Villeneuve. 1937. *Gregarella fabrearum* Chatton et Brachon Protiste parasite du Cilié *Fabrea salina* Henneguy. La Notion de dépolarisation chez les Flagellés et la conception des Apomastigines. Arch. zool. exp. gén., 78: N. et R., 216-37.

Chen, T. T., and R. M. Stabler. 1935. Further studies on the amoebae parasitic in opalinid ciliate protozoans. (Abstract). J. Parasit., 21: 428-29.

—— 1936. Further studies on the endamoeba parasitizing opalinid ciliates. Biol. Bull., 70: 72-77.

Cienkowsky, L. 1855a. Bemerkungen über Stein's Acinetenlehre. Bull. Acad. Imp. Sci. St.-Pétersb., phys-math., 13: 297-304. Also Quart. J. micr. Sci., 5: 96-103 (1857).

—— 1855b. Über Cystenbildung bei Infusorien. Z. wiss. Zool., 6: 301-6.

—— 1865. Beiträge zur Kenntnis der Monaden. Arch. mikr. Anat. 1: 203-32.

Claparéde, É., and J. Lachmann. 1857. Note sur la reproduction des Infusoires. Ann. Sci. nat., Zool., 4(8): 221-44.

—— 1858-59. Études sur les Infusoires et les Rhizopodes. vol. 1, Genève. Mém. Inst. nat. genev., 5, 6.

—— 1860-61. Études sur les Infusoires et les Rhizopodes. vol. 2, Genève. Mém. Inst. nat. genev., 7: 1-291.

Cleveland, L. R. 1928. Further observations and experiments on the symbiosis between termites and their intestinal Protozoa. Biol. Bull., 54: 231-37.

Cleveland, L. R., S. R. Hall, E. P. Sanders, and J. Collier. 1934. The wood-feeding roach *Cryptocercus*, its Protozoa, and the symbiosis between Protozoa and roach. Mem. Amer. Acad. Arts Sci., 17: i-x, 185-342.

Cohn, F. 1851. Beiträge zur Entwickelungsgeschichte der Infusorien. Z. wiss. Zool., 3: 257-79.

—— 1857. Über Fortpflanzung von *Nassula elegans* Ehr. Z. wiss. Zool., 9: 143-46.

—— 1858. Bemerkungen über Räderthiere. Z. wiss. Zool., 9: 284-94.

Collin, B. 1907. Note préliminaire sur quelques Acinétiens. Arch. zool. exp. gén. (4)7: N. et R., 93-103.

—— 1912. Étude monographique sur les Acinétiens. II. Morphologie, Physiologie, Systématique. Arch. zool. exp. gén., 51: 1-457.

Connell, F. H. 1930. The morphology and life cycle of *Oxymonas dimorpha* sp. nov., from *Neotermes simplicicornis* Banks. Univ. Cal. Publ. Zool., 36: 51-66.

——1932. *Gigantomonas lighti* sp. nov., a trichomonad flagellate from *Kalotermes (Paraneotermes) simplicicornis* Banks. Univ. Cal. Publ. Zool., 37: 153-88.

Cook, W. R. I. 1935. The genus *Lagenidium* Schenk, with special reference to *L. Rabenhorstii* Zopf and *L. entophytum* Zopf. Arch. Protistenk., 86: 58-89.

Cragg, F. W. 1919. A contribution to our knowledge of *Entamoeba coli*. Indian J. med. Res., 6: 462-84.

Craig, C. F. 1911. The parasitic amoebae of man. Philadelphia.

Crouch, H. B. 1933. Four new species of *Trichomonas* from the woodchuck (*Marmota monax* Linn.). J. Parasit., 19: 293-301.

Cunha, A. M. da, and J. Muniz. 1923. Parasitismo de *"Trichomonas"* pro "Chytridacae" do genero *"Sphaerita"* Dangeard. Brazil-med., 37: 19-20.

—— 1934. Observations sur un parasite des Flagellés du genre *Chilomastix*. C. R. Soc. Biol. Paris, 117: 208-10.

Dangeard, P. A. 1886a. Sur un nouveau genre de Chytridinés parasites des Rhizopodes et des Flagellates. Bull. Soc. bot. Fr., 33: 240-42.

—— 1886b. Recherches sur les organismes inférieurs. Ann. Sci. nat., (7) Bot., 4: 241-341.

—— 1889a. Recherches sur les Cryptomonadinae et les Euglenae Botaniste, 1: 1-38.

—— 1889b. Mémoire sur les Chytridinées. Botaniste, 1: 39-74.

—— 1895. Mémoire sur les parasites du noyau et du protoplasma. Botaniste, 4: 199-248.

—— 1896. Contribution a l'étude des Acrasiées. Le Botaniste, 5: 1-20.

—— 1900a. L'Organisation el le développement du *Colpodella pugnax*. Botaniste, 7: 5-31.

—— 1900b. Recherches sur la structure du *Polyphagus euglenae* Nowak. et sa reproduction sexuelle. Botaniste, 7: 213-61

—— 1902. Sur le caryophysème des Eugléniens. C. R. Acad. Sci. Paris, 134: 1365-66.

—— 1908. Sur un nouveau genre, parasite de Chrysomonadinées, le *Lecythodytes paradoxus*. C. R. Acad. Sci. Paris, 146: 1159-60.

—— 1910. Études sur le développement et la structure des organismes inférieurs. Botaniste, 11: 1-311.

Deflandre, G. 1928. Le Genre *Arcella* Ehrenberg. Morphologie- Biologie. Essai phylogénétique et systématique. Arch. Protistenk., 64: 152-287.

Dobell, C. 1919. The amoebae living in man. London.

Doflein, F. 1907. Studien zur Naturgeschichte der Protozoen. V. Amöbenstudien, Erster Teil. Arch. Protistenk., Suppl. 1: 250-93.

—— 1909. Lehrbuch der Protozoenkunde, 2d ed., Jena. Also 3d ed., 1911; 4th ed., 1916; 5th ed. (F. Doflein, and E. Reichenow), 1927-29.

Dogiel, V. 1906. Beiträge zur Kenntnis der Gregarinen. I. *Cystobia chiridotae* nov. sp. II. *Hyalosphaera gregarinicola* nov. gen. nova spec. Arch. Protistenk., 7: 106-30.

—— 1916. Researches on the parasitic Protozoa from the intestine of termites. I. Tetramitidae. Russk. zool. Zh., 1: 1-54.

—— 1917. Researches on the parasitic protozoa from the intestine of termites. II. Lophomonadidae. Sci. Res. Zool. Exped. Brit. E. Africa made by V. Dogiel and I. Sokolow in 1914, no. 10.

—— 1922a. Untersuchungen an parasitischen Protozoan aus dem Darmkanal der Termiten. III. Trichonymphidae. Arch. russ. Protist., 1: 172-234.

—— 1922b. Sur un nouveau genre de Metchnikovellidae. Ann. Inst. Pasteur, 36: 574-77.

—— 1929. Biologische Notizen über Darminfusorien der Huftiere. Arch. russ. protist., 8: 153-62.

Douvillé, N. 1930. Parasitisme ou commensalisme chez les Foraminifères. Les canaux chez les Nummulitides. Soc. geol. France, Livre Jubilaire 1830-1930, 1: 257-62.

Duboscq, O., and B. Collin. 1910. Sur la reproduction sexuée d'un Protiste parasite des Tintinnides. C. R. Acad. Sci. Paris, 151: 340-41.

Duboscq, O., and Grassé, P. 1925. Notes sur les Protistes parasites des Termites de France. IV. Appareil de Golgi, mitochondries et vesicules sous-flagellaires de *Pyrsonympha vertens* Leidy. C. R. Soc. Biol. Paris, 93: 345-48.

—— 1926. Les Schizophytes de *Devescovina hilli* n. sp. C. R. Soc. Biol. Paris, 94: 33-34.

—— 1927. Flagellés et Schizophytes de *Calotermes* (*Glytotermes*) *iridipennis* Frogg. Arch. Zool. exp. gén., 66: 452-96.

—— 1928. Note sur les Protistes parasites des termites de France. L'appareil parabasal de *Joenia annectens Grassi*. C. R. Soc. Biol., 99: 1118-20.

—— 1933. L'appareil parabasal des Flagellés. Arch zool. exp. gén., 73: 381-621.

—— 1934. Notes sur les Protistes parasites des Termites de France. VII. Sur les *Trimitus de Calotermes flavicollis* Rossi. VIII. Sur *Microrhopalodina inflata* (Grassi). Arch. zool. exp. gén., 75: 615-37.

—— 1937. Les Flagellés de l'*Acanthotermes ochraceus* Sjöst. du Sud-Algérien. C. R. Acad. Sci. Paris, 205: 574-76.

Duboscq, O., P.-P. Grassé, and M. Rose. 1937. Les Flagellés de l'*Anacanthotermes ochraceus* Sjöst. du Sud-Algérien. C. R. Acad. Sci. Paris, 205: 574-76.

Ehrenberg, C. G. 1838. Die Infusionsthierchen als vollkommene Organismen. Leipzig.

Elmassian, M. 1909. Une Nouvelle Coccidie et un nouveau parasite de la tanche, *Coccidium rouxi* nov. spec., *Zoomyxa legeri* nov. gen., nov. spec. Arch. zool. exp. gén., (5)2: 229-70.

Elpatiewsky, W. 1907. Zur Fortpflanzung von *Arcella vulgaris* Ehrbg. Arch. Protistenk., 10: 441-66.

Engelmann, T. W. 1862. Zur Naturgeschichte der Infusionsthiere. Z. wiss. Zool., 11: 347-93.

—— 1876. Über Entwickelung und Fortpflanzung von Infusorien. Morph. Jb., 1: 573-635.

Epstein, H. 1922. Über parasitische Infektion bei Darmamöben. Arch. russ. protist., 1: 46-81.

—— 1935. Bacterial infection in an amoeba. J. R. micr. Soc., 55: 86-94.

Fauré-Fremiet, E. 1909. Sur un cas de symbiose presente par un Infusoire cilié. C. R. Soc. Biol. Paris, 67: 113-14.

—— 1913. Sur les "Nématocystes" de *Polykrikos* et de *Campanella*. C. R. Soc. Biol. Paris, 75: 366-68.

Fitzpatrick, H. M. 1930. The lower fungi. Phycomycetes. New York.

Fiveiskaja, A. 1929. Einfluss der Kernparasiten der Infusorien auf den Stoffwechsel. Arch. Protistenk., 65: 275-98.

Foà, A. 1905. Due nuovi Flagellati parassiti. R. C. Accad. Lincei, (5)14: 542-46.

Focke, G. W. 1845. Ergebnisse ferneren Untersuchungen dere polygastrischen Infusorien. Amtlicher Bericht über die zweiundzwanzigste Versammlung deutscher Naturforscher und Ärztte in Bremen im September 1844. Verh. Ges. deutscher Naturf. Ärztte, 22: 109-10.

Fol, H. 1883. Sur le *Sticholonche zanclea* et un nouvel ordre de Rhizopodes. Mem. Inst. nat. genev., Nr. 15.

Fortner, H. 1934. Untersuchungen an *Pelomyxa palustris* Greeff. Studien zur Biologie und Physiologie des Tieres. I. Teil. Arch. Protistenk., 83: 381-464.

França, C. 1918. Observations sur les Trichonymphides. An. Fac. Med. Porto, (4) 2: 5-14.

Ganapati, P. N., and R. G. Aiyar. 1937. Life-history of a dicystid gregarine, *Lecudina brasili* n. sp., parasitic in the gut of *Lumbriconereis* sp. Arch. Protistenk., 89: 113-32.

Gassovsky, G. 1919. On the microfauna of the intestine of the horse. Trav. Soc. Nat. St.-Pétersbg. (Leningr.), 49: 20-37; 65-69.

Geitler, L. 1937. Über einen Pilzparasiten auf *Amoeba proteus* und über die polare Organization des Amöbenkörpers. Biol. Zbl., 57: 166-75.

Georgevitch, J. 1929. Sur la faune intestinale des Termites de Yougoslavie. C. R. Soc. Biol. Paris, 103: 325-28.

—— 1932. Recherches sur les Flagellés des Termites de Yougoslavie. Arch. zool. exp. gén., 74: 81-109.

—— 1936a. [Studies on a hyperparasite *Leismania esocis* nov. spec.] (In Serbian.) Ghlas Srpska Kralj. Akad. Belgrade, 172: 127-37.

—— 1936b. Ein neuer Hyperparasit, *Leishmania esocis* nov. spec. Arch. Protistenk., 88: 90-92.

Goetsch, W., and L. Scheuring. 1926. Parasitismus und Symbiose der Algengattung *Chlorella*. Z. Morph. Ökol. Tiere, 7: 220-53.

Gojdics, M. 1939. Some observations on *Euglena sanguinea* Ehrbg. Trans. Amer. micr. Soc., 58: 241-48.

Goldschmidt, R. 1907. Lebensgeschichte der Mastigamöben, *Mastigella vitrea* n. sp. und *Mastigina setosa* n. sp. Arch. Protistenk., (Suppl.) 1: 83-168.

Gonder, R. 1910. Ein Parasit von *Colpoda cucullus*. Arch. Protistenk., 18: 275-77.

Gönnert, R. 1935. Über Systematik, Morphologie, Entwicklungsgeschichte und Parasiten einiger Dendrosomidae nebst Beschreibung zweier neuer Suktorien. Arch. Protistenk., 86: 113-54.

Gould, L. J. 1894. Notes on the minute structure of *Pelomyxa palustris* (Greef). Quart. J. micr. Sci., 36: 295-306.

Grassé, P. P. 1926a. Sur la nature des côtes cuticulaires des *Polymastix* et *Lophomonas striata*. C. R. Soc. Biol. Paris, 94: 1014-15.

—— 1926b. Contribution a l'étude des Flagellés parasites. Arch. zool. exp. gén., 65: 345-602.

—— 1937. Sur un Flagellé termiticole "*Caduceia theobromae*" França. C. R. xiie Cong. Int. Zool.: 1324-29.

—— 1938. La Vêture schizophytique des Flagellés termiticoles: *Parajoenia*, *Caduceia* et *Pseudodevescovina*. Bull. Soc. zool. Fr., 63: 110-22.

Grassi, B., and Foà, A. 1904. Ricerche sulla riproduzione dei Flagellati. I. Processo di divisione delle Joenie e forme affini. Nota preliminare. R. C. Accad. Lincei, (5) 13, 2 sem.: 241-53.

—— 1911. Intorno ai Protozoi dei Termitidi. Nota preliminare. R. C. Accad. Lincei, (5) 20, 1 sem.: 725-41.

Greeff, R. 1866. Über einige in der Erde lebende Amöben und andere Rhizo-poden. Arch. mikr. Anat., 2: 299-311.

—— 1870. Unterschungen über den Bau und die Naturgeschichte der Vorticellen. Arch. Naturgesch., 1: 353-84.

—— 1874. *Pelomyxa palustris* (Pelobius), ein amöbenartiger Organismus des süssen Wassers. Arch. mikr. Anat., 10: 51-73.

Greenway, D. 1926. *Endolimax nana.* Arch. argent Enferm. Apar. dig., p. 174.

Gruber, A. 1884. Die Protozoen des Hafens von Genua. Nova Acta K. Leop. Carol., 46: 475-539.

—— 1892. Eine Mittheilung über Kernvermehrung und Schwärmerbildung bei Süsswasserrhizopoden. Ber. naturf. Ges. Freiburg i. B., 6: 114-18.

—— 1904. Über *Amoeba viridis* Leidy. Zool. Jb., (Suppl.) 7 (Festschr. Weissmann) : 67-76.

Günther, F. 1928. Über den Bau und die Lebensweise der Euglenen, besonders der Arten E. *terricola, geniculata, proxima, sanguinea* und *lucens* nov. spec. Arch. Protistenk., 60: 511-90.

Gwynne-Vaughan, H. C. I., and B. Barnes. 1937. The structure and develop-ment of the fungi. 2d ed., Cambridge.

Hafkine, M. W. 1890. Maladies infectieuses des Paramecies. Ann. Inst. Pasteur, 4: 148-62.

Hertwig, R. 1876. Über *Podophrya gemmipara* nebst Bemerkungen zum Bau und zur systematischen Stellung der Acineten. Gegenbaurs Jb., 1: 20-82.

—— 1879. Der Organismus der Radiolarien. Jena.

Hesse, E. 1909. Contribution à l'étude des Monocystidées des Oligochètes. Arch. zool. exp. gén., (5) 3: 27-301.

Hetherington, A. 1932. On the absence of physiological regeneration in *Stentor coeruleus.* Arch. Protistenk., 77: 58-63.

Hoelling, A. 1910. Die Kernverhältnisse von *Fusiformis termitidis.* Arch. Protistenk, 19: 239-45.

Hofker, J. 1927. The Foraminifera. Siboga Expedition. Leiden.

—— 1931. Studie über Tintinnoidea. Arch. Protistenk., 75: 315-402.

Hollande, A. 1938. *Bodo perforans* n. sp. Flagellé nouveau parasite externe du *Chilomonas paramaecium* Ehrenb. Arch. zool. exp. gén., 79: N. et R., 75-81.

Hovasse, R. 1923a. Les Peridiniens intracellulaire — zooxanthelles et *Syn-diniums* — chez les Radiolaires coloniaux. Remarques sur la reproduction des Radiolaires. Bull. Soc. zool. Fr., 48: 247-54.

—— 1923b. Sur les Peridiniens parasites des Radiolaires coloniaux. Bull. Soc. zool. Fr., 48: 337.

—— 1936. *Rhizophidium Beauchampi* sp. nov., Chytridinée parasite de la Volvocinée *Eudorina (Pleodorina) illinoisensis* (Kofoid). Ann. Protist., 5: 73-81.

Howland, R. B. 1928. A note on *Astasia captive* Beauch. Science, 68: 37.

Hsiung, Ta-Shih. 1928. Suctoria of the large intestine of the horse: *Allantosoma intestinalis* Gassovsky, *A. discorniger* sp. nov. and *A. brevicorniger* sp. nov. Iowa St. Coll. J. Sci., 3: 101-3.

Hudson, C. T., and P. H. Gosse. 1889. The rotifera; or wheel-animalcules, both British and foreign. London.

Issel, R. 1906. Intorno alla struttura ed alla biologia dell'infusorio *Trichodinopsis paradoxa* Clap. et Lachm. Ann. Mus. Stor. nat. Genova, (3)2: 334-57.

Ivanić, M. 1925. Zur Kenntnis der Agamogonieperiode einiger Amoebenparasiten. Zool. Anz., 63: 250-56.

—— 1934. Über einen Kernparasiten der roten Blutzellen beim grünen Frosche (*Rana esculenta* L.), *Erythrocytonucleophaga ranae* gen. nov., sp. nov. Zbl., Bakt., I Abt., 113: 1-6.

—— 1936. Über einen Protoplasmakörperparasiten von *Entamoeba histolytica* Schaudinn (*Entamoebophaga hominis* gen. nov. spec. nov.). Zbl. Bakt., I Abt., 138: 48-56.

Jahn, T. L. 1929. On certain parasites of *Phacus* and *Euglena* (Abstract). Anat. Rec., 44: 249-50.

—— 1933. On certain parasites of *Phacus* and *Euglena; Sphaerita phaci*, sp. nov. Arch. Protistenk., 79: 349-55.

Janicki, C. 1909. Über Kern und Kernteilung bei *Entamoeba blattae* Bütschli. Biol. Zbl., 29: 381-93.

—— 1915. Untersuchungen an parasitischen Flagellaten. II Teil: Die Gattungen *Devescovina, Parajoenia, Stephanonympha, Calonympha*. Ueber den Parabasalapparat. Über Kernkonstitution und Kernteilung. Z. wiss. Zool., 112: 573-691.

Jírovec, O. 1931a. Die Silberlinien bei den Pyrsomymphiden. Arch. Protistenk., 73: 47-55.

—— 1931b. Symbiose von Bakterien und *Trichonympha serbica*. Zbl. Bakt., I Abt., 123: 184-86.

—— 1933. Beobachtungen Über die Fauna des Rinderpansens. Z. Parasitenk., 5: 584-91.

Kahl, A. 1928. Die Infusorien (Ciliata) der Oldesloer Salzwasserstellen. Arch. Hydrobiol., 19: 50-123.

—— 1930. Urtiere oder Protozoa I: Wimpertiere oder Ciliata (Infusoria) I. Allgemeiner Teil und Prostomata, *in* Dahl: Die Tierwelt Deutschlands, 18 Teil, Jena.

—— 1931. *Ibid.*, II. Holotricha, 21 Teil.

—— 1932. *Ibid.*, III. Spirotricha, 25 Teil.

—— 1933. Ciliata libera et ectocommensalia, *in* Grimpe, G., und E. Wagler: Die Tierwelt der Nord-und Ostsee Lief. 23, Teil II. C3, 29-146, Leipzig.

—— 1935. *Ibid.*, IV. Peritricha und Chonotricha, 30 Teil.

Kalmus, H. 1928. Über den Bodenfauna der Moldau im Gebiete von Prag.

Ein Jahreszyklus. II. Protozoa, etc. Mit einem Anhang: Ökologische Beobachtungen und Versuche. Int. Rev. Hydrobiol., 19: 349-429.

Kent, W. S. 1880-82. A manual of the Infusoria. London.

Keppen, N. A., see Koeppen, N.

Kessel, J. F. 1923. On the genus *Councilmania*, budding intestinal amoebae parasitic in man and rodents. Univ. Cal. Pub. Zool., 20: 431-45.

—— 1924. The distinguishing characteristics of the parasitic amoebae of culture rats and mice. Univ. Cal. Publ. Zool., 20: 489-544.

Kirby, H., Jr. 1924. Morphology and mitosis of *Dinenympha fimbriata* sp. nov. Univ. Cal. Publ. Zool., 26: 199-220.

—— 1926a. On *Staurojoenina assimilis* sp. nov., an intestinal flagellate from the termite, *Kalotermes minor* Hagen. Univ. Cal. Publ. Zool., 29: 25-102.

—— 1926b. The intestinal flagellates of the termite, *Cryptotermes hermsi* Kirby. Univ. Cal. Publ. Zool., 29: 103-20.

—— 1927. Studies on some amoebae from the termite *Mirotermes,* with notes on some other Protozoa from the Termitidae. Quart. J. micr., Sci., 71: 189-222.

—— 1928. A species of *Proboscidiella* from *Kalotermes (Cryptotermes) dudleyi* Banks, a termite of Central America, with remarks on the oxymonad flagellates. Quart. J. micr. Sci., 72: 355-86.

—— 1929. *Snyderella* and *Coronympha,* two new genera of multinucleate flagellates from termites. Univ. Cal. Publ. Zool., 31: 417-32.

—— 1930. Trichomonad flagellates from termites I. *Tricercomitus* gen. nov., and *Hexamastix* Alexeieff. Univ. Cal. Publ. Zool., 33: 393-444.

—— 1932a. Flagellates of the genus *Trichonympha* in termites. Univ. Cal. Publ. Zool., 37: 349-476.

—— 1932b. Protozoa in termites of genus *Amitermes*. Parasitology, 24: 289-304.

—— 1934. Some ciliates from salt marshes in California. Arch. Protistenk., 82: 114-33.

—— 1936. Two polymastigote flagellates of the genra *Pseudodevescovina* and *Caduceia*. Quart. J. micr. Sci., 79: 309-35.

—— 1937. The devescovinid flagellate *Parajoenia grassii* from a Hawaiian termite. Univ. Cal. Publ. Zool., 41: 213-24.

—— 1938a. The devescovinid flagellates *Caduceia theobromae* França, *Pseudodevescovina ramosa* new species, and *Macrotrichomonas pulchra* Grassi. Univ. Cal. Publ. Zool., 43: 1-40.

—— 1938b. Polymastigote flagellates of the genus *Foaina* Janicki, and two new genera *Crucinympha* and *Bulanympha*. Quart. J. micr. Sci., 81: 1-25.

—— 1940. Microorganisms associated with the flagellates of termites. Pp. 407-8 *in* Third International Congress for Microbiology: Report of Proceedings. New York.

Klebs, G. 1883. Über die Organisation einiger Flagellaten-Gruppen und ihre Beziehungen zu Algen und Infusorien. Unters. Bot. Inst. Tübingen, 1: 233-62.

Koeppen, N. [Keppen, N. A.]. 1894. *Amoebophrya Sticholonchae* nov. gen. et sp. ("Corps spiral" de Fol) Zool. Anz., 17: 417-24.

——— 1889. *Hyalosaccus Ceratii,* nov. gen. et sp., parazit Dinoflagellat. Kiev. obshch. estest. Zap., 16: 89-135.

Kofoid, C. A., and O. Swezy. 1915. Mitosis and multiple fission in trichomonad flagellates. Proc. Amer. Acad. Arts. Sci., 51: 289-378.

Koidzumi, M. 1921. Studies on the intestinal Protozoa found in the termites of Japan. Parasitology, 13: 235-309.

Korotneff, A. 1880. Études sur les Rhizopodes. Arch. zool. exp. gén., (1), 8: 467-82.

——— 1891. Zoologische Paradoxen. Z. wiss. Zool., 51: 613-28.

Krüger, F. 1933. *Epistylis umbellaria* mit "Nesselkapseln." Zool. Anz. (Suppl.) 6. Verh. dtsch. zool. Ges., 35: 262-63.

Kuczynski, M. H. 1918. Über die Teilungsvorgänge verscheidener Trichomonaden und ihre Organization im allgemeinen. Arch. Protistenk., 39: 107-46.

Kudo, R. 1924. A biologic and taxonomic study of the microsporidia. III. Biol. Monogr., 9: 3-268.

Lachmann, C. F. J. 1856. Über die Organisation der Infusorien, besonders der Vorticellen. Arch. Anat. Physiol. wiss. Med., 1856: 340-98.

Lackey, J. B. 1936. Some fresh water Protozoa with blue chromatophores. Biol. Bull., 71: 492-97.

Lauterborn, R. 1916. Die sapropelische Lebewelt. Ein Beitrag zur Biologie des Faulschlammes natürlicher Gewässer. Verh. naturh.-med. Ver. Heidelberg, n. s., 13: 395-481.

Lavier, G. 1935a. Action, sur la biologie d'une Entamibe, due parasitism intranucléaire par une *Nucleophaga.* C. R. Soc. Biol. Paris, 118: 457-59.

——— 1935b. Sur une *Nucleophaga* parasite d'*Entamoeba ranarum.* Ann. Parasit. hum. comp., 13: 351-61.

Leger, L., and O. Duboscq. 1904. Notes sur les Infusoires endoparasites. Arch. zool. exp. gén., (4)2: 337-56.

——— 1909a. Sur une Microsporidie parasite d'une Grégarine. C. R. Acad. Sci. Paris, 148: 733-34.

——— 1909b. *Perezia lankesteriae* n. g., n. sp., Microsporidie parasite de *Lankesteria ascidiae* (Ray-Lank.). Arch. zool. exp. gén., (5)1: N. et R., 89-93.

——— 1909c. Études sur la sexualité chez les Grégarines. Arch. Protistenk., 17: 19-134.

Leidy, J. 1879. Fresh-water rhizopods of North America. U. S. Geol. Sur. Terr. Reps., 12: 1-324.

——— 1881. The parasites of the termites. J. Acad. nat. Sci. Philad., (2)8: 425-47.

Leiner, M. 1924. Das Glycogen in *Pelomyxa palustris* Greef, mit Beiträgen zur Kenntnis des Tieres. Arch. Protistenk., 47: 253-307.

Liebmann, H. 1936a. Auftreten, Verhalten und Bedeutung der Protozoen bei der Selbstreinigung stehenden Abwassers. Z. Hyg. InfektKr., 118: 29-63.

—— 1936b. Die Ciliatenfauna der Emscherbrunnen. Z. Hyg. InfektKr., 118: 555-73.

—— 1937. Bakteriensymbiose bei Faulschlammziliaten. Biol. Zbl., 57: 442-45.

Light, S. F. 1926. On *Metadevescovina debilis* gen. nov., sp. nov. Univ. Cal. Publ. Zool., 29: 141-57.

Lutz, A., and A. Splendore. 1908. Über Pebrine und verwandte Mikrosporidien. 2. Mitt. Zbl. Bakt., Abt. I, 46: 311-15.

Lwoff, A. 1925. Chytridinées parasites des Amibes de l'homme. Possibilité de leur utilisation comme moyen biologique de lutte contre la dysenterie amibienne. Bull. Soc. Path. exot., 18: 18-23.

Lynch, J. E., and A. E. Noble. 1931. Notes on the genus *Endosphaera* Engelman and on its occasional host *Opisthonecta henneguyi* Fauré-Fremiet. Univ. Cal. Publ. Zool., 36: 97-114.

Mackinnon, D. L. 1914. Observations on amoebae from the intestine of the crane-fly larva, *Tipula sp.* Arch. Protistenk., 32: 267-77.

Mackinnon, D. L., and H. N. Ray. 1931. Observations on dicystid gregarines from marine worms. Quart. J. micr. Sci., 74: 439-66.

Mainx, F. 1928. Beiträge zur Morphologie und Physiologie der Eugleninen. I. Teil Morphologische Beobachtungen, Methoden und Erfolge der Reinkultur. Arch. Protistenk., 60: 305-54.

Mangenot, G. 1934. Sur l'association d'une Rhodobactérie et d'un Infusoire. C. R. Soc. Biol. Paris, 117: 843-47.

Martin, C. H. 1909. Some observations on Acinetaria. I. The "Tinctinkörper" of Acinetaria and the conjugation of *Acineta papillifera*. II. The life-cycle of *Tachyblaston ephelotensis* (gen. et spec. nov.), with a possible identification of *Acinetopsis rara*, Robin. Quart. J. micr. Sci., 53: 351-89.

Mattes, O. 1924. Über Chytridineen im Plasma und Kern von *Amoeba sphaeronucleolus* und *Amoeba terricolo*. Arch. Protistenk., 47: 413-30.

Mayer, M. 1920. Zur Cystenbildung von *Trichomonas muris*. Arch. Protistenk., 40: 290-93.

Mecznikow, E., *see* Metschnikoff, É.

Mercier, L. 1907. Un parasite du noyau d'*Amobea blattae* Bütschli. C. R. Soc. Biol. Paris, 62: 1132-34. Also Réun. Biol., Nancy, 1907: 52-54.

—— 1910. Contribution a l'étude de l'Amibe de la Blatte (*Entamoeba blattae* Bütschli). Arch. Protistenk., 20: 143-75.

Metschnikoff, É [Mecznikow, E.]. 1864. Über die Gattung *Sphaerophrya*. Arch. Anat. Physiol. wiss. Med., 1864: 258-61.

—— 1892. Leçons sur la pathologie comparée de l'inflammation. Faite à l'Institut Pasteur en Avril et Mai, 1891. Paris.

Minden, M. von. 1915. Chytridiineae *in* Pilze I. Kryptogamenfl. Mark Brendenb., 5: 209-422.

Mitchell, J. B., Jr. 1928. Studies on the life history of a parasite of the Englenidae. Trans. Amer. micr. Soc., 47: 29-41.

Miyashita, Y. 1933. Drei neue parasitische Infusorien aus dem Darme einer Japanischen Süsswasseroligochate. Annot. zool. jap., 14: 127-31.

Molisch, H. 1903. Amoeben als Parasiten in *Volvox*. Ber. dtsch. bot. Ges., 21: 20-23.

Müller, J. 1856. Einige Beobachtungen an Infusorien. Monatsber. preuss. Akad. Wissensch., 1856: 389-93.

Myers, E. H. 1938. The present state of our knowledge concerning the life cycle of the Foraminifera. Proc. nat. Acad. Sci. Wash., 24: 10-17.

Nägler, K. 1910. Fakultativ parasitische Micrococcen in Amöben. Arch. Protistenk., 19: 246-54.

—— 1911a. Studien über Protozoen aus einem Almtümpel. I. *Amoeba hartmanni* n. sp. Anhang: Zur Centriolfrage. Arch. Protistenk., 22: 56-70.

—— 1911b. Studien über Protozoen aus einem Almtümpel. II. Parasitische Chytridiaceen in *Euglena sanguinea*. Arch. Protistenk., 23: 262-68.

Neresheimer, E. 1904. Über *Lohmannella catenata*. Z. wiss. Zool., 76: 137-66.

—— 1908. Die Mesozoen. Zool. Zbl., 15: 257-312.

Nöller, W. 1921. Über einige Wenig bekannte Darmprotozoen des Menschen und ihre nächsten Verwandten. Arch. Schiffs u. Tropenhyg., 25: 35-46.

—— 1922. Die wichtigsten parasitischen Protozoen des Menschen und der Tiere. I. Teil. Berlin.

Nowakowski, L. 1876. Beitrag zur Kenntnis der Chytridiaceen. II. *Polyphagus Euglenae*. Beitr. Biol. Pfl., ed. by Ferdinand Cohn, 2: 201-19.

Pascher, A. 1929. Studien über Symbiosen. I. Über einige Endosymbiosen von Blaualgen in Einzellern. Jb. wiss. Bot., 71: 386-462.

Pellissier, M. 1936. Sur certains constituants cytoplasmiques de l'Infusoire cilié *Trichodinopsis paradoxa* Clap. et Lach. Arch. zool. exp. gén., 78: N. et R., 32-36.

Penard, E. 1893. *Pelomyxa palustris* et quelques autres organismes inférieurs. Arch. Sci. phys. nat. (3), 29: 165-82.

—— 1902. Faune rhizopodique du bassin du Léman. Genève.

—— 1904. Héliozoaires d'eau douce. Genève.

—— 1905a. Sur les Sarcodinés du Loch Ness. Proc. roy. Soc. Edinb., 25: 593-608.

—— 1905b. Observations sur les Amibes à pellicule. Arch. Protistenk., 6: 175-206.

—— 1905c. Notes sur quelques Sarcodinés. Ire partie. Rev. suisse Zool., 13: 585-616.

—— 1908-9. Über ein bei *Acanthocystis turfacea* parasitisches Rotatorium. Mikrokosmos, 2: 135-43.

—— 1909. Sur quelques Mastigamibes des environs de Genève. Rev. suisse Zool., 17: 405-39.

—— 1912. Nouvelles Recherches sur les Amibes du groupe Terricola. Arch. Protistenk., 28: 78-140.

—— 1922. Études sur les Infusoires d'eau douce. Geneva.

Petschenko, B. de. 1911. *Drepanospira mülleri* n. g., n. sp. parasite des Parameciums; contribution a l'étude de la structure des Bacteries. Arch. Protistenk., 22: 248-98.

Pierantoni, O. 1936. La simbiosi fisiologica nei termitidi xilofagi e nei loro flagellati intestinali. Arch. zool. (ital.) Torino, 22: 135-71.

Pinto, C., and F. da Fonseca. 1926. *Trichomonas vitali* nova especie. Parasitismo das *Trichomonas* por *Sphaerita minor* Cunha et Muniz, 1923, e relacão das especies de Sphaeritas conhecidas. (In Portuguese). Bol. biol. Fac. Med. S. Paulo, 1926: 34-37.

Plate, L. 1886. Beiträge zur Naturgeschichte der Rotatorien. Jena. Z. Naturw., 19: 1-120.

—— 1887. Über einige ectoparasitische Rotatorien des Golfes von Neapel. Mitt. Zool. Sta. Neapel, 7: 234-63.

—— 1888. The genus *Acinetoides,* g. n., an intermediate form between the ciliated Infusoria and the Acinetae. Ann. Mag. Nat. Hist., (6) 2: 201-8. (Translation.) Original in Zool. Jahrb., Abt. Anat. Ont., 3: 135-43 (1888).

Powell, W. N. 1928. On the morphology of *Pyrsonympha* with a description of three new species from *Reticulitermes hesperus* Banks. Univ. Cal. Publ. Zool., 31: 179-200.

Powers, P. B. A. 1933. Studies on the ciliates from sea-urchins. I. General Taxonomy. Biol. Bull., 65: 106-21.

—— 1935. Studies on the ciliates of sea-urchins. A general survey of the infestations occurring in Tortugas echinoids. Publ. Carneg. Instn., 452: 293-326.

Prandtl, H. 1907. Der Entwicklungskreis von *Allogromia* sp. Arch. Protistenk., 9: 1-21.

Pringsheim, E. G. 1928. Physiologische Untersuchungen an *Paramecium bursaria.* Ein Beitrag zur Symbioseforschung. Arch. Protistenk., 64: 289-418.

Puymaly, A. de. 1927. Sur le *Sphaerita endogena* Dangeard, Chytridiaceé parasite des Euglènes. Bull. Soc. bot. Fr., 74: 472-76.

Raymond, G. 1901. Sur un Monadinazoosporeé parasite de *Chlamydomonas.* Microgr. prép., 9: 128-31.

Rieder, J. 1936. Biologische und ökologische Untersuchungen an Süsswasser-Suktorien. Arch. Naturgesch., 5: 137-214.

Robertson, M. 1905. *Pseudospora volvocis,* Cienkowski. Quart. Jour. micr. Sci., 49: 213-30.

Roskin, G. 1927. Zur Kenntnis der Gattung *Pseudospora* Cienkowski. Arch. Protistenk., 59: 350-68.

Ryder, J. A. 1893. The growth of *Eugleno viridis* when constrained principally to two dimensions of space. Contr. zool. Lab. Univ. Pa., 1: 37-50.

Sand, R. 1899. Étude monographique sur le groupe des Infusoires tentaculifères. Ann. Soc. belge Micr., 24: 57-189; 25: 7-205.

Sassuchin, D. N. 1928a. Zur Frage über die Parasiten der Protozoen. Parasiten von *Nyctotherus ovalis* Leidy. Arch. Protistenk., 64: 61-70.

—— 1928b. Zur Frage über die ecto- und entoparasitischen Protozoen der Froschkaulquappen. Arch. Protistenk., 64: 71-92.

—— 1931. Zum Studium du Darmprotozoenfauna der Nager im Süd-Osten RSFSR. I. Darmprotozoen des *Citellus pygmaeus* Pallas. Arch. Protistenk., 74: 417-28.

—— 1933. Materials on hyperparasitism in Protozoa (in Russian). Rev. Microbiol. Saratov, 12: 219-27.

—— 1934. Hyperparasitism in Protozoa. Quart. Rev. Biol., 9: 215-24.

Sassuchin, D. N., P. P. Popoff, W. A. Kudrjewzew, and W. P. Bogenko. 1930. Über parasitische Infektion bei Darmprotozoen. Arch. Protistenk., 71: 229-34.

Schereschewsky, Hélène. 1924. La Famille Metchnikovellidae (C. & M.) et la plaie qu'elle occupe dans le Systéme des Protistes (Russian with French summary). Arch. russ. Protist., 3(3-4): 137-45.

Scherffel, A. 1925a. Endophytische Phycomyceten-Parasiten der Bacillariaceen und einige neue Monadinen. Ein Beitrag zur Phylogenie der Oomyceten (Schröter). Arch. Protistenk., 52: 1-141.

—— 1925b. Zur Sexualität der Chytridineen. (Der Beiträge zur Kenntnis der Chytridineen. Teil I. Arch. Protistenk., 53: 1-58.

—— 1926a. Einiges über neue oder ungenügend bekannte Chytidineen. (Der Beiträge zur Kenntnis der Chytridineen. Teil II. Arch. Protistenk., 54: 167-260.

—— 1926b. Beiträge zur Kenntnis der Chytridineen. Teil III. Arch. Protistenk., 54: 510-28.

Schouten, G. B. 1937. *Nyctotherus ochoterenae* n. sp. y *Nyctotherus gamarrai* n. sp. Protozoarios parasitos de batracios (*Engystoma ovale bicolor* Schn. e *Hyla venulosa* Laur.). An. Inst. Biol. Univ. Méx., 8: 387-92.

Schubotz, H. 1908. *Pycnothrix monocystoides* nov. gen., nov. sp., ein neues ciliates Infusor aus dem Darm von *Procavia (Hyrax) capensis* (Pallas). Denkschr. med-naturw. Ges. Jena, 13: 1-18.

Schulze, F. E. 1875. Rhizopodienstudien. V. Arch. mikr. Anat., 11: 583-96.

Serbinow, J. L. 1907. Beiträge zur Kenntnis der Phycomyceten. Organisation u. Entwickelungsgeschichte einiger Chytridineen-Pilze (Chytridineae Schröter). Scripta bot. Petropol, 24: 1-173.

Skvortzow, B. W. 1927. Über einige Phycomycetes aus China. Arch. Protistenk., 57: 204-6.

Sokoloff, D. 1933. Algunas nuevas formas de flagelados del Valle de Mexico. An. Inst. Biol. Univ., Mex., 4: 197-206.

Stabler, R. M. 1933. On an amoeba parasitic in *Zelleriella* (Protozoa, Ciliata). J. Parasit., 20: 122.

Stabler, R. M., and T. T. Chen. 1936. Observations on an *Endamoeba* parasitizing opalinid ciliates. Biol. Bull., 70: 56-71.

Stein, F. (Ritter) von. 1849. Untersuchungen über die Entwickelung der Infusorien. Arch. Naturgesch., 15; 1: 92-148.

—— 1854. Die Infusionsthiere auf ihre Entwickelungsgeschichte untersucht. Leipzig.

—— 1859. Der Organismus der Infusionsthiere. I. Abtheilung. Algemeiner Theil und Naturgeschichte der hypotrichen Infusionsthiere. Leipzig. pp. 96-100.

—— 1867. Der Organismus der Infusionsthiere. II. Abtheilung. (1) Darstellung der neuesten Forschungsergebnisse über Bau, Fortpflanzung und Entwickelung der Infusionsthiere. (2) Naturgeschichte der heterotrichen Infusorien. Leipzig.

—— 1878. Der Organismus der Infusionsthiere. III. Abtheilung. Flagellaten. Leipzig.

—— 1883. Der Organismus der Infusionsthiere. III. Abtheilung. II. Hälfte die Naturgeschichte der Arthrodelen Flagellaten. Leipzig.

Stempell, W. 1909. Über *Nosema bombycis* Nägeli. Arch. Protistenk., 16: 281-358.

—— 1918. Über *Leptotheca coris* n. sp. und *Nosema marionis*. Mitt. zool. Inst. Univ. Münster, 1: 1-6.

—— 1919. Untersuchungen über *Leptotheca coris* n. sp. und das in dieser schmarotzende *Nosema marionis* Thel. Arch. Protistenk., 40: 113-57.

Stokes, A. C. 1884. A microscopical incident. Microscope, 4: 33-35.

Stubblefield, J. W. The morphology and life history of *Amphiacantha ovalis* and *Amphiacantha attenuata,* two new haplosporidian parasites of gregarines. Univ. Cal. Library, MS. 1937.

Studitsky, A. N. 1932. Über die Morphologie, Cytologie und Systematik von *Ptychostonum Chattoni* Rossolimo. Arch. Protistenk., 76: 188-216.

Sutherland, J. L. 1933. Protozoa from Australian termites. Quart. J. micr. Sci., 76: 145-73.

Swarczewsky, B. 1908. Über die Fortpflanzungerscheinungen bei *Arcella vulgaris* Ehrb. Arch. Protistenk., 12: 173-212.

Takagi, S. 1938. On a bacterial parasite with active motility inhabiting the the internal protoplasm of *Spirostomum ambiguum* Ehrenb. Annot. zool. Jap., 17: 170-78.

Thélohan, P. 1895. Recherches sur les Myxosporidies. Bull. Sci. Fr. Belg., 26: 100-394 (p. 360).

Tyzzer, E. E. 1920. Amoebae of the caeca of the common fowl and of the turkey.—*Entamoeba gallinarum*, sp. n. and *Pygolimax gregariniformis*, gen. et spec. nov. J. med. Res., 41: 199-209.

Veley, L. J. 1905. A further contribution to the study of *Pelomyxa palustris* (Greeff). J. Linn. Soc. (Zool.), 29: 374-95.

Wager, H. 1913. The life-history and cytology of *Polyphagus Euglenae*. Ann. Bot. Lond., 27: 173-202.

Walker, E. L. 1909. Sporulation in the parasitic Ciliata. Arch. Protistenk., 17: 297-306.

Wallengren, H. 1895. Studier öfver ciliata infusorier. Årsskr. Lunds Univ., 31: 1-77.

Wallich, G. C. 1863a. Further observations on an undescribed indigenous amoeba, with notices on remarkable forms of *Actinophrys* and *Difflugia*. Ann. Mag. Nat. Hist., (3) 11: 365-71.

—— 1863h. Further observations on *Amoeba villosa* and other indigenous rhizopods. Ann. Mag. Nat. Hist., (3) 11: 434-53.

Wenrich, D. H. 1921. The structure and division of *Trichomonas muris* (Hartmann). J. Morph., 36: 119-55.

—— 1932. The relation of the protozoan flagellate, *Retortomonas gryllotalpae* (Grassi, 1879) Stiles, 1902 to the species of the genus *Embadomonas* MacKinnon, 1911. Trans. Amer. micr. Soc., 51: 225-38.

—— 1937. Studies on *Iodamoeba bütschlii* with special reference to nuclear structure. Proc. Amer. phil. Soc., 77: 183-205.

—— 1940. Studies on the biology of *Dientamoeba fragilis*. Pp. 408-9 *in* Third International Congress for Microbiology: Report of Proceedings. New York.

Wenyon, C. M. 1907. Observations on Protozoa in the intestine of mice. Arch. Protistenk., suppl. 1: 169-201.

—— 1926. Protozoology, a manual for medical men, veterinarians and zoologists. New York.

Wesenberg-Lund, C. 1929. Rotatoria, Rotifera - Rädertierchen, *in* Kükenthal, W. and T. Krumbach: Handbuch der Zoologie. Berlin.

Wetzel, A. 1925. Zur Morphologie und Biologie von *Raphidocystis infestans* n. sp. einem temporar auf Ciliaten parasitierenden Heliozoon. Arch. Protistenk., 53: 135-82.

Winogradowa, T. 1936. *Sphaerita*, ein Parasit der Wiederkäuerinfusorien. Z. Parasitenk. 8: 356-58.

Yagiu, R. 1933. Studies on the ciliates from the intestine of *Anthocidaris crassispina* (A. Agassiz). I. *Cyclidium ozakii* sp. nov. & *Strobilidium rapulum* sp. nov. J. Sci. Hiroshima Univ., (B:I) 2:211-22.

Yakimoff, W. L. 1930. Zur Frage über Parasiten bei Protozoa. Arch. Protistenk., 72: 135-38.

—— 1931. Bolesni domaschnick schirootnich wisiroaemie prosteischmimi (Protozoa) (in Russian). (Cited by Sassuchin, 1934, without journal.)

Yakimoff, W. L., and A. P. Winnik. 1933. Die Ansteckung der Oocysten der Kaninchencoccidien durch Bakterien. Arch. Protistenk., 79: 131-32.

Yuan-Po, Li. 1928. *Entamoeba bobaci* n. sp. des tarabagans (*Marmota bobac*). Ann. Parasit. hum. comp., 6: 339-42.

Zacharias, O. 1909. Parasitische Amöben in *Volvox minor*. Arch. Hydrobiol., 5: 69-70.

Zeliff, C. C. 1930. A cytological study of *Oxymonas*, a flagellate, including the descriptions of new species. Amer. J. Hyg., 11: 714-39.

INDEX

Abbott, 972

Ablastin, 838; involved in immunity against trypanosomes, 856-62 *passim*

Accessory vacuoles, *see* Vacuoles, accessory

Accidental parasitism, defined, 895; *see also* Parasitism

Acclimatization, to experimental conditions, 507; and immunity, inherited, 717-21

Acids, effects on consistency, 55, 61; motor responses to, 333, 342

Actinophrys sol, 600

Actinopoda, sexual reproduction, 598 ff.

Active contraction *vs.* elastic shortening, 90, 91

Activities influencing longevity, 16

Adhesiveness (or stickiness), 71-77; relation to phagocytosis, 74; in animals that live in association with hosts, 930 ff., 944, 948, 949

Adolph, 65, 362, 363, 364, 424, 525

Aërobes, why die in absence of oxygen? 394

Aërobic respiration, *see* Respiration

Afridi, 849

Age changes; sexual immaturity and maturity, 714 f., 761

Aggregation, in light, 281, 297; in acid region, 342

Agranulocytes, 835

Albuminoid reserves, 160, 162

Alda Calleja, M. de, 876

Alexander, 58

Alexeieff, 160, 903, 907, 910, 970, 1033, 1042, 1050

Alkalies, effects on consistency, 55; motor responses to, 333, 342

Allelocatalytic growth, 533 ff.

Allison, 79

Alsup, 314

Alternating current, responses to, 310-14

Altmann's bioblast theory, 174, 175

Alverdes, 793

Amberson, 356, 363, 369

American Society of Parasitologists, Committee on Terminology, 891

Amoeba, properties of protoplasm as exhibited in, 46-50; surface precipitation reaction, 48; consistency, 51-61 *passim;* surface properties, 63-74 *passim;* specific gravity, 79, 80, 81; other properties, 82, 84, 90-97 *passim;* survey of functions having granular basis, 168-75 *passim;* structure, 272; responses to light, 272-80; to electricity, 305-20, 323; to chemicals, 333-41; dysentery and bacterial complications, 819-22

Amoebae, parasitic in a free-living ciliate, 1075

Amoebophrya Koeppen, 1089-93

Anaërobes, why are they anaërobes, 390-94

Anaërobic metabolism, measurement of, 385 f.

Anaërobiosis, occurrence of glycolysis and, 386-90

Ancestry, relation to conjugation, 615; necessity of diverse, 696

Ancistridae, 933

Ancistruma, fibrillar system, 228

Ancistrumidae, adaptation, 934-38, 939, 941

André, 900, 916, 934

Andrews, 927

Andrews and von Brand, 365, 368, 369, 485

Anentera, 12

Anesthetics, effect on O_2 consumption, 366

Aneurin, or thiamine (vitamin B_1), 490

Angerer, 48, 58

Animals: relationships between certain Protozoa and other animals, 890-1008 (*see also under* Relationship)

Anisogamy, term, 584

Anoplophrya circulans, 951, 952

Antibodies and antigens involved in immunity, 837-39, 873

Antiserum, 837

Antitoxin, 838

Anus, presence in Infusoria ascertained, 11

Apostomea, 956; adaptation, 957-60

Archer, 1094

logical and functional studies of, often separated, 112; segregation, 126, 132-38, 168, 171, 174, 178; segregation of neutral red by, 127; digestive, 129-32, 172, 178; excretory, 124, 144-50, 173, 178, 405, 440; carbohydrate, 154; basophilic, 160, 162, 163-65; metachromatic granules, 160, 162, 165; survey of functions having granular basis, in group of five Protozoa, 168-74; secretion, 168, 178; number of types, 169; failure to find general uniformity, 169, 174; comparison on basis of composition, 171; as permanent organelles and as temporary components, 174-77; active and passive, 175, 177; classification, 177-79; comparison with cells of the Metazoa, 179-81; unknown, 179; permanence and self-perpetuation, in Protozoa and Metazoa, 181; basal, in *Paramecium,* 195, 198, 200, 204, 226, 227; relationship between contractile vacuole and, 405 ff., 424, 435, 436, 438 f., 440; beta, 406, 435; osmiophilic, 438 f., 440

Granulocytes, 835

Grassé, 892, 893, 912, 918, 933, 1011, 1012, 1013, 1014, 1015, 1016, 1017, 1018, 1019, 1020, 1021, 1028, 1029, 1044

Grassé and Boissezon, 898

Grassi and Foà, 1028, 1046

Grave and Schmitt, 260, 343

Gray, 46, 548

Greeff, 201, 208, 213, 1025, 1026, 1041, 1042

Greeley, 54, 57, 306, 326

Green and Breazeale, 972

Greenleaf, W. E., 535

Greenway, 1045

Gregarines, gamete brood, 35; Bütschli's discovery re carbohydrate granules of, 111; Golgi bodies, 140

Gregory, 541, 543, 628, 629, 631, 796

Grieg, Von Rooyen, and Hendry, 876

Griffin, 23, 205, 221, 647

Griffiths, 422, 424

Gross, 589

Grosse-Allermann, 91

Growth, environmental conditions suitable for most rapid division and, 45; oxidation-reduction potential vs. respiration and, 394-96; food requirements and

other factors influencing growth of Protozoa in pure cultures, 475-516; specific factors, or vitamins, 489-93; stimulants, 493-95; in cultures as a population problem, 495-99; in relation to waste products, 499 f.; in relation to food concentration, 500 f.; in relation to pH of the medium, 501-3; oxygen relationships, 503 f.; in relation to temperature, 505 f.; in relation to light and darkness, 506 f.; methods for measurement of, 517-20; individual Protozoa, 520-26; colonial Protozoa, 526 f.; pedigree isolation culture and life cycles, 527-31; protozoan successions: nonlaboratory, 531 f.; laboratory, 532 f.; autocatalysis and allelocatalysis, 533-37; nutrition and, 537-44; population, 544-52; struggle for existence, 553 f.; literature cited, 554-64; studies of regeneration and, 772

Growth factor, term, 489

Gruber, A., 78, 82, 787, 788, 789, 797, 1043, 1053, 1075, 1083

Gruber, K., 47

Gruby and Delafond, 983

Günther, 1042

Guyer, 384

Gwynne-Vaughan and Barnes, 1047

Gymnostomes, association with host, 914

Haas, 917

Habenicht, 315

Hackett, 822, 823, 825

Haeckel, 201, 208, 1092

Hämmerling, 720

Hafkine, 1035, 1036, 1037

Hahnert, 61, 310, 324

Haldane, 356, 553

Haldane-Henderson, methods of gas analysis, 356

Hall, R. P., 135, 139, 143, 359, 360, 361, 362, 371, 407, 408, 435, 436, 437, 476, 478, 479, 481, 487, 489, 490, 494, 501, 502, 504, 507, 537, 542, 543; Food Requirements and Other Factors Influencing Growth of Protozoa in Pure Cultures (Chap. IX), 475-516

Hall, R. P., and Dunihue, F. W., 130, 437

Hall, R. P., and Elliot, A. M., 484, 492, 493

Hall, R. P., and Jahn, T. L., 907